PROCEEDINGS OF THE FOURTH COMPTON SYMPOSIUM

PROCEEDINGS OF THE FOURTH COMPTON SYMPOSIUM

Williamsburg, VA   April 1997

PART TWO: Papers and Presentations

EDITORS
Charles D. Dermer
Mark S. Strickman
James D. Kurfess
Naval Research Laboratory

AIP CONFERENCE
PROCEEDINGS 410

American Institute of Physics   Woodbury, New York

Authorization to photocopy items for internal or personal use, beyond the free copying permitted under the 1978 U.S. Copyright Law (see statement below), is granted by the American Institute of Physics for users registered with the Copyright Clearance Center (CCC) Transactional Reporting Service, provided that the base fee of $10.00 per copy is paid directly to CCC, 222 Rosewood Drive, Danvers, MA 01923. For those organizations that have been granted a photocopy license by CCC, a separate system of payment has been arranged. The fee code for users of the Transactional Reporting Service is: 1-56396-659-X/97 /$10.00.

© 1997 American Institute of Physics

Individual readers of this volume and nonprofit libraries, acting for them, are permitted to make fair use of the material in it, such as copying an article for use in teaching or research. Permission is granted to quote from this volume in scientific work with the customary acknowledgment of the source. To reprint a figure, table, or other excerpt requires the consent of one of the original authors and notification to AIP. Republication or systematic or multiple reproduction of any material in this volume is permitted only under license from AIP. Address inquiries to Office of Rights and Permissions, 500 Sunnyside Boulevard, Woodbury, NY 11797-2999; phone: 516-576-2268; fax: 516-576-2499; e-mail: rights@aip.org.

L.C. Catalog Card No. 97-77179
ISBN 1-56396-659-X (Set)
ISBN 1-56396-772-3 (Part One)
ISBN 1-56396-773-1 (Part Two)
ISSN 0094-243X
DOE CONF- 9704154

Printed in the United States of America

CONTENTS

Preface xxv
Prologue xxvii

PART ONE-THE COMPTON OBSERVATORY IN REVIEW

COMPTON STATUS AND FUTURE

Status and Future of the Compton Gamma Ray Observatory 3
    N. Gehrels and C. Shrader

SOLAR AND STELLAR GAMMA RAY ASTRONOMY

Solar and Stellar Gamma Ray Observations with Compton 17
    G. H. Share, R. J. Murphy, and J. Ryan

GALACTIC GAMMA RAY ASTRONOMY

Gamma-Ray Pulsars: The Compton Observatory Contribution
to the Study of Isolated Neutron Stars 39
    D. J. Thompson, A. K. Harding, W. Hermsen, and M. P. Ulmer
Recent Results from Observations of Accreting Pulsars 57
    M. H. Finger and T. A. Prince
Low-Mass X-Ray Binaries and Radiopulsars in Binary Systems 75
    M. Tavani and D. Barret
GRO J1744-28: The Bursting Pulsar 96
    C. Kouveliotou and J. van Paradijs
The Soft Gamma-Ray Repeaters 110
    I. A. Smith
Galactic Black Hole Binaries: High-Energy Radiation 122
    J. E. Grove, J. E. Grindlay, B. A. Harmon, X.-M. Hua, D. Kazanas,
    and M. McConnell
Galactic Black Hole Binaries: Multifrequency Connections 141
    S. N. Zhang, I. F. Mirabel, B. A. Harmon, R. A. Kroeger,
    L. F. Rodriguez, R. M. Hjellming, and M. P. Rupen
CGRO Studies of Supernovae and Classical Novae 163
    M. D. Leising
Supernova Remnants and Plerions in the Compton Gamma-Ray
Observatory Era 171
    O. C. de Jager and M. G. Baring
Diffuse Galactic Continuum Radiation 192
    S. D. Hunter, R. L. Kinzer, and A. W. Strong
Galactic $e^+ - e^-$ Annihilation Line Radiation 208
    D. M. Smith, W. R. Purcell, and M. Leventhal

Galactic Gamma-Ray Line Emission from Radioactive Isotopes 218
    R. Diehl and F. X. Timmes
Galactic Nuclear Deexcitation Gamma-Ray Lines......................... 249
    H. Bloemen and A. M. Bykov

EXTRAGALACTIC GAMMA RAY ASTRONOMY

High Energy Emission from Starburst Galaxies and Clusters 271
    Y. Rephaeli and C. D. Dermer
Seyferts and Radio Galaxies.. 283
    W. N. Johnson, A. A. Zdziarski, G. M. Madejski, W. S. Paciesas,
    H. Steinle, and Y.-C. Lin
Gamma-Ray Blazars 307
    R. C. Hartman, W. Collmar, C. von Montigny, and C. D. Dermer
Multiwavelength Campaigns.. 328
    C. R. Shrader and A. E. Wehrle
The Extragalactic Diffuse Gamma-Ray Emission 344
    P. Sreekumar, F. W. Stecker, and S. C. Kappadath

HIGH ENERGY GAMMA RAY ASTRONOMY

VHE and UHE Gamma-Ray Astronomy in the EGRET Era 361
    T. C. Weekes, F. Aharonian, D. J. Fegan, and T. Kifune

GAMMA RAY MYSTERIES

Pulsar Counterparts of Gamma-Ray Sources 387
    P. A. Caraveo and G. F. Bignami
On the Nature of the Unidentified EGRET Sources 394
    R. Mukherjee, I. A. Grenier, and D. J. Thompson
A Review of Gamma Ray Bursts....................................... 407
    C. Meegan, K. Hurley, A. Connors, B. Dingus, and S. Matz
Gamma-Ray Line Transients 418
    M. J. Harris
Constraints from Undetected Gamma-Ray Sources 436
    C. E. Fichtel and P. Sreekumar

HIGH ENERGY PHYSICS AND ASTROPHYSICS

Gamma Ray Implications for the Origin and the Acceleration
of Cosmic Rays.. 449
    R. Schlickeiser, M. Pohl, R. Ramaty, and J. G. Skibo
Spectral Signatures and Physics of Black Hole Accretion Disks............. 461
    E. Liang and R. Narayan

Comptonization Processes in Galactic and Extragalactic
High Energy Sources 477
    L. G. Titarchuk
Radiation Processes in Blazars 494
    M. Sikora

COMPTON INSTRUMENTS AND HISTORY

Overview of the Compton Observatory Instruments 509
    J. D. Kurfess, D. L. Bertsch, G. J. Fishman, and V. Schönfelder
The COMPTON Observatory: Reflections on its Origins and History 524
    D. A. Kniffen and N. Gehrels

PART TWO-PAPERS AND PRESENTATIONS

ISOLATED NEUTRON STARS

COMPTEL Gamma-Ray Study of the Crab Nebula 537
    R. D. van der Meulen, H. Bloemen, K. Bennett, W. Hermsen,
    L. Kuiper, R. P. Much, J. Ryan, V. Schönfelder, and A. Strong
5 Years of Crab Pulsar Observations with COMPTEL 542
    R. Much, K. Bennett, C. Winkler, R. Diehl, G. Lichti, V. Schönfelder,
    H. Steinle, A. Strong, M. Varendorff, W. Hermsen, L. Kuiper,
    R. van der Meulen, A. Connors, M. McConnell, J. Ryan, and R. Buccheri
The Infrared to Gamma-Ray Pulse Shape of the Crab Nebula Pulsar 547
    S. S. Eikenberry and G. G. Fazio
Observation of the Crab Pulsar with BeppoSAX: Study of the
Pulse Profile and Phase Resolved Spectroscopy 553
    G. Cusumano, D. Dal Fiume, S. Giarrusso, E. Massaro, T. Mineo,
    L. Nicastro, A. N. Parmar, and A. Segreto
The Spectrum of TeV Gamma Rays from the Crab Nebula 558
    J. P. Finley, S. Biller, P. J. Boyle, J. H. Buckley, A. Burdett,
    J. Bussóns Gordo, D. A. Carter-Lewis, M. A. Catanese, M. F. Cawley,
    D. J. Fegan, J. A. Gaidos, A. M. Hillas, F. Krennrich, R. C. Lamb,
    R. W. Lessard, C. Masterson, J. E. McEnery, G. Mohanty, J. Quinn,
    A. J. Rodgers, H. J. Rose, F. W. Samuelson, G. H. Sembroski,
    R. Srinivasan, T. C. Weekes, M. West, and A. Zweerink
First Stereoscopic Measurements at the Whipple Observatory 563
    F. Krennrich, for the Whipple Collaboration
The "COS-B/EGRET 1997" Geminga Ephemeris 568
    J. R. Mattox, J. P. Halpern, and P. A. Caraveo
On the Accurate Positioning of Geminga 573
    P. A. Caraveo, M. G. Lattanzi, G. Massone, R. Mignani, V. V. Makarov,
    M. A. C. Perryman, and G. F. Bignami
Studies of the Gamma Ray Pulsar Geminga 578
    S. Zhang, T. P. Li, M. Wu, W. Yu, X. J. Sun, L. M. Song, and F. J. Lu

Timing Analysis of Four Years of COMPTEL Data on PSR B1509-58 583
    A. Carramiñana, K. Bennett, W. Hermsen, L. Kuiper, V. Schönfelder,
    A. Connors, V. Kaspi, M. Bailes, and R. N. Manchester

Search for the Pulse of PSR B1823-13 in the COMPTEL and EGRET Databases 588
    A. Carramiñana, K. T. S. Brazier, K. Bennett, W. Hermsen,
    L. Kuiper, V. Schönfelder, and A. Lyne

Very High Energy Observations of PSR B1951+32 592
    R. Srinivasan, P. J. Boyle, J. H. Buckley, A. M. Burdett, J. Gordo,
    D. A. Carter-Lewis, M. Catanese, M. F. Cawley, E. Colombo, D. J. Fegan,
    J. P. Finley, J. A. Gaidos, A. M. Hillas, R. C. Lamb, F. Krennrich,
    R. W. Lessard, C. Masterson, J. E. McEnery, G. Mohanty, P. Moriarty,
    J. Quinn, A. J. Rodgers, H. J. Rose, F. W. Samuelson, G. H. Sembroski,
    T. C. Weekes, and J. Zweerink

A Candidate γ-Ray Pulsar in CTA 1 597
    K. T. S. Brazier, O. Reimer, G. Kanbach, and A. Carramiñana

Discovery of the Young, Energetic Radio Pulsar PSR J1105-6107 602
    V. M. Kaspi, M. Bailes, R. N. Manchester, B. W. Stappers, J. S. Sandhu,
    J. Navarro, and N. D'Amico

RXTE Observation of PSR1706-44 607
    A. Ray, A. K. Harding, and M. Strickman

VHE Gamma Rays from PSR B1706-44 612
    P. M. Chadwick, M. R. Dickinson, N. A. Dipper, J. Holder,
    T. R. Kendall, T. J. L. McComb, K. J. Orford, J. L. Osborne, S. M. Rayner,
    I. D. Roberts, S. E. Shaw, and K. E. Turver

RXTE Observations of the Anomalous Pulsar 4U0142+61 617
    S. Dieters, C. Wilson, M. Finger, M. Scott, and J. van Paradijs

Search for a Gamma-Ray Pulsar in the SNR RCW103 623
    M. Mori and K. Ebisawa

Search for X-ray Pulsation from Rotation-Powered Pulsars with *ASCA* 628
    Y. Saito, N. Kawai, T. Kamae, and S. Shibata

Gamma Ray Pulsar Luminosities 633
    M. A. McLaughlin, J. M. Cordes, and M. P. Ulmer

A New Class of Radio Quiet Pulsars 638
    M. G. Baring and A. K. Harding

The Pulse Profile of γ-ray Pulsars and the Emission Region Geometry 643
    E. Massaro and M. Litterio

Geometry of Pulsar X-Ray and Gamma-Ray Pulse Profiles 648
    A. K. Harding and A. Muslimov

A New Method for Statistical Study of Gamma-Ray Phase Curves of Radio Pulsars 653
    A. Chernenko

Evidence for Spontaneous Magnetic Field Decay in an Isolated Neutron Star 658
    J. C. L. Wang

NEUTRON STAR BINARIES

A Multi-Year Light Curve of Sco X-1 Based on BATSE SD Data and the Variability States of Sco X-1 665
    B. J. McNamara, T. E. Harrison, P. A. Mason, M. Templeton,
    C. W. Heikkila, T. Buckley, E. Galvan, and A. Silva

Comparison of the BATSE LAD and SD Light Curves of Sco X-1: 1991-1996 670
    T. E. Harrison, B. J. McNamara, P. A. Mason, and M. Templeton

Application of the Gabor Transform to BATSE Spectroscopy Detector Observations of Scorpius X-1 675
    P. A. Mason, M. Templeton, B. J. McNamara, T. E. Harrison, E. Galvan, and T. Buckley

High-Energy Transient Events From Scorpius X-1 and Cygnus X-1 679
    P. A. Mason, B. J. McNamara, and T. E. Harrison

Low-Energy Line Emission from Cygnus X-2 Observed by the BeppoSAX LECS 683
    E. Kuulkers, A. N. Parmar, A. Owens, T. Oosterbroek, and U. Lammers

BATSE Observations of the Second Outburst of GRO J1744-28 687
    P. Woods, C. Kouveliotou, J. van Paradijs, M. S. Briggs, K. Deal,
    C. A. Wilson, B. A. Harmon, G. J. Fishman, W. H. G. Lewin, and J. Kommers

Determination of Peak Fluxes and α for Bursts from GRO J1744-28 692
    T. E. Strohmayer, K. Jahoda, J. H. Swank, and M. J. Stark

Kilohertz Oscillations in 4U 0614+091 and Other LMXBs 697
    E. C. Ford, P. Kaaret, M. Tavani, D. Barret, P. Bloser, J. Grindlay,
    B. A. Harmon, W. S. Paciesas, and S. N. Zhang

General Relativity and Quasi-Periodic Oscillations 703
    P. Kaaret and E. C. Ford

Compact Hard X-ray Sources Near Galactic Longitude 20 708
    G. V. Jung, J. D. Kurfess, and W. R. Purcell

Hard and Soft X-ray Observations of Aquila X-1 713
    B. C. Rubin, B. A. Harmon, W. S. Paciesas, C. R. Robinson, and S. N. Zhang

Observation of X-Ray Bursters with the Beppo-SAX Wide Field Cameras 719
    A. Bazzano, M. Cocchi, L. Natalucci, P. Ubertini, J. Heise, J. in't Zand,
    J. M. Muller, and M. J. S. Smith

Aperiodic Variability of the X-ray Burster 1E1724-3045: First Results from RXTE/PCA 724
    J.-F. Olive, D. Barret, L. Boirin, J. Grindlay, P. Bloser, J. Swank, and A. Smale

New X-Ray Bursters with the WFCs on Board SAX: SAX J1750. 8-2900, GS 1826-24 and SLX 1735-269 729
    A. Bazzano, M. Cocchi, L. Natalucci, P. Ubertini, J. Heise,
    J. in't Zand, J. M. Muller, and M. J. S. Smith

Kilo-Hertz QPO and X-ray Bursts in 4U 1608-52 in
Low Intensity State 734
      W. Yu, S. N. Zhang, B. A. Harmon, W. S. Paciesas, C. R. Robinson,
      J. E. Grindlay, P. Bloser, D. Barret, E. C. Ford, M. Tavani, and P. Kaaret

Long-Term Observations of Her X-1 with BATSE 739
      R. B. Wilson, D. M. Scott, and M. H. Finger

RXTE Spectroscopy of Her X-1 744
      D. E. Gruber, W. A. Heindl, R. E. Rothschild, R. Staubert, M. Kunz,
      and D. M. Scott

Observations of Pulse Evolution in Her X-1 748
      D. M. Scott, R. B. Wilson, M. H. Finger, and D. A. Leahy

Evolution of the Orbital Period of Her X-1: Determination
of a New Ephemeris Using RXTE Data 753
      B. Stelzer, R. Staubert, J. Wilms, R. D. Geckeler, D. Gruber,
      and R. Rothschild

The Pulsed Light Curves of Her X-1 as Observed by BeppoSAX 758
      D. Dal Fiume, M. Orlandini, G. Cusumano, S. Del Sordo, M. Feroci,
      F. Frontera, T. Oosterbroek, E. Palazzi, A. N. Parmar, A. Santangelo,
      and A. Segreto

The 35 Day Cycle of Her X-1 and the Coronal Wind Model 763
      S. Schandl, R. Staubert, and M. König

CGRO/EGRET Observations of Centaurus X-3 768
      W. T. Vestrand, P. Sreekumar, and M. Mori

GRO J2058+42 X-Ray Observations 773
      C. A. Wilson, M. H. Finger, B. A. Harmon, R. B. Wilson,
      D. Chakrabarty, and T. Strohmayer

Observation of a Long Term Spin-up Trend in 4U1538-52 778
      B. C. Rubin, M. H. Finger, D. M. Scott, and R. B. Wilson

EGRET Observations of X-R Binaries 783
      B. B. Jones, Y. C. Lin, P. F. Michelson, P. L. Nolan, M. S. E. Roberts,
      and W. F. Tompkins

Observations of Vela X-1 with RXTE 788
      P. Kretschmar, I. Kreykenbohm, R. Staubert, J. Wilms, M. Maisack,
      E. Kendziorra, W. Heindl, D. Gruber, R. Rothschild, and J. E. Grove

BeppoSAX Observation of the X-ray Binary Pulsar Vela X-1 793
      M. Orlandini, D. Dal Fiume, L. Nicastro, S. Giarrusso, A. Segreto,
      S. Piraino, G. Cusumano, S. Del Sordo, M. Guainazzi, and L. Piro

New Radio Observations of Circinus X-1 798
      R. P. Fender

Orbit Determination for the Be/X-Ray Transient EXO 2030+375 803
      M. T. Stollberg, M. H. Finger, R. B. Wilson, D. M. Scott, D. J. Crary,
      and W. S. Paciesas

The Orbital Ephemeris and X-Ray Light Curve of Cyg X-3 808
      S. M. Matz

A Multiwavelength Study of Cygnus X-3 813
    M. L. McCollough, C. R. Robinson, S. N. Zhang, B. A. Harmon,
    W. S. Paciesas, R. M. Hjellming, M. Rupen, A. J. Mioduszewski,
    E. B. Waltman, R. S. Foster, F. D. Ghigo, G. G. Pooley, R. P. Fender,
    and W. Cui

Is There Any Evidence for a Massive Black Hole in Cyg X-3 818
    A. Mitra

Generation of Periodical Gamma Radiation in Binary System
with a Millisecond Pulsar 822
    M. A. Chernyakova and A. F. Illarionov

GALACTIC BLACK HOLE CANDIDATES

The MeV Spectrum of Cygnus X-1 as Observed with COMPTEL 829
    M. McConnell, K. Bennett, H. Bloemen, W. Collmar, W. Hermsen,
    L. Kuiper, R. Much, J. Ryan, V. Schönfelder, H. Steinle, A. Strong,
    and R. van Dijk

Five Years in the Life of Cygnus X-1: BATSE Long-Term Monitoring 834
    W. S. Paciesas, C. R. Robinson, M. L. McCollough, S. N. Zhang,
    B. A. Harmon, and C. A. Wilson

Spectral Evolution of Cyg X-1 During Its 1996 Soft State Transition 839
    S. N. Zhang, W. Cui, B. A. Harmon, and W. S. Paciesas

X-ray and γ-ray Spectra of Cyg X-1 in the Soft State 844
    M. Gierliński, A. A. Zdziarski, T. Dotani, K. Ebisawa, K. Jahoda,
    and W. N. Johnson

RXTE Observation of Cygnus X-1: Spectra and Timing 849
    J. Wilms, J. Dove, M. Nowak, and B. A. Vaughan

Spectral Variability of Cygnus X-1 in the Soft State 854
    W. Focke, J. Swank, B. Phlips, W. Heindl, and W. Cui

Modeling Cygnus X-1 γ_2 Spectra Observed by BATSE 858
    X.-M. Hua, J. C. Ling, and Wm. A. Wheaton

A Model for the High-Energy Emission of Cyg X-1 863
    I. V. Moskalenko, W. Collmar, and V. Schönfelder

A Thermal-Nonthermal Inverse Compton Model for Cyg X-1 868
    A. Crider, E. P. Liang, I. A. Smith, D. Lin, and M. Kusunose

Two Distinct States of Microquasars 1E1740-294 and GRS1758-258 873
    S. N. Zhang, B. A. Harmon, and E. P. Liang

Broad-Band Spectral Modeling of Cyg X-1 and 1E1740.7 878
    S. Sheth, E. Liang, M. Burger, C. Luo, A. Harmon, and S. N. Zhang

Observational Constraints on Annihilation Sites in 1E 1740.7-2942
and Nova Muscae 881
    I. V. Moskalenko and E. Jourdain

Two-Phase Spectral Modelling of 1E1740.7-2942 887
    O. Vilhu, J. Nevalainen, J. Poutanen, M. Gilfanov, P. Durouchoux,
    M. Vargas, R. Narayan, and A. Esin

Multi-Wavelength Monitoring of GRS 1915+105 892
    R. Bandyopadhyay, P. Martini, E. Gerard, P. A. Charles,
    R. M. Wagner, C. Shrader, T. Shahbaz, and I. F. Mirabel
The Hard X-Ray Spectrum of GRS 1915+105 897
    W. A. Heindl, P. Blanco, D. E. Gruber, M. Pelling,
    R. Rothschild, E. Morgan, and J. H. Swank
OSSE Upper Limit on Positron Annihilation from GRS 1915+105 902
    D. M. Smith, M. Leventhal, L. X. Cheng, J. Tueller, N. Gehrels,
    I. F. Mirabel, L. F. Rodriguez, and W. Purcell
RXTE Observations of GRS 1915+105 907
    J. Greiner, E. H. Morgan, and R. A. Remillard
Near-Infrared Observations of GRS 1915+105 912
    W. A. Mahoney, S. Corbel, Ph. Durouchoux, T. N. Gautier,
    J. C. Higdon, and P. Wallyn
Infrared Observations and Energetic Outburst of GRS 1915+105 917
    S. Chaty and I. F. Mirabel
ASCA Observations of Galactic Jet Systems 922
    T. Kotani, N. Kawai, M. Matsuoka, T. Dotani, H. Inoue, F. Nagase,
    Y. Tanaka, Y. Ueda, K. Yamaoka, W. Brinkmann, K. Ebisawa,
    T. Takeshima, N. E. White, A. Harmon, C. R. Robinson, S. N. Zhang,
    M. Tavani, and R. Foster
BATSE Observations of GX339-4 927
    B. C. Rubin, B. A. Harmon, W. S. Paciesas, C. R. Robinson,
    S. N. Zhang, and G. J. Fishman
Multiwavelength Observations of GX 339-4 932
    I. A. Smith, E. P. Liang, M. Moss, J. Dobrinskaya, R. P. Fender,
    Ph. Durouchoux, S. Corbel, R. Sood, A. V. Filippenko, and D. C. Leonard
Radio Observations of the Black Hole Candidate GX 339-4 937
    S. Corbel, R. P. Fender, Ph. Durouchoux, R. K. Sood, A. K. Tzioumis,
    R. E. Spencer, and D. Campbell-Wilson
Infrared Observations of the Ellipsoidal Light Variation in J0422+32 942
    D. M. Leeber, T. E. Harrison, and B. J. McNamara
Rapid X-ray Variability in GRO J0422+32 (Nova Per 1992) 947
    F. van der Hooft, C. Kouveliotou, J. van Paradijs, D. J. Crary,
    B. C. Rubin, M. H. Finger, B. A. Harmon, M. van der Klis,
    W. H. G. Lewin, and G. J. Fishman
BATSE Observations of Two Hard X-ray Outbursts from 4U 1630-47 952
    P. F. Bloser, J. E. Grindlay, D. Barret, S. N. Zhang, B. A. Harmon,
    G. J. Fishman, and W. S. Paciesas
Hard X-ray Observations of GRS 1009-45 with the SIGMA Telescope 957
    P. Goldoni, M. Vargas, A. Goldwurm, P. Laurent, E. Jourdain,
    J.-P. Roques, V. Borrel, L. Bouchet, M. Revnivtsev, E. Churazov,
    M. Gilfanov, R. Sunyaev, A. Dyachkov, N. Khavenson, N. Tserenin,
    and N. Kuleshova
**Relativistic Effects in the X-ray Spectra of the Black Hole Candidate
GS 2023+338** 962
    P. T. Życki, C. Done, and D. A. Smith

A Search for Gamma-ray Flares from Black-Hole Candidates
on Time Scales of ~1.5 Hours 967
    R. van Dijk, K. Bennett, H. Bloemen, R. Diehl, W. Hermsen,
    M. McConnell, J. Ryan, and V. Schönfelder
Physical Characteristics of the Spectral States of Galactic Black Holes 972
    J. Poutanen, J. H. Krolik, and F. Ryde
Temporal Characteristics of Compton Reflection from Accretion Disks 977
    W. T. Bridgman, C. D. Dermer, and J. G. Skibo
Temporal and Spectral Properties of Comptonized Radiation 982
    D. Kazanas, X.-M. Hua, and L. Titarchuk
Phase Difference and Coherence as Diagnostics of Accreting Sources 987
    X.-M. Hua, D. Kazanas, and L. Titarchuk
Global Spectra of Transonic Accretion Disks............................. 992
    E. Liang and C. Luo
Horizontal Branch Oscillations from Black Hole Candidates 995
    X. Chen, R. E. Taam, and J. H. Swank
Evolution of the Optically Thick Disk in Nova Muscae.................... 1000
    F. Melia and R. Misra

GALACTIC GAMMA-RAY LINE EMISSION

TGRS Results on the Spatial and Temporal Behavior
of the Galactic Center 511 keV Line.................................... 1007
    B. J. Teegarden, T. L. Cline, N. Gehrels, R. Ramaty, H. Seifert,
    M. Harris, D. Palmer, and K. H. Hurley
A BATSE Measurement of the Galactic Positron Annihilation Line 1012
    D. M. Smith, L. X. Cheng, M. Leventhal, J. Tueller, N. Gehrels,
    and G. Fishman
OSSE Constraints on the Galactic Positron Source Distribution 1017
    P. A. Milne and M. D. Leising
Is Positron Escape Seen in the Late-Time Light Curves
of Type Ia Supernovae?... 1022
    P. A. Milne, L.-S. The, and M. D. Leising
The Origin of the High-Energy Activity at the Galactic Center............. 1027
    F. Yusef-Zadeh, W. Purcell, and E. Gotthelf
The Galactic Center Lobe and its Interpretation 1034
    M. Pohl
Evidence for GeV Emission from the Galactic Center Fountain 1039
    D. H. Hartmann, D. D. Dixon, E. D. Kolaczyk, and J. Samimi
Positron Transport and Annihilation in Expanding Flows:
A Model for the High-Latitude Annihilation Feature..................... 1044
    C. D. Dermer and J. G. Skibo
Issues Concerning the Orion Gamma Ray Line Observations:
Overview and X-Ray Emission 1049
    R. Ramaty, B. Kozlovsky, and V. Tatischeff

Constraints from Pion Production on the Spectral Hardness
of the Low Energy Cosmic Rays in Orion 1054
    V. Tatischeff, R. Ramaty, and N. Mandzhavidze
Gamma-Ray Lines from OB Associations at $Z=Z_\odot$ and $Z=2Z_\odot$ 1059
    E. Parizot, J. Paul, and M. Cassé
On the Origin of the Orion Energetic Particles 1064
    E. M. G. Parizot
On the Origin of 3 to 7 MeV γ-Ray Excess in the Direction of Orion 1069
    V. A. Dogiel, M. J. Freyberg, G. E. Morfill, and V. Schönfelder
COMPTEL Spectral Study of the Inner Galaxy 1074
    H. Bloemen, A. M. Bykov, R. Diehl, W. Hermsen, R. van der Meulen,
    V. Schönfelder, and A. W. Strong
OSSE Results on Galactic γ-Ray Line Emission......................... 1079
    M. J. Harris, W. R. Purcell, K. McNaron-Brown, R. J. Murphy,
    J. E. Grove, W. N. Johnson, R. L. Kinzer, J. D. Kurfess, G. H. Share,
    and G. V. Jung
Reassessment of the ^{56}Co emission from SN 1991T 1084
    D. J. Morris, K. Bennett, H. Bloemen, R. Diehl, W. Hermsen,
    G. G. Lichti, M. L. McConnell, J. M. Ryan, and V. Schönfelder
RXTE Observations of Cas A 1089
    R. E. Rothschild, R. E. Lingenfelter, W. A. Heindl, P. R. Blanco,
    M. R. Pelling, D. E. Gruber, G. E. Allen, K. Jahoda, J. H. Swank,
    S. E. Woosley, K. Nomoto, and J. C. Higdon
Fluctuation Analysis of OSSE Measurements of the
1.275 MeV Line of ^{22}Na 1094
    M. J. Harris
COMPTEL All-Sky Imaging at 2.2 MeV 1099
    M. McConnell, S. Fletcher, K. Bennett, H. Bloemen, R. Diehl,
    W. Hermsen, J. Ryan, V. Schönfelder, A. Strong, and R. van Dijk
^{26}Al Constraints from the COMPTEL/OSSE/SMM Data 1104
    R. Diehl, M. D. Leising, J. Knödlseder, and U. Oberlack
^{26}Al and the COMPTEL ^{60}Fe Data....................................... 1109
    R. Diehl, U. Wessolowski, U. Oberlack, H. Bloemen, R. Georgii,
    A. Iyudin, J. Knödlseder, G. Lichti, W. Hermsen, D. Morris, J. Ryan,
    V. Schönfelder, A. Strong, P. von Ballmoos, and C. Winkler
Models for COMPTEL ^{26}Al Data 1114
    R. Diehl, U. Oberlack, J. Knödlseder, H. Bloemen, W. Hermsen,
    D. Morris, J. Ryan, V. Schönfelder, A. Strong, P. von Ballmoos,
    and C. Winkler
γ-Ray Emitting Radionuclide Production in A Multidimensional
Supernovae Model 1119
    G. Bazán and D. Arnett
Predictions of Gamma-Ray Emission from Classical Novae
and their Detectability by CGRO 1125
    M. Hernanz, J. Gómez-Gomar, J. José, and J. Isern
New Studies of Nuclear Decay γ-rays From Novae 1130
    S. Starrfield, J. W. Truran, M. C. Wiescher, and W. M. Sparks

SUPERNOVA REMNANTS AND COSMIC RAYS, DIFFUSE GAMMA-RAY CONTINUUM RADIATION, AND UNIDENTIFIED SOURCES

CTA 1 Supernova Remnant: A High Energy Gamma-Ray Source? 1137
    D. Bhattacharya, A. Akyüz, G. Case, D. Dixon, and A. Zych

Constraints on Cosmic-Ray Origin from TeV Gamma-Ray Observations of Supernova Remnants 1142
    R. W. Lessard, P. J. Boyle, S. M. Bradbury, J. H. Buckley,
    A. C. Burdett, J. Bussóns Gordo, D. A. Carter-Lewis, M. Catanese,
    M. F. Cawley, D. J. Fegan, J. P. Finley, J. A. Gaidos, A. M. Hillas,
    F. Krennrich, R. C. Lamb, C. Masterson, J. E. McEnery,
    G. Mohanty, J. Quinn, A. J. Rodgers, H. J. Rose, F. W. Samuelson,
    G. H. Sembroski, R. Srinivasan, T. C. Weekes, and J. Zweerink

Hard X-ray Emission from Cassiopeia A SNR 1147
    L.-S. The, M. D. Leising, D. H. Hartmann, J. D. Kurfess, P. Blanco,
    and D. Bhattacharya

Nonthermal SNR Emission 1152
    S. J. Sturner, J. G. Skibo, C. D. Dermer, and J. R. Mattox

Gamma-Rays from Supernova Remnants: Signatures of Non-Linear Shock Acceleration 1157
    M. G. Baring, D. C. Ellison, S. J. Reynolds, and I. A. Grenier

Modelling Cosmic Rays and Gamma Rays in the Galaxy 1162
    A. W. Strong and I. V. Moskalenko

Production of Beryllium and Boron by Spallation in Supernova Ejecta 1167
    D. Majmudar and J. H. Applegate

Gamma Rays and Cosmic Rays from Supernova Explosions and Young Pulsars in the Past 1172
    L. I. Dorman

Angle Distribution and Time Variation of Gamma Ray Flux from Solar and Stellar Winds, 1. Generation of Flare Energetic Particles 1178
    L. I. Dorman

Angle Distribution and Time Variation of Gamma Ray Flux from Solar and Stellar Winds, 2. Generation of Galactic Cosmic Rays 1183
    L. I. Dorman

Diffuse High-Energy Gamma-Ray Emission in Monoceros 1188
    S. W. Digel, I. A. Grenier, S. D. Hunter, T. M. Dame,
    and P. Thaddeus

Diffuse 50 keV to 10 MeV Gamma-Ray Emission from the Inner Galactic Ridge 1193
    R. L. Kinzer, W. R. Purcell, and J. D. Kurfess

Diffuse Galactic Continuum Emission: Recent Studies using COMPTEL Data 1198
    A. W. Strong, R. Diehl, V. Schönfelder, K. Bennett, M. McConnell,
    and J. Ryan

Galactic Diffuse γ-ray Emission at TeV Energies
and the Ultra-High Energy Cosmic Rays................................ 1203
    G. A. Medina Tanco and E. M. de Gouveia Dal Pino
The Diffuse Galactic Continuum Observed with EGRET:
Where's the Bump?... 1208
    J. Skibo
The Pulsar Contribution to the Diffuse Galactic γ-ray Emission........... 1213
    M. Pohl, G. Kanbach, S. D. Hunter, and B. B. Jones
The Total Cosmic Diffuse Gamma-Ray Spectrum from 9 to 30 MeV
Measured with COMPTEL 1218
    S. C. Kappadath, J. Ryan, K. Bennett, H. Bloemen, R. Diehl,
    W. Hermsen, M. McConnell, V. Schönfelder, M. Varendorff,
    G. Weidenspointner, and C. Winkler
The Cosmic γ-Ray Background from Supernovae 1223
    K. Watanabe, D. H. Hartmann, M. D. Leising, L.-S. The, G. H. Share,
    and R. L. Kinzer
The γ-Ray and Neutrino Background and Cosmic
Chemical Evolution 1228
    D. H. Hartmann, K. Watanabe, M. D. Leising, L.-S. The,
    and S. E. Woosley
The Contribution of Blazars to the Extragalactic Diffuse γ-ray
Background... 1233
    A. Mücke and M. Pohl
Absorption of High Energy Gamma Rays by Interactions
with Extragalactic Starlight Photons at High Redshifts................... 1238
    M. H. Salamon and F. W. Stecker
Further COMPTEL Observations of the Region Around GRO J1753+57:
Are There Several MeV Sources Present? 1243
    O. R. Williams, K. Bennett, R. Much, V. Schönfelder, W. Collmar,
    H. Bloemen, J. J. Blom, W. Hermsen, and J. Ryan
Temporal and Spectral Studies of Unidentified EGRET High
Latitude Sources... 1248
    O. Reimer, D. L. Bertsch, B. L. Dingus, J. A. Esposito, R. C. Hartman,
    S. D. Hunter, B. B. Jones, G. Kanbach, D. A. Kniffen, Y. C. Lin,
    H. A. Mayer-Hasselwander, C. von Montigny, P. L. Nolan, P. Sreekumar,
    D. J. Thompson, and W. F. Tompkins
Discovery of a Non-Blazar Gamma-ray Transient in the Galactic Plane 1253
    M. Tavani, R. Mukherjee, J. R. Mattox, J. Halpern, D. J. Thompson,
    G. Kanbach, W. Hermsen, S. N. Zhang, and R. S. Foster
Searches for Short-Term Variability of EGRET Sources
in the Galactic Anticenter.. 1257
    D. J. Thompson, S. D. Bloom, J. A. Esposito, D. A. Kniffen,
    and C. von Montigny
Short Time-scale Gamma-Ray Variability of Blazars and EGRET
Unidentified Sources 1262
    S. D. Bloom, D. J. Thompson, R. C. Hartman, and C. von Montigny

Optical Identification of EGRET Source Counterparts 1267
    A. Carramiñana, J. Guichard, K. T. S. Brazier, G. Kanbach,
    and O. Reimer
**Possible Identification of Unidentified EGRET Sources
with Wolf-Rayet Stars** 1271
    R. K. Kaul and A. K. Mitra
Accreting Isolated Black Holes and the Unidentified EGRET Sources 1275
    C. D. Dermer

SEYFERT AND RADIO GALAXIES

Multi-Year BATSE Earth Occultation Monitoring of NGC4151 1283
    A. Parsons, N. Gehrels, W. Paciesas, A. Harmon, G. Fishman,
    C. Wilson, and S. N. Zhang
Broad-Band Continuum and Variability of NGC 5548 1288
    P. Magdziarz, O. Blaes, A. A. Zdziarski, W. N. Johnson,
    and D. A. Smith
**Detection of a High Energy Break in the Seyfert Galaxy
MCG+8−11−11** 1293
    P. Grandi, F. Haardt, G. Ghisellini, J. E. Grove, L. Maraschi,
    and C. M. Urry
**Compton Gamma-Ray Observatory Observations of the Nearest
Active Galaxy Centaurus A** 1298
    H. Steinle, K. Bennett, H. Bloemen, W. Collmar, R. Diehl,
    W. Hermsen, G. G. Lichti, D. Morris, V. Schönfelder, A. W. Strong,
    and O. R. Williams
An Anisotropic Illumination Model of Seyfert I Galaxies 1303
    P. O. Petrucci, G. Henri, J. Malzac, and E. Jourdain
Scattered Emission and the $X-\gamma$ Spectra of Seyfert Galaxies 1308
    J. Chiang, C. D. Dermer, and J. G. Skibo
Pair Models Revivified for High Energy Emission of AGNs 1313
    G. Henri and P. O. Petrucci
Big Blue Bump and Transient Active Regions in Seyfert Galaxies 1318
    S. Nayakshin and F. Melia
Magnetic Flares and the Observed $\tau_T \sim 1$ in Seyfert Galaxies 1323
    S. Nayakshin and F. Melia
Physical Constraints for the Active Regions in Seyfert Galaxies 1328
    S. Nayakshin and F. Melia
Are Gamma-ray Bursts related to Active Galactic Nuclei? 1333
    J. Gorosabel and A. J. Castro-Tirado

BLAZARS

Evidence for γ-Ray Flares in 3C 279 and PKS 1622-297 at ~ 10 MeV 1341
    W. Collmar, V. Schönfelder, H. Bloemen, J. J. Blom, W. Hermsen,
    M. McConnell, J. G. Stacy, K. Bennett, and O. R. Williams

EGRET Observations of PKS 0528+134 from 1991 to 1997 1346
R. Mukherjee, D. L. Bertsch, S. D. Bloom, B. L. Dingus,
J. A. Esposito, R. C. Hartman, S. D. Hunter, G. Kanbach,
D. A. Kniffen, A. Kraus, T. P. Krichbaum, Y. C. Lin, W. A. Mahoney,
A. P. Marscher, H. A. Mayer-Hasselwander, P. F. Michelson,
C. von Montigny, A. Mücke, P. L. Nolan, M. Pohl, O. Reimer,
E. Schneid, P. Sreekumar, H. Teräsranta, D. J. Thompson, M. Tornikoski,
E. Valtaoja, S. Wagner, and A. Witzel

Imaging Analysis of PKS0528+134 During Its Flare with A Direct Demodulation Technique 1351
S. Zhang, T. P. Li, M. Wu, and W. Yu

First Results of an All-Sky Search for MeV-Emission from Active Galaxies with COMPTEL 1356
J. G. Stacy, J. M. Ryan, W. Collmar, V. Schönfelder, H. Steinle,
A. W. Strong, H. Bloemen, J. J. Blom, W. Hermsen, O. R. Williams,
and M. Maisack

Variability Time Scales in the Gamma-ray Blazars Using Structure Function Analysis 1361
G. Nandikotkur, P. Sreekumar, and D. A. Carter-Lewis

A Spectral Study of Gamma-ray Emitting AGN 1366
M. Pohl, R. C. Hartman, P. Sreekumar, and B. B. Jones

EGRET Observations of PKS 2005-489 1371
Y. C. Lin, D. L. Bertsch, S. D. Bloom, B. L. Dingus, J. A. Esposito,
S. D. Hunter, B. B. Jones, G. Kanbach, D. A. Kniffen,
H. A. Mayer-Hasselwander, P. F. Michelson, C. von Montigny,
R. Mukherjee, A. Mücke, P. L. Nolan, M. K. Pohl, O. L. Reimer,
E. J. Schneid, P. Sreekumar, D. J. Thompson, and W. F. Tompkins

Whipple Observations of BL Lac Objects at E>300 GeV 1376
M. Catanese, P. J. Boyle, J. H. Buckley, A. M. Burdett, J. Bussóns Gordo,
D. A. Carter-Lewis, M. F. Cawley, D. J. Fegan, J. P. Finley, J. A. Gaidos,
A. M. Hillas, F. Krennrich, R. C. Lamb, R. W. Lessard, C. Masterson,
J. E. McEnery, G. Mohanty, J. Quinn, A. J. Rodgers, H. J. Rose,
F. W. Samuelson, G. H. Sembroski, R. Srinivasan, T. C. Weekes,
and J. Zweerink

Multiwavelength Observations of Markarian 421 1381
J. H. Buckley, P. Boyle, A. Burdett, J. Bussóns Gordo, D. A. Carter-Lewis,
M. Catanese, M. F. Cawley, D. J. Fegan, J. P. Finley, J. A. Gaidos,
A. M. Hillas, F. Krennrich, R. C. Lamb, R. W. Lessard, C. Masterson,
J. McEnery, G. Mohanty, J. Quinn, A. Rodgers, H. J. Rose, F. Samuelson,
G. H. Sembroski, R. Srinivasan, T. C. Weekes, and J. Zweerink

Observation of Strong Variability in the X-Ray Emission from Markarian 421 Correlated with the May 1996 TeV Flare 1386
M. Schubnell

The Energy Spectrum of Mrk 421 1391
F. Krennrich, for the Whipple Collaboration

Study of the Temporal and Spectral Characteristics of
TeV Gamma Radiation from Mkn 501 During a State of
High Activity by the HEGRA IACT Array 1397
    F. Aharonian, A. Akhperjanian, J. Barrio, K. Bernlöhr, J. Beteta,
    S. Bradbury, J. Contreras, J. Cortina, A. Daum, T. Deckers, E. Feigl,
    J. Fernandez, V. Fonseca, A. Fraß, B. Funk, J. Gonzalez, V. Haustein,
    G. Heinzelmann, M. Hemberger, G. Hermann, M. Heß, A. Heusler,
    W. Hofmann, I. Holl, D. Horns, R. Kankanian, O. Kirstein, C. Köhler,
    A. Konopelko, H. Kornmayer, D. Kranich, H. Krawczynski, H. Lampeitl,
    A. Lindner, E. Lorenz, N. Magnussen, H. Meyer, R. Mirzoyan, H. Möller,
    A. Moralejo, L. Padilla, M. Panter, D. Petry, R. Plaga, J. Prahl, C. Prosch,
    G. Pühlhofer, G. Rauterberg, W. Rhode, R. Rivero, A. Röhring, V. Sahakian,
    M. Samorski, J. Sanchez, D. Schmele, T. Schmidt, W. Stamm, M. Ulrich,
    H. Völk, S. Westerhoff, B. Wiebel-Sooth, C. A. Wiedner, M. Willmer,
    and H. Wirth (HEGRA collaboration)

Multiwavelength Observations of a Flare from Markarikan 501 1402
    M. Catanese, S. M. Bradbury, A. C. Breslin, J. H. Buckley,
    D. A. Carter-Lewis, M. F. Cawley, C. D. Dermer, D. J. Fegan,
    J. P. Finley, J. A. Gaidos, A. M. Hillas, W. N. Johnson, F. Krennrich,
    R. C. Lamb, R. W. Lessard, D. J. Macomb, J. E. McEnery, P. Moriarty,
    J. Quinn, A. J. Rodgers, H. J. Rose, F. W. Samuelson, G. H. Sembroski,
    R. Srinivasan, T. C. Weekes, and J. Zweerink

Recent Observations of γ-rays above 1.5 TeV from Mkn 501
with the HEGRA 5 m Air Čerenkov Telescope 1407
    D. Kranich, E. Lorenz, and D. Petry for the HEGRA Collaboration

BeppoSAX Monitoring of the BL Lac Mkn 501 1412
    E. Pian, G. Vacanti, G. Tagliaferri, G. Ghisellini, L. Maraschi,
    A. Treves, C. M. Urry, F. Fiore, P. Giommi, E. Palazzi, L. Chiappetti,
    and R. M. Sambruna

Multiwavelength Observations of the February 1996 High-Energy
Flare in the Blazar 3C 279 1417
    A. E. Wehrle, E. Pian, C. M. Urry, L. Maraschi, G. Ghisellini,
    R. C. Hartman, G. M. Madejski, F. Makino, A. P. Marscher,
    I. M. McHardy, J. R. Webb, G. S. Aldering, M. F. Aller, H. D. Aller,
    D. E. Backman, T. J. Balonek, P. Boltwood, J. Bonnell, J. Caplinger,
    A. Celotti, W. Collmar, J. Dalton, A. Drucker, R. Falomo, C. E. Fichtel,
    W. Freudling, W. K. Gear, N. Gonzalez-Perez, P. Hall, H. Inoue,
    W. N. Johnson, M. R. Kidger, R. I. Kollgaard, Y. Kondo, J. Kurfess,
    A. J. Lawson, B. McCollum, K. McNaron-Brown, D. Nair, S. Penton,
    J. E. Pesce, M. Pohl, C. M. Raiteri, M. Renda, E. I. Robson,
    R. M. Sambruna, A. F. Schirmer, C. Shrader, M. Sikora, A. Sillanpää,
    P. S. Smith, J. A. Stevens, J. Stocke, L. O. Takalo, H. Teräsranta,
    D. J. Thompson, R. Thompson, M. Tornikoski, G. Tosti, P. Turcotte,
    A. Treves, S. C. Unwin, E. Valtaoja, M. Villata, S. J. Wagner, W. Xu,
    and A. C. Zook

Radio to γ-Ray Observations of 3C 454.3: 1993–1995 1423
  M. F. Aller, A. P. Marscher, R. C. Hartman, H. D. Aller, M. C. Aller,
  T. J. Balonek, M. C. Begelman, M. Chiaberge, S. D. Clements,
  W. Collmar, G. De Francesco, W. K. Gear, M. Georganopoulos,
  G. Ghisellini, I. S. Glass, J. N. González-Pérez, P. Heinämäki,
  M. Herter, E. J. Hooper, P. A. Hughes, W. N. Johnson, S. Katajainen,
  M. R. Kidger, A. Kraus, L. Lanteri, G. F. Lawrence, G. G. Lichti,
  Y. C. Lin, G. M. Madejski, K. McNaron-Brown, E. M. Moore,
  R. Mukherjee, A. D. Nair, K. Nilsson, A. Peila, D. B. Pierkowski,
  M. Pohl, T. Pursimo, C. M. Raiteri, W. Reich, E. I. Robson, A. Sillanpää,
  M. Sikora, A. G. Smith, H. Steppe, J. Stevens, L. O. Takalo, H. Teräsranta,
  M. Tornikoski, E. Valtaoja, C. von Montigny, M. Villata, S. Wagner,
  R. Wichmann, and A. Witzel

Multi-Wavelength Radio Monitoring of EGRET Sources and Candidates. 1428
  P. G. Edwards, J. E. J. Lovell, R. C. Hartman, M. Tornikoski,
  M. Lainela, P. M. McCulloch, B. M. Gaensler, and R. W. Hunstead

VLBI Observations of Southern Hemisphere Gamma-Ray Loud and Quiet AGN. 1433
  S. J. Tingay, D. W. Murphy, P. G. Edwards, M. E. Costa,
  P. M. McCulloch, J. E. J. Lovell, D. L. Jauncey, J. E. Reynolds,
  A. K. Tzioumis, R. A. Preston, D. L. Meier, D. L. Jones,
  and G. D. Nicolson

VLBA Monitoring of Three Gamma-Ray Bright Blazars: AO 0235+164, 1633+382 (4C 38. 41) & 2230+114 (CTA 102) 1437
  W. Xu, A. E. Wehrle, and A. P. Marscher

Coordinated Millimeter-Wave Observations of Bright, Variable Gamma-ray Blazars with the Haystack Radio Telescope 1442
  J. G. Stacy, W. T. Vestrand, and R. B. Phillips

The Burst Activity of Millimeter Wavelengths Compared to Gamma-Activity of AGN 1447
  H. Teräsranta

Relationships Between Radio and Gamma-ray Properties in Active Galactic Nuclei. 1452
  A. Lähteenmäki, H. Teräsranta, K. Wiik, and E. Valtaoja

Fast Variations of Gamma-Ray Emission in Blazars 1457
  S. J. Wagner, C. von Montigny, and M. Herter

A z = 2.1 Quasar as the Optical Counterpart of the MeV Source GRO J1753+57 1462
  A. Carramiñana, V. Chavushyan, and J. Guichard

ASCA Observations of Blazars and Multiband Analysis 1467
  T. Takahashi, H. Kubo, G. Madejski, M. Tashiro, and F. Makino

Spectral Modelling of Gamma-ray Blazars 1473
  M. Böttcher, H. Mause, and R. Schlickeiser

Modelling the Rapid Variability of Blazar Emission 1478
  J. G. Kirk and A. Mastichiadis

OVERVIEWS, SURVEYS, AND MISCELLANEOUS

BeppoSAX Overview 1485
    L. Piro, on behalf of the BeppoSAX team
Initial Results from the High Energy Experiment *PDS*
Aboard *BeppoSAX* 1493
    F. Frontera, D. Dal Fiume, E. Costa, M. Feroci, M. Orlandini,
    L. Nicastro, E. Palazzi, G. Zavattini, and P. Giommi
The CFA BATSE Image Search (CBIS) as Used for a Galactic
Plane Survey 1498
    D. Barret, J. E. Grindlay, P. F. Bloser, G. P. Monnelly, B. A. Harmon,
    C. R. Robinson, and S. N. Zhang
TeV Gamma Ray Emission from Southern Sky Objects
and CANGAROO Project 1507
    T. Kifune, S. A. Dazeley, P. G. Edwards, T. Hara, Y. Hayami,
    S. Kamei, R. Kita, T. Konishi, A. Masaike, Y. Matsubara, Y. Matsuoka,
    Y. Mizumoto, M. Mori, H. Muraishi, Y. Muraki, T. Naito, K. Nishijima,
    S. Ogio, J. R. Patterson, M. D. Roberts, G. P. Rowell, T. Sako,
    K. Sakurazawa, R. Susukita, A. Suzuki, R. Suzuki, T. Tamura, T. Tanimori,
    G. J. Thornton, S. Yanagita, T. Yoshida, and T. Yoshikoshi
Saturated Compton Scattering Models for the Soft Gamma-Ray
Repeater Bursts 1512
    I. A. Smith, E. P. Liang, A. Crider, D. Lin, and M. Kusunose
The GRB 970111 Error Box 19-Hours After the High Energy Event 1516
    A. J. Castro-Tirado, J. Gorosabel, N. Masetti, C. Bartolini,
    A. Guarnieri, A. Piccioni, J. Heidt, T. Seitz, E. Thommes, C. Wolf,
    E. Costa, M. Feroci, F. Frontera, D. Dal Fiume, L. Nicastro, E. Palazzi,
    and N. Lund
The Duration-Photon Energy Relation in Gamma-Ray Bursts
and its Interpretations 1520
    D. Kazanas, L. G. Titarchuk, and X.-M. Hua

FUTURE MISSIONS AND INSTRUMENTATION

IBIS: The Imaging Gamma-Ray Telescope on Board INTEGRAL 1527
    P. Ubertini, on behalf of the IBIS Consortium
SPI: A High Resolution Imaging Spectrometer for INTEGRAL 1535
    B. J. Teegarden, J. Naya, H. Seifert, S. Sturner, G. Vedrenne,
    P. Mandrou, P. von Ballmoos, J.-P. Roques, P. Jean, F. Albernhe,
    V. Borrel, V. Schonfelder, G. G. Lichti, R. Diehl, R. Georgii, P. Durouchoux,
    B. Cordier, N. Diallo, J. Matteson, R. Lin, F. Sanchez, P. Caraveo,
    P. Leleux, G. K. Skinner, and P. Connell
The Spectral Line Imaging Capabilities of the
SPI Germanium Spectrometer on INTEGRAL 1544
    G. K. Skinner, P. H. Connell, J. Naya, H. Seifert, S. Sturner,
    B. J. Teegarden, and A. W. Strong

The IBIS View of the Galactic Centre: INTEGRAL's Imager
Observations Simulations 1549
    P. Goldoni, A. Goldwurm, P. Laurent, and F. Lebrun

Can the INTEGRAL-Spectrometer SPI detect γ-ray Lines
From Local Galaxies? 1554
    R. Georgii, R. Diehl, G. G. Lichti, and V. Schönfelder

Contribution of Passive Materials to the Background Lines
of the Spectrometer of *INTEGRAL* (SPI) 1559
    N. Diallo, B. Cordier, M. Collin, and F. Albernhe

MGEANT—A Generic Multi-Purpose Monte-Carlo
Simulation Package for Gamma-Ray Experiments 1567
    H. Seifert, J. E. Naya, S. J. Sturner, and B. J. Teegarden

A Small Scan Angle-Dependent Background Systematic
in Non-Standard OSSE Observations 1572
    J. D. Kurfess, K. McNaron-Brown, W. R. Purcell, R. L. Kinzer,
    and W. N. Johnson

A Time Dependent Model for the Activation of COMPTEL 1577
    M. Varendorff, U. Oberlack, G. Weidenspointner, R. Diehl, R. van Dijk,
    M. McConnell, and J. Ryan

Statistical Analysis of COMPTEL Maximum Likelihood-Ratio
Distributions: Evidence for a Signal from Previously Undetected AGN 1582
    O. R. Williams, K. Bennett, R. Much, V. Schönfelder, J. J. Blom,
    and J. Ryan

Improved COMPTEL 10-30 MeV Event Selections for
Point Sources from Inflight Data 1587
    W. Collmar, U. Wessolowski, V. Schönfelder, G. Weidenspointner,
    C. Kappadath, M. McConnell, and K. Bennett

Earth Occultation Technique with EGRET Calorimeter Data
Above 1 MeV 1592
    B. L. Dingus, D. L. Bertsch, and E. J. Schneid

Maximum-Entropy Analysis of EGRET Data 1596
    M. Pohl for the EGRET collaboration, and A. W. Strong

Non-Parametric Estimates of High Energy Gamma-ray Source
Distributions 1601
    D. D. Dixon, E. D. Kolaczyk, J. Samimi, and M. A. Saunders

Development of Gas Micro-Structure Detectors for
Gamma-Ray Astronomy 1606
    S. D. Hunter, S. V. Belolipetskiy, D. L. Bertsch, J. R. Catelli,
    H. Crawford, W. M. Daniels, P. Deines-Jones, J. A. Esposito, H. Fenker,
    B. Gossan, R. C. Hartman, J. B. Hutchins, J. F. Krizmanic, V. Lindenstruth,
    M. D. Martin, J. W. Mitchell, W. K. Pitts, J. H. Simrall, P. Sreekumar,
    R. E. Streitmatter, D. J. Thompson, G. Visser, and K. M. Walsh

The Design of a 17 m Air Cerenkov Telescope for VHE
Gamma Ray Astronomy above 20 GeV 1611
    E. Lorenz for the MAGIC Telescope Design Group

Monte Carlo Simulations of the Timing Structure of Cherenkov
Wavefronts of Sub-100 GeV Gamma Ray Air Showers 1616
    D. R. Peaper, C. L. Gottbrath, M. P. Kertzman, and G. H. Sembroski

The University of Durham Mark 6 VHE Gamma Ray Telescope 1621
    P. M. Chadwick, M. R. Dickinson, N. A. Dipper, J. Holder,
    T. R. Kendall, T. J. L. McComb, K. J. Orford, S. M. Rayner,
    I. D. Roberts, S. E. Shaw, and K. E. Turver

Solar Tower Atmospheric Cherenkov Effect Experiment (STACEE) for Ground Based Gamma Ray Astronomy 1626
    D. Bhattacharya, M. C. Chantell, P. Coppi, C. E. Covault, M. Dragovan,
    D. T. Gregorich, D. S. Hanna, R. Mukherjee, R. A. Ong, S. Oser,
    K. Ragan, O. T. Tümer, and D. A. Williams

On the Potential of the HEGRA IACT Array 1631
    F. A. Aharonian (HEGRA collaboration)

A Site for Čerenkov Astronomy in the White Mountains of California 1636
    J. R. Mattox and S. P. Ahlen

Simulation of HEAO 3 Background 1642
    B. L. Graham, B. F. Phlips, R. A. Kroeger, and J. D. Kurfess

Activation of Gamma Detectors by 1.2 GeV Protons 1647
    J. L. Ferrero, C. Roldán, I. Arocas, R. Blázquez, B. Cordier, J. P. Leray,
    F. Albernhe, and V. Borrel

Participant List 1653
Author Index 1665

Preface

Over 300 scientists met at *The Fourth Compton Symposium* in Williamsburg, Virginia on April 27-30, 1997 to discuss the latest developments in gamma-ray astronomy. This meeting was hosted by the Naval Research Laboratory and the Compton Gamma Ray Observatory Science Support Center, and is the fourth in a series of conferences devoted to *Compton Observatory* science. It followed *The Violent Universe Workshop*, held on April 25-27, 1997, which provided a forum to instruct and excite educators and the public about gamma-ray astronomy.

Six years after the launch of the *Gamma Ray Observatory* on April 5, 1991, new results and discoveries continue to pour forth. This document reviews the scientific achievements stemming from the *Compton Gamma Ray Observatory* and chronicles ongoing research in the field of gamma-ray astronomy as of mid-1997. Papers summarizing related developments from X-ray astronomy missions such as the *Rossi X-ray Timing Explorer*, *Beppo-SAX* and *ASCA*, from ground-based high-energy gamma-ray observatories such as *Whipple* and *HEGRA*, and from multiwavelength campaigns correlated with gamma-ray observations are also included.

The Scientific Organizing Committee for *The Fourth Compton Symposium* consisted of J. D. Kurfess (NRL, Chair), C. D. Dermer (NRL), C. E. Fichtel (GSFC), G. J. Fishman (MSFC), N. Gehrels (GSFC), I. A. Grenier (Saclay), J. E. Grindlay (CfA), K. C. Hurley (UCB SSL), C. Kouveliotou (USRA), R. C. Lamb (ISU and CIT), M. D. Leising (Clemson), P. F. Michelson (Stanford), R. E. Rothschild (UCSD), J. M. Ryan (UNH), V. Schönfelder (MPE), R. Sunyaev (IKI), P. von Ballmoos (CESR), and A. E. Wehrle (IPAC). The Local Organizing Committee consisted of M. Strickman (NRL, Co-Chair), C. R. Shrader (GROSSC, Co-Chair), S. Barnes (GROSSC), J. E. Grove (NRL), R. C. Hartman (GSFC), W. N. Johnson (NRL), J. P. Norris (GSFC), T. Obrebski (NRL), and E. Pentecost (USRA). We would like to thank everyone for their hard work and planning which contributed to the success of this meeting. Special thanks go to Tina Obrebski and Liz Pentecost for their excellent administrative support, and to Mali Friedman for help with the proceedings. We gratefully acknowledge TRW, Inc., Ball Aerospace and Technologies Corporation, and Universities Space Research Association for generously supporting this symposium.

The growth in the size of the gamma-ray astronomy community compelled us, as in the past two symposia, to exclude gamma-ray burst and solar contributions. Even so, we received more than 330 abstracts, which exceeded our most optimistic expectations. Our wish to avoid parallel sessions unfortunately meant that time for oral contributions and poster viewing was severely limited. Judging from the size of the proceedings, the lesson here seems to be that gamma-ray astronomy is too vast a subject and three days is too short a time to survey adequately the richness and variety of the high-energy universe.

The success of the *Fourth Compton Symposium* left little doubt about the necessity for a *Fifth Compton Gamma Ray Symposium*, and attendees at the symposium banquet enthusiastically endorsed another such meeting. The banquet was held on a beautiful spring evening at the Sherwood Forest Plantation, the residence of John Tyler after he left office as the tenth president of the United States. It also provided an opportunity for the *Compton* commmunity to pay well-derserved tribute to Frank McDonald and Don Kniffen, who were so instrumental in the genesis, development, and success of the *Compton Gamma Ray Observatory*.

Charles D. Dermer
Mark S. Strickman
James D. Kurfess

Prologue

Revolution is an overused word, but nothing short can describe the impact that the *Compton Gamma Ray Observatory* has had on the field of gamma-ray astronomy. It has been such a short time, relatively speaking, that this subject went from the speculative imaginings of Philip Morrison [1] to a discipline too great for any single individual to master. As a result of the *Compton Observatory*, gamma-ray astronomy has completed a process of "subfield-specialization" [2]: for example, traditional subfields such as gamma-ray burst astronomy and high-energy solar astronomy are now large enough to support separate communities; TeV astronomy has been energized by the *Compton Observatory* results; blazar physics has been reinvented; and gamma-ray astronomy plays an ever larger role in cosmic-ray studies.

The discoveries of the *Compton Observatory* have been hardly less important to the mainstream of astronomical research than to the field of gamma-ray astronomy, and no well-versed research astronomer can afford to be ignorant of the knowledge that this mission has provided about the universe. These proceedings provide a record of gamma-ray astronomy six years after the launch of the *Compton Observatory*. The papers presented at the *Fourth Compton Symposium* are assembled in the second volume of the proceedings, entitled "Papers and Presentations." In addition to this volume, it was agreed by the Scientific Organizing Committee to organize a review of the scientific achievements of the *Compton* mission, authored by the central figures in the various topics that comprise gamma-ray astronomy.

The outcome of this endeavor is Volume One of the proceedings, entitled "The Compton Observatory in Review." The goal here is to provide a pithy summary of the new science provided specifically by the Compton Observatory, either alone or in consortium with other ground-based or space-based observatories. The motivation for producing such a volume arises from the frustrating experience of trying to derive an overall understanding of the results of past missions in the absence of an astronomical *Baedeker*. Volume One is therefore intended less for gamma-ray astronomers, who know where to find the primary literature sources, than for astronomers from other disciplines who would like a technically accurate and relatively concise summary of the *Compton* results. Try as we might, no single volume could still hope to be complete. Topics such as the remarkable gamma-ray flashes from terrestrial thunderstorms [3] and the elusive and poorly understood Cygnus X-3 [4] are not covered here.

The need for a review volume was also brought home by a number of less felicitous circumstances, including the downward trajectory of the *Compton Observatory* budget and, most worrisome of all, the departure of many of the mission scientists. In a large number of cases, scientists are moving from one project to another, as for European scientists moving to *INTEGRAL* and US scientists to *GLAST*. That is all well and good, and it is hoped that we are using this opportunity to distill their expertise before they set their minds on another challenge. Far more distressing is the departure of younger scientists who see, oftentimes more clearly than their tenured seniors, that the prospect of a US gamma-ray astronomy mission in the 100 keV - 30 MeV range is quite bleak. This is a simple matter of physics versus policy. It takes a space-based telescope of substantial area and cross section to stop a gamma ray, which means a massive and expensive payload, contrary to the central tenet of the "faster, cheaper, better" philosophy.

Yet in spite of this management policy, the *Compton Gamma Ray Observatory* and the *Hubble Space Telescope*, each hardly cheap or fast, have consistently scored high on the Senior Review science-per-dollar assessment. We should therefore take this opportunity to celebrate the great discoveries and new science emerging from the *Compton Gamma Ray Observatory*. The success of the *Compton* mission attests first and foremost to its outstanding design, to the skill of the technicians who built it, and to the ingenuity and hard work of the *Compton* scientists. Its success furthermore demonstrates the variety and unpredictability of the gamma-ray universe, the value of multiwavelength campaigns to monitor sources over the entire electromagnetic spectrum, and the need for long observing periods to detect weak gamma-ray fluxes.

Next year's great discovery may well be travelling at the speed of light toward Earth if we are only there to see it.

¡*Viva Compton!*

Charles D. Dermer
September, 1997

[1] Morrison, P., *Il Nuovo Cimento* **7**, 858 (1958).
[2] Ziman, J., *Knowing everthing about nothing: Specialization and change in scientific careers* (New York: Cambridge University Press) (1987).
[3] Fishman, G. J., et al., *Science* **264**, 1313 (1994).
[4] Matz, S. M., these proceedings.

ISOLATED NEUTRON STARS

COMPTEL gamma-ray study of the Crab nebula

R.D. van der Meulen[1,2], H. Bloemen[1,2], K. Bennett[3],
W. Hermsen[1], L. Kuiper[1], R.P. Much[3], J. Ryan[4],
V. Schönfelder[5], A. Strong[5]

[1] *SRON-Utrecht, Sorbonnelaan 2, 3584 CA Utrecht, The Netherlands*
[2] *Leiden Observatory, P.O. Box 9513, 2300 RA Leiden, The Netherlands*
[3] *Astrophysics Division, ESTEC, P.O. Box 299, 2200 AG Noordwijk, The Netherlands*
[4] *Space Science Center, Univ. of New Hampshire, Durham, NH 03824, U.S.A.*
[5] *Max-Planck Institut für Extraterrestrische Physik, P.O. Box 1603, 85740, Garching, Germany*

Abstract. We re-examine the Crab nebula, using five years of COMPTEL observations. A 30 bin 0.78–30 MeV spectrum is presented. It shows a spectral break at the edge of the COMPTEL energy range, and connects well to the EGRET spectrum, probably reflecting electron energy losses in the synchrotron emission scenario. Such a smooth continuum model alone may not be sufficient to explain the observations. A weak bump in the spectrum at 1–2 MeV may be present. No evidence for distinct line emission is seen.

INTRODUCTION

The Crab nebula is an extensively studied 943 year old SNR. The pulsar wind from the central, isolated pulsar supplies high-energy particles that synchrotron radiate while gyrating in the magnetic field of the nebula, but in situ acceleration has to occur as well.

The radio-to-γ-ray spectrum consists of a mix of different power-law components, which seem to indicate three electron populations. The continuum emission up to $\sim 1.8 \times 10^{13}$ Hz is associated with the optically visible nebula. The break may reflect the energy losses for electrons as old as the SNR; the average field of the optical nebula can then be calculated to be $\sim 3 \times 10^{-4}$ G. Consequently, the electrons responsible for the torus of X-ray emission, the spectrum of which breaks somewhere between 60 and 150 keV [1], should have a lifetime of ~ 1 month. The inner edge of the nebula lies at 0.08 pc, so it is clear that these electrons have been accelerated in or near the nebula.

CP410, *Proceedings of the Fourth Compton Symposium*
edited by C. D. Dermer, M. S. Strickman, and J. D. Kurfess
© 1997 The American Institute of Physics 1-56396-659-X/97/$10.00

De Jager et al. [2] discuss evidence for a third spectral break at ~26 MeV, based on early observations by COMPTEL and EGRET, which is addressed below. This would indicate another population of energetic electrons, with lifetimes of ~1 day. We present further evidence for this spectral break using more COMPTEL data, obtained during the first five years of the mission, and a more recent EGRET spectrum.

OBSERVATIONS AND DATA ANALYSIS

The Crab region is very well exposed by COMPTEL. Observations 0 (data obtained during the verification phase), 1.0, 31, 36.0, 36.5, 39.0, 213, 221, 310, 321.1, 321.5, 337, 412, 413, 419.1, 419.5, 420, 426, 502 were pointed within 30° of Crab, and are used in this analysis.

The Crab nebula is disentangled from the central pulsar by selecting events in the off-pulse phase 0.525–0.915, as defined by Nolan et al. [3], assuming there is no pulsar emission in the off-pulse phase. Our analysis thus gives 39% of the flux, which is then normalized to the full period.

We applied a maximum-likelihood method [4] to obtain flux estimates and to construct spectra with narrow energy bins equal to twice the FWHM of the COMPTEL energy resolution (5–10%). At high energies ($>$ 10 MeV) the limited number of events requires a larger binning, resulting in a total of 30 statistically independent bins. The likelihood analysis requires, for each energy interval separately, a careful estimate of the instrumental and isotropic background. In this analysis we applied a filter technique to the data space [5]. This filter eliminates, to first order, any source signature present. An iterative process then further corrects the background for the smeared-out source signature in each iteration.

In this work we use point spread functions (PSFs) from analytical modeling based on single-detector calibrations. PSFs from Monte Carlo simulations of the instrument are preferred, but not available yet for the narrow energy bins used here. A globally somewhat softer spectrum is expected with the simulated response (about 0.1 in spectral index).

RESULTS

Fig. 1 shows a power-per-decade spectrum of the off-pulse fluxes. Our data points are in good agreement with the COMPTEL spectrum presented by Much et al. [6], which was based on observations obtained during the first two years of the mission and contains 7 data points only. Statistical error bars are shown. The absolute calibration uncertainty is conservatively estimated to be 30%. Measurements from GRIS [7] and EGRET [8] (April 1991 – Aug. 1994) are added to place the COMPTEL points in a broader perspective.

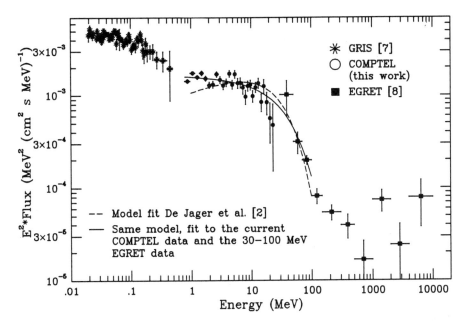

FIGURE 1. COMPTEL spectrum of the Crab nebula, together with GRIS low-energy data [7] and EGRET high-energy data [8]. Up to 10 MeV, the bin width of the COMPTEL data points is equal to twice the FWHM of the instrumental energy resolution. Between 10 and 30 MeV, broader energy bins were chosen. The two COMPTEL upper limits at the highest energies are not shown.

De Jager et al. [2] have modeled emission of the Crab nebula. They used an inverse-Compton (IC) component and a synchrotron component with an exponential cut-off to fit COMPTEL [6] and EGRET data (both April 1991 – May 1993):

$$dN/dE = K_s(E/3.5\text{MeV})^{-\Gamma_s} \cdot \exp(-E/E_0) + K_{\text{IC}}(E/1000\text{MeV})^{-\Gamma_{\text{IC}}}.$$

Their best-fit result is shown as a dashed line in Fig. 1 (limited to the 1–100 MeV range). Fig. 1 shows that the spectral break represented in the model begins in the COMPTEL domain. We have fit our data and the first 3 EGRET points to the first term of the equation. We ignore the IC component of the model, which is valid because the IC influence on the used data points is negligible. For the fit parameters we find: the energy cut-off $E_0 = 41 \pm 3$ MeV, $K_s = (1.29 \pm 0.04) \times 10^{-4}$ cm^{-2} s^{-1} MeV^{-1}, and $\Gamma_s = 2.02 \pm 0.03$, with a reduced χ^2 of 1.9 for 28 degrees of freedom (dof). The 1σ errors on the fit parameters have been calculated by increasing the χ^2_{min} with 3.5 [9]. The result can be seen in Fig. 1 (solid line). The cut-off energy is somewhat sensitive to the EGRET data points. Leaving all EGRET points out gives fit parameters $E_0 = 25 \pm 3$ MeV, $K_s = (1.39 \pm 0.04) \times 10^{-4}$ cm^{-2} s^{-1} MeV^{-1}, and $\Gamma_s = 1.96 \pm 0.03$, with a reduced χ^2 of 2.0 for 25 dof. For comparison, De Jager et al. [2], who included the COMPTEL calibration uncertainty in their fit but excluded the < 1 MeV COMPTEL point, found: $E_0 = 26^{+26}_{-9}$ MeV, $K_s = 1.25 \times 10^{-4}$ cm^{-2} s^{-1} MeV^{-1}, and $\Gamma_s = 1.74 \pm 0.42$.

The fit is not perfect, as indicated by the χ^2 value obtained above, but no obvious systematic trends can be seen. The deviations from the model fit are below the 3σ level in the individual energy bins. Monte Carlo simulations show similar deviations, although mostly at higher energies [10].

In order to obtain the cleanest data, we selected only 39% (the unpulsed fraction) of the available nebula events. However, as the emission from the nebula dominates the total emission from the Crab, we have used the set of all available events (pulsar contaminated) for our analysis as well. The result is shown in Fig. 2. A fit of the 0.78–10 MeV data points to a simple power-law spectrum gives a reduced χ^2_{18} of 4.5, which is a clear indication that the spectrum is more complicated. In particular, a feature at 1–2 MeV, for which some evidence could in fact already be seen in Fig. 1, seems more pronounced in the total Crab spectrum. We cannot exclude, however, that it is associated with the pulsar, if real. See also Much et al. in these proceedings. The BATSE (35–1700 keV) Cycle 1–3 total Crab spectrum [11], although higher than the GRIS and COMPTEL points [12], shows a deviation from a simple power-law spectrum, which can be interpreted as a rise near 1 MeV as well.

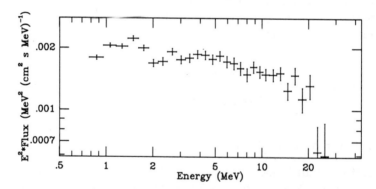

FIGURE 2. COMPTEL spectrum of the total Crab emission. Up to 10 MeV, the bin width is equal to twice the FWHM of the instrumental energy resolution. Between 10 and 30 MeV, broader energy bins were chosen. The highest energy upper limit is not shown.

CONCLUSIONS

COMPTEL observations from the first five years of the mission were combined to obtain and study a fine binned spectrum of the Crab supernova nebula. A reasonable fit to the off-pulse fluxes can be obtained with the synchrotron component of the model from De Jager et al. [2], giving a break energy of about 25–40 MeV. No significant evidence for line emission is found. An indication for a bump at 1–2 MeV cannot readily be explained by known systematic and statistical errors.

ACKNOWLEDGEMENTS

The COMPTEL project is supported by the German government through DARA grant 50 QV 90968, by NASA under contract NAS5-26645 and by NWO. R.D. van der Meulen is supported by the Netherlands Foundation for Research in Astronomy (NFRA) with financial aid from the Netherlands Organization for Scientific Research (NWO).

REFERENCES

1. Bartlett, L.M.: 1994, Ph.D. thesis "High Resolution Gamma-Ray spectroscopy of the Crab", Univ. of Maryland
2. De Jager, O.C., Harding, A.K., Michelson, P.F. et al., 1996, ApJ 457, 253
3. Nolan, P.L., Arzoumanian, Z., Bertsch, D.L. et al., 1993, ApJ 409, 697
4. De Boer, H. et al., 1992, In: Data Analysis in Astronomy IV, eds. Di Gesù et al., Plenum Press, New York, vol. 59, 241
5. Bloemen, H., Hermsen, W., Swanenburg, B.N. et al., 1994b, ApJS 92, 419
6. Much, R.P., Bennett, K., Buccheri, R. et al., 1995b, Adv. Space Res. 15(5), 81
7. Bartlett, L.M. et al., 1993, In: The Second Compton Symposium, eds. C.E. Fichtel, N. Gehrels, J.P. Norris, (AIP: New York), vol. 304
8. Fierro, J.M.: 1996, Ph.D. thesis "Observations of Spin-Powered Pulsars with the EGRET Gamma Ray Telescope", Stanford University
9. Lampton, M., Margon, B., Bowyer, S., 1976, ApJ 208, 177
10. Van der Meulen, R.D., Bloemen, H. et al., 1997, A&A, submitted
11. Ling, J.C. et al., 1997, ApJS, submitted
12. Much, R.P, Harmon, B.A., Nolan, P. et al., 1996, A&AS 120, 703

5 years of Crab Pulsar observations with COMPTEL

R.Much[4], K. Bennett[4], C. Winkler[4], R. Diehl[1], G. Lichti[1], V. Schönfelder[1], H. Steinle[1], A. Strong[1], M. Varendorff[1], W. Hermsen[2], L. Kuiper[2], R. van der Meulen[2], A. Connors[3], M. McConnell[3], J. Ryan[3], R. Buccheri[5]

[1] *Max-Planck Institut für Extraterrestrische Physik, P.O. Box 1603, 85740 Garching, F.R.G.*
[2] *SRON-Utrecht, Sorbonnelaan 2, NL-3584 CA Utrecht, the Netherlands*
[3] *Space Science Center, Univ. of New Hampshire, Durham NH 03824, U.S.A.*
[4] *Astrophysics Division, ESTEC, P.O. Box 299, NL-2200 AG Noordwijk, the Netherlands*
[5] *IFCAI/CNR, Piazza G. Verdi 6, 90139 Palermo, Italy*

Abstract. Using the COMPTEL data of the first 5 years of the CGRO mission we have derived the average pulsed spectrum of the Crab Pulsar, as well as phase-resolved spectra of the pulsed emission of the Crab Pulsar. The spectra in the COMPTEL energy range (0.75 to 30 MeV) are compared to those in the neighbouring energy bands. The pulsed flux has been examined for its stability. Pulsed lightcurves have been derived for different energy intervals. Preliminary results of this analysis are presented.

INTRODUCTION

For our analysis of COMPTEL data we have selected all Crab viewing periods (VPs) out of the first 5 years of the CGRO operation, where the Crab Nebula was within 30° of the pointing direction. In total there are 19 viewing periods, namely: 0, 1, 31, 36, 36.5, 39, 213, 221, 310, 321.1 321.5, 337, 412, 413, 419.1, 419.5, 420, 426, 502. This adds up to a total exposure of the Crab Nebula of $1.09 \cdot 10^9$ s cm^2 in the 1-3 MeV energy band.

ANALYSIS

The timing analysis was performed with the pulsar analysis subsystem of the COMPTEL Processing and Analysis Software System (COMPASS) using contemporary Crab radio observations (Arzoumanian et al. 1992). Lightcurves

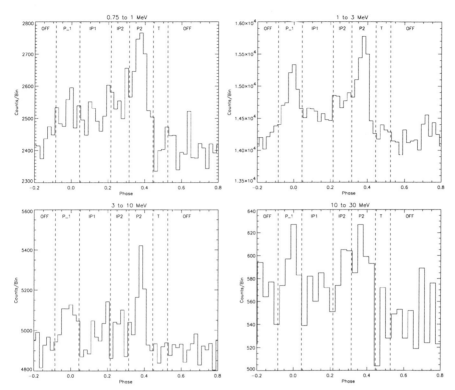

FIGURE 1. The Crab lightcurves of the combined observations for the four standard energy intervals.

were created for the combined observations by folding the photon arrival times with the pulsar period once they were transformed to the Solar System Barycentre.

We used a maximum likelihood ratio method to calculate the source flux with its associated statistical uncertainty (Bloemen et al. 1994). COMPTEL cannot resolve the Crab pulsar from the surrounding nebula. Therefore we used phase selection to disentangle the unpulsed and pulsed emission assuming that the unpulsed emission remains constant over the full pulsar period.

The pulsed emission Φ_{Δ_p} was determined by calculating the total (i.e. pulsed plus unpulsed) flux $(\Phi_{total})_{\Delta_p}$ in a pulsed phase interval Δ_p with subsequent subtraction of the unpulsed component $\Phi_{\Delta_{up}}$:

$$\Phi_{\Delta_p} = (\Phi_{total})_{\Delta_p} - \Phi_{\Delta_{up}} \cdot \frac{\Delta_p}{\Delta_{up}}$$

where Δ_p/Δ_{up} is the ratio of the width of the pulsed phase to the width of the unpulsed phase interval.

Phase-resolved photon spectra of the pulsed emission were calculated for the

Phase	Phase Interval	I_0 $10^{-4} ph/cm^2/MeV$	α	χ^2 (5 dof)
P1	0.915-0.445	2.59±0.41	-2.22±0.18	7.2
IP1	0.045-0.215	1.89±0.45	-2.65±0.35	1.7
IP2	0.215-0.315	2.73±0.50	-2.63±0.35	3.1
P2	0.315-0.445	4.57±0.43	-2.31±0.13	3.8
tot. pulsed	0.445-0.525	2.52±0.28	-2.37±0.15	1.6

TABLE 1. Results of powerlaw fitting of the 5-year averaged spectra. All fluxes are instantaneous.

smaller phase intervals indicated in Fig. 1. The phase definitions used earlier by Nolan et al. (1993) and Much et al. (1995) were adopted.

LIGHTCURVES

Figure 1 shows the lightcurves we obtained for the combined observations in the four COMPTEL standard energy intervals (0.75 to 1 MeV, 1 to 3 MeV, 3 to 10 MeV and 10 MeV to 30 MeV). The lightcurves show two peaks with a phase separation of 0.4. Compared to energies above 50 MeV (Nolan et al., 1993), there is relatively significant interpulse emission. The peak intensities change within the COMPTEL energy band relative to each other. Towards higher energies the intensity of the first peak increases relative to the second peak. This is to be expected considering the shape of the lightcurve measured above 30 MeV by EGRET, where the first peak dominates.

SPECTRAL ANALYSIS

All derived COMPTEL spectra (both total pulsed and phase-resolved), when fitted alone, can be described with a simple powerlaw, $I_0/1.65$ MeV$\cdot E^\alpha$. Our fit results are listed in Table 1.

Total Pulsed Spectrum

The total pulsed emission for the COMPTEL standard energy intervals was calculated for each VP. For the interval with the best statistics (1-3 MeV) the Crab pulsar's emission is shown as a funtion of time (TJD) in Fig. 2. The flux is compatible with a constant flux, except for the combination of VP 412+413, where we only received an upper limit. Further analysis of these VPs are required.

The spectrum of the total pulsed emission is shown in Fig. 3 as measured by

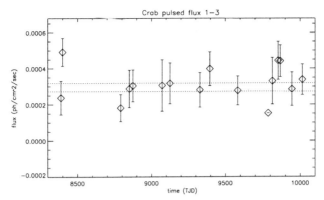

FIGURE 2. Instantaneous flux of the Crab pulsar in the 1-3 MeV energy band measured in the different VPs. The two horizontal lines indicate the ± 1σ uncertainty of the average flux in the entire period. The data point without error-bar is an upper limit (sum of VPs 412+413).

OSSE, COMPTEL and EGRET. However the OSSE, EGRET and COMPTEL spectra are not contemporaneous. Nevertheless the total pulsed spectra as measured by OSSE, COMPTEL and EGRET connect smoothly.

Phase-Resolved Spectra

The phase-resolved spectra for the first peak (P1), first and second interpulse (IP1 and IP2) and for the second peak (P2) were calculated. We compared the COMPTEL spectrum with the EGRET spectra obtained by Nolan et al (1993). The data have to be interpreted carefully, as the EGRET data are not contemporaneous with the COMPTEL data. However the constancy of the COMPTEL total pulsed flux (Fig 2.) and the smooth connection of the total pulsed spectrum (Fig 3.) suggest that a comparison is valid.
Although our results are preliminary (e.g. the P1 spectrum requires further analysis), the following conclusions can already be drawn from the phase-resolved spectra.

- the COMPTEL P2 spectrum smoothly connects to EGRET

- the IP1 and IP2 spectra clearly require a change in slope between the COMPTEL and the EGRET data

SUMMARY

- In the COMPTEL energy range the ratio of the intensities of the two main peaks of the lightcurve changes, namely, with increasing energy the first peak becomes more dominant relative to the second peak.

- The total pulsed emission can be described by a single powerlaw
- Combining the COMPTEL with the EGRET phase-resolved spectrum shows that
  - the phase-resolved spectrum of the second pulse connects smoothly to EGRET
  - the IP1 and IP2 spectra require a change in slope between the COMPTEL and the EGRET data

ACKNOWLEDGMENTS

The COMPTEL project is supported by the German government through DARA grant 50 QV 90968, by NASA under contract NAS5-26645 and by the Netherlands Organisation for Scientific Research (NWO).

REFERENCES

1. Arzoumanian Z., et al., 1992, GRO/radio timing data base, Princeton University
2. Bloemen, H., et al., 1994, A&AS 92, 419
3. Much, R., et al., 1995, A&A, 299, 435
4. Nolan, P., et al., 1993, ApJ, 409, 697
5. Ulmer, M., et al., 1994, ApJ, 432, 228

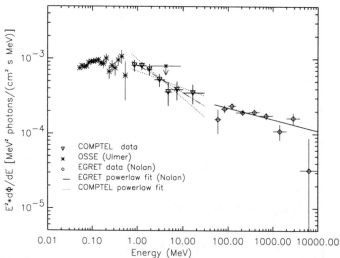

FIGURE 3. Instantaneous spectrum of the Crab total pulsed emission of the combined observations. The OSSE and EGRET spectra shown are not contemporaneous. Nevertheless the spectra of the three instruments connect smoothly.

The Infrared to Gamma-Ray Pulse Shape of the Crab Nebula Pulsar

Stephen S. Eikenberry and Giovanni G. Fazio

Harvard-Smithsonian Center for Astrophysics
60 Garden St.
Cambridge, MA 02138

Abstract.
We analyze the pulse shape of the Crab Nebula pulsar in the near-infrared, optical, ultraviolet, X-ray, and gamma-ray bands. We find that the phase separation between the two peaks of the pulse profile decreases nearly continuously as a function of energy over 7 decades of energy. We find that the differences between the energy dependences of the leading and trailing edge half-width half-maxima of both peaks found by Eikenberry et al. (1996a) also continue over 7 decades of energy. We show that the cusped shape of Peak 2 reverses direction between the infrared/optical and X-ray/gamma-ray bands, while the cusped shape of Peak 1 shows weak evidence of reversing direction between the X-ray and gamma-ray bands. These and many other pulse shape parameters are not predicted by current pulsar emission models, and offer new challenges for the development of such models.

I DATA

We selected the pulse profiles from a variety of sources. For the γ-ray band, we chose the OSSE pulse profile (Ulmer et al., 1994) from the Compton Gamma-Ray Observatory (CGRO) for its combination of energy coverage and high signal-to-noise. The X-ray pulse profile comes from the ROSAT HRI (Eikenberry and Fazio, 1997). The UV and optical pulse profiles come from Hubble Space Telescope High-Speed Photometer observations of the Crab Nebula pulsar (Percival et al., 1993), while the near-infrared pulse profiles come from the Solid-State Photomultiplier photometer on the MMT (Eikenberry et al., 1997b; Eikenberry, Fazio and Ransom, 1996).

II PEAK-TO-PEAK PHASE SEPARATION

Percival *et al.* (1993) were among the first to show that the phase separation between Peak 1 and Peak 2 changes with energy, and since then many authors have measured this effect in an attempt to understand its relation to and impact on the emission mechanism (i.e. Ransom *et al.* 1994; Ramanamurthy, 1994; Ulmer *et al.*, 1994). However, the techniques used to measure the phase separation have varied from author to author and from energy band to energy band, occasionally resulting in method-dependent biases. Here, we present a measurement of the peak-to-peak phase separation as a function of energy using the same techniques in all energy bands. The resulting separations and their uncertainties are presented in Figure 1.

The peak-to-peak phase separation appears to be a more or less smooth function of energy from IR to γ-ray energies. The separation decreases with energy over the range from 0.9 eV to 10^6 eV. However, there is some evidence of a turnover or break in this trend at $E = 0.7$ eV ($\lambda = 1.65\mu$m). While, Ulmer *et al.* (1994) have shown that the peak-to-peak separation shows no change with energy over the range ~ 50 keV to ~ 500 keV, their uncertainties of ~ 0.01 in phase are large enough to hide the effects that we see here.

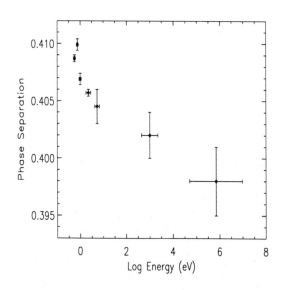

FIGURE 1. Peak-to-peak phase separation versus energy

III HALF-WIDTH HALF-MAXIMA

Eikenberry et al. (1996) found differences in the energy dependences of the peak half-widths for the leading and trailing edges of the IR-UV pulse profile peaks of the Crab Nebula pulsar. We perform similar analyses here to investigate whether such differences are consistently present across this larger energy range and to determine their form over this range. We present the HWHM measurements and their uncertainties for Peak 1 in Figure 2.

The HWHM results show a range of interesting characteristics in their energy dependences. First, we note that the HWHM energy dependences of the leading edge differs from the trailing edge. Second, we note that the difference between leading and trailing edge HWHM energy dependence is clearly visible in Peak 1 for the X-ray and γ-ray data points alone, confirming that these differences do indeed persist over the entire energy range. Finally, in Figure 2(a), we see that the leading edge half-width shows a distinct maximum in the IR at E=0.9 eV, which is either a peak in a smooth curve from 0.5 eV to 10^6 eV, or a break between 2 curves covering this range. The trailing edge HWHM (Figure 2(b)), on the other hand, shows a minimum in the UV (5.4 eV). Again, this may either be part of a smooth curve over the energy range, or evidence of a break between two curves.

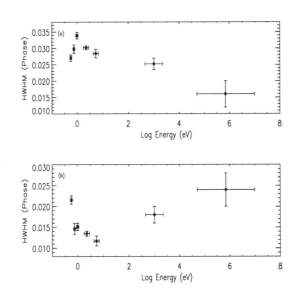

FIGURE 2. Peak half-width half-maximum versus energy for (a) Peak 1 leading edge, (b) Peak 1 trailing edge

IV PEAK ASYMMETRY INDEX

As noted in Eikenberry and Fazio (1997), we see evidence of a reversal in the cusped shape of Peak 2, from a short rise and long fall in the IR profile, to a long rise and short fall in the X-ray and γ-ray bands. In order to quantify this behavior, we introduce a new parameter which we call the "peak asymmetry index". The peak asymmetry index (PAI) is defined to be the logarithm of the ratio of the leading edge half-width to the trailing edge half-width, or

$$PAI = log_{10} \frac{HWHM(lead)}{HWHM(trail)}. \qquad (1)$$

As can be seen, a change in sign in the PAI means a reversal in the relative steepness of the leading and trailing slopes of the peak - from a faster rise than fall to a faster fall than rise, or vice versa. We present the PAIs for Peaks 1 and 2 in Figure 3. The Peak 1 PAI appears to be constant from 0.75 eV to 5.4 eV, with a significantly lower value at 0.5 eV and an apparently continuous decline from 5.4 eV to 10^6 eV. Note that all of the values are positive, except for the γ-ray data point. However, this point lies only $\sim 1.2\sigma$ below 0, and thus fails to provide convincing evidence of a shape reversal in Peak 1. On the other hand, such a reversal is clearly evident in the Peak 2 PAI. The IR-UV data points are all below 0 or consistent with negative values, while the X-ray and γ-ray data points are clearly positive. Thus we see that even some of the

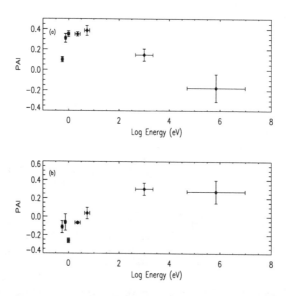

FIGURE 3. Peak asymmetry index (PAI) versus energy for (a) Peak 1, (b) Peak 2

large-scale properties of the "canonical" Crab Nebula pulsar pulse shape, such as a faster rise than fall in Peak 2, are in fact energy-dependent.

V CONCLUSION

Below we present a "scorecard" which shows how several emission models fare in comparison with the observations (for a detailed discussion see Eikenberry and Fazio, 1997). The models we compare are the Polar Cap Gap model (Daugherty and Harding, 1996), the 2-Gap Outer Gap model (Cheng, Ho, and Ruderman, 1986a,b), and the 1-Gap Outer Gap (Chiang and Romani, 1994; Romani and Yadigaroglu, 1995).

Emission Model Scorecard

Parameter	PC	2-Gap	1-Gap
Pk. Phase Sep.	Y	Y	?
HWHM Egy. Dep.	?	N	?
Rapid Spectral Var.	?	N	Y
Pk2 Reversal	N(?)	N(?)	?
IR Turnovers	?	?	?
Pk1/Pk2 Asymmetry	N	Y	Y

Key:

- Y - Yes, this feature is expected from the model
- N - No, this feature does not fit with model expectations
- ? - This feature is not predicted by the model, but it does not necessarily contradict the model

REFERENCES

1. Cheng, K.S., Ho C., and Ruderman, M. 1986a, ApJ, 300, 500
2. Cheng, K.S., Ho C., and Ruderman, M. 1986b, ApJ, 300, 500
3. Chiang, J. and Romani, R.W. 1994, ApJ, 436, 754
4. Daugherty, J.K. and Harding, A.K. 1996, ApJ, 458, 278
5. Eikenberry, S.S., et al. 1997, ApJ, 477, 465 .
6. Eikenberry, S.S. and Fazio, G.G. 1997, ApJ, 476, 281 .
7. Eikenberry, S.S., et al., 1996, ApJ, 467, L85 .
8. Percival, J.W., et al. 1993, ApJ, 407, 276
9. Ramanamurthy, P.V. 1994, A&A, 284, L13
10. Romani, R.W. and Yadigaroglu, I.-A. 1995, ApJ, 438, 314
11. Ulmer, M.P., et al. 1994, ApJ, 432, 228

Observation of the Crab Pulsar with BeppoSAX: study of the pulse profile and phase resolved spectroscopy

G. Cusumano*, D. Dal Fiume[†], S. Giarrusso*,
E. Massaro[§], T. Mineo*, L. Nicastro[†], A.N. Parmar[‡], A. Segreto*

*Istituto di Fisica Cosmica e Appl. Informatica, CNR, Palermo
[†] Istituto TeSRE, CNR, Bologna
[§] Ist. Astr., Unità GIFCO Roma 1, Univ. "La Sapienza", Roma
[‡] Astrophysics Division, SSD-ESA, ESTEC, Noordwijk

Abstract. The Crab Pulsar (PSR B0531+21) was observed by the four Narrow Field Instruments on board the italian-dutch satellite BeppoSAX in August and September 1996, during the Science Verification Phase. The fine time resolution (15 μs) and the high statistics of the data provided phase histograms of very good quality, well suited for phase resolved spectroscopy over the entire energy band (0.1-300 keV) covered by BeppoSAX. In this contribution we present preliminary results of the phase resolved analysis. Moreover, we also carried out a detailed analysis of the behaviour of the P2/P1 ratio with the energy.

INTRODUCTION

The Crab pulsar (PSR B0531+21) is the most observed spin powered pulsar since about thirty years from radio to γ rays. Many problems, however, concerning the physics and the geometry of the emission are still unresolved. The broad energy band and the high throughput of the BeppoSAX instruments allow a detailed study of the pulse profile and an accurate spectral analysis, useful to understand the physical processes occurring in the neutron star magnetosphere. The fine time resolution and the high statistics provided phase histograms of very good quality, well suited for phase resolved spectroscopy and pulse shape analysis. In this contribution we present some preliminary results of the Beppo-SAX observation of this source carried out during the Science Verification Phase.

INSTRUMENTS AND OBSERVATIONS

The italian-dutch satellite BeppoSAX observed the Crab pulsar (PSR B 0535+21) with all the four Narrow Field instrument (NFI) in 1996 August-September during the Science Verification Phase. These instruments are: the Low Energy Concentrator Spectrometer (LECS) operating in 0.1-10 keV range (Parmar et al. 1997), the Medium Energy Concentrator Spectrometer (MECS) which consists of three units operating in the 1-10 keV range [2], the High Pressure Gas Scintillation Proportional Counter (HPGSPC) operating between 4 and 120 keV [3] and the Phoswich Detector System (PDS), with four detection units, operating in the 15-300 keV energy band [4]. A description of the mission and the payload is given by Boella et al. [5]. The total exposure times were $\sim$9690 s for the LECS, $\sim$33481 s for the MECS (September 6–7), $\sim$12487 s for HPGSPC and $\sim$14130 s for the PDS(August 31 – September 1). Data reduction was performed with the standard procedures and selection criteria to avoid disturbances due the South Atlantic Anomaly crossing and to the sun and bright Earth radiation.

We computed the spectra of the LECS and MECS data considering only spatially selected events within circular regions with diameters of 16 and 10 arcmin, respectively; the PDS and HPGSPC net signals were obtained by subtracting the on-off collimator position count rates. Response matrices available to all observers after the 31th of December were used.

X-RAY PULSE PROFILES

The pulse profiles of PSR B 0531+21 in six energy bands from 0.1-300 keV with a phase resolution from 300 (0.11 ms) to 50 bins are shown in Fig. 1. These histograms were obtained by folding the photon arrival times, after conversion to the Solar System Barycenter, with the pulsar period derived from the radio ephemeris provided by A.G. Lyne and R.S.Pritchard (Jodrell Bank Crab Pulsar Timing Results). We took the zero phase at the centre of the main peak (P1).

The energy dependence of the well known double peaked structure is very clearly evident in this profile series with a very high statistical significance. The increase of the second peak (P2) intensity with respect to that of P1 is a well known phenomenon, (see Massaro, Feroci and Matt 1997 [6] for a compilation of all published X- and γ-ray pulse profiles). Notice also that the same trend is also well apparent in the interpeak (Ip) region. We evaluated for the intensity ratios between the two main peaks (P2/P1) and the interpeak to P1 (Ip/P1) for several profiles in different energy ranges. The off-pulse level, defined as the mean value in the phase interval (0.47,0.77) was subtracted to every bin content before computing the ratios. The phase boundaries of P1 and P2 and Ip were (-0.05,0.05), (0.27,0.47), (0.05–0.27), respectively. The

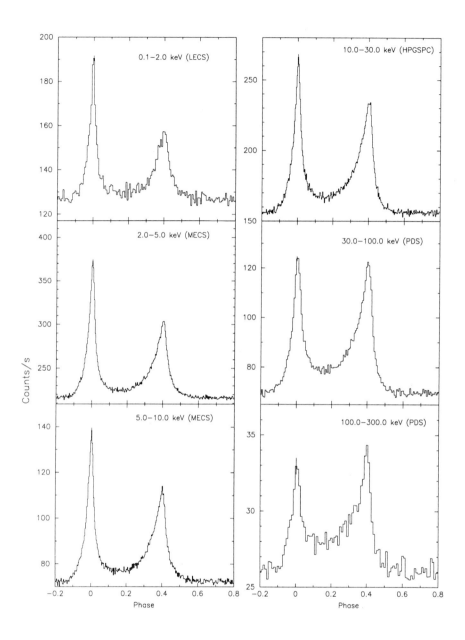

FIGURE 1. Phase histograms of PSR 0531+21 (Crab) observed with the NFI of Beppo-SAX in six energy ranges from 0.1 to 300 keV.

resulting values of P2/P1 and Ip/P1 as a function of the photon energy are plotted in Fig. 2.

SPECTRAL ANALYSIS

The results of the spectral analysis are still preliminary because the intercalibration of the NFI is not complete. We obtained the spectra for eachone of the three phase intervals given above, after subtraction of the mean off–pulse spectrum. Spectral shapes were modeled with a single power law absorbed at low energies: the column density N_H was estimated using only the LECS data

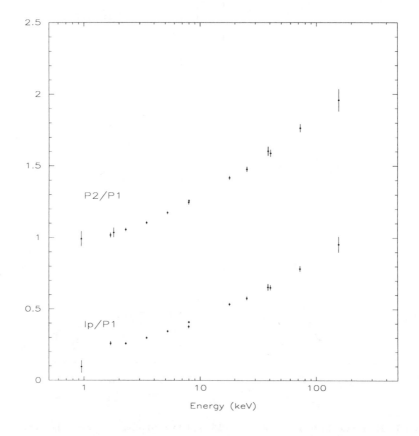

FIGURE 2. The P2/P1 (upper data) and Ip/P1 (lower data) ratios *vs* Log E (keV) measured by all Beppo-SAX NFI

and resulted equal to $3.334\pm0.002\ 10^{21}$ cm^{-2}, this value was after used to fit the MECS spectra. The spectral indices of P1, P2 and Ip in the MECS range (2 – 10 keV) were 1.80 ± 0.02, 1.68 ± 0.04 and 1.53 ± 0.15, respectively, whereas that of the total pulsed signal was 1.70 ± 0.021. The uncertainties correspond to 90% confidence level for a single parameter; the reduced χ^2 values ranged from 0.90 (P2) to 1.17 (P1) with 189 d.o.f.. All these results agree very well with the previous OSO–8 measures [7]. The fits in the PDS energy range (30 – 300 keV) give the following values of the spectral indices 2.02 ± 0.02 (P1, $\chi_r^2=0.95$, 48 d.o.f.), 1.88 ± 0.02 (P2, $\chi_r^2=1.72$), 1.72 ± 0.05 (Ip, $\chi_r^2=0.96$) and 1.88 ± 0.02 (Total, $\chi_r^2=1.09$). Notice that, while the PDS spectral index of P1 is practically coincident with that found by OSSE [8], that of P2 is softer of 0.12 and that of Ip of 0.21. These discrepancies can be partially explained by the different phase interval chioce, but it is more likely that they depend on the different energy ranges because the spectra of the various components seem, in fact, to show a smooth stepeening with increasing energy.

REFERENCES

1. Parmar,A.N. et al., 1997, AA 320.
2. Boella,G. et al., 1997, AA 320.
3. Manzo,G. et al., 1997, AA 320.
4. Frontera,F. et al., 1997, AA 320.
5. Boella,G. et al., 1997, AA 320.
6. Massaro,E., Feroci,M., Matt.,G., 1997, AAS in the press.
7. Pravdo,S.H., Serlemitsos,P.J., 1981, ApJ 246, 484.
8. Ulmer,M.P. et al., 1994, ApJ, 432, 228.

The Spectrum of TeV Gamma Rays from the Crab Nebula

J.P.Finley[*], S.Biller[†], P.J.Boyle[‡], J.H.Buckley[||], A.Burdett[†],
J.Bussons Gordo[‡], D.A.Carter-Lewis[¶], M.A.Catanese[¶],
M.F.Cawley[§], D.J.Fegan[‡], J.A.Gaidos[*], A.M.Hillas[†],
F.Krennrich[¶], R.C.Lamb[**], R.W.Lessard[*], C.Masterson[‡],
J.E.McEnery[‡], G.Mohanty[¶], J.Quinn[‡], A.J.Rodgers[†],
H.J.Rose[†], F.W.Samuelson[¶], G.H.Sembroski[*], R.Srinivasan[*],
T.C.Weekes[||], M.West,[†], J.A.Zweerink[¶]

[*] *Purdue University, U.S.A.,* [†] *University of Leeds, United Kingdom,* [‡] *University College, Dublin, Ireland,* [||] *Whipple Observatory, Harvard-Smithsonian CfA, U.S.A.,* [¶] *Iowa State University, U.S.A.,* [§] *St.Patrick's College, Maynooth, Ireland,* [**] *Space Radiation Lab, Caltech, U.S.A.*

Abstract. The Crab Nebula has become established as the standard candle for TeV gamma-ray astronomy. No evidence for variability has been seen. The spectrum of gamma rays from the Crab Nebula has been measured in the energy range 500 GeV to 8 TeV at the Whipple Observatory. Two methods of analysis involving independent Monte Carlo simulations and two databases of observations (1988-89 and 1995-96) were used and give close agreement. Using the complete spectrum of the Crab Nebula, the spectrum of relativistic electrons is deduced and the spectrum of the resulting inverse Compton gamma-ray emission is in good agreement with the measured spectrum if the ambient magnetic field is $\sim$ 25-30 nT.

I INTRODUCTION

The Crab Nebula has become a standard candle in TeV astronomy; it has been detected by many groups and its integral flux appears constant. The Crab Nebula is also well on the way to becoming a standard candle with regard to TeV spectral content. Synchrotron emission from the Crab covers a remarkably broad energy range terminating at about 10^8 eV, where a new component attributed to inverse Compton scattering begins. It is this component that we detect. The TeV spectrum is sensitive to the primary electron spectrum, the nebular magnetic field and the spatial distribution of electrons and magnetic field within the nebula.

In this paper we briefly describe two methods for extracting TeV spectra, compare results from the Whipple Observatory Imaging Cherenkov Telescope for the 1988/89 and 1995/96 observing seasons and comment on implications for the physics of the nebula. The methods are described in detail in "Paper I," [8], and the resulting TeV spectrum is put into context of other observations with implications for the physics of the nebula in "Paper II," [6]. The latter paper also compares our Crab TeV spectrum with those of other groups.

II METHODS

A straightforward approach to the determination of TeV spectra was developed at Iowa State University. Several components are required. First, a method of distinguishing gamma-ray images from background cosmic-ray images. The standard method is to use "supercuts" [10]. More than 99% of the background is rejected by the appropriate choice of the "supercuts" parameters. However, this procedure results in a strong bias against the images of higher energy gamma rays which tend to be longer and broader and hence more cosmic-ray like. A modified procedure [7] incorporating the total brightness (*size*) of the image has been employed to circumvent this bias. A second component is an estimate of the energy of each gamma-ray image with (a) good resolution and (b) negligible bias. We obtained a resolution of $\Delta E/E \sim 0.36$ with negligible bias by using a second order polynomial in *size* and *distance* as described in Paper I. The energy resolution function is, to a good approximation, Gaussian in the variable $\log(E)$.

In data taken for spectral analysis, each on-source observation is followed by an "off-source" observation covering the same range of azimuth and elevation angles. The images for both types of observation are selected for gamma-like events and estimated energies from corresponding observations are histogrammed and the difference ascribed to gamma rays from the source. This data then constitutes the observed energy spectrum and analysis methods are applied to extract the underlying source spectrum [7].

A different approach with emphasis on verifiability was developed at the University of Leeds. There are two aspects to this approach; a method for selecting images likely to have been initiated by cosmic gamma rays, and a method for determining the primary gamma-ray energy spectrum from the observed *size* spectrum [6,13]. The selection criterion is a "cluster" or "spherical" method in which a single parameter is used to characterize the gamma-ray-like nature of an image and correlations between image parameters are incorporated naturally.

In order to extract a spectrum, a *size* histogram is then computed for on-source and off-source observations and the difference histogram is ascribed to gamma rays. A simulated *size* spectrum is then computed. A weight is given to each simulated gamma ray and, by adjusting these weights, the spectrum is varied so that its *size* spectrum matches that of the difference histogram [8].

III RESULTS

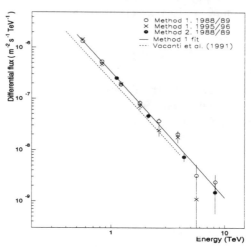

FIGURE 1. The Crab spectrum in the range 0.3 to about 8 TeV extracted using Methods 1 (open circles) and 2 (solid circles) for the Whipple 1988/89 database and using Method 1 (x's) for the 1995/96 database are shown above. Also shown is a fit to the combined Method 1 results (solid line) as well as an earlier spectrum (dashed line) taken from Vacanti *et al.* (1991).

The spectra obtained from the 1988/89 and 1995/96 seasons using Method 1 and 1988/89 season using Method 2 are in good agreement (see Figure 1). We have tested the sensitivity of the results to uncertainties in the Monte Carlo simulations and have found that the results are relatively robust [8]. The combined TeV fluxes for both seasons from Method 1 are well fit with a simple power law spectrum in which the differential flux ($J(E)$) is given by:

$$(3.3 \pm .2 \pm .7)\, 10^{-7} \left(\frac{E}{\text{TeV}}\right)^{-2.45 \pm .08 \pm .05} \text{m}^{-2}\,\text{s}^{-1}\,\text{TeV}^{-1} \quad (1)$$

where the first error is statistical and the second is our estimated systematic error. A simple extrapolation of this fit to lower energies, however, yields a flux far in excess of that observed by EGRET (see Figure 2). A quadratic fit of $\log(J)$ vs. $\log(E)$ to our data and a point representing the average EGRET flux at 2 GeV [9] is:

$$J(E) = (3.25)\, 10^{-7}\, (E/\text{TeV})^{-2.44-0.135\log_{10}(E)}\ \text{m}^{-2}\,\text{s}^{-1}\,\text{TeV}^{-1}. \qquad (2)$$

IV COMMENTS ON INTERPRETATION

Most of the early inverse Compton models for TeV gamma rays assumed a constant magnetic field in the principal source region where these are produced [4,11] whereas more recent models [3,1] incorporate hydrodynamic plasma/field flow making the calculations more complex and the results probably more realistic. Here, we make the simple assumption that the field is constant. The broad synchrotron emission band apparently extends up to 10^8 eV and is boosted to higher energies via inverse Compton scattering. The scattering giving rise to TeV gamma rays occurs in the Klein-Nishina rather than in the Thomson scattering regime. This implies that the electrons giving rise to our detected gamma rays must have energies in the range of 2-10 TeV and the corresponding scattered photons would mostly have energies of about 0.005 to 0.3 eV. This conclusion is only very weakly model dependent (see Paper II). Since electrons with energies of a few TeV generate synchrotron radiation at about 0.4 keV in a field of about 25 nT (see next paragraph), the Einstein Observatory X-ray images [5] of the Crab Nebula also reveal that part of the nebula which is emitting TeV gamma rays.

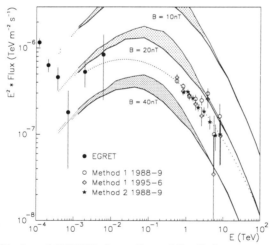

FIGURE 2. Whipple and EGRET observations of the Crab gamma-ray spectrum. The dotted line is a fit to the WHipple points and a single 2 GeV flux value. Full-line curves are predicted inverse Compton fluxes for 3 different assumed B fields. The shaded regions reflect uncertainties arising from the UV to soft X-ray region of the synchrotron spectrum

For assumed magnetic field values and the observed synchrotron flux, it is possible to deduce the spectrum of the primary electrons, presumably generated in the shock at the termination of the pulsar wind [2]. From the ambient photon density and the deduced electron spectrum, the TeV flux can be calculated, and results for B fields of 10, 20 and 40 nT are shown in Figure 2. As can be seen from the figure, the effective B field must lie between 20 and 40 nT with 27 nT falling very near the TeV data. Since even more energetic electrons have a short radiative lifetime, they should exist only near the terminus of the pulsar wind shock. Hence, measurement of the TeV spectrum over a wider energy range may probe spatial variations in the nebular magnetic field (see Paper II).

This work is supported by grants from the US DOE and NASA, by PPARC in the UK and by Forbairt in Ireland.

REFERENCES

1. Aharonian, F.A. and Atoyan A.M, Astropart. Phys. **3**, 275 (1995)
2. Coroniti, F.V. and Kennel, C.F., in "The Crab Nebula and related supernova remnants," ed. M.C.Kafatos and R.B.C. Henry, CUP, p 25 (1985).
3. De Jager, O.C., and Harding, A.K., Ap. J., **396**, 161 (1992).
4. Gould, R.J., Phys. Rev. Lett., **15**, 577 (1965).
5. Harnden, F.R.,Jr. and Seward, F.D., Ap. J. **283** 279 (1984).
6. Hillas, A.M., et al, "Paper II," to be submitted to Ap. J. (1997).
7. Mohanty, G., PhD Thesis Iowa State University (1995).
8. Mohanty, G., et al., "Paper I," Astroparticle Physics, submitted (1997).
9. Nolan, P.L. et al., Ap. J., **409**, 697 (1993).
10. Punch, M., et al., Proc. 22nd ICRC Dublin **1** 464 (1991).
11. Rieke, G.H., and Weekes, T.C., Ap. J., **155**. 429 (1969).
12. Vacanti, G., et al., Ap. J. **377**, 467 (1991).
13. West, M., PhD Thesis University of Leeds (1994).

First Stereoscopic Measurements at the Whipple Observatory

Frank Krennrich

Physics & Astronomy Department, Iowa State University, Ames, IA, 50011
for the Whipple Collaboration[1]

Abstract.
Observations of photons at $E \geq 550$ GeV from the Crab Nebula are presented. In contrast to previous observations these data have been taken with a system of two imaging atmospheric Čerenkov telescopes. The Whipple Observatory 10m and 8m γ-ray telescopes have been used to provide a stereoscopic view of air showers and therefore, the data presented here contain a more complete measurement of air shower parameters. The arrival direction of primary γ-rays has been reconstructed unambiguously using the stereoscopic view of two telescopes with an accuracy of $\sigma = 0.14°$. The relevance for future multiple-telescope installations operating in this energy range will be discussed.

INTRODUCTION

The ground-based imaging atmospheric Čerenkov technique has succeeded to detect TeV-photons from several galactic and extragalactic TeV γ-ray sources [1]. At lower energies results from the Energetic Gamma-Ray Experiment Telescope (EGRET) on the Compton GRO have established the field of GeV photon astronomy through the detection of 129 sources [2]. However, the present observations at GeV energies do not yet provide answers to many basic questions regarding the origin of the observed radiation. It is expected that

---

[1] D. A. Carter-Lewis, M.A. Catanese, G. Mohanty, *Iowa State University*; C. W. Akerlof, M. S. Schubnell, *Laboratory of Physics, University of Michigan, Ann Arbor, MI 48109-1120* ; J. H. Buckley, K.H. Harris, T. C. Weekes, *Fred Lawrence Whipple Observatory, Harvard-Smithsonian CfA, P.O. Box 97, Amado, AZ 85645-0097*; M. F. Cawley, *Physics Department, St.Patrick's College, Maynooth, County Kildare, Ireland*; J. Bussóns-Gordo, V. Connaughton, D. J. Fegan, J. Quinn, *Physics Department, University College, Dublin 4, Ireland* ; J. P. Finley, J. A. Gaidos, G. H. Sembroski, C. Wilson, *Department of Physics, Purdue University, West Lafayette, IN 47907*; A. M. Hillas, A. J. Rodgers, H.J. Rose, *Department of Physics, University of Leeds, Leeds, LS2 9JT, Yorkshire, England, UK*; R. C. Lamb *Space Radiation Laboratory, California Institute of Technology*; M. J. Lang, *Department of Physics, University College, Galway, Ireland*

next generation GeV satellite experiments such as GLAST [3] and ground-based counterparts such as VERITAS [4], MAGIC [5] and STACEE [6] at energies E $\geq$ 50 GeV will address most unsolved questions about relativistic processes in those astrophysical objects more efficiently.

Because of their huge collection areas ($10^4 - 10^5 m^2$), the energy region above 50 GeV remains the domain of ground-based Čerenkov telescopes. The imaging atmospheric Čerenkov technique employs large optical light collectors that focus the Čerenkov light emitted by secondary particles of an air shower onto an array of fast photomultipliers covering a 3°-5° field of view. The two-dimensional image of the Čerenkov light distribution in the focal plane is used to separate γ-ray induced showers from a background of cosmic-ray induced showers. Čerenkov images generated by γ-rays are characterized by narrower widths and lengths compared to cosmic-ray images, and for a point source are aligned parallel to the telescope optic axis, resulting in small angles between the image major axis and the line joining the image centroid and the center of the field of view (the 'alpha' parameter) [7].

Although substantial improvements in sensitivity have been made by second generation Čerenkov detectors such as Whipple [8] and HEGRA [9], new efforts are underway to improve the technique further to lower the energy threshold and increase the sensitivity. High resolution focal plane detectors are employed in the CAT telescope [10] and will be used in the so-called GRANITE III project [11] of the Whipple Collaboration. A different approach is to measure γ-ray showers with a system of two or more telescopes as described by [12] and [13] . The stereoscopic view of γ-ray showers could potentially improve the atmospheric imaging technique regarding:

- Sensitivity
- Accuracy of Measuring Air Shower Parameters
- Angular Resolution
- Energy Resolution

Stereo detectors may also play an important role in lowering the energy threshold ([14], [13]). A large scale imaging atmospheric Čerenkov detector with 9 telescopes operating at E$\geq$50 GeV (VERITAS) has been proposed and is motivated by the reasons given above [4]. The Whipple collaboration has pursued the stereo approach with a 10m-8m stereo system to test the performance in the few hundred GeV region and first results are presented here.

STEREO DETECTOR

The stereo detector used for the observations described here consists of the 10m and an 8m telescope (formerly called the 11m telescope and put in operation in December 1995) located on the southern ridge of Mount Hopkins near Amado, Arizona, at a latitude of 31° 41' and an altitude of 2300 meter. It is important to point out that the performance of the stereo technique depends on the particular configuration and the distance spacing of the telescopes used. The 8m telescope (146m apart from the 10m telescope) does not match the optical quality of the 10m reflector. The point spread function of the 8m reflector is broader by a factor of 2 than for the 10m. However, since the point spread function of the 8m telescope (0.22°) is still smaller than the pixel spacing in the focal plane (0.25°), the somewhat reduced optical resolution causes a marginal degradation of the imaging quality of the instrument. The dispersive mirror reflectivity of the 8m telescope has its primary reflectivity in the range 330 - 470 nm. The energy threshold of the 10m (250 GeV) and 8m (550 GeV) telescope system has been estimated using Monte Carlo simulations and is 550 GeV for coincident γ-ray showers. A more detailed description of the stereo detector can be found in [15].

OBSERVATIONS

The data presented here consists of observations of the Crab Nebula. Since the γ-ray emission from the Crab Nebula is steady and bright, it has become the standard candle in ground-based γ-ray astronomy, and is ideal to test the performance of a new stereo system. Sixteen ON-source measurements (28 minutes each) under clear sky conditions have been recorded during October and November 1996. Since we are testing a new technique, corresponding OFF-source runs (28 minutes each) at the same elevation and azimuth region as the ON-source runs have been carried out to provide a reliable background estimate. The total live time during the ON-source observations is 5.4 hours. The observations were carried out with both telescopes operating independently. For each telescope, the contemporaneous data have been recorded individually and marked with an absolute time accuracy of 0.25 ms, provided by two GPS clocks.

ANALYSIS

Air shower coincidences are found in the off-line analysis by matching the time of the events in both telescopes, accepting them as coincident if their recorded times differ by less than 0.5 ms.

TABLE 1. Selection criteria used for stereo events

Parameter	Lower Bound 10m	Upper Bound 10m	Lower Bound 8m	Upper Bound 8m
width:	0.07°	0.17°	-	-
length:	0.16°	0.34°	-	-
distance:	0.51°	1.1°	0.30°	0.85°
size:	200 d.c.	-	100 d.c.	-

Further selection criteria shown in Table 1 have been applied.

RESULTS

We have derived the arrival direction of air showers for 5.4 hours of Crab Nebula observations using the intersection of the major axes of the 10m and the 8m images. The intersection point in a common coordinate system corresponds to a point of origin on the sky. Figure 1 shows the results of the two-dimensional analysis of the arrival direction for coincident, candidate γ-ray events from stereoscopic observations of the Crab Nebula.

The gray scale in the histogram indicates the number of excess events (ON-source - OFF-source) passing the γ-ray selection criteria of Table 1 and a cut on the angular distance Θ from the putative source position. The search for an excess has been carried out on a $3° \times 3°$ wide grid system corresponding to the telescopes field of view on a sky. The grid system is divided into cell sizes of $0.1°$. The cut on Θ corresponds to a radial search window determined by the angular resolution of the stereo detector (see also [16]). Assuming a uniform background the optimum significance is obtained by using a search window of 1.59 σ (assuming a two dimensional Gaussian function) which keeps 72 % of the γ-ray events. From Monte Carlo simulations is has been found that $\sigma = 0.14°$ [15]. The contours show the likelihood ratio test statistic in

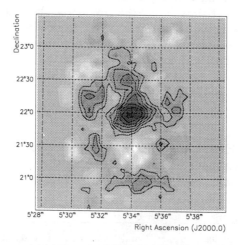

FIGURE 1. The two-dimensional reconstruction of the arrival direction of air showers showing the γ-ray sky in the direction of the Crab Nebula. It can be seen that a clear signal (7.9 σ) is present in the data. The peak is consistent with the direction of the Crab Nebula indicated by the cross.

1σ steps based on the method of Li and Ma [17] as derived from the number of counts ON-source and OFF-source. For this analysis the significance is 7.9 σ (85 excess events). The excess is consistent with the source position as indicated by the cross in Figure 1. The angular resolution for the two-dimensional analysis is consistent with $\sigma = 0.14°$, determined by varying the bin size to obtain the maximum significance.

CONCLUSIONS

We have tested the angular resolution that can be achieved with a system of two medium resolution imaging Čerenkov telescopes. It has been demonstrated that by using the intersection of the major image axes of the Čerenkov light images of γ-ray induced air showers an angular accuracy of $\sigma = 0.14°$ can be reached. Extrapolating this result for a telescope array like VERITAS where 9 telescopes with a pixel spacing of $0.12°$ - $0.23°$, an accuracy of $\sigma = 0.02°$ - $0.04°$ could be obtained. A limit to this extrapolation occurs due to the shower development itself [18] and is at about $0.02°$.

REFERENCES

1. Weekes T.C., et al., *Review paper, Proc. of 4th Compton Symp.*, this volume (1997a).
2. Thompson D.J., et al., *ApJS* **101**, 259 (1995).
3. Gehrels N., et al., *Proc. of 4th Compton Symp.*, this volume (1997).
4. Weekes T.C., et al., *Proc. of 4th Compton Symp.*, this volume (1997b).
5. Lorenz E., et al., *Proc. of 4th Compton Symp.*, this volume (1997).
6. Ong R., et al., *in Towards a Major Atmospheric Čerenkov Detector-IV,* , ed. M. Cresti, 261 (1995).
7. Reynolds P.T., et al., *ApJ* **404**, 206 (1993).
8. Cawley M.F., et al., *Exp. Astr.* **1**, 173 (1990).
9. Panter M., et al., *Proc. of 24th Int. Cosmic Ray Conf., Rome* **1**, 958 (1995).
10. Punch M., et al., *in Towards a Major Atmospheric Čerenkov Detector-IV,* , ed. M. Cresti, 356 (1995).
11. Lamb R.C., et al., *in Towards a Major Atmospheric Čerenkov Detector-IV,* , ed. M. Cresti, 386 (1995).
12. Aharonian F.A., et al., *in Towards a Major Atmospheric Čerenkov Detector-II,* ed. R.C. Lamb, 81 (1993).
13. Hillas A.M., *Space Science Reviews* **75**, 17 (1996).
14. Krennrich F., & Lamb R.C., *Exp. Astr.* **6**, 285 (1995).
15. Krennrich F., et al., *Astroparticle Physics* (in preparation).
16. Akerlof C.W., et al., *ApJ* **377**, L97 (1991).
17. Li T., & Ma Y., *ApJ* **272**, 317 (1983).
18. Hillas A.M., *Very High Energy Gamma Ray Astronomy, Crimea* 134 (1989).

The "COS–B/EGRET 1997" Geminga Ephemeris

J.R. Mattox*, J.P. Halpern[†], P.A. Caraveo[#]

Department of Astronomy, Boston University, Boston, MA 02215
[†]*Department of Astronomy, Columbia University, New York, NY 10027*
[#]*Istituto di Fisica Cosmica del CNR, Via Bassini, 15, 20133 Milano, Italy*

Abstract. We derive an ephemeris for the Geminga pulsar using all available EGRET, COS–B, and SAS–2 data spanning 24 yr. A cubic ephemeris predicts the rotational phase of Geminga with errors smaller than 50 milliperiods. The braking index obtained is 17±1. If Geminga continues to rotate without glitch as it has for at least 23 yr, we expect this ephemeris to continue to describe the phase with an error less than 100 milliperiods until the year 2008. Statistically significant timing residuals are detected in the EGRET data with an RMS value ~20 milliperiods. The addition of a sinusoidal term to the ephemeris reduces the residuals to an RMS value of less than 6 milliperiods. This sinusoid corresponds to a planet of mass $1.7/\sin i\ M_\oplus$ orbiting Geminga at a radius of 3.3 AU, although these residuals could be caused by timing noise rather than a planet.

INTRODUCTION

Although the periodicity of Geminga was initially found in ROSAT X-ray data (Halpern & Holt 1992), much more precise timing can be done with EGRET because the ROSAT exposures are short, the soft X-ray peaks are broad, and their modulation is shallow. An ephemeris for the rotation of Geminga based on EGRET observations spanning 2.1 yr was published by Mattox et al. (1994). Subsequently, an ephemeris for observations spanning 3.9 yr was published (Mattox, Halpern, & Caraveo 1996).

New observations now extend the baseline of EGRET observations to 5.9 yr. This long baseline allows the rotation parameters of Geminga to be sufficiently constrained so that the rotation phase during EGRET observations can be compared to the phase during COS–B observations. We thus obtain a cubic ephemeris which describes the rotation of Geminga from the beginning of SAS–2 observations (1973.0) to the end of the most recent EGRET observation (1997.2). The details of the derivation of this ephemeris are provided by Mattox, Halpern, & Caraveo (1997).

I THE DERIVATION OF THE EPHEMERIS

As described by Mattox et al. (1994, 1996), the ephemeris parameters are estimated as the values which give the largest value of the Z_{10}^2 statistic (i.e., the most non-uniform light curve). The recent determination of the time dependent position of Geminga to within $\sim$40 mas (Caraveo et al. 1997) is an order of magnitude better than required for this timing analysis. The position at the epoch of an HST observation and the proper motion given in Table 1 are used for our analysis.

We initially analyzed the EGRET data alone. With the new observations (concluding on 18 March 1997), the EGRET observations now span 5.9 yr. The downhill simplex method was used to simultaneously estimate the f and $\dot{f}$ which produced a maximum Z_{10}^2 statistic for various $\ddot{f}$. The cubic ephemeris thus obtained for EGRET with $T_0 = $ JD 2448750.5 is: $f = 4.21766909394(3)$, $\dot{f} = -1.95226(1) \times 10^{-13}$, and $\ddot{f} = 8.0(8) \times 10^{-25}$. The 95% confidence uncertainty of the last digit of each parameter is indicated by the digit in parenthesis. The corresponding braking index, $\eta = f\ddot{f}/\dot{f}^2 = 89 \pm 9$, is much higher than the value of $\eta = 3$ expected for magnetic dipole radiation. This discrepancy is discussed below.

The same analysis was done for the COS−B data alone. With $T_0 = $ JD 2443946.5, the cubic ephemeris is: $f = 4.2177501227(1)$, $\dot{f} = -1.95239(2) \times 10^{-13}$, and $\ddot{f} = 4.5(2.0) \times 10^{-25}$. This implies a braking index, $\eta = 50 \pm 20$, consistent with that of Hermsen et al. (1992). Because the accurate position and proper motion of Caraveo et al. (1997) were used for this analysis, we can reject the hypothesis of Bisnovatyi-Kogan & Postnov (1993) that the large value of $\ddot{f}$ reported by Hermsen et al. (1992) is due to proper motion. Another explanation is proposed in §4.

We note that the COS−B and EGRET values for $\ddot{f}$ are inconsistent. Furthermore, the value obtained from the two measurements of $\dot{f}$,

$$\ddot{f} = \frac{\dot{f}_{\text{EGRET}} - \dot{f}_{\text{COS-B}}}{T_{0_{\text{EGRET}}} - T_{0_{\text{COS-B}}}} = 3.1(5) \times 10^{-26}, \qquad (1)$$

although consistent with a braking index of 3, is not consistent with either the COS−B value for $\ddot{f}$ nor with the EGRET value. These discrepancies can be attributed to timing noise. However, they make the search for a timing solution which would coherently connect COS−B and EGRET problematic. Notwithstanding, having coherent solutions for 6.7 yr of COS−B data and 5.9 yr of EGRET data, it seemed plausible that one could find a coherent solution which would bridge the 9.0 yr between these observations. In order to compare the rotational phase of Geminga during the COS−B observations to the EGRET phase, we have updated the COS−B barycenter from that of the MIT PEP 740 solar system ephemeris to that of the JPL DE 200 ephemeris as described by Mattox et al. (1997).

In the search for a coherent COS−B/EGRET timing solution, the downhill simplex method was used to estimate $f, \dot{f}$, and $\ddot{f}$. An epoch near the center of the COS−B/EGRET time-span was chosen. The search was initiated at ∼1000 different initial values of these three parameters to attempt to find a global maximum in the midst of hundreds of local maxima. After several days of processing, a solution was found which gave a value of Z_{10}^2, 3283, which is distinctly larger than the next best value, 3043. The cubic ephemeris thus obtained for COS−B/EGRET with $T_0 = JD2446600$ is:

$$f = 4.217705363090(13)$$
$$\dot{f} = -1.9521717(12) \times 10^{-13}$$
$$\ddot{f} = 1.48(3) \times 10^{-25} \qquad (2)$$

The 95% confidence uncertainty of the last digit of each parameter is indicated by the digit in parenthesis.

Unlike the local maxima, equation 2 produces a light curve for each observation which is shown by Mattox et al. (1997) to be constant to within 50 milliperiods in phase and invariant in shape after allowing for variable instrument response and statistical fluctuation. Also, the 81 SAS−2 Geminga events (E > 100 MeV) yield a phase which is consistent with COS−B and EGRET when epoch folded with equation 2. This consistency in phase leads Mattox et al. (1997) to conclude that it is very unlikely that a glitch of the Geminga pulsar larger than $\Delta\nu/\nu < 5 \times 10^{-10}$ has occured between 1973 and 1996.6. Equation 2 is therefore thought to be an accurate description of the rotation of Geminga from the time of the SAS−2 observation through the EGRET observations. Because of this long timing interval, the parameters of this cubic ephemeris are much more precisely determined than for previous Geminga ephemerides.

However, Mattox et al. (1997) find evidence of deviation from this cubic ephemeris with a significance of 5×10^{-7}. They find that the timing residuals, shown in Figure 1, are well represented by a sinusoid with a amplitude of 6.2±0.9 ms and a period of 5.1±0.1 yr. This modulation is consistent with that expected from a planet of mass $1.7/\sin i\ M_\oplus$ orbiting Geminga at a radius of 3.3 AU. However, with EGRET observations covering only ∼1 potential period, and with only modest precision in the determination of the values of the timing residuals, it is possible that this is simply timing noise. The values of $f, \dot{f}$, and $\ddot{f}$ in Table 1 where optimized along with the orbital parameters. They differ from those in equation 2 by less than the 1σ uncertainty of each parameter, reflecting the fact that the orbital term is a very minor modification of the timing solution.

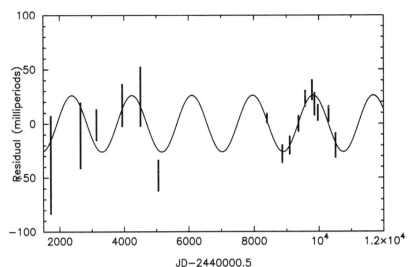

FIGURE 1. The Geminga timing residuals for SAS−2, COS−B, and EGRET. The timing residuals relative to the cubic ephemeris of Table 1 excluding the sinusoidal term are show with error bars which demarcate the 68% confidence ranges.. The sinusoidal term in Table 1 is shown with the continuous line.

Epoch, $T_0 = 2446600$ JD
(1986 June 18.5 Barycentric Dynamical Time)
Frequency at epoch, $f = 4.217705363081(13)$ Hz
Frequency derivative, $\dot{f} = -1.9521712(12) \times 10^{-13}$ Hz s^{-1}
2nd frequency derivative, $\ddot{f} = 1.49(3) \times 10^{-25}$ Hz s^{-2}
Position at JD 2449794, $\alpha_{2000} = 6^h\ 33^m\ 54^s.153$, $\delta_{2000} = +17°\ 46'\ 12''.91$
Proper motion: 169 mas yr^{-1} at position angle 54°
($\mu_{\alpha_{2000}} = 138$ mas yr^{-1}, $\mu_{\delta_{2000}} = 97$ mas yr^{-1})
Possible binary term:
Projected semi-major axis, $a_1 \sin i = 6.2(9)$ light ms
Orbital period, $P_b = 5.1(1)$ yr
Epoch of periastron passage, $T_b = $ JD 2449360(20)
Longitude of periastron, $\omega = 90°$
Eccentricity, $e = 0.0(4)$
Assuming $M_1 = 1.4\ M_\odot$:
$a_2 = 3.31(4)$ AU
$M_2 \sin i = 1.7(2)\ M_\oplus$
Peak one occurs 0.556(2) of a rotation after T_0.

TABLE 1. The "COS−B/EGRET 1997" ephemeris for Geminga. The digit in parenthesis following the derived parameters is the ∼95% confidence uncertainty of the last digit. Peak one precedes the strongest emission bridge (see Figure 1 of Mattox et al. 1997).

II DISCUSSION

It is now clear why the optimal cubic ephemeris for the EGRET data alone implies a braking index of 90. The large second derivative is a partial fit to the timing residuals of Figure 1. The extended timing baseline which COS−B provides allows a braking index of 90 to be ruled out. We expect that the large second derivative in the COS−B data alone is also caused by this effect. The braking index of 17±1 implied by the coherent solution probably reflects timing noise as well.

If the apparent sinusoidal modulation is timing noise, Mattox et al. (1997) derive a timing noise activity parameter of −0.54. If it is a planet, an upper limit on the activity parameter is −1.2. Either is feasible for a pulsar with the period and period derivative of Geminga (Cordes 1993). If further timing observations establish the reality of the Geminga planet, it would be the second confirmed pulsar planet (joining ms pulsar PSR B1257+12, Wolszczan 1994), and the first confirmed planet around a "slow" pulsar.

J. Mattox acknowledges support from NASA Grants NAG 5-3384 and NAG 5-3806, and J. Halpern from NAG 5-2051.

REFERENCES

1. Bisnovatyi-Kogan, G. S., & Postnov, K. A., 1993, Nature, 366, 663.
2. Caraveo, P. A., Lattanzi, M. G., Massone, G., Mignani, R., Makarov, V. V., Perryman, M. A. C., & Bignami, G. F. 1997, ApJL, submitted.
3. Cordes, J. M. 1993, in "Planets Around Pulsars," ed. J. Phillips, S. Thorsett, & S. Kulkarni, ASP Conf Ser, 36, 43.
4. Halpern, J. P., & Holt, S. S. 1992, Nature 357, 222.
5. Hermsen, W. et al. 1992, IAU Circ., No. 5541.
6. Mattox, J. R., et al. 1994, Proc 2nd *CGRO* Symp, AIP Conf. Proc. #304, 77.
7. Mattox, J.R., Halpern, J.P., Caraveo, P.A., 1996, A&A S, 120, C77.
8. Mattox, J.R., Halpern, J.P., Caraveo, P.A., 1997, ApJ, in press.
9. Wolszczan, A. 1994, Science, 264, 538.

On the Accurate Positioning of Geminga

P.A. Caraveo[1], M.G. Lattanzi [2], G. Massone [2], R. Mignani [3],
V.V. Makarov [4], M.A.C. Perryman[5] and G.F. Bignami [6,1]

[1] *Istituto di Fisica Cosmica del CNR, Via Bassini, 15, 20133 Milano, Italy*
[2] *Osservatorio Astronomico di Torino, 10025 Pino Torinese, Italy*
[3] *Max-Plack-Institute für Extraterrestrische Physik, Garching, Germany*
[4] *Copenhagen University Observatory, 1350 Copenhagen K, Denmark*
[5] *Astrophysics Division, ESTEC, 2200AG Noordwijk, The Netherlands*
[6] *Agenzia Spaziale Italiana, Via di Villa Patrizi 13, Roma, Italy*

Abstract. Accuracy in the absolute position in the sky is one of the limiting factors for pulsar timing, and timing parameters have a direct impact on the understanding of the physics of Isolated Neutron Stars (INS).This is particularly true for Geminga, the only example of radio silent neutron star known so far. We have combined the Hipparcos and Tycho catalogues, ground-based astrometric data,and Hubble Space Telescope (HST) Wide Field Planetary Camera (WFPC2) images, to yield for the $m_v = 25.5$ optical counterpart of Geminga a ~ 40 *mas* (per coordinate) uncertainty. Such a positional accuracy, unprecedented for the optical position of an object this faint, is needed to combine in phase γ-ray photons collected over more than 20 years, i.e. over 2.5 billions of the star's revolutions.

POSITION VS. TIMING

To properly analyze the timing signature of a rapidly varying source, such as a pulsar, the photon arrival times must be corrected for the motion of the detector, be it on the ground or onboard a satellite. Arrival times are corrected for the light travel time they would have needed to reach the barycentre of the solar system and the procedure, routinely applied before any time analysis at any wavelength, is called barycentrization. Such a correction is maximum for sources located near the ecliptic plane, while its magnitude decreases for sources at high ecliptic latitude. Any error on the source coordinates would affect the accuracy of the corrected arrival times and thus limit the precision potentially achievable for the measure of the source timing parameters. Indeed, the position of a radio pulsar is obtained by minimizing its timing residuals over a reasonably long time span. However, a number of factors, such as the pulsar period (fast pulsars are positioned with higher accuracy than slow ones), timing noise and glitching activity, come into play to limit the positional accuracy achievable with radio data.

The Importance of Optical Observations

Precise optical positioning of pulsar counterparts can give a significant contribution to improve the accuracy of the timing analysis and to reduce uncertainties in the determination of the pulsar ν and $\dot{\nu}$. This is a crucial step to nail down the pulsar frequency second derivative, aiming at the measurement of its braking index $n = \ddot{\nu}\nu/\dot{\nu}^2$. This quantity, expected to be 3 for magnetic dipole braking, has been measured so far for young objects such as Crab (Lyne et al. 1988), PSR 0540-69 (Guiffes et al.1992), PSR 1509-58 (Kaspi et al.,1994), and PSR 0833-45 (Lyne et al. 1996).
Table 1 lists the pulsar parameters relevant to the braking index measurement, namely: the value of the period and its derivative, the accuracy achieved in the measure of such derivative as well as on the objects' coordinates, the characteristic age, the frequency second derivative and the braking index itself. Table 1 also shows the importance of accurate optical positioning for the determination the timing parameters of pulsars affected by significant timing noise, such as the Crab and Vela pulsars, or too faint to be accurately positioned in radio, such as PSR 0540-69. Indeed, the only object relying on radio coordinates is PSR 1509-58, the optical identification of which is tentative (Caraveo et al., 1994).

The Case of Geminga

Optical positioning is all the more important for Geminga, so far a unique example of radio quiet neutron star (Caraveo, Bignami and Trümper, 1996) The radio silence of the source makes it impossible to access the information on its position and distance as inferred from the optimization of the radio timing parameters and from the dispersion measure of the radio pulses. Thus, to measure position, distance and timing parameters of Geminga all other branches of astronomy had to be exploited in a 20 y long chase, recently summarized by Bignami and Caraveo (1996).
Our knowledge of the Geminga pulsar is now good enough to warrant, "honoris causa", inclusion in the radio pulsar catalogue of Taylor, Manchester and Lyne (1993). Indeed, the radio silent Geminga stands out among normally behaving radio pulsars for the remarkable accuracy achieved in the measure of its parameters. Inspection of the pulsar catalogue shows that the period derivate of Geminga is known with an accuracy greater than that of the Crab (see Table 1). This is mainly due to the very stable behaviour of this 10^5 y

TABLE 1. Pulsars with measured braking index $n = \nu\ddot{\nu}/\dot{\nu}^2$.

Name	P (s)	$\dot{P}(s/s)$	$\Delta\dot{P}(s/s)$	$\Delta\alpha(sec)$	$\Delta\delta('')$
PSR0531+21	0.033	4.2 10^{-13}	2 10^{-19}	0.005	0.06[1]
PSR1509-58	0.151	1.5 10^{-12}	1 10^{-19}	0.089	1
PSR0540-69	0.050	4.8 10^{-13}	8 10^{-19}	0.029	0.49[2]
PSR0833-45	0.089	1.2 10^{-13}	2 10^{-17}	0.019	0.29[3]
Geminga	0.237	1.1 10^{-14}	1.4 10^{-19}	0.003[4]	0.040[4]

Name	Age (y)	n	$\ddot{\nu}$
PSR0531+21	1,260	2.5 ± 0.01	1.2 10^{-20}
PSR1509-58	1,545	2.8 ± 0.01	1.9 10^{-21}
PSR0540-69	1,670	2.0 ± 0.2	3.7 10^{-21}
PSR0833-45	11,350	1.6 ± 0.3	3.5 10^{-23}
Geminga	3.5 10^5	??	??

[1] McNamara, 1971; [2] Caraveo et al, 1992; [3] Manchester et al, 1978; [4] present work

old neutron star which does not seem to be affected by the glitching activity typical of younger objects, nor by significant timing noise. Indeed, Mattox, Halpern and Caraveo (1996) claim that, during the first 3 years of EGRET coverage, every pulsar revolution can be accounted for. Thus, the γ-ray photons, collected in week-long observing periods taken several month apart over a span of years, can be aligned in phase to form the very spiky light curve seen in high energy γ-rays. Had Geminga been a glitching pulsar, like Vela or Crab, it would have been impossible to obtain a satisfactory (and accurate) long-term solution. Indeed, the growing high-energy coverage offers the possibility to further exploit the stability of Geminga to refine the knowledge of the source timing parameters, including its second period derivative and, thus, its braking index. However, to take full advantage of the potentialities offered by γ-ray astronomy, a very accurate value of the source absolute coordinates is required to barycentrize the arrival time of each photon. This procedure is particularly critical for Geminga, which is close to the ecliptic equator, where the barycentric correction is maximum. Mattox et al. (1996) have shown that the uncertainty in the source absolute positioning can induce an error in the barycentric correction up to 2.3 δ_e msec, where δ_e is the source positional uncertainty (in arcsec) projected on the plane of the ecliptic. Thus, with a sensitivity to phase errors of 10^{-2} over a period of 237 msec, the positional accuracy of 1", presently available, is the limiting factor for Geminga's photons timing analysis.
An improvement of one order of magnitude in absolute position is mandatory to phase together SAS-2, COS-B and EGRET data.

ABSOLUTE COORDINATES

The highest resolution images of Geminga have been obtained with the Planetary Camera onboard HST (see Bignami et al, 1996 for a reproduction of the image). However, the HST ability to provide absolute coordinates is limited by the accuracy of the Guide Star Catalogue. To transform the PC coordinates, measured very precisely in the HST images, into absolute ones, we have developed an ad hoc precedure based on the newly available the Hipparcos reference frame (ESA,1997). However, to tie Geminga into the Hipparcos system, one has to overcome the 16-18 magnitude gap between the bright stars used as a primary reference and our $m_v = 25.5$ target. Moreover, one has to link astrometric images covering the field of view of ~ 1 square degree to those obtained by the PC instrument, with a 14,000 times smaller field of view. The principle of our method, fully described in Caraveo et al, (1997), is to cover the field of interest with images of increasingly smaller field of view and deeper limiting magnitude. This was accomplished with a program of dedicated astrometric measurements carried out at the Torino Observatory which, in conjunction with Hipparcos and HST data, yielded for Geminga a position (epoch=1995.21)
$\alpha = 6^h 33^m 54.1530^s$, $\delta = 17°46'12.909"$.
with a standard error in each coordinate of 40 mas.

CONSEQUENCES

Among the pulsars seen as high energy emitters, Geminga's positional accuracy is the best so far, better than that of the 9-mag brighter Crab pulsar. This improved position, in conjunction with the HST measure of the proper motion (Caraveo et al., 1996), will allow the accurate barycentrization of all the γ-ray data collected so far.
It will thus be possible to measure the source second derivative and, hence, its braking index. The task is a challenging one: for a braking index of 3, a $\ddot{\nu}$ of $2.7 \ 10^{-26} s^{-3}$ is expected. This value is three order of magnitude smaller than that recently measured for Vela and 6 order of magnitude smaller than that of the Crab (see Table 1). However, it appears to be within reach of the combined SAS-2, COS-B and EGRET data.
What makes Geminga suitable for the measurement of such a tiny value of the frequency second derivative? If we order known pulsars according to their increasing values of $\ddot{\nu}$ (expected for a braking index of 3), Geminga does not come out prominently. No less than 40 pulsars have hypothetical $\ddot{\nu}$ bigger than the value expected for Geminga. However, for all of them, these values are not measurable because the errors on ν and $\dot{\nu}$ are too big. What singles out Geminga is the possibility to reduce such errors by phasing together 20 years worth of γ-ray data, now that the source positional accuracy is not a

limiting factor. Thus, once again, Geminga appears to be in a special position amongst Isolated Neutron Stars. What matters most here is the source instrinsic stability, which made it worthwhile to devote a dedicated effort to the accurate measurement of its absolute position.

ACKNOWLEDGMENTS

We are grateful to the Hipparcos and Tycho collaborations who made the positions available to us prior to the official release of the catalogues.

REFERENCES

Bignami, G.F., and Caraveo, P.A. 1996, *A.R.A.A.* 34,331
Caraveo P.A., Bignami G.F., Mereghetti S., and Mombelli M. 1992, *Ap.J.* 395,L103
Caraveo P.A., Mereghetti S., Bignami G.F. 1994, *Ap.J.* 423,L125
Caraveo P.A., Bignami G.F., Trümper, J.A. 1996,*A & A. Review* 7, 209
Caraveo P.A., Bignami G.F., Mignani R., and Taff L.G. 1996, *Ap.J.* 461,L91
Caraveo P.A.et al 1997, ESA SP 402, in press
ESA 1997, *The Hipparcos and Tycho Catalogues* ESA SP 1200
Guiffes C, Finley JP, Ögelman H 1992, *Ap.J.* 394,581
Kaspi V.M., Manchester R.N., Siegman B., Johnston S., Lyne A.G. 1994, *Ap.J.* 422, L83
Lyne A.G., Pritchard R.S., Smith F. 1988, *M.N.R.A.S.*, 233,667
Lyne A.G., Pritchard R.S., Graham-Smith F., and Camilo F. 1996, *Nature* 381,497
Manchester R.N., et al. 1978, *M.N.R.A.S.* 184, 159
Mattox J.R., Halpern J.P. and Caraveo P.A. 1996, *A.& A. Suppl.* 120C,77
McNamara B.J. 1971, *P.A.S.P.* 83,491
Taylor,J., Manchester, R.M. Lyne A.G. 1993, *Ap.J.Suppl.* 88,529

Studies of the Gamma Ray Pulsar Geminga

S.Zhang ,T.P.Li,M.Wu,W.Yu,X.J.Sun,L.M.Song,F.J.Lu

*High Energy Astrophysics Lab.,Institute of High Energy Physics,
P.O.BOX 918-3,100039,Beijing,China*

Abstract. Ephemeris parameters of gamma ray pulsar Geminga have been revised using EGRET. Timing and imaging analyses show that the majority emission of Geminga are in phase region around the first peak in 10-30 MeV during COMPTEL VP1. No turn off of spectrum appear at energies below 30 MeV.

INTRODUCTION

After the confirmation by ROSAT[1] and EGRET[2] , gamma ray pulsar Geminga has been made full analysis using EGRET[3] and COMPTEL. Emission of Geminga in individual COMPTEL observations has only been found in VP0[4] . We have reanalyzed COMPTEL individual observations data using a direct demodulation method[5] . Timing analysis of Geminga has also been made with EGRET. We present in this paper the main results.

OBSERVATION AND ANALYSIS METHOD

From the beginning of the all-sky survey of CGRO Geminga has been observed several times by EGRET and COMPTEL during Phase I,II and III. Data from five observations(VP0.3-0.5, 1.0, 2.5, 221.0, 310.0) have been chosen for the analyses in this paper.

The direct demodulation method owns the ability of reconstructing the objects from incomplete and noisy data due to its principle of solving the modulation equations iteratively under physical constraints. This method has already been successfully applied to scan observation data of slat collimator telescopes, e.g. the scan imaging for CygX-1 by the balloon-borne hard X-ray collimatted telescope HAPI-4[6] and the reanalysing result for EXOSAT-ME galactic plane survey[7]. It can also be applied to analyse observation

TABLE 1. Used definition of phase intervals of Geminga.

Region	Symbol	Phase Interval
Peak1	P1	0.15-0.3
Interpulse1	I1	0.3-0.65
Interpulse2	I2	0.8-0.15
Peak2	P2	0.65-0.8

data from rotating modulation telescope[8], coded-mask aperture telescope[9] or COMPTEL.

RESULTS

1. Parameters revision of Geminga

Phase motions of Geminga were found from EGRET VP221 and VP310 relative to VP1 with the parameters (epoch $T_0=8400$ (TJD)) reported by Mayer-Hasselwander[10]. In our analysis, events were chosen with energies above 70MeV, within a cone of radius less than $\theta=5.85°[E_\gamma/100 \text{ MeV}]^{-0.534}$. The phase motions were calculated from the maximum of the following phase cross correlation function(CCF)

$$CCF(i) = \sum_{m=1}^{k} a_m b_{i+m} \qquad (1)$$

where k is the number of channels (k=200), a_m and b_{i+m} are the counts of the m and the i+m channel in two phases. The phase motions $\Delta\phi$ for VP221 and VP310 were obtained as 0.07±0.007 and 0.11±0.009. The phase motions have not appeared when the same data were processed using another set of parameters with more precise value of $\dot{f}$ and zero value of $\ddot{f}$ (epoch $T_0=8750(\text{TJD}))^{11}$. Thus indicates the $\Delta \dot{f}$ caused phase motions and the possible neglection of $\ddot{f}$. $\Delta \dot{f}$ and $\ddot{f}$ were then derived as $(3.48\pm0.26)\times10^{-17}$ Hz s^{-1} and $(-1.952252\pm0.000026)\times10^{-13}$ Hz s^{-1} (epoch $T_0 = 8400(\text{TJD})$). Although the revised parameters of Geminga have been adopted for the analyses in this paper, it is necessary to announce here the equality of the following results using parameters given by Mattox[11].

2. Timing and Imaging Analyses

COMPTEL VP1 VP221 and VP310 have been chosen for timing analysis. With the revised parameters Geminga light curve was got from EGRET VP1 and the phase intervals definition is listed in Table 1 and indicated in Figure 1(a).

Visible signal was found only in VP1 in energy range 10-30 MeV and the significance was about 2.1σ (for 10 channels). Figure 1(b) may indicate a

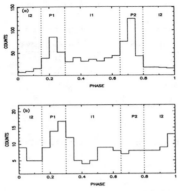

FIGURE 1. The Light Curve of Geminga Pulsar. (a) The phase histogram for the γ-ray emission from Geminga above 200 MeV observed by EGRET VP1. (b) The phase histogram for the γ-ray emission from Geminga in 10-30 MeV observed by COMPTEL VP1. Events satisfying $8°\leq\bar{\varphi}\leq 30°$ were used.

majority emission in phase region around peak1 for VP1. In order to confirm the result, we have made image analysis on the phase-resolved data of Geminga using a direct demodulation technique.

The image results of the phase-resolved data from COMPTEL VP1 VP221 and VP310 were similiar to that of timing analysis. Image of Geminga was only obtained in VP1 in phase region (0.15-0.3) in the energy range 10-30 MeV ($8°\leq\bar{\varphi}\leq 18°$)(see Figure 2). The result shows clearly the existance of Geminga (with flux of $(1.48\pm 0.41)\times 10^{-5}$ ph cm^{-2} s^{-1}) and the exactly resolved Crab nebula (with flux of $(0.74\pm 0.26)\times 10^{-5}$ ph cm^{-2} s^{-1}). The total flux of Crab nebula summed from four phase regions is about 7.63×10^{-5} ph cm^{-2} s^{-1}. The disappearance of PKS0528+134 in Figure 2 is due to its poor statistics in only 0.15 phase region and the excess of structure in this skymap may reflect the uncontinuous portion of the background in the observed data.

Approximately took the observed emission as the total emission, the energy spectrum of Geminga is shown in Figure 3. The data points at energies above 30 MeV and the fitted line in this figure were the results of EGRET[3]. The result shows no suppression at energies even below 30 MeV.

Variation of intensity ratio peak 2/peak 1 with energy for VP1 is shown in Figure 4. The data points above 30MeV which were the results of EGRET were fitted as $b(E/100MeV)^{\alpha}$ and the upper limit of peak 2/ peak1 obtained from COMPTEL VP1 is consistent with the fitted feature. Similiar fitting of EGRET data have also been done for VP0 (0.3-0.5) and VP2.5. The time evolution of the index α during these three subsequent observations indicates the variation factor of α for about 2.7±1.5 from VP0 to VP1.

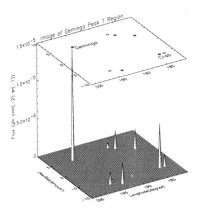

FIGURE 2. The direct demodulation map of Geminga phase region (0.15-0.3) for COMPTEL VP1.

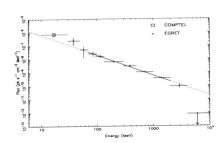

FIGURE 3. The spectrum of Geminga

DISCUSSION

Geminga has been analysed in details by Mayer-Hasselwander[3] with EGRET. Their result of no suppression of Geminga spectrum are confirmed by our analysis of COMPTEL. The upper limit of peak 2/peak 1 of Geminga for COMPTEL VP1 follows the trend extended from EGRET. The γ-ray emission region of Geminga observed by COMPTEL varied from interpulse 1 for VP0[4] to peak 1 for VP1, while the index α of peak 2/peak 1 of Geminga observed by EGRET in these two periods show a trend of increasing. Relationships may exist between these two phenomena.

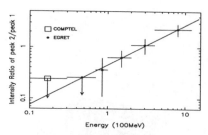

FIGURE 4. Variation of intensity ratio peak 2/peak 1 with energy for VP1

ACKNOWLEDGEMENT

The authors are grateful to Dr. W.Collmar of COMPTEL Team at MPE for providing the data, especially for his assistance in the understanding of the instrument and the data.

REFERENCES

1. Halpern J.P. and Holt S.S., *Nature* **357**, 222 (1992)
2. Bertsch D.L. et al., *Nature* **357**, 306 (1992)
3. Mayer-Hasselwander H.A. et al., *ApJ* **421**, 276 (1994)
4. Hermsen W., et al., *ApJSS* **92**, 559 (1994)
5. Li T.P. and Wu M., A direct method for spectral and image restoration, in: Worrall D.M., Biemesderter C. & Barnes J., eds., *Astronomical Data Analysis Software and System I., A.S.P. Conf. Ser.* **25**, 229(1992)
6. Lu Z.G. et al., *Nucl. Instr. and Meth. in Phys. Res. Sec. A* **362**, 551(1995)
7. Lu F.J., Li T.P., Sun X.J., Wu M. & Page C.G., *A&AS* **115**, 395(1996)
8. Chen Y., Li T.P. & Wu M., Direct demodulation technique for rotating modulation collimator imaging, *Exper. Astron.*, (1996)(submitted)
9. Li T.P., *Exper. Astron.* **6**, 63(1995)
10. Mayer-Hasselwander H.A. et al., *IAU Circ.* No.5649(1992)
11. Mattox J.P., et al., *Proc. AIP Conference* **304**, 77(1993)

Timing analysis of four years of COMPTEL data on PSR B1509-58

A. Carramiñana[1], K. Bennett[2], W. Hermsen[3], L. Kuiper[3],
V. Schönfelder[4], A. Connors[5], V. Kaspi[6], M. Bailes[7],
R.N. Manchester[8]

[1] *INAOE, Luis Enrique Erro 1, Tonantzintla 72840, MEXICO*[1]
[2] *Astrophysics Division, European Space Research and Technology Centre, Noordwijk, The Netherlands*
[3] *Space Research Organization Netherlands, Utrecht, The Netherlands*
[4] *Max-Planck Institute für Extraterrestrische Physik, Garching, Germany*
[5] *Space Science Center, University of New Hampshire, Durham NH, USA*
[6] *IPAC, California Institute of Technology, Pasadena CA 91125, USA*
[7] *Physics Department, University of Melbourne, Australia*
[8] *Australia Telescope National Facility, CSIRO, Epping, Australia*

Abstract. Time series analysis have been performed in search for the periodic signal of PSR B1509-58 in the COMPTEL database. We have marginal, and self-consistent, evidence for emission in the (1.0-3.0) MeV energy band. The pulse agrees both in phase and shape with that observed by OSSE and GINGA.
The lack of pulsations in the (3.0-10.0) MeV range is in conflict with an extrapolation of the γ-ray spectrum observed by OSSE, indicating a spectral break for photon energies above 3 MeV. This is consistent with γ-ray attenuation in the strong magnetic field at the vicinity of the neutron star.

THE YOUNG PULSAR PSR B1509-58

PSR B1509-58 is a young pulsar with a period of 150ms, somewhat higher than one would expect for its dynamical age, 1550 yrs. It has a particularly high inferred magnetic field at the surface of the neutron star, $B_* \simeq 3.08 \times 10^{13}$ G. The radio observations have also shown that this pulsar is highly stable, allowing a first measurement of $\ddot{\nu}$ in a pulsar [7].
PSR B1509-58 was originally found as a bright X-ray pulsar inside the X-ray nebula MSH 15-52 [12]. Its pulsations have been clearly seen in hard X-rays: GINGA showed a pulse with a ~ 0.25 offset in phase relative to the radio

---

[1]) supported through CONACYT grant 4142-E9404

pulse [8], and a hard spectral index of 1.3±0.05 throughout the (2-60) keV energy range; BATSE showed light curves from its lowest (20-40) keV energy range, up to (230-740) keV and, at marginal signal-to-noise ratio, even for ($E \geq 1.8$ MeV) [14], with a mean spectral index of 1.6±0.1 [15]; finally, OSSE has good signal-to-noise light curves up to 360 keV [13], with a pulse offset by 0.32 ± 0.02 with respect to the radio pulse an a spectral index of 1.68 ± 0.09.

In the medium γ-ray energy ranges, COMPTEL reported a detection in its lowest (0.75-1.0) MeV band during a single observation [2], this being the third pulsar found in the COMPTEL spectral window during the All-Sky Survey [5]. At high γ-ray energies, PSR B1509-58 has not been detected by EGRET [3], in spite of its relatively high sensitivity and success in pulsar detections.

COMPTEL DATA ON PSR B1509-58

Previous COMPTEL results were only based on three observations made during the first year of CGRO data. Here we report on timing analysis of four years of data, as show in Figure 1. COMPTEL data for timing analysis are selected assuming a perfect response of the instrument, i.e. a single scatter in the D1 layer and an absorption at D2 (for a description of the instrument see [11]). Most of the timing data available up to CGRO Viewing Period 424 were taken during Phase 3 and Cycle 4. The database comprises 918938 events in the (0.75-30) MeV range, spans 1376 days of which 130 correspond to the actual data, selected from observations where the pulsar was at $\leq 30°$ from the centre of the field of view. The effective exposure time, after correcting for Earth blockage in the COMPTEL data space, passages through the South Atlantic anomaly and other restrictions, amounts to about 20 days.

Due to the intense induced background and, for Compton telescopes, the azimuthal ambiguity of the arrival direction of γ-rays, the MeV regime is

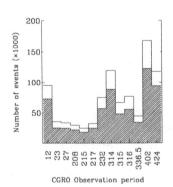

FIGURE 1. Distribution of COMPTEL timing events on each CGRO pointing. The shaded region represents (0.75-3.0)MeV data while the total is for (0.75-30.0)MeV.

TABLE 1. Radio ephemerides of PSR B1509-58 used in this work

MJD coverage	48522 - 49956
MJD epoch	49239.000000527
ν	$6.6324050404788\,\text{s}^{-1}$
$\dot{\nu}$	$-6.75457\times10^{-11}\,\text{s}^{-2}$
$\ddot{\nu}$	$1.96\times10^{-21}\,\text{s}^{-3}$

intrinsically difficult to explore. As a consequence, even for the MeV luminous Crab pulsar, signal-to-noise ratios are typically of a few percent.

RESULTS ON TIMING ANALYSIS

COMPTEL time series were phase folded with an ephemeris generated from radio data contemporaneous with the CGRO dataspan, having a coverage slightly larger than the COMPTEL data, namely from TJD8522 to TJD9956. The duration of the data is short enough not to require a $\ddot{\nu}$ term, the residuals between the radio data and the ephemeris being $\lesssim$ 4ms. The parameters are shown in Table 1. The phase folded events were tested under the null hypothesis using the Z_n^2 test, where n, the number of harmonics involved, was restricted to 1,3 or 5, following previous tests with Crab data.

The data were separated in the four standard COMPTEL energy ranges, namely (0.75-1.0)MeV, (1-3) MeV, (3-10)MeV and (10-30)MeV. The only individual observation showing any statistically significant effect (prob$\leq$ 1%) was the previously reported observation 23, in the (0.75-1.0) MeV range.

Timing analysis was also performed on the the combined observations in phase, still divided in energy bands. The results are shown on Table 2. The (1-3) MeV data, which comprises two thirds of all events, has $Z_1^2 = 9.41$, a value with 0.9% random probability. Adding the (0.75-1.0) MeV data barely changes the harmonic content. The (3-10)MeV and (10-30) MeV data show no deviation from noise, all the statistically significant effects being restricted to the low and medium COMPTEL energy bands.

The overall (0.75-3.0) MeV light curve, shown in Figure 2, peaks at radio phase 0.30 ± 0.06, as the OSSE light curve does. There is only a single broad pulse, similar to that observed by GINGA, BATSE and OSSE.

THE HIGH ENERGY SPECTRUM OF PSR B1509-58

Qualitative Spectral Behaviour

Proper fluxes and upper limits for pulsed emision have to be estimated through imaging analysis and will be shown in a future publication [1]. Timing

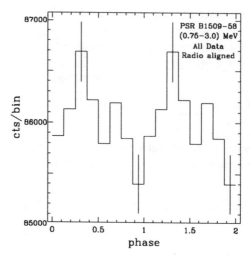

FIGURE 2. COMPTEL 8-bin light curve of PSR B1509-58 for (0.75-3.0) MeV data. The phase origin corresponds to the radio peak.

information can provide only a preliminary estimates of these values. However, using the Crab pulsar as a reference [10], one can predict the harmonic content expected in the PSR B1509-58 data, assuming the power-law spectrum seen by OSSE extends into the COMPTEL regime. For this purpose, Crab data of similar characteristics (∼total events, ∼data span, same data cuts) as those of PSR B1509-58 were generated. The Crab Z_n^2 values were translated into number of pulsed events, and, by comparing the Crab photon fluxes [10] and the OSSE extrapolation [9], we predicted Z_3^2 values for the PSR B1509-58 and compared with the observed values, as shown on Table 2.

TABLE 2. Predicted and measured Z_n^2 values for PSR B1509-58

Energy range MeV	Crab		PSR B1509-58			
	Events	Z_3^2	Events	OSSE extrapolated flux/Crab	Z_3^2 Predicted	Measured
(0.75-1.0)	78195	53.66	87069	0.215	2.76	11.27
(1.0-3.0)	520254	228.62	601058	0.23	13.97	13.52
(3.0-10)	193695	48.67	218714	0.75	30.91	4.25
(10-30)	12451	26.06	12097	0.875	19.38	4.13

While the observed (0.75-1.0)MeV and (1-3)MeV Z_3^2 values are in agreement with expectation, the (3-10)MeV and (10-30)MeV should show very significant effects corresponding to probabilities of 2.6×10^{-5} and 3.6×10^{-3} which are not observed. This indicates that the OSSE spectrum breaks before the (3-10)MeV COMPTEL energy range. Although the statistical significance of the effect in the (1-3)MeV band is weak, it is reinforced by the consistency of the COMPTEL light curve with that of other instruments. The indication is that the OSSE spectrum most probably extends into the (1-3)MeV energy range and breaks somewhere between 3MeV and 5MeV.

The spectral decay at a few MeV

PSR B1509-58 behaves differently than other γ-ray pulsars. It is the first pulsar releasing most of its energy at a few MeV, without comparable GeV emission. This is compatible with γ-ray production very close the neutron star surface followed by interactions in a strong magnetic field [4]. Magnetic photon processes become important when $(E_\gamma/mc^2)(B/B_{cr})$, approaches or exceeds unity ($B_{cr} \equiv m^2c^3/e\hbar$). For PSR B1509-58, $B_* \approx 0.7\,B_{cr}$ at the star surface, the pair production threshold is $\sim$0.7 MeV. γ-ray production at $\sim 1.9 R_*$ would raise it to $\sim$ 5MeV.

Recently Harding, Baring & Gonthier ([6]) have modelled in detail magnetic attenuation processes near the surface of the neutron star. The preferred mechanism is not magnetic pair production, $\gamma \rightarrow e^+e^-$, but photon splitting, $\gamma \rightarrow \gamma\gamma$, a no threshold process that can occur even if $(E_\gamma/mc^2)(B/B_{cr}) < 1$, i.e. the $\gamma \rightarrow e^+e^-$ threshold, and will be important if $B \gtrsim 0.3 B_{cr}$. The authors modelled splitting cascades for a polar cap model of PSR B1509-58, explaining the MeV cutoff and setting restrictions to physical parameters.

REFERENCES

1. Bennett K. et al. 1997, in preparation
2. Bennett K. et al. 1993, 23rd ICRC, 1, 172
3. Brazier K.T.S., Bertsch D.L., Fichtel C.E., et al. 1994, MNRAS 268, 517
4. Carramiñana A. & Bennett K. 1995, Adv. Space Res. 15, 565
5. Carramiñana A., Bennett K., Buccheri R., et al. 1995, A&A 304, 258
6. Harding A., Baring M. & Gonthier P.L. 1997, ApJ 476, 246
7. Kaspi V.M., Manchester R.N., Seigman B., et al., 1994, Ap.J. 422, L83
8. Kawai N., Okayasu R., Binkmann W., et al. 1991, ApJ 383, L65
9. Matz S.M., Ulmer M.P., Grabelsky D.A., et al. 1994, ApJ 434, 288
10. Much R., Bennett K., Buccheri R., et al. 1995, A&A 299, 435
11. Schönfelder V., et al. 1993, ApJS 86, 629
12. Seward F.D. & Harden F.R. 1982, ApJ 256, L45
13. Ulmer M.P., Matz S.M., Wilson R.B., et al. 1993, ApJ 417, 738
14. Wilson R.B., Finger M.H., Pendleton G.N., et al. 1993, *Isolated Pulsars*, eds. K.A. van Riper and C. Ho, Cambridge Univ. Press, p. 257
15. Wilson R.B. et al. 1993, AIP Conf Proc 280, 291

Search for the pulse of PSR B1823-13 in the COMPTEL and EGRET databases

A. Carramiñana[1], K.T.S. Brazier[2], K. Bennett[2], W. Hermsen[3], L. Kuiper[3], V. Schönfelder[4], A. Lyne[5]

[1] INAOE, Luis Enrique Erro 1, Tonantzintla 72840, MEXICO[1]
[2] Physics Department, Durham University, England
[3] Astrophysics Division, European Space Research and Technology Centre, Noordwijk, The Netherlands
[3] Space Research Organization Netherlands, Utrecht, The Netherlands
[4] Max-Planck Institute für Extraterrestrische Physik, Garching, Germany
[5] University of Manchester, Nuffield Radio Astronomy Laboratories, Jodrell Bank, Cheshire, England

Abstract. PSR B1823-13, one of the pulsars with high priority for gamma-ray pulsar searches, is positionally consistent with a well-known excess in the COMPTEL and EGRET maps. Although the pulsar has not been reported as a detection by COMPTEL or EGRET, previous non-statistical behaviour in COMPTEL CGRO Phase 1 data, together with a glitch occurring during Phase 2, encourage us to make new searches using the first three years of CGRO observations from 0.75 MeV up to ~ 10 GeV. We do not find any convincing evidence for the pulsed signature of PSR B1823-13 in these data.

INTRODUCTION

PSR B1823-13 is a fast (101ms), young ($P/2\dot{P} \simeq 21,400$ years) radio pulsar. Although it is not associated to an optical SNR, PSR B1823-13 is embedded in a compact X-ray nebula of 20" radius, contained in a larger region of diffuse X-ray emission [3]. There is evidence of unpulsed X-ray emission in the ROSAT data [1]. Given its energetics ($\dot{E}_{rot} \simeq 3 \times 10^{36}$ erg · s^{-1}) and distance (4 kpc), PSR B1823-13 is often quoted as a prime γ-ray pulsar candidate.

PSR B1823-13 is marginally coincident with the EGRET source 2EG J1825-1307. The pulsar ($l = 18.0, b = -0.69$) is $\sim 1\sigma$ away from the EGRET source location ($l = 18.38, b = -0.43$). The association between the pulsar and the

---

[1] supported through CONACYT grant 4142-E9404

the nature of the γ-ray source. The emission observed by EGRET does not correspond to a single point source, but is consistent with a strong source with either more sources around it or diffuse emission. The ($l = 18.0, b = -0.7$) region also corresponds to a strong excess in the COMPTEL maps, which can be modelled as a point source on top of diffuse emission. In any case, if the pulsar is a source, it is likely to suffer from an enhanced background, complicating the analysis.

Given the spatial coincidence and the characteristics of the pulsar, previous searches for pulsations were made during the respective all-sky survey EGRET [4] and COMPTEL [2] pulsar searches. While EGRET reported only upper limits, the COMPTEL data showed some hints of emission during validation period 5.0, in the highest (10-30)MeV energy range.

There are motivations to renew the search for the pulsed signature of this object in γ-ray data: a glitch occurred during CGRO phase 2, invalidating the phase 3 data analysis of EGRET data by [4]; also, PSR B1823-13 was targetted for pointed observations during phase 3, when it received $\sim$ 7.5 weeks of exposure, compared with $\sim$ 6 weeks during Phase 1 and $\sim$ 3.5 during phase 2. With this in mind, we analysed COMPTEL and EGRET phase 1 data using a pre-glitch ephemeris and phase 3 data with a post-glitch ephemeris. Phase 2 was put aside partly because of the lower exposure and partly because the time of the glitch is not well determined.

TIMING OF COMPTEL AND EGRET DATA

COMPTEL analysis and results

Phase 1 (CGRO observations 5, 7.5+13.0 and 20) and phase 3 (CGRO observations 324, 440 and 332) data were selected in the four standard COMPTEL energy ranges, (0.75-1.0)MeV, (1-3)MeV, (3-10)MeV and (10-30)MeV. We restricted the scatter angle of selected events to $4° \leq \bar{\Phi} \leq 36°$ (for a description of the COMPTEL instrument and data charateristics see [5]). The

FIGURE 1. EGRET image of the region around PSR B1823-13. Observation 5, $E_\gamma \geq 300$MeV. Each pixel is 0.5° per side. The pulsar is at the centre of the circle.

TABLE 1. Radio ephemerides of PSR B1823-13 used in this work

	Pre-glitch	Post-glitch
MJD coverage	48401-48935	49244-49625
MJD epoch	48668.000000343	49434.000001162
ν (Hz)	9.8569288561195	9.8564755503032
$\dot{\nu}$ (Hz^2)	-7.28140×10^{-12}	-7.31195×10^{-12}
$\ddot{\nu}$ (Hz^3)	1.58×10^{-22}	3.43×10^{-22}

event arrival times were phase folded using the ephemerides shown in Table 1 and a Z_3^2 test was performed for individual observations and for each CGRO phase.

In the case of COMPTEL data we note a couple of hints in individual observations (VP 5 and VP 332) and for phase 3, (1-3)MeV. The Z_3^2 are only marginally high and the actual result lacks statistical significance. The corresponding light curves are shown in Figure 2. Note their relative misalignment, which goes against a real pulsed signal.

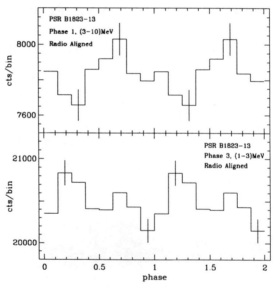

FIGURE 2. COMPTEL light curves for PSR B1823-13. Top is for Phase 1 data, (3-10) MeV; bottom is Phase 3 data, (1-3) MeV. The phase origin for both curves corresponds with the respective radio ephemerides.

EGRET analysis and results

EGRET photon events were divided in two energy bands: (30-100)MeV and ($E_\gamma > 100$MeV), phase folded using the same ephemerides as for the COMPTEL data 1, and tested with the Z_n^2 statistics. None of the individual observations, nor the phase 1 and phase 3 sets provided evidence of pulsed γ-ray emission.

TABLE 2. Z_3^2 values found per CGRO phase (COMPTEL & EGRET)

		COMPTEL				EGRET	
	Energy (MeV)	0.75-1.0	1.0-3.0	3.0-10	10-30	30-100	> 100
Phase 1	Events	31692	178790	62659	4964	5895	5679
	Z_3^2	0.68	5.43	11.96	5.33	9.53	3.32
Phase 3	Events	21276	163905	60253	4436	2947	3873
	Z_3^2	8.91	13.53	4.49	5.81	9.41	13.86

DISCUSSION

This analysis provided no evidence for pulsed γ-ray emission from PSR B1823-13. The association with the γ-ray source remains so far a positional coincidence and an actual physical link between the pulsar and the γ-ray source has not been proven. However, there are several indications that there might well be such a physical association. The X-ray data shows a point source inside a region of an extended emission [1], consistent with a synchrotron nebula being supplied of relativistic electrons by the pulsar [3]. PSR B1823-13 itself is similar to other γ-ray pulsars, namely Vela and PSR B1706-44. While the final proof resides in the finding of coherent pulsations, PSR B1823-13 may well just happen to be a steady source of MeV and GeV γ-rays, just as the Crab is a steady source of TeV photons.

REFERENCES

1. Finley J.P. & Ögelman H. 1993, IAUC 5787
2. Carramiñana A. et al, 1995, A&A 304, 258
3. Finley J.P., Srinivasan R., Park S., 1996, ApJ 466, 938
4. Nel H.I. et al. 1996, ApJ 465, 898
5. Schönfelder V., et al. 1993, ApJS 86, 629

Very High Energy Observations of PSR B1951+32

R. Srinivasan[*1], P.J. Boyle[†], J.H. Buckley[‡], A.M. Burdett[∥],
J. Gordo[†], D.A. Carter-Lewis[**], M. Catanese[**], M.F. Cawley[¶], E.
Colombo[§], D.J. Fegan[†], J.P. Finley[*], J.A. Gaidos[*], A.M. Hillas[∥],
R.C. Lamb[§§], F. Krennrich[**], R.W. Lessard[*], C. Masterson[†],
J.E. McEnery[†], G. Mohanty[**], P. Moriarty[††], J. Quinn[†],
A.J. Rodgers[∥], H.J. Rose[∥], F.W. Samuelson[**], G.H. Sembroski[*],
T.C. Weekes[‡], and J. Zweerink[**]

*Purdue University, Indiana 47906. † University College, Dublin, Ireland,
∥ University of Leeds, Yorkshire, UK, ‡ P.O. Box 97, Amado, AZ 85645,
¶ St. Patrick's College, Maynooth, Ireland, § CONAE, Argentina, †† RTC, Galway,
Ireland, §§ Caltech, CA 91125. ** Iowa State University, IA 50011

Abstract. PSR B1951+32 is a γ-ray pulsar detected by the *Energetic Gamma Ray Experiment Telescope* (EGRET) and identified with the 39.5 ms radio pulsar in the supernova remnant CTB 80. The EGRET data shows no evidence for a spectral turnover. Here we report on the first observations of PSR B1951+32 beyond 30 GeV. The observations were carried out with the 10m γ-ray telescope at the Whipple Observatory on Mt. Hopkins, Arizona. In 8.1 hours of observation we find no evidence for steady or periodic emission from PSR B1951+32 above $\sim$ 260 GeV. Flux upper limits are derived and compared with model extrapolations from lower energies and the predictions of emission models.

INTRODUCTION

The pursuit of Very High Energy (VHE) astrophysics has resulted in the discovery of five sources, of which three are associated with young spin-powered pulsars, namely, Crab Nebula [8], Vela pulsar [7] and PSR B1706-44 [4]. No

---

[1)] This research is supported by grants from the U.S. Department of Energy, NASA, PPARC in the UK and by Forbairt in Ireland. The authors wish to thank A. Lyne for providing the radio ephemeris of PSR B1951+32 and D.J. Thompson for providing the multiwavelength spectrum for PSR B1951+32

evidence has been found for periodic emission at these energies in these experiments.

PSR B1951+32 has been detected as a pulsating X-ray source [6] and as a high energy γ-ray pulsar at E $\geq$ 100 MeV at the radio period [5]. The best fit outer gap model of Zhang and Cheng (1997) suggests that PSR B1951+32 should emit detectable levels of TeV γ-rays (Figure 2). The multiwavelength spectrum of PSR B1951+32 (Figure 1b) indicates a maximum power per decade at energies consistent with a few GeV and still rising at 10 GeV. These factors make PSR B1951+32 a good candidate for observations with the Atmospheric Cherenkov technique above 100 GeV.

OBSERVATIONS

The observations of PSR B1951+32 reported here were acquired with the 10m reflector located at the Whipple Observatory on Mt. Hopkins in Arizona. A total of 14 *Tracking* runs and 4 *On/Off* pairs taken between 13th May, 1996 and 17th July, 1996 constitute the database for all subsequent discussion. The total *On* source observing time is 8.1 hrs. The radio position (J2000) of PSR B1951+32 ($\alpha = 19^h\ 52^m\ 58.25^s$, $\delta = 32°\ 52'\ 40.9''$) was assumed for the subsequent timing analysis.

ANALYSIS AND RESULTS

Standard Analysis

The event selection criteria are collectively called *Supercuts95* and a detailed description can be found elsewhere [1]. *Supercuts95* raises the effective energy threshold of the detector with its software trigger and *size* cuts. PSR B1951+32 appears to have a steep spectrum at EGRET energies and since the pulsar spectrum is expected to cut off, it behooves us to reduce the threshold of our analysis to search for a lower energy signal. The dominant background at lower energies is due to muons whose images appear in the camera as arcs and can be discriminated by a cut on their large *length/size* values. Hence the selection criteria used *Supercuts95* on images with sizes larger than 400 p.e. and *Smallcuts* (Table 1) for images with sizes less than 400 p.e.

No steady emission is detected from PSR B1951+32 and 3σ flux upper limits are displayed in Table 2. The effective area for *Supercuts95*, that was used to calculate the upper limit, was taken as $A_{eff} \sim 3.5 \times 10^8$ cm^2: the same area was used for the dataset that resulted from a combination of *Supercuts95* and *Smallcuts* although here there is more systematic uncertainty. The energy threshold was obtained from simulations and extrapolating the Crab Nebula γ-ray rate for each set of cuts used assuming a spectrum $\sim E^{-2.4}$.

TABLE 1. Parameter ranges for selecting γ-ray images

Parameter	Supercuts95	Smallcuts
length	0°16 - 0°30	unchanged
width	0°073 - 0°15	unchanged
distance	0°51 - 1°1	unchanged
alpha	< 15°	unchanged
max1	> 100 p.e.	45 p.e. - 100 p.e.
max2	> 80 p.e.	45 p.e. - 80 p.e.
size	$\geq$ 400 p.e.	0 - 400 p.e.
length/size	not used	< 7.5 × 10^{-4} °/p.e.

Periodic Analysis

The arrival times of the Cherenkov events were registered with a relative resolution of 0.1 μsec, transformed to the solar system barycenter and folded to produce the phases, ϕ_j, of the events modulo the pulse period. The ephemeris frequency parameters used were ν= 25.2963719901267 s^{-1} and $\dot{\nu}$=-3.73940×10^{-12}s^{-2}, at the epoch t_0=JD 2450177.5. This frequency was extrapolated 72 days to obtain a timing solution relevant to the epoch of observation. The datasets, however, were taken within the validity interval of the above ephemeris. An *optical* observation of the Crab pulsar with the 10m reflector yielded a clear detection of the Crab pulsar signal in phase with the radio pulse and verified the accuracy of the Whipple Observatory timing systems. No evidence of pulsed emission from PSR B1951+32 at the radio period exists. To calculate a pulsed flux upper limit we assumed the same pulse profile as seen at EGRET energies, i.e. with the phase range for the main pulse and secondary pulse as 0.12 - 0.22 and 0.48 - 0.74 respectively [5].

TABLE 2. Integral Flux Upper limits

	Steady Emission cm^{-2}s^{-1})	Periodic Emission (cm^{-2}s^{-1})	Threshold (GeV)
Supercuts95	0.97 × 10^{-11}	3.7× 10^{-12}	$\geq$ 370
Supercuts95 + Smallcuts	1.95× 10^{-11}	6.7× 10^{-12}	$\geq$ 260

DISCUSSION

PSR B1951+32 is surrounded by a compact nebula which gives a plerionic nature to the supernova remnant, CTB80. It is expected that for X-ray plerions, such as that associated with PSR B1951+32 where the density of nebular synchrotron photons is too low for synchrotron self Compton process (SSC) to take place, detectable VHE emission should be produced by the inverse

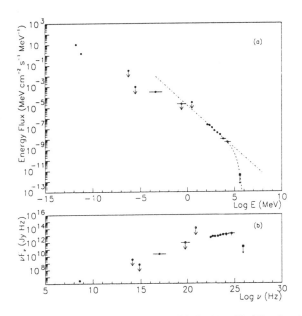

FIGURE 1. The pulsed energy spectrum of PSR B1951+32. The dot-dashed curve represents the power law fit to the EGRET points. The dashed curve is Equation 1 with a cut off energy $E_0 = 75$ GeV. The Whipple limit is indicated as a filled square at 370 GeV.

Compton scattering of the 2.7K cosmic microwave background by the same electrons radiating synchrotron X-ray photons. Interpreting the unpulsed X-ray emission form CTB80 as the synchrotron emission from a plerion, the estimated IC flux > 1 TeV is 6.6×10^{-13} TeV/cm^2/s/TeV [2].

To model the pulsed high energy spectrum, a function of the form

$$dN_\gamma/dE = KE^{-\Gamma}e^{(-E/E_0)} \quad (1)$$

was used where E is the photon energy, Γ is the photon spectral index and E_0 is the cut off energy. Equation 1 was used to extrapolate the EGRET spectrum to VHE energies constrained by the TeV upper limit reported here and indicates a cut off energy of $E_0 \leq 75$ GeV for pulsed emission (Figure 1a). The Whipple upper limit is compared with the outer gap model of Zhang and Cheng (Figure 2). This model includes the effect of geometry in the treatment of pulsed emission via a parameter $\alpha = r/r_L$, the radial distance to the synchrotron emitting region near the outer gap, r, as a function of the light cylinder radius r_L. Our pulsed upper limits are consistent with the outer gap model if $\alpha > 0.6$ implying an emission region far out in the magnetosphere.

PSR B1951+32 exhibits very similar spectral behavior and morphological features, such as an associated synchrotron nebula, to PSR B1706-44 [3]. If these factors are any indication of similar emission mechanisms in pulsars then

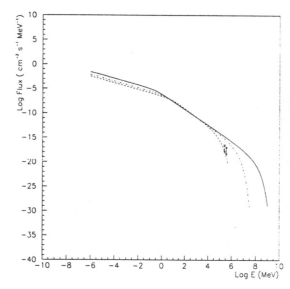

FIGURE 2. *Predicted pulsed γ-ray flux of PSR B1951+32 from the Zhang and Cheng outergap model. The solid, dot-dash and dashed curves correspond to α=0.5, 0.6, 0.7.*

the lack of unpulsed emission from PSR B1951+32 is puzzling considering that PSR B1706-44 was detected as a VHE source of unpulsed emission > 1 TeV [4]. Lack of pulsed emission indicates that the processes producing pulsed high energy photons over two decades of energy in the EGRET energy range somehow become ineffective over a decade of energy to result in a lack of VHE γ-rays. The low magnetic field of PSR B1951+32 relative to the average pulsar field implies that attenuation of γ-rays by magnetic absorption is not a likely explanation for the non-detection.

REFERENCES

1. Catanese, M. et al. 1995, Towards a Major Atmospheric Cherenkov Detector-IV, Padova, Italy, 335
2. De Jager, O.C. et al. 1995, in Proc. 24th ICRC (Rome), 2, 528
3. Finley, J.P. et al. 1997, ApJ, submitted
4. Kifune, T. et al. 1995, ApJ, 438, L91
5. Ramanamurthy, P.V., et al. 1995, ApJ, 447, L109
6. Safi-Harb, S., Oğelman, H., & Finley, J.P. 1995, ApJ, 439, 722
7. Takanori, Y. 1996, Ph.D. thesis, University of Tokyo
8. Vacanti, G., 1991, ApJ, 377, 467
9. Zhang, L. and Cheng, K.S. 1997, ApJ, submitted

A candidate γ-ray pulsar in CTA 1

K.T.S. Brazier[1], O. Reimer[2], G. Kanbach[2] & A. Carramiñana[3]

[1] Department of Physics, University of Durham, South Road, England, DH1 3LE
[2] Max-Planck-Institut für Extraterrestrische Physik, 85740 Garching, Germany
[3] Astrofísica, INAOE, Luis Enrique Erro 1, Tonantzintla, Puebla 72840, Mexico

Abstract. 2EG J0008+7307 is a prominent EGRET source 10.5 degrees from the Galactic Plane. We report the results of our study of this object, made as part of our on-going programme to identify individual EGRET point sources. We find that 2EG J0008+7307 has a constant flux and is located within the CTA 1 supernova remnant. It has recently been discovered that the centrally-concentrated X-ray emission from CTA 1 is non-thermal [14]. The 95% confidence contour for the EGRET source location includes only one point-like X-ray source, which is at the centre of the non-thermal X-ray emission. We propose that this object is a unknown γ-ray pulsar.

2EG J0008+7307

The high-energy γ-ray source 2EG J0008+7307 is a well-defined, isolated source 10.5 degrees from the Galactic plane. In a study of high-latitude EGRET sources, Nolan et al. [9] found that the source was variable (> 95% confidence) and had a spectral index of -1.58 ± 0.20; McLaughlin et al. [6] also placed the source above their variability threshold. Yet the object is unidentified. In this paper, we will present a detailed analysis of the EGRET data and our search for a counterpart at lower energies.

The optimal position for 2EG J0008+7307 is derived from > 1 GeV data, where the instrumental point spread function is much smaller than at lower energies. 2EG J0008+7307 is clearly visible above 1 GeV and we derive a position of l=119.87, b=10.52, with a 95% confidence radius of 15 arcmin. Using this position, the long term flux history of the source shown in Fig. 1 was produced, using 9 observations in which the source was < 35 degrees from the EGRET axis. This is a more restrictive selection than used by Nolan et al., but includes a longer time base. Following the method described by McLaughlin et al., we find a variability index $V = 0.55$, where $V \geq 1$ indicates variability. There is therefore no evidence that the source is variable.

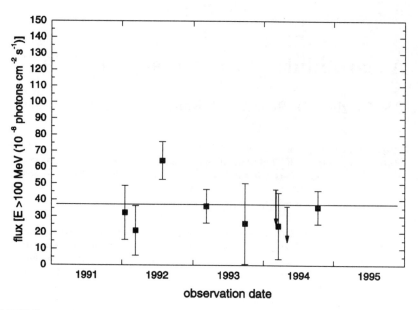

FIGURE 1. 1992–1995 EGRET flux history for 2EG J0008+7307. The horizontal line is a weighted mean of the individual fluxes.

The source spectrum was constructed from the three prime observations in which the source was within 20 degrees of the instrument axis. Between 70 MeV and 2 GeV the data are consistent with a power law of photon index of -1.58 ± 0.18, while above 2000 MeV the spectrum softens, falling below the extrapolated power law. The total flux above 100 MeV is $(46.4 \pm 6.2) \times 10^{-8}$ cm^{-2} s^{-1}.

IDENTIFYING 2EG J0008+7307

Nolan et al. [9] suggested that the most plausible counterpart to 2EG J0008+7307 was the brightest nearby X-ray source, an AGN identified by Seward, Schmidt and Slane [13]. From Figure 2 it is clear that this is now an unlikely counterpart, lying well outside the EGRET 95% contour. We also exclude QSO B0016+731, the nearest flat-spectrum radio source considered by Mattox et al. [5], on positional grounds. No other catalogued object has been proposed as the source of the γ-ray emission.

The supernova remnant, G119.5+10.2 (CTA 1) or an associated pulsar were discounted as counterparts by Nolan et al. because of the apparent variability of the EGRET flux. Since, however, we have not confirmed the variability, these options become viable.

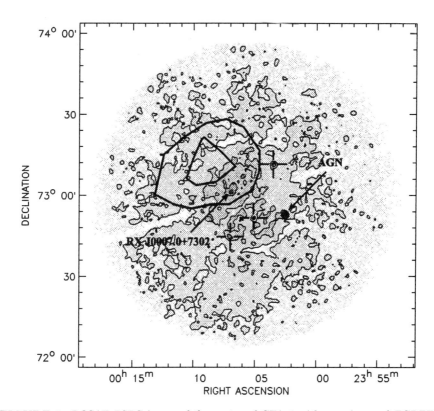

FIGURE 2. ROSAT PSPC image of the centre of CTA 1 with superimposed EGRET contours (68%, 95% confidence). The AGN identified by Seward et al. is the bright source towards the right of the field.

Observations of CTA 1

CTA 1 consists of an incomplete radio shell, ~ 100 arcmin in diameter, with a diffuse bar running north-south across its centre. Pineault et al. [10,11] explain the missing northwest part of the shell as expansion of the remnant into a lower density region of the interstellar medium. They derive a kinematic distance of 1.4 ± 0.3 kpc. The age of the remnant is estimated to be 10^4 years but could be as low as 5000 years [14].

CTA 1 has been observed by both Einstein and ROSAT [12,13]. The X-ray remnant is dominated by a central patch of diffuse emission; Slane et al. [14] have recently shown that this central X-ray emission is non-thermal, suggesting strongly that there is a compact object within the SNR. They propose that a weak point source at the heart of the diffuse emission is this object. Fainter, thermal X-ray emission from the outer parts of the SNR are also detected in CTA 1 [14].

Five X-ray point sources can be located within the SNR boundaries (Fig. 2). The area enclosed by the EGRET 95% confidence contour includes only one of these, denoted RX J0007.0+7302, the object noted by Slane et al. in the heart of the diffuse X-ray emission. From the X-ray observations alone, we of course cannot tell whether this object is connected with the EGRET source. To constrain the possible options, we obtained B, V, R and I images at the Guillermo Haro astrophysical observatory in northern Mexico. A $6' \times 10'$ field centred on the X-ray source position was exposed for a total of ~ 2 hours per filter. These frames were analysed with the IRAF software package, using POSS positions for reference stars.

The ROSAT uncertainty radius was assumed for this work to be $10''$, in order to allow for systematic errors caused by the non-thermal nebula, uncertainty in the satellite pointing, plus the statistical $6''$ uncertainty. No object is found in any of the images inside this $10''$ radius, with 3σ limits of B > 23.1, V > 22.5, R > 22.6 and I > 21.5. The nearest detection, $\sim 18''$ from the centre of the error box; is a $m_V = 17.2$ star of spectral type $\sim$K0 and with no signs of coronal activity.

Using a ROSAT adaptation [2] of the distributions of the X-ray to optical flux ratios for stars, AGN, galaxy clusters and BL Lacs detected in the Einstein Medium Sensitivity Survey [15], our upper limit of $m_I = 21.5$ for a source within the ROSAT error circle implies that an AGN is unlikely and a stellar source is ruled out. We require deeper optical and radio observations before we can rule out the possibility of a BL Lac.

If the K star is the counterpart to the X-ray object, then it is extremely X-ray luminous among stellar sources. This seems unlikely for the spectral type; given the lack of emission lines and the positional offset, we reject the star as the source of the X-ray flux.

DISCUSSION

2EG J0008+7307 has γ-ray properties found among identified sources only in pulsars: a hard spectrum with steepening above $E_\gamma \sim 2$ GeV, plus a non-variable flux. It is also coincident with the supernova remnant CTA 1. The γ-ray source is positioned in the northeast quadrant of the SNR, which includes the maximum of the non-thermal X-ray emission associated with the SNR. Only a handful of non-thermal X-ray nebulae are known, including those around the Crab and Vela pulsars, PSR 1951+32 and PSR 1509-58. Like Slane et al. [14], we propose that the point-like X-ray source at the heart of the CTA 1 emission is a pulsar that drives the diffuse X-radiation, and we further propose that it is the source of the γ-rays.

The EGRET source properties are consistent with this suggestion. For an assumed 1 steradian beam, the observed 100 MeV - 2 GeV flux corresponds to a luminosity of 4×10^{33} erg s^{-1}, inside the range seen in the six confirmed

EGRET pulsars. The γ/X-ray flux ratio is similar to the values seen in the Vela pulsar and Geminga. The spectrum of 2EG J0008+7307 is consistent with a young pulsar at the distance and age of the SNR. Unfortunately neither the X-ray nor the γ-ray source is bright enough for us to attempt to identify pulsations with current data, and no pulsed radio source has been found, either in recent surveys of SNRs and EGRET sources [1,8] or in a targeted observation of the X-ray point source position with the Jodrell Bank 76 m telescope. The upper limit of 0.3 mJy at 1412 MHz (J. Bell, private communication) means that, if 2EG J0008+7307 is a pulsar, then it is extremely radio-weak.

To take the identification further, we must confirm that RX J0007.0+7302 is not merely superimposed on the field by chance. To date, our optical observations have ruled out stellar and most AGN source categories. A faint BL Lac is still a possibility with current data; this will be tested by forthcoming observations.

A pulsar in CTA 1 would not be surprising, given that five radio pulsars and one radio-quiet pulsar are the only galactic EGRET sources identified so far [17]. We have previously identified 2EG J2020+40 as a candidate pulsar, again probably radio-quiet [3], and a number of papers have concluded that many of the unidentified EGRET sources near the Galactic plane could be pulsars, some of them radio-quiet (e.g. [5,18,4,7,16]. If confirmed, the proposed pulsar in CTA 1 will strongly support this prediction, with implications for other unidentified γ-ray sources and for our understanding of the pulsar population.

REFERENCES

1. Biggs J.D., Lyne A.G., 1996, MNRAS, 282, 691
2. Bower R.G., et al., 1996, MNRAS, 281, 59
3. Brazier K.T.S., et al., 1996, MNRAS, 281 1033
4. Kanbach G., et al., 1996, A&AS, 120, C461
5. Mattox J.R., et al., 1997, ApJ, 481, in press
6. McLaughlin M.A., et al., 1996, ApJ, 473, 763
7. Merck M., et al., 1996, A&AS, 120, C465
8. Nice D.J., Sayer R.W., 1997, ApJ, 476, 261
9. Nolan P.L., et al., 1996, ApJ, 459, 100
10. Pineault S., et al., 1993, AJ, 105, 1060
11. Pineault S., et al., 1997, A&A, submitted
12. Seward F.D., 1990, ApJS, 73, 781
13. Seward F.D., Schmidt B., Slane P., 1995, ApJ, 453, 284
14. Slane P., et al., 1997, ApJ, in press
15. Stocke J.T., et al., 1995, ApJS, 76, 813
16. Sturner S.J., Dermer C.D., A&A, 293, L17
17. Thompson D.J., et al., 1994, ApJ, 436, 229
18. Yadigaroglu I.-A., Romani R.W., 1997, ApJ, 476, 347

Discovery of the Young, Energetic Radio Pulsar PSR J1105−6107

V. M. Kaspi[1], M. Bailes[2], R. N. Manchester[3], B. W. Stappers[4], J. S. Sandhu[5], J. Navarro[6], N. D'Amico[7]

[1] *MIT, Department of Physics & Center for Space Research, Cambridge, MA 02139*
[2] *University of Melbourne, School of Physics, Parkville, Victoria 3052, Australia*
[3] *Australia Telescope National Facility, CSIRO, PO Box 76, Epping NSW 2121, Australia*
[4] *Mount Stromlo and Siding Spring Observatories, ANU, Private Bag, Weston Creek, ACT 2611, Australia*
[5] *California Institute of Technology, MS 105-24, Pasadena, CA 91125*
[6] *National Radio Astronomy Observatory, P.O. Box 0, Socorro, NM 87801*
[7] *Osservatorio Astronomico, via Zamboni 33, 40126 Bologna and Istituto di Radioastronomia, Via P. Gobetti 101, 40129 Bologna, Italy*

Abstract. We report on the discovery and follow-up timing observations of the 63 ms radio pulsar, PSR J1105−6107. The pulsar is young, having a characteristic age of only 63 kyr, and its spin-down luminosity, 2.5×10^{36} erg s^{-1}, is in the top 1% of all those known. Given its estimated distance of 7 kpc, PSR J1105−6107 is likely to be observable at high energies. Indeed, it is coincident with the known CGRO/EGRET source 2EG J1103−6106; we consider the possible association and conclude that it is likely. We also note the pulsar's apparent proximity to the supernova remnant G290.1−0.8 (MSH 11−6*I*A); an association requires that the pulsar's transverse velocity be ~650 km s^{-1} directed away from the remnant center, assuming a distance of 7 kpc and that the characteristic age is the true age.

INTRODUCTION

We have discovered a 63 ms radio pulsar, PSR J1105−6107, during a recent search for pulsars using the 64 m radio telescope at Parkes, NSW, Australia. Details of the discovery and follow-up timing observations have recently been published elsewhere [9]. Here we summarize the published results.

TABLE 1. Timing Parameters for PSR J1105−6107.

Parameter	Value
Right Ascension, α (J2000)	$11^h\ 05^m\ 26^s.07(7)$
Declination, δ (J2000)	$-61°\ 07'\ 52''.1(4)$
Galactic Latitude, l	$290°.4896(2)$
Galactic Longitude, b	$-0°.8465(1)$
Period, P	$0.063191252792(3)$ s
Period Derivative, $\dot{P}$	$15.80466(12) \times 10^{-15}$
Dispersion Measure, DM	$271.01(2)$ pc cm^{-3}
Epoch of Period	MJD 49545.0000
R.M.S. timing residual	6.2 ms
Surface Magnetic Field Strength, B	1.0×10^{12} G
Characteristic Age, τ_c	63,350 yr
Spin-Down Luminosity, $\dot{E}$	2.5×10^{36} erg s^{-1}

OBSERVATIONS AND RESULTS

PSR J1105−6107 was discovered in 1994 July in a search for radio pulsars at a central radio frequency of 1420 MHz [8]. Regular timing observations were begun not long after the pulsar's discovery, although some archival data were available from previous, unrelated observations. A total of 96 timing observations of PSR J1105−6107 were obtained between 1993 July 8 and 1996 July 4 at Parkes at central radio frequencies ranging from 660 to 2050 MHz. The average profile at 1650 MHz, shown in Figure 1, was obtained by aligning and summing numerous individual folded profiles. Arrival times were analyzed using the standard TEMPO pulsar timing software package [20] together with the JPL DE200 ephemeris [17]. Typical arrival time uncertainties were ~ 250 μs for ~ 10 min integrations with signal-to-noise ratio ~ 20 at frequencies above 1390 MHz. Arrival times at 660 MHz had uncertainties approximately twice as large. Timing parameters for PSR J1105−6107 can be found in Table 1. Timing residuals left after the subtraction of the best-fit model reveal significant timing noise, as expected for a young pulsar. The pulsar's small τ_c, as well as the large amount of timing noise, suggest that PSR J1105−6107 is an excellent candidate for glitches.

POSSIBLE ASSOCIATION WITH γ-RAY SOURCE 2EG J1103−6106

The Taylor & Cordes (1993) DM-distance model places PSR J1105−6107 at a distance of 7 kpc. Given its large spin-down luminosity, PSR J1105−6107 ranks 19th in a list of rotation-powered pulsars ordered by $\dot{E}/d^2$. Six of the seven top spots are held by known γ-ray pulsars (the seventh being the millisecond pulsar PSR J0437−4715), while most of the top 30 are known

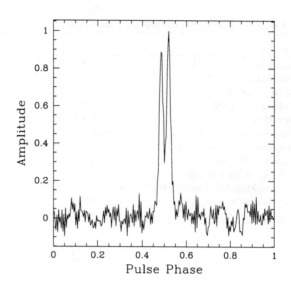

FIGURE 1. Pulse profile for PSR J1105−6107 at 1650 MHz.

X-ray sources. On this list, PSR J1105−6107 ranks 15 spots higher than the known X-ray and γ-ray pulsar PSR B1055−52 [3]. Thus, PSR J1105−6107 is a good candidate to be an observable high-energy emitter.

In fact, the radio timing position of PSR J1105−6107 (Table 1) lies well inside the 95% confidence $49' \times 32'$ error ellipse of the second EGRET catalog source 2EG J1103−6106 [22]. This γ-ray source was referred to in the first EGRET catalog as GRO J1110−60 [2], and is near, but outside, the error box of the second COS-B catalog source 2CG 288−00 [18]. Reported $E > 100$ MeV fluxes of 2EG J1103−6106 show no evidence for significant variability, consistent with its interpretation as a rotation-powered pulsar [22,13]. If the sources are associated, the estimated mean flux of 2EG J1103−6106 suggests that PSR J1105−6107 converts approximately 3% of its spin-down luminosity to high-energy γ-rays for a beaming angle of 1.0 sr, comparable to the efficiencies of the Vela pulsar and PSR B1706−44 [21,7].

There are 18 radio pulsars within 10° of the Galactic plane that have higher $\dot{E}/d^2$ than PSR B1055−52 for which pulsations have not yet been detected by EGRET, omitting millisecond pulsars. With the discovery of PSR J1105−6107, four of these lie within the 99% confidence contours of unidentified EGRET sources. By contrast, of 268 known radio pulsars within 10° of the Galactic plane whose energetics should be below the EGRET threshold for detection, only two lie within the 99% confidence contours of unidentified EGRET sources [4]. Assuming that this control group is spatially distributed like the young pulsars, using Poisson statistics, the probability for four coincidences among the 18 energetic pulsars is $\sim 1\times 10^{-5}$. Even conservatively accounting for the possibility that the control group is less concentrated near the Galactic plane (for example, by assigning it a significantly larger mean z-height), we find that the probability for four coincidences must be under $\sim 1\%$, although exact probabilities are difficult to estimate given the uncertainties in pulsar distances and spatial distributions, and in unidentified EGRET source properties. Even so, the evidence argues strongly that at least three of the four coincidences of high $\dot{E}/d^2$ pulsars with the unidentified EGRET sources are real. Furthermore, Yadigaroglu & Romani (1997) showed that most of the unidentified low-latitude EGRET sources such as 2EG J1103−6106 are likely to be young pulsars like PSR J1105−6107. We therefore conclude that the association between PSR J1105−6107 and 2EG J1103−6106 is likely. However, only the detection of γ-ray pulsations at the radio period will demonstrate the association unambiguously.

POSSIBLE ASSOCIATION WITH G290.1−0.8

PSR J1105−6107 is located near the Galactic supernova remnant G290.1−0.8, also known as MSH 11−61A [16]. Distance estimates to the two sources agree within the substantial uncertainties: the DM-distance model suggests a range 5−9 kpc, and Rho (1995) concluded from the neutral hydrogen absorption component of the remnant's X-ray spectrum that the distance is ~ 7 kpc. Independently, Rosado et al. (1997) argued that the remnant must be at ~ 7 kpc, on the basis of the kinematics of the optical emission. Similarly, age estimates agree, though both have substantial uncertainties. The pulsar's characteristic age, $P/2\dot{P}$, is 63 kr, but the true age depends on the birth spin period and the true braking index and may be anywhere between 63 and 250 kyr. Age estimates for the remnant depend strongly on its distance. Rho (1995) suggests the remnant's age is ~ 10 kyr. The resemblance of the optical emission to that of the Monoceros Ring, whose age is ~ 50 kyr [10], suggests a much larger age for G290.1−0.8. PSR J1105−6107 lies just over two remnant radii from the approximate remnant geometric center along a remnant axis of symmetry. For a distance of 7 kpc, and assuming the age of the system to be 63 kyr, the transverse velocity of the pulsar, if it is associated with the remnant, is ~ 650 km s^{-1}. This is larger than the mean pulsar transverse velocity [11], but much less than has been suggested for pulsars in other proposed associations (e.g. [5,12,1]), and well within the range of measured pulsar velocities [11].

An independent velocity estimate for PSR J1105−6107 is needed to test the association with the supernova remnant. A timing proper motion will not be forthcoming, given the large amount of timing noise exhibited by the pulsar. Also, its low flux density (~ 1.8 mJy at 1420 MHz) will make interferometric observations using currently available telescopes difficult, although pulse gating may improve the feasibility. The recent discovery of X-ray emission from PSR J1105−6107 [6] may indicate the presence of a pulsar-powered synchrotron nebula. Such a nebula could imply that the pulsar wind is being confined by interstellar medium ram pressure, as would be the case for a high-velocity pulsar.

ACKNOWLEDGEMENTS

We thank Joseph Fierro for helpful discussions regarding EGRET source coincidences with radio pulsars.

REFERENCES

1. Caraveo, P. A. 1993, ApJ, 415, L111
2. Fichtel, C. E. et al. 1994, ApJS, 94, 551
3. Fierro, J. M. et al. 1993, ApJ, 413, L27
4. Fierro, J. M. 1995, PhD thesis, Stanford University
5. Frail, D. A. & Kulkarni, S. R. 1991, Nature, 352, 785
6. Gotthelf, E. V. & Kaspi, V. M. 1997, ApJ, in preparation
7. Grenier, I. A., Hermsen, W., & Clear, J. 1988, A&A, 204, 117
8. Kaspi, V. M., Manchester, R. N., Johnston, S., Lyne, A. G., D'Amico, N. 1996, AJ, 111, 2028
9. Kaspi, V. M., Bailes, M., Manchester, R. N., Stappers, B. W., Sandhu, J. S., Navarro, J., D'Amico, N. 1997, ApJ, 485, 820
10. Leahy, D. A., Naranan, S., & Singh, K. P. 1986, Ap&SS, 119, 249L
11. Lyne, A. G. & Lorimer, D. R. 1994, Nature, 369, 127
12. Manchester, R. N., Kaspi, V. M., Johnston, S., Lyne, A. G., & D'Amico, N. 1991, MNRAS, 253, 7P
13. Ramanamurthy, P. V. et al. 1995, ApJ, 450, 791
14. Rho, J. 1995, PhD thesis, University of Maryland
15. Rosado, M., Ambrocio-Cruz, P., Le Coarer, E., & Marcelin, M. 1997, A&A, 315, 243
16. Shaver, P. A. & Goss, W. M. 1970, Aust. J. Phys. Astr. Supp., 14, 77
17. Standish, E. M. 1982, A&A, 114, 297
18. Swanenburg, B. N. et al. 1981, ApJ, 243, L69
19. Taylor, J. H. & Cordes, J. M. 1993, ApJ, 411, 674
20. Taylor, J. H. & Weisberg, J. M. 1989, ApJ, 345, 434
21. Thompson, D. J. et al. 1992, Nature, 359, 615
22. Thompson, D. J. et al. 1995, ApJS, 101, 259
23. Yadigaroglu, I. A. & Romani, R. W. 1997, ApJ, 476, 347

RXTE Observation of PSR1706-44

A. Ray[*][1], A.K. Harding[*], and M. Strickman[†]

[*]NASA/Goddard Space Flight Center Greenbelt, MD 20771
[†]Naval Research Laboratory, Space Science Division, Washington, D.C. 20375

Abstract.
We report on results of an observation with the Rossi X-Ray Timing Explorer of PSR1706-44, to search for pulsed X-ray emission. PSR1706-44 is a radio and high-energy gamma-ray pulsar (detected by EGRET), but no pulsed emission has been detected in the X-ray band. Since most of the other known gamma-ray pulsars emit pulsed X-rays, it is expected that PSR1706-44 would also be an X-ray pulsar. However, ROSAT detected a source at the pulsar position, but did not detect pulsations, giving a pulsed fraction upper limit of 18%. We report here results of a search for modulation at the pulsar period from our $\approx$ 90K second RXTE observation (out of the total approved time of 147 K second), carried out in November 1996 during a low state of the nearby X-ray binary 4U1705-44.

INTRODUCTION & MOTIVATION

X-ray emission from rotation powered pulsars can be a combination of varying amounts of three components: 1) a power law, 2) soft blackbody and 3) a hard component connected with heated polar caps. Blackbody-like emission seen at energies between 0.1 and 2 keV could result from either surface cooling of the neutrons star or polar cap heating by energetic particles or internal sources. A power-law component could result from non-thermal radiation of particles accelerated in the pulsar magnetosphere.

Measurements of the pulsed X-ray fluxes (especially in the yet unexplored spectral region above 10 keV, for most pulsars) can act as a calorimeter of the returned particle flux from either a polar cap (PC) or outer gap (OG) vacuum discharge accelerator. The returned particle flux is an indicator of the properties and mechanisms of the non-thermal (gamma-ray and power law X-ray) emission from the pulsar. In particular PC models predict a small percentage of total accelerated particle flux as the returned flux [1], whereas the OG models predict roughly equal number of particles accelerated towards the PC as flowing away [7], and therefore would give rise to proportional X-ray

---

[1]) On leave of absence from Tata Institute of Fundamental Research, Bombay 400005, India

and gamma-ray photon fluxes. These measurements for pulsars could serve as an important separator of the two classes of models.

At present, it is not clear how and where in the pulsar magnetosphere the pulsed non-thermal high energy emission originates. Polar cap models assume that particles are accelerated above the neutron star surface and that radiation results from a curvature radiation [4] or inverse-Compton [10] induced pair cascade in a strong magnetic field. Outer-gap models [3] assume that acceleration occurs along null charge surfaces in the outer magnetosphere and that the radiation results from curvature radiation/photon-photon pair production cascades [9]. The computed spectra of the pulsars in both polar cap [5] and outer-gap [6] models predict that the spectrum will extend below 50 keV. In polar cap models this emission is primarily power-law synchrotron emission that will terminate at the cyclotron energy of the local magnetic field, which could be anywhere from $\sim$ 10 keV at the neutron star surface down to $\sim$ 0.1 keV several stellar radii above the surface. A measurement of the pulsed emission below 50 keV would determine how far the spectrum of the non-thermal component extends. If a cutoff in the spectrum is observed, it might be a measure of the local field strength in the emission region and thus, a limit of the proximity of a polar cap cascade to the neutron star surface.

CHARACTERISTICS OF PSR 1706-44:

Studies of multi-waveband electromagnetic spectra of rotation powered (isolated) neutron stars have shown that most of the Gamma-ray pulsars also emit in pulsed X-rays. Of the seven gamma-ray pulsars detected by the Compton Gamma-ray Observatory, six are known to pulse in the X-ray band as well (see, [11]). The seventh gamma-ray pulsar PSR B1706-44, has been detected in the X-ray as a d.c. level source in the soft X-ray ROSAT band [2], although it has not shown any modulation in the X-ray up to a 18% limit. These pulsars because of their high spin down luminosities and their association with EGRET gamma-ray sources have probable pulsed X-ray fluxes. In the case of PSR 1706-44 there is not enough spectral information to classify the emission in a thermal vs non thermal model or whether any fraction is pulsed.

The Low Mass X-ray Binary 4U 1705-44 - a bright source is about half a degree away from the PSR 1706-44. In addition there is a faint source in soft X-ray bands at RA = 17 10 44 DEC=-44° 33' 28". These two sources are within the RXTE FWHM response circle of the XTE PCA detector. The observation of PSR 1706-44 was accorded as a RXTE Target of Opportunity (TOO) whenever the nearby bright LMXB would go into a low state (corresponding to 5 ct/s in the All Sky Monitor).

TABLE 1. Summary of RXTE Observations of PSR 1706-44

Observation Date	Beginning MJD	PCA/EDS Configuration	Total Live Time(s)
1996 Nov 9 & 11	50396.371	GoodXenon	33487
1996 Nov 10 & 11	50397.922	E_125us_64M_0-1s	52162

TABLE 2. Radio Pulsar Ephemerides used in Epoch Folding[a]

Epoch	ν_0 (Hz.)	$\dot{\nu}(s^{-2})$	$\ddot{\nu}(s^{-3})$	RMS (milliperiod)
50273.000000458	9.7598013942166	−8.85435D−12	1.47D−22	0.8

[a] Courtesy: V. Kaspi, M. Bailes N. Wang and R.N. Manchester; RA= 17 09 42.722, DEC= -44 29 8.44 (J2000), range of fit: MJD 50114-50433.

RXTE OBSERVATION & ANALYSIS COMPLETED TO DATE

About 96.9 ksec of observation has been completed so far out of the allocated 147 ksec of TOO observation. After excising the times for SAA passage and earth occultations etc., the live time of observation of this source is: 85649 s. The dates, instrument configurations, exposures etc., are listed in Table 1.

The science events (photon events) from the two separate PCA/EDS configurations were separately analyzed and fed into an epoch folding program "faseBin" adapted for RXTE observations, (developed by A. Rots (private communication) from the TEMPO software used in radio observations of pulsars). The radio pulsar ephemerides used in the epoch folding program - kindly supplied by V. Kaspi et al, are displayed in Table 2. The epoch folded phaseograms of these events are displayed in Figure 1 (the GoodXenon events). The phaseogram displayed here has the origin at the radio pulse arrival phase origin (i.e. in absolute phase with respect to the radio pulse).

There is no significant evidence for pulsed emission at the pulsar period. Currently we can only obtain the following 2 σ upper limit of the *pulsed* flux, based on the total observation of 85694 s, in the energy interval 1.5 keV < E_ν < 18.5 keV band [12] :

$$UL = \frac{2}{\Delta E} \frac{\sqrt{C_{tot}}}{(A_{eff}t)} \sqrt{\frac{\beta}{1-\beta}} = 9 \times 10^{-7} \text{photons cm}^{-2}\text{s}^{-1}\text{keV}^{-1}$$

here, we have taken the pulsar duty cycle $\beta = 1/2$ and the RXTE effective area A_{eff} to be 6000 cm^2. (The reduced χ^2 per degree of freedom is: 1.4 for 20 degrees of freedom, for the GoodXenon data). The corresponding upper limit for the νf_ν of the flux would be: 9×10^{-6}keV/cm^2/s/keV. These upper limits are shown together with the detections and upper limits of PSR 1706-44 in other energy bands in Figure 2.

The current RXTE upper limits seem to be consistent with an extension of the power law flux at EGRET energies to the relevant RXTE band. That is, it does not yet rule out a non-thermal photon flux extending from EGRET to RXTE bands. If with the rest of the allotted time on RXTE, we still fail to detect any pulsation in future, then the decreased upper limit to the pulsed flux may require a flattening from the power law extension of the EGRET spectrum. The upper limit to the modulations in the flux appears to be at the 0.4 % level (inclusive of the 'contaminating' emission from 4U 1705-44 and other sky and instrument backgrounds).

Acknowledgments:

We thank V. Kaspi, M. Bailes, N. Wang and R.N Manchester for sharing with us the radio ephemerides of PSR 1706-44 measured from ATNF. We

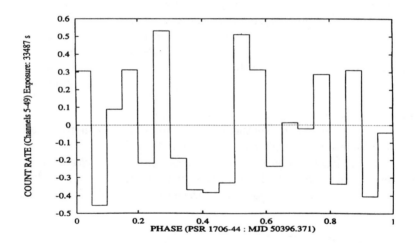

FIGURE 1. PSR1706-44 phaseogram of the GoodXenon data (channels 5 through 49 corresponding to 1.5-18 keV) epoch folded according to parameters listed in Table 2. The average count rate in the phase interval $0.0 < \phi < 1.0$ (marked as 0 on the y-axis) corresponds to 121.0 counts/s; the 1 σ error bar in the individual bins is typically 0.27 cts/s.

thank D.J. Thompson for discussions and for the compilation of data on this pulsar at other energy bands and Arnold Rots, Alan Smale, Tess Jaffe and Gail Rohrbach for discussion and help with data analysis at the XTE Guest Observation Facility. A.R. acknowledges support through a Senior Research Associateship of the National Research Council at NASA/Goddard.

REFERENCES

1. Arons, J. 1981, ApJ 248, 1099.
2. Becker, W., Brazier, K.T.S. and Trümper, J. 1995, A & A. 298, 528.
3. Cheng, K.S., Ho, C. and Ruderman, M.A. 1986, ApJ 300, 500.
4. Daugherty, J.K. and Harding, A.K. 1982, ApJ 252,337.
5. Daugherty, J.K. and Harding, A.K. 1996, ApJ 458, 278.
6. Ho, C. 1989, ApJ 342, 396.
7. Halpern, J. and Ruderman, M.A. 1993 ApJ 415, 286.
8. Nel, H. et al 1996 ApJ, 465, 898.
9. Ray, A. and Benford, G. 1981, Phys Rev D, 23, 2142.
10. Sturner, S.J. and Dermer, C.D. 1994, ApJ 420, L79.
11. Thompson, D.J. 1996, in Pulsars: Problems & Progress (eds S. Johnston et al) Astron Soc Pacific Ser. 105, 307.
12. Ulmer, M.P., Purcell, W.R., Wheaton, W.A. and Mahoney, W.A. 1991, ApJ 369, 485.

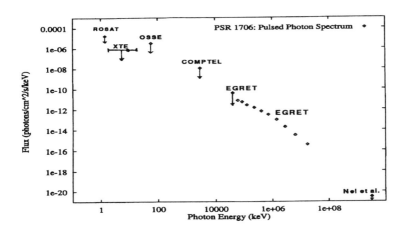

FIGURE 2. PSR 1706-44 pulsed photon spectrum, including upper limits, in the high energy bands.

VHE Gamma Rays from PSR B1706-44

P. M. Chadwick, M. R. Dickinson, N. A. Dipper, J. Holder,
T. R. Kendall, T. J. L. McComb, K. J. Orford, J. L. Osborne,
S. M. Rayner, I. D. Roberts, S. E. Shaw and K. E. Turver

Department of Physics, University of Durham, South Road, Durham DH1 3LE, UK

Abstract. Observations made with the University of Durham Mark 6 atmospheric Čerenkov telescope confirm that PSR B1706-44 is a VHE γ-ray emitter. The γ-ray flux has been measured at 250 GeV, and suggests a spectral slope that is consistent with extrapolation from the sub-10 GeV region. There is no indication from our dataset that the VHE γ-rays are pulsed, in contrast to the findings at < 20 GeV, which indicate that 100% of the flux is pulsed.

INTRODUCTION

PSR B1706-44 is a 102 ms radio pulsar which may be associated with the supernova remnant G343.1-2.3. The pulsar was discovered via its radio emission [1]. Observations of the 102 ms periodicity with the EGRET detector on *CGRO* [2,3] associated the *COS-B* gamma-ray source 2CG 342-02 [4] with this pulsar and have shown it to be one of only seven identified pulsars which emit high energy γ-rays.

A number of models of gamma-ray emission from pulsars suggest that PSR B1706-44, as well as the Crab, should be a source of VHE γ-rays (see, for example, [5,6]). The Crab nebula is a well established source of VHE gamma-rays, detectable from a few hundred GeV [7,8] to higher energies [9-11]. The Crab pulsar has also been detected in VHE γ-rays [12]. McAdam et al. (1993) [13] have suggested that PSR B1706-44 is associated with the shell-type supernova remnant G343.1-2.3 (but see [14] for an opposing view).

Kifune et al. (1995) [15] found evidence of gamma-ray emission above 1000 GeV from PSR B1706-44 but have detected no pulsations. We have observed this object with the University of Durham Mark 6 gamma-ray telescope to extend measurements closer to the energy range of the EGRET measurements.

TABLE 1. Observing log for observations of PSR B1706-44 with the University of Durham Mark 6 telescope at Narrabri during 1996.

Date (1996)	No. of scans	Date (1996)	No. of scans
May 11	3	July 9	7
May 12	4	July 11	4
May 21	5	July 15	4
May 22	2	July 18	5
May 24	6		

OBSERVATIONS

Observations of PSR B1706-44 were made on clear moonless nights at zenith angles 15° − 50° in May and June 1996 using the University of Durham Mark 6 VHE gamma ray telescope [16,17]. An observing log is shown in Table 1. Our method of observation is to make 15 minute exposures on and off source at similar zenith angles, hence providing identical samples of the cosmic ray background.

Data were selected for analysis if pairs of ON and OFF observations have count rates which differ by $\leq$ 2.5 σ, extensive environmental monitoring showed clear and stable skies, calibration data were available during the observation, and the pointing of the telescope was good. Event arrival times were measured using a well-established Rb oscillator-based clock, monitored by GPS, to a relative accuracy of $1\mu s$ and an absolute epoch of $10\mu s$.

RESULTS

The γ-ray signal in the data from PSR B1706-44 has been enhanced using parameters developed by the Whipple group, together with novel aspects of the response of the triggering detectors of the Mark 6 telescope to γ-ray and proton initiated events. These constitute an initial selection and are summarised in Table 2.

Data suitable for analysis comprise those events which are confined within the sensitive area of the camera and which contain sufficient information for reliable image analysis ($SIZE$ 200 − 50000 digital counts, where 1 digital count $\sim$ 3 photoelectrons). Image parameters are calculated using defined 'picture' and 'border' PMTs (see, e.g., [8]). Parameters 1, 2, 4 and 6 in Table 2 are defined in the conventional way. Parameter 3 (D_{dist}) is a new parameter and is defined as the separation in degrees of the centroids of the samples of light in the left and right detectors of the telescope. γ-ray events are expected to show images with a small separation between their centroids. Parameter 5

TABLE 2. Values of the parameters adopted for our analysis of PSR B1706-44 data. The parameters are defined in the text.

	Parameter	Initial Value
1	DISTANCE	$0.35° - 0.85°$
2	ECCENTRICITY	$0.35 - 0.65$
3	D_{dist}	$\leq 0.15°$
4	LENGTH	size dependent selection
5	CONCENTRATION	size dependent selection
6	ALPHA	$\leq 25°$

(*CONCENTRATION*) is a variant of the concentration parameter introduced by the Whipple group. Here it is defined as the fraction of the total light detected outside the 'picture' and 'border' PMTs; typical values for γ-ray events are smaller than for background proton events.

Events were selected as γ-ray candidates if they had characteristics within the limits shown in Table 2. The values of *LENGTH* were dependent on the *SIZE* of the image and varies from $0.35 - 0.4$ for *SIZE* ~ 300 to $0.45 - 0.55$ for *SIZE* ~ 10000. Similarly, *CONCENTRATION* varies systematically with size, from $0.5 - 0.7$ for *SIZE* ~ 300 to < 0.12 for *SIZE* ~ 10000.

Once selections 1 to 5 have been applied (see Table 3), the *ALPHA* distributions of the data are plotted for both ON and OFF source events — see Figure 1. There is an excess of events at small *ALPHA*, the expected γ-ray domain. Imposing a γ-ray selection of *ALPHA* $\leq 25°$ yields a γ-ray detection significant at the 6.9 σ level.

To demonstrate that this excess of gamma-ray like events originates from the source direction we have calculated the number of events passing the selections for a number of different assumed source positions. We show the results of this false source analysis in Figure 2 which clearly shows that the excess emanates from the direction of PSR B1706-44. Note that during these observations, the position of the source was not at the geometrical centre of the camera, which reduces the possibility that the observed excess is a result of bias arising from the geometry of the camera.

An initial investigation using simulations suggests that the maximum triggering probability for the Mark 6 telescope occurs at a γ-ray energy of 300 GeV [17]. Fluxes have been estimated on the basis of the number of excess

TABLE 3. Results using selections in Table 2

	ON	OFF	Excess	Significance
Total no. of events	217456	216318	1138	1.7 σ
Selected events (1-5)	6193	5671	522	4.8 σ
$ALPHA < 25°$	1989	1583	406	6.8 σ

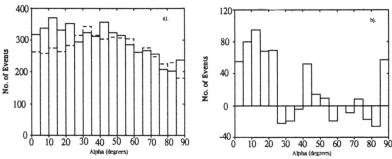

FIGURE 1. *Distributions of the pointing parameter ALPHA for events recorded during our 580 min of observations of PSR B1706-44. (a) shows the on-source (full line) and off-source (dotted line) distributions, while (b) shows their difference.*

events recorded with $ALPHA \leq 25°$ divided by the time taken to make the observations and the effective collecting area of the telescope. We estimate the γ-ray detection rate to be 0.7 ± 0.1 per minute. We then take into account an estimate of the efficiency with which we retain γ-rays in our dataset. Both the effective collecting area and the efficiency of γ-ray retention are factors which require further investigation. For the purposes of the calculation below we have assumed an effective collecting area of 3×10^8 cm^2 and a γ-ray retention factor of approximately 40%, which makes the incident γ-ray rate 1.8 ± 0.3 per minute. The integral flux from PSR B1706-44 is therefore estimated to be $(1.0 \pm 0.4) \times 10^{-10}$ cm^{-2} s^{-1} at E > 300 GeV.

A pulsar timing analysis, involving reduction of event times to the Solar System Barycentre and folding with a contemporary ephemeris [18] has been per-

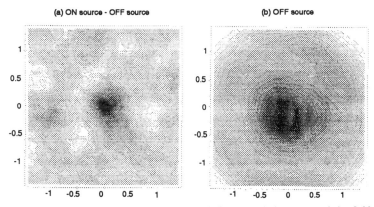

FIGURE 2. *False source plots for the region of sky around the centre of the field of view for on-source − off-source events (a) and off-source events (b). PSR B1706-44 is at the centre of these plots, and the axes are in units of degrees from the source position.*

formed. We have no evidence for periodicity from PSR B1706-44 in the small (580 min) dataset reported here. The 3σ flux limit is $\sim 4 \times 10^{-11}$ cm^{-2} s^{-1}.

DISCUSSION

Our observations of VHE γ-ray emission from PSR B1706-44 confirm the earlier reports by Kifune et al. (1995) that this object emits at $E \geq 300$ GeV.

Our dataset does not show any evidence for pulsations at the known pulsar frequency. The high-energy γ-ray observations [3] have shown evidence for periodicity, while the earlier VHE observations at > 1 TeV have shown no strong evidence [15]. It is of interest in this context to investigate the lower energy events (100 − 200 GeV) detected by the Mark 6 telescope, and such an analysis is underway.

ACKNOWLEDGEMENTS

We are grateful to the UK Particle Physics and Astronomy Research Council for support of the project and the University of Sydney for the lease of the Narrabri site. The Mark 6 telescope was designed and constructed with the assistance of the staff of the Physics Department, University of Durham.

REFERENCES

1. Johnston, S. et al., *Mon. Not. R. Astr. Soc.* **255**, 401 (1992).
2. Thompson, D. J. et al., *Nature* **359**, 615 (1992).
3. Thompson, D. J. et al., *Astrophys. J.* **465**, 385 (1996).
4. Swanenberg, B. N. et al., *Astrophys. J.* **243**, L69 (1981).
5. Cheng, K. S. and Ding, W. K. Y., *Astrophys. J.* **431**, 724 (1994).
6. Aharonian, F.A.. Atoyan, A.M. and Kifune, T., preprint (1997).
7. Weekes, T. C. et al., *Astrophys. J.* **342**, 379 (1989).
8. Vacanti, G. et al., *Astrophys. J.* **377**, 467 (1991).
9. Baillon, P. et al., *Proc. 23rd. Int. Cosmic Ray Conf.* vol 1, Calgary: University of Calgary, 271 (1993).
10. Tamimori, T. et al., *Astrophys. J.* **429**, L61 (1994).
11. Konopelko, A. et al., *Astroparticle Phys.* **4**, 199 (1996).
12. Dowthwaite, J.C. et al., *Astrophys. J.* **286**, L35 (1984).
13. McAdam, W. B., Osborne, J. L. and Parkinson, M. L., *Nature* **361**, 516 (1993).
14. Nicastro, L., Johnston, S. and Koribalski, B. *Astron. Astrophys.* **306**, L49 (1996).
15. Kifune, T. et al.,*Astrophys. J.* **438**, L91 (1995).
16. Armstrong, P. et al., in preparation.
17. Chadwick, P. M. et al., this conference.
18. Kaspi, V. M. et al., *GRO/radio timing database*, Princeton University, (1995).

RXTE Observations of the anomalous pulsar 4U0142+61

Stefan Dieters, Colleen Wilson, Mark Finger, Matt Scott and Jan van Paradijs

ES84 Space Sciences Building #4481
MSFC/NASA, Huntsville, Alabama. 35816

Abstract. We have observed the X-ray pulsar X0142+61 using RXTE in March 1996. This pulsar is one of a small group that are either isolated pulsars or members of very compact, low mass binaries. The pulse period is measured as 8.68814($\pm$9) sec with an upper limit of $\dot{P}/P \leq 8\times 10^{-10}$ s/s upon the short term spin up/down rate. A compilation of all historical measurements show an overall spin down trend with an episode of more rapid spin up and down between 1984 and 1996. A search for orbital modulations in the pulse arrival times yielded an upper limit ≤ 0.28 lt.sec. in $a_x \sin i$ over the interval 600 to 10^5 sec.

INTRODUCTION

Based upon their similar properties a small class of anomalous or "6 sec" X-ray pulsars has been identified [2,4,8]. These systems are 4U 0142+61, 1E 1048.1-5937, RX J 1838.4-0301, 1E 2259+589 and possibly also 4U 1626-67 (see pro [4] and con [8] arguments). These pulsars share: **(1)** a pulse period between 5.5 and 8.7 sec., **(2)** their pulse periods tend to increase with time, while most X-ray pulsars with similar periods tend to spin up, **(3)** in all but 4U 1626-67 no orbital periodicity has been found, **(4)** a low X-ray luminosity of 10^{34}–10^{36} ergs s^{-1} which is relatively stable on time-scales of days to years, **(5)** a high X-ray to optical luminosity ratio of $L_x/L_{opt} > 10$, **(6)** a very soft X-ray spectrum (except 4U 1626-67) with a photon index Γ=2–4, **(7)** they seem to be young objects because of their small galactic scale height (rms $\sim$100 pc) and association with either a molecular cloud or stellar remnant.

Their properties are consistent with them being either a low mass X-ray binary with an orbital period of a few hours or less and with a companion mass of at most a few tenths of a solar mass, or an isolated neutron star accreating from a molecular cloud or circumstellar debris [1]. They may even be the remnant of a Thorne-Zytkow object [8].

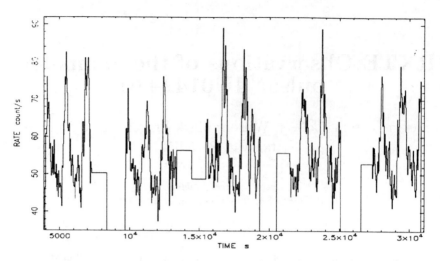

FIGURE 1. A section of the Standard 2 data for 4U 0142+61 and RX J0146.9+6121. The background as estimated using the *pca_bs* routine has been removed. The ~1500 sec modulation is due to RX J0146.9+6121.

In an attempt to detect an orbital modulation in the pulse arrival times or to put limits sufficiently stringent to eliminate any possible companion we observed 4U 0142+61 with RXTE. We present the results of our preliminary orbital period search.

DATA ANALYSIS

The observations were made on the 25th and 28-30th March 1996. The PCA was pointed directly at 4U 0142+61, but the field of view also included RX J0146.9+6121; a 24 min Be, X-ray pulsar system. We have concentrated the present analysis on the timing analysis on the PCA good xenon mode data on the 28-30 March. The PCA count rates (2-60 keV) were 160-275 cts/sec of which 55-125 cts/sec were background. See Fig 1.

To maximize the signal to noise ratio in the phase folds, only data from the top layer of the PCA and over the energy range (3.7-9.2 keV) were selected. The data was binned with a 125 ms time resolution and the times corrected to the solar system barycenter. The light curves were ported to numerous IDL routines written by the BATSE team.

Initialy the pulse period was determined by an epoch folding period search. No evidence for a period change over the 4 day span of the observations was found. Thus, the orbital period search could be limited to one that did not include any intrinsic spin period changes.

A template was created, using the period found with the initial period search, from the first 40 ksec. of data taken on the 28-30th March. This data is only interrupted by earth occultations. The data was then divided into 250 sec. (550 sec. bins were also used as a check) long segments. In each segment the data were phase folded with the best pulse period. The pulse profiles of each segment and the template were represented by the Fourier coeffiecents of their first three harmonics. This is done to suppress higher frequency (mainly counting) noise.

The phase of each segment with respect to the template was then calculated by cross-correlating the phase fold of each segment (as represented by its Fourier coeffiecents) and the template (also represented by it's Fourier coeffiecents). The errors on the phases were weighted by the amount of data in each segment. The time history was examined for cycle count errors and bad data points. The phases where searched for any periodicity (orbital) using the Lomb-Scargle periodogram (Press et al. 1986 and reference therein). The powers were normalized so that they followed an exponential distribution. The Lomb-Scargle periodogram is equivalent to a sinusoidal fit to the data so at each period a two parameter 99% confidence upper limit could be set to the amplitude. These upper limits were recast in terms of $a_x \sin i$. See Fig 3.

RESULTS

The light-curve with or without background subtraction, clearly shows a $\sim$25 min. variation. The period was measured using both Fourier and χ^2 periodograms as 23.45$\pm$0.2 min. This is slightly shorter than the $\sim$25 min measurement made in Aug 1984 [3]. This modulation was found to be much harder than the 8 sec. pulsations. ASCA observations [9] have shown that this modulation is associated with the Be source RX J0146.9+6121 and probably represents the very slow spin rate of the system's neutron star. Aside from this period and the pulse period of 4U 0142+61 no other periodicities were detected.

From the full orbital period search analysis, our pulse period estimate is P_{pulse}=8.688079(6) sec. A linear fit to the phase history set an upper limit of $\dot{P}/P \leq 8 \times 10^{-10}$ s/s upon the 'one-day' spin up/down rate on the 28th-30th March. Figure 2 combines period measurements using data collected with Einstein, EXOSAT, ASCA, ROSAT [9,3] and RXTE (this paper). Apparently this is the first time all the historical period determinations have been plotted together. The overall trend is one of spin down but between the 1984 and 1993 there must have been an episode of much more rapid spin up and spin down. Such a spin up/down episodes is reminiscent of most accreating X-ray pulsars.

As with previous studies we found that the relative pulse amplitude decreased with energy and that there is evidence for pulse shape evolution with

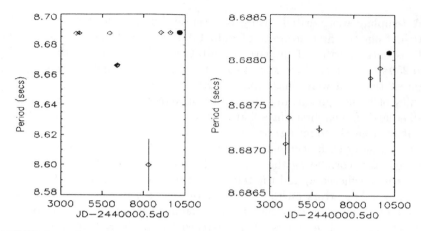

FIGURE 2. The historical pulse period history. Our measurement is indicated by a filled circle. The left-hand panel shows all historical measurements of the spin period while the right-hand panel is an enlargement showing the only the general spin down.

energy. In the 3.7–9.2 keV energy band used for the orbital period search the peak to peak amplitude is ∼11%. Over this energy range the average pulse profile is approximately sinusoidal in shape with slightly longer rise than fall. Such a pulse profile is well represented by the Fourier coeffiecents of its first three harmonics.

No periodicities exceeding the 99% confidence level was found in the phase shifts over the range 500–10^5 sec. The highest peak at ∼3000 sec. had a 17% probability of being due to noise alone. There is no evidence of a period near 24 min. Figure 3 shows the 99% confidence upper limits on the amplitude of a sinusoid at each period. For all periods the amplitude was consistent with being zero. There is a loss of search sensitivity towards both long and short periods and at the orbital period of RXTE. The limits are generaly lower than the limit of 0.37 on $a_x \sin i$ set over a similar period range by Israel *et al.* 1994. Lines of equal mass function are shown as a guide to the possible companions in this system.

CONCLUSIONS

By considering the mass radius relation for low mass stars [5,7] and assuming Roche lobe overflow our limits on $a_x \sin i$ eliminate hydrogen rich stars with $M_* \geq 0.2 M_\odot$ for most inclinations. Our limits are a factor of a few too high to eliminate less massive hydrogen rich stars and about a factor 10 too high to rule out He rich degenerate stars. We intend to extend our period search to

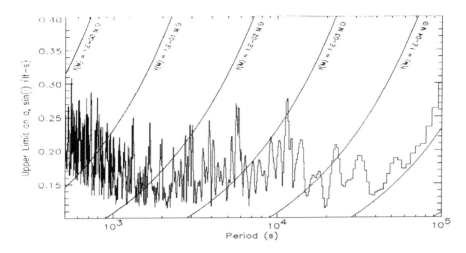

FIGURE 3. Upper limits to $a_x \sin i$

even shorter time-scales to cover the possible orbital periods of higher mass He degenerate stars. Future work will also include the analysis of the energy dependence in the pulse profile and the fitting of the X-ray spectrum.

REFERENCES

1. Corbet R.H.D., Smale A.P., Ozaki M., Koyama K., & Iwasawa K., ApJ, 443, 786 (1995).
2. Hellier C., MNRAS, 271, 21 (1994).
3. Israel G.L., Mereghetti S., Stella L., ApJ, 433, L25 (1994).
4. Mereghetti S., & Stella L., ApJ, 442, L17 (1995).
5. Paczyński B., & Sienkiewicz R., ApJ, 248, L27 (1981).
6. Press W.H., Flannery B.P, Teukolsky S.A., Vetterling W.T., in *Numerical Recipes*, Cambridge University Press, 1986, ch 12, pp 381-397.
7. Rappaport S., Nielson L.A., Ma C.P., & Joss P.C., ApJ, 322, 842 (1987).
8. van Paradijs J., Taam R.E., & van den Heuval E.P.J, A&A, 299, L41 (1995).
9. White N.E., Angelini L., Ebisawa K., Tanaka Y., & Gosh P., ApJ, 463, L83 (1996).

Search for a Gamma-ray Pulsar in the SNR RCW103

Masaki Mori* and Ken Ebisawa†

*Departemnt of Physics, Miyagi University of Education[1]
Aza Aoba, Aramaki, Sendai, Miyagi 980, Japan
†Laboratory for High Energy Astrophysics, NASA/Goddard Space Flight Center
and University Research Assocciation, Greenbelt, MD 20771, USA

INTRODUCTION

RCW103 (G332.4-0.4) is a supernova remnant with filaments which correspond well to the radio shell. Its distance is estmated to be 3.3 kpc from HI absorption. It includes an X-ray point source 1E161348-5055.1, whose location is marginally coincides with COS B source 2CG333+01. The GINGA satellite, which is sensitive for 2–30 keV, detected a coherent pulsation at 69 ms from the 2×4 degree^2 field-of-view including RCW103 and 2CG333+01 [1]. ASCA observation of RCW103 revealed the point source is bright above 3 keV, ruling out the possibility that X-ray emission is a soft thermal emission from neutron star surface. We suspect that 1E161348-5055.1, 2CG333+01 and the GINGA pulsar are all the same source, and it might be a gamma-ray pulsar with a character just between Crab and Vela pulsar. To verify this hypothesis, periodicity search has been performed for the EGRET data around the proposed 69 ms period.

OBSERVATION

The position of RCW103 as an X-ray point source is: $\alpha = 16^h17^m36^s.1$ and $\delta = -51°02'25".6$ (J2000) [2] or $\ell = 332°.43$ and $b = -0°.37$. Table 1 summarizes the EGRET observations of RCW103. Here aspect angle in degrees, exposure in $10^7 cm^2 s$ and number of events above 50 MeV within a 67% acceptance cone. $\sqrt{TS}$ is a significance as a point source. The total exposure is 4.32×10^8 cm^{2}s for > 50 MeV.

---

[1] E-mail: m-mori3@ipc.miyakyo-u.ac.jp

TABLE 1. EGRET observations of RCW103.

VP	Start date	End date	Aspect angle	Exposure	Events	$\sqrt{TS}$
0230	03/19/92	04/02/92	10.83	8.19	1076	1.3
0270	04/28/92	05/07/92	2.90	8.75	1160	0.9
2100	02/22/93	02/25/93	24.08	0.88	166	0.0
2140	03/29/93	04/01/93	24.08	1.05	167	1.2
2190	05/05/93	05/08/93	23.83	0.42	102	1.5
2260	06/19/93	06/29/93	23.18	3.58	716	3.2
2320	08/24/93	09/07/93	15.08	8.03	1096	0.0
3365	08/04/94	08/09/94	8.62	4.13	524	0.0
4020	10/18/94	10/25/94	22.60	1.82	462	0.2
4210	06/06/95	06/13/95	22.92	1.14	516	2.8
4220	06/13/95	06/20/95	22.98	1.69	558	0.3
4235	06/30/95	07/10/95	19.10	2.72	819	2.6

DC ANALYSIS

For each observation, maximum likelihood analysis [3] is used to test the excess in the direction of RCW103 above the Galactic diffuse background. The test statistic, $\sqrt{TS}$, is approximately equal to the statistical σ: in the second EGRET catalog [4] at least one detection with $\sqrt{TS} > 5$ is required for sources within $|b| < 10°$ to be claimed as a gamma-ray "source". The value of $\sqrt{TS}$ from the $E > 100$ MeV map for each viewing period is tabulated in the last column of Table 1. Thus RCW103 has been never detected as a "source" (see Fig. 1): nevertheless, one can still have a chance to detect its possible pulsation among possible contamination of source photons immersed in abundant background photons.

PERIODICITY SEARCH

The GINGA findings are summarized at first: observations were carried out between 4.9 and 5.5 March 1989 with the total exposure of ~ 15000 s and showed a significant pulsation at 69.3190 ± 0.0003 ms with a chance probability of 10^{-21} [1]. The period derivative was assumed to be zero, since the observation was too short to derive it. However, since the EGRET observations are far apart from the GINGA observations, we cannot ignore this quantity: supposing the pulsar in RCW103 is on a stage just between the Crab and Vela (see Table 2), we have searched for the range of $\dot{P} = 0 \sim 1 \times 10^{-13}$ or $\dot{f} = -\dot{P}/P_0^2 = (0 \sim -2) \times 10^{-11}$ Hz/s.

The program searches the frequency which gives the maximum non-uniformity with the H-test [5] within the given frequency range for each set of frequecy derivative. This test is proved to be powerful for light curves of unknown shape. In order to reduce the number of independent searches to

TABLE 2. Characteristics of some pulsars associated with SNRs.

Name	Period (ms)	$\dot{P}$	$\dot{f}$ (Hz/s)
Crab	33	4.1×10^{-13}	3.8×10^{-10}
PSR 1951+32	39.5	5.9×10^{-15}	3.8×10^{-12}
RCW103	69?	?	?
Vela	89	1.4×10^{-13}	1.8×10^{-11}

be realistic, only the viewing period showing some excess, though statistically not significant, were processed.

First, gamma-rays of energies greater than 50 MeV from within a cone of radius

$$\theta \leq 5.85°(E_\gamma/100\,\text{MeV})^{-0.534} \qquad (1)$$

which is expected to contain 67% of events from the point source are analyzed. These are standard criteria for EGRET pulsar analysis. The test periods are varied between 69.31 ms and 69.35 ms. Results for three viewing periods are summarized in Table 3. Here *steps* are number of searches in the f–$\dot{f}$ plane, H is the maximum value of H-test statistic during the search, $\mathcal{P}(H)$ is the probability of chance occurrence, f and $\dot{f}$ are those at H, and $\mathcal{P}$ is the overall

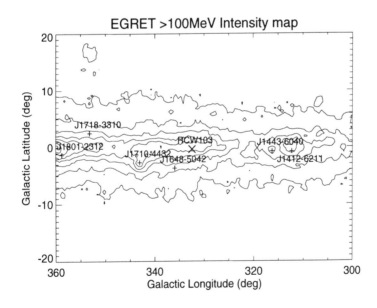

FIGURE 1. The gamma-ray sky around the RCW103 observed by EGRET (phase 1, > 100 MeV) shown by intensity contours. The EGRET second catalog sources are marked.

TABLE 3. Results of periodicity search using standard criteria

Vieing period	Events	Steps	Maximum nonuniformity				
			H	$\mathcal{P}(H)$	f (Hz)	$\dot{f}$ $(10^{-12}$Hz/s$)$	$\mathcal{P}$
2260	716	7001 × 500	37.6	9.9×10^{-7}	14.426522	-0.80	~ 1
4210	516	4840 × 1000	39.8	5.3×10^{-7}	14.421879	-2.14	~ 1
4235	819	7082 × 500	38.2	8.3×10^{-7}	14.424015	-5.72	~ 1

TABLE 4. Upper limits to the pulsed fraction of gamma-rays above 50 MeV assuming 10% and 50% duty cycles.

Vieing period	Upper limits ($cm^{-2}s^{-1}$, 95% C.L.)	
	10% duty cycle	50% duty cycle
2260	1.6×10^{-7}	4.8×10^{-7}
4210	5.0×10^{-7}	1.5×10^{-6}
4235	2.1×10^{-7}	6.3×10^{-7}

chance probability after multiplying the number of independent searches. One can see the maximum nonuniformity found is consistent with accidental and occurrs at significantly different f and $\dot{f}$ for each viewing period. The 95% C. L. upper limits derived from the maximum H found in the search [6] are tabulated in Table 4.

Second, in order to search for wider period range while keeping the computing time modest, gamma-rays coming within 2° of the source direction are further selected in addition to the above criteria: the test period ranges from 69.31 ms to 69.36 ms and the test period derivative varies in the range of $\dot{P} = (0 \sim 4) \times 10^{-13}$ or $\dot{f} = -\dot{P}/P_0^2 = (0 \sim -8) \times 10^{-11}$ Hz/s. This time periods are oversampled by a factor of 20 and the overall chance probability seems to be lower than the first search because of this "fine tuning" effect. In Table 5 the results are shown similarly to Table 3. There are two entries for each viewing period: upper entry is for $\dot{f} = (0 \sim -2) \times 10^{-11}$ Hz/s and lower entry for $\dot{f} = (-2 \sim -8) \times 10^{-11}$ Hz/s. Again we find no evidence for siginificant nonuniformity.

In conclusion, we found no evidence for pulsation near the 69 ms period in gamma-rays coming from the direction of RCW103 detected by EGRET.

One of the authors (M.M.) acknowldges Dave Bertsch for his assistance with the period search program.

REFERENCES

1. Aoki, T. , *IAU Circ.* 5588 (1992).
2. Tuohy, I. R. et al., *Astrophys. J.* **268**, 778 (1983).

TABLE 5. Results of periodicity search using the narrower angular window. There are two entries for each viewing period: upper entry is for $\dot{f} = (0 \sim -2) \times 10^{-11}$ Hz/s and lower entry for $\dot{f} = (-2 \sim -8) \times 10^{-11}$ Hz/s.

Vieing period	Events	Steps	\multicolumn{5}{c}{Maximum nonuniformity}				
			H	$\mathcal{P}(H)$	f (Hz)	$\dot{f}$ (10^{-12}Hz/s)	$\mathcal{P}$
2260	203	85512 × 1000	38.4	7.8×10^{-7}	14.423905	−11.30	∼ 1
2260	203	85512 × 600	45.4	1.2×10^{-7}	14.420185	−30.0	0.31
4210	93	58945 × 1000	48.2	6.1×10^{-8}	14.422130	−10.02	0.18
4210	93	58945 × 600	47.2	7.8×10^{-8}	14.427663	−40.7	0.14
4235	152	88379 × 1000	46.4	9.4×10^{-8}	14.425227	−4.04	0.42
4235	152	88379 × 600	43.2	2.1×10^{-7}	14.419325	−76.7	0.56

3. Mattox, J. et al., *Astrophys. J.*, **461**, 396 (1996).
4. Thompson, D. J. et al., *Astrophys. J. Suppl.* **101**, 259 (1995).
5. De Jager, O. C. et al., *Astron. Astrophys.* **221**, 180 (1989).
6. De Jager, O. C., *Astrophys. J.* **436**, 239 (1994).

Search for X-ray Pulsation from Rotation-Powered Pulsars with *ASCA*

Y. Saito[1], N. Kawai[2], T. Kamae[3], and S. Shibata[4]

[1] *Institute for Astronautical Science, Sagamihara, Kanagawa, 229, JAPAN*
[2] *Institute of Physical and Chemical Research (RIKEN) Wako, Saitama, 350-01, JAPAN*
[3] *University of Tokyo, Bunkyo-ku, Tokyo 113, JAPAN*
[4] *Yamagata University, Yamagata, Yamagata 990, JAPAN*

Abstract. We observed 8 pulsars with high spin down flux and searched for pulsation in the X-ray band (0.6-10 keV) with *ASCA*. We detected X-ray emission from all pulsars, including new detection in the 2-10 keV band for 3 pulsars. We also found pulsed emission from 3 pulsars including the first detection of millisecond pulsar PSR B1821−24 [1]. The pulse of PSR B1821−24 has sharp double peaks similar to that of the Crab pulsar and power-law spectrum which strongly suggest the magnetospheric origin.

We found an empirical relation between the pulsed luminosity $L_{X(pulse)}$ and the rotation energy loss $\dot{E}_{rot}$ in the X-ray band among all X-ray pulsars observed with *ASCA*. The pulsed luminosity is consistent with the relation $L_{X(pulse)} \propto \dot{E}_{rot}^{\frac{3}{2}}$, while those for the pulsed emission in γ-ray band $L_{\gamma(pulse)}$ is $L_{\gamma(pulse)} \propto \dot{E}_{rot}^{\frac{1}{2}}$. It should be noted that the empirical relation in the X-ray band holds also for the millisecond pulsar PSR B1821−24, which suggests same pulse emission mechanism works for millisecond pulsars.

INTRODUCTION

Thirty years have passed since the first discovery of the pulsar and we have not yet known the pulse emission mechanism. Difficulties in examining the pulsar models are due to following factors : small number of observed pulsars and unknown geometry such as inclination angle of magnetic and rotation axis. The large amount of the pulsed luminosity of magnetospheric origin is released in the X-ray to the γ-ray band. Thus, it is important to observe pulsars in the high energy band, while low sensitivity of past detectors were allowed to detect the pulsation from only a few number of pulsars.

After the launch of *CGRO*, 7 pulsars including middle-aged pulsars (10^5 yr) were identified as γ-ray pulsars (See review article in this proceedings). Based on the exploding of new parameter space of middle-aged pulsars, emission models made remarkable progresses recently, while they are still incomplete.

In the X-ray band, the launch of ASCA in 1993 opened a new window to observe the high energy pulsation in the hard X-ray (2-10 keV) band, which is suitable to distinguish non-thermal magnetospheric emission from thermal emission from neutron star surface.

ASCA has observed 19 pulsars until AO4. We analyzed for 8 pulsars and borrowed the results of 8 pulsars for comprehensive discussion. All pulsars observed have high spin down flux, while the spin parameters of individual pulsar are different which cover wide band of both the period and its derivative space : for example, rotation energy loss ranges 6 decades. This wide coverege enable us to find an empirical relation between the pulsed emission and the spin parameters regardless of the particular geometry. Here, we concentrate only on the results of pulsed emission, which makes clear difference from previous work based on the total emission (e.g. [2], [3]).

SEARCH FOR AN EMPIRICAL RELATION

We detected X-ray emission from all eight pulsars, including new detection in the 2-10 keV band for three pulsars, PSR B1821-24, PSR B1046-58, and PSR B1706-44, and spectral break around 2 keV for PSR J0437-4715. We summarized spectral parameters of a power-law model fitting in Table 1 and pulse shapes in Fig. 1. Interestingly, millisecond pulsar PSR B1821-24 shows a double peaked sharp pulse profile and non-thermal power law emission [1]. This is the first clear detection of pulsation of magnetospheric pulsation from millisecond pulsars.

We searched for a possible empirical relation between magnetospheric emission and spin parameters using the luminosity of pulsed X-ray emission above 2 keV ($L_{X(pulse)}$). We found that the pulsars line up when plotted $L_{X(pulse)}$ against the rotation energy loss ($\dot{E}_{rot}$) as shown in Fig. 2 (a). A solid line represents $\left(\frac{L_{X(pulse)}}{10^{34} \text{ erg/s}}\right) = \left(\frac{\dot{E}_{rot}}{10^{38} \text{ erg/s}}\right)^{\frac{3}{2}}$. Here, we assumed the beaming angle to be 1 sr for all pulsars. The photon indices do no correlate strongly to $\dot{E}_{rot}$

We also plotted the γ-ray (0.1-2 GeV) pulsed luminosity in Fig. 2 (b) with a solid line showing $\left(\frac{L_{\gamma(pulse)}}{5\times 10^{34} \text{ erg/s}}\right) = \left(\frac{\dot{E}_{rot}}{10^{38} \text{ erg/s}}\right)^{\frac{1}{2}}$.

DISCUSSION

Most pulsar theories deals with γ-ray pulsation and try to explain its energy spectrum of pulsed emission, time profile, and phases lag relative to the radio pulse. In our study, we focussed on the two statistical relation for $L_{X(pulse)}$ and $L_{\gamma(pulse)}$ and try to average out geometrical dependence.

TABLE 1. Spectral parameters of the pulsed flux and upper limits.

Name	d(kpc)	L_X[a]	Pulsed Frac [b]	$L_{X(pulse)}$[c]	Γ[d]
B1821−24	5.1	$6.4^{+0.7}_{-0.4}\times 10^{33}$	15	$9.4\pm 0.4 \times 10^{32}$	1.2 ± 0.5 (P1)
					2.9 ± 0.6 (P2)
J0437−4715	0.14	$1.2^{+0.2}_{-0.2}\times 10^{30}$	<100	$<1.2\times 10^{30}$	−
Crab [4] [5]	2.0	1.05×10^{37}	6.5	$6.8\pm 0.4 \times 10^{34}$	1.81 ± 0.02 (P1)
					1.84 ± 0.03 (P2)
					1.63 ± 0.03 (IP)
B1951+32[e]	2.5	$\approx 6.1\times 10^{33}$	<14.1	$<8.6\times 10^{32}$	−
B0540−69[e]	50	$\approx 8.4\times 10^{36}$	16	$\approx 1.3\times 10^{36}$	≈ 2.0
Vela	0.5	$1.91^{+0.02}_{-0.02}\times 10^{33}$	<3.6	$<6.8\times 10^{31}$	−
B1706−44	2.4	$6.7^{+2.1}_{-2.1}\times 10^{32}$	<81.0	$<5.4\times 10^{32}$	−
B1046−58	3.0	$9.1^{+1.1}_{-1.1}\times 10^{32}$	<66.0	$<6.0\times 10^{32}$	−
B1509−58	4.2	$4.0^{+0.5}_{-0.5}\times 10^{34}$	33.0	$6.3\pm 0.9 \times 10^{34}$	1.1 ± 0.1
B1055−52 [6]	1.5	$2.7^{+1.4}_{-1.4}\times 10^{30}$	<100	$<2.7\times 10^{30}$	−
B1929+10	0.25	$1.2^{+0.3}_{-0.3}\times 10^{30}$	<69	$<8.3\times 10^{29}$	−
B1610−50[e]	7.3	$\approx 2.3\times 10^{34}$	<100	$<2.3\times 10^{34}$	−
Geminga [9]	0.16	$\approx 6.2\times 10^{29}$	40	$\approx 3.6\times 10^{29}$	≈ 1.5[f]
B0950+08	0.13	$1^{+550}_{-0.5}\times 10^{29}$	<100	$<1.0\times 10^{29}$	−
B1853+01 [7]	3.1	$1.4^{+0.3}_{-0.3}\times 10^{33}$	<10	$<1.4\times 10^{32}$	−
B0656+14 [6]	0.5	−	<100	$<1.7\times 10^{31}$	−

[a] Total luminosity in the 2-10 keV band in unit of erg/s fitted by power-law.
[b] Pulsed fraction and upper limits in the 2-10 keV band in unit of %. Upper limits set at 99 % confidence level.
[c] Pulsed luminosity and upper limits in the 2-10 keV band in unit of erg/s. Upper limits are set at 99 % confidence level.
[d] Photon Index in the 0.7-10 keV band.
[e] These pulsars parameters are from private communication with M. Hirayama & K. Tamura
[f] Determined by interpolating between the *ASCA* and the *CGRO* EGRET luminosity.

It should be noted that the millisecond pulsar PSR B1821−24 is consistent with the relation. This indicate that the same pulse emission mechanism works also for weak field pulsars.

Correlation between $L_{X(pulse)}$ and the characteristic age (T) was not found for X-ray pulsation, while a correlation between L_γ has been found for some years based mostly on EGRET data (e.g. $\left(\frac{L_\gamma}{\dot{E}_{rot}}\right) \propto T^{1.17}$ in Thompson et al.(1994) [10], and $\left(\frac{L_\gamma}{\dot{E}_{rot}}\right) \propto T^{1.28}$ in Nel et al. (1996) [11]). This relation $\left(\frac{L_\gamma}{\dot{E}_{rot}}\right) \propto T^{-1}$ can be reinterpreted as $L_\gamma \propto \dot{E}_{rot}^{\frac{1}{2}}$, since $T = \dot{E}_{rot}^{-\frac{1}{2}} B^{-1}$ and the magnetic field (B) is similar for all EGRET samples. In our case, we detected pulsation from the millisecond pulsar, pulsar with much weaker magnetic field. Thus, we could clearly favored $\dot{E}_{rot}$ for this scaling relation.

Some emission models introduce the thermal emission of initial cooling to generate secondary pairs by $\gamma - \gamma$ collision of primary γ-ray photons (e.g.

Romani (1996) [12]). The above result shows that at least the X-ray emission mechanism is independent of the thermal emission of initial cooling which relate to the age of the pulsar.

Both $L_{X(pulse)}$ and $L_{\gamma(pulse)}$ are correlated with $\dot{E}_{rot}$, while they depend differently and X-ray luminosity depends much to $\dot{E}_{rot}$. We suspect that the population of the parent pairs to emit X-ray and γ-ray is different and the γ-rays are from primary pairs, and X-rays are from secondary pairs.

As for possible interpretation of these empirical relations, we found following conditions naturally explain the properties shown above.

1. the primary particles are generated in the gap (accelerating region) where acceleration voltage is regulated independent of $\dot{E}_{rot}$ and the resultant energy spectrum of the primary $e^{\pm}$ is same.

2. The number of the primary $e^{\pm}$ injected in the gap ($\dot{N}_{inject}$) is proportional to the Goldreich-Julian current $\propto \dot{E}_{rot}^{\frac{1}{2}}$

3. The primary particle emit γ-ray photons via the curvature radiation.

4. The primary γ-ray generate secondary $e^{\pm}$ which have same shape of energy spectrum independent of $\dot{E}_{rot}$.

5. The secondary $e^{\pm}$ emit X-rays as synchrotron radiation.

6. The emission regions of γ-rays and X-rays toward us are located near the light cylinder and their radial extents are proportional to the size of the light cylinder.

7. The secondary $e^{\pm}$ are re-accelerated outside of the light cylinder and loose some of the energy as the pulsar nebula.

REFERENCES

1. Saito, Y., et al., *ApJ Lett.*, **477**, 37, (1997)
2. Sewaed, F., D.,and Wang, Z., R., *ApJ*, **332**, 199, (1988)
3. Becker, W., and Trümper, J. Submitted to *A&A* (1997)
4. Makishima, K., et al., *PASJ*, **48**, 171 (1996)
5. Knight, F. K., *ApJ*, **260**, 538
6. Greiveldinger, C., et al., *ApJ Lett.*, **465**, 35 (1996)
7. Harrus, I. M., Hughes, J. P., and Helfand, D. J., *ApJ Lett.*, **464**, 161 (1996)
8. Saito, Y., et al., *ASCANEWS*, **5**, 34 (1995).
9. Arnaud, K. A. et al. "ASCA science highlights", ASCA Guest Observer Facility, http://heasarc.gsfc.nasa.gov/docs/asca/science/science.html
10. Thompson, D. J., et al., *ApJ*, **436**, 229 (1994)
11. Nel, H. J., et al., *ApJ*, **465**, 898 (1996)
12. Romani, R. W. *ApJ*, **470**, 469 (1996)

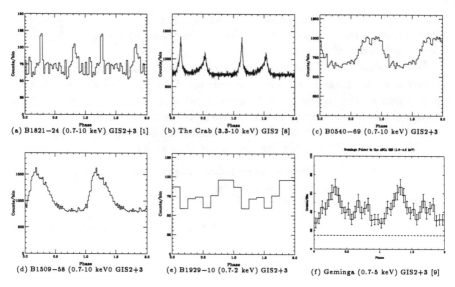

FIGURE 1. Pulse shapes of pulsars obtained with *ASCA*. Background is not subtracted. The pulse profile of PSR B0540-69 was suplied by M. Hirayama. Note, the pulse detected from PSR B1929+10 was lonly below 2 keV and may be thermal origin

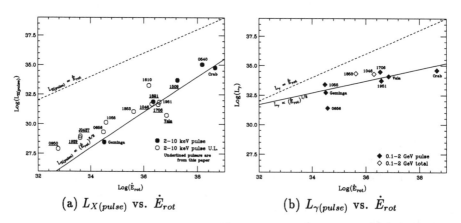

FIGURE 2. (a) Pulsed luminosity in the 2-10 keV band ($L_{X(pulse)}$) vs. rotation energy loss ($\dot{E}_{rot}$) and (b) that in 0.1-2 GeV band ($L_{\gamma(pulse)}$) vs. $\dot{E}_{rot}$ The solid angle for the pulsed emission is assumed to be 1 sr. For (a), filled circles represent the luminosity and open circles indicate the upper limit of 99 % confidence level. For (b), filled diamonds represent the luminosity of pulsed emission, and open diamonds indicate the total luminosity of the emission likely to be pulsed

Gamma Ray Pulsar Luminosities

M. A. McLaughlin[1], J. M. Cordes[1], & M. P. Ulmer[2]

[1] *Cornell University, Ithaca, NY, 14853*
[2] *Northwestern University, Evanston, IL, 60208*

Abstract. Gamma rays represent the dominant form of energy loss for several isolated pulsars and are most likely important in the energetics of many others. However, the small number of detected gamma ray pulsars makes modeling the gamma ray pulsar population and understanding gamma ray emission mechanisms difficult. We therefore apply a likelihood analysis to both detections and upper limits for pulsed gamma ray flux from known radio pulsars in order to constrain the luminosity law for gamma ray pulsars.

We have applied this analysis to pulsar detections and upper limits from both the OSSE and EGRET instruments. We find that the dependence of luminosity on period and magnetic field is very different in these two energy ranges, suggesting that different mechanisms are responsible for hard x-ray and high energy gamma ray emission. We use our luminosity law to predict which radio pulsars are likely to be strong emitters in the OSSE and EGRET energy ranges and discuss the implications of our results. Finally, we describe our ongoing analysis, in which OSSE and EGRET diffuse background measurements are used to further constrain the gamma ray luminosity law and provide important constraints on pulsar population parameters and the pulsar contribution to the diffuse gamma ray background.

INTRODUCTION

Although pulsars were discovered over 25 years ago, many problems in pulsar science remain unsolved. The emission mechanisms and magnetospheric physics of pulsars are not well understood. In particular, why some pulsars are bright gamma ray emitters and the relationship of the gamma ray beam to the radio beam remain undetermined. Furthermore, the galactic distribution of pulsars, especially gamma ray pulsars, is not well constrained.

Unfortunately, the small number of gamma ray pulsar detections makes modeling the gamma ray pulsar population and addressing the above issues difficult. For this reason, we have developed a likelihood analysis which employs all available information about gamma ray pulsars (detections, upper limits, and diffuse background measurements) to constrain their luminosity law and various pulsar population parameters, and estimate their contribution to the diffuse background.

We have applied our method to measurements taken with both the OSSE and the EGRET instruments. In this paper, we outline our method, present our preliminary results, discuss their implications, and describe our ongoing analysis.

THE METHOD

We model the pulsar gamma ray luminosity as

$$L = \gamma P^{-\alpha} B_{12}^{\beta} \tag{1}$$

where P is the period in seconds and B_{12} is the surface magnetic field in units of 10^{12} G. For each set of model parameters, we calculate the model flux as $F = L(\gamma, \beta, \alpha)/(\Omega D^2)$ for each pulsar with a detection or upper limit. In light of the wide pulse profiles of the detected gamma ray pulsars, we have set the solid angle into which the gamma rays are emitted, Ω, equal to 2π. In future analysis, this parameter will be varied as we include a beaming model. We calculate the distance D to each pulsar using the Taylor and Cordes model for Galactic electron density [10], except for pulsars with more available information. We account for distance errors by including a distance probability distribution function in our likelihood function.

We calculate the total likelihood of a model (i.e. one combination of parameters α, β, and γ) as $\mathcal{L}_{\text{tot}} = \mathcal{L}_d \mathcal{L}_u$, where the total likelihood for the detections, $\mathcal{L}_d$, is $\Pi_{i=1}^{N_d} \mathcal{L}_{i,d}$, and the total likelihood for the upper limits, $\mathcal{L}_u$, is $\Pi_{i=1}^{N_u} \mathcal{L}_{i,u}$. N_d and N_u are the number of detections and upper limits, respectively. Figure 1 shows $\mathcal{L}_{i,d}$ and $\mathcal{L}_{i,u}$, the likelihood functions for individual detections and upper limits. For detections, the likelihood peaks when model and detected fluxes are equal. The likelihood for upper limits is constant for all model fluxes less than the measured upper limit, and decreases for higher model fluxes.

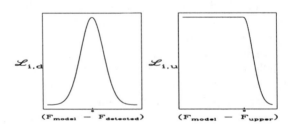

FIGURE 1. Left: likelihood function for detections; Right: likelihood function for upper limits.

THE DATA

In the OSSE energy range, our data consist of three detections (Crab, Vela, B1509-58) and 26 2σ pulsed flux upper limits [9,11]. At EGRET energies, we include all 6 pulsar detections (Crab, Vela, Geminga, B1055-52, B1706-44, B1951+32), and 3 σ upper limits on pulsed flux for 354 others [3,5,1]. For both energy ranges, we assume a photon spectrum that is power law, with a photon index of -2. However, this parameter will be allowed to vary in future analysis.

RESULTS

The luminosity law parameters yielding the highest likelihood in the **OSSE** energy range are $\boxed{\alpha = 8.3^{+0.7}_{-0.5},\ \beta = 7.6^{+0.8}_{-0.4},\ \log\gamma = 19.4^{+0.9}_{-1.5}.}$ The **EGRET** best parameters are $\boxed{\alpha = 1.8 \pm 0.1,\ \beta = 1.6 \pm 0.2,\ \log\gamma = 32.0 \pm 0.1.}$ For both cases, our analysis yields well defined likelihood maxima and unambiguous best values for parameters. Some example analysis outputs are shown in Figures 2 and 3.

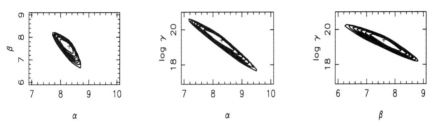

FIGURE 2. Contours of equal log likelihood for OSSE data as a function of parameter pairs. Crosses mark the maximum of the log likelihood.

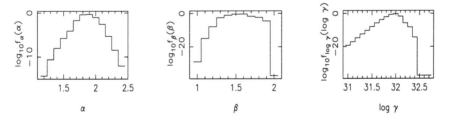

FIGURE 3. Marginalized probability distribution functions for EGRET data.

Our results are very different in the two energies ranges, suggesting that different mechanisms are responsible for pulsar hard x-ray and high-energy gamma ray emission. The OSSE results are roughly consistent with

ROSAT/ASCA detections, [6,2,8], which show stronger dependences on period and magnetic field than do EGRET results. Our luminosity models for both OSSE and EGRET differ from the standard $\dot{E}$ model [$\alpha = 4$, $\beta = 2$], and therefore produce a different ranking of pulsars as candidates for detection. While our model does not currently discriminate between polar cap and outer gap physics, we will attempt to do so by including a beaming model and orientation-angles derived from radio polarization data.

For this analysis, we included data on ALL pulsars with available upper limits. Separate analyses on millisecond pulsars and the general population of pulsars resulted in similar parameter values, suggesting that our model is an equally valid predictor of high energy flux for all pulsars. Using our model to calculate the predicted flux for all known pulsars in OSSE and EGRET energy ranges produces the following rankings.

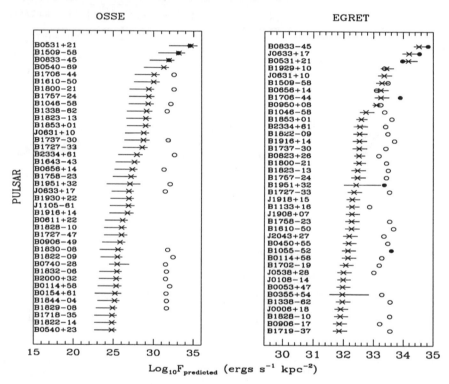

FIGURE 4. Crosses denote predicted gamma ray fluxes, with error bars. Solid dots mark detected fluxes, while open circles indicate upper limits.

We note that our current EGRET model underpredicts the fluxes of several detections, suggesting that more refinements must be made. Including a beaming model should enhance our analysis significantly. Our models will also improve as Cycle 7 observations and better timing ephemerides result in upper limits or detections for several of the highly ranked pulsars with no data (e.g. B0540-69, J0631+10, B1610-50) and more sensitive upper limits or detections for several others (e.g. B1929-10, B0656+14, B0950+08). In addition, including a diffuse term in the likelihood, so that $\mathcal{L}_{tot} = \mathcal{L}_{det}\mathcal{L}_{ul}\mathcal{L}_{dif}$, will further constrain the luminosity law and several other pulsar population parameters (i.e. initial spin period, magnetic field, braking index).

To calculate $\mathcal{L}_{dif}$ for a particular model, we first assume a spatial distribution for the pulsar population (currently a gaussian disk, exponential halo, and molecular ring). We may then calculate the total flux from pulsars in some direction given some combination of various pulsar population parameters (e.g. pulsar birthrate, inital spin period, magnetic field, braking index). Combining this with the measured diffuse flux at OSSE [7] an EGRET [4] energies, and allowing some maximum percentage of the total diffuse flux to be attributable to pulsars, we are able to constrain this percentage, further refine our luminosity model, and estimate best values for the various pulsar population parameters. For instance, Figure 5 illustrates the very strong dependence of the OSSE diffuse galactic center flux attributable to pulsars on initial spin period and magnetic field.

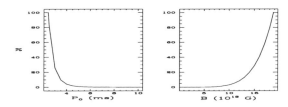

FIGURE 5. Percentage of OSSE diffuse flux attributable to unresolved pulsars as a function of initial spin period (left) and surface magnetic field (right).

REFERENCES

1. Arzoumanian et al., in preparation
2. Becker & Trumper, 1996, A&AS, 120, 69
3. Fierro et al., 1995, ApJ, 447, 807
4. Hunter et al., 1997, ApJ, 481, 205
5. Nel et al., 1996, ApJ, 465, 898
6. Ogelman, 1993, AdSpR, 131, 351
7. Purcell, 1997, in preparation
8. Saito, 1997, PhD. Thesis
9. Schroeder et al., 1995, ApJ, 450, 784
10. Taylor & Cordes, 1993, ApJ, 411, 674
11. Ulmer et al., 1997, in preparation

A New Class of Radio Quiet Pulsars

Matthew G. Baring[1] and Alice K. Harding

*Laboratory for High Energy Astrophysics, Code 661,
NASA Goddard Space Flight Center, Greenbelt, MD 20771*

Abstract. The complete absence of radio pulsars with periods exceeding a few seconds has lead to the popular notion of the existence of a high P *death line*. In the standard picture, beyond this boundary, pulsars with low spin rates cannot accelerate particles above the stellar surface to high enough energies to initiated pair cascades through curvature radiation, and the pair creation needed for radio emission is strongly suppressed. In this paper we postulate the existence of another pulsar "death line," corresponding to high magnetic fields B in the upper portion of the $\dot{P}$–P diagram, a domain where few radio pulsars are observed. The origin of this high B boundary, which occurs when B becomes comparable to or exceeds 10^{13} Gauss, is again due to the suppression of magnetic pair creation $\gamma \to e^+e^-$, but in this instance, primarily because of ineffective competition with the exotic QED process of magnetic photon splitting. This paper describes the origin, shape and position of the new "death line," above which pulsars are expected to be radio quiet, but perhaps still X-ray and γ-ray bright.

INTRODUCTION

Due to the broad range of period derivatives observed for isolated radio pulsars, the population spans over four decades in their estimated spin-down field strengths (e.g. Taylor, Manchester and Lyne 1993). However, none have inferred (dipolar) fields exceeding a few time 10^{13} Gauss, suggesting that there is an observational bias against observing high-field pulsars. This bias could be due to a complete absence of neutron stars with fields much above 10^{13} Gauss, or perhaps radio emission is somehow suppressed at such high field strengths, diminishing their observability. The former hypothesis has no intrinsic theoretical basis, and is contradicted by the suggestion (Duncan and Thompson 1992) that soft gamma repeaters have supercritical fields, above 10^{14} Gauss. Hence, it is of interest to examine the latter possibility, that high field pulsars do not produce radio emission; this is the focus of this paper.

---

[1] Compton Fellow, Universities Space Research Association

Magnetic one-photon pair production, $\gamma \to e^+e^-$, has traditionally been the only gamma-ray attenuation mechanism assumed to operate in polar cap models for radio (e.g. Sturrock, 1971) and gamma-ray pulsars (Daugherty & Harding 1982, 1996; Sturner & Dermer 1994), providing the means for both types of pulsars to radiate efficiently. Such an interaction can be prolific at pulsar field strengths, specifically when the photons move at a substantial angle $\theta_{\rm kB}$ to the local magnetic field. Pair creation has a threshold of $2m_ec^2/\sin\theta_{\rm kB}$. The exotic higher-order QED process of the splitting of photons in two, $\gamma \to \gamma\gamma$, will also operate in the high field regions near pulsar polar caps and until very recently, has not been included in polar cap model calculations. Magnetic photon splitting has recently become of interest in neutron star models of soft gamma repeaters (Baring 1995), and Harding, Baring and Gonthier (1997) have determined that splitting will play a prominent role in the formation of spectra for PSR1509-58, the gamma-ray pulsar having the lowest high-energy spectral turnover, around ~ 1 MeV.

The key property of photon splitting that renders it relevant to neutron star environs is that it has *no* threshold, and can therefore attenuate photons below the threshold for pair production, $\gamma \to e^+e^-$. Hence, when it becomes comparable to $\gamma \to e^+e^-$, it will diminish the production of secondary electrons and positrons in pair cascades. Since pairs are probably essential to the generation of radio emission (e.g. Sturrock 1971), such a "quenching" of pair creation can potentially provide a pulsar "death-line" at high field strengths; this phenomenon is the subject of this paper. While about a dozen radio pulsars have spin-down magnetic fields above 10^{13} Gauss, little attention was paid to $\gamma \to \gamma\gamma$ in pulsar contexts prior to the launch of the Compton Gamma-Ray Observatory (CGRO) in 1991 because until then, the three known gamma-ray pulsars had estimated field strengths of less than a few times 10^{12} Gauss. The detection of PSR1509-58 by the OSSE and COMPTEL experiments on CGRO provided the impetus to focus on high-field neutron star systems.

QUENCHING OF PAIR CREATION IN PULSARS

In polar cap models, pair cascades in radio and gamma-ray pulsars are initiated by relativistic electrons either via curvature radiation (e.g. Daugherty and Harding 1982), or by their resonant (magnetic) Compton scattering (e.g. Sturner and Dermer 1994) with thermal X-rays that emanate from the stellar surface. The cascades are perpetuated and amplified by synchrotron radiation interspersed with generations of pair creation. The nature of these three processes is well understood. For relativistic electrons with Lorentz factor γ, photons produced by these mechanisms are collimated to angles $\sim 1/\gamma$ to the direction of the electron's momentum. Furthermore, the produced radiation in each of these processes is highly polarized. The degree of polarization $\mathcal{P}$ of synchrotron and curvature emission (they are identical: see, for example,

Jackson 1975) is $(p+1)/(p+7/3)$ for power-law electrons $n_e(\gamma) \propto \gamma^{-p}$, and is in the 60%–70% range (e.g. see Bekefi 1966), favouring the production of photons in the $\perp$ state. Here the label $\parallel$ refers to the state with the photon's *electric* field vector parallel to the plane containing the magnetic field and the photon's momentum vector, while $\perp$ denotes the photon's electric field vector being normal to this plane. Likewise, it can be deduced from Herold's (1979) expression for resonant Compton scattering in the Thomson limit that the upscattered photons are predominantly in the $\perp$ state also, achieving $\mathcal{P} \sim 50\%$. Hence any pair creation in pulsars is primarily initiated by photons with polarization state $\perp$, thereby simplifying the considerations here.

In this paper we will assess how effective photon splitting is relative to magnetic pair creation in attenuating photons that are produced by these radiation mechanisms. In doing this, we propagated photons outwards from some point on or above the stellar surface, computing their attenuation probabilities for both of these processes. Of specific interest is the *escape energy*, ε_{esc}, the energy below which (for each mechanism) photons escape the neutron star magnetosphere without attenuation. We fully include the general relativistic effects of a Schwarzschild spacetime, and details of the geometry and propagation set-up are described at length in Harding, Baring and Gonthier (1997). In that work, which focused on the high-field test-case pulsar PSR1509-58, it was clearly demonstrated that pair creation is suppressed when photon splitting dominates it at higher field strengths. Also, generally, larger polar cap sizes favour the suppression of cascades and hence radio emission.

We computed the magnetic fields B_d for given polar cap angles (colatitudes) Θ for which the escape energies for splitting and pair creation were equal, so that for $B \gtrsim B_d$ pair creation is strongly suppressed by splitting. Since the multiplicity of pairs in a pulsar cascade rapidly becomes large in just a few generations, quenching is extremely abrupt and effective at high fields. Therefore, we expect a rapid decline in pair creation and hence also radio luminosity when B rises above B_d. Remembering that the polar cap size is coupled to the pulsar period P (in flat space time $\Theta \approx (2\pi/P)^{1/2} (R_{ns}/c)^{1/2}$, and we included general relativistic corrections to this), the so-defined (B, Θ) relationship becomes a critical curve on the P-$\dot{P}$ diagram. This curve delineates the phase spaces for radio-loud and radio-quiet pulsars, and examples are depicted in Figure 1, along with the latest population distribution from the Princeton pulsar catalogue. This boundary delineates a zone where pair creation is suppressed, like its long period counterpart. However, there is no pulsar evolution across the boundary (without field evolution): high field pulsars are born radio-quiet. Hence it is not a true death-line, just a border to the radio quiet region.

There are four examples of such boundaries in Figure 1 because the results differ according to the initial angles of photons with respect to B, and their original location. For photons that start out almost along the field, as in a curvature radiation-initiated cascades (the solid curves), $B_d \propto \Theta^{-1/3} \Rightarrow \dot{P} \propto$

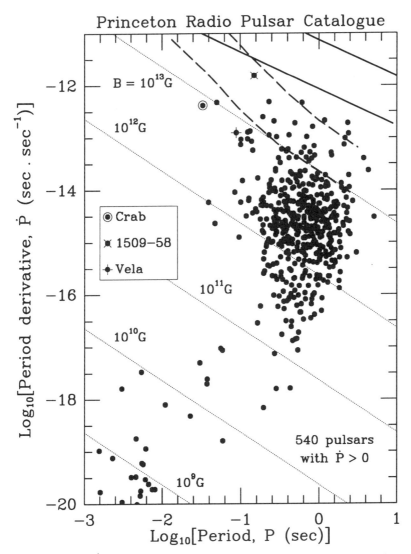

FIGURE 1. The P-$\dot{P}$ diagram for the latest Princeton Radio Pulsar catalogue (May 3, 1995: see also Taylor, Manchester and Lyne 1993) together with four possible high-field "death" lines (heavy solid and dashed curves), above which pulsars are radio-quiet. For the four cases depicted, the solid curves represent situations where photons that seed potential cascades (e.g. curvature radiation) are initially beamed very close to the field lines, while the dashed curves have such photons initially propagating at an angle of 0.57° to the local field. For both these scenarios, the lower curves are for emission from the stellar surface, and the upper ones are for photons originating half a stellar radius above the surface. The light dotted diagonal lines define contours of constant B, as labelled, and three gamma-ray pulsars in the diagram are marked as indicated in the inset.

$P^{-5/6}$. When photons initially have appreciable angles to **B**, as can be the case in resonant Compton-initiated (IC) cascades, photon splitting competes more effectively with pair creation for smaller polar cap sizes and the "death-line" drops to lower field strengths. Clearly the position of the line, which marks *surface* fields, strongly depends on the radius of photon origin since the physics of this problem couples to the magnitude of B. Hence there is, at present, significant uncertainty in the location of the radio-quiet boundary, principally because the location of the acceleration of primary electrons is not fully understood. Note that ground state pair creation also becomes prevalent for high B (Harding and Daugherty 1983), thereby aiding cascade cessation and lowering the radio-quiet boundary in the P-$\dot{P}$ diagram. Note also that there is marginal evidence for a drop in pulsar radio-luminosity when their fields exceed around 3×10^{13} Gauss, contrary to the slow increase with B seen for lower spin-down fields.

Clearly, when pair creation is suppressed and pulsars become radio quiet, they can still emit γ-rays prolifically, via the primary electrons and spectral reprocessing via splitting. Hence it is reasonable to conjecture that the radio quiet pulsars may actually be a class of objects formerly known as Gemingas. Motivations for searching for such γ-ray pulsars are therefore self-evident. Soft γ-ray observability is perhaps governed by $(B/P^2)^{1/2}$, favouring sources to the upper left of the P-$\dot{P}$ diagram, which should guide pulsar searches with OSSE and COMPTEL. On the other hand, hard γ-ray (i.e. EGRET, $>$ 100 MeV) observability implies no spectral cutoffs (as in PSR1509-58), and favours small Θ, in the upper right of the diagram (for high B). The very shapes of the "death-lines" in Figure 1 indicate that radio searches for high-field pulsars should focus on sub-second periods and high $\dot{P}$. In conclusion, in the polar cap model, the physics described here may well imply the existence of a radio-quiet pulsar population with high surface fields.

REFERENCES

1. Baring, M. G. 1995, *Ap. J. (Lett.)* **440**, L69
2. Bekefi, G. 1966 *Radiation Processes in Plasmas*, (Wiley & Sons, New York).
3. Daugherty, J. K. & Harding A. K. 1982, *Ap. J.* **252**, 337
4. Daugherty, J. K. & Harding A. K. 1996, *Ap. J.* **458**, 278
5. Duncan, R. C. & Thompson, C. 1992, *Ap. J. (Lett.)* **392**, L9
6. Harding, A. K., & Daugherty, J. K. 1983, in AIP Conf. Proc. 101, p. 194.
7. Harding, A. K., Baring, M. G. & Gonthier, P. L. 1997, *Ap. J.*, **476**, 246
8. Herold, H. 1979, *Phys. Rev. D* **19**, 2868
9. Jackson, J. D. 1975 *Classical Electrodynamics*, (Wiley & Sons, New York).
10. Sturner, S. J. & Dermer, C. D. 1994, *Ap. J. (Lett.)* **420**, L79
11. Sturrock, P. A. 1971, *Ap. J.* **164**, 529
12. Taylor, J. H., Manchester, R. N. & Lyne, A. G. 1993, *Ap. J. Suppl.* **88**, 529

The pulse profile of γ-ray pulsars and the emission region geometry

E. Massaro, M. Litterio

Ist. Astronomico, Unità GIFCO Roma 1, "La Sapienza", Roma

Abstract. γ-ray pulsars are grouped in two classes: "double-peaked" (D) and "single-peaked" (S) pulsars. D pulsars have phase separation between the peaks in the range 0.4–0.5, while S pulsars show broad peaks. We evaluated for two geometrical models of the emission region (polar cone and outer gap models) the probabilities to detect a D or a S pulsar, and the corresponding distribution of the peak separation and conclude that both these models are not able to explain the lack of D sources with separation less than 0.4.

INTRODUCTION

Rotation powered pulsars are the most important class of firmly identified galactic γ-ray sources. Six of them (Crab, Vela, Geminga, PSR B1055-52, PSR B1706-14, PSR B1951+32) have been detected by EGRET up to a few GeV [1]; furthermore, a possible evidence for pulsed emission at $E_\gamma > 50$ MeV has been recently reported for PSR B0656+14 [2]. Another pulsar, PSR B1509-58, has been observed by OSSE and BATSE but not seen by EGRET [3]. Ulmer [4] grouped the γ-ray pulsars in two types on the basis of the pulse shape: "double-peaked" (D) and "single peaked" (S) pulsars. The four D pulsars have phase separations δ between the two peaks in the range 0.4–0.5 and peak widths of about 0.1, while the three S pulsars show much broader peaks (about 0.3).

It is general opinion that the same spatial distribution of the emission can account for both types of pulse shapes, whose occurrence should depend on the viewing direction. It is not clear, however, if the active region is located over the polar cap region, within a few neutron star radii [5,6] or near the light cylinder as expected in the outer gap geometry [7].

In the present contribution we study the probability distributions of the peaks' phase separation for these two geometrical models. In particular, we will assume that a pulsar can be recognized of the D type only when $\delta > 0.25$;

for smaller values the two peaks could not be clearly distinguished and the source appears a S pulsar.

THE POLAR CAP MODEL

The emission pattern in the polar cap models is generally taken as an hollow cone whose axis coicides with the magnetic dipole axis. This model was considered in literature several years ago to explain the radio pulse shapes [8–10] and was also applied to the γ-ray emission by Massaro, Salvati and Buccheri (1979) [11]. Neglecting the pulse width, this geometry is defined by the following three angles: the two inclinations between the rotation axis and the magnetic axis α and the line of sight ζ, and the cone aperture ψ.

The plane α,ζ can be used to describe all the possible geometrical configurations deriving from the interplay between these angles; ψ divides this plane into 5 domains (Fig. 1), but only in two of them the pulsar can be observed: domain 1 in which the line of sight crosses only one cone (two peaks), and

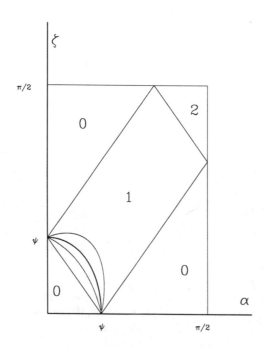

FIGURE 1. Domains in the plane α,ζ: 0 indicates the domains in which the pulsar is not seen. The curves for $\delta = 0.4$ and 0.5 (thick) are plotted

domain 2 in which both cones are seen (four peaks); in the domains indicated with 0 the line of sight is either always outside or always inside the emission cone. In the latter case a weakly modulated γ-ray emission could be observed, depending on the emission pattern thickness. The peak phase separation is given by

$$\cos(\delta\pi) = (\cos\psi - \cos\alpha \cos\zeta)/\sin\alpha \sin\zeta,$$

and the probability distribution of finding δ in a given interval, for random orientations of α and ζ can be computed via the double integration

$$P \propto \int_{\alpha_1}^{\alpha_2} \int_{\zeta_1(\alpha)}^{\zeta_2(\alpha)} d\alpha\, d\zeta\, \sin\zeta \sin\alpha \ .$$

Both the integration domains and the normalisation factors depend on the particular problem under consideration. We computed the integral in the domain delimited by the curves corresponding to a fixed value of δ, and normalised it over the entire domain 1. Numerical integrations were performed for different values of ψ in the interval ($5°,40°$): the expected fraction of observed pulsars with $\delta > 0.4$ resulted always between 0.15 and 0.18, almost independent by the value of ψ.

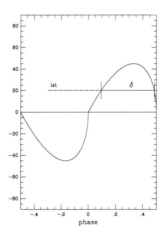

FIGURE 2. The analytical approximation of the Chiang–Romani outer gap emission pattern for $\alpha = 86°$. The line of sight trajectory (lst) for $\zeta=70°$ and the resulting peak sepration are shown.

PEAK SEPARATION IN THE OUTER GAP MODEL

The spatial distribution and the angular pattern of the γ-ray emission expected from the outer gap model has been computed by Chiang and Romani [7]. These authors considered only the case of a nearly perpendicular rotator ($\alpha = 86°$ - the value estimated by Rankin [12] for Crab), but the further computations by Romani and Yadigaroglu [13] showed that this pattern is not strongly dependent on the pulsar parameters and therefore it can considered representative of a rather wide interval of α (say, $\alpha > 75°$).

Our estimate of the probability to observe a peak separation δ was derived by means of a simple but accurate analytical approximation of the CR pattern (Fig. 2). The two γ-ray peaks correspond to the intersections of this curve with a horizontal line at $\pi/2 - \zeta$ and their phase separation is equal to the difference between two solutions of a resulting cubic equation.

Assuming a random distribution of ζ we computed the corresponding cumulative probability distribution of the peak separation $P(> \delta)$ (Fig. 3). We found $P(\delta > 0.4) = 0.40$, while the probability to find $0.25 < \delta < 0.40$

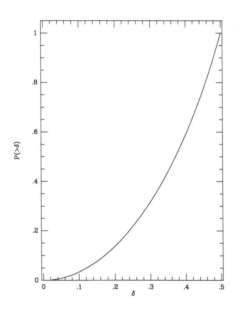

FIGURE 3. The cumulative probability distribution of δ computed for the pattern shown in Fig. 2.

is 0.39. We then expect that ∼50% of the observed D pulsars should have a peak separation less than 0.4. Furthermore, those with $\delta < 0.25$, which could be classified as S rather than D sources, are expected to be only ∼ 20 % of all detected pulsar.

CONCLUSIONS

Our analysis, based on the assumption that α and ζ are randomly distributed, has shown that the expected fraction of D pulsars with $0.25 < \delta < 0.40$ should range between 0.50 and 0.85, depending on the emission model. No γ-ray pulsar with a δ value in this interval has been detected and, therefore, other hypotheses must be invoked. The possibility that the cone, in which the emission is confined, is nearly aligned with the rotation axis [6] requires large values of the probability distribution of α only for values smaller or comparable to ψ, quite different from that due to a random orientation. A statistical analysis of the radio pulse width distribution [14] provided indications for a random distribution of α. Previous computations [15,16] based on the same physical processes, have shown that γ-ray emission is efficient also for highly inclined rotators. The inconsistency between the data and this model is still unsolved. The outer gap model is in a better agreement with the data and seems to be more favorite. Likely a distortion of the the field lines from the dipole geometry close to the light cylinder is sufficient to increase the probability of large peak separations.

REFERENCES

1. Nolan,P.L. et al., 1996, AAS 120, 61
2. Ramanamurthy,P.V. et al., 1996, ApJ 458,755
3. Ulmer, M. et al., 1993, ApJ 417, 738
4. Ulmer, M., 1994, ApJS 90, 789
5. Daugherty,J.K., and Harding,A.K., 1994, ApJ 429, 325
6. Daugherty,J.K., and Harding,A.K., 1996, ApJ 458, 278
7. Chiang,J., and Romani,R.W., 1994, ApJ 436, 754
8. Backer, D.C., 1976, ApJ 209, 895
9. Prószyński, M., 1979, AA 79, 8
10. Rankin, J.M., 1991, in "Neutron Stars: Theory and Observation", (J. Ventura and D. Pines eds.), p. 349
11. Massaro,E., Salvati,M., and Buccheri,R., 1979, MNRAS 189, 823
12. Rankin, J.M., 1990, ApJ, 352 247
13. Romani,R.W., and Yadigaroglu, ApJ 438, 314
14. Gil,J.A. and Han,J.L., 1996, ApJ 458, 265
15. Salvati,M., and Massaro,E., 1978, AA 67, 55
16. Massaro,E., and Salvati,M., 1979, AA 71, 51

Geometry of Pulsar X-Ray and Gamma-Ray Pulse Profiles

Alice K. Harding and Alex Muslimov[†]

NASA/GSFC Greenbelt MD 20771
[†]*NRC/NAS Resident Research Associate*

Abstract.
We model the thermal X-ray profiles of Geminga, Vela and PSR0656+14, which have also been detected as γ-ray pulsars, to constrain the phase space of obliquity and observer angles required to produce the observed X-ray pulse fractions and pulse widths. These geometrical constraints derived from the X-ray light curves are explored for various assumptions about surface temperature distribution and flux anisotropy caused by the magnetized atmosphere. We include curved spacetime effects on photon trajectories and magnetic field. The observed γ-ray pulse profiles are double peaked with phase separations of 0.4 - 0.5 between the peaks. Assuming that the γ-ray profiles are due to emission in a hollow cone centered on the magnetic pole, we derive the constraints on the phase space of obliquity and observer angles, for different γ-ray beam sizes, required to produce the observed γ-ray peak phase separations. We compare the constraints from the X-ray emission to those derived from the observed γ-ray pulse profiles, and find that the overlapping phase space allows both obliquity and observer angles to be smaller than $20 - 30^0$, implying γ-ray beam opening angles of less than $30 - 35^0$.

INTRODUCTION

The pulsars Geminga, Vela, and PSR 0654+14 have two-component X-ray spectra ([6], [4] [13]), with the soft component likely due to thermal emission from the whole stellar surface. We model the soft X- and γ-ray light curves for these pulsars surveying all possible obliquity and observer angles (Figure 1, Left). We take into account the effects of strong gravity on X-ray emission: the decrease in pulse fraction and increase in pulse width, caused by the fact that any intrinsic flux variations with pulsar rotation become smeared out by general-relativistic light bending (cf. [7]). The effect of a strong magnetic field on X-ray emission is twofold: 1) the anisotropization of the surface tem-

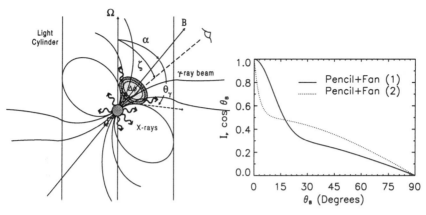

FIGURE 1. Left: Schematic illustration of polar cap γ-ray beam geometry, Right: Two different profiles assumed for the anisotropic distribution of thermal X-ray emission.

perature distribution (the transverse heat conductivity is strongly suppressed due to magnetization of electrons, which results in a surface temperature at the magnetic pole a factor of 1.5-2.5 higher than at the magnetic equator); 2) beaming of the thermal emission along the direction of the magnetic field in a strongly magnetized atmosphere (when $h\nu_c \gg kT$, where $\nu_c = eB/2\pi m_e c$ is the electron cyclotron frequency) resulting from a lower opacity along the magnetic field than in the transverse direction (see e.g. [5], and [8]).

DESCRIPTION OF THE MODEL

X-Ray Emission

We consider the thermal X-ray emission from the whole stellar surface, including the effect of relativistic light bending. The anisotropy of the surface temperature distribution, due to anisotropy of the electron thermal conduction in the NS crust in a strong magnetic field ($\sim 10^{12} - 10^{13}$ G), is described by an approximate formula (see [3]): $T(\Theta) = T_p(\cos^2\Theta + \chi^4 \sin^2\Theta)^{1/4}$, where T_p is a temperature at the magnetic pole, Θ is the angle between the local normal to the surface and tangent to the field line, $\chi = T_{eq}/T_p \approx 0.3 - 0.6$, and T_{eq} is a temperature at the magnetic equator. We assume a dipole magnetic field modified by the static part of the gravitational field. We also take into account the effect of anisotropic X-ray emission. In our calculations the emission pattern consists of a pencil component (peaked along the direction of the magnetic field) and a broad fan component. Plots (Figure 1, Right) of the function $I_\nu(\theta_B)\cos(\theta_B)$ (where I_ν is the intensity of radiation, and θ_B is the angle between the wave vector and direction of a magnetic field) are similar to those presented by [16] (see Fig. 2a and 2b). We assume a NS mass of 1.4 $M_\odot$ and radius of 10 km. We also assume that the surface effective temperature (at the magnetic pole) and polar value of the magnetic field strength are respectively 5×10^5 K (Geminga), 10^6 K (Vela and PSR 0654+14) and $\sim 3 \times 10^{12}$ G (Geminga), $\sim 6 \times 10^{12}$ G (Vela and PSR 0654+14).

Gamma-Ray Emission

Most of the observed γ-ray pulsars have double-peaked profiles with peak phase separations of 0.4 - 0.5 ([15]). The γ-ray emission in polar cap models ([1], [2]; [14]) is a hollow cone, with opening angle θ_γ, centered on the magnetic pole (Figure 1), producing either double-peaked or single-peaked pulse profiles depending on observer orientation. When $\theta_\gamma \sim \alpha$, an observer may see a broad double-peaked γ-ray pulse profile with the peak separation $\Delta\phi$ given by $\cos \Delta\phi = (\cos \theta_\gamma - \cos \alpha \cos \zeta)/(\sin \alpha \sin \zeta)$, where α and ζ are the angles of the magnetic axis and observer direction wrt the rotation axis. The γ-ray beam opening angle is determined approximately by the locus of the tangent to the outermost open field line:

$$\tan \theta_\gamma \simeq \frac{3\theta_{pc}(1 - \theta_{pc}^2 r/R)^{1/2}(r/R)^{1/2}}{3(1 - \theta_{pc}^2 r/R) - 1}, \quad (1)$$

where θ_{pc} is the polar cap half-angle, r is the radius of the emission region and R is the NS radius.

If $r > R$ and/or the polar cap half-angle is larger than the standard value, $\theta_{pc} \simeq (\Omega R/c)^{1/2}$, then θ_γ could be as large as 20^0 (Figure 1, Left). General relativistic effects on the photon trajectory and on the dipole magnetic field structure introduce small corrections to θ_γ and θ_{pc}.

MAIN RESULTS

The observed thermal X-ray profiles can be produced by an anisotropic emission pattern even for small obliquity angles. If the pulsed thermal soft X-ray emission for Geminga, Vela, and PSR 0654+14 is dominated by a beamed component (produced e.g. by the effect of anisotropic opacity of a magnetized atmosphere), then the model X-ray light curves for these pulsars agree very well with the observed pulsed fractions and pulse widths of their soft X-ray emission (at a median energy of 0.18 keV). The modulation which produces the pulsed fractions of $10\% - 20\%$ observed for these pulsars is primarily due to the anisotropic emission rather than the variation in the surface temperature, as found by Shibanov et al. (1996) [12]. In Figure 2, the calculated contours of constant opening angle θ_γ of the γ-ray beam for the measured $\Delta\phi = 0.5$ (Geminga) and 0.4 (Vela and PSR 0654+14), are superposed on the contours of constant X-ray pulse fraction. The simultaneous modelling of the soft (thermal) X-ray and γ-ray emission for Geminga, Vela, and PSR 0654+14 constrains the phase space for the possible obliquity and observer angles, and favors relatively small values for these angles, but γ-ray beam opening angles of at least 10^0 (Figure 2).

SUMMARY

- The possibility of beaming of the thermal X-ray emission in Geminga, Vela, and PSR 0654+14 provides a consistent explanation for their observed X-ray light curves and strongly supports the polar cap models for their γ-ray emission.

- For the anisotropic X-ray emission, the observed pulsed fraction and pulse width are much less sensitive to the effective temperature and are determined by the degree of a beaming.

- The obliquity and/or observer angle in Geminga, Vela, and PSR 0654+14 may be less than 30^0. The γ-ray opening angles must be at least 13^0 for Geminga and at least 5^0 for Vela and PSR 0654+14.

REFERENCES

1. Daugherty, J. K., & Harding A. K.: *Ap. J.* **429**, 325 (1994).
2. Daugherty, J. K., & Harding A. K.: *Ap. J.* **458**, 278 (1996).
3. Greenstein, G., & Hartke, G. J. *Ap. J.* **271**, 283 (1983).
4. Halpern, J. P., & Wang, Y.-H. *Ap. J.* **477**, 905 (1997).
5. Kaminker, A. D., Pavlov, G. G., & Shibanov, Yu. A. *Ap. & Sp. Sci.* **86**, 249 (1982).
6. Ögelman, H. in *The Lives of the Neutron Stars*, ed. A. Alpar, Ü. Kiziloglu, & J. van Paradijs, Dordrecht: Kluwer, 1995, p. 101.
7. Page, D. *Ap. J.* **442**, 273 (1995).
8. Pavlov, G. G., Shibanov, Yu. A., Ventura, J., & Zavlin, V. E. *Astron. & Astroph.* **289**, 837 (1994).
9. Pechenick, K. R., Ftaclas, C., & Cohen, J. M. *Ap. J.* **274**, 846 (1983).
10. Riffert, H., Mészáros, P. *Ap. J.* **325**, 207 (1988).
11. Shibanov, Yu., A., Zavlin, V. E., Pavlov, G. G., & Ventura J. *Astron. & Astroph.* **266**, 313 (1992).
12. Shibanov, Yu., A. et al., Proc. 17th Texas Symp. (Munich) (1996).
13. Strickman, M. S., Harding, A. K. & De Jager, O. C., these proceedings (1997).
14. Sturner, S. J. & Dermer, C. D.: *Ap. J.* **420**, L79 (1994).
15. Thompson, D. J. in *Pulsars: Problems and Progress* (IAU Coll. 160), ed. S. Johnston, M. A. Walker & M. Bailes, 1996, p. 307.
16. Zavlin, V. E., Shibanov, Yu. A., & Pavlov, G. G. *Astron. Lett.* **21**, 149 (1995).

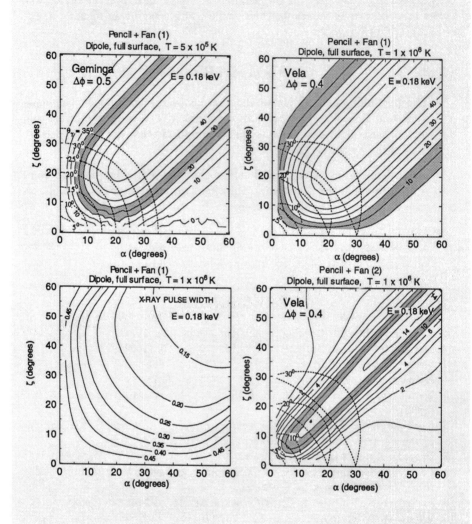

FIGURE 2. Upper-left, upper-right and lower-right panels: solid lines are contours of constant X-ray pulse fraction (%) with shaded contours denoting observed pulse fraction range, dashed lines are contours of constant γ-ray beam half-angle (degrees), $\Delta\phi$ is γ-ray pulse phase separation and E is X-ray energy; Lower left panel: contours of constant X-ray pulse width. Plots for Vela also apply to PSR0656+14.

A New Method for Statistical Study of Gamma-Ray Phase Curves of Radio Pulsars

Anton Chernenko

Space Research Institute,
Profsoyuznaya 84/32, 117810 Moscow, Russia

Abstract.
I analyze data on 6 radio pulsars in order to reveal systematic interrelation of their gamma and radio phase curves. The method used allows to rigorously prove a significant evolution effect that is the drift of the pulses in phase as the pulsar's period increases. On the other hand, no significant evolution with age is found. The revealed dependency of gamma-ray phase curves on period allows to construct a powerful detection algorithm that is at least an order of magnitude more powerful than those based on χ^2 and H-tests. The particular kind of the found evolution favors the outer gap model of gamma-ray emission, provided the emission region is closer to the surface in long period pulsars.

THE METHOD

The analysis consists of the following steps:

- A low phase resolution gamma-ray curves are built so that their zero phases are centered on the radio peaks. When there are two radio peaks, polarization measurements [4] help to choose the one marking the moment of minimum inclination of the line of sight to the magnetic axis.

- The background is subtracted under assumption that the minimal apparent count corresponds to the background one.

- Finally, normalized phase curves f_i (where i indicates the phase bin) are obtained by dividing the background-free data by the total number of counts.

In Fig.1 the phase curves of the pulsars are presented. One may see that for all pulsars $f_{[0.8-1.0]} \sim 0$. The distribution of $f_{[0.4-0.6]}$ is compact around ~ 0.4. The distributions of $f_{[0.0-0.2]}$ and $f_{[0.2-0.4]}$ spread from 0.0 to about 0.6 and are anti-correlated (linear correlation equals to -0.74).

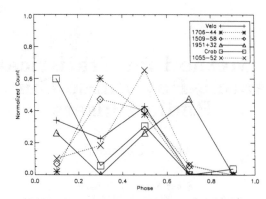

FIGURE 1. Gamma-ray phase curves of the 6 pulsars. The radio peaks are at phase 0.0.

EVOLUTION OF GAMMA-RAY PHASE CURVES

To reveal possible evolution of phase curves with period and/or age I introduce a crude phenomenological model of emission that is sketched in Fig.2. The model contains the following parameters: beginning of the first pulse ϕ_1, beginning of the second pulse ϕ_2, width of the pulses $\Delta\phi$, and total count of the first pulse c_1. Owing to normalization $c_2 = 1 - c_1$. By evolution model I define the one where ϕ_1, ϕ_2, and c_1 vary with pulsar period or age. The quality of the fit was tested using the global χ^2 value:

$$\chi^2 = \sum_{ij} \left(\frac{f_{ij} - \hat{f}_{ij}}{\delta f_{ij}} \right)^2 \tag{1}$$

where f_{ij} is the count accumulated in j-th phase interval of the j-th pulsar, $\hat{f}_{ij}$ is the model value and δf_{ij} is the Poisson error. Three models that are described below were tested. The results are presented in Table 1:

- No-evolution model. The model parameters are the same for all pulsars.

- Evolution with period. Parameters ϕ_1 and ϕ_2 are monotonous functions of period, having therefore different values for different pulsars.

- Evolution with age, ϕ_1 and ϕ_2 are monotonous functions of age.

In both evolution models the χ^2 values where not improved when the pulse amplitude c_1 were made also evolutional. It is apparent that the evolution with period is significant while the evolution with age is not. The evolution with period manifests itself primarily as the movement of the first pulse (from $\phi_1 = 0.05$ to 0.35). The second pulse moves about 3 times more slowly from ($\phi_2 = 0.38$ to 0.52). The constant width of the pulses $\Delta\phi = 0.07$. The first

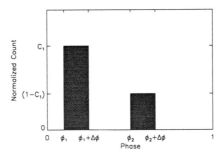

FIGURE 2. The phenomenological model of the gamma-ray phase curve. The radio peak is at phase 0.0.

TABLE 1.

Model	Best fit χ^2	D.O.F	F-test probability[a]
No evolution	360	20	—
Evolution with period	50	18	$5 \cdot 10^{-5}$
Evolution with age	197	18	0.28

[a] For comparison with the no-evolution model

pulse contains $c_1 = 0.6$ of the total count and the second one – the remaining 0.4. Since the χ^2 value remains large we are unable to assign errors to the best fit parameters. In Fig.3 I present the measured phase curves together with the predictions of the best fit model with evolution with period.

DETECTION OF FAINT PULSARS

The principle property of detection methods based on χ^2 and H-tests is that none particular pulse shape is assumed. Thus, a phase curve is tested against the uniform distribution.

On the contrary, I propose a test that employs the found systematics of distribution of gamma-rays in phase. The test is based on the postulate that given the pulsar period one can identify background and signal phase intervals. Then the null hypothesis will be that the actual number of counts accumulated over the signal interval is consistent with the prediction of the background model estimated using background interval.

Results of the previous section allows one to propose that for short period pulsars ($P < 0.1$s) the background interval shall be [0.2-0.4; 0.6-1.0] and signal interval be [0.0-0.2; 0.4-0.6], while for long period pulsars ($P > 0.1$s) they shall be: [0.0-0.2; 0.6-1.0] and [0.2-0.6], correspondingly.

Let C_B be the number of counts accumulated in the background phase interval, $\Delta\phi_B = 0.6$ – the total width of the backgrpound interval, C_S – the number of counts over the signal interval, and $\Delta\phi_B = 0.4$ – the total width of

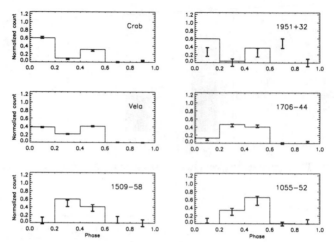

FIGURE 3. The observed phase curves f of individual pulsars are shown with the predictions $\hat{f}$ of the best fit "evolution with period" model (thin line).

signal interval. Then the probability of a fluctuation larger that C_S is:

$$\mathcal{P}\{C > C_S\} = 1 - \sum_{k=0}^{C_S} \int_0^\infty P(k,B)\rho(B)dB \qquad (2)$$

where B is the true unknown background rate. Poissonian probability to have k counts in the signal interval for given B is

$$P(k,B) = \frac{(B\Delta\phi_S)^k \exp(-B\Delta\phi_S)}{k!}, \qquad (3)$$

while posterior distribution of B given actual background count C_B is [3]

$$\rho(B) = \frac{\Delta\phi_B(B\Delta\phi_B)^{C_B} \exp(-B\Delta\phi_B)}{C_B!}. \qquad (4)$$

After substituting eqs. (3) and (4) and integration eq.(2) reduces to:

$$\mathcal{P}\{C > C_S\} = 1 - \frac{(\Delta\phi_B)^{C_B+1}}{C_B!} \sum_{k=0}^{C_S} \frac{(\Delta\phi_S)^k (C_B+k)!}{k!} \qquad (5)$$

It is interesting to compare the power of this test with that of χ^2 and H-tests, using the faintest pulsar PSR 1055-52 as the example. In this case $C_S = 59$, $C_B = 25$, and from eq.(5) we obtain the probability of chance as small as $2 \cdot 10^{-8}$. That is about 400 times smaller than that obtained in χ^2 test and about 30 times smaller than that from H-test.

Another field where the found systematics would help to increase sensitivity is the telescopic search of pulsars as points sources. In this case only those photons that fall to signal intervals should be used for the generation of the sky maps.

Finally, the photons collected from the directions of many undetected pulsars may be combined separately to background and signal intervals hopefully giving a significant collective detection.

IMPLICATION FOR EMISSION MODELS

One important theoretical inference from radio/gamma-ray phenomenology should be the geometry of gamma-ray emission. Thus far, two theoretical frameworks have been developed: polar cap model (e.g. [2]) and outer gap model (e.g. [1]).

Although the polar cap model can produce all apparent gamma-ray curves it has difficulties explaining why the radio peak is *always* located beyond the range occupied by the gamma-ray emission.

On the other hand, the outer gap geometry naturally predicts this effect. Besides that it predicts that evolution of the gap leads to a considerable change in position of the first gamma-ray pulse but not of the second one [5]. More specifically, the observed shift of the first pulse from 0.05 ($P = 0.033$) to 0.35 ($P = 0.197$) relative to the radio peaks implies that the outer limit of the 'active' gap moves from about $r = 0.9\ r_{LC}$ to about $r = 0.3\ r_{LC}$.

REFERENCES

1. Cheng K.S., Ho C, & Ruderman M.A., 1986, ApJ, 300, 522
2. Daugherty J.K. & Harding A.K., 1982, ApJ, 252, 337
3. Loredo T.J., in Feigelson E.D., Babu G.J., eds., Proc. of the Statistical Challenges in Modern Astronomy., Springer-Verlag, NY, p. 275 (1992)
4. Lyne A.G., and Manchester R.N., 1988, MNRAS, 237, 477
5. Smith F.G. 1986, MNRAS, 219, 729

Evidence for spontaneous magnetic field decay in an isolated neutron star

John C. L. Wang

Department of Astronomy, University of Maryland, College Park, MD 20742-2421

Abstract. The recently discovered source RX J0720.4-3125 is most likely an old isolated neutron star accreting from surrounding media. With this interpretation, we argue that if it was born as a canonical pulsar with spin period $P_i \lesssim 0.5$ s, the magnetic field on this star *must* have decayed since birth. [1] With a birth field $B_i \sim 10^{12}$ G, we find decay timescales $\gtrsim 10^7$ yrs for power law decay or $\gtrsim 10^8$ yrs for exponential decay. A measured period derivative $\dot{P} \lesssim 10^{-16}$ s s^{-1} would be consistent with an old accreting isolated neutron star.

OBSERVATIONS

The source RX J0720.4-3125 is an unidentified steady soft X-ray source seen by the Einstein IPC, EXOSAT LE, and the ROSAT PSPC and HRI instruments. The ROSAT PSPC count rate (0.1 – 2.4 keV) is 1.6 cts/s, and the best fit X-ray spectrum is given by a black body of $kT = 80$ eV with a hydrogen absorption column of $N_H = 1.3 \times 10^{20}$ cm^{-2} (Haberl et al. 1996, 1997). With this spectrum, the unabsorbed photon energy flux is $F_\nu(0.1 - 2.4 \text{ keV}) \approx 1.7 \times 10^{-11}$ erg cm^{-2} s^{-1} (K. Arnaud, priv. comm.), giving a source luminosity of $L_X \equiv L(0.1 - 2.4 \text{ keV}) = 1.9 \times 10^{31} d_{100}^2$ erg s^{-1}, where $d = 100 d_{100}$ pc is the distance to the source.

In all pointed ROSAT observations, there is a steady *periodic* modulation in the X-ray flux with an 8.39 s period (Haberl et al. 1996, 1997), which we interpret as the rotation period of the source. The position of the source is (Haberl et al. 1997; J2000) $\alpha =$ 7h, 20m, 24.90s; $\delta =$ -31° 25′ 51.3″ (with ±3″ uncertainty). The corresponding Galactic coordinates are $l = 244°$, $b = -8°$. The absence of an optical counterpart down to $V \sim 20.7$ places a lower limit on the X-ray to optical flux ratio of ~ 500 (Haberl et al. 1997).

From the hydrogen column density and the matter distribution of the interstellar medium along the line of sight (cf. Dickey and Lockman 1990; Welsh

---

[1] For a more detailed discussion, see Wang (1997)

et al. 1994, their Figure 3; Wang and Yu 1995), we obtain a rough distance estimate of ~ 100 pc. Requiring the hot spot area to be much less than the star's surface area for pulsations to be observed gives $d_{100} \ll 5.3R_6$, where $R = 10^6 R_6$ cm is the stellar radius. The conclusions regarding magnetic field decay are not sensitive to the distance estimate.

The observational evidence points consistently to a nearby isolated neutron star as the source. We assume throughout this paper that the neutron star has $M = 1.4 M_\odot$ and $R = 10$ km.

ARGUMENTS AGAINST A YOUNG ORIGIN

If the source is a young (age$\sim 10^4$–10^6 yrs) active pulsar, its magnetic dipole spin-down power $\dot{E}_R \gg L_X$ (e.g., Ögelman & Finley 1993), yielding $B_{12} \gg 140 d_{100} P_{8.39}^2$, where $P = 8.39 P_{8.39}$ s is the current observed pulsar period and $B = 10^{12} B_{12}$ G. This qualifies the source as a "magnetar" (Duncan & Thompson 1992). The young spin down age of such an object — $\tau_{sp} = \frac{P}{2\dot{P}} \ll 2 \times 10^5 d_{100}^{-2} P_{8.39}^{-2}$ yrs — together with its close proximity (~ 100 pc), however, argues against a young neutron star/pulsar origin. In addition, no radio emission has been detected from this object even though it would lie well above the (extrapolated) observed radio pulsar death line ($B_{12} > 15 P_{8.39}^2$; e.g., Chanmugam 1992).

A Geminga-type (γ-ray loud, radio-quiet) pulsar would have $\dot{E}_R \sim L_\gamma \gg L_X$. Using $L_\gamma/L_X \sim 10^3$ as for Geminga (e.g., Halpern & Holt 1992) gives $\tau_{sp} \sim 200$ yrs, which argues against this source being a Geminga-type pulsar. This is consistent with absence of EGRET detection (Haberl et al. 1997). Thus, this source *could not* have created the Local Bubble even though the line of sight to this source lies only $\sim 15°$ away from the center of an evacuated "tunnel" ($l \sim 230°$) which appears to define the Local Bubble in this direction (cf. Welsh et al. 1994). (For reference, the Geminga pulsar is on the other side of this tunnel with $l \approx 195°$.)

ACCRETING OLD NEUTRON STAR

For accretion to occur, we require $r_A < r_{co}$ (cf. Illarionov & Sunyaev 1975), where r_A is the Alfvénic radius where the energy density in the accretion flow balances the local magnetic pressure, and r_{co} is the corotation radius. This implies $B < B_{crit} \approx 10^{10} V_{20}^{-3/2} n_H^{1/2} P_{8.39}^{7/6}$, G for the present day surface field, where $V = (v^2 + c_s^2)^{1/2} = 20 V_{20}$ km s^{-1} with v being the star's speed relative to the ambient medium, and c_s and n_H are the sound speed and hydrogen number density of the medium, respectively. For *pulsed* emission to occur, we require $r_A \gg R$ (e.g., Shapiro & Teukolsky 1983). This implies $B \gg 10^5 V_{20}^{-3/2} n_H^{1/2}$, G. Unfortunately, it would not be possible to measure such a field directly through cyclotron (emission) line observations, since this requires $B_{12} \sim 1$ (Nelson et al. 1995).

The neutron star's spin evolution is divided into three phases: magnetic dipole, propeller, and accretion phases (cf. Blaes & Madau 1993; Lipunov et al. 1997; Wang 1997). The star enters the propeller phase when its period $P > P_0 \approx 4.4 \, V_{20}^{1/2} n_H^{-1/4} B_{12}^{1/2}$, s, and it enters the accretion phase when $P > P_a \approx 470 \, V_{20}^{9/7} n_H^{-3/7} B_{12}^{6/7}$ s. Since $B < B_{crit}$, we have $P_0 < P_{crit} \approx 0.42 \, V_{20}^{-1/4}$ s.

If the star's field *does not* decay, and its birth period $P_i < P_0 < P_{crit}$, then it will take $\gtrsim 10^{11}$ yrs — much longer than a Hubble time ($\sim 10^{10}$ yrs) — just to spin down to P_0. If its birth period $P_0 < P_i < P_{crit}$, it goes directly to the propeller phase but would still take $\gtrsim 10^{10}$ yrs to spin down to P_a.

Thus, if the star was born as a canonical isolated neutron star with $P_i < P_{crit}$, its field at birth *must* have been stronger than its present day field so that sufficient spin down can occur to allow accretion onto the star, that is, *the stellar magnetic field must have decayed.*

In general, the star's spin and magnetic field history is determined by P_i, B_i, the field decay law, and decay timescale t_d, though the dependence on P_i is very weak whenever $P_i \ll P_a$.

In Figure 1, we illustrate sample evolutionary tracks in B-P space for both a power law decay law (solid; e.g., Narayan & Ostriker 1990; Sang & Chanmugam 1987), and an exponential decay law (dot-dash; e.g., Ostriker & Gunn 1969). Spin down spans the magnetic dipole, propeller, and accretion phases. In both models, the star spins down to 8.39 s after a few billion years.

For $B_{i,12} \sim 1$, we find $t_d \gtrsim 10^7$ yrs for power law decay models (e.g., Urpin et al. 1994), while $t_d \gtrsim 10^8$ yrs for exponential decay models.

During the accretion phase, the spin down rate is $-\dot{\Omega}_{brk} \sim (\dot{I}/I)\Omega \sim (\frac{\dot{M}r_A^2}{MR^2})\Omega$, where I is the moment of inertia of the star plus magnetosphere system and $\dot{M}$ is the mass accretion rate (cf. Mestel 1990). Using $B < B_{crit}$ gives an upper bound to the current spin down rate; $\dot{P}|_{NOW} < 2 \times 10^{-16} n_H V_{20}^{-3} P_{8.39}^{7/3}$ (s s^{-1}). In both models shown in Figure 1, the 8.39 s period is reached shortly after the star enters the accretion phase, and $\dot{P}|_{NOW} \approx 10^{-16}$ s s^{-1}.

If magnetic field decays indefinitely, then it is evident from Figure 1 that P in the exponential decay model asymptotes after entering the accretion phase while P for the power law decay model continues to increase monotonically. This is because in exponential decay models, the field, and hence r_A (lever arm) decreases more rapidly at late times, whereas the field in power law decay models persists longer. Thus, if neutron star fields undergo exponential decay and the 8.39 s period in RX J0720.4-3125 is the star's asymptotic period, then we expect $\dot{P}|_{NOW} \ll 10^{-16}$ s s^{-1}.

Once the stellar field has decayed sufficiently to enable accretion, however, no further decay is required by the observations. Indeed, the decaying field may level out at late times to a steady finite long-lived value (e.g., Kulkarni 1986). If this is the case, then the late time behavior of $\dot{P}$ could be quite different (see Figure 1).

Regardless of how the field evolves during the accretion phase, however, the bound on $\dot{P}$ given above remains robust as long as the spin down rate during this phase is given by $\dot{\Omega}_{brk}$.

ACKNOWLEDGMENTS

We thank G. Hasinger for pointing out this source to us, and D. Hamilton, E. Ostriker, J. Stone, M. Wolfire, and S. Veilleux for valuable discussions. This work was supported in part by NASA grant NAG5-3836.

REFERENCES

1. Blaes, O. and Madau, P., *Ap. J.* **403**, 690 (1993).
2. Blaes, O., Warren, O., and Madau, P., *Ap. J.* **454**, 370 (1995).
3. Bondi, H., *MNRAS* **112**, 195 (1952).
4. Chanmugam, G., *ARA&A* **30**, 143 (1992).
5. Dickey, J. M. and Lockman, F. J., *ARA&A* **28**, 215 (1990).
6. Duncan, R. C. and Thompson, C., *Ap. J.* **392**, L9 (1992).
7. Haberl, F., Pietsch, W., Motch, C., Buckley, D. A. H., IAUC No. 6445 (1996).
8. Haberl, F. et al., *A&A*, in press (1997).
9. Halpern, J. P. & Holt, S. S., *Nature* **357**, 222 (1992).
10. Hunt, R., *MNRAS* **154**, 141 (1971).
11. Illarionov, A. F. and Sunyaev, R. A., *A&A* **39**, 185 (1975).
12. Kulkarni, S. R., *Ap. J.* **306**, L85 (1986).
13. Lipunov, V. M., Postnov, K. A., and Prokhorov, M. E., *Soviet Astronomy Review*, ed. R. A. Sunyaev, 1997, in press.
14. Mestel, L., *Basic Plasma Processes on the Sun*, eds. E. R. Priest and V. Krishan, Dordrecht: Kluwer, 1990, p. 67.
15. Narayan, R. and Ostriker, J. P., *Ap. J.* **352**, 222, (1990).
16. Nelson, R. W., Wang, J. C. L., Salpeter, E. E., and Wasserman, I., *Ap. J.* **438**, L99 (1995).
17. Ögelman, H. and Finley, J. P., *Ap. J.* **413**, L31 (1993).
18. Ostriker, J. P. and Gunn, J. E., *Ap. J.* **157**, 1395 (1969).
19. Sang, Y. and Chanmugam, G., *Ap. J.* **323**, L61 (1987).
20. Shapiro, S. L. and Teukolsky, S. A., *Black Holes, White Dwarfs, and Neutron Stars: The Physics of Compact Objects*, New York: Wiley, 1983.
21. Urpin, V. A., Chanmugam, G., and Sang, Y., *Ap. J.* **433**, 780 (1994).
22. Wang, Q. D. and Yu, K. C., *AJ* **109**, 698 (1995).
23. Wang, J. C. L. and Sutherland, R. S., *Proc. IAU Colloq. No. 163*, eds. D. T. Wickramasinghe et al., ASP Conf. Proc., 1997, in press.

24. Wang, J. C. L., *Ap. J.*, in press (1997).
25. Welsh, B. Y., Craig, N., Vedder, P. W., and Vallerga, J. V., *Ap. J.* **437**, 638 (1994).

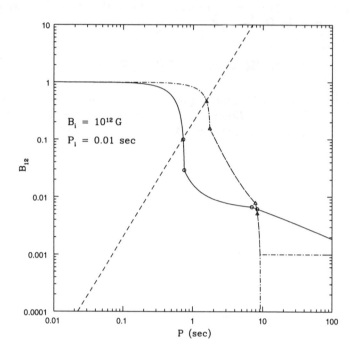

FIGURE 1. The spin period and magnetic field evolutionary track for a neutron star ($M = 1.4\,M_\odot$, $R = 10$ km) born with period $P_i = 0.01$ s and surface dipole field strength $B_i = 10^{12}\,G$. The star is assumed to be moving at $v = 20\,\mathrm{km\,s^{-1}}$ through a medium (solar abundance) with density $n_H = 1\,\mathrm{cm^{-3}}$ and $c_s = 10\,\mathrm{km\,s^{-1}}$. The star's field undergoes (solid curve) power law decay ($B(t) = B_i/(1+t/t_d)$) with $t_d = 3.8 \times 10^7$ yrs, or (dot-dashed curve) exponential decay ($B(t) = B_i \exp(-t/t_d)$) with $t_d = 4 \times 10^8$ yrs. Dashed line gives the observed radio pulsar death line ($B_{12} = 0.2P^2$). For power law decay, the star drops below the death line at $t \approx 3.5 \times 10^8$ yrs after its birth when $P = 0.73$ s. Propeller spindown begins at $t \approx 1.3 \times 10^9$ yrs when $P = P_0 = 0.76$ s. The star enters the accretion phase at $t \approx 5.7 \times 10^9$ yrs when $P = 6.96$ s. Spin down to the observed 8.39 s period occurs after 6.2×10^9 yrs. For exponential decay, the corresponding times and periods are (3.1×10^8 yrs, 1.56 s), (7.5×10^8 yrs, 1.74 s), (1.9×10^9 yrs, 7.86 s), and (2.1×10^9 yrs, 8.39 s), respectively. Each of these events is marked with an open circle on the curve for the power law decay model (solid) and an open triangle on the curve for the exponential decay model (dot-dash). The horizontal dot-dash line gives the evolutionary track for a star whose magnetic field decays exponentially until $B = B_{min} = 10^9\,G$ after which the field remains constant at this residual value.

NEUTRON STAR BINARIES

A Multi-Year Light Curve of Sco X-1 Based on BATSE SD Data and the Variability States of Sco X-1

B.J. McNamara, T.E. Harrison, P.A. Mason, M. Templeton, C.W. Heikkila, T. Buckley, E. Galvan, and A. Silva

Astronomy Department, New Mexico State University, Las Cruces, NM 88003

B. A. Harmon

ES84 NASA/Marshall Space Flight Center, Huntsville, AL 35812

Abstract. We have completed an internal calibration of 4.4 years of BATSE 10-20 keV SD data for Sco X-1. The resulting light curve shows variations on time scales from hours to weeks. A full description of the calibration used in arriving at this light curve has been submitted to the ApJ Supplement (McNamara et al. 1997). Fourier transform analysis of the resulting light curve found no significant periodicities on these time frames. Further period searches using short blocks of SD data, and the Gabor transform, is discussed in a companion paper by Mason et al. We have compared the BATSE SD Sco X-1 10-20 keV light curve to those produced by other missions/instruments and find that the amplitude of the variability of Sco X-1 increases with increasing energy. This result supports the suggestion of Mason et al. (1997a,b) that some positionally consistent catalogued gamma-ray bursts originate from Sco X-1 and Cyg X-1.

1. INTRODUCTION

Sco X-1 is the brightest non-solar persistent 1-10 keV X-ray source in the sky and is therefore observable using modest size instruments. Since Sco X-1 is the historical prototype of a class of binaries called low mass X-ray binaries (LMXBs) and is a Z source, it is an important object to study. LMXBs are binary systems in which a low mass secondary star transfers material, via an accretion disk, to a neutron star. Z sources are LMXBs that display part or all of a Z shaped pattern in a X-ray color-color plot of their emission (e.g. Hasinger and van der Klis 1989).

Multi-year high energy light curves have frequently been used to search for long duration emission cycles, often called super-periods, in X-ray binary systems (Priedhorsky and Holt 1987, Smale and Lochner 1992). In addition, these light curves sample the emission from these objects over extended periods of time and can therefore be used to quantify the amount of time they spend in their various activity states. Changes in these states may arise from variations in the mass accretion rate, opacity changes in the disk, or through the influence of third bodies. The time scales over which the emission changes associated with these phenomena occur are therefore relevant to an understanding of the physical conditions within both the accretion disk and neutron star environment.

The analysis of the long term, high energy, behavior of X-ray binaries has been hindered by a lack of suitable data bases. The BATSE SDs can be used to create multi-year light curves, at higher energies than was possible from Ariel 5 and Vela 5B or from the recently launched all-sky monitor of RXTE. The BATSE SD data therefore provide an additional, valuable tool, for the study of these systems.

2. THE BATSE SD DATA SET

Each BATSE SD consists of a NaI(TI) scintillator encased in a shielded frame with a 127 cm^2 beryllium entrance window. The energy sensitivity of a SD depends on its gain. At a gain setting of 4X, its lowest energy channel roughly covers the 10-20 keV energy range. When a high energy photon enters the NaI crystal, it liberates visible light. A photomultiplier, placed at the rear of the NaI crystal, detects this light and electronically converts it into a digital signal that is transmitted to Earth. As the CGRO orbits the Earth, the count rate of the SD pointing towards Sco X-1 dramatically changes as this source rises and sets with respect to the Earth's limb. These count changes are referred to as "Earth occultation steps" and their measured values are used to construct the 4.4 year Sco X-1 light curve presented in this paper.

Three adjustments must be made to the Earth occultation steps to place them into a consistent system. These corrections account for: 1) differing pointing angles between the detector normal and Sco X-1, 2) differing effective bandpasses of the SDs caused by different gain settings, and 3) sensitivity differences in the eight SDs.

3. THE 4.4 YEAR 10-20 KEV BATSE SD SCO X-1 LIGHT CURVE

A 4.4 year light curve of Sco X-1 consisting of 17,440 BATSE SD Earth occultation steps is presented in Figure 1. These steps are corrected for the three effects mentioned previously. The minimum time between points is about 40 minutes. It is clear from this figure that Sco X-1 varies on a number of timescales ranging from less than a day to over two weeks. An example of these emission changes is shown in Figure 2. A detailed discussion of the Sco

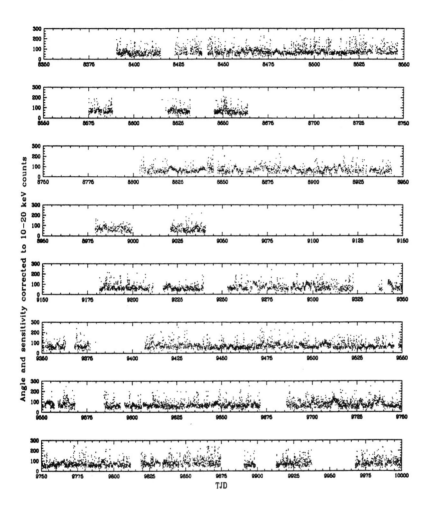

FIGURE 1. The 10-20 keV light curve of Sco X-1 based on BATSE SD step fluxes. Data gaps are caused by poor SD pointing angles to Sco X-1 or an inappropriate SD gain setting. The y axis units are counts per 2.048s, TJD refers to the truncated Julian date and is in units of days.

X-1 emission changes can be found in McNamara et al. 1997. In Figure 3, BATSE SD data is shown in comparison with coincident RXTE/ASM data. It is clear that the amplitude of the emission variability increases with energy. Balloon-borne instruments suggest that the increase in variability amplitude may extend to 300 keV (Haymes et al. 1972).

4. CONCLUSION

It is clear that Sco X-1 exhibits nonperiodic behavior on time scales extending from a fraction of a day to at least 20 days. This variability has a characteristic behavior, sharp rises are followed by equally quick declines. The longer term emission episodes (see Figure 2) seem to be extensions of those present on shorter timescales. In between these periods of activity, the 10-20 keV emission from Sco X-1 can be nearly constant. The presence of extended periods of nearly constant emission has been previously noted by Bradt et al. (1975). A comparison between BATSE SD, RXTE/ASM, and earlier balloon-based observations suggest that the amplitude of the Sco X-1 emission changes increases with energy. We postulate that at very high ener-

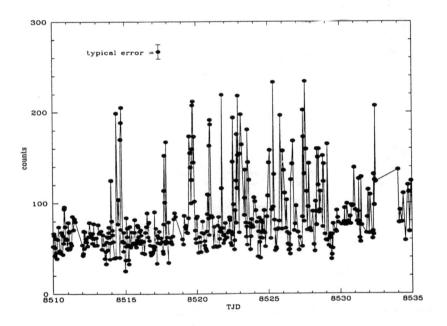

FIGURE 2. A 20 day active period in the 10-20 keV emission from Sco X-1. The activity consists of a prolonged series of short but intense emission changes.

gies (50-300keV) these changes can become intense enough to be identified as a gamma-ray burst. Mason et al. (1997a,b) provide a complete discussion of this hypothesis.

REFERENCES

Bradt, H. V., et al. 1975, ApJ, 197, 443.
Hasinger, G., & van der Klis, M. 1989, A&A, 225, 79.
Haymes, R.C., 1972, ApJ, 172, L47.
Mason, P. et al. 1997a, AJ (July issue).
Mason, P. et al. 1997b, (these proceedings).
McNamara, B. J., et al. 1997, ApJ Supplement, in press.
Priedhorsky, W. C, & Holt, S. S. 1987, Space Sci. Rev., 45, 291.
Smale, A. P., & Lochner, J. C. 1992, ApJ, 395, 582.

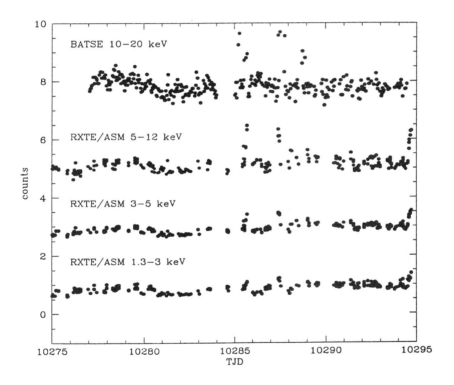

FIGURE 3. A comparsion of the emission variability from Sco X-1 at a variety of energies. The amplitude of the emission changes increases with energy.

Comparison of the BATSE LAD & SD Light Curves of Sco X-1: 1991 – 1996

T. E. Harrison, B. J. McNamara, P. A. Mason, M. Templeton

Astronomy Department, New Mexico State University, Las Cruces, NM 88003

B. A. Harmon

ES84 NASA/Marshall Space Flight Center, Huntsville, AL 35812

Abstract. Using recently calibrated Spectroscopy Detector (10 – 20 keV) data, we compare the high and low energy x-ray light curves of Sco X-1 since the beginning of the CGRO mission. The BATSE LAD data consists of daily averages of Earth occultation measurements. The BATSE SD Earth occultation measurements have been rebinned to the same temporal resolution as the LAD data. Both the LAD and SD data are morphologically similar, with long-lasting periods of low flux punctuated by short periods of very high flux. Additionally, the light curves show periods where Sco X-1 remained very bright in both energy regimes for several days. We also examine how the hardness evolves through individual high-flux episodes. We find that the amplitude of variability of Sco X-1 increases with increasing energy.

1. INTRODUCTION

Scorpius X-1 (Sco X-1) is the brightest, persistent cosmic X-ray source in the sky. It is a low-mass x-ray binary (LMXB) where the primary is believed to be a neutron star with a relatively weak magnetic field, and the secondary is a low-mass main sequence star. Sco X-1 is also a "Z source", a subset of LMXBs whose x-ray colors trace out a z-pattern in a high energy color-color plot (c.f., Hasinger & van der Klis, 1989). It is the brightest member of its class, and can be efficiently observed with modest instrumentation.

The low-energy (< 10 keV) behavior of Sco X-1 has been observed on many different time scales from years (e.g., Holt et al. 1976, Preidhorsky & Terrell 1983, Preidhorsky & Holt 1987, and Smale & Lochner 1992), to weeks (e.g., Canizares et al. 1975, Bradt et al. 1975, Mook et al. 1975, White et al. 1976, Willis et al. 1980, Illovasky 1980), to minutes (van der Klis et al. 1996).

High energy observations of Sco X-1 have not been as extensive. Haymes et al. (1972) reported (balloon-borne) observations of Sco X-1 that suggested a hard, high-energy tail of the Sco X-1 spectrum extending all the way to 300 keV. Later, more sensitive observations (Johnson et al. 1980, Rothschild et al. 1980), failed to detect this high-energy tail. Rothschild et al. concluded that to reconcile the different observations, Sco X-1 must be variable by a factor of 20 at the highest energies (50 − 300 keV).

To further examine the behavior of Sco X-1 at high energies, we combine two recently available data sets for Sco X-1: the BATSE Spectroscopy Detector 10 − 20 keV light curve (see McNamara et al., these proceedings), and the BATSE Large Area Detector (LAD) 20 − 50 keV Earth-occultation light curve. These two light curves cover a considerable fraction of the lifetime of the CGRO mission, allowing for extensive monitoring of Sco X-1 at energies above 10 keV. We present results below which suggest that the variability of Sco X-1 increases as you go to higher energies. This observation reconciles the result above, which suggested that Sco X-1 is highly variable at the highest energies. In fact, our new analysis suggests that Sco X-1 produces high energy transients which have been catalogued as Gamma-Ray Bursts (see the contribution by Mason et al., these proceedings).

2. DATA

The data extraction and calibration process for the Spectroscopy Detector (SD) 10 − 20 keV data are fully discussed in McNamara et al. (1997). While an internal calibration of the eight detectors for a variety of instrumental effects has been achieved, the conversion to flux units has not yet been finalized. In this poster we simply use previous observations in the 10 − 20 keV bandpass to provide a normalization for modeling purposes (note that the 10 − 20 keV light curve shows a hard lower limit to the flux level from Sco X-1, the "quiescent flux level", used for the normalization in what follows). The SD data is composed of Earth occultation measurements of which there are 10 to 20 per day. We have rebinned these data to construct a daily average for direct comparison to the available LAD data.

The LAD light curve is composed of daily averages of the Earth occultation fluxes for Sco X-1 in the 20 − 50 keV bandpass. These data are accessible on the WWW, and are discussed in a document by Harmon & Wilson available at HEASARC website. Sco X-1 is a weak source at high energies. To achieve a reasonable S/N, the multiple LAD Earth occultation measures of Sco X-1 obtained each day are averaged to produce a single daily measurement. The LAD data have been converted to flux units, but the calibration remains somewhat uncertain. We will again use previous high S/N observations in this bandpass (Ubertini et al. 1992) to provide a quiescent-state normalization.

3. THE HIGH ENERGY LIGHT CURVE OF SCO X-1

In Figure 1 we plot 150 days of data for Sco X-1. In the top panel of this plot is the SD (10 − 20 keV) light curve for Sco X-1. The middle panel is the LAD (20 − 50 keV) light curve, while the bottom panel is a "hardness" measurement–the LAD flux divided by the SD flux. It is obvious from these plots that the two light curves look similar–periods of high flux in the SD data (active periods) are usually seen in the LAD data.

The plot of the LAD/SD ratio shows that normally (in quiescence), the relative hardness (×100, note these are not absolute measures of hardness!) of Sco X-1 is about 1.5 to 2. For the entire 600 day light curve, the minimum

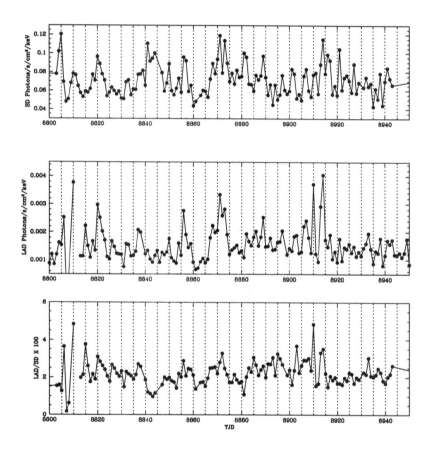

FIGURE 1. SD (top) and LAD (middle) light curves, and their ratio (×100,bottom), for Sco X-1 covering the period TJD8800-8950.

hardness was around 1.0, but during active states this ratio can reach 5, or higher. This factor of more than five in the observed hardness ratio is a demonstration that the range of variability at higher energies is much larger than seen at lower energies. Further proof of this fact is shown in Figure 20 of McNamara et al. (1997). For that plot, they constructed histograms of the various long-term light curves of Sco X-1 at a variety of energies, and from a variety of missions. The result is that the active state of Sco X-1 exhibits a larger variation in amplitude the higher the energy bandpass.

4. SPECTRAL MODELING

We now attempt to determine if we can reconcile the variations in the SD and LAD fluxes with commonly accepted models for Sco X-1. We have used XSPEC to produce the two commonly proposed models for Sco X-1 in this energy regime: a thermal bremstrahlung model (kT = 6.7 keV; Christian & Swank 1997), and a Comptonized accretion disk model (kT = 5.3 keV, τ = 5.36; Ubertini et al. 1992). The average Sco X-1 low-state values for the SD and LAD are 0.019 counts/cm^2/s/keV and 0.0047 photons/cm^2/s/keV, respectively. The observations by Ubertini et al. suggest that these fluxes should be 0.05 and 0.0015 photons/cm^2/s/keV, respectively. We have normalized the SD and LAD data in all of our figures using these values. This normalization may weaken some of our conclusions about the true spectral form of Sco X-1.

We fitted the two models to the data for TJD8843 & TJD8853 when Sco X-1 had low LAD and/or SD fluxes. For TJD8853 both model fits are adequate. The data for TJD8843 are not fitted as well using these two standard models, and suggest a cooler temperature model would be more appropriate for this particular day. We have also fit the interesting decline from a flare event covering the period TJD8820-8824. As shown in Figure 1, the decline at both energies was smooth and linear. The two standard models fit the lowest point in the decline (TJD8824) quite well, but not the flaring data and early decline. The thermal bremstrahlung model does a much poorer job of fitting these high-flux points when compared to the single temperature scaled Comptonized model. Our results suggest that hotter models are needed to fit the data during flaring episodes.

This result is reinforced when we compare the normal models fitted to active and quiescent periods, to model fits using higher temperatures. To get a Comptonized model to fit the higher flux period we had to adjust the temperature to 5.7 keV. The result was a reasonable fit. To get a thermal bremstrahlung model to fit the data, however, we had to raise its temperature to 12 keV! To our knowledge, no one has ever suggested such a model for this source, and we conclude that such a model is unreasonable. We find that even with this limited data set, a Comptonized model with a varying temperature provides a better fit for the various states of Sco X-1. There remains evidence, however, for deviations from this standard model, and additional observations with better spectral resolution are needed to fully understand what Sco X-1 is really doing at very high energies. RXTE has the energy coverage, resolution, and sensitivity needed to accomplish this task.

5. CONCLUSION

We have compared two recently available data sets for Sco X-1: the BATSE SD (10 − 20 keV) and LAD (20 − 50 keV) light curves during the first part of the CGRO mission. We have found that, in general, the light curves at the two energies track each other. There are many periods, however, when this is not true, and one bandpass has a higher flux than the other. We have found that the range in variability above the quiescent level depends on energy: as you go to higher energies the variations ("flares") get larger relative to the normal quiescent flux level. Modeling the two different fluxes during different periods shows that the Comptonized model works best for Sco X-1 outside of its quiescent phase (the thermal bremsstrahlung model works equally well at quiescence). Even with this limited data set, there is some evidence for deviations from the standard models. More data on the high energy behavior of Sco X-1 are needed to better understand the behavior of this object.

REFERENCES

Bradt, H. V., et al. 1975, ApJ, 197, 443.
Canizares, C. R., et al. 1975, ApJ, 197, 457.
Christian, D. J. & Swank, J. H. 1997, ApJS, 109, 177.
Hasinger, G., & van der Klis, M. 1989, A&A, 225, 79.
Haymes, R. C., 1972, ApJ, 172, L47.
Holt, S. S., et al. 1976, ApJ, 205, L27.
Illovasky, S. A., et al. 1980, MNRAS, 191, 81.
Johnson, W. N., 1980, ApJ, 238, 982.
McNamara, B. J., et al. 1997, ApJ Supplement, in press.
Mook, D. E., et al. 1975, ApJ, 197, 425.
Priedhorsky, W. C, & Terrell, J. 1983, ApJ, 273, 709.
Priedhorsky, W. C, & Holt, S. S. 1987, ApJ, 312, 743.
Rothschild, R. E., et al. 1980, Nature, 286, 786.
Smale, A. P., & Lochner, J. C. 1992, ApJ, 395, 582.
Ubertini, P., et al. 1992, ApJ, 386, 710.
van der Klis, M., et al. 1996, ApJ, 469, L1.
White, N. E., et al. 1976, MNRAS, 176, 91.
Willis, A. J., et al. 1980, ApJ, 237, 596.

Application of the Gabor Transform to BATSE Spectroscopy Detector Observations of Scorpius X-1

P. A. Mason, M. Templeton, B. J. McNamara, T. E. Harrison, E. Galvan, T. Buckley

Astronomy Department, New Mexico State University
Las Cruces, NM 88003

Abstract. We present an application of the Gabor transform to Scorpius X-1 data obtained using the CGRO BATSE Spectroscopy Detectors (SDs). Two types of data are obtained with the BATSE SDs. The first set consists of the high S/N step fluxes obtained using the Earth-occultation method (see McNamara et al. these proceedings). The second (much larger) set is the 2.048 second time resolution raw flux data obtained between rise and set of Sco X-1. It is this latter set of data for which the Gabor transform is applicable. In this poster we present initial progress towards analysis of the substantial amount (10^7 points) of 2.048 second time resolution data.

INTRODUCTION

Fourier transform analysis is a well known and powerful technique for revealing periodic behavior in time-series data. However, the power at a certain frequency is reduced when 1) the period does not persist through the entirety of the dataset (i.e. a transient period), 2) when the period in question is not strictly periodic (i.e. Quasi-Periodic Oscillations), or 3) when the period is changing secularly (e.g. the spin up or down of the primary star).

One approach to this problem utilizes the short-time Fourier transform. This method involves breaking up the data into short segments. A Fourier power spectrum is then calculated for each segment. Some time-dependent information about the power spectrum may then be then recognized, but the method has definite limitations. If the interval is broken up into too many pieces then frequency information is severely constrained (since the Nyquist frequency is reduced for the short segments). If too few segments are chosen only a limited amount of time-dependent information is gained. What is needed is a method which utilizes as much of the dataset as possible (to retain frequency information) while weighting the data going into the transform by a running window function. Hence, we employ the Gabor transform which

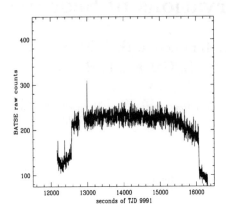

FIGURE 1. Raw BATSE SD data for an example of the common quiescent flux state is shown.

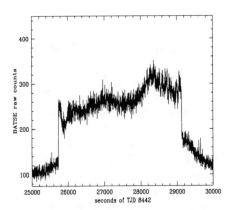

FIGURE 2. Raw BATSE SD data for an example of the rare active flux state is shown.

utilizes a running Gaussian window function.

THE GABOR TRANSFORM

The Gabor transform (Gabor 1947) remedies a limitation of Fourier power spectral analysis, namely the ability to detect the time-evolution of frequencies contained in a time-series. This is accomplished with the construction of a time-frequency power spectrum, which can be represented using a contour plot of power in the time-frequency plane. Conceptually it is a straightforward extension of the Fourier transform to two dimensions. See Boyd et al. (1995) for a derivation of the Gabor transform.

Examples of raw BATSE SD flux data are shown for a quiescent flux state and an active flux state in figures 1 and 2 respectively. The 2.048 second time resolution data are extracted by truncating the raw data at the "steps" and then fitting the light curve with a spline function derived from just a few points. The spline function is then subtracted from the raw data to remove trends caused mainly by the variation in the X-ray background. The mean is also subtracted from the data. In figure 3, the Gabor transform results are shown for a quiescent state (left) and an active state (right) In the top panels the time-frequency power spectra are shown. The detrended data are shown in the bottom panels. Upper and lower panels have identical scales for the time axis. The frequency range is that defined by the Nyquist frequency.

The active states most often display power only at low frequencies. Power at these frequencies is temporally short lived, often lasting just 2 or 3 cycles. Power increases at these low frequencies when the light curve is seen to modulate strongly. Sometimes there is significant power evident at higher frequencies in the data obtained during active states. Significant power at higher frequencies is more often observed during quiescent periods (e.g. Figure 3 (left), where a power spike is seen at 1200 seconds and a frequency of 0.13 Hz). In the Gabor Transform this feature has a bit of a fish-tail appearance, with power near 0.15 Hz just before the power spike and down to about 0.08 Hz just after the power spike. This indicates that the frequency evolves across the detectability of this feature. This project has only begun to reveal the wealth of detail in the 2.048 second resolution BATSE SD light curve since only about 20 of the thousands of similar light curves have been analyzed. We thank Karen Schaefer (STSci) for helpful discussions on properties of the Gabor transform.

REFERENCES

1. Boyd, P. 1995, *ApJ*, **323**, L131.
2. Gabor, 1946, *J. Inst. Elect. Eng.*, **93**, 429.
3. McNamara, B.J., et al. 1997, these proceedings

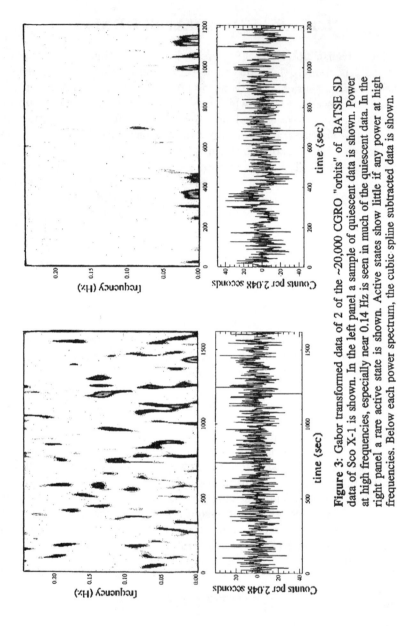

Figure 3: Gabor transformed data of 2 of the ~20,000 CGRO "orbits" of BATSE SD data of Sco X-1 is shown. In the left panel a sample of quiescent data is shown. Power at high frequencies, especially near 0.14 Hz is seen in much of the quiescent data. In the right panel a rare active state is shown. Active states show little if any power at high frequencies. Below each power spectrum, the cubic spline subtracted data is shown.

High-Energy Transient Events From Scorpius X-1 and Cygnus X-1

Paul A. Mason, Bernard J. McNamara, Thomas E. Harrison

Astronomy Department, New Mexico State University
Las Cruces, NM 88003

Abstract. We present evidence that the X-ray binaries Scorpius X-1 and Cygnus X-1 are responsible for high-energy transient events which have been cataloged in the CGRO BATSE 3B Gamma-ray burst (GRB) catalog as classical GRBs. We locate bursts in the 3B catalog which are positionally consistent with these binaries and compare the observed burst distribution properties with those found from an ensemble of randomly generated 3B catalogs and from repeater models. Two 3B cataloged GRBs that are likely high-energy transient events from Scorpius X-1; based on the fact that they occurred on the days with the two brightest "daily averages" in the Scorpius X-1 LAD light curve. We find that GRBs are emitted from Sco X-1 at the 99% confidence level.

Numerous GRBs are emitted by Cygnus X-1 and removed by the BATSE team. We find that the some of the bursts remaining in the BATSE 3B catalog are probably high-energy transient events from Cygnus X-1 based on the anomalously low V/V_{max} for the region near Cygnus X-1 as determined from ensembles of mock 3B catalogs.

COMPARISON OF MOCK CATALOGS TO THE BATSE DATA OF SCO X-1

Mock 3B catalogs were produced by generating GRBs at random sky positions and assigning them random times (during the interval covering the 3B catalog). During this process, temporal and sky coverage biases are incorporated. The temporal bias reflects the fact that the amount of sky coverage has varied over the lifetime of CGRO. In the 1B catalog (Fishman et al. 1994) the average sky coverage was 34%. Tape recorder malfunctions reduced the coverage to 28% for the 2B − 1B bursts. During the interval corresponding to the 3B − 2B GRBs the only major loss of live time was due to Earth-occultation and passages through the South Atlantic Anomaly, resulting in an average sky coverage of 41%. The relative sky coverage bias reflects the fact that the equatorial poles are more often within the field of view of the CGRO than is the celestial equator. Fishman et al. (1994) provides a table of sky coverage as a function of declination that is applicable to the entire 3B catalog. The temporal and sky coverage biases are imposed upon the isotropic parent distribution by rejecting a percentage of GRBs as a function of time and declination respectively. Once a mock catalog containing 1122 GRBs is generated, the BATSE error circles are randomly reordered so that the mock catalogs all have the same error distribution as the real 3B catalog. We find that there is an excess of 3B GRBs consistent with Sco X-1 at the 96% confidence level (see Mason, McNamara, and Harrison 1997a).

BATSE Large Area Detector (LAD: 20-50 keV) and Spectroscopy Detector (SD: 10-20 keV) data points are obtained each time Sco X-1 appears or disappears from behind the limb of the Earth (see McNamara et al. 1997). Each GRB consistent with Sco X-1 is examined to see if it temporally corresponds to an unusual daily average flux level. We find that the two highest LAD daily averages, both correspond to the occurrence of a GRB consistent with Sco X-1. The individual LAD dwells on each of these days were all at high LAD flux levels. Of 25,000 properly biased mock catalogs only 9 or 0.0036% contained GRBs that occurred on the two highest daily averages in the LAD data and were consistent with Sco X-1. Only 1.58% of these catalogs had consistent GRBs that occurred on either one or both of the two highest Sco X-1 LAD days. The fact that there is a significant surplus of GRBs which are consistent with the position of Sco X-1 along with the extremely unlikely occurrence of GRBs on the two highest LAD daily averages suggests that at least one of these GRBs (and probably both) originate from Sco X-1. To obtain an upper limit to the probably that this is due to random chance, we calculated the mean of the LAD daily averages of those GRBs which are consistent with Sco X-1. It was found that less than 1% of the mock catalogs had both an excess of GRBs consistent with Sco X-1 and a mean LAD daily average that matched or exceeded this value. Therefore at the 99% confidence level, we conclude that GRBs are emitted during high LAD flux states of Sco X-1.

REPEATING BURSTER MODELS FOR SCO X-1

To test the possibility that Sco X-1 is a *repeating burster* a series of properly biased mock catalogs were generated that included a single repeating GRB so that among the 1122 cataloged GRBs one or more is emitted by a burster at the position of Sco X-1. An ensemble of 25,000 synthetic catalogs were produced for each repeater model. In Figure 1, we present results of simulated 3B catalogs where a Sco X-1 repeating GRB source is modeled so that it will produce a GRB exactly twice, but only at or above a certain LAD flux cutoff level. In order for the mock catalog distribution to resemble the observed distribution the LAD cutoff level must be high (0.35-0.42 photons cm^{-2} s^{-1}) so that the GRBs will be produced only when the LAD flux is at its highest levels. We predict that future GRBs will be detected positionally consistent with Scorpius X-1 and at high (20-50 keV) flux levels.

BATSE Triggers for Cygnus X-1

The spatial distribution of GRBs appears to be inhomogeneous. Evidence for inhomogeneity of GRBs rests with a determination of the sample's $<V/V_{max}>$ (Schmidt 1968; Schmidt, Higdon, and Hueter 1988). Here V is defined as the volume occupied by the detected bursts and V_{max} is the volume containing all bursts to the burst detection limit assuming that the bursts may be regarded as approximate standard candles. If GRBs homogeneously occupy the full volume V_{max}, and the space is Euclidean, then the average V/V_{max} would be 0.5. If rather, we are observing past the edge of the GRB distribution then $<V/V_{max}>$ is less than 0.5. V/V_{max} information is avail-

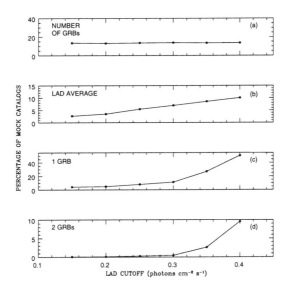

FIGURE 1. Mock catalog results for models which have a repeating GRB source at the position of Scorpius X-1 which repeats exactly twice, but only at or above various LAD flux levels is shown. The percentage of mock catalogs with (a) the number of GRBs greater than or equal to the observed number, (b) a LAD flux level higher than or equal to the observed level for those GRBs consistent with Sco X-1, (c) one or more consistent GRBs occurring on the highest LAD flux daily averages, and (d) the occurrence of 2 consistent GRBs on the 2 highest LAD flux days are shown. In order for these statistics to resemble the observed quantities, a repeater model with 2 or more repetitions which occur only at high (0.35-0.40 photons cm^2 s^{-1}) LAD flux levels is required.

able for 657 of the 1122 BATSE 3B bursts. The average V/V_{max} for the 657 BATSE 3B bursts is $< V/V_{max} > = 0.328 \pm 0.0113$. We find that $< V/V_{max} >$ for the region within 30^0 of Cygnus X-1 is anomalous at the 94% confidence level indicating that some of the 3B GRBs are from Cyg X-1 (see Mason, McNamara, and Harrison 1997b). Of the $\sim$ 1800 BATSE 1B catalog triggered events, 36 occurred as Cyg X-1 appeared from behind the limb of the Earth. There is no question that these events are not classical GRBs. An additional 33 BATSE 1B GRB triggers were classified as Cyg X-1 events by the BATSE team (Meegan et al. 1993). These 33 bursts are shown along with the Cyg X-1 Lad light curve in Figure 2 The method used to differentiate between classical GRBs and Cyg X-1 events involves a Bayesian statistical analysis that relies upon two assumptions (1) that Cyg X-1 outbursts would be localized near Cyg X-1 and (2) that Cyg X-1 only produces weak GRBs. We argue that the failure of the Baysian analysis is due to assumption (2). That is, some strong Cyg X-1 events remain in the 3B catalog and some of weak "classical" GRBs have been improperly attributed to Cyg X-1.

REFERENCES

1. Fishman et al. 1994, *ApJS*, **92**, 229.
2. Mason, P.A., McNamara, B.J., and Harrison, T.E. 1997a, submitted to ApJ Letters
3. Mason, P.A., McNamara, B.J., and Harrison, T.E. 1997b, *AJ*, **114**, 238.
4. McNamara et al. 1997, submitted to ApJS
5. Meegan, C.A., et al. 1993, in *Compton Gamma-Ray Observatory, AIP Conference Proceedings* No. 280, p 1117.

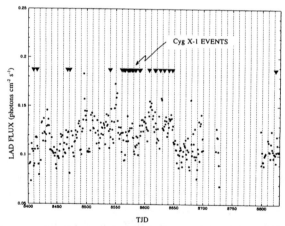

FIGURE 2. A comparison between the LAD flux and BATSE GRB triggers from Cygnus X-1 is shown. Note that the events often occur at low to moderate flux levels.

Low-energy line emission from Cygnus X-2 observed by the BeppoSAX LECS

Erik Kuulkers[*,†], Arvind N. Parmar[*], Alan Owens[*], Tim Oosterbroek[*] and Uwe Lammers[*]

[*] *ESA/ESTEC, Astrophysics Division (SA)*
P.O. Box 299, 2200 AG Noordwijk, The Netherlands
[†] *Astrophysics, University of Oxford, Nuclear and Astrophysics Laboratory,*
Keble Road, Oxford, OX1 3RH, UK

Abstract. We present a 0.2–10 keV spectrum of the low-mass X-ray binary Cygnus X-2 obtained using the Low Energy Concentrator Spectrometer on-board *BeppoSAX*. The spectrum can be described by a cut-off power-law model with absorption of $(2.28\pm0.07)\times10^{21}$ atoms cm^{-2}, a power-law index of 0.78±0.02 and a cut-off energy of 4.30±0.08 keV (68% confidence errors), except at energies near ∼1 keV where excess emission is present. This can be modeled by a broad Gaussian line feature with an energy of 1.02±0.04 keV, a full width half-maximum of 0.47±0.07 keV and an equivalent width of 74^{+25}_{-11} eV. This result confirms earlier reports of line emission near 1 keV and shows the intensity and structure of the feature to be variable. For a more detailed account of these observations we refer to Kuulkers et al. [4].

THE SOURCE

Cygnus X-2 is a bright persistent low-mass X-ray binary, whose X-ray spectrum has been studied from ∼0.1 to several hundred keV. Together with some of the other bright persistent low-mass binary X-ray sources it is classified as a "Z" source (Hasinger & van der Klis [2]). Z sources are thought to be accreting material at near-Eddington rates via an accretion disk onto a neutron star (e.g. Hasinger et al. [3]).

THE INSTRUMENT

The Low-Energy Concentrator Spectrometer (LECS) is sensitive in the energy range 0.1–10 keV (Parmar et al. [7]). Its unique design utilizes a driftless

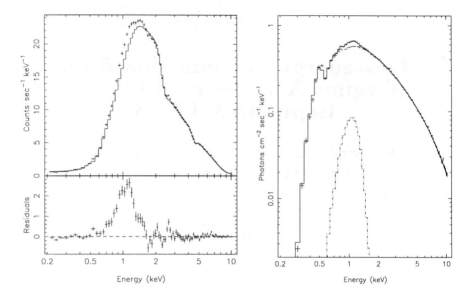

FIGURE 1. The observed LECS spectrum together with the best-fit cut-off power-law model with the normalization of the 1.02 keV emission feature set to zero (upper left panel). The lower left panel shows the residuals. The right panel shows the inferred photon spectrum and the model prediction including the 1.02 keV feature

gas scintillation proportional counter to make the lowest energies accessible with a good energy resolution while providing 16 μs time resolution and moderate spatial resolution. The LECS has a circular field of view of 37' diameter and a 0.1–10 keV background counting rate of 9.7×10^{-5} arcmin^{-2} s^{-1}.

Good data were selected from intervals when the minimum elevation angle above the Earth's limb was >4° and when the instrument configuration was nominal using the SAXLEDAS 1.4.0 data analysis package (Lammers [5]). The LECS was only operated during satellite night-time giving a total on-source exposure of 7.1 ks. A spectrum was extracted centered on the mean source position using the standard LECS extraction radius of 8' and the appropriate response matrix was generated. Background subtraction was performed using a standard blank field 46 ks exposure, but is not critical for such a bright source. The spectrum was rebinned to have at least 20 counts per channel. Channels below 0.2 keV were discarded since the source was absorbed below this energy.

THE OBSERVATIONS

BeppoSAX observed Cyg X-2 between 1996 July 23 00:19 and 18:18 UTC during Performance Verification. The mean LECS count rate observed from Cyg X-2 is $67\,\mathrm{s}^{-1}$, which corresponds to an observed 1–10 keV flux of $7.4\times10^{-9}\,\mathrm{erg\,cm^{-2}\,s^{-1}}$.

The spectrum can be described by a cut-off power-law model with absorption of $(2.28\pm0.07)\times10^{21}\,\mathrm{atoms\,cm^{-2}}$, a power-law index of 0.78 ± 0.02 and a cut-off energy of $4.30\pm0.08\,\mathrm{keV}$ (68% confidence errors), except at energies near $\sim 1\,\mathrm{keV}$ where excess emission is present (see Figure 1). This can be modeled by a broad Gaussian line feature with an energy of $1.02\pm0.04\,\mathrm{keV}$, a full width half-maximum of $0.47\pm0.07\,\mathrm{keV}$ and an equivalent width of $74^{+25}_{-11}\,\mathrm{eV}$ (see also Figure 1, right panel). There is no evidence for Gaussian line emission near 6.7 keV, with a 3σ upper limit on the equivalent width of 62 eV for a narrow line.

THE INTERPRETATION

Our results confirm earlier reports of line emission near 1 keV (e.g. Vrtilek et al. [10], Smale et al. [8]). The intensity and structure of the feature appears to be variable compared to these previous reports. The feature is most probably due to a combination of unresolved Fe L-shell and Ne K-shell line emission (e.g. Vrtilek et al. [11]).

Line emission at energies near 1 keV is not unique to Cyg X-2. Similar features have been observed in other low-mass X-ray binaries, e.g. in the Z source Sco X-1 (see Vrtilek et al. [11]) and in the accreting pulsars Her X-1 (e.g. McCray et al. [6]) and 4U 1626−67 (e.g. Angelini et al. [1]).

Inspection of the *Rossi* X-ray Timing Explorer All Sky Monitor (ASM) light curves (see Wijnands, Kuulkers & Smale [12]) indicates that the *BeppoSAX* observation occured 5–10 days before Cyg X-2 entered a probable low intensity state, suggesting that the source was transitioning from a medium to a low intensity state. We note that a similar transition may have been seen by Vrtilek et al. [9] (their "medium state B"). Vrtilek et al. [9] fit the 1–20 keV *Einstein* Monitor Proportional Counter (MPC) spectra obtained during this state with a thermal bremsstrahlung model of $kT\sim 7.0$–$8.7\,\mathrm{keV}$ and $N_\mathrm{H}\sim 2.3$–$3.5\times10^{21}\,\mathrm{H\,atoms\,cm^{-2}}$. A comparison with the best-fit parameters obtained when the same model is fit to the 1–10 keV LECS spectrum of $kT\sim 9.2\,\mathrm{keV}$ and $N_\mathrm{H}\sim 3.5\times10^{21}\,\mathrm{cm^{-2}}$ indicates that the spectral shape was similar on both occasions.

REFERENCES

1. Angelini L., White N.E., Nagase F., et al., *ApJ* **449**, L41 (1995).
2. Hasinger G., van der Klis M., *A&A* **225**, 79 (1989).
3. Hasinger G., van der Klis M., Ebisawa K., Dotani T., Mitsuda K., *A&A* **235**, 131 (1990).
4. Kuulkers E., Parmar A.N., Oosterbroek T., Owens A., Lammers U., *A&A* **323**, L29 (1997).
5. Lammers U., *The SAX/LECS Data Analysis System - Software User Manual*, ESA/SSD, SAX/LEDA/0010 (1997).
6. McCray R.A, Shull J.M., Boynton P.E., et al., *ApJ* **262**, 301 (1982).
7. Parmar A.N., Martin D.D.E., Bavdaz M., et al., *A&AS* **122**, 309 (1997).
8. Smale A.P., Angelini L., White N.E., Mitsuda K., Dotani T., *BAAS* **185**, 1484 (1994).
9. Vrtilek S.D., Kahn S.M., Grindlay J.E., Helfand D.J., Seward F.D., *ApJ* **307**, 698 (1986).
10. Vrtilek S.D., Swank J.H., Kallman T.R., *ApJ* **326**, 186 (1988).
11. Vrtilek S.D., McClintock J.E., Seward F.D., Kahn S.M., Wargelin B.J., *ApJS* **76**, 1127 (1991).
12. Wijnands R.A.D., Kuulkers E., Smale A.P., *ApJ* **473**, L45 (1996).

BATSE Observations of the Second Outburst of GRO J1744-28

Peter Woods[1], Chryssa Kouveliotou[2], Jan van Paradijs[1,3],
Michael S. Briggs[1], Kim Deal[1], C. A. Wilson[4], B. A. Harmon[4],
G. J. Fishman[4], W. H. G. Lewin[5] and J. Kommers[5]

[1] *Dept. of Physics, University of Alabama in Huntsville*
[2] *Universities Space Research Association*
[3] *Astonomical Institute "Anton Pannekoek", University of Amsterdam*
[4] *NASA Marshall Space Flight Center*
[5] *Dept. of Physics and Center for Space Research, Massachusetts Institute of Technology*

Abstract.
On 2 December 1996, exactly one year after its discovery, hard X-ray bursts were again detected with BATSE from the Bursting Pulsar, GRO J1744-28. Similar to the first outburst, a flurry of bursts were observed on the first day, the rate of which dropped dramatically to a near constant level for the next several months. From 2 December 1996 to 10 April 1997, BATSE observed over 2400 bursts. After April 10th, burst intensities dropped below the instrument's threshold. We present preliminary results of spectral analysis performed on 1663 bursts together with a comparison of these results to the burst properties of the first outburst.

INTRODUCTION

The only known bursting and pulsating X-ray source was discovered on 2 December 1995 (Fishman et al. 1995), when a series of hard X-ray bursts from the galactic center were detected with the BATSE instrument. A hard X-ray pulsar was also discovered in this region one month later (Paciesas et al. 1996, Finger et al. 1996). Shortly thereafter, it was realized that the burst source and pulsar were one in the same (Kouveliotou et al. 1996). The Bursting Pulsar, as it has come to be known, reached a maximum luminosity in late January 1996 and became undetectable with BATSE during May of that year. Surprisingly, the source went into outburst again precisely one year from the onset of the first outburst. We discuss burst properties of this second outburst and draw comparisons to the first.

OBSERVED BURST PROPERTIES

The typical burst light curve from the Bursting Pulsar as observed with BATSE (30 - 100 keV) lasts 10 sec, with a faster rise than decay. In a small percentage of events (~5%) concentrated during the first few days of the outburst, extended tails of emission are seen tens of seconds beyond the peak of the burst. Superposed on the burst envelope are pulsations from the rotating neutron star, as well as sporadic dips in intensity of unknown origin. These temporal characteristics are present in bursts from both the first and second outbursts. Detailed temporal analyses of events from the first outburst have been discussed by Strickman et al. (1996), Giles et al. (1996), Stark et al. (1996) and Koshut et al. (1997).

We calculated a corrected burst rate for the source using the livetime of the source (i.e. the time it is observable by BATSE). Over the duration of the outburst, two factors played an important role in determining livetime: Earth occultation modulated by the orbital precession of the CGRO spacecraft and data telemetry gaps. From the CGRO orbital parameters, we determined how long the source was visible versus how long it was occulted by the earth. As the plane of the CGRO orbit precesses due to Earth oblateness, the source latitude with respect to the CGRO orbital plane changes, causing occultation time intervals to change as well. The second correction is for data telemetry gaps which constitute about 20% of the data coverage for any given day. Accounting for these effects on livetime of the source, we calculated the fractional portion of the day the source was visible for each day of the outburst. Dividing the observed burst rate by this fraction yields the corrected burst rate for the source (Figure 1b). Figure 1a shows (with the same scale) the corrected burst rate history of the first outburst. The only significant difference between the two is the duration of each outburst. As shown later, the average burst intensities of the second outburst were less than those of the first, which could explain why they fell below the detection threshold sooner.

The spectra of bursts from GRO J1744-28 are best fit by a simple optically thin thermal Bremsstrahlung (OTTB) model, where the photon flux is given by

$$\frac{dN}{dE} \propto \frac{1}{E} e^{-\frac{E}{kT}} \qquad (1)$$

Over the energy range 30 - 100 keV, we fit the spectra of 1663 bursts using the following procedure. We subtracted the background flux by fitting a polynomial of order no greater than four to about 100 seconds of preburst and postburst data and interpolating between. A source interval was selected from which the bin with the largest count rate was chosen. This bin was then fit using the OTTB model yielding the peak flux (2.048 sec time bin) for all events and kT for about 57% of the events. Some of the events were too weak to determine kT, so a fixed kT of 7.5 keV was used in these cases. For

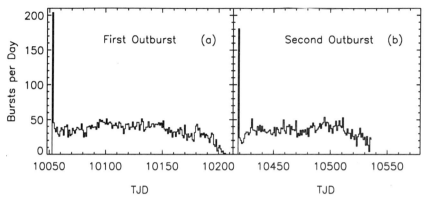

Figure 1 : Burst rate for the first and second outbursts, corrected for exposure time.

clarity, we chose to plot daily averages of burst peak flux and kT through the outburst (Figures 2b and 2d). The error bars here represent the statistical errors of each individual event as well as the error in the mean due to sample variance. Plotted in figures 2a and 2c are the same measurements for the first outburst (Briggs et al. 1997).

It is evident from Figures 2a and b that the second outburst was dimmer by a factor of two than the first. Not shown here are burst fluences which were also calculated for each event and show similar differences between outbursts. Variations are also apparent in the kT measurements between outbursts. The kT appears smaller with greater variability during the second outburst, in contrast to the first where it was larger and remained fairly constant.

To better understand the burst mechanism, we calculated the α parameter. The parameter α is defined as the ratio of the bolometric burst fluence divided by the bolometric persistent fluence over a given time interval (Lewin, van Paradijs and Taam, p 191). In addition to calculating burst fluences, average persistent fluxes were also measured during the outburst from occultation steps (Deal et al. 1997). Using these two pieces of information, one can calculate α. The spectral form of the persistent and burst emission are the same for this source (Giles et al. 1996, Strickman et al. 1996), so α can be calculated by comparing discrete wavelength regions of the spectrum. We calculated α on a daily basis for the energy range 30 – 100 keV. Due to the large number of bursts on the first day of each outburst, α was extremely low. For the first day of the second outburst, we found a 2σ upper limit of 6.5. Details of α during the first outburst can be found in Briggs et al. (1997). Past the first day, α rose discontinuously to ~ 20 and continued to rise slowly thereafter, reaching a maximum of ~ 55 in late January 1997. For the remainder of the outburst, α did not change significantly.

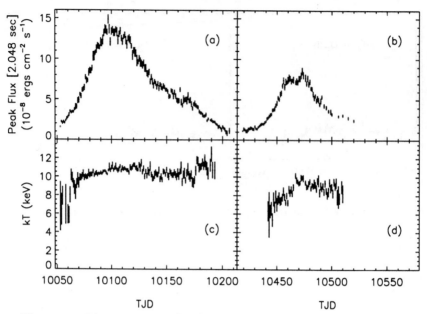

Figure 2 : The peak fluxes (top) and spectral kT (bottom) of the bursts from GRO J1744-28 during the first (left) and second (right) outbursts. All points above represent daily averages.

DISCUSSION

We have shown that aside from a difference in intensity for burst and persistent emission fluxes, and some modest changes in kT, the two outbursts of the Bursting Pulsar are nearly identical. The time scale for each outburst is roughly the same despite the significant difference in luminosities. They both yield low values of α especially on the first day suggesting an accretion instability as the burst mechanism. It is striking that the two outbursts show such similar burst durations and recurrence patterns, in particular a very high burst rate during the first day (hundreds per day), and a sudden decrease in that rate to ~ 35 per day.

ACKNOWLEDGEMENTS

This work was made possible by NASA grants NAG 5-3003 and NAG 5-2735.

REFERENCES

1. Briggs, M.S., Kouveliotou, C.K., van Paradijs, J., Woods, P., Deal, K., Lewin, W. & Kommers, J., in preparation (1997)
2. Deal, K.J., Harmon, B.A., Kouveliotou, C., van Paradijs, J., Wilson, C.A., Zhang, S.N. & Fishman, G.J., *4th Compton Symposium*, **19-10**, (1997)
3. Finger, M.H., et al., *IAU Circular*, **6285**, (1996)
4. Fishman, G.J., et al., *IAU Circular*, **6272**, (1995)
5. Giles, A.B., Swank, J.H., Jahoda, K., Zhang, Strohmayer, T., Stark, M.J. & Morgan, E.H., *ApJ*, **469**, L25 (1996)
6. Koshut, T., Kouveliotou, C., van Paradijs, J., Woods, P., Finger, M.H., Briggs, M.S., Fishman, G.J., Lewin, W. & Kommers, J., in preparation (1997)
7. Kouveliotou, C., et al., *IAU Circular*, **6286**, (1996)
8. Kouveliotou, C., van Paradijs, J., Fishman, G.J., Briggs, M.S., Kommers, J., Harmon, B.A., Meegan, C.A. & Lewin, W.H.G., *Nature*, **379**, 799 (1996)
9. Lewin, W.H.G., Rutledge, R.E., Kommers, J.M., van Paradijs, J. & Kouveliotou, C., *ApJ*, **462**, L39 (1996)
10. Lewin, W.H.G., van Paradijs, J. & Taam, R.E., in *X-ray Binaries*, W.H.G. Lewin et al. (Eds), Cambridge Univ. Press, pp 190-196 (1995)
11. Paciesas, W.S., et al., *IAU Circular*, **6284**, (1996)
12. Stark, M.J., Baykal, A., Strohmayer, T. & Swank, J.H., *ApJ*, **470**, L109 (1996)
13. Strickman, M.S., Dermer, C.D., Grove, J.E., Johnson, W.N., Jung, G.V., Kurfess, J.D., Philips, B.F., Share, G.H., Sturner, S.J., Messina, D.C. & Matz, S.M., *ApJ*, **464**, L131 (1996)

Determination of Peak Fluxes and α for Bursts from GRO J1744-28

Tod E. Strohmayer[1], Keith Jahoda[2], Jean H. Swank[2] and Michael J. Stark[3]

Laboratory for High Energy Astrophysics
NASA/GSFC Greenbelt, MD 20771
[1] *Universities Space Research Association:* [2] *NASA/GSFC*
[3] *University of Maryland, College Park*

Abstract. The Rossi X-ray Timing Explorer (RXTE) has been monitoring the Bursting Pulsar, GRO J1744-28, since Jan. 1996. Large deadtimes in the proportional counter array (PCA) have made measurements of the peak flux during bursts problematic, therefore the value α, of the ratio of time averaged persistent luminosity to time averaged burst luminosity, has not been well determined. The value of α has important implications for understanding the nature of the bursts. Some RXTE observations begun after the recent reemergence of GRO J1744-28 in Dec. 1996 were performed with the source offset from the axis of peak PCA response. This reduces the deadtime in the PCA and allows a more precise determination of burst peak fluxes and thus α. These observations provide an understanding of the deadtime processes during periods when the source was viewed at peak response. Between 17 - 24 January 1997 we estimate that α was ≈ 34. Based on similarity of the overall lightcurve and countrate, this value is also appropriate for the period near 1 March 1996. We also present improved estimates of the peak luminosity during bursts.

INTRODUCTION

The bursting pulsar, GRO J1744-28, was discovered in December, 1995 with BATSE [1]. The source is a low mass X-ray binary with a 2.14 Hz spin frequency and produces powerful X-ray bursts which are likely caused by an accretion disk instability [2]. Measurements of the peak burst luminosities and total energies have been difficult due to both detector deadtime problems in the RXTE-PCA and the lack of spectral sensitivity below 20 keV in BATSE. Hence our understanding of the burst energetics with respect to the persistent X-ray emission has been incomplete. Here we describe new observations which overcome these uncertainties and provide a clearer understanding of the burst energetics and the implications for the burst process.

DATA

RXTE has performed regular observations of J1744-28 since 18 January 1996, when the source was first far enough from the sun to be observed by RXTE. Since 3 November 1996, J1744-28 has been observed in a weekly, peer approved, monitoring campaign. The monitoring campaign was coupled with a Target of Opportunity proposal for more frequent observations should the source once again go into outburst. Observations were not possible from mid November, 1996 through mid January, 1997 due to the RXTE sun constraint. Both bursting and pulsed emission from J1744-28 recurred during this period when it was within 30° of the sun [3,4]. RXTE observations resumed on 17 January 1997, and the 2-60 keV count rate was comparable to the Crab Nebula. This flux is comparable to the brightness of J1744-28 on 1 March 1996 [5]. We also noted the presence of frequent bursting activity, with peak apparent counting rates reaching 150,000 counts s^{-1}, a level at which there are substantial and poorly understood dead time effects in the PCA [5,6]. Because the source had unexpectedly brightened, the target of opportunity proposal was triggered, and observations were made approximately daily over the following two weeks. Observations of the galactic center region, primarily designed to search for and locate bursting activity from other transient sources have also been made [7]; the pointing direction choosen for these observations is about 0.7° away from J1744-28. The peak counting rate during the bursts, observed in the off-axis position is 50,000 - 80,000 counts s^{-1} and the peak deadtime, now dominated by good X-ray events, is < 25%. We use these off axis bursts to estimate the true peak fluxes for on axis bursts, all of the observations to estimate the frequency of the bursts, and thus derive α, the ratio of the time averaged persistent flux to the time averaged burst flux. In this work we restrict our attention to the bursts observed during a one week period in 1997 January. A total of 22 bursts were detected in 884 minutes of exposure, yielding an average rate over the week of 1 burst every 40 minutes. A pair of bursts characteristic of those observed are displayed in Figure 1.

ANALYSIS

We have examined nine bursts observed when RXTE was offset from J1744-28. Despite the pointing offset, J1744-28 dominates the observed counting rate; the non bursting rate is 4500 s^{-1}. During bursts the observed good count rate reaches 60,000 - 80,000 s^{-1}. At this level, deadtime, while important, is < 25% so we correct the observed count rate for deadtime, and then for collimator transmission. This overcomes the uncertainties in the large deadtime corrections required for bursts observed on axis, and allows a better estimate of the intrinsic burst fluxes. Details of our deadtime correction follow. We analyzed the bursts with 0.125 sec resolution, taking advantage of the Standard 1 data which counts every event in the detector exactly once, allowing an estimate of the deadtime at this time scale. The important rates are the "good" count rate from each of the 5 Proportional Counter Units (PCU), the sum of the events with a propane flag, the sum of events with a Very Large Event flag, and the sum of the "remaining counts". The typical deadtime for any event is 10 μsec except that after the VLE events processing is inhibited for 150 μsec. The events are spread over five detectors, so the dead time is approximated as $t_d = 2.0 \times 10^{-6}(r_G + r_{Vp} + r_{rem}) + 3.0 \times 10^{-5} r_{VLE}$, where r_G is the sum of the 5 good rates, r_{Vp} is the summed propane rate, r_{rem} is the summed remaining rate, r_{VLE} is the summed VLE rate, and t_d is the fractional deadtime. The deadtime is dominated by the substantial increase in r_G. For the largest burst in this sample r_G ranges from 4800 s^{-1} outside the burst to a peak of 82,000 s^{-1}, corresponding to a deadtime range of 1 - 16 %. In addition, r_{rem} ranges from 4000 to 21,000 s^{-1} (deadtime range of 1 - 4.2 %), r_{Vp} from 600 to 5800 s^{-1} (deadtime range of 0.1 - 1.2 %), and in r_{VLE} the burst is not detected against the constant rate of 600 s^{-1} (deadtime of 1.8 %). The total deadtime ranges from a typical value of 4 % to a peak value of 23 % at the peak of this burst. The deadtime is dominated by good events; for the on-axis bursts (not used for estimating fluences here) r_{rem} becomes comparable to r_G and the interpretation of deadtime is more complex. Figure 2 shows the rates which enter the deadtime calculation, and the derived deadtime, for the burst discussed.

Using the derived deadtime, we correct the burst counting rate by a factor of $(1.0 - t_d)^{-1}$ and by a constant factor of 3.36 to account for the collimator transmission at the off axis position. We subtract a constant value which is the measured corrected count rate for a 5 second interval before each burst, and

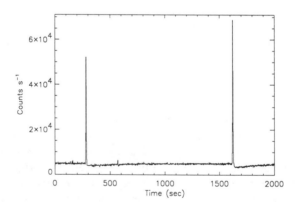

FIGURE 1. Bursts from GRO J1744-28 observed with the PCA. Note the dips in the persistent countrate after the bursts.

integrate the resulting lightcurve over a 25 second interval to obtain fluences. Using only preburst data to estimate the baseline avoids the dips often noted after these bursts [8,5]. Because these bursts observed off axis did not saturate the PCA, we can measure the burst to burst variability, which is seen to be about 20 % relative to the average for the extreme cases. To estimate α we take the average burst flux to be 923866 counts, the average interval to be 40 minutes and the average persistent flux to be 13,000 s^{-1}. Combining these values gives a value for α of about 34. Our calculation assumes that the spectrum of the bursts is the same as the spectrum of the persistent emission. Both the burst and nonburst emission can be fit with an absorbed powerlaw with an exponential cutoff above some energy. The detailed parameters are somewhat different for the bursting and non bursting emission, but the ratio of counts to flux varies by less than 5 % in the 2-10 keV band and only by about 10 % in the 2-60 keV band. We reject the hypothesis that the bursts have the 2-3 keV thermal spectrum characteristic of Type I bursts. The above is an average value, specific values computed using only data prior to a given burst yield values as low as 18 - 20. The spectral fits yield a conversion factor of flux to luminosity, assuming isotropic emission and a distance of 8 kpc, of 3.5×10^{34} ergs s^{-1} (pca-ct/sec)$^{-1}$. This implies a peak luminosity during bursts observed in 1997 of about 1.2×10^{40} ergs s^{-1}, or about a factor

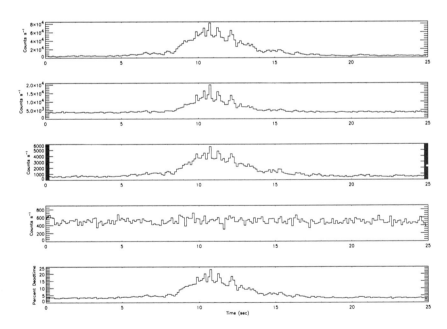

FIGURE 2. Counting rates which enter the deadtime correction. The counting rates from top to bottom are; r_G, r_{rem}, r_{Vp}, r_{VLE}, and the bottom panel gives the percent deadtime.

of 100 above the implied isotropic Eddington luminosity for a neutron star. Our estimate of α is necessarily uncertain at the 10 % level as we averaged a variable burst rate over a week. We cannot make a strong statement as to whether burst fluence is anticorrelated with the time between bursts, as is often the case for the Rapid Burster [8]. The time between bursts is greater than half the maximum unocculted viewing period, leading to uncertainty in the time between adjacent bursts (and providing data only on the short half of the distribution). The burst from Jan 24:01:26:50, which is only 15 minutes after its predecessor, is only 1.5 % below the mean fluence in this small sample.

DISCUSSION

Bursts observed with RXTE-PCA during the first outburst in Jan. 1996 were significantly affected by deadtime in the PCA. Some of these bursts showed evidence of strong modulation within the bursts, but saturation of the detectors made this difficult to quantify. Our analysis of off-axis bursts confirms that modulations of the flux within the bursts can occur with amplitudes greater than 1/2 of the peak countrate on timescales as short as the pulse period. It has been argued that the similarity of persistent and burst spectra, the similarities to bursts from the Rapid Burster, and the low values for α inferred from BATSE data when the burst recurrence rate was high in Dec. 1995 all point toward an identification of the bursts as type II, ie. due to release of gravitational potential energy via accretion [8]. The value of α inferred for Jan. 1997 is not incompatible with a thermonuclear origin based on the energetics alone, but it would require that hydrogen burning be the primary source of energy release. However, at the very high accretion rates inferred from the persistent emission (about 4×10^{-8} M_{sun} yr^{-1} ie, well above the isotropic Eddington limit) it is very unlikely that the hydrogen would burn unstably [9]. Thus our estimates of α favor a type II origin for these bursts.

REFERENCES

1. Fishman, G. et al. 1996, IAUC, 6290
2. Cannizzo, J. K. 1996, ApJ, 466, L31
3. Kouvelioutou, C. *et al.* 1997, IAUC 6350
4. Finger, M. J. *et al.* 1997, IAUC, 6350
5. Giles, A. B. *et al.* 1996, ApJ, 469, L25
6. Jahoda, K. *et al.* 1996, Proc, SPIE 2808, EUV, X-ray, and Gamma-ray Instrumentation for Space Astronomy VII, O. H. W. Siegmund and M Gurmin, eds, pg. 59
7. Strohmayer, T. E., Lee, U. and Jahoda, K. 1996, IAUC, 6484
8. Lewin, W. H. G. *et al.* 1996, ApJ, 462, L39
9. Bildsten, L., & Brown, E. F. 1997, ApJ, 477, 897

Kilohertz Oscillations in 4U 0614+091 and Other LMXBs

E.C. Ford[1], P. Kaaret[1], M. Tavani[1], D. Barret[2], P. Bloser[3], J. Grindlay[3], B.A. Harmon[4], W.S. Paciesas[4,5], S.N. Zhang[4,6]

*1:Department of Physics and Columbia Astrophysics Lab,
Columbia University, 538 W. 120th Street, New York, NY 10027
2:Centre d'Etude Spatiale des Rayonnements (CESR),
9 Avenue du Colonel Roche BP4346, 31028 Toulouse Cedex 4, FRANCE
3:Harvard Smithsonian Center for Astrophysics,
60 Garden Street, Cambridge, MA 02138
4:NASA/Marshall Space Flight Center, ES 84, Huntsville, AL 35812
5:University of Alabama in Huntsville, Department of Physics, Huntsville, AL 35899
6:Universities Space Research Association/MSFC, ES 84, Huntsville, AL 35812*

Abstract. Strong oscillations near 1000 Hz have been discovered in a number of low mass x-ray binaries with NASA's Rossi X-Ray Timing Explorer. Their behavior should ultimately lead us to a better understanding of the accretion dynamics and the fundamental properties of the neutron star itself. Here we summarize the known properties of fast QPOs, focusing on results from the source 4U 0614+091. We highlight some of the aspects yet to be explored.

PROPERTIES OF THE FAST OSCILLATIONS

Fast quasi-periodic oscillations (QPOs) from x-ray binaries were discovered with RXTE almost immediately after its launch. In observations to date, these signals, ranging in frequency from 330 to 1210 Hz, are known in at least ten low mass x-ray binaries. The great interest in fast QPOs is borne from the exciting revelations they potentially offer. They open the possibility of measuring fundamental properties of neutron stars in low mass x-ray binaries: their spin periods (Strohmayer et al. 1996, Ford et al. 1997a), their masses (Kaaret, Ford, & Chen 1997; Zhang, Strohmayer, & Swank 1997) and constraints on their radii (Miller, Lamb & Psaltis 1997). The exciting possibility of probing strong field gravity may also be realized with the fast QPOs, produced in the inner-most regions of the accretion disk near the neutron star (Kaaret, Ford, & Chen 1997, Kaaret & Ford 1997).

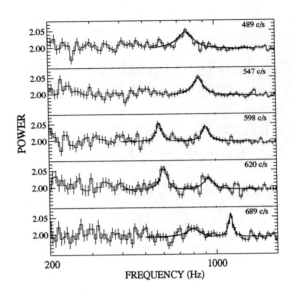

FIGURE 1. High Frequency power density spectra for 4U 0614+091. Power density spectra of 4U 0614+091 for five intervals beginning UTC 4/25/96 0:10:23, 4/24 19:47:27, 4/25 4:58:23, and 8/6 20:52:01. The total count rates are shown for each observation. Lorenztian fits to the QPO features are plotted. The power is normalized according to Leahy et al. (1983).

An example is shown in Figure 1 of some of the QPOs detected in the low mass x-ray binary 4U 0614+091. One, and at times two, narrow features appear in the power density spectra. The centroid frequencies of these features are variable and change with the intensity of the source.

There are currently eleven sources known to exhibit fast QPOs, referred to collectively as 'kilohertz QPOs'. Table 1 summarizes present observational results, and an updated list is maintained at http://www.astro.columbia.edu/~eric/qpos.html. A review can be found in van der Klis (1997b). The first reported fast QPOs discovered with RXTE were from Sco X-1 (van der Klis et al. 1996). Numerous other low mass x-ray binaries will be observed with RXTE, for example GX 349+2 and X1556-605.

The fast QPOs so far detected are broadly similar. The main defining aspect of the QPO, its frequency, is roughly the same in all sources. There are often two QPOs simultaneously present in a given source, and the difference between their frequencies is roughly the same for all sources: 250–400 Hz. The QPOs share other major features as well; they are all relatively narrow, with Q values (ν/FWHM) of 20–200, and are suprisingly strong; the RMS fraction exceeds 20% at high energies in some of the lower luminosity sources (Berger et al. 1996, Ford et al. 1997a).

Though the oscillations are very similar, the sources themselves vary widely. Table 1 implicates both major varieties of LMXB: the Z sources and the Atoll sources. The sources exhibiting these QPOs have strikingly different luminosities; the Z source Sco X-1, for example, is two orders of magnitude brighter than 4U 0614+091. The binaries may even have very different evolutionary paths. 4U 1820-30, for example, has a degenerate companion, a fact inferred from its very short orbital period (Stella, Priedhorsky & White 1987).

How representative are the detections in Table 1? There is an obvious bias against detecting QPOs in sources which have not been observed. For well-probed sources, though, the high frequency regime is well covered; with detectable RMS fractions of approximately 2% up to typically 4096 Hz. The

Source	Type	L	QPO	Q	RMS	$\Delta\nu$	Burst	Reference
X0614+091	A	1.7	500–1145	5–20	6–12[a]	323 ± 4	–	2,4,3
			550–800	5–20	4–8			IAU 6426,6428
X1608-52	A	2.1	567–890	10–75	6–8[b]	–	no QPOs	1,12,16,17 IAU 6336
X1820-30	A	4.8	1066	30	3.2	275 ± 8	–	5
			544–802	30–50	4.1			IAU 6507
X1735-444	A	5.4	1149	40	3	–	–	IAU 6447
X1728-34	A	11	988–1058	up to 120	5–7	360 ± 12	363	7
			640–716		8–6			IAU 6320, 6387
Aql X-1	–	≤ 17	750				549	9
X1636-536	A	19	1147–1200	20–50	5–11[c]	273 ± 9	581	15,13
			840–922	20–50	6–12			IAU 6428, 6541
KS1731-260	–	48	1170–1207	40 to 70	5–4	260 ± 10	524	6,14
			903	45	4			IAU 6437
GX 5-1	Z	210	567–895	2–18	6.7–2.0	325	–	IAU 6511
			325–448	–	–			
Sco X-1	Z	220	850–1075	8–20	2.5–1.0	230–310	–	10,11
			550–840					IAU 6319, 6424
GX 17+2	Z	240	988	~3 to 30	~3 to 5	306 ± 5	–	IAU 6565
			880–682					
Gal. Center	–	–	–	–	2–4	–	589	8, IAU 6484

TABLE 1. L is the 1–20 keV luminosity ($\times 10^{36}$ erg s^{-1}) estimated from the XTE/ASM lightcurves, using distances from van Paradijs & White (1995). Source types are indicated (A=atoll, Z=Z-source). $\Delta\nu$ is the frequency difference between the higher frequency QPO and the lower frequency QPO. 'Burst Frequency' is the frequency of the QPO detected during x-ray bursts. The RMS fraction increases with energy in [a]Ford et al. (1997b), [b]Berger et al. (1996) and [c]Zhang et al. (1996). The source near the galactic center may be MXB 1743-29 (Strohmayer et al. 1997). *References:* IAU refers to an IAU circular with the associated number. [1]Berger et al. (1996) [2]Ford et al. (1997a), [3]Ford et al. (1997b), [4]Mendez et al. (1997) [5]Smale, Zhang & White (1997) [6]Smith et al. (1997) [7]Strohmayer et al. (1996) [8]Strohmayer et al. (1997) [9]Swank et al. (1997) [10]van der Klis, et al. (1996) [11]van der Klis et al. (1997) [12]Vaughan et al. (1997) [13]Wijnands et al. (1997a) [14]Wijnands et al. (1997b) [15]Zhang et al. (1996) [16]Yu, W. et al. (1997a) [17]Yu, W. et al. (1997b)

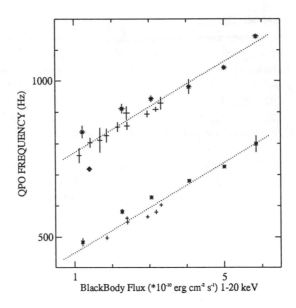

FIGURE 2. QPO frequency versus flux of the blackbody spectral component in 4U 0614+091. The plusses are data from an observation in April, and the asterisks are from August. A power law fit has a slope of $\alpha = 0.29 \pm 0.01$. The fluxes are from a spectral fit which include a power law and a blackbody component. The data point at 719 Hz (diamond) is from Mendez et al. (1997).

detection of lower frequency QPO features, however, may be selected against, since lower frequency QPOs are now known to be associated with weaker source fluxes. The presence and properties of the QPOs also seem to depend strongly on the state of the source when the observation was made. In X1820-30, for example, only the lower frequency of the two QPOs is detected except when the source is observed at its highest fluxes (Smale, Zhang & White 1997). In X0614+091, the opposite seems to be true; below a certainly flux only the higher frequency QPO is present (Ford et al. 1997a).

The QPO frequency tends to increase as the count rate of the variable source increases as is suggested by the plots in Figure 1. The details of this are borne out in a more careful study of the variations of the QPO frequency with respect to the energy spectra of the source (Ford et al. 1997b). The frequency of the QPO correlates well with the flux of a soft, blackbody component of the spectra as shown in Figure 2.

KEPLERIAN OSCILLATIONS AND THE BEAT FREQUENCY MODEL

The broadly similar properties of the QPOs in all of the sources points to a common mechanism for their generation. In particular, the presence and behavior of the double QPO suggests a beat frequency mechanism (see for example Ford et al. 1997b for discussion). The higher frequency QPO arises from Keplerian motion in the inner accretion disk, and a decreasing inner disk radius at higher fluxes causes the QPO frequency to increase. The correlation of frequency with flux (Figure 2) supports this view (Ford et al. 1997b).

A number of outstanding issues, however, are raised by the observational evidence. How are high frequency QPOs near 1000 Hz related to the well known low frequency QPOs at about 10 Hz? The 10 Hz, horizontal branch oscillations, are exhibited simultaneously with the fast QPOs in the Z-sources GX 5-1, GX 17+2, and Sco X-1, and seem to be related. In Sco X-1, the frequencies of the 10 Hz QPO and the 1000 Hz QPO are correlated (van der Klis 1996). This is awkward, as the horizontal branch oscillations have always been understood in terms of a model with a disrupted disk. Two frequencies are needed. Can the disk be disrupted twice?

Current observations also suggest the simplest beat frequency model will need refinement to explain all the fast QPOs phenomena. The model predicts that the difference in frequency between the two QPOs is always constant; the frequency difference of the QPOs in Sco X-1, however, changes from 230 to 310 Hz (van der Klis et al. 1997). Perhaps the effects of radiation are more important in Sco X-1 where the luminosity is about 100 times larger than in 4U 0614+091. There is an additional hint of a luminosity dependence from 4U 1636-536. In this intermediate luminosity source, the x-ray bursts exhibit QPOs whose frequency is in slight disagreement (at the 2σ level) with the expected value of twice the difference frequency (Wijnands et al. 1997a, Zhang et al. 1997). Higher luminosity sources may also produce weaker QPOs. Sco X-1, for example, produces QPOs with an RMS fraction 10 times smaller than those in X0614+091 ($\sim$1.5% vs. 10% when corrected for deadtime).

Luminosity, then, seems to be the most crucial parameter. Within a single source, the luminosity of at least one spectral component has a strong influences on the frequency of the QPO (Figure 2). And though the QPOs are roughly similar between sources, their somewhat different behaviors may be linked to luminosity.

In observing the QPOs produced at very high luminosities, one is also observing activity very close to the neutron star, as higher luminosities seem to correspond to small inner disk radii. At the highest luminosities one might observe the QPOs at the marginally stable orbit predicted in general relativity, which is at a radius $6r_g$ in a Schwarzschild metric (Kaaret, Ford, & Chen 1997). There is already indications for such an effect in at least two sources

(Kaaret, Ford, & Chen 1997) and further observations should provide even clearer evidence.

REFERENCES

1. Berger, M. et al. ApJL, 469, L13-16 (1996)
2. Ford, E.C. et al. ApJL, 475, L123-126 (1997a).
3. Ford, E.C. et al. ApJL, in press (1997b).
4. Kaaret, P., Ford, E.C. & Chen, K., ApJL, 480, L27 (1997)
5. Kaaret, P., & Ford, E.C. Science, 276, 1386-1391 (1997)
6. Leahy, D.A. et al. ApJL, 266, 160 (1983)
7. Mendez, M. et al. ApJL, accepted (1997)
8. Miller, C., Lamb, F. & Psaltis, D. ApJL, submitted (1997)
9. Smale, A.P., Zhang, W. & White, N.E. ApJL, 483, L119-122 (1997)
10. Smith, D.A. et al. ApJL, 479, L137-140 (1997)
11. Stella, L., Priedhorsky, W. & White, N.E. ApJ, 312, L17 (1987)
12. Strohmayer, T. et al. ApJL, 469, L9-12 (1996)
13. Strohmayer, T. et al. preprint (1997)
14. Swank, J., these proceedings (1997)
15. van der Klis, M. et al. ApJL, 469, L1-3 (1996)
16. van der Klis, M. et al. ApJL, 483, L97-100 (1997)
17. van der Klis, M., to appear in Proceedings of the Wise Observatory 25th Anniversary Symposium (1997b)
18. Vaughan, B.A. et al. ApJL, 483, L115-118 (1997)
19. Wijnands, R.A.D. et al. ApJL, 479, L141-144 (1997a)
20. Wijnands, R.A.D. et al. ApJL, 482, L65-68 (1997b)
21. Zhang, W. et al. ApJL, 469, L17-19 (1996)
22. Zhang, W. et al. IAU Circular No. 6541 (1997)
23. Zhang, W., Strohmayer, T. & Swank, J. ApJL, 482, L167-170 (1997)
24. Yu, W. et al., these proceedings (1997a)
25. Yu, W. et al., ApJL, submitted (1997b)

General Relativity and Quasi-Periodic Oscillations

Philip Kaaret and Eric C. Ford

Columbia Astrophysics Laboratory
Columbia University
538 West 1020th Street
New York, NY 10027

Abstract.
Quasi-periodic oscillations (QPOs) at frequencies near 1000 Hz have been detected from a number of neutron star x-ray binaries using RXTE. These oscillations are likely associated with orbital motion at the inner edge of accretion disks near neutron stars with low magnetic fields and, thus, give us information about the behavior of the accretion disk in the strong gravitational field near the neutron star. We discuss implications of general relativity for the interpretation of the fast QPOs.

INTRODUCTION

There are two important radii that determine the behavior of the accretion flow very close to a weakly magnetized neutron star. The first is the radius of the neutron star itself. This radius is determined by the mass of the star and the equation of state of nuclear matter. Although myriad efforts have been made to measure neutron star radii, our knowledge of this fundamental parameter is good only to about a factor of 2. The radius of a $1.4 M_\odot$ neutron star is roughly 10 km.

The second radius is that of the marginally stable orbit. In general relativity, stable orbits do not exist arbitrarily close to a massive compact object. Within a certain radius, relativistic corrections cause the shape of the potential energy curve to deviate from the $1/r$ form familiar in Newtonian mechanics and preclude the existence of stable particle orbits [1]. This radius, the marginally stable orbit, occurs at three Schwarzschild radii ($r_{ms} = 6GM/c^2$) for a non-rotating object of mass M and is moved inward if the object is rotating. For typical neutron star spin periods, rotation changes the radius by less than about 20%. The marginally stable orbit for a non-rotating, $1.4 M_\odot$ neutron star has a radius of 12.4 km.

A crucial question in the study of accretion onto weakly magnetized neutron stars is whether the neutron star surface lies inside or outside the marginally stable orbit [2]. If the surface lies outside the marginally stable orbit, then the accretion disk will terminate in a boundary layer at the surface. If the surface lies inside the marginally stable orbit, then the disk will terminate at the marginally stable orbit and there will be a gap between the inner edge of the disk and the surface - an 'accretion gap' [3].

Recently, observations with the *Rossi X-Ray Timing Explorer* [4] have revealed oscillations in the x-ray emission of accreting neutron stars at frequencies near 1000 Hz [5-8]. Matter orbiting a $1.4 M_\odot$ neutron star at a radius of 16.8 km will have a Keplerian orbital frequency of 1000 Hz. In the following, we argue that these fast oscillations are, indeed, related to orbital motion very close to the neutron star and, thus, provide a direct probe of the innermost part of the accretion flow. Further, we present evidence that the accretion disk around certain neutron stars is terminated at the marginally stable orbit. Finally, we discuss constraints on the equation of state of nuclear matter which can be derived from these observations.

QUASI-PERIODIC OSCILLATIONS

Quasi-periodic oscillations (QPOs) at frequencies near 1000 Hz have now been detected from more than 10 sources, see [9,10] and references therein. Here, we concentrate on the atoll sources [11]. The behavior of the fast QPOs from the various atoll sources are basically similar. There are two fast QPOs which sometimes appear simultaneously in the persistent emission. In the atoll sources, when two QPOs appear simultaneously, the frequency difference remains constant even though the QPO frequencies vary.

This phenomenology is naturally interpreted with a beat-frequency model [12]. There are three frequencies in such a model: the Keplerian orbital frequency at the inner edge of the accretion disk ν_K, the spin frequency of the neutron star ν_S, and the beat frequency which is the difference between these two $\nu_B = \nu_K - \nu_S$. If we identify the higher QPO frequency in the persistent emission as the Keplerian frequency and the lower frequency as the beat frequency, then the constant difference is simply the spin frequency of the star. This picture is strikingly confirmed by observations of 4U 1728-34. In x-ray bursts from this source, a single QPO is detected with a frequency which is constant to within a few percent and is equal to the frequency difference of the QPOs in the persistent emission [6]. The burst QPO is then direct detection of the neutron star spin frequency.

The QPO frequency in the persistent emission is strongly positively correlated with the x-ray count rate from the source (with two exceptions discussed in the next section). This behavior is expected in the beat-frequency model, if x-ray count rate is an indicator of the mass accretion rate in the disk. A robust correlation has been demonstrated between the QPO frequency and the flux, F_{BB}, of a blackbody component of the x-ray spectrum in 4U 0614+091 [13,14]. The correlation is well described by a power law relation $\nu_K \propto F_{BB}^\alpha$ with $\alpha = 0.27 - 0.37$. This behavior is expected in the magnetospheric beat-frequency model since the inner edge of the disk moves inward, and therefore the QPO frequency increases, as the mass accretion rate increases. The power law relation and the numerical value of the exponent are in good agreement with the predictions of the model [12].

MARGINALLY STABLE ORBIT

In two sources, 4U 1608-52 and 4U 1636-536, a QPO sometimes appears in the persistent emission which has a frequency which is not correlated with the x-ray count rate or inferred mass accretion rate of the source. In 4U 1608-52, the QPO frequency changes by less than 10% as the count rate changes by a factor of 2 [7]. The constancy of the QPO frequency seems inconsistent with the strong dependence predicted by the beat-frequency model. However, the marginally stable orbit provides a natural explanation of this behavior in the context of the beat frequency model [15].

As the mass accretion rate increases, the inner edge of the disk moves inward. At a certain mass accretion rate, the inner disk radius will move in to the marginally stable orbit, see Fig. 1. If the mass accretion rate increases further, the inner disk radius will not move as it did before since the inner disk cannot move significantly past the marginally stable orbit. Thus, the transition from a QPO frequency correlated with x-ray count rate to a QPO frequency independent of x-ray count rate is a simple consequence of where the disk is disrupted relative to the marginally stable orbit [15].

Additional support for this picture comes from 4U 1636-536 where both mass accretion rate dependent and independent QPOs have been observed

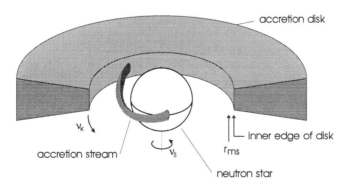

FIGURE 1. Accretion flow near the marginally stable orbit.

[16]. The mass accretion rate is inferred from the position of the source on an x-ray color-color diagram. Only one of the two QPO behaviors appears at any given time, which is consistent with this picture. In addition, the QPO frequencies observed in the two distinct behaviors are very similar, as would be expected in this picture. If the QPOs with the two distinct behaviors are produced by distinct mechanisms, the similarity of the frequencies would be an unexpected coincidence. One obvious interpretation, that the mass accretion rate independent QPO frequency is the neutron star spin frequency, is excluded by the extent of the frequency variations. For 4U 1636-536, the constant QPO frequency seen in persistent emission differs from the QPO frequency seen in x-ray bursts, again arguing against interpretation of the persistent emission QPO as due to the neutron star spin [17].

DISCUSSION

The behavior of the fast QPOs in sources similar to 4U 1728-34 is strong evidence for a beat-frequency model of QPO production. Above, we identified the higher frequency QPO with the Keplerian orbital frequency at the inner edge of the disk. This is the typical assumption made in beat-frequency models [12,18,19]. However, this assumption is more restrictive than is needed to explain the observed QPO behavior.

The beat-frequency model is applicable as long as the higher frequency QPO is generated by some rotational or orbital motion in the inner part of the accretion disk. Any orbital or rotational motion is allowed as long as the frequency of the motion monotonically increases as the inner disk radius decreases. It is possible that the orbital motion in the disk is sub-Keplerian due to radiation pressure; particularly, since the luminosities in the atoll sources may approach 10% of Eddington. It is also possible that the higher frequency QPO is generated by a traveling sound wave. However, in order to generate the beat-frequency, the wave must produce a density perturbation which rotates around the disk. Standing waves will not suffice.

Even with this less restrictive assumption, the observation of QPOs with both mass accretion rate dependent and independent frequencies is evidence that when the QPO frequency is independent of mass accretion rate the accretion disk is disrupted close to the marginally stable orbit. The marginally stable orbit provides a natural explanation for the two distinct behaviors from a single QPO production mechanism and also for the narrow range of maximum QPO frequencies observed [20]. Thus, the fast x-ray oscillations are the first experimental evidence for the existence of the marginally stable orbit [15].

We can now answer the question posed earlier: does the neutron star surface lie inside or outside the marginally stable orbit? At least for some neutron stars (those in 4U 1608-52 and 4U 1636-536), the neutron star surface lies inside the marginally stable orbit.

REFERENCES

1. Misner, C.W., Thorne, K.S., and Wheeler, J.A., *Gravitation* San Francisco: W.H. Freeman, 1970.
2. Kluźniak, W., and Wagoner, R.V., *Astrophys. J.* **297**, 548 (1985).
3. Kluźniak, W., Michelson, P., and Wagoner, R.V., *Astrophys. J.* **358**, 538 (1990).
4. Bradt, H.V., Rothschild, R.E., and Swank, J.H., *Astr. Astrophys. Suppl.* **97** 355 (1993).
5. van der Klis, M. et al., *Astrophys. J.* **469**, L1 (1996).
6. Strohmayer, T. et al., *Astrophys. J.* **469**, L9 (1996).
7. Berger, M. et al. *Astrophys. J.* **469**, L13 (1996).
8. Zhang, W. et al. *Astrophys. J.* **469**, L17 (1996).
9. Kaaret, P., and Ford, E.C., *Science* **276**, 1386 (1997).
10. Ford, E.C. et al. these proceedings.
11. Hasinger, G., and van der Klis, M., *Astron. Astrophys.* **255**, 79 (1989).
12. Alpar, M.A., and Shaham, J., *Nature* **316**, 239 (1985).
13. Ford, E. et al., *Astrophys. J.* **475**, L123 (1997).
14. Ford, E.C. et al., *Astrophys. J.* in press (1997).
15. Kaaret, P., Ford, E.C., and Chen, K., *Astrophys. J.* **480**, L27 (1997).
16. Wijnands et al., *Astrophys. J.* **479**, L141 (1997).
17. Zhang, W. et al., *IAU Circular* **6541** (1997).
18. Lamb, F.K. et al., *Nature* **317**, 681 (1985).
19. Miller, M.C., Lamb, F.K., and Psaltis, D., *Astrophys. J.* submitted (1997).
20. Zhang, W., Strohmayer, T.E., and Swank, J.H., *Astrophys. J.* **482**, L67 (1997).

Compact Hard X-ray Sources Near Galactic Longitude 20.

Gregory V. Jung, [1] J. D. Kurfess, [2], and W.R. Purcell [3]

Abstract. OSSE survey observations were conducted with scans crossing the galactic plane at longitudes $l^{II} = +16$ and $l^{II} = +20$. These data indicate the presence of three or more compact sources of hard X-ray emission. One of these is identified to be the source 4U 1812-12 a type I burster. Such semi-persistent hard X-ray emissions in the galactic LMXRB population may be responsible for a significant portion of the 'distributed' hard X-ray spectrum of the galactic center, as recently measured with the joint SIGMA/OSSE observations. For 4U 1812-12 we find a spectrum well-described from 50 keV to 200 keV with a power-law form of index $\approx$ -2.6 or with a thermal bremmstrahlung model (OTTB) of kt $\approx$ 55 keV.

INTRODUCTION

Hard emission from galactic binaries may produce a substantial portion of the distributed galactic emission of hard X-rays above 50 keV. This assertion is probable *a priori* as these are the most common class of galactic X-ray emitter, but until recently it was not supported by observation. Although several binaries are known for their X-ray emission at energies below 10 keV, only 11 of the 63 galactic sources of the HEAO A-4 survey had detectable emission above 80 keV [10]. None of these are members of the "ordinary" LMXRB's such as the type-I burst sources or the bright bulge sources comprising the bulk of objects in the catalogs sensitive in the 2-10 keV band. However, recent hard X-ray observations have demonstrated that at least a few of the type I X-ray bursters are capable of producing hard X-ray emission in transient events lasting a few or several days. The imaging telescope SIGMA first identified the source MXB 1728-34 with a hard X-ray outburst in the Spring of 1992 [4] which was also observed by the non-imaging BATSE and OSSE instruments [6], [8]. SIGMA also caught single hard emission episodes from the burster sources KS 1731-260 [1], A1742-294 [3], as well as more persistent hard emissions from sources in globular clusters Terzan 1 [5] and Terzan 2 [2]. Data from BATSE earth occultations have turned up hard emissions from the bursters Aquila X-1 [7] and from 4U 1608-52 [14]. The OSSE data discussed here show convincing evidence of another object believed to be a type I burster, the source 4U 1812-12.

---

[1] Universities Space Research Association, Washington, D.C.
[2] Naval Research Laboratory, Washington, D.C.
[3] Northwestern University, Evanston, Illinois

The OSSE instrument has a collimated rectangular field-of-view of 11.4° × 3.7° (FWHM). This was chosen to be small enough to deal with most source confusion issues but large enough to be sensitive to diffuse emission sources. In the spring of 1992 coordinated OSSE/SIGMA observations measured the distributed hard X-ray emission from the galactic center [11]. Observations from the imaging telescope SIGMA were used, with a sensitivity limit of $\approx$ 30 mCrab, to estimate the compact source contribution to the OSSE spectrum from the field $\{(11.4 > l^{II} > -11), (3.6 > b^{II} > -3.6)\}$. The shape of the corrected spectrum is roughly the shape of the spectrum of the summed sources: a power-law of index -2.6 in the 50-200 keV band, consistent with a thin thermal bremmstrahlung fit of about 45-55 keV. This hard X-ray component lies well above the extrapolation from the low-energy gamma emission above 600 keV [9], which is dominated by bremmstrahlung of galactic electrons in the interstellar medium. If the hard X-ray ($\alpha \approx -2.6$) spectrum were postulated to be an extension of the higher energy galactic bremmstrahlung, the energy requirements needed to overcome coulomb losses would be significant [13].

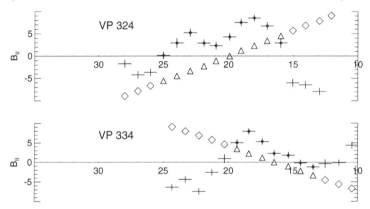

FIGURE 1. The sky positions of the two observations are represented in these two panels with symbols designating the center of each field-of-view, in (l^{II}, b^{II}) coordinates. The triangular symbols mark the positions of the designated "source" region. Background corrections for the source positions were obtained from the exposures taken at the positions marked by diamonds, while the correction for the background positions used the observations in the source region. The 40-180 keV difference count-rate profiles are shown in arbitrary units, aligned in galactic longitude.

OBSERVATIONS AND ANALYSIS

The OSSE galactic plane observation strategy

In these observations the OSSE instrument used a 'scanning' procedure to survey the galactic plane. The method simulates a traditional scan survey with a series of observations fields across the galactic plane, with the OSSE collimator aligned at an oblique angle. The averaged final exposure is nearly uniform along the OSSE scan line and within the span of the positioning sequences. The observation is designed to measure distributed emission from the galactic ridge as well as that of isolated sources. Each of the four OSSE detectors is assigned a sequence of eight orientations along the spacecraft XZ plane, where data are accumulated in 2-minute intervals. The selected positions alternate between a designated 'source' field in the center of the range (typically $\pm$ 10 degrees) and 'background' regions towards both ends of the range. The timing is thus similar to the more usual OSSE observing sequence of an isolated source. The interleaved source and background spectra are processed into a series of difference spectra, whereby a quadratic fit to three or four background sets are subtracted from the source sets, and spectra from the source positions are used in a symmetrical manner to correct the spectra for the background positions, and each set of spectra is considered in the analysis.

These observations were obtained from two scans that crossed the galactic plane at $l^{II} = 20$ and at $l^{II} = 16$, conducted in 1994 April (VP324) and in 1994 July (VP334). The two observations are illustrated with the maps in figure 1 that show the pointing positions along the scan path. The profile of the count-rate differences in the 40-180 keV band are shown on the same plots.

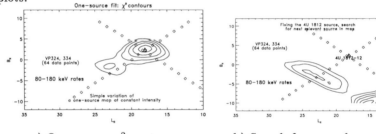

a) One-source χ^2 contours   b) Search for second source

FIGURE 2. Maps of the region in the χ^2 contours for a single source (a) and for a second source (b), after the effect of 4U 1812-12 is removed.

We analyse the data with trial skymap models using compact sources and a component confined to the galactic plane. A least-squares fit determines intensities of the components, and the overall χ^2 value is computed to judge the relative suitability. It can be seen in figure 1 that the rates of VP324 indicate the presence of at least two compact sources in the region $25 > L_{II} > 15$, while only one source is so evidently active in the scan of VP334 (1994 July). The residuals of the fit to this (six source + galactic ridge) map are shown in figure 3 for the 40-180 keV band. The histograms overlaying the residuals (plotted with error bars) are the background-subtracted rate profiles. The reduced χ^2 of this model is about 0.94 for the 40-180 keV band.

Contours of the dependence of the χ^2 value for a single-source map is shown in figure 2a. After accounting for the second source, significant only in the first period (VP324), the minimum χ^2 for variation of the position of the main source is centered at the actual position of the burster 4U 1812-12.

Additional map components

After the source 4U 1812-12 is accounted for, the χ^2 map of fig 2a indicates two additional sources, both exclusive to only one scan and thus they cannot be localized well in two dimensions. Further iteration of the procedure yields a "best-fit" skymap model with three more weak compact sources (total of six) and a galactic 'ridge'. The combined influence of the extra sources is statistically significant however they are individually only weakly significant. Moreover, there is a systematic dependence of the detector rate on absolute scan angle that is not well characterised. The residuals of the fit to this (six source + galactic ridge) map are shown in figure 3 for the 40-180 keV band. The histograms overlaying the residuals (plotted with error bars) are the background-subtracted rate profiles. The reduced χ^2 of this model is about 0.94 for the 40-180 keV band.

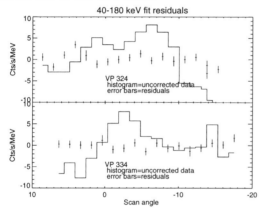

FIGURE 3. Residuals (40-180 keV) for the fullest fit found consistent with the data.

Source Spectra

A spectrum of each skymap components can be obtained with the same least-square fit on a channel-by-channel basis. The spectrum is only significant at lower energies for which a simplified response matrix is appropriate. The forward-folded spectrum of 4U 1812 is plotted in figure 4 with the fitted power-law model of index $\alpha \approx -2.6$.

CONCLUSIONS

From two week-long scans by OSSE across the galactic plane near (l^{II}, b^{II}) = (20,0) we find evidence for hard X-ray emission above 50 keV from up to six compact sources; only one of these can be located well enough to identified a counterpart, the burster 4U 1812-12. The spectrum of this source is similar to the persistent hard X-ray spectra found in a growing class of LMXRBs with long episodes of hard emission. These are generally prolific sources of lower temperature thermal X-ray emissions but only recently have they been shown to be capable of a sustained emission at higher energies. Such hard emissions are probably non-thermal in origin, possibly due to magnetic turbulence at the inner accretion disk boundary of a weakly magnetized neutron star [12]. An estimate of the hard emissions from this class is important for the interpretation of the extended hard X-ray emission from our galaxy.

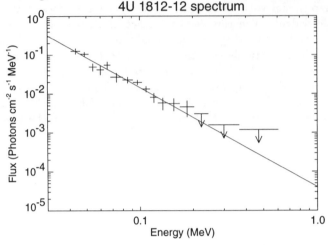

FIGURE 4. Hard X-ray spectrum of persistent emission from 4U 1812-12. The model plotted through the data is a power law of index -2.6.

REFERENCES

1. Barret, D., et al. 1992, ApJ 394, 615
2. Barret, D., et al. 1991, ApJ 379 L21
3. Churazov, E., et al, 1995, ApJ 443,341
4. Claret, A., et al, 1994, ApJ 423,436
5. Borrel, V., et al, 1996, ApJ 462,754
6. Harmon, B.A., et al, 1993 AIP conf. proc 304, 456
7. Harmon, B.A., et al, 1996, A&A supp. 120, 197
8. Jung, G.V., et al, 1993 AIP conf. proc 304, 426
9. Kinzer, R.L., Kurfess, J.D., Purcell, W.R. 1997, These proceedings
10. Levine, A.M., et al, 1984 ApJS 54,581
11. Purcell, W.R., et al, 1996, A&AS 120, 389
12. Tavani, M. and Liang, E. 1996, A&AS 120, 133
13. Skibo, J.G., et al. 1997, ApJ 483, L95
14. Zhang, S.N., et al. 1996, A&A supp. 120, 279

Hard and Soft X-ray Observations of Aquila X-1

Brad C. Rubin[1,2], B.A. Harmon[3], W.S. Paciesas[2,3], C.R. Robinson[3,4], S.N. Zhang[3,4]

[1] *RIKEN Institute, Wako-shi, Saitama 351-01 Japan*
[2] *Department of Physics, University of Alabama in Huntsville, Huntsville, AL 35899*
[3] *NASA/Marshall Space Flight Center, Huntsville, AL 35812*
[4] *Universities Space Research Association*

Abstract.
  Aquila X-1 is one of the most active of the soft x-ray transients (SXTs). Soft x-ray outbursts can produce emission close to the Eddington limit in the 1-10 keV band and are typically observed about once a year. Between becoming operational in the spring of 1991, and the beginning of 1996, BATSE observed at least three episodes of hard x-ray emission lasting 50 days or longer, during which the 20-100 keV flux exceeded 50 mCrab [1]. XTE-ASM has observed two outbursts in the 2-10 keV range since becoming operational in early 1996. One of these outbursts was also observed by BATSE. The cause of the hard x-ray emission, and its relation to the soft x-ray emission, is still unknown.

INTRODUCTION

Aquila X-1 is known to be a neutron star in a 19 hour binary orbit with a 20th magnitude K star, which typically brightens by two magnitudes during outbursts. X-ray bursts have been detected during the declining phase of outbursts [2]. Thirty soft x-ray outbursts were observed between 1966 and 1992 [3]. Salient features of this long term light curve include: (1) An unstable long term (quasi)-periodicity in which an underlying, nearly (but not exactly) periodic outburst pattern is reset, for example, from a ~ 122 to ~ 308 day period, sometime in the early to mid 80s. (2) A correlation between outburst peak flux (and total fluence) with time elapsed since the previous outburst, and with the outburst duration.

Hard x-ray emission from Aquila X-1 was first observed by BATSE in the summer of 1991 [1]. The spectrum could be fit as a power law with a photon index of ~ 2.5 between 20 and 100 keV, but emission at higher energies was not detectable, due to the low intensity of the hard x-ray emission (~ 50 mCrab). How high in energy the power law extends is still an open question.

OBSERVATIONS

Figure 1 shows the BATSE flux history prior to TJD 9800 (April 1995) obtained using the Earth occultation technique, together with information from other experiments on the soft x-ray and optical state. There appear to be at least three distinct outbursts (8450-8550, 8700-8800, and 9350-9600) and one interval which may contain several outbursts, each of shorter duration (9000-9300). The optical state is always correlated with the soft x-ray state when both types of observation are available simultaneously. This state is usually, but not always correlated with the hard x-ray state.

Since early 1996 XTE-ASM has monitored the soft x-ray flux. Figure 2(a) shows both light curves. A weak outburst, at about the 30 mCrab level in both instruments, was observed from 10250-10325. A later strong outburst was observed only in the soft x-ray band, at a level of about 400 mCrab.

Figure 2(b) compares the soft and hard x-rays during the weak outburst

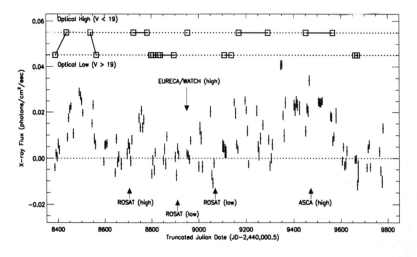

FIGURE 1. BATSE light curve for Aql X-1 in the 20-100 keV band, prior to April 1995. Photon fluxes are given as five-day averages with only the statistical error shown. The optical state is shown above the BATSE lightcurve. The optical high state is defined here when the optical counterpart brightens above a visual magnitude of about V=19. Epochs of ROSAT, EURECA/WATCH and ASCA observations of Aql X-1, and the soft X-ray intensity state relative to the quiescent level are also indicated. The X-ray low and high states are defined by luminosities typically smaller or larger than ~ 10^{33} erg/s, respectively.

Outburst	Previous (days)	Duration (days)	Peak Flux (mCrab)	Energy Log(ergs)
S1	$> 100, < 315$	80	40	42.6
S2	230	55	400	43.2

TABLE 1. Time and energy scales of soft x-ray outbursts observed by XTE-ASM. Previous is the time between the end of the previous and the begininning of the current outburst; Peak Flux is the maximum flux observed; and Energy is the Log of the total energy released during the outburst. These numbers are estimated from count rates and are approximate.

observed by both instruments. Though it is not possible to draw firm conclusions, especially from the BATSE data, the hard x-ray emission appears to continue for some time after the soft x-ray decline. Such a shift to harder emission as the intensity decreases was noted by Einstein [4].

For the two outbursts observed by the ASM, we have estimated the time elapsed since the beginning of the previous outburst, and the outburst durations, peak intensities, and total energy. These results are shown in Table 1, where the first outburst is labeled S1 and the second one S2. A distance of 1.7 kpc has been assumed. The peak luminosity of S1 is, very approximately, 6×10^{35} ergs/s and of S2, 8×10^{36} ergs/s. For S1, lower and upper limits to the time elapsed since the previous outburst can be set by noting (1) that an optical high state began on TJD 9915 and (2) that the ASM began returning reliable data around TJD 10150. Together with the outburst start date of TJD 10230, this implies that the time elapsed since the next previous outburst must have been in the range of $\sim$ 100-315 days. If the observed optical high was associated with the beginning of the next previous outburst, then the upper limit would be the correct value.

DISCUSSION

Observations of Aquila X-1 and other x-ray bursters indicate that the hard x-ray emission occurs only at low luminosities. It has been suggested that above a critical luminosity, Compton cooling by soft photons prevents significant hard x-ray emission [5]. S1, for example, has a peak luminosity of $\sim 6 \times 10^{35}$ ergs/s, which is at or below the critical luminosity inferred from other observations. Proposed emission mechanisms for the hard x-ray emission have been motivated by the power law tail and the existence of a critical luminosity. The well known interpretation of hard tails in black hole candidates as being due to comptonization of soft photons may or may not be correct for weakly magnetized neutron stars, and nonthermal mechanisms have been proposed which would generate a hard power law tail [7], [8]. These tails could extend to higher energies than would comptonized spectra. Future observations of the hard spectrum above 100 keV are needed to test these models.

The thermal disk instability model predicts that accretion disks around black holes and neutron stars are thermally and viscously unstable below a certain critical time averaged x-ray luminosity which depends on the orbital period [6]. If this is the case, the relatively frequent outbursts of Aquila X-1 may be attributed to the fact that it is close to the stability line in a plot of x-ray luminosity versus orbital period. However, at present no model successfully explains either the recurrence pattern of bright soft x-ray outbursts, or the cause of the hard x-ray emission.

The correlation between total energy released in an outburst and time to the previous outburst noted in [3] is consistent with a disk instability picture in which the excess mass accumulated in the accretion disk at a constant accretion rate between outbursts accounts for the energy released through accretion onto the neutron star during outbursts. The thirty outbursts analyzed in [3] for which this correlation, and the correlation bewteen outburst fluence (or peak flux) with duration also held, all had peak fluxes greater than 200 mCrab. From Table 1, the estimated peak flux of S1 is 40 mCrab, so using this outburst, it is possible to examine whether these correlations also hold for lower peak intensity outbursts. Note that only the peak flux, and not the fluence of S1, is substantially smaller, by a factor of 5, then other outbursts which have been observed from Aquila X-1.

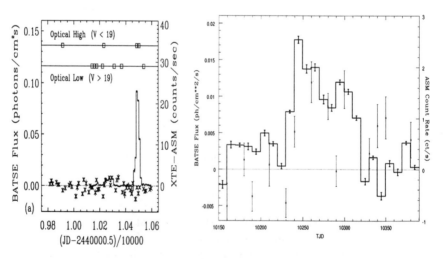

FIGURE 2. BATSE (20-100 keV, crosses) and XTE-ASM (2-10 keV, histogrammed) lightcurves and the optical state. (a) March 1995 - May 1997. (b) Detail of lightcurves in (a). 1 Crab ~ 75 ASM counts/sec.

Comparing the numbers in Table 1 with Figure 5 of [3] shows that S2 is consistent with the rough correlation between either fluence or peak flux and duration found for the other outbursts plotted there, but that S1 is clearly underluminous with respect to that correlation by at least one order of magnitude. Making the same comparison for fluence or peak flux and time to the previous outburst (T_p) (Figure 4 of [3]) is more difficult because the value of T_p is less certain. Using 230 days for the time elapsed prior to S2 (since S1) makes S2 consistent with these correlations. Using the largest possible value of 315 days for S1 would make it underluminous for this relation, although somewhat smaller values of T_p, which could be possible if an outburst were missed, would be consistent.

These observations, together with a clustering of previously observed outbursts in the 200-400 mCrab range raise the possibility that there may be a gap in the outburst peak luminosity distribution. Outbursts with peak luminosities below the gap (of which S1 is an example) do do not follow the energy-time correlations of those above the gap. These lower peak flux outbursts, which may not have been noticed previously due to instrumental sensitivity limitations, may also generate hard x-ray emission, as the BATSE data probably indicate. Evidently continued high sensitivity monitoring is needed to assess the outburst distributions.

REFERENCES

1. Harmon, B. A., et al. 1996, *Astronomy & Astrophysics*, 120, 197.
2. Koyama, K., 1981, *ApJ*, 247, L27
3. Kitamoto, S., 1993, *ApJ*, 403, 315
4. Czerny, M., Czerny, B., & Grindlay, J. 1987, *ApJ*, 312, 122
5. Barret, D., & Vedrenne, G. 1994, *ApJS*, 92, 505
6. van Paradijs, J. 1996, *ApJ*, 464, L39
7. Kluzniak, W., et al. 1988, Nature, 336, 558 *ApJ*, 403, 315
8. Tavani, M. & Liang, E. 1993 in *AIP Conf. Proceedings 280, 1st Compton Symposium*, p. 428

Observation of X-Ray Bursters with the Beppo-SAX Wide Field Cameras

A. Bazzano[1], M. Cocchi[1], L. Natalucci[1], P. Ubertini[1]
J. Heise[2], J.in 't Zand[2], J.M. Muller[2,3], M.J.S. Smith[3]

[1] *Istituto di Astrofisica Spaziale, C.P., 67, 0044 Frascati (Roma), Italy*
[2] *Space Research Organisation Netherlands, Sorbonnelaan 2, 3584 CA, Utrecht, the Netherlands*
[3] *BeppoSAX, Science Data Center and SOC, NuovaTelespazio, Roma, Italy*

Abstract. A deep observation of the Galactic Bulge has been performed with the Wide Field Cameras (WFCs) on board the currently operating satellite BeppoSAX. The observation consists of two different campaigns during October 1996 and March-April 1997 for a total exposure time of 500 ksec. During these observations a series of numerous X-ray bursts occurred, originating from sources already known as bursting sources, namely: KS 1731-26, H1724-30, GX 354-0, A1742-294, SLX 1744-299, 4U1820-30 and the peculiar object GRO 1744-28. Besides these ones, bursts have been observed for the first time from the new discovered source, SAX J1750.8-2900, and from the direction of two already known sources: GS 1826-24 and SLX 1735-269. Preliminary results on bursting sources are presented.

INTRODUCTION

The Galactic Center region has been observed and exploited successfully both at low and high energy during the 1990's with the imaging instruments on board SPACELAB 2 , MIR-KVANT and GRANAT missions. A number of new sources has been revealed and the monitoring of previously known ones allowed for a better understanding of their nature (for review papers see Skinner at al., 1989, Pavlinsky et al., 1994 and Vargas et al., 1996) In particular many new X-ray bursters have been detected, some of them showing hard X-ray persistent emission (Barret et al., 1996 and references therein). Most of the identified X-ray bursters show clustering in the Galactic Center direction. In fact, on a total of 40 bursting sources so far detected, 22 (55%) are concentrated within 15 degree from the Sgr A* position as can be inferred using the X-ray binary catalogue by van Paradijs, 1995.

The WFCs with their imaging capability and large field of view, offer the opportunity to deeply investigate and study the X-ray behaviour of the many

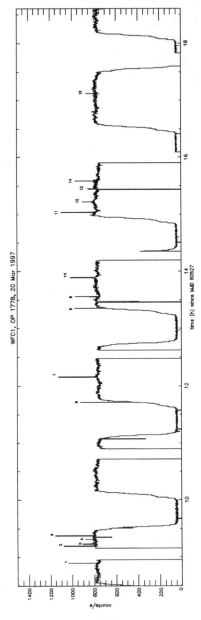

FIGURE 1.

sources belonging to different classes of X-ray emitters and to unambiguously identify the objects originating X-ray bursts.

OBSERVATION

The design of the Wide Field Cameras on board the BeppoSAX satellite, is based on the coded mask principle. Each separate unit consists of a mask and a position sensitive proportional counter. The field of view is restricted to 20 x 20 degree Full Width Half Maximum by means a stainless steel structure holding the mask at 0.7 m from detector. The position sensitive detectors, effective area 520 cm^2 each, combining a Beryllium window and a filling gas mixture pressure of 2.1 bar allow for an operative energy range of 1.8-30 keV. The energy resolution is 18% at 6 keV, the angular resolution is 5 arcmin, the position accuracy is better than 1 arcmin and the sensitivity is a few milliCrab in 10^4 s depending on the total flux of sources in the field of view.

The two identical WFCs point 180 degree away from each other, both perpendicular to the sun vector and to the Narrow Field Instruments (NFI). Two scientific modes are foreseen: a normal (imaging) mode and a high time resolution mode. For details on WFCs see Jager et al 1997.

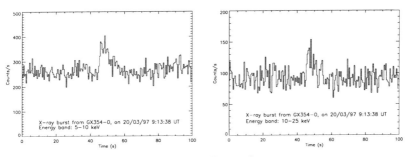

FIGURE 2. a, b.

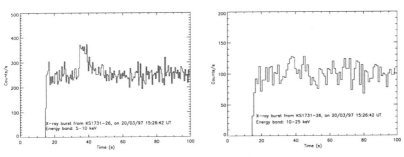

FIGURE 3. a, b.

During the period March-April 1997 a deep exposure (400 ks) on the Galactic Center was performed. At least 30 sources were detected and many bursts occurred (at least 115 from GRO1744-28, 20 from GX354-0, 8 from KS1731-26, 6 from GS 1826-24, 5 from H1724-307, 3 from A1742-294, 3 from the newly discovered SAX J1750.8-2900, 2 from SLX 1744-299 and 4U1820-30). In this paper preliminary results on bursts originated from previously known X-ray sources are reported. In figure 1 the time history of the WFC1 total count rates in the 2-30 keV energy band during observing period March 20.3536 to 20.7770 U.T., containing 15 identified bursts, is shown. X-ray bursts appear in this figure as a rapid and sudden increase to a peak well distinct above the background level. The rate is due to the integrated contribution from all the sources in the field of view, the diffuse emission from X-ray background, the diffuse emission from the galactic center itself and the intrinsic instrumental background. The imaging property of WFCs allows for an accurate determination of the arrival direction of the excess counts in the time profile and, in turn, the association with bright point-like sources in the sky. In this way burst number 3,13,15 and 7 have been identified as originating from GX354-0, KS 1731-26, A1742-294 and GRO 1744-28 respectively.

Time profile of bursts in two different energy bands (5-10 and 10-25 keV) from GX354-0, KS1731-26, A1742-29 and GRO1744-28 are shown in figure 2a

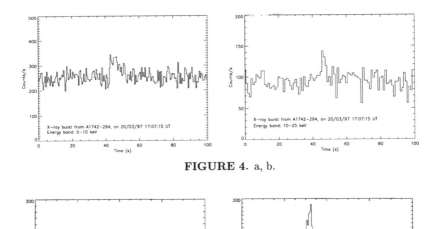

FIGURE 4. a, b.

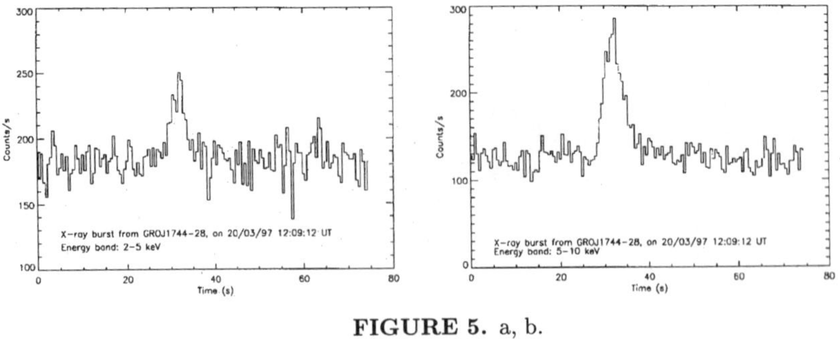

FIGURE 5. a, b.

and 2b, 3a-b, 4a-b and 5a-b.

RESULTS

The results of a preliminary analysis of the bursts profiles for each of the known sources can be summarised as follows: GX354-0 peak intensity $\sim$0.7 Crab (at least 5 times the quiescent state), fast rise, duration of burst about 15 s, no precursor, no strong difference in the peak intensity in the low and high energy band.

KS 1731-26 burst peak intensity $\sim$0.5 Crab (2 times the quiescent state), fast rise(1s), duration of burst about 10 s, no precursor, most of the peak intensity released in the 5-10 keV band.

A1742-29 burst peak intensity $\sim$0.4 Crab(at least 10 times the quiescent state), fast rise, duration of burst about 10 s, no precursor, no strong difference in the peak intensity in low and high energy band. The high energy band shows a delay in the peak rise.

The burst from GRO 1744-28 on March 20 at 12:08 30 has a peak intensity of $\sim$2.5 Crab (i.e. about 10 times the non-bursting or quiescent level), about 10 s duration, smoother rise in comparison to other bursts, slower decline and the flux seems higher in the energy range 5-10 keV. The peak intensity is maintained for a few second as also shown in burst seen with the PCA/RXTE (Giles et al., 1996). Many bursts have been collected from this source. A detailed analysis will be performed to search for miniburst and microburst as the ones reported during PCA observations and to look for any possible periodicity in the bursts occurrence.

REFERENCES

1. Barret, D., et al., *Astr & Astroph. Suppl.* 120, 4., 121 (1996).
2. Giles, A.B., et al., *Ap.J. Letters* 469, L25 (1996).
3. Jager R. et al., *Astr. & Astroph.* in press (1997).
4. Pavlinsky M.N., Grebenev, S.A. and Sunyaev, R.A. , *Ap.J.* 425, 110 (1994).
5. Skinner, G.K., 1989, in *"The Centre of the Galaxy"* M. Morris ed. , 567 (1989).
6. van Paradijs, J., in *X-Ray Binaries* ed. W.H.G. Lewin, J. van Paradijs, &E.P.J. van den Heuvel, Cambridge University Press, 214 (1995).
7. Vargas, M., et al., ESA SP-382, 129 (1996).

Aperiodic variability of the X-ray burster 1E1724-3045 : First results from RXTE/PCA

J.-F. Olive[1], D. Barret[1], L. Boirin[1],
J. Grindlay[2], P. Bloser[2], J. Swank[3] & A. Smale[3]

[1] *Centre d'Etude Spatiale des Rayonnements, CESR-CNRS, Toulouse, FRANCE*
[2] *Harvard Smithsonian Center for Astrophysics, Cambridge, USA*
[3] *Goddard Space Flight Center, Greenbelt, USA*

Abstract.
The X-ray burster 1E1724-3045, located in the globular cluster Terzan 2 was observed on November 1996 for 100 ksec with the PCA and HEXTE experiments aboard the *Rossi X-ray Timing Explorer*. During the observation, the PCA count rate was about 470 cts/s in the 2-20 keV range (background substracted). We have characterized the aperiodic variability of the source over a wide frequency band ($2\ 10^{-3}$ - 40 Hz). The 2-20 keV emission of this source shows a ligh level of noise variability (RMS $\sim$ 25 %). The Fourier spectrum can be modelled in terms of the shot noise model with two characteristic lifetimes of 670 and 16 msec although a continuous distribution of shots lifetimes can not be excluded. These two components account for about the same contibution to the total RMS of the source ($\sim$ 10 %). In addition to these components, a broad and asymetric peaked noise feature, located at 0.8 Hz, appears in the spectrum. This QPO-like feature accounts for a $\sim$ 5 % of the RMS. Both the lifetimes of the shots and the QPO frequency seems independent on the energy but the integrated RMS of all three components increases between the 2-5 keV and 5-20 keV range. We show that 1E1724-3045 has striking timing similarities with the black hole candidate GRO J0422+32 (Nova Persei 1992).

INTRODUCTION

The X-ray burster 1E1724-3045, close to the galactic center, is located at a distance of about 7 kpc, in the globular cluster Terzan 2. Discovered in the early seventies with the UHURU satellite (Forman et al. 1978), it has been further observed by several X-ray instruments such as EXOSAT, TTM (see Olive et al. 1997 for a review of existing observations), but so far no detailed timing analysis of its persistent X-ray emission has been reported.

In this paper, we focus on the aperiodic variability of 1E1724-3045 which is an important clue to address the nature of this source. In particular, we wish to detail the power density spectrum from 2×10^{-3} to 40 Hz; a frequency range where we have discovered a broad low frequency QPO-like feature. The results of our analysis on kilo-Hertz QPOs in the power density spectrum are to appear in Olive et al., 1997.

OBSERVATIONS AND ANALYSIS

The RXTE observations took place on November 4,5 and 8 1996, for a total of about 100 ksec of usefull time. For this observation, we choose a 16 msec time resolution mode for the PCA data, in 64 energy channels.

Preliminary spectral analysis of the RXTE data reveals that the source was in its hard state during all our observation and that the X-ray spectrum can be approximatively fitted by a power law of index ~ 2 extending up to at least 50 keV in the PCA data. More detailled analysis in underway. Nevertheless, the present analysis is sufficient to estimate the source flux to be 1.3×10^{-9} ergs cm^{-2} s^{-1} in the 2-20 keV range.

Each continuous set of data (lasting ~ 2000 sec.) was divided into M segments of $N = 16384$ bins of $\delta t = 0.012207$ second duration. A Fast Fourier Transform algorithm was used to convert each data segment into frequency space. The final power spectrum of one segment was obtained by taking the ensemble average over all M segments. The power spectra were subsequently normalized to fractional RMS densities according to Belloni & Hasinger (1990). Then, all power spectra were rebinned logarithmically in order to increase the statistical weight of individual frequency bins.

TABLE 1. Modelling of the Mean Power Spectrum in Various Energy Bands. Obs Id. : 1) cover the whole XTE pointing, 2) 4-5 Nov. 97, 3) 8 Nov. 97

Obs Id.	E (keV)	τ_1 msec.	τ_2 msec.	RMS_1 (%)	RMS_2 (%)	RMS_{QPO} (%)	RMS_{total} (%)
1	02-05	686 ± 16	15.2 ± 0.4	8.6	9.2	4.4	22.2
1	05-20	665 ± 13	16.9 ± 0.2	12.4	12.0	5.3	29.6
1	02-20	677 ± 13	16.5 ± 0.1	10.5	10.8	5.3	26.6
2	02-05	679 ± 21	15.1 ± 0.5	8.6	9.2	4.3	22.1
3	02-05	693 ± 24	15.6 ± 0.6	8.7	9.4	4.3	22.4
2	05-20	650 ± 19	16.7 ± 0.3	11.7	11.8	5.8	29.4
3	05-20	751 ± 13	17.1 ± 0.3	12.3	12.2	5.5	30.0
2	02-20	657 ± 18	16.2 ± 0.2	10.5	10.7	5.2	26.4
3	02-20	702 ± 20	16.6 ± 0.2	10.6	10.9	5.3	26.9

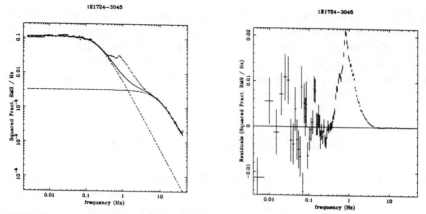

FIGURE 1. Left : All observation-averaged power spectra for 1E1724-3045 in the 2-20 keV energy range. Right : Residual noise power after the two-shot model is substracted

As an example, figure 1 shows the normalized power spectral density function in the 2-20 keV band for the entire XTE pointing. This average spectrum shows a flat top below a low frequency cut off around 10^{-1} Hz, a strong noise component peaking at 0.8 Hz (FWHM $\sim$ 0.6 Hz), and a second cut off around 8 Hz. If one excludes the 0.3-3 Hz frequency range, the spectra can be modelled with the sum of two Lorentzians, with amplitudes K_1 and K_2 :

$$P(f) = \frac{K_1}{1+(\frac{f}{f_1})^2} + \frac{K_2}{1+(\frac{f}{f_2})^2} \qquad (1)$$

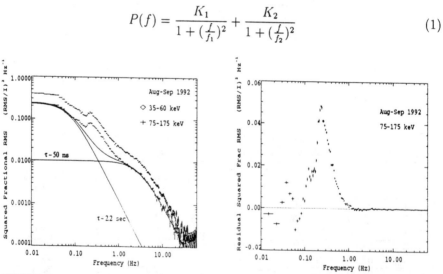

FIGURE 2. Left : Normalized power spectra for GRO J0422+32 in 30-60 and 75-175 keV bands. Right : Residual noise power in the 75-175 keV band after the two-shot model is substracted (from Grove et al. 1993)

This model can describe the two cut off and the $1/f^2$ behavior above several Hz. In the well known "shot noise" model, the characteristic frequencies f_1 and f_2 are related to the lifetimes of randomly occuring exponential shots ($\tau_1 = \frac{1}{2\pi f_1}$ and $\tau_2 = \frac{1}{2\pi f_2}$).

The profile of the residual power noise obtained after subtracting the model to the data (see Figure 2) is a broad and asymetric peak extending from 0.3 to 3 Hz approximatively.

This fitting procedure has been used for various combinations of time intervals and energy range. We obtain similar Fourier spectra (two lorentzians components and the broad peak) both for the 2-5 and 5-20 keV time series. The fitted parameters are reported in Table 1, together with the integrated fractional RMS for the two Lorentzians (RMS_1 and RMS_2) and for the residuals (RMS_{QPO}). The total integrated fractional RMS increases with the energy range from $\sim$ 22% (2-5 keV) to $\sim$ 29% (5-20 keV).

DISCUSSION

Both the high level of fractional RMS ($\sim$ 20-30 % RMS total) and the Fourier spectral shape strongly support the classification of 1E1724-3045 as an "Atoll source" in its "Island" state.

The fact that the Fourier spectrum of 1E1724-3045 can be decomposed into two Lorentzian functions suggests that the shots appear on two characteristics time scales (16 and 670 msec). Nevertheless one can not exclude a continuous distribution of shot time scales between these two extreme values. In the intermediate range between the two breaks, we have found a broad structure, peaking at 0.8 Hz. If the shot time scales were continuously distributed, the amplitude of the residuals would be lesser although its shape would be similar.

The Fourier spectrum of 1E1724-3045 can be compared to the one of the hard X-ray transient source and black hole candidate GRO J0422+32 (XN Per 92) as observed at higher energies with OSSE (see Fig. 2, from Grove et al. 1993). The two Fourier spectral shapes are identical : two Lorentzians (τ_1=50 msec and τ_2=2200 msec) and a broad assymetric peak at 0.23 Hz for the black hole candidate. The two spectra are not only identical in shape but one can remark that all characteristics times can be scaled by a factor of about 3.3 ($\tau_1^{GRO}/\tau_1^{1E} = 3.12$, $\tau_2^{GRO}/\tau_2^{1E} = 3.25$, $\nu_{QPO}^{1E}/\tau_{QPO}^{GRO} = 3.45$). This may suggest a simple scaling of shot frequency with compact object mass if the shots arise at similar gravitational radii.

Low frequency QPO simultaneously with a high level of RMS is not an exclusive property of black holes. Our result extends the growing list of similarities between Atoll sources and black hole systems (Van der Klis, 1995).

REFERENCES

1. Belloni T, Hasinger G., *Astron. Astrophys.*, **230**, 103 (1990).
2. Forman, W., et al., *Ap. J. Suppl.*, **38**, 357 (1978).
3. Grove J.E et al., in "The Second Compton Symposium", AIP. Conf. Proc. No 304, Eds: C.E. Fichtel, N. Gehrels, J. P. Norris, 192 (1993).
4. Olive J.F et al., in preparation (1997).
5. Van der Klis M., Chapter 6 in "In X-ray Binaries", Eds (W.H.G. Lewin, J. van Paradijs and E.P.J. Van den Heuvel) Cambridge University Press (1995).

New X-Ray Bursters with the WFCs on board SAX: SAX J1750.8-2900, GS 1826-24 and SLX 1735-269

A. Bazzano[1], M. Cocchi[1], L. Natalucci[1], P. Ubertini[1]
J. Heise[2], J. in 't Zand[2], J. M. Muller[2,3], M.J.S. Smith[3]

[1] *Istituto di Astrofisica Spaziale, C.P., 67, 0044 Frascati (Roma), Italy*
[2] *Space Research Organisation Netherlands, Sorbonnelaan 2, 3584 CA, Utrecht, the Netherlands*
[3] *BeppoSAX, Science Data Center and SOC, NuovaTelespazio, Roma, Italy*

Abstract. A deep observation of the Galactic Bulge has been performed with the Wide Field Cameras (WFCs) on board the currently operating satellite BeppoSAX. The observation consists of 5 sets of exposures, each lasting 100 ks performed in the period fall 1996 and spring 1997. During these observations a series of numerous X-ray bursts occurred originating from already known bursting sources. Furthermore, X-ray emitters showing bursting behaviour have been detected for the first time. Preliminary results are presented on a new source, SAX J 1750.8-2900 and on two previously known sources GS 1826-238 and SLX1735-269. These objects were tentatively associated by other authors with Black Hole Candidates on the basis of their spectral behaviour and lack of bursts detection.

INTRODUCTION

During the last 10 years the Galactic Center region has been observed and explored successfully both at low and high energy from a few keV up to 400 keV (for a review see Skinner 1993, Pavlinsky et al. 1994, Goldwurm et al. 1994). Whenever the galactic plane and in particular the central radian of the galaxy was surveyed at X and gamma-ray wavelengths, a number of new sources, possibly associated with black hole and neutron stars, have been revealed. Still controversial is the identification of the compact object in some of the newly discovered Spacelab 2, Ginga, GRANAT and GRO sources (e.g. Pavlinsky et al., 1992, Barret et al. 1996, Grebenev et al., 1996). Monitoring of such sources is necessary to look for spectral and time signatures capable to discern the nature of the compact object. The wide field of view of the

WFCs on board BeppoSAX satellite coupled with imaging capability allows for observation of sources in the Galactic Center region previously inaccessible to other instruments. These measurements will in turn permit to identify unambiguously the X-ray emitter and to use both spectral and timing properties of the central emitting object up to 30 keV to search for association.

OBSERVATION

The WFCs

The design of the two Wide Field Cameras (WFCs) on board the BeppoSax satellite is based on the coded mask principle and each separate unit consists of a mask and a position sensitive proportional counter. The field of view is restricted to 20 x 20 degree Full Width Half Maximum by means a stainless steel structure holding the mask at 0.7 m from detector. The position sensitive detectors, effective area 520 cm^2 each, combining a Beryllium window and a Xenon based gas mixture pressure of 2.1 bar allow for an operative energy range of 1.8-30 keV. The energy resolution is 18% at 6 keV, the angular resolution is 5 arcmin, the position accuracy is better than 1 arcmin and the sensitivity is a few milliCrab in 10^4 s depending on the total flux of sources in the field of view. The two identical WFCs point 180 degree away from each other, both perpendicular to the sun vector and to the Narrow Field Instruments (NFI). Two scientific modes are foreseen: a normal (imaging) mode and a high time resolution mode. For details on WFCs see Jager et al. 1997.

Detection of X-ray Bursts

SAX J1750.8-2900:

The new BeppoSAX source was detected since the beginning of the run started on March 18.032 U.T. and the estimated intensity was ranging between 30-50 mCrab in the energy range 2-8.6 keV. This new source was monitored carefully along all orbits to look for any transient behaviour. On March 20.38790 a flux in excess of the previous detected one was observed from the monitored position not corresponding to any known source. The event lasted for about 30 s and the peak intensity was of the order of 0.7 Crab, i.e 15-20 times higher than the persistent flux (Bazzano et al.1997).

Again, we detected bursts from the same position about 6 hours later (20.63408) and on March 31.41514. Peak intensity flux corresponded to about 0.75 and 0.65 Crab respectively. The first bursts is shown in figure 1 with different ones while in figure 2 the burst profile in the low and high energy band are reported. From figures we can derive that duration of burst is of the order of 15 s, and burst is characterized by a smooth rise and decline. Furthermore we suggest most of the peak flux is released in the 5-10 keV band but a single

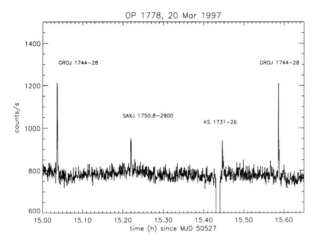

FIGURE 1.

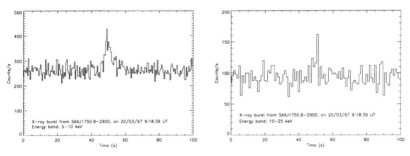

FIGURE 2. a, b.

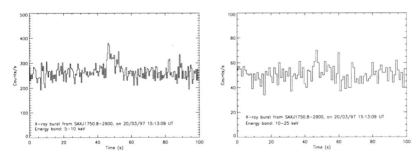

FIGURE 3. a, b.

bin peak is detectable at higher energies with some delay. The second burst profile in the two energy band is shown in figure 3 and a second peak after 10 s is visible that seems to be related mostly with the 5-10 keV energy band. Analysis of spectra of all bursts as well as the persistent emission is ongoing.

GS 1826-238:

During the 3rd shift of the Galactic Center observation, X-ray Bursts have been detected for the first time from the GINGA Transient GS 1826-238 (Makino et al. 1988). These bursts occurred on March 30.96169, 31.69363 and 31.95521. Their peak intensity were of the order of 0.35, 0.52 and 0.48 Crab unit respectively while the persistent flux was about 12 mCrab in the 2-8 keV band (Ubertini et al., 1997). The source was also active during last run performed on April 13-15 when 3 bursts were detected. The bursts from this source are all similar in shape showing fast rise time, ($\sim$1s), exponential decay with typical time of 150 s and spectral softening. In figure 4 a burst

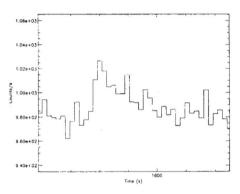

FIGURE 4.

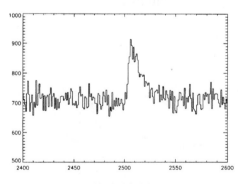

FIGURE 5.

profile is shown.

SLX 1735-269:

During the last run, at U.T. 13.668, a burst was detected from the direction of the high energy source SLX 1735-269 firstly reported by Skinner et al., 1987 and later monitored by the Sigma experiment on board the GRANAT satellite. The burst lasted less than 30 s and was detected at an intensity level of about 0.9 Crab in comparison to a quiescent emission of about 10 mCrab (Bazzano et al., 1997). The burst profile is shown on figure 5.

The nature of SLX 1735-269 and GS 1826-238:

Both SLX 1735-269 and GS 1826-238 were in the past tentatively associated with black hole candidates because of their observed spectral behaviour and the lack of any timing signature indicating a different origin (Tanaka, 1989 for GS1826-238 and Grebenev et al., 1996 for SLX1735-269). The bursts detection with the WFCs rule out this hypothesis since type I bursts, detected with BeppoSAX, are an unequivocal signature that the compact object is a neutron star. In fact, type I bursts are due to thermonuclear flashes within accreted material on the surface of a neutron star and usually show spectral softening during decays associated with the energy dependence of the burst profile (for a recent review see Lewin et al., 1995).

REFERENCES

1. Barret, D., Mc Clintock J.M. and Grindlay J.E., *Ap. J.* 472 (1996).
2. Bazzano, A. et al., IAU Circ. 6668 (1997).
3. Goldwurm, A., et al., *Nature* 371, 589 (1994).
4. Grebenev, S.A., Pavlinsky, M.N. and Sunyaev, R.A., ESA SP-382,183 (1996).
5. Jager R. et al., *Astr. & Astroph.* in press (1997).
6. Lewin, H.G.W. et al., in *X-Ray Binaries* ed. W.H.G. Lewin, J. van Paradijs, &E.P.J. van den Heuvel, Cambridge University Press, 175 (1995).
7. Makino, F. et al., IAU Cir. n. 4653 (1988).
8. Pavlinsky, M.N., Grebenev, S.A. and Sunyaev, R.A., *Soviet Astronomy Letters* 18, 217 (1992).
9. Pavlinsky, M.N., Grebenev, S.A. and Sunyaev, R.A., *Ap.J.* 425, 110 (1994).
10. Skinner, G.K. et al., *Nature* 330, 544 (1987).
11. Skinner, G.K., *Astr. & Astroph Suppl.* 97, 149 (1993).
12. Ubertini, P. et al., IAU Circ. 6611 (1997).
13. van Paradijs, J., in *X-Ray Binaries* ed. W.H.G. Lewin, J. van Paradijs, &E.P.J. van den Heuvel, Cambridge University Press, 214 (1995).

Kilo-Hertz QPO and X-ray Bursts in 4U 1608-52 in Low Intensity State

W. Yu[†], S. N. Zhang[**], B. A. Harmon[*], W. S. Paciesas[*‡], C. R. Robinson[**], J. E. Grindlay[°], P. Bloser[°], D. Barret[▷], E. C. Ford[#], M. Tavani[#], P. Kaaret[#]

[*] *Marshall Space Flight Center, Huntsville, AL 35810*
[*] *Universities Space Research Association*
[†] *CRHEAL, Institute of High Energy Physics, Beijing, 100039, China*
[‡] *University of Alabama in Huntsville, Huntsville, AL 35889*
[°] *Harvard-Smithsonian Center for Astrophysics, 60 Garden Street, Cambridge, MA 02138*
[▷] *Centre d'Etude Spatiale des Rayonnements, CNRS-UPS, France*
[#] *Columbia Astrophysics Laboratory, Columbia University, New York, NY 10027*

Abstract. We present the results from RXTE/PCA observations of 4U 1608-52 in its island state on March 15, 18 and 22 of 1996. Three type I X-ray bursts were detected in one RXTE orbit on March 22. We observed QPO features peaking at 567-800 Hz on March 15 and 22, with source fractional rms amplitude of 13%-17% and widths of 78-180 Hz in the power density spectra averaged over each spacecraft orbit. The rms amplitudes of these QPOs are positively correlated with the photon energy. The three X-ray bursts, with burst intervals of 16 and 8 minutes, have a duration of 16s. The blackbody emission region of the smallest X-ray burst among the three suggest it was a local nuclear burning. We also discuss a type I X-ray burst candidate in the observation.

INTRODUCTION

4U 1608-52 is a transient X-ray burster with recurrence intervals from 80 days to about 2 years [1] [2]. It was classified as an atoll source based on the correlated X-ray spectral variability and High-Frequency- Noise (HFN) [3] [4]. 4U 1608-52 is known to show spectral and timing characteristics similar to those in blackhole candidates (BHCs) in the low state [5]. When in a low luminosity state ($L_x \leq 10^{37}$erg/s), its energy spectrum above 2 keV is usually dominated by a power-law component [4] [6] [7], which is gradually cut off above $\sim 60 - 70$ keV [8].

An outburst was observed from 4U 1608-52 with RXTE in early 1996 [9]. A few RXTE/PCA pointings on March 3, 6, 9 and 12 show kilo-hertz

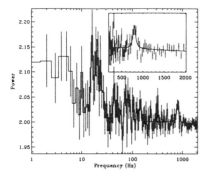
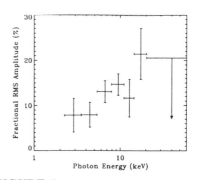

FIGURE 1. The QPO peak at 800 Hz in the first orbit on March 15. The PDS is complex and a broad peak at 20 Hz is also visible.

FIGURE 2. Average fractional *rms* amplitude of QPO as a function of photon energy. We average the results obtained from *Event Mode* data analysis of the 6 orbits.

QPO at 690-890 Hz [10]. Here we present our analysis results obtained from RXTE/PCA observations on March 15, 18 and 22 with source luminosity $\sim 10^{36}$ erg/s.

OBSERVATIONS AND DATA ANALYSIS

The X-ray monitoring with RXTE/ASM shows that our observations on March 15, 18 and 22 were taken near the end of the outburst when source fluxes were below 50 mCrab. The persistent X-ray flux (2-20 keV) was $(4.6 - 11) \times 10^{-10}$ erg/s/cm^2. The count rate ranged between 190-450 cps (1-60 keV) with source intensity below previous RXTE observations [10]. Three X-ray bursts were observed in one orbit on March 22.

Power Density Spectra (PDS)

By analyzing the *Event Mode* data with background subtraction, we have found that the Power Density Spectra (PDSs) were dominated by HFN with rms 10%-14% in our observations. We detect kilo-hertz QPO features at 567-800 Hz in 6 orbits, with source fractional rms amplitude of 13%-17% and widths of 78-180 Hz. As an example, Fig. 1 is the PDS obtained from the 1-30 keV *Event Mode* data averaged over the first orbit on March 15. We also investigate the energy dependence of the QPO features. A positive correlation between the average QPO rms amplitude and the photon energy is observed, as shown in Fig. 2. No kilo-hertz pulsations have been detected at 99% confidence level in the X-ray bursts in our preliminary searching. The details of the timing analysis will be reported elsewhere [11].

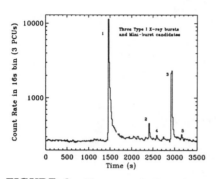
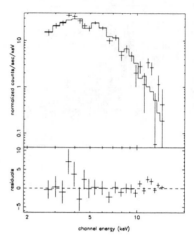

FIGURE 3. Three type I X-ray bursts (marked as 1-3) and mini-burst candidates (marked as 4-5). The time resolution is 16s and the count rates were obtained from 3 PCUs.

FIGURE 4. Energy spectrum of burst 2 with $kT_{BB} = 1.2$ keV and $\chi^2/21(dof) = 1.49$.

X-ray Bursts and Burst-like Variations

Three X-ray bursts were observed in the second orbit on March 22, 1996. They all have profiles consistent with a fast rise and exponential decay, and show blackbody-type energy spectra, indicating that they are type I X-ray bursts. Fig. 3 is the light curve with these bursts. The lightcurve is obtained from 3 PCUs and with 16 s resolution. The properties of these bursts are listed in Table 1. The average BB radii were obtained from spectral fitting to a BB model with $N_H = 1.5 \times 10^{22} cm^{-2}$.

The burst durations indicate that they belong to the "slow" class of bursts usually observed in 4U 1608-52 in its low state [12]. The intervals between

TABLE 1. Properties of Type I X-ray Bursts

Burst No.	Peak Count Rate (cps)[a]	Duration (s)[b]	BB Radius (km)[c]
1	16500	16.7 ± 0.1	~ 13
2	240	15.4 ± 1.0	~ 4
3	4150	16.7 ± 0.3	~ 18

[a] from 3 PCUs of RXTE/PCA
[b] defined as the ratio between the integrated burst counts and the burst peak count rate in PCA band
[c] from results of spectral fitting to a blackbody(BB) model with a correction assuming that the ratio between the color temperature and the effective temperature is 1.5

the bursts were about 16 min and 8.5 min, among the shortest now known [1]. The ratio between the average persistent flux and the average burst flux of burst 3, regarding the relative time interval of 24 minutes between burst 1 and 3, was smaller than 6. In Fig. 4, we show the average energy spectrum of the burst 2 with the best fit BB model with $N_H = 1.5 \times 10^{22} cm^{-2}$. The burst BB temperature was 1.2 ± 0.3 keV. The derived BB emission radius is smaller than the radii obtained for burst 1 and 3 (Table 1.).

Marked as 4 and 5 in Fig. 3 were burst-like intensity variations which were more than 4σ above the persistent intensity level in the light curve with the 10 s resolution during the entire observations. The second one was stronger. Its peak count rate in 3 PCUs was 90 ± 18 cps. Its profile can be fitted to a model with a linear rise and an exponential decay (see Fig.5). The derived rise time (from 10% to 90% of peak flux) and decay time (e-folding time) are 7.6 ± 1.3 s and 12.9 ± 2.5 s respectively. For the burst 2, the rise time and decay time are 8.2 ± 1.2 s and 13.1 ± 1.5 s. So they have similar profiles. In addition, the spectrum of this burst-like variation is consistent with the BB model, but a power law model can not be excluded. Fitting its spectra to a BB model yield a blackbody temperature of 1.2 ± 0.1 keV (Fig.6). Its corrected radius of BB emission region is 2.2 km. The hardness ratio (5.4-20.4 keV/2.2-5.4 keV) during the decay was also similar to that in burst 2.

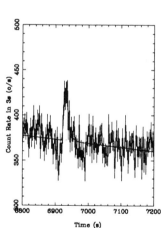
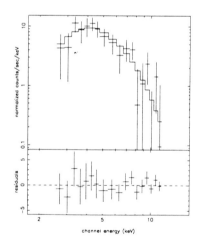

FIGURE 5. Time profile of one of the burst candidates (5). A model composed of a linear rise and an exponential decay was fit to the data.

FIGURE 6. The Energy spectrum of the burst candidate (5) with $kT_{BB} = 1.2$ keV and $\chi^2/16(dof) = 0.51$.

DISCUSSION AND CONCLUSION

Strong HFN components in PDSs and low X-ray luminosity (2-20 keV) of $(0.7-1.7) \times 10^{36}$erg/s assuming a distance of 3.6 kpc suggest 4U 1608-52 was in its island state [3]. The QPO features in our observations show a similar relation between QPO rms amplitude and photon energy to those observed in a higher intensity state [10]. The burst 3 cannot be explained as being produced by replenishing sufficient fuel through accretion within the short interval between burst 3 and burst 1. This suggest that a portion of the nuclear fuel had survived the previous bursts [1]. The burst 2 had a BB emitting radius smaller than those of the two stronger bursts and the size of the canonical neutron star radius. One burst-like intensity variation, which could not be observed with EXOSAT Medium Energy (ME) Experiment (estimated EXOSAT ME peak count rate (1-20 keV) is 12.4 c/s), also shows signatures of a type I X-ray burst with an even smaller BB emitting radius. This strongly supports the local nuclear burning interpretation. Our results suggest that these small bursts or variations, only observed associated with the strong bursts in our observations, may be attributed to the mixing mechanism during the post burst phase of the strong bursts [13].

We appreciate various assistances from the members of RXTE/GOF. WY thanks Prof. R. E. Taam of Northwestern University and Dr. William Zhang of GSFC for many helpful suggestions. WY would like to acknowledge the support from the National Natural Science Foundation of China.

REFERENCES

1. Lewin W.H.G, van Paradijs J., and Taam R.E., *Space Sci.Rev.* **62**, 223 (1993)
2. Lochner J.C., Roussel-Dupré D. *ApJ* **435**, 840 (1994)
3. Hasinger G., van def Klis M., *A&A* **225**, 79 (1989)
4. Yoshida K., et al., *PASJ* **45**, 605 (1993)
5. van der Klis M. *ApJS* **92**, 511 (1994)
6. Mitsuda K., et al. *PASJ* **36**, 741 (1989)
7. Penninx W., Damen E., et al. *A&A* **208**, 146 (1989)
8. Zhang S.N., et al. *A&A* **120**, 279
9. Marshall F.E., Angelini L. *IAU Circ.*6331
10. Berger M. et al., *ApJ* **469**, L13
11. Yu W. et al., submitted to *ApJ*, 1997
12. Murakami T., et al. *ApJ* **240**, L143
13. Woosley S.E., and Weaver T.A. *High Energy Transients in Astrophysics, AIP Conf.Proc. 115*, New York: AIP Press, 1985, pp. 273.

Long-term Observations of Her X-1 with BATSE

Robert B. Wilson*, D. Matt Scott[†], and Mark H. Finger[†]

Space Sciences Laboratory, Marshall Space Flight Center, Huntsville,AL 35812
[†] *USRA, Marshall Space Flight Center, Huntsville,AL 35812*

Abstract. Pulsed emission from Her X-1 has been observed by BATSE during each Main High state throughout the CGRO mission. This long observation set by a single instrument provides new information on long-term behavior of the Her X-1 system. The pulsed emission varies by more than a factor of 3 between different Main High states. Frequency and flux histories do not show a simple relationship between the source intensity and spin behavior, but do show that the source is normally in spinup, and spindown episodes occur at all source flux levels. Orbital analyses are presented, including determination of the orbital period derivative, which is in agreement with Deeter et al. (1991).

INTRODUCTION

The low-mass x-ray binary system Her X-1 has been observed frequently by many observers since its discovery in 1971 [3]. Except during the extended observations of previous all-sky monitors (*UHURU, Ariel 5, HEAO A-4*), determination of pulse frequency and source luminosity were not sufficiently frequent to fully characterize the behavior of the source. The typical 20 – 50 keV source flux observed by BATSE is $\sim$130 mCrab, sufficient to obtain high quality pulse profiles (Figure 1) and intensity measurements (Figure 2). With 6 years of BATSE observations in hand we can better determine normal and anomalous behavior of this source. The source is known to undergo large variations in peak flux in the Main High state - it was in an extended low state during part of 1983 [2,4]. BATSE has observed large (factor of $\sim$3) changes in Main High peak source flux in 20 - 50 keV during the mission.

Her X-1 has been observed to mostly spin up between consecutive Main High states. Previous measurements with pointed instruments were too sparse to characterize the occurrence rate of spindown episodes, which were first observed by *UHURU*. We find that spindown episodes with $\dot{\nu}$ greater than $0.2\ pHz \cdot s^{-1}$ in magnitude (as measured by shifts in pulse frequency between Main High states) have occurred at least 5 times since April 1991 and are

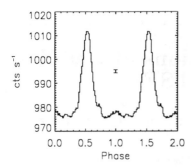
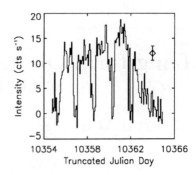

FIGURE 1. Pulse profile for Her X-1 for TJD 9586 - 9593. The counting rate is the sum of background and source contributions. A typical ±1σ uncertainty is shown at phase 1.0

FIGURE 2. Aspect-corrected pulsed intensity history for the Main High state on TJD 10354-10364. A typical ±1σ uncertainty is shown at TJD 10364.

not confined to times of low source flux. Using data available through 1989, Deeter et al. (1991) found that the orbital period of Her X-1 was decreasing at a rate of $(2.25 \pm 0.27) \cdot 10^{-8} days \cdot yr^{-1}$. BATSE observations (combined with the previous data) give the value shown in Table 1, which has an improved uncertainty, and is consistent with [1].

OBSERVATIONS

For detailed studies of Her X-1 with BATSE folded-on-board data is required. These observations are normally scheduled only during the Main High portions of the 35 day cycle, and consist of 8 - 16 s readouts of epoch-folded 16-energy-channel count-rate histograms (with 64 phase-bin resolution) of selected detectors which view the source direction. Substantial data coverage exists for about 80% of the Main High states since *CGRO* became operational. All states have been observed using the aliased signal of the source using the DISCLA 1.024 s data type.

ANALYSES

Orbital Determination

All data collected during one *CGRO* orbit are barycentered and epoch-folded to produce one pulse profile. In the Fourier domain a best-fit correlation

TABLE 1. Orbital Parameters for Her X-1

$T_{\pi/2}$	JD 2449420.6734653(45)
P_{orb}	1.7001674461(96) $days$
$a_x sin(i)$	13.18549(32) s
$\dot{P}_{orb}$	$-1.80(08) \cdot 10^{-8} days \cdot yr^{-1}$

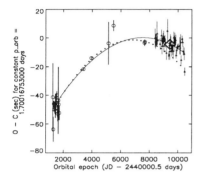

FIGURE 3. Observed minus computed epochs ($T_{\pi/2}$), relative to a constant orbital period model. The solid curve is the best quadratic fit, using all data. The large diamond is the mean BATSE epoch. The dashed curve is a fit excluding the BATSE data.

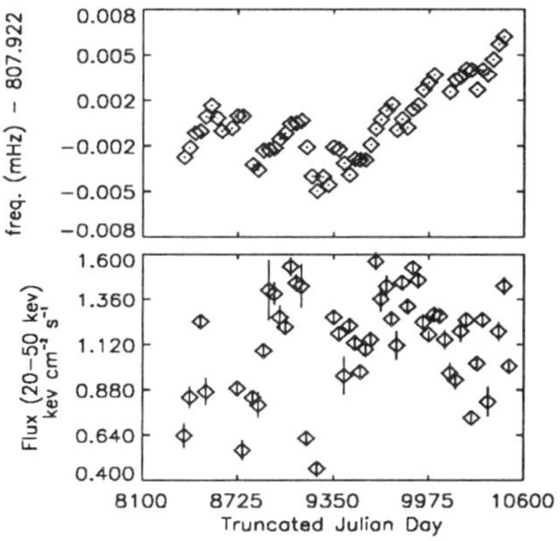

FIGURE 4. History of pulse frequency and Main High peak pulsed flux over the CGRO mission. Flux measurements cannot be made for all intervals, either due to lack of suitable scheduled data or low source intensity.

with a master pulse template is obtained, giving a phase and amplitude. For each Main High interval a χ^2 minimization fit to the pulse phases is performed, with free parameters $T_{\pi/2}, a_x sin(i)$, and the pulse frequency and frequency derivative. The fit was then repeated with $a_x sin(i)$ held constant at the mean of the Main High values, to obtain a separate determination of orbital epoch for each Main High state. The mean $T_{\pi/2}$ and P_{orb} were found by a fit to the orbital epochs. The values obtained are shown in Table 1.

Using the same procedure as Deeter et al. (1991), residual orbital epochs have been compared to expected values for a constant orbital period. Historical and BATSE data are shown in Figure 3. The period is not consistent with a constant value, as also found by Deeter et al. (1991). Fitting the residuals to a quadratic, a value is obtained for $\dot{P}_{orb}$, as shown in Table 1. The fit is not formally satisfactory, with a reduced χ^2 value of about 2 for either the historical data alone, or all of the data. The value we obtain has a smaller uncertainty than [1], and is in agreement with that measurement.

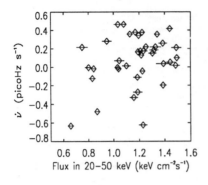 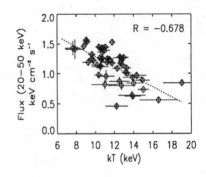

FIGURE 5. Scatterplot of peak Main High flux versus mean frequency derivative (defined as the frequency difference between successive Main High states).

FIGURE 6. Scatterplot of measured flux versus modeled kT, using an optically thin thermal bremstrahlung model. Values are from the *peaks* of each measureable Main High state.

Flux & Pulsar Spin Behavior

Using an on-pulse/offpulse analysis, optically thin thermal bremsstrahlung fits have been made to the average source counting rate over each Her X-1 orbit, excluding eclipse intervals and the prior 0.5 d intervals, to exclude pre-eclipse dips. The intervals at the peak intensity of each Main High state are selected and plotted in Figure 4, lower panel. The pulse frequencies obtained above are shown in the upper panel. Although in [5] we found a correlation between times of spindown and low source flux, that trend has not continued. As shown in Figure 5 there are times of spindown even when the source 20 – 50 keV flux is greater than 1.2 $keV cm^{-2} s^{-1}$. The system spins up between $\sim 75\%$ of the states measured, with the other cases evenly distributed relative to flux level when spindown occurs. Both spinup and spindown are observed at times of low flux.

We note a correlation of the measured flux versus the modeled kT value as shown in Figure 6. These are data selected only from the peaks of each Main High state. The increasing hardness at low flux levels is what would be expected if absorption by varying amounts of obscuring or scattering material was the cause of the flux variations between Main High states. The trend shown here is *not* shown during the traversal of any particular Main High state, as shown in Figure 7. If absorption and/or scattering is the cause of this effect, it is not due to the same material which controls the onset and end of each Main High envelope.

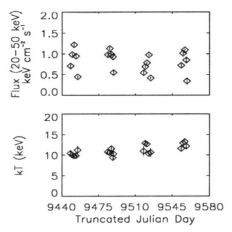

FIGURE 7. Flux and kT history for entire Main High states during TJD 9440 - 9580. Each flux point is averaged over 1 orbit of the system, excluding eclipse intervals, and the 0.5 day interval prior to the eclipse when pre-eclipse dips may occur.

DISCUSSION

The period derivative of Her X-1 is consistent with a constant value using all of the available observations, but with reduced χ^2 values of about 2, suggesting that the mass loss (and angular momentum loss) from the system may be variable on timescales we cannot measure.

There is not a simple relation between the brightness of a Main High state and the neutron star spin behavior, indicating that the hard x-ray flux is *not* simply related to the mass accretion rate. If we are viewing the source through obscuring material even at the peak of the Main High state, we would not observe a simple correlation of flux and spinup rate. It is also possible that variations in the source spectrum may alter the hard x-ray flux even for nearly constant mass accretion rate. Changes of the mass accretion rate on timescales less than 35 d also could explain these observations.

REFERENCES

1. Deeter et al., *ApJ* **383**, 324 (1991).
2. Jones, C. A., Forman, W., and Liller, W., *ApJ* **182**, L189 (1973).
3. Tananbaum, H. et al., *ApJ* **174**, L143 (1972).
4. Parmar, A. N. et al., *Nature* **313**, 118 (1985).
5. Wilson, R. B. et al., *Second Compton Symposium*, New York: AIP, 451 (1994).

RXTE Spectroscopy of Her X-1

D. E. Gruber*, W. A. Heindl*, R. E. Rothschild*, R. Staubert[†], M. Kunz[†] and D. M. Scott[‡]

* Center for Astrophysics and Space Science, UCSD, La Jolla CA, USA
[†]Institut für Astronomie und Astrophysik - Univ. of Tübingen, Germany
[‡]Marshall Space Flight Center, Huntsville, AL, USA

Abstract.
The accreting x-ray pulsar, Hercules X-1, emits a spectrum with a localized feature at about 35 keV which has been observed repeatedly and is generally interpreted as an absorption line resulting from transitions between quantized levels in an intense magnetic field at the neutron star's polar cap. It thus affords a diagnostic of plasmas under such unusual conditions, as well as providing a key to the structure of the magnetosphere and to the mass flow. The Rossi X-Ray Timing Explorer has observed Her X-1 three times to date, with 10000 seconds for the shortest observation. This rich data set permits spectral analysis to a new level of precision. We describe some of the more interesting results of this analysis, with particular emphasis on the stability of the line and continuum over time.

I INTRODUCTION

Detection of cycloton lines has been claimed for nine accreting x-ray pulsars [1], but repeated measurements have been made only for Her X-1. The large area and wide energy range of the Rossi X-Ray Timing Explorer (RXTE) have made it possible for the first time to pursue detailed spectroscopy on this class of objects, as well as studies of variability. We report results of an initial study of Her X-1, in which we find remarkable stability of the line centroid.

II OBSERVATIONS AND ANALYSIS

Three RXTE observations were used, summarized in Table 1. The third set was selected from a monitoring campaign spanning September 14 to Oct 25 1996. The selected observations were all obtained during the "High-On" phase, lasting about 10 days, of the 35-day cycle. About two thirds of the total data of observations two and three was available for this analysis.

TABLE 1. Observations and their exposures.

	Observation Time (UT)	Live Time (ks)
1	1996 Feb. 1	10
2	1996 Jul. 26,27	43
3	1996 Sep. 30, Oct. 2-8	28

The standard suite, sometimes the development versions, of XTE Ftools was used to accumulate phase-average spectra for each of the three data sets. A substantial live-time correction was applied to the HEXTE data and background was directly measured from the 16-second rocking. For PCA data the released background model produced an obvious underestimate of the detector background at higher energies, so in analysis the PCA data was limited to energies less than 27 keV.

The HEXTE response matrices did not produce noticeable systematics in the fits, but the PCA data showed clearly 1-2 percent ripples in the residuals to fits, which we attribute to errors in the matrix. The effect of these systematics on the fit results was reduced by applying a 1 percent systematic error to each channel with xspec, and by eliminating PCA channels below 4 keV.

Since the goal was to measure stability of the spectral shape and to permit comparison with earlier results, the traditional power-law continuum with high-energy cutoff model was employed in fitting with xspec, even though this model has a discontinuous first derivative at the cut-off energy and can thus introduce an artifact, an apparent absorption line, at this energy.

III RESULTS

Table 2 displays aspects of the best fits of the data, conditioned as described above, to the model of a power law continuum with exponential cutoff with single Gaussian absorption line. No convinving evidence was obtained for a harmonic overtone of the line.

The numbers in parentheses are the 90 percent confidence limits from the joint fit, and apply to the final decimal places of the best-fit value. The reduced chi-squares are very poor, but the dominant contribution appears to be from the inaccuracy of the PCA response matrix. The next largest appears to be from incorrect PCA background subtraction above 20 keV. A more interesting pattern consists of low residuals on the low-energy side of the 39 keV line, and high residuals on the other side. Clearly, the subtractive gaussian line profile incorrectly describes the data, and the next attempt should be with a multiplicative line. Such a form makes more physical sense anyway, because it is likely that the line is a resonance absorption of the continuum.

TABLE 2. Results of Spectral Fitting

parameter	Feb	July	Oct	mean	RMS
continuum					
norm(2-10)	2353(2)	1671(1)	1128(1)		
index	.931(1)	.788(1)	.870(1)	.863	.072
cutoff	17.95(5)	17.83(2)	17.53(3)	17.77	.22
fold	15.41(20)	15.04(6)	12.72(9)	14.4	1.5
Fe line					
norm	.00629(8)	.00488(3)	.00395(4)		
abs. line					
norm	.0153(10)	.0162(4)	.0047(4)		
centroid	39.27(36)	38.68(15)	38.74(32)	38.90	.32
sigma	7.46(39)	8.43(16)	6.54(35)	7.48	.94
chi-square	2.8	9.5	4.5		

The formal errors of the fits are so small that it makes sense to compare the shape parameters of the three observations in terms of their dispersions, expressed as RMS. One sees that three of the five parameters have RMS about 10 percent of the mean value, while two, the cutoff and line centroid, are stable to one percent. The chance probability that the true RMS for these quantities is near 10 percent, like the other parameters, seems very small, of the order of $(0.04)^3$.

The power law index, the e-folding energy, and the 38 keV line width are thus seen to be variable on a 10 percent scale on timescales of months. This is about the level seen for variability of the index from previous experiments e. g. [2], although intercalibration errors has certainly contributed in the past. The fold energy is more variable than was seen in the HEAO-1 data [3], which had a number of looks and good statistics. The sensitive measurement of the line width and its variability is new to this set of observations. Previous scintillator measurements (see [2,3]) were severely limited by detector resolution comparable to or larger than the line width. Although having better resolution than scintillators, existing observations with proportional counters [4] and germanium [5] were limited by counting statistics.

The very stable values for the cut energy and line centroid are new and surprising. The value of about 18 keV is different from that obtained in scintillator experiments, e. g. [3] 20 keV or higher, or even with proportional counters, where a value of 20 keV is also reported [6] (get a better number). The present measurements should be considered superior because they have very good spectal leverage with good statistics to both sides of the cutoff energy. The measurement of the line centroid also differs from earlier scintillator measurements, which centered near 35 keV [2]. Again, the present result should be more reliable because of the combination of better statistics and much better detector resolution. The stability of the values of these spectral parameters in the three observations is surprising in light of the variability of the other parameters, and the factor of two variability of the overall x-ray flux level.

IV INTERPRETATION

The 38-keV line is widely regarded to result from resonance absorption of electrons in a 3 x 10**12 Gauss magnetic field [7]. In fact, the line centroid is usually accepted as providing a direct measure of the field strength at the site of emission above the polar cap of the neutron star[8]. Likewise, the cutoff energy has been interpreted [9] as resulting from the onset of opacity in an extreme magnetic field of this order. If both are accepted as measures of the magnetic field strength, then the average magnetic field in the region of emission into the line of sight remains remarkably constant for the three observations, and is also stable to the order of one percent. Since the magnetic field is anchored on the neutron star, the average height of emission of the line photons is constrained. This one percent RMS corresponds to a 100 km constraint on the average height of emission for a dipole field, and 50 km for quadrupole.

References

1. Makashima, K., et al., 1992, in "Frontiers of X-Ray Astronomy", eds Tanaka and Koyama, (Tokyo:Universal), 23.
2. Kunz, M., unpublished dissertation, 1995, Univ. Tübingen.
3. Soong, Y., et al., 1990, Ap. J., 348, 641.
4. Pravdo, S. H., et al., 1979, M. N. R. A. S., 188, 5.
5. Tueller, J., et al., 1984, Ap. J., 279, 177.
6. White, N. E., Swank, J. H., and Holt, S. S., 1983, Ap. J., 270, 711.
7. Voges. W., et al., 1982, Ap. J., 263, 803.
8. Trümper, et al., 1978, Ap. J. (Letters), 219, L105.
9. Boldt, E. A., et al., 1976, Astron. Astrophys., 50, 161.

Observations of Pulse Evolution in Her X-1

D. Matthew Scott[†], Robert B. Wilson[‡], Mark H. Finger[†,*], & Denis A. Leahy[*]

[†] *University Space Research Association.*
[*] *CGRO Science Support Center/Goddard Space Flight Center.*
[‡] *Space Sciences Laboratory, NASA/Marshall Space Flight Center.*
[*] *Dept. of Physics & Astronomy, University of Calgary.*

Abstract. We present recent *RXTE* observations of pulse evolution in Her X-1 and compare the evolution pattern to that observed earlier with *Ginga*, *HEAO 1* and *Uhuru*. The pulse profile evolves continuously during the $\sim$ 5-day long Short High state of the 35-day cycle and shows dramatic and systematic changes near the end of the $\sim$ 10-day long Main High state.

INTRODUCTION

Her X-1/HZ Her is a unique accretion-powered pulsar system exhibiting a great wealth of phenomena. This eclipsing system contains a 1.24 second period pulsar in a 1.7 day circular orbit with its optical companion HZ Her. In addition, the system displays a longer 35-day cycle that was first discovered as a repeating pattern of High and Low X-ray flux states. A Main High and Short High state, lasting about ten and five days each, respectively occur once per 35-day cycle and are separated by ten day long Low states. X-ray pulsations are visible during the HIGH states but cease during the Low states. The pulse profile exhibits significant and systematic changes during each type of High state. We present detailed observations of the profile changes including the most complete view of the Short High pulse evolution to date.

OBSERVATIONS

The *RXTE* 2-60 keV lightcurve for Her X-1 is displayed in figure 1 showing a consecutive Short-Main High sequence from September-October 1996. No background subtraction has been applied. Simultaneous *BATSE* observations

of the 20-70 keV pulsed emission during the Main High arc shown in the bottom panel. Note the rapid drop in the hard X-ray pulsed emission near 35-day phase 0.2 while the total flux is dropping much more slowly. A time sequence of average pulse profiles in the energy band 2-60 keV is displayed in figure 2. The pulses were formed by folding the Standard 1 data (0.125 sec time resolution) over spans of 1000-2500 seconds. The Main High folding period was determined from *BATSE* monitoring and the Short High was folded using an average period from adjacent Main Highs. Arrival times were converted to the solar system barycenter and then corrected to the neutron star by removing the light travel time delay across the binary orbit prior to folding. Some phase drift is apparent in the Short High pulses showing that a refined period can be achieved. The relative phase alignment of the all the displayed Main and Short High pulses has been determined from pulse phase ephemeris extrapolations [1].

A 2-37 keV lightcurve for Her X-1 from *Ginga* observations is displayed in figure 3 showing a consecutive Main-Short High state sequence from April-May 1989. Background and aspect corrections have been applied. A time sequence of pulse profiles in 1.0-4.6 keV and 9.3-14 keV energy bands are shown in figure 4. The insets on the first panel display the 35-day phase (with $P_{35} = 20.5 P_{orb}$ for all figures). The last four pulses of the Main High occupy a span of only 6.7 hours. The *Ginga* observations are described in [1, 2].

Figure 5 shows a sequence of *Uhuru* pulses from [3] from a single Main High. Each pulse is folded data during a single orbit of Her X-1 denoted by the number in the inset. The second column shows a sequence of *HEAO 1* pulses from [4] with 35-day phase shown in the inset.

PHENOMENOLOGY OF THE PULSE EVOLUTION

Inspection of the figures reveals a pulse evolution pattern that is systematic across both High state types and has been remarkably stable in form over many years of observation. Most profile changes during the Main High occur during the last few days when the flux is declining significantly whereas the profile changes most rapidly near the beginning of the Short High. Smaller random profile changes are sometimes observed that may be related to accretion rate variations.

The first major change in the Main High profile pulse is a deepening near pulse phase 0.25 that starts within the first few days after Turn-On. Otherwise the profile is quite constant in shape for roughly seven days and then the main pulse and interpulse both begin to decay away preferentially on their leading edges at a rate which accelerates during the Main High flux decline. In both *RXTE* and *Ginga* observations rapid changes in the profile occured on timescales of a few hours. Finally a quasisinusoidal profile is left which then disappears after another day or so.

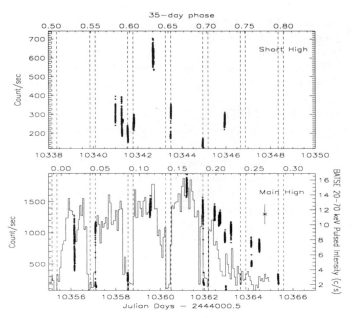

FIGURE 1. The 2-60 keV lightcurve of Her X-1 observed with *RXTE* (diamonds). Eclipse interval marked by dashed lines. *BATSE* pulsed emission lightcurve shown at bottom (histogram) with typical error bar on right.

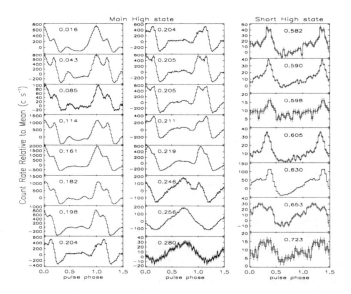

FIGURE 2. *RXTE* observation of pulse evolution in Her X-1 (2-60 keV). The 35-day phase is shown in the inset.

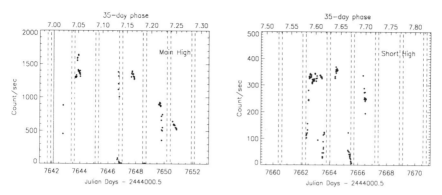

FIGURE 3. The 2-37 keV lightcurve observed with *Ginga*. Eclipse interval marked by dashed lines.

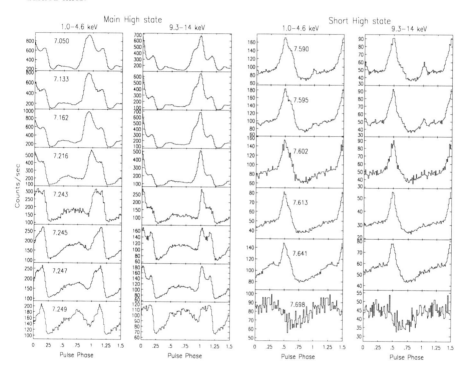

FIGURE 4. *Ginga* observation of pulse evolution in Her X-1 in two energy channels. The 35-day phase is shown in the inset.

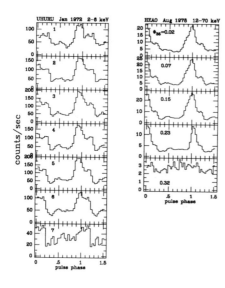

FIGURE 5. *Uhuru* and *HEAO 1* observations of Main High pulse evolution.

The Short High pulse profile initially appears with either one or two peaks. A larger, softer peak near phase 0.5 is always present that corresponds to the interpulse in the Main High pulse. A second harder peak appears at phase 0.0 in the *Ginga* but not the *RXTE* Short High observations. An earlier Exosat observation also showed the second harder peak which was initially larger than the softer peak[6]. In both the *Ginga* and Exosat observations the hard peak decayed away in the less than one day leaving only the softer peak. The softer peak takes 3-4 days to decay away and apparently narrows in the process, finally leaving only the underlying quasisinusoidal profile.

The Short High quasisinusoidal profile has about 50% the flux of the Main High quasisinusoid and is shifted in phase by 0.5.

The persistant, systematic and apparently repeating evolution of the pulse profile in Her X-1 is a unique phenemonena among the accretion powered pulsars. Some pulsars, such as EXO 2030+375, have shown systematic pulse profile changes, but these have all been associated with transient changes in the X-ray luminosity. The Her X-1 profile evolution is clearly a different type of phenomena and may play a key role in furthering our understanding of the pulse emission process.

REFERENCES

1. Deeter J., Scott D. M., Boynton P., Miyamoto S., Kitamoto S., Takahama S. and Nagase F. (1997) submitted to *Ap.J.* (in 1993)
2. Scott D. M., (1993) PhD Thesis, University of Washington
3. Joss P.C., Fechner W.B., Forman W. & Jones C. (1978), *Ap.J.*, 225, 994
4. Soong Y., Gruber D.E., Peterson L.E. and Rothschild R.E. (1990) *Ap.J.*, 348, 634
5. Kahabka P. 1989, Proc. of the 23rd ESLAB Symp. on Two-Topics in X-ray Astronomy, ed. Hunt J. & Battrick B. (The Netherlands: ESA Publications Division), 447

Evolution of the Orbital Period of Her X–1: Determination of a New Ephemeris using RXTE Data

B. Stelzer, R. Staubert, J. Wilms and R. D. Geckeler

Institut für Astronomie und Astrophysik – Astronomie, Univ. of Tübingen, Germany

D. Gruber and R. Rothschild

Center for Astrophysics and Space Sciences (CASS), UCSD, USA

Abstract. Her X–1 was observed by RXTE in July 1996 during the MAIN HIGH state of its 35 day cycle. Using data from the Proportional Counter Array (PCA) we redetermine the orbital parameters of the binary system. The analysis based on pulse-timing measurements yields new estimates for the orbital elements and an accurate spin period for the time of observation. By comparing our results with previous observations of Her X–1 we are able to report a new orbital ephemeris including an improved value for the decrease in orbital period.

INTRODUCTION AND DATA BASE

Her X–1 is a Low Mass X-ray Binary consisting of a neutron star orbiting a $2\,M_\odot$ companion. Due to the various periodic phenomena it exhibits in the X-ray range, Her X–1 has been observed by many satellite instruments since its discovery in 1972 [1]. These periodicities are the spin period of the neutron star (1.24 sec), the orbital period (1.7 days), the accretion disk precessional period (35 days) and a dip period (1.62 days). The latter is a beat period between 1.7 days and 35 days.

Her X–1 was observed by the Proportional Counter Array (PCA) on board the Rossi X-ray Timing Explorer (RXTE) on the 26th and 27th of July, 1996. The total observing time was 50 ksec. During the observation Her X–1 was in the MAIN HIGH state of its 35 day cycle. Pre-eclipse dips are present in part of the observation. No data were taken during X-ray eclipse. The time resolution we used for the pulse timing analysis is 10 ms.

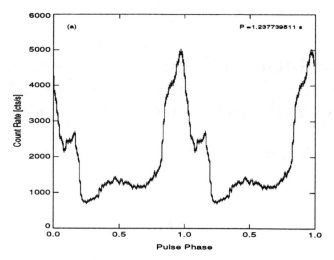

FIGURE 1. Pulse Profile at Orbital Phase 0.6

ORBITAL PARAMETERS AND SPIN PERIOD

As a first step in the data analysis the photon arrival times have been transformed to the solar system barycenter. The first fifty 1.24 s pulses of the observation were then folded with a trial spin period in order to establish a reference pulse profile. The rest of the data has been divided into groups of ten spin periods respectively and folded with the trial period to build mean pulse profiles. A typical profile (100 phase bins) is displayed in Fig. 1.

The mean pulse profiles exhibit phase offsets with respect to the reference pulse due to their time delay caused by the orbital motion of Her X-1.

These offsets were determined by a minimum χ^2 technique: sliding the profiles over the reference profile in 100 steps and finding the minimum χ^2, corresponding to the best match, by a quadratic fit to the 100 χ^2 values.

The offsets are shown in the upper part of Fig. 2 (a). In this plot we show our final result, that is the data have been folded with our best value for the spin period, P_{spin}= 1.237739511 s. The way in which this value was obtained is explained below. Gaps in the data stream are due to Earth occultations of the source during individual satellite orbits. The data of 17 RXTE orbits were used, covering 82% of the binary orbit. A typical RXTE orbit contributed about 230 phase offset values.

The measured phase offsets were then fit by a cosine function with four free parameters (B, $a \cdot \sin i$, P_{orb}, $T_{\frac{\pi}{2}}$):

$$\Delta t_{shift} = B + a \cdot \sin i \cdot \cos\left(\frac{2 \cdot \pi}{P_{orb}} \cdot (t - T_{\frac{\pi}{2}})\right)$$

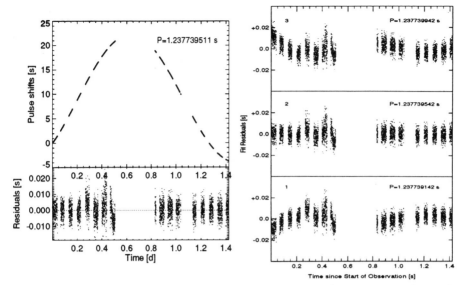

FIGURE 2. (a) Pulse Arrival Time Delays with Respect to the Reference Pulse at the Beginning of the RXTE-Observation and Residuals of a Cosine Fit. (b) Fit Residuals of the Pulse Arrival Time Delay for Three Different Trial Spin Periods

We have applied the above method to twelve distinct trial spin periods in steps of 10^{-7}s. The resulting residuals for three of the twelve spin periods are shown in Fig. 2 (b). They are definitely more structured than the residuals for the spin period of Fig. 2 (a). The mean secular change of the spin period ($10\frac{ns}{d}$ [5]) is small enough to be neglected in this analysis.

The χ^2-values of the twelve fits as a function of the spin period match a cubic function (see Fig. 3) whose minimum determines the final value for the spin period, $P_{spin} = 1.237739511(19)$ s[1]. The trial periods of Fig. 2 (b) are labeled by numbers 1–3.

TABLE 1. Orbital Parameters determined by this RXTE observation

P_{spin}	1.237739511(19)	s
$T_{\frac{\pi}{2}}$	50290.659249(15)	MJD(TDB)[2]
P_{orb}	1.699606(63)	days
$a \cdot \sin i$	13.1826(4)	lt-s

---

[1] values given in parenthesis are 1 σ uncertainties refering to the two last digits
[2] MJD(TDB)=JD(TDB)-2400000.5
where TDB is the barycentric dynamical time

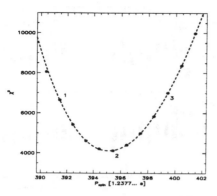

FIGURE 3. χ^2 as a function of P_{spin}

The uncertainty of ± 19 ns for P_{spin} is the 68 % confidence interval for joint fits with five free parameters: B, a· sin i, P_{orb}, $T_{\frac{\pi}{2}}$ and P_{spin} (using χ^2_{min}+5.86).
The fit to the time delays was repeated for this value of P_{spin} in order to determine the best orbital parameters. The best fit parameters yield an updated value for the orbital epoch $T_{\frac{\pi}{2}}$, as well as for the orbital period P_{orb} and the radius of the orbit a· sin i (see Tab. 1).

ORBITAL EPHEMERIS

Using our new value for $T_{\frac{\pi}{2}}$ and the existing data from previous missions ([3,2], and [4]) it is possible to find an improved value for the secular change of the orbital period, $\dot{P}_{orb}$.

TABLE 2. Best Fit Parameters for a Quadratic Ephemeris

T_0	45477.484951(17)	MJD(TDB)^2
P_{orb}	1.7001676337(55)	days
$\dot{P}_{orb}$	$(-1.71 \pm 0.19) \cdot 10^{-8}$	d/yr

A fit with a linear ephemeris results in no acceptable description of the data, with $\chi^2_{red} = 16.34$ as compared to $\chi^2_{red} = 2.74$ for the quadratic ephemeris. Fig. 4 shows the deviations of the timings for $T_{\frac{\pi}{2}}$ from the linear component of our quadratic best fit ephemeris (dotted line). The best fit parameters are given in Tab. 2 together with their uncertainties.

We find a secular decrease of the orbital period of Her X-1 of $\dot{P}_{orb}=(-1.71 \pm 0.19) \cdot 10^{-8}$ d/yr.

It might be worth noting that despite the good fit for the quadratic ephemeris a sudden change in orbital period near MJD 45477 (Orbit No.0 in Fig. 4) cannot be ruled out since a combination of separate linear fits to the data before and after Orbit No. 0 yields $\chi^2_{red} = 1.40$. The period change in this case is $2.84 \cdot 10^{-7}$d.

RESULTS

The new RXTE observation extends the baseline for the determination of the orbital ephemeris of Her X-1 . As a result we present a new value for

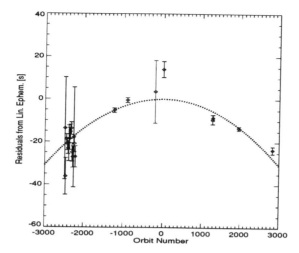

FIGURE 4. Residuals of $T_{\frac{\pi}{2}}$ after Subtracting the Linear Part of the Quadratic Bestfit Ephemeris. The first OSO-8 point, the HEAO-1 point and the third UHURU point are omitted in our analysis, the former two for its possible systematic errors, the latter because of its huge uncertainty.

the change in orbital period: $\dot{P}_{orb} = (-1.71 \pm 0.19) \cdot 10^{-8}$ d/yr, which — in comparison to the value given by [3] — has a smaller uncertainty and its absolute value is lower by about two standard deviations. We note that there is no problem with a change in orbital period of this order: the coronal wind model [5], [6] naturally provides the necessary loss of angular momentum.

ACKNOWLEDGEMENTS

This research was supported by DARA under grant 50 OO 96058 and NASA under grant NA G5 - 3272.

REFERENCES

1. H. Tananbaum et al. 1972, ApJ, 174,L143
2. J. E. Deeter et al. 1981, ApJ, 247,1003–1012
3. John E. Deeter et al. 1991, ApJ, 383,324–329
4. R. B. Wilson et al. 1994, AIP conference proceedings 304,235
5. S. Schandl et al. 1997, Proceed. 4th Compton Symposium, Williamsburg
6. S. Horn 1992, Doctoral thesis Ludwig-Maximilians-Universität München

The pulsed light curves of Her X-1 as observed by BeppoSAX

D. Dal Fiume[1], M. Orlandini[1], G. Cusumano[2],
S. Del Sordo[2], M. Feroci[3], F. Frontera[1,4],
T. Oosterbroek[5], E. Palazzi[1], A. N. Parmar[5],
A. Santangelo[2], A. Segreto[2]

[1] *Istituto TESRE/CNR, via Gobetti 101, 40129 Bologna, Italy*
[2] *Istituto IFCAI/CNR, via La Malfa 153, 90146 Palermo, Italy*
[3] *Istituto Astrofisica Spaziale/CNR, via Fermi 21, 00044 Frascati, Italy*
[4] *Dipartimento di Fisica, Università di Ferrara, via Paradiso, 1, 44100 Ferrara, Italy*
[5] *Astrophysics Division, Space Science department of ESA, ESTEC, P.O. Box 299, 2200 AG Noordwijk, The Netherlands*

Abstract. We report on the timing analysis of the observation of the X-ray binary pulsar Her X-1 performed during the BeppoSAX Science Verification Phase. The observation covered more that two full orbital cycles near the maximum of the main-on in the 35 day cycle of Her X-1. We present the pulse profiles from 0.1 to 100 keV. Major changes are present below 1 keV, where the appearance of a broad peak is interpreted as re-processing from the inner part of the accretion disk, and above 10 keV, where the pulse profile is less structured and the main peak is appreciably harder. The hardness ratios show complex changes with pulse phase at different energies.

INTRODUCTION

Her X-1 is one of the most observed and best studied sources in the X-ray sky. This eclipsing binary pulsar (orbital period 1.7 days, pulsation period 1.23 sec) was one of the first X-ray sources in its class to be discovered [5,15]. It shows a 35 day period on-off cycle in which a main-on and a short-on are present. The flux from the source varies roughly a factor of three between the main-on and the short-on [6]. Low flux level emission was also detected between the two on of the 35 day cycle.

The pulsed light curves observed during main-on show a broad, structured single peak from 2 to 100 keV [14]. There is evidence that the pulse shape varies during the 35 day cycle both in the low energy band below 10 keV and in the hard X-rays above 20 keV. In 1-30 keV a double peaked pulse shape was observed with EXOSAT during the short-on state [16]. Soong et al. [14] measured with HEAO-1 a change in the 12-70 keV pulse shape during the main on, even if they conclude that the changes they observe are more likely related to the source intensity than to the phase of the 35 day cycle.

OBSERVATION

BeppoSAX is a program of the Italian Space Agency (ASI) with participation of the Netherlands Agency for Aerospace Programs (NIVR). It is composed by four co-aligned Narrow Field Instruments (NFIs) [1], operating in the energy ranges 0.1-10 keV (LECS) [13], 1-10 keV (MECS) [2], 3-120 keV (HPGSPC) [8] and 15-300 keV (PDS) [4]. Perpendicular to the NFI axis there are two Wide Field Cameras [7], with a 40° × 40° field of view.

During the BeppoSAX Science Verification Phase (SVP) Her X-1 was observed from 1996 07 24 00:34:46 UT to 1996 07 27 11:54:46 UT. Data were telemetred in direct mode for all the four NFIs, with information on arrival time, energy and position for each photon. We report on the pulsed light curves observed during a fraction of the out-of-eclipse phase of the binary orbit.

The data were recorded in single-event mode: each detected photon was tagged with its arrival time in the detectors. The arrival times were corrected to the solar system barycentre and folded at the pulsar period of 1.2377396 s [11]. A correction for the Her X-1 orbital motion was also included.

RESULTS

The folded light curves as observed by BeppoSAX are shown in Figure 1. Two clear transitions are seen: one at approximately 1 keV and the other at approximately 10 keV.

The 1 keV transition goes from a broad sinusoidal pulse shape (Figure 1, panel (a)) to a more peaked pulse shape that remains almost unchanged up to 10 keV. The transition is accompanied by a phase shift of $\sim 250°$.

The 10 keV transition goes from a structured single peak, with a prominent shoulder on its trailing edge, to an almost perfectly triangular pulse shape. The pedestal observed between phase 0.9 and phase 1.4 in Fig. 1 remains visible at least up to 40 keV. This second transition occurs exactly at the energy where the Her X-1 X-ray spectrum begins to deviate consistently from a simple power law shape [3].

These two transition are even more evident in Fig. 2, in which we show the ratios between light curves in different energy bands. The hardness ratios also show some complex changes in the 2-10 keV energy interval. These changes are percentually small, but are easily detected with high significance.

The transition at ~1 keV is likely to be due to a change from a reflected to a directly observed pulse beam [9,11]. As discussed in [11], the phase shift of ~250° observed between the soft and the hard pulse shape suggests that the reflection zone is the inner part of a tilted accretion disk.

The pulse width decreases with energy above 2 keV, mainly due to the suppression at higher energies, above 10 keV, of the shoulder on the trailing edge of the peak. This variation of the pulse shape is likely due to a variation in the intrinsic beaming pattern with energy, as predicted by emission models (e.g. [10,17]). However it is not straightforward to obtain the observed asymmetric shapes directly from the models and from simple pencil or fan beams.

An attempt to introduce some degree of geometrical complexity in the emission region was done by Panchenko and Postnov [12], reproducing qualitatively the Her X–1 pulse shape. However our detailed and simultaneous measurement in a broad energy interval shows that any geometrically complex model must explain also the energy dependence of the pulse shapes. This can be done only by taking into account the spatial anisotropy and energy dependence of the intrinsic beaming pattern from the emission zone.

Acknowledgements. The authors wish to thank the BeppoSAX Scientific Data Center staff for their support during the observation and data analysis. This research has been funded in part by the Italian Space Agency.

REFERENCES

1. Boella G., Butler R.C., et al. 1997a, A&AS, 122, 299
2. Boella G., Chiappetti L., et al. 1997b, A&AS, 122, 327
3. Dal Fiume D., Orlandini M., et al. 1997, in preparation
4. Frontera F., Costa E., et al. 1997, A&AS, 122, 357
5. Giacconi R., Gursky H., et al. 1973, ApJ, 184, 227
6. Gorecki A., Levine A., et al. 1982, Apj, 256, 234
7. Jager R., Mels W.A., et al. 1997, A&AS, in press
8. Manzo G., Giarrusso S., et al. 1997, A&AS, 122, 341
9. McCray R.A., Shull J.M., Boynton P.E., et al., 1982, ApJ 262, 301
10. Mészáros P., Nagel W. 1985 ApJ, 299, 138.
11. Oosterbroek T., et al. 1997, A&A, in press
12. Panchenko I. E., Postnov K. A. 1994, A&A, 286, 497
13. Parmar A.N., Martin D.D.E., et al. 1997, A&AS, 122, 309
14. Soong Y., Gruber D.E., et al. 1990, ApJ, 348, 634
15. Tananbaum H., Gursky H., et al. 1972, Apj, 174, L143
16. Trümper J., Kahabka P., et al. 1986, ApJ, 300, L63
17. Yahel R. Z., 1980, A&A, 90, 26

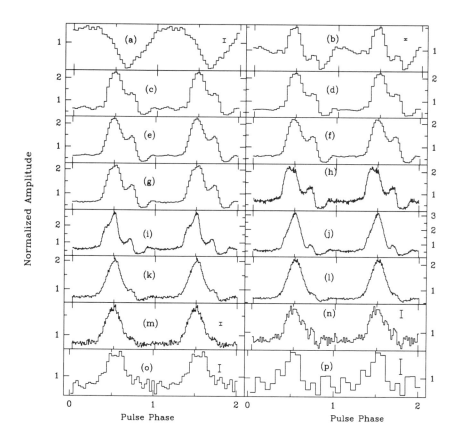

FIGURE 1.
Folded light curves of Her X-1: (a) 0.1–0.4 keV, LECS; (b) 0.4–1.6 keV, LECS; (c) 1.6–2.4 keV, LECS; (d) 2.4–10 keV, LECS; (e) 2–4 keV, MECS; (f) 4–6 keV, MECS; (g) 6–10 keV, MECS; (h) 4–8 keV, HPGSPC; (i) 8–15 keV, HPGSPC; (j) 15–30 keV, HPGSPC; (k) 13–20 keV, PDS; (l) 20–30 keV PDS; (m) 30–40 keV, PDS; (n) 40–50 keV, PDS; (o) 50–70 keV, PDS; (p) 70–100 keV, PDS. Where not indicated, the error bar is smaller than the symbol size.

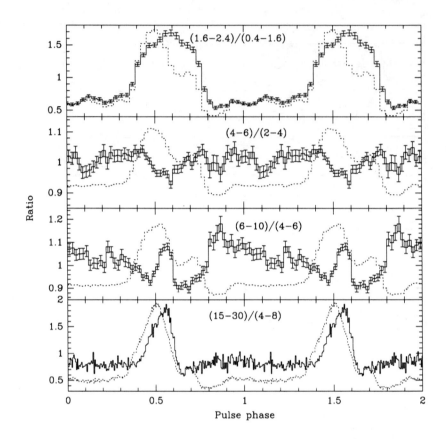

FIGURE 2.
Hardness ratios between folded light curves in different energy bands. In each panel the dashed line shows the measured light curve in the higher energy band for reference.

The 35 day cycle of Her X-1 and the coronal wind model

S. Schandl, R. Staubert, M. König

Institut für Astronomie und Astrophysik - Astronomie
University of Tübingen, Germany

Abstract. We present the longterm behaviour of the pulse period and the 35 day precession including historical data as well as new results from GRO/BATSE and the All Sky Monitor onboard RXTE. We find a strong correlation showing shorter precession periods during spin-down phases. Spin-down corresponds to smaller accretion rates (Gosh and Lamb 1979). With reference to the coronal wind model we propose now that the shape of the warped accretion disk depends on the accretion rate. And the precessional period depends on the disk shape. The effect is of the right order to explain the observed correlation.

INTRODUCTION

The LMXB Her X-1/Hz Her shows periodicities on very different time scales. The spin period of the neutron star is 1.24 sec, the orbit lasts 1.7 days and the 35 day cycle corresponds to the precession of the warped accretion disk in the tidal field of the companion. The binary system shows a high inclination of more than 80°, and therefore the disk covers the central neutron star for the observer temporarily during the 35 day precession resulting in strong variations of the X-ray signal. The underlying clock is not very accurate but we will connect its temporal behaviour now with changes in the mass transfer rate. This is supported by a strong correlation between the duration of the precession cycle and changes in the spin-up rate where the latter is due to variations of the mass transfer rate (Gosh and Lamb 1979). Our suggestions are derived from the coronal wind model (Schandl and Meyer 1994, Schandl 1996) describing a mechanism for the warped shape of the accretion disk.

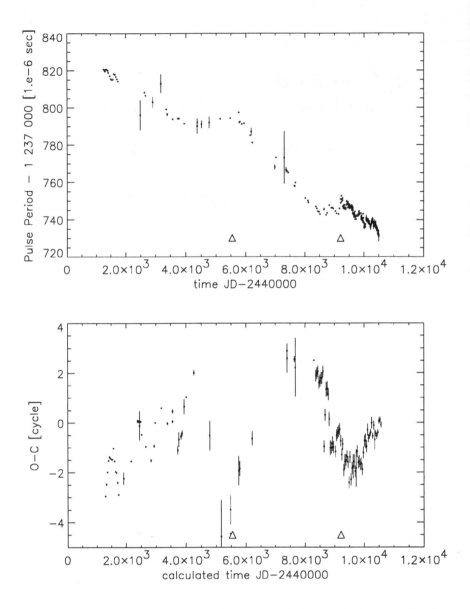

FIGURE 1. The upper panel shows the pulse period evolution. In the lower we plot the turn-on data where O-C is the difference between observed and calculated turn-on time. The calculated time follows the epoch of Staubert et al. (1983) for the 31st turn-on: T_{31} = JD 2442410.349 d. We used a turn-on period of 34.853 d equal to 20.5 P_{orbit} with P_{orbit} = 1.700167788 d (Deeter et al. 1981). – The triangles mark the times of observed extended low periods (Parmar et al. 1985 and Vrtilek et al. 1994).

DATA BASE

The Turn-on Data

Turn-ons, defined by a sharp rise in the X-ray signal, are observed since the discovery of Her X-1 in 1972 (Tananbaum et al. 1972). Our historical data set is based on Staubert et al. (1983) and Kunz (1996). Additionally, we included turn-ons derived from the occultation data of the GRO/BATSE instrument taken from the public GRO data archive at NASA/GSFC. Recently, also data from the All Sky Monitor onboard RXTE yield turn-on times. These data are taken from the HEASARC archive at NASA/GSFC. Details of the determination of the turn-on times will be given in Schandl et al. (1997).

The Spin Period Data

The historical data are taken from Nagase (1989), Sunyaev et al. (1988) and Kunz (1996). We included also the pulse periods determined by BATSE of the public pulsar data of the GRO data archive and one value of a pointed observation of RXTE (Stelzer et al. 1997).

OBSERVATIONAL RESULTS

The temporal evolution of the pulse period and of the turn-on times are strongly correlated. This had earlier been noted on a shorter set of data by Bochkarev et al. (1988). During constant spin-up the precession cycle length is also constant. During times of spin-down (following a reduced mass accretion rate, Gosh and Lamb 1979) the precession of the warped disk is faster (reducing O-C).

At the end of both spin-down phases the disk precession gets slower. At these times the extended low in 1984 observed by EXOSAT (Parmar et al. 1985) and the low of the multiwavelength campaign 9 years later (Vrtilek et al. 1994) occur. This supports our model where an increasing mass accretion rate yields a more tilted accretion disk covering the X-ray source permanently.

Additionally to the longterm changes, the pulse period shows quasiperiodic oscillations on a time scale of 400 - 600 days. A correlation with the turn-on behaviour will be considered.

THE CORONAL WIND MODEL

Heating of the accretion disk surface by X-ray illumination yields a hot corona. This is hydrostatically layered above the inner disk regions and escapes from the system at outer radii where its inner energy gets larger than the gravitational potential. Above the surfaces of a warped disk such a coronal wind is asymmetrical. The repulsive forces of the leaving matter yield therefore a torque which does not vanish. This torque acts on the inclination of each disk ring as well as on the precessional period.

The stationary torque equation includes viscosity, tidal forces and coronal wind torque. For the boundary condition torques from outside are excluded. The resulting solution for the disk shape is shown in Schandl and Meyer (1994), Schandl (1996) (see Fig. 2).

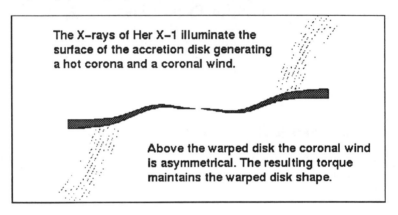

FIGURE 2. A cut through the warped disk as it results from the coronal wind model (Schandl and Meyer 1994)

APPLICATION OF THE CORONAL WIND MODEL TO HER X-1 DATA

To explain the observed correlation between faster disk precession and spin-down (Fig. 1) or rather a reduced accretion rate (Gosh and Lamb 1979) we now consider the influence of the accretion rate on the disk precession. The precession itself depends on the geometry of the disk shape which results from the torque equation. The acting torques have different functional dependencies on the accretion rate: The smoothing viscous torque is proportional to the surface density $\Sigma \propto \dot{M}^{3/4}$, while the warping wind torque is proportional to the irradiated X-ray luminosity $L \propto \dot{M}$. This causes changes in the disk shape and the precessional period when the accretion rate changes!

A reduced accretion rate results in a generally less inclined disk shape. The flatter disk geometry adjusts itself to keep the balance between mass loss rate in the wind and matter accreted by the neutron star nearly constant. As a result the ratio of the disk inclination at outer radii (the wind region) to the inclination of the inner disk increases. This causes a faster disk precession, as observed. For details see Schandl et al. (1997).

While the disk adjusts to the changing accretion rate its shape may evolve far from the equilibrium state. This may explain the higher inclinations during increases of the accretion rate yielding the observed extended lows.

CONCLUSIONS

We have shown that changes in the accretion rate act on the spin period of the neutron star as well as on the precessional period of the accretion disk. This causes the observed correlation between the turn-on times and the pulse period evolution of Her X-1. - Finally, the variety in the behaviour of Her X-1 is caused by the complex structure of its tilted and twisted accretion disk.

ACKNOWLEDGEMENT

This work was supported by DARA under grant No. 50OR9205.

REFERENCES

Bochkarev, N.G., Lyutyi, V.M., Sheffer, E.K., Voloshina, I.B., 1981, Sov. Astron. Lett. 14(6), 421
Deeter J.E., Boynton P.E., Pravdo S.H., 1981, ApJ 247, 1003 Gosh P., Lamb F.K., 1979, ApJ 234, 296
Kunz M., 1996, PHD Thesis, University of Tübingen, Germany
Nagase F., 1989, PASJ 41, 1
Parmar A.N., Pietsch W., McKechnie S., et al., 1985, Nat 313, 119
Schandl S., 1996, A&A, 307, 95
Schandl S., Meyer F., 1994, A&A 289, 149
Schandl S., Staubert R., König M., 1997, to be submitted
Staubert R., Bezler M., Kendziorra E., 1983, A&A 117, 215
Stelzer B., Staubert R., Wilms J., et al., 1997, see poster presentation 'Evolution of the Orbital Period of Her X-1', this conference
Sunyaev R., Gilfanov M., Churazov E., Loznikov V., Efremov V., 1988, SvAL 14, 418
Tananbaum H., Gursky H., Kellog E.M., et al., 1972, ApJ 174, L143
Vrtilek S.D., Mihara T., Primini F.A., et al., 1994, ApJ 436, L9

CGRO/EGRET Observations of Centaurus X-3

W. Thomas Vestrand*, P. Sreekumar[†], and Masaki Mori[‡]

Space Science Center, University of New Hampshire, USA
[†]*LHEA/NASA Goddard Space Flight Center, Greenbelt, MD, USA*
[‡]*Department of Physics, Miyagi University of Education, Sendai, Miyagi 980, Japan*

Abstract.
We review evidence for the first detection of GeV gamma-rays from the massive x-ray binary system Centaurus X-3. The CGRO/EGRET detection of an outburst of gamma-ray emission from the direction of Cen X-3 occurred during an interval of rapid spindown by the x-ray pulsar. Our phase analysis of photon arrival times, employing the contemporaneous x-ray pulse observations by CGRO/BATSE, indicates modulation of the gamma-ray emission at the pulsar spin frequency with a significance level higher than 99.5%. Straightforward interpretation of the observations suggests that accreting pulsars may constitute a class of highly variable GeV gamma-ray emission.

INTRODUCTION

Numerous authors have proposed mechanisms for particle acceleration and gamma-ray production by X-Ray Binary systems (XRBs). Among the acceleration mechanisms that have been proposed are pulsar acceleration [3,25], shock acceleration at an accretion shock front [11,16], shock acceleration at a pulsar wind termination shock [1], plasma turbulence excited by the accretion flow [15], and a number of electrodynamic mechanisms [6,18,7]. Likely gamma-ray production mechanisms include Compton scattering of photons from the companion star by energetic electrons and pion-decay emission arising from energetic proton interactions with the atmosphere of the companion star, accretion disk, accretion column, or accretion wake [25,11,20]. There are therefore many potential scenarios for generating very energetic gamma-rays in the XRB environment, but, until recently, there has been very little data to support the idea that energetic particles are accelerated in XRBs.

The Centaurus X-3 system has played an important role in the development of our understanding of galactic x-ray binary sources. Timing analysis of the UHURU x-ray observations for the luminous Cen X-3 source revealed the first evidence for coherent x-ray pulsations from an object in a binary system [14,23]. It was quickly understood that the pulsed x-ray emission could be generated by the accretion of matter from a companion star onto a rotating neutron star and led to the adoption of binary star models as the fundamental model for galactic x-ray sources [22,19]. Since then, observations of the pulsar in Cen X-3 have provided important insights about accretion processes and how magnetized neutron stars interact with accretion disks [13].

Here we review gamma-ray observations of the Cen X-3 system that suggest it may once again play a key role in furthering our understanding of high-energy processes in XRBs.

GAMMA RAY EMISSION FROM CEN X-3

Ground-based observations with atmospheric Cerenkov telescopes suggest that gamma rays with TeV energies are emitted by the Cen X-3 system. The Durham Group has reported detections, made with their Mk 3 telescope in Narrabri, Australia, of a variable >250 GeV gamma-ray flux from Cen X-3 that is modulated at the 4.8 second pulsar spin period [5]. The Potchefstroom group in South Africa has also reported detections of spin-modulated TeV gamma rays from Cen X-3 [21]. Barring obscuration by photon-photon pair production within the XRB [24,2], one would expect Cen X-3 to also be a source of sporadic GeV gamma-ray emission.

Such considerations prompted our Cycle 4 Guest Investigator program to search for GeV gamma-ray emission from Cen X-3 with CGRO/EGRET. During Cycle 4 Cen X-3 was within the EGRET field-of-view for two weeks (18 October-1 November 1994: VP 402.0 and VP 402.5). Our likelihood analysis of those EGRET measurements, which employs the standard software and techniques developed by the instrument team, revealed the presence of a significant test statistic peak ($\sim 5\sigma$) at $l = 292.20°$ and $b = 0.48°$ with both a shape and size that is consistent with a point source located at the position of Cen X-3 [26]. Our subsequent analysis of archival data indicated that the source is not present in likelihood maps derived from earlier EGRET measurements of the region. This failure to detect the emission in earlier observations indicates that the excess is not an artifact generated by an error in the diffuse radiation model. A careful consideration of gamma-ray blazar properties also indicates that generation of the outburst by a background blazar is a very unlikely possibility [26]. The simplest hypothesis is that the gamma-ray excess was generated by an outburst of GeV emission from Cen X-3— a luminous XRB ($L_x \sim 10^{38}$ergs/s) and suspected TeV source.

All EGRET observations of the Cen X-3 region gathered through Cycle 5, with viewing angles $< 25°$, are listed in table 1. Only one other viewing period, VP 14.0, showed evidence ($\sim 2.3\sigma$) for emission from Cen X-3 that exceeded the 95% confidence level. Comparison of the flux measured dur-

TABLE 1. CGRO/EGRET Observations of Centaurus X-3

Viewing Period	Duration	View Angle (degrees)	Counts[a]	Exposure ($cm^2 s$)	Flux >100 MeV[b] ($\gamma\ cm^{-2} s^{-1}$)
14.0[c]	91/11/14–91/11/27	7.5	69.9±32.3	28.22×10^7	$24.8 \pm 11.4 \times 10^{-8}$
32.0	92/06/25–92/07/01	23.8	<21.8	3.20×10^7	$<68.3 \times 10^{-8}$
230.0	93/07/27–93/08/02	16.0	<19.0	3.39×10^7	$<56.1 \times 10^{-8}$
314.0, 315.0	94/01/03–94/01/22	12.2, 12.2	<48.9	21.96×10^7	$<22.3 \times 10^{-8}$
402.0, 402.5	94/10/18–94/10/30	18.6, 14.8	74.6±18.4	8.15×10^7	$91.6 \pm 22.6 \times 10^{-8}$
522.0	96/06/11–96/06/13	6.6	<18.7	2.78×10^7	$<67.3 \times 10^{-8}$
531.0	96/10/03–96/10/14	8.3	<22.1	8.20×10^7	$<27.0 \times 10^{-8}$
phase 1,2,3	—	—	<184.5	88.23×10^7	$<20.9 \times 10^{-8}$

[a] For non-detections, a 95% limit on the number of possible source counts is quoted.
[b] All flux upper limits are at the 95%-confidence level.
[c] The "detection" quoted for VP 14.0 only has a 2.3σ significance.

ing VP 402.0 and VP 402.5 with the limit placed by joint analysis of all the data accumulated in the first three phases of the mission allows us to reject the constant gamma-ray flux hypothesis at the 99.96% confidence level. Furthermore, comparison with the measurements from VP 314.0 and VP 315.0 indicates variability on a ten month timescale at a confidence level of >99.9%.

To derive the phase-averaged Cen X-3 photon spectrum during the outburst, exposure and count maps were accumulated for 8 *EGRET* energy bands between 70 MeV and 10 GeV. The counts associated with Cen X-3 in each energy band were then derived by modeling the observed count maps with three components: an isotropic diffuse component, a Galactic diffuse component, and a point source at the known position of Cen X-3. Employing the standard *EGRET* spectral fitting procedures, we found the best-fitting power-law is given by an index $\alpha = 1.81 \pm 0.37$ and an integral flux in the range 0.1-10 GeV of $9.2(\pm 2.3) \times 10^{-7}$ photons $cm^{-2}\ s^{-1}$. Extrapolation of these best fitting EGRET spectral parameters predict a flux for >250 GeV gamma-rays of $1.3 \times 10^{-10} cm^{-2} s^{-1}$; a value that is comparable to the peak flux of $\sim 3 \times 10^{-10}$ measured above 250 GeV [5].

At x-ray energies the Cen X-3 emission displays 4.8-second modulation associated with spin of the accreting pulsar. The TeV gamma-ray measurements also suggest the presence of a 4.8 second spin modulation. We therefore searched the Cycle 4 dataset for evidence of spin modulation at GeV energies. After correcting the arrival times for motion of EGRET relative to Cen X-3 and for motion of the pulsar within the binary system, the arrival times were epoch folded according to the x-ray pulse ephemeris. To model the pulse frequency drift during the GeV gamma-ray outburst, we least-squares fit the functional form $\nu(t) = \nu_o + \dot{\nu}_o (t-t_o) + \frac{1}{2}\ddot{\nu}_o(t-t_o)^2$ to the *BATSE* pulse frequency measurements (Finger 1996, private communication)) for a 37 day interval that brackets the EGRET observation. The best fitting function parameters are given by: $\nu_o = 207.62829(\pm 0.00018)$ mHz, $\dot{\nu}_o = -3.004(\pm 0.385) \times 10^{-9}$ mHz

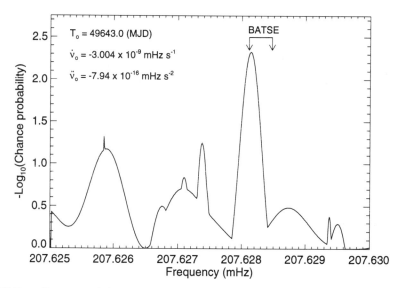

FIGURE 1. H- test statistical estimate of the probability that the gamma-ray emission is modulated at frequencies in the vicinity of the x-ray pulse frequency. The range plotted for the BATSE frequency denotes the formal error derived from our least squares fit to the pulse frequency history.

s^{-1}, $\ddot{\nu}_o = -7.94(\pm 2.49) \times 10^{-16}$ mHz s^{-2}, and $t_o = 49643.0$ MJD.

To test for spin modulation of the GeV gamma-ray emission, we employed the H-test statistic, which makes no assumptions about binning or the shape of the light curve [9]. Test statistics were calculated for modulation at frequency values within one independent frequency step for the accumulation (i.e., (14 days)$^{-1}$=0.00083 mHz) of the BATSE derived ν_o value, while the pulse frequency drift terms, $\dot{\nu}_o$ and $\ddot{\nu}_o$, were fixed at the BATSE derived values. The data for October 1994 data show a test statistic peak at 207.62812($\pm$0.00014) mHz with an amplitude that indicates a chance probability of only 0.47% that the event arrival times are consistent with a flat phasogram. When we apply the Z_2^2 test [4] to the data, we again find a peak at a frequency which is consistent with the x-ray pulse frequency and an amplitude that corresponds to a chance probability for consistency with a flat phase distribution of 0.16%. Since the modulation search was limited to a range about the BATSE value which is the independent frequency step size for the accumulation, the detection significance was not devalued for independent frequency trials. We did, however, search a broader frequency range, 207.625-207.630 mHz, for other peaks and found none with comparable amplitude. Analysis of control datasets for four other locations in the field of view showed no significant peaks in the BATSE pulsation frequency range. Furthermore, removal of the binary ephemeris correction from the Cen X-3 data eliminates the frequency peak. The statistical tests therefore indicate gamma-ray modulation at the Cen X-3 pulsar spin frequency with a significance level comparable to a 3σ detection.

The EGRET observations of Cen X-3 indicate that the GeV gamma-ray flux can vary significantly on a timescale of ten months and arguably represent the first evidence for variable GeV emission from any XRB. Production of the outburst of spin-modulated gamma rays requires, at least, sporadic acceleration of particles to GeV energies within the binary system. The pulsed x-ray flux from Cen X-3 was weak during the GeV flux outburst and the pulsar was undergoing an interval of rapid spin down. However, relationship between the gamma-ray emission and the variable x-ray emission is unclear. Simultaneous multiwavelength observations of Cen X-3 at x-ray through TeV gamma-ray energies are clearly needed if one is going to fully understand particle acceleration and interaction within the system.

REFERENCES

1. Arons, J. and Tavani, M., *Ap. J.* **403**, 249 (1993).
2. Bednarek, W., *A&A* **278**, 307 (1993).
3. Bignami, G. et al., (1977).
4. Buccheri, R., et al. *A&A* **128**, 245 (1983).
5. Carramiñana, A. et al., *Timing Neutron Stars*, Dordrecht:Kluwer, 1989, p. 369.
6. Chanmugam, G. and Brecher, K., *Nature* **313**, 767 (1985).
7. Cheng, K.S. and Ruderman, M., *Ap. J.* **371**, 187 (1991).
8. Day, C.S., and Stevens, I.R., *Ap. J.*, **403**, 322 (1993).
9. deJager, O.C., et al. *A&A* **221**, 180 (1989).
10. Eichler, D. and Vestrand, W.T., *Nature* **307**, 613 (1984).
11. Eichler, D. and Vestrand, W.T., *Nature* **318**, 345 (1985).
12. Finger, M.H., Wilson, R.B., and Fishman, G.J., *The Second Compton Symposium*, New York:AIP, 1994, p. 304.
13. Finger, M.H. and Prince, T. A., these proceedings (1997).
14. Giaconni, R. et al., *Ap. J.* **167**, L67 (1971).
15. Katz, J.I. and Smith, I.A., *Ap. J.* **326**, 733 (1988).
16. Kazanas, D. and Ellison, D.C., *Nature* **319**, 380 (1986).
17. Krzemiński, W., *Ap. J.* **192**, L135 (1974)
18. Kluzniak, W. et al., *Nature* **336**, 558 (1988).
19. Lamb, F.K., Pethick, C.J., and Pines, D., *Ap. J.*, **184**, 271 (1973).
20. Hillas, M., *Proc. 19th Int. Cosmic Ray Conference* **3**, 455 (1985).
21. North, A.R., et al., *Proc. 21th Int. Cosmic Ray Conference* **2**, 274 (1990).
22. Pringle, J.E. and Rees, M., *A&A* **21**, 1 (1972).
23. Schreier, E., et al., *Ap. J.* **172**, L79 (1972).
24. Vestrand, W.T., *Ap. J.* **271**, 304 (1983).
25. Vestrand, W.T., and Eichler, D., *Ap. J.* **261**, 251 (1982).
26. Vestrand, W.T., Sreekumar, P., and Mori, M., *Ap. J.* **484**, L49 (1997).

GRO J2058+42 X-Ray Observations

Colleen A. Wilson*, M.H. Finger*, B.A. Harmon*, R.B. Wilson*, D. Chakrabarty[†], and T. Strohmayer[††]

NASA Marshall Space Flight Center, Huntsville, AL 35812
[†] *Center for Space Research, Massachusetts Institute of Technology, Cambridge, MA 02139*
[††] *USRA, Laboratory for High Energy Astrophysics, NASA/GSFC, Greenbelt, MD*

Abstract. GRO J2058+42, a transient 198 second x-ray pulsar, was discovered by BATSE during a giant outburst in 1995 September-October. The total flux peaked at about 300 mCrab (20-50 keV) as measured by Earth occultation. The pulse period decreased from 198 s to 196 s during the 46-day outburst. BATSE has observed five additional weak outbursts from GRO J2058+42 with peak pulsed fluxes of about 15 mCrab (20-50 keV) that were spaced by about 110 days. The periodicity of the outbursts allowed a ToO observation by the *Rossi X-ray Timing Explorer (RXTE)* Proportional Counter Array (PCA) which localized the source to a 4' error circle. We present pulse frequency history, flux histories from pulsed and Earth occultation data, and representative pulse profiles.

INTRODUCTION

A periodic signal at about 198 seconds was observed in the BATSE data starting on 14 September, 1995. At the same time a new source was brightening in the Earth Occultation data. A location (figure 1) was determined from the pulsed data and the non-pulsed data with an error box of about 4°×1° [1]. A *CGRO* target of opportunity was declared and the spacecraft was reoriented to allow scans of the region by OSSE. The error box (figure 1) was improved to 30'×60' [2]. The total flux (figure 2) peaked at about 300 mCrab (20-100 keV) and the pulsed flux (figure 2) peaked at 140 mCrab (20-50keV) on 27 September 1995. The bright outburst continued until 30 October 1995. A search of archival BATSE data from April 1991 until the bright outburst showed no previous outbursts.

An analysis of data following the bright outburst initially revealed three much weaker outbursts lasting about two weeks with pulsed fluxes peaking at 10-15 mCrab (20-50 keV). These outbursts were spaced by about 110 days (figure 2). From this regular spacing, the peak of the next outburst was predicted to be on November 28, 1996. A public target of opportunity scan of

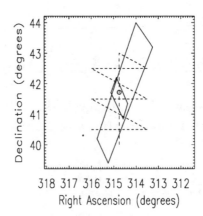

FIGURE 1. History of localization of GRO J2058+42. The large error box was determined from BATSE pulsed and non-pulsed data [1]. An OSSE scan of the BATSE error box yielded the smaller error box [2]. The $RXTE$ PCA scan followed the path shown by the dashed line. The final location by $RXTE$, a 4' error circle, is shown by the small circle [3].

the error box was performed on November 28 with the $RXTE$ PCA yielding a 90% confidence 4' error circle (figure 1) centered on (J2000) R.A. = $20^h\ 59^m.0$, Decl. = +41°43' [3]. (See Wilson et al. in preparation for a description of the PCA observation.) BATSE was also detecting the source from November 23-December 1, 1996. An additional outburst was detected by BATSE about 110 days later (March 16-20, 1997). An archival search of $ROSAT$ data found no sources within the error circle (J. Greiner, 1997, private communication). No optical counterpart has been found to date.

OBSERVATIONS

Figure 2 shows histories of spin frequency, pulsed intensity, and total (pulsed +unpulsed) flux determined at 4-day intervals. Pulse profiles were determined from the BATSE 20-50 keV DISCLA (1.024 second) data for a range of trial frequencies to search for a signal from GRO J2058+42. To generate each profile, the data are fit with a background model plus a sixth order Fourier expansion in pulse phase model for short intervals. In each four day span, the pulse profiles from the short intervals are shifted in phase and summed for a range of trial frequency offsets from the pulse phase model. The best fit frequency is determined by the Z_6^2 test which measures the significance of the first 6 Fourier amplitudes. For GRO J2058+42, root-mean-square (RMS) pulsed fluxes are estimated from the best fit pulse profile as $F_{RMS} = \left[\int_0^1 (F(\phi) - \bar{F})^2 d\phi\right]^{1/2}$ where $F(\phi)$ is the flux of the pulse profile at phase ϕ, $0 \leq \phi \leq 1$, and

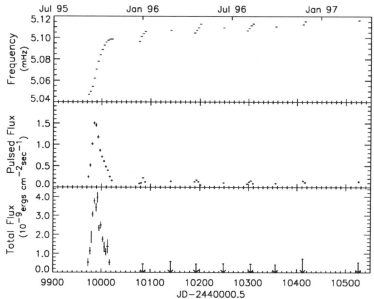

FIGURE 2. GRO J2058+42 frequency, pulsed flux (20-50 keV), and total flux (20-50 keV) determined at 4-day intervals. The spin frequencies have been barycentered, but have not been orbitally corrected since the orbit is unknown. Upper limits (3σ) are shown only for the peak total flux of the weak outbursts.

$\bar{F} = \int_0^1 F(\phi)d\phi$ is the average flux. The pulsed fluxes in the center panel of figure 2 were determined at 4-day intervals, assuming an exponential spectrum with an e-folding energy of 20 keV. See [4] for a detailed description of the pulse frequency and pulsed flux estimation techniques. The bottom panel of figure 2 shows the total 20-50 keV flux as measured by BATSE Earth occultation. The Earth occultation method measures the intensity of a known source by calculating the difference in total count rate in source facing detectors just before and just after source occultation by the Earth. This method measures the total flux from a source, including pulsed and unpulsed components [5]. The fluxes were determined at 4-day intervals by fitting count rates with an assumed exponential spectrum with an e-folding energy of 20 keV. Upper limits (3σ) for 4-day sums are shown for the peak fluxes of the weak outbursts.

GRO J2058+42 experienced a large spin up during its initial 46 day outburst (figure 2). The pulse period changed by about 2 seconds across this outburst, corresponding to an average $P/\dot{P}$ of 12 years. Measurements of the spin-up rate give a peak value of $2.48 \pm .02 \times 10^{-11}$ Hz s^{-1} (not orbitally corrected) which is comparable to peak spin-up rates seen by BATSE in giant outbursts from known Be transients [4]. The large spin-up rate and brightness of this outburst compared to later outbursts leads us to classify the initial outburst from GRO J2058+42 as a giant outburst [6]. The bright initial outburst is followed by a series of much weaker outbursts, too dim to be unambiguously detected in Earth occultation data. The weak outbursts appear to alternate in duration and intensity (figure 2), producing brighter outbursts every 110 days with dimmer outbursts halfway between. The All-Sky Monitor (ASM) on the *RXTE* shows outbursts of approximately equal intensity and duration every 55 days [7]. No known systematic effects would remove or weaken outbursts in the BATSE data at 55 day intervals. However, the ASM measures total (pulsed+unpulsed) flux, while the BATSE measurements are of pulsed flux only, hence energy dependent changes in pulse fraction could cause the observed differences. See Wilson et al. in preparation for a full description of BATSE and ASM observations.

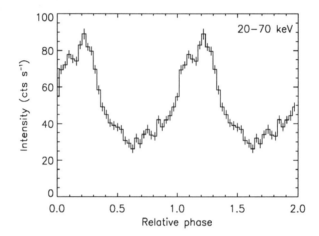

FIGURE 3. A pulse profile for 20-70 keV from September 23-27, 1995, near the peak of the giant outburst. The average count rate from occultation data for September 23-27,1995 has been added to the count rate plotted, but has not been included in the errors.

Pulse profiles for four energy bands, 20-30, 30-40, 40-50, and 50-70 keV were generated by epoch-folding four day intervals of BATSE CONT (2.048 second) data into 32 phase bins using a quadratic phase model, with no orbital model. Figure 3 shows a pulse profile for 20-70 keV from near the peak of the giant outburst of GRO J2058+42. Epoch-folded subsets of the September 23-27 interval also show a dip at phase ≈ 0.15, just preceding the peak of the profile emission. Our analysis has revealed preliminary evidence for profile variations during the giant outburst (See Wilson et al. in preparation for details). The pulse fraction appears to be constant across the giant outburst. An exponential photon spectrum was well fit to the total flux as measured by earth occultation and the pulsed flux, with consistent best fit e-folding energies of 15-22 keV. No spectral trends with intensity were seen in the BATSE data from the giant outburst.

DISCUSSION

Long term studies of Be/X-ray binaries have demonstrated that giant outbursts followed by or intermixed with a series of periodic normal outbursts appear to be typical behavior for Be/X-ray binaries [4]. GRO J2058+42 is most likely a Be/X-ray pulsar because it exhibits transient outbursts recurring with presumably an orbital period of 110 days and it has both giant and normal outbursts. The odd-even effects in the intensity of the normal (weak) outbursts seen by BATSE suggest that we are seeing two outbursts per 110 day orbit. An inclined neutron star orbit out of the equatorial plane of the companion, which could produce two outbursts per orbit, has often been suggested as a mechanism for outbursts in Be/X-ray transients [8] . However, two outbursts per orbit have not been seen in other Be transients to date.

A Corbet diagram shows the spin period of accreting neutron stars versus the binary orbital period [9], [10]. On this diagram, the spin-periods of the Be transients are strongly correlated with their orbital periods. Two other types of systems, neutrons stars orbiting Roche lobe filling supergiants and neutron stars orbiting underfilled Roche lobe supergiants, fall into regions on the diagram. An orbital period of 110 days places GRO J2058+42 along the orbital period vs. spin period correlation for Be/X-ray binaries.

REFERENCES

1. Wilson, C. A. et al. **IAUC 6238** (1995)
2. Grove, J. E. **IAUC 6239** (1995).
3. Wilson, C. A., Strohmayer, T., & Chakrabarty, D. **IAUC 6514** (1996)
4. Bildsten, L. et al. *ApJS* (1997), accepted.
5. Harmon, B. A., et al., in *Compton Observatory Science Workshop*, ed. C. R. Schrader, N. Gehrels, & B. Denis, *NASA Conf. Publ.* **3137**, 69 (1992).
6. Stella, L., White, N. E., & Rosner, R. *ApJ* **308**, 669 (1986).
7. Corbet, R., Peele, A., & Remillard, R. **IAUC 6556** (1997).
8. Priedhorsky, W. C. & Holt, S. S. *Space Science Reviews* **45**, 291 (1987).
9. Corbet, R. H. D. *MNRAS* **220**, 1047 (1986).
10. Waters, L. B. F. M. & van Kerkwijk, M. H. *A&A* **223**, 196 (1989).

Observation of a Long Term Spin-up Trend in 4U1538-52

Brad C. Rubin[1,2], M.H. Finger[3,4,5], D.M. Scott[3,4], R.B. Wilson[3]

[1] *RIKEN Institute, Wako-shi, Saitama 351-01 Japan*
[2] *Department of Physics, University of Alabama in Huntsville, Huntsville, AL 35899*
[3] *NASA/Marshall Space Flight Center, Huntsville, AL 35812*
[4] *Universities Space Research Association*
[5] *CGRO Science Support Center*

Abstract.

The high mass x-ray pulsar 4U1538-52 has been continually monitored by the BATSE (Burst and Transient Source Experiment) on CGRO (Compton Gamma-ray Observatory). The frequency history reveals that a reversal of the long term trend in the accretion torque, from spinning down to spinning up, has occurred, probably in 1988. This is the first time a long term change is known to have occurred for this source. The magnitude of the average $|\dot\nu/\nu| \sim 10^{-11} s^{-1}$ over an interval of ~ 5 years is similar during spin-down and spin-up. Shorter term pulse frequency variations of either sign are also observed. The power density spectrum of fluctuations in angular acceleration is consistent with white noise on timescales from 16 to 1600 days. It can be well fit with a power law with power law index 0.10 ± 0.21 and white noise strength $(7.6 \pm 1.6) \times 10^{-21} (Hz/s)^2 Hz^{-1}$.

INTRODUCTION

The pulse period of the eclipsing binary X-ray pulsar 4U1538-52 (pulse period ~ 530 seconds, orbital period ~ 3.7 days) were sporadically observed between 1976 and 1988 by *Ariel* [8], OSO-8 [1], EXOSAT [14], HEAO A-2 [17], *Tenma* [12], and *Ginga* [6]. Archival results from *Uhuru*, HEAO A-1, and EXOSAT have also been analyzed [5]. Optical observations of the companion, QV Nor [7] [13], have helped to establish that it is a 14.5 mag B0 supergiant which is not overfilling its Roche lobe.

The sporadic pulse period observations from these experiments indicated a long term spin-down trend, lasting at least 10 years, with rapid and probably random pulse period variations on shorter time scales. Here we report observing, for the first time, a long term spin-up trend. This trend has continued for at least four years, with occasional shorter term reversals. The nearly continuous observations made by BATSE also make possible an analysis of the torque variation process. This can be accomplished through a measurement of the power density spectrum of the intrinsic fluctuations in angular acceleration. The results of these observations are similar to results which have previously been obtained for Vela X-1 [3] [11]. Further details about the present work, including a determination of orbital epochs and a new limit on the rate of orbital period change can be found in [16].

OBSERVATIONS

We have performed an epoch-folding analysis of the pulsed signal, obtaining the pulse profile, pulse frequency, and pulsed intensity for 4U1538-52 using background-subtracted BATSE LAD Discriminator data in the 20-50 keV band. Background due to diffuse cosmic, atmospheric and other hard x-ray sources has been subtracted using a global model which is updated every two CGRO orbits (about three hours) [15].

Background subtracted (count) data for each detector with normal less than 90 degrees to the source were weighted by the square of the cosine of the angle between the source and the normal and summed over detectors. This form is a very good approximation to the detector response, and represents the response better than a projected area weighting. These data were fit to a sixth order harmonic expansion in the pulse phase during intervals between source rise and source set (~ 4000 seconds) during one CGRO orbit. A solar system barycenter ephemeris and preliminary binary ephemeris from [12] were used in constructing the phase model. Further details about the epoch-folding method appear in [2].

To observe a detectable signal, it is necessary to sum these harmonic amplitudes coherently over time intervals longer than 4000 seconds. We have found that 16 days is a sufficiently long detection interval for most of the mission data. The mean pulse frequency for each 16 day interval is obtained by varying the pulse phases in each set of harmonic amplitudes according to a set of search frequencies and locating a maximum in the significance of the coherently summed amplitudes.

The pulse frequencies are plotted as a pulse period history in Figure 1a and shown together with all data from previous missions in Figure 1b. There are some intervals in the BATSE data during which the source is undetectable. The largest of these intervals (about TJD 8640-8720) occurs during an interval of sparse data coverage due to tape recorder malfunction.

The mean pulse profile is converted to a pulse template by normalizing it to a root mean square of 1.0, as shown in Figure 2(a). Intervals during which the source is in binary eclipse were removed before constructing this profile. The profile is similar to the profiles observed by Ginga in the 20-40 keV band [4]. The weakness of the secondary peak is partly due to its virtual disappearance around 20 keV because of the phase dependent cyclotron absorption line present there.

The source intensity is determined by correlating this template with the pulse profile during each segment. The pulsed flux history during the 1600 day span of our observations is shown in Figure 2(b), where each data point is a sum over the uneclipsed data during a 16 day interval. The average pulsed flux as a function of orbital phase is obtained from a sum of the data in each orbital phase bin, and is shown in Figure 2(c). In this analysis we have assumed that the pulse profile shape does not vary significantly. From Figure 2(b) we see that the average flux varies by factors of 2 to 4 and that there does not appear to be a correlation between the flux and the spin frequency changes in Figure 1(a). The mean pulsed rms energy flux of all of the uneclipsed data in the 20-50 keV energy band is 1.6×10^{-10} ergs/cm^2/s. This corresponds to a luminosity of 5×10^{35} ergs/s for an assumed source distance of 5.5 kpc. Figure 2(c) shows clear evidence for eclipse of the source at the expected binary phase, and little significant intensity variation away from the eclipse. The eclipse duration is consistent with 0.59 ± 0.01 days measured by [6].

POWER SPECTRUM OF THE FLUCTUATIONS

If the pulse frequency variations are either non-deterministic in nature and/or are unresolved by measurement then statistical methods must be used to characterize these variations. Orthonormal polynomial power density spectral estimators have been developed for the purpose of interpreting pulse frequency variations in Vela X-1 [9] [10] [11]. Here the same method is used to characterize the fluctuations in angular acceleration of the neutron star in 4U1538-52.

The power density spectrum we have obtained from the BATSE data is

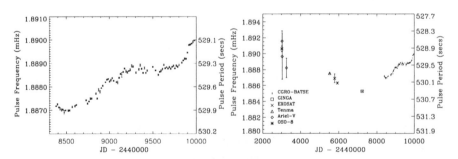

FIGURE 1. Pulse Frequency History. (a) BATSE data. Each data point represents one 16 day interval. Gaps indicate that the source was not detectable during a 16 day interval. The large gap around JD2448700 is associated with sparse data coverage. (b) All known data.

shown as a function of frequency (inverse timescale, not to be confused with pulse frequency), in Figure 3. Quadratic estimators have been applied to the pulse frequencies. At high frequencies, excess power caused by pulse shape variations may be expected, as has been observed in Vela X-1. However, frequencies high enough to observe this upturn are not being probed. Instead, the spectrum appears to be independent of frequency ($\propto f^0$) over the range of accessible timescales.

We find a power law slope $\alpha = 0.10 \pm 0.21$ and a mean rms noise level $S = (7.6 \pm 1.6) \times 10^{-21}$ $(\text{Hz/s})^2 \text{Hz}^{-1}$ with a $\chi^2 = 3.1$ for 4 degrees of freedom. This result implies that there are significant pulse frequency variations on all timescales accessible to the BATSE observations and that these variations are consistent with a white noise model in the pulse frequency derivative. The observed noise level for 4U1538-52 is of the same order of magnitude as that found for Vela X-1, $(1.6 \pm 0.5) \times 10^{-20} (\text{Hz/s})^2 \text{Hz}^{-1}$ [3].

REFERENCES

1. Becker, R. H., et al. 1977 *ApJ*, 216, L11
2. Bildsten, L., et al. 1997, to appear in *ApJS 113*
3. Boynton, E. P., 1984, *ApJ*, 283, L53
4. Clark, G. W., et al. 1990, *ApJ*, 353, 274
5. Cominsky, L. R., & Moraes, F. 1991, *ApJ*, 370, 670
6. Corbet, R. H. D., Woo, J. W., & Nagase, F. 1993, *Astronomy & Astrophysics*, 276, 52
7. Crampton, D., Hutchings, J. B., & Cowley, A. P. 1978, *ApJ*, 225, L63
8. Davison P. J., Watson M. G., & Pye J. P. 1977 *MNRAS*, 181, 73P
9. Deeter, J. E., & Boynton, P. E. 1982, *ApJ*, 261, 337
10. Deeter, J. E., 1984, *ApJ*, 281, 482
11. Deeter, J. E., et al. 1989, *ApJ*, 336, 376

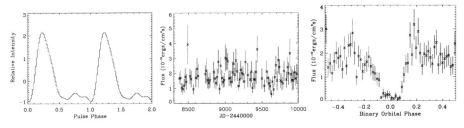

FIGURE 2. (a) Mean Pulse Profile. The average is over all uneclipsed data. The area is normalized to 1.0 and errors are at the 1% level. (b) Pulsed Flux History. Each data point is the average of 16 days of uneclipsed data. Missing points indicate 16 day intervals during which the source was not detected. (c) Mean pulsed flux as a function of orbital phase. Each data point is an average of 1/64 of the orbital phase.

12. Makishima, K., et al. 1987, *ApJ*, 314, 619
13. Reynolds, A. P., Bell, S. A., & Hilditch, R. W. 1992, *MNRAS*, 256, 631
14. Robba, N. R., et al. 1992, *ApJ*, 401, 685
15. Rubin, B. C., et al. 1996 *Astronomy & Astrophysics Supplements*, 120, 687
16. Rubin, B. C., et al. 1997, to appear in *ApJ*, 488
17. White, N. E., Swank, J. H., & Holt, S. S. 1983, *ApJ*, 270, 711

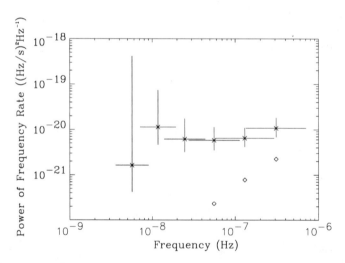

FIGURE 3. Power density spectrum of the pulsed frequency derivative as a function of frequency. Horizontal error bars represent the $\pm 1\sigma$ interval in the frequency response of each estimator. Vertical error bars represent the 90% confidence interval in the power estimate. Diamonds show the expected level of pulse shape noise. For data points with no noise estimate, it was too small to appear on the plot.

EGRET Observations of X-Ray Binaries

B.B. Jones, Y.C. Lin, P.F. Michelson, P.L. Nolan,
M.S.E. Roberts, and W.F. Tompkins

Stanford University

Abstract. Several known low- and high- mass X-ray binaries with orbital periods between tens of minutes and several days were examined in EGRET data. Instrument thresholds for orbital modulation were calculated, and appropriate observations were analyzed with new software which uses maximum likelihood methods for signal detection, fully accounting for variations in EGRET exposure, as well as the structured gamma-ray background and the energy-dependent point-spread function. While some sources were not detected by EGRET at all, several were marginally detected with less than 5 σ significance. However, due to significant background and low statistics, the observations were unable to constrain the modulation fraction of any of the sources.

INTRODUCTION

The search for orbital modulation of X-ray binary flux in gamma rays has been hampered by the time scales involved. Most orbital periods are between tens of minutes and several hours. This is comparable to, or longer than, the typical time scale of exposure changes in EGRET; thus any attempt to resolve orbital flux modulation must take explicit account of exposure changes. Meanwhile, gamma-ray fluxes of X-ray binaries are low enough that direct observation of a single period of modulation is impossible. As with pulsar analyses, epoch folding offers a way to improve the signal-to-noise ratio and increase chances for the detection of orbital flux modulation. However, not only must the arrival times of photons be epoch folded, but also the changing exposure must be folded at the same period and phase to find accurate fluxes.

METHODS

In order to fully integrate all observational information, a maximum likelihood method was used to search for orbital flux modulation. This method also allows detection thresholds to be calculated. Because simultaneous X-ray and gamma-ray ephemerides are not usually available, a small area of parameter space (in p and $\dot{p}$) is searched for each source.

Detection Method

The maximum likelihood method that we have implemented epoch folds photons and exposure and includes point-spread function and structured background information. The details of the structured background are taken from Bertsch et al. [1]. This information is stored and used in a way which facilitates searching over a range of period and period derivative parameters. The algorithm does a likelihood comparison of a light-curve model consisting of an m-bin flux model versus a null model. These m fluxes are allowed to vary independently to maximize the likelihood of the given data, which is then compared to the null likelihood—the likelihood of the data under the best fit flat (i.e. 1-bin) model. If the likelihood of the best light-curve model is $\mathcal{L}_{LC}$, and the likelihood of the best null model is $\mathcal{L}_{null}$, then we can define

$$C = 2(\ln \mathcal{L}_{LC} - \ln \mathcal{L}_{null}) \qquad (1)$$

Wilks' Theorem shows that C is distributed as $\chi^2(m-1)$ in the null model [2]. The distribution of C in the null model was measured using known non-periodic data, and was indeed $\chi^2(m-1)$.

Threshold Calculation

The results of such a search must take into account the thresholds for detection of orbital flux modulation. The exact value of the threshold is a function of the structure of the background, various instrument parameters, and the shape and duty cycle of the orbital light curve. While it may be hoped that some X-ray binaries have duty cycles as short as 20%, it is more likely that gamma emission is fairly constant, then drops sharply but briefly during eclipse [3]. A sinusoid assumption is a compromise between these two extremes. For the purposes of threshold calculation, the light curve is assumed to be sinusoidal, and the total number of photons (source + background) is assumed to be large enough that the errors are Gaussian. This assumption has been verified in almost every case. The expectation value of the maximum likelihood is then calculated as a function of total photons per bin and modulation fraction. The 99% threshold is the locus of points in parameter space where one-half of all possible data sets would yield a detection with 99% significance, assuming a typical number of 2000 trials are used during the search process.

RESULTS

Most X-ray binaries [3] have source-to-background ratios so low that even if their orbital modulation fraction were 100%, the modulation would be undetectable. It would still be possible to detect modulation in such sources if the duty cycle were sufficiently short, but under the standard model of X-ray binary emission, this is unlikely. For candidate sources that have parameters near threshold, a maximum likelihood period search is done. The sinusoidal assumption is dropped in favor of a five independent bin light curve model. The choice of five bins was made to maximize flexibility in the model while retaining sufficient numbers of photons in each bin. Photons are barycenter corrected and epoch folded with trial periods in a small range (± 10-20%) of the known X-ray orbital periods. 27 observations of 16 promising low-mass X-ray binaries (LMXBs) and 7 observations of 4 promising high-mass X-ray binaries (HMXBs) yielded no periodic signal detections significant at the 99% level. The high-mass X-ray binary results are displayed in Table 1, and the low-mass X-ray binary results are displayed in Table 2.

CONCLUSION

The EGRET data was analyzed for orbital periodic signals from X-ray binaries. In most cases, the source-to-background ratio was low enough to place any possible sinusoidal variation below threshold; that is, Poisson fluctuations in the background were large enough to overwhelm the expected signal. Several sources were analyzed despite a low signal-to-noise ratio, in case they were to display variation with a very short duty cycle. Only Cyg X-3, which has been extensively studied in gamma rays [4-6], was bright enough to have a non-negligible chance of being detected, assuming a sinusoidal light curve. No evidence for variation was found in any of the sources.

REFERENCES

1. Bertsch, D.L., et al. *ApJ* **416**, 587 (1993).
2. Eadie, W.T., Drijard, D., James, F. E., Roos, M., & Sadoulet, B. *Statistical Methods in Experimental Physics* Amsterdam: North-Holland, 1971.
3. White, N.E., Nagase, F., and Parmar, A.N., *X-Ray Binaries*, ed. H.G. Lewin, J. van Paradijs, and E.P.J van den Heuvel, Cambridge: Cambridge University Press, 1995.
4. Lamb, R.C., et al. *ApJ* **212**, L63 (1977).
5. Michelson, P.F., et al. *ApJ* **401**, 724 (1992).
6. Mori, M. et al. *ApJ* **476**, 842 (1997).

TABLE 1. High mass X-ray binaries. For each source, the viewing period of the observation is listed, along with the average number of photons in each phase bin, the modulation fraction that would be measured if all the source flux were modulated, and the threshold modulation fraction that would yield a 99% significance detection in half of all possible data sets.

Name	VP	Photons/bin	Max Modulation	Threshold
X0532–664 = LMC X-4	6.0	44	16.1%	50%
	17.0	52	13.6%	45%
	224.0	12	37.3%	80%
X0538–641 = LMC X-3	6.0	42	7.9%	50%
X0540–697 = LMC X-1	17.0	55	24.0%	43%
	6.0	46	10.9%	47%
X1119–603 = Cen X-3	14.0	164	3.7%	27%
	402.5	31	18.6%	57%
	402.0	27	19.3%	57%
	208.0	19	21.5%	70%
	215.0	8	22.3%	>80%
X1538–522 = QV Nor	516.1	20	1.2%	70%
X1700–377 = HD153919	508.0	36	5.4%	54%
X1956+350 = Cyg X-1	318.1	60	7.3%	44%
	601.1	36	15.8%	54%
X2030+407 = Cyg X-3	203.0	476	13.3%	15%
	2.0	296	14.4%	21%
	7.1	133	10.8%	30%
	212.0	243	14.8%	23%
	303.2	77	16.1%	38%
	328.0	67	27.9%	42%
	331.0	30	26.4%	57%
	331.5	48	15.9%	47%
	333.0	64	3.6%	42%
	601.1	32	26.1%	57%
	34.0	54	5.1%	47%
	p12[a]	1039	10.6%	12%

[a] Combined data from Phases 1 and 2

TABLE 2. Low mass X-ray binaries. For each source, the viewing period of the observation is listed, along with the average number of photons in each phase bin, the modulation fraction that would be measured if all the source flux were modulated, and the threshold modulation fraction that would yield a 99% significance detection in half of all possible data sets.

Name	VP	Photons/bin	Max Modulation	Threshold
X0543−682 = Cal 83	329.0	8	14.4%	>80%
X0547−711 = Cal 87	224.0	13	33.2%	75%
	17.0	57	13.6%	45%
X1124−685 = N'Mus 91	230.0	22	14.6%	65%
X1323−619 = 4U1323-62	23.0	63	4.7%	40%
X1455−314 = Cen X-4	217.0	14	15.3%	75%
X1625−490	23.0	94	7.2%	35%
	529.5	76	9.0%	40%
XB1636−536	27.0	135	1.8%	30%
X1656+354 = Her X-1	9.2	37	6.7%	50%
XB1658−298	232.0	226	2.0%	25%
	5.0	438	1.2%	∼10%
X1659−487 = GX339-04	336.5	98	2.6%	35%
	270.0	136	5.3%	30%
	226.0	127	2.1%	30%
	323.0	178	2.1%	27%
	210.0	31	4.4%	60%
	214.0	38	7.6%	52%
	5.0	246	2.2%	22%
	219.0	12	18.8%	75%
	302.3	68	2.2%	40%
	423.0	54	1.7%	45%
X1735−444	226.0	183	0.4%	27%
X1755−338	226.0	183	0.7%	27%
	229.0	18	4.9%	68%
X1820−303	323.0	260	0.4%	17%
X1822−371	508.0	47	2.5%	47%
	529.5	46	3.1%	47%
X1908+005 = Aql x-1	43.0	24	14.7%	60%
X1957+115 = 4U1957+11	331.5	42	3.1%	50%
X2023+338 = V404 Cyg	303.2	62	13.4%	42%
X2127+119 = AC211	19.0	46	1.6%	47%

Observations of Vela X–1 with RXTE

P. Kretschmar*, I. Kreykenbohm*, R. Staubert*, J. Wilms*,
M. Maisack*, E. Kendziorra*, W. Heindl†, D. Gruber†,
R. Rothschild† and J. E. Grove‡

*Institut für Astronomie und Astrophysik – Astronomie, Univ. of Tübingen, Germany
†Center for Astrophysics and Space Science, UCSD, La Jolla, CA, USA
‡E.O. Hulburt Center for Space Research, NRL, Washington DC, USA

Abstract. We present results from observations of Vela X–1 obtained in February 1996 with *RXTE*. We have fitted phase averaged spectra with several models and compared our results with prior results obtained with *HEXE* and *Ginga*. No acceptable fit is possible with the models normally applied to X-ray binaries, instead we found a best fit using a power law with a Fermi-Dirac cutoff modified by one or two cyclotron absorption lines at $\sim 25\,\mathrm{keV}$ and $\sim 58\,\mathrm{keV}$. In phase resolved spectral analysis we find that the cyclotron absorption is detectable only in the pulse, and is deeper in the first stronger pulse.

INTRODUCTION

Vela X–1 is an eclipsing high mass X-ray binary with an orbital period of 8.94 days and a spin period of approximately 283 seconds. The partner is the massive B0.5 star HD77581.

Vela X–1 has been observed and studied in depth in the past years with most X-ray instruments. The observations with *RXTE* offer the chance to observe over a broad energy range (5–200 keV) with a large detector area. The spectrum of Vela X–1 has in the past been successfully described by the 'standard spectrum' of X-ray pulsars: a power law with exponential cutoff. Cyclotron absorption features have been reported from observations with *HEXE* at $\sim 50\,\mathrm{keV}$ [1], while *Ginga* observed a spectral feature at $\sim 25\,\mathrm{keV}$ [4]. These results have been supported by the analysis of the combined data from *HEXE* and *TTM* [2,3] and in the detailed analysis of *Ginga* data [5].

RXTE observed Vela X–1 five times with a total of about 5 ksec observation time each, uniformly spaced with orbital phase (see Table 1). Observations #2 to #5 have been analyzed.

TABLE 1. Orbital phase of the 5 *RXTE* observations of Vela X-1.

Obs. No.	#1	#2	#3	#4	#5
Date	20.2.97	22.2.97	23.2.97	25.2.97	27.2.97
Orbital phase	0.097–0.108	0.311–0.324	0.492–0.509	0.701–0.717	0.885–0.895

APPLIED MODELS

We have tried to fit a large number of spectral models to the data, including bremsstrahlung and comptonized spectra. All spectral fits were done including the effects of photoelectric absorption and an additive iron line. Due to problems with the background model and the calibration of the *PCA* the iron line parameters found are inconclusive. To avoid unphysical results, they were constrained to $E \equiv 6.4\,\mathrm{keV}$ and $\sigma \leq 0.4$.

Using the most common model for X-ray binaries, a power law with high energy cutoff (HECUT) we always found features that could be interpreted as absorption lines around the cutoff energy. The same effect has been observed in the analysis of *RXTE* spectra of Her X-1 [6] where such a feature has never been reported before. This led us to suspect an artificial feature caused by the sharp turnover introduced by HECUT (see Fig. 1)

As alternatives we used two models with continous turnover. The first is a power law with a so-called "Fermi-Dirac" cutoff (FDCO), first mentioned by Tanaka [7]; the second is the NPEX model, which contains the sum of two powerlaws multiplied by an exponential factor. This model was used for the analysis of *Ginga* data by Mihara [5]. Table 2 gives the analytical expressions for these models.

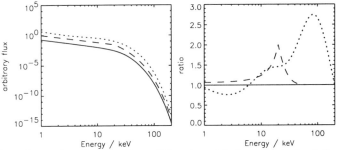

FIGURE 1. Comparison between three different models for a spectral cutoff: power law with Fermi-Dirac cutoff (——), power law with high energy cutoff (– – –) and NPEX model (· · ·). For clarity, the graphs of the latter two models in the left panel have been shifted by factors of 5 and 25, respectively. The right panel displays the ratios of HECUT, respective NPEX to FDCO

TABLE 2. Analytical expressions for the three models discussed in the text.

HECUT	FDCO	NPEX
$E^{-\alpha} \times \begin{cases} 1 & ; E < E_{\text{cut}} \\ e^{-\frac{E-E_{\text{cut}}}{E_{\text{fold}}}} & ; E > E_{\text{cut}} \end{cases}$	$E^{-\alpha} \times \left(\dfrac{1}{e^{\frac{E-E_{\text{cut}}}{E_{\text{fold}}}} + 1} \right)$	$A \times (E^{-\alpha_1} + B \times E^{+\alpha_2})$ $\times e^{-E/kT}$

PHASE AVERAGED SPECTRA

After excluding other models we fitted the phase averaged spectra with the three models explained in the previous section. The HECUT model is not able to describe the data, with significant residuals suggesting broad line-like features. The FDCO spectrum, which avoids the problems of a sharp onset of the cutoff, gives less marked residuals but still fails to describe the data. Similar results are obtained with the NPEX model.

With all continua a good fit is achieved by including cyclotron absorption line features. Our fits show that the usual assumption of the energy of the second harmonic cyclotron absorption feature being twice the energy of the first does not result in satisfactory modeling of the data. To obtain a good fit we had to introduce a free scaling factor $f > 2$ with $E_2 \equiv f \cdot E_1$ for the line centroid energies and correspondingly for the line widths.

Comparing the different continua, we found that the FDCO model usually gives a slight improvement over the HECUT model and achieves fits as good or better as the NPEX model, which has one parameter more. Therefore we concentrated on the FDCO model. We note that with the NPEX continuum our results are in very good agreement with those obtained from *Ginga* observations [5] as long as we limit the energy range to 5–50 keV.

TABLE 3. Results of fits for the phase averaged spectra using an FDCO continuum with cyclotron lines.

Component	Parameter	obs #2	obs #3	obs #4	obs #5
Absorption	n_H	$0^{+2.6}_{-0.0}$	$0.8^{+3.4}_{-0.8}$	$22.0^{+1.5}_{-1.5}$	$25.7^{+1.9}_{-1.8}$
Power law	α	$0.81^{+0.09}_{-0.03}$	$1.33^{+0.13}_{-0.10}$	$1.60^{+0.04}_{-0.03}$	$0.73^{+0.03}_{-0.03}$
FD-cutoff	E_{cut}	$26.7^{+1.0}_{-2.0}$	$34.9^{+2.2}_{-2.5}$	$37.2^{+2.8}_{-3.5}$	$23.4^{+2.5}_{-6.5}$
	E_{fold}	$8.5^{+0.2}_{-0.2}$	$8.3^{+0.8}_{-0.9}$	$8.3^{+1.5}_{-1.5}$	$10.9^{+0.5}_{-0.3}$
1st Cyclo. line	E_{cyc}	$24.1^{+0.4}_{-0.5}$	$22.7^{+1.3}_{-1.2}$	$23.1^{+2.0}_{-1.6}$	$24.2^{+0.4}_{-0.5}$
	width (fixed)	5.0	5.0	5.0	5.0
	depth	$0.22^{+0.04}_{-0.04}$	$0.11^{+0.06}_{-0.04}$	$0.13^{+0.06}_{-0.05}$	$0.18^{+0.05}_{-0.04}$
2nd Cyclo. line	depth	$0.29^{+0.9}_{-\infty}$	—	—	$0.18^{+0.05}_{-0.04}$
	factor f	$2.43^{+\infty}_{-0.4}$	—	—	$2.25^{+0.17}_{-0.09}$
χ^2_{red}		0.56	0.61	0.43	0.60

Table 3 gives the fit parameters for the four observations. Because of the limited statistical quality of our data we could not determine the width and the depth of the lines independently. We settled for a fixed width of 5.0 keV for the fundamental line, to be able to compare the line depths found. This value is consistent with the best-fit values found when we left the parameter free.

For observations #3 and #4 no significant second line was found, the line parameters were effectively unconstrained without limiting the unknown energy of the second line. Therefore we decided to present the results for a fit with just one cyclotron line.

PHASE RESOLVED SPECTRA

We present preliminary results for observation #2 obtained with privately written software. The data were split into 3 phase segments of equal width (see Fig. 2) and fit results were obtained for the phase resolved spectra using the FDCO model.

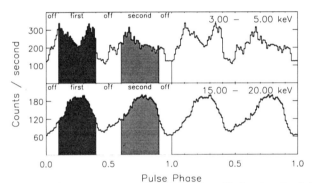

FIGURE 2. Pulse profiles in two energy bands for obs. #2 showing the definition of the phase segments used for the analysis.

In the spectrum of the first pulse we find indications for two lines but the uncertainities are too large to detect the second harmonic with confidence. For the second peak only the first harmonic is detected and the line features vanish alltogether between the two peaks. The parameters given in Table 4 are therefore for a FDCO model with two, one and no cyclotron line features for the first pulse, second pulse and off-pulse phase segments respectively. Figure 3 shows the spectra of the three phase segments fitted by a FDCO model without line features.

The parameters of the continuum also vary with the pulse phase. Note that the values given are for the FDCO model and are not directly comparable with the values derived for the more common HECUT spectrum.

TABLE 4. Results of fitting FDCO model spectra to the phase segments defined in Fig. 2.

Component	Parameter	first pulse	second pulse	off-pulse
Absorption	n_H	2^{+2}_{-2}	0^{+1}_{-0}	$0.4^{+4.0}_{-0.4}$
Power law	α	$0.8^{+0.15}_{-0.2}$	$0.66^{+0.04}_{-0.03}$	$0.9^{+0.2}_{-0.1}$
FD-cutoff	E_{cut}	$27^{+2.5}_{-6}$	$28.4^{+0.9}_{-0.9}$	$17.3^{+5}_{-4.5}$
	E_{fold}	$10.5^{+1}_{-0.5}$	$7.0^{+0.3}_{-0.3}$	$9.7^{+1}_{-0.5}$
1st Cyclo. line	E_{cyc}	$24.6^{+0.6}_{-0.6}$	$24.5^{+1.7}_{-1}$	—
	width (fixed)	5.0	5.0	—
	depth	$0.3^{+0.03}_{-0.05}$	$0.2^{+0.25}_{-0.05}$	—
2nd Cyclo. line	depth	$0.95^{+\infty}_{-0.7}$	—	—
	factor f	$2.37^{+\infty}_{-0.2}$	—	—
χ^2_{red}		0.65	0.55	0.56

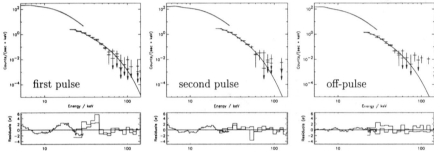

FIGURE 3. The pulse phase resolved spectra of observation #2 fitted by an FDCO continuum without the inclusion of cyclotron line features.

Acknowledgements This research was supported by DARA under grant 50 OO 96058 and NASA under grant NAG5-3272.

REFERENCES

1. Kendziorra, E., Mony, B., Kretschmar, P., et al., 1992, in [8], 51
2. Kretschmar, P., Pan, H. C., Kendziorra, E., et al., 1996, A&AS, 120, C175
3. Kretschmar, P., Pan, H. C., Kendziorra, E., et al., 1997, A&A, in press
4. Makishima, K., Mihara, T., Nagase, F., Murakami, T., 1992, in [8], 23
5. Mihara, T. 1995, *Ph.D. thesis*, University of Tokyo
6. Staubert, R., *priv. comm.*
7. Tanaka, Y. 1986, in D. Mihalas, K. Winkler (eds.), Radiation Hydrodynamics in Stars and Compact Objects, 198, Springer, Berlin
8. Tanaka, Y., Koyama, K., (eds.) 1992, Frontiers of X-Ray Astronomy, Frontiers Science Series, 2

BeppoSAX observation of the X–ray binary pulsar Vela X–1

M. Orlandini[1], D. Dal Fiume[1], L. Nicastro[1],
S. Giarrusso[2], A. Segreto[2], S. Piraino[2],
G. Cusumano[2], S. Del Sordo[2], M. Guainazzi[3], L. Piro[4]

[1] *Istituto TESRE/CNR, via Gobetti 101, 40129 Bologna, Italy*
[2] *Istituto IFCAI/CNR, via La Malfa 153, 90146 Palermo, Italy*
[3] *SAX/Scientific Data Center/Nuova Telespazio, via Corcolle 19, 00131 Rome, Italy*
[4] *Istituto Astrofisica Spaziale/CNR, via Fermi 21, 00044 Frascati, Italy*

Abstract. We report on the spectral (pulse averaged) and timing analysis of the ~ 20 ksec observation of the X-ray binary pulsar Vela X–1 performed during the BeppoSAX Science Verification Phase. The source was observed in two different intensity states: the low state is probably due to an erratic intensity dip and shows a decrease of a factor ~ 2 in intensity, and a factor 10 in N_H. We have not been able to fit the 2–100 keV continuum spectrum with the standard (for an X-ray pulsar) power law modified by a high energy cutoff because of the flattening of the spectrum in ~ 10–30 keV. The timing analysis confirms previous results: the pulse profile changes from a five-peak structure for energies less than 15 keV, to a simpler two-peak shape at higher energies. The Fourier analysis shows a very complex harmonic component: up to 23 harmonics are clearly visible in the power spectrum, with a dominant first harmonic for low energy data, and a second one as the more prominent for energies greater than 15 keV. The aperiodic component in the Vela X–1 power spectrum presents a *knee* at about 1 Hz. The pulse period, corrected for binary motion, is 283.206 ± 0.001 sec.

INTRODUCTION

Vela X–1 is the prototype of the class of accreting high-mass X–ray binary pulsars, with a spectrum that was fit by a power law modified at high energy by an exponential cutoff [17]. Moreover, the spectrum shows line features: a ~ 6.4 keV Iron fluorescence emission line, and a cyclotron resonance feature (CRF) at ~ 55 keV [14]. The pulse-averaged spectrum depends on the 8.96 day orbital period [8]: episodes of strong absorption are explained in terms of accretion [2] or photoionization [10] wakes.

The pulse period history of the 283 sec X-ray binary pulsar Vela X-1 shows a typical wavy behavior [12]. The reversal of spin-up and spin-down has been explained in term of a (temporary) accretion disk, acting as reservoir for the momentum transfer [1].

OBSERVATION

BeppoSAX is a program of the Italian Space Agency (ASI) with participation of the Netherlands Agency for Aerospace Programs (NIVR). It is composed by four co-aligned Narrow Field Instruments (NFIs) [3], operating in the energy ranges 0.1–10 keV (LECS, not operative during this observation) [15], 1–10 keV (MECS) [4], 3–120 keV (HPGSPC) [11] and 15–300 keV (PDS) [7]. Perpendicular to the NFI axis there are two Wide Field Cameras [9], with a $40° \times 40°$ field of view.

During the BeppoSAX Science Verification Phase (SVP) Vela X-1 was observed on 1996 July 14 from 06:01 to 20:54 UT, corresponding to orbital phases 0.28–0.35 [6]. All the instruments were operated in direct modes, which provide information on each photon. The two mechanical-collimated instruments were operated in rocking mode with a 96 sec stay time for HPGSPC and 50 sec stay time for PDS. The first part of the observation, about 40% of the total, was characterized by a lower flux and a higher absorption — the 6–40 keV flux passed from 0.34 to 0.64 photons cm^{-2} sec^{-1} in the two states. This is probably due to the passage of the neutron star through clumpy circumstellar material [16].

SPECTRAL ANALYSIS

In Fig. 1 we show the results of the spectral fit to the 2–100 keV Vela X-1 spectrum with the usual (for an X-ray pulsar) power law modified by a high-energy cutoff [17]. We added to the model an Iron emission line and a CRF at ~ 55 keV. The spectral results are summarized in Table 1, from which it is evident that this model is not able to adequately describe the complex Vela X-1 continuum, and in particular the flattening of the spectrum in the 10–30 keV. A fit with *two* power laws modified by an exponential cutoff — the so-called NPEX model — is able to describe this flattening [14], although a more detailed model is needed. It is noteworthy that it is in this energy range that the pulse shape changes dramatically from a five to a two peak structure.

TIMING ANALYSIS

Each photon arrival time was corrected to the solar system barycenter in order to decouple effects due to Earth and satellite motion. We then defined a

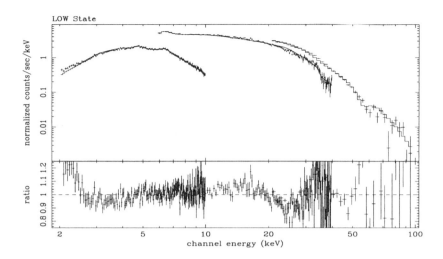

FIGURE 1. BeppoSAX wide-band 2-100 keV spectrum of Vela X-1 during the LOW state. The continuum has been modelled by a power law modified by a high energy cutoff [17]. The inclusion of an Iron emission line and a CRF at ~ 55 keV was not able to describe the data, in particular the flattening in the 10-30 keV range.

TABLE 1. Fit results to the Vela X-1 spectra for both the two intensity states. The continuum has been modelled with a power law modified by a high energy cutoff [17]. All quoted errors represent 90% confidence level for a single parameter.

	LOW State	HIGH State
wabs (cm^{-2})	7.86 ± 0.09	1.63 ± 0.03
PhoIndex	1.23 ± 0.07	1.12 ± 0.04
cutoffE (keV)	24.3 ± 0.1	24.74 ± 0.07
foldE (keV)	13.4 ± 0.1	12.11 ± 0.06
LineE (keV)	6.44 ± 0.02	6.47 ± 0.02
Sigma (keV)	0.21 ± 0.04	0.20 ± 0.02
Ecyc (keV)	55.4 ± 0.5	55.9 ± 0.3
Width (keV)	10 (fixed)	10 (fixed)
Depth	2 (fixed)	5.3 ± 0.5
χ^2_{dof} (dof)	2.9 (334)	9.0 (335)

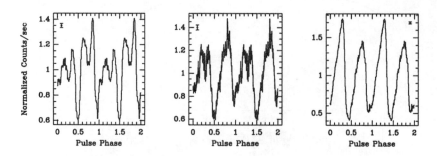

FIGURE 2. Vela X-1 pulse profiles as observed by the three operative NFIs aboard BeppoSAX. From left to right: 2-10 keV (MECS3), 6-40 keV (HPGSPC), 20-100 keV (PDS).

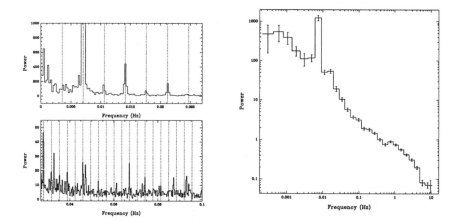

FIGURE 3. Results of the Fourier analysis performed on the PDS data. On the left we show the power spectrum in linear scale, in order to better enhance the effects due to the coherent component (harmonics are shown with dotted lines). The second harmonic is the most prominent because of the double-peak structure of the pulse profile at this energy. On the right we show the same data in a logarithmic scale, in order to evidence the aperiodic component. Note the *knee* at ~ 1 Hz.

fiducial point in the pulse profile — the minimum between the two main peaks — and determined the arrival times corresponding to these points. Those points were then fit with a line corrected for the Doppler delays due to the orbital motion (we assumed the ephemeris given by [6]). We find a pulse period of 283.206 ± 0.001 sec for $T_o = 2,450,288.5$ JD. In Fig. 2 we show the pulse profile as determined from the three operative NFIs instruments. It is noteworthy the dramatic change of pulse shape from low (five peaks) to high (two peaks) energy, with the transition clearly visible in the HPGSPC data.

We also performed a Fourier analysis on the data, and the results are shown in Fig. 3 for the PDS data. From the point of view of the coherent component, we find that the second harmonic is more pronounced than the first because in this energy range the pulse is double-peaked. For low energy data this behavior is reversed. For what concerns the aperiodic component, we find a sort of *knee* at about 1 Hz, indicating a ~ 1 sec time scale typical of processes that occur at the magnetospheric limit. This result is very similar to that obtained by EXOSAT [5], in which a bump at ~ 1 Hz and a high frequency cutoff was observed.

Acknowledgements. The authors wish to thank the BeppoSAX Scientific Data Center staff for their support during the observation and data analysis. This research has been funded in part by the Italian Space Agency.

REFERENCES

1. Anzer, U., Börner, G., & Monaghan, J.J. 1987, A&A, 176, 235
2. Blondin, J.M., Stevens, I.R., & Kallman, T.R. 1991, ApJ, 371, 684
3. Boella, G., Butler, R.C., et al. 1997a, A&AS, 122, 299
4. Boella, G., Chiappetti, L., et al. 1997b, A&AS, 122, 327
5. Dal Fiume, D., et al. 1992, in *Frontier Objects in Astrophysics and Particle Physics*, eds. Giovannelli F. & Mannocchi G., SIF Conference Proceedings, Bologna, p.207
6. Deeter, J.E., Boynton, P.E., et al. 1987, AJ, 93, 877
7. Frontera, F., Costa, E., et al. 1997, A&AS, 122, 357
8. Haberl, F., & White, N.E. 1990, ApJ, 361, 225
9. Jager, R., Mels, W.A., et al. 1997, A&AS, in press
10. Kaper, L., Hammerschlag-Hensberge, G., & Zuiderwijk, E.J. 1994, A&A, 300, 446
11. Manzo, G., Giarrusso, S., et al. 1997, A&AS, 122, 341
12. Nagase, F. 1989, PASJ, 41, 1
13. Orlandini, M. 1993, MNRAS, 264, 181
14. Orlandini, M., Dal Fiume, D., et al. 1997, A&A, submitted
15. Parmar, A.N., Martin, D.D.E., et al. 1997, A&AS, 122, 309
16. Sato, N., Hayakawa, S., et al. 1986, PASJ, 38, 731
17. White, N.E., Swank, J.H., & Holt, S.S. 1983, ApJ, 270, 711

New radio observations of Circinus X-1

R. P. Fender

Astronomy Centre, University of Sussex, Falmer, Brighton BN1 9QH, UK

Abstract. New radio observations of the radio-jet X-ray binary Circinus X-1 over nearly an entire 16.6-day orbit are presented. The source continues to undergo radio flaring in the phase interval 0.0 – 0.2 and appears to be brightening since observations in the early 1990s. The radio flux density is well correlated with simultaneous soft X-ray monitoring from the XTE ASM, including a secondary flare event around phases 0.6 – 0.8 observed at both energies.

INTRODUCTION

Circinus X-1 is a highly unusual radio-bright southern X-ray binary. Every 16.6 days the source undergoes X-ray, infrared and radio outbursts (e.g. Glass 1994; Haynes et al. 1978). This is interpreted as heightened accretion during periastron passage of a compact object in an elliptical orbit around its companion star. There are few good spectroscopic observations due in part to the presence of two confusing sources within 2 arcsec of the optical counterpart (Moneti 1992), and the nature of the companion star remains uncertain.

Cir X-1 is embedded within a synchrotron nebula which trails back towards the nearby SNR G321.9-0.3. Synthesis mapping at 6 cm with ATCA has resolved jet-like structures within this nebula, originating at the location of the binary and curving back towards G321.9-0.3 (Stewart et al. 1993) – this suggests the source may be a runaway X-ray binary with an origin in the SNR. The radio brightness of the source at cm wavelengths declined significantly from the late 1970s to the early 1990s, with Haynes et al. (1978) recording flux densities in excess of 1 Jy, while Stewart et al. (1993) only measured a few mJy.

Here we present new radio monitoring of Cir X-1, over most of an orbital period, in 1996 July, including a comparison with simultaneous soft X-ray monitoring with the XTE ASM.

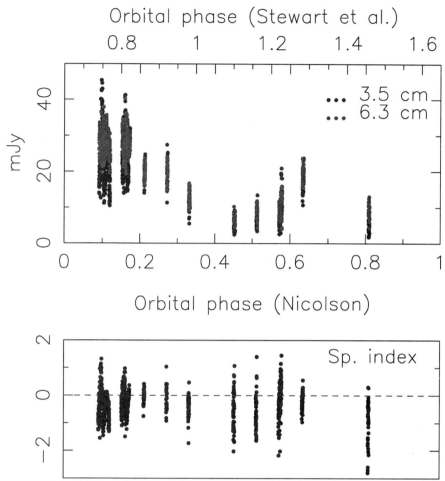

FIGURE 1. ATCA observations, simultaneously at 6.3 & 3.5 cm, of Cir X-1 between 1996 July 1 – 14. Orbital phase as calculated both by the linear ephemeris of Stewart et al. and the quadratric ephemeris of Nicolson is indicated – clearly the Nicolson ephemeris is more accurate. Flux density variations within a single observation are confused by resolution and u-v effects, but the day-to-day variations are real and the source is clearly brightest in the phase interval 0.0 – 0.2. The spectral index, plotted in the lower panel, is consistent with nonthermal synchrotron emission.

OBSERVATIONS

Cir X-1 was observed with the Australia Telescope Compact Array (ATCA) on fourteen days of its 16.6 day orbit between 1996 July 1 – 14. Observations were made simultaneously at wavelengths of 6.3 & 3.5 cm, with the array in high resolution 6D configuration (6 km maximum baseline). The effective resolution of the array in this configuration is ~ 2 and ~ 1 arcsec at 6.3 & 3.5 cm respectively.

The first two runs on Cir X-1 were of duration ~ 12 and ~ 10 hr respectively, in order to map the source at high resolution. However, while there is some evidence of jet-like structure in the resultant maps at both wavelengths, uncertainty as to the contribution both from intrinsic source variability and the surrounding nebula render such features unreliable. The lack of much of the synchrotron nebula in the maps suggests that it is of relatively low surface brightness, with little structure on small (few arcsec) angular scales.

RESULTS

The light curve

Fig 1 shows the radio light curve from Cir X-1 over the entire set of observations. As described above, it is hard to differentiate intrinsic source variability from resolution and u-v coverage effects, and so apparent variability on short timescales should not be taken too seriously without further careful analysis. However, the large day-to-day changes, with a drop in flux density by a factor of ~ 3 are indeed significant. The flux density of the source, at all phases, while well below that reported in the 1970s, appears to have risen considerably since the observations of Stewart et al. (1993) in the early 1990s.

The spectral index ($\alpha = \Delta \log S_\nu / \Delta \log \nu$) of the radio emission, plotted in the lower panel of Fig 1, is consistent with optically thin synchrotron emission with α around -0.5.

A note on the ephemerides ..

Stewart et al. (1993) discuss two ephemerides for Cir X-1 : a quadratic ephemeris from Nicolson (private communication), and a simplified linear ephemeris of their own. The orbital phase as calculated from both ephemerides is shown in Figure 1; from this it is apparent that the linear ephemeris of Stewart et al. is in error in its prediction of the time of outburst, and the Nicolson ephemeris must be considered more reliable. However, given that the XTE ASM has by now observed more than thirty periodic outbursts of the system, a newly refined X-ray ephemeris may be called for.

Comparison with X-ray monitoring

Fig 2 compares the radio light curve at 6.3 cm with the 2-12 keV lightcurve obtained by the XTE ASM. A clear correlation exists, with flaring actvity

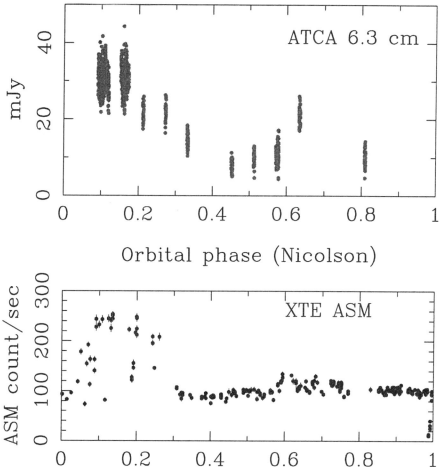

FIGURE 2. A comparison of the ATCA 6.3 cm light curve with simultaneous soft X-ray monitoring with the XTE ASM. The activity in the two bands is clearly correlated at all phases. The major flaring, presumably occurring around the time of periastron passage of the compact object, is obvious. Note also however, the secondary flaring, both in X-rays and radio emission, around phase 0.6 – 0.8.

in both energy regimes occuring between phases 0.0 – 0.2 followed by a subsequent decline. This supports the picture of some enhanced accretion and related particle acceleration/ejection occurring around the time of periastron passage.

Note also the lesser flaring, again at **both** energies, in the phase interval 0.6 – 0.8 (near apastron). Such a secondary radio outburst half an orbit after the primary flare has not previously been reported.

CONCLUSIONS

Cir X-1 has been observed over most of its 16.6-day orbit simultaneously at 6.3 & 3.5 cm, and in the 2-12 keV energy range with the XTE ASM. The source is clearly continuing to undergo radio flaring around phase 0.0 – 0.2 (from the quadratic ephemeris of Nicolson), and may have begun brightening since observations in the early 1990s. There is also a clear coupling between the soft X-ray activity (reflecting the state of the accretion process) and the radio brightness (probably reflecting the ejection of synchrotron-emitting material). In addition, secondary radio flaring around phases 0.6 – 0.8, which is correlated with the X-ray behaviour, has been discovered.

Cir X-1 is the only X-ray binary for which there is both strong evidence for radio jets and direct evidence that the compact object is a neutron star (from Type I X-ray bursts). It is a key system in our understanding of radio emission, in particular jets, from X-ray binaries. Future radio observations, including flux monitoring, mapping and accurate proper motion measurements, will be of great importance.

Acknowledgments

I thank Jim Caswell, George Nicolson, Bob Sault, Tasso Tzioumis and Kinwah Wu for useful discussions, and Karen Southwell and Vince McIntyre for assistance with the observations. The ATCA is funded by the Commonwealth of Australia for operation as a National Facility managed by CSIRO. I acknowledge quick-look results provided by the ASM/XTE team.

REFERENCES

1. Moneti A., 1992, ApJ, 260, L7
2. Glass I.S., 1994, MNRAS, 268, 742
3. Stewart R.T., Caswell J.L., Haynes R.F., Nelson G.J., 1993, MNRAS, 261, 593
4. Haynes R.F., et al., 1978, MNRAS, 185, 661

Orbit Determination for the Be/X-Ray Transient EXO 2030+375

Mark T. Stollberg*, Mark H. Finger[†,*], Robert B. Wilson[‡], D. Matthew Scott[†], David J. Crary[†], & William S. Paciesas*

*Department of Physics, University of Alabama in Huntsville.
[†] University Space Research Association.
*CGRO Science Support Center/Goddard Space Flight Center.
[‡] Space Sciences Laboratory, NASA/Marshall Space Flight Center.

Abstract. The Be/X-ray binary transient pulsar EXO 2030+375 has been observed during twenty-two outbursts over four years (1991-1994) using the large area detectors (LADs) of the Burst and Transient Source Experiment (BATSE) on the Compton Gamma-Ray Observatory (CGRO). Thirteen outbursts between 1992 February and 1993 August occurred regularly at ~46 day intervals, close to the orbital period determined using EXOSAT data (Parmar et al. 1989). EXOSAT discovered this pulsar during a "giant" outburst in 1985 May (Parmar et al. 1985). All BATSE outbursts were "normal" type. Pulse phases derived from the thirteen consecutive outbursts were fit to two different models to determine a binary orbit. A summary of the results are presented here.

INTRODUCTION

The Be/X-ray transient pulsar EXO 2030+375 was observed by BATSE during twenty-two outbursts from 1991 May to 1994 June. The flux history (20-50 keV) and spin frequency history for the source from the BATSE observations appear in Figures 1 and 2. The vertical dashed lines in both figures bracket thirteen outbursts that recurred at consecutive ~46 day intervals. Each of these outbursts lasted 7-15 days. Preliminary results for these data were presented in [3] and [4]. The epoch-folded pulse profile was double peaked and showed no significant variations either with flux or energy. One-day spectral fits over all thirteen outbursts were performed using an exponential model with a variable kT parameter. The weighted average kT value was found to be 20.2 ± 0.3 keV. Here we summarize results from a timing analysis of the 13 consecutive outbursts.

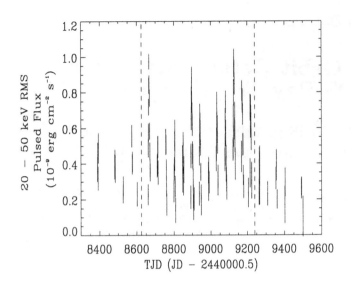

FIGURE 1. Flux history (20–50 keV) for EXO 2030+375 for the BATSE observations.

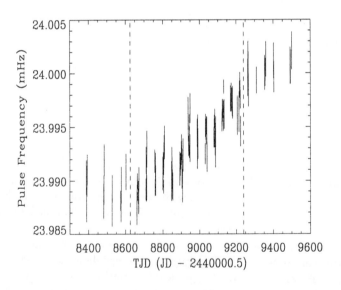

FIGURE 2. Spin frequency history for EXO 2030+375 for the BATSE observations.

PULSE TIMING

The spin frequencies at mid–outburst for the thirteen outbursts were linearly fit to obtain global values of ν and $\dot{\nu}$ for a preliminary phase model that was quadratic in t, the emission time of pulses from the pulsar. The BATSE DISCLA data (20–50 keV) were folded using this phase model and the orbital parameters from [4].

Phase offsets with respect to the preliminary ephemeris were obtained using a template matching procedure in Fourier space between the harmonic representation of the pulse profile for each folded single day of data and that of a scaled and shifted master template extracted from the first of the thirteen outbursts, Truncated Julian Days (TJDs) 8660–8672. The procedure was equivalent to a cross–correlation between the profile and the template [5] but allowed weighting of the harmonics by their errors. Monte Carlo simulations using the master template showed that this method could find accurate phase offsets down to a signal–to–noise ratio of 2.5 (defined here as the ratio of the daily pulsed flux value to its error) before substantial departures from the expected Gaussian distribution occurred. Therefore, a 2.5 signal–to–noise cutoff for the data was established and days with a lower value were discarded.

Cygnus X-1 was a source that contributed noise to the BATSE observations of EXO 2030+375. To account for the presence of this source, the ten Fourier harmonics used to construct the daily pulse profiles were identified in daily power spectra for EXO 2030+375 and the average power obtained for 21 mHz intervals centered on each harmonic. The errors for each harmonic were multiplied by the square root of the average Leahy normalized power divided by two.

Deeter [5] has suggested that retaining harmonics in the master template dominated by noise would degrade the pulse phase estimation. Therefore only the first 7 harmonics were retained. A comparison via a χ^2 minimization was performed between the first outburst template and three other templates drawn respectively from all of the outbursts, only the strong outbursts, and only the weak outbursts to determine if a better master template could be used. The lowest χ^2 value was obtained for the all–outburst template and so it was selected as the new master template.

The provisional phase model was corrected further by fitting a second degree polynomial to the total phase (phase model + phase offsets) for each outburst to obtain improved values of ν and $\dot{\nu}$. The DISCLA data were refolded using these corrections.

ORBIT FITTING

The orbit fit from [4] had an unacceptable reduced χ^2 of 3.39, leading us to try to improve the fit. As a first attempt, a polynomial was used to model the

TABLE 1. EXO 2030+375 Orbit Parameters from the Two Models.

Parameter	Polynomial Model	Gaussian Model
χ^2/dof	183.14/116	121.25/102
P_{orb}	46.017 ± 0.003 days	46.01 ± 0.02 days
e	0.37 ± 0.02	0.37 ± 0.02
$a_x \sin i$	257 ± 14	264 ± 21 lt-sec
ω	223.0 ± 1.9 degrees	223.4 ± 3.9 degrees
T_p	JD 2448936.9 ± 0.2	JD 2448937.0 ± 0.3
$f(M)$	8.42 ± 1.34	9.12 ± 2.17 $M_\odot$
σ		3.98 ± 1.22 days

phase contribution from the pulsar for all thirteen outbursts. The model used the difference between the pulse emission times in the pulsar rest frame and a phase epoch T_0 near the middle of the data. The coefficients for each order of the polynomial were estimated and held constant over the fit. All of the orbit parameters which affect the delay between emission and arrival times were allowed to vary for each polynomial order. Phases could not be connected between outbursts because of long periods when no X-rays were seen, so non-integer cycle slips were determined in the fitting process. Goodness of fit was evaluated for each polynomial order using a χ^2 minimization. Beginning with polynomial order 3, fits were made for each succeeding order until order 15 was reached. Fits beyond order 15 proved unmanageable. The resulting χ^2 value and orbit parameters for the 15th order fit are presented in Table 1.

A second model was tried that concentrated the spin-up torque for each outburst during the times when hard X-ray flux was observed by BATSE. This torque model was of the form

$$\dot{\nu}(t) = \phi''(t) = \dot{\nu}_0 + \sum_{i=1}^{13} B_i e^{-(t-\tau_i)^2/2\sigma_i^2} . \qquad (1)$$

$\dot{\nu}_0$ represented the torque between outbursts while the summation was a Gaussian contribution to the frequency derivative during each outburst with a peak time τ_i and half-width σ_i. A polynomial representation for $\dot{\nu}_0$ was tried but was found unnecessary. The B_i coefficients are a direct measure of the peak torque. Gaussian fits to the observed X-ray flux for each outburst determined preliminary values of τ_i and showed that all of the σ_i's were consistent with a single half-width σ, which was fixed in the final fit. All of the A_j's, B_i's, τ_i's, σ, as well as all orbit parameters were allowed to vary for the fit. The resulting χ^2 value, the orbit parameters, and the Gaussian half width from the fit are shown in Table 1. Figure 3 shows the derived torque model for all thirteen outbursts.

The error bars at each peak are those of the Gaussian amplitudes determined by the fit. The best fit solution shows spin-up occurring during each outburst and a slight spin-down occurring between outbursts. The mean peak spin-up

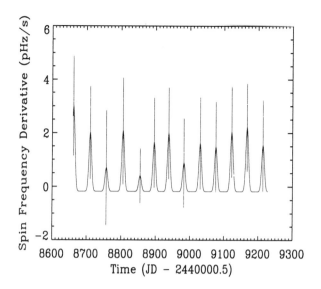

FIGURE 3. Torque values for EXO 2030+375 resulting from Gaussian orbit fit model. The uncertainties are highly correlated.

rate during outbursts above the inter–outburst level has a significance of only 1.1σ. Since this model does account for the physically expected behavior, the resulting orbital parameters and errors should be preferred to those from the polynomial model. A full presentation of this analysis is given in [6].

REFERENCES

1. Parmar, A. N., White, N. E., Stella, L., Izzo, C., & Ferro, P. 1989, ApJ, 338, 359
2. Parmar, A. N., Stella, L., Ferri, P., & White, N. E. 1985, IAU Circ., No. 4066
3. Stollberg, M. T., Pendleton, G. N., Paciesas, W. S., Finger, M. H., Fishman, G. J., Wilson, R. B., Meegan, C. A., Harmon, B. A., & Wilson, C. A. 1993, in Compton Gamma-Ray Observatory, eds. M. Friedlander, N. Gehrels, & D. J. Macomb, (New York: AIP Press), 371
4. Stollberg, M. T., Paciesas, W. S., Finger, M. H., Fishman, G. J., Wilson, R. B., Harmon, B. A., & Wilson, C. A. 1994, in The Evolution of X-Ray Binaries, eds. S. S. Holt & C. S. Day, (New York: AIP Press), 255
5. Deeter, J. E. 1981, Ph. D. dissertation, University of Washington
6. Stollberg, M. T., Finger, M. H., Wilson, R. B., Scott, D. M., Crary, D. J., & Paciesas, W. S. 1997, ApJ, in preparation

The Orbital Ephemeris and X-Ray Light Curve of Cyg X-3

Steven M. Matz*

*Northwestern University
Dept. of Physics and Astronomy
Dearborn Observatory
Evanston Illinois 60208-2900

Abstract.
The orbital dynamics of Cyg X-3 are a key to understanding this enigmatic X-ray binary. Recent observations by the RXTE ASM and the OSSE instrument on GRO enable us to extend the baseline of arrival time measurements and test earlier models of orbital period evolution. We derive new quadratic and cubic ephemerides from the soft X-ray data (including ASM). We find a significant shift between the predicted soft X-ray phase and the light curve phase measured by OSSE from $\sim$ 44 to 130 keV. Some of the apparent phase shift may be caused by a difference in light curve shape.

INTRODUCTION

Cygnus X-3 is a unique and poorly understood X-ray source which may represent a very short-lived, transitional state of binary evolution. A study of its dynamics (via the orbital ephemeris), and local environment (via the light curve and spectrum) may have broad implications for X-ray binaries in general. The period (4.8 hr, presumably orbital) is characteristic of low mass X-ray binaries. However, there is evidence that the companion is actually a high-mass Wolf-Rayet star [1]. The mass loss from the stellar wind of such a star could explain the large *positive* $\dot{P}$ measured in Cyg X-3, which is inconsistent with mass transfer via Roche lobe overflow [2]. The light curve, observed in IR and X-rays, is asymmetric with a non-zero eclipse, resulting, in most models, from interactions in a dense cocoon or wind around the system [3–6]. If the asymmetry is instead caused by an elliptical orbit, the light curve shape is expected to change significantly over time due to apsidal motion [7] which could also produce some or all of the apparent period increase [8]. Therefore, measurements of the light curve shape are directly related to the orbital dynamics.

OBSERVATION SUMMARY

OSSE [9] observed Cyg X-3 for $\sim$ 65 days in 7 intervals between 1991 May 30 and 1994 July 12. We use the 2-min background-subtracted $\sim$ 50 keV to 10 MeV spectra produced by the standard OSSE spectral analysis routines for these 7 observations. There was no significant flux contribution from Cyg X-1 in any of the source or background pointings. Some results of the earlier OSSE observations have been described elsewhere [10].

For the ASM we used the background subtracted 2-10 keV Cyg X-3 counting rates from the quick-look results provided by the ASM/RXTE team over the WWW. The data analyzed here cover 1996 Feb 22 – 1997 Apr 11.

TABLE 1. ASM Arrival Times

Interval (HJD − 2440000.)	Cycle	Arr. Time (HJD − 2440000.)	σ_T (days)
10135.8–10235.4	46251.0000	10185.6237	1.0×10^{-3}
10236.0–10335.9	46753.0000	10285.8687	1.1×10^{-3}
10335.9–10435.8	47254.0000	10385.9134	1.4×10^{-3}
10435.8–10535.7	47754.0000	10485.7588	1.2×10^{-3}

ASM ARRIVAL TIME DERIVATION

To correct for large, long term changes in the source DC flux level the ASM 2-10 keV Cyg X-3 rates were first "cleaned" by fitting a line to successive 10-day segments of the data and subtracting the fit. The linear fit removes trends on time scales much longer than an orbital period. This process did not significantly change the arrival times derived from the data.

The cleaned data were divided into approximately 100-day intervals. Long intervals are needed to average over the cycle-to-cycle light curve (LC) variations and to completely sample the Cyg X-3 orbit. The data in each interval were then corrected to the SSB and epoch-folded (using a constant period ephemeris) to produce orbital light curves. We fit the resulting light curves with the EXOSAT template [11] to determine the relative phase of the LC minimum. This phase is then used to calculate an arrival time for a cycle near the middle of the observation interval [12]. The arrival times are shown in Table 1 in heliocentric JD (HJD) − 2440000.

Figure 1 shows differences (residuals) between the arrival times predicted by a constant period ephemeris and those observed by the ASM and by other soft X-ray instruments going back to 1970 ([13] and references therein, plus [14]). Overplotted are fits with quadratic and cubic ephemeris models.

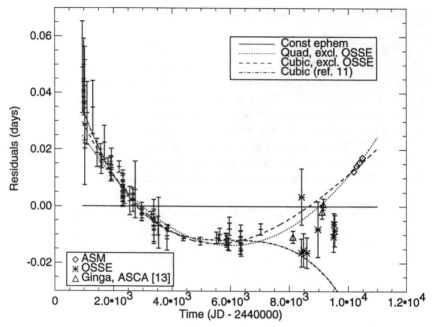

FIGURE 1. Arrival time residuals for OSSE, ASM, and historical data. The errors on the ASM points are approximately equal to the symbol size. Also shown are the best-fit quadratic and cubic ephemerides. An extrapolation of an older cubic fit [11] with a larger negative $\ddot{P}$ is shown for comparison.

ORBITAL EPHEMERIS

The ASM arrival times were combined with those from earlier soft X-ray observations and fit with both quadratic and cubic ephemeris models (Figure 1). The resulting best-fit parameters are shown in Table 2. Neither quadratic nor cubic models provide a statistically acceptable fit to all the data. Possibly the errors in earlier data have been underestimated due to cycle-to-cycle variations or other systematic effects [13]. No single data set drives the bad χ^2.

The cubic term is still statistically significant, but the fitted magnitude has decreased with time: from $\ddot{P} = 1.2 \times 10^{-10}$ [11] to 4.1×10^{-11} yr^{-1} in this work. The wind model predicts essentially zero $\ddot{P}$. Future ASM data should provide a definitive test of the reality of the non-zero $\ddot{P}$.

OSSE ARRIVAL TIME DERIVATION

Following the same procedure used for the ASM data, arrival times were determined for the each of the 7 OSSE observations of Cyg X–3 from 1991 to

TABLE 2. PRELIMINARY Best-Fit Quadratic and Cubic Ephemerides

	$T_n = T_0 + P_0 n + c_0 n^2$		$T_n = T_0 + P_0 n + c_0 n^2 + dn^3$
T_0	$2440949.89185 \pm 0.00088$ HJD	T_0	2440949.8967 ± 0.0014 HJD
P_0	$0.199684393 \pm (8.9 \times 10^{-8})$ d	P_0	$0.199683305 \pm (2.6 \times 10^{-7})$ d
c_0	$(6.06 \pm 0.17) \times 10^{-11}$ d	c_0	$(1.17 \pm 0.13) \times 10^{-10}$ d
		d	$(-7.54 \pm 1.71) \times 10^{-16}$ d
χ^2 (dof)	168.42 (84)	χ^2 (dof)	136.37 (83)
Conf. level	1.3×10^{-7}	Conf. level	2.0×10^{-4}

1994 using light curves of 2-min background-subtracted counting rates ~ 44–130 keV. The arrival times for the 7 observations were compared with the values predicted using the soft X-ray ephemerides (Table 2). The OSSE points fall systematically below the predicted curves on the residual plot.

The OSSE phase minimum (~ 44–130 keV) is significantly (6–7σ) earlier than the predicted soft X-ray (1–10 keV) minimum for both quadratic and cubic ephemerides ($\Delta t \sim -20 \pm 3$ min, $\Delta\phi \sim -0.065 \pm 0.010$). The data are consistent with a constant offset over more than 3 years of observations.

ORBITAL LIGHT CURVE ANALYSIS

Light curves (Figure 2) were produced from the ASM and OSSE data using the quadratic ephemeris in Table 2. The plotted errors in the ASM data reflect only the reported statistical errors. The light curves were fit with the EXOSAT soft X-ray template [11], varying phase, amplitude, and DC intensity to determine the phase of minimum and the consistency of the overall shape with the template.

Both the OSSE and ASM light curves are statistically *inconsistent* with the canonical X-ray template [11]. Qualitatively, however, the template reasonably describes the ASM data. The large χ^2 is due to the fact that the actual observed variations in each bin are larger that the statistical errors due to underlying source fluctuations.

Evolution of the light curve shape with time is expected if the apparent orbital period change is due to apsidal motion [7,8]. There is no evidence for this in the ASM observations.

The OSSE LC appears somewhat more symmetric than the template, with a faster rise, but this is difficult to constrain with current data. The differences in LC shape may contribute to the apparent phase shift with energy.

REFERENCES

1. van Kerkwijk, M. H. et al., *Nature*, **355**, 703 (1992).

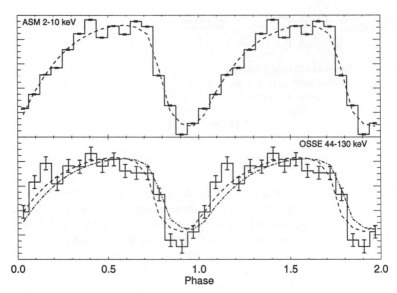

FIGURE 2. ASM (top) and OSSE (bottom) light curves with best-fit X-ray templates overplotted (dashed line). The dashed-dotted line in the OSSE plot shows the best fit to the OSSE data with the ASM phase. The shift between the two energies is apparent.

2. Molnar, L. A., *ApJ*, **331**, L25 (1988).
3. Pringle, J. E., *Nature*, **247**, 21 (1974).
4. Milgrom, M. and Pines, D., *ApJ*, **220**, 272 (1978).
5. White, N. E. and Holt, S. S., *ApJ*, **257**, 318 (1982).
6. Willingale, R., King, A. R., and Pounds, K. A., *MNRAS*, **215**, 295 (1985).
7. Ghosh, P., Elsner, R. F., Weisskopf, M. C., and Sutherland, P. G., *ApJ*, **251**, 230 (1981).
8. Elsner, R. F., Ghosh, P., Darbro, W., Weisskopf, M. C., Sutherland, P. G., and Grindlay, J. E., *ApJ*, **239**, 335 (1980).
9. Johnson, W. N., et al., *ApJS*, **86**, 693 (1993).
10. Matz, S. M., Grabelsky, D. A., Purcell, W. R., Ulmer, M. P., Johnson, W. N., Kinzer, R. L., Kurfess, J. D., and Strickman, M. S., in *The Evolution of X-Ray Binaries (AIP Conf. Proc. 308)*, Holt, S. S. and Day, C. S., editors, 263 (AIP, New York, 1994).
11. van der Klis, M. and Bonnet-Bidaud, J. M., *A&A*, **214**, 203 (1989).
12. Matz, S. M., Fender, R. P., Bell Burnell, S. J., Grove, J. E., and Strickman, M. S., *A&AS*, **120**, 235 (1996).
13. Kitamoto, S. et al., *PASJ*, **47**, 233 (1995).
14. Bonnet-Bidaud, J. M. and van der Klis, M., in *Cataclysmic Variables and Low-Mass X-Ray Binaries*, Lamb, D. Q. and Patterson, J., editors, 147 (D. Reidel, Dordrecht, 1985).

A Multiwavelength Study of Cygnus X-3

M. L. McCollough[1,2], C. R. Robinson[1,2], S. N. Zhang[1,2],
B. A. Harmon[2], W. S. Paciesas[2,3], R. M. Hjellming[4], M. Rupen[4],
A. J. Mioduszewski[5], E. B. Waltman[6], R. S. Foster[6],
F. D. Ghigo[7], G. G. Pooley[8], R. P. Fender[9], and W. Cui[10]

[1] *Universities Space Research Association, Huntsville, AL 35806*
[2] *ES84 NASA/Marshall Space Flight Center, Huntsville, AL 35812*
[3] *University of Alabama in Huntsville, Huntsville, AL 35899*
[4] *National Radio Astronomy Observatory, Socorro, NM 87801*
[5] *JIVE/National Radio Astronomy Observatory, Socorro, NM 87801*
[6] *Naval Research Laboratory, Washington, D.C. 20375*
[7] *National Radio Astronomy Observatory, Green Bank, WV 24944*
[8] *Mullard Radio Astronomy Observatory, Cambridge, United Kingdom*
[9] *University of Sussex, Brighton BN1 9QH, United Kingdom*
[10] *Massachusetts Institute of Technology, Cambridge, MA 02139*

Abstract. We present a global comparison of long term observations of the hard X-ray (20–100 keV), soft X-ray (1.5–12 keV), infrared (1–2 μm) and radio (2.25, 8.3 and 15 GHz) bands for the unusual X-ray binary Cygnus X-3. Data were obtained in the hard X-ray band from CGRO/BATSE, in the soft X-ray band from RXTE/ASM, in the radio band from the Green Bank Interferometer and Ryle Telescope and in the infrared band from various ground based observatories. Radio flares, quenched radio states and quiescent radio emission can all be associated with changes in the hard and soft X-ray intensity. The injection of plasma into the radio jet is directly related to changes in the hard and soft X-ray emission. The infrared observations are examined in the context of these findings.

INTRODUCTION

Cyg X-3 represents one of the most unusual X-ray binaries ever to have been observed (see [2] for a review). It does not fit well into any of the established classes of X-ray binaries. Its orbital period, 4.8 hours, is typical of a low mass X-ray binary, but infrared observations [9] indicate that the mass donating star may be a high mass Wolf-Rayet star. In addition, Cyg X-3 undergoes giant radio outbursts and there is evidence of jet-like structures moving away from Cyg X-3 at 0.3–0.9 the speed of light [4,3,8].

In the radio there are three distinct states [10–12]. **Quiescent:** The radio flux stays around $\sim$ 100 mJy and shows little variability. **Minor Flaring:** Cyg X-3 undergoes a series of small (< 1 Jy) flares. Between flares, the radio flux may drop below the flux values seen during quiescent radio emission. **Major Flaring:** The radio undergoes a single or multiple large flares (> 1 Jy). Preceding major flares the radio can undergo periods of *quenched emission* (very low level of radio emission $\sim$ 10–20 mJy) for an extended period of a few days to a few weeks.

OBSERVATIONS

Long term monitoring observations have been obtained in hard X-ray, radio, and soft X-ray to better understand this source. In addition, various IR observations have been collected for comparison. *Hard X-Ray Observations:* BATSE Earth occultation observations were used to obtain a hard X-ray (HXR) light curve for Cyg X-3. The data were fit with a power law with a fixed spectral index of –3.0 to determine a flux in the 20 – 100 keV band. *Radio Observations:* Cyg X-3 was observed by the Green Bank Interferometer (GBI) at 2.25 and 8.3 GHz on a daily basis. The average integration times for individual scans is approximately 10 minutes with up to 12 scans made on the source per day. Cyg X-3 was also observed on a regular basis by the Ryle Telescope at 15 GHz . Typically several 5-minute observations were made each day. *Soft X-Ray Observations:* Cyg X-3 is routinely monitored by the ASM, on RXTE, in the 1.5–12 keV energy band. Single dwell data were averaged for comparison with the observations at other wavelength bands. *IR Observations:* IR K band photometry of Cyg X-3, made with UKIRT and TCS (Telescopio "Carlos Sánchez"), were obtained for comparion with the other observations.

RELATIONSHIPS BETWEEN THE HARD X-RAY, RADIO, AND SOFT X-RAY BANDS

A recent study [6] compared the 20–100 keV HXR emission detected from Cyg X-3 by BATSE with the radio data from GBI. Correlation studies between the two bands (using the Spearman rank, Kendall τ, and Pearson r tests) showed important relationships. *(a)* During the quiescent radio state the HXR flux anticorrelates with the radio. It is during this time that the HXR reaches its highest level (see Fig. 1). *(b)* During flaring states (major and minor) in the radio, the HXR flux evolves from an anticorrelation to a correlation with the radio. In particular, for major radio flares and the quenched radio emission which precedes them the correlation is strong (see Fig. 1). *(c)* The onset of large radio flares (> 5 Jy) can be predicted by noting when the HXR flux drops below the BATSE detection limit for several days.

As part of this work we have compared the HXR flux to the soft X-ray (SXR) flux observed by the ASM/RXTE (see Fig. 2). We find that the HXR flux anticorrelates with the SXR flux (using the same statistical tests we had used for the HXR and radio). This anticorrelation holds both in the low and high SXR states that we observed. (See [13] and references therein for a discussion of these SXR states.) This type of pivoting between the HXR and SXR is strikingly similar to the behavior observed in Cyg X-1 [5,14]. We also confirm the result of Watanabe et al. [13] that the radio flaring periods occur during the high SXR states (see Fig. 2).

HARD X-RAY AND IR COMPARISON

We have also examined the relationship between the HXR and the IR. We collected various IR observations (K band) that overlap the time of the HXR measurements made by BATSE. In Fig. 3 is a plot of the HXR versus the IR (K band) flux for daily averages of the data. There is a tentative indication of a possible anticorrelation between the HXR and the IR.

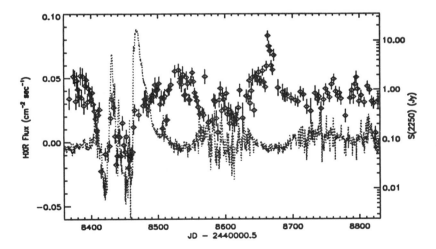

FIGURE 1. Two day averages of the HXR data (diamonds with error bars) with the 2.25 GHz GBI data (dashed line) overlayed.

RELATIVISTIC JET

In February 1997 (TJD 10483) a large radio flare (10 Jy) was observed by GBI and Ryle after a period of quenched radio emission. This flare was detected by BATSE in the HXR and showed a strong correlation with the radio. The flare in the HXR was preceded by a long period of very low HXR flux (below the BATSE one day detection limit). The flare triggered a VLBA observation to obtain high resolution radio images of Cyg X-3 during the major

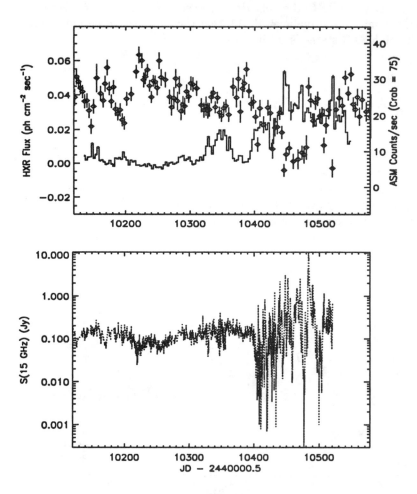

FIGURE 2. *Top panel:* Three day averages of the HXR data (diamonds with error bars) and the ASM data (histogram with error bars). *Bottom panel:* Ryle 15 GHz data over the same time period.

flare. The first image [7] showed a striking curved and one-sided jet. From a comparison with a second image a proper motion of 5.5 milliarcseconds/day was deduced from the motion of features in the jet. The jet was found, assuming a twin-jet model and a distance of 10 kpc, to have a velocity of 0.8–0.9c with an inclination close to our line of sight [8]. Thus major flares in Cyg X-3 appears to result in high velocity jets similar to those in the superluminal sources GROJ1655-40 and GRS1915+105.

REFERENCES

1. Becklin, E. E. et al. 1972, *Nature Phys. Sci.*, **239**, 134.
2. Bonnet-Bidaud, J. M. & Chardin, G. 1988, *Physics Reports*, **170**, 326.
3. Geldzahler, B. J. et al. 1983, *Ap. J.*, **273**, L65.
4. Gregory, P. C. et al. 1972, *Nature Phys. Sci.*,**239**, 114.
5. Liang. E. P. & Nolan, P. L. 1984 *Space Sci. Rev.*,**38**, 353.
6. McCollough, M. L. et al. 1997, in *"Transparent Universe"*, ESA, **SP-382**, p.265.
7. Mioduszewski, A. J. et al. 1997, Proc. of IAU Colloqium 164, in press.
8. Mioduszewski, A. J. et al. 1997, in preparation.
9. van Kerkwijk, M. H. et al. 1992, *Nature*, **355**, 703.
10. Waltman, E. B. et al. 1994, *Astron. J.*, **108**, 179.
11. Waltman, E. B. et al. 1995, *Astron. J.*, **110**, 290.
12. Waltman, E. B. et al. 1996, *Astron. J.*, **112**, 2690.
13. Watanabe, H. et al. 1994, *Ap. J.*, **433**, 350.
14. Zhang, S. N., et al. 1997, *Ap. J.*, **477**, L95.

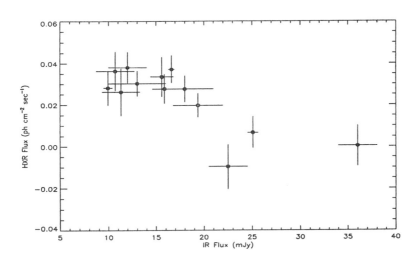

FIGURE 3. Daily averages of the HXR data versus the IR data (K band measurements).

Is There Any Evidence for a Massive Black Hole in Cyg X-3

Abhas Mitra

Nuclear Research Laboratory, Bhabha Atomic Research Center, Mumbai- 400 085, India

Abstract. There has been a recent suggestion that Cyg X-3 contains a black hole (BH) of mass $M_1 \sim 17 M_\odot$. This interpretation is closely linked to a previous claim that Cyg X-3 contains a Wolf-Rayet star of mass $M_2 \sim 10 M_\odot$. The latter interpretation would imply that the X-ray source in this close binary ($P = 4.8$ hr) is enshrouded by a relatively cool superstrong wind with $\dot{M} \geq 10^{-5} M_\odot$ yr^{-1}. It can be shown that such a wind would be completely opaque to the low energy X-rays observed from the source and hence the Wolf-Rayet hypothesis can not be a correct one. By pursuing the same argument it also follows that the compact object must be of modest mass ruling out the existence of a massive BH in Cyg X-3. And a more probable scenario for Cyg X-3 would be one where the compact object is a canonical neutron star (NS) and the companion is an extremely low mass dwarf, $M_2 \sim 10^{-2} M_\odot$ much like $PSR 1957 + 20$.

I INTRODUCTION

In a recent paper, Schmutz, Geballe & Schild (1996, henceforth SGS) have claimed to have presented evidence for the existence of a $\sim 17 M_\odot$ BH in Cyg X-3. This conclusion is based primarily on the previous interpretation of van Kerkwijk et al. (1992, VK) that the companion of Cyg X-3 is a moderately massive Wolf-Rayet star, $M_2 \sim 10 M_\odot$. SGS have claimed to have measured the value of the mass function of the binary as $f(m) = 2.3 M_\odot$ by using the numerical relationship

$$f(m) = 1.035 \times 10^{-7} K^3 P (1-e^2)^{3/2} = \frac{M_1^3 \sin^3 i}{(M_1 + M_2)^2} \quad (1)$$

where P is the binary period in days, e is the eccentricity of the binary, i is the angle of inclination of the orbit and K is the measured orbital Doppler velocity amplitude (in Km s^{-1}). SGS obtained $M_1 = 17 M_\odot$ by taking a value of $e = 0$, $P = 0.2$, $i = 50°$, $K = 480 \pm 20$ Km recent paper (Mitra 1996, henceforth M96), it has been discussed in considerable detail why this Wolf-Rayet interpretation is unlikely to be true for Cyg X-3.

M96 attempted to show that had Cyg X-3 really contained a Wolf-Rayet star with a wind much stronger than $10^{-7} M_\odot\ yr^{-1}$, it would be opaque to low energy X-rays. And massive Wolf-Rayet stars have wind stronger than $\dot{M} \geq 10^{-5} M_\odot\ yr^{-1}$. Following the analytical work on photoionization opacities of a cosmic plasma, it was shown in M96 that for $2-8$ keV X-rays passing through a wind, the photoelectric absorption cross-section approximately varies as $\sigma \sim 5 \times 10^{-22}(E/1\ \text{keV})^{-3}\ \text{cm}^2$ per H atom. Simultaneously the thick wind offers a very large column density (l) to the central X-ray source which results in a very large absorption optical thickness $\tau \gg 1$ for the low energy X-rays.

And it is important to note that the fact that the Wolf-Rayet wind would be absolutely opaque to soft X-rays for a close X-ray binary with $P = 4.8$ hr only *has been verified independently by A.C. Fabian by making use of the photoionization code CLOUDY* (M96).

Nonetheless, in M96, it was assumed that the compact object was of the order of $\sim 1.4 M_\odot$ and let us try to adjudge the probable changes in the former interpretation in case the value of $(M_1 + M_2)$ were $\sim 30 M_\odot$. Since the semimajor axis of the binary changes only modestly in this process, $a \sim (M_1 + M_2)^{1/3}$, the value of l is lowered approximately by a factor of three for a given value of $\dot{M}$ and mean wind speed v. On the other hand, the average value of the ionization parameter, ξ, (Tarter, Tucker & Salpeter 1969) remains approximately constant:

$$\xi = \frac{L_x}{n r_x^2} = \frac{4\pi v m r^2 L_x}{\dot{M} r_x^2} \qquad (2)$$

where L_x is the X-ray luminosity, n is the atomic number density, r is the distance measured from the centre of the companion and r_x is the distance measured from the center of the compact object. For a mean value of $\frac{r}{r_x} \sim 1$, we can see that the value of ξ does not change significantly. In M96, the value of v was taken to be $1000 Km\ s^{-1}$. Even if the value of v is higher, say, $2500 Km^{-1}$, for the extremely high value of $\dot{M} > 10^{-5} M_\odot\ yr^{-1}$ required by SGS, we would still have a low value of $\xi \sim 10^2$ as was originally found in M96.

Consequently, the value of the electron temperature of the wind continues to be very low $\sim 10^4 K$. Since, for a cosmic plasma, the previously referred value of the photoelectric absorption cross-section $\sigma \sim 5 \times 10^{-22}(E/1\ \text{keV})^{-3}$ cm^2 per H-atom holds good over a wide range of temperature $\sim 10^4 - 10^6$ K, all that happens now is that the final optical depths would be reduced by a factor of $3 \times 2.5 \sim 7.5$. This means that if the claim of SGS were true, still we would have values of τ crudely ranging between $40 - 400$ depending on the value of i. Thus, the actual value of $\dot{M}$ must be much lower than what is implied by SGS. And simply this fact rather than any presumed value of M_1 would try to constrain the value of M_2.

Therefore, we believe that the conclusion of M96 that $q \equiv M_2/M_1 \ll 1$ for Cyg X-3 remains valid for a wide range of values of M_1 and M_2.

A REANALYSIS

There are several implicit assumptions behind the philosophy of determining the mass function of a the companion(s) of a binary. One assumption is that the region from where the probe line or signal is emitted behaves like a point source and track the orbital motion. In otherwords the radial extent of the region $\Delta r \ll a$, the semimajor axis of the binary. This condition is approximately satisfied for optical lines emitted from the surface (photosphere) of even a massive star orbiting a wide binary ($P \gg 1$ d), or for emission from a compact object like a neutron star in case of binary (radio) pulsars or X-ray binaries even when the value of P is only a few hr. And if the IR lines observed by SGS really emanate from the surface of the companion this condition might be valid. Even then, for $q \ll 1$, we obtain from equn. (1) $f(m) \sim M_1 \sin^3 i$. So far, the best observational limit on the inclination angle is that due to White & Holt (1982) and which was obtained from the X-ray studies of Cyg X-3: $i \leq 70°$. Therefore, a value of $f(m) = 2.3\ M_\odot$ could actually imply $M_1 \approx 2.7\ M_\odot$ contrary to the estimate of SGS, $M_1 \approx 17\ M_\odot$! The compact object of Cyg X-3 is likely to be spinning rapidly and therefore a value of $M_1 \approx 2.7\ M_\odot$ is well below upper mass limit of a similar neutron star ($\sim 3.8\ M_\odot$) (Friedman & Ipser 1987). This range of a value of M_1 is in broad agreement with the general idea that the compact object in Cyg X-3 is radiating at a near Eddington rate (Mitra 1992a).

The above conclusion has been reached by considering the interpretation of SGS that the measured velocity dispersion could be really ascribed to the orbital motion. However, following some clarification due to R. Gies, now we point out that this very interpretation is unlikely to be true because of the following reasons. In their paper, SGS show a light curve in their Fig. 1 and a radial velocity curve in their Fig. 5. We see that the IR flux minimum (which must occur at an orbital conjunction phase) occurs the epoch as the radial velocity extremum, which, if it were, the result of orbital motion, must occur at a quadrature phase. Thus *it is unlikely that SGS have measured the orbital motion*. Instead, it appears much more probable that that the features SGS have measured are related to the structures in the wind outflow. In fact, their light curve and line profiles resemble the models for wind outflow given by van Kerkwijk et al. 1996), and thus it seems extremely likely that the SGS velocity curve reflects changes in the orientation of structure in the wind and not orbital motion. Note also that in any case, the observed IR lines are supposed to be produced far off in the wind in a region which is relatively cool ($\Delta r \gg r$?). If this is true, the entire work of SGS is rendered irrelevant for the use of mass-function formula (1).

Let us remind here that if the companion is a He-rich dwarf and then strong X-ray irradiation due to the compact object may ablate the companion and generate an evaporative wind either from the companion or from the accretion disk or from both. And in such a case, we would not be hard put to explain how we can have a W-R star with a radius apparently larger than the narrow Roche lobe of CYg X-3 as the companion. Note also that this latter difficulty is usually explaies away by assuming that the strong emission lines are emitted from a region much larger than the orbit size. If it were really so, again, we are led to the conclusion that the observations of SGS do not refer to the orbital motion at all.

II CONCLUSION

We find that even if we take the measured value of SGS at its face ($K = 480 \pm 20$ km/s), the value of the compact object in Cyg X-3 could be as low as $2.7 M_\odot$ and which invalidates the claim for finding evidence for a massive $\sim 17\ M_\odot$ blackhole in Cyg X-3. Yet more importantly, we find that the measurement of SGS may not at all be ascribed to orbital motions rendering it irrelevant for determination of the mass function(s) of the binary. Once, we accept that the the compact object is of modest mass and use the fact that $q \ll 1$, the analysis of M96 would strongly suggest the companion in this binary is a very low mass object $M_2 \sim 0.01 M_\odot$ and, presumably, a He-rich white dwarf ablated in response to the various bombardment of various radiations emanating from the compact object (Mitra 1992b). This estimate would be in agreement with the idea that Cyg X-3 could be an immediate predecessor of a system like PRS 1957+20. Then the evolutionary status of Cyg X-3 suggested by the present work and M96 would be quite similar to the one discussed by Tavani, Ruderman, & Shaham (1989).

REFERENCES

1. Friedman, J.L.& Isper, J.R., *ApJ* **314**, 59 (1987)
2. Mitra, A., *ApJ*, **384**, 255 (1992a)
3. Mitra, A. *ApJ*, **390**, 345 (1992b)
4. Mitra, A., *MNRAS*, **280**, 953 (1996)
5. Schmutz, W., Geballe, T.R., & Schild, H. , *A & A*, **311**, L25 (1996)
6. Tarter, C.B., Tucker, W.H. & Salpeter, E.E., *ApJ* **156**, 943 (1969)
7. Tavani, M., Ruderman, M., & Shaham, J., *ApJ* **342**, L31 (1989)
8. van Kerkwijk, H.M., et al., *Nature* **355**, 703 (1992)
9. van Kerkwijk, H.M. et al., *A&A* **314**, 521 (1996)
10. White, N.E. & Holt S.S., *ApJ*, bf 257, 318 (1982)

Generation of periodical gamma radiation in binary system with a millisecond pulsar

M.A.Chernyakova, A.F.Illarionov

Astro Space Center, P. N. Lebedev Physical Institute,
84/32 Profsoyuznaya St., Moscow 117810, Russia

Abstract. Consider binary system with a millisecond pulsar ejecting relativistic particles and an optical star emitting soft photons with the energy $\omega \simeq 1-10$ eV. These low energy photons are scattered by the relativistic electrons and positrons of the pulsar's wind. The scattered photons forms a wide spectrum from X-ray band up to $\varepsilon \simeq 1 - 1000$ GeV. The luminosity of gamma radiation $L_\gamma = L_\gamma(\psi)$ depends heavily on the angle ψ between the directions to the optical star and to the observer from the pulsar, even when the pulsar wind is isotropic. During the orbital motion this angle varies periodically giving rise to the periodical change of the observed intensity of the gamma radiation and its spectrum. We have calculated the spectral shape $L_\gamma(\varepsilon, \psi)$ of the scattered hard photons. Under the assumption that the energy losses of the relativistic particles are small we receive analytical formulas. We apply our results to the binary system PSR B1259-63 and show that if the wind from the Be star is accounted for then it is possible to reproduce the observed spectrum.

INTRODUCTION

The principal accessible store of energy in a pulsar is its rotational energy, which is liberated at a rate $L_p \sim 10^{32} \div 10^{36}$ erg/s. The bulk of the pulsar luminosity L_p is transferred to the pulsar wind which consists of electrons, positrons and probably heavy ions. The Lorentz factor of the relativistic particles in the wind may vary in range $\gamma \sim 10 - 10^6$. Consider the case of binary with a pulsar ejecting relativistic particles and an optical star emitting soft photons in optic and UV band with the energy $\omega \simeq 1 - 10$ eV (see fig.1). These low-energy photons are scattered by the pulsar wind relativistic electrons and positrons. The energy of the photon after the inverse Compton scattering is very high - $\varepsilon_{max} \sim \omega\gamma^2$ in the Thomson limit ($\omega\gamma \ll mc^2$) and $\varepsilon_{max} \sim mc^2\gamma$ in the opposite case (m is a mass of an electron, c is a speed of light). The scattered photons forms a wide spectrum from X-ray band up to $\varepsilon \simeq 1-1000$ GeV.

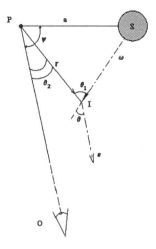

FIGURE 1. The geometry under consideration. P is a pulsar, S is an optical star, I is a point of interaction of the electrons and photons, O is an observer. Note that these points are not in one plane in general case.

ANISOTROPY OF THE ENERGY RADIATED FROM THE BINARY SYSTEM

Lets calculate the ψ-dependence of the luminosity of the hard radiation emitted from the binary system $L_\gamma(\psi)$ in the Thomson limit ($\omega\gamma/mc^2 \ll 1$). The luminosity of scattered hard photons is equal to the particle energy losses in the course of the particle motion from the pulsar to the infinity. The total energy losses of the electron moving along the radial trajectory at an angle ψ to the line PS in the radiation field of the optical star are (Chernyakova&Illarionov,1997) $mc^2\gamma_0 \left[1 - (1 + K\Phi(\psi))^{-1}\right]$, where

$$\Phi(\psi) = \frac{3(\pi - \psi)}{2\sin\psi} - 2 - \frac{\cos\psi}{2}, \qquad (1)$$

$$K = \sigma_T L_* \gamma_0 \big/ (4\pi a m c^3), \qquad (2)$$

γ_0 is the initial Lorentz factor of the electron, σ_T - is the Thomson cross-section. Parameter K represents the efficiency of the pulsar luminosity transformation to the energy of gamma radiation.

As the bulk of the energy losses of the relativistic electron transfers to the scattered photons moving in the direction of the electron movement the angular dependence of the energy radiated from the binary system $L_\gamma(\psi)$ agree with the angular dependence of the energy losses of the electrons. Thus the total energy losses of the isotropic flow of all relativistic particles in the direction of the observer located at an angle ψ to the line PS in a unit solid angle in the case of monoenergetic pulsar wind is

$$L_\gamma(\psi) \approx \frac{L_p}{4\pi}\left(1 - \frac{1}{1+K\Phi(\psi)}\right). \qquad (3)$$

This is the energy transferred to the gamma band. Equation (3) is valid for the arbitrary value of K but it has no use for ψ less then R_*/a and bigger then $\pi - R_*/a$ (R_* - is a radius of the optical star) as the optical star is not a point source. If ψ is not too small then it follows from (3) that under the condition $K \ll 1$ the energy decrease is small and $L_\gamma(\psi) = \frac{L_p}{4\pi}K\Phi(\psi)$. From (3) it follows that the total energy transferred to the gamma band under the assumption that $K \ll 1$ is :

$$L_\gamma = \left(\frac{3}{8}\pi^2 - 2\right) L_p K. \qquad (4)$$

THE SPECTRAL AND THE ANGULAR DEPENDENCE OF THE OUTGOING RADIATION

Lets find the spectral and angular dependence of the outgoing hard radiation going beyond the Thomson limit under the assumption that the optical star emits a black body radiation with temperature T. Then the luminosity of the scattered quanta moving at an angle ψ to the line PS within a unit solid angle in a unit of time in the energy range from ε to $\varepsilon + d\varepsilon$ in the case of monoenergetic pulsar wind is:

$$L_\gamma(\varepsilon,\gamma,\psi) = \frac{45\sigma_T N_{e\pm} L_*}{128\pi^6 \gamma^2 c T^2} \frac{\varepsilon F_{bb}(\varepsilon,\gamma,\psi)}{a}, \qquad (5)$$

where $N_{e\pm}$ is the number of particles leaving the pulsar per second (see the definition of $F_{bb}(\varepsilon,\gamma,\psi)$ and the derivation of this formula in the paper of Chernyakova&Illarionov, 1997).

During the orbital motion ψ changes periodically. In Figure 2 the dependence of F_{bb} on ε for different positions of the companions in the circular orbit and for different values of γ is shown.

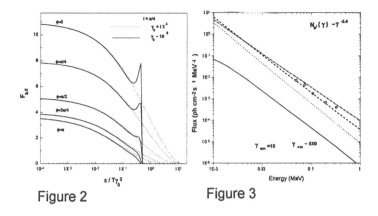

Figure 2   Figure 3

FIGURE 2. The dependence of F_{bb} on ε for different positions of the companions in the circular orbit for $\gamma = 10^4$ and $\gamma = 10^6$. i is an inclination angle, φ is a true anomaly. There is simple relation between ψ and φ $\cos\psi = \sin i \cos\varphi$.

FIGURE 3. The observed spectrum of the system PSR B1259-63. OSSE spectrum of emission from 1994 January 3-23(black dots)(Grove et al. 1995) and schematic extrapolations to the power-law spectra of $ASCA$ observations from 1993 December 28(dashed line), 1994 January 10 (dotted line), and 1994 January 26 (dashed-dotted line)(Kaspi et al. 1995) are shown. The solid line shows the result of a free pulsar wind model.

COMPARISON WITH THE OBSERVATIONS

The only known binary system which contains radio pulsar and emits non pulsed X-ray radiation is PSR B1259-63 system. The PSR B1259-63 system contains the radio pulsar with the spin period $P = 47.76$ ms, rotating around the massive Be star SS 2883 along the highly eccentric orbit ($e = 0.87$) with an orbital period 3.4 years. The distance between the companions at periastron is about 10^{13}cm. Spin-down luminosity of the pulsar is $L_p \simeq 9 \times 10^{35}$erg/s. The luminosity of the Be star is $L_* = 2.2 \times 10^{38}$erg/s (Johnston et al. 1994). The distance from the Earth to the system is about $D = 2$kpc.

The review of the different discussed models of the origin of the non pulsed X-ray spectrum from the system can be found in the paper of Tavani&Arons 1997. We applied the results of our paper to this system. For receiving the observed photon power law spectrum with an index about 1.7 in the energy range 2-1000 keV we take the power law distribution of the electrons and positrons in the pulsar wind $\frac{dN_{e^\pm}}{d\gamma} = C\gamma^{-2.4}L_p/mc^2$, $C = \left[0.4 \Big/ \left(\gamma_{\min}^{-0.4} - \gamma_{\max}^{-0.4}\right)\right]$.

Taken the average of formula (5) over the electrons spectrum in a range $10 < \gamma < 500$ we obtain the X-ray spectrum resulting from our model (see Fig.3 solid line). As it can be seen from the Figure 3 while the form of the spectrum is quite close to the observed one the intensity of the radiation resulted from our model is less then the observed one by a factor of 30.

This discrepancy is due to the fact that we used formulas, derived under the assumption of the undisturbed free flow of the pulsar wind which is not correct in the system PSR B1259-63 where the strong mass outflow from the Be star presents. Writing the mass rate of the polar wind of the Be star as $\dot{M} = (10^{-8} M_\odot/yr)\, \dot{M}_{-8}$ and the wind velocity as $v = 10^8 v_8$ cm/s we find the ratio of the impulses of the two winds $L_p/cv\, \dot{M} = 0.45/v_8\, \dot{M}_{-8}$. Thus for the Be star polar wind typical parameters the centrally located shock between the pulsar and the star due to the interaction between the two winds seems to appear and thus the free flow of the pulsar wind will be disturbed. The big difference between the values of the velocities of the particles from the different sides of the tangential discontinuity leads to the growth of the instabilities and the two winds will be macroscopically mixed between the shocks. The results of numerical calculations of Igumenshchev (1997) verified such a picture. Then the heavy non relativistic wind slows down the volumes filled by the relativistic electrons and positrons and they acquire essentially non relativistic hydrodynamic drift velocity v_d along the shock while the energy of electrons and positrons does not changes significantly. With the decrease of the hydrodynamic velocity of the relativistic plasma the time which it spends near the optical star increases in c/v_d times. The idea of small drift velocity as applied to the system LSI 61°303 was discussed by Maraschi&Treves (1981). The effective transformation parameter $K_{eff} \sim \frac{c}{v_d} K$ thus can be large enough to overcome the discrepancy between the simple theory and observations.

ACKNOWLEDGMENTS

This work is supported in part by the RFBR grant 97-02-16975 and the Cariplo Foundation for Scientific Research.

REFERENCES

1. Chernyakova M.A., Illarionov A.F., *MNRAS*, in press (1997).
2. Grove J.E., Tavani M., Purcell W.R., Johnson W.N., Kurfess J.D., Strickman M.S., Arons J., *ApJ*, **447**, L113 (1995).
3. Igumenshchev I.V., private communication (1997).
4. Johnston S., Manchester R.N., Lyne A.G., Nicastro L., Spyromilo J., *MNRAS* **268**, 430 (1994).
5. Kaspi V.M., Tavani M., Nagase F., Hirayma M., Hoshino M., Aoki T., Kawai N., Arons J., *ApJ*, **453**, 424 (1995).
6. Maraschi L., Treves A., *MNRAS*, **194**, 1p (1981).
7. Tavani M., Arons J., *ApJ*, **477**, 439 (1997).

GALACTIC BLACK HOLE CANDIDATES

The MeV Spectrum of Cygnus X-1 as Observed with COMPTEL

M. McConnell*, K. Bennett[‡], H. Bloemen[||], W. Collmar[¶], W. Hermsen[||], L. Kuiper[||], R. Much[‡], J. Ryan*, V. Schönfelder[¶], H. Steinle[¶], A. Strong[¶], and R. van Dijk[‡]

University of New Hampshire, Durham, NH
[‡]*Astrophysics Division, ESTEC, Noordwijk, The Netherlands*
[||]*SRON-Utrecht, Utrecht, The Netherlands*
[¶]*Max Planck Institute (MPE), Garching, Germany*

Abstract. The COMPTEL experiment on the Compton Gamma-Ray Observatory (CGRO) has observed the Cygnus region on several occasions since launch. These data represent the most sensitive observations to date of Cygnus X-1 in the 0.75-30 MeV range. The spectrum shows significant evidence for emission extending out to several MeV. These data alone suggest a need to modify the thermal Comptonization models or to incorporate some type of non-thermal emission mechanism. Here we report on the results of an analysis of selected COMPTEL data collected during the first three years of the CGRO mission. These data are then compared with contemporaneous data from both BATSE-EBOP and OSSE. Given a lack of consistency between the OSSE and BATSE-EBOP spectra, it is difficult to draw firm conclusions regarding the exact shape of the spectrum near 1 MeV. A few general conclusions can, however, be drawn from these data.

INTRODUCTION

It has become increasingly apparent over the last several years that the standard thermal Comptonization model [1] does not provide an adequate description of the broad-band spectrum of Cyg X-1. Several modifications to the standard model have been proposed that seek to provide a better fit to the data. For example, modifications to the standard model have been developed which expand the range of allowable parameter space [2-4]. Other models have pursued alternative geometries that can also lead to improvements in the model. These include the incorporation of Compton backscatter radiation from a cooler optically-thick accretion disk [5,6] or models based on a thermally stratified geometry [7,8]. Still other theorists have proposed schemes which are based on nonthermal acceleration processes [9]. All of these models have their merits. Unfortunately, given the quality of the available data, it is difficult to determine a clearly favored candidate to account for the observed spectrum near 1 MeV.

OBSERVATIONS AND DATA ANALYSIS

To date, COMPTEL has obtained numerous observations of the Cygnus region. Most of the high-quality (i.e., near on-axis) observations took place during the first three years of the mission. Here, we have selected a subset of these data for analysis (Table 1). The choice of observations was dictated by the availability of contemporaneous OSSE data. These data, along with contemporaneous results from BATSE, can, in principle, be used to assemble an improved picture of the spectrum near 1 MeV. In all cases except VP 318.1, the BATSE 45-140 keV flux level (as derived from Earth occultation analysis) is fairly constant at ~ 0.1 photons cm^{-2} sec^{-1}. During VP 318.1, the hard X-ray flux was lower by about a factor of five.

The analysis of COMPTEL data involves generating a series of images, one for each energy interval of interest. Assumptions regarding the spectral shape are incorporated into the point-spread-functions (PSFs) used in the analysis of each image. Flux values derived from each image are used to compile a spectrum of the source. The resulting spectrum is then compared versus that assumed for the PSF generation to insure a consistent analysis. The COMPTEL image analysis for Cyg X-1 is complicated by the fact that we are looking in the galactic plane. Images generated with COMPTEL data generally show some level of spatial structure, much of which is believed to result from galactic diffuse emission. In the present case, the spatial analysis of each energy

TABLE 1. List of *COMPTEL* observations used in the present analysis. The effective exposure is a measure of the equivalent on-axis exposure (measured in days), taking into account earth occultations, data gaps, etc.

Viewing Period	Start Date	Start TJD	End Date	End TJD	Viewing Angle	Effective Exposure
2.0	30-May-1991	8406	8-Jun-1991	8415	1.7°	3.65
7.0	8-Aug-1991	8476	15-Aug-1991	8483	11.2°	2.72
203.0	1-Dec-1992	8957	8-Dec-1992	8964	7.0°	1.75
203.3	8-Dec-1992	8964	15-Dec-1992	8971	7.0°	1.75
203.6	15-Dec-1992	8971	22-Dec-1992	8978	7.0°	1.69
212.0	9-Mar-1993	9055	23-Mar-1993	9069	15.4°	2.71
318.1	1-Feb-1994	9384	8-Feb-1994	9391	4.5°	1.78
328.0	24-May-1994	9496	31-May-1994	9503	7.0°	1.56
331.0	7-Jun-1994	9510	10-Jun-1994	9513	7.0°	0.95
331.5	14-Jun-1994	9517	18-Jun-1994	9521	7.0°	1.34
333.0	5-Jul-1994	9538	12-Jul-1994	9545	7.0°	1.86

interval was performed independently using a variety of spatial distributions. These included models for the expected distribution of the galactic diffuse emission (based on the known gas distributions) and also empirical modeling using a superposition of one or more sources. Models for PSR 1951+32 (located 2.6° away from Cyg X-1) were also included in the analysis. This pulsar has been detected by EGRET and there is evidence (based on a joint timing and spatial analysis) for it in the COMPTEL data as well [10]. Variations in the derived flux using different spatial models provided a handle on the systematic uncertainties in the analysis.

The COMPTEL spectrum for Cyg X-1 shows clear evidence for emission extending out to at least 2 MeV. These data alone can be modeled as a power law spectrum with a photon index of -3.7. Good fits can also be obtained using Comptonization models (with electron temperatures in the range of 450–700 keV), but the extrapolation of these fits to lower energies is quite poor. Such values for the electron temperature are much higher than those derived from fits at lower energies. The COMPTEL data alone, therefore, suggest some inadequacy in the ability of the Comptonization models to fit the broad-band spectrum.

The COMPTEL spectrum (accumulated from the viewing periods listed in Table 1) is shown along with contemporaneous BATSE-EBOP and OSSE data in Figure 1. (BATSE-EBOP refers to an analysis of BATSE data using the JPL Enahnced BATSE Occultation Package [11].) This comparison demonstrates an inherent difficulty in precisely determining the shape of the spectrum near 1 MeV. In particular, the BATSE-EBOP data shows a clear trend towards higher flux levels, while the OSSE spectrum shows a clear trend toward lower flux levels. We are presently investigating the possibility that both trends (especially near 1 MeV) may be due to the presence of additional sources of emission in the Cygnus region. The COMPTEL analysis, for example, requires spatial modeling of several features in order to obtain a reliable spectrum. The emission in the region can be modeled either as a collection of (two or three) point sources or as a distribution which follows the general gas distribution within the galaxy. In either case, such emissions may have an impact on the spectra derived from both BATSE-EBOP and OSSE data. For example, the BATSE-EBOP spectrum is derived from an analysis which involves modeling various BATSE background components along with a large number of point sources distributed throughout the sky [11]. No provision is made for any diffuse emissions, which may be far more important at 1 MeV than at 100 keV. Any galactic diffuse emission, if present, would therefore be included in the derived spectrum for Cyg X-1. Under the hypothesis of spatially distributed emission in the Cygnus region, the BATSE-EBOP spectrum would therefore tend to show unrealistically high flux levels. Likewise, the OSSE background subtraction process (on-source / off-source) would result in a background level that would be too high, leading to a reduced flux level in the derived Cyg X-1 spectrum. This type of trend is precisely what we observe. Whether or not it can explain the observed trends quantitatively (particularly near 1 MeV) is currently being investigated. The importance of this result is that observations with COMPTEL can be used to map out the spatial structure of emission near 1 MeV and to more precisely pin down the 1 MeV flux level. It may be that accurate spectral measurements near 1 MeV will require more detailed knowledge of the spatial distribution of the emission, information which only COMPTEL can easily provide.

DISCUSSION

Given the spectra presented in Figure 1, it is difficult to draw any firm conclusions about the nature of the spectrum near 1 MeV. As noted above, the COMPTEL data alone seems to further corraborate the conclusion that

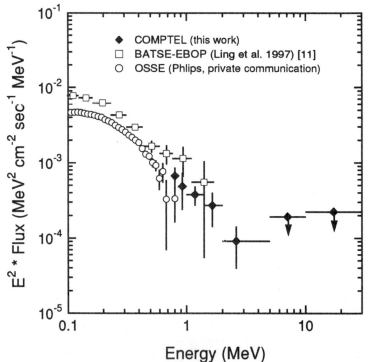

FIGURE 1. Contemporaneous spectra of Cyg X-1 as derived from COMPTEL, BATSE-EBOP and OSSE. Some OSSE upper limits have been removed for the sake of clarity. Error bars are 1σ. Upper limits are 2σ. Errors on COMPTEL data include estimates of systematic uncertainties due to spatial modeling of the COMPTEL images.

standard thermal Comptonization models may be inadequate in describing the observed spectrum – Comptonization models fall off far too rapidly near 1 MeV. A comparsion with BATSE-EBOP and OSSE data is difficult to make, given the discrepancies between the OSSE and BATSE-EBOP spectra. Independent fits to the BATSE/COMPTEL and the OSSE/COMPTEL spectra both lead to two conclusions: 1) the standard Comptonization models [1,2], when fit over the full range of the data, both fall off too rapidly to account for the flux near 1 MeV; and 2) improved fits with the Comptonization models can be obtained by limiting the fit to energies above 300 keV. These conclusions are both broadly consistent with those models which introduce Compton reflection. The reflection component in these models contributes only to energies below $\sim$ 300 keV, so that the fits we obtain at energies above 300 keV may be a more realistic estimate of the electron temperature. On the other hand, stochastic acccleration models [9] predict a hard tail feature near 1 MeV which may also be capable of modeling the COMPTEL results. Unfortunately, more detailed conclusions must await a resolution of the discrepancies between all of these data at energies below 1 MeV.

ACKNOWLEDGEMENTS

The *COMPTEL* project is supported by NASA under contract NAS5-26645, by the Deutsche Agentur für Raumfahrtgelenheiten (DARA) under grant 50 QV90968 and by the Netherlands Organization for Scientific Research NWO. We would also like to acknowledge both B. Phlips for providing the contemporaneous OSSE spectrum and J. Ling for providing the contemporaneous BATSE-EBOP spectrum.

REFERENCES

1. Sunyaev, R. A. & Titarchuk, L. G. 1980, A&A, 86, 121.
2. Titarchuk, L. 1994, ApJ, 434, 570.
3. Hua, X.M. & Titarchuk, L. 1995, ApJ, 449, 188.
4. Skibo, J.G., et al. 1995, ApJ, 446, 86.
5. Haardt, F., Done, C., Matt, G., & Fabian, A. C. 1993, ApJ Letters, 411, L95.
6. Wilms, J., Dove, J.B., Maisack, M., and Staubert, R. 1996, A&A Supp., 120, C159.
7. Skibo, J.G. & Dermer, C.D. 1995, ApJ Letters, 455, L25.
8. Ling, J.C., et al., 1997, ApJ, in press.
9. Li, H., Kusunose, M., & Liang, E.P. 1996, ApJ Letters, 460, L29.
10. Hermsen, W., et al. 1997, Proc. 2nd INTEGRAL Workshop "The Transparent Universe", ESA SP-382, p. 287.
11. Ling, J.C., et al., 1997, ApJ Supp, submitted.

Five Years in the Life of Cygnus X-1: BATSE Long-Term Monitoring

W. S. Paciesas*[#], C. R. Robinson[b#], M. L. McCollough[b#],
S. N. Zhang[b#], B. A. Harmon[#] and C. A. Wilson[#]

University of Alabama in Huntsville, AL 35899
[b] Universities Space Research Association
[#] NASA/Marshall Space Flight Center, Huntsville, AL 35812

Abstract. The hard X-ray emission from Cygnus X-1 has been monitored continually by BATSE since the launch of CGRO in April 1991. We present the hard X-ray intensity and spectral history of the source covering a period of more than five years. Power spectral analysis shows a significant peak at the binary orbital period. The 20–100 keV orbital light curve is roughly sinusoidal with a minimum near superior conjunction of the X-ray source and an rms modulation fraction of approximately 1.7%. No longer-term periodicities are evident in the power spectrum. We compare our results with other observations and discuss the implications for models of the source geometry.

INTRODUCTION

Cyg X-1, the prototypical galactic black hole, is one of the most intensively studied objects in the X-ray sky. Nevertheless, our understanding of its detailed nature has been slow to evolve, due in large part to the strong but rather chaotic variability of its X-ray emission. Long-term monitoring in soft X-rays has been performed by several instruments, including the all sky monitors on Ariel 5 [4], Vela 5B [11], and Ginga [5]. These have identified two main periodic components in Cyg X-1, the 5.6 day binary orbital period and a less well defined period of ~300 days [11], the cause of which is speculative.

The now well-known soft ("high") and hard ("low") emission states were also identified from soft X-ray data. Such states are now known to be common features of black holes, although not all black hole candidate systems have shown both states (e.g., GS 2023+33, LMC X-1).

With CGRO/BATSE we have now accumulated more than 5.5 years of monitoring of Cyg X-1 in hard X-rays. Moreover, since the launch of RXTE we now have wide band long-term monitoring of Cyg X-1. This enabled us to obtain the most comprehensive observations yet of a complete soft state episode [14]. We report here exclusively on the BATSE long-term monitoring, including a broad overview of the intensity and spectral variability and our search for periodic or quasi-periodic components.

OBSERVATIONS

Data obtained between 21 Apr 1991 and 24 Sep 1996 (TJD 8367–10350) were processed using the standard BATSE Earth occultation software. Fluxes in the energy range 20–100 keV were calculated by fitting standard spectral models to the 16-channel count spectra, either from individual occultation steps or summed over one day. Two models were used: a single power-law and an optically thin thermal bremsstrahlung (OTTB). In general, our conclusions do not depend significantly on the choice of model spectrum.

Figure 1 shows the long-term intensity and spectral history of Cyg X-1 in the 20–100 keV energy range using one-day integrations. The two previously known soft state episodes are clearly visible around TJDs 9350–9410 and 10220–10300. During the remaining time, the source intensity in the hard state fluctuated rather randomly, staying mostly between 0.2 and 0.35 ph cm^{-2} s^{-1}. The hard state spectral index remained relatively steady around a value near -1.85, although extended periods with slightly softer spectra occur, e.g., TJD 8950–9030. Flares above the 0.35 ph cm^{-2} s^{-1} level typically last only a few days and show no spectral differentiation, whereas dips below the 0.2 ph cm^{-2} s^{-1} level typically last a week or so, and may or may not show spectral softening (cf. the intensity dips around TJDs 9510 and 9610).

Figure 2 shows a plot of the spectral index vs. flux for the same data set. The predominance of the -1.85 spectral index over a wide range of intensities is obvious, as is the trend toward softer spectral indices at low intensity. However, the hard state can persist down to intensities as low as 0.15 ph cm^{-2} s^{-1} and spectra as soft as $\alpha \simeq -2.2$ can be present at essentially any intensity. Spectra with $\alpha \lesssim -2.2$ are mainly confined to flux levels below ~ 0.15 ph cm^{-2} s^{-1}, which appear to occur only during the low energy "high" states. We do not see well-defined intensity/spectral states corresponding to the three-state classification scheme outlined by Ling et al. [6].

To search for periodic and quasi-periodic signals, we used fluxes deconvolved in a similar manner, but at the resolution of individual occultation steps. We present results using the OTTB model, which produced slightly more robust fits. The unevenly sampled power density spectrum (PDS) (Figure 3) is rather flat at low frequencies, falls off roughly as $1/f$ above ~ 0.005 cycle/day (200 day period), and reaches the Poisson noise level at $f \simeq 0.4$. To estimate the significance of peaks in the data, we first fit the data with a combination of a constant power Z_0 at low frequencies and a power-law $a_0 f^{a_1}$ above a break frequency f_c. The resulting parameters were $Z_0 = 148$, $a_0 = 0.333$,

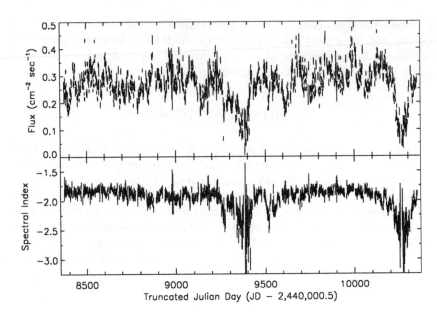

FIGURE 1. Long-term hard X-ray intensity and spectral history of Cyg X-1 from BATSE observations. The upper panel shows the integrated photon flux in the 20–100 keV band derived by fitting a power-law model to one-day average count spectra. The lower panel shows the corresponding photon number spectral index.

$a_1 = -1.117$, and $f_c = 0.00425$. After dividing by the red noise fit, the maximum of the power spectrum is at $f = 0.178606$, consistent with the binary orbit $f = 0.178580$ [2]. Treating this as an *a priori* interesting frequency, the probability of a chance fluctuation is 1.4×10^{-7}. If we ignore the *a priori* argument and scale by the number of independent trial frequencies, the chance probability is 5.4×10^{-5}. Our data show no evidence for a peak around $f = 0.0033$ (300 day period); however, our sensitivity below ~ 0.01 Hz is limited by the red noise.

Figure 4 shows the data folded at the orbital period. The modulation is roughly sinusoidal, with a minimum near phase 0 (supergiant companion nearest the observer). The best-fitting sine function has a minimum at phase 0.025 ± 0.008 and peak-to-peak amplitude 0.0094 ± 0.0004 ph cm^{-2} s^{-1} (statistical errors only), which corresponds to 3.8% of the average intensity. The rms scatter about the mean is $\sim 1.7\%$.

DISCUSSION

Detection of the 5.6 day orbital variation in hard X-rays was first reported by Ling et al. [7], who found a peak-to-peak amplitude of $\sim 6\%$ in 50–140

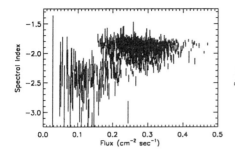
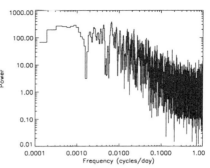

FIGURE 2. Correlation plot of spectral index vs. photon number flux using the same data as in Figure 1. For clarity, the flux errors are not shown.

FIGURE 3. The power density spectrum of Cyg X-1 hard X-ray flux, computed from single occultation step measurements using the Scargle algorithm [13] for unevenly sampled data.

keV using $\sim$120 days of data from the HEAO-3 gamma ray spectrometer. Priedhorsky et al. [12] reported a marginal detection of $10 \pm 4\%$ in 17–33 keV from $\sim$70 days of observations with WATCH/Eureca. Phlips et al. [10], using $\sim$120 days of CGRO/OSSE data, did not find a significant variation in 60–140 keV, with an upper limit to the rms fraction of 5%. Our result is consistent with the OSSE upper limit, and marginally consistent with the earlier observations.

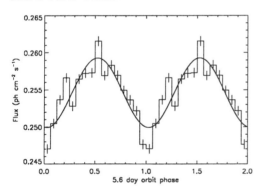

FIGURE 4. Cyg X-1 single occultation step data folded at the 5.6 day binary orbital period. Two cycles are shown for clarity. The best-fitting sine function is superimposed. Error bars represent the statistical error in the mean for each phase bin. Phase 0 corresponds to the time when the supergiant companion is closest to our line of sight.

Since the low energies show obvious absorption dips near phase 0 [4,8,9], it is natural to consider absorption as responsible for the variation we see. The decrease in our data centered roughly around phase 0 cannot be due to absorption by cold matter because the column density required would cause a total eclipse of the soft X-rays, which is not observed. Electron scattering by highly ionized material would require a maximum column density $\sim 6 \times 10^{24}$ cm^{-2}. If the material is in a stellar wind, the nearly sinusoidal shape we observe implies that this material is spread over a large portion of the orbit. This is inconsistent with the much lower column density ($\simeq 3 \times 10^{23}$ cm^{-2}) estimated for such a wind from 9–12 keV data [12].

An alternative possibility is a variable reflection component from the accretion disk or the companion star. Done et al. [1, also see ref. [3]] showed that the Cyg X-1 spectrum can be fit with a model involving reflection from an ionized accretion disk, similar to models for active galactic nuclei. In these fits, the reflection component represents ∼30% of the flux in the 20–100 keV range, so that our results could be explained by a variation of 5–10% in reflectivity as a function of phase.

SUMMARY

BATSE has observed Cyg X-1 continually for more than 5.5 years. The hard X-ray light curve is dominated by red noise that has a flat power spectrum below a frequency of ∼0.004 cycle/day (periods $\simeq$ 250 days) and falls off roughly as $1/f$ at higher frequencies. Periodic variability is detected at the binary orbital period, with an rms modulation of ∼1.7% and a minimum flux at the time of superior conjunction of the supergiant companion (phase 0). We find no evidence for the previously reported long-term period of ∼300 days.

The 20–100 keV spectrum of Cyg X-1 appears to have a spectral hardness limit around a power-law index $\alpha \simeq -1.8$. BATSE has observed such a spectrum over a flux range of at least a factor of three, from ∼0.15 to $\gtrsim$0.45 ph cm^{-2} s^{-1}. However, softer spectra can be present at any observed flux level. Below ∼0.15 ph cm^{-2} s^{-1}, only soft spectra ($\alpha \lesssim -2.25$) associated with the soft (high) X-ray state have so far been seen.

REFERENCES

1. Done, C., et al. 1992, *ApJ* **395**, 275
2. Gies, D.R., & Bolton, C.T. 1982, *ApJ* **260**, 240
3. Haardt, F., et al. 1993, *ApJ* **411**, L95
4. Holt, S.S., et al. 1979, *ApJ* **233**, 344
5. Kitamoto, S., et al. 1997, preprint
6. Ling, J.C., et al. 1983, *ApJ* **275**, 307
7. Ling, J.C., et al. 1990, *Proc. 21st ICRC, Adelaide*, Vol. 1, p. 197
8. Mason, K.O., et al. 1974, *ApJ* **192**, L65
9. Murdin, P. 1975, in *X-Ray Binaries*, NASA SP-389, p. 425
10. Phlips, B.F., et al. 1996, *ApJ* **465**, 907
11. Priedhorsky, W.C., et al. 1983, *ApJ* **270**, 233
12. Priedhorsky, W.C., et al. 1995, *A&A* **300**, 415
13. Scargle, J.D. 1982, *ApJ* **263**, 835
14. Zhang, S.N., et al. 1997, *ApJ* **477**, L95

Spectral Evolution of Cyg X-1 During Its 1996 Soft State Transition

S. N. Zhang**, Wei Cui†, B. A. Harmon* W.S. Paciesas*‡

Marshall Space Flight Center, Huntsville, AL 35810
* Universities Space Research Association
† *Center for Space Research, Massachusetts Institute of Technology, Cambridge, MA 02139*
‡ *University of Alabama in Huntsville, Huntsville, AL 35899*

Abstract. We report the broad band, near continuous spectral evolution of Cyg X-1 during its 1996 soft state transition, observed with ASM/RXTE (1.3-12 keV) and BATSE/CGRO (20-2000 keV). The spectra above 20 keV can be well described by a power law with an exponential cutoff. During the hard to soft state transition the power law became steeper and the cutoff energy became higher, and vise versa during the soft to hard state transition. In the middle of the soft state, there is no evidence for any spectral break up to at least 250 keV.

INTRODUCTION

Cyg X-1 is the first known black hole candidate, with a compact object mass estimated to be in excess of 3 $M_\odot$ [16,2], but more likely to be 10-20 $M_\odot$ [8]. Despite extensive observational and theoretical studies over the past 30 years, many properties of the system, especially the mass accretion conditions near the assumed black hole and the hard X-ray production mechanism, still remain unknown.

One of the most important observational characteristics of Cyg X-1 is its state transitions, between its regular hard/low state and the soft/high state. Until 1996, all observations of these state transitions [15,9,11] were restricted to narrow energy bands with rather poor temporal coverage. The most comprehensive observations of a state transition occurred in 1996 during a complete hard to soft [3,4,20] to hard state [21,20] transition cycle. Except for the near continuous monitoring observations with ASM/RXTE and BATSE/CGRO throughout the entire transition episode [18], all other observations reported so far were taken during the soft state of the source. These include the detailed spectral and timing studies with RXTE PCA and HEXTE [1,5,6], a high spectral resolution study with ASCA [7], broad band studies with SAX [14] and other pointed pbservations with OSSE/CGRO and RXTE [13]. In this paper we report the details of the broad band (1.2-300 keV) spectral evolution of Cyg X-1 throughout its 1996 soft state transition, observed with ASM/RXTE and BATSE/CGRO.

DATA ANALYSIS AND RESULTS

We use the data from the near continuous monitoring with ASM/RXTE and BATSE/CGRO. The entire state transition episode is divided into 15 smaller segments according to the individual CGRO pointing periods. Within each pointing period, the BATSE detector count rates (8-10 channels between 20-300 keV) are fit with a simple spectral model consisting of a power-law with an exponential cutoff (see Table 1 for details), using the known BATSE detector responses. Other simple spectral models, such as a straight power-law, or the optically thin and thermal bremsstrahlung model, produce generally much poorer fits to the data. Because the correct spectral model is unknown and the derived photon spectrum is model dependent, we need to estimate the systematic errors introduced by assuming a spectral model. We use the differences between the spectrum derived from the straight power-law model and that obtained with the more flexible power-law with an exponential cutoff model as the estimated systematic errors. Since the main differences between these two spectral models are the high energy tail above the spectral break, and this part of the spectrum contributes the most to the non-photoelectric detections, the estimated systematic errors should be taken as the upper limits to the real systematic errors.

The three channel (1.3-3.0, 3.0-4.8 and 4.8-12 keV) ASM detector count rates are first converted into relative Crab flux units by using the corresponding average count rates from the Crab Nebula. Then the ASM count rates are converted into the real flux units, using the Crab Nebula spectrum detected with BATSE (a power-law with a photon index of -2.0 and an integrated flux of 0.28 $ph/cm^2/s$ between 20-100 keV). Since the Cyg X-1 spectrum does not deviate signficantly in this energy range from the straight power-law spectrum of the Crab Nebula, such conversion does not produce signficant systematic errors. These results are shown in Table 1 and Figure 2. For clarity, only the best fit spectral model (without errors) and the statistic errors for the data points are shown in the plots.

From these first ever continuous observations of a complete episode of state transitions from Cyg X-1, several conclusions can be drawn: (a) the observations presented here covered two distinct spectral states, i.e., the soft state between $\sim$10220 and $\sim$10310 and the hard state before and after that period; (b) the hard state power-law photon index is -1.8 to -1.4 and cutoff between $\sim$120 and $\sim$150 keV; (c) in the soft state, the power-law became much steeper (index -2.5 to -2.0), without obvious cutoff up to $\geq$250 keV; (d) segment 'A' covers an intermediate state between the hard and the soft state; (e) extrapolations of the BATSE power-law into the ASM enery range (1.3-12 keV) do not always match the ASM measurements, especially at the beginning of the hard to soft and soft to hard state transitions; this is due to the observation that the state transitions seems to occur more abruptly in the soft band than in the hard band.

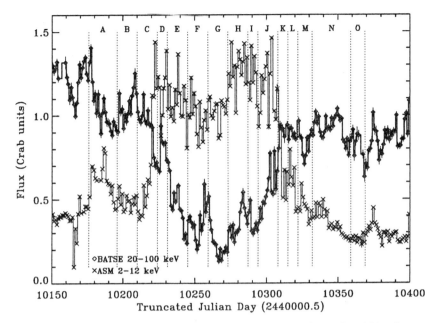

FIGURE 1. ASM and BATSE light curves of Cyg X-1 throughout the 1996 soft state transition.

DISCUSSION

The physical mechanism responsible for the observed state transitions in Cyg X-1 is still not understood. Many of the earlier models assumed that the soft state transition was caused by a sudden increase of the mass accretion rate through the inner disk boundary located at three times the Schwarzschild radius of the black hole. However the observed bolometric luminosity and the inferred mass accretion rate variations during the transitions seem to rule out this possibility [18]. The Lightman-Eardley instability [10] also seems unlikely to work here, since it would require a higher luminosity in the hard state, again contrary to the observeations [18]. The recently developed Advection Dominated Accretion Flow model [12] allows the transition to occur with only a marginal increase of mass accretion rate over a threshold value. This model, however, requires usually a much larger inner disk boundary change between the soft and hard states than inferred from the observations [18]. The observed inner disk boundary movements have been proposed to be due to the rotation direction reversal of the accretion disk surrounding a mildly spinning black hole [17]. In all the above mentioned models, the apparently different transition times in the soft and hard bands are not explained naturally. Therefore a self-consistent model is still needed to explain all these key observations of the Cyg X-1 state transitions.

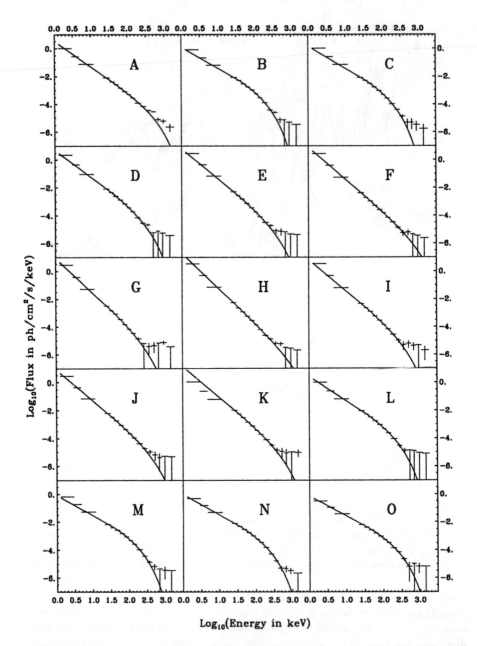

FIGURE 2. ASM and BATSE spectra of Cyg X-1 throughout the 1996 soft state transition, during each CGRO pointing period. The corresponding periods and the model parameters are shown in table 1.

	Time	Index α	Energy E_{cutoff}
A	10176-10196	-1.82 ± 0.02	354.7 ± 28.3
B	10196-10210	-1.50 ± 0.03	123.9 ± 6.0
C	10210-10224	-1.57 ± 0.03	118.3 ± 6.2
D	10224-10231	-1.95 ± 0.05	192.6 ± 23.5
E	10231-10245	-2.19 ± 0.06	251.8 ± 56.9
F	10245-10259	-2.30 ± 0.08	>250
G	10259-10273	-2.43 ± 0.15	>250
H	10273-10287	-2.51 ± 0.09	>250
I	10287-10294	-2.24 ± 0.13	>250
J	10294-10308	-2.20 ± 0.05	299.2 ± 68.6
K	10308-10315	-2.20 ± 0.06	>250
L	10315-10322	-1.75 ± 0.05	149.6 ± 16.3
M	10322-10332	-1.46 ± 0.04	119.9 ± 6.7
N	10332-10359	-1.48 ± 0.02	149.2 ± 5.0
O	10359-10369	-1.43 ± 0.06	157.1 ± 16.4

TABLE 1. Model parameters of the spectra shown in figure 2. Only the BATSE data are used for the spectral deconvolution. The spectral model consisted of a power law with an exponential cutoff, i.e., $F(E) = A_0(\frac{E}{E_p})^\alpha M(E)$ (where $M(E) = \exp(\frac{E_{\text{cutoff}}-E}{E_{\text{fold}}})$ if $E \geq E_{\text{cutoff}}$; $M(E)=1$ if $E < E_{\text{cutoff}}$). Both E_p and E_{fold} are chosen to be 40 keV.

REFERENCES

1. Belloni, T. et al. 1996, ApJ, **472**, L107
2. Bolton, C.T. 1972, Nature, **235**, 271
3. Cui, W. 1996, IAUC, 6404
4. Cui, W. et al. 1996, IAUC, 6439
5. Cui, W. et al. 1997, ApJ, **474**, L57
6. Cui, W., Zhang, S.N., Focke, W. & Swank, J. H. 1997, ApJ, *484*, 383
7. Dotani, T. et al. 1997, ApJ, **485**, L87
8. Gies, D.R. & Bolton, C.T. 1986, ApJ, **304**, 371
9. , S.S. et al. 1976, ApJ, **203**, L63
10. Lightman, A.P. & Eardley, D.M., 1974, ApJ, **187**, L1
11. Ling, J.C. et al. 1983, ApJ, **275**, 307
12. Narayan, R. 1996, in *Accretion phenomena and Related Outflows*, Port Douglas, Australia, ASP Conf. Series
13. Phlips, B, et al. 1997, these proceedings
14. Piro, L. et al. 1996, **IAUC**, 6431
15. Tananbaum, H. et al. 1972, ApJ, **177**, L4
16. Webster, B.L. & Murdin, P. 1972, Nature, **235**, 37
17. Zhang, S.N., Cui, W. & Chen, W. 1997, **482**, L155
18. Zhang, S.N., et al. 1997, ApJ **477**, L95
19. Zhang, S.N., et al. 1997, ApJ, to be submitted.
20. Zhang, S.N., et al. 1996, IAUC, 6405
21. Zhang, S.N., et al. 1996, IAUC, 6447
22. Zhang, S.N., et al. 1996, IAUC, 6462

X-ray and γ-ray spectra of Cyg X-1 in the soft state

Marek Gierliński[*], Andrzej A. Zdziarski[†], Tadayasu Dotani[‡], Ken Ebisawa[§], Keith Jahoda[§] and W. Neil Johnson[¶]

[*] *Jagiellonian University Observatory, Cracow, Poland*
[†] *Copernicus Astronomical Center, Warsaw, Poland*
[‡] *Institute of Space and Astronautical Science, Sagamihara, Japan*
[§] *NASA/GSFC, Greenbelt, USA*
[¶] *E. O. Hulburt Center for Space Research, Naval Research Laboratory, Washington, USA*

Abstract.
We present X-ray/γ-ray observations of Cyg X-1 in the soft state during 1996 May–June. We analyze *ASCA*, *RXTE* and OSSE data. The spectrum consists of soft X-ray blackbody emission of an optically thick accretion disk in the vicinity of a black hole and a power law with an energy index $\alpha \sim 1.2$–1.5 extending to at least several hundred keV. In the spectra, we find the presence of strong Compton reflection, which probably comes from the disk.

INTRODUCTION

Cyg X-1, the primary black-hole candidate, undergoes transitions between two spectral states: the hard ('low') state in which its X-ray spectrum is hard ($\alpha \sim 0.6$) and extends up to several hundred keV, and the soft ('high') state dominated by a strong soft X-ray emission together with a much softer power law ($\alpha \sim 1.5$) tail. Usually, Cyg X-1 remains in the hard state. The last transition to the soft state began around 1996 May 16 when the soft X-ray flux started to increase. The source remained in the soft state until about August 11 (Zhang et al. 1997).

THE DATA

In this paper, we analyze three groups of observations of Cyg X-1 in 1996. An *RXTE* observation on May 23 shows the object still undergoing a transition between the states (Fig. 1). The remaining data sets are from a simultaneous *ASCA/RXTE* observation on May 30 and from six simultaneous

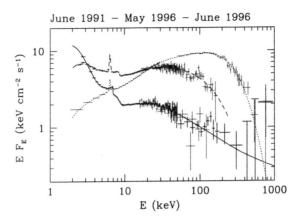

FIGURE 1. The spectral states of Cyg X-1. The solid curve is a fit to the soft state observed by *RXTE* and OSSE on 1996 June 17. The dashed curve is a fit to an intermediate state observed by *RXTE* on 1996 May 23. For comparison, we also show a *Ginga*/OSSE spectrum in the hard state (on 1991 June 6) with the model represented by the dotted curve (see Gierliński at al. 1997). The data have been rebinned for clarity.

RXTE/OSSE observations on June 17-18, when the object was in the soft state.

The *RXTE* data come from the public archive. A 2% systematic error has been added to each PCA channel to represent calibration uncertainties. Since dead-time effects of the HEXTE clusters are not yet fully understood, we allowed a free relative normalization of the HEXTE data with respect to the PCA data.

The *CGRO*/OSSE observation in the soft state on June 14–25 (from 50 keV to 1 MeV) overlaps with the *RXTE* observations on June 17 and 18. We have extracted six OSSE data sets near-simultaneous with the *RXTE* observations. In order to get better statistics, we have increased each OSSE data interval to include 30 minutes on either side of the corresponding *RXTE* interval. The spectrum is shown in Fig. 1, which also shows a spectrum from the hard state for comparison.

ASCA observed Cyg X-1 from May 30 5:30 (UT) through May 31 3:20 (Dotani at al. 1997, hereafter D97). The GIS observation was made in the standard PH mode. The SIS data suffer from heavy photon pile-up and thus are not usable. We have selected 1488 live seconds of the *ASCA* data near-simultaneous with the corresponding *RXTE* observation.

RESULTS

First, we fit the simultaneous *RXTE*/OSSE soft-state data of June 17–18. At soft X-rays, the spectra are dominated by a black-body component. The OSSE data show the emission extending up to at least 800 keV. Our basic model consists of a soft X-ray blackbody disk model and a power law. The power-law energy index, α, varies between ~ 1.3 and ~ 1.5. We do not observe any high-energy cutoff in the power law, which suggests a non-thermal origin of the emission. We note that the power-law component comes probably from Comptonization of the soft photons and should be therefore cut off at low energies. However, this cutoff would occur only well below 1 keV and its neglect in the model does not affect our conclusions.

Between ~ 10–200 keV, the observed spectrum systematically departs from the power law and forms a smooth hump. There is also a broad absorption feature above 7 keV, which can be attributed to an Fe Kα absorption edge. Considering both effects, we conclude that Compton reflection from cold matter takes place in the soft state of Cyg X-1. The covering factor of the reflector, $\Omega/2\pi$, varies in the range 0.6-0.7. The reflector can be identified with the optically thick disk also responsible for the soft blackbody.

The iron edge is smeared, which we attribute to Doppler and strong-gravity effects in the vicinity of the black hole. Therefore, we consider models in the Schwarzschild metric with the accretion disk inclined at angle $35°$ and reflection taking place between $R_{in} = 3R_g$ and $R_{out} = 15R_g$ (where $R_g = 2GM/c^2$). In this paper, we use a model of angle-dependent Compton reflection (Magdziarz & Zdziarski 1995) convolved with the relativistic line profile (Fabian at al. 1989). We note that this is only an approximation of the real disk reflection and does not take into account all angular effects near the black hole.

We then study the *ASCA/RXTE* data of May 30. We apply a conservative lower limit of 4 keV to the PCA data, and use the *ASCA* data in the range of 0.7–10 keV. First, we use a blackbody disk emission model taking into account general relativity (Hanawa 1989) together with a power law with relativistic reflection (as for the June data above). This, however, yields an unacceptable χ^2 of about 1100/577 d.o.f. (see the residuals in the upper panel of Fig. 2). The fit can be significantly improved by an additional weak high-energy tail on top of the disk spectrum. We model this tail as due to thermal Comptonization of a 300 eV blackbody in a plasma with $kT \approx 5$ keV and $\tau \approx 3$ (shown by the long-dashed curve on Figure 3). The resulting χ^2 is 680/575 d.o.f. The additional component can be interpreted as Compton radiation of an intermediate layer between the optically thick disk and an optically thin corona. We note that a similar reduction of χ^2 can be (instead of adding the Comptonization tail) obtained by breaking the power law to a softer one below a few keV. A similar softening of the power law at low energies was observed by *ASCA* in the hard state (Ebisawa et al. 1996).

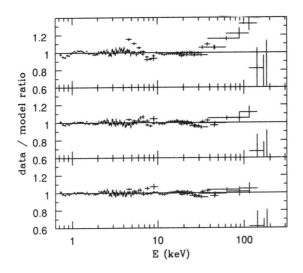

FIGURE 2. The data-to-model ratios for the 1996 May 30 observation. The upper panel corresponds to the model including power law, reflection and disk emission, the middle panel includes a Comptonization tail to the disk emission, and the lower panel includes the tail and an Fe line.

The fit can be further improved by adding an Fe Kα line. For that, we use a relativistic disk line (Fabian at al. 1989) with the same parameters as for the Compton reflection. The obtained χ^2 is 641/573 d.o.f., i.e., the presence of line is statistically significant at a very high confidence level. Figure 2 shows the residuals corresponding to the above models. The model spectrum is presented in Fig. 3.

The spectrum is absorbed by $N_H = 0.47 \pm 0.01$. The fitted power-law index is $\alpha = 1.23 \pm 0.03$. We assume no cutoff in the power law (as implied by the June 14-25 OSSE data). The covering factor of the reflector is $\Omega/2\pi = 0.55 \pm 0.1$, and the reflecting medium is ionized with ionization parameter $\xi \equiv L/nr^2 = 430^{+160}_{-130}$ erg cm s^{-1}, corresponding to Fe XXI-XXIV as the most abundant species.

The best fit value of the rest-frame line energy, $E_{\rm Fe} = 6.54^{+0.11}_{-0.15}$ keV, strongly depends on the assumed disk inclination angle. Since the Auger resonant destruction strongly suppresses Kα emission following photoionization of Fe XVII-XXIII, we should expect fluorescence from lithium-like Fe XXIV at ~ 6.7 keV. On the basis of our model, this line energy is consistent with the disk inclination angle of 30°. Notwithstanding, we stress that due to approximate character of the model and uncertainties in the PCA response matrices, the line parameters obtained here might be inaccurate.

If we use the same blackbody disk model as D97 assuming $T_{\rm col}/T_{\rm eff} = 1.7$, we obtain a black hole mass of 21 ± 2 M$_\odot$ and an accretion rate of $6.5 \pm 0.4 \times$

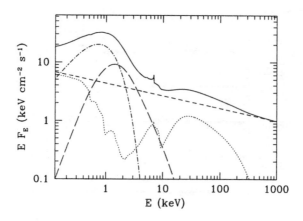

FIGURE 3. The model spectrum of Cyg X-1 in the soft state based on the simultaneous *ASCA/RXTE* observations of 1996 May 30. The short-dashed, dotted, dash-dotted, and long-dashed curves show the power law, the Compton reflection, the blackbody-disk emission, and the thermal Comptonization components, respectively. Solid line is the sum of all components. In order to show the details of the soft X-ray emission, the interstellar absorption was removed from the model.

10^{17} g s^{-1}. That mass is significantly higher than 12^{+3}_{-1} M$_\odot$ found by D97. The discrepancy is explained by influence of the additional Comptonization component used here. On the other hand, we obtain $M_x \approx 16$ M$_\odot$ by assuming $T_{\rm col}/T_{\rm eff} = 1.5$. We stress that the ratio of $T_{\rm col}/T_{\rm eff}$ is very uncertain due to theoretical difficulties. In particular, the standard Shakura-Sunyaev solution is unstable at $\dot{m} \equiv \dot{M}c^2/L_{\rm Edd} \sim 1$. In Cyg X-1 the total disk luminosity in the soft state is $L_{\rm disk} \approx 5 \times 10^{37}$ erg s^{-1} (about 5 times higher than in the hard state), which corresponds to $\sim 0.05 L_{\rm Edd}$. Assuming the emission efficiency of 0.08, we find $\dot{m} \sim 1$.

REFERENCES

1. Dotani T., et al. *ApJL*, in press (1997).
2. Ebisawa K., at al. *ApJ* **467**, 419 (1996).
3. Fabian A. C., at al. *MNRAS* **238** 729 (1989).
4. Gierliński M., at al. *MNRAS*, in press (1997).
5. Magdziarz P. and Zdziarski A., *MNRAS* **273**, 837 (1995).
6. Zhang S. N., at al. *ApJ* **477**, L95 (1997).

RXTE Observation of Cygnus X-1: Spectra and Timing

J. Wilms*, J. Dove[†], M. Nowak[†], B. A. Vaughan[‡]

*IAA Tübingen, Astronomie, Waldhäuser Str. 64, D-72076 Tübingen
[†] JILA, University of Colorado, Campus Box 440, Boulder, CO 80309-0440
[‡] Space Radiation Laboratory, Caltech, Pasadena, CA 91125

Abstract. We present preliminary results from the analysis of an RXTE observation of Cygnus X-1 in the hard state. We show that the observed X-ray spectrum can be explained with a model for an accretion disk corona, in which a hot sphere is situated inside of a cold accretion disk. Accretion disk corona models with a slab-geometry do not successfully fit the data. We also present the observed temporal properties of Cyg X-1, i.e. the coherence-function and the time-lags, and discuss the constraints these temporal properties impose for the modeled accretion geometry in Cyg X-1.

INTRODUCTION

Cygnus X-1 is one of the most firmly established persistent galactic black hole candidates (BHCs). Its X-ray spectrum in the hard state can be described by a power-law with a photon-index $\Gamma \approx (1.5\text{–}1.7)$, modified by an exponential cutoff with a folding energy $E_{\text{fold}} \approx 150\,\text{keV}$ [1,2] and reprocessing features below 10 keV. This spectral form has generally been interpreted as being due to an accretion disk corona (ADC).

As it has been previously pointed out, the standard "slab-like" ADC has problems explaining the high-energy radiation of BHCs, since 1. the strength of the observed reflection component is too small to be consistent with a slab geometry [2], 2. the observed temporal characteristics of Cyg X-1 cannot be explained with a pure slab-like ADC [3], and, 3., numerical models show that the spectral parameters of Cyg X-1 are inconsistent with those predicted for slab-like ADCs [4,5].

In this paper, we use spectra simulated with a non-linear Monte Carlo code to model the observed spectrum of Cyg X-1. In addition, we present the global temporal characteristics of Cyg X-1. The results are based on 10 ksec of data from Cyg X-1 obtained with the *Rossi X-ray Timing Explorer* (RXTE) on 1996 October 23. The data were reduced with ftools 4.0, using version 2.6.1

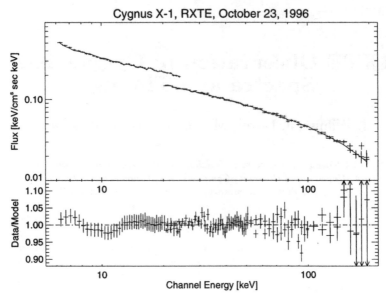

FIGURE 1. Spectrum of the best-fit sphere+disk model, and the ratio between the data and the best-fit model. The jump between the low-energy and the high-energy spectra is due to the uncertainty in the absolute effective areas of the instruments. The relative normalization between the HEXTE clusters and PCA is 0.70 in this fit.

of the PCA response matrix, version 1.5 of the PCA background-estimator program, and the HEXTE response matrices dated March 20, 1997. Due to the uncertainy in the response matrix we model the spectrum between 6 and 200 keV only and add a systematic error of 2% to the PCA data.

SPECTRAL FITTING

Spectral fitting of the RXTE data shows that the spectrum can be well described with a model of the form $KE^{-\Gamma}\exp(-E/E_\mathrm{f})$, where $\Gamma = 1.4$ and $E_\mathrm{f} = 150\,\mathrm{keV}$, *plus* an additional black-body with a temperature of $(1.0 \pm 0.1)\,\mathrm{keV}$. Since we are ignoring the spectrum below 6 keV, the constraint on the temperature of the soft-excess is fairly weak. There is evidence only for a weak reflection component in the spectrum, with the covering factor being $< 0.1 \cdot \Omega/2\pi$.

We have computed grids of spectra for self-consistent ADC models using a non-linear Monte Carlo scheme based on the code of Stern et al. [7] and further described in [4,5]. We have written interpolation routines that enable us to fit the observational data with the data reduction software XSPEC [8]. The data and subroutines are available upon request.

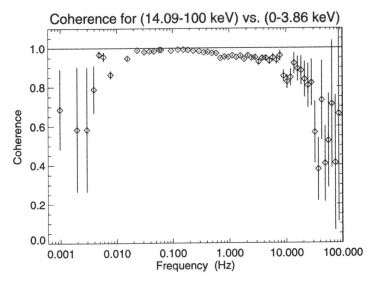

FIGURE 2. Coherence between the hard and soft energies.

Modeling with spectra from a standard slab-corona, where the accretion disk is embedded between two hot ADCs, does not result in an acceptable fit. Due to the large covering fraction of the cold disk, a high amount of the coronal radiation is reprocessed in the cold disk, resulting in a large flux of thermal radiation which can efficiently Compton-cool the corona. Therefore, the temperature of the slab ADCs cannot get high enough to produce a spectrum as hard as that observed. Slab-models also have too strong of a soft-excess [4]. In contrast, the sphere+disk model, in which a spherical corona is situated at the center of a cold accretion disk, is able to explain the observed spectrum. In Fig. 1 we show the best-fit obtained with the sphere+disk model, as well as the ratio between the data and the best-fit. The seed optical depth of the corona is $\tau_e = 2.1 \pm 0.1$ and the temperature of the corona is $T = 66 \pm 3\,\text{keV}$. The deviation between the model and the data is always less than 2%, which is an indication that the model is describing the data quite well. Our formal χ^2_{red}-value for this fit is 1.0. For a more detailed discussion, see [6]

TEMPORAL PROPERTIES

Although there is good evidence from the above spectral observations that the accretion geometry in Cyg X-1 differs from a simple slab, the results do not rule out other geometries in which the covering fraction of the cold material is smaller than 30%. In order to further constrain the geometry, the temporal properties of the source should be considered. For Comptonization

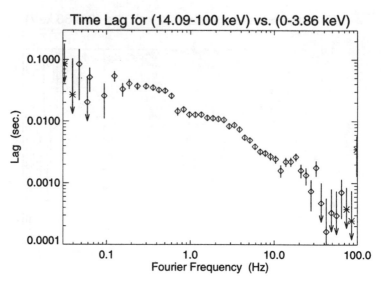

FIGURE 3. Time lag measured between the hard and the soft energy band. Stars indicate a negative time lag, i.e. the hard band leads the soft band. The minimum observed time lags of $< 1\,\mathrm{ms}$ indicates a maximum coronal size of $< \mathcal{O}(50GM/c^2)$ for $M = 10 M_\odot$. The maximum observed time lag constrains the dynamics of the corona.

the hard photons emerging from the source will have, on average, undergone more scatterings than the low energy photons, resulting in a time-lag between hard and soft X-rays. The exact form of this time-lag can, in principle, be used to further constrain the accretion geometry of the system. Only a geometry in which both, the spectral *and* the temporal data, can be explained, should be considered a valid candidate geometry for Cyg X-1 [9,10].

Much of the information on the geometry can be obtained by interpreting the phase information of the observed light curve [9,10]. In Fig. 2 we display the coherence between the energy band above 14.9 keV and the energy band below 5 keV. The coherence-function for the frequency f is defined as [9]

$$\gamma^2(f) = \frac{|<C(f)>|^2}{<|S_1(f)|^2> <|S_2(f)|^2>}$$

where $C(f)$ is the cross-spectrum between the intrinsic signals $S_i(f) = X_i(f) - N_i(f)$ where $X_i(f)$ is the observed signal and $N_i(f)$ is the Poisson noise-component ("observational rms noise"). The measured coherence of unity from 0.01 Hz to 10 Hz indicates a remarkable stability in the timing properties of the signal over the whole spectrum — an indication that either there is a single global source for the observed fluctuations, or that there is a global coherent response from the Compton corona, or both [3].

Similar information is also contained in Fig. 3, where the time-lag between the same energy bands is shown. In regions where the coherence function is unity, i.e., 0.01 Hz to 10 Hz, the time-lag can be well determined. Above $\approx$10 Hz the coherence is not preserved and, Poisson noise begins to dominate; therefore, the time-lag varies erratically.

Summary

We have, for the first time, fit the RXTE spectrum of Cyg X-1 using self-consistent ADC models. We have shown that the slab-geometry models do not explain the spectrum of the object, while the sphere+disk model, having a smaller covering factor, *can* describe the observed spectrum. In addition, we have presented the measured coherence function and time-lags, which can be used to further constrain the accretion geometry. We emphasize that only the direct application of the models to the observational data can show whether a given geometry is or is not applicable.

We acknowledge helpful discussions with M. Begelman, D. Gruber, I. Kreykenbohm, K. Pottschmidt, Ch. Reynolds, and R. Staubert. This work has been financed by NSF grants AST95-29175, INT95-13899, NASA Grant NAG5-2025, NAG5-3225, NAG5-3310, DARA grant 50 OR 92054, and by a travel grant to J.W. from the DAAD.

REFERENCES

1. K. Ebisawa, 1997, Adv. Space Res. **19**, 5.
2. M. Gierliński et al., 1997, MNRAS, in press.
3. M. A. Nowak, B. A. Vaughan, J. Dove, and J. Wilms, 1997, IAU Coll. 163, in press
4. J. B. Dove, J. Wilms, and M. C. Begelman, 1997, ApJ **487**, in press (October 1, 1997, issue).
5. J. B. Dove, J. Wilms, M. G. Maisack, and M. C. Begelman, 1997, ApJ **487** in press (October 1, 1997, issue).
6. J. B. Dove, J. Wilms, M. A. Nowak, B. A. Vaughan, and M. C. Begelman, 1997, ApJ (Letters), submitted
7. B. E. Stern, M. C. Begelman, M. Sikora, and R. Svensson, MNRAS **272**, 291 (1995).
8. K. A. Arnaud, 1996, in ADASS V, J. H. Jacoby and J. Barnes (eds.), ASP Conf. Ser., 101, San Francisco: Astron. Soc. Pacific., 17
9. B. A. Vaughan and M. A. Nowak, 1997, ApJ **474**, L43.
10. M. A. Nowak and B. A. Vaughan, 1996, MNRAS **280**, 227.

Spectral Variability of Cygnus X-1 in the Soft State

Warren Focke[1,2], Jean Swank[1], Bernard Phlips[3], William Heindl[4], and Wei Cui[5]

[1] Code 662, Goddard Space Flight Center, Greenbelt, MD 20771
[2] Department of Physics, University of Maryland, College Park, MD 20741
[3] Universities Space Research Association, Washington, DC 20024
[4] University of California, San Diego, CA 92093
[5] Massachusetts Intstitute of Technology, Cambridge, MA 02139

Abstract. Cygnus X-1 made a transition to the relatively rare soft state in 1996 May, and stayed in this state until 1996 August. A number of Target of Opportunity (TOO) observations of the source were made during this time with the Rossi X-ray Timing Explorer (RXTE). The hardness-intensity diagram showed similarity to that of GRS 1915-105 in recent observations. Time-dependent model fits to the spectrum show that the photon index of the hard tail varies significantly on minute timescales.

INTRODUCTION

Cygnus X-1 made a transition to the relatively rare soft (X-ray high, γ-ray low) state in 1996 May, and stayed in this state until 1996 August [1]. A number of Target of Opportunity (TOO) observations of the source during this time were made with the Rossi X-ray Timing Explorer (RXTE). The photon spectrum in this state is characterized by a soft, approximately thermal, portion and a steep (photon index around 2) powerlaw tail extending to high energies [2].

Spectral data from the Proportional Counter Array (PCA) and the High Energy Timing Experiment (HEXTE) were gathered into spectra covering time intervals from 16 to 128 s. These spectra were analyzed both by performing model fits and by examining the hardness-intensity diagram.

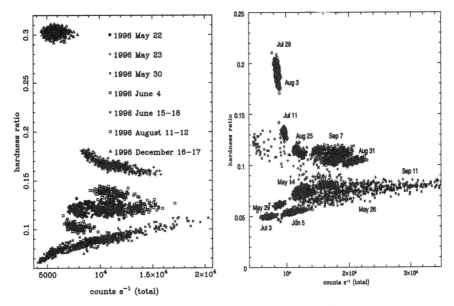

FIGURE 1. Left panel: Cygnus X-1 hardness-intensity diagram. The horizontal grouping in the upper left corner of the plot is hard state data. Note that the first observations in the TOO are in the middle of the lower region, the source then moves to the upper horizontal branch, and is on the lower horizontal branch during the true soft state. Also note that there are three different groupings of points in the central region from three orbits in the data from 1996 August, on the return from the soft state to the hard state. Right panel: GRS 1915+105 hardness-intensity diagram. This diagram seems to lack the horizontal area in the upper left corresponding to the hard state in Cygnus X-1.

COLOR STUDIES

The hardness-intensity diagram (HID) is a plot of the ratio of count rates in two different energy bands against total rate. Figure 1 (left panel) shows a plot of the ratio of the 11–30.5 keV rate to the 2–11 keV rate against the 2–45 keV rate. Each data point represents 16 s. The horizontal grouping in the upper left corner of the plot is hard state data from 1996 December. The source moves within the groupings representing individual orbits on fairly short (around a minute) timescales, comparable to the variations seen in the model fits in this paper. Variations within orbits in the soft state typically are about half the length of the lower horizontal branch. The plot shows remarkable similarity to a plot of the same quantities for the superluminal source GRS 1915+105 presented in [3] (Figure 1 (right panel)). This is suggestive of similarities in the accretion processes occurring in these sources during the times observed.

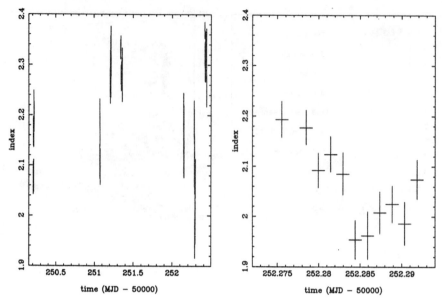

FIGURE 2. Cygnus X-1 photon index vs. time. Left Panel: Data from 1996 June 16 – 1996 June 18. Each point represents a 128 s integration. PCA and HEXTE data above 15 keV were fit to an absorbed powerlaw, and the photon spectral index plotted against time. Right Panel: A blowup of one orbit (second from the right) from the left panel.

MODEL FITS

Model fits were performed to determine the variability of the high-energy powerlaw tail. The spectra were extracted in time intervals which were multiples of 32 s in order to include full periods of the background rocking of the HEXTE. The portions of the spectra above 15 keV were fit to a powerlaw subject to interstellar absorption (N_h was fixed at 6.0×10^{21}). Significant variations on the shortest timescales observable were seen. The photon index varied by as much as 10% in 100 s, with larger variations up to 20% (total range 1.95–2.35) between orbits. Figure 2 shows the photon index plotted against time.

DISCUSSION

Many theories of the high energy emissions from this and other suspected black hole sources involve two components: low temperature thermal emission from an optically thick plasma comprising the main body of the accretion disk, and a hard tail formed by Comptonization of a portion of the thermal emission in a hotter, optically thinner, plasma [4], [5], [6], [7]. In a pure thermal model, the slope of the powerlaw tail is determined by the optical depth and temperature of the Comptonizing medium [8]. In some cases, the effects of bulk motion may dominate over thermal motion in determining the slope [9]. It should be noted that, for energies above those present in the seed spectrum and significantly less than the temperature of the Comptonizing medium, a powerlaw is produced independently of the energy distribution of the seed spectrum (usually thermal) or the Comptonizing medium (such as whether the main source of energy in the Comptonizing medium is thermal or bulk motion). Studies of the variability of this slope may thus reveal properties of the accretion system. For example, the fact that it is observed to vary on minute timescales places an upper bound near a light-minute on the size of the Comptonizing region.

REFERENCES

1. Zhang, S. N., Cui, W., Harmon, B. A., Paciesas, W. S., Remillard, R. E., and van Paradijs, J. 1997 The Astrophysical Journal Letters, 477, L95
2. Cui, W., Heindl, W. A., Rothschild, R. E., Zhang, S. N., Jahoda, K., and Focke, W. 1997, The Astrophysical Journal, 474, L57
3. Chen, X., Swank, J., and Taam, R. 1997, The Astrophysical Journal Letters, 477, L41
4. Skibo, J. G., and Dermer, C. D. 1995, The Astrophysical Journal Letters, 455, L25
5. Chakrabarti, S. K., and Titarchuk, L. G. 1995, The Astrophysical Journal, 455, 623
6. Narayan, R. 1996, The Astrophysical Journal, 462, 136
7. Czerny, B., Witt, H. J., and Życki, P. T. 1996, Acta Astronomica, 46, 9
8. Sunyaev, R. A., and Titarchuk, L. G. 1980, Astronomy and Astrophysics, 86, 121
9. Titarchuk, L. G., Mastichiadis, A., and Kylafis, N. D. 1996, Astronomy and Astrophysics Supplement Series, 120, 171

Modeling Cygnus X-1 γ_2 Spectra Observed by BATSE

Xin-Min Hua*, James C. Ling[†] and Wm. A. Wheaton[†]

*LHEA, Goddard Space Flight Center, Greenbelt MD 20771
[†]JPL, California Institute of Technology, Pasadena, CA 91109

Abstract. The γ_2 spectra of Cygnus X-1 from the BATSE earth occultation observations between 25 keV and 1.8 MeV has two components: a Comptonized part seen below 300 keV, and a high-energy tail in the $0.3 - 2$ MeV range. The spectra was interpreted in terms of an interacting two-zone model, consisting of a high temperature core embedded in a corona with relatively lower temperature. In this study, we perform a comprehensive search in the 4-dimensional parameter space of the two-zone model. We give the estimation of the best fitting parameters of the model and their errors. The results are compared with the one-zone model, both analytical and numerical, traditionally used in spectral data fitting. It is found that the two-zone model yields significantly better fits to the observation, an indication that there exists a high-energy core in the cloud surrounding the black hole candidate.

SINGLE-TEMPERATURE MODEL

Ling et al. [4] reported the average Cygnus X-1 spectrum observed by BATSE during the 22-day period TJD 9226-9250 (August 1993), when the source was at the γ_2 level. We first use the single-temperature Comptonization model to fit the data. The model simulates the Comptonization in a spherical plasma cloud with Thomson optical depth τ_0 and electron temperature T_e using Monte Carlo method described in Hua & Titarchuk [2]. The resulting spectra from clouds with a matrix of τ_0 and T_e values are compared with the observed data listed in Table 2 of Ling et al. [4]. The best-fit spectrum is obtained from Monte Carlo calculation following 4 million photons in the cloud with $\tau_0 = 1.35$ and $kT_e = 80$ keV, with $\chi^2 = 28.1$ for 11 degrees of freedom, or reduced $\chi^2_\nu = 2.55$. The soft source photons at black-body temperature 0.5 keV is injected radially into the spherical plasma cloud. The observed Cygnus X-1 spectrum together with the best-fit spectrum calculated from the single-temperature Comptonization model are plotted in Figure 1. The fluxes measured by the COMPTEL experiment [3] in August 1991 (data points with circles) are also plotted.

For comparison, the best fitting parameters obtained by analytical method [2] are $\tau_0 = 2.68$ and $kT_e = 53.1$ keV, with $\chi^2_\nu = 4.3$ [4]. Because the spectra is obtained from Monte Carlo method, there is statistical fluctuation in the χ^2 value. In order to estimate the uncertainty of the χ^2 values caused by this fluctuation, we performed several runs, each with the same number of source photons (4 million) and the same cloud parameters but different initial random number seeds. It was found that the standard deviation was about 2.7, about 10% of the χ^2 value in this case.

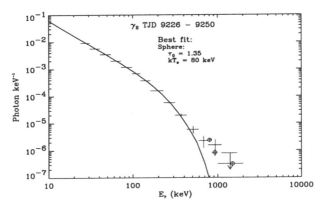

FIGURE 1.

In determining the best-fit spectrum displayed in Figure 1, we performed a search over a large area of the 2-dimensional $kT_e - \tau_0$ parameter space. We did Monte Carlo calculations for the grid points with $\Delta\tau_0 = 0.05$ and $\Delta(kT_e) = 5$ keV in the area. The spectrum obtained for each points was then compared to the same observation. The contours of $\chi^2 = 22, 33, 44$ resulting from the fittings are shown in Figure 2, corresponding to the χ^2_ν levels 2, 3 and 4 respectively. Because of the statistical fluctuation of the χ^2 values and the finite grid intervals, the minimum-curvature-surface method was used to interpolate between the grid points. It is seen that the best-fit parameters fall within a narrow "valley" roughly following the curve $\tau_0(kT_e)^{1.4} = 626$ (dashed line) and the minimum χ^2_ν value is ~ 2. The ranges of parameters for $\chi^2_\nu \leq 3$ are $\tau_0 = 1.275 - 1.650$ and $kT_e = 70 - 83$ keV.

TWO-TEMPERATURE MODEL

The spectra from the single-temperature Comptonization model have $\chi^2_\nu \gtrsim 2$. The best-fit spectrum, while fitting the observation data below 300 keV nicely, cuts off sharply above it and fails to account for the high-energy tail observed

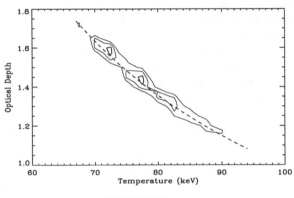

FIGURE 2.

for this state. In order to improve the fitting, we adopt a two-temperature concentric spherical geometry configuration which is both conceptually simple and intuitively realistic for the analysis of the spectrum [4]. The model was first proposed by Skibo and Dermer [5] to interpret the X-ray spectral hardening at high energies observed in AGNs. It consists of a high temperature "core" embedded in a cooler "corona". For simplicity, spherical symmetry and equal electron densities in the core and corona are assumed. Monte Carlo method is used to simulate the photon scatterings in the two-zone medium [1], taking into account the possible criss-crossing of photons between the two regions. As inputs, 0.5 keV black-body temperature soft source photons are radially injected into the system at the outer surface of the corona. This corona-core model is characterized by four parameters, the temperature kT_1 and optical depth τ_1 of the corona and those of the core, kT_2 and τ_2. Some of the photons can be scattered into the core, raised to higher energies than in the corona. These photons are expected to account for the observed high-energy tail above 300 keV.

In order to determine the best-fit configuration and estimate the uncertainty of the spectral fitting in the 4-dimensional parameter space $(\tau_1, kT_1, \tau_2, kT_2)$, we performed a comprehensive search of the parameter space at the grid points with $\Delta\tau_1 = 0.1, \Delta\tau_2 = 0.04, \Delta(kT_1) = 5$ keV and $\Delta(kT_2) = 10$ keV. For each of the grid point, energy spectrum was calculated using Monte Carlo method by following 4 million source photons. The calculated spectrum was then fitted to the observational data. The resulting χ^2 values, obtained for the corona optical depth $\tau_1 = 1.3$ and a series values of core optical depth τ_2, are displayed in Figure 3 as the cross sections of the contours projected in the 2-dimensional $kT_1 - kT_2$ plane. The figure displays the contours at levels $\chi^2 = 9.0$ and 18.0, obtained by interpolation using the minimum-curvature-surface method, corresponding to $\chi^2_\nu = 1$ and 2 for 9 degrees of freedom. We plotted the similar contours for τ_1 values between 0.9 and 1.7 and found that for $\tau_1 < 0.11$ or

> 1.6, no χ^2 value below 18 was found. Because of the statistical uncertainty and the coarse grid, the contours are not very well defined. Nevertheless they clearly indicate the region in the parameter space yielding the best fit to the observations. It is seen that for the two-temperature model, χ_ν^2 can be $\lesssim 1$ over a wide range of parameters, while for single-temperature model, the χ_ν^2 value is $\gtrsim 2$.

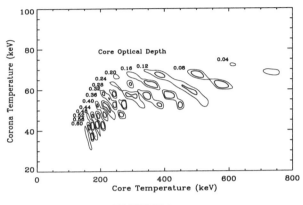

FIGURE 3.

The regions of $\chi_\nu^2 \lesssim 1$ given by Figure 3 and similar figures for $\tau_1 = 1.1$ and 1.5 are plotted in Figure 4. The areas enclosed by the thicker curves correspond to $\tau_1 = 1.1, 1.3$ and 1.5 respectively. The thinner lines roughly indicate the τ_2 values. In summary, the 4-dimensional parameter space range ($\chi_\nu^2 \lesssim 1$) is given by: $\tau_1 \approx 1.1 - 1.6$, $kT_1 \approx 40 - 80$ keV, $\tau_2 \approx 0.08 - 0.56$ and $kT_2 \approx 160 - 710$ keV.

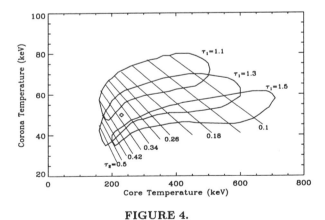

FIGURE 4.

One of the best-fit spectrum from the two-temperature model is plotted in Figure 5. Its parameters are: $\tau_1 = 1.3$, $kT_1 = 50$ keV, $\tau_2 = 0.36$ and $kT_2 = 230$ keV with $\chi_\nu^2 = 0.97$, corresponding to the point indicated by the diamond in Figure 4. Also plotted (dotted line) is the best-fit spectrum from the one-temperature model given in Figure 1.

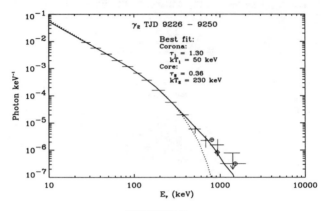

FIGURE 5.

CONCLUSION

For the Cygnus X-1 γ_2 spectrum observed by BATSE during the period TJD 9226-9250, the single-temperature model parameters for the best-fit spectrum are $\tau_0 = 1.35$, $kT_e = 80$ keV and $\chi_\nu^2 = 2.6$. Taking into account the statistical uncertainty of the Monte Carlo calculation, the minimum χ_ν^2 value is ~ 2. The parameters for analytical model are $\tau_0 = 2.68$, $kT_e = 53.1$ keV and $\chi_\nu^2 = 4.3$. On the other hand, the two-temperature model can significantly improve the fit. The spectra calculated based on such a model can fit the same observation with $\chi_\nu^2 \lesssim 1$ over a wide range of parameters: $\tau_1 \approx 1.1 - 1.6$, $kT_1 \approx 40 - 80$ keV, $\tau_2 \approx 0.08 - 0.56$ and $kT_2 \approx 160 - 710$ keV. The presence of the central high-energy core provides a reasonable explanation of the energy spectrum at energies over 300 keV.

REFERENCES

1. Hua, X.-M., *Computers in Physics*, in press (1997).
2. Hua, X.-M. and Titarchuk, L., *Ap. J.*, **449**, 188 (1995).
3. McConnell, M.L. et al., *Ap. J.* **424**, 933 (1994).
4. Ling, J.C., Wheaton, Wm.A., Wallyn, P., Mahoney, W.A., Paciesas, W.S., Harmon, B.A., Fishman, G.J., Zhang, S.N. and Hua, X.-M., *Ap.J.*, in press (1997).
5. Skibo, J.G. and Dermer, C.D., *Ap.J.*, **455**, L25, (1995).

A model for the high-energy emission of Cyg X-1

Igor V. Moskalenko[†*], Werner Collmar[†], Volker Schönfelder[†]

[†] *Max-Planck-Institut für extraterrestrische Physik, D-85740 Garching, Germany*
[*] *Institute for Nuclear Physics, Moscow State University, 119 899 Moscow, Russia*

Abstract. We construct a model of Cyg X-1 which describes self-consistently its emission from soft X-rays to MeV γ-rays. Instead of a compact pair-dominated γ-ray emitting region, we consider a hot optically thin and spatially extended proton-dominated cloud surrounding the whole accretion disc. The γ-ray emission is due to bremsstrahlung, Comptonization, and positron annihilation, while the corona-disc model is retained for the X-ray emission. We show that the Cyg X-1 spectrum accumulated by OSSE, BATSE, and COMPTEL in 1991–95, as well as the HEAO-3 γ_1 and γ_2 spectra can be well fitted by our model (see [1] for details). The derived parameters are in qualitative agreement with the picture in which the spectral changes are governed by the mass flow rate in the accretion disc. In this context, the hot outer corona could be treated as the advection-dominated flow co-existing with a standard thin accretion disc.

INTRODUCTION

Cyg X-1 is believed to be powered by accretion through an accretion disc. Its X-ray spectrum indicates the existence of a hot X-ray emitting and a cold reflecting gas. The soft blackbody component is thought to be thermal emission from an optically thick and cool accretion disc [2,3]. The hard X-ray part $\gtrsim 10$ keV with a break at ~ 150 keV has been attributed to thermal disc emission Comptonized by a corona with a temperature of ~ 100 keV [4,5]. A broad hump peaking at ~ 20 keV [6], an iron Kα emission line at ~ 6.2 keV [7], and a strong iron K-edge [8,9] have been interpreted as signatures of Compton reflection of hard X-rays off cold accreting material. There have also been reports of a hard component extending into the MeV region. Most famous was the so-called 'MeV bump' observed at a 5σ level during the HEAO-3 mission [10]. For a discussion of the pre-CGRO data see [11].

The average X-ray flux of Cyg X-1 shows a two-modal behavior. Most of its time it spends in a so-called 'low' state where the soft X-ray luminosity is low. There are occasional periods of 'high' state emission. Remarkable is

TABLE 1. Luminosity of Cyg X-1.

Energy band	Luminosity, 10^{36} erg/s
≥ 0.02 MeV	26
0.02–0.2 MeV	20.5
0.2–1 MeV	4.8
≥ 1 MeV	0.6

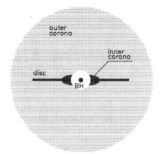

FIGURE 1. A schematic view.

the anticorrelation between the soft and hard X-ray components [5], which is clearly seen during transitions between the two states.

Since its launch in 1991 CGRO has observed Cyg X-1 several times. The COMPTEL spectrum shows significant emission out to several MeV, which however, remained always more than an order of magnitude below the MeV bump reported by HEAO-3. The spectrum accumulated between '91 and '95 by the COMPTEL [12], BATSE [13], and OSSE [14] instruments is shown in Fig. 2. Although the OSSE and BATSE normalizations are different, the spectral shape is very similar. Table 1 shows the average luminosity of Cyg X-1 (at 2.5 kpc) as derived from the BATSE-COMPTEL spectrum.

THE MODEL

The existence of a compact pair-dominated core around a BH in Cyg X-1 is probably ruled out by the CGRO observations. A signature of such a core would be a bump [15,16] similar to the one reported by HEAO-3. However no evidence for such a bump was found in the CGRO data [12,14]. Additionally, the luminosity of Cyg X-1 above 0.5 MeV, though small, exceeds substantially the Eddington luminosity for pairs, which is ~ 2000 times lower than for a hydrogen plasma. Also the hard MeV tail can not be explained by Comptonization in a corona ($kT \sim 100$ keV) and thus another mechanism is required.

We consider the proton-dominated optically thin solution [17], $\Theta \equiv \frac{kT}{m_e c^2} \lesssim 1$, where the γ-ray emission is attributed to a spatially extended cloud surrounding the whole accretion disc (Fig. 1), the outer corona, which emits via bremsstrahlung, Comptonization, and positron annihilation, and analyze possible consequences of that. We adopt a standard model for X-rays, where the hard X-rays are produced by Comptonization of the soft X-ray emission in an inner corona, and the soft X-rays are a composition of the local blackbody emission from the disc and the reflection component. The optical depth of the outer corona has to be so small that the disc and inner corona emission is only slightly reprocessed in it.

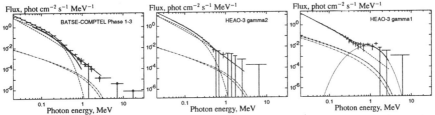

FIGURE 2. *Left panel:* the Cyg X-1 spectrum based on the CGRO Phase 1–3 observations (◇: the COMPTEL data [12], +: the BATSE data [13]). The thick solid line is the best fit to the time average OSSE spectrum [14]. *Central and right panels:* the HEAO-3 γ_2, γ_1 spectra [10]. In all panels the thin solid lines represent our model fit for the parameter sets I. The spectral components shown are the annihilation line (dotted line), ee-, e^+e^--, ep-bremsstrahlung (dash-dot), dashed lines: the Comptonized spectra from the i- and o-corona (shown up to 3 MeV, where it agrees with simulations, and also significant data points are available).

The spectral modelling has been carried out with the ee- and e^+e^--bremsstrahlung emissivities given by numerical fits of [18,19]. For the ep-bremssrahlung and annihilation emissivities we use the integration formulas of [18,20]. To calculate the effect of Compton scattering we follow the model of [21], which generally agrees well with Monte Carlo simulations except at high temperatures, $\Theta \sim 1$, and small optical depth. But it still provides a correct spectral index. We found that a power-law with a cutoff, $\propto \left[\frac{E_0}{E}\right]^{\alpha+1} (1 - e^{-kT/E})$, where α is determined by the equation of [22], gives a reasonable agreement with Monte Carlo simulations up to ~ 3 MeV. The chosen normalization provides a correct value of the amplification factor [23]. The emission of the accretion disc which is reprocessed by the inner corona was taken monoenergetic, $E_0 = 1.6kT_{bb}$, where $kT_{bb} \approx 0.13$ keV is the effective temperature of the soft excess [3]. The intensity of the narrow annihilation line from the disc plane can be estimated by $I_a \simeq \frac{n_+ c}{4} \frac{R_d^2}{D^2} \cos i_d$, where n_+ is the e^+ number density in the outer corona, c the speed of light, R_d the disc radius, D the distance, and i_d ($\approx 40°$) the inclination angle of the disc plane.

The fitting parameters are: kT_i, τ_i, and kT_o, τ_o, the temperature and optical depth of the inner (i) and outer (o) coronae (spheres), L^*_{soft}, the luminosity of the *disc* effectively Comptonized by the i-corona, L_{soft}, the *total* effective luminosity of the central source in soft X-rays illuminating the o-corona, R, the o-corona radius, and, $Z = n_+/n_p$, the positron-to-proton ratio in it.

The bremsstrahlung and annihilation photon fluxes from the outer corona are proportional to $R^3 n_p^2$. Thus, if the annihilation contributes significantly, there is a continuum of solutions given by an equation $R^3 n_p^2 Z(1+Z) = \frac{R\tau_o^2 Z(1+Z)}{\sigma_T^2 (1+2Z)^2} = const$, at kT_o, τ_o fixed, where $\frac{Z(1+Z)}{(1+2Z)^2}$ varies slowly for $Z \gtrsim 0.5$. For a negligible positron fraction the continuum of solutions is defined by $\tau_o = Rn_p \sigma_T = const$, where $R \leq R_{\max}$, and $R_{\max}$ is fixed from fitting.

TABLE 2. The best fit model parameters.

Parameters	CGRO Phase 1–3		HEAO-3: γ_2-state		γ_1-state	
	I	II	I	II	I[a]	II
Soft X-ray luminosity, $L_{\rm soft}$ (10^{36} erg s^{-1})	9.7	10.6	10.6	10.7	9.8	7.9
i-corona temperature, kT_i (keV)	75.9	73.9	95	94.9	–	93.0
i-corona optical depth, τ_i	2.30	2.40	1.41	1.42	–	1.44
$L^*_{\rm soft}$, 10^{36} erg s^{-1}	0.86	0.83	1.96	1.95	–	0.51
o-corona temperature, kT_o (keV)	430	479	450	448	346	361
o-corona optical depth, τ_o	0.05	0.037	0.056	0.056	0.12	0.10
o-corona radius, R (10^8 cm)[b]	$\lesssim 100$	$\lesssim 100$	$\lesssim 100$	150	391	812
Positron-to-proton ratio, Z[b]	0	0	0	1.00	1.0	0.5
Proton number density, n_p (10^{10} cm^{-3})[b]	–	–	–	187	154	93
Accretion disc radius, R_d (10^8 cm)	–	–	–	1	1	1
I_a, 10^{-5} photons cm^{-2} s^{-1}	0	0	0	0.18	0.15	0.04
χ^2_ν	3.1	3.1	1.4	1.4	0.9	0.9

[a] The inner corona is small or even absent at all; [b] For R, n_p, Z dependence see text.

RESULTS AND DISCUSSION

The observed spectra of Cyg X-1 are shown in Fig. 2 together with our model calculations. The parameters obtained from spectral fitting are listed in Table 2. For comparison two parameter sets with the same χ^2_ν are shown, however the first one (I) seems to be more physical.

The average BATSE-COMPTEL spectrum corresponds probably to the normal state of Cyg X-1. Only two components contribute: the Comptonized emission from the inner and outer coronae. The parameters obtained for the HEAO-3 γ_2 state are similar, though the upper limits at $E_\gamma \gtrsim 1$ MeV allow some positron fraction (II). For the HEAO-3 γ_1 'bump' spectrum the outer corona size is several times larger, while the inner corona is small or even absent at all (set I). The non-negligible positron fraction (for R, n_p, Z dependence see above) is too high to be produced in the optically thin outer corona [17]. Therefore, we suggest a positron production mechanism, which could sometimes operate in the inner disc. The radiation pressure would necessarily cause a pair wind, which serves as energy input into the o-corona thereby increasing its radius.

The small luminosity of the disc which is Comptonized by the inner corona, $L^*_{\rm soft} \approx 10^{36}$ erg/s, probably implies a geometry where only the inner part of the disc is effectively covered by the corona. Otherwise, if the corona forms a disc-like structure where the intensity depends on the inclination angle, then it should cover almost all of the X-ray emitting area of the disc.

The soft (< 10 keV) X-ray luminosity of Cyg X-1 is $\sim 8.5 \times 10^{36}$ erg/s on average [5], while during the HEAO-3 γ_1, γ_2 states it was even lower [10]. Taking into account that for hard X-ray photons the Comptonization efficiency in the hot plasma drops substantially [21] while the number of photons decreases as well, the obtained values, $L_{\rm soft} \approx 10^{37}$ erg/s, match the data.

No pairs are required to reproduce the spectrum of Cyg X-1 in its normal state. If one takes $I_a \approx 4.4 \times 10^{-4}$ photons cm^{-2} s^{-1} [24] for the annihilation line flux in the γ_1 state, it allows the accretion disc radius to be estimated to $R_d \sim 1.7 \times 10^9$ cm (set I). The allowed upper limit derived from optical measurements is $R_d \approx 6 \times 10^9$ cm $(M/10M_\odot)$ [5].

The obtained parameters are consistent with a picture where the spectral changes are governed by the mass accretion rate $\dot{M}$ [25]. The γ_1 state probably corresponds to a smaller $\dot{M}$ compared to the normal state. In this context, the γ-ray luminosity should anticorrelate with the hard X-ray luminosity. The extended hot outer corona can be treated as the advection-dominated accretion flow (ADAF) co-existing with a cool optically thick disc, though in contrast to a standard ADAF [25], the electrons here are hot and the protons are cold. This is possible since cooling via bremsstrahlung and Coulomb ep-collisions at low density is unimportant while small optical depth prevents from effective Compton cooling. The adequate heating could be provided by the electron thermal conduction from a region with nearly virial ion temperature.

Useful discussions with N.Shakura, R.Narayan, L.Titarchuk, and M.Gilfanov are greatly acknowledged. We are particularly grateful to M.McConnell for providing us with the combined spectra of Cyg X-1 prior to publication, and E.Churazov for Monte Carlo simulations of Comptonization in $\Theta \sim 1$, $\tau \approx 0.1 - 0.05$ plasma.

REFERENCES

1. Moskalenko I. V., Collmar W., Schönfelder V., in preparation (1997)
2. Pringle J. E., *ARA&A* **19**, 137 (1981)
3. Bałucińska-Church M., et al., *A&A* **302**, L5 (1995)
4. Sunyaev R. A., Titarchuk L. G., *A&A* **86**, 121 (1980)
5. Liang E. P., Nolan P. L., *Spa. Sci. Rev.* **38**, 353 (1984)
6. Done C., et al., *ApJ* **395**, 275 (1992)
7. Barr P., White N., Page C. G., *MNRAS* **216**, 65p (1985)
8. Inoue H., in *Proc. 23rd ESLAB Symp.*, Noordwijk: ESA, 1989, **2**, p.783
9. Tanaka Y., *Lect. Notes in Phys.* **385**, 98 (1991)
10. Ling J. C., et al., *ApJ* **321**, L117 (1987)
11. Owens A., McConnell M. L., *Comments Astrophys.* **16**, 205 (1992)
12. McConnell M., et al., in *AIP Proc.*, 1997, this Symposium
13. Ling J. C., et al., *ApJS*, submitted (1997)
14. Phlips B. F., et al., *ApJ* **465**, 907 (1996)
15. Liang E. P., Dermer C. D., *ApJ* **325**, L39 (1988)
16. Liang E. P., *A&A* **227**, 447 (1990)
17. Svensson R., *MNRAS* **209**, 175 (1984)
18. Stepney S., Guilbert P. W., *MNRAS* **204**, 1269 (1983)
19. Haug E., *A&A* **178**, 292 (1987)
20. Dermer C. D., *ApJ* **280**, 328 (1984)
21. Hua X.-M., Titarchuk L., *ApJ* **449**, 188 (1995)
22. Titarchuk L., Lyubarskij Yu., *ApJ* **450**, 876 (1995)
23. Dermer C. D., Liang E. P., Canfield E., *ApJ* **369**, 410 (1991)
24. Ling J. C., Wheaton Wm. A., *ApJ* **343**, L57 (1989)
25. Narayan R., *ApJ* **462**, 136 (1996)

A Thermal-Nonthermal Inverse Compton Model for Cyg X-1

A. Crider*, E. P. Liang*, I. A. Smith*
D. Lin* and M. Kusunose[†]

*Department of Space Physics and Astronomy, Rice University
Houston, TX 77005-1892
[†] Kwansei University, Uegahara Ichiban-cho
Nishinomiya 662, Japan

Abstract. Using Monte Carlo methods to simulate the inverse Compton scattering of soft photons, we model the spectrum of the Galactic black hole candidate Cyg X-1, which shows evidence of a nonthermal tail extending beyond a few hundred keV. We assume an *ad hoc* sphere of leptons, whose energy distribution consists of a Maxwellian plus a high energy power-law tail, and inject 0.5 keV blackbody photons. The spectral data is used to constrain the nonthermal plasma fraction and the power-law index assuming a reasonable Maxwellian temperature and Thomson depth. A small but non-negligible fraction of nonthermal leptons is needed to explain the power-law tail.

INTRODUCTION

Cygnus X-1 is the brightest and most studied Galactic black hole candidate. Its high-energy spectrum has been extensively modeled in attempts to determine what processes produce the continuum emission. While several models debate the detailed geometry of the system [1–4], many agree that the 10-200 keV portion of the spectrum is a result of inverse Compton scattering of soft X-ray photons. Early modeling attempts assumed a single temperature Maxwellian plasma [5]. However such simple models do not adequately fit the observations [6,7]. More complex models, such as one with two plasma regions of different temperature, increase the goodness-of-fit substantially [3].

When Cyg X-1 is in the normal (γ_2) state, a high-energy power-law is observed in the γ-ray continuum [8]. This is typically attributed to some nonthermal process. It is possible that the power-law is due to inverse Compton scattering off of a high-energy power-law tail in the plasma energy distribution. If so, both the emission above and below 300 keV could originate from the same plasma.

FITTING THE CYG X-1 SPECTRUM

We wish to see if the high-energy spectrum of Cyg X-1 can be explained simply by Comptonization from a single plasma region with a Maxwellian+power-law energy distribution. We simulated Comptonized spectra with a Monte Carlo code based on the algorithm of Pozdnyakov, Sobol, and Syunyaev [9]. We compared the results to combined BATSE and COMPTEL data from 1991 [8,10]. Though this data has already been unfolded through the detector response using an assumed model, it serves to give us an approximate solution.

The parameters that define the shapes of our simulated spectra are the temperature of the thermal leptons (electrons and pairs) T_e, the Thomson depth τ_T, the fraction of nonthermal leptons ξ, and the energy index of the nonthermal leptons p. Generating and interpolating over a 4-dimensional parameter space is computationally intensive and for this study we fixed both $T_e = 65$ keV and $\tau_T = 2.45$. A value of T_e very close to this was determined for this data assuming single temperature analytic Comptonization [8]. The parameter τ_T was determined from the spectral index α of the 30 to 70 keV spectra, as suggested by Pozdnyakov, Sobol and Syunyaev [9], where the equations

$$\gamma = \frac{\pi^2}{3} \frac{mc^2}{(\tau_T + \frac{2}{3})^2 T_e} \qquad (1)$$

and

$$\alpha = -\frac{3}{2} + \sqrt{\frac{9}{4} + \gamma} \qquad (2)$$

are solved for τ_T. This left two free spectral shape parameters, ξ and p. We calculated a 6×6 grid of simulated spectra, with ξ ranging logarithmically from 0.25% to 8.0% and p ranging from 3.25 to 4.5. These grid points were chosen based on our experience with this code applied to other astrophysical phenomena [11,12]. We iterated our code until the statistical signal-to-noise within the range of 30 keV to 2020 keV was less than 10 in each of the 22 bins. Bins above 500 keV were smoothed in a manner similar to that in Pozdnyakov, Sobol and Syunyaev [9]. A simple spline algorithm allowed us make our model continuous so we could evaluate a χ^2 between our simulation and the discrete data.

Figure 1 shows a reasonable fit of our code to the BATSE and COMPTEL data, where $\xi = 0.5\%$ and p = 3.5. To see what range of the parameter space is acceptable, we next examined χ^2 over the entire grid. In Figure 2, we plot the confidence levels of fits on this grid. Contours are drawn here assuming a polynomial interpolation between grid points.. This shows that for a fixed $T_e = 65$ keV and $\tau_T = 2.45$, there is a 68.3% (1σ) confidence that ξ lies between 0.25% and 1.0% and p is between 3.25 and 4.25.

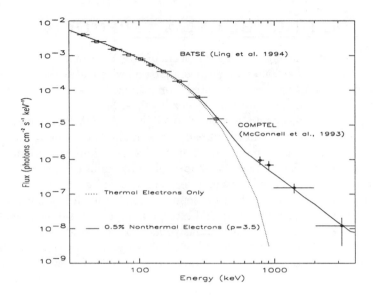

FIGURE 1. Fit of Maxwellian+nonthermal Comptonization model to unfolded spectrum of Cygnus X-1 where $T_e = 65$ keV, $\tau_T = 2.45$, $\xi = 0.5\%$, and $p = 3.5$. Also plotted for comparison is a fit of the same model with no nonthermal electrons or pairs.

We remind the reader that this is a crude way to test a model since we are comparing to already unfolded data. However, by showing a reasonable match between our simulated spectra and the unfolded spectra, we can be confident that more a rigorous procedure would also work.

SUMMARY AND CAVEATS

We find that the X-ray/γ-ray spectra of the normal state of Cygnus X-1 can roughly be reproduced by Comptonization of 0.5 keV blackbody photons through a combination thermal-nonthermal plasma. For T_e=65 keV and the Thomson depth τ_T=2.45, we determine that the fraction of nonthermal leptons $\xi = 0.5\%^{+0.5\%}_{-0.25\%}$ and the energy index of the nonthermal leptons $p = 3.5^{+0.75}_{-0.25}$.

While these results place first-order limits on the nonthermal lepton distribution, several modifications to these procedures would be necessary in order to reliably determine the allowed parameter space. The current Monte Carlo code does not require self-consistency between the lepton distribution and the radiative cooling. This obviously must be corrected to produce physically meaningful results. It is also necessary to generate a four-dimensional grid of simulated spectra to allow all four shape parameters to vary. This grid should also extend to lower values of ξ and p since our current grid does nec-

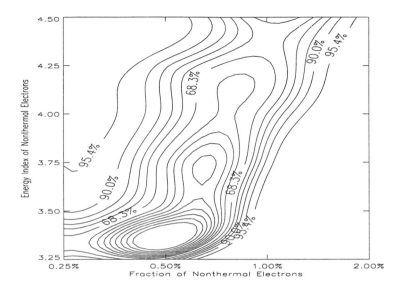

FIGURE 2. Confidence level contours in $\xi - p$ parameter space for fits of model to Cyg X-1 spectrum. Confidence levels are calculated from goodness-of-fit statistic χ^2 assuming 11 degrees of freedom (14 data points - 3 fit parameters). See Press et al. 1993 for details.

essarily exclude solutions in this portion of the parameter space. Finally, any future work would require folding our simulated spectra through the detector responses to allow direct comparison with the γ-ray count data.

ACKNOWLEDGEMENTS

Funding for this work is provided through a NASA GSRP Fellowship from Marshall Space Flight Center and NASA Grant NAG5-3824.

REFERENCES

1. Skibo, J.G & Dermer, C.D., *ApJ* **455**, L25 (1995).
2. Chakrabati, S.K., & Titarchuk L.G., *ApJ* **455**, 623 (1995).
3. Ling, J.C. et al., *ApJ* in press (1997).
4. Moskalenko, I.V., Collmar, W., & Schönfelder, V., *this volume* (1998).
5. Sunyaev R.A. & L.G. Titarchuk, *A&A* **86**, 121 (1980).
6. Titarchuk, L.G. *ApJ* **434**, 570 (1994).
7. Grabelsky, D.A., Matz, S.M., Purcell, W.R., et al., *AIP Conf. Proc.* **280**, 345 (1993).
8. Ling, J.C et al., *AIP Conf. Proc.* **304**, 220 (1994).

9. Pozdnyakov, L.A., Sobol, I.M., & Syunyaev R.A., *Soviet Scientific Reviews, Section E : Astrophysics and Space Physics Reviews* **2**, 189 (1983).
10. McConnell, M. et al., *ApJ* **424**, 933 (1994).
11. Liang, E.P., Kusunose, M., Smith, I.A. & Crider A., *ApJ* **479**, L35 (1997).
12. Smith, I.A., Liang, E.P., Crider, A., Lin, D., & Kusunose, M., *this volume* (1998).
13. Press, W.H., Teukolsky, S.A., Vetterling, W.T., & B.F. Flannery, *Numerical Recipes in C, 2nd. ed.*, Cambridge : Cambridge Press, 1988, ch. 15.6, pp. 689-699.

Two Distinct States of Microquasars 1E1740-294 and GRS1758-258

S. N. Zhang**, B. A. Harmon* and E. P. Liang[#]

ES-84, Marshall Space Flight Center, Huntsville, AL 35810
Universities Space Research Association
[!]*Department of Space Phys. & Astron., Rice University, Houston, TX 77251*

Abstract. We report the monitoring results over a ~5 year period of the two radio and hard X-ray sources 1E1740-294 and GRS1758-258 with BATSE/CGRO since 1991. Two hard X-ray flux and spectral states are found. The hard X-ray high flux state in 1E1740-294 can be represented by a hard power-law with an exponential cutoff, commonly seen in black hole candidates in their hard state. In its hard X-ray low flux state, its energy spectrum becomes a steeper power law without break up to 200-300 keV, usually seen in the soft state of black hole candidates and in the two galactic superluminal jet sources GRS1915+105 and GROJ1655-40. We also discuss the possible nature of the two systems.

INTRODUCTION

The hard X-ray sources 1E1740-294 and GRS1758-258 have been called "microquasars" because of the radio lobes associated with the central high energy and radio sources [15]. They are also considered to be black hole candidates due to their similar high energy spectra to Cyg X-1 (see ref. [4] for a review and references therein). Many of their properties, however, remain a mystery. Searches for their optical and infrared counterparts have resulted in upper limits of $I = 21$ and $K = 17$, respectively. Considering the detected column density, massive companions are unlikely [17,3]. Thus the two systems may be low mass X-ray binaries; although stable long term hard X- ray emission is never observed from any other low mass black hole binary systems (see ref. [31] and references therein).

Recent RXTE observations have shown that their high energy power density spectra are also similar Cyg X-1 [19]. The long term SIGMA/GRANAT observations of the sources revealed rather large flux variations on time scales of months and longer [12]. Correlated radio and hard X-ray flux variations were also reported in 1E1740-294 [15]. In this paper we report the two-state behavior of the two microquasars, i.e., the hard X-ray high and low flux states, observed from five years of monitoring with the BATSE/CGRO. The different energy spectra of 1E1740-294 in the two states are also obtained. We also discuss the possible nature of the two systems.

DATA ANALYSIS AND RESULTS

We use the earth occultation analysis technique [8,9] to produce the light curves of both 1E1740-294 and GRS1758-258 from the near continuous Large Area Detector (LAD) data. Since both sources are located in rather crowded sky regions, source confusion is a major problem for obtaining reliable intensity information. The earth occultation imaging technique [22–24] has been used to search for bright transient sources in these regions. In figures 1 (left) and (right) we show the region near 1E1740-294 before and after the initial outburst of the bursting pulsar [11] GROJ1744-28. In fact its persistent X-ray counterpart was discovered with the figure 1 (right) [20,10]. We also minimize the contamination from the galactic ridge diffuse emission by excluding data when the limb of the earth is nearly parallel to the galactic plane. This leaves periodic gaps in the derived light curves. This means that we cannot study long periodic variations in these sources around the satellite orbital precession period of around 52 days and harmonics of this frequency.

The derived detector count rate histories are extracted for 16 detector energy channels between 20-2000 keV. However, no statistically significant detection can be seen above 300 keV from both sources. Using the known detector response, their energy spectra can be deconvolved. Between 20-300 keV, there are typically 8-10 energy channels. Due to the limited statistics, no detailed spectral analysis is carried out for GRS1758-258. For their light curve generation, we fit their count rate spectra only to a power-law model, with the spectral index fixed at its typical value of -1.8, in order to make the fitting procedure converge consistently. Using other spectral models, such as the optically thin and thermal bremsstrahlung model does not produce significantly different light curves.

In figure 2 (left) and (right) we show the long term light curves of both sources between 20-100 keV. Each data point represents an integration time of 15 and 50 days for 1E1740-294 and GRS1758-258, respectively. From the light curves, we can see clearly that each of the two sources have two distinct flux levels, or states, and that they spend the majority of their lives in the hard X-ray high flux state. The time scale for the sources to move between the states is on the order several months to a year.

In figure 3 we show two energy spectra of 1E1740-294, representative of the two states. In the hard X-ray high flux state, the energy spectrum is described with a hard power-law between 20-100 keV (photon spectral index of about -1.7) with a gradual cut-off at higher energies ($E_{cutoff} \sim 3kT_{electron} \sim 130$ keV). In the hard X-ray low flux state, the energy spectrum is consistent with a steeper power-law (photon spectral index of about -2.6) up to 200-300 keV. At around 20 and 300 keV, the two spectra have pivot points. So the hard X-ray luminosity difference between the two states seems to be due to their different spectral shape between 20-300 keV.

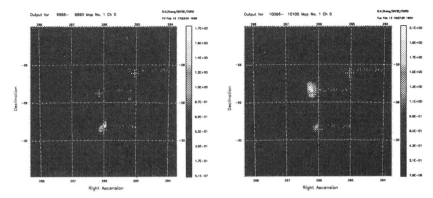

FIGURE 1. *Left:* BATSE image of the Galactic center region before the initial outburst of the bursting pulsar GRO J1744-28, between 20-100 keV. *Right:* BATSE *discovery* image of the persistent counterpart of the bursting pulsar GRO J1744-28, between 20-100 keV.

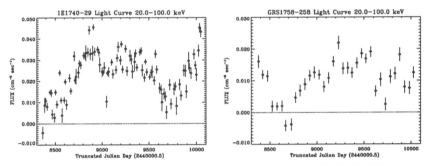

FIGURE 2. Hard X-ray light curves of 1E1740-29 (*left*) and GRS1758-258 (*right.*)

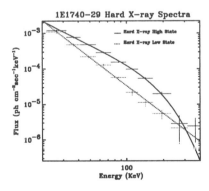

FIGURE 3. Energy spectra of 1E1740-29. The hard X-ray high (solid line) and low (dotted line) state spectra are averaged between TJD 8740–8854 and 8435–8658, respectively. The solid and dotted curves are the Sunyaev-Titarchuk Thermal Comptonization model ($kT_{electron} = 44$ keV and $F_{Comptonization} = 2.7$ and a simple power-law model (photon spectral index = -2.6).

875

DISCUSSION

The main result reported in this paper is the two state behavior of the two sources in the hard X-ray energy band. Part of our light curves also overlap with the long term monitoring data with SIGMA/GRANAT [4] and are consistent with the previous results. Suc two state behavior is very similar to Cyg X-1, whose regular state is the hard/low state, corresponding to the high hard X-ray flux state of the two microquasars. Occasionally Cyg X-1 enters into the soft/high state, during which its hard X-ray spectrum above 20 keV also becomes a steep power law without any detectable break [13,29]. Previous observations have revealed their other similarities with Cyg X-1, i.e., in high energy spectra [12] and power density spectra [19] during the hard X-ray high flux state, correlated hard X-ray and radio flux variations [15,30], and the persistent nature of their hard X-ray and radio emission [31].

Because Cyg X-1 is a much closer and brighter sources, it has been observed with much more details than the two microquasars. For example, in its hard/low state, an ultra-soft component with a blackbody temperature of ~ 0.1 keV [1,7] is observed. In its soft/high state, the luminosity and temperature of the soft component increases significantly [5,6]. Its radio emission transition was found to be simultaneous with the X-ray state transition [21,30]. A blackbody component has been detected with different temperature and luminosity in the soft and hard states, and is perhaps the only evidence of the existence of the accretion disk in Cyg X-1.

If the similarities between the two microquasars and Cyg X-1 extend beyond the two state behavior in the hard X-ray band, we then also expect the existence of blackbody components in their low energy spectra. Therefore any positive detection of a blackbody component would be important for understanding the mass accretion conditions near the black hole and the nature of the systems. The existence of an optically thick accretion disk would suggest that they are binary systems, since other possible accretion modes, for example from the interstellar medium or molecule clould would not likely form an optically thick accretion disk. Mereghetti, Belloni & Goldwurm have reported a soft excess in GRS1758-258 [14], which may indeed be the expected blackbody component. Further they also found an anti-correlation of the soft excess with the hard X-ray flux. This provides further support to the similar nature of the system to Cyg X-1. For 1E1740-294, it is very difficult to detect the expected blackbody component due to the very high column density from its direction (8×10^{22} cm^{-2} [17]).

Is it possible that these systems contain neutron stars instead of the assumed black holes? This is unlikely, since (a) none of the two neutron star signatures, i.e., pulsations and type I X-ray bursts, have been ever detected from them; (b) hard X-ray emission at this level has never been seen from any known accreting neutron star systems [2], although a similar energy spectra (hard X-ray power-law with a high energy cutoff) with a much lower luminosity was detected from a neutron star X-ray burster binary system 4U1608-52 [25]; and (c) the two hard X-ray state behavior has also never been seen from any known neutron star systems. In Cyg X-1, the state transition has been proposed to be due to the disk rotation direction reversal [16,28]. The location of the last stable orbit changes significantly when the reversal happens, if the black hole spins rapidly. In a neutron star system, the neutron star angular momentum is too small to influence significantly the location of its last stable orbit, even if the magnetic field of the neutron star is weak enough to allow the disk to extend down to its last stable orbit. Nevertheless, we cannot exclude firmly the possibility that they may harbor neutron stars, without dynamical mass estimates of their compact objects.

REFERENCES

1. Balucinska-Church, M. *et al.* 1995, *A&A*, **302**, L5
2. Barret, D., McClintock, J.E. & Grindlay, J.E. 1996, *ApJ*, **473**, 963
3. Churazov, E., Gilfanov, M. & Sunyaev, R. 1996, *ApJ*, **464**, L71
4. Chen, W., Gehrels, N. & Leventhal, M. 1994, *ApJ*, **426**, 586
5. Cui, W. *et al.* 1996, *ApJ*, **474**, L57
6. Dotani, T. *et al.* 1997, *ApJ*, **485**, L87
7. Ebisawa, K. *et al.* 1996, *ApJ*, **467**, 419
8. Harmon, B.A. *et al.* 1992, *AIP* 280, 314
9. Harmon, B.A. *et al.* 1993, *AIP* 304, 456
10. Hoverston, P. *USA Today*, 1996, Feb. 29, page 1
11. Kouveliotou, C. *et al.* 1996, *Nature*, **379**, 799
12. Kuznetsov, S. *et al.* 1995, "Proceedings of the Wurzburg X-ray conference"
13. Liang, E.P. & Nolan, P.L. 1984, *Sp. Sci. Rev.*, **38**, 353
14. Mereghetti, S., Belloni, T. & Goldwurm, A. 1994, *ApJ*, **433**, L21
15. Mirabel, I.F. *et al.* 1992, *Nature*, **358**, 215
16. Shapiro, S.L. & Lightman, A.P. 1976, *ApJ*, **204**, 555
17. Sheth, S. *et al.* 1996, *ApJ*, **468**, 755
18. Smith, D.M. *et al.* 1997, *ApJ*, in press.
19. Paciesas, W.S. *et al.* 1996, *IAUC*, 6284
20. Tananbaum, H. *et al.* 1977, *ApJ*, **177**, L5
21. Zhang, S.N. *et al.* 1993, *Nature*, **366**, 245
22. Zhang, S.N. *et al.* 1994, *IEEE Tran. NS.*, 41, 1313
23. Zhang, S.N. *et al.* 1995, *Exp. Astr.*, **6**, 57.
24. Zhang, S.N., *et al.* 1996, *A&A* **120**, 279
25. Zhang, S.N. *et al.* 1997, *ApJ*, **477**, L95
26. Zhang, S.N., *et al.* 1997, *ApJ* **479**, 381
27. Zhang, S.N., Cui, W. & Chen, W. 1997, **482**, L155
28. Zhang, S.N., *et al.* 1997, *ApJ*, **477**, L95
29. Zhang, S.N., *et al.* 1997, *ApJ*, to be submitted.
30. Zhang, S.N., *et al.* 1997, these proceedings.

Broad-Band Spectral Modeling of Cyg X-1 and 1E1740.7

S. Sheth*, E. Liang[†], M. Burger[†], C. Luo[†]
A. Harmon[‡], S. N. Zhang[‡]

Harvard University, Cambridge, MA 02138
[†]*Rice University, Houston, TX 77251*
[‡]*Marshall Space Flight Center, Huntsville, AL*

Abstract. We present the simultaneous ASCA and BATSE spectra of the "low-hard" state of Cyg-X1 and 1E1740.7 for several epochs in 1993 and 1994. We model the broad-band spectral data with the Sunyaev-Titarchuk (ST) inverse Compton model. The best-fit parameters are consistent with a temperature range of 29 keV - 57 keV and an energy index of 0.47 - 0.84. These correspond to Komponeet parameter y of 1.2 - 2.5 and Thomson depth of 1.6 - 3.3, consistent with past observations.

CYG X-1

The ASCA X-ray data for Cyg X-1 are best fit with a power law plus a soft blackbody. This blackbody temperature is usually 0.3 - 0.5 keV except on Nov. 23, 1994, when it shot up to 1.75 keV. We note that the high energy coninuum for that observation also corresponds to the highest Thomson depth (3.3) and lowest ST temperature (29 keV). Hence there may be some hint of anticorrelation between the soft photon source and the inverse Compton temperature. The best-fit Sunyaev-Titarchuk [1] parameters are shown in Table 1, and the data from Oct. 23, 1993 is shown in Figure 1 as an example of a broad-band spectrum.

1E1740.7

The ASCA data from 1E1740 is best fit with a simple power law with hydrogen column absorption [2]. Our model did not incorporate this low-energy absorption, and thus only data above 7 keV were used. Table 2 shows the best-fit data, and Figure F:ssheth:2 shows a spectrum from the PV phase (Nov. 93).

TABLE 1. Cyg X-1 Sunyaev-Titarchuk Best-fit Parameters

Date[a]	kT (keV)	Energy Index	Reduced χ^2
23 Oct 93	41.5 ±6.6	0.78 ±2.4E − 02	1.31
11 Nov 93	38.1 ±2.7	0.74 ±1.3E − 02	1.83
12 Nov 93	41.6 ±3.7	0.72 ±1.5E − 02	1.75
23 Nov 94	38.2 ±1.1	0.62 ±7.2E − 03	3.24
24 Nov 94	55.3 ±7.8	0.84 ±1.9E − 02	1.48
25 Nov 94	38.1 ±1.3	0.58 ±1.1E − 02	2.57
26 Nov 94	45.9 ±2.1	0.71 ±8.3E − 03	1.53

[a] ASCA data raised by factor of 1.2 to account for instrumental differences between ASCA and BATSE

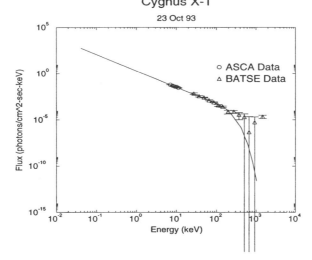

FIGURE 1. ASCA and BATSE Cyg X-1 data from 23 Oct 93, including best-fit model

REFERENCES

1. Sunyaev, R. and Titarchuk, L., *Ast. Ap.* **86**. 121 (1980).
2. Sheth, S. et al., *Ap. J.* **468**, 755 (1996).

TABLE 2. 1E1740 Sunyaev-Titarchuk Best-fit Parameters

Epoch[a]	kT (keV)	Energy Index	Reduced χ^2
PV	40.4 ±14.4	0.48 ±6.3$E-02$	0.61

[a] ASCA data raised by factor of 1.2 to account for instrumental differences between ASCA and BATSE

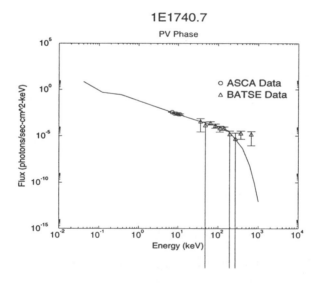

FIGURE 2. ASCA and BATSE 1E1740.7 data from Nov. 93, including best-fit model

Observational constraints on annihilation sites in 1E 1740.7−2942 and Nova Muscae

Igor V. Moskalenko* and Elisabeth Jourdain[†]

Max-Planck-Institut für extraterrestrische Physik, D-85740 Garching, Germany
[†] *Centre d'Etude Spatiale des Rayonnements, 31028 Toulouse Cedex, France*

Abstract. The region of the Galactic center contains several sources which demonstrate their activity at various wavelengths and particularly above several hundred keV [1]. Escape of positrons from such a source or several sources into the interstellar medium, where they slow down and annihilate, can account for the 511 keV narrow line observed from this direction. 1E 1740.7-2942 object has been proposed as the most likely candidate to be responsible for this variable source of positrons [2]. Besides, Nova Muscae shows a spectrum which is consistent with Comptonization by a thermal plasma $kT_e \lesssim 100$ keV in its hard X-ray part, while a relatively narrow annihilation line observed by SIGMA on Jan. 20–21, 1991 implies that positrons annihilate in a much colder medium [3,4].

We estimate the electron number density and the size of the emitting regions suggesting that annihilation features observed by SIGMA from Nova Muscae and 1E 1740.7-2942 are due to the positron slowing down and annihilation in thermal plasma. We show that in the case of Nova Muscae the observed radiation is coming from a pair plasma stream ($n_{e^+} \approx n_{e^-}$) rather than from a gas cloud. We argue that two models are probably relevant to the 1E source: annihilation in (hydrogen) plasma $n_{e^+} \lesssim n_{e^-}$ at rest, and annihilation in the pair plasma stream, which involves matter from the source environment.

OBSERVATIONS

Observations with the SIGMA telescope have revealed annihilation features in the vicinity of ~ 500 keV in spectra of two Galactic black hole candidates (Fig. 1): 1E 1740.7–2942 (1E 1740) and Nova Muscae (NM). Three times a broad excess in the 200–500 keV region was observed in the 1E 1740 emission spectra [6–9]. The features detected on Oct. 13–14, 1990 and Sept. 19–20, 1992 showed similar fluxes and line widths, the lifetime was restricted by 1–3 days. In Oct. 1991 an excess at high energies was observed during 19 days and was not so intensive as two others: the average flux was $(1.9\pm0.6)\times10^{-3}$ phot cm^{-2}

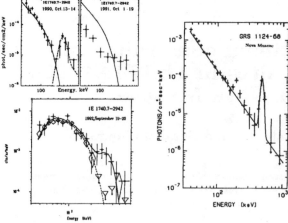

FIGURE 1. Energy spectra of 1E 1740 [1,6–9] and NM [4] observed by SIGMA are shown together with fits of the authors. For Sept. 1992 flare shown is counts s^{-1} keV^{-1}. The dashed line in the upper left panel shows the annihilation line shape for Gaussian-like injection of energetic particles into the thermal plasma of $kT = 35$ keV for $E/A = 20$. The line is shifted left to approach the data.

s^{-1} in the 300–600 keV region. On Jan. 20–21, 1991, the NM spectrum showed a clear emission feature near 500 keV with the intrinsic line width $\lesssim 58$ keV [4,5]. Meanwhile, during all periods of observation the hard X-ray emission, $\lesssim 200$ keV, was found to be consistent with the same law. Observations of NM after the X-ray flare are well fitted by a power-law of index $2.4 - 2.5$, the spectrum of 1E 1740 is well described by Sunyaev-Titarchuk model [10] with $kT \approx 35 - 60$ keV, $\tau \approx 1.1 - 1.9$. The observational data [4–8] are summarized in the first part of Table 1.

ANALYSIS AND DISCUSSION

The spectral features observed by SIGMA are, commonly believed, related to e^+e^--annihilation. Relatively small line widths imply that the temperature of the emitting region is quite low, $kT \approx 35 - 45$ keV for 1E 1740 and $4 - 5$ keV for NM. Since the hard X-ray spectra showed no changes, most probably that e^+e^--pairs produced somewhere close to the central object were injected into surrounding space where they cool and annihilate. Radiation pressure of a near-Eddington source alone can accelerate e^+e^--plasma up to the bulk Lorentz factor of $\gamma_0 \sim 2 - 5$ [11], while Comptonization by the emergent radiation field could provide a mechanism for cooling the plasma which further annihilate 'in flight' (for a discussion see [3]). If there is enough matter around a source, then particles slow down due to Coulomb collisions and annihilate in the medium. We explore further this last possibility by checking whether the inferred parameters of the emitting region are consistent with those obtained by other ways (for details see [12]). We assume single and short particle ejection on a timescale of hours; since the ejection would probably impact on the whole spectrum, longer spectral changes would be observable.

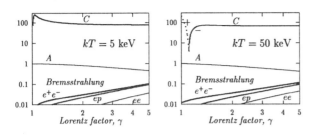

FIGURE 2. The dimensionless annihilation rate (A), and energy losses due to bremsstrahlung and Coulomb scattering (C) in hydrogen plasmas [12].

The relevant energy loss rates $|\frac{d\gamma}{dt}|$ and annihilation rate per one positron are shown in Fig. 2 in units $n_e \pi r_e^2$. Annihilation rate is small in comparison with the relaxation rate, thus most of positrons annihilate after their distribution approaches the steady-state one.

Suggesting that the energetic particles slow down due to Coulomb scattering in the surrounding matter, one can estimate its (electron) number density $n_- \approx \frac{\gamma_0 - 1}{\pi r_e^2 c \Delta_i} |\frac{d\gamma}{dt}|^{-1}$, where γ_0 is the initial Lorentz factor of the plasma stream, c is the light speed, and Δ_i is the time scale of the annihilation line appearance. Taking a reasonable value for the bulk Lorentz factor, $\gamma_0 \approx 3$, one can obtain $n_- \approx 2.2 \times 10^7$ cm^{-3} $(\Delta_i/2$ days$)^{-1}$, and $n_- \approx 1.5 \times 10^8$ cm^{-3} $(\Delta_i/5$ hr$)^{-1}$.

If the particles were injected into the medium only once, then the annihilation time scale is $\Delta_d \approx \frac{1}{\pi r_e^2 c n_- A}$. It yields one more estimate of the number density in the emitting region $n_- \approx \frac{1}{\pi r_e^2 c \Delta_d A} \approx 1.55 \times 10^9$ cm^{-3} $(\Delta_d/1$ day$)^{-1}$, where we put $A = A(1) \approx 1$ (Fig. 2). Total duration of this state is $\Delta_d \approx 18 - 70$ hr, and $\Delta_d \leq 10$ days, that gives $n_- \approx (5-20) \times 10^8$ cm^{-3} and $n_- \approx 1.5 \times 10^8$ cm^{-3} $(\Delta_d/10$ days$)^{-1}$, correspondingly. The values obtained restrict the electron number density in the volume where particles slow down and annihilate (Table 1).

The time scales $\Delta_{i,d}$ are connected $\Delta_d/\Delta_i = \frac{1}{A(\gamma_0-1)}|\frac{d\gamma}{dt}|$, which is supported also by 1992 Sept. 19–20 observation. Therefore, the annihilation rise time on 1990 Oct. 13–14 should be $\Delta_i \approx 1 - 2$ hr for consistency.

The size of the emitting region λ can be estimated from a relation $2n_+\lambda^3 \approx \Delta_d L_{500}$ if we assume $n_+ \leq n_-$ for the positron number density. It gives $\lambda \gtrsim 1.34 \times 10^{13}$ cm $(\Delta_d/1$ day$)^{2/3} \approx (1.1 - 2.7) \times 10^{13}$ cm and $\lambda \gtrsim 1.3 \times 10^{13}$ cm $(\Delta_d/10$ days$)^{2/3}$, while the obvious upper limits are $\lambda < c\Delta_i \approx 2.2 \times 10^{14}$ cm $(\Delta_i/2$ hr$)$ and $\lambda \leq 5 \times 10^{14}$ cm.

The column density of the medium where particles slow down should exceed the value $N \sim \lambda n_-$, viz. $N \gtrsim 2 \times 10^{21}$ cm^{-2} $(\Delta_d/10$ days$)^{-1/3}$, and 2.1×10^{22} cm^{-2} $(\Delta_d/1$ day$)^{-1/3} \lesssim N < c\Delta_i n_- \approx 1.1 \times 10^{23}$ cm^{-2}. The column density of the gas cloud measured along the line of sight, where 1E 1740 embedded, is high enough $N \approx 3 \times 10^{23}$ cm^{-2} [13,14]. Note that ASCA measurements give the column density *to this source* of $\approx 8 \times 10^{22}$ cm^{-2} [15]. For NM the corresponding value is $N \sim 10^{21}$ cm^{-2} [16], less or marginally close to the obtained lower limit. If, on contrary, one suggests $n_+ \ll n_-$ it yields a

TABLE 1. Observational data and parameters of the emitting region.

	1E 1740.7−2942		Nova Muscae
	1990 Oct.13–14	1992 Sep.19–20	
Annihilation rise time, Δ_i	$\lesssim 2$ days[a]	few hours	~ 5 hr
Annihilation lifetime, Δ_d	18–70 hr	27–75 hr	$\lesssim 10$ days
Annihil. line flux, F_{500} (phot cm^{-2}s^{-1})	10^{-2}	4.3×10^{-3}	6×10^{-3}
Total line flux[b], L_{500} (photons s^{-1})	8.6×10^{43}	3.7×10^{43}	7.2×10^{41}
Line width, W (keV)	240	180	40
Column density, N (cm^{-2})	$\sim 10^{23}$		$\sim 10^{21}$
Plasma temperature, kT_e (keV)	35 − 45		3 − 4
Coulomb energy loss rate, $|d\gamma/dt|$	70		100
Annihilation rate, A	1		1
Electron number density, n_- (cm^{-3})	$(5-20) \times 10^8$		1.5×10^8
Size of the emitting region, λ (cm)	$(1.1-20) \times 10^{13}$		$(1.3-50) \times 10^{13}$

[a] Our estimation is 1–2 hr;   [b]1E 1740: for 8.5 kpc distance; NM: for 1 kpc distance.

condition $N \gg 2 \times 10^{21}$ cm^{-2} $(\Delta_d/10 \text{ days})^{-1/3}$, which considerably exceeds the measured value.

These estimations imply that the 500 keV emission observed from NM was coming from e^+e^--plasma jet $(n_+ \approx n_-)$ rather than from particles injected into a gas cloud[1] $(n_+ \ll n_-)$, therefore, particles have to annihilate 'in flight' producing a relatively narrow line shifted dependently on the jet orientation. If so, then our estimation of n_- from annihilation time scale gives the average electron/positron number density in the jet, fixing the total volume as $\lambda^3 \sim 2 \times 10^{39}$ cm^3 $(\Delta_d/10 \text{ days})^2$. The reported 6%–7% redshift [4,5] supports probably the annihilation-in-jet hypothesis, although its statistical significance is small. The large size of the emitting region and a small width of the line, both except the gravitational origin of the redshift.

For the emitting region in 1E 1740 our estimations give $n_- \gtrsim n_+$. Two events, Oct. 1990 and Sept. 1992, have shown similar parameters, which are consistent with single particle injection into the thermal (hydrogen) plasma. The redshift of the line $\sim 25\%$ [6–8] implies that positrons probably annihilate in a plasma stream moving away from the observer, since the size of the emitting region is too large and rules out its gravitational nature. A natural explanation of this controversial picture is that the plasma stream captures matter from the source environment and annihilation occurs in a moving plasma volume. The value of n_- obtained is then the average electron number density in the jet, $\lambda^3 \gtrsim 2.4 \times 10^{39}$ cm^3 $(\Delta_d/1 \text{ day})^2$ is its total volume, and the jet length should be of the order of $\sim 0.2c\Delta_d \approx 5.2 \times 10^{14}$ cm $(\Delta_d/1 \text{ day})$.

While some part of the pair plasma annihilates near 1E 1740 producing the broad line, the remainder could escape into a molecular cloud, which was found to be associated with this source [13,14]. Taking $\sim 10^5$ cm^{-3} for the average

---

[1] A possibility that NM lies in front of a large gas cloud can not be totally excluded. In this case, particles could be injected into this cloud, away from the observer.

number density of the cloud one can obtain for the slowing down time scale[2] $\Delta_i \lesssim 1$ year, the same as was obtained in [2]. The size of the turbulent region caused by propagation of a dense jet should also be of the order of 1 ly. It agrees well with the length 2–4 ly (15–30 arcsec at 8.5 kpc) of a double-sided radio jet symmetrical about 1E 1740 [17].

If the lines from 1E 1740 (Fig. 1) were produced by continuous injection of energetic particles, then observations of the narrow 511 keV line emission from the Galactic center allows to put an upper limit on the particle escape rate into the interstellar medium. Taking $\tau_0 = 1$ year for the positron lifetime in 10^5 cm^{-3} dense molecular cloud [2], and suggesting one hard state of $\Delta_d \gtrsim 2$ days long per period τ_0, one can obtain an escape rate $E/A \approx \frac{F_{511}}{F_{500}} \frac{\tau_0}{\Delta_d} \lesssim 20$, where we took $F_{511} \approx 10^{-3}$ phot cm^{-2} s^{-1} for the narrow line intensity [18], and $F_{500} = 10^{-2}$ phot cm^{-2} s^{-1} (Table 1). This is consistent with 1990 Oct. 13–14 and 1992 Sept. 19–20 spectra; the dashed line in Fig. 1 shows the annihilation line shape for Gaussian-like particle injection, $\sim \exp[-(\gamma - 4)^2]$, into the thermal plasma of $kT = 35$ keV for $E/A = 20$ [12]. The longest hard state (19 days, Oct. 1991) with the average flux of $F_{500} \approx 2 \times 10^{-3}$ phot cm^{-2} s^{-1} places the upper limit at almost the same level $E/A \approx 10$.

I.M. is grateful to the SIGMA team of CESR for hospitality and facilities.

REFERENCES

1. Churazov E., et al., *ApJS* **92**, 381 (1994)
2. Ramaty R., et al., *ApJ* **392**, L63 (1992)
3. Gilfanov M., et al., *Soviet Astron. Lett.* **17**, 437 (1991)
4. Goldwurm A., et al., *ApJ* **389**, L79 (1992)
5. Sunyaev R. A., et al., *ApJ* **389**, L75 (1992)
6. Bouchet L., et al., *ApJ* **383**, L45 (1991)
7. Sunyaev R. A., et al., *ApJ* **383**, L49 (1991)
8. Cordier B., et al., *A&A* **275**, L1 (1993)
9. Churazov E., et al., *ApJ* **407**, 752 (1993)
10. Sunyaev R. A., Titarchuk L. G., *A&A* **86**, 121 (1980)
11. Kovner I., *A&A* **141**, 341 (1984)
12. Moskalenko I. V., Jourdain E., *A&A* **in press** (1997) – astro-ph/9702071
13. Bally J., Leventhal M., *Nature* **353**, 234 (1991)
14. Mirabel I. F., et al., *A&A* **251**, L43 (1991)
15. Sheth S., et al., *ApJ* **468**, 755 (1996)
16. Greiner J., et al., in *Proc. Workshop on Nova Muscae*, ed. S. Brandt, Lyngby: Danish Space Research Inst., 1991, p.79
17. Mirabel I. F., et al., *Nature* **358**, 215 (1992)
18. Mahoney W. A., Ling J. C., Wheaton Wm. A., *ApJS* **92**, 387 (1994)

---

[2] The annihilation lifetime Δ_d was obtained for thermal plasma and is not valid for the cold medium where positrons mostly annihilate in the bound (positronium) state.

Two-phase spectral modelling of 1E1740.7-2942

Osmi Vilhu[1], Jukka Nevalainen[1], Juri Poutanen[2], Marat Gilfanov[3], Philippe Durouchoux[4], Marielle Vargas[4], Ramesh Narayan[5] and Ann Esin[5]

[1] *Observatory, FIN-00014 University of Helsinki, Finland*
[2] *Uppsala Observatory, Uppsala, Sweden*
[3] *Space Research Institute, 117810 Moscow, Russia*
[4] *C.E.Saclay, 91191 Gif-sur Yvette, France*
[5] *Harvard-Smithsonian Center for Astrophysics, Cambridge, MA 02138, USA*

Abstract.
Combined ASCA and SIGMA data of 1E1740.7-2942 during its standard state (September 1993 and 1994) were fitted with two-phase models (ISMBB [8,9] and ADAF [7,3]). In ISMBB's, the radius of the spherical hot (T_e = 150 - 200 KeV) corona lies between 200 - 250 km where it joins the classical inner disc. The disc radiates 40 % of the total luminosity (with $\dot M = 0.017 \dot M_{Edd}$ of $10M_\odot$). ADAF's need an extra component to reproduce the soft part of spectrum. However, the origin of the soft excess remains somewhat uncertain, although special care was taken in the background elimination.

INTRODUCTION

1E1740.7-2942 is the famous jet-source (micro-quasar, Great Annihilator) close to the Galactic Center, observable in the cm/mm and X/Gamma-ray wavelengths [11,1,5,6,10,2,12]. During the last few years, new tools for modelling of X-ray data have been developed. These include two-phase sombrero models ISMBB's [8,9] and ADAF's (advection dominated accretion flows [7,3]). While ADAF's are fixed basically by $\dot M$ (mass transfer rate), ISMBB's have more freedom to fit observations.

We used ASCA archive data and SIGMA hard X-ray observations (Sept 1993 and 1994) to explore 1E1740.7-2942 with the help of these models (see Figure 1). Simultaneous BATSE results were used for comparison. The nearby background in ASCA images was carefully analysed and subtracted.

ISMBB AND ADAF MODELS

The hot phase of ISMBB was handled as a pure spherical thermal pair plasma (defined by T_e and τ_e), although the background plasma (protons) and non-thermal electrons can be included in the treatment. To solve self-consistently the pair balance, the energy balance and the radiative transfer equations, the iterative scattering method (ISM) was used ([8]). The source of soft photons is the classical disc black body radiation with radial temperature-dependence $T(r) = T_{bb}(r/R_{in})^{-3/4}$. The disc is allowed to reach inside R_{in} (where T is assumed constant), but all our fits converged to the case where the radius of the hot phase equals to R_{in}. The radiatively heated reprocessed radiation is included in the disc black body, increasing the disc luminosity typically by 20 %. The solution was found running the models and the input data under XSPEC of the XANADU software (see Figure 2 and Table 1).

Several ADAF models were computed around the intermediate state [3] with $\alpha = 0.3$ (viscosity parameter), $\beta = 0.5$ (magnetic pressure parameter), m = 10 (mass in solar units) and $\dot{m} = 0.11$ (mass transfer rate in Eddington units with efficiency 0.1, $\approx \dot{m}_{crit}$). These models are too luminous by a factor of 4 which can be easily accounted for by slightly reducing the parameters (the luminosity is proportional to $\alpha^2 m \dot{m}$). The $\dot{m}$-value from ISMBB-fits is also smaller.

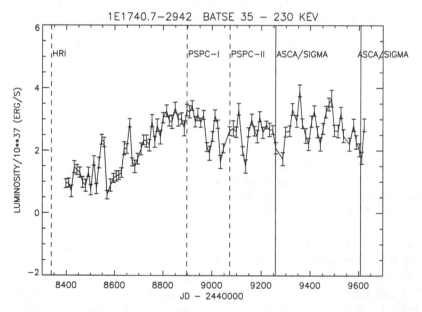

FIGURE 1. BATSE light curve of 1E1740.7-2942 showing the dates of the observations discussed in the text.

TABLE 1. ISMBB and ADAF fits to the standard state of 1E1740.7-2942

PARAM	ISMBB 94	ISMBB 93	ADAF-1 94	ADAF-1 93
N_H (10^{22} cm^{-2})	12.3	12.3	15.7	16.8
kT_{bb} (KeV)	0.25	0.23	≈ 0.2	≈ 0.2
T_e (KeV)	200	150	≈ 110	≈ 110
τ_e	0.7	1.3	≈ 2.1	≈ 2.1
cos(incl)	0.95	0.95	0.87	0.87
R_{in}[a]	7	8	10	10
L_{total} (10^{37} erg/s)	4.5	5.6	3.5	4.4
L_{disc}/L_h	0.82	0.55		
red. χ^2	1.13	1.12	1.17	1.16

[a] $R_{in} = Rc = R_{TR}$, in R_{Sch} units of $10M_\odot$

Figure 3 shows the fit with a model having transition radius $R_{TR} = R_{in} = 10$ (in Schwarzchild units). For a smaller radius the spectrum is too soft, while for a larger one the disc spectrum lies totally outside the ASCA-range (see Figure 4 where the models are compared). Even the model with $R_{TR} = 10$ needs an extra component, in Figure 3 the excess was modelled by a thin plasma (Raymond-Smith) with emission measure of 1.1×10^{62} cm^{-3} and T = 0.28 KeV. Using the Sedov-model, we can speculate that a shocked young (≤ 1000 years) supernova remnant, expanding in a dense ($\geq 10^{2.5}$ cm^{-3}) medium and originated in a relatively quiet explosion ($E_T \leq 10^{48.5}$ erg), can explain these values and the small size (point source in the ROSAT HRI-image).

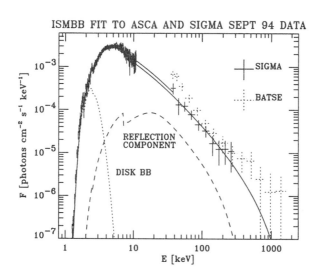

FIGURE 2. ISMBB fit to the September 1994 data. BATSE results are overplotted.

CONCLUSIONS

The standard state spectrum of 1E1740.7-2942 can be fitted with the ISMBB models. The disc has an inner radius of 7 - 8 Schwarzschild radii of $10M_\odot$ black hole radiating with $L_{disc} = 2.2\times10^{37}$ erg/s, corresponding to the mass transfer rate of 2.3×10^{17} g/s $= 0.017\dot{M}_{Edd}$ ($\dot{M}_{Edd} = 10L_{Edd}/c^2 = 1.39\times10^{18}M/M_\odot$ g/s, using efficiency of 0.1). Inside R_{in}, the spherical hot corona ($T_e = 150 - 200$ keV) emits Comptonized radiation with $L_h = 2.7\times10^{37}$ erg/s.

The ADAF models can reproduce the overall spectral shape and luminosity, provided that the total mass, the mass transfer rate and the viscosity parameters are properly selected. For small values of R_{in}, the ADAF hot flows are somewhat too cold and the disc spectrum too shallow (see Figure 4). However, it is possible that the soft excess has some other origin we missed (a local excess in the background, circumstellar scattering dust, young SNR in a dense medium, ...), but this can be decided only with future observations having a higher spatial resolution (like AXAF).

An important test for the nature of the soft excess may be the observed ROSAT count rates [4]. The ISMBB models of Table 1 predict 2.5 times higher and 0.8 times lower PSPC and HRI count rates, respectively. If the disc is removed, the predicted count rates are reduced by a factor of 3. However, if an anticorrelation between the soft and hard luminosities exists, a part of the difference can be due to a real variability (see Figure 1).

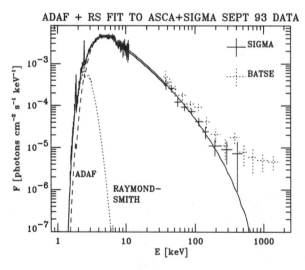

FIGURE 3. ADAF model ($R_{TR} = 10R_{Sch}$) fit to the September 1993 data. BATSE results are overplotted.

REFERENCES

1. Bally J. and Leventhal M. 1991, Nature 353, 234.
2. Churazov E., Gilfanov M. and Sunyaev R. 1996, ApJ 464, L71.
3. Esin A.A, McClintock J.E. and Narayan R. 1997, astro-ph/9705237.
4. Heindl W.A., Prince T.A. and Grunsfeld J.M. 1994, ApJ 430, 829.
5. Mirabel I.F. et al. 1991, Astron.Astrophys. 251, L43.
6. Mirabel I.F. et al. 1992, Nature 358, 215.
7. Narayan R. 1996, ApJ 462, 136.
8. Poutanen J. and Svensson R. 1996, ApJ 470, 249.
9. Poutanen J. 1997, private comm.
10. Sheth S., Liang E., Luo C., 1996. ApJ 468, 755.
11. Sunyaev R. et al. 1991, ApJ 383, L49.
12. Vilhu O. et al. 1997, Proc. 2nd INTEGRAL workshop 'The Transparent Universe', ESA SP-382, p.221, astro-ph 9612194.

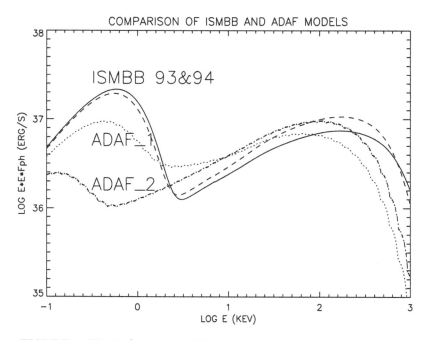

FIGURE 4. ISMBB (1993 and 1994) and ADAF ($R_{TR} = 10$ and 100) models.

Multi-Wavelength Monitoring of GRS 1915+105

R. Bandyopadhyay[1], P. Martini[2], E. Gerard[3], P.A. Charles[1], R.M. Wagner[2], C. Shrader[4], T. Shahbaz[1] and I.F. Mirabel[5,6]

[1] *University of Oxford;* [2] *Ohio State University;* [3] *Observatoire de Paris*
[4] *NASA/Goddard Space Flight Center;* [5] *Service d'Astrophysique, Saclay*
[6] *Instituto de Astronomia y Fisica del Espacio, Argentina*

Abstract. Since its discovery in 1992, the superluminal X-ray transient GRS 1915+105 has been extensively observed in an attempt to understand its behaviour. We present here preliminary results from a multi-wavelength campaign undertaken from July to September 1996. This study includes X-ray data from the RXTE All Sky Monitor and BATSE, two-frequency data from the Nancay radio telescope, and infrared photometry from the 1.8m Perkins telescope at Lowell Observatory. The K-band data presented herein provide the first long-term well-sampled IR light curve of GRS 1915+105. We compare the various light curves, searching for correlations in the behaviour of the source at differing wavelengths and for possible periodicities.

INTRODUCTION AND OBSERVATIONS

The X-ray transient GRS 1915+105 was discovered by the GRANAT satellite in 1992 [1]. VLA observations at the time of outburst led to the discovery of relativistic ejections of plasma clouds with apparent superluminal motions, possibly a smaller-s cale analogue to the jets observed in active galactic nuclei and quasars [2]. GRS 1915+105 has a highly variable spectral index and often exceeds the Eddington luminosity limit for a neutron star, indicating that the system may contain a black hole. Radio observations show that it is at a kinematic distance $D = 12.5 \pm 1.5$ kpc from the Sun; combined with measurements of the hydrogen column density along the line of sight, a visual extinction of $A_V = 26.5 \pm 1$ mag has been determined [3]. Variable radio and IR counterparts to the X-ray source have been found [4]. However, the optical counterpart has only been detected at $I = 23.4$ [5].

On the basis of its spectral morphology and the calculated absolute K magnitude, Castro-Tirado et al. [6] suggested GRS 1915+105 to be a low-mass X-ray binary (LMXB). However, its long term X-ray behaviour has shown erratic burts with recurrent peaks of similar intensity (maximum $L_x = 3 \times 10^{38}$ ergs/s; [3]), unlike typical X-ray light curves of LMXBs [7]. K-band spectroscopy of the counterpart revealed several prominent emission features but no detectable absorption features which would be indicative of a late-type secondary. On the contrary, the K-band spectrum is very similar to those of high-mass X-ray binaries (HMXB) with a Be star as the mass-losing component [8]. In addition, the absolute magnitude and colours of GRS 1915+105 are strikingly similar to other HMXBs, most notably the LMC Be/X-ray binary (XRB) A0538-66 [9].

JHK photometry of the IR counterpart has revealed both short- and long-term variability, but no periodicity has been seen [3]. In order to search for the long-term periodicity ($\sim$15-45 days) expected in an eccentric HMXB, a high-resolution ($\sim$1 day) light curve with a baseline of several months is necessary. At X-ray and radio wavelengths, GRS 1915+105 shows fluctuations of variable amplitude on a variety of timescales. As in the IR, no consistent periodicity has been found. However, some X-ray and radio observations have shown a quasi-periodicity of $\sim$30 days [10].

We present multi-wavelength light curves of GRS 1915+105 from the period July to September 1996, including K-band photometry obtained with the Ohio State Infrared Imager/Spectrometer (OSIRIS) on the 1.8m Perkins telescope at Lowell Observatory, X-ray data from the RXTE All-Sky Monitor (2-10 keV) and BATSE (20-200keV), and 1414 MHz and 3310 MHz observations from the Nancay radio telescope. This study is the first to include long-term well-sampled IR photometry as well as X-ray and radio information in a multi-wavelength study of GRS 1915+105.

THE INFRARED LIGHT CURVE

The five three-month light curves of GRS 1915+105 we obtained are shown in Figure 1; the K-band curve appears in the top panel. GRS 1915+105 exhibits variability from $K \sim$12.5 to $\sim$13.5, with maxima at UT dates 3 July, 7 August, and 15 September. A visual inspection of the curve between the latter two maxima indicates an apparent 40-day modulation. An attempt to search for a regular periodicity on this timescale was unsuccessful, as a 40-day period appears inconsistent with the $\sim$30 day s eparation of the first two maxima. However, we note that it is possible that our observations began *after* a true peak; therefore a $\sim$40-day cycle cannot be ruled out. The shape of the curve and the long timescale for the variations are intrigu ingly similar to several known Be/XRB systems with eccentric orbits and long periods (generally $\gtrsim$ 20 days; [11]). The IR photometric characteristics of GRS 1915+105 are especially similar to the LMC Be/XRB A0538-66 (X0535-668). Both sources show $\gtrsim$ 1 magnitude variability at K [8], and both reach $L_x \sim 10^{39}$ erg/s at outburst peak [12,13].

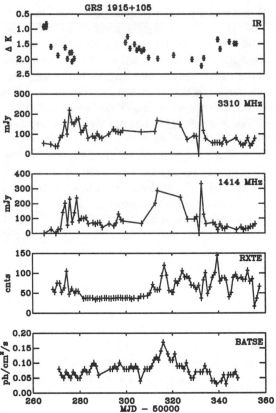

FIGURE 1. Multi-wavelength light curves of GRS 1915+105 from July-September 1996. Note the simultaneous flare of days 310-327 which appears in the radio and X-ray data, but which corresponds to a minimum in the IR light curve.

The inclination of GRS 1915+105 is well constrained to be $i = 70\pm2°$ from the angle of the jet emission [2]. We therefore expect IR variability on the orbital period due purely to either ellipsoidal modulations and/or X-ray heating of the companion star; however, the amplitude of these variations is partially dependent on the spectral type of the mass-donating star. Substantial orbital flux variations ($\sim$1 mag) are generally expected from high-i LMXBs, where the IR light curve is dominated by X-ray heating [14]. In HMXBs, more moderate ellipsoidal variability ($\sim$10-20%) is the norm [15] . We note, however, that the Be/XRB A0538-66 shows >1 mag variability at K [9]. In addition, the first maximum in our light curve is $\sim$0.3 mag brighter than the two subsequent maxima, indicating possible short-term variability such as that seen by Chaty et al.. It therefore seems likely that the IR variability in our K-band light curve does not result from a single cause, but from several different processes, perhaps with an underlying ~40-day orbital modulation.

In addition to orbital variability, there are a number of possible sources of IR emission in GRS 1915+105 which may account for the observed changes in magnitude. These include (1) IR emission from the accretion disk [6], (2) free-free emission from an X-ray driven wind [16], (3) time-variable Doppler-broadened spectral line emission from ions in the relativistic jets [8], (4) thermal dust reverberation of energetic outbursts [12], and (5) synchrotron emission from relativistic jets [17]. Of these possibilities, (3), (4), and (5) are related to ejection events and would therefore produce changes in the IR magnitude correlated with jet activity. Finally, enhanced IR emission could be produced as a result of advective accretion in the inner accretion disk.

RADIO AND X-RAY LIGHT CURVES

Radio and X-ray light curves for July-September 1996 appear in the lower four panels of Figure 1. While the two radio curves show very similar variability, the RXTE and BATSE data do not show similar behaviour, indicating strong spectral variations. The most striking feature of these light curves occurs at day 310 (15 August), when a flare occurs at all four wavelengths. This flare peaks at day 311 and then falls off over ~15 days in the hard X-rays and radio, while the soft X-rays continue to show large amplitude fluctuations.

On the basis of the similarity in rise, duration, and decay times of the hard X-ray/radio flare during days 310-327 to the April 1994 ejection event reported by Rodriguez et al. [18] and the suspected ejection event in August 1995 discussed by Foster et al. [19], it seems likely that the flare in our data indicates that an ejection event occurred at this time (August 1996). In the April 1994 and August 1995 ejections of GRS 1915+105, the hard X-ray flux peaks prior to the radio, while in our data, the radio and hard X-ray peaks appear coincident. Our data are not necessarily inconsistent with this pattern, as it is possible that the actual radio peak during the observed flare occurred during the gap in our radio coverage (days 311-325). The pattern of the appearance of a hard X-ray peak followed within a few days by a radio peak, with subsequent correlated decays, may be a signature of jet ejection in GRS 1915+105 (as suggested by Harmon et al. [7]). It seems likely, however, that this is not the only such signature. As yet undefined are the mechanisms which are necessary to cause the changes in the hard X-ray/radio behaviour surrounding ejection events, and hence altering the ejection signatures of GRS 1915+105 at various times. What does seem likely is that ejection events do not occur if X-ray (hard or soft) active states are not accompanied by radio emission levels $\gtrsim 100$ mJy [8,20].

CONCLUSIONS

(i) There is evidence for a long-term periodic IR modulation on the order of 30-40 days. The qualitative characteristics of this modulation are similar to those expected for a Be/XRB with an eccentric orbit. However, it is likely that the IR variability arises from a combination of causes, with short-term IR emission superimposed on an underlying orbital period.

(ii) By comparing our X-ray/radio data to previously observed ejection events, we believe that an ejection event took place in August 1996, when we see a large simultaneous outburst in the radio and X-rays. It is interesting to note that during this event the IR emission is at a low level, indicating that IR variations resulting from jet ejection may be minimal in this instance.

(iii) In general agreement with Harmon et al. [7], we surmise that the pattern of a hard X-ray flare rapidly followed by a radio flare, with subsequent correlated decays, is a signature of jet ejection in GRS 1915+105; however, the X-ray activity must be accompanied by radio emission levels $\gtrsim 100$ mJy. We also note that this is probably not the only such hallmark of ejection events.

REFERENCES

1. Castro-Tirado A.J. et al., 1994, ApJS, 92, 469.
2. Mirabel I.F. & Rodríguez L.F., 1994, Nature 371, 46.
3. Chaty S. et al., 1996, A&A, 310, 825.
4. Mirabel I.F. et al., 1994, A&A, 282, L17.
5. Boer M., Greiner J., & Motch C., 1996, A&A, 305, 835.
6. Castro-Tirado A.J., Geballe T.R. & Lund N. 1996, ApJ, 461, L99.
7. Harmon B.A. et al., 1996, ApJ, 477, L85.
8. Mirabel I.F. et al., 1997, ApJ, 477, L45.
9. Allen D.A., 1984, MNRAS 207, 45p.
10. Mirabel I.F. & Rodríguez L.F., 1996, *Solar and Astrophysical Magnetohydrodynamic Flows*, ed. K. C. Tsinganos. Kluwer Academic Publishers, p. 683.
11. van den Heuvel E.P.J. & Rappaport S., 1987, in *Physics of Be Stars*, Proc. IAU Colloq. No. 92, eds. A. Slettebak & T.P. Snow, Cambridge University Press.
12. Mirabel I.F. et al., 1996, ApJ, 472, L111.
13. Charles P.A. et al., 1983, MNRAS, 202, 657.
14. White N.E., 1989, in *Theory of Accretion Disks*, eds. F. Meyer, W.J. Duschl, J. Frank, & E. Meyer-Hofmeister. Kluwer Academic Publishers, p. 269.
15. van Paradijs J. & McClintock J.E., 1995, in *X-ray Binaries*, eds. Lewin W.H.G, van Paradijs J. and van den Heuvel E.P.J., Cambridge University Press, p. 101.
16. van Paradijs J. et al., 1994, ApJ, 429, L19.
17. Sams B.J., Eckart A., & Sunyaev R., 1996, Nature 382, 47.
18. Rodríguez L.F. et al., 1995, ApJS, 101, 173.
19. Foster R.S et al., 1996, ApJ, 467, L81.
20. Greiner J., Morgan E.H., & Remillard R.A., 1996, ApJ, 473, L107.

The Hard X-Ray Spectrum of GRS1915+105

W.A. Heindl[†], P. Blanco[†], D.E. Gruber[†], M. Pelling[†], R. Rothschild[†], E. Morgan[††], and J.H. Swank[†††]

[†] *Center for Astrophysics and Space Sciences, University of California, San Diego*
[††] *Massachusetts Institute of Technology*
[†††] *NASA/Goddard Space Flight Center*

Abstract.
We have fit 2-250 keV spectra of the Galactic microquasar GRS 1915+105 obtained with the *Rossi X-ray Timing Explorer* (*RXTE*). Two individual observations, from 1996 August and 1997 January, as well as 15-250 keV data averaged over multiple observations made during 1997 January through March are analyzed. We find that when GRS 1915+105 was bright ($L_{2-10keV} \sim 10^{39}$ ergs s^{-1}) and variable on day-long timescales, its spectrum was consistent with the standard black hole X-ray high (soft) state. When the source was dimmer ($L_{2-10keV} \sim 10^{38}$ ergs s^{-1}) and steady, its spectrum was well described by the black hole X-ray low (hard) state. We confirm that these X-ray states are associated with the gamma-ray "power-law" and "breaking" states identified by Grove et al. [3].

I INTRODUCTION

Along with GRO J1655-40, the transient X–ray source GRS 1915+105 is one of two known superluminal radio jet sources in the Galaxy. GRS 1915+105 was discovered in 1992 August with the *GRANAT*/WATCH all-sky X-ray monitor [1]. The 1992 outburst, as seen by *CGRO*/BATSE, lasted until 1993 September and was followed by outbursts from 1993 December to 1994 April and 1995 April to the present [5]. Observations made with the VLA from 1994 March through April revealed a pair of radio condensations moving away from a compact radio core at apparent velocities greater than the speed of light [9]. Mirabel *et al.*, determined that GRS 1915+105 lies in the Sagittarius arm of the Milky Way at a distance of 12.5 ± 1.5 kpc [7].

Along with the overall light curve, we discuss here the 2-250 keV spectrum from 2 observations – 1996 August 29 and 1997 January 29. We also present a summed 15-250 keV spectrum for the period 1997 January 7 through March 27.

TABLE 1. Observations and their exposures.

	Observation Time (UT)	Live Time (s) PCA	HEXTE
1a	1996 Aug. 29 11:33–11:50	960	548
1b	1996 Aug. 29 12:35–13:26	2448	1643
1c	1996 Aug. 29 13:26–15:02	2512	1652
2	1997 Jan. 29 20:56–01:07	9840	5708
3	1997 Jan. 01 – Mar. 27	——	48467

II OBSERVATIONS AND ANALYSIS

The *RXTE* pointed instruments – the Proportional Counter Array (PCA) [6] and the High Energy X-ray Timing Experiment (HEXTE) [4] – have been aimed at GRS 1915+105 at least once a week since 1996 April, amounting to more than 100 observations to date. In addition, the All Sky Monitor (ASM) observes the source several times daily in the band 2-10 keV.

Table 1 gives the details of the observations discussed here. For observations 1 and 2, *RXTE* pointed at the source during 3 consecutive spacecraft orbits ($\sim$ 90 min each). Because of spectral variations from one orbit to the next during observation 1, the data from each orbit have been fit individually (designated "a", "b", and "c"). Only the HEXTE data are discussed for observation 3 which is the sum of 13 pointings carried out in 1997 January through March. Observation 2 is a subset of these data.

Because of the steeply falling source spectrum and the relative effective areas of the PCA and HEXTE, the statistical uncertainties of the PCA data are much smaller than those of the HEXTE data. In order to allow the HEXTE data to constrain spectral fits, we added 2% systematic errors to the PCA data. Before combining the 13 individual pointings of observation 3, we fit each in order to verify that no significant spectral variations were present.

All the spectra require both a soft and a hard component to form a reasonable representation. As a general fitting template, we used a disk black body component plus a broken power law with an exponential cutoff at high energies. We also allowed an iron line, but no fit significantly required its presence. In some cases, other choices for the soft component (e.g. thermal bremsstrahlung) were possible, but the disk black body spectrum worked well for all spectra.

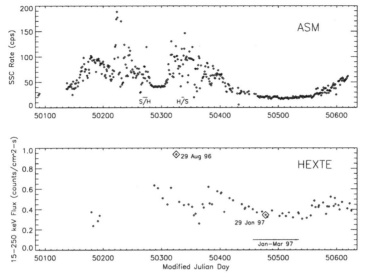

FIGURE 1. The light curve of GRS 1915+105 from the ASM and HEXTE on *RXTE*. Soft/Hard branch transitions from PCA observations [2] are indicated on the ASM plot.

III RESULTS AND DISCUSSION

A Light Curve

Figure 1 shows both the ASM and HEXTE light curves since 1996 January. The ASM data are daily average rates, while the HEXTE data are averaged over the duration of the observation made that day (typically a few thousand seconds). Therefore, the data are contemporaneous but not simultaneous. On day-long timescales the source exhibits two characteristic behaviors. When the ASM rate is above ∼50 counts/s (e.g. MJD 50320–50400), the soft emission varies by up to a factor of two. Meanwhile, at low fluxes, the source variations are less than ∼10%. It is also clear that the soft emission is typically more variable than the hard emission. However, an exception to this is seen during MJD 50282–50309 when the > 15 keV flux decreased by 30% as the soft flux remained essentially constant.

Chen, Swank, and Taam, analyzing PCA data from 1996 April – September, found that the emission is confined to two branches on a plot of X–ray hardness ratio versus counting rate [2]. As a function of counting rate, the PCA hardness (ratio of 11-30.5keV to 2-11keV count rates) falls either on a hard branch, which rises steeply at low counting rates, or a soft branch, whose hardness is only weakly dependent on rate. From late May to early July,

TABLE 2. Spectral Fits.

	N_h ($\times 10^{22}$cm^{-2})	γ_1	E_b (keV)	γ_2	E_{cut} (keV)	E_{fold} (keV)	T_{in} (keV)
1a	$3.3^{\pm 0.7}$	$2.5^{\pm 0.2}$	$15.7^{\pm 0.9}$	$3.24^{\pm 0.04}$	—	—	$1.47^{\pm 0.05}$
1b	$4.1^{\pm 0.7}$	$2.6^{\pm 0.2}$	$15.5^{\pm 0.6}$	$3.40^{\pm 0.02}$	—	—	$1.64^{\pm 0.05}$
1c	$4.4^{\pm 0.7}$	$2.6^{\pm 0.2}$	$15.6^{\pm 0.7}$	$3.38^{\pm 0.03}$	—	—	$1.62^{\pm 0.05}$
2	$2.6^{\pm 0.7}$	$1.9^{\pm 0.2}$	—	—	$16^{\pm 5}$	$60^{\pm 15}$	$1.53^{\pm 0.08}$
3	—	$2.10^{\pm 0.04}$	—	—	$24.9^{\pm 01.3}$	$88^{\pm 4}$	1.53 (fixed)

pointed observations always found the source on the soft branch. Between 3 and 11 July (MJD 50267 and 50275), it transitioned to the hard branch where it apparently remained until at least 7 September (MJD 50333). On 11 September, it was again found on the soft branch [2]. One might expect these transitions to correspond to the transitions between stable and highly variable emission seen in the ASM. However, the transition from the soft to hard branch preceded the shift to stable emission by several days, and the reverse transition came more than 30 days after GRS 1915+105 once again became highly variable (see fig. 1). It seems, therefore, that the mechanisms behind the pronounced variability and the hard/soft branch transitions are not the same.

B Spectra

Figure 2 shows the best fit input photon spectra for observations 1c and 2 respectively. The best fit model parameters are given in Table 2. We chose these observations to search for spectral differences between the variable and stable flux states. Figure 1 shows that observation 1 occurred during a highly variable state, while observations 2 and 3 are from a stable emission period. The observation 1 spectra are the brightest, having $L_{2-10keV} \sim 10^{39}$ ergs s^{-1} compared with $\sim 10^{38}$ ergs s^{-1} for observation 2. In both the variable and stable states, the disk black body has an inner temperature of $\sim$1.5 keV. It is the hardness ratio (here, 15-200 keV flux divided by 2-15 keV flux) and the shape of the underlying power law that distinguish the spectra. In observations 1a-c, the hardness had a value of $\sim$0.2, while in observation 2 it was $\sim$0.6. This indicates that the source was in X-ray high (soft) and low (hard) states for observations 1 and 2 respectively. Further, the nature of the power law components are distinct. The observation 1 spectra are well fit at energies above 15 keV by a simple, unbroken power law with a photon index of 3.2-3.4. This steep power law would overpredict the 2-15 keV flux and therefore must break to a harder slope (index $\sim$2.5) at low energies (see fig. 2). In observations 2 and 3, a hard power law (index $\sim$2) with an exponential cutoff at high energies is required. A broken power law gives an unacceptable fit.

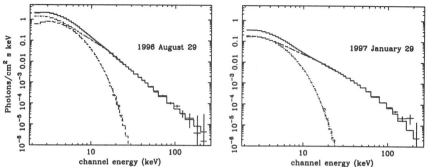

FIGURE 2. Unfolded photon spectra for observations 1c and 2 respectively. See table 2 for spectral parameters.

Grove et al. found that the black hole transients observed by OSSE can be divided into two classes by their gamma-ray spectra [3]. The "breaking spectra" class shows a hard but exponentially cut off power law spectrum, while the "power-law spectra" class has a soft but unbroken power law. Some black hole candidates (e.g. Cyg X-1) have shown both gamma-ray states, and Grove et al. associate the "breaking" gamma-ray state with the X-ray low (hard) state and the "power-law" gamma-ray state with the X-ray high (soft) state [3]. The observations discussed here confirm this association for GRS 1915+105.

We have seen that GRS 1915+105 shows both the high (soft) and low (hard) X-ray states typical of black hole candidates and that these X-ray states have corresponding changes in spectral shape at energies above 15 keV. It is most variable, at least on long timescales, below 10 keV. Meanwhile, the transitions between variable and stable emission, as seen in the ASM, are not contemporaneous with the spectral transitions. This suggests that these two emission properties are somewhat independent.

REFERENCES

1. Castro-Tirado, A., et al., *ApJS* **92**, 469, (1994).
2. Chen, X., Swank, J.H., and Taam R.E., *ApJL*, **477**, 41 (1997). (1997).
3. Grove, J.E., et al., *ApJ*, submitted, (1997).
4. Gruber, D.E., et al., *A&AS*, **120**, 641, (1996).
5. Harmon, B. A., et al., *ApJL*, **477**, 85, (1997).
6. Jahoda, K., et al., *SPIE*, 2808, 59 (1996).
7. Mirabel, I.F., et al., *A&A*, **282**, L17 (1994)
8. Mirabel, I.F., et al., *ApJL*, **477**, 45 (1997).
9. Mirabel, I.F., and Rodriguez, L.F., *Nature*, **371**, 46, (1994).

OSSE Upper Limit on Positron Annihilation from GRS 1915+105

D. M. Smith[1], M. Leventhal[2], L. X. Cheng[2], J. Tueller[3], N. Gehrels[3], I. F. Mirabel[4], L. F. Rodriguez[5], and W. Purcell[6]

[1] *Space Sciences Laboratory, University of California, Berkeley, CA 94720*
[2] *Department of Astronomy, University of Maryland, College Park*
[3] *NASA-Goddard Space Flight Center*
[4] *CEA-CEN, Saclay, France*
[5] *Instituto de Astronomia, UNAM, Mexico*
[6] *Dept. of Physics and Astronomy, Northwestern University*

Abstract. Black hole candidates with radio jets are a plausible source of the Galactic positrons which produce the observed 511 keV line. The most spectacular of these objects is GRS 1915+105, which we have observed with OSSE to look for annihilation in the interstellar medium around it; this would give a long-term average of its positron production over many thousands of years. We find a 3σ upper limit of 0.96×10^{-4} photons s^{-1} cm^{-2} on a narrow 511 keV line, which implies that the product of the fraction of synchrotron-producing particles which are positrons and the average number of blob ejections per year is less than 0.74.

I INTRODUCTION

It is not known what objects contribute most to the Galactic positron annihilation radiation. Recently, OSSE has mapped out a component of the annihilation radiation comparable in size and shape to the Galactic bulge (about 5° across), as well as a component along the Galactic plane and a component at positive Galactic latitude [15,3]. The bulge stands out from the plane more than in the COMPTEL map of 1809 keV radiation from decaying ^{26}Al. Therefore a major component of positron production is not correlated with the population of ^{26}Al-producing objects. This bulge component may be due to positrons expelled in relativistic jets from stellar-mass black holes. Radio jets have recently been discovered in several Galactic black hole candidates [10,11], and positrons appear in some of the models of jet production.

The longest-studied Galactic black-hole candidate with relativistic radio jets is 1E 1740.7-2942, about 1° from the Galactic center [10]. The high luminosity of the jets combined with their limited penetration into the interstellar medium implies that they are composed primarily of e^+e^- pairs [2]. Another Galactic bulge source, GRS 1758-258, also shows radio jets and similar hard x-ray behavior to 1E 1740.7-2942.

On three occasions, the SIGMA gamma-ray imager on the *GRANAT* spacecraft has reported bright, transient emission features from 1E 1740.7-2942 which are consistent with broadened, redshifted annihilation emission, as might be expected from annihilation at the inner edge of an accretion disk [4]. Another such transient [4] was reported by SIGMA from Nova Muscae 1991 (GS 1124-683), a confirmed black-hole candidate with a known mass function [16]. Other instruments have failed to confirm the existence of such transients [7,9,19,20,8].

Nova Muscae has come and gone, and 1E 1740.7-2942 and GRS 1758-258 are relatively faint easily confused with the rest of the Galactic bulge. The best candidate to study positron production would be far from the bulge, either emit persistently or at least repeatedly with a high luminosity, and show clear radio jets.

II GRS 1915+105

GRS 1915+105 meets all these criteria. It went into hard x-ray outburst in May 1992 and was discovered in August 1992 by WATCH/*GRANAT* [1]. The earlier light curve was analyzed later in BATSE data [6]. Unlike a "classical" x-ray nova, it has since undergone repeated outbursts lasting tens to hundreds of days. Radio jets associated with the source were identified by two of us [11], showing blobs associated with the x-ray outbursts and having apparent superluminal motion. The distance is estimated as 12.5 ±1.5 kpc [11], making its hard x-ray luminosity in outburst about 10 times the highest of 1E 1740.7-2942 and Cyg X-1.

Two infrared objects (IRAS 19124+1106 and IRAS 19132+1035, also visible at 20 cm) lie about 1/4 ° (60 pc) from GRS 1915+105 in opposite directions [17]. The three sources are remarkably colinear, and the distances from GRS 1915+105 to the two IRAS sources are the same to better than ∼2%. The orientation of the three sources (position angle 157°) is close to the angle of the radio jets of GRS 1915+105. Thus, although the infrared objects resemble H II regions due to late O-type stars, their positions suggest an association with GRS 1915+105, possibly as termination shocks of the jets.

OSSE's field of view is large enough to encompass all three sources and hundreds of parsecs in each direction, so that the annihilation site of any positrons created should be included regardless of the relevance of the IRAS sources. Since positron annihilation in the interstellar medium occurs over thousands to hundreds of thousands of years [5], we are searching for an annihilation signal related to the *long-term integrated activity* of this source, rather than to the recent outbursts since 1992.

III OBSERVATIONS

From 6-20 February 1992, OSSE observed at longitude 40° on the Galactic plane [14]. GRS 1915+105 would then have been at a collimator response of about 55%. The flux for a source at GRS 1915+105 was $(0.90 \pm 0.54) \times 10^{-4}$ ph cm^{-2} s^{-1}. Since this observation was intended to look for Galactic plane flux, however, the collimator was parallel to the plane and the background fields were out of the plane. Thus, even if a statistically significant flux had been detected, it might well have been due to a plane component, not GRS 1915+105. Another observation with the collimator parallel to the plane was made in 1995.

For our observation, OSSE viewed GRS 1915+105 from 15-29 October 1996. The collimators were centered on the source and oriented within 30° of perpendicular to the Galactic plane. This minimized the contribution to our 511 keV signal from the Galactic plane by viewing as little of the plane as possible, and by placing the background fields in the plane as well.

The spectrum was fit with a power law, a narrow 511 keV line, and an orthopositronium continuum. Only the power law was needed, giving $\chi^2 = 1.05$ when used alone. The best-fit power law index was (-3.09 ± 0.02), the line flux $(-0.41 \pm 0.32) \times 10^{-4}$ ph cm^{-2} s^{-1}, and the orthopositronium continuum flux $(0.27 \pm 0.73) \times 10^{-4}$ ph cm^{-2} s^{-1}.

IV DISCUSSION

In the absence of a detected signal, we constrain the rate of positron production in the source using our upper limit.

To estimate the energy in electrons (plus positrons) needed to produce the radio emission from the blobs ejected on 19 March 1994, one first calculates the radio flux density and size of a blob in its own frame of reference, taking into account relativistic and projection effects [12]. Then, assuming that the synchrotron system has the minimum possible energy [13] and that all electrons are relativistic, there is 3×10^{46} ergs of energy in electrons and a magnetic field of about 50 mG [12].

To convert the energy in electrons to a number of particles, we assume a power-law distribution. The radio emission early in the March 1994 ejection had a spectral index $\alpha = -0.49$, which gives a particle number index $p = 2\alpha - 1 = -1.98$ for optically thin synchrotron emission. This index was measured between 4.885 and 14.965 GHz [17], but may be applicable from 1.407 GHz all the way into the K band in the infrared, if data from different ejection events are combined [17,18]. Using all the data in these references, we could take values for α from -0.49 to -0.84, values for the high-frequency cutoff from 240 GHz to 2.2 μm in the infrared, and values for the low-frequency cutoff from 1.407 to 4.885 GHz. The cutoffs on either end are really cutoffs of the observations rather than the data. The high-frequency cutoff is less important because the particle index p is steeper than -1. Using different combinations of these values gives a mean energy per particle which runs from about $100mc^2$ to $1000mc^2$. We take $1000mc^2$ as the most conservative estimate (that which implies the fewest positrons assuming a pair plasma). Then we have 4×10^{49} particles.

Positron annihilation in the interstellar medium takes place over thousands of years [5], so the simplest assumption is that active periods come and go on a timescale shorter than the annihilation time. Then the annihilation process is in a steady state, with the annihilation rate equal to the production rate. If the average number of ejections per year is N, and the fraction of synchrotron-producing particles which are positrons is ϵ, then the annihilation rate in steady state is:

$$\frac{4 \times 10^{49} N\epsilon}{1 \text{year}} = 1.3 \times 10^{42} N\epsilon \text{ annihilations/s},$$

and the flux of 511 keV photons at Earth is

$$\frac{2 \frac{\text{photons}}{\text{decay}} \, 1.3 \times 10^{42} N\epsilon}{4\pi (12.5 \text{kpc})^2} = 1.3 \times 10^{-4} N\epsilon \text{ ph cm}^{-2} \text{ s}^{-1}.$$

Since our 3σ upper limit on annihilation flux is 0.96×10^{-4} ph cm^{-2} s^{-1}, we constrain $N\epsilon$ to be < 0.74. If the blobs are completely pair plasma ($\epsilon = 0.5$), then $N < 1.5$.

This result includes three assumptions worth noting explicitly:

- We have assumed that GRS 1915+105 has been in a state where ejections can occur (i.e. not evolved significantly) for a time longer than the positron decay time in the interstellar medium ($\sim 10^5$ yr).

- We have also assumed that the positronium fraction of the decays is small. Experience with the Galactic center 511 keV line suggests this would not be the case. Assuming instead a positronium fraction of 0.9, each decay produces an average of 0.65 line photons and 2.025 positronium continuum photons; using an optimum combination of the line and orthopositronium continuum limits, we find $N\epsilon < 1.3$.

- We have taken the most conservative particle spectrum (i.e. with fewer, more energetic particles) consistent with the available data. Simultaneous, sensitive observations at the beginning of a large ejection from the longer radio wavelengths to the infrared and even visible would define the average particle spectrum more completely, and probably make our limits more restrictive without additional gamma-ray observations.

This research was supported by NASA grant NAG5-3812. We thank Tom Bridgman and the Compton Observatory Science Support Center for their help.

REFERENCES

1. Castro-Tirado, A. et al. 1994, ApJ Suppl., 92, 469
2. Chen, W., Gehrels, N., & Leventhal, M. 1994, ApJ, 426, 586
3. Cheng, L. X. et al. 1997, ApJ, 481, 43
4. Gilfanov, M. et al. 1994, ApJ Suppl., 92, 411
5. Guessoum, N. et al. 1991, ApJ, 378, 170
6. Harmon, B. A. et al. 1994, AIP Conf. Proc., 304, 210
7. Harris, M. J., Share, G. H., & Leising, M. D. 1994, ApJ, 433, 87
8. Harris, M. J. 1997, these proceedings
9. Jung, G. V. et al. 1995, A&A, 295, L23
10. Mirabel, I. F. et al. 1992, Nature, 358, 215
11. Mirabel, I. F. & Rodriguez, L. F. 1994, Nature, 371, 46
12. Mirabel, I. F. and Rodriguez, L. F., in *Proceedings of the 17th Texas Symposium on Relativistic Astrophysics*, in Annals of the New York Academy of Sciences, 1995.
13. Pacholczyk, A. G. 1970, *Radio Astrophysics*, (San Francisco:Freeman)
14. Purcell, W. R. et al. 1993, ApJ, 413, L85
15. Purcell, W. R. et al. 1997, in *Proceedings of the 2nd Integral Workshop*, ESA SP-382, p. 67
16. Remillard, R. A. et al. 1992, ApJ 399, L145
17. Rodriguez, L. F. & Mirabel, I. F. 1995, in *Superluminal Radio Sources* ed. M. Cohen & K. I. Kellerman (Washington: Nat. Acad. Sci.)
18. Sams, B. J. et al. 1996, Nature, 382, 47
19. Smith, D. M. et al. 1996, ApJ, 458, 576
20. Smith, D. M. et al. 1996, ApJ, 471, 783

RXTE Observations of GRS 1915+105

J. Greiner*, E.H. Morgan**, R.A. Remillard**

*Astrophysical Institute Potsdam, 14482 Potsdam, Germany
**Center for Space Research, MIT, Cambridge, MA 02139, USA

Abstract. We report on extensive X-ray observations of the galactic superluminal motion source GRS 1915+105 with the RXTE satellite over the last year. More than 130 RXTE pointings have been performed on roughly a weekly basis.

GRS 1915+105 displays drastic X-ray intensity variations on a variety of time scales ranging from sub-seconds to days. In general, the intensity changes are accompanied by spectral changes on the same timescale. Three types of bursts with typical durations between 10–100 sec have been identified which have drastically different spectral properties and seem to occur in a fixed sequence. One of the most intense bursts has a bolometric X-ray luminosity of $\approx 5 \times 10^{39}$ erg/s during the 25 sec maximum-intensity part.

INTRODUCTION

GRS 1915+105 was discovered on 1992 August 15 with the WATCH detectors on *Granat* (Castro-Tirado et al. 1992). A comparison of the BATSE (> 25 keV) flux with that of ROSAT (1–2.4 keV) fluxes obtained during regularly performed pointings has shown that GRS 1915+105 has been active all the time, even during times of BATSE non-detections (Greiner et al. 1997). A variable radio source was found with the VLA (Mirabel et al. 1993a) inside the ±10″ X-ray error circle (Greiner 1993), which later was discovered to exhibit radio structures travelling at apparently superluminal speed (Mirabel & Rodriguez 1994) making GRS 1915+105 the first superluminal source in the Galaxy. Until then, apparent superluminal motion was only observed in AGN, the central engines of which are generally believed to be massive black holes. This similarity suggests that GRS 1915+105 harbors a stellar-sized black hole.

The X-ray spectrum as seen with ROSAT (Greiner 1993) and ASCA (Nagase et al. 1994) is strongly absorbed ($N_H \approx 5 \times 10^{22}$ cm^{-2}) consistent with the location in the galactic plane at 12.5 kpc distance (Mirabel & Rodriguez 1994).

POINTED RXTE OBSERVATIONS

The first two RXTE observations were performed on April 6 and 9, 1996. The surprising results of these observations and those of the daily RXTE ASM dwells triggered an unique sequence of RXTE pointings on GRS 1915+105 on a roughly weekly time scale. These data are publicly available and a number of papers have already appeared dealing with various aspects of the extremely rich variety of X-ray properties of GRS 1915+105: (Greiner, Morgan and Remillard 1996; Chen, Swank and Taam 1997; Belloni et al. 1997; Morgan, Remillard and Greiner 1997; Taam, Chen and Swank 1997).

Temporal characteristics

The X-ray light curves reveal a variety of features, one of which is large amplitude intensity variations. We identify the following properties in the light curves of GRS 1915+105:
- In 15 of the pointed PCA observations, we find large, eclipse-like dips in the X-ray flux, which we call sputters. During these sputters the flux drops from $\approx$2–3 Crab to a momentary lull at about 100–500 mCrab and then shoots up again. The spectrum softens dramatically during the sputters, thus arguing against absorption effects.
- On some occasions we see extremely large amplitude oscillations with an amplitude of nearly 3 Crab and periods of 30–100s.
- Between the episodes of large-amplitude variations the X-ray flux variations are more regular, developing into clearly visible quasi-periodic oscillations which seem to be stable over several days.
- The combination of the intense QPOs and the high throughput of the PCA enabled phase tracking of individual oscillations: the QPO arrival phase (relative to the mean frequency) exhibits a random walk with no correlation between the amplitude and the time between subsequent events. Furthermore, the mean 'QPO-folded' profiles are roughly sinusoidal with increased amplitude at higher energy, and with a distinct phase lag of $\approx$0.03 between 3 and 15 keV.

Spectral characteristics

We have started a comprehensive spectral investigation using PCA as well as HEXTE data of well-defined time stretches which are selected according to their different shapes in the lightcurve. While the work is still in progress we note the following, more general properties:
- The spectra are complex and rapidly variable. Single component spectra like pure power law, bremsstrahlung, synchrotron or comptonisation models do not fit these spectra. In general, the spectra are composed of at least two components: one soft component extending up to about

15–20 keV, and a flat, hard component extending up to 200 keV which is well represented by a power law of photon index 2.5–3.5. The hardness ratios demonstrate that the spectrum varies on timescales of seconds!

- The soft component can be well described by either an exponentially cut-off power law, a bremsstrahlung model, or disk blackbody models. Given typical X-ray luminosities during most of the April to November 1996 activity state of 10^{39} erg/s, and an upper limit for the size of the emission region defined by the observed spectral changes on timescales of seconds, the bremsstrahlung model is physically excluded (emissivity is too low). We therefore use a multicolor disk blackbody spectrum (DISK in XSPEC).

- The spectrum of the X-ray emission during the lulls is softer than during the high-intensity states, i.e. these lulls are not caused by absorption or any low-energy cut-off. We have selected photons (for individual layers and single PCA units) at different time intervals corresponding to these two intensity states. The gross energy distribution of the high-intensity emission can be described by a disk blackbody temperature of 1.9 keV and a power law with photon index $\alpha = -2.6$. The spectrum during the lulls is well represented by a disk blackbody with a temperature of 1.1 keV plus a power law of photon index $\alpha = -2.3$.

- As can be inferred from the hardness ratios, there are no major spectral changes during the decay phase between lulls before the onset of the large-amplitude oscillations. But during these oscillations the temperature oscillates on the same time scale as the intensity. The power law model has a photon index of –2.6, and nothing can be said on rapid variability of this component due to low statistics on these 1–2 sec timescales.

- We identify three burst-like events with completely different spectral behaviour (see Fig. 1). First, bursts with a typical disk temperature of ≈1.1 keV and little changes along the burst (though a slight, smooth softening seems possible; see Fig. 1 at t=110–120 sec). However, at the end of these *type 1* bursts the spectrum changes abruptly and starts hardening. Second, bursts with generally longer timescale and larger amplitude (see lower panel of Fig. 1 at t=150–280). They show a smooth hardening despite more erratic intensity variations on top of the general flare profile. Third, bursts with a temperature variation proportional to the intensity (see lower panel of Fig. 1 at t=290–300, 325–335 or 370–380 sec) and a maximum temperature of ≈2 keV (*three* bursts). In general, we find the following sequence: one type 1 burst preceding a major flare (type 2 burst), followed by a series of alternating type 1 and type 3 bursts. We have identified at least 6 such sequences during the October 13–25, 1996 period. It is interesting to note that the type 1 bursts are similar to those occuring during or at the end of prolonged lulls like those of May 26, 1996 (see Fig. 4 of Greiner *et al.* 1996) and April 6 (ibid. Fig. 3) or Oct. 7, 1996 (Fig. 1 of Belloni *et al.* 1997).

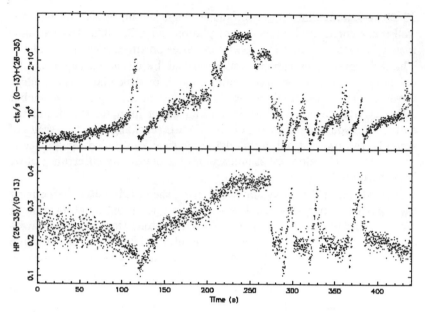

FIGURE 1. Part of the lightcurve as observed on October 13, 1996. The lower panel shows the countrate (hardness) ratio of channels 26–35 (10.6–12.2 keV) versus channels 0–13 (3.1–5.3 keV) while the top panel shows the summed countrate of these two bands at 0.2 sec resolution (see text for more details).

- The major flare (type 2 burst) on October 13, 1996 (Fig. 1) is the most intense emission we have detected from GRS 1915+105 so far. The nearly flat-top main peak has a duration of nearly 30 sec., and integrating over the 2 keV disk blackbody plus the −2.6 power law up to 100 keV results in an unabsorbed luminosity (at the adopted distance of 12.5 kpc) of 5×10^{39} erg/s.

DISCUSSION

The lack of coherent pulsations, the strong variability during the high-intensity states on time scales well below one second and the impossibility of a bremsstrahlung interpretation indicate that the emission originates in an accretion disk.

The drastic intensity variations were interpreted as an inherent accretion instability, rather than absorption effects, since there was spectral softening during these dips. The repetitive, sharp variations and their hierarchy of time scales are entirely unrelated to the phenomenology of absorption dips (Greiner et al. 1996). The nature of these astonishing X-ray instabilities is currently a mystery though attempts have been made to both interprete these as accretion disk instabilities leading to an infall of parts of the inner accretion disk (Greiner et al. 1996, Belloni et al. 1997) and relate them to radio flares (Greiner et al. 1996, Pooley & Fender 1997).

At photon energies above 10 keV, the high amplitudes and sharp profiles of the QPOs are inconsistent with any scenario in which the phase delay is caused by scattering effects. Alternatively, it appears that the origin of the hard X-ray spectrum itself (i.e. the creation of energetic electrons in the inverse Compton model) is functioning in a quasiperiodic manner. These results fundamentally link X-ray QPOs with the most luminous component of the X-ray spectrum in GRS 1915+105.

Taam et al. (1997) have investigated PCA data of October 15, 1996 and found that type 1 and type 3 bursts always occur together. Our finding confirms this behaviour with the addition that such sequences are always preceded by one pair of type 1 and 2 burst.

The maximum (during a flare on Oct. 13, 1996) unabsorbed X-ray luminosity is 5×10^{39} erg/s. This is well above the Eddington luminosity for a neutron star with any reasonable mass, suggesting that the system contains a black hole. Assuming that the emission is near (but slightly below) the Eddington luminosity, the inferred mass (35 $M_\odot$) is compatible with that derived from the stable 67 Hz QPO (Morgan et al. 1997).

Acknowledgements: JG is supported by the German Bundesministerium für Bildung, Wissenschaft und Forschung (BMBW/DARA) under contract Nos. 50 QQ 9602 3 and is grateful to DFG for a substantial travel grant (KON 1088/1997 and GR 1350/6-1).

REFERENCES

[1] Belloni T., Mendez M., King A.R., van der Klis M., van Paradijs J., 1997, ApJ 479, L145
[2] Castro-Tirado A.J., Brandt S., Lund N., 1992, IAU Circ. 5590
[3] Chen X., Swank J.H., Taam R.E., 1997, ApJ 477, L41
[4] Greiner J., 1993, IAU Circ. 5786
[5] Greiner J., Morgan E.H., Remillard R.A., 1996, ApJ 473, L107
[6] Greiner J., Harmon, B.A., Paciesas W.S., Morgan E.H., Remillard R.A., 1997, in *Accretion Phenomena and Associated Outflows*, Proc. of IAU Coll. 163, Port Douglas, July 1996, PASA (in press)
[7] Mirabel I.F., Rodriguez L.F., Marti J., Teyssier R., Paul J., Auriere M., 1993a, IAU Circ. 5773
[8] Mirabel I.F., Rodriguez L.F., 1994, Nat. 371, 46
[9] Morgan E.H., Remillard R.A., Greiner J., 1997, ApJ (in press)
[10] Nagase F., Inoue H., Kotani T, Ueda Y., 1994, IAUC 6094
[11] Pooley G.G., Fender R.P., 1997, MNRAS (subm.)
[12] Taam R., Chen X., Swank J., 1997, ApJ (in press)

Near-Infrared Observations of GRS 1915+105

William A. Mahoney[1], Stephane Corbel[2], Ph. Durouchoux[2],
Thomas N. Gautier[1], J. C. Higdon[3], and Pierre Wallyn[2]

[1] *Jet Propulsion Laboratory 169-327, 4800 Oak Grove Drive,
Pasadena, CA 91109*

[2] *DAPNIA, Service d'Astrophysique, CE Saclay,
91191 Gif sur Yvette Cedex, France*

[3] *Joint Science Department, The Claremont Colleges,
Claremont, CA 91711*

Abstract.
During 9 July 1995 and 28 April 1996 we carried out J, H, and K_s observations of the X-ray transient and Galactic black hole candidate GRS 1915+105 using the Cassegrain IR Camera on the 5-meter telescope at Mt. Palomar. Between the two observations the infrared intensity increased by approximately one magnitude to one of the highest levels yet seen, again confirming the highly variable nature of the emission. Our second observation was made during a time when the source was being continuously monitored by *CGRO* (20-100 keV), *RXTE* (2-12 keV), and the Ryle Telescope (15 GHz), allowing simultaneous multiwavelength spectra covering a very broad energy range.

I INTRODUCTION

The X-ray transient and black hole candidate GRS 1915+105 was the first Galactic object found to exhibit apparent superluminal motion. It was discovered in 1992 by *Granat*/WATCH [1] and was later identified with a radio source ejecting material in high-velocity jets [2]. GRS 1915+105 has been described [2,3] as a Galactic 'microquasar' with the potential of explaining analogous processes that power active galactic nuclei by taking advantage of a much closer object that varies on much faster timescales. Based on its time variability and a hard X-ray spectrum extending to at least 100 keV, it has been suggested that GRS 1915+105 is an accreting black hole in a binary system [4].

GRS 1915+105 is located in the Galactic plane at an estimated distance of

12.5 kpc, making it one of the brightest X-ray sources in the Galaxy. VLA observations of its radio counterpart indicate mass ejections in opposite directions with velocities of 0.92c [2]. Its spectrum is highly reddened ($A_v \sim 20$) [5] indicating little likelihood of seeing an optical counterpart. However, a clear and highly variable counterpart has been observed at most other energies.

In the hard X-ray range (20-100 keV), the source is being monitored continuously by BATSE using Earth occultation techniques [4]. Its emission is characterized mainly by intense outbursts separated by brief quiescent periods. Since February 1996, GRS 1915+105 has also been followed by the All Sky Monitor (ASM) and it has been observed over 30 times by the Proportional Counter Array (PCA), both on *RXTE* [6]. A recurrent QPO near 67 Hz has been seen, supporting the identification with an accreting black hole.

Radio observations with the VLA [7] and the Ryle Telescope [8] have identified a remarkable variety of intensity changes ranging from isolated outbursts to nearly periodic oscillations with timescales near 30 minutes. Similar variations have also been seen in the K-band, again with periods near 30 minutes [9]. Based on its time variability and spectral similarity to SS 433, it is believed that much of the infrared emission from GRS 1915+105 originates in the accretion disk [10].

As most X-ray binaries are highly variable [11], it is critical to carry out frequent multiwavelength observations, especially during periods of unusual activity. For two years we have been making near- and mid-infrared observations of gamma-ray source counterparts using the 5-meter telescope on Mt. Palomar. X-ray monitoring of GRS 1915+105 by the ASM [12] showed the source was in a highly active state during April 1996, motivating us to include it in our observations of 28 April 1996. We found that by April 1996 the emission in the J, H, and K_s bands had increased significantly compared to our observation the previous year.

II OBSERVATIONS AND DATA ANALYSIS

During the nights of 9 July 1995 (1125-1148 UT) and 28 April 1996 (1115-1127 UT) we observed GRS 1915+105 in the near-infrared band using the Cassegrain IR Camera on the Mt. Palomar 5-meter Hale telescope. On both occasions the instrument was configured with a 256 × 256 pixel InSb array which images a field-of-view of about 32″ × 32″. Observations to a sensitivity level of about 0.1 mJy can be made within a minute in the broad J (1.25 μm), H (1.65 μm), and K_s (2.15 μm) bands. During both nights, the sky was clear and the seeing was good although somewhat variable. On both occasions we included observations of a number of the UKIRT faint JHK standard stars [13] to allow photometric measurements accurate to a few percent.

A typical observational sequence with a given filter consisted of four images with GRS 1915+105 centered in each of the four quadrants of the field-of-view. This sequence facilitated sky subtraction and was repeated for each of the three wavelength bands. Data analysis was carried out using the IRAF package and included subtractions of the read noise, the dark current, and the sky background, as well as flat fielding and image shifting and combining to yield a mosaic image. The final J and K_s mosaics obtained on 28 April 1996 are shown in Figure 1 with GRS 1915+105 near the center of each image. The highly reddened nature of the source can be clearly seen in Figure 1 by noting the much higher intensity in K_s compared to J relative to nearby stars.

A similar analysis of the photometric calibration standards gave intensities consistent to about 3% for both nights. This calibration was verified by comparing the intensities of nearby stars between our two observations and with other measurements. For example, our K magnitudes of 13.22 ± 0.04 and 13.39 ± 0.04 for stars A and B (Figure 1), respectively, agree with measurements by Fender et al. (1997) to within 5% [9]. Our photometric magnitudes of GRS 1915+105 are summarized in Table 1.

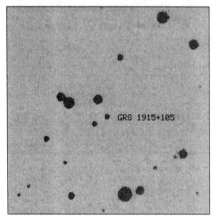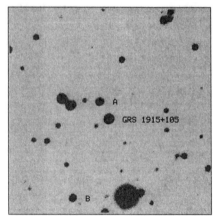

FIGURE 1. Field of approximately $1' \times 1'$ in J (left) and K_s (right) with GRS 1915+105 just below the center of each image. Because of the high extinction, the relative intensity of GRS 1915+105 is much higher in K_s than in J.

In Figure 2 our observations are compared with previously reported near-infrared magnitudes of GRS 1915+105 [3,10,14]. The highly variable nature

Date	J	H	K_s	$J - K_s$
9 July 1995	16.84 ± 0.07	14.44 ± 0.05	12.92 ± 0.04	3.92 ± 0.08
28 April 1996	15.95 ± 0.07	13.83 ± 0.05	12.46 ± 0.04	3.49 ± 0.08

TABLE 1. Near-infrared intensities of GRS 1915+105 (magnitudes).

of the source is clearly evident. Between our first observation in July 1995 and the second in April 1996, the intensity increased by nearly a magnitude to one of the brightest levels yet seen. In fact, the J magnitude is the brightest ever reported. The highly reddened nature of the object is evident and is consistent with an extinction of $A_v \sim 20$. Finally, comparison of the two spectra (Table

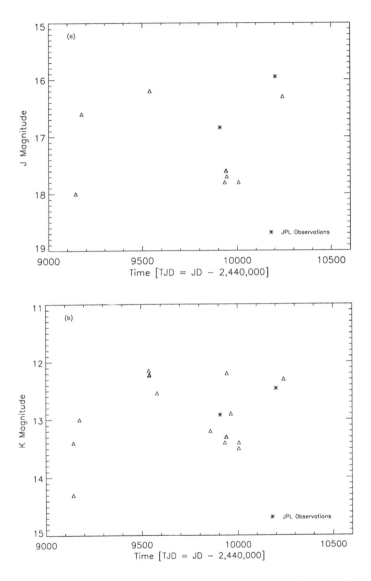

FIGURE 2. Comparison of JPL observations of GRS 1915+105 with previously reported measurements [3, 10, 14] in the (a) J and (b) K_s bands.

1) shows that between July 1995 and April 1996 the J intensity increased faster than the K_s intensity indicating that the spectrum became bluer.

III SUMMARY

We observed the X-ray transient GRS 1915+105 on 9 July 1995 and again on 28 April 1996 in the near-infrared using the Cassegrain IR Camera on the 5-meter Hale telescope. During both nights the source was clearly visible in the J, H, and K_s bands with an intensity that had increased by nearly a magnitude between the two observations to one of the brightest levels yet reported. We also found that the spectrum became bluer as the intensity increased. The April 1996 observation was made during a period of simultaneous monitoring of the source by *CGRO*, *RXTE*, and the Ryle radio telescope which will allow simultaneous multiwavelength spectral comparisons.

IV ACKNOWLEDGMENTS

We thank K. Matthews for his invaluable assistance in helping us operate the 5-meter Cassegrain IR Camera and A. Castro-Tirado for sharing with us data from previous near-infrared observations of GRS 1915+105 and nearby stars. The research described in this paper was carried out by the Jet Propulsion Laboratory, California Institute of Technology, under contract to the National Aeronautics and Space Administration.

REFERENCES

1. Castro-Tirado, A. J., et al., *ApJS*, **92**, 469-472 (1994).
2. Mirabel, I. F., and Rodríguez, L. F., *Nature*, **371**, 46-48 (1994).
3. Sams, B., Eckart, A., and Sunyaev, R., *IAUC* 6455 (1996).
4. Harmon, B. A., et al., *ApJ*, **477**, L85-L89 (1997).
5. Castro-Tirado, A. J., et al., *ApJ*, **461**, L99-L101 (1996).
6. Morgan, E. H., Remillard, R. A., and Greiner, J., *ApJ*, **482**, 993-1010 (1997).
7. Rodríguez, L. F., and Mirabel, I. F., *ApJ*, **474**, L123-L125 (1997).
8. Pooley, G. G., and Fender, R. P., *MNRAS*, in press (1997).
9. Fender, R. P., et al., *MNRAS*, in press (1997).
10. Chaty, S., et al., *A&A*, **310**, 825-830 (1996).
11. Lewin, W. H. G., van Paradijs, J., and van den Heuvel, E. P. J., editors, *X-Ray Binaries*, (Cambridge: Cambridge Univ. Press) (1995).
12. Morgan, E., Remillard, R., and Greiner, J., *IAUC* 6392 (1996).
13. Courteau, S., *A Compilation of the UKIRT Faint JHK Standards*, NOAO internal publication (1994).
14. Mirabel, I. F., et al., *ApJ*, **472**, L111-L114 (1996).

Infrared Observations and Energetic Outburst of GRS 1915+105

Sylvain Chaty and I. Felix Mirabel

Service d'Astrophysique, Centre d'études de Saclay,
CEA/DSM/DAPNIA/SAp, F-91 191 Gif-sur-Yvette Cedex, France

Abstract. Many near-infrared wavelengths observations, carried out since 1993 on the galactic superluminal source of relativistic ejections GRS 1915+105, have yielded the following important results:

1) The infrared counterpart of the microquasar GRS 1915 + 105 exhibits various variations in the 1.2 – 2.2 μm band: the strongest are of $\sim$ 1 magnitude in a few hours and of $\sim$ 2 magnitudes over longer intervals of time [2].

2) The strikingly similar infrared properties of GRS 1915+105 and of SS 433 suggest that GRS 1915+105 consists of a collapsed object (neutron star or black hole) with a thick accretion disk in a high mass and luminous binary system [2].

3) During an intense and long-term X-ray outburst of GRS 1915 + 105 in 1995 August, we observed the time-delayed reverberation of the radio flare and ejection event in the infrared wavelengths, the observed spectrum of the enhanced infrared emission suggesting th e appearance of a warm dust component [6].

4) A near-infrared jet, ejected by GRS 1915 + 105, discovered by Sams et al. (1996), was not seen by us, 17 days later. We derive some implications about the radiative lifetime of the electrons involved in this jet emission [1].

INFRARED OBSERVATIONS OF GRS 1915+105

Mirabel et al. [5] showed that there is no visual counterpart of GRS 1915 + 105 brighter than $R = 21$ magnitudes, and we observed with the NTT on 9 July 1994 a faint counterpart at ~ 1 μm, consistent with the $I = 23.4$ magnitudes (Bor et al. 1996). We carried out infrared observations of GRS 1915 + 105 at the ESO with the MPI 2.2 m telescope on 4–5 June 1993 and 5–8 July 1994 with the IRAC2(b) camera [2], in the J (1.25 μm), H (1.65 μm) and K (2.2 μm) bands, wi th a typical seeing of 1.2 arcsec, and at the 3.6 m CFHT on 16 August 1994, with the Redeye camera, with a typical seeing of 0.6 arcsec. The variations in the J, H and K bands (Fig. 1) show that GRS 1915+105 exhibits strong shor t-term variability in intervals of less than 24 hours, as well as strong long-term variability over intervals from one month to one year. Indeed, the luminosity of GRS 1915 + 105 increased by nearly 1 magnitude in H and K between the ! nights of 4 and 5 June 1993, and between 4 June 1993 and 5 July 1994 there was a change of nearly 2 magnitudes in J, 2.5 magnitudes in H, and 2.1 magnitudes in K. The infrared colors change with luminosity. The rapid increase of 1 magnitude observed in an interval of 24 hours in June 1993 could result from occultation, and it is also interesting to note that this rapid variation of the infrared luminosity occurred in a period when the source was strong and showing rapid variations of luminosity in the 8–60 keV en ergy band observed by WATCH (Sazonov et al. 1994), and in the 20–100 keV energy band observed by BATSE (Paciesas et al. 1995).

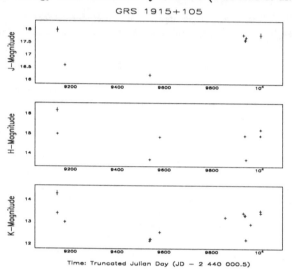

FIGURE 1. Time variation of J (1.25 μm), H (1.65 μm) and K (2.2 μm) bands luminosities, of the source GRS 1915 + 105 from 4 June 1993 to 17 October 1995. Note the short and long time scales variations.

GRS 1915+105 AND SS 433

From the apparent magnitudes we derived the absolute magnitudes, corrected from interstellar extinction, using a visual absorption of $A_v = 26.5 \pm 1$ magnitudes, and the kinematic distance $D = 12.5 \pm 1.5$ kpc [2]. The absorptio ns in the J, H and K bands are $A_J = 7.1 \pm 0.2$, $A_H = 4.1 \pm 0.2$ and $A_K = 3.0 \pm 0.1$ magnitudes respectively. The infrared emission of GRS 1915 + 105 cannot arise only in the photosphere of the second ary star, because of the shape of the spectrum, which cannot be reproduced by photospheric emission from any stellar type, and because of the rapid variations in luminosity and energy distribution. Therefore, besides the photospheric emission from the sec ondary, there must be an additional source of infrared emission in GRS 1915+105. The energy distributions of the most well studied galactic X-ray sources are shown in Fig. 2. To derive the absolute magnitudes of SS 433 we! assumed the kinematic distance of 4.2 ± 0.5 kpc (van Gorkom et al. 1982) and a visual absorption $A_V = 7.25 \pm 0.25$ magnitudes (Mc Alary et al. 1980). Besides the time variations, the infrared absolute magnitudes and colors of GRS 1915 + 105 are strikingly similar to the classic source of relativistic jets SS 433. This similarity in the observed infrared properties suggests that SS 433 and GRS 1915 + 105 are similar systems. Therefore, by analogy with SS 433, GRS 1915 + 105 could be a collapsed object with a thick accretion disk in a hot and luminous, high mass binary [2].

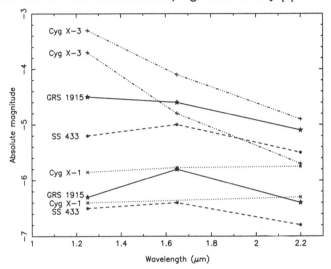

FIGURE 2. Infrared energy distributions of GRS 1915 + 105, SS 433, Cyg X-1 and Cyg X-3 for the periods of minimum and maximum luminosity. Note the similar infrared properties between GRS 1915 + 105 and SS 433.

INFRARED BURST OF GRS 1915+105

During an intense X-ray and radio outburst of GRS 1915 + 105 in 1995 August, we observed with the VLA a pair of bright radio-emitting clouds emerging from the compact radio core in opposite directions, and at relativistic speeds. At near-infrared wavelengths, we discovered an infrared outburst (Fig. 3) [6]. Due to the time-delayed reverberation of this sudden and major radio flare and ejection event, the infrared outburst was detected between two and five days after the radio outburst. There fore, the cause of the infrared response to this impulsive event must be at ≥ 2 light-days from the compact object. The source became redder by J-K = 1.2 magnitudes, and brightened by ~ 1 magnitude in K ($\sim 10^3$ L$_\odot$). The $1.0 - 2.5$ μm continuum rising to the red suggests the appearance of a warm dust emitting component of mass $10^{-10} M_\odot$, and temperature 2300 K. The thermal energy reradiated from heated dust in the near-infrare! d was $\sim 10\%$ of the mean X-ray luminosity of the source, or 0.1% of the typical kinetic energy in the bulk motion of the relativistic ejecta in GRS 1915+105 ($\sim 3 \times 10^{42}$ erg s^{-1}). At a distance of 500 a.u. (3 light-days) and with an X-ray lu minosity of 10^{37} erg s^{-1} we expect an equilibrium dust temperature of only ~ 100 K. We then believe that the observed near-infrared emission could come from small grains out of equilibrium with the X-ray field, and that most of the dust is radiati ng at lower temperatures.

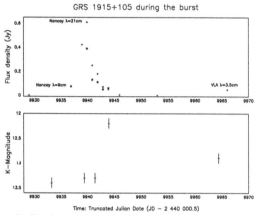

FIGURE 3. Top: Radio observations of GRS 1915+105 around the 1995 August outburst and ejection event, observed with the VLA at $\lambda = 3.5$ cm and with the Nanay radiotelescope at $\lambda = 9$ cm and $\lambda = 20$ cm. Bottom: Infrared K magnitudes. Note the time delay of the infrared brightening relative to the time of peak radio emission.

INFRARED JET OF GRS 1915+105

Sams et al. [7] observed the counterpart of GRS 1915+105 in the K band. These observations were done on 18–21 July 1995. The images show that GRS 1915 + 105 appears extended to the South-West: this jet has a total near-infrared magnitude of $K = 13.9$ magnitude, is separated from the central source by 0.3", and its characteristics are consistent to that observed in a radio outburst of the source [4]. On 16–17 October 1995, Eikenberry & Fazio [3], observing GRS 1915 + 105 in the K band w ith the KPNO 2.1 m telescope, found no evidence for extended structure, with an upper limit of $K > 16.4$ magnitude. To understand what happened between these two dates, it was necessary to analyze images that were taken in this period.

We have reanalyzed the i mages of Mirabel et al. [6], synthesizing the better point-spread-function (PSF) as possible, and subtracting the scaled PSF to the infrared counterpart of GRS 1915 + 105 [1]. The result is that we can not see any pr! esence of jet structure (Fig. 4). We also searched for possible extended emission around the source GRS 1915 + 105 by a multifrequency analysis, thanks to a wavelet approach, and the result is identical. The infer limit in magnitude that we can deri ve from our observations is $mag_K > 17.5$ magnitude. Therefore, the infrared flux decreased by a factor of ≥ 28, in a temporal delay between the observations of Sams et al. (1996) and those of Mirabel et al. (1996) of $\Delta t = 16.7$ days. Th en, the $\frac{1}{e}$ radiative timescale of the electrons involved in this jet emission would be $\tau \sim 5$ days.

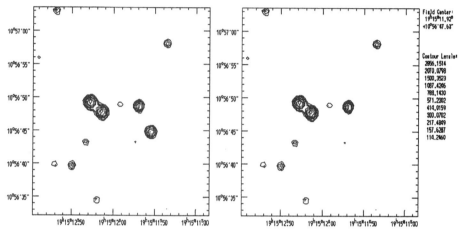

FIGURE 4. Left: original image of the field of view around GRS 1915 + 105. Right: image with the PSF of GRS 1915 + 105 being subtracted. We can not see any presence of jet structure, up to a magnitude in the K band of $m_K > 17.5$ magnitude.

REFERENCES

1. Chaty S., Mirabel I.F., 1997, Proceedings of the XXXIInd Rencontres de Moriond, "Very High Energy Phenomena in the Universe", in press
2. Chaty S., Mirabel I.F., Duc P.-A. et al., 1996, A&A, 310, 825
3. Eikenberry S.S., Fazio G.G., 1997, ApJL, 475, L53
4. Mirabel I.F., Rodríguez L.F., 1994, Nat, 371, 46
5. Mirabel I.F., Duc P.-A., Rodríguez L.F. et al., 1994, A&A, 282, L17
6. Mirabel I.F., Rodríguez L.F., Chaty S. et al., 1996, ApJ, 472, L111
7. Sams B.J., Eckart A., Sunyaev R., 1996, Nat, 382, 47

ASCA Observations of Galactic Jet Systems

T.Kotani[0], N.Kawai[0], M.Matsuoka[0], T.Dotani[1], H.Inoue[1],
F.Nagase[1], Y.Tanaka[1], Y.Ueda[1], K.Yamaoka[1], W.Brinkmann[2],
K.Ebisawa[3], T.Takeshima[3], N.E.White[3], A.Harmon[4],
C.R.Robinson[4], S.N.Zhang[4], M.Tavani[5], R.Foster[6]

[0]*RIKEN, Hirosawa 2-1, Wako, Saitama 351-01, Japan*
[1]*ISAS, Yoshinodai 3-1-1, Sagamihara, Kanagawa 229, Japan*
[2]*MPIE, Giessenbachstrasse, D-85740 Garching, Germany*
[3]*NASA/GSFC, Greenbelt, MD 20771, USA*
[4]*NASA/MSFC, ES-81, Huntsville, AL 35812, USA*
[5]*Columbia Univ., New York, NY 10027 USA*
[6]*NRL, Code 7210, Washington, DC 20375, USA*

Abstract. Recent studies with ASCA have shown very complicated, strange iron K features in the spectra of galactic jet systems. SS 433, the "classic" jet, was found to have pairs of Doppler-shifted lines, contrary to the previous belief that the receding X-ray jet is short and hidden behind the accretion disk. The transient jets, GRS 1915+105 and GRO J1655-40, show spectral dips, which have never been observed in any other source and are interpreted as absorption lines or Doppler-shifted absorption edges. If they are resonant absorption lines of helium-like iron, they would be the evidence of highly ionized, anisotropically distributed plasma near the jet engine. These features peculiar to galactic jet systems are expected to be explained in terms of the nature of the sources and the jet-formation mechanisms.

Since ASCA was proved to be an excellent tool for diagnostics of jets, observation campaigns of the jet systems were planned and performed. SS 433 was observed about thirty times in the three years of the campaign, covering the phase space of the 162.5-day precession and the 13-day orbital motion. The extracted physics of the system, such as X-ray-jet length ten times longer than previous estimations, jet kinetic luminosity exceeding 10^{40} erg s^{-1}, etc., draw a highly energetic and stormy, new picture of SS 433. The transient jets, GRS 1915+105 and GRO J1655-40, were also observed several times. GRS 1915+105 was found to be active in ASCA band even months after onsets of outburst. Violent variations were not seen. GRO J1655-40 was observed to be transit between high and low states, and the low state is consistent to occultation of a component.

We review ASCA Observations of galactic jet systems and present some topics from recent progresses.

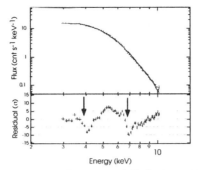

FIGURE 1. Spectrum of GRS 1915+105
Solid curve is an absorbed power law model with an exponential cut-off. An iron and a calcium K-feature are indicated in the residual.

I TRANSIENT JETS

A GRO J1655−40

GRO J1655−40, one of the two transient super-luminal jets, was observed with ASCA five times; 23 August 1994 [6], 27 September 1994 [13], 15 August 1995 [7], and 23 March 1996, and 25 February 1997. When the source was in low or high states, the 2-10 keV continuum was roughly expressed either by an absorbed power law with an exponential cut-off, or by an absorbed disk blackbody [5,14]. In the spectral residuals, resonant $K\alpha$ absorption lines of helium-like iron appeared. This is the first detection of iron $K\alpha$ absorption lines from X-ray binaries, and a strong evidence of highly ionized plasma in a non-spherical distribution around the jet engine. For more details, see Ueda [14].

B GRS 1915+105

The other transient jet, GRS1915+105, was observed four times so far; 27 September 1994 [13], 20 April 1995 [4,5], 23 October 1996, and 25 April 1997. In the first and second observation, the source was found to be bright in the ASCA band. The 2-10 keV continuum was roughly expressed by the same models as GRO J1655−40. In the spectral residuals, iron and calcium K features appeared. They were interpreted as a Doppler-shifted K-absorption edge, or a $K\alpha$ absorption line of hydrogenic ion. The spectrum of the first observation is shown in figure 1.

If they are absorption lines, such K-absorption lines are considered to be peculiar to transient super-luminal jets, and probably connected to the jet formation mechanism. Alternatively, if they are Doppler-shifted edges, the absorbing matter must be the ejecta itself, since the Doppler-shift parameter suggests a relativistic velocity of the absorber. Together with the proper motion [12], the parameters of the jet would be determined precisely. Assuming the ionization stage of the absorber to be helium-like, velocity of the jet, inclination, and distance to the source would be $v = 0.8585 \pm 0.0072$ c, $67.90° \pm 0.95°$, and $D = 11.6$ kpc, respectively.

II SS 433

The "classic" jet system SS 433 was one of the most frequently observed target of ASCA. The observation campaign began in 1993, and ended in 1996. About thirty observations were performed, the total exposure time exceeded 500 ks, and the total number of ground/space observatories involved in the campaign was twenty seven. Various phases of the 162.5-day precession and the 13-day orbital motion were sampled.

The first surprising was the discovery of the Doppler-shifted emission-line pairs [8]. It had been generally accepted that the X-ray jet is too short compared to the size of the accretion disk, and that only proceeding one is visible. However, the ASCA data shown that both X-ray jets are visible. Therefore, the X-ray jet must be longer than previous estimations of $\lesssim 10^{12}$ cm [15,17]. The physics of the jet, such as kinetic luminosity, mass outflow rate, initial density, and mass of the compact star, should be revised according to the new X-ray jet length.

For a precise determination of the X-ray jet length, an observation covering an eclipse was performed [10]. Since ASCA can resolve spectra into red-shifted and blue-shifted component, we can know how and which jet is occulted by the companion star as eclipse progresses. The geometry of the system was modeled and parameterized as shown in figure 2. The jet was assumed to be occulted by the accretion disk or the companion star. The temperature structure of the X-ray jet was numerically calculated based upon a jet model including the cooling effects of radiation and expansion [9]. The emission-line light curve obtained in March 1996 [10] was fitted with the model. The relative dimensions of X-ray jet length l_X, companion star radius R_C, and orbital separation a were determined as

$$l_X/a = 8.0^{+1.8}_{-1.0}, \quad R_C/a = 0.51^{+0.11}_{-0.03}. \tag{1}$$

From these values, the X-ray jet length was successfully determined [1] as

$$l_X = 20.1^{+4.8}_{-3.6} \times 10^{12} \times \left(\frac{K}{112 \text{ km s}^{-1}}\right) \text{ cm}, \tag{2}$$

where K is the Doppler modulation [3]. It should be stressed again that this value is ten times longer than the previous estimations.

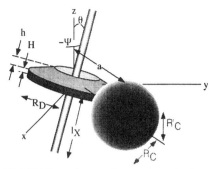

FIGURE 2. Assumed geometry of eclipse

Since the X-ray jet length is a function of initial electron density n_{e0} [11], it was determined to be

$$n_{e0} = 2.50^{+0.61}_{-0.53} \times 10^{12} \times \left(\frac{K}{112 \text{ km s}^{-1}}\right)^{-1.1} \text{ cm}^{-3}. \qquad (3)$$

The observed flux and the emissivity derived from this density constrain the mass outflow rate per jet $\dot{m}$ as

$$\dot{m} = 4.1 \pm 0.3 \times 10^{-6} \times \left(\frac{\Theta}{2.5°}\right)^{1/3} \left(\frac{K}{112 \text{ km s}^{-1}}\right)^{0.37} \text{ M}_\odot \text{ yr}^{-1}, \qquad (4)$$

where Θ is opening half angle of the jet. This corresponds to the kinetic luminosity per jet L_K of

$$L_K = 7.8 \pm 0.6 \times 10^{39} \times \left(\frac{\Theta}{2.5°}\right)^{1/3} \left(\frac{K}{112 \text{ km s}^{-1}}\right)^{0.37} \text{ erg s}^{-1}. \qquad (5)$$

These values are of a single jet, thus must be multiplied by 2 for the total bipolar flow. The resultant total mass outflow rate is comparable to 10^{-5} M$_\odot$, and the kinetic luminosity reaches 10^{40} erg s^{-1}.

III DISCUSSION

In table 1, the specification of the galactic jets are shown. The averaged mass outflow of SS 433 far exceeds that of transient jets. To power the jet, the compact object must accrete comparable or more matter than the jet, and much more matter is considered to be ejected from the system as a disk wind in super-critical accretion regime. Therefore, a large rate of mass loss, say 10^{-4} M$_\odot$ yr^{-1}, is expected from the companion star.

Other evidences found in ASCA data of SS 433 show that the accretion disk precesses according to the jet axis [11]. That supports the slaved disk

TABLE 1. Specification of galactic jets

Type	v/c	$2\dot{m}_{max}$ g s^{-1}	$2\dot{m}_{mean}$ M$_\odot$ yr^{-1}	$2L_{K,max}$ erg s^{-1}	Engine
SS 433	0.26	5×10^{20}	8×10^{-6}	1.6×10^{40}	?
transient jets	0.92	$\sim 10^{20}$ [a]	$\sim 10^{-7}$ [a]	$\sim 10^{41}$ [a]	Kerr black hole?

[a] Assuming baryonic ejecta [12].

scenario, *i.e.*, the precession is caused by the companion star. These evidences suggest that the abnormality of SS 433 lies in the environment, such as the companion star, or the orbit, rather than in the compact object itself.

On the other hand, a Kerr black hole was proposed for the engine of the transient jets [16]. If the angular momenta of these sources are essential, then the jet formation mechanisms would be very different between the two classes of galactic jets, transient jets and SS 433.

REFERENCES

1. Brinkmann W., Kawai N., and Matsuoka M., *Astron. Astrophys.* **218**, L13 (1989).
2. Crampton D., Hutchings J.B., *Astrophys. J.* **251**, 604 (1981).
3. D'Odorico S., Oosterloo T., Zwitter T., and Calvani M., *Natur.* **353**, 329 (1991).
4. Ebisawa K., White N.E., Kotani T., and Harmon A., *IAUC* **6171** (1995).
5. Ebisawa K., *X-Ray Imaging and Spectroscopy of Cosmic Hot Plasmas*, Tokyo: Universal Academy Press, 1997, 427
6. Inoue H., Nagase F., Ishida M., Sonobe T., and Ueda Y., *IAUC* **6063** (1994).
7. Inoue H., Nagase F., and Ueda Y., *IAUC* **6210** (1995).
8. Kotani T., Kawai N., Aoki T., Doty J., Matsuoka M., Mitsuda K., Nagase F., Ricker G. et al., *Publ. Astron. Soc. Japan* **46**, L147 (1994).
9. Kotani T., Kawai N., Matsuoka M., and Brinkmann W., *Publ. Astron. Soc. Japan* **48**, 619 (1996).
10. Kotani T., Kawai N., Matsuoka M., and Brinkmann W., *Accretion Phenomena and Related Outflows, IAU Colloquium 163*, San Francisco: Astronomical Society of the Pacific, 1997.
11. Kotani T., *Doctral Thesis*, University of Tokyo, 1997.
12. Mirabel I.F., Rodríguez L.F., *Nature* **371**, 46 (1994).
13. Nagase F., Inoue H., Kotani T., and Ueda Y., *IAUC* **6094** (1994).
14. Ueda Y., Inoue H., Tanaka Y., Ebisawa K., Nagase F., Kotani T., and Gehrels N., submitted to *Astrophys. J.* (1997).
15. Watson M.G., Stewart G.C., Brinkmann W., King A.R., *Mon. Not. R. Astr. Soc.* **222**, 261 (1986).
16. Zhang S.N., Cui W., and Chen W., *Astrophys. J. Lett.* **482**, L155 (1997).
17. Zwitter T., Calvani M., *Mon. Not. R. Astr. Soc.* **236**, 581 (1989).

BATSE Observations of GX339-4

Brad C. Rubin[1,2], B.A. Harmon[3], W.S. Paciesas[2,3], C.R. Robinson[3,4], S.N. Zhang[3,4], G.J. Fishman[3]

[1] RIKEN Institute, Wako-shi, Saitama 351-01 Japan
[2] Department of Physics, University of Alabama in Huntsville, Huntsville, AL 35899
[3] NASA/Marshall Space Flight Center, Huntsville, AL 35812
[4] Universities Space Research Association

Abstract.
Between the summer of 1991 and the fall of 1996, BATSE observed eight outbursts from the black hole candidate GX339-4. They occurred in a declining sequence with two types of outbursts. A rough linear correlation exists between the fluence emitted during an outburst and the time elapsed between the end of the previous outburst and the beginning of the current one. The light curves of the earlier, more intense, outbursts (except for the second one) can be modeled by a fast exponential (time constant $\sim$ 10 days) followed by a slower exponential ($\sim$ 100 days) on the rise and a fast exponential decay ($\sim$ 5 days) on the fall. The later, weaker, outbursts are modeled with a single rising time constant ($\sim$ 20 days) and a longer decay on the fall ($\sim$ 50 days). These observations can be used to constrain models of the behavior of the accretion disk surrounding the compact object.

INTRODUCTION

GX339-4 is usually considered a black hole candidate (BHC) due to the similarity of its x-ray spectral and timing states to dynamically established BHC such as Cygnus X-1. However, there is still considerable uncertainty about the mass of the compact object in GX339-4 [1]. Recently there has been a report of a radio jet from this source [2].

Most x-ray observations of GX339-4 have focused on determining which x-ray spectral state the source is in, and describing the properties of those states (for a recent review see [3]). Here we will present new data from six recent outbursts observed by CGRO-BATSE. Our analysis will also include two earlier outbursts observed by BATSE [4]. For more information on the present work, see [5]. The eight outbursts observed by BATSE can be roughly divided into two types based on the outburst fluence, light curve, spectral evolution, and recurrence pattern. In these respects, the first four outbursts appear to be different from the last four.

More important than this classification, however, is the observation that the pattern of these outbursts is one of decreasing total energy release, and that outburst fluence in the 20-300 keV band is correlated with the time elapsed since the previous outburst [5].

OBSERVATIONS

Figure 1(a) shows the BATSE flux history obtained using the Earth occultation technique and an optically thin thermal bremsstrahlung (OTTB) spectral model fit to the observed count rates in the 20-300 keV band. The first two outbursts have been published previously [4] wherein there is a description of

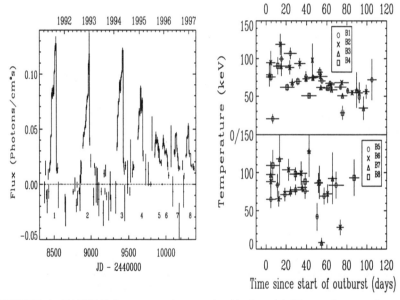

FIGURE 1. BATSE light curve and spectral evolution. (a) Photon flux as a function of time in ten day bins in the 20-300 keV energy band as determined from an OTTB fit to Earth occultation data. Vertical error bars are statistical only. Horizontal bars show data integration intervals. The data are histogrammed when an outburst is in progress. The numbers near the bottom of the plot label each of the outbursts B1-B8 (the B is omitted on the plot). (b) The spectral evolution of each outburst is shown as a plot of OTTB model temperature in the 20-300 keV band as a function of time since the beginning of the outburst. The plot symbol corresponding to each outburst is shown in the inset. Horizontal bars indicate data integration intervals. The vertical error bars are statistical only.

the observational techniques. Here each data point has a longer integration interval (nominally 10 days).

Figure 1(a) shows eight outbursts separated by intervals during which the source was not detected above the $\sim$ 30 mCrab threshold for ten day integrations. We will label these outbursts B1-B8. As discussed in [4], the light curve and spectral evolution of B1 and B2 are quite similar. B3 is also similar to B1 and B2, not only in its profile, but also in that the time between the peak of this burst and that of B2 is roughly the same as the time between B1 and B2, ($\sim$ 450 days). The peak flux of B4 is about 2/3 the peak flux of the first three outbursts, and it occurs at close to half the period ($\sim$ 240 days) established by the first three outbursts. The peak of B5 also comes approximately 210 days after that of B4. In contrast, B6-B8 peak at intervals which show no evidence of being related to the earlier (quasi-)period.

Three spectral models: OTTB, photon power law (PL), and Sunyaev-Titarchuk comptonization (ST) were fit to the outburst data over 20-300 keV. An OSSE observation near the peak of B1 was consistent with an OTTB model (kT $\simeq$ 70 keV) over the full energy range in which the source was detected (up to 400 keV), but is inconsistent with PL and a marginal fit at best to ST above 200 keV [6]. During the bright portions of B1-B4, PL gives a poor (often unacceptable) fit to the BATSE data between 20 and 300 keV, while the OTTB and ST models are both adequate, and fit the data equally well. These results, however, imply that it is sufficient to interpret BATSE data with an OTTB. Although an OTTB model also works for spectra obtained during B5-B8, there are also times when PL gives an adequate (sometimes equally good) fit to the data. During these times, the possibility that the spectrum is a power law which also extends to higher energies can not be ruled out.

The spectral evolution during each outburst is presented in a plot of OTTB fit temperature versus time in Figure 1(b). During outbursts B1-B4 the temperature gradually declined from its peak. In contrast, the temperature evolution of B5-B8 has no rising or falling trends.

ANALYSIS OF THE OUTBURST PATTERN

We have examined the possibility of correlations among the outburst fluence, peak flux, duration, and time between outbursts.

Figure 2(a) shows that peak flux is roughly linearly correlated with the total outburst fluence, as first reported in [7]. The correlation extends over a factor of 3.5 in each parameter. Peak outburst luminosities depend on source distance, which has been estimated to be between 1.3 kpc [8] and 4 kpc [9]. The peak luminosities range from $2.2 \times 10^{35} d_{kpc}^2$ ergs/s for B5 to $7.6 \times 10^{35} d_{kpc}^2$ ergs/s for B3 where d_{kpc} is the source distance in kpc.

In Figure 2(b) the time between outbursts is plotted versus outburst fluence in two different ways. The crosses show the time elapsed since the previous outburst (T_{pi}, the time between the end of the previous burst B(i-1) and the start time of Bi). B1 is not included because no burst preceding it could have been observed by BATSE; however its occurrence is implicit in the value of T_{p2}. The squares show the time between the end of the current outburst and the start of the next one (T_{ni}). The dashed line shows the result of a straight line fit to the T_{pi} and the dotted line to the T_{ni}. An approximate linear correlation between T_{pi} and fluence is observed, with the possible exception of B5, which appears underluminous for this relation. In contrast, while the general trend is similar for T_{ni}, the deviations of B1 and B3 from the trend are large. Thus, taking T_{pi} as the underlying variable correlated with outburst energy release is a better description. This correlation implies that the time averaged luminosity, $\overline{L} = 1.6 \times 10^{35} d_{kpc}^2$ ergs/s, is roughly constant. Outburst durations show no clear trend when plotted against fluence.

OUTBURST TIMESCALES

In order to obtain a better understanding of the outburst timescales, and to facilitate comparisons to models, we have examined the outburst light curves at a resolution of one day. Spectral fits in which flux is the only free parameter were used to obtain a flux estimate for each day of data. In each fit, the temperature is held fixed at the temperature determined from the corresponding

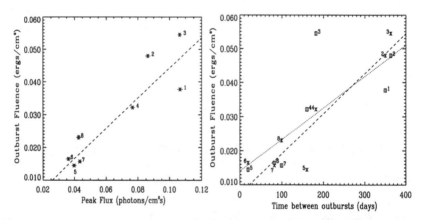

FIGURE 2. Outburst parameters vs. outburst fluence in the 20-300 keV band. In each of these plots, the number near each data point labels the corresponding outburst (B1-B8). The error bars are statistical only. (a) Outburst fluence versus peak flux. The dashed line shows the best fit straight line to the plotted data. (b) Outburst fluence versus time to the previous outburst (T_{pi}, crosses) and time to the next outburst (T_{ni}, squares). The dashed line is the best fit straight line of the fluence to T_{pi} and the dotted line to T_{ni}.

Outburst	Previous	Duration	Rise 1	Rise 2	$\chi^2_{\nu rise}(\nu)$	Decay	$\chi^2_{\nu decay}(\nu)$
B1	-	100	12.7 ± 1.5	153 ± 52	1.6 (65)	5.7 ± 0.5	1.5 (5)
B2	350	95	-	108 ± 25	0.9 (39)	3.4 ± 0.5	0.8 (8)
B3	363	93	7.6 ± 1.5	104 ± 12	3.8 (73)	8.9 ± 0.9	0.6 (10)
B4	182	71	21 ± 11	56 ± 12	1.4 (37)	5.8 ± 0.8	0.5 (6)
B5	160	86	19 ± 13	-	0.8 (8)	56 ± 14	1.0 (36)
B6	19	69	18 ± 7	-	1.0 (8)	55 ± 24	1.3 (22)
B7	82	61	13.5 ± 2.6	-	0.7 (16)	12.6 ± 3.6	0.8 (11)
B8	99	79	47 ± 6	-	1.3 (32)	64 ± 18	0.7 (13)

TABLE 1. Timescales during outburst sequence. The second column (labeled previous) is the time between the end of the previous and the begininning of the current outburst. Duration is the duration of the outburst. Rise 1 is the first e-folding rise time and Rise 2 the second e-folding rise time determined from an exponential model fit (with variable break time, when both timescales are present) on the rising portion of the outburst. Decay is the e-folding decay time from an exponential model fit on the falling side of the outburst. All times are in days. The reduced χ^2 and corresponding number of degrees of freedom, ν, are shown for each fit.

ten day spectral fit. We have attempted to model the first four outbursts (except for the second one) with an initial fast exponential rise followed by a second, much slower, rise. The last four outbursts (and the second one) have only a single rise, and each outburst has a single decay. Information about rise and decay times of all of the outbursts is shown in Table 1. In each case the rise (including variable break time) and decay intervals were fit separately.

For comparison power law models have been fit to rise/decay portions of the light curve. F-test probabilities that the power law model is preferred range from 3% to 85% and are typically 30%. Only on the B1 fall and B3 rise is this above 50%. Exponential models are thus marginally preferred.

REFERENCES

1. Callanan, P. J., et al. 1992, *MNRAS*, 259, 395
2. Fender, R. P., et al. 1997, *MNRAS*, 286, L29
3. Tanaka, Y., Lewin, W., in *X-Ray Binaries*, Lewin, W., van Paradijs, J., van den Heuvel, E., editors, Cambridge University Press, 1995, p. 126.
4. Harmon, B. A., et al. 1994, *ApJ*, 425, L17
5. Rubin, B. C., et al. 1997, submitted to *ApJ Letters*
6. Grabelsky, D. A., et al. 1995, *ApJ*, 441, 800
7. Robinson C. R., et al. 1996, in *Proceedings of the Second Integral Workshop 'The Transparent Universe'*, ESA SP-382, 249
8. Predehl, P., et al. 1991, *Astronomy and Astrophysics*, 246, L40
9. Cowley, A. P., Crampton, D., & Hutchings, J. B. 1987, *AJ*, 92, 195
10. Liubarskii, Y. E. & Shakura, N. 1987, *Soviet Astronomy Letters*, 13, 917

Multiwavelength Observations of GX 339–4

I. A. Smith*, E. P. Liang*, M. Moss*, J. Dobrinskaya*,
R. P. Fender[†], Ph. Durouchoux[‡], S. Corbel[‡], R. Sood[||],
A. V. Filippenko[¶], D. C. Leonard[¶]

*Department of Space Physics and Astronomy, Rice University, MS-108,
6100 South Main, Houston, TX 77005-1892
[†]Astronomy Centre, University of Sussex, Falmer, Brighton, BN1 9QH, U. K.
[‡]DAPNIA, Service d'Astrophysique, CE Saclay, 91191 Gif sur Yvette Cedex, France
[||]School of Physics, ADFA, Northcott Drive, Canberra ACT 2600, Australia
[¶]Department of Astronomy, University of California, Berkeley, CA 94720-3411

Abstract. We discuss our recent observations of the Galactic black hole candidate GX 339–4. The source was in a "soft X-ray off" state, and we found that this was similar to previous very hard states. A possible radio jet was found 1996 July 11 – 14. During this time, there were no dramatic changes in the X-ray or gamma-ray flux or hardness. Our 1996 May 12 Keck observations found that the optical emission was surprisingly bright and still dominated by the highly variable accretion disk, in spite of the fact that RXTE did not see a detectable flux. RXTE observations on 1996 July 26 revealed a 0.35 Hz QPO.

INTRODUCTION

Most Galactic black hole candidates exhibit at least two distinct spectral states (see Liang 1997 for a review). In the hard state (= soft X-ray low state) the spectrum from $\sim$ keV to a few hundred keV is a hard power law (photon index 1.5 ± 0.5) with an exponential cutoff. This can be interpreted as inverse Comptonization of soft photons. In the soft state (often, but not always, accompanied by the soft X-ray high state), the spectrum above ~ 10 keV is a steep power law (photon index > 2.2) with no detectable cutoff out to $\sim$ MeV. GX 339–4 is unusual in that it is a persistent source, being detected by X-ray telescopes most of the time, but it also has nova-like flaring states. It also has a super high state, and a "soft X-ray off" state. The definition of the latter has traditionally depended on the sensitivity of the particular instrument making the measurement, but is typically < 100 mCrab in the soft X-rays.

In this paper, we outline our recent observations of GX 339–4 that are part of our ongoing multiwavelength campaign. Figure 1 shows the long term evolution of the RXTE flux. The source remained in a "soft X-ray off" state for all the observations discussed here. Pointed RXTE observations of GX 339–4 were made 1996 July 26, and a 0.35 Hz QPO was found [4].

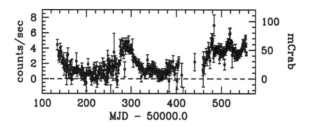

FIGURE 1. Evolution of the RXTE ASM 2−10 keV flux. 1997 January 1 = MJD 50449. These quick-look results were provided by the ASM/RXTE team.

RADIO OBSERVATIONS

The radio counterpart to GX 339–4 was discovered in 1994 by the Molonglo Radio Observatory, Australia (MOST) at 843 MHz [12]. Further monitoring has found that the radio emission is variable, with a flux density $\lesssim$ 10 mJy [5]. No correlation between the radio and X-ray emission has been found [12].

High resolution 3.5 cm radio observations with the Australia Telescope Compact Array (ATCA) detected a possible jet-like feature in 1996 July 11–14 (Figure 3 of Fender et al. 1997). An ATCA observation on 1997 February 3 may have a small extension in the direction opposite from this jet, but no strongly significant confirmation of the jets has so far been found [1]. Figure 2 shows the daily radio spectra in 1996 July. The spectrum is approximately

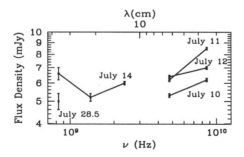

FIGURE 2. Preliminary daily radio spectra in 1996 July. All the data points are from ATCA, except those at 36 cm from MOST (Dick Hunstead, private communication).

flat, and shows a significant variability. However, the spectral shape and amplitude were not anomalous during the time of the possible radio jet.

OSSE OBSERVATIONS

OSSE observed GX 339–4 as a ToO during an outburst for 1 week beginning 1991 September 5. The OSSE flux was $\sim$ 300 mCrab [8]. A second 1 week observation was carried out beginning 1991 November 7, when the source was $\sim$ 40 times weaker [8].

We observed the source with OSSE 1996 July 9–23. During this time, the RXTE ASM flux was generally rising, with a flux $\sim$ 50 mCrab. Figure 3 shows that the OSSE flux was also generally rising, with no significant variation in the hardness ratio. This is very interesting, since it was during the first week that the radio source had the possible jet-like feature. One might have expected that the physics behind the jet formation could have led to a significant change in the higher energy emission. But since the radio flux was not unusually bright during this time, it is possible that the energy release in this case was relatively small. Further multiwavelength observations during

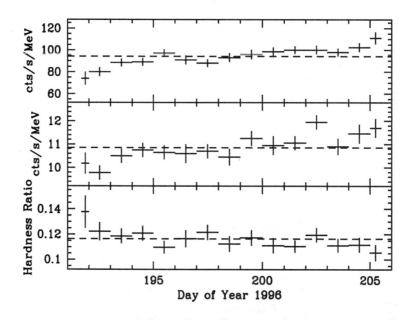

FIGURE 3. Daily fluxes for our 1996 July 9–23 OSSE observation of GX 339–4. Top panel: 0.05 – 0.07 MeV. Middle panel: 0.07 – 0.27 MeV. Bottom panel: hardness ratio (0.07 – 0.27 MeV) / (0.05 – 0.07 MeV). The possible radio jet-like feature was detected in observations on 1996 July 11-14 [6].

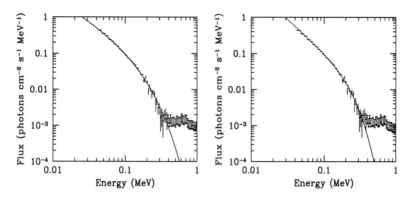

FIGURE 4. 1996 July OSSE observation of GX 339–4. Left panel: best fit PE model has power law index 1.15±0.07, $kT = 97±6$ keV, and flux 0.094±0.001 photons cm^{-2}s^{-1}MeV^{-1} at 100 keV. Right panel: best fit ST model has optical depth $\tau = 2.5 \pm 0.1$, $kT = 47 \pm 1$ keV, and flux 0.093 ± 0.001 photons cm^{-2}s^{-1}MeV^{-1} at 100 keV.

radio jet or radio flaring events will be very important. In other black hole candidates, violent changes in the high energy emission may or may not be associated with large radio flares, e.g. GRO J1655–40 [10,13].

Figure 4 shows the best power-law × exponential (PE) fit to our complete data set. The flux normalization at 100 keV is a factor of 2 lower than that found by Grabelsky et al. [8] in the "low hard" state, while our power law index is slightly steeper, and kT higher. However, it is apparent that our "off" state is similar to previous very hard states. A Sunyaev-Titarchuk (ST) function gave an equally good fit to our OSSE data (Figure 4). This differs from Grabelsky et al. [8], who found that the ST spectrum dropped too rapidly at high energies to give good fits. However, Compton scattering models could explain these harder spectra if a small fraction of non-thermal particles is added to the thermal population, as we have shown for Cygnus X-1 [3]. Extrapolating back to the X-ray region, the ST model that best fits our 1996 July OSSE data also agrees well with the RXTE ASM flux. The PE model that best fits the OSSE data is a factor ~ 3 too low in the X-ray region.

KECK OBSERVATIONS

The accretion disk has always dominated the optical emission, preventing a direct observation of the companion star. On 1996 May 12 we made Keck observations when the RXTE ASM did not detect a significant X-ray flux and it was hoped that the accretion disk would be in optical quiescence. However, the optical emission was still dominated by the highly variable accretion disk. Figure 5 shows two consecutive 400 s spectra. The data are not fully photometric but give V $\sim$ 17 mag, the brightest flux ever found for the "soft

X-ray off" state, which normally has V ∼ 18 − 20 mag. The relative scale of the two curves is accurate, and multiplying the brighter one by a factor of 0.942 we find that the general shape did not change markedly between the two observations.

The dominant emission line is from Hα. Its equivalent width (EW) of ∼ 6.5 Å is similar to that in previous GX 339–4 observations [9,2], but is small compared to other black holes, e.g. Nova Oph 1977 had EW = 85 Å during the same Keck run [7]. There is an indication that the line is double peaked, with $\Delta v \sim 250$ km s^{-1}.

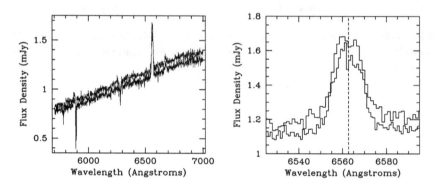

FIGURE 5. 1996 May 12 Keck observations of GX 339–4. Left panel: whole spectra. Right panel: zoom of Hα. Dashed line shows the rest wavelength.

This work was supported by NASA grant NAG 5-1547 at Rice University, and NSF grant AST-9417213 at UC Berkeley.

REFERENCES

1. Corbel, S., et al. 1997, these proceedings
2. Cowley, A. P., Crampton, D., & Hutchings, J. B., 1987, AJ, 92, 195
3. Crider, A., et al. 1997, these proceedings
4. Dobrinskaya, J., et al. 1997, to appear
5. Durouchoux, Ph., et al. 1997, ApJ, in press
6. Fender, R. P., et al. 1997, MNRAS, 286, L29
7. Filippenko, A. V., et al. 1997, PASP, 109, 461
8. Grabelsky, D. A., et al. 1995, ApJ, 441, 800
9. Grindlay, J. E. 1979, ApJ, 232, L33
10. Harmon, B. A., et al. 1995, Nature, 374, 703
11. Liang, E. P. 1997, Physics Reports, in press
12. Sood, R., et al. 1997, 'The Transparent Universe', ESA SP-382, 201
13. Tavani, M., et al. 1996, ApJ, 473, L103

Radio Observations of the Black Hole Candidate GX 339-4

S. Corbel[1], R.P. Fender[2], P. Durouchoux[1], R.K. Sood[3], A.K. Tzioumis[4], R.E. Spencer[5] and D. Campbell-Wilson[6]

[1] *CEA Saclay, DAPNIA-SAp, 91191 Gif sur Yvette Cedex, France*
[2] *Astronomy Centre, University of Sussex, Falmer, Brighton BN1 9QH, UK*
[3] *Australian Defence Force Academy, UNSW, Canberra, ACT 2600, Australia*
[4] *Australia Telescope National Facility, CSIRO, PO Box 76, Epping 2121, NSW, Australia*
[5] *University of Manchester, Jodrell Bank, Macclesfield, Cheshire SK 11 9DL, UK*
[6] *School of Physics, University of Sydney, Sydney, NSW 2006, Australia*

Abstract. The black hole candidate GX 339-4 was first detected as a variable radio source by Sood & Campbell - Wilson [14] in May 1994 with the Molonglo Observatory Synthesis Telescope (MOST). Since then, several observations have been obtained with the Australian Telescope Compact Array (ATCA) in order to study the radio behavior of this source in relation to its soft and hard X-ray activity. We present new results of high resolution radio observations performed with the ATCA in order to study the jet-like feature observed in GX 339-4 by Fender et al [3]. From the ATCA lightcurve at 8640 MHz, we find evidence of quenched radio emission from GX 339-4.

INTRODUCTION

Since its discovery in 1973 with the OSO-7 satellite [5], GX 339-4 has been extensively studied in optical, X-rays and gamma rays. The optical counterpart was identified with a 17th magnitude blue star by Doxsey et al. [2], which was found to be very variable (15 to 20 mag.). Based on its bimodal spectral behavior and its fast timing properties, GX 339-4 has been classified as a black hole candidate in a low mass system. GX 339-4 displays all three spectral modes (low state, high state and very high state) typical of black hole candidates based on the flux in the 2-10 keV band. Recently, Méndez et al. [6] have found evidence for an intermediate state between the low state and the hard state. The X-ray spectral behavior of GX 339-4 is quite similar to that observed in the well known black hole candidate Cyg X-1.

Some anti-correlation between X-ray states and optical brightness has been reported by Motch et al. [9]. The companion star is very faint in the off state (or very low state), brightens during the low (hard) state and is at an intermediate level during the high (soft) state. A brightness modulation of 14.8 hr, interpreted as the orbital period of the binary system, has been reported [1]. But due to the strong optical emission from the accretion disk, the orbital parameters of GX 339–4 have not yet been established in order to clearly demonstrate that it is a black hole binary. X-ray and optical quasi periodic oscillations have been observed at different frequencies, most of them during low state [13]. Its distance estimates range from 1.3 to 4 kpc, with a favorable distance of 1.3 kpc from X-ray halo measurement [10].

A variable radio source was first detected by Sood & Campbell–Wilson [14] in May 1994 with the MOST and since then, several observations have been performed in order to study the radio behavior of this source. High resolution radio observations by Fender et al [3] in July 1996 revealed the existence of a possible jet like feature issuing from the core of GX 339–4. Here we will concentrate on the observations conducted in February 1997.

OBSERVATIONS

Radio observations of GX 339–4 began on May 25 1994 and are continuing at present. Data were collected from the Molonglo Observatory Synthesis Telescope (MOST) and from the Australian Telescope Compact Array (ATCA) at a total of five frequencies: 843 MHz at the MOST, 1.5, 2.4, 4.7 and 8.6 GHz at the ATCA. The ATCA is a synthesis telescope consisting of 6 parabolic dishes, 22 m in diameter, mounted on an East-West array. At the time of our observations, the array was in the 6 km configuration with a maximum baseline of $\sim$ 6 km. The nominal angular resolution was $\sim$ 1 arcsec at 3.5 cm. Amplitude calibration was performed using B1934-638, while B1646-50 was used as the phase calibrator. The source was offset from the center and the distance between the target and the phase calibrator was small (2.8 degrees). We used a cycle of 8 min on source and 2 min on the phase calibrator in order to remove any phase error that could introduce artifacts in the imaging process. The data were edited and calibrated using the software package Miriad.

SEARCH FOR RADIO JETS

Miyamoto et al. [8] predicted the existence of a jet in GX 339–4 in its very high state. High resolution radio observations at 3.5 cm with ATCA in July 1996 revealed a possible extension to the West of the source [3], which was detected at a level of $\sim$ 6 mJy. While phase errors cannot be ruled out as the origin for this feature, it is interesting to note that just before these observations, the source's radio emission had been undetectable at an upper limit of 1 mJy from MOST. This is very reminiscent of the behaviour of Cygnus X-3, which undergoes periods of quenched radio emission prior to jet-producing outbursts [15].

In order to confirm the existence of this feature, we performed 3 observations spaced by one week in February 1997 (3, 10 and 17) in a similar configuration. A very weak extension was detected in the 3.5 cm map (Figure 1) of Feb 3 around the core of GX 339-4 on the opposite side of the feature observed by Fender et al. [3]. The axis of the elongation is consistent with an SE-NW orientation. There is no evidence for a counterjet, that could point to a relativistic plasmoid ejection. However, the source is compatible with a point source for the other two observations (Feb. 10 and 17). Due to the weakness of this feature, these observations are not enough to confirm unambiguously the presence of a jet-like feature in GX 339-4, even if the jets might have disappeared between two observations, because they are transient in nature in most X-ray binaries. It is worth noting that all our ATCA observations were performed during an X-ray low state.

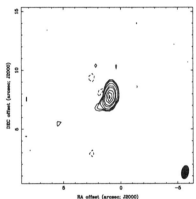

FIGURE 1. 3.5 cm map of GX 339-4 obtained with ATCA on 1997 Feb 3. Contours are -3, 3, 4, 6, 10, 20, 30, 50, 70 and 90 times r.m.s. noise of 90 μJy beam^{-1}. The solid ellipse represents the synthesised beam (restored beam of $1.17'' \times 0.66''$). The position (J2000) of the core is α: 17 h 02 m 49.3 s and δ: -48°47'31.0"

RADIO LIGHT CURVE AND SPECTRA

The radio lightcurve of GX 339-4 observed with ATCA at 8640 MHz is displayed in Figure 2. GX 339-4 is clearly detected as a variable radio source with variations from week to week, but no flaring activity (sudden increase from an essentially quiescent level) has yet been detected, such as that seen in Cyg X-3 [15], GRO J1655-40 [4] or GRS 1915+105 [12]. This is supported by data from the MOST (Hunstead, 1997, private communication). However, we should emphasize that the lightcurve is too sparsely sampled at present to

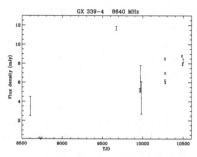

FIGURE 2. ATCA lightcurve of GX 339–4 at 8640 MHz

allow us to characterize the time variability with confidence. A slight variation at 3.5 cm ($\sim 10\%$) is also present on a timescale of one day.

We have analysed archival data obtained with ATCA and have found that the source has been detected on December 13, 1991 at a level of ~ 3 mJy. However, a more interesting result is the non-detection at 3.5 and 6.3 cm (with a one sigma upper limit of 0.2 mJy) on April 21, 1992, at a time when the source was very weak in hard X-rays. This is very similar again to the quenched radio emission observed in Cyg X-3. This is the first time radio emission was not detected from GX 339–4 at the level of this strong upper limit.

The radio spectrum of GX 339–4 obtained on 1997 February 17 from 843 MHz (36 cm) to 8640 MHz (3 cm) is presented in Figure 3. Observations at 843 MHz were performed with the MOST. Using the convention $S(\nu) \propto \nu^\alpha$, the spectrum could be fitted with a single powerlaw with an average spectral index of +0.24 if we excluded the MOST datapoint, but the latter seems to indicate a curvature in the spectrum and could suggest the presence of another emission component. The inverted spectrum component indicates a compact core which is partially optically thick, similar to previous observation [3].

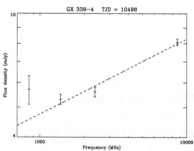

FIGURE 3. Radio spectra of GX 339–4 from 843 to 8640 MHz. The dotted line indicates the best fit to the ATCA datapoints.

DISCUSSION

Among the radio emitting X-ray binaries, GX 339-4 is a peculiar source. Its radio behavior is quite similar to that of Cyg X-1 (variable and flat radio spectra), but with very different temporal hard X-ray activity. Strong outbursts have been detected with BATSE [11] in hard X-rays from GX 339-4. The superluminal source, GRO J1655-40, exhibits a hard X-ray behavior "similar" to that of GX 339-4, with the main difference being that GX 339-4 doesn't show radio flaring and superluminal radio ejection. Recently, Mioduszewski et al. [7] have shown some evidence that Cyg X-3 might also be a superluminal source. The detection of quenched radio emission (if confirmed) from GX 339-4 prior to radio ejection could be a missing link between these sources. The long term radio behavior of GX 339-4 could link the jet sources (and transient radio sources) to Cyg X-1 (persistent radio source) and Cyg X-3. All our radio observations have been performed during the low state of the source, and a more complete coverage of the various X-ray states should provide new clues for the understanding of GX 339-4.

REFERENCES

1. Callanan P.J. et al., 1992, MNRAS, 259, 395
2. Doxsey, R. et al., 1979, ApJ, 228, L27
3. Fender, R.P. et al., 1997, MNRAS, 286, L29
4. Hellming, R.M. and Rupen M.P., 1995, Nature, 375, 464
5. Market, T.H. et al., 1973, ApJ, 184, L67
6. Méndez, M. and van Der Klis, M., 1997, ApJ, 479, 926
7. Mioduszewski, A. et al, 1997, Proc. IAU 164, in press
8. Miyamoto, S. and Kitamoto, S., 1991, ApJ, 374, 741
9. Motch, C. et al. 1985, Sp. Sc. Rev., 40, 219
10. Predehl, P. et al., 1991, A&A, 246, L40
11. Robinson, C.R. et al., 1997, Proc. 2nd INTEGRAL workshop, ESA SP-382
12. Rodríguez, L.F. et al., 1995, ApJS, 101, 173
13. Steiman-Cameron, T.Y. et al., 1997, ApJ, in press
14. Sood, R.K. and Campbell-Wilson, D., 1994, IAU Circ. 6006
15. Waltman, E. et al, 1996, AJ, 112, 2690

ACKNOWLEDGMENTS: The authors would like to thank Dick Hunstead and Mike Nowak for useful discussions and Vince McIntyre and Marc Elmouttie for helpful assistance during the ATCA observations. S. Corbel acknowledges support from Le Comité National Français d'Astronomie and the local organizers of the Compton symposium. The Australia Telescope is funded by the Commonwealth of Australia for operation as a National Facility managed by CSIRO. The MOST is supported by the Australian Research Council and the Science Foundation for Physics within the University of Sydney.

Infrared Observations of the Ellipsoidal Light Variation in J0422+32

Dawn M. Leeber, Thomas E. Harrison, Bernard J. McNamara

Astronomy Department, New Mexico State University, Las Cruces, NM 88003

Abstract. We have obtained K-band photometry of J0422+32 using the ARC 3.5 m telescope in January 1997 to search for the ellipsoidal variations expected from the secondary star. J0422+32 is a soft x-ray transient that erupted in August of 1992, containing a compact object primary and a late-type secondary star. Both a neutron star and a black hole have been proposed as the primary component for this system. The inclination of the orbit, however, is not known well enough to be able to decide between the two. We explore the use of infrared observations, where the secondary star is the dominant luminosity source, to measure the amplitude of the ellipsoidal variation of the system. The GENSYN light curve synthesis program is used in conjunction with these variations to estimate the orbital inclination, and hence the mass of the binary system.

INTRODUCTION

J0422+32 (= V518 Per) is a soft x-ray transient (SXT), and possible black hole x-ray nova (BXHN), which had its first recorded outburst in August of 1992. Details of the outburst can be found in Kato et al. (1995), and Shrader et al. (1994). The optical and x-ray light curves of J0422+32 and two other SXTs (V616 Mon, and GU Mus) are presented in Figure 1 of King et al. (1996). J0422+32's behavior at optical and x-ray energies is very similar to these other SXTs. The x-ray flux from SXTs declines exponentially following maximum, and all three systems exhibited a secondary maximum. Both GU Mus and V616 Mon have primaries with masses higher than the upper limits for neutron stars (Remillard et al. 1992, Charles & Casares 1995), and are therefore considered to be black holes.

Because the x-ray light curve of J0422+32 resembled the other two black hole systems, it too was expected to have a black hole primary. But there remains uncertainty in this assignment. Orosz & Bailyn (1995) estimate a primary mass of < 3 $M_\odot$ for this system, while Fileppenko et al. (1995) estimate a primary mass that lies between 3.23 and 3.91 $M_\odot$. The difficulty with mass determinations is assigning a velocity semi-amplitude to the primary star (in the case of a black hole system, it is especially difficult!). Another equally important factor in the determination of the mass of the system is the orbital inclination. This parameter, except for eclipsing systems, is very difficult to ascertain. Many of the parameters needed for the calculation of the system mass for J0422+32 appear to have well-determined values, and with a good measurement of the inclination, we should be able to calculate an accurate mass of the binary system. We explore the technique of modeling the infrared ellipsoidal light curve of the secondary star to determine the system inclination for J0422+32.

INFRARED OBSERVATIONS OF J0422+32

To attempt to measure the ellipsoidal light variations in J0422+32, we obtained time on the Apache Point Observatory 3.5 m telescope with the infrared camera GRIM II. The observing run consisted of 4.9 hours on 21 January 1997. Observations were obtained in the K' infrared band (2.12 μm) to avoid contamination from other sources in the system (including hot spots on the secondary star, the accretion disk, etc.). Differential photometry of several field stars was used for flux measurement and calibration. The resulting light curve for the January 1997 run is shown in Figures 1 & 2.

The period of J0422+32 has been determined to be 5.1 hours, though in the earliest observations, a period of 10.2 hours could not be excluded (Callanan et al. 1995). Thus, our run of 4.9 hrs covered more than 90% of an orbit, and should have contained either two minima or two maxima. The observed light curve, however, shows no evidence for the expected smooth sinusoid with a period of 5.1 hours. Shortly after our run, Iyudin and Haberl et al. reported that RXTE ASM observations showed that J0422+32 had been in a mini-outburst starting in November 1996 and lasting into January 1997. Thus, with an additional luminosity source, and x-ray heating from the outburst, the light curve would not be expected to be a simple sinusoid.

There have been earlier reports of ellipsoidal light variations being detected in the optical for J0422+32 (Callanan et al. 1995, Casares et al. 1995, Chevalier et al. 1996), but these certainly cannot have been due to the secondary star. When comparing the IR data and the most recent optical data for J0422+32, and fitting a 3500 K blackbody curve to the infrared data point with appropriate reddening, we note that the optical data is well above the extrapolation of this blackbody. This shows that the 20-30% variation expected from the secondary star at these wavelengths would have been drowned out by the additional luminosity in the system, and that the source of the sinusoidal light variations seen at those times was not coming from the minimum-light photosphere of the secondary (the heated face of the secondary might have been responsible for those variations).

ELLIPSOIDAL LIGHT CURVES IN THE INFRARED

The secondary stars in BXHN are cool, low-mass main sequence stars that fill their Roche lobes. Because the secondary stars are cool (T ≈ 3500 K), they emit most of their luminosity in the near-infrared. Due to the great distances to BXHN, the secondary stars are very faint at all wavelengths, but much more

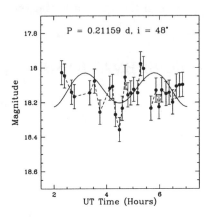

FIGURE 1. 5.1 hour period model plotted over the light curve. Error bars are 1σ. The GENSYN modeling program has been used to model the secondary component of the system. There is poor agreement between the two curves, no matter how they are phased.

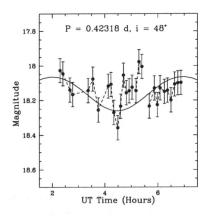

FIGURE 2. 10.2 hour period model plotted over the light curve. Error bars are 1σ. While this fit is not perfect, it fits the data better than the 5.1 hour period. The deviations from the 10.2 hour period are most likely due to contamination from other luminosity sources in the system at the time of observation.

so at optical wavelengths. For example, J0422+32 (in observations detailed above) had a mean magnitude of K=18.2. With a color of $V - K = 4.11$ for the M2V secondary, the optical brightness of the secondary star (if this star was the sole source of luminosity in the system, and interstellar extinction was neglected) would be V=22.3. As discussed in King et al. (1996), the visual extinction to J0422+32 is $A_V = 0.75$, thus at minimum, the J0422+32 system should have V > 23. It is difficult to do precision photometry on objects this faint with available telescopes. With K=18.2, however, photometry of adequate precision ($\approx$ 5-10%) can be obtained with 4 m class telescopes.

The use of ellipsoidal light variations to determine the parameters of binary star systems has a rich history (see the references in Wilson 1994). Infrared ellipsoidal light variations have even been employed in understanding cataclysmic variables (Szkody et al. 1988). Light curve modeling of binary systems has reached a sophisticated level. Computer programs by Mochnacki and collaborators (GENSYN, GDDSYN) and by Wilson and Devinney (WLD) calculate the light curves of real stars, in real binary systems. The distribution of light over the surface of the star (accounting for limb darkening, gravity darkening, and irradiation/reflection) are fully accounted for. The shape and temperature variations in the photosphere of the Roche lobe filling star are fully modeled. The physics employed by these programs is well understood, as are the values of various input parameters. Thus, accurate models of the ellipsoidal light curve are produced, allowing for determinations of parameters such as the inclination of the binary system.

Because of our need to adapt the code to infrared wavelengths, and the avail-

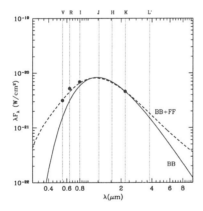

FIGURE 3. Our data at K and most recent optical data taken from the literature for J0422+32. The optical (V,R,I) ellipsiodal variations seen could not have been due to the minimum-light photosphere of the secondary star in the system, (M2V, $T_{BB} = 3500K$, solid curve) since they would have been diluted by the additional luminosity being generated in the system. The dashed curve is 75% BB and 25% free-free emission.

ability of the source code, we will be using the GENSYN program to model the infrared ellipsoidal light curves of BXHN. In Figure 1 we overlay a model of the 5.1 hour infrared light curve for a system with identical parameters to J0422+32 as determined by Filippenko et al. (1995), at an inclination of 48° onto our obsevations, assuming that the secondary star is the sole source of light in the system. Figure 2 shows the 10.2 hour model. The amplitude of the maximum to minimum is very dependent on the inclination of the system, and photometry good to 5 - 10% can easily determine the inclination (and period!) of such a system. Given adequate data, the technique of modeling the infrared light curves of BXHN is very powerful for determining system parameters.

FUTURE WORK AND CONCLUSIONS

Infrared ellipsoidal light curve modeling promises to be a valuable technique for determing the inclination of BXHN systems. For this to work, however, the system needs to be close to minimum light. Unfortunately, this was not the case in our observations. To allow the GENSYN program to accurately model the light curve of BXHN that are not near minimum light (i.e., when another source of light exists in the system other than the secondary) will require fully adapting the software to calculate the proper infrared flux distribution from the binary star system. We are currently doing this. Thus, we may yet be able to understand the light curve observed. A longer observational run is needed when the system is truly at, or at least close to, minimum light. Also, to increase the S/N, we will observe in the J infrared band where the spectral energy distribution of the $M2V$ secondary star peaks. We hope to conduct these observations during Autumn 1997.

REFERENCES

Casares, J., et al. 1995, MNRAS, 276, L35
Callanan, P. J., et al. 1995, ApJ, 441, 786
Charles, P. A., & Casares, J. 1995, IAU Circ., No. 6193
Chevalier, C., & Ilovaisky, S. A. 1996, A&A, 312, 105
Filippenko, A. V., Matheson, T., & Ho, L. C. 1995, ApJ, 455, 614
Iyudin, A., & Haberl, F. 1997, IAU Circ., No. 6605
Kato, T., Mineshige, S., & Hirata, R. 1995, PASJ, 47, 31
King, N L., Harrison, T. E., & McNamara, B.J. 1996, AJ, 111, 1675
Orosz, J., & Bailyn, C. D. 1995, AJ Let., 446, L59
Remillard, R., McClintock, J., & Bailyn, C. 1992, ApJ, 399, L145
Shrader, C. R., et al. 1994, ApJ, 434, 698
Szkody, P., & Feinswog, L. 1988, ApJ, 334, 422
Wilson, R. E. 1994, PASP, 106, 921

Rapid X-ray variability in GRO J0422+32 (Nova Per 1992)

F. van der Hooft[a], C. Kouveliotou[b,c], J. van Paradijs[a,d], D.J. Crary[c,e], B.C. Rubin[f], M.H. Finger[b,c], B.A. Harmon[c], M. van der Klis[a], W.H.G. Lewin[g], and G.J. Fishman[c]

[a] *Astronomical Institute "Anton Pannekoek", University of Amsterdam and Center for High Energy Astrophysics, Kruislaan 403, NL-1098 SJ Amsterdam, The Netherlands*
[b] *Universities Space Research Association, Huntsville, AL 35806, USA*
[c] *NASA/Marshall Space Flight Center, Huntsville, AL 35812, USA*
[d] *Department of Physics, University of Alabama in Huntsville, Huntsville, AL 35899, USA*
[e] *NAS/NRC Research Associate, NASA Code ES-84, Marshall Space Flight Center, Huntsville, AL 35812, USA*
[f] *RIKEN Institute, Wako-shi, Saitama 351-01, Japan*
[g] *Massachusetts Institute of Technology, 37-627 Cambridge, MA 02139, USA*

Abstract. We present the results of an analysis of the time variability of the soft X-ray transient GRO J0422+32 (Nova Per 1992) observed with BATSE. Our analysis covers the entire ∼ 200 day outburst, beginning with the first detection of the source on 1992 August 5. We obtained power density spectra (PDS) in the 20–100 keV energy band covering the frequency interval 0.002–0.488 Hz. The PDSs of GRO J0422+32 show a peak of quasi-periodic oscillations (QPOs) with a centroid frequency of ∼ 200 mHz, during the first 120 days of the outburst.

INTRODUCTION

The soft X-ray transient (SXT) GRO J0422+32 (X-ray Nova Persei 1992) was detected with the Burst And Transient Source Experiment (BATSE) on board the Compton Gamma Ray Observatory on 1992 August 5 (Truncated Julian Day [TJD] 8839) [12]. Initially, the source intensity of GRO J0422+32 increased rapidly, reaching a maximum flux of ∼ 3 Crab (20–300 keV) within three days after its first detection and remained at this level for the following days [8]. Hereafter, the X-ray intensity (40-150 keV) of the source decreased exponentially with a decay time of 43.6 days [23]. About 139 days (TJD 8978) after the first detection of the source, the X-ray flux reached a secondary maximum, after which the X-ray source continued to decrease at approximately the same rate. BATSE detected GRO J0422+32 above its 3σ 1-day detection treshold of 0.1 Crab (20-300 keV) for ~ 200 days following the start of the outburst. The daily averaged flux history of GRO J0422+32 in the 40–150 keV energy band, as determined using the Earth occultation technique [9], is displayed in Figure 1.

The spectrum of GRO J0422+32 was hard with a photon power-law index of about 2, detected up to 600 keV with OSSE [7] and SIGMA [14]; during X-ray maximum the source was detected up to a level of 1-2 MeV with COMPTEL [20]. Timing analysis of the hard X-ray data revealed quasi-periodic oscillations (QPOs) centered at 0.03 and 0.2 Hz (20-300 keV, [11]) and 0.3 Hz (40-150 keV, [23]). These QPOs were confirmed by ROSAT observations in the 0.1-2.4 keV energy band [13]. The observed hard X-ray spectrum and strong X-ray variability on short time scales resemble the properties of dynamically proven black hole candidates (BHCs) [16,17]; see however, [21], which led to the suggestion that GRO J0422+32 is a BHC too.

Shortly after the first X-ray detection, the optical counterpart of GRO J0422+32 was identified by Castro-Tirado et al. [1] and Wagner et al. [24] at a peak magnitude of V=13.2. The optical light curve showed a very slow decay and the presence of several mini-outbursts before it finally reached quiescence at V=22.35, ~ 800 days after the first detection of the X-ray source [6]. ¿From spectroscopic observations, Filippenko et al. (1995) [5] determined the orbital period and mass function to be 5.08 ± 0.01 hrs and 1.21 ± 0.06 M$_\odot$, respectively. The quiescent optical light curve of GRO J0422+32 is double-waved at the orbital period [2,3]. The orbital inclination was estimated from its ~ 0.03 mag semi-amplitude (I band) to be $\lesssim 45°$ [2]. This value for the orbital inclination implies a mass $\gtrsim 3.4$ M$_\odot$ for the compact object in GRO J0422+32, i.e. slightly above the (theoretical) maximum mass of a neutron star.

Here we report on the temporal analysis of data obtained with BATSE of the soft X-ray transient GRO J0422+32 during TJD 8839–9040.

TIME SERIES ANALYSIS

We have used 1.024 s time resolution count rate data from the large area detectors (four broad energy channels) and applied an empirical model [15] to subtract the signal due to the X-ray/γ-ray background. This model describes the background by a harmonic expansion in orbital phase (with parameters determined from the observed background variations), and includes the risings and settings of the brightest X-ray sources in the sky. It uses eight orbital harmonic terms, and its parameters were updated every three hours.

For our analysis, we considered uninterrupted data segments of 512 successive time bins (of 1.024 s each) on which we performed Fast Fourier Transforms (FFTs) covering the frequency interval 0.002–0.488 Hz. Per day, we obtained typically 35 of such segments while the source was above the Earth horizon. For each data segment and for each of the eight detectors separately, we cal-

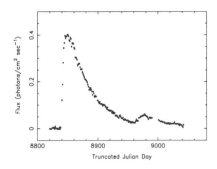
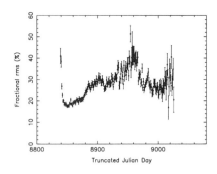

FIGURE 1. *(Left:)* Daily averaged flux history of GRO J0422+32 in the 40–150 keV energy band, as determined using the Earth occultation technique. The first detection of the source was on TJD 8839 (1992 August 5). *(Right:)* Fractional rms amplitudes (20–100 keV), determined from the daily averaged PDS, of the interval 0.01–0.48 Hz.

culated and coherently summed the FFTs of the lowest two energy channels (20–50, 50–100 keV). For those detectors which had the source within 60 degrees of the normal, these FFTs were again coherently summed (weighted by the ratio of the source to the total count rates) and converted to Power Density Spectra (PDS). The PDS were normalized such that the power density is given in units of $(\text{rms/mean})^2$ Hz^{-1} [22] and averaged over an entire day.

In Figure 2 we present four 20-day averaged PDSs covering TJD 8841–8860, 8861–8880, 8961–8980, and 8981–9000. At the start of the X-ray outburst, the break frequency of the PDSs of GRO J0422+32 is near 0.03 Hz. Above the break the PDSs decrease as a power law with superimposed a peak at ~ 200 mHz indicative of QPOs in the time series. This typical shape is observed in all PDSs of GRO J0422+32 obtained during the first 100 days of its outburst. The lower frequency QPOs at 0.03 Hz, reported by [11] are not detected. At the secondary X-ray maximum (TJD 8978), the break frequency in the PDSs has increased towards 0.07 Hz and the QPOs are absent. Such a shape is reminiscent of the PDSs of Cyg X-1 [4].

The fractional rms amplitudes were determined by integrating the single-day averaged PDSs of GRO J0422+32 over the frequency interval 0.01–0.48 Hz. The history of these daily fractional rms amplitudes, covering the interval TJD 8839–9040, are presented in Fig. 1. During the rise to the first maximum in X-ray intensity, the fractional rms amplitude decreased monotonically, followed by a gradual increase during the exponential decline of the light curve. The largest fractional rms amplitude was obtained near TJD 8960, shortly before the onset of the secondary maximum in X-rays. At the secondary maximum, the fractional rms amplitudes again reach a local minimum.

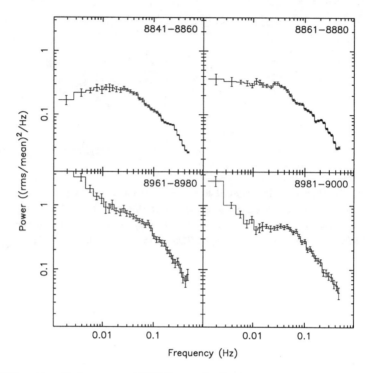

FIGURE 2. Four 20-day averaged PDS (20–100 keV), obtained during the first and second X-ray maximum of GRO J0422+32: TJD 8841–8860 *(top, left)*, TJD 8861–8880 *(top, right)*, TJD 8961–8980 *(bottom, left)*, and TJD 8981–9000 *(bottom, right)*.

DISCUSSION

GRO J0422+32 is a dynamically proven BHC of mass $\gtrsim 3.4$ $M_\odot$ [2,5], i.e. slightly above the maximum mass of a neutron star. Low-frequency (0.04–0.8 Hz) QPOs have been observed in the BHCs Cyg X-1, LMC X-1, GX 339-4 and GRO J1719-24 [18,22]. These QPOs were observed while the sources were in their so-called low state, with the exception of LMC X-1 where a 0.08 Hz QPO was found while an ultrasoft component dominated the energy spectrum, showing that the source was in the high state. ROSAT did not detect such an ultrasoft excess in the X-ray spectrum of GRO J0422+32 ∼ 40 days after the start of its outbursts [13]. Combined with the flat power law hard X-ray spectrum extending to several hundred keV and the strong rapid aperiodic variability, this indicates that GRO J0422+32 was also in the low state, although very bright, when the low-frequency QPOs were detected. Our observations support the suggestion that low-frequency QPOs may be a common property of BHCs in the low state [22].

ACKNOWLEDGMENTS

FvdH acknowledges support by the Netherlands Foundation for Research in Astronomy with financial aid from the Netherlands Organisation for Scientific Research (NWO) under contract number 782-376-011. FvdH also thanks the "Leids Kerkhoven–Bosscha Fonds" for a travel grant.

REFERENCES

1. Castro-Tirado, A.J. et al. 1993, A&A, 276, L37
2. Callanan, P.J. et al. 1996, ApJ, 461, 351
3. Chevalier, S. & Ilovaisky, S.A. 1996, A&A, 312, 105
4. Crary, D.J. et al. 1996, A&A Supp., 120, 153
5. Filippenko, A.V. et al. 1995, ApJ, 455, 614
6. Garcia, M.R. et al. 1996, ApJ, 460, 932
7. Grove, J.E. et al. 1994, in AIP Conf. Proc. 304, The Second Compton Symposium, 192
8. Harmon, B.A. et al. 1992, IAU Circ., 5584
9. Harmon, B. A. et al. 1993, AIP Conf. Proc. 280, 1st Compton Gamma Ray Observatory Symp., ed. M. Friedlander, N. Gehrels & D. Macomb (New York: AIP), 345
10. Kato, T. et al. 1995, PASJ, 47, 31
11. Kouveliotou, C. et al. 1993, in AIP Conf. Proc. 280, Proc. Compton Gamma Ray Observatory, ed. M. Friedlander, N. Gehrels, & D.J. Macomb (New York: AIP), 319
12. Paciesas, W.S. et al. 1992, IAU Circ, 5580
13. Pietsch, W. et al. 1993, A&A, 273, L11
14. Roques, J.P. et al. 1994, ApJS, 92, 451
15. Rubin, B.C. et al. 1996, A&A Supp., 120, 687
16. Sunyaev, R. et al. 1991, A&A, 247, L29
17. Tanaka, Y., & Lewin, W.H.G. 1995, in: X-ray Binaries, eds. W.H.G. Lewin, J. van Paradijs, & E.P.J. van den Heuvel (Cambridge: Cambridge University Press), 126
18. van der Hooft, F. et al. 1996, ApJL, 458, L75
19. van der Klis, M. 1995, in: The lives of the neutron stars, eds. M.A. Alpar, Ü. Kiziloğlu, & J. van Paradijs (Dordrecht: Kluwer Academic Publishers), 301
20. van Dijk, R. et al. 1995, A&A, 296, L33
21. van Paradijs, J. & van der Klis, M. 1994, A&A, 281, L17
22. van der Klis, M. 1995, in: X-ray Binaries, eds. W.H.G. Lewin, J. van Paradijs, & E.P.J. van den Heuvel (Cambridge: Cambridge University Press), 252
23. Vikhlinin, A. et al. 1995, ApJ, 441, 779
24. Wagner, R.M. et al. 1992, IAU Circ, 5589

BATSE Observations of Two Hard X-ray Outbursts from 4U 1630-47

P. F. Bloser[†], J. E. Grindlay[†], D. Barret[*], S. N. Zhang[‡°], B. A. Harmon[‡], G. J. Fishman[‡], W. S. Paciesas[‡⋆]

[†]*Harvard-Smithsonian Center for Astrophysics, Cambridge, MA 02138*
[*]*Centre d'Etude Spatiale des Rayonnements, CNRS-UPS, France*
[‡]*Marshall Space Flight Center, Huntsville, AL 35810*
[°]*Universities Space Research Association*
[⋆]*University of Alabama at Huntsville, Huntsville, AL 35889*

Abstract. Analyzing data from the BATSE instrument on the Compton Gamma Ray Observatory, we have found evidence of two hard X-ray outbursts from the ultra-soft transient and black hole candidate 4U 1630-47. Both outbursts are nearly simultaneous with observations at lower energies by other instruments. In July 1994, the source reached a hard X-ray (20-100 keV) flux of $\sim$ 100 mCrab, with a possible brief flare of $\sim$ 200 mCrab. BATSE occultation images confirm that this emission originated from 4U 1630-47. This outburst corresponds to an archival detection of the source by *ASCA*, as reported by Parmar et al. [9]. In June 1996, during the extended outburst observed by RXTE [6], the 20-100 keV flux also reached $\sim$ 100 mCrab. The origin of this emission was verified by BATSE images as well. Kuulkers et al. [5] present RXTE/ASM data from the same period of time. We compare the BATSE data with the *ASCA* and RXTE/ASM results in order to push our knowledge of this source's spectral characteristics into the hard X-ray range.

INTRODUCTION

4U 1630-47 is an ultra-soft X-ray transient (with no optical counterpart) whose early outbursts observed by VELA 5B, *Uhuru*, OSO 7, and *Ariel V* suggested a 600-day periodicity [4]. This periodicity is not exact, however [10], and the outburst lightcurves have displayed a wide variety of behavior [5]. An EXOSAT observation in 1984 suggested a spectral model of the form $E^{-\Gamma_S}\exp(-E/kT)+E^{-\Gamma_H}$ [8]. Due to its ultra-soft spectrum with a hard tail, the source is considered a black hole candidate (BHC) [11].

We have been making use of BATSE data to conduct a survey of hard (20-100 keV) X-ray transients in the Galaxy [2]. This survey includes searching for new sources and studying the hard X-ray characteristics of Low Mass X-ray Binaries (LMXBs) already known. Employing the Earth occultation technique [3] and occultation imaging technique [12], Bloser et al. [1] made the first hard X-ray detections of the recurrent ultra-soft transient 4U 1630-47.

PREVIOUS BATSE OBSERVATIONS OF 4U 1630-47

4U 1630-47 was detected by BATSE at about the 3.5σ level from TJD 8380-8410 (May 1991) [1]. The flux level reached $\sim 70 - 100$ mCrab and the spectrum could be fit by a power law with photon index $\alpha = 2.6 \pm 0.6$. This detection is not consistent with the previously-established 600-day periodicity of the source. 4U 1630-47 was also detected by BATSE at the $\sim 4\sigma$ level at a flux of ~ 15 mCrab from TJD 9450-9470 (April 1994). This sensitivity is possible with exceptional viewing geometry if no interfering sources are present. The spectrum could be fit by a power law with $\alpha = 2.32 \pm 0.5$. This detection is consistent with the previously-established ephemeris (but see discussion). BATSE did NOT detect 4U 1630-47 in September 1992, during which time Parmar et al. [7] report a detection of the source in archival ROSAT data. The 2σ upper limit on the flux for BATSE was ~ 70 mCrab for five day flux averages. (BATSE's sensitivity varies greatly depending on the limb geometry and presence of interfering sources.)

RECENTLY REPORTED ARCHIVAL DETECTIONS OF 4U 1630-47

Parmar et al. [9] and Kuulkers et al. [5] report the discovery of three archival detections of 4U 1630-47 that fall during or near times observed by BATSE. *Ginga* ASM data indicate that 4U 1630-47 underwent a long (870 days) outburst from 17 October 1988 to 4 March 1991 [5]. The light curve can be described as symmetric and parabolic, with a superimposed pattern of flares and minima occuring about every 220 days. The reported end of this outburst is about 60 days before the May 1991 BATSE detection [1], and so the BATSE detection is most likely associated with this same activity. Parmar et al. [9] report an archival detection of 4U 1630-47 by the *ASCA* Gas Imaging Spectrometers on 3 September 1994 (TJD 9598). They are able to fit the spectrum with the same cutoff power law model used for the 1984 EXOSAT observation, but without the high energy power law component, as the *ASCA* GIS give few counts above 8 keV. The best-fit parameters are quite different than those found by EXOSAT, but fixing Γ_S at 2 as found by EXOSAT also gives an acceptable fit. They then find $kT = 0.78 \pm 0.02$ keV, also consistent with EXOSAT. The 1-50 keV luminosity (for an assumed distance of 10 kpc) is $2.0 - 2.4 \times 10^{37}$ ergs s^{-1}, about a factor of 4 lower than that of the EXOSAT observation, and so Parmar et al. [9] estimate that the outburst began ∼ 75 days before the *ASCA* observation. The April 1994 BATSE detection [1] occured about 120 days before the *ASCA* detection, and about 40 days before the estimated start of the outburst. The ASM on RXTE detected an outburst from 4U 1630-47 beginning on 11 March 1996 (TJD 10153 [6]). Kuulkers et al. [5] present an analysis of these data. The light curve shows a fast rise, a long plateau phase lasting about 100 days, and an exponential decay with a decay time of ∼ 15 days. The source became rapidly harder during the rise, and then gradually harder (with large scatter) during the plateau phase, starting after about TJD 10200. The hardness (defined as the ratio of the count rates in the 5-12 keV and 1-5 keV bands) reached a maximum of about 1.5 around TJD 10270. Based on the estimated outburst times for these archival outbursts plus other, previously reported outbursts, Parmar et al. [9] and Kuulkers et al. [5] suggest that the outburst recurrence time for 4U 1630-47 changed from ∼ 600 to ∼ 690 days between the 1984 and 1987 outbursts.

NEW BATSE OBSERVATIONS OF 4U 1630-47

We have searched BATSE data at the times of the reported *ASCA* and RXTE/ASM detections in an effort to obtain simultaneous soft and hard X-ray measurements. Motivated by the *ASCA* detection on 3 September 1994 (TJD 9598) estimated to have begun on TJD 9523 ± 50 [9], we analyzed data from TJD 9470-9620; the light curve is shown in Figure 1. 4U 1630-47 is clearly detected at a level of ∼ 50 − 100 mCrab, with a possible brief flare of ∼ 200 mCrab (100 mCrab ≈ 0.03 photons cm^{-2} s^{-1}, 20-100 keV). An occultation image of the region is shown in Figure 2, in which a clear excess is evident. We produced five occultation images of the region at slightly offset positions and the source was clearly detected in all five. This procedure greatly increases confidence in detections of weak sources. Spectral fits to the BATSE data were made with both power law and optically thin thermal bremsstrahlung (OTTB) models. Best fit parameters were power law index $\alpha_{PL} = 2.56 \pm 0.21$ with $\chi^2_\nu = 0.86$, and $kT = 43.3 \pm 7.3$ keV with $\chi^2_\nu = 0.86$, respectively.

Motivated by the clear RXTE/ASM detection of 4U 1630-47 from 11 March 1996 to 3 October 1996 (TJD 10153-10359) [5], we analyzed the BATSE data from TJD 10100-10400. The source is clearly detected at a level of ∼ 100 mCrab, with a brief flare of ∼ 200 mCrab (see Figure 3). Flux from the nearby sources GX 339-4 and GRO J1655-40, which were bright at this time, has been removed. This period of emission corresponds to the portion of the RXTE/ASM light curve showing increased hardness ratio during the plateau phase [5]. An occultation image is shown in Figure 4. Best fit spectral parameters were $\alpha_{PL} = 2.32 \pm 0.25$ ($\chi^2_\nu = 1.4$) and $kT = 56.3 \pm 17.6$ keV ($\chi^2_\nu = 1.03$).

A summary of all BATSE detections of 4U 1630-47 is given in Table 1.

TABLE 1. Summary of BATSE detections of 4U 1630-47. A distance of 10 kpc is assumed. 1 mCrab ~ 0.03 photons cm^{-2} s^{-1}

Date	TJD	Flux$_{20-100}$ mCrab	α_{PL}	Luminosity$_{20-100}$ ergs s^{-1}	Reference
May 1991	8380-8410	70	2.6 ± 0.6	1.7×10^{37}	[1]
March/April 1994	9450-9480	15	2.3 ± 0.5	3.7×10^{36}	[1]
June/July 1994	9520-9570	90	2.6 ± 0.2	2.2×10^{37}	This paper
May/June 1996	10230-10300	100	2.3 ± 0.3	2.4×10^{37}	This paper

DISCUSSION AND CONCLUSIONS

BATSE has made two more convincing detections of 4U 1630-47 in the hard X-ray band. The hard X-ray detections further strengthen the case that 4U 1630-47 is a black hole. Due to the lack of detailed spectral information from either the *Ginga* or RXTE ASMs and the lack of high energy response in the *ASCA* GIS, it is difficult to compare these BATSE observations with the lower energy detections reported by Parmar et al. [9] and Kuulkers et al. [5]. Indeed, the *ASCA* detection occurs about 15 days after the outburst seen by BATSE, and occultation images made for this time do not confirm emission from 4U 1630-47. Thus the detections are not truly simultaneous. The 1996 outburst as seen by RXTE/ASM shows that the spectral characteristics of the source can change considerably during an outburst, and so comparing non-simultaneous observations is not useful. The RXTE/PCA public data from the 1996 outburst should be more valuable in this regard. A key question is, does 4U 1630-47 display quasi-persistent/weak outburst behavior? Parmar

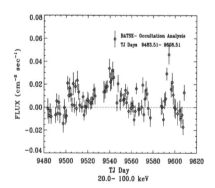
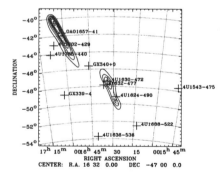

FIGURE 1. The light curve (20-100 keV) of 4U 1630-47 from May to September 1994 (TJD 9480 = 8 May 1994). Each point is averaged over one day.

FIGURE 2. Occultation image centered on the position of 4U 1630-47 using data from TJD 9535-9549. Emission due to OAO 1657-41 or 4U 1700-37 is seen in the upper left.

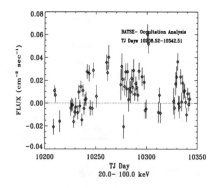 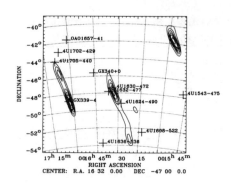

FIGURE 3. The light curve (20-100 keV) of 4U 1630-47 from April to September 1996 (TJD 10200 = 27 April 1996). Each point is averaged over one day.

FIGURE 4. Occultation image centered on the position of 4U 1630-47 using data from TJD 10270-10284. GX 339-4 is also visible.

et al. [9] and Kuulkers et al. [5] report several weak detections at times not corresponding to either the old or revised ephemeres for 4U 1630-47. Perhaps the weak detections of the source by BATSE reported by Bloser et al. [1] (see Table 1) are a similar form of "quiescent" or weak transient emission. 4U 1630-47 has shown a rich variety of behavior. BATSE, with it continuous all-sky monitoring, is a valuable tool for the study of such a source.

REFERENCES

1. Bloser, P., et al., *A&A Sup.* **120**, 191 (1997).
2. Grindlay, J., et al., *A&A Sup.* **120**, 145 (1997).
3. Harmon, B. A., et al., in "The Compton Observatory Science Workshop," NASA CP-3137, 1992, ed. C. Shrader, N. Gehrels, & B. Dennis, p. 69.
4. Jones, C., et al., *ApJ* **210**, L9 (1976).
5. Kuulkers, E. et al., *MNRAS*, submitted.
6. Marshall, F., *IAU Circular* 6389 (1996).
7. Parmar, A., Angelini, L., & White, N., *ApJ* **452**, L129 (1995).
8. Parmar, A., Stella, L., & White, N., *ApJ* **304**, 664 (1986).
9. Parmar, A., et al., *A&A*, in press.
10. Priedhorsky, W., *Ap. Space Sci.* **126**, 89 (1986).
11. White, N., & Marshall, F., *ApJ* **281**, 354 (1984).
12. Zhang, S. N., et al., *Nature* **366**, 245 (1993).

Hard X-ray observations of GRS 1009-45 with the SIGMA telescope

P. Goldoni[1], M. Vargas[1], A. Goldwurm[1], P. Laurent[1], E. Jourdain[2], J.-P. Roques[2], V. Borrel[2], L. Bouchet[2], M. Revnivtsev[3], E. Churazov[3,4], M.Gilfanov[3,4], R. Sunyaev[3,4], A.Dyachkov[3], N. Khavenson[3], N. Tserenin[3], N. Kuleshova[3]

[1] *CEA/DSM/DAPNIA, SAp, CEA-Saclay, F-91191 Gif-sur-Yvette, France*
[2] *Centre d'Etude Spatiale des Rayonnements, 9 Avenue du Colonel Roche, BP 4346, 31029 Toulouse Cedex France*
[3] *Space Research Institute, Profsouznaya 84/32, Moscow 117810, Russia*
[4] *Max-Planck-Institut für Astrophysik, Karl-Schwarzschild-Str. 1, 85740 Garching bei Munchen, Germany*

Abstract.
We report on hard X-ray observations of X-ray Nova Velorum 1993 (GRS 1009-45) performed with the SIGMA coded mask X-ray telescope in January 1994. The source was clearly detected with a flux of about 60 mCrab in the 40-150 keV energy band during the two observations with a hard spectrum ($\alpha \sim -2.1$) extending up to $\sim$ 150 keV. These observations confirm the duration of the activity of the source in hard X-rays over 100 days after the first maximum and suggest a spectral hardening which has already been observed in Nova Muscae 1991. These and other characteristics found in these observations strengthen the case for this Nova to be a Black Hole candidate similar to Nova Muscae.

INTRODUCTION

X-ray Novae so far have given the best evidence of the existence of galactic black holes from their dynamical mass determinations. Among them GRS 1009-45 (Nova Velorum 1993) is one of the least known. It was detected by GRANAT/WATCH between the 12th and the 13th September 1993 [8]. It appeared firstly in the 20-60 keV band at 0.18 Crab reaching shortly after 0.7-0.8 Crab in the 8-60 keV total band of the instrument. BATSE detected the source at the same level between 20 and 500 keV with a power law with photon index $\alpha \sim -2.5$. About a month later the TTM team established that the source spectrum was similar to the ones of GS2000+25 and Nova Muscae 1991 (GRS1124-68), both dynamically proven Black Hole candidates

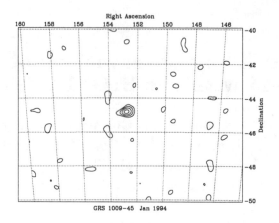

FIGURE 1. Image of the sky region observed by SIGMA

[7]. In the 2-200 keV band, the object presented a soft component with a black body spectrum with kT~0.52 ±0.03 keV which dominated between 2 and 10 keV. At higher energies the spectrum was well approximated by a power law $\alpha \sim -2.53 \pm 0.05$ [16].

) A week later ASCA confirmed this result giving a blackbody spectrum with kT ~ 0.6 keV and F_x ~ 0.8 Crab (1-10 keV) [17]. The absorption column density of this observation is 3 (-0.14,+0.22) × 10^{21} cm^{-2} (Ueda Y., private communication). This value points toward a relatively nearby source with a distance of a few kiloparsec.

The optical identification was performed by [1]: they identified the source with a V~14.6 late-G/early K type star. Very recently [14] it has been reported the discovery of an orbital, ellipsoidal modulation with an amplitude of 0.23 magnitudes and a period of 6.86 ± 0.12 hrs. No mass function was however derived as there is no available radial velocity curve for the secondary. High energy (>20 keV) monitoring of the source have been performed by the BATSE which produced a light curve and spectral indices for the source in the 20-100 keV band during about 150 days following the first maximum [10]. The overall form of the light curve is similar to the Nova Muscae 1991 one, with two later maxima in mid-October (~ 30 days after the primary) and in early December (~ 85 days after the primary) even if the Nova Muscae maxima occurred later [10]. The spectrum softens during the primary rise reaching $\alpha \sim -2.5$. The December maximum appears to be harder with $\alpha \sim -2.1$

and a similar behavior is seen in January observations.

OBSERVATIONS, DATA ANALYSIS AND RESULTS

The French coded mask telescope SIGMA provides high resolution images in the hard-X/soft γ-ray band from 30 keV to 1300 keV, with a typical angular resolution of 15' and a 20 hour exposure sensitivity of $\sim$ 26 mCrab in the 40-150 keV band [11]. Launched the 1st December 1989 onboard the Russian Granat Space observatory, SIGMA detected a fair number of galactic X-ray novae during its seven year lifetime [4,13,19,21,22]. This class of sources is one of the primary targets of this telescope due to its high precision localization capability unprecedented in this energy band.

The observations we report were performed on the 11th and 12th of January 1994 (JD 2449364-5). The campaign consisted of two distinct observation sessions with the telescope pointed in the source direction, both performed in spectral-imaging mode [11], for a total effective time of $\sim$ 35 hours.

In figure 1 is shown the source in the 40-75 keV energy band, GRS 1009-45 appears at a 6σ confidence level. To estimate the best source position, we selected data in the 60-120 keV energy band thus making a compromise between the higher statistics of the low energy band and the narrower instrument PSF at higher energies. A least square fit taking into account the instrumental PSF gives a source position at 10h 11m 26.4s($\pm$ 4'), $-$ 44 ° 49' 20"($\pm$ 4'), in complete agreement with the optical counterpart position [1,2].

Table 1 shows the flux intensities of the source during the two observing sessions. The second observing session displays a considerable rise in the hard band flux from $\sim$ 40 mCrab to $\sim$ 100 mCrab. However, the flux in the soft band is constant within the errors, and the hardness ratios of the two observing sessions are respectively HR(691) = 0.69 $\pm$ 0.4 and HR(692) = 1.53 $\pm$ 0.5. So there is no measurable spectral change of the source between the two observations. Given the relative faintness of the source, we consider this difference as not significative and the source as constant during the whole observation.

We thus summed the spectra obtained during the two observations. and performed a spectral fit to the data using a power law model. The average spectrum is well fitted with a power law with spectral index $\alpha = -$ 1.9 $\pm$0.4 and an integrated energy flux $F_x \sim$ 6.9 ($\pm$ 1.7) $\times$ 10^{-10} ergs cm^{-2} s^{-1} in the 40-150 keV energy band. The resulting spectrum is shown in Figure 2 along with a power law best fit.

The distance to the source has been estimated with optical observations by [2] as being between 1.5 and 4.5 kpc. On the basis of the hard X-ray outburst luminosity it has been proposed a distance of 2.6 kpc with a 100 % uncertainty [22] fully compatible with the optical one. The previously quoted hydrogen absorption column density agrees with these estimations. Adopting these

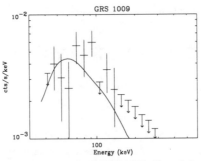

FIGURE 2. X-ray spectrum of GRS 1009-45 as detected by SIGMA.

TABLE 1. Detected flux of GRS 1009-45. Errors quoted are 1σ errors.

Obs. Session	Energy Band	Fluxes (mCrab)	Errors
691	40-75 keV	57	14.7
691	75-150 keV	39.4	20.4
692	40-75 keV	63.9	15.8
692	75-150 keV	97.8	22.7

estimates we obtain an X-ray luminosity L_X(40-150 keV) $\sim 10^{35}$-10^{36} ergs^{-1}. This hard X-ray luminosity is similar to the Nova Muscae 1991 one as detected by SIGMA about 150 days after the outburst, L_x(40-150 keV) $\sim 10^{35}$ erg s^{-1}. [6,15].

DISCUSSION

X-ray Nova Velorum 1993 is a soft X-ray transient which displayed during its outburst a fair number of characteristics in common with well known Black Hole Candidates. TTM and ASCA observations during the outburst showed that it displayed (so-called) ultrasoft X-ray emission along with an hard power law spectrum extending up to $\sim$ 500 keV. This spectrum is thought to be a good Black Hole indicator even if it is not considered to be conclusive [18]. Our observation showed that the source emitted hard X-rays up to $\sim$ 150 keV more than 100 days after the primary outburst.

With this observation we monitored Nova Velorum 1993 during the final phases of its third outburst. Its luminosity was similar to the Nova Muscae 1991 one as detected with SIGMA about 150 days after its primary outburst. The hard X-ray spectral index that we detected, $\alpha = -1.9 \pm 0.4$, is somewhat harder than the spectral index measured during the primary outburst ($\alpha = -2.5 \pm 0.03$). It is compatible with the spectral index measured by BATSE near the beginning of the second maximum ($\alpha = -2.12 \pm 0.07$, [10]).

A similar phenomenon possibly appeared also in Nova Muscae 1991 for which a harder x-ray spectrum was reported [10] during the second outburst. A similar average spectral hardening was detected by the ART-P and SIGMA instruments on board the GRANAT observatory [3].

We conclude that the hard X-ray emission of Nova Velorum 1993 looks similar to the ones of well established Black Hole candidates, especially Nova Muscae 1991.

We have shown with the result of SIGMA observations that GRS1009-45 was active over 100 keV about 120 days after the primary outburst. Its hard X-ray luminosity was comparable to the one of Nova Muscae 1991 $\sim$ 150 days after the outburst. Its hard X-ray spectrum was on average somewhat harder than the one detected during the primary outburst.

If we put our results together with all the other X-ray characteristics displayed by this source during its active phase, mainly the outburst ultrasoft spectrum and the 20-100 keV light curve, we believe that our observations strengthen the case for GRS 1009-45 to be a Black Hole X-ray Nova.

REFERENCES

1. Della Valle M. & Benetti S., IAUC **5890** (1993)
2. Della Valle M. et al., 1997 *A&A*, **318**, 179
3. Gil'fanov M. et al., 1993 *A&AS*, **97**, 303
4. Gil'fanov M. et al., in "The lives of the Neutron Stars" A.Alpar, Kiziloglu, U., van Paradijs, J. (1993)
5. Goldwurm A. et al., 1992, *ApJ*, **389**, L79
6. Goldwurm A. et al., 1993, *ApJS*, **97**, 293
7. Kaniowsky A., Borozdin K., Sunyaev R.,IAUC **5878** (1993)
8. Lapshov I., Sazonov S., Sunyaev, R., IAUC **5864** (1993)
9. Lapshov I. et al., 1994 *Ast. Lett* 20, 205
10. Paciesas W.S. et al., in "The Gamma ray sky as seen with Compton GRO and SIGMA" M.Signore, P.Salati,G.Vedrenne eds., Kluwer Academic Publisher, The Netherlands (1995)
11. Paul J. et al. Adv. Space Res., **11**, (8)2, (1991)
12. Remillard R., McClintock J., Baylin C. *ApJ*, **399**, L145(1992)
13. Roques J.P. et al. *ApJS*, **92**, 451(1994)
14. Shahbaz T. et al. *MNRAS*, **282**, L47(1996)
15. Sunyaev R. et al. *ApJ*, **L389**, 75(1992)
16. Sunyaev R. et al. Astronomy Letters **20**, 777 (1994)
17. Tanaka Y., IAUC **5888** (1993)
18. Tanaka Y. & Shibazaki N. *ARA&A*, **34**, 607(1996)
19. Trudolyubov et al. PAZh **22**, 740 (1996)
20. Tsunemi H. et al. *ApJ*, **337**, L81(1988)
21. Vargas M. et al. *A&A*, **313**, 828(1996)
22. Vargas M. et al. *ApJ*, **476L**, 23(1997)

Relativistic effects in the X-ray spectra of the Black Hole Candidate GS 2023+338

P. T. Życki*, C. Done* and D. A. Smith[†]

Department of Physics, University of Durham, South Road, Durham DH1 3LE, U.K.
[†]*Department of Physics and Astronomy, University of Leicester, Leicester, U.K.*

Abstract.
We present results of spectral analysis of *Ginga* data obtained during the decline phase after the 1989 outburst of GS 2023+338 (V404 Cyg). Our analysis includes detailed modelling of the effects of X-ray reflection/reprocessing. We have found that (1) the contribution of the reprocessed component (both continuum and line) corresponds to the solid angle of the reprocessor as seen from the X-ray source of $\Omega \approx (0.4 - 0.5) \times 2\pi$, (2) the reprocessed component (both line and continuum) is broadened ("smeared") by kinematic and relativistic effects, as expected from the accretion disk reflection. We discuss the constraints these results give on various possible system geometries.

INTRODUCTION

Some of the strongest evidence for the existence of accretion disks around black holes has come from X-ray observations of the relativistically smeared iron Kα line profile in Active Galactic Nuclei (e.g. [14]). This fluorescence line is produced by hard X–ray illumination of the accreting material, and the combination of high orbital velocities and strong gravity in the vicinity of a black hole gives the line a characteristically skewed, broad profile [3]. A reflected continuum should also accompany this line (e.g. [8]) and the amplitude and shape of both reprocessed components give constraints on the solid angle, Ω, subtended by the accreting material, its inclination, elemental abundance and ionization state. The Black Hole Candidates (BHC) show many X–ray spectral similarities to AGN, plausibly because both involve the same physical processes of disk accretion onto a black hole [15,4].

Many Soft X-ray Transients (SXT) are known to be BHC (see [13] for recent review). A number of these systems were observed by *Ginga* and the spectra obtained are amongst the best in which to investigate the overall effects of

X-ray reprocessing.

In this contribution we present results of spectral analysis of data obtained during the 1989 outburst of GS 2023+338. We have selected two data sets where there is little short time-scale spectral variability, obtained on June 20 and July 19-20 (one and two months after the outburst). We analyse these using models of X-ray reprocessing that consistently connect the properties of the iron Kα line with the properties of the reflected continuum. We show that the reflected component is present in the spectra and that it is smeared by kinematic and relativistic effects as expected from an accretion disk reflection.

MODEL

Main model components are a primary power law and its Compton reflection from a possibly ionized medium including consistently the iron Kα line emission [1,16,9]. Ionization state is parameterised by the ionization parameter, $\xi \equiv L_X/nr^2$. We assume elemental abundances as in [10] but we let the iron abundance, [Fe], be free.

The reprocessed component can then be "smeared" to simulate the relativistic and kinematic effects of disk emission (e.g. [12]). The model is parameterised by the inner and outer radius of the disk and the form of the irradiation emissivity. We fix the form of the emissivity, $F_{irr}(r) \propto r^{-3}$ as expected from coronal illumination, fix R_{out} at $10^4 R_g$ ($R_g \equiv \sqrt{GM/c^2}$) and fit R_{in}.

The soft X-ray excess is modelled as the multi-temperature accretion disk spectrum. We assume that the inclination of the system is $i = 56°$ [11].

RESULTS OF MODEL FITTING

We begin with the simplest possibility, a power law spectrum with narrow line at 6.4 keV. We than add the reflected continuum, assuming first that it is un-ionized and non-smeared. In the next step we fit ξ and [Fe] and finally we introduce the effect of smearing.

For the June 20th data set the simplest model (model 0 in Table 1) gives $\chi_\nu^2 = 1.9$. Adding the reprocessed component is highly significant even in the simplest version (model A), $\chi^2 = 19.9/25$ d.o.f. ($\chi_\nu^2 = 0.80$). We note that the normalization of the reflected component is significantly smaller than 1, $f \equiv \Omega/2\pi = 0.46 \pm 0.04$.

The fit can be improved by allowing for ionized reflection (model B) but the ionization is rather weak. Allowing the iron abundance to be free does not improve the fit, however introducing the effect of smearing is highly significant ($\Delta\chi^2 = 8.9$).

TABLE 1. Results of model fitting

model	Γ	ξ	$[Fe]^a$	f	R_{in} (R_g)	EW (eV)	χ^2/d.o.f.
June data							
0	1.34 ± 0.02	–	–	–	–	90^{+23}_{-14}	46.8/25
A	1.66 ± 0.01	0(f)	1(f)	0.46 ± 0.04	–	–	19.9/25
B	$1.66^{+0.01}_{-0.03}$	0.1^{+4}	1(f)	$0.45^{+0.04}_{-0.06}$	–	–	15.2/24
C	1.64 ± 0.04	0(f)	1.2 ± 0.35	0.43 ± 0.06	–	–	19.1/24
D	1.65 ± 0.03	0.1^{+4}	$1.0^{+0.4}_{-0.2}$	0.44 ± 0.06	–	–	14.9/23
E	1.64 ± 0.04	0^{+5}	$1.5^{+0.7}_{-0.5}$	0.47 ± 0.06	25^{+26}_{-11}	–	6.0/22
July data							
0	1.45 ± 0.25	–	–	–	–	93 ± 14	57/25
A	1.72 ± 0.015	0(f)	1(f)	0.40 ± 0.05	–	–	34.6/25
B	1.68 ± 0.04	13^{+25}_{-12}	1(f)	0.32 ± 0.08	–	–	22.9/24
C	1.68 ± 0.03	0(f)	$1.7^{+0.8}_{-0.5}$	0.35 ± 0.05	–	–	26.9/24
D	$1.68^{+0.04}_{-0.06}$	0.1^{+40}	1.5 ± 0.6	$0.35^{+0.05}_{-0.10}$	–	–	21.5/23
E	$1.67^{+0.04}_{-0.05}$	0^{+20}	$2^{+1}_{-0.8}$	$0.37^{+0.05}_{-0.08}$	35^{+110}_{-21}	–	15.4/22

A similar progression in quality of the fit is given by the July spectrum (Table 1; Figure 1), where again the relativistic smearing effects are significantly present in the data.

DISCUSSION

The results of spectral modelling clearly show that the reprocessed component is present in these data, and that it is smeared as expected from relativistic effects in an accretion disk/black hole system. The reflected fraction for both continuum and line is roughly $\sim 0.5\times$ that expected from an isotropically illuminated flat disk. Thus the line is *not* depleted by Auger ionisation masked by relativistic smearing [12] and the Auger ionisation is ruled out by the low ionisation state of the disk. Similarly low covering fraction is also commonly seen in the persistent BHC: Cyg X–1 and GX 339-4 [1,4,15] so is *not* some time dependent effect of disk evolution in transient systems, but rather represents a significant geometrical constraint. Another constraint comes from the fact that the observed relativistic smearing is less than that expected from a central point source illuminating a flat disk which extends down to $6R_g$.

One possible geometry is a spherical or flattened X-ray source centered on the black hole. A spherical source with $\tau_T \gg 1$ gives $f = 0.5$ naturally [1]. However, here the source is probably marginally optically thin with $\tau_T \sim 0.5-2$. This can still give a reduction in f since there is a hole in the inner disk at $r \leq R_{in}$, so some of the photons escape without illuminating the disk. The relativistic smearing constraints then imply that the the source emissivity is less steep than $\propto r^{-3}$ and/or the inner disk disk truncates at $R_{in} > 6\ R_g$ and/or the outer disk flares. There are few real physical constraints on the source

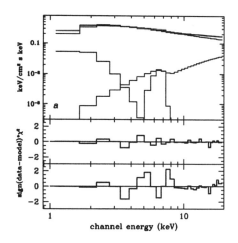
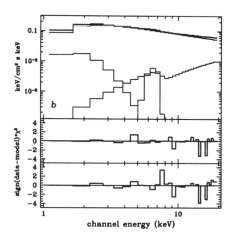

FIGURE 1. Best fit model spectrum (model E) (upper panels), their χ^2 residuals (middle panels) and the χ^2 residuals for models D i.e. without kinematic smearing. Panels *a* are for the June 20th data while panels *b* for the July data.

emissivity. Local release of gravitational energy in a Keplerian disk gives a luminosity $\propto r^{-3}$, but non–local mechanisms could also operate, possibly "flattening" the dependence. Outside of source the illumination is $\propto r^{-3}$ but within the spherical source the illumination produces an emissivity of $\propto r^{-2}$. This can reduce the most strongly smeared components to the observed level, though truncation of the inner disk radius may also be required. The geometry might then be in accord with recently proposed scenario of SXT evolution after outbursts [2] although the overall behavior of the source does not seem to follow the proposed scenario.

Small active regions on the disk perhaps powered by magnetic reconnection give another possible geometry [6]. However, these models give $f \sim 1$ unless the height of the reconnecting regions, h_{rec}, is large compared to the size of the disk. This would require that the reconnecting regions be concentrated along the inner disk radius R_{in}, and that $h_{\text{rec}} \sim R_{\text{in}}$, which would also go some way to satisfying the smearing constraints.

Another geometry that has been proposed is a continuous corona overlying the accretion disk [5]. A fraction $\sim 1 - \exp(-\tau/\cos i) \sim 0.5$ of the reflected spectrum is itself Compton scattered by the X–ray emitting corona, and loses its characteristic spectral shape [7]. The energy generation would be expected to be $\propto r^{-3}$, so the smearing constraints are a problem unless the inner disk truncates, or the outer disk flares.

Additional independent constraints on the geometry are provided by low normalization of the soft component and weak ionization of the reflecting medium. They both support the idea of the disk being truncated at $\sim 30\,R_g$ as then both the expected temperature of Shakura–Sunyaev disk emission component (for $\dot{m} = 0.01\,M_{\rm Edd}$ and $M = 10\,M_\odot$), $T \approx 0.15\,{\rm keV}$ and the ionization parameter, $\xi \sim 1$, would be in agreement with our results. However, low ξ may also mean that the disk is much denser than the SS solution due to e.g. coronal dissipation of energy.

CONCLUSIONS

We have performed spectral analysis of *Ginga* data of soft X-ray transient GS 2023+338 (V404 Cyg) obtained one and two months after its 1989 outburst. We have found that

- the Compton reflected continuum and iron fluorescent Kα line are present in the spectrum,
- the properties of both reprocessed components (normalizations, ionization parameters) are in agreement,
- the data require both reprocessed components to be broadened and smeared,
- the smearing is consistent with being due to reflection from a Keplerian disk,
- both the normalization of the reflected continuum and the amount of smearing constrain the geometry.

REFERENCES

1. Done, C. et al., *Ap. J.* **395**, 275 (1992).
2. Esin, A. A., McClintock, J. E., and Narayan, R. *Ap. J.* submitted, (1997).
3. Fabian, A. C. et al., *M.N.R.A.S.* **238**, 729 (1989).
4. Gierliński, M. et al., *M.N.R.A.S.* in press (1997).
5. Haardt, F., and Maraschi, L., *Ap. J.* **413**, 507 (1993).
6. Haardt, F., Maraschi, L., and Ghisellini, G., *Ap. J. Lett.* **432**, L95 (1994).
7. Haardt, F. et al., *Ap. J. Lett.* **411**, L95 (1993).
8. Lightman, A. P., and White, T. R., *Ap. J.* **335**, 57 (1988).
9. Magdziarz, P., and Zdziarski, A. A., *M.N.R.A.S.* **273**, 837 (1995).
10. Morrison, R., and McCammon, D., *Ap. J.* **270**, 119 (1983).
11. Pavlenko, E. P. et al., *M.N.R.A.S.* **281**, 109 (1996).
12. Ross, R. R., Fabian, A. C., and Brandt, W. N., *M.N.R.A.S.* **278**, 1082 (1996).
13. Tanaka, Y., and Shibazaki, N., *Annu. Rev. Astron. Astrophys.* **34**, 607 (1996).
14. Tanaka, Y. et al., *Nature.* **375**, 659 (1995).
15. Ueda, Y., Ebisawa, K., and Done C., *Publ. Astron. Soc. Jpn.* **46**, 107 (1994).
16. Życki, P. T., and Czerny, B., *M.N.R.A.S.* **266**, 653 (1994).

A search for gamma-ray flares from black-hole candidates on time scales of ~ 1.5 hours

R. van Dijk[4], K. Bennett[4], H. Bloemen[2], R. Diehl[1],
W. Hermsen[2], M. McConnell[3], J. Ryan[3], V. Schönfelder[1]

[1] *Max-Planck Institut für Extraterrestrische Physik, P.O. Box 1603, 85740 Garching, F.R.G.*
[2] *SRON-Utrecht, Sorbonnelaan 2, NL-3584 CA Utrecht, the Netherlands*
[3] *Space Science Center, Univ. of New Hampshire, Durham NH 03824, U.S.A.*
[4] *Astrophysics Division, ESTEC, P.O. Box 299, NL-2200 AG Noordwijk, the Netherlands*

Abstract. Strong short-lived flares from black-hole candidates have been detected in the hard X-ray regime and possibly also at γ-ray energies. Here we present a search for short-lived flares in the 0.75–30 MeV COMPTEL data. No flares are found during the 5 viewing periods considered, with typical upper limits of a few times the Crab flux.

INTRODUCTION

Galactic black-hole candidates (BHCs), which are binaries suspected to harbour a black hole, have been extensively studied during the last years. Most noticably, several X-ray transients have been the subject of correlated multi-wavelength campaigns, which yielded a wealth of observational data. At γ-ray energies, however, the amount of data gathered so far is limited. Although there is not much doubt that accretion onto a compact object is the main source of energy in BHCs, the detailed physics of such configurations are not yet completely clear. One of the basic uncertainties that still remains is whether or not the ions and electrons in the accretion flow near the compact object decouple to form a two-temperature plasma. If so, non-thermal processes such as π^0-decay and nuclear de-excitations may produce an appreciable γ-ray flux. The γ-ray domain is therefore important for assessing the possible existence of such two-temperature plasmas.

Gamma-ray flares from BHCs have been reported on several occasions in the past. Ling et al. (1987) observed a large ~ 1 MeV flare from Cyg X-1

VP	Start-end	l_Z	b_Z	Target BHCs
20.0	8658–8672	39.7	0.8	GRS 1915+105
36.5	8846–8854	168.2	-9.5	GRO J0422+32
318.0	9384–9391	68.5	-0.4	GRS 1915+105
336.5	9568–9573	340.4	2.9	GRO J1655-40
522.5	10248-10259	65.8	2.7	Cyg X-1

TABLE 1. The viewing periods (VPs) used in this analysis. Of these, VP 36.5, 336.5 and 522.5 were TOOs resulting from hard X-ray outbursts of GRO J0422+32, GRO J1655-40 and Cyg X-1 respectively. The other VPs were included to obtain results for GRS 1915+105 (BATSE detected hard X-ray flaring during TJDs 8748-8762 [Paciesas et al. 1996]). *Column 1*: the CGRO VP number; *column 2*: the start and end day [TJD ≡ JD−2440000.5]; *columns 3 and 4*: the Galactic longitude and latitude of the pointing direction; *column 5*: the main BHCs searched for in the field-of-view.

with a total flux of 1.6×10^{-2} photons cm^{-2} s^{-1}. Long-term observations with SMM have shown that MeV flares of this strength cannot be a frequent phenomenon (Harris et al. 1993). More recently, Boggs et al. (1996) reported on measurements with the balloon-borne HIREGS of a strong 1-10 MeV flare from GRO J1655-40, one of the Galactic superluminal BHCs. A search for MeV flares on time scales of two days in ~ 5 years of COMPTEL data did not reveal more evidence for occurrences of flares from GRO J1655-40 (van Dijk et al. 1997). The flare observed with HIREGS, however, lasted much shorter than two days ($\lesssim$ 1 hour; S. Boggs, private communication). In addition, BATSE has seen strong 20–100 keV flares from GRS 1915+105, the other Galactic superluminal source, on time scales of only 15 minutes (Paciesas et al. 1996). This prompted us to search for short-lived gamma-ray flares from black-hole candidates in the COMPTEL data, the results of which are presented here.

OBSERVATIONS AND ANALYSIS

The CGRO viewing periods used are listed in Table 1. The observations were divided into segments of ~ 1.55 hours, the duration of 1 orbit of the CGRO satellite. Due to excluded time periods such as SAA passages and incomplete real-time telemetry coverage, the actual lengths of the segments varied from ~ 1.5×10^3 to 5.2×10^3 seconds (left plot in Fig. 1). The number of events in each orbit is further influenced by event selections that effectively remove the Earth from the field-of-view (right plot in Fig. 1; see Section 2.3.3 in van Dijk 1996). For each of the time intervals separately, the exposure, geometry and event matrices were generated (Schönfelder et al. 1993). Instead of the standard energy selections, we used the 0.75–1.25, 1.25–3.0, 3.0–8.0 and 8.0–30.0 MeV energy ranges which yield a more balanced distribution of events

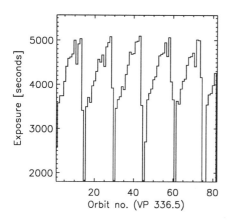 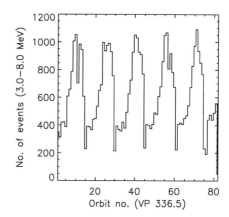

FIGURE 1. The exposure (left) and the number of 3–8 MeV events (right) for each orbit of the CGRO satellite during VP 336.5.

over energy. The analysis is based on the maximum likelihood ratio (MLR) method, which yields for each location within the search region an MLR value and a flux with error (de Boer et al. 1992). The square root of the MLR value is the detection significance for a source at that location. Here we typically used a circle of radius 20° as search region.

THE BACKGROUND MODEL

An often used data-space model for the instrumental-background radiation used in the analysis of COMPTEL data is obtained by applying a smoothing to the observed events in the 3-dimensional data space. Since COMPTEL data are usually dominated by instrumental background, this technique works quite well. Here, however, the total number of events in data space due to instrumental background is typically only a few hundred up to a few thousand and can be as low as ~ 50. Not only the sparseness of the data space but also the large relative contribution from any sources that might be detected were suspected to cause problems for this background model. We tested the applicability of the smoothing-based background model to sparse data spaces by adding simulated sources to the data spaces for the orbits in VP 336.5 and analysing them with the MLR method. In addition, we also tested a simple background model based on the geometry function, for which the scalings of the third data-space dimension (the scattering angle $\bar{\varphi}$) are taken from the observed event distributions. We conclude that the simple geometry-based background model works better for sparse data spaces than the usual smoothing-based background model. This can be inferred from Fig. 2, which shows the flux losses for both background models as a function of energy range

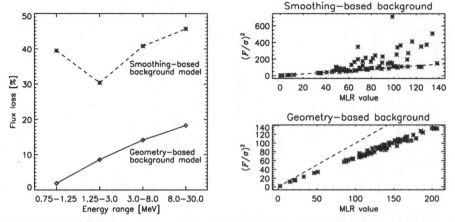

FIGURE 2. Results from the analysis of the data spaces for VP 336.5 with simulated sources added. Left: the average loss of source flux for each background model. Right: $(F/\sigma)^2$ for the 3-8 MeV energy range, where F is the flux and σ the error on the flux, versus the MLR value. In a proper MLR analysis (i.e., with an independent background model), these should be the same to first approximation (dashed lines).

(left plot) and the behaviour of the MLR values for different source strengths for the 3–8 MeV energy range (right plot; other energy ranges yield similar results). For the geometry-based background model, the flux losses are less and the MLR values do not scatter as much, although they do require to be calibrated.

RESULTS

The data for each CGRO orbit for the observations listed in Table 1 were analysed using a search region of 20° around the pointing direction. This way, we obtained N_i MLR maps and flux maps for each observation, with N_i the number of CGRO orbits during VP i. At the location of the black-hole candidates (Table 1), none of the detection significances were larger than $\sim 3\sigma$ (in the absence of trials, an MLR value of 9 formally indicates a 3σ detection for a known source). In Fig. 3, we show a typical example of the MLR values and fluxes obtained. For the other locations in the sky regions searched, the maximum MLR values correspond to detection significances of $\sim 3.8\sigma$. Given the large number of trials (# energy ranges × # orbits × # locations × # VPs) these are consistent with being statistical fluctuations.

Using least-squares fits to the N_i fluxes obtained for each location, we searched for time variability in the fluxes during the observations. No significant time variability was found. In cases when the small probability indicated a possible time variable signal, further analysis revealed that these small probabilities

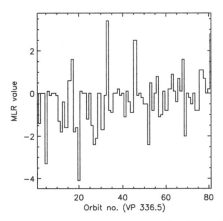 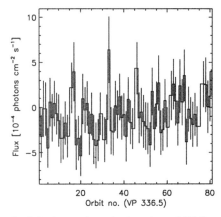

FIGURE 3. The MLR values (left) and fluxes (right) obtained at the location of GRO J1655-40 as a function of CGRO orbit number during VP 336.5.

were driven by significant negative fluxes. These could usually be explained by an incorrect instrumental-background model due to limited statistics.

Given the number of trials performed, we conclude that we did not detect any significant flaring emission in the COMPTEL data considered here. Typical upper limits during the CGRO orbits in these observations, in units of the Crab flux, are 10, 4, 4 and 8 in the 0.75–1.25, 1.25–3.0, 3.0–8.0 and 8.0–30.0 MeV energy ranges. Between 1 MeV and 10 MeV, these upper limits are below the fluxes reported for the flare from GRO J1655-40 observed with HIREGS in January 1995 and indicate that such strong short-time-scale flaring cannot be a phenomenon occurring frequently.

REFERENCES

1. Boggs S. et al., 1996, presentation at the High Energy Astrophysics Division (HEAD) meeting, San Diego, U.S.A.
2. de Boer H. et al., 1992, in *Data Analysis in Astronomy IV*, eds. V. Di Gesù et al. (New York: Plenum Press), Physical Sciences, Vol. 59, 241
3. Harris M.J. et al., 1993, ApJ 416, 601
4. Ling J.C. et al., 1987, ApJ 321, L117
5. Paciesas W.S. et al., 1996, A&AS 120, 205
6. Schönfelder et al., 1993, ApJS 86, 657
7. van Dijk R., 1996, PhD thesis, University of Amsterdam (available at ftp://astro.estec.esa.nl/pub/sciproj/rvdijk/thesis/thesis.html)
8. van Dijk R. et al., 1997, in *Proceedings 2nd INTEGRAL Workshop 'The Transparent Universe'* (16–20 Sep. 1996, St. Malo, France), ESA SP-382, 225

Physical Characteristics of the Spectral States of Galactic Black Holes

Juri Poutanen*, Julian H. Krolik[†] and Felix Ryde[‡]

*Uppsala Observatory, Box 515, SE-75120 Uppsala, Sweden
[†]Johns Hopkins University, Baltimore, MD 21218
[‡]Stockholm Observatory, SE-13336 Saltsöbaden, Sweden

Abstract. Using simple analytical estimates we show how the physical parameters characterizing different spectral states of the galactic black hole candidates can be determined using spectral data presently available.

SPECTRAL STATES OF GBHC

Galactic black hole candidates (GBHC) radiate in one of several spectral states, and some of them switch suddenly from one state to another. These states can be typified by their extremes: a "hard" state (HS, also called "low", because of relatively low flux in standard X-ray 2 – 10 keV band), and a "soft" state (SS, a "high" state with relatively strong 2 – 10 keV flux). The broad band spectra in both states can be described as the sum of a blackbody and a power-law with an exponential cut-off. The black body component (probably from the optically thick accretion disk) is more prominent in the SS, when it has a temperature of 0.3 – 1 keV. The lower temperature of the black body in the HS makes it difficult to detect, due to the interstellar absorption. The power-law energy index, α, is 1.0 – 1.5 in the SS, and roughly 0.3 – 0.7 in the HS [1,2]. Recent OSSE observations reveal that the cut-off energy, E_c, of the power-law is correlated with the spectral state; the power-law turns over at $\sim$ 100 keV in the HS, and $E_c \gtrsim 200$ keV in the SS [3,4]. There are also indications of that the amplitude of the Compton reflection "bump" increases when spectrum steepens [2].

The physical nature of the existence of the different spectral states and spectral transitions is not yet completely understood, although a number of models has been proposed (e.g. [5–8]). Recent progress on the theoretical side (we now understand much better how thermal Comptonization works, when the seed photons are produced mainly by reprocessing a part of the hard X-ray output, [11–13]) and the existence of broad band simultaneous spectral data for some of the sources (e.g., [4,9,10]) give us an opportunity to use the observed spectral characteristics in these states to determine the geometry and energy dissipation distribution in an accreting black hole system. The goal of the present investigation is to infer physical parameters of GBHC purely on the basis of *radiation* physics, this later can be used to guide efforts to obtain *dynamical* explanations for the changes in spectral state. A more detailed discussion can be found in [14].

ANALYTICAL ARGUMENTS

It is natural to attribute the two components with which the spectra are fitted to two physically related regions: an optically thick (quasi-thermal) accretion disk, which is responsible for the blackbody component, and an optically thin hot region (corona), which radiates the hard X-rays. The intrinsic dissipation rates in the "disk" and "corona" are L_s^{intr} and L_h, respectively. The size of the region over which the "disk" radiates most of its energy is R_s, and the size of the corona is R_h.

The hard X-rays are assumed to be produced by thermal Comptonization (e.g. [15]) of the seed photons that are partly created locally (by thermal bremsstrahlung or cyclo-synchrotron radiation [16]) and partly in the quasi-thermal region. The "soft" luminosity, L_s is partly due to local energy dissipation and partly due to reradiation of hard X-rays, created in the "corona".

The shape of the Comptonized spectrum produced by the "corona" may be described by two parameters: the power-law slope α, and the exponential cut-off energy E_c. Also two parameters (the effective temperature T_s and L_s^{obs}) define the soft part of the radiation. The relative ratio of the observed hard luminosity to the observed soft luminosity, $L_h^{\text{obs}}/L_s^{\text{obs}}$, and the magnitude of the reflection bump (described by the parameter, C, the fraction of solid angle that the optically thick region occupies around the "corona") complete the set of observables.

These phenomenological parameters are determined by two dimensional quantities, the total dissipation rate and R_s, and four dimensionless physical parameters: the ratio L_s^{intr}/L_h; the Compton optical depth of the "corona" τ_T; the fraction D of the light emitted by the thermal region which passes through the "corona"; and the ratio S of intrinsic seed photon production in the "corona" to the seed photon luminosity injected from outside. Another dimensionless parameter, the compactness $l_h \equiv L_h \sigma_T/(m_e c^3 R_h)$, may be used to determine the relative importance of $e^\pm$ pairs in the corona. In this context it is also useful to distinguish the net lepton Compton optical depth τ_p from the total Compton optical depth (including pairs), τ_T.

We will show how all these parameters, as well as several others of physical interest, may be inferred from observable quantities.

Some of the physical parameters of the system may be derived (or at least constrained) almost directly from observables. For example, the electron temperature in the corona (measured in electron rest mass units) is very closely related to the cut-off energy of the hard component:

$$\Theta \simeq f_x E_c/(m_e c^2), \tag{1}$$

where $f_x \sim 0.7$. Similarly,

$$L_h \simeq L_h^{\text{obs}}/\{1 - C[1-a]\}, \tag{2}$$

where a is the albedo for Compton reflection. The intrinsic disk luminosity is

$$L_s^{\text{intr}} \simeq L_s^{\text{obs}} - CL_h[1-a]. \tag{3}$$

Taking the local disk emission to be approximately black body, disk's inner radius is

$$R_s \simeq \left\{L_s^{\text{obs}}/4\pi\sigma T_s^4\right\}^{1/2}, \tag{4}$$

where T_s is the effective temperature at the inner edge. A number of correction factors (accounting for difference between color and local effective temperature [17] and incorporating the general relativistic corrections [18] etc.) were omitted in these formulae.

We next employ the two following analytic scaling approximations for thermal Comptonization spectra found by [12]:

$$D(1+S) = 0.15\alpha^4 L_h/[L_s^{\text{intr}} + CL_h(1-a)] \tag{5}$$

and

$$\tau_T = 0.16/(\alpha\Theta). \tag{6}$$

The first expresses how the power-law hardens as the heating rate in the corona increases relative to the seed photon luminosity; the second expresses the trade-off in cooling power between increasing optical depth and increasing temperature.

Finally, both C and D may be written in terms of R_h and R_s. If the corona is a sphere centered on the black hole, and the disk is an annulus of inner radius R_s and infinitesimal vertical thickness,

$$C \simeq R_h/(6R_s) \quad \text{and} \quad D \simeq (1/16)(R_h/R_s)^3. \tag{7}$$

The coefficient in the definition of C is most accurate when the emissivity is uniform within the sphere and it has order unity optical depth; the coefficient in the definition of D assumes the disk surface brightness is $\propto r^{-3}$ and its intensity distribution is isotropic. The relationship between R_h/R_s and C or D is somewhat different in other geometries, but the scaling tends to be similar. Equations (7) may be regarded as giving *independent* estimates of R_h/R_s. If C is used, this ratio is constrained by the observed magnitude of the Compton reflection bump; if D is used, it is determined by the observed hard and soft luminosities, and the slope of the hard X-ray power-law. The corona is assumed roughly spherical.

RESULTS AND CONCLUSIONS

The method described above can be applied to find physical parameters of the system $\Theta, \tau_T, R_s, D(1+S), R_h/R_s, L_s^{\text{intr}}/L_h^{\text{obs}}$ and l_h from the set of observables $\alpha, E_c, T_s, L_s^{\text{obs}}, C$ and $L_h^{\text{obs}}/L_s^{\text{obs}}$. Unfortunately, existing data are good enough only for a small number of objects. This method applied to Cyg X-1 (see [14]) shows that the inner edge of the cold disk shrinks by roughly a factor of 5 between the hard and soft states (from $\sim 40 r_g$ to $\sim 8 r_g$, assuming 10 $M_\odot$ black hole), while the corona shrinks in radial scale by only factor of two. In the SS, corona covers a sizable portion of the inner disk (see Fig. 1 and [8,9]).

Despite the change in coronal size and luminosity, the compactness of the corona almost does not change during the hard-soft transition. However, its optical depth drops by an order of magnitude from $\sim 1-2$ to ~ 0.1. Electron temperature of the corona is ~ 100 keV in the HS and probably is higher in the SS. In the HS the disk receives only a minority of the dissipation, $L_s^{\text{intr}}/L_h \sim 0.1$, but is the site of most of the heat release in the SS, $L_s^{\text{intr}}/L_h \sim 3$.

Detailed calculations based on the method by [19], that allow us to solve for the energy and electron-positron pair balance together with the self-consistent treatment of the Comptonization in the corona, confirm the conclusions made from simple analytical arguments. We also are able to show that in the case of thermal electrons there are very few $e^{\pm}$ pairs in the HS, but they can be comparable in number to the net electrons in the SS. In contrast the amount of pairs in the SS can be significant. This means that the fall in τ_p from the HS to the SS is greater than the fall in τ_T.

We close by noting that simultaneous broad band observations from soft

FIGURE 1. Geometry of the accretion flows around black holes in the hard (top) and soft (bottom) states.

X-rays up to gamma-rays are necessary in order to make strong conclusions regarding the geometry and physical conditions in the accretion flows around black hole.

REFERENCES

1. Tanaka, Y., and Levin, W. H. G., "Black-hole binaries" in *X-ray Binaries*, Cambridge: Cambridge University Press, 1995, pp. 126-174.
2. Ebisawa, K., Ueda, Y., Inoue, H., Tanaka, Y., and White, N. E., *ApJ* **467**, 419 (1996)
3. Phlips, B. F., Jung, G. V., Leising, M. D. et al., *ApJ* **465**, 907 (1996)
4. Grove J. E. et al., "Two gamma-ray spectral classes of black hole transients" in *The Transparent Universe*, Proc. 2nd INTEGRAL Workshop, ESA SP-382, 1997, pp. 197-200
5. Ichimaru, S., *ApJ* **214**, 840 (1977)
6. Chen, X., and Tamm, R. E., *ApJ* **466**, 404 (1996)
7. Ebisawa, K., Titarchuk, L., and Chakrabarti, S. K., *PASJ* **48**, 59 (1996)
8. Esin, A. A., McClintock, J. E., and Narayan, R., *ApJ* submitted (astro-ph/9705237)
9. Gierliński, M., Zdziarski, A. A., Done, C., Johnson, W. N., Ebisawa, K., Ueda, Y., Haardt, F., Phlips, B. F., *MNRAS*, in press
10. Zdziarski, A. A., Johnson, W. N., Poutanen, J., Magdziarz, P., and Gierliński, M., "X-rays and gamma-rays from accretion flows onto black holes in Seyferts and X-ray binaries," in *The Transparent Universe*, Proc. 2nd INTEGRAL Workshop, ESA SP-382, 1997, pp. 373-380.
11. Ghisellini, G., and Haardt, F., *ApJ* **432**, L95 (1994)
12. Pietrini, P., and Krolik, J. H., *ApJ* **447**, 526 (1995)
13. Stern, B. E., Poutanen, J., Svensson, R., Sikora, M., Begelman, M. C., *ApJ* **449**, L13 (1995)
14. Poutanen, J., Krolik, J. H., and Ryde, F., *MNRAS*, submitted
15. Shapiro, S. L., Lightman, A. P., Eardley, D. N., *ApJ* **204**, 187 (1976)
16. Narayan, R., and Yi, I., *ApJ* **452**, 710 (1995)
17. Shimura, T., and Takahara F., *ApJ* **445**, 780 (1995)
18. Zhang, S. N., Cui, W., and Chen, W., *ApJ* **482**, L155 (1997)
19. Poutanen, J., and Svensson, R., *ApJ* **470**, 249 (1996)

TEMPORAL CHARACTERISTICS OF COMPTON REFLECTION FROM ACCRETION DISKS

W. T. Bridgman[†], C. D. Dermer[‡], and J. G. Skibo[‡]

[†]*Goddard Space Flight Center/USRA*
Code 660.1, Greenbelt, MD 20771
[‡]*Naval Research Laboratory*
Code 7653, Washington, DC 20375-5352

Abstract. We treat the problem of time-dependent Compton reflection produced by a point source of radiation illuminating an infinite optically-thick disk of cold solar composition material. The problem is treated using a Monte-Carlo radiation transport simulation in the optically thick limit. The reflection produces a broad feature in the power spectrum at a frequency $f \sim c/h$, where h is the height of the photon source above the disk.

MOTIVATION

Reflection, as it applies to accreting systems such as black hole candidates [3-5] and AGN [12,6,7], has been studied extensively in the spectral domain [10] but very few studies explore its effects in the temporal domain. Initial analysis of analytic reflection models suggested presence of broad features in the power spectra with centroid frequencies $\sim c/h$ and widths $\sim c/2h$, where h is the height of the source above the disk. These features have some similarities to the low-frequency quasi-periodic oscillations (QPOs) reported for a number of black hole candidates.

We wish to determine what kinds of temporal signatures of reflection can be expected in more realistic accreting physical systems. This will provide additional evidence to validate or invalidate the reflection hypothesis. If the reflection hypothesis is valid, what physical parameters of the system can be determined from its temporal signatures?

METHODOLOGY

Attempts to extend the initial analytic models to more realistic physical scenarios were not successful. We therefore resorted to a Monte-Carlo analysis of the scattering process. In constructing the model, we have six major input assumptions: 1) a power law point-source spectrum, E^{-2}, 2) the source located at some height (which we will define as unity) above an optically thick disk, 3) disk composition is solar [8], 4) photoelectric absorption is included [1], 5) the iron fluorescense line is included [9,11], and 6) the light travel time into the disk is negligible.

The flux falls steeply at long times after the initial flash (roughly as t^{-2}) so we expect few photons to be scattered at large time lags. To improve the statistics at these large lag times (and therefore at the low frequencies of interest) we chose logarithmic time binning. This created another problem since Fourier transforms (and power spectra) require linear time binning. The resulting Monte-Carlo light curves are illustrated in Figure 1 for a range of observer angles. Near edge-on, the light curves exhibit a gentle rollover in power-law index from about zero to ~ 2.0. For a face-on observer, the power-law index starts in excess of 2.0 and then decreases to ~ 2.0. This same basic behavior by angle is preserved over all the energy bands examined (1-10 keV, 10-20 keV, 20-50 keV, 50-100 keV and 100-200 keV)

To get around the binning problem, we decided to fit the Monte-Carlo light curves with an analytic function, then resample the fitted function to obtain the linear time binning. Spline-curve fitting methods for this were unsuccessful. Through analysis of the Monte-Carlo light curves, we experimented with different analytic fitting functions. The best results were obtained using a function we call Disk Echo Model #3 which has the form:

$$flux(t) = \begin{cases} 0, & t < \tau \\ f_0\left(\frac{\tau}{t}\right)^q \exp\left\{-\frac{b}{c}\left[\left(\frac{\tau}{t}\right)^c - 1\right]\right\}, & t \geq \tau \end{cases} \quad (1)$$

where f_0 is the maximum flux; τ is the turn-on time, which is the same as the time of maximum flux for this model; q is the asymptotic power-law index; b is the difference between the asymptotic power-law index and the index at the turn-on time; and c is a time scaling index. We use the height above the disk to set our distance and time scale so the time, t, is in units of the height, h. If $h = 3 \times 10^5$ kilometers = 1 light-second, then the frequency unit is Hertz.

Once fitted for a given Monte-Carlo result, the function is then linearly resampled to produce the light curves for generation of power and lag spectra. When generating the linearly resampled curve, we include the contribution of the direct (unreflected) photons in the zero time delay bin, weighting them by their relative intensity. This is important since it is the correlation between the direct and reflected photon arrival times which create the QPO-like features. Samples of the resulting power spectra (Figure 2) and the auto-correlation functions (Figure 3) are presented below.

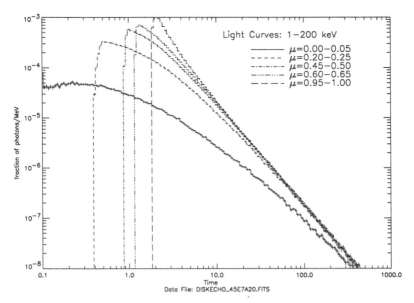

FIGURE 1. Sample light curves at different observer angles. We look at the entire energy band from 1-200 keV with observer angles ranging from $\mu = cos\theta = 0.0$ (the disk appears edge-on to the observer) to $\mu = cos\theta = 1.0$ (the disk appears face-on to the observer).

RESULTS & IMPLICATIONS

The frequency of the QPO-like feature is directly related to the distance of the emission region from the reflection region and the angle of the observer relative to the disk. It can therefore provide a sense of the scale of the system. The reflection-induced QPOs are not integrally spaced (as was predicted in the original analytic models). Their frequency and relative amplitude increase with increasing observer angle.

If reflection is indeed responsible for some of the spectral features observed in black hole candidates and AGN, then a temporal signature such as that described above should also be visible. Some recent temporal studies of GX 339-4 and GRO 1009-45 with ASCA and RXTE show some evidence of this behavior [2].

PROBLEMS AND AREAS FOR MORE WORK

The method selected for this analysis had some excellent successes but also exhibited some notable problems that need to be addressed. The discrete time binning can create large fluctuations in the frequency domain. This has also created a problem for performing lag analysis. The steep rise to maximum in

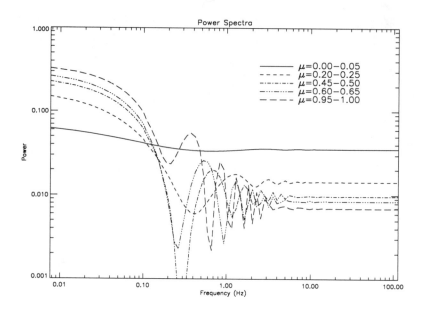

FIGURE 2. Power spectra by observer angle for energy band 3 (20-50keV). Logarithmic frequency rebinning has smoothed out higher order harmonics. The QPO-like features are most prominent when the disk is near edge-on to the observer. The power a high frequencies is related to the shot rate and the amplitude of the initial flash.

the light curves is only partially due to the angular binning so a more accurate model than Disk Echo Model #3 is desirable.

In addition to the problems created by the analysis, there are a number of improvements which can be made in the model. In our Monte-Carlo model, the disk was constructed with a discrete surface. What kind of effects would a density profile or atmosphere/corona create? Also, since the disk is optically thick and geometrically thin in the model examined, the time delays due to the travel of photons into the disk have been ignored. However, this effect may become important in the case of an optically thin disk or a disk with a substantial corona.

REFERENCES

1. Balucinska-Church, M. & McCammon, D. *ApJ* **400**, 699 (1992).
2. Dobrinskaya, J.Y. and Liang E., *in these proceedings* (1997).
3. Done, C.; Mulcahaey, J.S.; Mushotzky, R.F.; and Arnaud, K.A. *ApJ* **395**,275 (1992).
4. George, I.M. & Fabian, A.C. *MNRAS* **249**, 352–367 (1991).

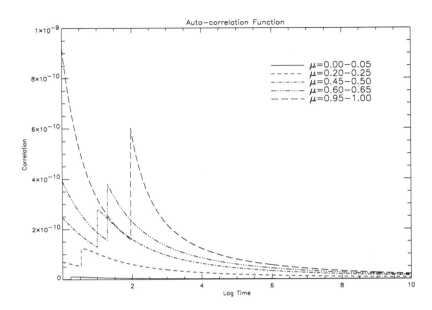

FIGURE 3. Auto-correlation function by observer angle for energy band 3 (20-50keV). The second peak at positive lags is due to the time gap between the initial and reflected photons.

5. Haardt, F.; Done, C.; Matt, G. & Fabian, A.C. *ApJ Letters* **411**, L95–L98 (1993).
6. Krolik, J.H. *Testing the AGN Paradigm: Proceedings of the 2nd Annual topical Astrophysics Conference.* pp.473–485 (1992).
7. Maisack, M.; Yaqoob, T.; Staubert, R.; Kendziorra, E. *Testing the AGN Paradigm: Proceedings of the 2nd Annual topical Astrophysics Conference.* pp. 362–365 (1992).
8. Morrison, R. & McCammon, D. *ApJ* **270**, 119 (1983).
9. Pounds, K. A.; Nandra, K.; Stewart, G.C.; George, I.M.; Fabian, A.C. *Nature* **344**, 132 (1990).
10. Tanaka, Y. *Iron Line Diagnostics in X-Ray Sources*, eds. A. Treves, G. C. Perola, & L. Stella, p. 98 (1991).
11. Zdziarski, A. A., Ghisellini, G., George, I. M., Svensson, R., Fabian, A. C., & Done, C. *ApJ* **363**, L1 (1990)
12. Zdziarski, A.A., *Advances in Space Research* **15**, (5)27–(5)35 (1995).

Temporal and Spectral Properties of Comptonized Radiation

Demosthenes Kazanas, Xin-Min Hua and Lev Titarchuk

LHEA, NASA/GSFC Code 661, Greenbelt, MD 20771

Abstract. We have found relations between the temporal and spectral properties of radiation Comptonized in an extended atmosphere associated with compact accreting sources. We demonstrate that the fluctuation power spectrum density (PSD) imposes constraints on the atmosphere scale and profile. Furthermore, we indicate that the slope and low frequency break of the PSD are related to the Thomson depth τ_0 of the atmosphere and the radius of its physical size respectively. Since the energy spectrum of the escaping radiation depends also on τ_0 (and the electron temperature kT_e), the relation between spectral and temporal properties follows. This relation allows for the first time an estimate of the accreting matter Thomson depth τ_0 independent of arguments involving Comptonization.

INTRODUCTION

The analysis of the energy spectra of the compact accreting objects has established the process of Comptonization by hot electrons as the main process by which these spectra are produced (see e.g. [7]). However, in order to completely determine the physical parameters of the systems, additional information is needed. These arguments have motivated the study of time variability of these sources.

The study of time variability of the sources, in particular Cyg X-1, has yielded a number of unexpected results: (a) The fluctuation power spectral densities (hereafter PSD) are generally power laws of indices $s \sim 1 - 1.5$ in the variability frequency ω (see e.g. [8]). The values of s are significantly flatter than those expected from exponential shots ($s \geq 2$). Furthermore, the turnover frequency, ω_c, is much smaller than those associated with the dynamics responsible for the emission. For Cyg X-1, $\omega_c \simeq 0.05$ Hz, a far cry from the kHz frequencies expected on the basis of dynamical considerations. (b) The studies [5] of the time lags between hard and soft photons as a function of the ω, has indicated that these lags decrease with increasing ω.

THE EXTENDED ATMOSPHERE MODEL

Motivated by the discrepancy of the above systematics, we present an model for the time dependent spectral formation of Comptonized radiation which can reproduce the basic observational features described above.

Our model assumes spherical accretion. We assume, in addition, that the accreting component behaves like a hot "atmosphere" of constant electron temperature $T_e \sim 50$ keV and, more importantly, with a density profile $n(r) \propto 1/r$ in radius, extending to $r_c \simeq 10^4 \times$ Schwarzschild radius. We assume that there is a source of soft photons within the spherical shock boundary of radius r_{sh}. The electron density is considered to be constant, n_+, inside this boundary and $n_+ r_{sh}/4r$ in the atmosphere. The physical size of this cloud is determined by the total optical depth τ_0.

We have calculated the response of this configuration to an impulsive input of soft photons within the radius of the shock [4]. The calculations were carried out by a Monte Carlo method described in [2]. Figures 1a and 1b show the resulting light curves at different energy bands for clouds with $\tau_0 = 1, 2$ and 3, $n_+ = 1.6 \times 10^{17}$ cm^{-3}, $r_{sh} = \tau_0 \times 10^{-4}$ l. s. and outer radius of the atmosphere $r_c = 5 \times 10^3 r_{sh} \simeq \tau_0$ 0.5 light seconds. It is apparent that their shapes have the form of power laws over the time range 10^{-3} s to ~ 1 s, followed by an exponential cutoff at times of order a few seconds. The exponential light curves from a cloud with uniform density $n = 2 \times 10^{14}$ cm^{-3}, radius $r = 1.5 \times 10^{10}$ cm ($\simeq 0.5$ l.s.) the same electron temperature T_e and $\tau_0 = 2$ are also shown (dotted curves).

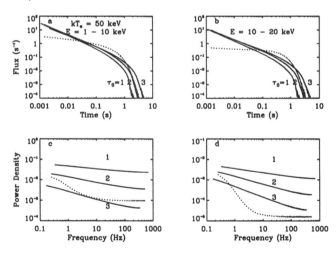

FIGURE 1.

The power law form of the light curves is the result of photons scattering in the extended atmosphere whose optical path has a logarithmic radial dependence. The slope of the the light curve for the uniform sphere of small scattering depth is close to zero (dotted curve) and that of the atmosphere considered here is close to -1 [4]. Therefore, measuring these light curves, e.g. by computing the resulting PSDs, can provide *a tool for uncovering the density profile of the atmosphere.*

As the optical depth of the atmosphere increases, the power law indices of the light curves evolve to smaller values. The flattening of the light curves is indicated in Figures 1a and 1b. This change in the light curve slopes reflects directly on the resulting PSD spectra, shown in Figures 1c and 1d. The flattening of the light curves leads to steepening of the PSD spectra, which are, however, flatter than those corresponding to a uniform source. Assuming that the temperature of the extended "atmosphere" is constant, the increase in optical depth will also lead to a harder energy spectrum for the emerging photons. Therefore, *there should exist well defined correlations between the slopes of the escaping photon spectra and those of the PSD.*

In Figure 2 we plot the energy spectrum (solid curve) resulting from the extended atmosphere with $kT_e = 50$ keV and $\tau_0 = 3$. It is almost identical to that from a uniform plasma cloud with the same electron temperature T_e but a different optical thickness $\tau_0 = 2$ (dashed curve). For comparison, we also plot the energy spectrum from a uniform cloud with the same temperature and $\tau_0 = 3$ (dotted curve). This spectrum is harder than the other two. This is because for the same total optical thickness ($\tau_0 = 3$), photons in the uniform cloud are harder to escape than in the $1/r$ atmosphere. This example shows clearly that *spectroscopic analysis alone can not provide complete information about the source structure* and to get a complete picture of the spectral formation it is necessary to analyze in conjunction the temporal as well as the spectral data.

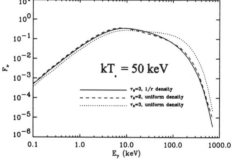

FIGURE 2.

Since it is believed that Comptonization is the main mechanism for the production of X-rays, measurements of the time lags between two different energies in the X-ray band should give us a direct estimate of the density of the region in which Comptonization takes place. The time lags are expected be of the order of msec for galactic objects for the uniform density hot clouds, considered exclusively to date in the literature. In addition, the time lags between soft and hard photons are constant and independent of the Fourier frequency ω [5], [3]. For the non-uniform density profile of the extended atmosphere discussed herein, the shape of the phase lag as a function of the variability frequency ω also changes. In Figure 3, we plot the phase lags of photons in the 10 – 20 keV band relative to those in 1 – 10 keV band, for the four cases depicted in Figure 1.

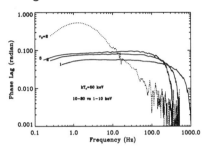

FIGURE 3.

As in Figure 1, the dotted curve corresponds the uniform cloud and its shape is similar to those displayed in [3], except that here the curve extends to higher frequencies with a roughly constant slope instead of a sharp drop, simply because in present analysis we use much finer time bins. The phase lags corresponding to the $1/r$ atmosphere configuration (solid curves) are flat across the entire range of frequencies. This type of phase lag is hinted in the GINGA observations of Cyg X-1 [5] and also seen in the recent observations of the high state of Cyg X-1 [1].

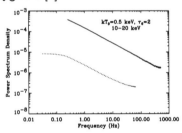

FIGURE 4.

The PSD depends strongly on the energy of the source photons [3]. In Figure 1, the source photons have a blackbody temperature 2 keV. In Figure 4, we present the PSD curve (solid curve) in the energy range 10 – 20 keV resulting from the same configuration as that of $\tau_0 = 2$ in Figure 1, but with the source photons at blackbody temperature $kT_0 = 0.5$ keV. It is seen that PSD with lower source photon energy is steeper.

In Figure 4, we also present one more PSD (dotted) curve, corresponding to the same light curve as the solid one but for different time resolution. For the solid curve the light curve is calculated over a range of 4 seconds in 4096 bins. For the dotted curve, the light curve is calculated over 32 seconds in the same number of bins. It is seen that the PSD turns flat for frequencies below $\omega_c \sim 0.25$ Hz. The time scale corresponding to $1/\omega_c = 4$ seconds is actually the time scale of the light curve beyond which it drops virtually to zero. Thus we have found a possible physical meaning for the shoulder frequency ω_c in the PSD curves, commonly seen in the PSD of many sources, namely, it indicates the time scale of the light curve, or the size of the extended $1/r$ atmosphere. The "white noise" below ω_c reflects the average frequency of the shots while the power-law above ω_c reflects the time structure within one single shot.

DISCUSSION

Much of the present discussion relies on the specific form of the density $n(r)$ as a function of radius r. Clearly, this is not the free-falling solution customarily used in association with spherically symmetric accretion (e.g. [6]), so a few comments are in order. If the presence of the extended atmosphere is due to the effects of preheating, then at the edge of this atmosphere, one would expect that the random and rotational velocities of matter to be comparable and that its subsequent evolution in radius to be predominantly governed by the removal of its angular momentum. The agent responsible for this process is considered herein to be the interaction of the fluid with the photons produced near the Schwarzschild radius. The fluid interacts and transfers its angular momentum to photons. [4] shows that, the specific density profile we have assumed guarantees that the removal of the angular momentum of the accreting matter by the photons can in fact proceed on dynamical time scales and thus preserve the assumed density profile. Most likely, this process is possible only for the density profile prescribed above ($n(r) \propto 1/r$), since this is the only profile which allows for significant photon scattering, and hence removal of angular momentum, from a large range of radii.

REFERENCES

1. Cui, W. et al. 1997, ApJ, 484 in press
2. Hua, X.-M. 1997 Computers in Physics, in press
3. Hua, X.-M. & Titarchuk, L. 1996 ApJ, 469, 280
4. Kazanas, D., Hua, X.-M. & Titarchuk, L. 1997 ApJ, 480, 735
5. Miyamoto, S. et al., 1988, Nature, 336, 450
6. Narayan, R. & Yi, I., 1994, ApJ, 428, L13
7. Sunyaev, R. A. & Titarchuk, L.G., 1980, A&A, 86, 121
8. van der Klis, M. 1995, in: X-ray Binaries, eds. W. Lewin, J. Van Paradijs and E. Van Den Heuvel (University Press, Cambridge) p. 252

Phase Difference and Coherence as Diagnostics of Accreting Sources

Xin-Min Hua, Demosthenes Kazanas and Lev Titarchuk

LHEA, NASA/GSFC Code 661, Greenbelt, MD 20771

Abstract. We present calculations of the time lags and the coherence function of X-ray photons for a novel model of radiation emission from accretion powered, high-energy sources. Our model involves only Comptonization of soft photons injected near the compact object in an extended but non-uniform atmosphere around the compact object. We show that this model produces time lags between the hard and soft bands of the X-ray spectrum which increase with Fourier period, in agreement with recent observations; it also produces a coherence function equal to one over a wide range of frequencies if the system parameters do not have significant changes, also in agreement with the limited existing observations. We explore various conditions that could affect coherence functions. We indicate that measurements of these statistical quantities could provide diagnostics of the radial structure of the density of this class of sources.

INTRODUCTION

It appears strange that the X-ray fluctuation power spectral densities (PSD) of accreting compact sources contain most of their power at frequencies $\omega \lesssim 1$ Hz, far removed from the kHz frequencies expected [6]. More seriously, Miyamoto et al. ([8], [9]) studied the time lags between the soft and hard photons in the X-ray light curves of Cyg X-1, using the GINGA data. It was shown that the hard time lags increase roughly linearly with the Fourier period P from $P \lesssim 0.1$ sec to $P \sim 10$ sec. These lags are very hard to understand in a model where the X-ray emission is due to soft photon Comptonization in the vicinity of the compact object. In such a model, they should simply reflect the photon scattering time in the region or $\simeq$ msec.

More recently, Nowak & Vaughan [11] and Vaughan & Nowak [12] have brought attention to another statistic, namely the coherence of the X-ray light curves. These authors computed the coherence function for the GINGA data of Cyg X-1 and GX 339-4 and found it to be equal to one for both sources over the frequency range 0.1 - 10 Hz. The coherence function for Cyg X-1 has also been computed with the more recent data of RXTE [2] and it was found to be equal to one up to frequency $\simeq 20$ Hz. this fact is considered to be quite surprising, since most of the models have coherence functions substantially smaller one.

Motivated by the discrepancy between the expected and the observed variability behavior of accreting compact sources, Kazanas et al. [6], proposed that the Comptonization process takes place in an non-uniform "atmosphere" which extends over several decades in radius. It was shown that this model can account for the form of the observed PSDs, the energy spectra and at the same time predicts a correlation between the slopes of the PSD and the energy spectra.

TIME AND PHASE DIFFERENCES

The model [6] considers Comptonization in a cloud of constant temperature but non-uniform density of profile $n(r) \propto r^{-1}$, which extends over several decades in radial coordinate from the compact source, r. In order to explore the Comptonization in clouds with density configurations such as these, we developed a Monte Carlo method which can treat photon propagation and Compton scattering in inhomogeneous media [3].

The parameters of the calculations were so chosen as to provide qualitative agreement with the spectrum of Cyg X-1 obtained by BATSE [7] at its γ_0 state. The spectrum is consistent with Comptonization in an electron cloud of temperature $\sim$ 110 keV and Thomson optical depth $\sim$ 0.45. We employ a spherical model with temperature $kT_e = 100$ keV. The source has a radius $r_2 \approx 1.5$ light seconds and consists of a central core of radius r_1 and an extended "atmosphere" with density profile $n(r) = n_1 r_1/r$ for $r_1 < r < r_2$. For the region $r < r_1$ we assume that it has a uniform density $n = n_1$, and that a soft photon source of blackbody spectrum at temperature 0.2 keV is located at its center.

The density gradient of the extended atmosphere can significantly affect the resulting photon spectrum [6]. We found that a total optical depth $\tau_0 = 1$ for our non-uniform configuration produces as good a fit to the BATSE data as the uniform one used by Ling et al. [7]. Furthermore, we assume three values for r_1: 2.4×10^{-2}, 2.4×10^{-3}, 2.4×10^{-4} light seconds. These conditions suffice to determine the density n_1.

With the above configurations, we calculated the energies and arriving times of the photons emerging from the cloud. The photons are collected in the energy bands $2-6.5$ and $13.1-60$ keV. Based on the light curves so obtained, we calculated the phase and time lags of the higher energy band with respect to lower one for clouds with the three values of r_1 [4]. The resulting time lags (solid curves) as well as the corresponding phase lags (dotted curves) as a function of Fourier frequency are shown in Figure 1. The time lags are power-laws of indices $\lesssim 1$, in rough agreement with observations of [2] and [8], [9].

The time lags as a function of Fourier frequency depends on the specific form of the radial profile of density of the extended "atmosphere". We also display

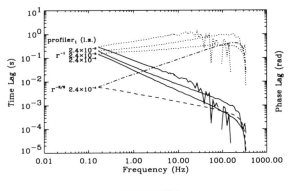

FIGURE 1.

in the same figure the phase and time lags resulting from an atmosphere with density profile $n(r) \propto r^{-3/2}$. It is evident that, *the observations of frequency dependent hard X-ray lags not only argues in favor of the presence of the extended atmosphere, but also point to their study as a means for probing its detailed density profile* [4].

THE COHERENCE FUNCTIONS

We assumed that the configuration changes in our model during an observation would lead to loss of coherence [4]. To simulate this evolution, we used the light curves in the energy bands $2 - 6.5$ and $13.1 - 60$ keV resulting from two configurations represented by different parameters of our model. We produced a group of eight light curves for each energy band of each configuration, obtained with different initial random number seeds. The difference between each light curve and the average of the eight curves in the same group is taken as the noise. For the two energy bands, labeled 1 and 2 respectively, we used the light curves and their respective noises to compute the coherence function defined in Equation (2) of [12]. The power spectra $|S_i|^2$ in this equation should be understood as being noise-corrected. The averages in the equation are taken over the 16 pairs of "measurements".

We first computed the coherence function of our model by averaging over 16 pair light curves from two identical configurations. The coherence function was 1 across the entire frequency range, as expected. We then computed coherence functions for a variety of configuration evolutions with the results presented in Figure 2. It is seen that changes in the parameters do result in loss of coherence. The four thick curves represent the evolution of the configuration starting from one of those described in the above section, namely that with $r_1 = 2.4 \times 10^{-3}$ light seconds to one of the following configurations:

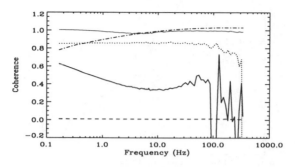

FIGURE 2.

(a) One with n_1 and τ_0 increased by a factor of 5 with the density profile unchanged (solid curve). (b) One with electron temperature decreased from 100 keV to 50 keV (dotted curve). (c) Configurations with different energy of initial soft photons. The dashed curve is obtained between the original initial configuration ($kT_e = 0.2$ keV) and one with source photons of a single energy $E_0 = 13.1$ keV. The dash-dotted curve indicates the final configuration has source photons at the blackbody temperature $kT_0 = 4$ keV.

The thin curve in Figure 2 represents an evolution between the configuration with $r_1 = 2.4 \times 10^{-2}$ l. s. to that with $r_1 = 2.4 \times 10^{-4}$ l. s. The coherence is virtually unity over the entire range of frequencies under consideration, although the the configuration changes. The cause of this is: The light curves represent two independent time series, say q and r. We rewrite the Equation (10) in [12] in terms of the ratios of PSD in the two energy bands $\alpha_1 = |Q_1|/|R_1|$ and $\alpha_2 = |Q_2|/|R_2|$.

$$\gamma_I^2 = \frac{1 + \alpha_1^2 \alpha_2^2 + 2\alpha_1 \alpha_2 \cos(\delta\theta_r - \delta\theta_q)}{(1 + \alpha_1^2)(1 + \alpha_2^2)}. \tag{1}$$

The ratios α_1 and α_2 are plotted in Figure 3 as thin solid and dashed curves respectively. For $\omega \lesssim 1$ Hz, $\alpha_1 \approx \alpha_2 \approx 1$. From Figure 1, The difference in the phase lag is small (Figure 1) $\delta\theta_r - \delta\theta_q \lesssim 10°$. Equation (1) then suggests that the coherence should be $\gamma_I^2 \simeq 1$. At higher frequencies, however, $\alpha_1 \neq \alpha_2$ and $\delta\theta_r - \delta\theta_q$ could be as great as 50° (see Figure 1). But both α_1 and $\alpha_2 \ll 1$, yielding $\gamma_I^2 \simeq 1$. Therefore the coherence at high frequencies is only superficial, since neither of the conditions outlined in [12] are satisfied. In a similar way, one can examine the true causes for the coherences or lack thereof as displayed in Figure 2.

Therefore, The measurements of coherence close to one, such as those in [12], may indicate the constant state of the responsible mechanism over the observation time or changes in the system to which the coherence statistic is insensitive.

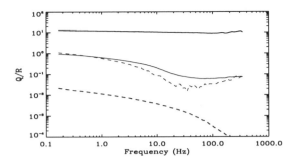

FIGURE 3.

CONCLUSIONS

1. Our model leads to the hard photons lag increasing with the Fourier period. This behavior is distinctly different from that of Compton scattering in clouds of uniform density but in agreement with observations. Furthermore, the observed lags are consistent with a density profile $n(r) \propto r^{-1}$. Thus measurements of the lag dependence on frequency could be used in the deconvolution of the density profile of the atmosphere from observations.

2. Our model produces coherence close to one if the parameters of the system remain constant over long time scales (hours). The coherence function can be reduced to less than one by changing the macroscopic parameters of the Comptonization cloud and/or the energy of the source photons. However, the inverse of this statement is not true.

REFERENCES

1. Cui, W. et al. 1997a, ApJ, 474, L57, also astro-ph/9610072
2. Cui, W. et al. 1997b, ApJ, submitted
3. Hua, X.-M. 1997 Computers in Physics, submitted
4. Hua, X.-M., Kazanas & Titarchuk, L. 1997 ApJ, 482, L57
5. Hua, X.-M. & Titarchuk, L. 1996 ApJ, 469, 280
6. Kazanas, D., Hua, X.-M. & Titarchuk, L. 1997 ApJ, 480, 735
7. Ling, J. C. et al., 1997, ApJ, in press
8. Miyamoto, S. et al., 1988, Nature, 336, 450
9. Miyamoto, S. et al., 1991, ApJ, 383, 784
10. Nowak, M. A., 1994, ApJ, 422, 688
11. Nowak, M. A., & Vaughan, B. A., 1996, MNRAS, 280, 227
12. Vaughan, B. A., & Nowak, M. A., 1997, ApJ, 474, L43 (VN97)

Global Spectra of Transonic Accretion Disks

E. Liang[†], C. Luo[†]

[†]*Rice University, MS108, Houston, TX 77005-1892*

Abstract. We compute the global spectral output of transonic disk solutions of Luo and Liang. We show that the spectrum is sensitive to both the angular momentum and viscosity of the disk. We discuss how the hard and soft spectral states of Galactic black holes may be caused by changes in these parameters.

I THE MODEL

Using the global structure of the transonic disk solutions obtained by Luo and Liang [1], we construct the global disk spectral output by dividing the solution into finite radial zones. We first compute the spectral flux from each ring and then sum over all rings with the area of each ring as a weighting factor. The Luo and Liang solution gives only the ion temperature, density and scale height. Hence we estimate the electron temperature assuming Coulomb coupling between ions and electrons. When the effective absorption depth is large we assume a local blackbody spectrum with the temperature fixed by the flux. When the effective absorption depth is less than unity we compute the local spectrum assuming Comptonization and use the Sunyaev-Titarchuk formula [2].

II RESULTS

Here we illustrate spectral output for sample values of the disk parameters. Figures 1 and 2 compare the spectrum of a low angular momentum J disk with that of a high (i.e. close to Keplerian) angular momentum disk. Note that the low-J disk spectrum is unsaturated, whereas the high-J disk spectrum consists of a large blackbody component plus a saturated Compton component. Such a spectrum has not been observed for Galactic black holes.

For a given J and accretion rate, it turns out that the disk structure is also sensitive to the viscosity alpha. As alpha is decreased from unity the

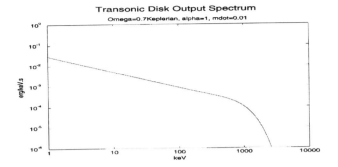

FIGURE 1. Global transonic disk output for J=0.7 Keplerian, alpha=1, and accretion rate=0.01 Eddington

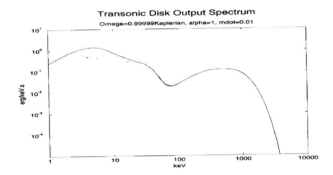

FIGURE 2. Same as Fig.1 except for J=0.99999 Keplerian

hot optically thin inner region grows in size and gets thinner, while the outer optically thick disk grows thicker. As a result the disk spectrum transitions from a single-component Comptonized spectrum (Fig.3) to a two-component spectrum with a blackbody component plus a steep power-law (Fig.4). This strongly resembles the transition from the low-hard state to the soft-high state of many Galactic black holes [3].

III ACKNOWLEDGEMENTS

This work was supported in part by NASA grant NAG5-3824.

REFERENCES

1. Luo, C. and Liang, E., *Ap. J.* to appear (1997).
2. Sunyaev, R. and Titarchuk, L., *Ast. Ap.* **86**, 121 (1980).

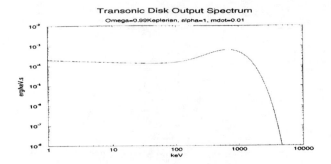

FIGURE 3. Same as Fig.1 except for J=0.99 Keplerian

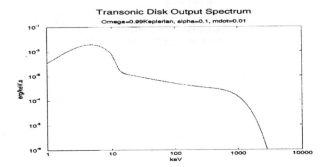

FIGURE 4. Same as Fig.3 except for alpha=0.1

3. Liang, E., *Phys. Rep.* to appear (1997).

Horizontal Branch Oscillations from Black Hole Candidates

Xingming Chen[1], Ronald E. Taam[2] and Jean H. Swank[3]

[1] *UCO/Lick Observatory, UCSC, Santa Cruz, CA 95064*
[2] *Department of Physics and Astronomy, Northwestern University, Evanston, IL 60208*
[3] *GSFC/NASA, Greenbelt, MD 20771*

Abstract. We study the $\sim 1 - 10$ Hz quasi-periodic oscillations (QPOs) from black hole candidate X-ray sources (BHCs), in particular, GRS 1915+105. This type of QPO exists only when the source is on the hard, horizontal branch on the color intensity diagram, and their frequency increase with the intensity of the source. Similar behavior at 20-60 Hz has been seen for the horizontal branch oscillations (HBOs) from bright neutron star X-ray sources. We therefore suggest that these QPOs may have a similar origin. Since a magnetosphere and stellar surface are absent in BHCs, it is suggested that for these QPOs, oscillations in an accretion disk may be more relevant to the phenomena.

INTRODUCTION

The black hole candidate X-ray source GRS 1915+105 is well known for its super-luminal radio jets [16]. The recent X-ray observations obtained from the Rossi X-ray Timing Explorer (RXTE) has revealed unprecedented rich phenomena, such as dips, rapid bursts, and various types of QPOs [6,4,17,2,20,19]. With the data observed from April to October 1996, we discovered that the $\sim 0.5 - 6$Hz QPO is present only when the source is located on the hard branch on the color-intensity diagram [4]. Here the color is defined as the ratio of (2-11)keV/(11-30.5)keV. It was also found that the frequency of the QPOs increases as the intensity of the source increases. These properties are very similar to those of another BHC, Nova Muscae 1991 (GS 1124-68). In this paper, we report on observations in 1997, and show that, these properties persist through various changes. We further point out that, when the mass difference between the central objects is considered, these QPOs are similar to QPOs observed in bright neutron star X-ray sources, i.e., the horizontal branch oscillations (HBOs) of $\sim 20 - 60$Hz.

THE QPO PROPERTIES FROM GRS 1915+105

From January to April 1997, GRS 1915+105 was relatively quiet in comparison to its highly active period in 1996, but its behavior was similar to that between July and September 1996, when most of the $\sim 0.5 - 6$ Hz QPOs in question here were present and the source was on the hard spectral branch. We note that, during May 1996 this source exhibited QPOs with frequencies $\lesssim 0.1$Hz. These lower frequency QPOs have different properties and will not be addressed here.

By using the same analysis procedures [4], we have examined whether this source has a different spectral and timing behavior. Four new observations are plotted on the color-intensity diagram in Figure 1. It is seen that, the hardness ratio is very large (> 0.2), and much above the soft branch. It does not lie exactly on the old hard branch however. The power spectrum reveals a strong QPO feature with a weak harmonic. The higher the source intensity, the higher the QPO frequency and the weaker the harmonic. The lower frequency noise steepens as the harmonic component weakens (see Figure 2).

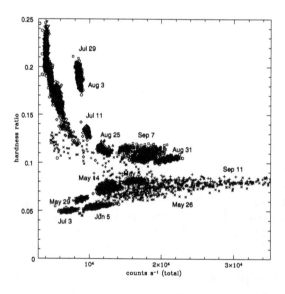

Figure 1

The Hardness-intensity Diagram of GRS1915+105 (see Chen et al. 1997). The four new observations are located on the upper-left section of the plot

The QPO frequency and count rate relation along with the old data is plotted in Figure 3. The more recent data depart from the data described previously, however, the positive relation between the QPO frequency and source intensity still remains.

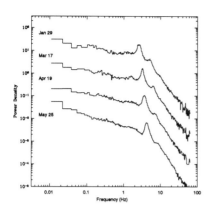
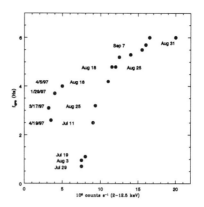

Figure 2
The power spectrum density

Figure 3: The QPO frequency and source intensity relation

COMPARISON TO HBOS

The horizontal branch oscillations (HBOs) are observed in the bright neutron star X-ray sources. These sources exhibit a Z-shaped curve in the so-called color-color diagram or color-intensity diagram. On the upper horizontal branch, QPOs of frequency $\sim 20-60$ Hz are seen and the QPO frequency has a positive relation with the source intensity. On the middle normal branch, QPOs of frequency $\sim 5-10$ Hz are independent of the source intensity.

The $\sim 20-60$ Hz HBOs of Z-sources have strikingly similar properties to the $\sim 1-10$ Hz QPOs from BHCs. In particular, both have a positive QPO frequency and source intensity relation, and a negative (or flat) spectral color (i.e., the hardness ratio) and intensity relation. In fact, the frequencies of the QPOs are in a similar range, considering the mass difference of the central objects and assuming that the QPO frequency scales with M^{-1}. That is,

$$M_{\text{Z-source}} \sim 1-2 M_\odot, \quad M_{\text{GRS1915+105}} > 10 M_\odot, \quad M_{\text{NovaMuscae}} \sim 7 M_\odot,$$

and therefore QPOs of frequency $0.5-10$Hz from GRS 1915+105 correspond to QPOs of frequency $3-8$Hz from Nova Muscae, and correspond to QPOs of frequency $20-60$Hz from the Z-sources.

The above comparisons suggest that these QPOs could have a similar origin independent of the nature of the central object. We note however that the energy spectra can be quite different for different sources or for the same source at different times. The energy spectra of the Z-sources on the hard-horizontal branch [18,7] are rather different from those of GRS 1915+105 on the hard branch [8]. Thus the QPO frequency - source intensity relation depends on the measure of the intensity (e.g. counts in an energy band or luminosity of a component). It is not clear at this point how to make this comparison. It is possible that with suitable identifications the relations are similar.

CONCLUSION

The most widely used model for HBOs from Z-sources is the magnetospheric beat frequency modulated accretion model [1,10]. In this model, the neutron star is assumed to have a magnetic field sufficiently weak that the accretion flow is disrupted only within several neutron star radii from its surface. At the magnetospheric boundary, inhomogeneities in the region are channeled to the magnetic poles at the frequency corresponding to the difference between the Keplerian frequency at the magnetosphere and the neutron star spin frequency. This mechanism leads to a modulation of the mass accretion rate onto the neutron star at the beat frequency.

The recent simultaneous detection of both HBOs and kiloHz QPOs in Sco X-1 [21], however, suggests that the accretion disk is not truncated or disrupted at the magnetopause as suggested in the magnetospheric beat frequency model. Detection of nearly coherent oscillations in some bursts from several atoll sources appears to affirm the presence of a neutron star with rotation period in the relevant range. But it is not clear that these nearly coherent oscillations are caused by the magnetic field channeling of the flow, since in atoll sources the magnetic field is believed very weak. Even if the field is strong enough to channel the flow, it remains difficult to explain the absence of the coherent period in the persistent flux of these atoll sources. Furthermore, as described in this paper, QPOs are observed over a wide range of frequencies in BHCs which do not have a magnetosphere or stellar surface to interact with the accretion flow. Given that the phenomenology of the $0.5 - 10Hz$ QPOs from BHCs are similar to HBOs from Z-sources, one should consider the possibility of a nonmagnetospheric mechanism for such variability.

Due to the difficulties encountered in the magnetospheric beat frequency model, a nonmagnetic model based upon oscillations trapped in the inner region of an accretion disk may be viable. One proposal involving a disk instability model is based on the axisymmetric approximation ($m = 0$) and has been applied to the low frequency oscillations (1-10 Hz) in BHCs [3] with the very rapid fluctuations in the model possibly related to the high frequency oscillations seen in GRS 1915+105 (at 67 Hz, [17]). Such models may equally apply to neutron star systems as well, provided that the neutron star is smaller than 3 Schwarzschild radii (r_g). In this model the variability is attributed to the action of inertial acoustic waves in the disk. Nonlinear time dependent calculations [12,13,3,14,15] reveal that the instabilities are global and take the form of rapid oscillations modulated further on longer timescales. The HBOs may be related to these slower modulations although the relation between QPO frequency and source intensity will need to be established.

Models involving oscillating modes in the non-axisymmetric approximation, particularly, the $m = 1$ mode have also been considered [11,9]. In that case, there are two fundamental oscillation frequencies which are related to local azimuthal rotation frequency, the corresponding epicyclic and vertical oscillation frequencies, and the sound speed. It can be shown that [5], if the instability occurs in the region near $3r_g$, then the higher frequency mode is closely comparable to the kiloHz QPOs. However, the corresponding lower frequency mode is about 100 − 200Hz and is too high to be consistent with the HBOs (assuming a reasonable scale height of the disk, say 10% of the radius). This region is special since the flow becomes supersonic inside $3r_g$. However, nature may choose another site far away from it (say $10r_g$). In that case the lower frequency mode can be comparable to the HBO frequency (although the higher frequency mode is then lower than the kiloHz QPOs). Further theoretical work is necessary before one can attribute the origin of the HBOs from both neutron stars and BHCs to the same mechanism.

REFERENCES

1. Alpar, M. A., & Shaham, J. 1985, Nature, 316, 239
2. Belloni, T., et al. 1997, ApJ, 479, L145
3. Chen, X., & Taam, R. E. 1995, ApJ, 441, 354
4. Chen, X., Swank, J. H., & Taam, R. E. 1997, ApJ, 477, L41
5. Chen, X., Swank, J. H., Taam, R. E. & Zhang, W. 1997, in preparation
6. Greiner, J., Morgan, E. H., & Remillard, R. A. 1996, ApJ, 473, L107
7. Hasinger, G., et al. 1990, A&A, 235, 131
8. Heindl, W. et al. 1997, preprint
9. Ipser, J. R. 1996, ApJ, 458, 508
10. Lamb,F.K., Shibazaki,N., Alpar,M.A., & Shaham,J. 1985, Nature, 317, 681
11. Kato, S. 1993, PASJ, 45, 219
12. Matsumoto, R., Kato, S., & Honma, F. 1988, in Physics of Neutron Stars and Black Holes, ed. Y. Tanaka (Tokyo: Universal Academy Press), 155
13. Matsumoto, R., Kato, S., & Honma, F. 1989, in Theory of Accretion Disks, ed. F.Meyer, W.Duschl, J.Frank, & E.Meyer-Hofmeister (Dordrecht:Kluwer), 167
14. Milsom, J. A. & Taam, R. E. 1996, MNRAS, 283, 919
15. Milsom, J. A. & Taam, R. E. 1997, MNRAS, 286, 358
16. Mirabel, I. F., & Rodriguez, L. F. 1994, Nature, , 371, 46
17. Morgan, E. H., & Remillard, R. A., & Greiner, J. 1997, ApJ, 482, 993
18. Shulz, N. S., Hasinger, G., & Trumper, J. 1989, A&A, 225, 48
19. Swank, J. H., Chen, X. & Taam, R. E. 1997, ApJ, in press
20. Taam, R. E., Chen, X. & Swank, J. H. 1997, ApJ, 485, in press
21. van der Klis, M., Wijnands, R., Horne, K., & Chen, W. 1997, ApJ, 481, L97

Evolution of the Optically Thick Disk in Nova Muscae

Fulvio Melia[*,1] and Ranjeev Misra[†]

[*] *Physics Dept. & Steward Observatory, The University of Arizona, Tucson AZ 85721*
[†] *Inter-University Centre for Astronomy & Astrophysics, Pune, India*

Abstract. We here model the soft X-ray flux from the black hole X-ray nova, GS 1124-68 (Nova Muscae, GRS 1124-68) as emission from an optically-thick accretion disk with Comptonization. We demonstrate that examining the disk's radial dependence on the accretion rate separately in the various spectral states allows for a variation in the location of the disk's inner edge. By extension, we infer the presence of a hot inner region, which can account for the hard component and thus produce a consistency with currently viable models. Our results suggest that the disk's inner edge has a radius whose dependence on the accretion rate is a power-law, and that the slope of this function changes between the different spectral states. We show, moreover, that the accretion rate does not appear to be a direct measure of the total luminosity. This implies that a certain fraction of the released gravitational energy is probably advected inwards through the event horizon.

INTRODUCTION

Nova Muscae's spectral evolution was quite similar to that of other black hole X-ray novae. The soft X-rays dominated the bolometric luminosity over the first 120 day period, during which the hard X-ray power-law contributed less than $\approx 40\%$ of the total power in the first 50 days (ultra soft-state) and $< 10\%$ near the secondary maximum of the soft X-ray light curve (soft-state). However, the hard component was dominant (at the $\approx 80\%$ level) 140 days after the outburst, giving rise to the hard state.

Ebisawa et al. [1] fitted the soft X-ray flux using a multi-color disk model [2]. They concluded that the inner radius of the optically thick disk does not vary significantly with $\dot{M}$, implying that it extends down to the last stable orbit during the entire evolution of the nova. This result contrasts with the prediction of several accretion disk models [3,4,5,6], in which the inner disk region becomes hot and emits the observed hard X-rays. The paradox arises because the hard X-rays emitted during the ultra soft state indicate that such a hot region is probably present. Here, we inspect the data more closely and demonstrate that examining the disk's radial dependence on $\dot{M}$ separately in the various spectral states does in fact allow for a variation in the location of the disk's inner edge.

THE OPTICALLY THICK DISK EVOLUTION

Unlike earlier work, we here pay attention to the variation of the inner radius $r_i \equiv R_i/R_g$ (where R_g is the Schwarzschild radius) with time separately in the individual spectral states, as demarcated by the dashed lines in Figure 1: Region I is the ultra-soft state; II, the intermediate state between the ultra-soft and soft states; III, the soft state, and IV, the intermediate state between the soft and hard states. We emphasize that the dependence on how the temperature is determined is not critical for this discussion. That is, the dependence of the ratio T_{wp}/T_{eff} on the disk parameters does not by itself bring out the variation in R_i, where T_{wp} is the inferred Wien peak temperature and T_{eff} is the effective temperature.

[1)] Presidential Young Investigator.

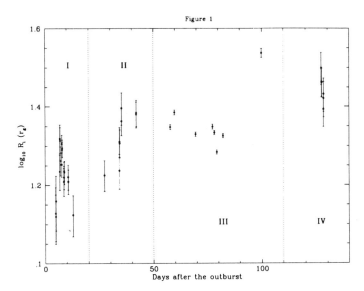

Figure 1

In Figures 2 and 3 we show the variation of R_i as a function of the flux and the accretion rate $\dot{M}$, respectively. The solid line in Figure 3 is the radius versus $\dot{M}$ where the Compton y parameter equals one. To the right of this curve, $y > 1$ and the effects of Comptonization must be taken into account. When viewed state by state, there appears to be a significant correlation between R_i and $\dot{M}$. Figure 4 shows the variation of $\dot{M}$ and the soft X-ray flux with time. We see that the accretion rate inferred with our analysis is in fact not proportional to the total flux, but rather is dependent on R_i and the soft flux.

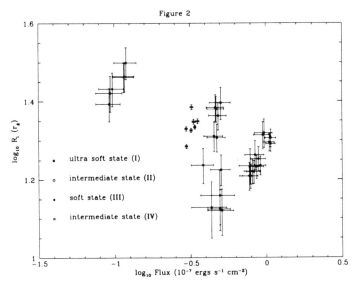

Figure 2

A hard X-ray power-law was detected even when Nova Muscae was in the ultra-soft state. Since the disk is expected to be hot for $R < R_i$ (i.e., inside of the optically thick region), the presence of the hard X-ray component may be taken as an indirect confirmation that R_i is variable, since the inner edge would then presumably not extend down to the last stable orbit at $R \approx 3R_g$. In Figure 5 we plot the ratio of the observed hard X-ray flux (1 – 20 keV) (taken from [7]) to the rate of gravitational energy release in this inner region, versus $\dot{M}$. In this plot, we assume that the inner edge of the hot disk (i.e., R_*) lies at $3R_g$.

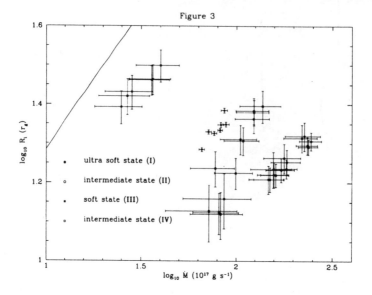

Figure 3

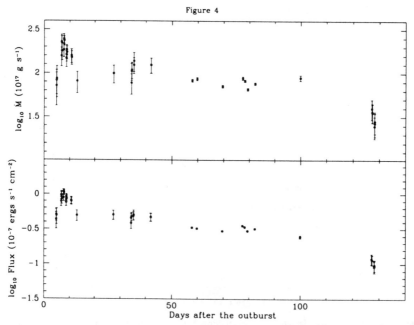

Figure 4

During the initial peak (i.e., the ultra soft state), this ratio is inversely proportional to the accretion rate—an expected result since the fraction of energy advected should increase with $\dot{M}$ [4,6]. The ratio continues to decrease as the accretion rate increases even in state II. Finally, in the next sequence of states, the ratio again increases with decreasing accretion rate as expected. Since there is a shift in this ratio between the two major states, it appears that the physical configuration of the hot disk is different during the ultra-soft (before 20 days) and soft states.

Figures 6 and 7 show the variation in R_i with $\dot{M}$ for the first two spectral states individually. The results for the later transitions show equally tight correlations. The solid lines represent the best fits to the data. For the ultra soft state (I), the straight line has a slope $s = .385^{+.20}_{-.19}$ (χ^2per d.o.f $= .02$), where the range corresponds to fits with a χ^2 at most double its minimum value. For the rest of the states we get: Intermediate state (II), $s = .83^{+.38}_{-.40}$ (χ^2per d.o.f $= .05$); Soft state (III), $s = .56^{+.31}_{-.33}$ (χ^2per d.o.f $= .74$); Intermediate state (IV), $s = .44^{+.09}_{-.10}$ (χ^2per d.o.f $= .003$).

CONCLUSIONS

In the standard model [3], the transition from optically thick to thin occurs secularly when radiation pressure dominates over that due to the gas. This mechanism predicts that the inner disk radius is $R_i \propto \dot{M}^{0.76}$, which does not appear to be consistent with the slopes found for states I and IV, where the χ^2 is small. It may be consistent with the behavior of states II (the intermediate state) and III (the soft state).

In an advection dominated disk model [4], the disk configures into a cold optically thick state, when $\dot{M}$ exceeds some critical value. This model predicts that R_i should be inversely proportional to $\dot{M}$. But this too appears to be contrary to the results presented here.

In the transition disk model [6], the optically thick disk terminates when the effective optical depth is no longer much greater than one, hence rendering the optically thick disk solution inconsistent. For a radiation pressure dominated disk, τ_{eff} is proportional to $\dot{M}^{-2} R_i^{2.9}$. The radius at which $\tau_{eff} \to 1$ is therefore $\propto \dot{M}^{0.69}$. Although this trend is significantly better, this dependence is again not fully consistent with the slopes found in states I and IV.

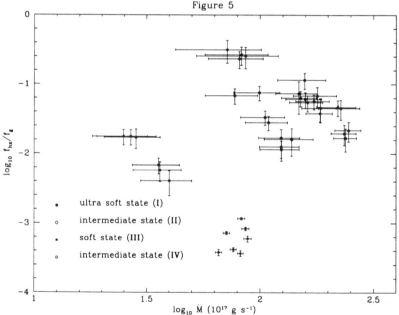

Figure 5

It appears that we still do not have a full understanding of the transition from an optically thick to an optically thin disk. The hard X-rays probably arise from the innermost part of the accretion disk, though the manner in which the disk makes this transition is not yet fully understood. In addition, the disk structure appears to be different in the ultra-soft and soft states.

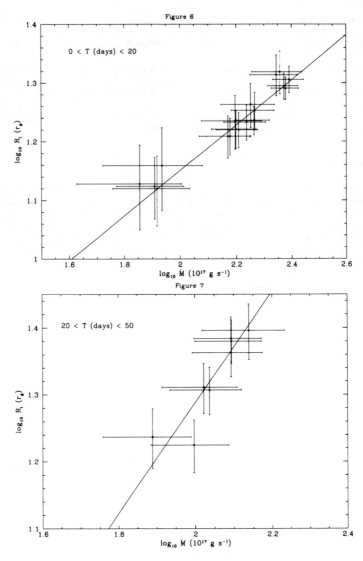

Figure 6

Figure 7

This work was supported in part by NASA grant NAG 5-3075.

REFERENCES

1. Ebisawa, K. et al., *PASJ* **46**, 375 (1994).
2. Shimura et al., *Ap.J.* **445**, 780 (1995).
3. Shapiro, S.L. et al., *Ap.J.* **204**, 187 (1976).
4. Narayan, R. et al., *Ap.J.* **452**, 710 (1995).
5. Ebisawa, K. et al., *PASJ* **48**, 59 (1996).
6. Misra, R. & Melia, F., *Ap.J.* **465**, 869 (1996).
7. Ebisawa, K. et al., *Ap.J.* **403**, 684 (1993).

GALACTIC GAMMA-RAY LINE EMISSION

TGRS Results on the Spatial and Temporal Behavior of the Galactic Center 511 keV Line

B. J. Teegarden[1], T. L. Cline[1], N. Gehrels[1], R. Ramaty[1], H. Seifert[2], M. Harris[2], D. Palmer[2], K. H. Hurley[3]

[1]*NASA, Goddard Space Flight Center, Greenbelt, MD 20771, USA*
[2]*NASA/USRA, Goddard Space Flight Center, Greenbelt, MD 20771, USA*
[3]*Univ. of California, Berkeley, CA 94720, USA*

Abstract. The TGRS (Transient Gamma-Ray Spectrometer) experiment is a high-resolution germanium detector launched on the WIND satellite on Nov. 1, 1994. Although primarily intended to study gamma-ray bursts and solar flares, TGRS also has the capability of studying steady sources near the ecliptic plane (including the Galactic Center). We present here preliminary results on the narrow 511 keV annihilation line from the general direction of the Galactic Center obtained from the TGRS occultation mode. Data are presented for approximately 2 years beginning 1 Jan. 1995. We detect the narrow annihilation line from the galactic center with an average flux of $((1.10\pm.05) \times 10^{-3}$ ph cm^{-2} s$^{-1})$. The data are consistent with a broadened source at the Galactic Center with a width $\sim 25°$ in ecliptic longitude. No evidence for temporal variability on time scales longer than 1 month was found. Our results are in reasonable agreement with the OSSE 511 keV data, but there are some hints of differences in the spatial distribution at the 2σ level.

INTRODUCTION

In this paper we present a new observation of the narrow annihilation line made by the TGRS experiment on the WIND satellite. Due to its high background and small collecting area TGRS is less sensitive than most earlier instruments. However, this is offset by the fact that it is collecting data from the Galactic Center region nearly 100% of the time. We present here the results of the first two years of TGRS operation. For earlier presentations of TGRS results see [1,2].

TGRS was launched on the WIND spacecraft on 1 Nov. 1994. The primary purpose of the WIND mission is to sample conditions in interplanetary space in vicinity of the earth's magnetosphere. Although in earth orbit, WIND spends

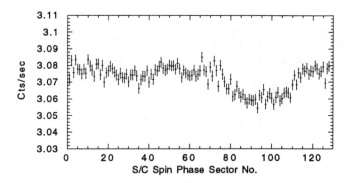

FIGURE 1. TGRS 505-515 keV counting rate as a function of spacecraft spin phase (1 sector = 360/128=2.81°). Data are accumulated from 1 Jan 1995 to 1 Jan 1997.

the great majority of the time in interplanetary space. As a consequence the environment is generally quite stable (background levels, temperatures, etc.) which lends itself quite well to the long data accumulations necessary for the analysis described in this paper. TGRS is an unshielded high-resolution germanium detector passively cooled to a temperature of 85K. Its energy range is $\sim$ 20 - 8000 keV, and its energy resolution at launch was $\sim$ 2.7 keV FWHM at 500 keV. The main axis of the detector is parallel to the spin axis of the spacecraft which is normal to the plane of the ecliptic. A 1-cm thick lead occulter located just outside of the radiative cooler approximately in the plane of the detector subtends an angle of 90° with respect to the detector. As the spacecraft spins (period = 3 s.), the occulter sweeps out a region 16° (FWHM) wide which is offset from the ecliptic plane by an angle of 4.5° so that it passes directly through the Galactic Center. The Crab nebula is included in this occultation region. The on-board processor accumulates data in 4 commandable energy windows (64 channels ea.) synchronized with the spin of the spacecraft. Each spacecraft rotation is divided into 128 sectors.

OBSERVATIONS

The 128 channel sector data is precisely gain corrected using instrumental background lines of known energies. Because of the stability of the TGRS background and gain it has been possible to make very long (2 yr) accumulations and to derive clean spectra and spatial distributions from them. Fig. 1 shows a 2 year accumulation of the counting rate as a function of spin phase (sector no.) in a narrow window (505-515 keV) about the 511 keV line. A clear dip is seen centered near sector no. 95 as well as a shallower dip centered

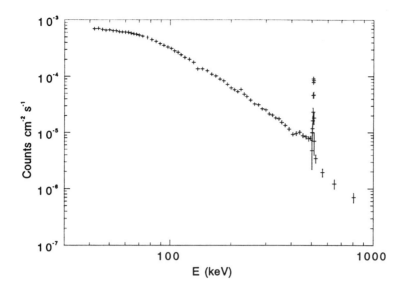

FIGURE 2. TGRS 40-1000 keV GC count spectrum derived from modulation of the GC flux by the on-board occulter. Data are accumulated from 1 Jan 1995 to 1 Jan 1997.

near sector 30. These are due, respectively, to the galactic center (GC) and to the Crab. The GC dip is much more pronounced since it is dominated by narrow 511 keV line emission. At these relatively high energies (> 500 keV) the GC and the Crab are the only two significant sources in the region scanned by the TGRS occulter, whereas at lower energies the situation is much more complicated with significant source confusion.

TGRS accumulates sector data like that shown in Fig. 1 as a function of energy over the range 20-1000 keV. The data are binned on board in a quasi-logarithmic fashion except for a 64 keV wide window centered about the 511 keV line, which is more finely binned (1 keV/channel) to be able to study this narrow line in more detail. We have generated sector data as a function of energy for each of the instrument energy channels. A simple fit (2 sources) has been carried out in each of these energy bins. From these fits count spectra have been derived for the Crab and the GC. The Crab result is consistent with the canonical Crab spectrum. The result of the channel-by-channel GC fit is given in Fig. 2. The narrow 511 keV line is clearly seen as well as a positronium-like continuum below the line. The positronium fraction will be presented in a later paper, but is generally consistent with prior values (f $\sim$ 90%). Also, more detailed information on the (resolved) line width will be given.

Temporal variability of the GC 511 keV emission has been suggested by

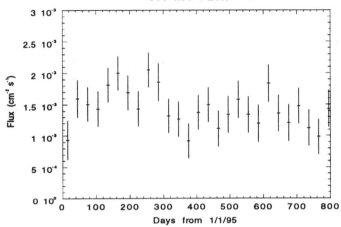

FIGURE 3. TGRS 505-515 count rate as a function of spacecraft spin phase. Data are accumulated from 1 Jan 1995 to 1 Jul 1996.

some of the balloon flight experiments in the 1980's (see Teegarden [3] for a review). However, more recent experiments have not found any evidence for such variability [4,5]. To search for time variation we have divided the TGRS data into 30 day periods and calculated the flux in the 505-515 keV window using the same method described previously. Note that in this window there will be a small contribution from the continuum under the line, but this is estimated to be < 15% of the total rate. The results of this analysis are shown in Fig. 3. No clear evidence is seen for any temporal variability.

Although TGRS is not well suited to mapping the spatial distribution of the 511 keV emission, with the accumulation of more than 2 years of data, it has now become possible to learn something about the gross features of this distribution. Again we use the same representation of the data as shown in Fig. 1. We fit these data with two different broadened distributions (in ecliptic longitude), a rectangle and a gaussian. The results of these fits are given in Table 1. In both cases the centroid of the emission is consistent with the center of our galaxy (to within 1.8 σ). Also, in both cases the width in ecliptic longitude is $\sim$ 25°. To compare with the recent OSSE results we have taken the OSSE 511 keV map [6] of the GC region and folded it through the TGRS occulter response function, producing a sector plot similar to Fig. 1. We then fit this data with the rectangular and gaussian distributions in the same way as was done with the TGRS data. The results of these fits are

TABLE 1. TGRS Galactic center 511 kev model fits and comparison with OSSE.

Model	Centroid (ecliptic long.)	Width	χ^2 (122 dof)
Point Source	266.5° ± 0.8°	------	148.3
Rectangle	266.3° ± 1.2°	27.1° (+7.6°/-5.8°)	137.1
Gaussian	264.6° ± 1.2°	25.4° (+6.0°/-5.8°)	135.6
OSSE (folded through TGRS response)	264.0°	~15°	

Note: Ecliptic longitude of galactic center = 266.8°.

also given in Table 1. As expected, the centroid of the OSSE emission lies above the galactic plane (ecliptic longitude = 264.0°). This is due to the high latitude 511 keV emission in the OSSE map (the "annihilation fountain"). However, the disagreement with the TGRS data is < 2σ depending on the assumed model. The ~ 15° width of the OSSE distribution is narrower than TGRS distribution, but again only at the 2σ level. This is a preliminary result and systematic errors remain to be carefully evaluated. The OSSE method of measuring the spatial distribution of the 511 keV involves taking differences of pointings separated typically by 10°-15°. This technique is rather insensitive to broad smooth regions of emission having an extent > 10°. TGRS, on the other hand, would be sensitive to such emission. At the 2σ level it is not yet clear whether there is any significant difference between the TGRS and OSSE results. However, TGRS continues to accumulate data from the GC region with essentially 100% duty cycle. Eventually, it should be possible to determine whether such a broadened component exists.

REFERENCES

1. Teegarden, B. J., et al., *Astron. & Astrophys.*, **120**, 283 (1996).
2. Teegarden, B. J., et al., *ApJ*, **463**, L75 (1996).
3. Teegarden, B. J., *ApJ Supp*, **92**, 363, (1993).
4. Share, G. et al., *Astrophys. & Space Sci.*, **231**(1), 161 (1995).
5. Purcell, W. et al., *Proceedings of Second Compton Symposium*, (New York: AIP), 403 (1994).
6. Purcell, W. et al., these proceedings (1997).

A BATSE Measurement of the Galactic Positron Annihilation Line

D. M. Smith[1], L. X. Cheng[2], M. Leventhal[2], J. Tueller[3], N. Gehrels[3], G. Fishman[4]

[1] *Space Sciences Laboratory, University of California, Berkeley, CA 94720*
[2] *Department of Astronomy, University of Maryland, College Park*
[3] *NASA-Goddard Space Flight Center*
[4] *NASA-Marshall Space Flight Center*

Abstract.
A specialized background subtraction technique has produced a measurement of the total 511 keV positron-annihilation flux from the inner Galaxy with the BATSE LADs: $(1.49 \pm 0.02) \times 10^{-3}$ ph cm^{-2} s^{-1}. The positronium fraction is found to be (0.91 ± 0.02) assuming a power law for the underlying continuum. This result is reproduced consistently in four consecutive yearly data sets. We compare it to other results from wide-field-of-view instruments and discuss future plans for the analysis.

Early in the study of 511 keV positron annihilation radiation from the Galactic center, it was noticed that instruments with larger fields of view tended to measure higher fluxes [1,2]. This implied that there was either diffuse emission over large scales ($\gg 10°$), multiple point-like sources, or both. Several review papers [6,13,10] discuss the evolution of our ideas about the distribution of the 511 keV emission.

Our analysis uses 5-minute, 128-channel spectra from the 8 BATSE Large Area Detectors (LADs) [3]. Spectra from the two LADs pointing closest to the Galactic center are selected for inclusion in our measurement if the Earth's limb is $> 45°$ from the Galactic center, and if they were not taken: 1) in an orbit following a passage of the South Atlantic Anomaly (SAA), or 2) at a time of high geomagnetic activity, or 3) during a time of unusually high cosmic ray count rates (from the plastic Charged Particle Detectors (CPDs) which cover the LADs).

Background spectra are constructed using a database which is assembled from another subset of the data, mutually exclusive with the first, in which no point within 45° of the Galactic center can be visible to the detector. The same exclusions for SAA, geomagnetic, and cosmic ray criteria are used. If the Crab is visible, its expected count spectrum is subtracted. Spectra including Nova Persei 1992 during its outburst were excluded.

The background spectra are binned into a two-dimensional library by two parameters: the angle from the Earth's center to the detector axis and the CPD rate for the same detector. For each source spectrum we do a two-dimensional interpolation among the bins in the library based on its value of these two parameters to create its corresponding background spectrum.

A series of corrections are then made to the background spectrum based on the differences between various other parameters related to the source and background spectra. A difference in each parameter is related to a differential spectrum ("template") [9]. The templates are generated by comparing many pairs of spectra within the background set. Templates are applied based on the following parameters; no further parameters showed significant correlations with spectral differences:

- The sum of all 8 CPD rates
- SAA doses (from OSSE data) convolved with the half life of ^{24}Na
- The dot product of the detector axis and magnetic East
- The detector/Earth angle during the last SAA transit
- The LAD rate above 1 MeV (which has a minor contribution from the sky itself; see below)
- The CPD rate convolved with the half life of ^{128}I
- The spacecraft altitude
- The dot product of the detector axis and the gradient of the geomagnetic cutoff

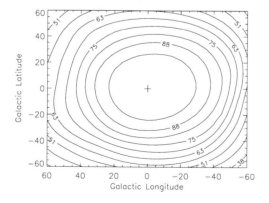

FIGURE 1. Averaged LAD response, normalized to 100% at the peak.

We analyzed four time periods separately, each nearly a year: 06/23/91 - 05/23/92, 05/23/92 - 04/23/93, 12/29/93 - 02/12/95, and 02/12/95 - 03/28/96. The gap in 1993 is due to a series of orbital reboosts, which make the background less precisely predictable (since the Galactic 511 keV signal is ~2% of the background 511 keV signal, this is critical). The background library was also created for each year separately, but in order to get adequate statistics for the templates, they could only be made twice, in two-year intervals.

We sum the background-subtracted spectra for each year and also their response on the sky. Figure 1 shows the response summed over 4 years and normalized to 100% at the center; the FWHM is ~ 130°. The LADs, although geometrically thin, still have a very flat angular response at 511 keV because they are also nearly optically thin: the reduction of projected area at large off-axis angles is compensated for by the increase in projected distance through the crystal, which gives greater stopping power.

We find the 511 keV line flux by folding model spectra forward through the LAD response matrices. We added one step to the forward-folding process to simulate the application of template #5 above, since a small count rate above > 1 MeV is produced in the LADs by the Galactic flux. We used a spectral model consisting of a power law continuum, orthopositronium continuum, and narrow 511 keV line, for a total of 4 free parameters. The spatial model is a point source at the Galactic center; from Figure 1, it is apparent that any distribution which is concentrated within 30° or so of the Galactic center will give nearly the same result.

Figure 2 shows the results of the fit to all 4 years of data summed together; the results for the individual years are similar (see below). The data are

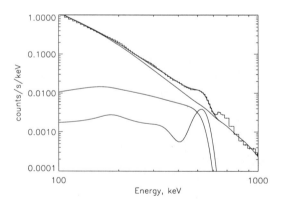

FIGURE 2. Averaged count spectrum, with 3 components and sum of the model fit. The feature at 650 keV is an instrumental artifact.

shown along with curves for the three components of the fit and their sum. The feature from 600-800 keV is a sporadically occurring instrumental artifact; the fit was done using data from 100-560 keV and 850-1000 keV to avoid it.

Figure 3 shows the fit line flux and positronium fraction (fraction of positrons decaying via positronium) as a function of time. The error bars shown are derived from the scatter of these four points, assuming the actual quantities are constant; they therefore include both statistical errors and that portion of the systematic errors which can vary from year to year. To look for variability with these data, the error bars (which are dominated by systematic errors) should be estimated by a bootstrap method; this is under development. The error shown for the average of the flux measurements, 2×10^{-5} ph cm^{-2} s^{-1}, corresponds to a count rate of 3×10^{-4} of the background line. Keen observers will recall that the positronium values presented in our talk were significantly lower; what has changed is the inclusion of the data from 850-1000 keV, which allow the underlying continuum to be better constrained.

The average of the points in Figure 3, $(1.49 \pm 0.02) \times 10^{-3}$ ph cm^{-2} s^{-1}, is smaller than the fluxes previously reported by instruments with wide fields of view [4,7]. The most extensive measurements were made with the *Solar Maximum Mission* Gamma-Ray Spectrometer (SMM/GRS) [4], which also had a 130° FWHM field of view. Their derived flux for a point source (the same spatial model used here), was $(2.7 + 0.6, -0.7) \times 10^{-3}$ ph cm^{-2} s^{-1}. Since their error bars are described as being the total allowed range of values considering their systematic errors, our values do not agree, despite the similar fields of view. The data analysis techniques differ, however, and it is possible that with the right spatial distribution the results can be made to agree. The SMM/GRS value assuming a distributed source in the Galactic plane is $(2.3 + 0.5, -0.8) \times 10^{-3}$ ph cm^{-2} s^{-1} for the inner radian. It is the nature of our

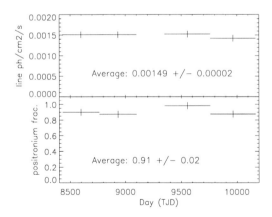

FIGURE 3. Fit 511 keV line flux and positronium fraction as a function of time.

analysis that any broadening of the distribution will cause our value to rise; at first, because flux is put in a place where the relative detector response is <100%, and, when the distribution is broadened beyond 45° radius, because some flux begins to appear in our background spectra.

In conclusion, finding a distribution which brings SMM/GRS into agreement with BATSE (as well as OSSE, the Transient Gamma-Ray Spectrometer (TGRS) on *Wind,* and other instruments [8,11,12]) may help us understand the distribution of annihilation on angular scales much larger than those so far probed by OSSE. We will pursue a number of spatial distributions with the BATSE data, as well as repeating the analysis with different spectral forms for the underlying continuum; this tends to alter the positronium fraction results somewhat but the line flux hardly at all [5]. Finally, we will do a bootstrap analysis of our errors and examine possible sources of systematic error which do not vary from year to year, such as uncertainties in the response matrix.

This research was supported by NASA grants NAG5-2380 and NAG5-3599. We thank Mark Finger, Geoff Pendleton, Rob Preece, Tom Bridgman, and the Compton Observatory Science Support Center for their help.

REFERENCES

1. Albernhe, F. et al. 1981, A&A, 94, 214
2. Dunphy, P. P., Chupp, E. L., & Forrest, D. J. 1983, in *Positron-Electron Pairs in Astrophysics,* ed. M. L. Burns, A. K. Harding, and R. Ramaty (New York, AIP), p. 237
3. Fishman, G. et al. 1992, in *The Compton Observatory Science Workshop,* ed. C. R. Shrader et al., NASA CP-3137, p. 26
4. Harris, M. J. et al. 1990, ApJ 362, 135
5. Kinzer, R. L. et al. 1996, A&AS, 120, 317
6. Lingenfelter, R. E. & Ramaty, R. 1989, ApJ, 383, 686
7. Neil, M. et al. 1990, ApJ, 356, L21
8. Purcell, W. et al. 1997, ApJ, submitted
9. Smith, D. M. et al. 1996, ApJ, 471, 783
10. Smith, D. M., Purcell, W. R. & Leventhal, M., these proceedings
11. Teegarden, B. J. et al. 1996, ApJ, 463, L75
12. Teegarden, B. J. et al., these proceedings
13. Tueller, J. 1993, in *Compton Gamma-Ray Observatory* , ed. M. Friedlander, N. Gehrels, and D. J. Macomb (New York, AIP), p. 97

OSSE Constraints on the Galactic Positron Source Distribution

Peter A. Milne & Mark D. Leising

Department of Physics and Astronomy, Clemson University, Clemson SC 29634-1911

Abstract. A detailed measurement of the distribution of 511 keV line radiation will be necessary to sort out the actual contributions of the many plausible positron sources. The OSSE instrument has measured well the flux in certain regions where it has been pointed long or often, but because of its small offset background subtraction technique and highly non-uniform exposure to date, it does not yet provide detailed constraints on the source distribution. Qualitatively and quantitatively different galactic distributions are consistent with the current data. In our opinion, claims of quantitative information about specific components, and of the existence of unforseen features are premature.

I INTRODUCTION

Mapping the galactic distribution of the 511 keV annihilation radiation has been a primary goal of the Oriented Scintillation Spectrometer Experiment (OSSE) on CGRO since its launch in 1991. To date, during some 50 CGRO viewing periods, 330 different source pointings have been made in the region within ninety degrees in longitude and forty-five degrees in latitude of the galactic center direction [1]. OSSE has no intrinsic imaging capability, so its angular information is limited to the variation in the flux in its nearly triangular field-of-view, which at 500 keV is 3.8° x 11.4° FWHM. Each source field is also coupled to two (or more) background fields, typically offset 10-12 degrees.

Two types of techniques have been used to infer the distribution of positron sources from the OSSE 511 keV data; a priori model-fitting and model-independent methods. Both can identify distributions consistent with the data, and both are biased. Because of the extremely low signal to noise ratio of these data, direct inversion techniques can be applied only with extreme regularization of some kind. These techniques generally show a large bias due to nonuniform exposure. The hope is that a variety of techniques will all indicate certain essential features of the sky distribution required for consis-

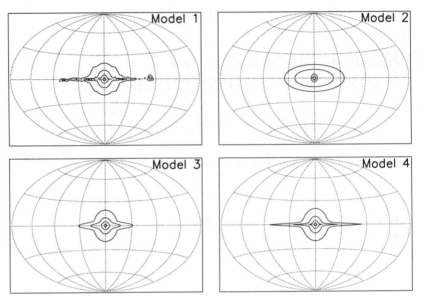

FIGURE 1. Acceptable Model Fits to 330 OSSE pointings at 511 keV. Model #1 is the map of COBE/DIRBE at 60μ and the $R^{1/4}$ bulge of de Vaucouleurs with $R_{eff} = 1$ kpc [4] [2] . Model #2 is a 60x20 gaussian and a 5x5 gaussian. Model #3 is a 60x5 gaussian and a $R^{1/4}$ bulge. Model #4 is the light from Higdon-Fowler and a $R^{1/4}$ bulge [3]. Flux contours are $(1, 3.3, 10, 33) \times 10^{-3} \gamma$ cm^{-2} s^{-1} St^{-1}.

tency with the data. Here we apply only a limited set of possible analyses to illustrate some of the range of sky distributions currently allowed by the data.

II GEOMETRIC AND ASTRONOMICAL MODELS

Since the earliest 511 keV observations, it has been understood that the emission is galactic in origin. Thus, the galactic distributions of various objects and sky maps at other wavelengths have been tested as potential maps. We have also tested various geometrical models, e.g., simple two-dimensional gaussians centered on the galactic center. Many of these distributions can be ruled out as dominant components of the annihilation radiation. Figure 1 shows examples of two-component models that are reasonably consistent with the current dataset. These models were chosen to illustrate the range of acceptable models.

The common features of all acceptable fits we have found are a bulge component and a disk component. The relative strengths of the two components varies greatly among models. One way to quantify the relative contributions of the two components is the ratio of the bulge flux to the disk flux (B/D). Table 1 shows this ratio, the χ^2 of the fit, and the total flux of each model. The $R^{\frac{1}{4}}$ bulge has a sharper central peak and broader wings than the 5 x 5 gaussian. For model #3 versus model #2 this translates to the bulge being able to provide some of the necessary extended emission (in l and b) as well as most of the galactic center emission.

Each single component model was also tested with a galactic center point source added. All such combinations have lower probabilities than combinations with an extended bulge included. This suggests that the central component is indeed extended and not a point source. For reference, the best single component model fit is the $R^{1/4}$ bulge which fits with a $\chi^2 = 374$ (Probability = 4 %). Thus it is clear that we can derive only limited details about the disk component(s).

TABLE 1. Characteristics of models for 330 OSSE pointings.

Model	χ^2	Prob.(%)	B/D	Total Flux (x$10^{-3}\gamma$ cm^{-2} s^{-1})
Dirbe60 +$R^{1/4}$	360	10.9	2.8	3.1
60x20 + 5x5	360	13.1	0.14	3.6
60x5 + $R^{1/4}$	360	11.0	4.7	2.6
HF Light + $R^{1/4}$	361	10.1	2.4	2.9

III COMPTEL 1.8 MEV MAP

The COMPTEL 1.8 MeV map is dominated by the decay of ^{26}Al, in which 82% of the time a positron is emitted. Thus the ^{26}Al map is a physically required component of the 511 keV map, if ^{26}Al positrons annihilate before propagating far. As a global tracer of current stellar activity, it should also reflect the distribution of other massive star/supernova positron contributors. The COMPTEL map plus a point source yields a poor fit, but combined with a $R^{1/4}$ bulge it provides a reasonable fit, having a $\chi^2 = 365$ (Probability = 7.8 %). The 511 flux associated with the ^{26}Al map is comparable to the 1.809 MeV flux - more than that provided by the ^{26}Al positrons by a factor 2→3. The enhanced number of positrons are presumably from ^{44}Ti ejected by the same core collapse supernovae, though additional sources may be required. A fraction of the galactic bulge flux is not associated with current star formation activity, thus other sources, perhaps Type Ia supernovae, are needed.

IV TWO MODEL-INDEPENDENT TECHNIQUES

To show just how non-unique solutions to the OSSE data are, we illustrate two other model-independent techniques. First, we test the significance of point sources at all positions in a 1° grid. The most significant point is put

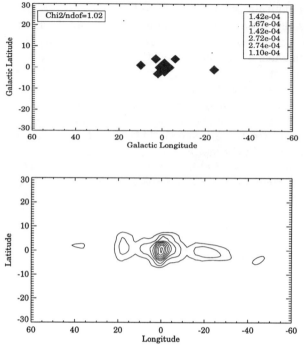

FIGURE 2. Sky maps consistent with the OSSE 511 keV line data. Upper: Iteratively selected point sources. Point significance is coded by simple size and flux by grayscale level. Lower: Eighth iteration of our adaptation of the Richardson-Lucy technique, starting from a 30° × 60° gaussian. This map has χ^2 per data point (the number of degrees of freedom is not clear) of 1.09.

in place with the best-fit flux, and the procedure is repeated, simultaneously refitting all points each time one is added. After the eighth point is added, as shown in Fig. 2, the reduced χ^2 is 1.09 and no other point is more significant than 3σ. We do not argue that this is the true map, but it produces an adequate fit. No "fountain" is apparent, although some asymmetry might be suggested. We also apply the Richardson-Lucy iterative technique that uses the ratios of the measurements to current model predictions of each and projects them onto the sky weighted by the respective responses. We can not describe this in detail here, but this procedure is adapted to the OSSE data, because negative measurements are physically allowed (as positive sources in background fields.) This technique is biased, although uniformly so, to the exposure map. It can not put flux where OSSE has not looked. Our current implementation can give sky distributions, even somewhat different ones depending on the initial iterate, etc., consistent with the data, as shown in Fig. 2. Only very low flux contours extend toward the putative fountain.

In no sense would we argue that any distribution shown here is "best". Each has its own biases, in some cases extreme ones, and each is consistent with the data at a reasonable level. We would place highly publicized renditions of other such techniques along with these. Unless one has strong prior information about the 511 keV sky, all should be regarded as possible representations of the actual sky map.

V SUMMARY

Quite different realizations of the 511 keV sky are consistent with the OSSE data. Model-independent techniques are useful for giving impressions of the data, but probably can not be interpreted literally for quantitative detail. We must continue to develop better physical understanding of the sources of the positrons. The basic ingredients of the models as we now understand them are, 1) A disk from core-collapse SN, traced adequately by COMPTEL ^{26}Al measurements, but scaled up to include (at least) ^{44}Ti positrons from the same events, 2) possibly ^{56}Co and ^{44}Ti positrons from disk SN Ia traced by intermediate disk populations, 3) Bulge SN Ia traced by e.g., starlight, and 4) Possible black hole/jet source contributions in both disk and bulge. Unfortunately, current constraints on the structure of the disk are weak at best. The major question regarding SN Ia ^{56}Co contributions is whether its positrons can really escape. This escape might actually be visible in the very late time SN Ia light curves (Milne et al. 1997 [5]). All distributions, but especially the bulge, might be modified for the smearing of MeV positron propagation.

We have not discussed fits including the high-quality data from other wide field instruments such as SMM and TGRS. These see larger fluxes than OSSE so modelling or imaging must put substantial flux somewhere where OSSE does not efficiently detect it. This can be anywhere OSSE has not looked, or it can be low surface brightness emission extended over scales of 10° or more anywhere. Thus this emission is currently essentially unconstrained. A possible additional component is diffuse, high-latitude emission. This might result from very nearly regions, and could be related to suggested extended emission in ^{26}Al in COMPTEL/SMM combined data [6].

REFERENCES

1. Purcell, W.R., et al., *ApJ*, in press.
2. de Vaucouleurs, G.,*Ann. d'Astrophys.* 11, 247.
3. Higdon, J.C., Fowler, W.A., *ApJ* 317, 710.
4. Weiland, J.L., et al., *ApJ* 425, L81.
5. Milne, P.A.,The, L.-S.,Leising, M.D., these proceedings.
6. Diehl, R. et al., these proceedings.

Is Positron Escape Seen In the Late-time Light Curves of Type Ia Supernovae?

Peter A. Milne, Lih-Sin The, Mark D. Leising

Department of Physics and Astronomy, Clemson University, Clemson SC 29634-1911

Abstract. At times later than 200 days, Type Ia SN light curves are dominated by the kinetic energy deposition from positrons created in the decay of ^{56}Co. In this paper, the transport of positrons after emission are simulated, for deflagration and delayed detonation models, assuming various configurations of the magnetic field. We find the light curve due to positron kinetic energy has a different shape for a radially combed magnetic field than it does for a tangled field that traps positrons. The radial field light curves fit the observed light curves better than do the trapping field light curves for the four SNe used in this study. The radial field permits a much larger fraction of the positrons to escape, perhaps enough to explain a large percentage of the 511 keV annihilation radiation from the Galactic plane observed by OSSE.

INTRODUCTION

Type Ia supernovae possess two qualities that make them attractive as a potential source of positrons; the large amount of positron-producing ^{56}Ni synthesized in the explosion, and the occurrence of Type Ia SNe in bulge populations. Their spectra and light curves show evidence of the ^{56}Ni $\rightarrow$ ^{56}Co $\rightarrow$ ^{56}Fe decays. In 19% of the ^{56}Co $\rightarrow$ ^{56}Fe decays, a positron is produced. The large number of positrons produced in a single SN combined with realistic Type Ia SN rates show that they can easily account for the 511 keV annihilation flux observed by SMM and other detectors if enough can escape from the SN into the ISM. All models of the galactic distribution of 511 keV annihilation emission that fit the OSSE observations contain a bulge component. Supernova searches in other spiral galaxies confirm that Ia SNe occur in the bulge populations [1].

Heterogeneity does exist within the Ia paradigm. Different models are created to fit the light curves and spectra of individual SNe. The mass of ^{56}Ni produced in models varies from 0.1 -0.9 M$_\odot$ [2]. The amount and location of

^{56}Ni in the ejecta has an obvious effect upon the fraction of positrons that escape. The magnetic field geometry is a less obvious, but even more important determinant of positron escape. If the field is tangled such that positrons are eternally trapped within the ejecta, the fraction of positrons that escape is zero. A small fraction may survive non-thermally in the expanding ejecta ($\leq 10^{-3}$ survival fraction). If the homologous expansion stretches the field lines to the extent that they are essentially radial, then many more positrons may escape (5% -15%). This subject was explained in detail by Chan and Lingenfelter (1993) [3]. While extreme SN Ia models vary in nickel mass by a factor of less than 10, the ratio between positron survival in the radial field geometry and positron survival in the knotted field geometry can vary by a factor of 200. Thus, the field geometry is the dominant determinant of positron survival. The problem is to determine which field geometry approximates the SN ejecta the best.

LIGHT CURVES

Six times as much energy goes into the γ-rays ($\simeq 3.6$ MeV per decay) as goes into the kinetic energy of the positron ($\simeq 0.632$ MeV) [3] For positron KE deposition to play a significant role, the gamma-ray opacity must drop to below one-sixth the effective opacity of the positron KE deposition. The gamma-ray transport was treated with a Monte Carlo simulation which estimated the fraction ($f_{dep,\gamma}$) of the decay energy that is deposited via Compton scattering and photoelectric absorption processes. $f_{dep,\gamma}$ is published for a number of SN models, including W7, in The, Bridgeman, Clayton (1994) [4]. The radial positron KE deposition was calculated by emitting positrons isotropically relative to the field lines and slowing them via ionization and excitation, inverse Compton scattering, synchrotron emission, and bremsstrahlung emission. Ionization and excitation dominates at all but the highest energies. The trapped positron KE deposition was calculated by emitting the positrons in the center of their respective zones and confining them to evolve co-located with the ejecta.

Figure 1 shows the various contributions to the bolometric light curve of the SN Ia model W7 [5]. The opacity to γ-rays drops quickly, so by 200^d only a small fraction of the γ-rays interact with the ejecta, and the KE deposited by the positrons as they slow down dominates the light curve. Escaping positrons take some of their KE with them creating a deficit in the light curve versus the "In Situ" approximation. Trapped, non-thermal positrons will store energy temporarily, giving it back at later times. The radial curve falls off more sharply than does the In Situ curve; the trapped curve crosses over above the In Situ curve and remains slightly above that curve.

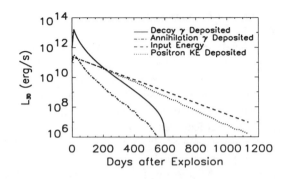

FIGURE 1. Contributions to the W7 light curve. The γ and e^+ KE deposition are in units of 10^{30} erg/s. The "input energy" is the KE input to the ejecta by β^+ decay. The "In Situ" approximation is the assumption that all of the KE is deposited on-site, instantaneously. The energy deposited by γ-rays produced by positron annihilation in the ejecta is not significant.

OBSERVATIONS

A few SNe have occurred close enough that photometry was taken at 350+ days. A model is judged to satisfactorily represent the observed supernova if it reproduces the early-time spectrum and the multi-band light curve. The test of the field geometry is achieved by using models which fit at early time and determining whether the predictions of either of the two field geometries fit the late-time B band light curve. The B band is assumed to be proportional to the bolometric light curve during this phase. The bases of this assumption are two-fold. First, the observed light curves do not show color evolution in this phase [6]. Second, once the nebular phase begins, the continuum falls off and the B band is dominated by florescence lines [7]. The lines should scale with the KE deposition until very late times ($\sim 1000^d$) at which time freeze out becomes an issue [8]. Thus, the phase to observe positron driven light curves is between 400 and 1000 days.

The light curves are fit on a supernova-by-supernova basis. Four SNe were analyzed in this preliminary study. The first two, SN 1972E [9] and SN 1992A [10] were considered "typical" SNe and are thus fit with the deflagration model W7 to demonstrate the technique. The second two, SN 1990N [11] and SN 1991T [8] were peculiar and are fit by late detonation models created specifically to fit their early spectra and light curves, W7DN and W7DT [12]. The lines are scaled to fit the data at 200 days. The fits are shown in figure 2.

SN 1972E, SN 1992A and SN 1990N follow linear declines in magnitude after 200 days, and the slopes of these declines are better fit by the radial models. Other models have been suggested as better representatives of these

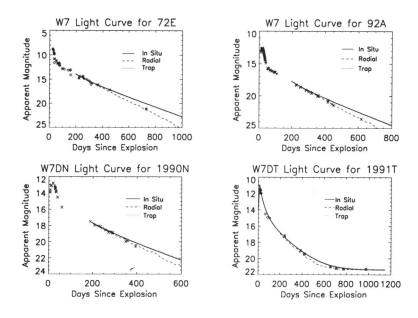

FIGURE 2. Model fits to four SNe observed at late-times. See text for references

individual SNe, in future, these other models will be used to strengthen the argument for these SNe. To fit W7DT to SN 1991T, we included a component with constant flux to account for the additional luminosity that is provided by light echoes, as identified by Schmidt, et.al.(1994) [8]. The existence of a light echo provides a luminosity source to offset the declining ^{56}Co input. Although it may still be possible to accurately model a light curve that contains a light echo, at present we caution against their use to discriminate between positron trapping and escape.

SURVIVAL FRACTIONS AND 511 KEV FLUX

TABLE 1. Positron survival 3 years post-explosion.

Model Name	Radial %	Radial #	Trapped %	Trapped #
W7	5.5	1.3×10^{53}	2.0×10^{-2}	4.8×10^{50}
W7DN	10.4	2.6×10^{53}	5.2×10^{-2}	1.3×10^{51}
W7DT	11.7	3.6×10^{53}	5.9×10^{-2}	1.8×10^{51}

Table 1 shows the survival fraction of positrons emitted in the SN models W7, W7DN and W7DT for the two field geometries. The values are for 3 years (99.99% of the positrons have been emitted by then). Positrons that escape and/or survive until three years after the supernova explosion exist in an environment that is very diffuse. This is due to two facts. First, the rapid expansion of the supernova creates a tenuous bubble, and second, Type Ia SNe tend to occur in a very low density phase of the ISM. This tenuous environment means that a positron can survive so long that the contributions from many supernovae can combine to produce a diffuse positron annihilation radiation source. Numerically, if the galactic 511 keV flux is 2.3 x 10^{-3} γ cm^{-2} s^{-1} [13], then a Type Ia SN rate of 0.2 -1 SN Ia $[100\ y]^{-1}$ would produce the necessary number of positrons, for a survival fraction of 5% from the SN model W7.

SUMMARY

This paper motivates the use of late-time light curves as a probe of positron transport in type Ia SN ejecta. A few SNe have been observed well enough to begin to discriminate between radial and trapping field geometries. Preliminary observations suggest that the field lines are radially combed, but future observations are required to solve this problem.

Determining the positron escape from typical Type Ia SNe does not solve the 511 keV emission problem by itself. The issues of the galactic distribution of past Ia SNe as well as of the ISM transport and the annihilation of positrons must also be addressed. Nonetheless, it appears that Type Ia SNe must be considered a potentially large contributor to the 511 keV emission.

REFERENCES

1. Branch, D., *ApJ*, 000 (1996).
2. Hoeflich, P., Khokhlov, A., *ApJ* **457**, 500 (1996).
3. Chan, K.-W., Lingenfelter, R.E., *ApJ* **405**, 614 (1993).
4. The, L.-S., Bridgman, W.T., Clayton, D.D., *ApJS* **93**, 531 (1994).
5. Nomoto, K., Thielemann, F.-K., Yokoi, K., *ApJ* **286**, 644 (1984).
6. Turatto, M., Cappellaro, E., Barbon, R., Della Valle, M., Ortolani, S., Rossino, L., *AJ* **100**, 771 (1990).
7. Colgate, S.A., Fryer, C.L., Hand, K.P., in *Thermonuclear Supernovae*, Dordrecht: Kluwer Academic Publishers, pp. 273 (1997).
8. Schmidt,B.P., Kirshner,R.P., Leibundgut, B., Wells, L.A., Porter, A.C., Ruiz-Lapuente, P., Challis, P., Filippenko, A.V., *ApJ* **434**, L19 (1994).
9. Kirshner, R.P., Oke, J.B., *ApJ* **200**, 574 (1975).
10. Suntzeff, N.B., *Unpublished*,(1997).
11. Lira, P., *Master's Thesis: Univ. Of Chile*, (1995).
12. Yamaoka, H., Nomoto, K., Shigeyama, T., Thielemann, F.-K., *ApJ* **393**, L55 (1992).
13. Share, G.J., Leising, M.D., Messina, D.C., Purcell, W.R., *ApJ* **358**, L45 (1990).

The Origin of the High-Energy Activity at the Galactic Center

Farhad Yusef-Zadeh*, William Purcell* & Eric Gotthelf[†]

Dept. Physics and Astronomy, Northwestern University, Evanston, Ill. 60208
[†]*NASA/GSFC, Greenbelt, MD 20771*

Abstract. Recent X-ray and gamma-ray observations of the Galactic center region by the ASCA and CGRO/EGRET instruments show evidence of 2–10 keV and > 1 GeV continuum emission as well as 6.7 and 6.4 keV line emission from the inner 0.2° of the Galactic center. This region is also known to host a bright nonthermal radio continuum source Sgr A East and a dense molecular cloud M–0.02–0.07 known as the 50 km s^{-1} cloud. The oval-shaped nonthermal Sgr A East is physically interacting with M–0.02–0.07 at the Galactic center. A comparison between the distribution of ionized, synchrotron and neutral gas suggests a self-consistent interpretation of the high-energy activity at the Galactic center. Our preliminary analysis of the data suggest a shock model of cosmic ray acceleration at the site of the interaction to explain the enhanced GeV γ-ray emission. We also address a number of issues related to the spatial correlation of the diffuse radio and X-ray emitting gas as well as to the origin of the fluorescent 6.4 and 6.7 keV emission at the Galactic center.

INTRODUCTION

Radio View of the Galactic Center Radio continuum observations of the inner 15' of the Galactic center show two prominent radio continuum structures known as the Sgr A complex and the filamentary continuum Arc. The Sgr A Complex consists of Sgr A East and its halo as well as Sgr A* and its thermal orbiting gas Sgr A West. Sgr A* is unique and considered by many to be a massive black hole with a mass of $10^6 M_\odot$ at the Galactic center. The oval-shaped structure known as Sgr A East is thought to be the remnant of an explosion located just behind the Galactic center (Yusef-Zadeh and Morris 1987; Pedlar et al. 1989), A number of authors (Khokhlov & Melia 1996; Mezger et al. 1989; Yusef-Zadeh and Morris 1987) question the interpretation of Sgr A East as a standard SNR. Khokhlov & Melia (1996) have considered that Sgr A East is the remnant of star that is tidally disrupted by a massive black hole, presumably by Sgr A*. The explosion energy is estimated to be an order of magnitude more than the energy released by a typical supernova. There is also considerable evidence that this explosion occurred inside the dense molecular cloud M–0.02–0.07, thus depositing more than 4×10^{52} ergs in the ISM (Mezger et al. 1989). Recent discovery of OH(1720MHz) masers at the interface of the 50 km s^{-1} molecular cloud and Sgr A East showed conclusively that these two are physically interacting with each other (Yusef-Zadeh et al. 1996).

On the largest scale, there is a diffuse 7–10' halo of nonthermal continuum emission surrounding the oval-shaped radio structure Sgr A East. The spectrum of the halo tends to be steeper than Sgr A East and is primarily nonthermal with the energy spectral index of >3. The optical depth toward Sgr A East and the halo at low frequencies lead Pedlar et al. (1989) to consider a mixture of both thermal and nonthermal gas but displaced to the front side of Sgr A East.

The Arc is a nonthermal filamentary source located near $l \approx 0.18°$ and runs in the direction perpendicular to the Galactic plane. The filaments are linearly polarized showing evidence that they are tracing magnetic field lines and emitting synchrotron radiation. The spectrum of the filaments is unusual in that it is flatter than typical nonthermal features. The energy spectrum of relativistic particles have a spectral index of ≈ 1.6 in the radio wavelengths. A number of Galactic center molecular clouds appear to outline the linear filaments, prompting the hypothesis that the filaments and clouds are physically interacting with each other, in which case the field strength is estimated to be at least 1 mG in order for the filaments to resist deflection at points of interaction with the clouds (Yusef-Zadeh and Morris 1987; Serabyn and Morris 1994).

X-Ray View of the Galactic Center Recent ASCA observations of the Galactic center showed conclusively the evidence for diffuse X-ray emission arising from the inner 15' of the Galactic center (Koyama et al. 1996). The continuum radiation is accounted for by thermal plasma having temperature of 10 keV. The strongest continuum radiation from the inner 30' in the energy band between 0.7 and 10 keV arises from within the shell of Sgr A East and is somewhat elongated along the Galactic plane. The electron density and the thermal energy of thermal gas within the shell of Sgr A East are estimated to be 6 cm^{-3} and 3×10^{50} ergs, respectively (Koyama et al. 1996). Weak and diffuse emission is also seen corresponding to the radio halo as well along the Galactic plane both in the positive and negative longitudes, but with low surface brightness. This weak emission extends over as far as 80 pc (33') on either side of the Galactic center. The strongest diffuse emission beyond the Sgr A complex arises from the positive longitude side outlined by the nonthermal radio filaments of the Arc. Figure 1 shows X-ray contours superimposed on the radio image displaying the Sgr A East shell, its halo and the filaments in the Arc. The electron density and the thermal energy of the hot gas outside the Sgr A complex are estimated to be about 0.3–0.4 cm^{-3} and 0.5–1×10^{53} ergs, respectively (Koyama et al. 1996).

One of the more fascinating aspect of ASCA observations of the Galactic center is the evidence of 6.4 keV emission peaking on two molecular clouds in the region between the Arc and the Sgr A complex and in Sgr B2. This fluorescent Kα line emission results from the K-shell photoionization of iron atoms.

Gamma-Ray View of the Galactic Center Recent report of high energy (30 MeV – 30 GeV) continuum emission from the Galactic center based on EGRET observations (Mattox 1997) indicate a source with a luminosity of 5×10^{36} erg s^{-1} at the distance of the galactic center. The source is situated within 0.2^0 of the Galactic center and could be either compact or diffuse within 100 pc of the Galactic center. The energy spectrum of this source is fit by a power law having an index of 1.7 which is harder than typical EGRET sources in the Galactic plane. This source has been considered to be associated with the Arc (Pohl 1997), or with Sgr A*, or possibly to have a diffuse origin (Thompson et al. 1996; Mattox 1997).

DISCUSSION

Sgr A East as the Source of High-Energy Activity The hypothesis that we are considering involves Sgr A East, the most energetic source in the Galactic center region. This source has been considered to be due to an unusual explosion, perhaps a Seyfert-like activity as seen in the nucleus of spiral galaxies (Pedlar et al. 1989). A number of observations indicate that Sgr A East is unusual and more energetic than a typical supernova, having released more than 5×10^{52} ergs into the Galactic center region (e.g. Mezger et al. 1989).

Since the interaction of Sgr A East with the 50 km s^{-1} molecular cloud is well established, the high-energy cosmic rays responsible for radio, X-ray and γ-ray emission could be generated at the site of the interaction of Sgr A East and its molecular cloud before diffusing out along the Galactic plane. In this hypothesis, the EGRET source is considered to be diffuse and the γ-ray spectrum is due to accelerated cosmic rays at the site of the interaction of the explosive event with the giant molecular cloud. The western edge of the 50 km s^{-1} giant molecular cloud is interacting with the Sgr A East shell whereas the eastern edge of the cloud appears to be outlined by the nonthermal linear filaments that cross the Galactic plane near $l \approx 0.2^0$. Figure 2 shows the overall distribution of ^{13}CO gas between 30 and 50 km s^{-1} (Bally et al. 1988) which appears to be correlated with the hot X-ray emitting gas displayed as contours in Figure 1. In addition, the distribution of 6.4 keV emission from the inner 30' of the Galactic center, as discussed below, appears to coincide with the peaks of two molecular gas clouds (Koyama et al. 1996) in M-0.02-0.07. These morphological correlations strongly suggest that Sgr A East, the 50 km s^{-1} molecular cloud, the hot X-ray emitting gas and the 6.4 keV emission are physically associated implying that Sgr A East is responsible for the high-energy activity at the Galactic center.

There is considerable evidence that nonthermal particles at high energies produce synchrotron emission from Sgr A East and the Arc as well as the continuum EGRET γ-ray source (2EGJ1746-2852) seen at the Galactic center. There is also conclusive evidence that shocks exit at the interface of the 50 km s^{-1} molecular cloud and Sgr A East. Thus, it is plausible that the low-energy cosmic ray particles must exist from extrapolation of high-energy nonthermal particles. In fact some of the brightest supernova remnants interacting with molecular clouds and showing OH maser shocks at 1720 MHz appear to have EGRET counterparts (Esposito et al. 1996).

The eastern edge of the Sgr A East lies a high density molecular core with a density of $1-2\times10^6$ cm^{-3} and a string of HII regions excited by massive stars (e.g. Serabyn et al. 1992). Both these features are associated with sites of star formation in the 50 km s^{-1} cloud. The cluster of hot stars and the cavity of ionized gas surrounded by the circumnuclear ring at the Galactic center begs the question of the association of these features with another site of star formation within the 50 km s^{-1} cloud. We sketch a scenario in which the cluster of hot stars IRS 16 and the circumnuclear ring of molecular gas are remnants of an episode of star formation that took place in the 50 km s^{-1} cloud. In this picture, the stars and the gas are captured at the Galactic center as the giant molecular cloud passed the Galactic center and is presently behind Sgr A West.

The Nature of 10^7-10^8 K Emission The 50 km s^{-1} molecular is distributed along the positive longitude side of the plane but immediately behind the Galactic center. The front side of the cloud is assumed to face Sgr A* and the shell of Sgr A East outlines the compressed molecular gas as a result of the shock wave driving into the the dense medium of M–0.02–0.07. The halo of Sgr A East seen both in the radio and X-ray wavelengths is a secondary manifestation of the explosion. The hot gas and the accelerated cosmic rays that are produced at the site of the shock leak preferentially toward the Galactic center because Sgr A East is bounded by the giant molecular cloud. The X-ray emitting gas is most concentrated within the shell of Sgr A East resembling the structure of composite supernova remnants. A strong morphological correlation between the nonthermal radio features, the Arc and Sgr A East and its halo, and the hot and cold thermal emitting gas as noted in contours of Figures 1 and 2 indicate strongly that the X-ray gas has a nonthermal component and is mixed with thermal gas at a temperature of 10 keV. In fact, there is evidence of a hard tail in the diffuse X-ray spectrum of the inner region of the Galaxy based on Ginga observations (Yamasaki et al. 1996).

The hot plasma at this temperature can not be confined by the gravitational potential in the Galactic center, thus an alternative mechanism is required to constrain the hot gas in the region. The morphology of X-ray and radio emitting gas suggests that the X-ray gas follows the magnetized filaments before reaching the Arc. Under the assumption that the magnetic pressure of the Arc is strong enough to confine much of the hot plasma, the estimated field strength has to be greater than 0.5 mG for an electron density of 0.3 cm^{-3} and a temperature of 10 keV.

The Nature of the 6.4 keV Emission Koyama et al. (1996) argue that the hot plasma can not account for exciting the 6.4 keV line emission since the observed X-ray luminosity is an order of magnitude less than the observed 6.4 keV line emission. Recent theoretical work by Borkowski and Szymkowiak (1997), however show that thermal electrons with few to 10 keV energies can penetrate interstellar dust grains and produce fluorescent Kα emission through the K shell ionization. The advantage of this mechanism over the excitation of Kα of Fe in the gas phase is that the dust grains are abundant in the content of their heavy atoms. These authors consider a non-equilibrium ionization model to account for X-ray spectra of hot plasma mixed in with neutral atoms of heavy elements. The ionization parameter τ is the product of electron density n_e and time t. In the case when τ is less than 10^{13} cm^{-3} s all ions are in equilibrium and the contribution of Fe Kα emission from dust grains is minimal. However, for the case when the plasma is underionized before the final equilibrium state, the Fe Kα emission could be totally due to dust grains. Since the observed 6.4 keV and 6.7 keV line intensities arise from the dense molecular clouds outside the Sgr A East complex, it is plausible that the thermal electrons of the plasma with a 10 keV temperature is exciting the Fe Kα of dust grains in selected clouds where the non-equilibrium ionization model is applicable.

We believe that the two prominent clouds toward Sgr A East and Sgr A West (the circumnuclear disk) at the Galactic center, namely the 50 km s^{-1} and the circumnuclear neutral gas orbiting Sgr A*, do not show any 6.4 keV line emission because the ionization parameter is close to 10^{13} cm^{-3} s. With the expansion time scale of 5×10^4 yrs over 80 pcs and the electron density of 6 cm^{-3} in Sgr A East, $\tau=10^{-13}$ cm^{-3}. However, the region where the Fe Kα is seen, the electron density of the hot plasma is much lower than 6 cm^{-3} and therefore the non-equilibrium ionization model is applicable.

Ginga results have shown a hard tail in the spectrum of the inner degree of the Galactic center (Yamauchi et al. 1990). The hard tail beyond 10 keV can be important in explaining the nature of 6.4 keV emission and the strong correlation seen between CO and 2-10 keV spatial distributions and in determining whether the low-energy cosmic ray electrons are responsible for heating of molecular clouds near the Galactic center.

Summary A number of different observations indicate strongly the co-existence of nonthermal and thermal gas in the inner 50 pc of the Galaxy. This mixture of gas can be accounted for in a self-consistent fashion by assuming that the center of the high-energy activity is due to the unusual explosion of Sgr A East which is expanding inside the 50 km s^{-1} giant molecular cloud. In this picture, the high-energy ($\approx$GeV) cosmic rays accelerated by the explosion are responsible for the γ-ray emission as detected by CGRO/EGRET, the nonthermal radio continuum emission from Sgr A East and its halo, and for illumination of the magnetic field lines of the Arc. On the other hand, the low-energy ($<$ 1 MeV) cosmic-ray particles are considered to heat the giant 50 km s^{-1} molecular cloud and produce the observed diffuse X-ray emission, the 6.4 keV and the hard tail observed in the Ginga spectrum. The strong

X-ray emission arising simultaneously from the Sgr A East shell and from the interior of the shell shows a resemblance to the centrally peaked morphology of composite SNR's such as W44 (e.g. Rho et al. 1994). A more detailed account of this model will be given elsewhere.

REFERENCES

1. Bally, J., et al., *ApJ*, **324**, 223 (1988).
2. Borkowski, K. & Szymkowiak, A.E., *ApJ*, **477**, L49 (1997).
3. Esposito, J. A., et al., *ApJ*, **461**, 820 (1996).
4. Khokhlov, A. & Melia, F., *ApJ*, **457**. L61 (1995).
5. Koyama K., et al., *PASJ*, **48**, 249 (1996).
6. Mattox, J. R., *GCNEWS*, vol. *4*, (Feb. 1997).
7. Mezger, P. G., et al., *A.A.*, **209**, 337 (1989).
8. Pedlar, A., et al., *ApJ*, **342**, 769 (1989).
9. Pohl, M., *A.A.* **317**, 441 (1997).
10. Rho, J. et al. *ApJ* **430**, 757 (1994).
11. Serabyn, E., et al., *ApJ*, **395**, 166 (1992).
12. Serabyn, E. & Morris, M., *ApJ*, **424**, L91 (1994).
13. Thompson, D.J. et al., *ApJS*, **207**, 227 (1996).
14. Yamasaki, N.Y. et al., *A.A.S*, **120**, 393 (1996).
15. Yamauchi, S., et al., *ApJ*, **365**, 532 (1990).
16. Yusef-Zadeh, F. Ph.D. thesis, Columbia University (1986).
17. Yusef-Zadeh, F., et al., *ApJ*, **466**, L25 (1996).
18. Yusef-Zadeh, F. & Morris, M., *ApJ*, **320**, 545 (1987).

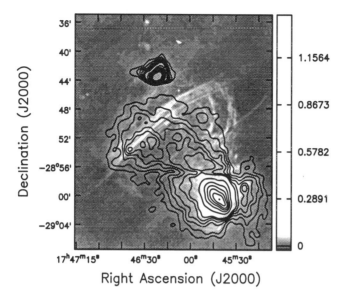

FIGURE 1. Radiograph of the λ20cm continuum image superimposed on X-ray contours in the energy range between 0.7 and 10 keV with a resolution of 1'. The ASCA data was kindly provided by Dr. Koyama and Dr. Maeda.

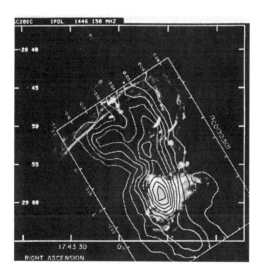

FIGURE 2. Using the 1950 coordinates, this figure is simialr to Figure 1 except that the contours represent the distribution of ^{13}CO emission in the velocity range between 30 and 50 km s^{-1} having similar resolution to that of X-ray data shown in Figure 1 (Yusef-Zadeh 1986; Bally et al. 1988).

The Galactic Center Lobe and its interpretation

M. Pohl[1,2]

1 MPI für Extraterrestrische Physik, Postfach 1603, 85740 Garching, Germany;
2 Danish Space Research Institute, Juliane Maries Vej 30, 2100 København Ø, Denmark

Abstract. In this paper I discuss the radio structures in the galactic center region and their current interpretation in the context of possible counterparts for the γ-ray sources in this part of the sky.

The EGRET source 2EG J1746-2852 is likely to be identified with the galactic center arc. For the extended source of e^+/e^- annihilation emission (Purcell, this volume) we do not find an obvious counterpart, though its location with respect to the galactic center lobe may indicate a connection.

I THE GALACTIC CENTER REGION AT RADIO FREQUENCIES

In Fig.1 we show the inner region around the galactic center at 20cm wavelength [1]. In this image the galactic plane runs roughly from the upper left corner to the middle of the bottom line. The Sgr A complex at (265.6,-29.0) is overexposed with our choice of grey scale. The main component of Sgr A is the extremely compact discrete source Sgr A*, which shows a similar behaviour like active galactic nuclei although its luminosity is orders of magnitudes less. There is dynamical evidence that Sgr A* harbors a black hole of $2 \cdot 10^6 \ M_\odot$ [2,3]. It is interesting to note that the linear size of Sgr A* is only 100 times the Schwartzschild radius of such a black hole [4].

Within 50 pc of Sgr A we find the arc, a magnetic structure which is oriented perpendicular to the galactic plane. It seems to be connected to the Sgr A complex by the so-called bridge. However, recombination lines detected from the bridge indicate the existence of thermal material while the detection of polarized radio emission from the arc and the absence of recombination lines prove the dominance of synchrotron radiation from this region. A number of thin filaments are superimposed on the diffuse emission of the arc. Both the filaments and the diffuse emission from the arc show inverted radio spectra with $\alpha \simeq 0.3$ ($S \propto \nu^\alpha$). The arc extends northward and southward to plumes

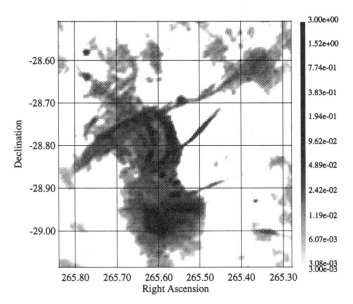

FIGURE 1. The inner region around the galactic center in a 20cm VLA image (Yusef-Zadeh and Morris 1988) in units of Jy/beam. The beam size is 30.5 ×26 arcsec.

with high degree of polarization but steeper spectrum. These extensions may be understood as due to ageing, initially monoenergetic electrons, which have been transported out by diffusion and convection from their sources in the arc [5]. With the same model both the flux and the spectrum of the EGRET source 2EG J1746-2852 [6,7] can be explained as inverse Compton emission of the young high-energy electrons on the far-infrared photons from the molecular cloud complex M0.20-0.033, which runs into the arc [8]. Thus the arc seems to be the source of a plasma outflow perpendicular to the galactic plane into plumes of roughly 1° size.

To find structures on larger scales it is necessary to filter out the thermal and non-thermal background emission in the single-dish surveys. By applying such a method [9] to the 408 MHz allsky survey [10] a jet-like structure has been found to be associated with the galactic center [11]. An image of this jet is shown in Fig.2. At 408 MHz the feature can only be traced down to 8° latitude. Data at 1408 MHz indicate the existence of two ridges, which can be followed to 1°-2° latitude and which connect the jet to the galactic center region [11]. When located at the Galactic Center, the projected length of the structure would be around 4 kpc. The spectral index is around $\alpha \simeq -0.6$ and the integrated luminosity (0.1 GHz to 100 GHz) is 10^{36} ergs/sec.

In a 10.5 GHz survey of the galactic plane a giant Ω-shaped loop, the Galactic Center Lobe, has been found [12]. This lobe is shown in 2.695 GHz data [13] in Fig.3. It has a flat radio spectrum between 1.4 GHz and 10.5 GHz

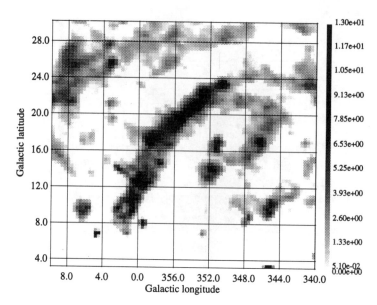

FIGURE 2. The Galactic Center Jet in the 408 MHz survey (Haslam et al. 1982) after background filtering (Sofue and Reich 1979). The scale is Kelvin brightness temperature.

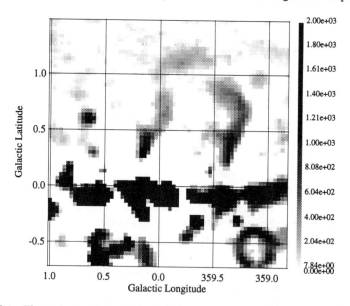

FIGURE 3. The Galactic Center Lobe in the 11cm galactic plane survey (Reich et al. 1990) after background filtering. The scale is milliKelvin brightness temperature. One ridge of the Ω-shaped structure coincides with the northern extension of the arc while the western ridge seems to point at the Sgr C complex.

in most parts. The eastern ridge of the Galactic Center Lobe is located at the northern extension of the arc. A comparison of the intensity and spectral changes in the northern and southern extension of the arc shows additional emission in the north, which indicates that the extensions of the arc and the Galactic Center Lobe are two different structures, which are seen in projection [5].

Although the outstanding appearance of the Galactic Center Lobe suggests a high-speed ejection, there is not much proof for such a behaviour. Recently it has been pointed out [14], that both ridges of the Galactic Center Lobe are associated with spurs of molecular gas based on ^{13}CO data. The gas coinciding with the eastern ridge has a velocity of around +100 km/sec and that at the western ridge has a velocity of -150 km/sec. So the two clumps appear to be symmetric with respect to the center, but slightly shifted towards negative velocities by about 20 km/sec. The symmetric appearance of the gas clumps suggests a physical association between the molecular gas and the radio structure and further indicates a rotation of both at ~ 100 km/sec. The total mass of the molecular gas is around $3 \cdot 10^5$ $M_\odot$ and the kinetic energy is of order 10^{53} ergs.

The whole structure is probably not the result of an explosion, because in this case the rotation would be small due to angular momentum conservation. The most promising scenario at the moment is that of a cylindrical outflow. Here disk gas would be vertically accelerated along the poloidal magnetic field [15]. The outflow velocity would be of the order of the Alfvén velocity. It is unclear whether the Galactic Center Lobe is physically connected to the large-scale jet that is shown in Fig.2.

II DISCUSSION

A recent reanalysis of OSSE, SMM, TGRS, and balloon data has revealed an extended feature in the 0.511 MeV e^+/e^- annihilation radiation centered at $l \simeq -2°$, $b \simeq 10°$ with a flux of $\sim 5 \cdot 10^{-4}$ ph./cm^2/sec [16,17] (see also Purcell, this volume). This finding has been interpreted as a positron carrying fountain in the galactic center region [18]. Here I discuss this idea in the context of the radio structures.

It is obvious, that the annihilation source has probably little to do with the large-scale jet at the galactic center (see Fig.2), because it is located a few degrees offset. But if the source is the result of a fountain, then the base of the fountain may be physically connected to the Galactic Center Lobe (see Fig.3). In this case the density of ionized gas in the fountain would have to small ($n_e < 1$ cm^{-3}) to prevent detectable thermal bremsstrahlung outside the Lobe. The radio emission from the Galactic Center Lobe would have to be synchrotron emission of electrons which are produced with a spectrum of $Q \propto E^{-1}$ or harder up to a few GeV, possibly by stochastic acceleration processes. The spatial and spectral distribution of the synchrotron emission of such electrons would harmonize with the data for the Galactic Center Lobe only if there is little convection or the fountain has a large opening angle (see Appendix of [8]). An important implication however would be that the western ridge of the Lobe is polarized.

A serious constraint is the density of neutral material in the Lobe. If the positrons would interact with the molecular material their life time would be around 30000 years, and consequently the convection velocity would have to be > 5000 km/sec to let the bulk of 0.511 MeV photons be produced above the Galactic Center Lobe. This requirement can only be relaxed if it is assumed that the positrons are produced with energies higher than 50 MeV. The energetic electrons which are produced in the arc and convect outward are too few in number by more than four orders of magnitude to produce observable 0.511 MeV emission, even if they all were positrons. Also the plasma streaming in the arc is symmetric with plumes in the northern and in southern direction, whereas the excess of positrons appears only on one side of the galactic plane.

To summarize: though the location of the annihilation source with respect to the Galactic Center Lobe may indicate a connection, the corresponding physics does not seem to allow a simple relation in the context of a large-scale plasma outflow. More detailed modelling of the fountain scenario is definitely required.

REFERENCES

1. Yusef-Zadeh F., Morris M., 1988, ApJ, 329, 729
2. Eckart A., Genzel R., 1996, Nature, 383, 415
3. Eckart A., Genzel R., 1997, MNRAS, 284, 576
4. Mezger P.G., Duschl W.J., Zylka R., 1996, ARA&A, 7, 289
5. Pohl M., Reich W., Schlickeiser R., 1992, A&A, 262, 441
6. Thompson D.J. et al., ApJS, 101, 259
7. Merck M., et al., 1996, A&AS, 120, C465
8. Pohl M., 1997, A&A, 317, 441
9. Sofue Y., Reich W., 1979, A&AS, 38, 251
10. Haslam C.G.T. et al., 1982, A&AS, 47, 1
11. Sofue Y., Reich W., Reich P., 1989, ApJ, 341, L47
12. Sofue Y., Handa T., 1984, Nature, 310, 568
13. Reich W. et al., 1990, A&AS 85, 633
14. Sofue Y., 1996, ApJ, 459, L69
15. Uchida Y., Shibata K., Sofue Y., 1985, Nature, 317, 699
16. Purcell W.R. et al., 1997, ApJ, submitted
17. Cheng L.X. et al., 1997, ApJ, 481, L43
18. Dermer C.D., Skibo J.G., 1997, ApJ Letters, in press

Evidence for GeV emission from the Galactic Center Fountain

D. H. Hartmann[*], D. D. Dixon[**], E. D. Kolaczyk[†], J. Samimi[††]

*Department of Physics and Astronomy
Clemson University, Clemson, SC 29634
**Institute of Geophysics and Planetary Physics
University of California Riverside, CA 92521
†Department of Statistics
University of Chicago, Chicago IL
††Sharif University
Tehran, Iran

Abstract. The region near the Galactic center may have experienced recurrent episodes of injection of energy in excess of $\sim 10^{55}$ ergs due to repeated starbursts involving more than $\sim 10^4$ supernovae. This hypothesis can be tested by measurements of γ-ray lines produced by the decay of radioactive isotopes and positron annihilation, or by searches for pulsars produced during starbursts. Recent OSSE observations of 511 keV emission extending above the Galactic center led to the suggestion of a starburst driven fountain from the Galactic center [1]. We present EGRET observations that might support this picture.

THE GALACTIC CENTER FOUNTAIN

The center of the Milky Way may have experienced a series of explosive events [2] [3] [4]. Large scale X-ray structures, such as the north polar spur, might be explained as propagating shocks induced by these explosions [5], although other arguments suggest the north polar spur is a rather local feature. To understand the angular distribution of the X-ray features, the shock model requires an impulsive energy release of $\sim 3 \times 10^{56}$ ergs about $\sim 1.5 \times 10^7$ yrs ago. A massive black hole at the Galactic center could have released this energy, but an alternative scenario is the energy deposition from a large number of supernovae.

A typical Type II supernova injects about $1-2 \times 10^{51}$ ergs of kinetic energy into the ISM, and contributions from pre-supernova winds may double this amount. Thus, a total of $\sim 10^5$ supernovae is needed to attain a total energy of 3×10^{56} ergs. The duration of the starburst should be $< 10^6$ yrs in order to yield a propagating shock that matches the observed X-ray features [5]. Activities observed near the Galactic center are manifest on various spatial scales, with perhaps the most dominant feature being the expanding molecular ring. At a Galactocentric distance of ~ 200 pc, the expanding molecular ring might contain as much as 10^{55} ergs of kinetic energy [6]. Enhanced 6.7 keV line emission was detected by the GINGA satellite at a distance compatible with being associated with the ring structure [7] [8]. If this line emission originates from a hot and tenuous plasma, then the X-ray observations suggest $\sim 10^{54}$ ergs of thermal energy were injected less than $\sim 10^6$ yrs ago [9].

On smaller scales, the Galactic superbubble G359.1–0.5 suggests that a starburst of $\sim 10^{2-3}$ supernovae may have occured within the past few million years [10]. Observations of stellar populations within ~ 1 pc of the Galactic center argue in favor of starburst models involving $\sim 4 \times 10^5$ $M_\odot$ of gas (implying $\sim 10^{2-3}$ supernovae) between 5 – 9 Myr ago [11]. Emission features from He I surveys of the central region also implies that at least several tens of massive stars were born within a few parsecs of the center in the last $\sim 10^6$ yrs [12] [13].

Other arguments support more extensive or intense starburst episodes. The total mass interior to 1 pc exceeds 10^6 $M_\odot$ [14]. If a significant fraction of this total mass is due to mass segregation of compact stellar remnants initially formed within the inner 10–100 pc, then the associated number of neutron stars could exceed $\sim 10^6$, for a Salpeter IMF [6]. However, the velocity that neutron stars are apparently born with may allow most of them to escape the central region. Thus, neutron stars would not contribute significantly to the mass interior to 1 pc. If the neutron stars were not produced in steady state but in a series of starbursts, one might consider the production of $\sim 10^3$ neutron stars in bursts separated $\sim 10^7$ yrs. Such starbursts would also inject over 10^{54} ergs of kinetic energy into the ISM. As noted above, however, the expanding molecular ring imply that the last energy deposition was a factor 10–100 larger [4]. Hydrodynamic simulations of gas–star systems near galactic centers suggest that starbursts which produce $> 10^5$ supernovae could occur quasi-periodically every $\sim 10^8$ yrs [3]. Bursts of this magnitude would be expected to severely influence the gas dynamics near the center, and (to a lesser extent) the disk and the halo through the influence of the propagating shock wave.

New evidence for a recent starburst in the inner Galaxy comes from the 511 keV mapping by OSSE [15] [16]. The global map can be decomposed into two components, a disk and a bulge. In addition, the data require a "hot spot" at $l \sim -4°$ and $b \sim 7°$. This positive latitude enhancement was interpreted by Dermer & Skibo [1] as the result of a recent starburst ($\sim 10^6$ yrs ago) involving $\sim 10^4$ supernovae. The resulting positrons lose energy and annihilate as they are convected upward with the gas flow. In this picture one also expects the coproduction of ^{26}Al (visible on a timescale of 10^6 yrs) and cosmic rays (CRs). Shocks would also produce a non-thermal population of electrons which might produce a radio afterglow. In fact, a 4 kpc long jet-like radio feature emanating from the Galactic center region has been detected at 408 MHz [17], and is commonly known as the Galactic Center Spur (GCS). If this $\sim$ 200 pc wide chimney indeed convects radiactive debris and CRs into the halo, we might also expect some emission in the GeV regime due to interactions of the CRs with the gas in the chimney.

DETECTION OF GEV EMISSION

The data analyzed are coadded EGRET observations through VP 429.0, selecting only events within 30 degrees of the detector zenith. The analysis method is a 2D variant [18] of the TIPSH algorithm for denoising Poisson data. In the particularly TIPSH method employed here, we specify a null hypothesis, consisting of a predicted distribution of counts/pixel in the data set. Here, we have used a hypothesis consisting of the predicted Galactic [19] and extragalactic [20] diffuse emission. TIPSH works by comparing the Haar wavelet coefficients of the data with the distribution of coefficients implied by a Poisson distribution whose pixel means (and variances) are described by the null hypothesis. Those coefficents which fall below some user prescribed significance cutoff are considered statistically consistent with the null hypothesis, and discarded. At the end, the non-zero wavelet coefficients describe the portion of the data which is "different" than the null hypothesis, within the statistics of the observations, along with some (preferably small) number of false detections (non-zero coefficients due solely to noise). For the analysis described here, we selected our significance threshhold such that the error rate was about 2%.

For the region under discussion, the denoised residual (significant differences w.r.t. null hypothesis) is shown in Figure 1, overplotted on filled contours showing the 511 keV model [16]. The Galactic plane flux has been truncated to show the feature of interest, which is an apparent northward extension of the Galactic emission, at $1° - 2°$ longitude, extending up to about $15°$ latitude as seen in this plot. Though we have not yet derived a spectrum for this feature, similar analysis in the 1-2 GeV band shows no evidence for a similar feature. In the 2-4 GeV band, there does appear to be enhanced emission in this region. However, it is confused with other nearby emission, and is not nearly so distinct. A reasonable conclusion is that the spectrum of this feature is fairly hard, and distinctly different than that of Galactic cosmic-ray induced emission. Further model fitting is required to verify this. Note that this feature coincides with the "jet" seen in the radio band. Also in Figure 1 are filled contours showing the most recent OSSE 511 keV model fit [16]. Though there is an apparent latitude offset in the positive latitude feature, it can be shown that due to a strong exposure-related systematic in the OSSE observations [21], a 511 keV feature corresponding to the EGRET/408MHz

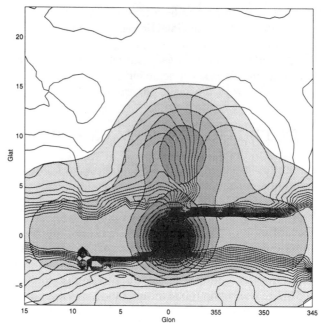

FIGURE 1. Flux contours from EGRET counts. The jet-like feature is reasonably well aligned with the galactic center spur seen in 408 MHz maps. The filled contours represent the 511 keV model fit described in [16]. The apparent offset between the 511 keV and GeV features is potentially due to exposure systematics in the OSSE observations.

"jet" would be consistent with the observed 511 keV maps.

A key question is the reality of the observed features, since when using a non-parametric estimation scheme such as TIPSH, one is always concerned with artifacts. Unfortunately, it is usually difficult (if not impossible) to assign a quantitative "significance" to a feature in a non-parametric estimate. At the time of this writing, we have not yet devised a method to accomplish this. An alternative approach would be to perform some sort of model fitting, but this has its own set of pitfalls, and must be accomplished with care. Future analyses will attempt to address this, hopefully using physically motivated models.

CONCLUSIONS

Whether driven by bursts of star formation or processes that occur near a massive black hole, the numerous activities going on near the Galactic center are hidden, for the most part, from optical observations. In the gamma-ray band evidence for starburst activity is harder to hide. Hartmann, Timmes, and

Diehl [22] discussed the possibility that the production of ^{26}Al in supernovae may lead to a detectable afterglow at 1.809 MeV, and Hartmann [23] suggested that such starbursts might be detectable through an excess of radio pulsars. The detection of 1.8 MeV emission from a galactic center starburst might be accomplished by the INTEGRAL mission. The recent OSSE observations of 511 keV emission above the galactic center [15] [16] were interpreted as the result of a major galactic center starburst driving a positron fountain into the halo [1]. We present EGRET GeV observations that perhaps support this picture, but new gamma-ray missions such as GLAST will be required to verify these observations. The Galactic center is one of the most dynamical regions of our Galaxy, and high energy gamma rays may be the best tool for studying its starburst history.

This work was supported by NASA Grant NAG5-3666 (DDD). DH acknowledges support from the Compton GI program.

REFERENCES

1. Dermer, C. D., and Skibo, J. G., ApJ, in press (1997).
2. Oort, J., ARA&A, **15**, 295 (1977).
3. Loose, H. H., Krügel, E., & Tutukov, A., A&A, **105**, 342 (1982).
4. Sofue, Y., in *The Center of the Galaxy*, ed. M. Morris, Kluwer, p. 213 (1989).
5. Sofue, Y., ApJ, **431**, L91 (1994).
6. Morris, M., ApJ, **408**, 496 (1993).
7. Koyama, K., *et al.*, Nature, **339**, 603 (1989).
8. Koyama, K., *et al.*, Nature, **343**, 148 (1990).
9. Yamauchi, S., *et al.*, ApJ, **365**, 532 (1990).
10. Uchida, K. I., *et al.*, ApJ, **398**, 128 (1992).
11. Tamblyn, P., & Rieke, G. H., ApJ, **414**, 573 (1993).
12. Krabbe, A., Genzel, R., Drapatz, S., & Rotaciuc, V., ApJ, **382**, L19 (1991).
13. Rieke, G. H., & Rieke, M. J., in *The Nuclei of Normal Galaxies*, eds. R. Genzel and A. I. Harris, (Kluwer Acad. Publ., 1994)
14. Genzel, R., Hollenbach, D., & Townes, C. H., Rep. Prog. Physics, **57**, 417 (1994).
15. Cheng, L. X. *et al.*, ApJ, **481**, L43 (1997)
16. Purcell, W. R., *et al.*, ApJ, in press (1997).
17. Sofue, Y., Reich, W., & Reich, P., ApJ, **341**, L47 (1989).
18. Dixon, D. D. *et al.*, these proceedings (1997).
19. Hunter, S. D. *et al.*, ApJ, **481**, 205 (1997).
20. Sreekumar, P. *et al.*, ApJ, submitted (1997).
21. Dixon, D. D., in preparation (1997).
22. Hartmann, D., Timmes, F. X., & Diehl, R. D., PASP Proc. 112, eds. A. Burkert, D. Hartmann, and S. Majewski (1996).
23. Hartmann, D. H., ApJ, **447**, 646 (1995).

Positron Transport and Annihilation in Expanding Flows: A Model for the High-Latitude Annihilation Feature

Charles D. Dermer and Jeffrey G. Skibo

*E. O. Hulburt Center for Space Research,
Code 7653, Naval Research Laboratory, Washington, DC 20375*

Abstract. Positron sources in the vicinity of the Galactic Center can account for the recently discovered high-latitude 0.511 MeV annihilation radiation if the positrons are convected away from the disk of the Galaxy and annihilate during transport. Here we describe our treatment for positron energy loss and thermalization, transport, and annihilation. The concept of the Maxwell-Boltzmann length (a generalized stopping distance) is outlined, and a model map of the distribution of annihilation radiation is presented for idealized flows.

INTRODUCTION

Recent analyses [1–3] of 0.511 MeV e^+-e^- annihilation data obtained with the Oriented Scintillation Spectrometer Experiment (OSSE) on the *Compton Gamma Ray Observatory* and other satellite and balloon experiments reveal a feature in the annihilation radiation pattern of our galaxy centered at $\ell \sim -2°$ and $b \sim 10°$ with a flux of $\sim 5 \times 10^{-4}$ 0.511 MeV ph cm^{-2} s^{-1}. The annihilation emission component is $\approx$ 5-10° north of the galactic plane in the general direction defined by the axis of the galactic center lobe [4–6]. If near the galactic center at distance $8d_8$ kpc, then positron sources are producing $\gtrsim 4 \cdot 10^{42} d_8^2$ e^+ s^{-1} which annihilate $\approx$ 1-2 kpc above the galactic plane.

We [7] recently proposed that the origin of the high latitude annihilation radiation feature is due to an episode of starbust activity taking place within a few hundred pc of the Galactic Center. The positrons are produced in the decay of freshly synthesized elements – prinicipally by the ^{56}Ni→^{56}Co →^{56}Fe and the ^{44}Ti→^{44}Sc→^{44}Ca decay chains – and are convected upward in a fountain of hot gas. The origin of the positrons and the normal matter annihilation target are thus explained by a single mechanism. Recent starburst activity is also indicated by X-ray observations of hot gas [8,9].

Here we describe in more detail the equations that are solved to calculate the high-latitude annihilation radiation patterns.

POSITRON ENERGY-LOSS AND TRANSPORT

In the model proposed in [7], positrons are injected in hot thermal gas which expands and rises away from the Galactic Center midplane region. To avoid a hydrodynamical fluid calculation (see [10] for a detailed treatment), it was assumed that the gas rises in a fountain geometry which has the shape of a cone cut off at one end. If the upward speed of the gas is prescribed, then the density of the medium and the time between injection and arrival at a given height z above the galactic midplane is exactly determined for gas whose number density depends solely on z. For simplicity, it was assumed in [7] that the gas rises with constant speed.

The instantaneous energy-loss rate of a positron with Lorentz factor γ in a fully ionized plasma with proton number density $n_p(z)$ at height z is given by

$$-\dot{\gamma}_{\rm tot}(z) = g_1 n_p(z) c\sigma_T \ln\Lambda/\beta + 8\alpha_f r_e^2 c g_3 n_p(z) \gamma \ln(2\gamma) + \dot{\gamma}_{\rm adi} +$$

$$+ \frac{4}{3} c\sigma_T [u_{\rm B}(z) + u_{\rm ph}(z)] \beta^2 \gamma^2 , \qquad (1)$$

where the terms on the right-hand-side of equation (1) represent Coulomb, bremsstrahlung, adiabatic, and synchrotron and Thomson losses, respectively. The symbols have their usual definitions, and $g_1 = 1 + 2Y + 8Z$ and $g_3 = 1 + 3Y + 36Z$, where $Y = n_\alpha/n_p$ is the α-particle to proton ratio, and Z is the ratio of the sum of the number densities of C, N, and O to n_p. The quantities $u_B(z)$ and $u_{\rm ph}(z)$ are the dimensionless magnetic field and photon energy densities in units of the electron rest mass energy.

The expression for the adiabatic energy-loss rate involves a number of subtleties. The expansion of the flow is due to the thermal gas pressure if we are dealing with a high-beta plasma (see [11] for the opposite limit). The nonthermal particles are coupled to the thermal gas if there is even a very weak magnetic field. Interactions with magnetic scattering centers which are expanding with the flow gives an energy-loss rate $\propto (\gamma - 1)\dot{V}/V$, where $V(t)$ is the fluid volume. If the nonthermal particles are simply entrained in the flow by a large-scale magnetic field, then there would be no energy loss but only a density change. On the other hand, if the scattering centers are due to magnetohydrodynamic turbulence characterized by a spectrum of plasma waves (see, e.g., [12]), then both energy losses and gains for different nonthermal particle energies could result from stochastic gyroresonant wave-particle interactions in the expanding flow. Noting these different possibilities, we let $-\dot{\gamma}_{\rm adi} = (\gamma - \gamma^{-1})\dot{V}/(3V)$, which has the correct form for the energy-loss rate due to expanding scattering centers and which reduces to the correct expression when the positrons become part of the thermal pool.

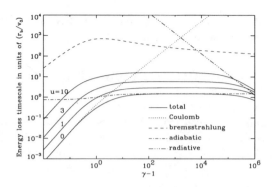

FIGURE 1. Time scale for energy loss of positrons with kinetic enegy $m_e c^2(\gamma - 1)$ at different heights $z = u r_b / \tan \chi$, where $\tan \chi = 0.1$, with separate contributions shown for $u = 0$.

Figure 1 shows the time scale for positron energy loss in units of (r_b/v_0), which characterizes the time scale for fluid rising with speed v_0 to cross a region of size r_b. Here we take $v_0 = 100$ km s^{-1} and $r_b = 100$ pc, giving a crossing time scale of $\cong 1$ million years. In Fig. 1, the proton density at the base of the fountain is $n_p(0) = 0.13$ cm^{-3} corresponding to the injection of 1 $M_\odot$ per century into the base of the fountain. The fountain opening angle $\chi = 5.7°$, implying that $\dot{V}/V = 2v_0 \tan \chi /(r_b + 2z \tan \chi)$. The quantity $u = z \tan \chi / r_b$ is the dimensionless height above the base of the fountain; thus $u = 1$ corresponds to a height of 1 kpc. A constant magnetic field of $B = 3$ μG is used in the calculation of the synchrotron energy-loss rate, and we set the Thomson losses to zero in Fig. 1.

As shown in Fig. 1, Coulomb losses dominate at mildly relativistic energies. The two-body Coulomb process is therefore the most important thermalization process to consider for β^+ injection from radionuclei decay [13] and for positrons produced by $\gamma\gamma \to e^+$-e^- processes in hot plasmas near black holes. Adiabatic losses could be important for sufficiently dilute thermal particle densities or large opening angles of the fountain. For the given magnetic field strength, synchrotron losses play a role only for very large values of γ. At such energies, diffusion could strongly affect the mode of transport, and a treatment following the Lerche/Schlickeiser model [14] is required.

THERMALIZATION AND ANNIHILATION

Positrons are thermalized principally through Coulomb interactions, and then annihilate with thermal electrons to produce the measured 0.511 MeV annihilation photons. Because of the motion of the hot gas, the location where positrons become effectively thermalized represents a mapping from

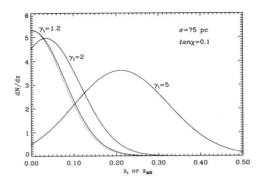

FIGURE 2. Mapping between nonthermal positron injection distribution and thermalization injection distribution. Dotted curve gives the nonthermal injection distribution with height z, and solid curves give the z-distribution of thermalized positrons which were initially injected with Lorentz factor γ_i.

the location where a positron is injected with Lorentz factor γ_i to another location where the positron becomes part of thermal distribution. We call the distance between injection and thermalization the Maxwell-Boltzmann length. This quantity is a nonlinear function of the location and positron injection energy. Figure 2 illustrates the the mapping between the spatial-distribution of the positron injection function and the thermalization injection distribution. The former is approximated by a Gaussian function centered at the Galactic midplane with a 75 pc standard deviation. The Maxwell-Boltzmann length is calculated in FIgure 2 by assuming that positrons lose energy through Coulomb processes only.

MODEL ANNIHILATION FOUNTAIN MAPS

Given the physics of the fountain outlined above, we calculated the spatial distribution of annihilation radiation [7]. The resultant annihilation emissivity is added to a Galactic disk and bulge distribution fitted to the measured [3] annihilation emissivity. A grayscale map of the annihilation emissivity is superimposed over radio contours [5] of the Galactic Center region in Figure 3. A fountain of hot, pair-laden gas that is driven into the galactic halo by starburst activity in the Galactic Center region can explain the high-latitude annihilation feature.

Acknowledgements: We thank Dr. Ronald Murphy for help in producing the maps of the annhihilation fountain. This work was supported by the NASA Guest Investigator Program and the Office of Naval Research.

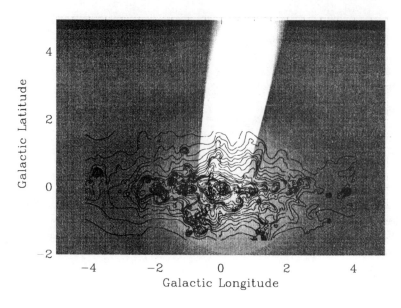

FIGURE 3. Grayscale map of the annihilation emissivity superimposed over a radio contour image of the Galaxy.

REFERENCES

1. Cheng, L. X. et al. ApJ **481**, L43 (1997)
2. Purcell, W. R. et al. 1997a, *The Transparent Universe*, Proceedings of the 2nd INTEGRAL Workshop, ESA SP-382, p. 67
3. Purcell, W. R. et al. 1997b, *ApJ*, in press
4. Pohl, M., Reich, W., and Schlickeiser, R., *A&A*, **262**, 441 (1992)
5. Mezger, P. G., Duschl, W. J., and Zylka, R., *A&A* **7**, 289 (1996)
6. Pohl, M., these proceedings (1997)
7. Dermer, C. D., and Skibo, J. G., *ApJ Letters*, in press (1997)
8. Koyama, K. et al., *Nature* **339**, 603 (1989)
9. Yamauchi, S. et al., *ApJ* **365**, 532 (1990)
10. Suchkov, A. A., Balsara, D. S., Heckman, T. M., Leitherer, C., *ApJ* **430**, 511 (1994).
11. Goldshmidt, O., and Rephaeli, Y., *ApJ* **444**, 113 (1995).
12. Dermer, C. D., Miller, J. A., and Li, H., *ApJ* **456**, 106 (1996)
13. Chan, K. W. and Lingenfelter, R. E. 1993, ApJ, 405, 614
14. Lerche, I. and Schlickeiser, R., *ApJ* **239**, 1089 (1980)

Issues Concerning the Orion Gamma Ray Line Observations: Overview and X-Ray Emission

R. Ramaty*, B. Kozlovsky† and V. Tatischeff**

*Goddard Space Flight Center, Greenbelt, MD 20771
†School of Physics and Astronomy, Tel Aviv University, Israel
**Goddard Space Flight Center, Greenbelt, MD 20771[1]

Abstract. We review the major implications of the COMPTEL gamma ray line observations of Orion. We present new calculations of the expected X-ray line and continuum emissions that accompany the observed gamma ray lines. We show that a serious conflict may exist between the predicted oxygen K_α line emission and upper limits from ROSAT. This conflict is alleviated, but not fully resolved, if the gamma rays are produced in the warm, partially ionized medium at cloud boundaries.

INTRODUCTION

We briefly review the main implications of the gamma ray line emission observed from Orion with COMPTEL [1,2] as discussed by one of us (RR) in an invited talk at the 4th COMPTON Symposium. More details can be found in [3-7]. We then provide new results on the X-ray production by the same accelerated nuclei which produce the observed gamma ray lines. Dogiel et al. [8] stated that the 0.5-2 keV bremsstrahlung, produced by knock-on electrons due to the ions which produce the gamma ray line emission in Orion, will exceed upper limits set with ROSAT. We show that the contribution of the knock-on process to the total X-ray production by energetic ions is negligible in comparison with inverse bremsstrahlung and that the X-ray emission from both these continuum processes is not inconsistent with the ROSAT data. However, in the 0.5-1 keV region the dominant X-ray production mechanism is line emission following electron capture onto fast O nuclei. We find that a serious conflict may exist between the ROSAT upper limits and the predicted X-ray line emission.

---

[1] NAS/NRC Research Associate

IMPLICATIONS OF THE GAMMA RAY LINES

The 3-7 MeV emission observed with COMPTEL is most likely nuclear line emission produced in accelerated ion interactions. However upper limits at other wavelengths, as well as these data themselves, place severe constraints on the nature of the accelerated particles:

(i) The near absence of accelerated protons and α particles. This follows from the gamma ray spectrum [2], which shows no evidence for narrow line emission [7], and is supported by the 1-3 MeV upper limits and by the energetics [3].

(ii) The suppression of the accelerated particle Ne-Fe relative to C-O. This follows [4] from the 1-3 MeV upper limits [2]. The suppression of the protons, α particles and Ne-Fe could be understood if the ejecta of supernovae, the winds of massive stars [9,10,4], or the pick-up ions from the breakup of dust [4], were the sources of the accelerated particles. The preferential acceleration of C and O, the main progenitors of the Galactic B and Be, provides a link to the origin of these elements [10,11].

(iii) The low flux of relativistic electrons implied by the upper limits on the MeV continuum. In analogy with gradual solar flares [3], this result supports suggestions that the acceleration is by shocks [9,12]. The entire 3-7 MeV emission cannot be relativistic electron bremsstrahlung [8] because of the structured nature of the observed spectrum.

(iv) A hard energy spectrum with a high energy cutoff for the accelerated particles. A harder spectrum implies a smaller power deposition by accelerated particles into the ambient medium [3,4], while the high energy cutoff, $E_0 \lesssim 100$ MeV/nucleon [13], follows from the absence of enhanced high energy (>30 MeV) gamma ray emission [14]. The assumed spectrum is of the form $E^{-1.5}e^{-E/E_0}$ [4]. Similar spectra are expected from acceleration by an ensemble of shocks [15].

(v) Possible anisotropic interactions of the accelerated particles. This follows from the $\sim$4 MeV feature [2] which could be due to redshifted broad 4.44 MeV line emission produced by particles interacting in a sharp ambient medium density gradient [6,7]. This could happen if particles trapped in the Orion-Eridanus bubble bombarded the Orion clouds [16]. Gamma ray line production at cloud boundaries, but not in their interiors, alleviates the difficulties caused by the ionization that accompanies the gamma ray line production. The accelerated particles could have ionized $\sim 2 \times 10^4 M_\odot$ in 10^5 years [5], a small fraction of the total available mass. Line splitting in an isotropic interaction model [17] cannot account for the observed spectrum because it predicts [7] a peak at $\sim$5 MeV which is not observed.

(vi) An extended gamma ray emitting region in Orion. This follows from the COMPTEL observations, which have imaged the emission [2], and the apparent discrepancy between the OSSE [18] and COMPTEL data which can be best understood if the emitting region is spatially extended.

X-RAY EMISSION

Energetic ions produce continuum X-ray emission via bremsstrahlung from both primary and secondary electrons [19]. The primary electrons are ambient electrons interacting with the energetic ions. The secondary electrons are knock-on electrons which subsequently produce bremsstrahlung by interacting with the ambient atoms. The primary bremsstrahlung is also referred to as inverse bremsstrahlung [20]. In addition to the continuum, the energetic ions also produce line emission due to atomic transitions (mostly 2p→1s) in both the accelerated ions and the ambient atoms. In the accelerated ions the transitions follow electron capture onto bare or singly charged ions or collisional excitations of the 1s electrons [21,22]. In the ambient atoms the transitions follow the creation of K-shell vacancies by accelerated particles. The creation of K-shell vacancies by X-rays is the standard mechanism for the production of fluorescent K_α line emission in astrophysics. The creation of such vacancies by fast particles is also possible [23].

We evaluated the inverse bremsstrahlung using the approach of Boldt and Serlemitsos [20]. For the secondary electron bremsstrahlung we first calculated the production of knock-on electrons using the cross section given by Chu [24] with corrections at high energies [25,26]. We then calculated the bremsstrahlung using equation 3BN of Koch and Motz [27] with appropriate electron energy losses. All the calculations were carried out in a thick target with a neutral ambient medium.

In Figure 1 we compare our calculated continuum X-ray production cross sections with experimental data [24]. Their relatively high energy proton beam

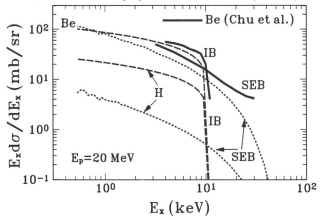

FIGURE 1. Continuum X-ray production cross sections by a 20 MeV proton beam. The Chu et al. data (solid curves) are for X-rays observed from a Be target at 90° to the beam. The calculations, for both Be and H targets, are angle averaged. IB - inverse bremsstrahlung; SEB - secondary electron bremsstrahlung.

allowed Chu et al. [24] to separate the inverse bremsstrahlung (IB) from the secondary electron bremsstrahlung (SEB). The dashed curves represent our angle-averaged calculated cross sections. We see that for Be the curves follow the data quite well, the discrepancy of less than a factor of 2 being due to the angle averaging and possible contaminations in the experiment, for example Compton scattering of gamma rays in the Be target. We also note that for the Be target the contributions of IB and SEB are approximately equal. With increasing target atomic number, the SEB becomes dominant relative to the IB, because the knock-on electrons produce bremsstrahlung efficiently off the high Z target nuclei. This, however, is not the case for a H target, for which IB dominates, except at the highest X-ray energies (Figure 1).

We evaluate the K_α line emission for energetic O ions capturing electrons from a neutral ambient medium with solar composition using the photon multiplicities of Pravdo and Boldt [21]. The most relevant accelerated ion for X-rays between 0.5 and 2 keV is O, whose K_α lines are at 0.65 keV (capture on fully stripped O) and 0.56 keV (capture on singly charged O). We ignore X-ray line production due to K-shell vacancy creation in ambient O by the accelerated particles. We estimate that only for a partially ionized medium will this process make a non-negligible contribution.

The results are shown in Figure 2. These thick target calculations, normalized to the observed 3-7 MeV gamma ray emission, are for a neutral ambient medium with solar composition, and accelerated particle energy spectra $E^{-1.5}e^{-E/E_0}$ and composition He:C:O=0.92:0.69:1, with almost vanishing abundances for other elements (WC [4]). N_H is the H column density towards Orion. The 'Total' curves, for three values of N_H, include IB, SEB and K_α line emission. The SEB contribution is shown with no absorption applied. The ROSAT upper limit [8] already includes nominal absorption. SEB indeed makes a negligible contribution. We repeated the SEB calculation of [8] using the same formalism and input data. The result agrees with our more accurate approach but is lower than that of [8] by about an order of magnitude.

We see that a serious conflict may exist between the predicted line emission and the data. In contrast to the continuum, the line emission could be suppressed by assuming that: the ambient medium is ionized; the current epoch accelerated particle spectrum is strongly cut off below several MeV/nucl (the bulk of the line emission is produced by ~2 MeV/nucl O nuclei); or there is much stronger absorption than that assumed in [8]. If H and He, but not O, are fully ionized, based on [21], we can expect a reduction of the predicted line emission by about an order of magnitude. A further reduction would result if the ambient O is in grains. We do not expect that O is fully ionized at the cloud boundaries, the most likely site for gamma ray line production. By increasing the accelerated proton and α particle abundances relative to O we would decrease the X-ray line emission relative to the continuum. But in this case the continuum would become inconsistent with the ROSAT upper limits.

REFERENCES

1. Bloemen, H. et al., *A&A* **281**, L5 (1994).
2. Bloemen, H. et al., *ApJ* **475**, L25 (1997).
3. Ramaty, R., Kozlovsky, B., and Lingenfelter, R. E., *ApJ* **438**, L21 (1995).
4. Ramaty, R., Kozlovsky, B., and Lingenfelter, R. E., *ApJ* **456**, 525 (1996).
5. Ramaty, R., *A&A Suppl.* **120**, C373, (1996).
6. Ramaty, R., Kozlovsky, B., and Lingenfelter, R. E., *Proc. 2nd INTEGRAL Workshop* ESA SP-382, 75 (1997).
7. Kozlovsky, B., Ramaty, R., and Lingenfelter, R. E., *ApJ* **484**, in press (1997).
8. Dogiel, V. A. et al., *24th Internat. Cosmic Ray Conf. Papers*, in press (1997).
9. Bykov, A. M., and Bloemen, H., *A&A* **283**, L1, (1994).
10. Cassé, M., Lehoucq, R., and Vangioni-Flam, E., *Nature* **373**, 318 (1995).
11. Ramaty, R., Kozlovsky, B., and Lingenfelter, R. E., *ApJ* **488**,in press (1997).
12. Nath, B. B., and Biermann, P. L. 1994, *MNRAS* **270**, L33 (1994).
13. Tatischeff, V., Ramaty, R., and Mandzhavidze, N., this volume.
14. Digel, S. W., Hunter, S. D., and Mukherjee, R., *ApJ* **441**, 270, (1995).
15. Bykov, A. M., *Space Science Revs.* **74**, 397, (1995).
16. Parizot, E., this volume.
17. Bykov, A. M., Bozhokin, S. V., and Bloemen, H., *A&A* **307**, L37, (1996).
18. Murphy, R. J. et al., *ApJ* **478**, **990**, (1996).
19. Anholt, R., *Rev. Mod. Phys.* **57**, 995 (1985).
20. Boldt, E. A., and Serlemitsos, P., *ApJ* **157**, 557 (1969).
21. Pravdo, S. H., and Boldt, E. A., *ApJ* **200**, 727 (1975).
22. Bussard, R. W., Ramaty, R., and Omidvar, K., *ApJ* **220**, 353 (1978).
23. Cahill, T. A., *Ann. Rev. Nucl. Part. Sci.* **30**, 211 (1980).
24. Chu, T. C. et al. *Phys. Rev. A* **24**, 1720 (1981).
25. Rudd, M. E. et al. *Phys. Rev.* **151**, 20 (1966).
26. Toburen, L. H., and Wilson, W. E., *Phys. Rev. A* **5**, 247 (1972).
27. Koch, H. W., and Motz, J. W., *Rev. Mod. Phys.* **31**, 920 (1959).

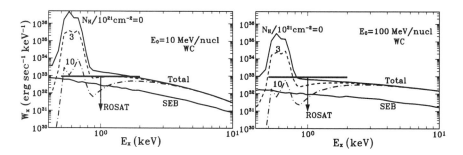

FIGURE 2. Differential X-ray luminosities at Orion.

Constraints from Pion Production on the Spectral Hardness of the Low Energy Cosmic Rays in Orion

Vincent Tatischeff[1], Reuven Ramaty, and Natalie Mandzhavidze[2]

Goddard Space Flight Center, Greenbelt, MD 20771

Abstract. The gamma ray line emission observed with COMPTEL from Orion is attributed to the interaction within the molecular cloud of fast nuclei resulting from diffusive shock acceleration of a seed source. We have carried out detailed calculations of pion production by such low energy cosmic rays enriched in heavy elements. In the light of the EGRET observations of the molecular complex, we show that the characteristic energy of the accelerated particles spectrum can be quite large, but should not exceed 100 MeV/nucleon in order to not overproduce the high energy gamma ray emission.

INTRODUCTION

The gamma ray excess observed from the Orion molecular cloud complex with the COMPTEL instrument has been attributed to nuclear deexcitation lines from accelerated ^{12}C and ^{16}O, themselves excited by nuclear collision [1-3]. This line emission provides evidence for the presence of low energy cosmic rays with high C/H and O/H abundance ratios in Orion. However, both the physical origin of the accelerated material, as well as the nature of the acceleration mechanism remain unknown. In particular, as a recent analysis of the COMPTEL data suggests that the nuclear interactions responsible for the gamma ray emission might be anisotropic [4], the hardness of the accelerated particle spectrum becomes poorly constrained by the width of the observed broad ^{12}C and ^{16}O gamma ray lines.

The EGRET observations of gamma ray emission above 30 MeV from Orion, which show no evidence for high energy gamma ray counterpart from pion decay [5], give interesting constraints on the energy spectrum of the accelerated

---

[1] NAS/NRC Research Associate
[2] University Space Research Associate

material. We have investigated in detail the pion production from the interaction of accelerated heavy ions with the ambient medium.

THE MODEL

One interesting aspect of heavy ion collisions at intermediate energies is the observation of a significant pion production cross section at "subthreshold beam energies", i.e. below the free nucleon-nucleon threshold at 290 MeV. We take this effect into account carefully in order to predict the pion-production from low energy cosmic rays enriched in heavy elements. We compile, from a number of accelerator measurements, five reference inclusive π-production cross sections, namely $p + p \to \pi^0 + X$, $p + \alpha \to \pi^0 + X$, $p + {}^{12}C \to \pi^0 + X$, ${}^{12}C + {}^{12}C \to \pi^0 + X$ and $p + p \to \pi^+ + X$, and determine the other needed cross sections by simple rescaling models.

The "pionic bremsstrahlung" model provides an elegant description of the collective pion-production mechanism below 290 MeV/nucleon [6] and we use it to describe the reaction kinematics below this threshold. Above it, we use the isobar model, which assumes that pions are formed through the excitation of isobaric resonances [7].

The pion production is followed by gamma ray emission from the following processes:

$$\pi^0 \to 2\gamma \quad T_{1/2} = 1.78 \cdot 10^{-16} s \tag{1a}$$

$$\left.\begin{array}{l}\pi^+ \to \mu^+ + \nu_\mu \\ \pi^- \to \mu^- + \bar{\nu}_\mu\end{array}\right\} \quad T_{1/2} = 2.55 \cdot 10^{-8} s \tag{1b}$$

$$\left.\begin{array}{l}\mu^+ \to e^+ + \nu_e + \bar{\nu}_\mu \\ \mu^- \to e^- + \bar{\nu}_e + \nu_\mu\end{array}\right\} \quad T_{1/2} = 2.20 \cdot 10^{-6} s, \tag{1c}$$

where the electrons and positrons produce gamma rays by bremsstrahlung and annihilation in flight. The lepton-production kinematics is calculated following Okun [8]. We use the prescription of Murphy, Dermer and Ramaty [9] for the production of bremsstrahlung and annihilation radiation.

We assume a thick target interaction model; an accelerated particle source spectrum resulting from diffusive strong shock acceleration [10],

$$\frac{dN_i}{dt}(E) = K_i E^{-1.5} e^{-E/E_0} \tag{2}$$

where the K_i's are proportional to the accelerated particle abundances; and the following accelerated particle compositions: cosmic ray source, ejecta of a 60 M$_\odot$ supernova, pick-up ions resulting from the breakup of interstellar dust, winds of Wolf-Rayet stars of spectral type WC [11]. The results are shown in Figure 1. We see that for the same E_0, the cosmic ray source provides a

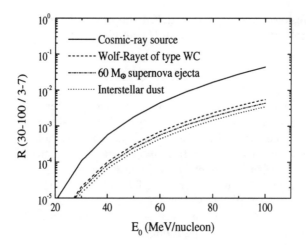

FIGURE 1. Ratio of the number of photons in the 30-100 MeV energy range produced by pion decay to the number of photons in the 3-7 MeV energy range resulting from nuclear deexcitations, as a function of E_0 for the shock acceleration spectrum of eq.(2). The 3-7 MeV emission is calculated as in [11].

much higher 30-100 MeV to 3-7 MeV ratio than the three other sources. This is a consequence of its lower ^{12}C and ^{16}O abundances relative to protons and α-particles.

RESULTS

The >30 MeV EGRET data are determined through a fit of the observed EGRET intensity map by a combination of H I (atomic hydrogen) and CO (molecular hydrogen) maps, plus an isotropic term for the extragalactic emission [5]. The comparison of the COMPTEL data [2] with the EGRET data thus requires that we specify the location of the emission produced by the low energy cosmic rays. The COMPTEL 3-7 MeV map, however, is not accurate enough to reveal whether this emission is coming from the molecular clouds or from the atomic hydrogen in their vicinity. Therefore, we consider three different cases, namely that the 3-7 MeV emission comes from the molecular clouds only, from the atomic hydrogen only, or from both the atomic and the molecular hydrogen. In latter case, the EGRET data are determined by adding the CO and H I components.

We fit both the EGRET and the COMPTEL data by the function:

$$Q_\gamma = \alpha \cdot Q_\gamma[GCR] + \beta \cdot Q_\gamma[LECR(E_0)] \qquad (3)$$

where $Q_\gamma[GCR]$ is the diffuse gamma ray emission in the solar neighborhood due to the Galactic cosmic rays [12], $Q_\gamma[LECR(E_0)]$ is the gamma ray emis-

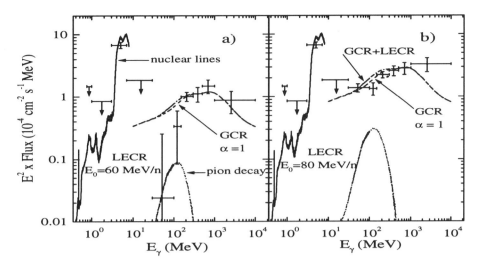

FIGURE 2. Illustration of the fit of the gamma ray emission from Orion with the local diffuse emission and the emission due to a low energy cosmic rays with Wolf-Rayet composition; (a) $E_0=60$ MeV/nucleon and the EGRET data (above 30 MeV) are determined by CO model fits; (b) $E_0=80$ MeV/nucleon and the EGRET data are determined by atomic hydrogen model fits.

sion (nuclear lines and pion decay) due to low energy cosmic rays enriched in heavy elements, α is the ratio of the high energy cosmic rays density in Orion to the high energy cosmic rays density in the solar neighborhood, and

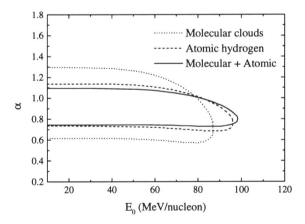

FIGURE 3. 99% confidence level contours of the fit of the gamma ray emission from Orion as functions of E_0 and α, the Galactic cosmic rays density in Orion normalized to the local cosmic rays density.

β is a scale factor for the low energy component, namely the product of the ambient medium density to the cosmic rays density. Thus we have three free parameters: α, β and E_0.

If the low energy cosmic rays irradiate the molecular clouds only or the atomic gas only, the best fits are provided for $\alpha=0.955$ with a reduced χ^2 of 2.15, and $\alpha=0.935$ with a reduced χ^2 of 1.62, respectively. If the low energy cosmic rays irradiate both the molecular and the atomic gas, the best fit is provided for $\alpha=0.92$ with a reduced χ^2 of 2.30. In all three cases, the best fits imply no significant pion production by the low energy component, which corresponds to $E_0 \lesssim 50$ MeV/nucleon. However, as can be seen from the 99% confidence level contours in Figure 3, E_0 can be as high as 85, 95 and 100 MeV/nucleon for the three cases respectively.

We also see that in the case of only molecular clouds being irradiated (Figure 2a), even without the contribution of the low energy component, the Galactic diffuse gamma ray emission exceeds the observations below about 150 MeV. This could suggest that not all high energy cosmic rays penetrate into the densest parts of the clouds.

In conclusion, as already pointed out by Digel, Hunter and Mukherjee [5], the high energy cosmic ray flux in Orion is the same, within the errors, as the local cosmic ray flux ($\alpha \cong 1$). Concerning the low energy component, the characteristic energy E_0 of the shock acceleration spectrum should not exceed 100 MeV/nucleon in order not to overproduce the high energy gamma ray emission. Even though this limit is higher that previous limits inferred from line width considerations (which are quite uncertain), it constitutes an important constraint on the acceleration mechanism responsible for the low energy component in Orion. We wish to thank Seth Digel for his essential help on the EGRET data.

REFERENCES

1. Bloemen, H., et al., *A&A* **281**, L5 (1994).
2. Bloemen, H., et al., *ApJ* **475**, L25 (1997).
3. Ramaty, R., *A&A Suppl.* **120**, C373 (1996).
4. Kozlovsky, B., Ramaty, R., and Lingenfelter, R. E., *ApJ* **484**, in press (1997).
5. Digel, S. W., Hunter, S. D., and Mukherjee, R., *ApJ* **441**, 270 (1995).
6. Vasak, D., Müller, B., and Greiner, W., *J. Phys.* **G11**, 1309 (1985).
7. Stecker, F. W., *Astrophys. Space Sci.* **6**, 377 (1970).
8. Okun, L. B., *Leptons and Quarks*, Moscow, Nauka, 1981, p. 24.
9. Murphy, R. J., Dermer, C. D., and Ramaty, R., *ApJ Suppl.* **63**, 721 (1987).
10. Ellison, D. C., and Ramaty, R., *ApJ* **298**, 400 (1985).
11. Ramaty, R., Kozlovsky, B., and Lingenfelter, R. E., *ApJ* **456**, 525 (1996).
12. Skibo, J. G., *Ph.D. thesis*, University of Maryland (1993).

Gamma-Ray Lines from OB Associations at $Z = Z_\odot$ and $Z = 2Z_\odot$

E. Parizot[*], J. Paul[*] and M. Cassé[*,†]

[*]*DAPNIA/Service d'astrophysique, CEA-Saclay, 91191 Gif-sur-Yvette, France*
[†]*Institut d'Astrophysique de Paris, CNRS, 98bis bd Arago, 75014 Paris, France*

Abstract. We calculate the gamma-ray line emission induced by energetic particles (EPs) in molecular clouds coupled with OB associations, assuming that the EPs originate from the winds of massive stars. Their composition reflects the average wind composition of an individual Wolf-Rayet star or of a whole OB association. We consider two initial stellar metallicities (solar and twice solar), and show that the corresponding gamma-ray line ratios are different, and that the lines are narrower at twice solar metallicity, which is relevant to the inner Galaxy. Addressing the Orion emission detected by COMPTEL, we show that our average wind compositions satisfy the available constraints. We analyse the most relevant present and future observables (broad band ratios, ^{12}C and ^{16}O line widths, ^{12}C*/^{16}O* line ratio) and discuss the differences between our results and those obtained with a WC composition based only on the extreme late phase wind, as used up to now. In particular, we predict a flux of 1–$2\,10^{-5}$ ph cm^{-2} s^{-1} at ~ 0.450 MeV from the Orion complex, due to the de-excitation of the fusion-spallation products ^{7}Li and ^{7}Be. Such a flux is below the current OSSE upper limit, but above the expected sensitivity of *INTEGRAL*.

INTRODUCTION

The OB associations release large amounts of energy in the ISM, from both stellar winds and supernovae explosions. This results in the formation of large bubbles (or superbubbles) within or in the vicinity of the parent molecular cloud, filled with a tenuous, hot, metal-rich plasma. Such bubbles have been proposed as natural sites of particle acceleration up to energies of a few tens or hundreds of MeV, because of the presence of multiple shocks and magnetic turbulence [1,2]. A well known example is the Orion-Eridanus bubble originated from the Orion OB1 association [3–5]. Its total energy is estimated around $2\,10^{52}$ erg [6]. It has been proposed that particles accelerated within this bubble are responsible for the gamma-ray emission detected by COMPTEL in Orion, attributed to the de-excitation of ^{12}C and ^{16}O nuclei [7–10]. In this context, we assume that the energetic particles (EPs) interacting in the

Orion molecular clouds have a composition which reflects that of the winds of massive stars supplying the acceleration process within the bubble. We therefore calculate the average wind composition of Wolf-Rayet (WR) stellar winds, using the evolutionary models of the Geneva group [11,12] in two different cases. First, we assume that the metallicity of the surrounding medium and that of the ZAMS stars are solar, as is approximately the case in Orion. Second, we investigate the case of a twice solar metallicity, which is relevant to the emission from the Galactic ring at $\sim$ 5 kpc, since most of the known metal rich massive stars are located in this region.

THE MODEL

The average composition of the wind ejecta of a massive star is calculated from the total ejected mass of each isotope as :

$$M_i = \int_{\text{whole life}} \dot{M}(t) X_i(t) \mathrm{d}t \tag{1}$$

where $\dot{M}(t)$ and $X_i(t)$ are respectively the stellar mass loss rate and the mass fraction of element i in the wind at time t.

We have estimated in this way the mean-wind composition of stars with ZAMS masses $M = 40, 60, 85$ and 120 M$_\odot$, for different models (C, D and E; see references [11,12]). Some of these EP compositions are shown in Table 1, where we also show, for comparison, the WC composition used by Ramaty and co-workers, based on the extreme late phase wind of a WC star, which corresponds to a negligible ejected mass (we denote this composition by late-WC). As can be seen, mean-wind and late-WC compositions are quite different : our compositions are much richer in H and He, and have a higher C/O abundance ratio. The consequences of these features are analysed below.

In the context of particle acceleration from the hot plasma filling a bubble caused by the activity of a whole OB association, the mean-wind compositions of individual stars are probably not appropriate. Rather, we have to consider the global average composition of all the stellar winds in the association. To obtain such 'mean-OB' compositions, we weight the contribution of each star according to its total mass loss and to its probability of occurence among the association, which is given by the IMF. For Orion, we take an IMF index of x=1.7 [13]. Some of the obtained compositions are shown in Table 1. It can be seen in particular that the H and He abundances are higher for associations with twice solar initial metallicity. This is because the WR phase begins earlier in the stellar evolution, at a stage when the core helium has not yet burned completely. Similarly, more carbon is saved from burning into oxygen, which results in a higher C/O abundance ratio.

Concerning the EP energy spectrum, we use the generic injection spectrum adopted by different authors [14–16] : $Q_i(E) = K_i (E/E_0)^{-1.5} \exp(-E/E_0)$,

TABLE 1. Mean wind and mean OB compositions assumed for the EPs

isot.	40 M$_\odot$/C	60 M$_\odot$/C	60 M$_\odot$/D	60 M$_\odot$/E	late-WC	OB/0.02	OB/0.04
^{1}H	8.03e+01	5.76e+01	1.48e+02	1.61e+02	1.00	5.32e+01	1.21e+02
^{4}He	2.52e+01	1.91e+01	7.98e+01	6.42e+01	9.20e-01	1.71e+01	5.63e+01
^{12}C	1.59	2.57e+00	5.66e+00	4.78e+00	6.90e-01	1.88e+00	3.79e+00
^{13}C	1.15e-03	6.88e-04	2.34e-03	2.09e-03	0.00	6.69e-04	4.29e-03
^{14}N	1.08e-01	4.60e-02	2.00e-01	2.11e-01	2.40e-03	5.90e-02	3.56e-01
^{16}O	1.00	1.00	1.00	1.00	1.00	1.00	1.00
^{20}Ne	1.56e-02	1.30e-02	3.97e-02	3.53e-02	2.30e-03	1.15e-02	5.95e-02
^{22}Ne	4.52e-02	7.66e-02	2.21e-01	1.83e-01	2.00e-02	5.53e-02	2.78e-01
^{24}Mg	4.71e-03	3.94e-03	1.20e-02	1.07e-02	6.30e-04	3.50e-03	1.81e-02
^{25}Mg	5.94e-04	4.97e-04	1.52e-03	1.35e-03	0.00	4.41e-04	2.28e-03
^{26}Mg	5.49e-04	1.67e-03	1.49e-03	1.49e-03	5.32e-03	4.87e-04	2.52e-03
^{27}Al	4.71e-04	3.95e-04	1.20e-03	1.07e-03	6.70e-05	3.50e-04	1.81e-03
^{28}Si	5.12e-03	4.29e-03	1.31e-02	1.17e-02	6.90e-04	3.80e-03	1.97e-02
^{32}S	2.71e-03	2.27e-03	6.93e-03	6.19e-03	3.30e-04	2.02e-03	1.04e-02
^{40}Ca	3.29e-04	2.75e-04	8.40e-04	7.49e-04	5.00e-05	2.44e-04	1.26e-03
^{56}Fe	4.45e-03	3.72e-03	1.14e-02	1.01e-02	9.10e-04	3.30e-03	1.71e-02

where the coefficients K_i are proportional to the abundances by number of each isotope i. We choose the overall normalisation so as to reproduce the detected flux from Orion in the range 3-7 MeV. This spectrum allows one to explore a variety of phenomenologically possible spectra by varying one single parameter, namely the 'break energy' E_0.

We distinguish between 'direct reactions' - in which the projectile is light (p, α...) and the target is heavy (^{12}C, ^{16}O...) - leading to narrow line emission, and 'inverse reactions' (heavy projectile and light target) leading to broad line emission (because the excited nuclei de-excite in flight, at a high velocity).

RESULTS

We show in Figure 1a the integrated fluxes in the (0.2–1 MeV) and (1–3 MeV) bands, calculated with some of our mean-wind compositions. These fluxes are normalised to the Orion flux in the (3–7 MeV) band. It appears that the fluxes are higher (with respect to the 3–7 MeV flux) at twice solar metallicity, by a factor of 1.5–3. Accordingly, we predict of variation of the gamma-ray line spectrum with Galactic longitude, which is of interest for the INTEGRAL mission. Of particular importance is the (1–3 MeV)/(3–7 MeV) band ratio, since the COMPTEL observations set a 2σ upper limit of 0.13 on the ratio R defined as : $R = 2 (\int_{1\,\mathrm{MeV}}^{3\,\mathrm{MeV}} E_\gamma^2 Q(E_\gamma) dE_\gamma) / (\int_{3\,\mathrm{MeV}}^{7\,\mathrm{MeV}} E_\gamma^2 Q(E_\gamma) dE_\gamma)$ [15,17].

In Figure 1b, we show this ratio as a function of the injection break energy E_0 for different EP compositions. As already demonstrated by some authors

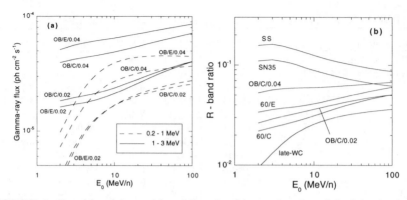

FIGURE 1. Broad band fluxes (a) and R ratios (b) as a function of the injection break energy E_0 for different EP compositions, normalised to the Orion flux in the (3–7 MeV) band. The labels refer to the model used for the massive star evolution (C or E) and to the initial stellar metallicity (0.02 = solar; 0.04 = twice solar).

(e.g. [17]), compositions such as SS or CRS can be excluded on the basis of this observational constraint. We confirm their conclusions and obtain similar results for any of our common test compositions. Concerning our mean-wind and mean-OB compositions, they are all consistent with the COMPTEL data, which is an important conclusion in itself.

The $^{12}C^*/^{16}O^*$ gamma-ray line ratio is shown in Figure 2a for various compositions. This observable appears to be very discriminating, and should be quite easily measured by INTEGRAL's spectrometer SPI. From the COMPTEL data, we estimated that $2 \leq\, ^{12}C^*/^{16}O^* \leq 4$. If this rough guess estimate is right, then one can exclude EP compositions like SS, GR (dust grains), SN35 (supernova ejecta from a 35 $M_\odot$ progenitor), but also late-WC. Conversely, our mean-wind compositions provide a more adequate line ratio. Although our estimate is admittedly uncertain, we point out that whatever the value of the $^{12}C^*/^{16}O^*$ line ratio will prove to be, its measurement will allow one to distinguish between mean-wind and late-WC compositions, and will provide a strong argument to exclude (or favour) compositions such as SS, GR or SN35.

Figure 2b shows the ratio of broad and narrow line components (inverse and direct processes), which provides an indication on the line profiles to be observed by a high resolution spectrometer. We find that the ^{16}O line is significantly narrower than the ^{12}C line, which makes it easier to detect by *INTEGRAL*. The ^{16}O narrow line emission even dominates at twice solar metallicity. Both C and O lines are expected to be narrower in the central radian.

We have also calculated the flux in the ^{7}Li-^{7}Be feature at ~ 0.450 MeV using our mean-wind compositions. This feature results from the de-excitation of ^{7}Li (478 keV) and ^{7}Be (439 keV) nuclei, following $\alpha + \alpha$ reactions. Because of

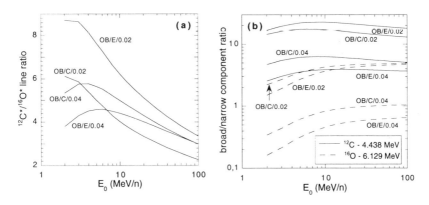

FIGURE 2. $^{12}C^*/^{16}O^*$ line ratio (a) and broad-to-narrow line component ratio (b) as a function of the injection break energy E_0 for different EP compositions, normalised to the Orion flux in the (3–7 MeV) band. The labels as in Figure 1.

the significative abundance of ^{4}He, we obtain much higher fluxes than with the late-WC composition, which is another distinctive feature. We predict a flux of 1–$2\,10^{-5}$ ph cm^{-2} s^{-1} from the Orion complex, which should be measured by *INTEGRAL*.

REFERENCES

1. Bykov, A.M. and Toptygin, I.N., *Sov. Phys. JETP* **71**, 702 (1990).
2. Bykov, A.M. and Fleishman, G.D., *MNRAS* **255**, 269 (1992).
3. Reynolds, R.J., and Ogden, P.M., *ApJ* **229**, 942 (1979).
4. Cowie, L.L., Songaila, A. and York, D.G., *ApJ* **230**, 469 (1979).
5. Burrows, D.N., Singh, K.P., Nousek, J.A., Garmire, G.P. and Good, J., *ApJ* **406**, 97 (1993).
6. Brown, A.G.A., Hartmann, D., and Burton, W.B., *A&A* **300**, 903 (1995).
7. Bykov, A.M., and Bloemen, H., *A&A* **281**, L1 (1994).
8. Bloemen, H., et al., *A&A* **281**, L5 (1994).
9. Parizot, E.M.G., this volume.
10. Parizot, E.M.G., *A&A* submitted (1997).
11. Schaller, G., Schaerer, D., Meynet, G. and Maeder, A., *A&AS* **96**, 269 (1992).
12. Meynet, G., Maeder, A., Schaller, G., Schaerer, D. and Charbonnel, C., *A&AS* **103**, 97 (1994).
13. Brown, A.G.A., de Geus, E.J., and de Zeeuw, P.T., *A&A* **289**, 101 (1994).
14. Ramaty, R., Kozlovsky, B., and Lingenfelter, R.E., *ApJ* **438**, L21 (1995).
15. Ramaty, R., *A&A* **120**, 373 (1996).
16. Parizot, E.M.G., Cassé, M. and Vangioni-Flam, E., *A&A* submitted (1997).
17. Ramaty, R., Kozlovsky, B., and Lingenfelter, R.E., *ApJ* **456**, 525 (1996).

On the Origin of the Orion Energetic Particles

Etienne M.G. Parizot

CEA/DSM/DAPNIA/Service d'astrophysique, CEA-Saclay, 91191 Gif-sur-Yvette, France

Abstract. We present a series of arguments leading to a model for the Orion gamma-ray emission. In particular, we address the crucial question of the origin of the energetic particles interacting with the molecular cloud complex. We show that it is very unlikely that these particles are accelerated within the clouds themselves, and propose that they originate from the Orion-Eridanus superbubble, located in front of the Orion clouds. Generalisation to other OB association-molecular cloud pairs is possible.

INTRODUCTION

The instrument COMPTEL has detected a gamma-ray emission with unexpected intensity in Orion [1,2]. The emission map shows that the source is extended, and coincides roughly with the CO observations tracing the molecular clouds Orion A and B, as well as Mon R2. The emission consists of a $(1.01 \pm 0.14)\,10^{-4}\,\mathrm{ph\,cm^{-2}s^{-1}}$ flux in the 3–7 MeV band, with only upper limits at other COMPTEL energies, i.e. in the bands 0.75–1 MeV, 1–3 MeV and 7–30 MeV [2]. This emission is thought to be due to nuclear interactions between energetic particles (EPs) and the Orion cloud material (e.g. [3–5]), leading principally to ^{12}C and ^{16}O de-excitation lines. This interpretation constitutes the 'standard paradigm' for the Orion gamma-ray emission, and we shall adopt it here. Many questions, however, remain to be answered. For example : where do the EPs come from ? What are they powered by ? How are they accelerated ? What is their chemical composition and energy spectrum ? What is their confining volume ? Are there other similar sources in the Galaxy ?

Some of these questions have already been discussed in the light of the various observational constraints, but a general self-consistent model is still lacking. We shall concentrate here on the question of the origin of the Orion EPs, and show that it provides an interesting guide toward the study of the other questions.

WHERE DO THE EPS COME FROM ?

EPs vs. Galactic Cosmic Rays

We should first point out that the EPs are exceptionally rich in C and O. This is required by the observations, according to three independent arguments : energetics, line profiles and line ratios. Indeed, the EPs responsible for the Orion emission lose energy through ionisation losses at a rate which depends on the EP composition. Now given the gamma-ray flux detected by COMPTEL, the energy deposition rate would be unreasonably high if the EPs were not enriched in C and O, because other nuclei contribute to the energy losses without producing gamma-rays in the 3–7 MeV band [3,4]. Second, the gamma-ray lines are apparently broad, which indicates that most of the flux comes from 'inverse reactions' in which energetic C and O nuclei are excited on ambient H and He, and de-excite in flight. This requires an enrichment of the EPs in C and O, with respect to the target composition (approximately solar). Finally, the absence of gamma-ray lines detected in the 1–3 MeV band allows one to quantify the overabundance of C and O in the EPs, with respect to other metals [4], all of which have their lines in this band.

A second crucial information can be drawn from the observations : the energy spectrum of the Orion EPs has to be cut at a few tens of MeV. Otherwise, i) the C and O de-excitation lines would be too broad, and ii) the most energetic EPs would produce π^0 pions, and in turn a gamma-ray continuum around 1 GeV with a flux much higher than observed by EGRET [6,7].

=46inally, given the EP spectrum and composition and the Orion gamma-ray flux, one can estimate the energy density of the EPs in Orion. It is found to be at least 10 times higher than that of the GCRs [8,9].

In conclusion, we argue that the EPs interacting in Orion represent an energetic component distinct from the ordinary Galactic cosmic rays (GCRs), having a different composition (much richer in C and O), characteristic energy (a few tens of MeV instead of 1 GeV), and energy density. The Orion EPs are thus specific, and this raises the crucial question of their origin.

Internal Source ?

Let us first investigate the case when the EPs are accelerated within the clouds themselves. In a medium as dense as Orion ($< n > \sim 100\,\mathrm{cm}^{-3}$), because of ionisation energy losses, C and O nuclei of a few tens of MeV/n spend only a few 10^3 years with an energy above the excitation thresholds. As a consequence, they cannot produce gamma-ray lines further than a distance $L \sim (Dt)^{1/2} \sim 1$ pc from their acceleration site (D is the diffusion coefficient in the Orion clouds, say $10^{26}\,\mathrm{cm}^2\,\mathrm{s}^{-1}$). Even if the EPs are continuously reaccelerated so that they never slow down to energies below the nuclear thresholds, their life-time is limited by the nuclear destruction time scale, i.e. $\sim 10^4$ yr for C and O and the Orion density. This gives a diffusion length $L \sim 3$ pc, which is still very small as compared to the linear size of the gamma-ray source (~ 45 pc).

Consequently, the internal source hypothesis requires many sources, and the above time scales indicate that all of these sources have to be recent (less than a few 10^3 years, or 10^4 if re-acceleration takes place). Now we know from observation that there were no SN explosions in the past 10^4 years in Orion - especially not several of them ! - and there are no currently active Wolf-Rayet (WR) stars. Furthermore, there is not enough power in the magnetic turbulence to re-accelerate the EPs at a rate of at least $2\,10^{38}$ erg s^{-1} [4]. All these difficulties make any internal source model unlikely.

External Source ?

In the case of an external source, there is no need anymore for a very energetic process taking place now. It suffices that a lot of energy was released in the past, accelerating EPs within a large reservoir, and that these EPs keep on diffusing (at least in part) toward the Orion clouds where they interact. In order that the EPs survive long enough, the reservoir has to have a low density. Now such a reservoir exists close to Orion : it is the Orion-Eridanus superbubble.

Another appeal of an external source model is that it accounts naturally for the extension of the gamma-ray source. Indeed, for a large enough reservoir, the clouds are irradiated over their whole surface. Moreover, the above argument based upon the diffusion length of the EPs inside the dense molecular complex is still valid, and leads here to the conclusion that the EPs do not penetrate deep into the clouds. The irradiation is only superficial, which provides a natural solution to specific problems related to the Orion ionisation and astrochemistry [10,11].

We further argue that an external source model may provide an explanation of the small scale anti-correlation between the gamma-ray emission and the CO contours tracing the Orion clouds, as seen on the superimposed map given in reference [2]. Indeed, the gamma-rays would originate from the clouds surface, while the CO column density is proportional to the clouds volume along the line of sight. Now surface and volume are, by essence, anti-correlated, especially in a very filamentary and clumpy region, as is the case in Orion [12].

=46inally, line profile considerations favour an interaction geometry in which the Orion EPs irradiate the target (the Orion clouds) from in front of them, suppressing the blue wings of the expected (splitted) broad lines, and thus leading to an apparent redshift [13]. This is in remarkable agreement with the Orion-Eridanus model, since the Orion clouds are actually located just behind the superbubble (e.g. [15]).

THE ORION-ERIDANUS MODEL

The Orion-Eridanus superbubble has been first observed in H_α in 1974 [14], showing shell structures extending from the Barnard's loop, around the Orion A and B clouds, down to 50 degrees below the Galactic plane, in Eridanus. It consists of a cavity filled with hot ionised gas, some 350 pc away from the sun, surrounded by an expanding shell of neutral hydrogen, most certainly related to the strong stellar wind and SN activity of the Orion OB1 association [15–18].

We thus dispose of a lot of energy ($\sim 2\,10^{52}$ erg) at the right place (close to, and in front of the Orion clouds). Moreover, superbubbles related to the intense activity of OB associations are known to provide a very efficient acceleration mechanism [19], with an energy spectrum similar to that required in Orion by phenomenological analysis, that is a hard spectrum with a strong cut-off at a few MeV/n [20,9]. The acceleration is due to large scale magnetic field fluctuations caused by interacting shocks associated with the activity of massive stars (strong WR winds and supernovae) in the bubble [19]. A particularity of this acceleration mechanism is that it involves many secondary weak shocks, and therefore predicts a very low electron-to-proton acceleration ratio [3,20]. This can explain the absence of observed continuum Bremsstrahlung gamma-rays from primary electrons.

=46inally, we have shown that the winds of the massive stars in a typical OB association have an average composition compatible with the Orion gamma-ray emission [21,22]. In particular, it satisfies the (1–3 MeV)/(3–7 MeV) band ratio upper limit. Now the Orion-Eridanus superbubble is filled with these wind materials, so that our model also provides a natural explaination for the strong enhancement of the C and O abundances in the Orion EPs.

Quantitative and detailed study of the energetics in the context of the Orion-Eridanus model will be found in reference [9], further supporting the model.

In conclusion, we have shown that many independent arguments (diffusion length, time scales, energetics, source extension, line profile, ionisation, EP composition...) converge toward an external source model, and more specifically, toward the Orion-Eridanus model. This identification allows us to generalise the Orion case to the whole Galaxy. Indeed, according to our model, the Orion emission results from the intense activity of an OB association inside or close to a dense molecular cloud complex. Now such OB association–cloud pairs are common in the Galaxy [23]. About 20% of the estimated 200 Orion-like objects (i.e. with same mass and luminosity) present in the Galaxy would then produce a diffuse emission between 3 and 7 MeV with a flux $\sim 2\,10^{-5}$ ph cm^{-2}s^{-1} if the sources are, say, uniformly distributed along a ring at 5 kpc from the Galactic center [9]. This may account for the diffuse emission observed by COMPTEL [24].

REFERENCES

1. Bloemen, H., et al., *A&A* **281**, L5 (1994).
2. Bloemen, H., et al., *ApJ* **475**, L25 (1997).
3. Bykov, A.M., and Bloemen, H., *A&A* **281**, L1 (1994).
4. Ramaty, R., Kozlovsky, B., and Lingenfelter, R.E., *ApJ* **438**, L21 (1995).
5. Cassé, M., Lehoucq, R., and Vangioni-Flam, E., *Nature* **373**, 318 (1995).
6. Digel, S.W., Hunter, S.D., and Mukherjee, R., *ApJ* **441**, 270 (1995).
7. Ramaty, R., Kozlovsky, B., and Tatischeff, V., *4th COMPTON Symposium Proc.* (1997).
8. Ramaty, R., *A&A* **120**, 373 (1996).
9. Parizot, E.M.G., *A&A* submitted (1997).
10. Ramaty, R., Kozlovsky, B., and Lingenfelter, R.E., *ApJ* **456**, 525 (1996).
11. Dogiel, V.A., *Nuovo Cimento* **19 C**, 671 (1996).
12. Gentzel, R., and Stutzki, J., *ARA&A* **27**, 41 (1989).
13. Ramaty, R., Kozlovsky, B., and Lingenfelter, R.E., in *ESA SP-382 "The Transparent Universe", ESA Publications Division*, p. 75 (1997).
14. Sivan, J.P., *A&AS* **16**, 163 (1974).
15. Burrows, D.N., Singh, K.P., Nousek, J.A., Garmire, G.P. and Good, J., *ApJ* **406**, 97 (1993).
16. Reynolds, R.J., and Ogden, P.M., *ApJ* **229**, 942 (1979).
17. Brown, A.G.A., de Geus, E.J., and de Zeeuw, P.T., *A&A* **289**, 101 (1994).
18. Brown, A.G.A., Hartmann, D., and Burton, W.B., *A&A* **300**, 903 (1995).
19. Bykov, A.M. and Fleishman, G.D., *MNRAS* **255**, 269 (1992).
20. Bykov, A.M., *Space Sci. Rev.* **74**, 397 (1995).
21. Parizot, E.M.G., Cassé, M. and Vangioni-Flam, E., *A&A* submitted (1997).
22. Parizot, E.M.G., Paul, J., and Cassé, M., this volume.
23. Williams, J.P., and McKee, C.F., *ApJ* **476**, 166 (1997).
24. Bloemen, H., et al., this volume.

On the origin of 3 to 7 MeV γ-ray excess in the direction of Orion

V.A.Dogiel[1], M.J.Freyberg[2], G.E.Morfill[2], V.Schönfelder[2]

[1] *Dept.of Theor.Physics, P.N.Lebedev Institute, 117924 Moscow, Russia*
[2] *Max-Planck-Institut für extraterrestrische Physik, 85740 Garching, Germany*

Abstract. COMPTEL has detected strong 3 to 7 MeV γ-ray emission from the Orion complex with no significant evidence for emission outside this range. We show that measurements of the diffuse X-ray emission from Orion can constrain the sources of the γ-ray emission, if the excess is due to γ-ray lines produced by low-energy cosmic ray nuclei. We also discuss an interpretation of the excess as continuum bremsstrahlung emission of relativistic electrons.

INTRODUCTION

COMPTEL has detected strong 3 to 7 MeV flux from the Orion complex with no significant evidence for emission outside this range (Bloemen et al. 1994, 1997). The emission is supposed to be due to γ-ray lines from excited ^{12}C and ^{16}O nuclei generated in the medium of the complex by low energy (tens MeV) cosmic rays whose density is higher than in the surrounding space (Bykov et al. 1996, Ramaty et al. 1996).

The luminosity of fast nuclei needed to produce the Orion γ-ray excess is about $(3-8) \cdot 10^{38}$ erg s^{-1} (see Bykov and Bloemen 1994; Bykov et al. 1996, Ramaty et al. 1996). These estimates of the energy flux depend on the chemical composition of the medium and on the position of the high energy break in the spectrum of accelerated nuclei (see Parizot et al. 1997).

As one can see, only a small fraction of the total energy of accelerated nuclei is transformed into the energy of γ-ray photons, since the efficiency of γ-ray photon production is low in this case and therefore one can expect a significant emission in other ranges of the electromagnetic spectrum. We shall show that the line emission should be accompanied by production of secondary knock-on electrons which generate bremsstrahlung radiation in the X-ray energy range. Detecting this flux of X-ray emission from Orion would confirm the γ-ray origin of the excess. We shall show that this model may have some problems and in this respect we discuss also other possibilities to explain its origin.

X-RAY FLUX OF KNOCK-ON ELECTRONS

In the approximation of the thick target model the spectrum of particles is described by the equation $\frac{d}{dE}\left(\frac{dE}{dt}N(E)\right) = Q(E)$ where N is the density of charged particles with energy E. Q is the production rate of sources and (dE/dt) is the rate of energy losses in the volume. In our case nuclei and electrons lose their energy by ionization. The used approximation of the thick target model means that the nuclei lose all their energy in Coulomb interactions (by ionization). In the collisions the electrons of the medium which are initially at rest acquire some energy W from a fast nucleus. As a result a large number of fast electrons are generated in the medium in which the γ-ray lines are produced. The energy of these secondary electrons produced by $< 20 - 30$ MeV n^{-1} nuclei, lies in the range between several keV and several tens of keV (see Hayakawa, 1969). These electrons lose their energy by ionization but at the same time they also produce a flux of bremsstrahlung X-ray photons.

The differential energy flux of bremsstrahlung photons $\frac{dF_x}{dE_x}$ determined in the framework of the thick target model is described by the equation

$$\frac{dF_x}{dE_x} = E_x \int_{E_x}^{W_m} dW_k \sigma_{br} v_e n_e \left(\frac{dW_k}{dt}\right)_i^{-1} \int_{W_k}^{W_m} dW \int_{W\frac{M_n}{4m}}^{E_m} dE_k \sigma_s v_n n_e \left(\frac{dE_k}{dt}\right)_i^{-1} \int_{E_k}^{E_m} dE Q_n(E)$$

Here E and W denote the kinetic energy of fast nuclei and secondary electrons, respectively, whose velocities and the rest masses are v_n and v_e, and M_n and m. The cross-sections σ_s and σ_{br} describe the processes of secondary electron and of bremsstrahlung X-ray production in a medium with electron density n_e. The terms dE/dt_i and dW/dt_i describe ionization loss rates of nuclei and electrons. One can see that in the case of the γ-ray line origin of the excess the flux of bremsstrahlung X-ray emission from Orion in the energy range $0.5 - 2$ keV is significant, $\sim (2-6) \cdot 10^{33}$ erg s^{-1} depending on the value of the nucleus luminosity. Thus it should be observable by X-ray telescopes and it should also be distinguishable from the thermal X-ray flux which depends exponentially on the energy of photons.

ROSAT PSPC ANALYSIS OF THE ORION REGION

Analysis of ROSAT PSPC data in the energy range $0.5 - 2$ keV shows that there is nonthermal X-ray emission (point-like sources are excluded) in the direction $l \sim 209°$ and $b \sim -19°$ close to the Orion Nebula. The size of the emitting area is, however, rather small – a few square degrees only i.e. much less than the area of the assumed emitting region of γ-ray lines (about hundred square degrees). The total luminosity of the nonthermal X-rays is about 10^{32} erg s^{-1}. Estimating the nonthermal X-ray flux over the whole γ-ray emitting region we obtain an upper limit of $\sim 1.3 \cdot 10^{33}$ erg s^{-1} in the energy range

0.5 − 2 keV. The results of ROSAT data analysis are shown in Figs.1 and 2. The analysis shows a lack of correlation of γ- and X-ray emission (that is expected in the γ-ray line model), in some regions even an anti-correlation is observed (see Fig.1).

We tried to discuss possible implications of this discrepancy in terms of intrinsic absorption in addition to normal interstellar absorption of the X-radiation by matter in the Orion complex. This leads us to the conclusion that a significant absorption is expected only, if the region emitting γ- and X-rays is rather small ($\sim$ 1 pc, just what is expected in the model of Morfill and Meyer (1981)). In this case the extended Orion γ-ray excess as seen by COMPTEL results from the superposition of several point-like sources.

EXCESS AS BREMSSTRAHLUNG CONTINUUM GENERATED BY CR ELECTRONS

Below we investigate another possible excess origin. We assume that the excess spectrum is continuous and the emission is generated by high energy electrons. If we look at the large error bars in the energy distribution of the Orion excess and at the preliminary character of the analysis we see that the data of Bloemen et al. (1997) do not exclude this interpretation. The spectrum is hard at low energies ($< 3\,\mathrm{MeV}$) and soft at high energies ($> 7\,\mathrm{MeV}$), and can be approximated by a broken power-law $dL_\gamma/dE_\gamma \propto E_\gamma^{-\alpha}$, where $\alpha < 1$ for $E_\gamma < 3\,\mathrm{MeV}$ and $\alpha > 3$ for $E_\gamma > 7\,\mathrm{MeV}$. In the framework of our model we assume that the intensity of the bremsstrahlung γ-radiation is larger than that of the γ-ray line emission. This is the case, if the efficiency of nucleus acceleration η_n is small in comparison with that of electrons, η_e, i.e. $\eta_e > 3 \cdot 10^{-4} \eta_i$. The production spectrum of electrons $Q_e(E_e)$ for the case of Fermi II acceleration has the form $Q_e(E_e) = K E_e^{-1}$ (K is constant) with a break at energies of about $E_b \approx 20\,\mathrm{MeV}$ in the spectrum to satisfy the data. In the framework of the thick target model the bremsstrahlung radiation of these electrons is described by the equation

$$\frac{dL_\gamma}{dE_\gamma} = \int_{E_\gamma}^{E_b} dE_e \sigma_{br} n_a c \left(\frac{dE_e}{dt}\right)_i^{-1} \int_{E_e}^{E_b} Q(E) dE$$

where σ_{br} is the bremsstrahlung cross-section, dE/dt_i is the rate of ionization losses, n_a is the background gas density. The spectrum of the bremsstrahlung power $F_\gamma \sim E_\gamma^2 dL_\gamma/dE_\gamma$ and the measurements of COMPTEL are shown in Fig.3. One can see that the bremsstrahlung model describes the Orion excess data rather well above 3 MeV, though there is a severe problem below 3 MeV. The calculations show that in this model an energy luminosity of the electrons of about $\sim 10^{36}$ erg s^{-1} is required to produce the observed γ-ray flux from Orion, i.e. much less than in the γ-ray line model.

CONCLUSION

It is shown that a significant flux of soft X-rays ($\sim (2-6) \cdot 10^{33}$ erg s^{-1} in the energy range 0.5 - 2 keV) can be generated by secondary knock-on electrons, if the interpretation of the 3 to 7 MeV excess as being due to γ-ray lines is correct. Analysis of ROSAT PSPC data has yielded an upper limit about a factor 2 - 5 lower than this estimate, $\sim 1.3 \cdot 10^{33}$ erg s^{-1} between 0.5 - 2.0 keV for the entire spatial region covered by the COMPTEL low-energy γ-ray excess. The difference between the calculated luminosity and the one estimated from ROSAT data is not large enough, however, to draw a final conclusion in favor or against the γ-ray line model.

In addition, the X-ray analysis of ROSAT data, does not show such a spatial correlation between the X-ray emission and the γ-ray line intensity in the investigated part of Orion. This conclusion (a lack of correlation) may be problematic for the γ-ray line model of the Orion excess. The results may be explained either by additional absorption in small dense clumps of gas with a large gas column density or by absorption in regions with abnormal high abundance of C and O. Both these assumptions lead us to the conclusion that the line emission is generated in small regions which could not be resolved by COMPTEL as point-like sources and thus several localized γ-ray emission regions may mimic the spatially extended Orion excess (which is not excluded from the COMPTEL data, see Bloemen et al., 1997).

Concerning the bremsstrahlung model we conclude that this model does not have the energy problem since much less energy is necessary to generate the observed γ-ray emission. On the other hand this model has a problem to reproduce the spectral shape of the excess.

Acknowledgements: VAD is grateful to his colleagues from the Max-Planck-Institut für extraterrestrische Physik (Garching) for helpful and fruitful discussions.

REFERENCES

Bloemen H. et al., *A&A*, 281, L5 (1994)
Bloemen H. et al., *ApJ Letters*, 475, L25 (1997)
Bykov A.M., Bloemen H., *A&A*, 283, L1 (1994)
Bykov A.M. et al., *A&A*, 307, L37 (1996)
Hayakawa S., *Cosmic Ray Physics*, Wiley-Interscience (1969)
Morfill G.E., Meyer P., *Proc 17th ICRC*, 1, 116 (1981)
Parizot E.M.G., Casse M., Vangioni-Flam E., *A&A*, submitted (1997).
Ramaty R. et al., *ApJ Suppl*, 40, 487 (1979)
Ramaty et al., *ApJ*, 456, 525 (1996)

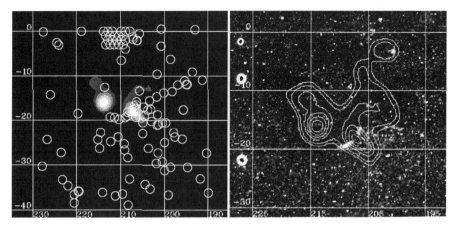

FIGURE 1. *Left:*) COMPTEL grey-scale image in galactic coordinates with ROSAT PSPC pointed observations indicated by circles (114' diameter). These pointed observations are used for spectral analysis as they have sufficient exposure but incomplete sky coverage. *Right:* ROSAT PSPC All-Sky Survey image (0.5 − 2 keV) in galactic coordinates with COMPTEL contours. It does not show a general correlation of γ- and X-ray emission. Point sources have not been removed from this image.

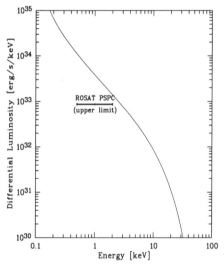

FIGURE 2. Differential luminosity spectrum $E_x \frac{dL_x}{dE_x}$ of X-rays generated by knock-on electrons in Orion (solid line). Additionally, the upper limit derived from ROSAT PSPC data is presented. Note, this value has been already corrected for $0 \, \text{g/cm}^2$ absorbing column density.

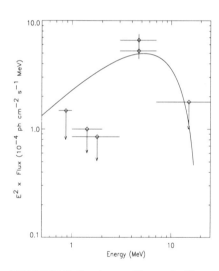

FIGURE 3. Spectrum of bremsstrahlung γ-ray emission from Orion and the flux measurements of COMPTEL.

COMPTEL Spectral Study of the Inner Galaxy

H. Bloemen[*][§], A.M. Bykov[†][*], R. Diehl[‡], W. Hermsen[*],
R. van der Meulen[*][§], V. Schönfelder[‡], and A.W. Strong[‡]

[*]*SRON-Utrecht, Utrecht, The Netherlands*
[§]*Leiden Observatory, Leiden, The Netherlands*
[†]*A.F. Ioffe Institute, St. Petersburg, Russia*
[‡]*MPI für extraterrestrische Physik, Garching, Germany*

Abstract. We present first steps towards fine-resolution COMPTEL spectra of the galactic γ-ray glow at 0.75–30 MeV, combining all data from the first 5 years of the mission. The spectra are not fully deconvolved yet. We show that there is evidence for deviations from a smooth continuum spectrum in the inner-Galaxy emission. Two broad spectral excesses seem to be present, namely below the ^{26}Al 1.8 MeV line down to ~ 1.3 MeV and at 3–7 MeV, each containing a flux of the order 10^{-4} γ cm^{-2} s^{-1} rad^{-1}. We find that the excess emission appears to have a rather wide latitude distribution. The spectral features suggest the presence of nuclear deexcitation lines as a possible explanation.

INTRODUCTION

Prior to CGRO, only very little spectral information on the 1–30 MeV emission from the inner Galaxy was available and any imaging information was basically lacking [4]. COMPTEL has provided first maps, but spectral analyses have hitherto been limited to 3 or 4 broad energy bands and focussed on diffuse emission from cosmic-ray interactions in the interstellar medium [7–10]. We present here first attempts to apply a much finer spectral binning, using all observations that were obtained in the period May 1991 to June 1996. A more extensive report is in preparation.

The finer binning enables us to search for the possible presence of collective γ-ray line signals from large ensembles of sources and to set further constraints on the origin of the continuum emission. The relative contributions of diffuse bremsstrahlung and inverse-Compton (IC) radiation are not well established yet and a significant contribution from unresolved point sources cannot be excluded. With regard to γ-ray lines, in addition to the ^{26}Al 1.809 MeV line, candidates are the nuclear decay lines from ^{22}Na (1.275 MeV) and ^{60}Fe (1.17

and 1.33 MeV), the 2.22 MeV line from neutron capture onto hydrogen, and nuclear deexcitation lines following nuclear interactions, for which evidence may have been seen already from the Orion complex at 3–7 MeV [2]. Candidate deexcitation lines [6] are from ^{12}C (4.44 MeV) and ^{16}O (6.13, 6.92, and 7.12 MeV) in the 3–7 MeV regime and from heavier nuclei such as ^{20}Ne (1.63 MeV), ^{22}Ne (1.28 MeV), ^{24}Mg (1.37 MeV), and ^{28}Si (1.78 MeV) at 1–2 MeV.

ANALYSIS METHOD

COMPTEL contains two detector layers in which the energy deposits and the interaction locations of incident photons are determined. These define for each photon the measured energy, the Compton scatter angle $\bar{\varphi}$, and the scatter direction [we use galactic coordinates (ℓ, b)]. For a selected energy range, the basic data space is 3-dimensional, $(\ell, b, \bar{\varphi})$. The signature of a point source, the point spread function (PSF), is cone-shaped in this data cube. Detailed PSF knowledge was obtained from pre-launch calibration measurements, from simulations of the instrument, and from analytical modelling based on single-detector calibrations. Simulated PSFs are preferred and used here, but these are not available yet for the narrow energy bins in Fig. 3.

We apply our standard likelihood analysis procedures, e.g. [3,1], but in a 2-D data space, obtained by integrating the 3-D one over galactic longitude. This is done for practical purposes. As a consequence, we are less sensitive to intensity variations in ℓ-direction, but we demonstrate that this should not influence our findings significantly. We apply forward folding, i.e. adopt an intensity distribution, fold it through the instrument response (given a certain input spectrum), integrate over ℓ, and then apply a likelihood fitting procedure. This includes a background model. A sufficiently accurate independent background estimate in the $(\ell, b, \bar{\varphi})$ data space cannot be obtained yet. But we have successfully applied a filtering technique [1], which is based on the fact that the (ℓ, b) structure of the data cube is largely due to known geometry effects. We use here this technique, but replace the associated response modification by a selfconsistent iterative approach with a likelihood convergence criterion.

We can thus fully deconvolve smooth *continuum* spectra by analysing the $(b, \bar{\varphi})$ data space for each *individual* energy bin of the spectrum under construction. In the case of γ-ray *line* emission, however, fully accounting for the instrument response would require the use of an additional data-space dimension, E_γ, because the measured photon energy will be below the nominal line energy for a significant fraction of the events, mainly due to energy losses in the lower detector layer. An analysis tool capable of performing the complex deconvolution procedure for many energy bins is not available yet. Restricting our analysis to seperate energy bands implies that the non-diagonal response elements of a γ-ray line (Compton tail and escape peaks) can lead to a noticable spill-over effect for strong lines, complicating the search for other (weak)

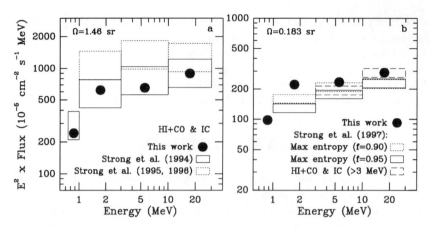

FIGURE 1. a) COMPTEL broad-band spectra of the inner Galaxy ($|\ell| < 60°$; $|b| < 20°$), as obtained by Strong et al. [7–9] and as obtained with the analysis technique applied in this paper, fitting the same HI and CO observations and IC intensity model (IC not used in [7]). b) Comparison with the most recent results of Strong et al. [10] ($|\ell| < 30°$; $|b| < 5°$).

lines at the low-energy side. We make, therefore, one important exception: *we account for the low-energy response tail from the strong 1.8 MeV line.* This can be done in a consistent manner by including the best-fit 1.8 MeV sky model (determined at 1.7–1.9 MeV) as a fixed component (with the appropriate response function) in the model fitting at lower energies.

RESULTS

We show here the impact of gradually reducing the spectral bin sizes. First, for comparison with COMPTEL studies by Strong et al. [7–10], we present broad-band spectra in Fig. 1, obtained by fitting simultaneously a map of the total gas column-density distribution (from HI and CO observations) and an IC intensity model. The result of the early work by Strong et al. [7] agrees with ours, but the more recent estimates [8,9] in Fig. 1a are too high at 1–10 MeV. The most recent values by Strong et al. [10] (Fig. 1b) are lower again at 1–10 MeV. These variations are due to inadequate handlings of the instrumental background, in particular in [8,9].

Fig. 2 shows our results for 12 narrower energy bins. The spectrum on the left-hand side was derived with the same model intensity distributions as in Fig. 1. The spectrum on the right-hand side was obtained under the assumption of a constant intensity distribution over the longitude range $\ell = -60°$ to $60°$, consisting of a narrow (FWHM=2°) and a broad (FWHM=20°) component, which were fitted simultaneously. The results are very similar. We have fully removed here the 1.8 MeV line by excluding all events between 1.7 and 1.9 MeV from the 1.5–2 MeV bin (and the 1.8 MeV response tail

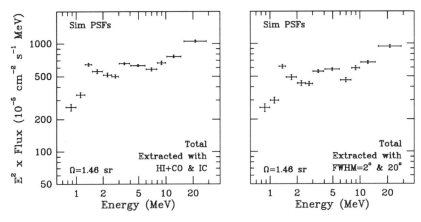

FIGURE 2. Spectra of the inner Galaxy ($|\ell| < 60°$; $|b| < 20°$) for two different types of model intensity distributions. *Left:* From fitting diffuse-emission models (HI+CO and IC). *Right:* From models with Gaussian latitude distributions of FWHM=2° and 20° (see text). All events between 1.7 and 1.9 MeV are excluded from the 1.5–2 MeV bin and the response tail of the 1.8 MeV line is modelled out. Statistical $\pm 1\sigma$ error bars are shown.

is modelled out as described above). So the plotted values in this energy bin actually cover 1.5–1.7 + 1.9–2 MeV. Fig. 3 shows a more preliminary spectrum (see caption) with 23 narrow energy bands. In order to study the spectra of the narrow and broad components separately (Fig. 4), we had to restrict ourselves to the 12 broader energy bins to limit the statistical errors.

The results presented here clearly suggest that spectral structure is present in the integrated inner-Galaxy emission, with excesses at about 1.3–1.7 and 3–7 MeV. We have verified through simulations (including the complete background structure) that such features are not artifacts introduced by our analysis technique. We have also studied the impact of differing PSFs and $\bar{\varphi}$ selections; the spectral features turn out to be robust. We have learned from these tests that a total flux uncertainty up to $\sim 30\%$ cannot be excluded, largely due to the PSFs (which can be improved upon in future work). Fig. 4 shows that the excess emission appears to have a rather wide latitude distribution, but the extent needs further study. Wide-latitude emission was seen previously [9]. Fig. 4 illustrates that we should be able to separate spatially the diffuse bremsstrahlung and IC model components as a function of energy, which can provide detailed individual spectra and is thus an important test of the origin of the continuum emission. Such an analysis is in progress.

Our main conclusion is that we start to find spectral features in the emission from the inner Galaxy. The details should be considered preliminary. Such features were not seen so far by e.g. SMM and OSSE, but this may not be surprising because the excesses are broad and relatively weak, containing fluxes of typically 10^{-4} γ cm^{-2} s^{-1}rad^{-1} (below SMM and OSSE upper limits [5]), and the emission region appears to extend well out of the galactic plane, which

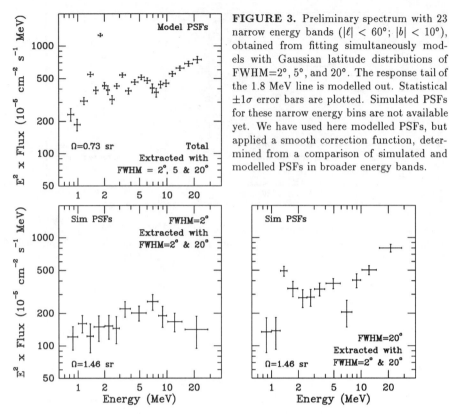

FIGURE 3. Preliminary spectrum with 23 narrow energy bands ($|\ell| < 60°$; $|b| < 10°$), obtained from fitting simultaneously models with Gaussian latitude distributions of FWHM=$2°$, $5°$, and $20°$. The response tail of the 1.8 MeV line is modelled out. Statistical $\pm 1\sigma$ error bars are plotted. Simulated PSFs for these narrow energy bins are not available yet. We have used here modelled PSFs, but applied a smooth correction function, determined from a comparison of simulated and modelled PSFs in broader energy bands.

FIGURE 4. Spectra of the narrow and broad components separately ($|\ell| < 60°$; $|b| < 20°$). This separation may of course not properly represent the true sky. All 1.7–1.9 MeV events are excluded and the 1.8 MeV tail is modelled out. Statistical $\pm 1\sigma$ error bars are shown.

is problematic for OSSE. An interpretation in terms of nuclear deexcitation lines seems most plausible and is currently being investigated. We cannot set meaningful constraints yet on ^{22}Na and ^{60}Fe decay lines.

REFERENCES

1. Bloemen H. et al., 1994, ApJS 92, 419
2. Bloemen H. et al., 1994, A&A 281, L5; 1997, ApJ 475, L25
3. de Boer H. et al., 1992, in Data analysis in astron., ed. V. Di Gesù et al., 241
4. Gehrels N. & Tueller J., 1993, ApJ 407, 597
5. Harris M.J. et al. 1995, ApJ 448, 157; 1996, A&A Suppl. Ser. 120, C343
6. Ramaty R., Kozlovsky B., Lingenfelter R.E., 1979, ApJS 40, 487
7. Strong A.W. et al., 1994, A&A 292, 82
8. Strong A.W. et al., 1995, Proc. 24th Int. Cosmic Ray Conf. 3, 234
9. Strong A.W. et al., 1996, A&A Suppl. Ser. 120, C381
10. Strong A.W. et al., 1997, these proceedings

OSSE RESULTS ON GALACTIC γ-RAY LINE EMISSION

M. J. Harris[1], W. R. Purcell [2], K. McNaron-Brown[3], R. J. Murphy [3], J. E. Grove [3], W. N. Johnson[3], R. L. Kinzer[3], J. D. Kurfess[3], G. H. Share[3], G. V. Jung[4]

Abstract.
We report progress in an ongoing search for γ-ray line emission from the Galaxy using OSSE. Four lines are of interest. The 4.4 and 6.1 MeV lines from de-excitation of ^{12}C and ^{16}O excited by cosmic-ray impacts, which were detected by COMPTEL from the Orion molecular cloud, have been searched for at several longitudes in the Galactic plane, without success. Constraints can be placed on one specific model of a point-source origin of the lines. The other two lines, the 1.17 and 1.33 MeV lines from ^{60}Fe, may be present at the limit of OSSE's sensitivity at positive longitudes in the central radian of the Galaxy.

I INTRODUCTION

This paper is a sequel to one presented at the Third Compton Symposium in which the method used, the spectra, the systematic errors, and the preliminary results were described[1]. Here we present updated results including data from the most recent available OSSE targets. The principle of the method is to measure the intensities of the relevant lines in individual OSSE fields of view (FOVs), and to search for correlations between the line strength and the exposure of the FOV to an expected Galactic distribution. The lines involved are those of ^{60}Fe (1.17 and 1.33 MeV), ^{22}Na (1.275 MeV), ^{26}Al (1.875 MeV, included as a check on ^{60}Fe), and the de-excitation lines from ^{12}C and ^{16}O excited by cosmic-ray impacts (4.4 and 6.1 MeV). The cosmic-ray induced lines may be broad (~ 1 MeV FWHM) or narrow (~ 100 keV), depending on whether the accelerated particles are predominantly heavy nuclei or protons. While ^{22}Na is expected to follow the distribution of Galactic novae, the other lines are expected to follow a thin Galactic disk distribution. The detection

---

[1] USRA/GVSP, Code 661, NASA/Goddard Spaceflight Center, Greenbelt, MD 20771.
[2] Northwestern University, 2145 Sheridan Road, Evanston, IL 60208-3112
[3] Mail Code 7650, NRL, Washington DC 20375
[4] USRA, Suite 801, 300 D Street S.W., Washington DC 20024

Species	Source	Energy MeV	Flux per OSSE FOV in the plane γ cm^{-2} s^{-1}	Flux or 3σ upper limit from model plane γ cm^{-2} s^{-1} rad^{-1}
^{26}Al	central rad	1.809	$4.0 \pm 0.7 \times 10^{-5}$	$2.37 \pm 0.44 \times 10^{-4}$
^{60}Fe[a]	$-3° \leq l \leq 25°$	1.17, 1.33	$2.7 \pm 1.2 \times 10^{-5}$	$1.62 \pm 0.69 \times 10^{-4}$
^{60}Fe[a]	central rad	1.17, 1.33	$1.1 \pm 0.8 \times 10^{-5}$	$\leq 1.49 \times 10^{-4}$
Cosmic-ray[b]	central rad	4.44, 6.13 broad	$2.0 \pm 1.1 \times 10^{-5}$	$\leq 2.03 \times 10^{-4}$
Cosmic-ray[b]	central rad	4.44, 6.13 narrow	$1.8 \pm 7.1 \times 10^{-6}$	$\leq 1.26 \times 10^{-4}$
^{22}Na	central rad	1.275	$0.0 \pm 1.2 \times 10^{-5}$	$\leq 2.13 \times 10^{-4}$
^{22}Na[c]	spheroid	1.275	$0.5 \pm 1.5 \times 10^{-5}$	$\leq 9.1 \times 10^{-5}$

[a] Flux in each of two lines.
[b] Summed flux in two lines.
[c] The OSSE FOV flux is for a pointing $l = 0°, b = 0°$. The model flux is calculated for the central 5° of the model spheroid.

TABLE 1. Line fluxes from the Galactic plane

by COMPTEL of broad cosmic-ray induced lines from the Orion molecular cloud[2] has however introduced a complicating factor into the distribution of these lines (§2).

The updated results for the central radian of the Galaxy, from 59 OSSE viewing periods containing 283 FOVs, are shown in Table 1. Additional results for other longitudes, where there is little OSSE exposure, are given in §2.

II COSMIC-RAY 4.4 AND 6.1 MEV LINES

Table 1 shows no significant evidence for either broad or narrow 4.4 and 6.1 MeV line emission in the central radian of the Galaxy, which had been expected to be the strongest source. In the light of the surprising COMPTEL detection of broad lines from Orion, a search was made for other sources in the database of OSSE Galactic plane observations. We also scheduled OSSE observations dedicated to the Orion region. Preliminary OSSE results for the central region of Orion (between Orion A and B)[3] showed that the COMPTEL emission could not be due to a point source here. Our updated analysis of FOVs covering a wider area yields the same conclusion for the whole molecular cloud region[4]. Our measurement of the total flux from the region mapped by COMPTEL[5] is $1.5 \pm 1.0 \times 10^{-4}$ photon cm^{-2} s^{-1}, which is not sensitive enough to confirm the COMPTEL value $1.28 \pm 0.15 \times 10^{-4}$ photon cm^{-2} s^{-1}. We also searched for a line at 0.511 MeV predicted to arise from the annihilation of e^+ produced by the cosmic-ray interactions in Orion[6]. Our result $-7 \pm 9 \times 10^{-5}$ photon cm^{-2} s^{-1} from the area of line emission mapped by COMPTEL cannot constrain the prediction, which is at a level $\sim 2 \times 10^{-5}$ photon cm^{-2} s^{-1}.

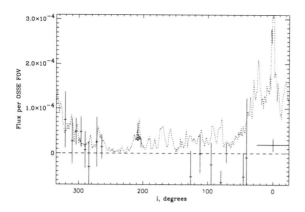

FIGURE 1. OSSE measurements of summed intensity of broad 4.4 and 6.1 MeV lines at several galactic longitudes. Dashed line — distribution of CO intensity with longitude[7], normalized to the COMPTEL line intensity in Orion in an OSSE FOV.

However we can exclude the possibility of point source emission at this level from the central region (Orion A).

Our results for the Galactic plane (including Orion and the central radian) are shown in Fig. 1. The overall Galactic CO distribution[7] is superimposed, normalized to the measured COMPTEL point at Orion. We made no detections of any source at any longitude. It is clear from our upper limit for the central radian that sources of the strength of Orion do not follow the molecular cloud mass distribution towards the Galactic center; it may however follow CO-like features such as spiral arm tangents at other longitudes.

It is therefore difficult to test diffuse models of the Orion emission by finding examples elsewhere in the Galaxy, since there is no obvious indicator such as CO mass telling one why Orion is special, and where else to look. Point source models have been proposed[8], and are easier to test using observations in other parts of the Galaxy. The OSSE data can be used to test one particular model which was proposed before the discovery of the Orion emission[9,10]. This model arose from the discovery in optical spectra of enhanced Li in the binary companions of several black hole candidates. The discoverers attributed this to the production of Li on the stellar surface by accelerated charged particles. The objects involved are A0620-00, V404 Cyg, Nova Mus 1991, and Cen X-4. OSSE data are available for the first three. The results of fitting both broad and narrow 4.4 and 6.1 MeV lines to the OSSE spectra are shown in Table 2.

We conclude that there is no steady flux from these objects down to 3σ upper limits of 6×10^{-5} photon cm^{-2} s^{-1} for broad lines, and 4×10^{-5} photon cm^{-2} s^{-1} for narrow lines. All of these objects undergo outbursts, so it is possible that Li production and line emission may occur during the outbursts.

Source	Broad line flux $\times 10^5$ γ cm^{-2} s^{-1}	Narrow line flux $\times 10^5$ γ cm^{-2} s^{-1}
A0620-00	0.2 ± 5.1	-0.3 ± 3.4
V404 Cyg	0.8 ± 2.6	-1.1 ± 1.7
Nova Mus	5.4 ± 4.0	1.8 ± 2.4
Sum	1.9 ± 2.0	-0.2 ± 1.3

TABLE 2. 4.4 and 6.1 MeV line fluxes from black hole candidates with Li-rich secondaries

III ^{60}FE LINES AT 1.17 AND 1.33 MEV

The result reported in ref. 1 — that the two lines of ^{60}Fe appear at about the 2σ level from positive longitudes in the Galactic center — is slightly improved in the new results, to a level 2.3σ (Table 1). In Fig. 2 we compare this result with preliminary results presented at the Symposium[11,12] and with the earlier *SMM* result[13]. All results for the emission *from the central radian* are in agreement, since they are all essentially upper limits. The OSSE result for the central radian, however, can be broken down into measurements at positive and negative longitudes, due to OSSE's superior spatial resolution.

The results in Fig. 2 are expressed as ratios of the ^{60}Fe line flux to the ^{26}Al 1.809 MeV line flux (fixed at the value measured by COMPTEL, 3.0×10^{-4} photon cm^{-2} s^{-1} rad^{-1}). This is because the sites of nucleosynthesis of ^{26}Al and ^{60}Fe are expected to be basically the same, i.e. core collapse supernovae. Theoretical predictions may therefore be expressed as a single line flux ratio which should be valid everywhere in the Galaxy; the best recent value for this ratio is 0.16[14]. The OSSE measurement at positive longitude is in disagreement with this value; it exceeds the theoretical prediction by a factor of ~ 3. This may indicate a new type of source for ^{60}Fe, or else a difference in the massive-star population (and hence the ^{60}Fe/^{26}Al ratio) between positive and negative longitudes.

Acknowledgements. This research was supported by NASA grant DPR S-67026F.

REFERENCES

[1] M.J. Harris et al., A&AS, 120, 343 (1996).
[2] H. Bloemen et al., A&A, 281, L5 (1994).
[3] R.J. Murphy et al., ApJ, 473, 990 (1996).
[4] M.J. Harris et al., A&A in press (1997).
[5] H. Bloemen et al., ApJ, 475, L25 (1997).

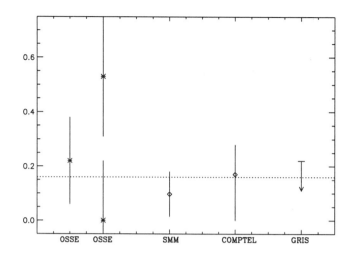

FIGURE 2. Existing measurements and upper limits on the flux ratio between ^{26}Al and ^{60}Fe (single line) from the central radian of the Galactic plane[11-13]. The OSSE measurement is resolved into results from $l > 0°$ (upper) and $l < 0$ (lower). Dashed line - theoretical line flux ratio.

[6] R. Ramaty, B. Kozlovsky, & R.E. Lingenfelter, ApJ, 438, L21 (1995).
[7] T.M. Dame et al., ApJ, 322, 706 (1987).
[8] J.A. Miller & C.D. Dermer, A&A, 298, L13 (1995).
[9] E.L. Martin et al., Nature, 358, 129 (1992).
[10] R. Rebolo et al., Mem.S.A.It., 66, 437 (1995).
[11] J.E. Naya et al., these proceedings.
[12] R. Diehl et al., these proceedings.
[13] M.D. Leising & G.H. Share, ApJ, 424, 200 (1994).
[14] F.X. Timmes et al., ApJ, 464, 332 (1996).

Reassessment of the ^{56}Co emission from SN 1991T

D. J. Morris*, K. Bennett‡, H. Bloemen§, R. Diehl†, W. Hermsen§,
G. G. Lichti†, M. L. McConnell*, J. M. Ryan* and V. Schönfelder†

*EOS/Space Science Center, University of New Hampshire, Durham NH 03824-3525, USA
†Max-Planck-Institut für extraterrestrische Physik, Postfach 1603, 85740 Garching Germany
§SRON-Utrecht, Sorbonnelaan 2, 3584 CA Utrecht, The Netherlands
‡Astrophysics Division, ESTEC, 2200 AG Noordwijk, The Netherlands

Abstract. The detection of ^{56}Co emission from SN 1991T has been previously reported at a level near the COMPTEL sensitivity threshold. The spectral analysis method, fitting the count spectrum to a background model plus a ^{56}Co emission template, is subject to possible systematic effects which had not been thoroughly studied at that time. To better evaluate the significance of that ~3.3σ detection, the same method has been applied to a grid of points with 5° spacing, out to 35° from the pointing direction, in each of 103 observing periods from phases 1 through 3. A dozen instances were found with a ^{56}Co signal as significant as that for either of the two observations of SN 1991T alone (~2σ). Nothing was found as significant as the combined observations of SN 1991T. The strongest instrumental background artifact in the vicinity of the two principal ^{56}Co lines, attributed to ^{27}Mg, falls between the ^{56}Co lines. It fills in the valley between those lines, and so will obscure real ^{56}Co emission rather than producing false ^{56}Co sources. Fortunately, this artifact was weak up to the time of the reboost during phase 3. Thus, it is very unlikely that the reported emission from SN 1991T was a statistical fluctuation or instrumental artifact. But, since the flux was so near the detection threshold, little can be said about the gamma-ray light curve of the supernova, the relative strengths of the ^{56}Co lines, or the line widths.

INTRODUCTION

The type Ia supernova SN 1991T was discovered on April 13, 1991, just 8 days after the launch of CGRO, in the spiral galaxy NGC 4527 on the edge of the Virgo cluster. The peak magnitude of SN 1991T was near V=11.5, about 0.5 magnitude brighter than a typical type Ia in the Virgo cluster. This, together with certain spectral peculiarities [1], suggested that an unusually large mass of ^{56}Ni (the short-lived progenitor of ^{56}Co) was produced, making SN 1991T intrinsically brighter than the typical type Ia in both the optical and γ-ray bands. Given its relative proximity, about 13.5 Mpc [2], SN 1991T was an obvious target for CGRO.

It was observed by COMPTEL during two 14-day periods beginning 66 days (obs. 3) and 176 days (obs. 11) after the supernova explosion.

The initial analysis of the SN 1991T observations [3] employed background-subtracted count spectra, and maximum likelihood imaging for energy windows around each of the two principal ^{56}Co lines, at 847 keV and 1238 keV. Those efforts produced only upper limits on the line fluxes. The choice of an appropriate background spectrum is a difficult problem in a standard spectral analysis. The instrumental background changes with time, the spacecraft pointing direction and across the field of view, often confounding efforts to find the "right" background. A second difficulty is that the statistical fluctuations in background spectra contribute to the uncertainties in background-corrected spectra.

As an alternative, a new analysis method was developed which avoids selecting a measured background spectrum. Rather, this method fits the measured spectrum directly with model components, exploiting the accumulated knowledge of the instrumental background and models of the instrumental response. Evidence was found of ^{56}Co emission in both SN 1991T observations, though it was statistically significant ($> 3\sigma$) only for the combined observations [4]. A consistent detection ($\sim 3\sigma$) was obtained with maximum likelihood imaging by combining events in the two energy windows, as well as the two observations, for the analysis and utilizing an improved point-spread function.

Though these results were encouraging, an investigation of possible systematic errors in the spectral fitting was needed to provide confidence in the detection of ^{56}Co. This paper presents an assessment of such systematic effects.

THE SPECTRAL-FITTING METHOD

The function which is used to model the instrumental background is

$$f_B = a_0\left(1 - e^{-b(E-E_0)}\right)e^{-\alpha E} + a_K \exp\left[-0.5\left(\frac{E-E_K}{\sigma_K}\right)^2\right] + a_D \exp\left[-0.5\left(\frac{E-E_D}{\sigma_D}\right)^2\right] \quad (1)$$

The first term, with four free parameters, is a smooth underlying background with a threshold at E_0 and an exponential decline at high energies. The two remaining terms are Gaussians to model a feature near 1500 keV, due in part to the decay of ^{40}K in various spacecraft components, and the 2223-keV line from deuterium formation in the COMPTEL D1-detector liquid scintillator.

Figure 1 shows a typical but particularly pertinent example of a fit to this background model. The spectrum contains events in obs. 3 and 11 within 20° of SN 1991T. The spectrum is fit from threshold (~720 keV) to 2600 keV. Figure 1b shows the deviations from the fit. Significant deviations are apparent, particularly around 1500-1700 keV, but in the vicinity of the ^{56}Co lines the fit is quite good ($\chi^2 = 10.9$ for 14 points at 800-1500 keV).

When searching for ^{56}Co emission the fit is restricted to the energy range from threshold to 1800 keV. This eliminates the three free parameters for the deu-

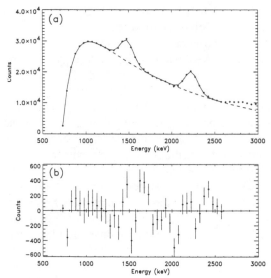

FIGURE 1. (a) Spectrum of events from obs. 3 and 11 combined, within 20° of SN 1991T. The solid line is the fit to the background approximation (1) with the dashed line showing the first term in (1) alone. (b) The residuals of the fitted spectrum.

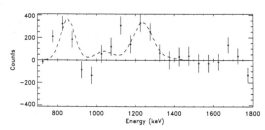

FIGURE 2. Deviations from the background model in a fit including the ^{56}Co template (dashed line) to events within 3° of SN 1991T.

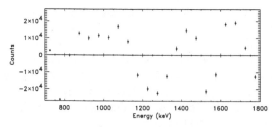

FIGURE 3. Deviations from the background model in a fit to all events within 20° of the pointing direction in observations through February 1997.

terium line and hopefully provides a better background fit in the vicinity of the ^{56}Co lines. After fitting to the background model alone, an emission template, derived from Monte Carlo simulations of the telescope response to the ^{56}Co lines, is included in the fit. If inclusion of the template significantly improves the goodness of the fit, as measured by the χ^2 statistic [5], the presence of ^{56}Co emission can be inferred. Figure 2 shows deviations from the background model for a fit with the ^{56}Co emission template to combined data from the two SN 1991T observations, within 3° of SN 1991T. The dashed line shows the fitted ^{56}Co template.

ASSESSMENT OF SYSTEMATIC ERRORS

Two approaches have been used to assess systematic errors in the spectral fitting analysis. Persistent deviations of the instrumental background from the model (1) have been sought by fitting spectra for large volumes of data. An empirical assessment of the significance of the ^{56}Co emission has been done by searching a large number of COMPTEL observations for ^{56}Co emission.

Figure 3 shows the residuals for a fit with the background

model (1) to a spectrum of events within 20° of the pointing directions, summed over all observations through February 1997. There is an interval with positive residuals between the energies of the principal ^{56}Co lines. This feature is probably due to ^{27}Mg activation (^{27}Al(n,p)^{27}Mg) in passive materials near the D1 modules. The ^{27}Mg nuclei beta decay with a halflife of 9.5 min, emitting either an 844-keV or 1014-keV photon. Most ^{27}Mg background events result from the absorption of the decay photon in the D2 detector together with the scattering of a bremsstrahlung photon in the D1 detector, depositing at least 50 keV (the D1 energy threshold). Because the ^{27}Mg events fill the valley between the principal ^{56}Co lines, they obscure real ^{56}Co emission rather than producing false ^{56}Co sources. Fortunately the ^{27}Mg feature was weak before the spacecraft reboost in November 1993.

To search for ^{56}Co sources, software was developed to simultaneously accumulate spectra for many points in the field of view and to fit the spectra with the background model (1) plus an emission template. This in effect allows mapping of ^{56}Co emission. We produced 103 such maps on a 5°×5° grid with a field of 70°×70° and spectra accumulating events within 3° of each grid point. The maps cover all observations through phase 3 (to September 1994).

Local maxima in ^{56}Co flux more significant than that at SN 1991T in either one of the two individual observations (~2σ) were found at 12 positions in 11 maps. This is much smaller than the number that should be expected by chance. However, all those excesses were in the 79 maps for observations preceding the reboost and they were concentrated toward the center of the field of view: 5 of the 12 were within 10° of the center and 11 were within 20°. The dearth of ^{56}Co excesses in the later observations is probably due to increased ^{27}Mg activation. The distribution of ^{27}Mg background events across the field of view may well lead to the concentration of ^{56}Co excesses near the center. This question can be investigated through Monte Carlo simulations of ^{27}Mg events.

For a 10°-radius field of view and 79 maps about 1000 points were tested, with about 45 deviations >2σ expected, half of them positive. However, the grid spacing was not so large that spectra at neighboring points are independent, so the discovery of only 5 local maxima exceeding 2σ may well be consistent with statistical expectations.

None of the excesses found in the ^{56}Co search was as significant as that at SN 1991T in the combined observations 3 and 11. No two of those excesses were found at the same point on the sky, so excesses in different observations cannot be combined to reveal a source more significant than SN 1991T.

OTHER CONFIDENCE TESTS FOR SN 1991T

Figure 4 is a spectral-analysis map of the ^{56}Co emission from SN 1991T on a 1°×1° grid, which can be compared directly with the maximum likelihood map in ref. 4. The two analysis methods produce very similar maps, with SN 1991T near the 1σ location contour, and the most likely source position about 1.5° to the northeast, away from the nearby bright continuum source 3C273. Independent imaging analysis with the 'software collimation' method [6], though less sensitive, also shows a 2σ excess at SN 1991T.

By varying the line weighting for the emission template, we have tested for absorption of the ^{56}Co γ-rays. The best fit is obtained with about a 20% reduction in the strength of the 847-keV line, relative to the 1238-keV line, corresponding to absorption in 10-20 $g\ cm^{-2}$ of Fe, but this fit is only marginally better than that for the production ratios (no absorption). Eliminating either of the principal lines entirely reduces the significance of the fit by about 1σ.

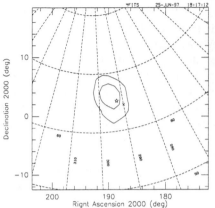

FIGURE 4. Map of the ^{56}Co emission in obs. 3 and 11 showing 1- and 2-σ source location contours. The positions of SN 1991T (star) and 3C273 (×) are indicated.

CONCLUSION

There is no evidence of systematic effects that mimic ^{56}Co emission in the spectral fitting analysis. Rather, the most persistent background feature in the vicinity of the principal ^{56}Co lines obscures ^{56}Co emission. A search for excesses of ^{56}Co flux found fewer than would be expected by chance, and none as significant as that at SN 1991T.

This empirical assessment of the significance of ^{56}Co emission shows that the formal statistical significance of 3.3σ does not overstate confidence in the source at SN 1991T. That confidence is bolstered by the coincidence of the emission in space and time with the expected source, the drop in significance if either of the principal lines is omitted from the emission template, and the good agreement between the spectral analysis and maximum likelihood mapping.

REFERENCES

1. Phillips, M. M. *et al.*, *Astron. J.*, **103**, 1632 (1992).
2. Tully, R. B., *Nearby Galaxies Catalog*, Cambridge: Cambridge University Press, 1988.
3. Lichti, G. G. *et al.*, *Astron. Astrophys.*, **292**, 569 (1994).
4. Morris, D. J. *et al.*, in *Seventeenth Texas Symposium on Relativistic Astrophysics and Cosmology*, ed. Böhringer, H. *et al.*, New York: New York Academy of Science 1995, pp. 397-400.
5. Eadie, W.T., Drijard, D., James, F. E., Roos, M. and Sadoulet, B., *Statistical Methods in Experimental Physics*, Amsterdam: North-Holland 1971, pp.192-199.
6. Diehl, R. *et al.*, *Astron. Astrophys. Suppl.*, **97**, 181 (1993).

RXTE Observations of Cas A

R. E. Rothschild*, R.E. Lingenfelter*, W.A. Heindl*, P.R. Blanco*, M.R. Pelling*, D.E. Gruber*, G.E. Allen[†], K. Jahoda[†], J.H. Swank[†], S.E. Woosley[‡], K. Nomoto**, and J.C. Higdon[°]

*University of California San Diego, La Jolla, CA 92093
[†]Goddard Space Flight Center, Greenbelt, MD 20771
[‡]University of California Santa Cruz, Santa Cruz, CA 95064
**University of Tokyo, 113 Tokyo, Japan
[°]The Claremont Colleges, Claremont, CA 91711

Abstract. The exciting detection by the COMPTEL instrument of the 1157 keV ^{44}Ti line from the supernova remnant Cas A sets important new constraints on supernova dynamics and nucleosynthesis. The ^{44}Ti decay also produces x-ray lines at 68 and 78 keV, whose flux should be essentially the same as that of the gamma ray line. The revised COMPTEL flux of 4×10^{-5} cm^{-2}s^{-1} is very near the sensitivity limit for line detection by the HEXTE instrument on RXTE. We report on the results from two RXTE observations — 20 ks during In Orbit Checkout in January 1996 and 200 ks in April 1996. We also find a strong continuum emission suggesting cosmic ray electron acceleration in the remnant.

INTRODUCTION

The Cas A supernova remnant is the remains of the explosion of a massive star 317 years ago (if indeed John Flamsteed observed it in 1680) at a distance of 3.4 kpc. Optical observations [1,2] suggest that Cas A is the remnant of an SNIb, or SNII from a WN Wolf-Rayet progenitor with a initial mass >25 M$_\odot$. X-ray imaging of Cas A reveals an outer, weak ring of emission assumed to be the result of the expanding shock wave interacting with the interstellar medium or wind material, and a brighter inner ring representing the reverse shock impinging upon the ejecta [3]. Spectral measurements clearly demonstrate that thermal equilibrium has not been attained and that relatively simple one- or two-temperature thermal models cannot describe the data [4]. At gamma-ray energies the detection of 1157 keV line photons [5] has provided an estimate of ^{44}Ti production in the supernova explosion which is at odds with the low apparent brightness of the supernova event [6]. As the youngest known supernova remnant, Cas A affords us the opportunity to to study both the nucleosynthesis involved and the dynamic evolution of the forward and reverse shocks as they impinge in the interstellar medium and the slower moving material within the remnant, respectively.

Radioactive ^{44}Ti $\rightarrow$ ^{44}Sc $\rightarrow$ ^{44}Ca produced by explosive silicon burning and the freeze out from nuclear statistical equilibrium in supernovae is thought to be the primary source of galactic ^{44}Ca. The detection of gamma-ray line emission from the decay of ^{44}Ti with a mean-life of about 78 to 96 years, and its short-lived daughter ^{44}Sc, should ultimately enable us to discover the sites of the most recent supernovae in our Galaxy. These supernovae have escaped detection because of the large optical extinction by dust in the inner Galaxy, and even recent radio searches with the VLA have failed to discover any very young (<100 years old) supernova remnants in our Galaxy suggesting that they may not turn on in radio until they are older.

PREVIOUS OBSERVATIONS

The principal gamma-ray lines from the ^{44}Ti decay chain to ^{44}Sc are at 67.87 and 78.32 keV [7] with essentially equal intensities, and at 1157.0 keV 100% of the time from the much shorter lived (mean-life of 5.7 hours) ^{44}Sc. Searches for these lines by the gamma-ray spectrometers on HEAO-3 and SMM, prior to GRO, set upper limits on the line fluxes at 67.87 and 78.32 keV of 2×10^{-4} photons/cm^{2}s [8] and at 1157 keV of 8×10^{-5} to 2×10^{-4} photons/cm^{2}s, depending on Galactic longitude [9], at the 99% confidence level from any potential "point source" supernova remnant in the Galactic disk. The SMM limit of 8×10^{-5} photons/cm^{2}s from any remnant in the direction of the Galactic center also sets [9] a 90% confidence limit on the typical ^{44}Ti yield of $\leq 1.0 \times 10^{-4}$ M$_\odot$ for any class of supernovae whose frequency was greater than 1 per 100 years. Even this, however, was still consistent with the ^{44}Ti yields calculated from most current supernova models, and recent estimates of the Galactic rates of SNIa, SNIb and SNII of 1.1 h^2, 1.2 h^2 and 6.1 h^2 supernovae per 100 years, respectively. These range for SNIa from about 1.8×10^{-5} M$_\odot$ for deflagration model W7 [10] to $(0.2$ to $4.0) \times 10^{-3}$ M$_\odot$ for sub-Chandrasekhar mass models [11]; $(3$ to $8) \times 10^{-5}$ M$_\odot$ for SNIb models [12]; and $(0.14$ to $2.3) \times 10^{-4}$ M$_\odot$ for SNII for the 11 to 40 M$_\odot$ models [11].

The HEAO-3 limits together with Monte Carlo simulations of Galactic supernovae also suggest [8] a limit on the total Galactic production of ^{44}Ti of $\leq 1.0 \times 10^{-3}$ M$_\odot$ per 100 years, which is consistent with that which would be expected from the present rate of Galactic ^{56}Fe production rate of $\sim (0.8 \pm 0.6)$ M$_\odot$ per 100 years [13], if the diffuse Galactic 511 keV positron annihilation radiation comes from positrons resulting from the decay of ^{56}Ni, ^{44}Ti, and ^{26}Al, produced in supernovae. This ^{56}Fe production rate implies a corresponding Galactic ^{44}Ti production rate of $\sim (1.0 \pm 0.7) \times 10^{-3}$ M$_\odot$ per 100 years, assuming that mean Galactic mass fraction X_{44}/X_{56} has the solar value of 1.23×10^{-3}.

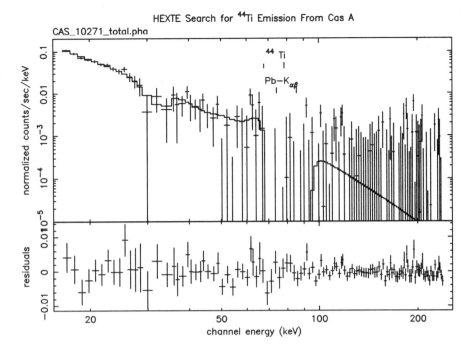

FIGURE 1. HEXTE background subtracted data for Cas A compared to best-fit model of the continuum plus ^{44}Ti lines at 67.9 and 78.3 keV. The model also includes oversubtracted $K_{\alpha\beta}$ lines of Pb from the collimator and Te and Xe from activation of Iodine. Positions of the ^{44}Ti and Pb-$K_{\alpha\beta}$ lines are shown.

The detection [14] of 1157 keV line emission from Cas A by the COMPTEL imaging gamma-ray spectrometer on GRO with a line flux of $(4.2\pm0.9)\times10^{-5}$ photons/cm^{2}s implies an initial ^{44}Ti mass of $(1.0\pm0.2)\times10^{-4}$ M$_\odot$, assuming a remnant distance of 3.4 kpc, a supernova date of 1680 A.D., and a ^{44}Ti mean-life of 96 yr. Taking the shorter mean-life of 78.2 yr would increase the initial ^{44}Ti mass by a factor of 2. Such masses are consistent or slightly higher than those that would be expected for Cas A, if it was either a massive SNIb or a SNII supernova.

The OSSE observations [15] of Cas A, searching for all three nuclear lines expected from the ^{44}Ti decay chain, yielded a best-fit of $0.40^{+2.26}_{-2.77}\times10^{-5}$ photons/cm^{2}s for each line, with a 99% upper limit of 5.7×10^{-5} photons/cm^{2}s. The OSSE and COMPTEL results are consistent at 15% confidence for a flux of 3.5×10^{-5} photons/cm^{2}s.

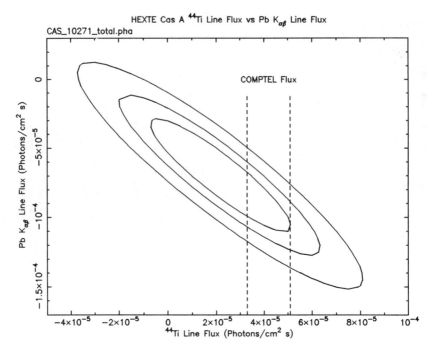

FIGURE 2. Chi-square contours (1-,2-,and 3-σ) for the line strengths of the ^{44}Ti and Pb fluorescence lines from fitting the HEXTE data. The dotted lines give the $\pm 1\sigma$ range of values from the COMPTEL analysis of the 1157 keV line.

ANALYSIS OF HEXTE OBSERVATIONS

Cas A has been observed by RXTE on two occasions: 1) during In-Orbit Checkout (IOC) on January 20, 1996 and 2) from March 31 to April 17, 1996 as our AO-1 proposal observation. The IOC observation provided 16 ks of "very clean" data, and the AO-1 observation yielded 128 ks of "very clean" data. Both data sets were fitted with a model containing a power law component plus lines at 68 and 78 keV that were constrained to have identical fluxes. The width of these lines was set at 2.5% of their energy to represent the expected Doppler broadening due to the ejecta outflow. Residuals to the fit revealed excess deviations ($\sim$1-2% of background) at the positions of the strong background lines due to fluorescence of the Pb collimators (74.2 and 85.4 keV) and to a lesser extent at 30 keV due to activation of Iodine. Thus, a simultaneous fit was performed that included the ^{44}Ti lines, the Pb lines, the 30 keV line, and a power law continuum. Figure 1 shows the result of this fit to the AO-1 observation.

Both observations yielded consistent power law indices ($\Gamma = 3.21 \pm 0.13$). The power law component may be an indication of synchrotron emission, suggesting that cosmic ray electron acceleration is taking place in Cas A at energies in excess of 10^{13} eV [4]. Figure 2 displays the χ^2 contours for the joint variation of the ^{44}Ti lines and the lead lines for the AO-1 observation. The best fit flux for the ^{44}Ti lines is $(2.19\pm1.95)\times10^{-5}$ photons/cm^{2}s. A similar analysis of the IOC data with 12.5% of the AO-1 livetime has a best fit flux of $(0.27\pm3.36)\times10^{-5}$ photons/cm^{2}s. These translate into 99% upper limits of $\leq 8.0\times10^{-5}$ photons/cm^{2}s and $\leq 10.3\times10^{-5}$ photons/cm^{2}s for the AO-1 and IOC observations, respectively.

CONCLUSIONS

The RXTE observations of Cas A have detected a power law component that dominates the flux above 10 keV and may be due to synchrotron emission. This in turn may be a direct indication that cosmic ray acceleration is taking place in Cas A at high energies. The 67.9 and 78.4 keV lines from ^{44}Ti decay are marginally consistent with that seen by COMPTEL for the 1157 keV line from the subsequent decay of ^{44}Sc and the OSSE 99% confidence upper limits. We are currently analyzing the second 200 ks observation of Cas A by RXTE during the AO-2 cycle.

REFERENCES

1. Jansen, F. et al., *Astrophys. J.* **331**, 949 (1988).
2. Fesen, R. A., & Becker, R. H., *Astrophys. J.* **371**, 621 (1991).
3. Tsunemi, H. et al., *Astrophys. J.* **306**, 248 (1986).
4. Allen, G. E. et al., *Astrophys. J.* (in press).
5. Iyudin, A. F., et al., *Astron. & Astrophys.* **284**, L1 (1994).
6. The, L.-S., et al., *Astrophys. J.* **444**, 244 (1995).
7. Wesselborg & Alburger, *Nuc. Instr. & Meth. Phys. Res.* **A302**, 89 (1991).
8. Mahoney, W., Ling, J., Wheaton, W., & Higdon, J., *Astrophys. J.* **387**, 314 (1992).
9. Leising, M. D., & Share, G. H., *Astrophys. J.* **424**, 200 (1994).
10. Thielemann, F.-K., Nomoto, K., & Yokoi, K., *Astron. Astrophys.* **158**, 17 (1986).
11. Woosley, S. E., & Weaver, T. A., *Astrophys. J. Supp.* **101**, 181 (1995).
12. Woosley, S. E., Langer, N., & Weaver, T. A., *Astrophys. J.* **448**, 315 (1995).
13. Chan, K.W. and Lingenfelter, R.E., *Astrophys. J.* **405**, 614 (1993).
14. Schonfelder, V. et al., *Astron. & Astrophys. Supp.* **120**, 13 (1996).
15. The, L.-S., et al., *Astron. & Astrophys. Supp.* **120**, 357 (1996).

FLUCTUATION ANALYSIS OF OSSE MEASUREMENTS OF THE 1.275 MeV LINE OF ^{22}Na

Michael J. Harris

USRA/GVSP, Code 661, NASA/Goddard Space Flight Center, Greenbelt, MD 20771.

Abstract.
We show that the time-scales on which OSSE makes and repeats observations make possible a very sensitive search for γ-ray lines which are constrained to vary on compatible time-scales. The method resembles that by which the amplitude of single shot-like events is determined from the fluctuations in the sum of many events, even though individual shots are too weak to detect. The variability of the 1.275 MeV line from decay of ^{22}Na in novae near the Galactic center (GC) is suitable for this analysis. Applying it to OSSE measurements of this line during 1991–1995, we obtain a value for the characteristic increase in line flux due to a single nova which is significant at a level $\sim 2\sigma$.

I INTRODUCTION

Since 1991 we have systematically measured the strengths of nucleosynthetic and cosmic-ray induced γ-ray lines in OSSE spectra from fields of view (FOVs) around the Galactic center (GC). Time series of line strengths have been obtained for the 1.275 MeV line of ^{22}Na, the 1.809 MeV line of ^{26}Al, and the cosmic-ray 4.4 and 6.1 MeV lines, among others[1].

The time series of ^{22}Na line strengths is shown in Fig. 1a. There is no evidence for a positive average flux in these data, down to a 3σ upper limit of 9.7×10^{-5} photon cm^2 s^{-1}. However a more sophisticated analysis can yield important information about the sources of ^{22}Na, by considering the fluctuations of the measurements about the average.

The case of ^{22}Na produced in Galactic novae is particularly favorable for such an analysis. The 1.275 MeV line light-curve is very strongly constrained by the intrinsic properties of novae and of its β-decay. Nova envelopes become transparent to it on time-scales ~ 10 d, and its flux must decay with the exponential fall-off of the ^{22}Na abundance (half-life 2.6 yr). These time-scales are very well suited to observations by OSSE, which nominally last 14 d, and which are repeated (for the GC) at intervals ranging from 14–1400 d. The nova rate in the $\sim 5°$ around the GC is probably ~ 10 per year[2], of which ~ 2 may be of the neon-rich subtype which produces the most ^{22}Na. As seen by OSSE, the 1.275 MeV light-curve thus consists of 'instantaneous' increases at random epochs, followed by long exponential decays which are sampled frequently (but very irregularly) by OSSE's repeated observations.

The method described here resembles the analysis of 'shot noise' in stochastic light-curves (as applied for example to Cyg X-1[3]). In this approach, the characteristic properties of an individual shot are extracted from the stochastic fluctuations, even though an individual shot increment may be far too weak to be detected. We will not consider the theory of shot noise at all here. Instead, in §2 we present a simple analytic treatment which illustrates the capability of the method. In §4 we describe an autocorrelation function (ACF) obtained from shot noise theory and illustrate how it is employed.

II ANALYTIC TREATMENT

Consider two observations separated by an interval t. If the nova rate is R the probability of one or more novae occurring during time t is $\sum_n P_n = \sum (Rt)^n/n! e^{Rt} = 1 - e^{-Rt}$ for $n \geq 1$ assuming Poisson statistics. The expected increase in flux between the two epochs is therefore $\sum_n P_n \phi$, where ϕ is the average flux per event.

Let x_0 be the pre-existing ^{22}Na abundance at the time of the first observation. By the time of the second observation, it will have decayed by a factor $(1 - e^{-t/\tau})$. However, by hypothesis, there has been a flux $\sum_n P_n \phi$ added during time t. We assume that the increase occurs at the mid-point of the interval. By the time of the second observation, this quantity will have decayed by a factor $(1 - e^{-0.5t/\tau})$. The expected change ('step') in flux between the two observations is

$$\Delta \phi = \sum_n P_n \phi (1 - e^{-0.5t/\tau}) - x_0 (1 - e^{-t/\tau}). \qquad (1)$$

We make the further assumption that the ^{22}Na abundance mid-way between the two observations is equal to the long-term average abundance, which enables us to eliminate x_0. The long-term average is $R\phi\tau$, so that

$$x_0 e^{-0.5t/\tau} + \sum_n P_n \phi = R\phi\tau. \qquad (2)$$

Eliminating x_0 between (1) and (2) yields the expected step in flux

$$\Delta\phi = \sum_n P_n \phi e^{0.5t/\tau} - R\phi\tau(e^{0.5t/\tau} - e^{-0.5t/\tau}). \qquad (3)$$

The distribution of flux step with t is clearly not random. The 148 measurements in Fig. 1a were arranged in pairs, for each of which the flux step $\Delta\phi$

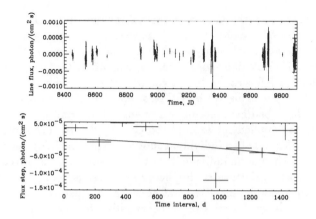

FIGURE 1. (a) 148 OSSE flux measurements of the 1.275 MeV ^{22}Na line 1991–1994. (b) Data points—7942 measurements of the flux step between the 148 observations in (a) plotted against the interval between observations, combined into 150 d bins. Full line—flux step as a function of interval predicted by Eq. (3), fit to the data.

and time interval t were calculated. Least-squares fitting of the two variables R and ϕ to (3)[1] gave $\phi = 2.3 \pm 0.6 \times 10^{-5}$ photon cm^{-2} s^{-1} per nova and $R = 0.85^{+0.15}_{-0.10}$ novae yr^{-1}. The fit is illustrated in Fig. 1b.

The quoted errors are unrealistic; they do not take account of the high degree of correlation due to multiple use of the same data points. Note also that the systematic departures from the fitted curve are highly significant, probably due to the stochastic nature of the underlying light-curve. We estimated the effect of the correlations by a Monte Carlo simulation of the data points in Fig. 1a within their errors. We found that the true error in ϕ was underestimated by almost a factor 2. The significance of our analytic result quoted above is therefore about 2σ. The effect of the stochasticity remains to be dealt with by Monte Carlo simulations of the ACF described in §3.

We checked the result by performing the same fits to the time series of ^{26}Al 1.809 MeV and cosmic-ray 4.4 and 6.1 MeV line strengths measured by ref. 1, and to a time series of ^{22}Na 1.275 MeV line fluxes measured from over 200 OSSE spectra of AGNs. In none of these cases was there a significant fit with Eq. (3).

III AUTOCORRELATION FUNCTION ANALYSIS

By comparison with standard shot noise analyses, our data suffer from the problem of being sampled at very irregular intervals (Fig. 1a). An ACF for

---

[1] The 148 OSSE FOVs were not identical, although overlapping. The expected values of $\Delta\phi$ were therefore weighted by the fractional overlap of each pair of FOVs.

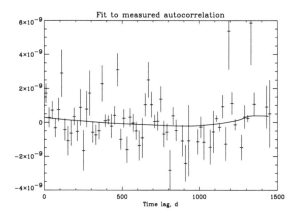

FIGURE 2. Data points—autocorrelation function (4) of the OSSE 1.275 line measurements in Fig. 1a. Solid line–Mean autocorrelation function of ~ 100 simulated light-curves having $r = 0.85$ yr^{-1}, fit to the data.

use with such data has the form

$$y_j = \frac{\sum_i (\phi_i - \overline{\phi})(\phi_{i+j} - \overline{\phi}) w_i w_{i+j}}{\sum_i w_i w_{i+j}} \quad (4)$$

(Finger 1994, private communication), where i labels uniform intervals, some containing data ϕ_i and some not, and the weights w_i are set to 1 for intervals containing data and to 0 for empty intervals.

The significance of this ACF obtained from a given data set must be obtained by comparing it to the function expected from the hypothesis. We did this by generating a mean ACF from $\sim 10^4$ simulations of the 1.275 MeV light-curve from randomly occurring novae at the GC. The ACF computed from the OSSE data in Fig. 1, despite its irregularity, yielded a significant correlation when fitted to the mean ACF of a subset of these simulations (Fig. 3). The amplitude of this correlation for the subset having $R = 0.85$ yr^{-1} is $A = 8.6 \pm 3.3 \times 10^{-10}$ (photon/(cm^2 s))^2. The amplitude ϕ of the input exponentially-decaying shot is related to this by $A = \phi^2 N_n$, where N_n is the number of shots, which for our sample with nova rate $R = 0.85$ yr^{-1} averaged 3.0. We derive a mean flux increase per nova $1.7 \pm 0.6 \times 10^{-5}$ photon cm^{-2} s^{-1}, in good agreement with the analytic result in §2.

IV ASTROPHYSICAL IMPLICATIONS

It appears to be possible to measure the 1.275 MeV line flux from an average neon nova down to a level $\simeq 10^{-5}$ photon cm^{-2} s^{-1} at the distance of the GC. The corresponding mass of ^{22}Na produced in an average neon nova is

$\sim 2\times 10^{-7} M_\odot$. This is comparable to the upper limits $\sim 4\times 10^{-8} M_\odot$ obtained from COMPTEL observations of nearby neon novae[4]. Predicted ^{22}Na masses from models of accretion onto the most massive ONeMg-rich white dwarfs[5,6] ($> 1.3 M_\odot$) range from 7.5–30 $\times 10^{-8} M_\odot$. These massive objects are presumably the 'average' contributors of ^{22}Na towards the GC, as detected by our analysis.

Acknowledgements. I am very grateful to Dr. W. T. Bridgeman for guidance on the use of ACFs. Supported by NASA grant DPR S-10987C.

REFERENCES

[1] M.J. Harris et al, A&AS, 120, 343 (1996).
[2] M.D. Leising & D.D. Clayton, ApJ, 295, 591 (1995).
[3] W.T. Bridgman et al., in *The Second Compton Symposium*, ed. C.E. Fichtel, N. Gehrels, & J. P. Norris (New York: AIP), 225 (1994).
[4] A.F. Iyudin et al., A&A, 300, 422 (1994).
[5] S. Starrfield et al., ApJ, 391, L71 (1992).
[6] M. Politano et al., ApJ, 448, 807 (1995).

COMPTEL All-Sky Imaging at 2.2 MeV

M. McConnell*, S. Fletcher[†], K. Bennett[‡], H. Bloemen[||], R. Diehl[¶], W. Hermsen[||], J. Ryan*, V. Schönfelder[¶], A. Strong[¶], and R. van Dijk[‡]

*University of New Hampshire, Durham, NH
[†]Los Alamos National Laboratory, Los Alamos, NM
[‡]Astrophysics Division, ESTEC, Noordwijk, The Netherlands
[||]SRON-Utrecht, Utrecht, The Netherlands
[¶]Max Planck Institute (MPE), Garching, Germany

Abstract. It is now generally accepted that accretion of matter onto a compact object (white dwarf, neutron star or black hole) is one of the most efficient processes in the universe for producing high energy radiations. Measurements of the γ-ray emission will provide a potentially valuable means for furthering our understanding of the accretion process. Here we focus on neutron capture processes, which can be expected in any situation where energetic neutrons may be produced and where the liberated neutrons will interact with matter before they decay (where they have a chance of undergoing some type of neutron capture). Line emission at 2.2 MeV, resulting from neutron capture on hydrogen, is believed to be the most important neutron capture emission. Observations of this line in particular would provide a probe of neutron production processes (i.e., the energetic particle interactions) within the accretion flow. Here we report on the results of our effort to image the full sky at 2.2 MeV using data from the *COMPTEL* experiment on the *Compton Gamma-Ray Observatory* (*CGRO*).

INTRODUCTION

The possibility of observing γ-ray lines from the radiative capture of neutrons has been recognized for some time [1]. Although several capture lines are possible, by far the most dominant line is expected to be that from neutron capture on hydrogen (producing a line at 2.223 MeV). There are several scenarios which might produce 2.2 MeV line emission in accreting compact sources (neutron stars or black holes). These include: 1) neutron capture within the accretion flow; 2) neutron capture in the atmosphere of a neutron star; 3) neutron escape from the accretion flow followed by capture in the compact object's companion star; and 4) neutron capture in a situation where a beam of accelerated particles impinges on the companion star (in analogy to solar flares).

The gravitational potential energy released from accretion of matter onto the surface of a compact object can lead to ion temperatures approaching 100 MeV ($T_i \sim 10^{12} K$), which subject heavier nuclei to breakup by spallation reactions. Some of the liberated neutrons might be captured on protons within the accretion flow itself, thus generating a 2.2 MeV line signature. It has been shown that, under most conditions, the neutrons are more likely to escape the production region rather than be captured [2]. Furthermore, neutron capture in the hot accreting plasma would lead to an extremely broad emission line [3] that might be difficult to observe. It therefore seems unlikely that any detectable 2.2 MeV line emission would be generated from *within* the accretion flow.

Matter accreting onto a neutron star has large enough kinetic energy to excite or destroy nuclei. Neutrons liberated by these reactions (principally by the spallation of ^{4}He), once thermalized, will either recombine radiatively with a proton (to produce a 2.223 MeV photon) or non-radiatively with ^{3}He. This problem has been studied in detail [4]. Predicted flux levels are as high as $\sim 2 \times 10^{-5}$ cm^{-2} s^{-1}. This level of emission is near the present all-sky sensitivity limit of *COMPTEL* observations collected over the first five years of the CGRO mission. Enhanced levels of 2.2 MeV emission might be expected from sources where the accreting material contains an unusually high abundance of heavier elements (as would be expected for a highly evolved massive companion). There are at least three cases where such heavy element enhancements may exist: 4U1916-05, 4U1626-67, and 4U1820-30 [5]. In this scenario, a 2.223 MeV neutron capture line would be gravitationally red-shifted to an energy as low as 1.76 MeV.

Neutrons that are produced within the accretion flow are not confined by any magnetic fields. Consequently, they are free to leave the production region provided they can escape the gravitational well of the compact object. Some fraction of the escaping neutrons may then interact in the atmosphere of the companion star. The thermalization of the interacting neutrons, and the subsequent capture by ambient protons, would lead to a γ-ray line at 2.223 MeV. Various considerations (e.g., the neutron decay time and the solid angle for interaction with the companion) suggest that close binaries are more probable sources of observable 2.2 MeV emission. In this scenario, the 2.2 line flux will originate on the side of the companion star irradiated by the neutron flux, i.e., the side of the companion that faces the compact object. Therefore, the 2.2 MeV flux will most likely be modulated by the binary period, with peak flux near the X-ray maximum. This process has been discussed in the context of Cyg X-1 [2], with predicted flux levels as high as $\sim 10^{-5}$ cm^{-2} s^{-1}.

The detection of VHE photons (E $> 10^{12}$ eV) has been reported from various accreting sources, including Cyg X-3, Vel X-1 and Her X-1, suggesting the presence of very energetic proton beams. If these beams interact with the companion star, we can, by direct analogy with solar flares, expect some emergent 2.2 MeV flux. Again, this would be a narrow, unshifted line at 2.223 MeV. As in the previous scenario, this line would also vary in intensity with orbital phase. The resulting 2.2 MeV line flux has been estimated, assuming that the protons are accelerated isotropically near the compact object [6]. The peak flux predicted for Cyg X-3 ($\sim 10^{-4}$ cm^{-2} s^{-1}) is well within the range of detectable emission with *COMPTEL*.

To date, the most sensitive search for 2.2 MeV line emission was that carried out using *SMM* data [7]. Their survey was constrained (by the nature of the *SMM* mission) to a region along the ecliptic plane. They set a 3σ upper limit of 1.0×10^{-4} cm^{-2} s^{-1} on the *steady* emission from the Galactic center and from Sco X-1. Upper limits on the 2.2 MeV line emission from Cyg X-1 were in the range of $(1.2-2.2) \times 10^{-4}$ cm^{-2} s^{-1}, according to different models of the emission process. The 3σ upper limit to the phase-averaged steady emission from Cyg X-3 was set at 1.2×10^{-4} cm^{-2} s^{-1}.

OBSERVATIONS AND DATA ANALYSIS

The *COMPTEL* experiment is ideally suited for studies of the 2.2 MeV line from a variety of sources. The wide field-of-view imaging capability of *COMPTEL* provides for continuing exposure to a number of sources and provides the first-ever all-sky survey at these energies. Despite the presence of a major background line at 2.2 MeV (resulting from neutron capture within the upper layer of liquid scintillators), the *COMPTEL* experiment maintains excellent sensitivity at this energy. This analysis incorporates all available *COMPTEL* data from the first five years of the *CGRO* mission. Specifically, we have used data from *CGRO* viewing periods 1.0 through 523.0, with the exception of viewing 2.5, when *COMPTEL* was operated in a special solar mode.

The *COMPTEL* data analysis typically is carried out in a 3-d dataspace defined by the direction of the photon scatter vector, specified by the angles χ and ψ, and by the derived Compton scatter angle, specified by the angle $\bar{\phi}$ [8]. In this case, all-sky images were generated using a procedure analagous to that which has been successfully employed in studies of the diffuse galactic 1.8 MeV emission [9,10]. This approach is based on independent background estimates at adajacent energies. More specifically, we rely on a background estimate that consists of separate empirical modeling of the distributions for χ and ψ (a 2-d distribution) and for $\bar{\phi}$ (a 1-d distribution). A broad energy band (1–10 MeV) that excludes the line interval (2.110–2.336 MeV) provides information on the (χ,ψ) distribution. The $\bar{\phi}$ distribution is derived directly from the data in the line interval (2.110–2.336 MeV). The resulting background model incorporates an estimate of the instrumental background along with the effect of any continuum sources within the FoV. Only sources of mono-energetic line emission (which exhibit a somewhat different scatter direction distribution) will remain in the resulting images. This approach has been validated for the 2.2 MeV line interval using data from the Crab (where we have no reason to expect such a line signature) and using solar flare data (where such a line signature is clearly present). The validation results were as expected. No signature from the Crab was detected, wheras a significant solar flare signature was detected at a level consistent with other, independent, measurements of the 2.2 MeV line flux.

RESULTS

Using the background estimate described above, we have generated all-sky maps with two different imaging algorithms. These include a maximum entropy algorithm and a maximum likelihood algorithm. The maps generated with these two different methods are similar in appearance. The all-sky map generated with the maximum entropy algorithm is shown in Figure 1. In general, the sky at 2.2 MeV is relatively featureless. For example, there is no evidence for any diffuse galactic emission at this energy. There is, however, evidence for emission at $(\ell, b) = (300°, -30°)$. With a peak likelihood value of 32.0, and given that the all-sky map represents about 500 independent trials,

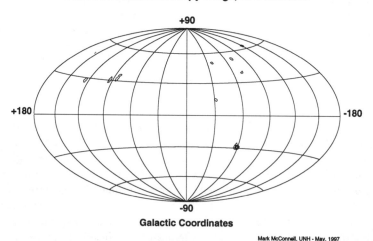

FIGURE 1. *COMPTEL* 2.2 MeV all-sky map derived using a maximum entropy imaging method. The only significant source is a point-like feature near $(\ell, b) = (300°, -30°)$, for which there is no obvious counterpart. This map appears nearly identical to a maximum likelihood map having a likelihood threshold value of 15.

this corresponds to a significance of $\sim 3.7\sigma$. There are no obvious counterparts (such as an X-ray binary) that are consistent with the emission models discussed above. We continue to search for a counterpart of this feature.

We used the X-ray binary catalog of van Paradijs [11] to search for emission from particular source candidates. None of the catalogued sources showed any sign of detectable emission. Flux limits (at the 3σ level) are typically in the range of $(1-2) \times 10^{-5}$ cm^{-2} sec^{-1}. Typical (3σ) upper limits include Cyg X-3 ($< 1.8 \times 10^{-5}$ cm^{-2} sec^{-1}), Sco X-1 ($< 2.5 \times 10^{-5}$ cm^{-2} sec^{-1}), 4U 1916-05 ($< 1.8 \times 10^{-5}$ cm^{-2} sec^{-1}), 4U 1626-67 ($< 2.5 \times 10^{-5}$ cm^{-2} sec^{-1}), and 4U 1820-30 ($< 1.6 \times 10^{-5}$ cm^{-2} sec^{-1}). For Cygnus X-1, we set a 3σ upper limit of 2.3×10^{-5} cm^{-2} sec^{-1}, which is about one order-of-magnitude below the limit set by Harris and Share (1991). This result, in conjunction with the model of Geussom and Dermer (1988), can be used to place constraints in the fraction of escaping neutrons that are captured by the companion star. For an assumed ion temperature (T_i) of 20 MeV, the data imply that less than 25% of the escaping neutrons are captured by the companion star. Further insight may be come from a phase-resolved analysis in progress.

ACKNOWLEDGEMENTS

The *COMPTEL* project is supported by NASA under contract NAS5-26645, by the Deutsche Agentur für Raumfahrtgelenheiten (DARA) under grant 50 QV90968 and by the Netherlands Organization for Scientific Research NWO.

REFERENCES

1. Fichtel, C.E., and Trombka, J.I. 1981, *Gamma-Ray Astrophysics*, NASA SP-453.
2. Guessom, N. & Dermer, C.D. 1988, in *Nuclear Spectroscopy of Astrophysical Sources* (AIP Conf. Proc. 107), ed. N. Gehrels and G.H. Share (New York: AIP), p. 332.
3. Aharonian, F.A. and Sunyaev, R.A., 1984, *MNRAS*, 210, 257.
4. Bildsten, L., Salpeter, E.E., & Wasserman, I. 1993, *ApJ*, **408**, 615.
5. Bildsten, L. 1991, in *Gamma-Ray Line Astrophysics* (AIP Conf. Proc. 232), ed. P. Durouchoux & N. Prantzos (New York: AIP), p. 401.
6. Vestrand, W.T. 1989, in the *Proceedings of the Gamma Ray Observatory Workshop*, ed. N. Johnson, p. 4-274.
7. Harris, M.J. & Share, G.H. 1991, *ApJ*, **381**, 439.
8. Schönfelder, V., et al. 1994, *ApJS*, **86**, 629.
9. Diehl, R., et al. 1995, *A&A*, **298**, 445.
10. Knödelseder, J., et al., 1996, *SPIE Conf. Proc.*, 2806, 386.
11. van Paradijs, J., 1995, in *X-Ray Binaries* (New York: Cambridge Univ. Press), p. 536.

^{26}Al Constraints from COMPTEL/OSSE/SMM Data

R. Diehl[1], M.D. Leising[2], J. Knödlseder[3], U. Oberlack[1],

[1] Max-Planck-Institut für extraterrestrische Physik, 85740 Garching, Germany
[2] Clemson University, Clemson, SC 29631, USA
[3] Centre d'Etude Spatiale des Rayonnements, 31028 Toulouse Cedex, France

Abstract. ^{26}Al emission at 1.809 MeV has been measured by several different instruments. The constraints set by each are very different, due to differing sky exposures and background estimation methods. We report on a first combined analysis of data from the COMPTEL, OSSE, and SMM instruments. The COMPTEL 1.8 MeV image as deconvolved with the Maximum-Entropy algorithm was tested for its flux normalization and its fine structure in the inner Galaxy, using data from SMM and OSSE. We find consistency of the OSSE data with COMPTEL's image structure, in particular the choice of the 'stopping point' of the iterative deconvolution method. The SMM data suggest substantially more total flux than that given by the COMPTEL map. We discuss possible explanations, in particular large-scale emission outside the plane of the Galaxy, and outline future work on combined data from the three instruments.

INTRODUCTION

Observations of 1.809 MeV gamma-rays from radioactive decay of ^{26}Al by more than eight different experiments have provided a unique view of recent nucleosynthesis in the Galaxy. Production of ^{26}Al with its ~1 million year decay time is expected to occur as a by-product of nucleosynthesis in supernovae and novae, as well as in the interior of massive stars. The measurement of ^{26}Al yields and spatial distribution sets constraints on the nuclear burning conditions at those sites, and addresses the detailed theoretical models of the candidate sources [4].

The COMPTEL images show the plane of the Galaxy as the prominent source, with irregularities along the plane, in particular with enhancements in the Vela and Cygnus regions far from the inner Galaxy. Yet, indications are that the COMPTEL image fails to record some of the 1.809 MeV flux that other non-imaging instruments have measured. We investigate such differences among instruments, directly comparing sky distributions in the re-

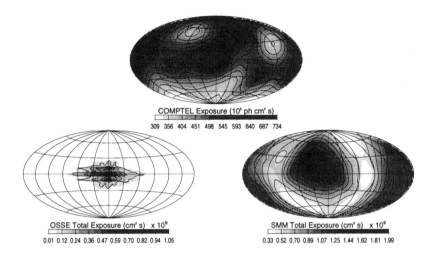

FIGURE 1. Exposure of the sky at 1.8 MeV in Galactic coordinates for the three measurements used in this study: COMPTEL (top) covers the entire sky with degree-type spatial resolution, OSSE (left) exposes the inner Galaxy region with similar spatial resolution, SMM (right) emphasizes a wide band around the ecliptic with differences selected for a particular point (here the Galactic center at all times), with practically only large-scale structure sensitivity.

spective data spaces, and in particular seek celestial/astronomical origins of these differences. All experiments are severly limited by various types of instrumental background. In each case one is forced to employ a differential measurement technique of some kind, determining its background from independent but representative measurements. An analysis of data from different instruments might allow some insight into the celestial signature at 1.809 MeV beyond those available from the individual instruments. We combine data from 3 instruments for this study, and report on the status of this ongoing work.

MEASUREMENTS

All three instruments have comparable spectral resolution, and spatial resolution in the range from ∼degrees (COMPTEL and OSSE) to ∼160 degrees (SMM), for fields of view from ∼50 deg^2 (OSSE) through 1 sr (COMPTEL) to ∼3 sr (SMM). The overall signal depends on model assumptions, and ranges from ∼7σ (OSSE) through above 30σ (COMPTEL and SMM). Underlying background is derived differently in all three cases: COMPTEL utilizes its imaging information to interpolate the background under the 1.809 MeV line

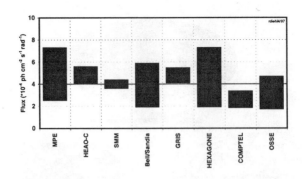

FIGURE 2. Flux measurements in the 1.809 MeV line by eight different instruments, evaluated for the inner Galaxy regime. The apparent differences may hold information about the spatial distribution of the true 1.809 MeV emission

from adjacent energy intervals per 2-week observation, OSSE performs spectral on/off background subtraction on a 2-minute time scale, and SMM splits each earth orbit into an on/off pair determined by earth occultation of a selected point to remove slowly varying background and uses source modulation from its annual scan of the ecliptic. The sky exposure of the three instruments overlaps mostly in the central region of the Galaxy, where most of the 1.809 MeV emission appears to arise; yet, exposures are sufficiently different (see Figure 1) to allow further conclusions on the spatial distribution of the emission. Figure 2 illustrates the 1.809 MeV measurements, if converted to a flux from the inner radian of the Galaxy, for comparison. The displayed uncertainties reflect the sensitivity to the assumed sky distribution.

COMPTEL: The COMPTEL imaging telescope consists of two layers of scintillation detectors, which record a Compton scatter interaction through a coincidence event in both layers [6]. The imaging information is held in event parameters 'Compton scatter angle φ' (as derived from the measured energy deposits in the upper and lower detector plane) and 'Scattered Photon Direction Angles (χ, ψ)' (as derived from the interaction locations). In this data space, a point source event distribution describes a hollow cone. Imaging analysis corresponds to finding these response cones in data space, constructing a sky map that matches the measured data, given a model for distribution of background in data space. Background dominates above a celestial 1.809 MeV event fraction of $\sim 2\%$. The spatial structure of background is modelled from simultaneous measurements at adjacent energy intervals. The COMPTEL map [3] is constructed by an iterative deconvolution process utilizing the maximum-entropy constraint. Starting from a flat map which is defined to hold an assumed total sky flux at 1.809 MeV, subsequent iterations re-distribute this flux with more and more contrast, improving the fit with iteration up to the regime of super-resolution, where the map is also fitted to statistical fluctuations in the data, and thus approaches a (positivity constrained) maximum-likelihood map. The reconstructed flux within a given region depends on the effective signal-to-background, and is more suppressed in early iterations when surface brightness or exposure are low. Additional biases arise from the background normalization: pre-setting of the global scale of the background contributions from spectral analysis is problematic when spectral shapes are different in the individual observations; also, the assumed absence of celestial 1.809 MeV glow from galactic latitudes above 60° generates a bias. Normally we use COMPTEL model-independent flux results as best determined from high iterations of maximum entropy maps. He we test the range of map iterations for both structure and overall flux against SMM and OSSE data as a means to vary the 1.809 MeV spatial distribution model in a continuous and systematic way.

OSSE The OSSE instrument aboard the Compton Observatory features two pairs of scintillation detectors with a typical energy resolution of $\sim 8\%$ (FHWM) [2]. Tungsten collimators limit the field- of-view to an approximately rectangular area of 3.8*11.4 degrees, detector pairs are rocked at 2 minute intervals on/off source with an angular separation ('scan angle') of typically 10° towards the long end of the field of view, thus defining 'source' and 'background' datasets. For each of the four detectors, four off-pointings bracketing an on-pointing are weighted with an analytical orbital variation function, and accumulated to provide the background for each particular 2-minute 'on' measurement. The background-subtracted measurements are accumulated over a typically two-week observation period, and fitted wth a spectral model including the 1.809 MeV line. Evaluating the exposure of the sky for each pointing, the measured flux can be fitted for an assumed spatial distribution of the emission in the sky. Typical conditions yield a background of 75 counts s^{-1} in the 1.8 MeV line, compared with a celestial signal of 0.02 counts s^{-1} from the central Galaxy; this translates into a signal-to-background ratio of 0.03%. The OSSE background subtraction eliminates part of the celestial signal at 1.809 MeV due to the partial transparency of the collimator: the angular response of OSSE still retains $\sim 25\%$ of its peak value near its main field of view, and does not fall off below 10% even very far from the pointing direction. The effective modulation of the observed 1.809 MeV flux is relatively low. All data from observations along the Galactic plane have been used for 1.809 MeV analysis, which comprises 223 pointings, most of which are clustered around the Galacic Center.

SMM The Gamma Ray Spectrometer aboard the Solar Maximum Mission Spacecraft was a set of seven NaI scintillation detectors continuously pointed towards the sun [5]. Its active shield defined an aperture of $\sim 160°$ (FWHM) at 1.8 MeV. Annual transits tracking the Sun along the plane of the ecliptic provide the basic signal modulation. Our analysis of SMM data subdivides the measurement of each orbit around the earth, defining an 'on-source' part of the orbit where the pivot point on the sky is visible, versus an 'off-source' part of the orbit where this pivot point is occulted by the earth. The difference spectra ('on'-'off', applying proper normalization with time) are accumulated in time bins of, e.g., 10 days, and fitted with a spectral model including the 1.809 MeV feature. The effective exposure of the detector is then evaluated for an assumed spatial distribution of the emission. Fitting the expected excess signal profile with the exposure, and a correction term for the time dependence of prompt activation, provides a flux value for this assumed distribution. The typical background rate in the 1.8 MeV line was ~10 counts s^{-1} versus peak source rates of 0.04 counts s^{-1}, corresponding to a signal-to-background ratio of 0.4%. More than 9 years of SMM data have been analyzed.

DISCUSSION

Fitting SMM data, we observe that all iterations of the COMPTEL maximum-entropy deconvolution process provide equally good, but not adequate, fits – but always with substantially larger flux than measured by COMPTEL. The SMM/COMPTEL flux difference is not reduced when using later (more structured) image iterations. The OSSE fits of the different iterations of the COMPTEL map are acceptable except for the first 2-3 iterations. The OSSE-fitted intensity of the COMPTEL map however varies with iteration; the best match of the OSSE-derived 1.8 MeV flux with the COMPTEL value is obtained for iterations 5-7. This can be regarded as independent confirmation of the COMPTEL experimenters' choice of a most realistic image from the possible iterations. Also, this supports the basic structure displayed in the inner Galaxy region of the COMPTEL map. The SMM data suggest a larger flux than what is found in the COMPTEL map, by a factor 1.5-2.0. With the exposure maps and biases discussed above, we suggest that the 'missing' flux can be placed preferentially in nearby regions, leading to large-scale low surface-brightness emission which could be suppressed by the COMPTEL high-latitude background normalization. Although a Galaxy-wide diffuse radioactivity is established, such additional local/nearby emission provides an interesting revival of early theories for a diffuse 1.809 MeV glow [1].

REFERENCES

1. Dearborn, D.S.P. & Blake, J.B. 1989, ApJ, 288, L21
2. Johnson N. et al. 1993, ApJS, 86, 693
3. Oberlack U., et al. 1996, A&AS, 120, 4, 311
4. Prantzos N.& Diehl R. 1996, Phys.Rep. 267(1), 1
5. Share G., et al. 1985, ApJ, 292, L61
6. Schönfelder V. et al. 1993, ApJS, 86, 657

^{26}Al and the COMPTEL ^{60}Fe Data

R. Diehl[1], U. Wessolowski[1], U. Oberlack[1], H. Bloemen[2],
R. Georgii[1], A. Iyudin[1], J. Knödlseder[5], G. Lichti[1],
W. Hermsen[2], D. Morris[3], J. Ryan[3], V. Schönfelder[1],
A. Strong[1], P. von Ballmoos[5], C. Winkler[4]

[1] *Max-Planck-Institut für extraterrestrische Physik, 85740 Garching, Germany*
[2] *SRON-Utrecht, Sorbonnelaan 2, 3584 CA Utrecht, The Netherlands*
[3] *Space Science Center, University of New Hampshire, Durham NH 03824, U.S.A.*
[4] *Astrophysics Division, ESTEC, ESA, 2200 AG Noordwijk, The Netherlands*
[5] *Centre d'Etude Spatiale des Rayonnements, 31028 Toulouse Cedex, France*

Abstract. Nucleosynthesis models predict the production of ^{60}Fe by the same massive stars which are responsible for ^{26}Al synthesis. With a radioactive decay time similar to ^{26}Al, the gamma-ray line emission at 1.173 and 1.332 MeV is predicted to be $\sim$16% of the 1.809 MeV ^{26}Al line intensity, from the same source regions. We investigate with COMPTEL all-sky data from CGRO Phases 1-5 whether this source of ^{60}Fe can be detected, using the known spectral signature plus the spatial distribution as imaged with COMPTEL ^{26}Al measurements. Uncertainties in spatial signature of the instrumental and continuum background limit the sensitivity, such that only an upper limit of $\sim$ 44% (2σ) is quoted at this time.

INTRODUCTION

The detailed study of 1.809 MeV emission has become a source of constraints on models of nucleosynthesis in the Galaxy (see review [5]). COMPTEL imaging data provide the best constraints on the spatial distribution of the emission down to the few-degree scale, while large field-of-view instruments contribute large-scale integrated flux constraints (like the SMM result, see [2] and Diehl et al., these proceedings). Nucleosynthesis models predict the production of ^{60}Fe by the same massive stars which are responsible for ^{26}Al synthesis. In models of nucleosynthesis in massive stars, both during their main sequence evolution and during the core collapse supernova, it was found that ^{26}Al production occurs in hydrostatic H burning and explosive Ne burning in the O/Ne shell, while ^{60}Fe is produced in He shell burning and at the base of the O/Ne shell [6]. Neutrino spallation enhances the production of ^{26}Al, while being negligible for ^{60}Fe production. Both ^{26}Al and ^{60}Fe are produced mainly during the presupernova evolution, between mass coordinates $\sim 3-6$ M$_\odot$ for a 25 M$_\odot$ star: they originate in approximately the same region of the star, in spite of those differences in detail processes. The core-collapse supernova production of these two isotopes should therefore have similar spatial distributions after the explosion of the star, on a Galactic scale. With a radioactive decay time similar to ^{26}Al, the gamma-ray line emission at 1.173 and 1.332 MeV is predicted to be $\sim 16\%$ of the 1.809 MeV ^{26}Al line intensity, from the same source regions [7]. There may be additional ^{60}Fe production from type I supernovae superimposed [8], but also other sources may contribute to the observed ^{26}Al. Yet, from earlier results, it was expected that massive stars are among the most plausible sources of ^{26}Al in the Galaxy [5], ejected into interstellar medium through the wind phases and / or supernova events, while AGB stars and novae could dominate the observed emission only under somewhat extreme assumptions.

This paper attempts to find the signal of the ^{26}Al-correlated ^{60}Fe contribution, using COMPTEL data from 5 years of observation.

COMPTEL ^{26}Al AND ^{60}Fe DATA

The COMPTEL all-sky data from 5 years of observations have been translated into an image in the 1.809 MeV gamma-ray line, using the Maximum Entropy deconvolution method and a background model based on simultaneous measurements in adjacent energy bands [4]. The image shows dominant emission from the plane of the Galaxy, yet also contains irregularities and features at intermediate latitudes, which may have a more local origin (≤ 1 kpc).

This map serves as the basis of our test, constituting one of the assumed ^{26}Al source distributions, derived without *a priori* assumptions on the physical nature of the source. Additionally, we also use model distributions based on tracers, which were found to correlate with or describe the ^{26}Al data of COMPTEL adequately [1]. The background modelling which underlies the 1.8 MeV image is based on independent data taken simultaneously at adjacent energies [4]. We employ a similar approach here, with some adjustments for the two-line signal and the 1.809 MeV correlation. The model exploits the signatures in the different dimensions of the COMPTEL imaging data space from different energy regimes: the energy-dependent signature in the calculated Compton scatter angle φ is obtained at the line energy intervals themselves. The signature of the background and of continuum emission from the sky in the χ and ψ imaging dataspace domains were obtained from the 0.85-10 MeV regime altogether, excluding the line intervals itself and the 1.8 MeV interval; this ensures that the analysis is sensitive only to line emission above the celestial continuum, yet does not suppress ^{26}Al correlated signal.

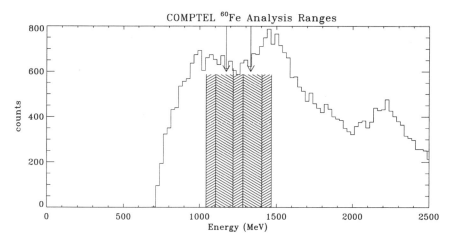

FIGURE 1. COMPTEL choice of energy bands for ^{60}Fe analysis

Our choice of energy bands is shown in Fig. 1. We chose the line intervals being 1.5 σ broad at the high and low ends, centered at the 1.332 and 1.173 MeV decay lines from ^{60}Fe, and 1 σ broad towards the regime between the lines. Thus we leave still a narrow band between the lines for (future) energy dependence interpolation of the background model; outer adjacent background bands for this purpose are of same width. We account for the double line structure adding the responses of both lines in the composite ^{60}Fe signal band. We must exclude the 1.8 MeV line interval from our background dataset, since we are searching for signal correlated to the signal in this band. We verify the consistency and integrity of this analysis by simulations and by analysis in the individual line energy intervals at 1.332 and 1.173 MeV. Adding the expected counts from a 20% (of ^{26}Al) intensity to our background model recovers this signal with a variance of 4% from Poisson samples of the composite simulated dataset. This variance is consistent with the statistical uncertainty derived in our fits of different models to the real measurement. We also perform MEM imaging deconvolution from the same data in the individual energy bands of both lines, confirming the absence of correlated map features, but also the presence of more spurious map features in the 1.173 MeV band, compared to the 1.332 MeV band. This is probably largely due to the more intense variation of the instrumental background spectrum in the 1 MeV regime, from shortlived instrumental activation during SAA transits. We note that our background approach becomes worse as we approach the detector thresholds, and adjacent bands are unevenly populated across the detectors. Also, this energy band includes possible emission from celestial sources in the ^{44}Ti and ^{22}Na γ-ray lines at 1.157 and 1.275 MeV.

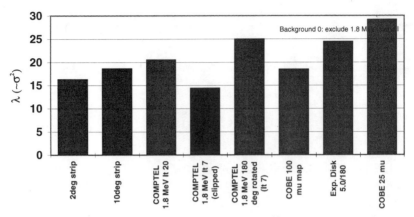

FIGURE 2. COMPTEL model fit results on ^{60}Fe data from 5 years

RESULTS

We test the detection significance of a variety of different distributions in our ^{60}Fe line energy bands. For all varieties that encode some sort of Galactic signal, we detect emission at a formal significance of 4σ and above (Fig.2). The models based on the COMPTEL 1.8 MeV data do not stand out particularly. In fact, an inverted version of the 1.8 MeV map even fits somewhat better than the original. We therefore conclude that we do not detect ^{60}Fe emission correlated to ^{26}Al with our analysis. We interpret the detected 'signal' as artifact of background imperfections.

Deriving upper limits, we have to make choices on how to interpret the apparent Galactic signal. Even though it could be Galactic emission altogether, we cannot exclude that Galactic models compensate for systematic inadequacies of our background model in the fit. On the other hand, the distribution we are testing may be inappropriate for ^{60}Fe emission, and the scatter in our different detected emission levels may be interpreted as indicating this systematic uncertainty (note that we include quite improbable geometrical slab models of 2 and 10° latitude in our test, as well as the bulge-like COBE 25μ map and disk-like emission models). We then derive variances from our systematic model uncertainty from the variances of the detected emission levels, and also determine the statistical uncertainty from the measured counts in the line intervals.

Upper limit values can be derived now with two different interpretations of our results. We may report that our ^{26}Al derived model does not yield any excess emission above other models, and we may add to the statistical uncertainty a systematical uncertainty, which is based on the variance of the result for different models. This yields an upper limit of 28% ^{60}Fe emission (expressed in percentage of ^{26}Al emission). More conservatively, we include the average level of our detected signal in our systematic uncertainty, hence derive the one-σ value from the linear sum of the statistical uncertainty plus the RMS of the average signal level and their variances for our models used. This yields our preferred upper limit value of 44% of ^{26}Al, or $1.2\ 10^{-4}$ ph cm^{-2}s^{-1}rad^{-1} in either line.

CONCLUSIONS

First analysis of ^{60}Fe emission with COMPTEL data aims at detection of a signal that is spatially correlated with the COMPTEL 1.8 MeV result for ^{26}Al. No correlated signal is found. Uncertainties in this analysis arise from background uncertainty in the 1 MeV regime, and from possible ^{60}Fe signal with a spatial distribution different than the tested ^{26}Al map. The derived upper limit for ^{26}Al-correlated ^{60}Fe emission is about 2-3 times higher than theoretical estimates, indicating that the expected emission is at the sensitivity level of this study. Improved background modeling over this first study is in progress, and can be expected to improve our sensitivity to a level required for the ^{60}Fe objective.

ACKNOWLEDGMENTS

The COMPTEL project is supported by the German government through DARA grant 50 QV 90968, by NASA under contract NAS5-26645, and by the Netherlands Organisation for Scientific Research NWO. J. Knödlseder is supported by the European Community through grant number ERBFMBICT 950387.

REFERENCES

1. Diehl R., et al., 1997, these proceedings
2. Leising M. D., Share G. H., 1994, ApJ, 424, 200
3. Oberlack U., et al., 1997, these proceedings
4. Oberlack U., et al., 1997, A&A, in preparation
5. Prantzos N.& Diehl R. 1996, Phys.Rep. 267(1), 1
6. Timmes, F. X., et al., 1995, ApJ, 449, 204
7. Woosley S.E., & Timmes F.X., 1997, ESA-SP-382, 21
8. Woosley S.E., 1997, ApJ, 476, 801

Models for COMPTEL ^{26}Al Data

R. Diehl[1], U. Oberlack[1], J. Knödlseder[5], H. Bloemen[2], W. Hermsen[2], D. Morris[3], J. Ryan[3], V. Schönfelder[1], A. Strong[1], P. von Ballmoos[5], C. Winkler[4]

[1] *Max–Planck–Institut für extraterrestrische Physik, 85740 Garching, Germany*
[2] *SRON-Utrecht, Sorbonnelaan 2, 3584 CA Utrecht, The Netherlands*
[3] *Space Science Center, University of New Hampshire, Durham NH 03824, U.S.A.*
[4] *Astrophysics Division, ESTEC, ESA, 2200 AG Noordwijk, The Netherlands*
[5] *Centre d'Etude Spatiale des Rayonnements, 31028 Toulouse Cedex, France*

Abstract. The all-sky data from CGRO Phases 1-5 provide the most detailed constraints to date on the ^{26}Al distribution in the Galaxy. With improved modelling of the instrumental background, different source distribution models can be distinguished. We derive large-scale parameters from first-order models, i.e., a characteristic scale height of 130 pc, and a Galactocentric scale radius of 5 kpc. Spiral structure of the ^{26}Al source distribution improves the fits, as molecular gas provides a better description than geometrical models in the inner Galaxy. We note that infrared emission from dust also matches the ^{26}Al emission quite well. Hints for contributions from more local and localized ^{26}Al components such as Loop I and the Gould Belt are compatible with a massive star origin of ^{26}Al.

INTRODUCTION

The detailed study of 1.809 MeV emission has become a source of constraints on models of nucleosynthesis in the Galaxy (see review [18]). COMPTEL imaging data provide the best constraints on the spatial distribution of the emission down to the few-degree scale, while large field-of-view instruments contribute large-scale integrated flux constraints (like the SMM result [19] [8]) or details on the spectral shape of the line (like the GRIS result [14]). In order to exploit the COMPTEL measurements, we follow two complementary approaches: (1) unbiased imaging deconvolution, through the maximum entropy method, providing a result free of astrophysical bias yet subject to exposure- and intensity dependent distortions [17] [16], and (2) estimation of parameters of astrophysically plausible spatial model distributions, providing a reliable answer to a well-posed astrophysical question, yet depending on the validity and significance of the hypothesis [6] [7] [13]. This paper reports the latest results on the latter approach.

From earlier results, it was expected that massive stars are among the most plausible sources of ^{26}Al in the Galaxy [18], ejected into interstellar medium through the wind phases and / or supernova events, while AGB stars and novae could dominate the observed emission only under somewhat extreme assumptions. Therefore recent research on ^{26}Al modelling concentrates on finding optimum tracers of massive stars in the Galaxy [7].

Massive stars in the nearby region of the Galaxy have been detected through a variety of observables, such as their optical and UV brightness and unique spectral lines from the stellar photosphere, coronal X rays, and emission over a wide range of frequencies from their wind-blown bubbles and their interaction with the ambient gas. Even if stars cannot be detected directly due to obscuration, regions of high space density of massive stars have been identified indirectly, through, e.g., HII regions, Hα emission, infrared emission from massive protostars and circumstellar dust, and supernova remnants visible in radio through X rays. If we can establish ^{26}Al emission to be correlated with those observables (excluding incomplete sample regions), we may be able to invert the argument: With its million-year decay time, gamma-rays from ^{26}Al decay measure the massive-star formation history averaged over a time scale of several million years, thus extending the time scale beyond that accessible by other means.

COMPTEL MODELING RESULTS

The COMPTEL all-sky data from 5 years of observations have been translated into an image in the 1.809 MeV gamma-ray line, using the Maximum Entropy deconvolution method and a background model based on simultaneous measurements in adjacent energy bands [16]. The image shows dominant emission from the plane of the Galaxy, yet also contains irregularities and features at intermediate latitudes, which may have a more local origin (≤ 1 kpc). Our modelling proceeds as follows: we account for the Galaxy-wide emission to first order, and derive global parameters which characterize the large-scale emission. Then we investigate correlation with formation sites of massive stars in the nearby region of the Galaxy, attempting to improve this first-order model. We fit a set of plausible large-scale models to COMPTEL data: geometrical disk models with exponential scale height and Galactocentric scale radius dependencies [6], spiral-arm models derived from free electron measurements with the exponential scale height as a parameter [21], molecular gas data as measured in CO [3], and stellar light- through dust-dominated measurements of the whole Galaxy in the infrared [20]. Globally, we find that exponential disks are among the best generic models of our measurement, insignificantly inferior to the best-fitting spiral model based on free electrons.

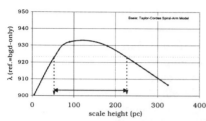
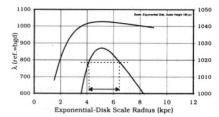

FIGURE 1. COMPTEL 1.8 MeV emission scale radius (right, with expanded peak region) and scale height (left), ordinates are fitted likelihood ratio for two-parametric models.

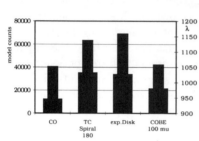
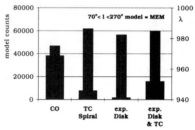

FIGURE 2. COMPTEL 1.8 MeV model comparisons, extracted from the entire plane of the Galaxy (left) and in the inner Galaxy (right; see text). (left ordinate / narrow bars: fitted counts in model; right ordinate / wide bars: likelihood ratio, $\simeq \sigma^2$).

The large-scale emission can be characterized by a (1/e) Galactic scale height of 130 ($^{+100}_{-75}$) pc, and a Galactocentric (1/e) scale radius of 5 ($^{+1.5}_{-0.8}$) kpc ($R_\odot$=8.5 kpc), with features from spiral structure (Fig.1). The associated total mass is 2.1 (± 0.1) $M_\odot$. The scale radius was derived from exponential-disk models, and may be biased by up to 1.5 kpc toward high values due to structures such as the Cygnus / Vela / Carina regions. For determination of the scale height, we use the slightly-better fitting spiral-arm model, and optimize the fit through scale height variation. Note also here that this assumes a generic scale height value for the Galactic plane emission, averaging specific features along the plane. Comparing other models based on large-scale surveys in observables that may plausibly relate to ^{26}Al emission (Fig.2), we note that neither molecular gas nor warm dust can represent the contrast between inner Galaxy and outer disk regime, nor account for the particularly bright regions in Cygnus, Vela, and Carina. Alleviating the contrast issue, we focus our model on the inner Galaxy, through a two-component model using the deconvolved Maximum-Entropy image as a model in the outer Galaxy (-90° $\leq l \leq$ 70°) in addition to our tracer model (Fig.2 right). This demonstrates that in the inner Galaxy the CO model provides a better fit than geometrical models, supporting the irregularity observed in the deconvolved COMPTEL 1.809 MeV image.

The emission features far outside the inner Galaxy direction, in the Cygnus, Vela, and Carina regions, are detected at significances $\geq 5\sigma$ (each region by itself). This suggests significant deviations from a large-scale symmetric and smooth ^{26}Al emission pattern. Those regions have been studied for candidate sources, each of them found to have special characteristics, which are compatible with a localized enhancement of star formation and ^{26}Al: the Cygnus region with the Cygnus superbubble as remnant of possibly 60 recent supernovae as well as several Wolf Rayet stars along the line of sight [5], the Vela region with the Vela supernova remnant and the closest Wolf Rayet star in the system 'γ Vel' [15], and the Carina region with η Car but also the largest space density of young open clusters in the Galaxy [?]. Detailed models for those regions will be addressed in a separate paper.

A specific search for correlation with catalogues from massive-star related objects was performed: neither Wolf Rayet star [22] or OB star catalogues [10], nor catalogued radio supernova remnants [11] significantly improve the description of COMPTEL data above our first-order models (neither Galaxy-wide, nor in the inner-Galaxy direction). We note that object catalogues are strongly biased in all cases, however: the completeness limit for massive-star catalogues does not extend beyond $\simeq$3kpc, and radio supernova remnants can only be detected in less confused directions. Therefore, since our main ^{26}Al emission arises from the inner Galaxy and is integrated over all source regions throughout the Galaxy, we cannot expect to find a predominant correlation between our measurement and these catalogues. Rather, ^{26}Al measurements may provide a more realistic map of massive stars. Yet, for nearby regions ($\leq$500pc), specifically those outside the general direction of the inner Galactic plane, additional emission at intermediate or high latitudes may be identified by modelling plausible candidate sources in more detail.

The Gould Belt has been found to describe a local structure of O stars deviating from the plane of the Galaxy, modelled by a disk with $\simeq 20°$ inclination to the plane [2]. The origin and precise inclination and center of the Gould Belt are subject to some discussion, depending on whether it is modeled based on O stars or HI gas structures. Our geometrical model encompasses the O star population of the belt, and indicates a slight improvement of our fit ($\simeq 2\sigma$), yet insufficient to claim detection. We note however that testing an "inverted Gould Belt" (i.e., a model belt reflected on the Galactic plane) yields an *anti*correlation of similar significance.

The closest OB association to the Sun, the Sco-Cen association, is related to one of the most prominent large scale radio structures on the sky, the North Polar Spur / Loop I, through supernova activity originating from the most massive stars of this association. The last supernova event may have occured as late as $\simeq 2\ 10^5$y ago, enriching the Loop I bubble with nucleosynthesis products including ^{26}Al. Being so nearby, this could result in low surface-brightness 1.809 MeV emission barely detectable by COMPTEL. Our search with a geometrical model of Loop I (depositing the ^{26}Al in a thin outer shell, due to the age of the supernova) reports a minor hint ($\simeq 1.5\sigma$) only.

^{26}Al production from the Sco-Cen association as a whole is under study: ages of stellar subgroups suggest separable features on the sky, detailed HI maps help to relate those to low-significance 1.8 MeV image structures [1] [4].

CONCLUSIONS

Radioactive ^{26}Al appears to trace nucleosynthesis activity of massive stars throughout the Galaxy. On largest scales, the emission at 1.809 MeV is characterized by a Galactocentric radius of 5 kpc and a disk scale height of 130 pc. This is slightly larger than the expected scale height from massive star distributions, which may be related to the ejection of ^{26}Al into bubbles blown by earlier stages of massive star evolution. In the inner Galaxy, we note support for our irregularly structured image from the fit improvement by the (also structured) molecular gas model, as compared to smooth geometrical models. We interpret this to support global correlation with massive stars. Further second-order model components such as nearby star catalogues or specific structures originating from massive nearby stars cannot be clearly detected.

REFERENCES

1. Blaauw A. 1964, Ann.Rev.Astr.Astroph. 2, 213
2. Comeron F., Torra J., Gomez A. 1994, A&A, 286, 789
3. Dame T.M., et al., ApJ, 322, 706
4. deGeus E.J. 1992, A&A, 262, 258
5. del Rio E., et al. 1996, A&A, 315, 237
6. Diehl R., et al. 1995, A&A, 298, 445
7. Diehl R., et al. 1996, A&AS, 120, 4, 321
8. Diehl R., et al. 1997, these proceedings
9. Garmany C.D., Stencel R.E. 1992, A&AS, 94, 214
10. Green D.A., 1996, http://mrao/cam.ac.uk/surveys/snrs/
11. Knödlseder J., et al. 1997, ESA-SP 382, 55
12. Knödlseder J., et al., 1996, A&AS, 120, 4, 327
13. Knödlseder J., et al. 1996, A&AS, 120, 4, 339
14. Naya J.E., et al. 1996, Nature 384, 44
15. Oberlack, U., et al. 1994, ApJ Suppl., 92, 443
16. Oberlack, U., et al. 1997, these proceedings
17. Oberlack U., et al. 1996, A&AS, 120, 4, 311
18. Prantzos N.& Diehl R. 1996, Phys.Rep. 267, 1, 1
19. Share G., et al. 1985, ApJ, 292, L61
20. Sodroski T.J. et al. 1994, ApJ, 428, 638
21. Taylor J.H., Cordes J. M. 1993, ApJ, 411, 674
22. vanderHucht K., et al., 1988, A&A 199, 217.

γ-Ray Emitting Radionuclide Production In A Multidimensional Supernovae Model

Grant Bazán[1] and David Arnett[2]

[1]*Lawrence Livermore National Laboratory, PO Box 808, L-170,
Livermore, CA 94550
bazan4@llnl.gov*
[2]*Steward Observatory, Tucson, AZ 85721
darnett@as.arizona.edu*

Abstract. We examine the effects of multidimensional hydrodynamics on γ-ray emitting radionuclide yields from massive star progenitor supernovae. Significant differences are expected between explosive nucleosynthesis product yields from 1- and 2-dimensional hydrodynamical models due to the rather high perturbation amplitudes in density (10%), temperature (5%), and neutron excess (50%) realized in 2-dimensional simulations of pre-core collapse stellar evolution. Shocks are introduced into these models of pre-collapse evolution and are further evolved with a 160 isotope network coupled to our PPM hydro code.

INTRODUCTION

Among the wealth of the data supplied by Supernova 1987A was the undeniable presence of hydrodynamical instabilities, whose influence extended from the hydrogen envelope to the explosive nucleosynthesis layers. Interaction with explosive nucleosynthesis products was evident in a number of measured quantities. The early emergence of the γ-ray light curve, as well as the time behavior of the bolometric light curve, is now understood as evidence of hydrodynamical mixing of ^{56}Ni from its explosive nucleosynthesis birthsite throughout at least 30% of the entire stellar envelope [1]. In addition, the best explanation for the 'Bochum Event', where fine structure appeared in the Hα line between days 20-100, is that a large clump ($\sim 10^{-3}$ M$_\odot$) of ^{56}Ni extended into the hydrogen envelope at high velocity ($\sim$ 4700 km/s). γ-ray deposition and the resulting ionization is then responsible for the fine structure [5,10]. Further evidence appeared in the velocity asymmetries of iron peak element emission lines [7,9].

Multidimensional hydrodynamic modelling of the pre-core collapse structure of a SN 1987A progenitor could provide the explanation for the instabilities, along with posing questions for the role of thermodynamic perturbations in explosive nucleosynthesis. In previous work, we have found that 2-dimensional hydrodynamic models of oxygen burning convective shells give rise to large perturbations in several state variables [3,4]. The perturbation amplitudes range from $\sim$ 5 - 10% in density and temperature to over 50% in neutron excess.

The importance of these structure defects is twofold. First, explosive nucleosynthesis product abundances are extremely sensitive to small changes in temperature and neutron excess. The presence of these perturbations should lead to an initial state of abundance inhomogeniety as a function of azimuthal position. Also, the density perturbations act as seeds for Richtmeyer-Meshkov instabilities induced by the passage of the forward moving supernova shock. Later, the already non-linear perturbations in the instability region would seed two bouts of Richtmeyer-Meshkov and Rayleigh-Taylor instabilities caused by the passage two reverse shocks. These reverse shocks appear due to the the forward moving ejecta encountering the rapid flatennings in the density profile at the H/He interface and the stellar surface. In this paper, we explore the early phases of this evolution with the explosive nucleosynthesis and Richtmeyer-Meshkov instabilities.

CALCULATIONS

This study couples a large nuclear network solver to a high order hydrodynamic code. Our calculations make use of the PROMETHEUS code, which employs the direct eulerian implementation of PPM [6], and also has a long history of studying supernova related instabilities [2,8]. It includes a stellar EOS from the TYCHO stellar evolution code and accounts for nuclear energy deposition. The nuclear network is composed of 160 nuclei from neutrons to ^{71}Ga and is solved via first order implicit, backwards differentiation and a preconditioned backwards conjugate gradient solver for the resulting matrix. Sub-cycling is allowed over the nuclear network when warrented.

The simulation begins from the structure of the previous oxygen burning shell convection simulations [4]. Namely, we have a 460 by 128 (r, θ) mesh from $\theta = \pi/4$ to $3\pi/4$ with logarithmically spaced radial zones extending from the Si core boundary to the H/He interface. To mimic a supernova core collapse, we allow the lower boundary to infall at 1000 km/s for 1 sec, so as to move the position of explosive oxygen burning out to less neutron-rich regions. A ballistic is then applied to the bottom to create an explosion of 1.5×10^{51} ergs. In practice, the hydrodynamics takes about 1/3 of the time of the network solution for the entire grid.

RESULTS

γ-ray emitting nuclei are synthesized in a variety of density, temperature, and neutron excess environments. The most important of those arising from massive star progenitor supernovae (^{26}Al, ^{44}Ti, ^{56,57}Ni) are no exception, and their spatial and velocity distributions reflect these birthsites.

^{26}Al

Our simulations show a very asymmetric distribution of ^{26}Al, which is only due to its formation in explosive carbon burning (T $\sim 2 \times 10^9$K). We note, however, that the ^{26}Al made in this fashion is only a part of the total ^{26}Al formed during earlier stages of hydrogen and carbon burning. In our convective oxygen simulations, significant amounts of ^{12}C are mixed inhomogeneously throughout the O shell from convective overshoot into the overlying layers. When the supernova shock wave encounters the asymmetric spatial distribution of ^{12}C, the result is a spatial distribution of ^{26}Al denoting former azimuthal maxima of ^{12}C (figure 1).

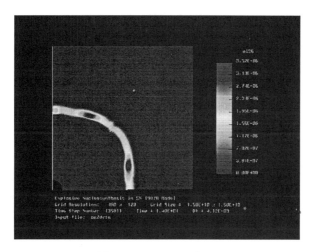

FIGURE 1. ^{26}Al distribution at t = 14 sec.

^{44}Ti

^{44}Ti is mostly produced in the α-rich freezeout. Here, this region occurs in the middle of large perturbations. However, since the abundances of α-nuclei are not very dependent on neutron excess, the only initial azimuthal asymmetries are the result of density and temperature inhomogeneities. The picture below is taken after roughly two R-M instability timescales, and so reflect the R-M instability in operation.

^{56}Ni

^{56}Ni is almost wholly produced in low η environments via explosive oxygen burning, and because it's an α-nucleus, its abundance displays a similar lack of dependence on initial neutron excess as ^{44}Ti. Figure 3 shows the spatial distribution roughly two R-M timescales after nucleosynthesis.

^{57}Ni

^{57}Ni is produced in complete and incomplete Si burning in moderate η environments. Because this region lies below the main density perturbations,

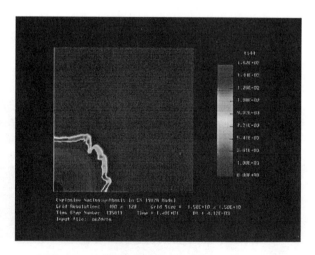

FIGURE 2. ^{44}Ti distribution at t = 14 sec

the actual yield will eventually depend on how much instabilities can dredge-up into the envelope. Because of these instabilities, one can place the 'mass cut' further out in mass than corresponding 1D models, but still expect a small fraction of underlying material rich in ^{57}Ni to be ejected (see figure 4).

REFERENCES

1. Arnett, W. D., Bahcall, J. N., Kirschner, R. P., and Wossley, S. E. *Ann. Rev. Astr. Astrophys.*, **27**, 629 (1989).
2. Arnett, W. D., Fryxell, B. A., and Müller, E. *Astrophys. J.*, **341**, L63 (1989).
3. Bazán, G. and Arnett, D. *Astrophys. J. Lett.*, **433**, L41 (1994).
4. Bazán, G. and Arnett, D. *Astrophys. J.*, submitted.
5. Chugai, N. N. *Sov. Astr. Lett.* **18**, 50 (1992).
6. Colella, P. and Woodward, P. R. *J. Comp. Phys.* **54**, 174 (1984).
7. Danziger, I. J., Bouchet, P., Fosbury, R. A. E., et al., *Supernova 1987A in the Large Magellanic Cloud*, Cambridge: Cambridge University Press, 1988, p. 37.
8. Fryxell, B. A., Müller, E. and Arnett, D. *Astrophys. J.*, **367**, 619 (1991).
9. Meikle, W. P. S., Allen, D. A., Spyromilio, J., and Varani, G.-F. *Mon. Not. Roy. Astr. Soc.* **238**, 193 (1989).
10. Utrobin, V. P., Chugai, N. N., and Andronova, A. A. *Astro. and Astrophys.* **295**, 129 (1995).

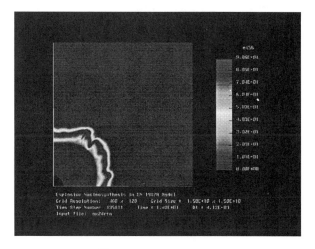

FIGURE 3. ^{56}Ni distribution at t = 14 sec.

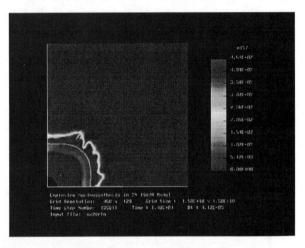

FIGURE 4. ^{57}Ni distribution at t = 14 sec.

Predictions of gamma-ray emission from classical novae and their detectability by CGRO

M. Hernanz[1], J. Gómez-Gomar[1], J. José[2] and J. Isern[1]

[1] *Institut d'Estudis Espacials de Catalunya (IEEC), CSIC Research Unit*
Edifici Nexus-104, C/Gran Capita 2-4, 08034 Barcelona, SPAIN
[2] *Departament de Física i Enginyeria Nuclear (UPC)*
Avda. Víctor Balaguer s/n, 08800 Vilanova i la Geltrú, SPAIN

Abstract. An implicit hydrodynamic code following the explosion of classical novae, from the accretion phase up to the final ejection of the envelope, has been coupled to a Monte-Carlo code able to simulate their gamma-ray emission. Carbon-oxygen (CO) and oxygen-neon (ONe) novae have been studied and their gamma-ray spectra have been obtained, as well as the gamma-ray light curves for the important lines ($e^- - e^+$ annihilation line at 511 keV, ^{7}Be decay-line at 478 keV and ^{22}Na decay-line at 1275 keV). The detectability of the emission by CGRO instruments has been analyzed. It is worth noticing that the γ-ray signature of a CO nova is different from that of an ONe one. In the CO case, the 478 keV line is very important, but lasts only for ~ 2 months. In the ONe case, the 1275 keV line is the dominant one, lasting for ~ 4 years. In both cases, the 511 keV line is the most intense line at the beginning, but its short duration (~ 2 days) makes it very difficult to be detected. It is shown that the negative results from the observations made by COMPTEL up to now are consistent with the theoretical predictions. Predictions of the future detectability by the INTEGRAL mission are also made.

INTRODUCTION

Nova explosions are caused by thermonuclear runaways on white dwarfs accreting hydrogen from a main sequence companion in a cataclysmic variable. The explosive burning of hydrogen on the top of a CO or an ONe degenerate core leads to the synthesis of some β^+ unstable nuclei, such as ^{13}N, ^{14}O, ^{15}O, ^{17}F and ^{18}F. These nuclei have short lifetimes (~ 1-2 minutes for ^{14}O, ^{15}O and ^{17}F, ~ 15 min for ^{13}N and ~ 3 hours for ^{18}F), and they emit a positron when they decay. These positrons annihilate with electrons leading to the emission of photons with energy less or equal to 511 keV. Furthermore, other medium- and long-lived radiocative nuclei are synthesized in classical novae: ^{7}Be (τ=77days), whichs emits a photon of 478 keV after an electron capture, ^{22}Na (τ=3.75yr) and ^{26}Al (τ=1.04×10^6 yr), which experience a β^+-decay emitting photons of 1275 and 1809 keV, respectively. Thus, classical novae are potential γ-ray emitters, as was pointed out in previous works (Clayton & Hoyle [1], Clayton [2], Leising & Clayton [3]). There are also some previous works concerning nucleosynthesis in classical novae (see Politano et al [4], Prialnik & Kovetz [5] and references therein for nucleosynthesis in ONe and CO novae, respectively, as well as Hernanz et al [6], José et al [7], José & Hernanz [8] and [9]). But to our knowledge there are no previous works which follow all the phases (from the accretion stage to the final explosion and mass ejection phases) and couple them to the study of the γ-ray emission of classical novae.

In this paper we have used realistic profiles of densities, velocities and chemical compositions (obtained by means of a hydrodynamical code) to determine the production and transfer of γ-rays (by means of a Monte-Carlo code) during nova explosions. In this way, we are able to do a detailed analysis of the γ-ray emission of classical novae and predict their detectability by instruments onboard the CGRO (see our previous paper by Hernanz et al [10] for our first results and their relation to the future mission INTEGRAL). We are mainly concerned by the emission and potential detectability of individual novae, related to the decay of the medium-lived nuclei ^{7}Be and ^{22}Na (besides of the short-lived nuclei ^{13}N and ^{18}F).

γ-RAY SPECTRA AND LIGHT CURVES OF CO AND ONE NOVAE

In table 1 we present the main properties of the ejecta of some of the most representative models we have computed (accretion rate $2\times10^{-10}M_\odot.yr^{-1}$). There is an important difference between CO and ONe novae: CO novae are important producers of ^{7}Be (thus line emission at 478 keV is expected during some days), whereas ONe novae are important producers of ^{22}Na (thus line emission at 1275 keV during the first years after the explosion is expected). This can clearly be seen by comparing the γ-ray spectra of a typical CO nova (CO1) and those of an ONe one (ONe2), shown in figure 1 for different epochs after the explosion. In all cases a continuum component, mainly below 511 keV, appears as well as some lines (511 keV in all cases and 478 keV and 1275 keV in CO and ONe novae, respectively). Two phases can be distinguished in the evolution of all models. During the early expansion, the ejected envelope is optically thick and there is an important contribution of the continuum below 511 keV, related to the comptonization of 511 keV photons. Later on, when the envelope becomes optically thin, absorption and comptonization become negligible; therefore, the intensity of the lines is exclusively determined by the total mass of the radioactive nuclei ^{7}Be (CO novae) and ^{22}Na (ONe novae). The

TABLE 1. Main properties of the ejecta one hour after peak temperature. Initial mass, total ejected mass and ejected mass of the most relevant radioactive nuclei are in $M_\odot$ and the mean kinetic energy of the ejecta, $<E_k>$, is in erg.g^{-1}

Model	M_{WD}	M_{ejec}	$<E_k>$	^{7}Be	^{13}N	^{18}F	^{22}Na
ONe1	1.15	1.8 10^{-5}	3.1 10^{16}	~ 0	5.5 10^{-9}	7.1 10^{-8}	9.8 10^{-10}
ONe2	1.25	1.6 10^{-5}	3.3 10^{16}	1.2 10^{-11}	2.9 10^{-8}	6.7 10^{-8}	1.6 10^{-9}
CO1	0.8	6.3 10^{-5}	8 10^{15}	7.8 10^{-11}	1.6 10^{-7}	1.7 10^{-7}	~ 0
CO2	1.15	1.4 10^{-5}	3.2 10^{16}	1.1 10^{-10}	1.3 10^{-8}	3.6 10^{-8}	~ 0

TABLE 2. Properties of the 511 keV line (columns 2 and 3) and of the 1275 and 478 keV lines (columns 5 and 6): time and flux (d=1kpc) at maximum.

Model	t [hours]	Flux$_{511}$ [photons/s/cm^2]	line	t [days]	Flux [photons/s/cm^2]
ONe1	6	1.3 10^{-2}	1275 keV	7.5	3.8 10^{-6}
ONe2	5	1.6 10^{-2}	1275 keV	6.5	6.1 10^{-6}
CO1	7.5	1.4 10^{-4}	478 keV	13	1.4 10^{-6}
CO2	6.5	7.3 10^{-3}	478 keV	5	2.6 10^{-6}

main properties of the lines are shown in table 2. It is worth noticing that fluxes are rather low, due to the small ejected masses (see table 1). This explains the null results of the observations of the 1275 keV emission by some novae made with COMPTEL onboard CGRO (Iyudin et al [12]). Our predicted fluxes are fully compatible with the upper limits obtained in that work. It is also worth noticing the important fluxes associated with the 511 keV line that we predict for all models (see table 2). This flux should be detected by the current instruments onboard the CGRO, provided that novae are observed very early after their explosion. A retrospective analysis of the BATSE data would perhaps provide important clues to this issue (see Fishman et al [13]).

We want to mention that there is a lack of agreement between all the theoretical models of novae (including ours) and some observations of ejected masses of classical novae, although large uncertainties affect sometimes the determination of observed ejected masses. For instance, our model ONe1 fits quite well the observed abundances of the neon nova QU Vul 1984, but observations of this nova give an ejected mass that can reach $\sim 10^{-3} M_\odot$ (Saizar et al 1992), almost two orders of magnitude larger than the theoretical one. Thus the flux of the 1275 keV line could be considerably larger, but no theoretical models are by now able to produce simultaneously such large ejected masses and neon in the ejecta.

In figure 2 we show the light curves of the 478 and 1275 keV lines for CO and ONe models, respectively. For the 478 keV line two different phases can be distinguished: during the first $\sim$1.5 days, the line is completely dominated by the continuum generated by ^{13}N and ^{18}F decays, whereas later on the line follows the typical light curve of the ^{7}Be radioactive decay, with τ=77 days. The 1275 keV line has a rise phase lasting $\sim$7days, followed by a decline related

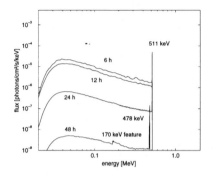 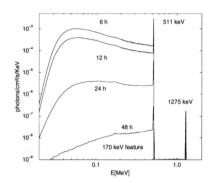

FIGURE 1. Evolution with time of the γ-ray spectrum of a CO nova of 0.8 $M_\odot$ (left) and an ONe nova of 1.25 $M_\odot$ (right).

to the decay of ^{22}Na. The small fluxes obtained make it impossible to detect this line with the current CGRO instruments, unless a very close explosion occurs (the same as for the future SPI instrument onboard INTEGRAL).

Concerning the light curve of the 511 keV line for all the models, as several isotopes, with different decay timescales, contribute to this emission, its temporal evolution is somewhat complex. Besides one very early maximum appearing in the CO novae (related to ^{13}N decay), one later maximum appears in all cases at $\sim$ 6 hours (see table 2), which is related to ^{18}F decay. The further development of the light curves is different in the CO and in the ONe cases: in the ONe case, as ^{22}Na emits a e$^+$ when it decays, the emission at 511 keV lasts for a longer time (as the ^{22}Na decay- timescale is much longer than the ^{18}F one). An early observation of a nova explosion, either of a CO or an ONe one, would provide a positive detection of the 511 keV line (an estimation of the detection distance of our novae by the future SPI instrument onboard INTEGRAL yields around 10 kpc for all of them except for the low-mass CO1 model).

CONCLUSIONS

We have developed a Monte-Carlo code in order to treat the production and transfer of γ-rays in nova envelopes, coupled to a hydrodynamical code that provides realistic profiles of all the relevant magnitudes. Thus a complete view of the nova explosion is obtained. The most relevant features of the γ-ray emission of classical novae are their intense 511 keV emission, lasting only some hours, the 478 keV emision in the case of CO novae, related to ^{7}Be decay,

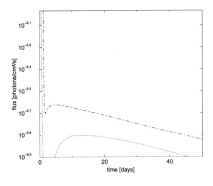
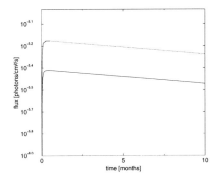

FIGURE 2. Left: Light curves of the 478 keV line (d=1 kpc) for CO novae (CO1: dotted, CO2: dot-dashed). Right: Light curves of the 1275 keV line (d=1 kpc) for ONe novae (ONe1: solid, ONe2: dotted)

and the 1275 keV emission in the case of ONe novae, related to ^{22}Na decay. The 1275 keV emission is the most long-lived one, but the fluxes obtained are too low to be detected for novae at typical distances, thus explaining the negative detections of the novae observed up to now by COMPTEL.

We thank for partial support the CICYT and DGICYT Projects ESP95-0091 and PB94-0827-C02-02.

REFERENCES

1. Clayton, D.D., and Hoyle, F. *ApJ* **187**, L101 (1974).
2. Clayton, D.D., *ApJ* **244**, L97 (1981).
3. Leising, M., and Clayton, D.D., *ApJ* **323**, 159 (1987).
4. Politano M., Starrfield S., Truran J.W., Weiss A., Sparks W.M., *ApJ* **448**, 807 (1995).
5. Prialnik D., Kovetz A., *ApJ* **477**, 356 (1997).
6. Hernanz M., José J., Coc A., Isern J., *ApJ Letters* **465**, L27 (1996).
7. José J., Hernanz M., Coc A., *ApJ Letters* **479**, L55 (1997).
8. José J., Hernanz M., *Nuclear Physics*, in press (1997).
9. José J., Hernanz M., in preparation.
10. Hernanz M., Gómez-Gomar, J., José, J., and Isern, J., *The Transparent Universe*, ESA SP-382, Noordwijk: ESA Publications Division, 1997, pp. 47-50.
11. Saizar, P., et al, *ApJ* **398**, 651 (1992).
12. Iyudin A.F., et al, *A&A* **300**, 422 (1995).
13. Fishman G.J., et al, *Gamma-Ray Line Astrophysics*, New York: AIP, 1991, pp. 190-192.

New Studies of Nuclear Decay γ-rays From Novae

S. Starrfield[1], J. W. Truran[2], M. C. Wiescher[3], W. M. Sparks[4]

[1] Department of Physics and Astronomy, Arizona State University, P.O. Box 871504, Tempe, AZ 85287-1504
[2] Department of Astronomy and Astrophysics and Enrico Fermi Institute, University of Chicago, Chicago, IL 60637
[3] Department of Physics, University of Notre Dame, Notre Dame, IN 46556
[4] XNH, Nuclear and Hydrodynamic Applications, MS F664, Los Alamos National Laboratory, Los Alamos, NM 87544

Abstract. The cause of the nova outburst is a thermonuclear runaway (TNR) in hydrogen rich material transferred by a companion onto a white dwarf. Studies of this phenomenon have shown that the TNR produces large concentrations of the short lived positron unstable isotopes of the CNO nuclei which are transported to the surface by convection so that *early in the outburst* we expect significant numbers of radioactive decays to occur at the surface. The resulting γ-ray emission may be detectable from *nearby* novae early in their outbursts. The TNR is also expected to produce substantial amounts of ^{7}Be and ^{22}Na. Their decays also yield potentially detectable levels of γ-ray emission for relatively nearby novae. We are also interested in the role played by novae in the production of the $\sim 2 M_\odot$ of ^{26}Al found in the galaxy. In order to improve our predictions of this phenomenon, we have performed a new set of calculations of TNR's on ONeMg and CO white dwarfs with an updated nuclear reaction network and opacities.

I INTRODUCTION

Novae occur in cataclysmic variable (CV) binary systems in which a Roche lobe filling secondary is losing hydrogen-rich material through the inner Lagrangian point onto a white dwarf (WD) primary. The accumulating shell of material on the WD mixes with material from the core and eventually becomes unstable to a thermonuclear runaway (TNR). The TNR causes an explosion and the accreted plus core material is ejected into space where it can be analyzed. Hydrodynamic simulations of this phenomenon reproduce the gross features of the nova outburst: the typical amount of mass ejected, the kinetic energies of the ejecta, and the optical light curves [9,13]. We note that, in addition to the γ-ray production, determining the elemental abundances in the ejecta, in combination with the hydrodynamic simulations, can provide information on the core composition for a number of different types of WDs.

Studies of the TNR have shown that hydrogen burning around the time of peak energy generation produces large concentrations of the short lived positron unstable isotopes of the CNO nuclei: ^{13}N ($\tau = 9.97$ m), ^{14}O ($\tau = 71$ s), ^{15}O ($\tau = 122$s), ^{17}F ($\tau = 66.6$ s), and ^{18}F ($\tau = 109.8$ m). These nuclei are transported to the surface by convection on time scales of a few seconds so that we expect significant numbers of radioactive decays to occur at the surface. In fact, the simulations [13,14] predict that the rates of nuclear energy generation at the surface can reach values exceeding 10^{15} erg gm^{-1} s^{-1}. The resulting γ-ray emission may be detectable from nearby novae early in their outbursts. We have described this process in some detail and will not repeat that discussion here [11].

The TNR also is expected to produce substantial amounts of ^{7}Be ($\tau = 53$ d), and ^{22}Na ($\tau = 2.6$ yr), and these nuclei form part of the material ejected into space by the explosion. Their decays also yield potentially detectable levels of γ-ray emission for relatively nearby novae. Nevertheless, no detections of these γ-ray episodes have occurred for any of the novae discovered in outburst during the lifetime of COMPTON GRO. We argue that this is the unfortunate consequence of the fact that the expected fluxes are at the limit of detectability with GRO for novae at distances of $\gtrsim 1$ kpc. This can certainly be remedied with the launch of future instruments with higher sensitivities.

A detection of a nova outburst in γ-rays would provide an important confirmation of the TNR theory of the outburst and, in addition, serve to impose critical constraints on the role of convection in the earliest stages of the runaway. In order to improve the accuracy of our predictions, we have begun a new set of calculations of TNR's on massive white dwarfs (both carbon-oxygen [CO] and oxygen-neon-magnesium [ONeMg]) in which we have updated the nuclear reaction network, with the inclusion of new and improved experimental and theoretical determinations of the nuclear reaction rates. We have also incorporated new and improved opacities from the OPAL carbon rich tables and have investigated the effects of changes in convective efficiency on the evolution [13]. Our results show that the changes in the reaction rates and opacities that we have introduced produce important changes with respect to our previous studies. For example, we find that less mass is accreted and ejected. This is primarily a consequence of the fact that the OPAL opacities are larger than those we previously used, which results in less mass being accreted onto the white dwarf.

Relevant to nucleosynthesis considerations, a smaller amount of ^{26}Al ($\tau = 7.4 \times 10^5$ yr) is produced, while the abundances of ^{31}P and ^{32}S increase by factors of more than two. This change is attributable to the increased proton-capture reaction rates for some of the intermediate mass nuclei near ^{26}Al and beyond, such that nuclear buildup to higher mass nuclei is enhanced. The calculations are discussed in detail in [13] so that here we only present the results with respect to γ-ray production from the decays of ^{7}Be and ^{22}Na plus the contribution of ^{26}Al to Galactic Nucleosynthesis.

TABLE 1. Predictions of the Amount of Radioactive Nuclei Ejected in CO Sequences

WD Mass	$0.6M_\odot$	$0.8M_\odot$	$1.00M_\odot$	$1.25M_\odot$
^{7}Be	4×10^{-8}	6×10^{-8}	6×10^{-8}	7×10^{-8}
^{22}Na	2×10^{-6}	2×10^{-6}	2×10^{-6}	4×10^{-6}
^{26}Al	4×10^{-7}	3×10^{-7}	4×10^{-6}	7×10^{-5}
$M_{ej}{}^a$	$3\times10^{-5}M_\odot$	$2\times10^{-5}M_\odot$	$4\times10^{-6}M_\odot$	$1\times10^{-6}M_\odot$

^{a}Mass extending past a radius of 10^{11}cm at the end of the evolution.

TABLE 2. Predictions of the Amount of Radioactive Nuclei Ejected in ONeMg Sequences

WD Mass	$1.0M_\odot$	$1.25M_\odot$	$1.35M_\odot$
^{7}Be	2×10^{-10}	6×10^{-8}	5×10^{-8}
^{22}Na	5×10^{-5}	3×10^{-4}	5×10^{-3}
^{26}Al	2×10^{-2}	1×10^{-3}	7×10^{-3}
M_{ej}	$1\times10^{-5}M_\odot$	$5\times10^{-6}M_\odot$	$6\times10^{-6}M_\odot$

II NUCLEOSYNTHESIS RESULTS

We have recently computed new simulations of TNR's on CO white dwarfs with masses ranging from $0.6M_\odot$ to $1.25M_\odot$ [14]. Results of other groups can be found in [3,1]. The details of the evolutionary sequences and nucleosynthesis results will be given in [14]. In Table 1 we present the abundance results for the three radioactive nuclei: ^{7}Be, ^{22}Na, and ^{26}Al.

It is clear that the CO sequences do not eject sufficient quantities of these nuclei for this class of novae to be important in questions of detectability by COMPTON GRO. It is also clear that this class of novae cannot contribute much ^{26}Al to the Galaxy. While the amount of ^{7}Be ejected by these novae is insufficient to produce a detectable signal in all but the very nearest novae, it is important to realize that our initial abundance of ^{3}He was solar and not enriched as might be the case in the low mass secondaries of these novae [12,2].

The results for the ONeMg sequences are more promising. While the amount of ^{7}Be is still too low to be interesting, the amounts of ^{22}Na and ^{26}Al are sufficiently large to be worth considering in the context of COMPTON. We also point out that the ejected mass fraction of ^{22}Na is an increasing function of WD mass while the ejected mass fraction of ^{26}Al is a decreasing function of WD mass. This implies that those novae that produce an interesting amount of ^{26}Al will have occurred on low mass WD's while those novae predicted to be detectable in 1.275MeV γ-rays by COMPTON will be the more massive novae.

We have, therefore, put together Table 3 which gives the predicted γ-ray emission, plus the contribution to Galactic ^{26}Al production, by four recent well studied novae. We caution the reader that this table contains both observational results and theoretical predictions. The first row gives the name of the nova (V838 Her, V1974 Cyg, QU Vul, or V705 Cas), the second through fourth rows give the year of outburst, relative speed class, and composition. The first three are ONeMg novae and the fourth, V705 Cas, is a CO nova that formed an optically thick dust shell during its outburst. The next three rows give the *estimated* mass of the WD (note colon), the estimated distance, and estimates (from observations) of the total amount of mass ejected during the outburst. These numbers are averages of a number of different determinations while the value for QU Vul is based on the arguments in [6] that the ejected mass of this nova has been underestimated by a large factor. The next two sets of data are our best estimate of the mass fraction of the particular isotope for the given WD mass. We also provide the γ-ray photon luminosity of the nova using the distance and isotopic abundance. The final set of rows is the estimated ejected mass fraction of ^{26}Al and the amount of ^{26}Al that should be found in the galaxy if roughly one third of the 35 novae per year [7] were novae of this type. The formulae for determining the γ-ray luminosity and ^{26}Al abundance can be found in [15,5].

The results presented in Table 3 show, as already stated, that none of these novae should have been detected in γ radiation from the ^{7}Be K-capture reaction. In contrast, given the detection limit of $\sim 10^{-4}$ photons cm^{-2} s^{-1} for the OSSE and COMPTEL instruments, V838 Her should have been detected but was not detected [8]. In addition, [4] searched the SMM data base for any signal from QU Vul. While their search was negative, their upper limit was $\sim 3 \times 10^{-4}$ photons cm^{-2} s^{-1}.

III CONCLUSIONS

We have computed new sets of evolutionary sequences for both CO and ONeMg novae using new reaction rates and opacities. These updates have produced interesting changes in the results of our computations (Starrfield et al. 1997). We find, for example, that the detection of ^{7}Be by COMPTON is hopeless for all reasonable estimates of the initial abundance of ^{3}He. This is true for both CO and ONeMg novae. In contrast at least two out of the recent ONeMg novae should have been detected by COMPTON or SMM. The fact that they were not detected requires further calculations to determine the cause of this discrepancy. Finally, two out of the recent ONeMg novae experienced relatively slow outbursts and are predicted to have ejected $\sim 10^{-4} M_\odot$ of material. If a significant fraction of novae in the Galaxy are of this type (30%), then novae far outproduce the $\sim 2 M_\odot$ of ^{26}Al observed in the galaxy.

The work reported in this paper was supported in part by NSF and NASA grants to the University of Chicago, Arizona State University, and Notre Dame University, and by the DOE.

TABLE 3. Theoretical Predictions of the Radioactive Element Production by Four Novae

Nova	V838 Her	V1974 Cyg	QU Vul	V705 Cas
Year	1991	1992	1984	1993
Speed Class	V. Fast	Fast	Moder.	Slow
Composition	ONeMg	ONeMg	ONeMg	CO
WD Mass	$1.35:M_\odot$	$1.25:M_\odot$	$1.1:M_\odot$	$0.8:M_\odot$
Distance	$\sim$3kpc	$\sim$2 kpc	$\sim$2 kpc	$\sim$2 kpc
Mass Ejected	$7 \times 10^{-5} M_\odot$	$5 \times 10^{-5} M_\odot$	$1 \times 10^{-3} M_\odot$	$1 \times 10^{-4} M_\odot$
^{22}Na	5×10^{-3}	3×10^{-4}	1×10^{-4}	2×10^{-6}
F_{22}(time=0)	2×10^{-4}	2×10^{-5}	1×10^{-4}	2×10^{-7}
^{7}Be	5×10^{-8}	6×10^{-8}	1×10^{-8}	6×10^{-8}
F_7(time=0)	8×10^{-8}	2×10^{-7}	5×10^{-8}	6×10^{-8}
^{26}Al	7×10^{-3}	1×10^{-3}	1×10^{-2}	3×10^{-7}
$M_\odot$ of ^{26}Al	$5 M_\odot$	$0.4 M_\odot$	$30 M_\odot$	$4 \times 10^{-4} M_\odot$

REFERENCES

1. Josè, J., Hernanz, M., and Coc, A. 1997, ApJ, 479, L55.
2. Hernanz, M., Josè, J., Coc, A., and Isern, J. 1996, ApJ, 465, L27.
3. Kovetz, A., and Prialnik, D. 1997, ApJ, 477, 356.
4. Leising, M. D., Share, G. H., Chupp, E. L., and Kanbach, G. 1988, ApJ, 328, 755.
5. Nofar, I., Shaviv, G., and Starrfield, S. 1991, ApJ, 369, 440.
6. Saizar, P., and Ferland, G. J. 1994, ApJ, 425, 755.
7. Shafter, A. 1997, ApJ, 487, in press.
8. Shrader, C., Starrfield, S., Truran, J. W., Leising, M. D., and Shore, S. N. 1994, Bull. AAS, 26, 1325.
9. Starrfield, S. 1995, in "Physical Processes in Astrophysics", ed. I. Roxburgh, and J. L. Masnou, (Springer: Heidelberg), 99.
10. Starrfield, S., Gehrz, R. D., and Truran, J. W. 1997, in "Astrophysical Implications of the Laboratory Study of Presolar Material", AIP Conference Proceedings #?, ed. T. Bernatowicz and E. Zinner, in press.
11. Starrfield, S., Politano, M., Truran, J. W., and Sparks, W. M. 1992, in "Proceedings of the Second COMPTON GRO conference", ed. C. Shrader, N. Gehrels, and B. Dennis (NASA CP 3137), 377.
12. Starrfield, S., Truran, J. W., Sparks, W. M., and Arnould, M. 1978, ApJ, 222,600.
13. Starrfield, S., Truran, J. W., Wiescher, M. C., and Sparks, W. M. 1997a, MNRAS, in press.
14. Starrfield, S., Truran, J. W., Wiescher, M. C., and Sparks, W. M. 1997b, in preparation.
15. Weiss, A., and Truran, J. W. 1990, A&A, 238, 178.

SUPERNOVA REMNANTS AND COSMIC RAYS, DIFFUSE GAMMA-RAY CONTINUUM RADIATION, AND UNIDENTIFIED SOURCES

CTA 1 SUPERNOVA REMNANT : A HIGH ENERGY GAMMA-RAY SOURCE?

D. Bhattacharya, A. Akyüz, G. Case, D. Dixon and A. Zych

Institute of Geophysics and Planetary Physics, University of California, Riverside, CA 92521

Abstract. We discuss the excess gamma-ray emission seen from the direction of CTA 1, a supernova remnant, in the energy range of 100 MeV to 10 GeV with the EGRET instrument. The X-ray emission peak of CTA 1 seen by ROSAT falls within the 95% EGRET likelihood contour giving rise to the possibility that a central source is responsible for the high energy emission. We consider the probability that pulsars are responsible for most of the EGRET SNR candidate source emission.

INTRODUCTION

The possibility that a number of EGRET source positions at low galactic latitudes are supernova remnants (SNR) is pointed out by [1] and [2]. Later, a detailed analysis of the EGRET data of 14 radio bright (with radio flux > 100 Jy at 1 GHz) SNR positions detected gamma rays from 5 of them: γ Cygni, IC 443, W28, W44 and Monoceros [3]. A recent analysis by [5] finds evidence of extended high energy gamma-ray emission associated with the Monoceros SNR. Two more EGRET sources are tentativeley connected to SNR G312.4-0.4 and MSH 11-61A [4].

Recent spatially resolved ASCA measurements of SN1006 [6] provide evidence for synchrotron emission in the range 0.5–10 keV from its rims. For a magnetic field of 6-10 μG, the ASCA measurements infer the radiating electron energies to be $\geq$ 200 TeV. Similar results have been obtained from ASCA measurements of SNR IC443 [7] indicating shock accelerations of electrons to 10 TeV energy. Because shock acceleration does not distinguish between electrons and protons, the protons are expected to be accelerated to such energies also. The acceleration of protons is treated elaborately in time dependent two fluid models; the production of high energy gamma rays due to this process has been given by [8]. Esposito et al. [3] explain the non-detection of the

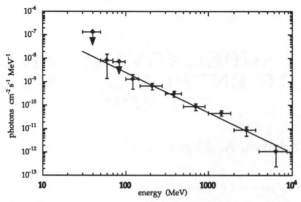

FIGURE 1. *Derived energy spectrum for 2EG 0008+7303.*

other selected SNRs as either the SNRs are still in the free expansion phase and have not built up a significant density of locally accelerated cosmic rays, or the SNRs are in a region without nearby dense clouds, or simply some of them are too distant.

If the five EGRET sources identified with SNR positions are cosmic-ray accelerators it is expected that they would accelerate particles up to the knee of the cosmic ray spectrum (at $\sim 3 \times 10^{15}$ eV) with spectral indices ~ 2.3. However, the CYGNUS air shower experiment (threshold $\sim 10^{14}$ eV) does not show any evidence of the resulting ultra-high energy gamma rays from these sources [9]. This implies the cosmic-ray spectra of these sources soften below 10^{14} eV. It is, however, also possible that cosmic-ray electrons are responsible for the EGRET source emission, in which case the sensitivities of air shower experiments are too low to detect the ultra-high energy emission [9]. Furthermore, none of the SNR candidate sources have shown any conclusive evidence of a spectral bump around 70 MeV that is expected from the decay of neutral pions, the prime signature of cosmic ray acceleration. There is a possibility that emission from one or more of these remnants is associated with a pulsar. Evidences for pulsar associations have been presented for W28 [10], W44 [11] and γ-Cygni [12]. In this paper we discuss the positional coincidence of another supernova remnant CTA 1 with the unidentified EGRET source 2EG J0008+7307 and possible mechanisms for the gamma-ray emission.

RESULTS AND DISCUSSION

Analyzing the data from the first two phases of EGRET observations, Thompson et al. [13] find a flux of 5.68×10^{-7} ph cm^{-2} s^{-1} above 100 MeV and a photon index of about 1.6 ± 0.2 for 2EG 0008+7307. We have reanalyzed the region adding VP 401 (regions only within 30° of EGRET pointing were used) using the standard Likelihood technique [14]. The intensity variations over different viewing periods are consistent with a constant value to

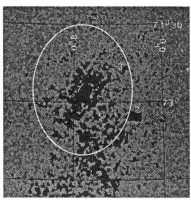

FIGURE 2. *95% confidence contour (for E > 300 MeV) of TS overlaid on the ROSAT X-ray contous of CTA 1.*

within 2σ for all observations. The best fit can be described by the following expression ($\chi_n^2 = 0.72$): $F(E) = (1.12 \pm 0.15) \times 10^{-10} (E/619)^{-1.71 \pm 0.11}$ ph cm^{-2} s^{-1} MeV^{-1}. The energy spectrum is shown in Figure 1. Assuming a 4π beaming fraction and a distance of 1.4 kpc we derive a luminosity of $\sim 7 \times 10^{34}$ ergs s^{-1} over the energy range (0.1-5) GeV.

To obtain a better source position we decided to use the data above 300 MeV. In Figure 2, we show the EGRET 95% confidence contour of the likelihood test statistic (TS) overlaid on the ROSAT X-ray contours obtained by Seward, Schmidt, & Slane [15]. The best fit position is $l = 119.75$, $b = 10.55$ in galactic coordinates and RA = $0^h 08^m$, Dec = $73°09'$ in equitorial.

Radio measurements show CTA 1 consisting of bright emission areas connected with a shell structure over only a part of the circumference of the object and weak emission that could be traced around the whole circular area of the SNR [16], [17]. ROSAT X-ray results find the emission from the center to be harder and considerably brighter than the rim [15]. The EGRET best fit position of the gamma-ray source coincides within 16 arc min of the X-ray emission peak associated with the central nebula seen by ROSAT, i.e., source 7. We could not find any obvious counterparts either to source 2 (considered to be an AGN by [15]) or 7 in the list of 1420 MHz compact radio sources within the remnant [17]. As the candidate AGN source was unambigously detected only in X-rays and at optical wavelengths, the chance of it being a blazar capable of producing high energy gamma rays is small. Furthermore, its location outside the 95% TS level prompts us to suggest that the gamma-ray emission is not associated with source 2.

If we consider that the gamma-ray emission is due to the accelerated particles then estimations of cosmic ray density can be made if the total amount of SNR matter, including the swept up material, is known. However, the predicted or assumed ISM density in this region varies: 0.04 [16], 1 [17], 0.017 [15] cm^{-3}. Using the ISM density of 0.04 cm^{-3} and a shock radius of 21.6

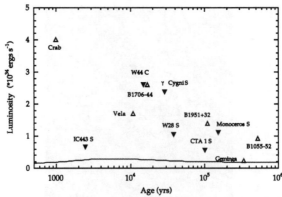

FIGURE 3. *Gamma-ray luminosities of the six EGRET pulsars and six SNR candidate sources as a function of their ages assuming that their radiation is beamed into 1 sr. The solid line is the theoretical SNR gamma-ray luminosity. S and C after SNR name denotes shell or composite remnant.*

pc [15] we estimate the mass of the CTA 1 remnant to be approximately 15 $M_\odot$. Using an expression given by [3] where gamma-ray intensity is the product of combined bremsstrahlung/π^o decay emissivity and the SNR matter, we find the expected intensity to be 4.7×10^{-11} ph cm^{-2} s^{-1}. This is a factor of 745 lower than the observed intensity. Hence, if the gamma ray emission arises solely in the nebula, the cosmic ray density would have to be 745 times larger than the local solar value. This number is much larger than the average cosmic ray density enhancement factor (~ 150) calculated for other EGRET candidate SNR sources [3].

The dynamic age of CTA 1 is estimated to be 1.5×10^4 yr. It is possibly in a pressure driven snowplow phase where the shell is radiatively cooling. From the blast wave characteristics, a kinetic energy of 3×10^{49} ergs was derived for the initial supernova debris [15]. This is considerably lower than 10^{51} ergs calculated for many other remnants. This makes the CTA 1 blast wave relatively weak, perhaps incapable of processing enough energy through shocks to produce gamma rays at a detectable level for EGRET.

In Figure 3, we show the gamma-ray luminosities of the six EGRET pulsars and six SNR candidate sources, including CTA 1, as a function of their ages. The pulsar luminosities are given over the range 0.1 to 5 GeV assuming a beaming angle of 1 sr [18]. We have also shown SNR luminosities reduced by a factor of 4π (for reasons stated below). Overall, the SNR candidate sources have higher average luminosities than the pulsars if we assume isotropic emission. According to [8] the total gamma-ray luminosity of a remnant increases very rapidly, almost two orders of magnitude, during the initial 1000 years after which it remains constant for the the rest of the remnant life ($\sim 10^6$ yr). (the solid line in Figure 3 indicates the theoretical luminosity). It is not clear to us whether such a trend is apparent in Figure 3; this could be due to the facts that i) we are dealing with only a few remnants, ii) the blast characteristics and the ISM densities are different, and iii) one or more of them could harbor a pulsar.

If we assume that each one of the SNR sources harbor a gamma-ray pulsar and its beaming fraction is 1 sr, we obtain luminosities as shown in 3. It is apparent that gamma-ray luminosities of all the sources become progressively weaker as they age and all the SNR sources, except for IC 443, follow the general pulsar luminosity vs. age dependence. It is possible that pulsar emission is responsible for most of the EGRET SNR sources - this conclusion is consistent with the recent work by [19] where they find that the spatial distribution and luminosities of EGRET unidentified sources are consistent with the proposition that all Galactic EGRET sources are pulsars. CTA 1's gamma-ray luminosity is consistent with the pulsar luminosity of its age group. This, together with the requirement for a very high cosmic ray energy density (in spite of a weak blast energy), the location of central X-ray source near the most likely gamma-ray position, high gamma- to X-ray flux ratio and relatively steady emission prompts us to suggest that the gamma-ray emission from CTA 1 is originating from a central source, probably a pulsar located at the center of the nebula.

We wish to thank the CGRO Science Support Center for providing the data and much of the image and spectral analysis tools.

REFERENCES

1. Sturner, S. J. & Dermer, C. D. 1995, *A&A*, **293**, L17.
2. Esposito J. A. et al.: 1994, *BAAS*, **26**, 970.
3. Esposito, J. A.et al.: 1996, *ApJ*, **461**, 820.
4. Sturner, S. J., Dermer, C. D. and Mattox, J. R.: 1996, *A&AS*, **120**, 445.
5. Jaffe, T. R. et al.: 1997, *ApJL*, in press.
6. Koyama, K. et al.: 1995, *Nature*, **378**, 255.
7. Keohane, J. W. et al.: 1997, *ApJ*, accepted for publication.
8. Drury, L. O'C, Aharonian, F. A., and Völk, H. J.: 1994, *A&A*, **287**, 959.
9. Allen, G. et al.: 1995, *ApJ*, **448**, L25.
10. Frail D. A., Kulkarni S. R., & Vashist, G.: 1993, *Nature*, **365**, 136.
11. Wolszczan, A., Cordes, J. M. and Dewey, R. J.: 1991, *ApJ*, **372**, L99.
12. Brazier, K. T. S. et al.: 1996, *MNRAS*, **281**, 1033.
13. Thompson, D. J. et al.: 1995, *ApJS*, **101**, 259
14. Mattox, J. R. et al.: 1996, *ApJ*, **461**, 396.
15. Seward, F. D., Schmidt, B. and Slane, P.: 1995, *ApJ*, **453**, 284.
16. Sieber W., Salter C. J., & Mayer C. J.: 1981, *A&A*, **103**, 393.
17. Pineault, S. et al.: 1993, *AJ*, **105**, 1060.
18. Thompson, D. J. et al.: 1994, *ApJ*, **436**, 229.
19. Yadigarglu, I.-A and Romani, R. W.: 1997, *ApJ*, **476**, 347.

Constraints on Cosmic-Ray Origin from TeV Gamma-Ray Observations of Supernova Remnants

R.W.Lessard[1], P.J.Boyle[2], S.M. Bradbury[4], J.H.Buckley[3],
A.C.Burdett[4], J.Bussóns Gordo[2], D.A.Carter-Lewis[5],
M.Catanese[5], M.F.Cawley[6], D.J.Fegan[2], J.P.Finley[1],
J.A.Gaidos[1], A.M.Hillas[4], F.Krennrich[5], R.C.Lamb[7],
C.Masterson[2], J.E.McEnery[2], G.Mohanty[5], J.Quinn[2],
A.J.Rodgers[4], H.J.Rose[4], F.W.Samuelson[5], G.H.Sembroski[1],
R.Srinivasan[1], T.C.Weekes[3] and J.Zweerink[5]

[1] *Dept. of Physics, Purdue University, West Lafayette, IN 47907-1396, U.S.A.* [1]
[2] *Dept. of Exper. Physics, University College Dublin, Belfield, Dublin 4, Ireland*
[3] *Whipple Obs., Harvard-Smithsonian CfA, P.O. Box 97, Amado, AZ 85645-0097, U.S.A.,*
[4] *Dept. of Physics, University of Leeds, Leeds, LS2 9JT, Yorkshire, U.K.,*
[5] *Dept. of Physics and Astronomy, Iowa State University, Ames, IA 50011-3160 USA,*
[6] *Dept. of Physics, Saint Patrick's College, Maynooth, Co. Kildare, Ireland,*
[7] *Space Radiation Lab., California Institute of Technology, Pasadena, CA 91125 USA*

Abstract. If supernova remnants (SNRs) are the sites of cosmic-ray acceleration, the associated nuclear interactions should result in observable fluxes of TeV gamma-rays from the nearest SNRs. Measurements of the gamma-ray flux from six nearby, radio-bright, SNRs have been made with the Whipple Observatory gamma-ray telescope. No significant emission has been detected and upper limits on the >300 GeV flux are reported. Three of these SNRs (IC443, gamma-Cygni and W44) are spatially coincident with low latitude unidentified sources detected with EGRET. These upper limits weaken the case for the simplest models of shock acceleration and energy dependent propagation.

INTRODUCTION

It is generally believed that cosmic rays with energies less than ~ 100 TeV originate in the galaxy and are accelerated in shock waves in shell-type SNRs.

---

[1] This research has been supported in part in the U.S. by the Department of Energy and NASA, Forbairt in Ireland and PPARC in the UK.

This hypothesis is supported by several strong arguments. First, supernova blast shocks are one of the few galactic sites capable of sustaining the galactic cosmic ray population against loss by escape, nuclear interactions and ionization energy loss assuming a SN rate of about 1 per 30 years and a 10% efficiency for converting the mechanical energy into relativistic particles. Second, models of diffuse shock acceleration provide a plausible mechanism for efficiently converting this explosion energy into accelerated particles with energies $\sim 10^{14} - 10^{15}$ eV and naturally give a power-law spectrum similar to that inferred from the cosmic ray data after correcting for energy dependent propagation effects. Finally, observations of non-thermal X-ray emission in SN1006 [11] and IC443 [10] suggest the presence of electrons accelerated to ~ 100 TeV and ~ 10 TeV respectively.

If SNRs are sites for cosmic ray production, there will be interactions between the accelerated particles and the local swept-up interstellar matter. Drury, Aharonian and Volk [6] (DAV) and Naito and Takahara [13] have calculated the expected gamma-ray flux from secondary pion production using the model of diffusive shock acceleration. The expected intensity (DAV) is

$$F(>E) = 9 \times 10^{-11} \left(\frac{E}{1\text{TeV}}\right)^{-1.1} \left(\frac{\theta E_{SN}}{10^{51}\text{erg}}\right) \left(\frac{d}{1\text{kpc}}\right)^{-2} \left(\frac{n}{1\text{cm}^{-3}}\right) \text{cm}^{-2}\text{s}^{-1}$$

(1)

where E is the photon energy, θ is the efficiency for converting the supernova explosion energy, E_{SN}, into accelerated particles, d is the distance to the SNR and n is the density of the local ISM.

OBSERVATIONS

We report on the results of observations of six nearby SNR (W44, W51, gamma-Cygni, W63, Tycho and IC443) by the Whipple Observatory's high energy gamma-ray telescope situated on Mount Hopkins in southern Arizona. The telescope [5] employs a 10 m diameter optical reflector to focus Čerenkov light from air showers onto an array of 109 photomultipliers covering a 3 degree field of view. By making use of distinctive differences in the lateral distribution of gamma-ray induced showers and hadronic induced showers, images can be selected as gamma-ray like based on their angular spread. The determination of the incident direction of the selected gamma-ray like events is accomplished by making use of the orientation, elongation and asymmetry of the image. Monte Carlo studies have shown that gamma-ray images are a) aligned towards their source position on the sky b) elongated in proportion to their impact parameter on the ground and c) have a cometary shape with their light distribution skewed towards their point of origin in the image plane. Results on the Crab Nebula indicate that the angular resolution function for

TABLE 1. Results of Observations.

Source Name	Pointing α, δ (1950)	Aperture Radius (deg)	ON Source Counts	OFF Source Counts	Total Time (min)	Upper Limit $\times 10^{-11}$ (cm^{-2}s^{-1})
W44	18:53:29, 01:14:57	0.55	450	426	360.1	3.0
W51	19:21:30, 14:00:00	0.68	619	559	468.0	3.6
γ-Cygni	20:18:59, 40:15:17	0.76	1040	1104	560.0	2.2
W63	20:17:15, 45:24:36	1.05	452	501	140.0	6.4
Tycho	00:22:28, 63:52:11	0.29	315	302	867.2	0.8
IC443	06:14:00, 22:30:00	0.64	1565	1522	1076.7	2.1

the telescope using this technique is a Gaussian with a standard deviation of 0.13 degrees [12]. A combination of Monte Carlo simulations and results on the Crab Nebula indicate that the energy threshold of the technique is 300 GeV and the effective collection area for a point source located at the center of the field of view is 2.1×10^8 cm^2 and is reduced for offset sources [12].

The analysis of data from extended sources involves binning the event arrival directions. We define the source region for the SNR by a circular aperture which matches the maximum extent of the radio shell [9] plus twice the width of the angular resolution function to account for the smearing of the edge of the remnant. The number of gamma-ray candidate events is obtained by subtracting the number of events in the OFF-source observations from the number of events in the ON-source observations.

RESULTS

The observations were made over three observing seasons, from 1993 to 1996. Of the six remnants observed, W 51 showed the greatest excess at an offset location which is spatially consistent with hard X-ray emission and peak radio intensity. More data are required to determine if this excess is signicant. No significant excess has been recorded for the other remnants and 99.9% confidence upper limits on the flux have been calculated (see Table 1). The upper limit assumes uniform emission from the remnant in the absence of *a priori* knowledge of a more defined emission region.

DISCUSSION

In Figure 1 the Whipple upper limits and EGRET data (gamma-Cygni and IC443 [7], W44 [8] and [15] for the remaining) are compared with an $E^{-2.1}$ extrapolation of the EGRET data using the contribution to the gamma-ray spectrum from secondary pion decay as derived by Buckley et al. [4] using the model of DAV. The upper dotted curve assumes a source spectrum of $E^{-2.1}$ and a reasonable maximum value of the product $E_{SN}\theta/d^2$ used in the model.

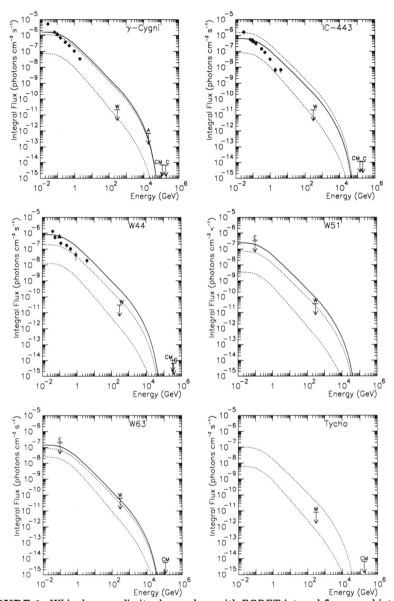

FIGURE 1. Whipple upper limits shown along with EGRET integral fluxes, and integral spectra. These are compared to extrapolations from the EGRET integral data points (solid curves), as well as a conservative estimate of the allowable range of fluxes from the model of DAV (dotted curves). Also shown are upper limits from CASA-MIA [2], CYGNUS [1], and AIROBIC [14].

The lower dotted curve assumes a source spectrum of $E^{-2.3}$ and a reasonable minimum value of the product $E_{SN}\theta/d^2$.

We interpret our results in the context of two hypotheses, (1) that the EGRET data gives evidence for acceleration of cosmic ray nuclei in SNR and that the observed gamma-ray emission comes not from primary electrons but from nuclear interactions of cosmic rays with ambient material or (2) that the EGRET flux is produced by some other mechanism.

Under the assumption that the contribution from electron bremsstrahlung and inverse Compton (IC) scattering are negligible, it is reasonable to compare the high energy gamma-ray upper limits to an extrapolation of the integral EGRET fluxes using the model by DAV. In the case of gamma-Cygni, IC443 and W44 the Whipple upper limits lie a factor of $\sim$ 25, 10 and 10 respectively below the extrapolation and require either a spectral break or a source spectrum steeper than $E^{-2.5}$ for gamma-Cygni and $E^{-2.4}$ for IC443.

Another plausible explanation for the results is that the EGRET flux is produced by high energy electrons accelerated in the vicinity of pulsars. If this is the case, then the Whipple upper limits must be compared with the *a priori* model predictions. There is enough uncertainty in the parameters of the SNR that the upper limits are not in strong conflict with these predictions, but it is still strange that in these objects which show strong evidence for interactions with molecular clouds in no case is there an observable TeV gamma-ray flux. Evidence of an X-ray point source embedded in gamma-Cygni [3] and IC443 [10] and the observation of a pulsar, B1853+01, in W44 [16], all provide support to a pulsar origin for the EGRET flux.

REFERENCES

1. Allen, G.E., et al., *ApJ*, **448**, L25 (1995).
2. Borione, A., et al., in *Proc. 24th Int. Cosmic Ray Conf. (Rome)*, **2**, 439 (1995).
3. Brazier, K.T., et al., *MNRAS*, **281**, 1033 (1996).
4. Buckley, J.H., et al., submitted to *AA*, (1997).
5. Cawley, M.F., et al., *Exper. Astr.*, **1**, 173 (1990).
6. Drury, L.O'C., et al., *AA*, **287**, 959 (1994).
7. Esposito, J.A., et al., *ApJ*, **461**, 820 (1996).
8. Fierro, J.M., *PhD Thesis, Stanford University*, (1995).
9. Green, D.A., *A Catalog of Galactic Supernova Remnants (1995 July version)*, Cambridge, UK, MRAO, (1995).
10. Keohane, J.W., et al., accepted by *ApJ*, (1997).
11. Koyama, K., et al., *Nature*, **378**, 255 (1995).
12. Lessard, R.W., *Ph.D Thesis, National University of Ireland*, (1997).
13. Naito, T. and Takahara, F., *J.Phys. G:Nucl.Part.Phys.*, **20**, 477 (1994).
14. Prosch, C., et al., *AA*, **314**, 275 (1996).
15. Thompson, D.J., et al., *ApJS*, **101**, 259 (1995).
16. Wolszczan, A., et al., *ApJ*, **372**, L99 (1991).

Hard X-ray Emission from Cassiopeia A SNR

Lih-Sin The*, Mark D. Leising*, Dieter H. Hartmann*,
James D. Kurfess†, Philip Blanco§, and Dipen Bhattacharya‡

* Department of Physics and Astronomy, Clemson University, Clemson, SC 29634-1911
† E.O. Hulburt Center for Space Research, Naval Research Laboratory,
Code 7650, Washington, DC 20375-5352
§ Center for Astrophysics and Space Sciences, University of California,
San Diego, CA 92093
‡ Institute of Geophysics and Planetary Sciences, University of California,
Riverside, CA 92521

Abstract. We report the results of extracting the hard X-ray continuum spectrum of Cas A SNR from RXTE/PCA Target of Opportunity (TOO) observations and CGRO/OSSE observations. The data can rule out the single thermal bremsstrahlung model for Cas A continuum between 2 and 150 keV. The single power law model gives a mediocre fit (~5%) to the data with a power-law index, $\Gamma = 2.94 \pm 0.02$. A model with two component (bremsstrahlung + bremsstrahlung or bremsstrahlung + power law) gives a good fit. The power law index is quite constrained suggesting that this continuum might not be the X-ray thermal bremmstrahlung from accelerated MeV electrons at shock fronts [1] which would have $\Gamma \simeq 2.26$. With several SNRs detected by ASCA showing a hard power-law nonthermal X-ray continuum, we expect a similar situation for Cas A SNR which has $\Gamma = 2.98 \pm 0.09$. We discuss the implication of the hardest nonthermal X-rays detected from Cas A to the synchrotron radiation model.

INTRODUCTION

X-ray observations of supernova remnants have been stimulated by recent reports of nonthermal power-law X-ray detections suggesting supernovae as sites of charged particle accelerations and sources of cosmic rays. The first strong evidence for charged particle acceleration near supernova shock fronts in the X-ray energy band is demonstrated by the morphological and spectral correlation between X-ray and radio emission from the bright NE and SW rims of SN 1006 [2]. The brightest radio emission regions show almost featureless power law X-ray spectra when compared with SN 1006 central region which is dominated by emission lines of highly ionized elements in a non-equilibrium

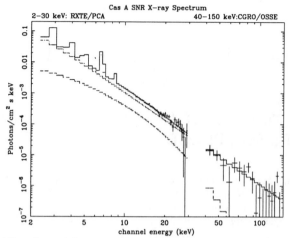

FIGURE 1. The thermal bremsstrahlung (kT=7.93 keV; dashed line) + power-law (Γ=2.98; dashed-dotted line) + 6 K X-ray line model produces a good fit (solid line) to the PCA & OSSE data (crosses) with a χ^2/dof=1.077, dof=62.

ionization thermal plasma. The nonthermal X-ray component has been modeled as a synchrotron emission from electrons accelerated to ~100 TeV within shock fronts [3–5]. In this model, the radio spectrum is produced by synchrotron radiation of electrons accelerated to ~GeV energies with the radio power-law index being less steep than the X-ray power-law index. Several similar evidences also have been demonstrated by ASCA measurement of RX J1713.7-3946 [6] and IC 443 [7], and RXTE and OSSE measurements of Cas A [8–10].

OSSE with a total accumulation time of 15×10^5 detector-seconds, detected a hard continuum between 40-150 keV from Cas A SNR at a 4σ confidence with a flux of $\sim 9 \times 10^{-7}$ γ cm^{-2} s^{-1} keV^{-1} at 100 keV [10]. The detection is the hardest X-ray detection from a SNR without plerionic source. However, the shape of the continuum has not been strongly constrained. The continuum can either be a bremsstrahlung with kT$\simeq$35 keV or a power law with $\Gamma \simeq 3.06$. In this paper, we use the 2-30 keV RXTE/PCA TOO and the 40-150 keV CGRO/OSSE Cas A data to better determine the shape of Cas A hard X-ray continuum.

RESULTS

The RXTE/PCA data used in this analysis is the TOO by RXTE on Aug. 2, 1996 with a total observing time of ~4000 sec. We fit the 2-30 keV PCA data simultaneously with the 40-150 keV OSSE data. The results of fitting several models with one or two continuum plus about six emission lines from

TABLE 1. Spectral fits to the 2-150 keV PCA & OSSE X-ray fluxes.

Continuum Model	1st-component T(keV) or Γ	1st-component Flux[a] at 1 keV	2nd-component T(keV) or Γ	2nd-component Flux[a] at 1 keV	χ^2/dof	dof	Prob.[b]
Thermal Bremss.	(8.14 ± 0.21)	$(1.24 \pm 0.04) \times 10^{-1}$			1.934	64	1×10^{-5}
Power-law	(2.94 ± 0.02)	(1.26 ± 0.07)			1.313	64	0.047
Cold + Hot Thermal Bremss	(4.27 ± 0.49)	$(2.03 \pm 0.30) \times 10^{-1}$	(22.8 ± 3.49)	$(2.85 \pm 0.61) \times 10^{-2}$	1.037	62	0.40
Thermal Bremss. + Power-law	(7.93 ± 1.36)	$(3.52 \pm 1.41) \times 10^{-2}$	(2.98 ± 0.09)	$(9.87 \pm 2.21) \times 10^{-1}$	1.078	62	0.31
Tenma's best fit[c]	3.76 fixed	0.483			46.20	74	0
EXOSAT's best fit[d]	0.65 fixed	1.23	3.74 fixed	4.58×10^{-1}	82.88	74	0

[a] Flux in units of γ cm^{-2} s^{-1} keV^{-1}.
[b] Confidence level that the model is consistent with the data.
[c] The single thermal bremsstrahlung + 7 K X-ray line model that best fits the 1.5-20 keV *Tenma* data.
[d] The two thermal bremsstrahlung + 6 K X-ray line model that best fits the 0.5-25 keV EXOSAT data.

Si, S, Ar, Ca, and Fe are shown in Table 1. The line widths are fixed to 100 eV and the hydrogen column density toward Cas A is fixed at 1×10^{22} H cm^{-2} [11,12]. We find that a single thermal bremsstrahlung model for the data can be ruled out. Nevertheless, a single power law model gives a mediocre fit. We also find the two models that best fit the EXOSAT [12] and *Tenma* [11] data cannot give acceptable fits to the PCA and OSSE data. A two component model, either the bremsstrahlung + bremsstrahlung or the bremsstrahlung + power law, each gives an equally good fit. In Figure 1 for the current interest in a nonthermal power law model, we show the result of the thermal bremsstrahlung + power-law model fit to the data. The power law continuum is detected with a higher confidence level and the power-law index is more tightly constrained than using the OSSE data alone [10]. The PCA and OSSE data is consistent only at a 3.8% level with the X-ray bremsstrahlung from accelerated MeV electrons at shock fronts as suggested by Asvarov et al. [1] which would have $\Gamma = \alpha + 1.5 = 2.26$ where α is the radio spectral index.

DISCUSSIONS

RXTE PCA + HEXTE in its AO-1 observing period detected a continuum between 10 and 60 keV [8,9]. The continuum is fitted well with a model of two Raymond-Smith thermal bremsstrahlung of kT = 0.7 keV and kT = 2.8 keV and a power law with $\Gamma \sim 2.4$ which exponentially steepens at e-fold 50 keV. The shape of the model is in good agreement with the result we find here.

The bremsstrahlung + bremsstrahlung model is expected in the reverse shock model. In this model, the hot temperature (T$\sim$23 keV) of the shocked circumstellar gas behind the blast-wave is consistent with the assumption that the blast-wave velocity is slightly larger than the observed median-filament velocity of 5500 km/s [13,14] and the circumstellar density n_H <1.5 cm^{-3}. The shocked gas temperature behind the reverse shock of T$\sim$4 keV suggests the swept up mass is $\sim$6 M$_\odot$ and $n_H \simeq 1.4$ cm^{-3} [14,15].

A summary of the detected nonthermal continuum from four SNRs is shown in Table 2. Cas A is of interest because from the 4 SNRs with measured nonthermal power-laws, it is the youngest, has the strongest magnetic field,

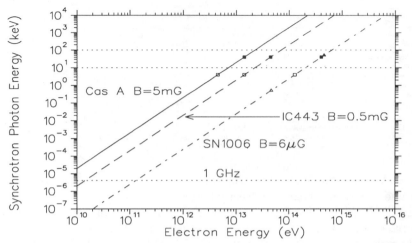

FIGURE 2. The synchrotron photon energy relation with the electron energy [2] for Cas A (solid line), IC 443 (dashed line), and SN 1006 (dashed-dotted line). The solid squares and the open squares mark the maximum electron energy constrained by the synchrotron energy loss (Eq.(1) of [4]) for $f^*R_J=1$ and $f^*R_J=10$, respectively. The solid triangles and the open triangles mark the maximum electron energy bounded by the age of the SNRs (Eq.(2) of [4]) for $f^*R_J=1$ and $f^*R_J=10$, respectively. This maximum energy is shown only for SN 1006 due to its low magnetic field. Synchrotron energy loss sets the maximum electron energy in Cas A and IC 443. The shock speed is assumed to be 5000 km/s in each case. f is the ratio of the electron mean free path to the electron gyroradius. R_J is the factor that contains the orientation of the shock relative to the magnetic field. The dotted lines shows some interesting synchrotron photon energies at 1 GHz, 10 keV, and 100 keV.

its detected nonthermal energy is the highest, and it has the steepest X-ray and radio power-law indexes, of four SNRs with measured nonthermal power law. In Figure 2, we show for Cas A, IC 443, and SN 1006 how close their detected hard X-ray energies to the maximum synchrotron energies. The synchrotron energy is related to the maximum electron energy which is limited by the synchrotron energy loss and the SNR's age. Cas A, as the brightest radio SNR source, has the most intense magnetic field and therefore it causes the large synchrotron energy loss which limits Cas A's maximum electron energy. Figure 2 shows that the hard X-ray emission detected from Cas A by OSSE is near the cutoff synchrotron energy or indirectly to the maximum electron energy. Further Cas A measurements by OSSE may detect the cutoff energy and hence would provide the evidence that the nonthermal power law emission is the synchrotron radiation from accelerated electrons at the shock fronts near the maximum energy. Another evidence of the synchrotron radiation being the process for the nonthermal emission is the steepening of spectral index from radio to X-ray energy. It has been suggested that the turnover or break

TABLE 2. Comparison of the measured nonthermal continuum

SNR	Age (yrs)	Best Estimate of B	α	Measured Power Law Energy (keV)	Γ
SN1006	971	3-10 μG	0.56	0.5 - 20	1.95
RX J1713.7-3946	≥1000	?	?	0.5 - 10	2.45
IC 443	1000	500 μG	<0.24	0.5 - 12	2.35
Cas A	317	1-5 mG	0.76	15-150	2.98

∼0.25 keV is observed in SN 1006 spectra [5]. We expect the turnover in Cas A spectra is ∼1-10 keV due to Cas A's strong magnetic field. However, in order to observe this turnover in Cas A spectra, an X-ray imaging detector is needed because Cas A's thermal bremsstrahlung contributes substantially near this energy.

REFERENCES

1. Asvarov, A.I., Guseinov, O.Kh., Dogel, V.A., Kasumov, F.K., *Soviet Astron.*, **33**, 532 (1989)
2. Koyama, K., Petre, R., Gotthelf, E.V., Hwang, U., Matsuura, M., Ozaki, M., and Holt, S.S., *Nature*, **378** (1995)
3. Ammosov, A.E., Ksenofontov, L.T., Nikolaev, V.S., and Petukhov, S. I., *Astron. Lett.* **20**, 157 (1994)
4. Reynolds, S.P., *ApJ*, **459**, L13 (1996)
5. Mastichiadis, A., *A & A*, **305** L53 (1996)
6. Koyama, K., Kinugasa, K., Matsuzaki, K., Nishiuchi, M., Sugizaki, M., Torii, K., Yahauchi, S., and Aschenbach, B. *PASJ*, **49**, in press (1997)
7. Keohane, J.W., Petre, R., Gotthelf, E.V., Ozaki, M., and Koyama, K., *ApJ*, in press (1997)
8. Rothschild, R., et al., *this proceeding* (1997)
9. Allen, G.E., et al., submitted to *ApJL* (1997)
10. The, L.-S., Leising, M.D., Kurfess, J.D., Johnson, W.N., Hartmann, D.H., Gehrels, N., Grove, J.E., and Purcell, W.R., *Astron. Astrophys. Suppl.* **120**, 357 (1996)
11. Tsunemi, H., Yamashita, K., Masai, K., Hayakawa, S., and Koyama, K., *ApJ*, **306**, 248 (1986)
12. Jansen, F., Smith, A., Bleeker, J.A.M., De Korte, P.A.J., Peacock, A., and White, M.E., *ApJ*, **331**, 949 (1988)
13. Woltjer, L., *Ann. Rev. Astr. Ap.*, **10**, 129 (1972)
14. McKee, C., *ApJ*, **188**, 335 (1974)
15. Borkowski, K.J., Szymkowiak, A.E., Blondin, J.M., and Sarazin, C.L., *ApJ*, **466**, 866 (1996)

Nonthermal SNR Emission

Steven J. Sturner[1,2], Jeffrey G. Skibo[3], Charles D. Dermer[3], & John R. Mattox[4]

[1] *Universities Space Research Association, Greenbelt, MD 20706*
[2] *NASA Goddard Space Flight Center, Code 661, Greenbelt, MD 20771*
[3] *Naval Research Laboratory, Code 7653, Washington, DC 20375*
[4] *Boston University, Boston, MA 02215*

Abstract. We model the temporally evolving nonthermal particle and photon spectra at different stages in the lifetime of a standard shell-type supernova remnant. A characteristic νF_ν spectrum of a SNR consists of a peak at radio-through-optical energies from nonthermal electron synchrotron emission and another high-energy gamma-ray peak due primarily to nonthermal electron bremsstrahlung, Compton scattering, and secondary pion production. We find that SNRs are capable of producing maximum gamma-ray luminosities $\gtrsim 10^{34}$ erg s^{-1} if the density of the local ISM is $\gtrsim 1$ cm^{-3}. This emission will persist for $\gtrsim 10^5$ years after the supernova explosion. This long gamma-ray lifetime suggests that SNRs with a wide range of ages could be gamma-ray sources and could constitute some of the unidentified EGRET sources.

INTRODUCTION

Although it has been decades [1–3] since it was proposed that Galactic cosmic rays are accelerated by supernova remnants (SNRs), the evidence that would prove this scenario has remained elusive until recently. Observations by the *ASCA* satellite have provided evidence for the acceleration of electrons by SNRs. *ASCA* observations [4] of the SNR SN1006 show a nonthermal, hard x-ray tail extending to energies above the thermal bremsstrahlung emission. It has been suggested that this hard tail is synchrotron emission from nonthermal electrons with energies near 100 TeV [5,6].

The EGRET instrument on the *Compton Gamma Ray Observatory* (*CGRO*) may provide a database from which to test for the acceleration of both cosmic ray protons and electrons by SNRs [7–9]. It has been argued that some of the 96 unidentified EGRET sources are SNRs [10–12]. There exist some compelling positional associations of EGRET sources with radio-bright, young, nearby, shell-type SNRs such as IC 443, W28, and γ Cygni. It has been suggested [13,14] that SNRs are favored to be detected if they are in the vicinity of dense interstellar clouds as these are.

Here we outline a model for nonthermal particle and photon production in temporally evolving SNR shocks, yielding multiwavelength photon spectra at different stages in the SNR lifetime. We urge the reader to see [15] for a more complete description of the model and our results.

THE MODEL

The model we have developed for the production of photons by shock accelerated electrons and protons can be broken into three parts: the production of energetic particles, the temporal evolution of the particle energy distributions, and the production of photons. We assume that the nonthermal primary electron and proton sources are produced through first-order Fermi acceleration in SNR shocks. Shock acceleration theory predicts a particle spectrum that is a power-law in momentum with spectral index ≈ 2 for strong SNR shocks which should have a compression ratio $r \approx 4$. The particle spectra are cut-off at high energies by three mechanisms (see, e.g., [5,6,16]): the finite age of the SNR, energy-loss processes such as electron synchrotron radiation, and free escape from the shock region when the particles are no longer effectively scattered by the MHD turbulence near the shock.

The energy gain rate for particles in a shock is [5,17]

$$\dot{E}_{\text{shock}}(t) = 100 \; \frac{B_{\text{ISM}} \; v_8^2(t)}{f R_{\text{J}}} \quad \text{MeV s}^{-1}, \qquad (1)$$

where B_{ISM} is the magnetic field strength of the local ISM into which the SNR is expanding, $v_8(t)$ is the shock velocity in units of 10^8 cm s^{-1}, $f \sim 10$ is the particle mean free path along the magnetic field in units of its gyro-radius, and $R_{\text{J}} \sim 1$ is a factor that accounts for the orientation of the shock relative to the magnetic field. If energy losses and free escape from the shock region can be neglected, the maximum energy of the accelerated particles can be found by simply integrating eqn. (1) over the age of the remnant. At high energies, synchrotron losses will restrict the maximum energy of electrons to the energy where the synchrotron energy loss rate equals the shock acceleration rate. The third possible maximum kinetic energy is the free escape energy, which can be calculated by setting the maximum wavelength of MHD turbulence, λ_{max}, equal to f times the particle gyro-radius. Reynolds [5] finds $\lambda_{\text{max}} \approx 10^{17}$ cm when fitting the x-ray data of Koyama et al. [4] of SN1006. In our model the cut-off energy for electrons is taken to be the minimum of these three maximum energies while proton energies are limited only by the age of the remnant and free escape from the shock region.

We assume that the particle injection rate is a constant per unit shock surface area until the SNR enters its radiative phase at which time the particle source is turned off. We assume that total kinetic energy injected in both protons and electrons over the lifetime of the remnant equals 10% of the initial bulk kinetic energy of the ejecta.

We then calculate the instantaneous differential particle densities, $n(E,t)$, throughout the lifetime of the remnant. This is done by numerically solving the Fokker-Planck equation in energy space, given by

$$\frac{\partial n(E,t)}{\partial t} = -\frac{\partial}{\partial E}\left[\dot{E}_{\text{tot}}\, n(E,t)\right] + \frac{1}{2}\frac{\partial}{\partial E^2}\left[D\, n(E,t)\right] + Q(E,t) - \frac{n(E,t)}{\tau_{\text{esc}}}, \quad (2)$$

where the terms on the right side represent, from left to right, systematic energy losses, Coulomb diffusion in energy space, the primary particle source function Q, and catastrophic energy losses due to pion-producing events.

The energy loss mechanisms for electrons which we consider are adiabatic losses, Coulomb collisions, synchrotron radiation, bremsstrahlung, and Compton scattering of the cosmic microwave background (CMB) as well as the diffuse Galactic IR and optical emission. For protons we consider adiabatic losses, Coulomb collisions, and inelestic pion-producing collisions.

The photon production mechanisms we consider are synchrotron radiation, bremsstrahlung, and Compton scattering for electrons. For protons, we only consider photons produced by the decay of neutral pions. We do not consider the radiation emitted by the secondary electrons and positrons which result from the decay of charged pions produced by proton-nucleon interactions. This emission will generally be negligible compared to that of the primary electrons [6].

RESULTS

In Figure 1 we show an example of our model results for the case of a Type 1a supernova expanding into an interstellar medium with density 1 cm^{-3} and a magnetic field strength of 5 μG. We note several characteristice of this emission. First, the spectral power consists of a peak at radio-through-optical energies from electron synchrotron emission and another at gamma-ray energies due primarily to electron bremsstrahlung and Compton emission. The dominance of the electron emission processes over neutral pion decay is a result of our assumption that equal amounts of kinetic energy are given to the electrons and the protons. From our results we find that at least 6 times as much kinetic energy must be injected in protons as electrons in order for the pion decay gamma-rays to dominate the electron bremsstrahlung emission given power-law indices ≈2.

There are several temporal features that should be noted. The emission peaks when the SNR enters the radiative phase (∼29,000 years in this case) at which time the particle sources are turned off. Once the particle sources are turned off, the x-ray synchrotron and gamma-ray Compton emission both

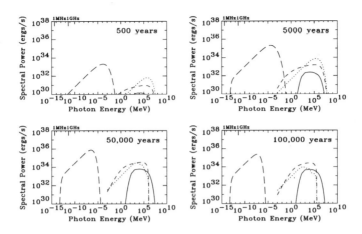

FIGURE 1. Temporal evolution of the nonthermal emission from a Type I SN expanding into an ISM of density 1 cm^{-3} and a magnetic field strength of 5 μG. Here long-dashed, short-dashed, dotted, and solid lines represent the synchrotron, bremsstrahlung, inverse Compton, and pion-decay emission, respectively

diminish rapidly as the number of electrons with energies > 1 TeV diminishes very quickly. This also results in the change in the spectral shape of the Compton emission. Initially the Compton emission is dominated by Comptonization of the CMB. As the electron spectrum decays with time, the energy of the most energetic electrons decreases so that Comptonization of the higher energy photon fields starts to dominate the Compton emission at the highest energies. Also note that as the x-ray synchrotron and Compton emission diminishes rapidly, the radio synchrotron, bremsstrahlung, and pion decay emission remain nearly constant beyond 100,000 years. This is due to the relatively long radiative lifetimes of the protons and the GeV electrons that produce this radiation.

In Figure 2 we compare the results of our model with the nonthermal spectrum from the SNR IC 443 [15,18–21] where we have assumed that the unidentified EGRET source 2EG J0618+2234 is associated with the SNR. Recent *ASCA* results have shown that much of the nonthermal x-ray emission from IC 443 is being emitted from a small region of the remnant which has an extremely flat radio spectral index, ~0.25 [18]. The model spectrum we show has two components, one representing the majority of the remnant where we have assumed a spectral index of 2.0 for the source particles (solid), and a second component with a particle spectral index of 1.5 (as indicated by the flat radio spectrum) representing the region that is emitting much of the observed nonthermal x-rays (dashed). The flat spectrum component has a much lower total injected kinetic energy than the steep spectrum source [0.12%

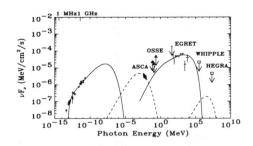

FIGURE 2. Comparison of our two component model with the multiwavelength nonthermal emission from IC 443/2EG J0618+2234.

(electrons+protons) as opposed to 8.6%] and the free escape energy is much higher ($\lambda_{max} = 5 \times 10^{18}$ cm compared to 10^{15} cm). This model provides a reasonable fit to the nonthermal emission from this source.

REFERENCES

1. Ginzburg, V. L., & Syrovatskii, S. I. 1964, The Origin of Cosmic Rays (New York: Macmillan)
2. Hayakawa, S. 1969, Cosmic Ray Physics (New York: Wiley-Interscience)
3. Pinkau, K. 1970, Phys. Rev. Let., 25, 603
4. Koyama, K., et al. 1995, Nature, 378, 255
5. Reynolds, S. P. 1996, ApJ, 459, L13
6. Mastichiadis, A. 1996, A&A, 305, L53
7. Fichtel, C. E. et al. 1994, ApJS, 94, 551
8. Thompson, D. J., et al. 1995, ApJS, 101, 259
9. Thompson, D. J., et al. 1996, ApJS, 107, 227
10. Sturner, S. J., & Dermer, C. D. 1994, A&A, 293, L17
11. Sturner, S. J., Dermer, C. D., & Mattox, J. R. 1996, A&AS, 120, 445
12. Esposito, J. A., Hunter, S. D., Kanbach, G., & Sreekumar, P. 1996, ApJ, 461, 820
13. Aharonian, F. A., Drury, L. O'C., & Völk, H. J. 1994, A&A, 285, 645
14. Grenier, I. A. 1995, Adv. Sp. Res., 15, 73
15. Sturner, S.J., Skibo, J. G., Dermer, C. D., & Mattox, J. R. 1997, ApJ, in press
16. Gaisser, T. K. 1990, Cosmic Rays and Particle Physics (Cambridge: Cambridge University Press)
17. Reynolds, S. P. 1995, 24th ICRC, 2, 17
18. Keohane, J. K., et. al. 1997, ApJ, in press
19. Erickson, W. C., & Mahoney, M. J. 1985, ApJ, 290, 596
20. Lessard, R. W., et al. 1995, 24th ICRC, 2, 475
21. Prosch, C., et al. 1995, 24th ICRC, 2, 405

Gamma-Rays from Supernova Remnants: Signatures of Non-Linear Shock Acceleration

Matthew G. Baring,*[1] Donald C. Ellison,[†]
Stephen J. Reynolds[†] and Isabelle A. Grenier[‡]

*LHEA, NASA Goddard Space Flight Center, Greenbelt, MD 20771, USA
[†]Dept. of Physics, North Carolina State University, Raleigh, NC 27695, USA
[‡]EUROPA-Université Paris, Paris VII and
DAPNIA/Service d'Astrophysique, CE-Saclay, Gif/Yvette, France

Abstract. Results from a study of non-linear effects in shock acceleration theory and their impact on the gamma-ray spectra of supernova remnants (SNRs) are presented. These non-linear effects describe the dynamical influence of the accelerated cosmic rays on the shocked plasma at the same time as addressing how the non-uniformities in the fluid flow force the distribution of the cosmic rays to deviate from pure power-laws. Such deviations are crucial to gamma-ray spectral determination. Our self-consistent Monte Carlo approach to shock acceleration, which is ideally suited to addressing SNR spectral issues, is used to predict ion and electron distributions that spawn neutral pion decay, bremsstrahlung and inverse Compton emission components for SNRs. The cessation of acceleration above critical energies in the 1 TeV - 10 TeV range caused by the spatial and temporal limitations of the expanding SNR shell yields gamma-ray spectral cutoffs: the resulting emission spectra are quite consistent with Whipple's TeV upper limits to those EGRET unidentified sources that have SNR associations.

INTRODUCTION

Supernova remnants have long been invoked as a principal source of galactic cosmic rays, created via the process of diffusive Fermi acceleration at their expanding shock fronts. They can also provide gamma-ray emission via the interaction of the cosmic ray population with the remnant environment; this concept was explored recently by Drury, Aharonian and Völk (1994). In their model, the gamma-ray luminosity is spawned by collisions between the cosmic rays and nuclei from the ambient SNR environment. In the more recent

---

[1]) Compton Fellow, Universities Space Research Association

models of Mastichiadis and De Jager (1996), Gaisser, Protheroe and Stanev (1997), and Sturner, et al. (1997), additional components are formed by ee and ep bremsstrahlung, and inverse Compton scattering involving shock-accelerated electrons interacting with the cosmic microwave background and also IR/optical emission (from dust/starlight).

With no definitive detections of gamma-rays from known supernova remnants, the motivation for modelling these "hypothetical" sources hinges on a handful of spatial associations of unidentified EGRET sources (Esposito et al. 1996) at moderately low galactic latitudes with well-studied radio and X-ray SNRs. These include IC 443, γ Cygni and W44, and most have neighbouring dense environments that seem necessary in order to provide sufficient gamma-ray luminosity to exceed EGRET's sensitivity threshold. The remnants associated with several of the EGRET unidentified sources show an apparent absence of TeV emission, as determined by Whipple (Lessard et al. 1995), which could be explained an intrinsic cutoff in the cosmic ray distribution generated by remnants' expanding shocks.

All of the above models invoke simple power-law accelerated particle populations. In this paper, we utilize the more sophisticated output of shock acceleration simulations (e.g. Jones and Ellison 1991), and address the issues of spectral curvature and the maximum energy of acceleration in the context of SNR gamma-ray emission. We use output from the fully non-linear, steady-state Monte Carlo simulations of Ellison, Baring and Jones (1996) to describe the accelerated particles in environments where they influence the dynamics of the SNR shell. Our results make clear predictions of what maximum energies of gamma-rays are expected and more accurate predictions of the level of TeV emission in these sources. Our model provides a prescription for defining realistic values of the non-thermal electron/proton abundance ratio.

SHOCK ACCELERATION AND γ-RAY SPECTRA

Shock acceleration is usually assumed in astrophysical models to generate power-law particle populations. This approximation omits the effect the accelerated particles have on the shock hydrodynamics. Such non-linear effects, well-documented in the reviews of Drury (1983) and Jones and Ellison (1991), have a feedback on the acceleration mechanism and its efficiency. The slope of the cosmic ray distributions depends purely on the compression ratio r of flow speeds either side of the shock, and this ratio ultimately depends on the shape of the ion distributions. Our kinematic Monte Carlo technique (see Ellison, Baring and Jones 1996) for simulation of Fermi acceleration in the non-linear regime is ideal for fully exploring particle populations, acceleration efficiencies and spectral properties. The simulation follows particle convection and diffusion in the shock environs using a simple scattering law: a particle's mean free path λ is proportional to its gyroradius r_g.

The maximum energy of Fermi-accelerated ions is of central importance to TeV observations. It can be determined in the Sedov phase (appropriate to sources like IC 443, γ Cygni and W44) by equating the acceleration time to the remnant age, producing values of the order of (Baring et al. 1997):

$$E_{\max} \sim 30.7 \frac{r-1}{r} \frac{Q}{\eta} \left(\frac{B_1}{3\mu G}\right) \left(\frac{n_1}{1\,\mathrm{cm}^{-3}}\right)^{-1/3} \mathcal{E}_{51}^{1/2} \text{ TeV} \quad (1)$$

where r is the velocity compression ratio, n_1 is the ISM density, B_1 is the field in units of Gauss, Qe is the particle's charge, $\mathcal{E}_{51}$ is the energy released in the supernova (in units of 10^{51} erg/sec) and $\eta = \lambda/r_g$. In the Sedov phase, where the peak luminosity is expected (see Drury, Aharonian and Völk 1994), $E_{\max}$ is actually a slowly increasing function of time. For young SNRs in the early Sedov phase (200–5000 yrs), typically age-limited acceleration terminates at around $1 - 10$ TeV. The diffusion scale of the SNR medium is κ/u for a diffusion coefficient of $\kappa = \lambda v/3$, which, noting Eq. (1), is generally, but not always, considerably smaller than the shock radius.

Typical simulation output is shown in Fig. 1, where the resulting ion and e^- distributions are exhibited. The pressure of the accelerated ions (p and He^{2+}) acts to slow down the fast-moving flow upstream of the shock, creating a

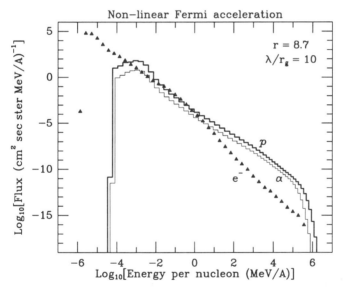

FIGURE 1. Proton, He^{2+} and e^- distributions resulting from the Monte Carlo shock acceleration simulation of acceleration at SNR shocks. The transition to relativistic energies produces a "bump" at around 1 GeV. The ion spectra are the two histograms, providing nearly all of the total energy density of the system. The triangles represent the e^- distribution (see next section for a discussion). The shock speed was 1950 km/sec and upstream parameters were a density of $n_1 = 1\,\mathrm{cm}^{-3}$ and field strength $B_1 = 3\mu$ G.

maximum compression ratio r that is much greater than the canonical "strong shock" value of 4. Such large r are realized at the largest scales, implying an increased "efficiency" of accelerating higher energy particles than those at lower energies. Hence upward curvature appears in the proton, α and electron distributions due to the non-linear effects: high energy particles have longer mean-free paths and therefore typically influence the flow on larger scale-lengths. The escape of particles gives a cessation of acceleration, in this case at TeV energies, forcing the compression ratio to increase to compensate in the flow hydrodynamics. These non-linear predictions of the spectrum differ (e.g. see Ellison, Baring and Jones 1996) significantly from the test-particle case, with deviations from power-law behaviour by as much as a factor of around 3 over four decades in particle energy: this is a significant influence on predictions of TeV fluxes in gamma-ray SNRs.

In applying the full non-linear Monte Carlo simulation to the prediction of SNR γ-ray emission, we model the π^0 decay component, bremsstrahlung and inverse Compton scattering of background radiation fields. Cosmic ray ions collide with nuclei in the cold ambient ISM to produce π^0 s, which subsequently decay to create two photons; the decay spectra were calculated much along the lines of the work of Dermer (1986). Bremsstrahlung and inverse Compton contributions were calculated along the standard lines used in Gaisser, Protheroe and Stanev (1997). Fig. 2 shows the emergent γ-ray spectra resulting from particle distributions generated by the Monte Carlo simulation. The non-linear modifications can alter the TeV/EGRET flux ratio by as much as a factor of 2-3 (Baring, Ellison and Grenier 1997). In this case, ions are accelerated out to around a TeV per nucleon, providing compatibility with the Whipple upper limits. Turnovers at such low energies are not predicted in the work of Gaisser, Protheroe and Stanev (1997), but are invoked by Mastichiadis and De Jager (1996), and Sturner et al. (1997). One implication of such low maximum energies is that γ-ray SNRs must be a γ-ray bright minority of the remnant population, with other SNRs being required to produce the bulk of cosmic rays out to the $10^{14} eV$ "knee."

The two levels of inverse Compton and bremsstrahlung emission depicted in Fig. 2 correspond to two alternative electron populations. These represent our modelling of two different physical situations, where electrons start efficiently resonating with Alfvén-whistler modes above 1 MeV (weaker continuum) and 1 GeV (stronger one). These choices span a range of possibilities that describe how electrons are injected into the Fermi process. Electrons "injected" at lower energies (i.e. 1 MeV) have shorter mean free paths at < 1 MeV than those "injected" at 1 GeV, and consequently steeper spectra. The 1 GeV case represents, more or less, an extreme upper bound (discussed in Baring et al. 1997) to the non-thermal electron population, and lower values like 1 MeV or below are favoured. Hence, e/p ratios considerably less than unity are expected, consistent with cosmic ray abundances (e.g. Müller et al. 1995) and modelling of the diffuse γ-ray background (Hunter et al. 1997). The shape of

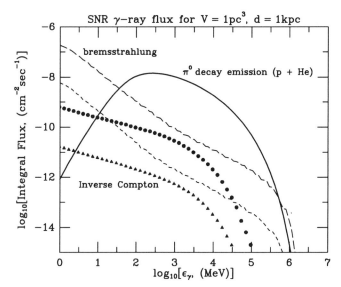

FIGURE 2. The γ-ray spectral components, integrated over shock acceleration-produced ion and electron distributions much like those in Fig. 2. Electrons were "injected" at two different energies (1 MeV and 1 GeV) producing two alternative distributions for each of the inverse Compton (points) and bremsstrahlung (broken lines) processes; higher fluxes corresponding to 1 GeV injection and the higher e/p ratio. Given our shock parameters ($u = 10^4$ km/sec, $n_1 = 1$ cm^{-3} and field strength $B_1 = 3\mu$ G), γ-ray spectra appropriate to remnants like IC443 and γ Cygni are an order of magnitude larger.

the γ-ray spectrum is strongly dependent on the e/p ratio above 1 GeV.

In conclusion, the non-linear effects of shock acceleration addressed here produce spectra with cutoffs that can comfortably accommodate Whipple's TeV upper limits to unidentified EGRET sources.

REFERENCES

1. Baring, M. G., Ellison, D. & Grenier, I. 1997, 2nd Integral Workshop, p. 81.
2. Baring, M. G., et al. 1997, *Astrophys. J.* in preparation.
3. Dermer, C. D. 1986, *Astron. Astr.* **157**, 223.
4. Drury, L. O'C., Aharonian, F., & Völk, H. 1994, *Astron. Astr.* **287**, 959.
5. Ellison, D. C, Baring, M. G. and Jones, F. 1996, *Astrophys. J.* **473**, 1029.
6. Esposito, J. A., et al. 1996, *Astrophys. J.* **461**, 820.
7. Gaisser, T. K., Protheroe, R. & Stanev, T. 1997, *Astrophys. J.* submitted.
8. Hunter, S. D., et al. 1997, *Astrophys. J.* **481**, 205.
9. Jones, F. C. & Ellison, D. C. 1991, *Space Sci. Rev.* **58**, 259.
10. Lessard, R. W., et al. 1995, *Proc. 24th ICRC (Rome)*, Vol 2, p. 475.
11. Müller, D., et al. 1995, *Proc. 24th ICRC (Rome)*, Vol 3, p. 13.
12. Sturner, S., et al. 1997, *Astrophys. J.* in press.

Modelling cosmic rays and gamma rays in the Galaxy

Andrew W. Strong* and Igor V. Moskalenko*†

*Max-Planck-Institut für extraterrestrische Physik, D-85740 Garching, Germany
†Institute for Nuclear Physics, Moscow State University, 119 899 Moscow, Russia

Abstract. An extensive program for the calculation of galactic cosmic-ray propagation has been developed. This is a continuation of the work described in [1]. The main motivation for developing this code [2] is the prediction of diffuse Galactic gamma rays for comparison with data from the CGRO instruments EGRET, COMPTEL, and OSSE. The basic spatial propagation mechanisms are (momentum-dependent) diffusion, convection, while in momentum space energy loss and diffusive reacceleration are treated. Primary and secondary nucleons, primary and secondary electrons, and secondary positrons are included. Fragmentation and energy losses are computed using realistic distributions for the interstellar gas and radiation fields.

INTRODUCTION

We are developing a model which aims to reproduce self-consistently observational data of many kinds related to cosmic-ray origin and propagation: direct measurements of nuclei, electrons and positrons, gamma rays, and synchrotron radiation. These data provide many independent constraints on any model and our approach is able to take advantage of this since it must be consistent with all types of observation. We emphasize also the use of realistic astrophysical input (e.g., for the gas distribution) as well as theoretical developments (e.g., reacceleration). The code is sufficiently general that new physical effects can be introduced as required. The basic procedure is first to obtain a set of propagation parameters which reproduce the cosmic ray B/C ratio, and the spectrum of secondary positrons; the same propagation conditions are then applied to primary electrons. Gamma-ray and synchrotron emission are then evaluated. Models both with and without reacceleration are considered. The models are three dimensional with cylindrical symmetry in the Galaxy, the basic coordinates being (R, z, p) where R is Galactocentric radius, z is the distance from the Galactic plane and p is the total particle momentum. The numerical solution of the transport equation is based on a

Crank-Nicholson implicit second-order scheme. In the models the propagation region is bounded by $z = z_h$ beyond which free escape is assumed. A value $z_h = 3$ kpc has been adopted since this is within the range which is consistent with studies of ^{10}Be/Be and synchrotron radiation. For a given z_h the diffusion coefficient as a function of momentum is determined by B/C for the case of no reacceleration; with reacceleration on the other hand it is the reacceleration strength (related to the Alfvén speed v_A) which is determined by B/C. Reacceleration provides a natural mechanism to reproduce the B/C ratio without an ad-hoc form for the diffusion coefficient [3,4]. The spatial diffusion coefficient for the case *without* reacceleration is $D = \beta D_0$ below rigidity ρ_0, $\beta D_0(\rho/\rho_0)^\delta$ above rigidity ρ_0, where β is the particle speed. The spatial diffusion coefficient *with* reacceleration assumes a Kolmogorov spectrum of weak MHD turbulence so $D = \beta D_0(\rho/\rho_0)^\delta$ with $\delta = 1/3$ for all rigidities. For this case the momentum-space diffusion coefficient is related to the spatial one [4]. The injection spectrum of nucleons is assumed to be a power law in momentum. The interstellar hydrogen distribution uses HI and CO surveys and information on the ionized component; the Helium fraction of the gas is taken as 0.11 by number. The interstellar radiation field for inverse Compton losses is based on stellar population models and IRAS and COBE data, plus the cosmic microwave background. Energy losses for electrons by ionization, Coulomb, bremsstrahlung, inverse Compton and synchrotron are included, and for nucleons by ionization and Coulomb interactions. The distribution of cosmic-ray sources is chosen to reproduce the cosmic-ray distribution determined by analysis of EGRET gamma-ray data [5]. The bremsstrahlung and inverse Compton gamma rays are computed self-consistently from the gas and radiation fields used for the propagation. The π^0-decay gamma rays are calculated explicitly from the proton and Helium spectra using [6]. The secondary nucleon and secondary $e^\pm$ source functions are computed from the propagated primary distribution and the gas distribution, and the anisotropic distributions of $e^\pm$ in the $\mu^\pm$ system was taken into account [7].

ILLUSTRATIVE RESULTS

Some results obtained are shown in the Figures. The energy dependence of the B/C ratio, and local *proton and Helium spectra* are shown in Figs. 1 and 2. The spectrum of *primary electrons* is shown in Fig. 6. The adopted electron injection spectrum has a power law index –2.1 up to 10 GeV; this is chosen using the constraints from synchrotron and from gamma rays. The electron spectrum is consistent with the direct measurements around 10 GeV where solar modulation is small and it also satisfies the constraints from $\frac{e^+}{e^-+e^+}$. Above 10 GeV a break is required in the injection spectrum to at least –2.4 for agreement with direct measurements (which may however not be necessary if local sources dominate the directly measured high-energy electron spectrum).

The synchrotron spectrum at high Galactic latitudes (Fig. 3) is important since its shape constrains the shape of the 1–10 GeV electron spectrum. An injection index –2.1 (without reacceleration) is the steepest which is allowed by the radio data over the range 38 to 1420 MHz. As illustrated, an index –2.4 as often used (e.g., [8]) gives a synchrotron spectrum which is too steep.

The modelled *gamma-ray* spectrum for the inner Galaxy, shown here for the case of no reacceleration (Fig. 4), fits well the COMPTEL [9] and EGRET [5] data between 1 MeV and 1 GeV beyond which there is the well-known excess not accounted for by π^0-decay with the standard nucleon spectrum [10,11]. The electron spectrum used here is increased by a factor 2 over that shown in Fig. 6; this is required to reproduce well the observed gamma intensities. It is still within the range allowed by the positron fraction (see below), and the shape accords well with the flatter electron injection spectrum required by synchrotron data (Fig. 3). Inverse Compton is dominant below 10 MeV, bremsstrahlung becomes important for 3–200 MeV. The lower bremsstrahlung combined with π^0-decay leads to a good fit to the flat spectrum observed in this range, in contrast to previous attempts to model the spectrum with a steeper bremsstrahlung spectrum [8]. The 10–30 MeV γ-ray longitude profile (Fig. 5) at low latitudes from this model can be compared with that from COMPTEL [9]. The e^+ *spectrum* (Fig. 6) and the *positron fraction* (Fig. 7) agree well with the most recent data compilation for 0.1–10 GeV [7].

This study indicates that it is possible to construct a model satisfying a wide range of observational constraints and provides a basis for future developments.

More details can be found on *http://www.gamma.mpe-garching.mpg.de/~aws/aws.html*

REFERENCES

1. Strong A.W., Youssefi G., *Proc. 24th ICRC* **3**, 48 (1995)
2. Strong A.W., Moskalenko I.V., Schönfelder V., *Proc. 25th ICRC*, OG 8.1.1 (1997)
3. Heinbach U., Simon M., *ApJ* **441**, 209 (1995)
4. Seo E.S., Ptuskin V.S., *ApJ* **431**, 705 (1994)
5. Strong A.W., Mattox J.R., *A&A* **308**, L21 (1996)
6. Dermer C.D., *A&A* **157**, 223 (1986)
7. Moskalenko I.V., Strong A.W., *ApJ*, submitted (1997)
8. Strong A.W. et al., *A&AS* **120**, C381 (1996)
9. Strong A.W., et al., in *AIP Proc.*, 1997, this Symposium
10. Hunter S.D., et al., *ApJ* **481**, 205 (1997)
11. Mori M., *ApJ* **478**, 225 (1997)
12. Webber W.R., et al., *ApJ* **457**, 435 (1996)
13. Seo E.S., et al., *ApJ* **378**, 763 (1991)
14. Engelmann J.J., et al., *A&A* **233**, 96 (1990)
15. Protheroe R.J., *ApJ* **254**, 391 (1982)
16. Barwick S.W., et al., *ApJL* **482**, L191 (1997)

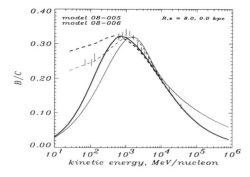

FIGURE 1. The energy dependence of the B/C ratio can be reproduced with $D_0 = 2.0 \times 10^{28}$ cm^{2}s^{-1}, $\delta = 0.6$, $\rho_0 = 3$ GV/c without reacceleration (thick line) and $D_0 = 4.2 \times 10^{28}$ cm^{2}s^{-1}, $v_A = 20$ km s^{-1} with reacceleration (thin line). Dashed lines are modulated to 500 MV. Data from [12].

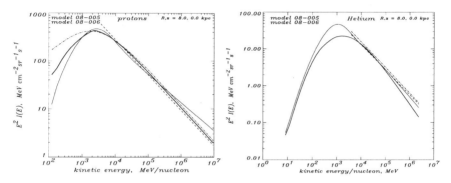

FIGURE 2. *Left panel:* the local proton spectrum for injection index 2.15 (thin solid line), 2.25 (thick solid line) without and with reacceleration respectively, compared with the measured 'interstellar' spectrum (dashed [13] and dashed-dot [11] lines). *Right panel:* the Helium spectrum with injection index 2.25 (thin solid line), 2.45 (thick solid line) without and with reacceleration respectively, compared with the measured 'interstellar' spectrum (dashed [13] and dashed-dot [14] lines).

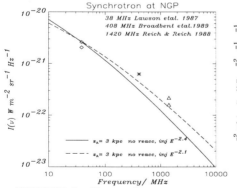

FIGURE 3. The synchrotron spectrum at the NGP, and predictions for electron injection indices –2.1 (dashed line) and –2.4 (solid line).

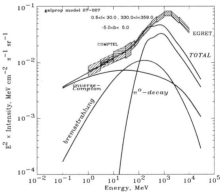

FIGURE 4. The γ-ray spectrum for the inner Galaxy, $330° < l < 30°, |b| < 5°$. EGRET data [5], COMPTEL data [9].

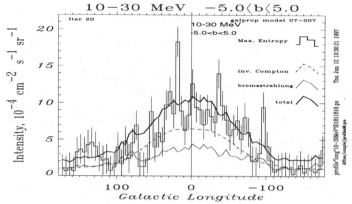

FIGURE 5. Longitude distribution of gamma rays in energy range 10-30 MeV. Histogram: COMPTEL [9], dashed: inverse Compton, thin line: bremsstrahlung, thick line: total.

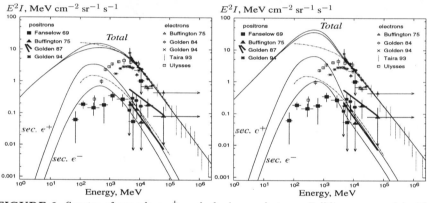

FIGURE 6. Spectra of secondary $e^{\pm}$, and of primary electrons. Full lines: our model with no reacceleration (left) and with reacceleration (right). Dash-dotted lines by Protheroe [15]: lower is his leaky-box prediction for e^+, upper is his adopted electron spectrum.

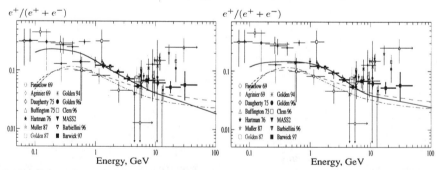

FIGURE 7. Positron fraction for model with no reacceleration (left panel) and with reacceleration (right panel). Dashed and dash-dotted lines by Protheroe [15]: his predictions of the leaky-box and diffusive halo models respectively. The data collection is taken from [16].

Production of Beryllium and Boron by Spallation in Supernova Ejecta

Deepa Majmudar[1], James H. Applegate[2]

1:Dept. of Physics, Columbia University, 538 W. 120th Street, New York, NY 10027
2:Dept. of Astronomy, Columbia University, 538 W. 120th Street, New York, NY 10027

Abstract.
The abundances of beryllium and boron have been measured in halo stars of metallicities as low as [Fe/H] =-3. The observations show that the ratios Be/Fe and B/Fe are independent of metallicity and approximately equal to their solar values over the entire range of observed metallicity. These observations are in contradiction with the predictions of simple models of beryllium and boron production by spallation in the interstellar medium of a well mixed galaxy. We propose that beryllium and boron are produced by spallation in the ejecta of type II supernovae. In our picture, protons and alpha particles are accelerated early in the supernova event and irradiate the heavy elements in the ejecta long before the ejecta mixes with the interstellar medium. We follow the propagation of the accelerated particles with a Monte-Carlo code and find that the energy per spallation reaction is about 5 GeV for a variety of initial particle spectra and ejecta compositions. Reproducing the observed Be/Fe and B/Fe ratios requires roughly 3×10^{47} ergs of accelerated protons and alphas. This is much less than the 10^{51} ergs available in a supernova explosion.

INTRODUCTION

Spallation reactions involving protons or alpha particles colliding with the nuclei of the abundant light elements carbon, nitrogen, and oxygen have long been recognized as important or dominant contributors to the production of the isotopes of lithium, beryllium, and boron (Reeves, Fowler & Hoyle 1970). Evidence for this picture includes the fact that the ratios of abundances of spallation products are equal to the ratios of the production cross sections, the large overabundance of spallation products in cosmic rays, and the fact that the observed cosmic ray flux irradiating a poulation I composition for the age of the galaxy produces the observed spallation to CNO abundance ratio. For a summary of these arguments see the review by Reeves (1982).

Spallation nucleosynthesis in the interstellar medium of a well mixed galaxy is very inefficient if the metallicity of the gas is low. In the simplest closed box model of galactic chemical evolution, the abundance of spallation products is proportional to the square of the iron abundance at low metallicity. This prediction is contradicted by the observations of beryllium and boron abundances in low metallicity stars (Duncan et al. 1992, Gilmore et al. 1992, Boesgaard 1996) which show that the Be/Fe and B/Fe ratios are independent of metallicity and approximately equal to their solar values for stars in the metallicity range -1>[Fe/H]>-3.

To account for these observations we propose that spallation nucleosynthesis took place in the supernova event itself. In our model, particles are accelerated and irradiate the CNO nuclei in the supernova ejecta long before the ejecta mixes with the interstellar medium of the early galaxy. We follow the propagation of the accelerated particles with a Monte-Carlo analysis and find that reproducing the observed spallation to iron ratio requires about 3×10^{47} ergs to go into accelerating particles. We also find, not surprisingly, that we produce the same isotopic ratios as are found in calculations which irradiate a solar composition interstellar medium.

MODEL DESCRIPTION & RESULTS

A supernova releases a large amount (10^{51}ergs) of energy as it explodes. A fraction of this energy may go into accelerating particles to very high energies. The type II supernova ejecta are rich in CNO nuclei which serve as targets for spallation by energetic protons and α particles that are accelerated in the expanding ejecta.

We have modelled the spallation production of Li, Be and B in supernova ejecta by writing a Monte-Carlo simulation. The program takes as its input the spectrum of accelerated particles (p, α) in supernova explosion and irradiates the ejecta (CNO) of specified composition and density with these particles.

In the simulation, the accelerated particles start out with initial energy as specified by the spectrum and have elastic, inelastic or spallation collisions based on the probability of that collision according to their relative cross sections and target abundance. The energy dependent cross-sections for these collisions are taken from the compilations by Read & Viola (1984) and Meyer (1971). The main energy loss processes for these accelerated particles are by spallation, ionization and elastic/inelastic collisions. Protons go through

TABLE 1. E_{sp} (GeV) for various E_c (MeV) and n, calculated for a 25M⊙ supernova composition

$E_c(MeV)=$	50	100	200	500
n=5	6.3	4.3	5.4	9.6
n=10	4.3	4.2	4.9	8.6

elastic and inelastic collisions with other protons and He nuclei in the ejecta and spallation reactions with CNO nuclei. In the case of a proton-proton or a proton-α elastic collision, the incident proton loses energy to the stationary target and accelerates the target proton or α particle. This creates a cascade of accelerated particles, each of which in turn goes through a sequence of collisions. The energy loss suffered by the incident proton in a proton-proton elastic collision is calculated kinematically using differential cross sections for elastic scattering (Meyer 1971). The proton-proton inelastic collision, p + p $\to \pi^0$ + p + p, produces π^0s which decay as $\pi^0 \to \gamma + \gamma$, producing gamma-ray flux. The remaining energy after π^0 production is shared between the two outgoing protons. The proton-α inelastic collision produces secondary particles (d,^{3}He) that are not relevent to our simulation and the incident proton is assumed to lose all its energy. Spallation between protons and CNO produces one of the light element isotopes according to their relative cross sections (p + CNO $\to$ ^{6}Li, ^{7}Li, ^{9}Be, ^{10}B, ^{11}B). Protons also lose energy by ionization in passing through the ejecta. These losses begin to dominate at lower energies. When the ionization energy loss is much greater ($\approx$ 50 times) than the energy loss by other collisions (spallation, elastic and inelastic), the proton is assumed to lose all its energy by ionization and doesn't suffer any more collisions. Similar treatment is applied to accelerated α particles. In this case, the spallation reactions are $\alpha + \alpha \to$ ^{6}Li, ^{7}Li and α + CNO $\to$ ^{6}Li, ^{7}Li, ^{9}Be, ^{10}B, ^{11}B.

The simulation is run for various accelerated particle spectra and compositions of irradiated material. The compositions of 15$M_\odot$, 25$M_\odot$ and 35$M_\odot$ supernova ejecta (Weaver & Woosley 1993) are used as irradiated material distributed uniformly in a sphere of radius of 10^{15}cm. The source spectrum of accelerated particles is described as constant (flat spectrum) up to a certain cutoff energy E_c and at energies $E > E_c$, a power law decrease by index n. We use various cutoff energies (E_c = 50 MeV-500 MeV/nucleon) and power law indices (n = 2-10) and compare the results.

We calculate the total number of light element isotopes (^{6}Li, ^{7}Li, ^{9}Be, ^{10}B, ^{11}B) produced by spallation in the irradiated ejecta, the total number of elastic

TABLE 2. Results for E_c = 100 MeV/nucleon, power law index n = 5 calculated for various supernova masses

	15$M_\odot$	25$M_\odot$	35$M_\odot$	Observed
B/Be	16.4	12.9	12.5	10-20[1]
^{11}B/^{10}B	2.6	2.5	2.5	4.0[2]
^{7}Li/^{6}Li	1.5	1.5	1.6	12.6[2]
^{6}Li/^{9}Be	3.7	3.0	2.7	3.7[2]
$N_{\pi 0}$	2.0x10^{47}	6.5x10^{46}	4.4x10^{46}	
E_{sp} (GeV)	8.7	4.3	3.9	

1. Duncan et al. (1992) 2. Cameron (1982)

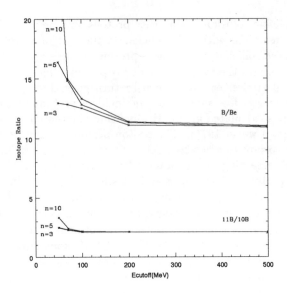

FIGURE 1. B/Be and ^{11}B/^{10}B for various cutoff energies (E_c) and power law index n using 25M⊙ supernova composition

and inelastic collisions and the energy needed per spallation (E_{sp}).

We find the energy per spallation (E_{sp}) to be in the range of 1-10 GeV, as shown in Table 1. The value of E_{sp} is higher in the case of $E_c = 500$ MeV/nucleon because the number of inelastic collisions is dramatically increased as the accelerated particle spectrum extends well beyond the 280 MeV threshold for inelastic collisions.

Using the Solar abundances (Cameron 1982), corrected for ^{7}Li to account for its production in the big bang nucleosynthesis, we get the total number of spallations per ^{56}Fe nucleus as 2.6×10^{-5}. The total amount of ^{56}Fe ejected from a supernova is taken as 0.07M⊙, as observed in SN1987A (Erickson et al. 1988). Therefore the total number of spallations per supernova required for light element production is 3.9×10^{49}. Using 5 GeV as the value of energy per spallation, we get the total energy required as 3×10^{47} ergs. A typical supernova releases about 10^{51} ergs, so only a small fraction of the total energy

TABLE 3. Number of π^0s produced for various E_c (MeV) and n, calculated for 25M⊙ supernova

	E_c=50	100	200	500
n=5	0	6.5×10^{46}	1.5×10^{48}	2.8×10^{49}
n=10	0	0	3.4×10^{46}	1.8×10^{49}

needs to be directed towards spallation reactions.

The variation of B/Be and ^{11}B/^{10}B ratios with cutoff energy E_c and power law index n are shown in Fig. 1. The isotopic ratios ^{11}B/^{10}B, ^{6}Li/^{9}Be, ^{7}Li/^{6}Li and B/Be, the number of π^0s produced as a result of p-p inelastic collision and the energy required per spallation (E_{sp}), calculated for supernova masses of 15M$\odot$, 25M$\odot$ and 35M$\odot$ with accelerated particle spectrum as $E_c = 100$MeV/nucleon, index $n = 5$ are shown in Table 2. The observed isotopic ratios are also shown for comparison.

The B/Be ratio is recently observed in several Halo Dwarfs (Duncan et al. 1992, Boesgaard 1996) and is found to be in the vicinity of 10, which is consistent with our calculations. The Solar ^{11}B/^{10}B ratio of 4 is higher than our calculated value of 2.5. There may be other sources of ^{11}B production, such as neutrino induced nucleosynthesis in a type II supernova (Woosley et al. 1990), giving us the high ^{11}B/^{10}B ratio.

The energy required per spallation decreases with increasing supernova mass because the H/CNO ratio decreases with increasing supernova mass and so there are more spallation collisions and fewer elastic/inelastic collisions, thus utilizing more of the energy for spallation. For the same reason the number of π^0s produced decreases with increasing supernova mass.

Table 3 shows the number of π^0s produced for various cutoff energies E_c and index n, calculated for a 25M$\odot$ supernova composition. The threshold for π^0 production is at 280 MeV, so for certain accelerated particle spectra we get no pion production. The π^0s subsequently decay producing two γ rays at energies centered around 70 MeV.

REFERENCES

1. Boesgaard, A., Formation of the Galactic Halo...Inside and Out, ASP Series, Vol. 92, 327 (1996)
2. Cameron, A. G. W., Essays in Nuclear Astrophysics, eds. C. Barnes, D. Clayton, and D. N. Schramm, (Cambridge University Press) (1982)
3. Duncan, D., Lambert, D., and Lemke, M., ApJ, 401, 584 (1992)
4. Erickson, E. F., Haas, M. R., Colgan S. W. J., Lord, S. D., Burton, M. G., Wolf, J., Hollenbach, D. J., Werner, M., ApJL, 330, L39 (1988)
5. Gilmore, G., Gustafsson, B., Edvardsson, B., and Nissen, P. E., Nature, 401, 584 (1992)
6. Meyer, J. P., A & A Suppl., 7, 417 (1971)
7. Read, S. M., and Viola, V. E. Atomic Data and Nuclear Data Tables, 31, 359 (1984)
8. Reeves, H., Fowler, W. & Hoyle, F., Nature, 226, 227 (1970)
9. Weaver, T.A., and Woosley, S.E., Physical Reports, 227, 65M (1993)
10. Woosley, S.E., Hartmann, D. H., Hoffman, R. D., and Haxton, W. C., ApJ, 356, 272 (1990)

GAMMA RAYS AND COSMIC RAYS FROM SUPERNOVA EXPLOSIONS AND YOUNG PULSARS IN THE PAST

LEV I. DORMAN

Technion, Haifa 32000, Israel and IZMIRAN, Moscow 142092, Russia
e-mail: lid@physics.technion.ac.il

Abstract. Gamma rays (GR) as well as cosmic rays (CR) from supernova (SN) explosions and young pulsars (YP) in the past produced in the Earth's atmosphere neutrons; the interaction of neutrons with atoms of nitrogen ^{14}N generates radiocarbon ^{14}C with decay time 5730 years. By using a method of coupling functions, taking into account vertical and global mixing and exchange processes of elements on the Earth, on the basis of experimental data of radiocarbon contents in annual rings of trees and on contents of some other cosmogenic radionuclides in dated samples became possible to estimate GR fluxes from SN explosions and YP in our Galaxy in the last few thousand years as well as CR fluxes from local SN explosions (up to about 100 pc from the Sun) in the last 150 thousand years.

INTRODUCTION. Konstantinov & Kocharov (1965) supposed that radiocarbon data contain information on many astrophysical phenomena in the past, including SN explosions. The problem is how to extract this information? For this purpose we developed special method of radiocarbon coupling functions (Dorman, 1976, 1977 and then in Dorman, 1997) by analogy with coupling functions for research of CR variations (Dorman, 1957, 1974). In Section 1 we shortly list main results of this method in aspect of SN explosions in the past, and then consider the expect effects and published radiocarbon data which show the existing of these effects in the past in GR (Section 2), and in CR (taking into account also ^{10}Be data, Section 3).

1. METHOD OF COUPLING FUNCTIONS

1.1. Elements mixing and coupling functions for CR and GR. As the first step we took into account the vertical mixing processes in the atmosphere and introduced local coupling function. In the second step we took into account the global elements mixing (with characteristic time about $1 \div 2$ years) and introduced global coupling function $W_g(R)$ for CR:

$$\delta Q_g(t)/Q_{go} = \int_0^\infty (\delta I(R,t)/I_o(R)) W_g(R) dR, \quad (1)$$

where $R = cp/Ze$ is rigidity of particles with momentum p and charge Ze, $Q_{go} \approx 2.4$ atom^{14}C/cm^2 sec is the global production rate and $I_o(R)$ - spectrum of galactic CR in minimum of solar activity,

$$W_g(R) = \begin{cases} akf_1(R,R_o)\left[1-\left(1-(R/R_o)^{1/2}\right)^{1/2}\right], & \text{if } R \leq R_o \\ akf_1(R,R_o), & \text{if } R > R_o. \end{cases} \quad (2)$$

In Eq. (2) $R_o = 14.9$ GV for the present time with magnetic moment of the Earth μ_o (for the past $R_o(-t) = R_o\mu(-t)/\mu_o$), $a = 4.076 + 1.865 \times 10^{-2}W$, $k = 1.104 + 6.614 \times 10^{-4}W$ (here W is sunspot number; a and k were calculated on basis of paper Lingenfelter, 1963), and

$$f_1(R,R_o) = f_2(R)\left(1 - ak\int_0^{R_o} f_2(R)\left(1-(R/R_o)^{1/2}\right)^{1/2}dR\right)^{-1}, \quad f_2(R) = R^{-(k+1)}\exp\left(-aR^{-k}\right).$$ The GR effect will be described by

$$\delta Q_{g\gamma}(t)/Q_{go} = \int_0^\infty I_\gamma(E_\gamma,t) W_{g\gamma}(E_\gamma) dE_\gamma, \quad (3)$$

where coupling function $W_{g\gamma}(E_\gamma) = M_\gamma(E_\gamma)/Q_{go}$ and $M_\gamma(E_\gamma)$ is the number of atoms ^{14}C produced in the atmosphere by one gamma quanta with energy E_γ (until now $W_{g\gamma}(E_\gamma)$ is not determined exactly; some rough estimations will be describe below, in Section 2).

1.2. Exchange of elements and radiocarbon contents. In the third step we took into account the exchange processes between planetary reservoirs ($N_A(t)$-radiocarbon contents in atmosphere, $N_F(t)$-in all other):

$$dN_A(t)/dt = Q_g(t) - \lambda N_A(t) - \lambda_{AF}N_A(t) + \lambda_{FA}N_F(t), \quad (4)$$
$$dN_F(t)/dt = -\lambda N_F(t) - \lambda_{FA}N_F(t) + \lambda_{AF}N_A(t), \quad (5)$$

where $\lambda = 1.21 \times 10^{-4}$ year^{-1} is the radioactive decay constant for ^{14}C, $\lambda_{AF} = 0.145$ year^{-1} is the probability for radiocarbon atoms to pass from reservoir A to reservoir F and $\lambda_{FA} = 1.66 \times 10^{-3}$ year^{-1} is the probability to pass in the opposite direction. The solution of (4), (5) for initial conditions $N_A(0) = N_{Ast} = Q_{go}(\lambda+\lambda_{FA})(\lambda\lambda_1)^{-1}$, $N_F(0) = N_{Fst} = Q_{go}\lambda_{AF}(\lambda\lambda_1)^{-1}$:

$$N_A(t) = \frac{\lambda_{FA}}{\lambda_{AF}+\lambda_{FA}}e^{-\lambda t}(I(t)+Q_{go}/\lambda) + \frac{\lambda_{AF}}{\lambda_{AF}+\lambda_{FA}}e^{-\lambda_1 t}(I_1(t)+Q_{go}/\lambda_1) \quad (6)$$
$$N_F(t) = \frac{\lambda_{AF}}{\lambda_{AF}+\lambda_{FA}}e^{-\lambda t}(I(t)+Q_{go}/\lambda) - \frac{\lambda_{AF}}{\lambda_{AF}+\lambda_{FA}}e^{-\lambda_1 t}(I_1(t)+Q_{go}/\lambda_1) \quad (7)$$

where $\lambda_1 = \lambda + \lambda_{AF} + \lambda_{FA}$, $I(t) = \int_0^t Q_g(\tau)e^{\lambda\tau}d\tau$, $I_1(t) = \int_0^t Q_g(\tau)e^{\lambda_1\tau}d\tau$.

1.3. Inverse problems. Important feature of the radiocarbon method is the possibility to solve inverse problems: by measurements of $N_A(t)$ to determine $Q_g(t)$, and then by Eq. (1) and (3) to estimate approximately

$\delta I(R,t)/I_o(R)$ and $I_\gamma(E_\gamma, t)$. Namely, any unknown variation of radiocarbon production rate can be present as $Q_g(t) = Q_{go}(1 + \sum A_j \cos\omega_j(t - t_j))$. After introducing in (6)-(7) and integrating, taking into account global mixing with characteristic time T_g (details see in Dorman, 1997b), we obtain $N_A(t) = N_{Ast}[1 + \sum A_j \beta_j \cos\omega_j(t - t_j - \gamma_j)]$, where

$$\beta_j = \frac{2\lambda\lambda_1}{\omega_j T_g(\lambda + \lambda_{FA})} \left[\frac{(\lambda + \lambda_{FA})^2 + \omega_j^2}{(\lambda^2 + \omega_j^2)(\lambda_1^2 + \omega_j^2)} \right]^{1/2} \sin(\omega_j T_g/2), \quad (8)$$

$$\gamma_j = T_g/2 + \omega_j^{-1} \arctan\left(\frac{\omega_j(\omega_j^2 + \lambda(\lambda + \lambda_{FA}) + \lambda_{FA}\lambda_1)}{\omega_j^2(\lambda + \lambda_{AF}) + \lambda\lambda_1(\lambda + \lambda_{FA})} \right). \quad (9)$$

So by data on radiocarbon contents in dated samples $N_A(t)$ we determine observed frequencies ω_j, amplitudes $A_j\beta_j$ and phases $t_j + \gamma_j$, then for each ω_j we estimate β_j according to (8) and γ_j according to (9), and finally the amplitudes A_j and phases t_j for $Q_g(t)$.

1.4. Application to GR and CR increases from SN and YP. It is easy to see from Eq. (8) that the reduction is very important for short time GR burst (e.g., for characteristic time 2 years and $T_g = 1.5$ years, $\beta_j = 9.55 \times 10^{-4}$ and $\gamma_j = 1.24$ years - it means that observed amplitude in contents must be multiplied on factor $\beta_j^{-1} = 1.05 \times 10^3$ to obtain real increasing in radiocarbon production rate). For CR effect from local SN explosion the characteristic time $\geq 10^3$ years and $Q_g(t)$ will have time profile similar to observed $N_A(t)$.

2. GR FROM SN AND YP IN THE PAST

2.1. Expected GR fluxes from YP. According to Berezinsky & Prilutsky (1977) the total number of accelerated particles in the expanding SN

$$N_p(E,t) = (\gamma - 1)(kL_o/E_o^2)(1 + t/\tau)^{-2}(t^3/t_a^2)(E/E_o)^{-(\gamma+1)}, \quad (10)$$

where L_o is the initial luminosity of the pulsar, k is the fraction of energy L_o transferred to accelerated particles, E_o is the low energy cut off in the spectrum of energetic particles, $\tau \geq 10^7$ sec is the time of the pulsar braking due to magnetic dipole radiation. The spectrum of emitted gamma quanta will have a maximum at ~ 70 MeV and have a form $\propto E_\gamma^{-\gamma}$ for higher energies. The total flux of GR per 1 cm^2 on the Earth from the shell of YP on the distance r in the time interval $t_\gamma \leq t \leq t_a$ (where t_γ is the time after explosion when SN shell becomes transparent for GR and t_a - when adiabatic energy losses begin to dominate over those due to nuclear collisions) will be

$$F_\gamma(r) \approx 60 (r/1\text{kpc})^{-2} (kL_o/10^{44}) \frac{\tau^2(t_a - t_\gamma)}{(\tau + t_\gamma)(\tau + t_a)}. \quad (11)$$

For $r = 1$ kpc and $kL_o = 10^{44}$ erg/sec it gives $F_\gamma \geq 2 \times 10^8$ quanta/cm² in the time interval ~ 1 year.

2.2. Production of neutrons and radiocarbon in the Earth's atmosphere by GR from YP. The production of secondary neutrons by gamma quanta was considered by Povinec & Tokar (1979) and it was shown that the main contributions are given by: 1) giant resonance (is dominant for $E_\gamma \leq 50$ MeV and gives about 88 % of generated neutrons); 2) quasideuteron disintegration (is dominant for $E_\gamma \approx 80 \div 300$ MeV and gives about 10 % of generated neutrons; 3) pion photoproduction (effective for $E_\gamma \geq 170$ MeV and gives about 2 % of generated neutrons). For SN explosion on distance 1 kpc it was obtained production by GR from π^0-decay ~ 4.3 atom^{14}C/cm^2 sec and by GR from electron bremsstrahlung ~ 2.6 atom^{14}C/cm^2 sec during $1 \div 2$ years. So the expected increase in radiocarbon production rate will be about 3 times in comparison with Q_{go}, but the expected amplitude in $\delta N_A(t)/N_{Ast}$ according to Eq. (8) will be about thousand times smaller.

2.3. Possible GR effect from SN 1006 AD. According to Murdin (1985), this SN was the brightest and closest (~ 1.3 kpc) of those observed historically. Precisely dated annual samples of sequoia wood from the Big Stump Grove were used by Damon et al. (1995) to measure the radiocarbon contents variation in period from 1003 to 1020 AD. It was found that at 1009 AD radiocarbon contents increased on about 0.8 % and then decreased during the next 9 years; obtained results are in agreement with theoretical curve calculated for 1.75 times increasing of radiocarbon production rate during 2 years. This requires a total energy in accelerated particles $\sim 10^{50}$ ergs.

3. CR FROM LOCAL SN AND YP IN THE PAST

3.1. Expected CR fluxes. Expected CR fluxes as well as expected anisotropies in present time from local SN explosions in the past (with age smaller than 5×10^4 years according to Nishimura at al:, 1979) were estimated in Dorman et al. (1983), Ghosh et al. (1983). In this case the expected variation of radiocarbon contents in dated samples will be determined by Eq. (6), where $Q_g(t)$ is determined by Eq. (1) and

$$\delta I(R,t) = I_{sn}(R)(2\pi D(t-t_{sn}))^{-3/2} \exp(-r^2/4D(t-t_{sn})). \quad (12)$$

In Eq. (14) τ_{sn} is the time of explosion and r is the distance from the Earth, $I_{sn}(R)$ is the spectrum of CR emitted from SN explosion and YP (we suppose that the main time of emission is much smaller than the time of propagation to the Earth so the source function can be described as $I_{sn}(R)\delta(t-t_{sn})$), D is the diffusion coefficient of CR in local part of Galaxy. We suppose that D is about the same as average for Galaxy, determined in Webber et al. (1992):

$$D = 0.6 \times 10^{28}(R/1 \text{ GV})^{0.6} \text{ cm}^2/\text{sec}. \quad (13)$$

By (12) and (13) we can determine the time t_{max} of maximum CR intensity increase on the Earth

$$t_{max} - t_{sn} = 8.3 \times 10^4 (r/100 \text{ pc})^2 (R/1 \text{ GV})^{-0.6} \text{ years}. \quad (14)$$

In Table 1, based on Nishimura at al. (1979) list of SNR, are shown expected t_{max} for $R_{ef} \sim 10$ GV and amplitude $(\delta N_A(t)/N_{Ast})_{max}$ for total energy in generated CR $\sim 3 \times 10^{49}$ ergs. Table 1 shows that only for Loop 1 we expect sufficient increase in radiocarbon production rate in the past with amplitude more than 100 % with maximum at ∼5 thousand years after explosion. For all other local SN listed in Table 1 the maximum will be reach only after several 10^5 years in future.

Table 1. Values of $t_{\max}$ and $(\delta N_A(t)/N_{Ast})_{\max}$ for local SNR.

SNR	$r/100$ pc	t_{sn}, years	$t_{\max}$, years	$(\delta N_A(t)/N_{Ast})_{\max}$
CTB	7	-3.2×10^4	$+99.0 \times 10^4$	6.0×10^{-4}
Cyg. Loop	6	-3.5×10^4	$+71.6 \times 10^4$	9.5×10^{-4}
HB 21	8	-2.3×10^4	$+131.1 \times 10^4$	4.0×10^{-4}
CTB 1	9	-4.7×10^4	$+164.1 \times 10^4$	2.8×10^{-4}
CTB 13	6	-3.2×10^4	$+71.9 \times 10^4$	9.5×10^{-4}
HB 9	8	-2.7×10^4	$+130.7 \times 10^4$	4.0×10^{-4}
S 149	7	-4.3×10^4	$+97.9 \times 10^4$	6.0×10^{-4}
Monoceros	6	-4.6×10^4	$+70.5 \times 10^4$	9.5×10^{-4}
Vela	4	-1.1×10^4	$+32.3 \times 10^4$	3.2×10^{-3}
Lupus Loop	4	-3.8×10^4	$+29.6 \times 10^4$	3.2×10^{-3}
Loop 1	0.5	-3×10^4	-2.48×10^4	1.64

3.2. Observation of CR effects from local SN. Using the data from several laboratories on radiocarbon contents in annual tree rings and in stalactites for 40,000 years (averaged for groups of 10 and 100 years) as well as ^{10}Be data from Greenland and Antarctic stations for 150,000 years Kocharov et al. (1990, 1991) analyzed four long-term (several thousand years) gradual CR intensity increases. The first event starts about 40 thousand years ago, and after ∼ 6 thousand years reaches a maximum with amplitude ∼ 100%, which requires a total energy in accelerated particles ∼ 10^{50} ergs. This result were confirmed by Cini Castagnoli et al. (1995) on the basis of ^{10}Be data in the Mediterranean sea sediments (it is important that during this event there was no sufficient enhancement in ^{18}O contents, that it can't be caused by changing in atmospheric circulation processes). The remnant of this SN can be Loop 1 (see Table 1). The other three increases of ^{10}Be contents were observed at ∼ 60, ∼ 95 and ∼ 140 thousand years ago (more than 50 %), but during these events were observed sufficient enhancements in ^{18}O contents and they need in additional analysis.

CONCLUSIONS. 1. By the method of radiocarbon coupling functions, taking into account mixing and exchange processes of elements on the Earth, is possible to solve inverse problem: on the basis of experimental data of radiocarbon contents to estimate GR fluxes from YP in our Galaxy.

2. Application of this method to CR effect from SN explosions and YP shows that inverse problem can be solved and exact information on diffusion coefficient and energetic characterictics can be obtain only for local SN (r not more than hundred pc from the Sun, see Table 1).

3. The total energy in CR estimated by GR effect for SN 1006 AD according

to Damon et al. (1995), and by CR effect for local SN (probably identified with SNR Loop 1) according to Kocharov et al. (1990, 1991) is $\sim 10^{50}$ ergs.

4. For more exact investigations of GR effects in radiocarbon necessary to determine more accurately coupling function $W_{g\gamma}(E_\gamma)$ for GR in Eq. (3) and sufficiently reduce the statistical errors for radiocarbon measurements in dated samples; the progress in this area will give possibility to obtain important information on GR effects of many hundred YP in our Galaxy in the past, and on CR effects of many local SN explosions.

Acknowledgements. I would like to thank J. Kurfess and M. Strickman for the kind possibility to take part in the 4-th Compton Symposium and interesting discussions. I thank also A. Dar, B. Mendoza, Yu. Ne'eman, L. Pitaevski, L. Pustil'nik and J.F. Valdes-Galicia for fruitful discussions. This research was partly supported by Physical Dep. of Technion (Haifa) and Instituto de Geofisica UNAM (Mexico).

REFERENCES

1. Berezinsky V.S. & Prilutsky O.F., *Proc. 15th ICRC, Plovdiv*, **1**, 122 (1977)
2. Cini Castagnoli G. et al., *Proc. 24th ICRC, Rome*, **4**, 1204 (1995).
3. Damon P.E. et al., *Proc. 24th ICRC, Rome*, **2**, 311 (1995).
4. Damon P.E. et al., *Radiocarbon*, **34**, No 2, 235 (1992).
5. Dorman L.I., *CR Variations*, Gostekhteorizdat, Moscow (1957).
6. Dorman L.I., *CR: Variations and Space Exploration*. Amsterdam (1974).
7. Dorman L.I., In *Astrophysics Phenomena and Radiocarbon*, ed. G.E. Kocharov, Metzniereba, Tbilisi, pp. 49-96 (1976).
8. Dorman L.I., *Proc. 15th ICRC, Plovdiv*, **4**, 369, 374, 378, 383, 387 (1977)
9. Dorman L.I., In *Towards the Millenium in Astrophysics*, ed. M.M. Shapiro and J.P. Wefel, World Sci. Publ. Co., Singapore (1997).
10. Dorman L.I. et al., *Astrophys. and Space Sci.*, **109**, 87 (1983).
11. Ghosh A. et al., *Proc. 18th ICRC, Bangalore*, **9**, 243 (1983).
12. Kocharov G.E. et al., *Proc. 21th ICRC, Adelaide*, **7**, 120 (1990).
13. Kocharov G.E. et al., *Proc. 22th ICRC, Dublin*, **2**, 388 (1991).
14. Konstantinov V.P. & Kocharov G.E., *Doklady AN SSSR*, **165**, 63 (1965).
15. Lingenfelter R.E., *Rev. of Geophysics*, **1**, No. 1, 35 (1963).
16. Murdin P., *Supernovae*, Cambridge Univ. Press, Cambridge (1985).
17. Nishimura J. et al., *Proc. 16th ICRC, Kyoto*, **1**, 488 (1979).
18. Povinec P. & Tokar S., *Proc. 16th ICRC, Kyoto*, **2**, 237 (1979).
19. Stuiver M. & Braziunas T.F., *The Holocene*, **3**, No 4, 289 (1993).
20. Webber W.R. et al., *ApJ*, **390**, 96 (1992).

ANGLE DISTRIBUTION AND TIME VARIATION OF GAMMA RAY FLUX FROM SOLAR AND STELLAR WINDS, 1. GENERATION BY FLARE ENERGETIC PARTICLES

LEV I. DORMAN

Technion, Haifa 32000, Israel and IZMIRAN, Moscow 142092, Russia
e-mail: lid@physics.technion.ac.il

Abstract. On the basis of data on flare energetic particle (FEP) generation and its propagation in solar (or stellar) wind we calculate the interaction of these particles with matter and estimate the angle distribution and time variation of generated gamma ray (GR) fluxes for local and distant observers.

INTRODUCTION

The generation of GR by interaction of FEP with solar or stellar wind matter shortly was considered in Dorman (1996). Here we will consider this problem in more details. Three factors are important: 1st- space- time distribution of FEP in the Heliosphere, their energetic spectrum and chemical composition; 2nd- wind matter distribution in space and its change during solar or stellar activity cycle; 3rd- FEP interaction with wind matter accompanied with GR generation.

1. Space-Time Distribution and Spectrum of FEP

The problem of FEP generation and propagation through the solar corona and in the interplanetary space as well as its energetic spectrum and chemical and isotopic composition (1st-factor) was reviewed in Dorman (1957, 1963a,b, 1974, 1978), Dorman & Miroshnichenko (1968), Dorman & Venkatesan (1993), Stoker (1995). We assume that for stellar FEP the situation will be similar.

1.1. *Approximation of isotropic diffusion from pointing instantaneous source.* In the first approximation according to numeral data of observations of many events for 5 solar cycles the time change of solar cosmic ray intensity and energy spectrum change can be described by the solution of isotropic diffusion (characterized by the diffusion coefficient $D_i(E_k)$) from some pointing instantaneous source $N_{oi}(E_k)\delta(r)\delta(t)$ of solar energetic particles of type i (protons, $\alpha-$ particles and heavier particles, electrons) generated in the event at point $r = 0$ at the moment $t = 0$:

$$N_i(E_k, r, t) = N_{oi}(E_k)\left(2\pi^{1/2}(D_i(E_k)t)^{3/2}\right)^{-1}\exp(-r^2/(4D_i(E_k)t)), (1)$$

where $N_{oi}(E_k)$ is the energy spectrum in source.

1.2. *Differential energy spectrum of FEP in the source.* According to numeral experimental data, the energetic spectrum of generated solar energetic particles in the source (see review in Dorman & Venkatesan, 1993) approximately can be described as

$$N_{oi}(E_k) \approx N_{oi}E_k^{-\gamma}, (2)$$

where γ increases with increasing of energy from about $0 \div 2$ at $E_k \leq 1$ GeV/nucleon to about $5 \div 6$ at $E_k \approx 5 - 10$ GeV/nucleon. Approximately

$$\gamma \approx \gamma_o + \ln(E_k/E_{ko}), (3)$$

where parameters γ_o and E_{ko} are different for individual events, but typically they are in intervals $2 \leq \gamma_o \leq 5$ and $2 \leq E_{ko} \leq 10$ GeV/nucleon. Parameter N_{oi} changes sufficiently from one event to other: for example, for events of February 23, 1956, November 15, 1960, July 18, 1961 and May 23, 1967 $N_{oi} \approx 10^{34} \div 10^{35}$, 3×10^{32}, 4×10^{31} and 10^{31} particles.(GeV/nucleon)$^{\gamma-1}$, accordingly. The total energy contained in FEP will be according to Eq. (2) and (3):

$$E_{\text{FEP}} = N_{io}\int_0^\infty E_k^{1-\gamma_o-\ln(E_k/E_{ko})}\, dE_k = N_{oi}\Psi(\gamma_o, E_{ko}) \text{ GeV}, (4)$$

where parameter $\Psi(\gamma_o, E_{ko}) = \int_{-\infty}^\infty E_{ko}^{2-\gamma_o-x}\exp(2x - \gamma_o x - x^2)\, dx$ for $\gamma_o = 2$ increases from 2.40 at $E_{ko} = 3$ up to 6.67 at $E_{ko} = 10$ and for $\gamma_o = 5$ decreases from 4.38 at $E_{ko} = 3$ to 2.00 at $E_{ko} = 10$ GeV/nucleon. For many great solar cosmic ray events E_{FEP} reaches $10^{31} \div 10^{32}$ ergs and for flares on some types of stars it can be several orders higher.

2. Space-Time Distribution of Wind Matter

The detail information on the 2nd-factor (space-time distribution of solar wind matter) for local Heliosphere was obtained recently by the mission of Ulysses. Important information for bigger distances (up to 50 AU) was obtained from missions Pioneer 10, 11, Voyager 3, 4, but only not far from the ecliptic plane. If we assume the model of Parker (1963) of radial wind expanding that

$$n(r, \theta, t) = n_1(\theta, t)u_1(\theta, t)r_1^2/(r^2 u(r, \theta, t)), (5)$$

where $n_1(\theta, t)$ and $u_1(\theta, t)$ are the matter density and solar wind speed at the helio-latitude θ on the distance $r_1 = 1\ AU$ from the Sun. The dependence $u(r, \theta, t)$ is determined by the interaction of solar wind with galactic cosmic rays and anomaly component of cosmic rays, with interstellar matter and interstellar magnetic field, by interaction with neutral atoms penetrating from interstellar space inside the Heliosphere, by the nonlinear processes caused by these interactions (see review in Dorman, 1995 and recent calculations in Le Roux and Fichtner, 1997). We assume that the space-time distribution of stellar wind matter can be described by equation similar to (5).

3. Generation of Gamma Radiation by FEP in the Heliosphere and space-time distribution of GR emissivity

3.1. *Generation of neutral pions in the Heliosphere by FEP.* According to Stecker (1971), Dermer (1986a,b) the neutral pion generation caused by nuclear interactions of energetic protons with hydrogen atoms through reaction $p + p \to \pi^\circ +$ anything will be determined by

$$F_{pH}^{\pi}(E_\pi, r, \theta, t) = 4\pi n(r, \theta, t) \times$$
$$\int_{E_{k\min}(E_\pi)}^{\infty} dE_k N_p(E_k, r, t) \langle \varsigma \sigma_\pi(E_k) \rangle (dN(E_k, E_\pi)/dE_\pi) dE_k, \quad (6)$$

where $E_{k\min}(E_\pi)$ and $\langle \varsigma \sigma_\pi(E_k) \rangle$ are the threshold energy and inclusive cross section for this reaction, $n(r, \theta, t)$ is determined by Eq. (5) and $N_p(E_k, r, t)$ by Eq. (1), and the distribution of generated pions in elementary act is normalized: $\int_0^\infty (dN(E_k, E_\pi)/dE_\pi) dE_\pi = 1$.

3.2. *Expected space-time distribution of GR emissivity in periods of FEP events.* Eq. (6), taking into account (1) and (5), gives according to Dermer (1986a,b) expected space-time distribution of GR emissivity:

$$F_{pH}^{\gamma}(E_\gamma, r, \theta, t) = 2 \int_{E_{\pi\min}(E_\gamma)}^{\infty} dE_\pi (E_\pi^2 - m_\pi^2 c^4)^{-1/2} F_{pH}^{\pi}(E_\pi, r, \theta, t) =$$
$$\left(\frac{3^{3/2} n_1(\theta,t) u_1(\theta,t)}{2^{-7/2} \pi^{-1/2} r^2 r_1 u(r,\theta,t)}\right) \int_{E_{\pi\min}(E_\gamma)}^{\infty} (E_\pi^2 - m_\pi^2 c^4)^{-1/2} dE_\pi \int_{E_{k\min}(E_\pi)}^{\infty} N_{op}(E_k) \times$$
$$\langle \varsigma \sigma_\pi(E_k) \rangle (t/t_1)^{-3/2} \exp\left(-3r^2 t_1/2r_1^2 t\right) dE_k, \quad (7)$$

where $E_{\pi\min}(E_\gamma) = E_\gamma + m_\pi^2 c^4/4E_\gamma$ and $t_1 = r_1^2/6D_p(E_k)$ is the time maximum of FEP flux on the Earth's orbit. In the first approximation the space distributions of gamma ray emissivity at different t/t_1 will be mainly determined by the function $r^{-2}(t/t_1)^{-3/2} \exp(-3r^2 t_1/2r_1^2 t)$, where t_1 corresponds to some effective E_k according to Eq. (7) in dependence of E_γ. The biggest gamma ray emissivity will be in the region $r \leq r_i = r_1(2t/3t_1)^{1/2}$, where the level of GR emissivity $\propto r^{-2}(t/t_1)^{-3/2}$. For example, in the inner region at $t = t_1 = 10^3 s$ ($r_i = \frac{2}{3} r_1 = 10^{13} cm$, $D_p(E_k) \approx 4 \times 10^{12} cm^2/sec$) for the powerful event with $E_{\text{FEP}} \approx 10^{32}$ ergs we expect emissivity $F_{pH}^{\gamma}(E_\gamma > 100\ MeV, r) \approx$

$10^8 r^{-2}$ photon.cm^{-3}.s^{-1} (on the distance 5 solar radii it gives $\sim 10^{-15}$ photon.cm^{-3}.s^{-1}, but at $t/t_1 = 0.1$ it gives $\sim 3 \times 10^{-14}$ photon.cm^{-3}.s^{-1}). Out of the inner region the level of emissivity fall $\propto r^{-2} \exp\left(-(r/r_i)^2\right)$.

4. Expected Variations of GR Fluxes from Solar and Stellar Winds

4.1. Local observer. Let us assume that the observer is inside the Heliosphere, on the distance r_{obs} from the Sun and helio-latitude θ_{obs}. The sight line of observation we can determine by the angle θ_{sl}, computed from the equatorial plane from anti-Sun direction to the North. In this case the expected angle distribution and time variation of GR flux will be

$$\Phi^\gamma_{pH}(E_\gamma, r_{obs}, \theta_{obs}, \theta_{sl}, t) =$$

$$\int_0^{L_{max}} F^\gamma_{pH}(E_\gamma, r(L, r_{obs}, \theta_{obs}, \theta_{sl}), \theta(L, r_{obs}, \theta_{obs}, \theta_{sl}), t) \, dL, \quad (8)$$

where the distribution of GR emissivity $F^\gamma_{pH}(E_\gamma, r, \theta, t)$ in the right hand of (8) is determined by Eq. (7), taking into account that for the Heliosphere of radius r_o

$$L_{max} = r_o \sin(\Delta\theta - \arcsin(r_{obs} \sin\Delta\theta/r_o))/\sin\Delta\theta, \quad \Delta\theta = \theta_{sl} - \theta_{obs},$$
$$r(L, r_{obs}, \theta_{obs}, \theta_{sl}) = (r^2_{obs} + L^2 + 2r_{obs}L\Delta\theta)^{1/2}, \quad \theta(L, r_{obs}, \theta_{obs}, \theta_{sl}) =$$
$$\theta_{obs} + \arccos\left((r^2_{obs} + r_{obs}L\Delta\theta)/\left(r_{obs}(r^2_{obs} + L^2 + 2r_{obs}L\Delta\theta)^{1/2}\right)\right). \quad (9)$$

For the spherically-symmetrical problem, taking into account that effective is only inner region $r \leq r_i = r_1 (2t/3t_1)^{1/2}$ where $F^\gamma_{pH} \propto r^{-2}$, we obtain

$$\Phi^\gamma_{pH}(E_\gamma, r_{obs}, \varphi, t) = F^\gamma_{pH}(E_\gamma, r_{obs} \sin\varphi, t)(\theta_{max} - \theta_{min}) r_{obs} \sin\varphi, \quad (10)$$

where φ is the angle between direction of observation and direction to the Sun, and $\theta_{max} = \arccos(r_{obs} \sin\varphi/r_i)$, $\theta_{min} = -\left(\frac{\pi}{2} - \varphi\right)$ if $r_{obs} \leq r_i$ and $\theta_{min} = -\theta_{max}$ if $r_{obs} \geq r_i$. Eq. (10) gives the lower limit of expected GR flux because here we not take into account the GR production in region $r \geq r_i$.

4.2. Distant observer. For distant observer ($r_{obs} \gg r_o$) we obtain

$$\Phi^\gamma_{pH}(E_\gamma, r_{obs}, t) = 2\pi r_{obs}^{-2} \int_{-\pi/2}^{+\pi/2} \cos\theta d\theta \int_0^{r_o} r^2 F^\gamma_{pH}(E_\gamma, r, \theta, t) \, dr, \quad (11)$$

where $F^\gamma_{pH}(E_\gamma, r, \theta, t)$ is determined by Eq. (7). For FEP propagation and GR generation is important only inner region with $r \leq r_i \ll r_o$, where we expect about constant wind speed; that

$$\Phi^\gamma_{pH}(E_\gamma, r_{obs}, t) \approx \Phi^\gamma_{pH}(E_\gamma, r_{obs}, t_1)(t_1/t), \quad (12)$$

where $t_1 = r_1^2/6D_p(E_k)$ is the time of maximum of FEP intensity on the distance $r_1 = 1$ AU from the star and we take into account that $\int_0^{r_o} \exp\left(-\frac{r^2}{4D_p(E_k)t}\right) dr \approx \int_0^\infty \exp\left(-\frac{r^2}{4D_p(E_k)t}\right) dr = \sqrt{\pi D_p(E_k) t}$. Eq. (12) gives a simple test to control a mode of FEP propagation in stellar wind.

Conclusions. 1. According to Eq. (8), (9) and (10), by observations of GR generated by FEP interaction with solar wind matter can be organized monitoring of FEP propagation and solar wind matter distribution as well as of the distribution of solar wind speed (by using also Eq. (5)) in the inner Heliosphere in periods of great solar cosmic ray events.

2. According to Eq. (11) and (12), observations of GR caused by stellar FEP interaction with stellar wind matter can give important information on stellar flare cosmic ray spectrum and mode of energetic particle propagation as well as on stellar wind matter distribution.

Acknowledgements. I would like to thank J. Kurfess and M. Strickman for the kind possibility to take part in the 4-th Compton Symposium and interesting discussions. I thank also A. Dar, B. Mendoza, Yu. Ne'eman, L. Pitaevski, L. Pustil'nik and J.F. Valdes-Galicia for fruitful discussions.

This research was partly supported by Physical Department of Technion (Haifa) and Instituto de Geofisica UNAM (Mexico).

REFERENCES

1. Dermer C.D., *A&A* **157**, 223 (1986a).
2. Dermer C.D., *ApJ* **307**, 47 (1986b).
3. Dorman L.I., *Cosmic Ray Variations*, Gostekhteorizdat, Moscow (1957).
4. Dorman L.I., *Cosmic Ray Variations and Space Research*, Nauka, Moscow (1963a).
5. Dorman L.I., *Astrophysical and Geophysical Aspects of Cosmic Rays*, in *Progress of Cosmic Ray and Elementary Particle Physics*, ed. J.G. Wilson and S.A.Wouthuysen, Vol. **7**, North-Holland Publ. Co., Amsterdam (1963b).
6. Dorman L.I., *Cosmic Rays of Solar Origin* (in Series "Summary of Science", Space Investigations, Vol.**12**). VINITI, Moscow (1978).
7. Dorman L.I., In *Currents in High Energy Astrophysics* (ed. M.M. Shapiro, R. Silberberg and J.P. Wefel), Kluwer Ac. Publ., Dordrecht /Boston /London, NATO ASI Series, **458**, 183, 193 (1995).
8. Dorman L.I., *A&ASS* **120**, No. 4, 427 (1996).
9. Dorman L.I. and Miroshnichenko L.I., *Solar Cosmic Rays*. Physmatgiz, Moscow (1968).
10. Dorman L.I. and Venkatesan D., *Space Sci. Rev.*, **64**, 183 (1993).
11. Le Roux J.A. and H. Fichtner, *ApJ* **477**, L115 (1997).
12. Parker E.N., *Interplanetary Dynamically Processes*, Intersci. Publ., New York-London (1963).
13. Stecker, F.W., *Cosmic Gamma Rays*, Mono Book Co, Baltimore (1971).
14. Stoker P.H., *Space Sci. Rev.*, **73**, 327 (1995).

ANGLE DISTRIBUTION AND TIME VARIATION OF GAMMA RAY FLUX FROM SOLAR AND STELLAR WINDS, 2. GENERATION BY GALACTIC COSMIC RAYS

LEV I. DORMAN

Technion, Haifa 32000, Israel and IZMIRAN, Moscow 142092, Russia
e-mail: lid@physics.technion.ac.il

Abstract. In our investigations of galactic cosmic rays (CR)- solar activity hysteresis phenomenon were obtained data on CR modulation parameters changing during solar cycle. Here we use these data to calculate the change of CR density distribution in the Heliosphere during solar cycle in dependence of particle rigidity. Then we calculate the expected gamma ray (GR) emissivity distribution and GR fluxes generated by interaction of modulated galactic CR with solar wind matter. It is shown that GR observations can give important additional information on solar wind distribution and CR modulation during solar cycle as well as on galactic CR modulation by stellar winds, stellar wind matter distribution and on stellar activity cycles.

INTRODUCTION

The generation of GR by interaction of galactic CR with solar wind matter shortly was considered in Dorman (1996). Here we will consider this problem in more details. GR generation in the interplanetary space by galactic CR is determined mainly by 3 factors: 1st- by space-time distribution of galactic CR in the Heliosphere; 2nd- by the solar wind matter distribution in space and its change during solar activity cycle; 3rd- by properties of galactic CR interaction with solar wind matter accompanied with GR generation. For 1st and 2nd factors are important nonlinear processes: influence of galactic CR pressure and kinetic stream instability on solar wind properties and on galactic CR propagation (review on CR nonlinear collective effects see in Ch.9 of Berezinsky et al., 1990 and in Dorman, 1995; recent analysis made in Le

Roux and Fichtner, 1997).

1. Galactic CR Space-Time Distribution

The problem of galactic CR propagation through the interplanetary space as well as modulation of its intensity and energetic spectrum (1st- factor) was reviewed in Dorman (1957, 1963a,b, 1974, 1975a,b, 1991), and with taking into account CR nonlinear processes in Dorman (1995). According to this research, the modulation of energy spectra of galactic CR and its space-time distribution in the Heliosphere in the frame of the spherically-symmetrical model can be described as

$$N(E_k, r, t) = N_o(E_k) \exp\left(-\frac{\gamma+2}{3} \int_r^{r_o} \frac{u(r,t)dr}{D(E_k,r,t)}\right), \quad (1)$$

where r_o is the radius of Heliosphere, $u(r,t)$ is the solar wind speed, and $D(E_k, r, t)$ is the diffusion coefficient. According to Dorman & Dorman (1967a,b), Dorman et al. (1997), the change of $D(E_k, r, t)$ during solar cycle can be determined from the hysteresis phenomenon, taking into account the time lag of processes in the interplanetary space relative to the Sun:

$$D(E_k, r, t) \approx D(E_k, r)_{\max} (W(t-r/u)/W_{\max})^{\left(\frac{1}{3} + \frac{2}{3}(W(t-r/u)/W_{\max})\right)}, \quad (2)$$

where W is the sunspot number and $W_{\max}$ is the sunspot number in maximum of solar activity. In (1) for protons the differential energy spectrum outside of the Heliosphere

$$N_o(E_k) = 2.2 \left(E_k + m_p c^2\right)^{-2.75}, \quad (3)$$

and for α−particles instead of 2.2 will be 0.07 and E_k is the kinetic energy per nucleon (Simpson, 1983). In the most part of Heliosphere $u(r,t)$ and $D(E_k, r, t)$ depend very weekly from r that instead of (1) we obtain

$$N(E_k, r, t) \approx N_o(E_k) \exp\left(-\frac{B(t)}{R\beta}\left(1 - \frac{r}{r_o}\right)\right), (4)$$

where $B(t)$ is the parameter of global modulation, $R = cp/Ze$ and $\beta = V/c$ are the particle rigidity (in GV) and particle velocity in units of light speed c. According to Dorman & Dorman (1967a,b), Zusmanovich (1986) and Belov et al. (1990), near the minimum of solar activity $B_{\min} \approx (0.3 \div 0.4)\ GV$. In the maximum of solar activity the modulation for positive and negative particles of CR will be different and will depend on the direction of the general magnetic field of the Sun (caused by drift effects): for protons and nuclei $B \approx 1.6 GV$ in cycles 1955-1965, 1975-1985 (and expected in 1995-2005) and $B \approx 1.2 GV$ in 1965-1975, 1985-1995.

2. Space-Time Distribution of Solar Wind Matter

The 2nd factor was described in Dorman (1997) and we will use Eq.(5) from this paper for determining solar wind matter distribution. We will use also the result of Le Roux & Fichtner (1997) for solar wind speed (taking into account the influence of galactic and anomaly CR pressure, and pickup protons on solar wind propagation):

$$u(r) \approx u_1 (1 - k(r/r_o)), \quad (5)$$

where u_1 is the speed on distance $r = 1$ AU, r_o is the distance to the terminal shock wave and parameter $k \approx 0.13 \div 0.45$ in dependence of subshock compression ratio (from 3.5 to 1.5) and from injection efficiency of pickup protons (from 0 to 0.9).

3. Expected GR from Interaction of Galactic CR with Solar and Stellar Wind Matter

3.1. *GR emissivity distribution.* In this case the result will be described (according to Stecker, 1971 and Dermer, 1986a,b) by the same equations (6) and (7) in Dorman (1997), only instead of flare energetic particle distribution function necessary to use space-time distribution of galactic CR, described by Eq. (1)-(4). In the first approximation the space-time distributions of GR emissivity will be mainly determined by the function

$$F^\gamma_{pH}(E_\gamma, r, t) = F^\gamma_{pH}(E_\gamma) \frac{n_1(\theta,t)}{n_o} \left(1 - \frac{kr}{r_o}\right)^{-1} \left(\frac{r_1}{r}\right)^2 \exp\left(-A(E_\gamma, t)\left(1 - \frac{r}{r_o}\right)\right) (6)$$

where $F^\gamma_{pH}(E_\gamma)$ is the emissivity spectrum from galactic CR protons out of the Heliosphere (as background emissivity from interstellar matter with density n_o according to Dermer, 1986a,b) and $n_1(\theta, t)$ is the density of solar or stellar wind on the helio-latitude θ on the distance $r_1 = 1$ AU from the star. In (6) $A(E_\gamma, t) = B(t)/(R\beta)_{ef}(E_\gamma)$ and $(R\beta)_{ef}(E_\gamma)$ is some effective value of $R\beta$ for particles responsible for GR generation with energy E_γ. According to Dermer (1986a,b) the expected GR emissivity from all particles in galactic CR will increase in about 1.45 times if we take into account also $\alpha-$ particles and heavier particles in galactic CR.

3.2. *Expected angle distribution and time variations of GR fluxes for local observer.* From (6) for the spherically-symmetrical problem, for observation on the distance r_{obs} in direction determined by the angle φ between direction of observation and direction to the Sun, we obtain GR flux

$$\Phi^\gamma_{pH}(E_\gamma, r_{obs}, \varphi, t) = F^\gamma_{pH}(E_\gamma)(n_1(t)/n_o) G(r_{obs}, \varphi, k, A), \quad (7)$$

where

$$G(r_{obs}, \varphi, k, A) = r_1^2 (r_{obs} \sin \varphi)^{-1} \times$$
$$\int_{\theta_{min}}^{\theta_{max}} \left(1 - \frac{kr_{obs} \sin \varphi}{r_o \cos \theta}\right)^{-1} \exp\left(-A(E_\gamma, t)\left(1 - \frac{r_{obs} \sin \varphi}{r_o \cos \theta}\right)\right) d\theta. \quad (8)$$

In (8) $\theta_{max} = \arccos(r_{obs} \sin \varphi / r_o)$, $\theta_{min} = -\left(\frac{\pi}{2} - \varphi\right)$ if $r_{obs} \leq r_o$ (important for present observations) and $\theta_{min} = -\theta_{max}$ if $r_{obs} \geq r_o$ (possible future observations). We calculate (8) numerically for $r_{obs} = r_1 = 1\ AU = 1.5 \times 10^{13}$ cm, $k = 0.13, 0.30, 0.45$, $\varphi = 2, 10, 45, 90, 178°$, and $A(E_\gamma, t) = 0$ (no modulation), $0.2, 0.4, 0.8, 1.6, 3.2, 6.4, 12.8$ (the case $E_\gamma \geq 100\ MeV$ corresponds to $R\beta \geq 2\ GV$ what means $A(E_\gamma, t) \leq 0.2$ in minimum of solar activity and $A(E_\gamma, t) \leq 0.8$ in maximum of solar activity). The dependence from k is sufficient only for $A(E_\gamma, t) \geq 6.4$, in other cases G is about the same for $k = 0.13, 0.30, 0.45$. In Table 1 we show values $G(r_{obs}, \varphi, k, A)$, sufficient for cases of generation gamma radiation with $E_\gamma \geq 100$ MeV ($A \leq 0.8$) and $E_\gamma < 100$ MeV (corresponds to $A \geq 0.8$).

Table 1. Parameter $G(r_{obs}, \varphi, k, A)$ (in cm) for $r_{obs} = 1$ AU, $k = 0.3$ in dependence of φ (in degrees) and of $A(E_\gamma, t)$ (characterized both effective E_γ and level of solar activity).

$\varphi \backslash A$	0	0.2	0.4	0.8	1.6	3.2
2	1.34E+15	1.09E+15	8.96E+14	6.01E+14	2.7E+14	5.5E+13
10	2.57E+14	2.10E+14	1.73E+14	1.16E+14	5.3E+13	1.1E+13
45	5.02E+13	4.14E+13	3.41E+13	2.31E+13	1.1E+13	2.3E+12
90	2.37E+13	1.96E+13	1.62E+13	1.11E+13	5.3E+12	1.2E+12
178	1.51E+13	1.25E+13	1.04E+13	7.19E+12	3.5E+12	8.7E+11

3.3. The expected time variations of GR flux for distant observer. Let us suppose that some observer is on the distance $r_{obs} \gg r_o$. In this case the expected time variations of GR flux for distant observer from interaction of galactic CR with stellar wind matter will be, taking into account (6):

$$\Phi^\gamma_{pH}(E_\gamma, r_{obs}, t) = 4\pi r^{-2}_{obs} \int_0^{r_o} r^2 F^\gamma_{pH}(E_\gamma, r, t)\, dr \approx 4\pi r^{-2}_{obs} F^\gamma_{pH}(E_\gamma) \frac{n_1(t)}{n_o} r_1^2 r_o \times$$
$$(\exp(-A))\left[\left(1 + \frac{A}{2} + \frac{A^2}{6} + .\right) + k\left(\frac{1}{2} + \frac{A}{3} + \frac{A^2}{8} + .\right) + k^2\left(\frac{1}{3} + \frac{A}{4} + .\right) + .\right]. \quad (9)$$

Table 2 shows values $J(k, A) = \Phi^\gamma_{pH}(E_\gamma, r_{obs}, t) r^2_{obs} \times \left(4\pi F^\gamma_{pH}(E_\gamma)(n_1(t)/n_o) r_1^2 r_o\right)^{-1}$ for $E_\gamma \geq 100$ MeV.

Table 2. Parameter $J(k, A)$ in dependence of k and A.

$k \backslash A$	0	0.1	0.2	0.3	0.4	0.6	0.8
0.13	1.071	1.020	0.972	0.927	0.884	0.804	0.730
0.30	1.180	1.126	1.075	1.026	0.980	0.893	0.813
0.45	1.292	1.235	1.180	1.128	1.078	0.984	0.897

CONCLUSIONS. 1. The space-time distribution of GR emissivity caused by interaction of galactic CR with solar and stellar wind matter is determined by Eq. (6).

2. Expected angle distribution and time variations of GR fluxes for local observer is determined by Eq. (7) and (8), and Table 1. These observations will give important information on galactic CR modulation in the Heliosphere and solar wind matter distribution.

3. The expected time variations of GR flux and energy spectrum for distant observer from stellar winds are determined by Eq. (9) and Table 2. These observations will give important information on galactic CR modulation in stellar winds, stellar activity cycles, and stellar wind matter distribution.

Acknowledgements. I would like to thank J. Kurfess and M. Strickman for the kind possibility to take part in the 4-th Compton Symposium and interesting discussions. I thank also A. Dar, B. Mendoza, Yu. Ne'eman, L. Pitaevski, L. Pustil'nik and J.F. Valdes-Galicia for fruitful discussions.

This research was partly supported by Physical Department of Technion (Haifa) and Instituto de Geofisica UNAM (Mexico).

REFERENCES

1. Belov A.V. et al., *Proc. 21 ICRC*, Adelaide, **6**, 52.
2. Berezinsky V.S., Bulanov S.V., Ginzburg V.L., Dogiel V.A., and Ptuskin V.S., *Cosmic Ray Astrophysics*, ed. V.L. Ginzburg, Physmatgiz, Moscow (1990).
3. Dermer C.D., *A&A* **157**, 223 (1986a).
4. Dermer C.D., *ApJ* **307**, 47 (1986b).
5. Dorman L.I., *Cosmic Ray Variations*, Gostekhteorizdat, Moscow (1957). English translation published in USA in 1958, Dep. of Defence, Ohio Base.
6. Dorman L.I., *Cosmic Ray Variations and Space Research*, Nauka, Moscow (1963a).
7. Dorman L.I., *Astrophysical and Geophysical Aspects of Cosmic Rays*, in *Progress of Cosmic Ray and Elementary Particle Physics*, ed. J.G. Wilson and S.A.Wouthuysen, Vol. **7**, North-Holland Publ. Co., Amsterdam (1963b).
8. Dorman L.I., *Cosmic Rays: Variations and Space Exploration*. North-Holland Publ.Co., Amsterdam (1974).
9. Dorman L.I., *Experimental and Theoretical Principles of Cosmic Ray Astrophysics*. Physmatgiz, Moscow (1975a).
10. Dorman L.I., *Variations of Galactic Cosmic Rays*. Moscow University (1975b).
11. Dorman L.I.,*Nuclear Phys. B*, **228**, 21 (1991).
12. Dorman L.I., In *Currents in High Energy Astrophysics* (ed. M.M. Shapiro et al.), Kluwer Ac. Publ., Dordrecht, NATO ASI Serie, **458**, 183, 193 (1995).
13. Dorman L.I., *A&ASS* **120**, No. 4, 427 (1996).
14. Dorman L.I., This issue, Paper 17-13 (1997).
15. Dorman I.V. & Dorman L.I., *J. Geophys. Res.* **72**, 1513 (1967a).
16. Dorman I.V. & Dorman L.I., *J. Atmosph. and Terr. Phys.* **29**, 429 (1967b).
17. Dorman L.I. et al., *Proc. 25th ICRC*, Durban, **2**, 69, 73; **7**, 341, 345 (1997).
18. Le Roux J.A. & Fichtner H., *ApJ* **477**, L115 (1997).
19. Parker E.N., *Interplanetary Dynamically Processes*, Intersci. Publ., New York-London (1963).
20. Simpson J.A., *Ann. Rev. Nucl. Particle Phys.*, **33**, 323 (1983).
21. Stecker, F.W., *Cosmic Gamma Rays*, Mono Book Co, Baltimore (1971).
22. Zusmanovich A.G., *Galactic Cosmic Rays in the Interplanetary Space*, Nauka, Alma-Ata (1986).

Diffuse High-Energy Gamma-Ray Emission in Monoceros

S. W. Digel*, I. A. Grenier[+], S. D. Hunter[†],
T. M. Dame**, and P. Thaddeus**

*Hughes STX, NASA/GSFC, Code 631
Greenbelt, MD 20771
[+]Université Paris 7 and Service d'Astrophysique, CE Saclay
[†]NASA/GSFC, Code 661
**Harvard-Smithsonian Center for Astrophysics

Abstract. We present a study of the diffuse gamma-ray emission observed by EGRET toward Monoceros in the outer Galaxy. The region studied, $l = 210-250°$, $b = -15-+20°$, includes the molecular clouds associated with Mon R2 and Maddalena's cloud as well as more distant clouds in the Perseus arm. This is the sector of the third quadrant best suited for study of variations of the gamma-ray emissivity and molecular cloud mass calibration in the outer Galaxy. The molecular mass calibrating ratio in Monoceros is found to be $N(H_2)/W_{CO} = (1.56 \pm 0.29) \times 10^{20}$ cm^{-2} (K km s^{-1})$^{-1}$ and the integral gamma-ray emissivity is $Q(E > 100 \text{ MeV}) = (1.92 \pm 0.14) \times 10^{-26}$ H-atom^{-1} s^{-1} sr^{-1}. The emissivity profile across the outer Galaxy is found to have a minimum between the local interstellar clouds and those in the Perseus arm.

INTRODUCTION

The advances provided by EGRET for the detection of diffuse, high-energy gamma-ray emission have made it possible to study the spatial and spectral variation of the emission with unprecedented detail. The object of such studies, aside from identifying point sources that would otherwise be confused with diffuse emission, has been to measure the high-energy cosmic-ray proton and electron densities at remote locations, and to calibrate the CO $J = 1 \to 0$ line as a tracer of interstellar molecular hydrogen.

A number of individual interstellar cloud complexes are resolved by EGRET and have been studied in the way described here (e.g., Ophiuchus [1], Orion [2], and Cepheus [3]). The local cloud complex in Cepheus is in the foreground of distant, massive clouds in the Perseus arm. In the Galactic longitude range of Cepheus, $l \sim 100-130°$, the Doppler shifts of the H I and CO emission lines from the interstellar medium are a strong function of Galactocentric distance. This consideration, and the good sensitivity of EGRET, permitted the local and distant components of the diffuse emission to be studied separately across the outer Galaxy. The approximate analog of the Cepheus region in the third quadrant is the complex of local interstellar clouds in Monoceros, which again lie in front of more distant clouds in the Perseus arm.

We undertook a study of the diffuse emission in Monoceros to compare the radial variations of cosmic-ray density and gas mass calibration in the third quadrant with that inferred in the second quadrant from the Cepheus study. Some results from this work [4] are presented here.

DATA

The gamma-ray dataset includes all relevant EGRET viewing periods through 510.0 and 510.5, which were pointed observations of Monoceros with EGRET in restricted field of view mode. Owing to the large field of view of EGRET and the proximity of the well-observed Geminga and Vela pulsars to either longitude extreme of the region of interest, more than 50 viewing periods are combined here. The intensities of the viewing periods were intercompared to check the relative calibrations. The total number of photons in the region of interest ($l = 210 - 250°$, $b = -15 - +20°$) is 19,814 for the representative energy range $E > 100$ MeV.

As tracers of the neutral interstellar gas, we use the 21-cm H I surveys of Weaver & Williams [5], Heiles & Habing [6], Cleary et al. [7], and Kerr et al. [8] and the 2.6-mm CO survey of Dame et al. [9]. For purposes of the analysis, integrated maps were constructed for four adjacent ranges of Galactocentric distance R across the outer Galaxy: 1. Monoceros local ($R < 10$ kpc); 2. Interarm ($R = 10 - 12.5$ kpc); 3. Perseus arm ($R = 12.5 - 16$ kpc); 4. Beyond Perseus arm ($R > 16$ kpc). Galactocentric distances were derived on the assumption of a flat rotation curve with $R_\odot = 8.5$ kpc and $V_\odot = 220$ km s^{-1}.

ANALYSIS

As is commonly done [1–3], we assume that the interstellar medium is transparent to high-energy gamma-rays, that the cosmic rays responsible for the gamma-ray production are uniform within each annulus and within the atomic and molecular gas, and that the $N(H_2)/W_{CO}$ proportionality (X) is constant within each annulus. Under these assumptions the observed gamma-ray intensity for a given energy range may be modeled as a linear combination of the $N(HI)$ and W_{CO} maps. Additional terms for the isotropic intensity and for four unidentified point sources in the field [10,11] must also be included; the column density of ionized hydrogen and the inverse Compton intensity are neglected here. The CO emission in the outer two annuli was not significant and not included in the model; for any given gamma-ray energy range, the final model thus had 11 free parameters. Coefficients of the linear combination, which are directly related to the gamma-ray emissivities and the X-ratios in the annuli, are determined from a maximum likelihood fit [12,13] to the EGRET observations.

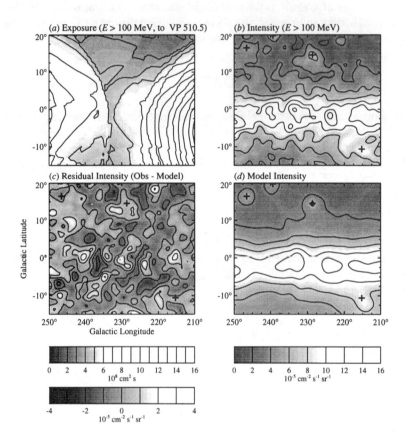

FIGURE 1. Observed and model intensities for $E > 100$ MeV. (a) The effective EGRET exposure, derived on the assumption of a gamma-ray spectral index of -2.1, is largest near the Geminga and Vela pulsars, outside the high and low longitude ranges of the region of interest. (b) Observed intensity, smoothed slightly to decrease fluctuations. The crosses mark the positions of the four point sources – all cataloged unidentified EGRET point sources [10,11] – in the model. (c) The residual intensity (observed minus model) shows no large-scale deviations from zero. (d) Model intensity, displayed with the same smoothing and contour levels as (b).

RESULTS

The model was fit to the EGRET observations for several gamma-ray energy ranges spanning 30-10,000 MeV. The observed and model gamma-ray intensities are compared in Figure 1 for $E > 100$ MeV. The residual map (Fig. 1c) has no positive fluctuations consistent with significant point sources, and has no apparent large-scale trends.

The integral gamma-ray emissivities for $E > 100$ MeV are compared in Figure 2 with the emissivities found in other large-area studies. In this energy range, the emissivity is approximately proportional to the density of high-energy cosmic-ray protons. Also shown in the figure is the profile of cosmic-ray density from the model of Hunter et al. [14] for the same longitude range, normalized at the Solar circle. The emissivity gradient in Monoceros, as the model of Hunter et al. predicts, is not monotonic with R. Rather, the emissivity has a minimum between the Solar circle and the Perseus arm. This variation may not be evident in the Strong et al. [15] and Strong & Mattox [16] emissivities, because they are averaged over essentially the entire outer Galaxy, which would tend to smear away such contrasts. The emissivity gradient found for the Cepheus longitude range [3], which would not be subject to such smearing, nevertheless does appear to be monotonic. This may be an effect of the different binning used – the Perseus arm is closer at lower longitudes and no 'Interarm' range of interstellar gas could be defined. Another possible interpretation is that the 1.8 kpc scale length for gas-cosmic ray

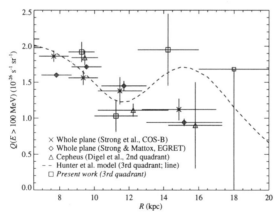

FIGURE 2. The derived emissivity gradient toward Monoceros for $E > 100$ MeV; one point is plotted for each annulus. For comparison, also shown are the profile of cosmic-ray enhancement for the third quadrant from the model of Hunter et al. [14], the gradient derived from an EGRET study of the Cepheus region in the second quadrant [3], and the whole-Galaxy averages of Strong et al. [15] and Strong & Mattox [16].

coupling in the model of Hunter et al. [14] spans the interarm gap in Cepheus but not the broader gap ($\sim$2.5 kpc) in Monoceros.

The X-ratio for the local annulus, the principal molecular clouds of which are Mon R2 and CMa OB1, is $N(H_2)/W_{CO} = (1.56 \pm 0.29) \times 10^{20}$ cm^{-2} (K km s^{-1})$^{-1}$. This is somewhat greater than found for other, closer, molecular clouds in recent studies, e.g., Ophiuchus 1.1 $\pm$ 0.2 [1], Orion 1.06 $\pm$ 0.14 [2], and Cepheus 0.92 $\pm$ 0.14 [3] in the same units. However, the clouds in Monoceros have Galactocentric distance several hundred parsecs greater than these others; the magnitude of the difference of X is consistent with a generally increasing trend found beyond the solar circle, which may be related to overall gradients of cloud temperature or abundances (e.g., [17–19]).

CONCLUSIONS

Our study of diffuse gamma-ray emission in Monoceros indicates that the density profile of high-energy protons may have a secondary maximum in the Perseus arm, where the cosmic-ray density is approximately the same as it is in the Solar vicinity. The X-ratio for the clouds in Monoceros is somewhat larger than for other local clouds, but they are significantly more distant, in the direction where there is already evidence that the ratio tends to increase.

REFERENCES

1. Hunter, S. D., et al., *ApJ* **436**, 216 (1994)
2. Digel, S. W., Hunter, S. D., & Mukherjee, R., *ApJ* **441**, 270 (1995).
3. Digel, S. W., et al., *ApJ* **463**, 609 (1996).
4. Digel, S. W., et al., *ApJ* in prep.
5. Weaver, H., & Williams, D. R. W., *A&AS* **8**, 1 (1973).
6. Heiles, C., & Habing, H. J., *A&AS* **14**, 1 (1974).
7. Cleary, M. N., Haslam, C. G. T., & Heiles, C., *A&AS* **36**, 95 (1979).
8. Kerr, F. J., et al., *A&AS* **66**, 373 (1986).
9. Dame, T. M., et al., *ApJ* **322**, 706 (1987).
10. Fichtel, C. E., et al., *ApJS* **94**, 551 (1994).
11. Thompson, D. J., et al., *ApJS* **107**, 227 (1996).
12. Pollock, A. M. T., et al., *A&A* **146**, 352 (1985).
13. Mattox, J. R., et al., *ApJ* **461**, 396 (1996).
14. Hunter, S. D. et al., *ApJ* **481**, 205 (1997).
15. Strong, A., et al., *A&A* **207**, 1 (1988).
16. Strong, A. & Mattox, J. R., *A&A* **308**, 21L (1996).
17. Mead, K. N., & Kutner, M. L., *ApJ* **330**, 399 (1988).
18. Digel, S., Bally, J., & Thaddeus, P., *ApJ* **357**, L29 (1990).
19. Sodroski, T. J., *ApJ* **366**, 95 (1991).

Diffuse 50 keV to 10 MeV Gamma-ray Emission from the Inner Galactic Ridge

R.L. Kinzer [1], W.R. Purcell [2], and J.D. Kurfess[1]

Abstract.
Measurements of the 50 keV to 10 MeV gamma-ray spectra from the inner Galactic ridge made by the OSSE are reported. Spectra are found to comprise a soft low-energy continuum component which can be represented by an exponentially absorbed power-law, a positronium continuum, a 0.511 MeV positron annihilation line, and an underlying continuum which is consistent with a power-law extension of the cosmic-ray induced spectrum observed above 30 MeV. The cosmic-ray component is measured from $\sim$ 200 keV to 10 MeV. This component appears to exhibit a constant spectral shape over the 0° to 60° longitude range, in agreement with measurements above 30 MeV. When account is taken of variable discrete source contributions and the energy-dependent collimator response, both the soft low-energy continuum and the cosmic-ray induced continuum longitude distributions are consistent with the Galactic CO distribution.

INTRODUCTION

The principal components of the central Galactic ridge spectrum in the 0.05 to 10 MeV range are: a variable soft low energy component, important below $\sim$ 300 keV, known to contain strong contributions from discrete sources; a positronium annihilation continuum; a 0.511 MeV positron annihilation line; a 1.809 MeV ^{26}Al line; and a component resulting from cosmic-rays interacting with the interstellar medium. The soft low-energy component has a latitude width of about 5° FWHM (full-width at half-maximum) (Purcell et al. 1996). The gamma-ray continuum above 30 MeV is principally from a narrow disk along the plane with an $\sim$ 80° wide, roughly flat, central maximum. This component has a similar spectrum above 30 MeV throughout the Galaxy.

OSSE has made six modestly deep observations of the Galactic plane with the long-axis of the collimator aligned with the plane. These observations are at $l_{II} = 0°, 25°, 40°$, and 58°.

EXPERIMENTAL RESULTS

---

[1] Naval Research Laboratory, Washington, D.C.
[2] Northwestern University, Evanston, Illinois

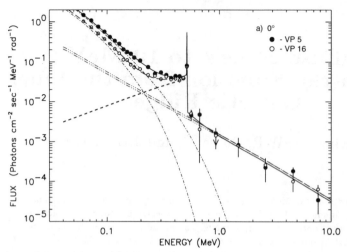

FIGURE 1. Spectra of 2 observations of the Galactic plane at $l_{II}=0°$, deconvolved using an assumed 5° FWHM Galactic ridge distribution. Best-fit composite models are shown by the solid line and by a long-dashed line. Model components are: dashed-dotted line - exponentially absorbed power-law component; dashed line - positronium continuum component and a 2.5 keV wide narrow line at 0.511 MeV; dashed-triple-dotted line - cosmic-ray interaction continuum model of Skibo (1993).

Accurate analyses of diffuse source observations with OSSE require response matrices which take detailed account of the response to photons from all angles in both the source and background fields. Measurements of the Galactic plane continuum spectra use background offset angles over twice as large as the nominal 4.5 degrees. For this reason, measurements of the Galactic plane spectrum above ~ 700 keV must take into consideration a subtle dependence of the observed count-rate spectra on the scan angle of the detector relative to the spacecraft (Kurfess et al. 1997). In the current analysis, a full treatment of these effects has been implemented.

Fig. 1 shows the measured photon spectra, along with the best-fit composite photon spectral models, for the two observations of the Galactic plane at $l_{II}=0°$. These photon spectra were obtained from the measured count-rate spectra using standard forward-folding techniques to unfold the detector response using assumed models for the spectral components. As indicated in the figure and discussed above, the spectra contain at least four different components. Fig. 2 shows these two spectra again, along with measurements at $l_{II}=25°$, 40°, and 58°; the composite best-fit model is shown for each observation.

Fig. 3 shows the longitude distribution of the soft low-energy component (100 keV intensity) along with the Galactic CO distribution shown by the

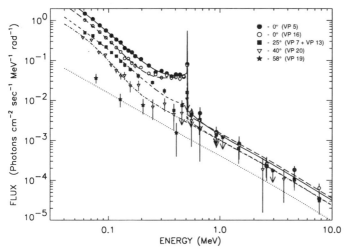

FIGURE 2. Observations and composite best fit models (as shown in Fig. 1) at 0° (filled circles, solid line for VP 5 and open circles, long-dashed line for VP 16), at 25° (filled squares, dashed line), at 40° (open triangles, dashed-dotted line), and 58° (filled stars, dotted line). At 58°, only the cosmic-ray interaction model of Skibo is used.

dashed line (Dame et al. 1987) as smoothed by the OSSE collimators (solid line). The smoothed CO distribution, which is normalized to the lowest intensity Galactic center observation (where variable source intensities are low), fits the soft component distribution reasonably well. The same comparison is shown in Fig. 3b for the cosmic-ray induced continuum. Here, the broad wings on the collimator response at high energies causes significant smoothing and spreading of the input CO distribution, resulting in a spatial distribution in agreement with that observed. Both continuum components are consistent with the same CO distribution, with the *apparent* differences resulting from

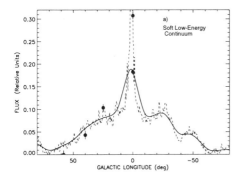

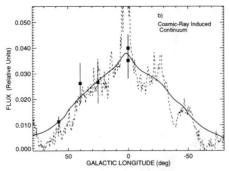

FIGURE 3. a). Longitude distribution of the soft low-energy continuum component along with the Galactic plane CO distribution as smoothed by the OSSE collimator for comparison. The curve is normalized to the lowest-intensity Galactic center observation where variable discrete source contributions were small. The un-smoothed CO distribution is shown by the dashed line. b). A similar distribution of the cosmic-ray induced component, along with the smoothed CO distribution (normalized to the average of the Galactic-center observations.

the differing collimator responses at low and high energies.

Fig. 4 again displays the composite OSSE spectrum of the plane toward the center (with the positronium annihilation components subtracted from the model and the data points), unfolded using an assumed Galactic plane latitude width of 5° FWHM. Also shown are source-subtracted OSSE data points obtained by Purcell et al. (1996) in simultaneous OSSE-SIGMA observations of the Galactic center region (also with the positronium continuum, as measured here, subtracted off). Measurements EGRET and COMPTEL, and earlier measurements by SAS 2 and COS B are also given.

DISCUSSION

The OSSE, COMPTEL, and EGRET instruments on the CGRO have provided measurements which give a consistent picture of the complex Galactic plane gamma-ray spectrum over the 50 keV to 50 GeV range. Within the 50 keV to 10 MeV OSSE observations, the Galactic plane spectrum displays the presence of four or more components which follow at least two different spatial dependences. Each component provides insight to the nature of one or more components of the inner-Galaxy interstellar medium.

Because of solar modulation, it is not possible to directly measure the Galactic cosmic-ray electron spectrum below ~ 1 GeV energies. The current observation of a soft gamma-ray plane which appears to be roughly twice as wide as that above a GeV and which is more peaked toward the center, roughly following the Galactic CO distribution, provides new constraints to these models.

The current results provide significant improvements in our understanding

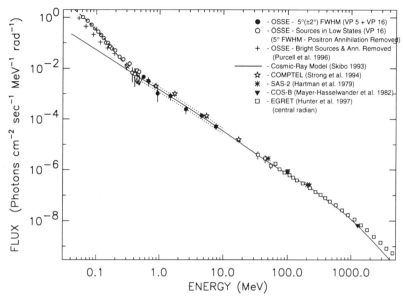

FIGURE 4. Comparison of the OSSE Galactic center observations (VP 5 & 16), with the annihilation components subtracted, and best-fit OSSE model for a 5° FWHM Galactic ridge, along with higher energy measurements. The parallel dotted lines are fits to the OSSE data unfolded assuming a 7° width (upper curve), and 3° width (lower curve), an over-estimate of the OSSE systematic error range in the 1 to 10 MeV region.

of both the intensity and spectral distributions of the cosmic-ray induced and the soft low-energy gamma-ray continuum components from the Galactic plane towards the center. OSSE has the inherent capability to greatly improve the quality of this understanding by performing deeper (4 to 8 week) observations in a program designed to study the Galactic plane continuum emissions.

REFERENCES

1. Dame, T. et al. 1987, ApJ, 322, 706
2. Kinzer, R. L., et al., J. 1996, Astron. Astrophys. Suppl., 120, 317
3. Kinzer, R. L., Purcell, W. R., & Kurfess, J. D. 1997, Ap J., (In preparation)
4. Kinzer, R. L., et al. 1997, Ap J., (Submitted)
5. Kurfess, J. D., et al. 1997,in The 4th Compton Symp. (this volume)
6. Purcell, W., et al. 1994, in The 2nd Compton Symp., (New York:AIP), 403
7. Purcell, W. R., et al. 1996, Astron. Astrophys. Suppl., 120, 389
8. Skibo, J. G. 1993, Ph.D. Dissertation, University of Maryland.

Diffuse Galactic Continuum Emission: Recent Studies using COMPTEL Data

A. W. Strong*, R. Diehl*, V. Schönfelder*,
K. Bennett[†], M. McConnell[‡], J. Ryan [‡]

* Max-Planck-Institut für extraterrestrische Physik, D-85740 Garching, Germany
[†] Astrophysics Division, ESTEC, 2200 AG Noordwijk, The Netherlands
[‡] University of New Hampshire, Durham NH 03824, U.S.A.

Abstract. COMPTEL full sky maximum entropy maps using 5 years of data have been produced using background estimates based on high-latitude observations. The Galactic diffuse emission can be studied using latitude and longitude profiles from these maps. Direct comparison of profiles with theoretical models is illustrated for the 10-30 MeV range. We demonstrate the presence of a broad latitude component, consistent with results from model-fitting studies. The method is also used to obtain model-independent broad-band spectra of the Galactic emission.

INTRODUCTION

The COMPTEL all-sky survey, containing over 5 years of mission data, has previously been used to generate whole-sky maps in the 1-30 MeV range [1–3]. These images are generated using the maximum entropy method. Up to now these maps have mainly been used qualitatively to give a large-scale panorama of the gamma-ray sky at these energies. It is however desirable to try to make quantitative use of maximum entropy images since the intensity distribution can be compared directly with models and with images at other energies from other experiments. This provides an alternative to model fitting and gives a more direct visualization of the information coming from the data; and it is a model-independent technique. This approach has been pursued and validated for the case of 1.8 MeV line emission [4,5,7], where the background is derived from independent measurements at adjacent energies. The high background of COMPTEL makes imaging of extended continuum structure a challenging task. Previous all-sky continuum imaging studies [1–3] have derived the background model by a smoothing procedure applied to the same data; while this produces visually useful images, their quantitative use is limited due to the loss of signal from the smoothing process. Another method [8] is to make use of the observational data at high Galactic latitude, where little Galactic emission is expected, to derive estimates of the instrumental background structure. This method has the advantage of being largely independent of the data being analysed, hence reducing loss of signal. It is thus well adapted to studying extended structures. Time-dependence of the background at energies below 10 MeV is difficult to incorporate however, and remains an uncertainty in this study. It is possible to handle these effects and this is foreseen in future work.

METHOD

COMPTEL data from CGRO Cycles 1-5 (199 observations) were used in the standard energy ranges 1-3, 3-10 and 10-30 MeV (the 'ON' dataset). The data from 23 high-latitude observations ($|b| > 50°$) were combined in the instrument coordinate system (the 'OFF' dataset); the background estimate in each of the ON datasets was obtained using a bicubic spline interpolation in the OFF dataset. These OFF counts were normalized to the ON dataset for each observation and then combined into one global background model for the entire survey. A corresponding global dataset was produced for the ON events and for the COMPTEL response. An additional factor in the range 0.90-0.95 is applied to the background estimate to ensure that ON exceeds OFF everywhere in the dataspace; this procedure has only a second-order effect on the intensity distribution. The COMPTEL full sky maximum entropy software [1] was then used to produce the images. In order to evaluate the technique, simulations using model images of the Galaxy were performed. While these simulations are idealized they show that large-scale structure is recovered reliably, including the correct intensity scale, provided high iterations of the maximum-entropy solution are used. These high iterations approach the maximum-likelihood solution and hence are over-structured on small scales, and must be averaged over several degrees in l or b to obtain reliable profiles of large-scale structure. Note that these are the scales in which we are interested for diffuse emission studies. An important problem is the assignment of zero level, since this is by definition not determined in such a method. This degree of freedom is left when comparing with models or deriving spectra.

RESULTS

Fig 1 shows latitude and longitude profiles for the inner galaxy, averaged over $|b|$ <5° for the longitude profile and $|l|$ <30° for the latitude profile. The error bars are estimates from the observed fluctuations in the profiles at intermediate galactic latitudes where the celestial emission is weak. For comparison, the profile produced by a fit *in the COMPTEL dataspace* to a combination of narrow (gas) and broad component (IC) models is also shown,

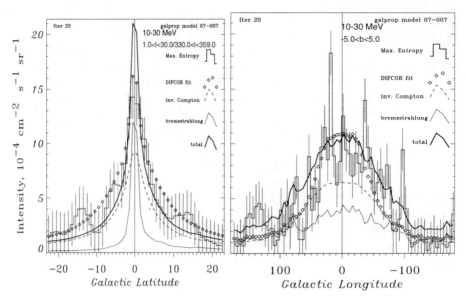

Fig 1. Latitude and longitude profiles of COMPTEL maximum entropy images are shown by the histograms. The intensity is averaged over $330° < l < 30°$ for the latitude profiles and $|b| < 5°$ for the longitude profiles. The profiles based on a two-component model fit to the data is also shown (diamonds). For comparison, emission predicted in [10] is shown, from bremsstrahlung (lower full line), inverse Compton (dashed line) and the total (full line).

the amplitudes of the two components being free parameters; this model-dependent profile agrees in overall shape with the maximum entropy profile. This is an important demonstration of the the reliability of the method, since it illustrates the degree of consistency between model-dependent and model-independent approaches. We also compare with a detailed physical model for the emission [9,10] in which a narrow bremsstrahlung and a broad inverse Compton component contribute roughly equally. This gives immediately an idea of how well the data supports such a model. The similarity in overall shape shows that such a model can reproduce the main features of both the latitude and longitude profiles. Clearly there is more structure present than in the model, due in part to unidentified point sources such as that at $l=18°$ (other peaks are due to the Vela pulsar at $l=262°$ and the Crab at $l=185°$). Study of such distributions shows clearly in what respects the model fails to fit the data and suggests where improvements can be made. For example the particular model illustrated gives intensities which are rather too high suggesting that a slight decrease in the normalization of the cosmic-ray electron spectrum would fit better the lower envelope, with the localized excesses attributed to point sources. The difficulty in determining the absolute zero level is still a limiting factor which prevents a unique solution at present.

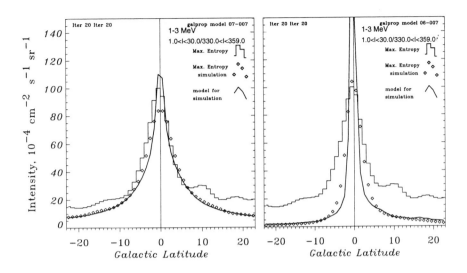

Fig. 2 Latitude profiles from maximum entropy images for 1-3 MeV, averaged over $330° < l < 30°$. The histogram is from COMPTEL data, the diamonds are simulations based on (a) a broad (IC-like) latitude distribution (b) a narrow (gas-like) latitude distribution. The continuous curve is the model distribution used for the simulations.

For the energy range 1-3 MeV, we have made simulations for narrow and broad latitude distributions (Fig 2). These demonstrate that we expect to be able to distinguish between these on the basis of the images. The COMPTEL latitude profile resembles rather better the simulated result for the broad profile. Quantitatively the wide profile yields a factor 0.73 smaller unweighted minimum χ^2 than the narrow one, for both zero level and amplitude free in each case. This provides support for the hypothesis that a broad component is important; it may plausibly be attributed to inverse Compton or maybe to another component [11].

It is also possible to use the intensity images to construct spectra, taking the zero level to be the mean intensity at high latitude ($|b| > 30°$). This gives the spectrum of the non-isotropic emission (Fig 3). The latitude range $|b| < 5°$ is chosen for this spectrum since this optimises the signal-to-noise ratio in this method. The result can be compared with that from model fitting, here illustrated by a free fit to gas/cosmic-ray interactions plus inverse Compton emission; for this method only points >3 MeV are shown since background temporal variations make it unreliable below this energy. Both maximum entropy and model fitting yield spectra which are harder than in [2]. The $|b| < 20°$ spectrum in [2] is unreliable because of inadequate handling of the instrumental background. The present results are consistent with the higher

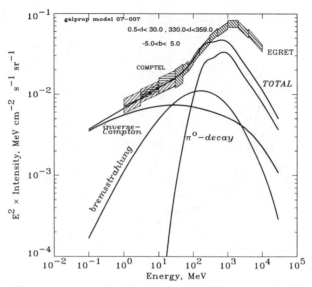

Fig. 3: The γ-ray spectrum for the inner Galaxy, 330° < l < 30°, |b| < 5°. ///: from COMPTEL maximum entropy images using two different background scaling parameters (0.90,0.95); ≡: free fit of COMPTEL data to model based on gas + inverse Compton. \\\ : EGRET data [12]. Curves show the bremsstrahlung, inverse Compton and π°-decay components and total for the physical model described in [10].

energy resolution spectra based on model fitting given in [11]. For comparison, a model from [9,10] is shown, which fits the COMPTEL-EGRET spectrum from 1 MeV to 300 MeV, beyond which the well known discrepancy with the canonical π°-decay spectrum [6] is evident.

REFERENCES

1. Strong A.W., *Experimental Astronomy* **6/4**, 97 (1995)
2. Strong A.W. et al., *A&AS* **120**, C381 (1996)
3. Strong A.W. et al., *ESA* **SP-382**, 533 (1997)
4. Diehl R, et al. Adv. Sp. Res **15**, 5, 123 (1995)
5. Diehl R. et al. A&A **298** , 445 (1995)
6. Hunter S.D. et al., ApJ **481**, 205
7. Oberlack U. et al. (1997) these Proceedings
8. Strong A.W., *A&A* **292**, 82 (1994)
9. Strong A.W., Moskalenko I.V., Schönfelder V., *25th ICRC*, OG8.1.1 (1997)
10. Strong A.W., Moskalenko I.V, in *AIP Proc.*, (1997), these Proceedings
11. Bloemen H. et al., in *AIP Proc.*, (1997), these Proceedings
12. Strong A.W., Mattox J.R., *A&A* **308**, L21 (1996)

Galactic Diffuse γ-ray Emission at TeV Energies and the Ultra-High Energy Cosmic Rays

G. A. Medina Tanco[*,†] and E. M. de Gouveia Dal Pino[*]

[*] Instituto Astronômico e Geofísico, Universidade de São Paulo [1], (04301-904) São Paulo - SP, Brasil

[†] Royal Greenwich Observatory, Cambridge CB3 0EZ, UK, gmt@ast.cam.ac.uk

Abstract. Using the cosmic ray (CR) data available in the energy interval ($10 - 2 \times 10^7$) GeV/particle, we have calculated the profile of the primary γ-ray spectrum produced by the interaction of these CR with thermal nuclei of the ISM. Normalized to the EGRET measurements, this allows an estimate of the galactic diffuse γ-ray background due to intermediate and high energy CR at TeV energies. On the other hand, over the last few years, several particles with energies above 10^{20} eV (beyond the Greisen-Zatsepin-Kuzmin cut-off) have been detected. These particles are very likely extragalactic protons originated at distances not greater than 30 – 50 Mpc [e.g., 1]. The propagation of these ultra-high energy protons (UHEP) through the intergalactic medium leads to the development of γ-ray cascades and an ultimate signature at TeV energies. To assess the statistical significance of this γ-ray signature by the UHEP, we have also simulated the development of electromagnetic cascades triggered by the decay of a 10^{19} eV π^o in the intergalactic medium after an UHEP collision with a cosmic microwave background photon.

THE γ-RAY SPECTRUM AT $10^{12} - 10^{15}$ EV

The γ-ray production mechanisms related to CR interactions with the ISM are well understood and have been described in detail by a number of authors [e.g., 2]. The processes which contribute to diffuse γ-ray production are: (i) bremsstrahlung; (ii) inverse Compton scattering; and (iii) nuclear interactions, but for the energies of interest in the present work, which are above 10 GeV, the latter is the most relevant mechanism.

In order to calculate the γ-ray differential spectrum, dN_γ/dE, from the observed CR differential spectrum, dJ_{cr}/dE, we have employed the *Lund*

---

[1] This work was partially sponsored by the Brazilian Agencies FAPESP and CNPq

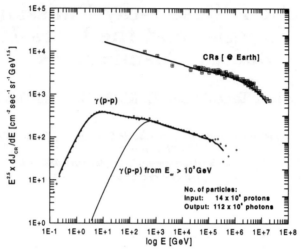

FIGURE 1. Observed CR spectrum (obtained from BASJE, TIBET, TACT, TUNKA, JACEE and Grigorov experiments), and the calculated background γ-ray spectrum at TeV energies.

Monte Carlo for Hadronic Processes routines (PYTHIA version 6.1, March 1997) [3]. The CR energy spectrum includes data from BASJE, TIBET, TACT, TUNKA, JACEE and Grigorov [4]. CR nuclei interact with the ISM thermal nuclei producing γ-ray through different processes (e.g., $\pi^o \to 2\gamma$; $\pi^+ \to \mu^+ \to e^+ \to \gamma$; $q\bar{q} \to g\gamma$; $f\bar{f} \to \gamma\gamma$; $qg \to q\gamma$; etc). The resulting γ-ray spectrum is depicted in Fig. 1.

The evaluation of the mean-free path of the γ-ray photons through the galactic background radiation field (stellar, IR, and cosmic microwave background (CMB) photons) shows that the primary γ-spectrum at $E_\gamma \simeq (10^{10} - 10^{15})$ eV is almost unaffected by interactions with background photons [5]. Therefore, the actual value of the galactic γ-background at these energies can be estimated by normalizing the γ-spectrum of figure 1 to the EGRET observations at 1-10 GeV.

ELECTROMAGNETIC CASCADES DUE TO UHEP

For the analysis of the electromagnetic cascading triggered by UHEP propagation through the intergalactic medium (IGM), two different situations must

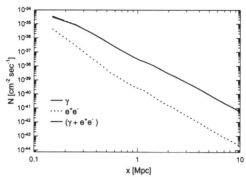

FIGURE 2. Total flux of γ-ray photons and electron-positron pairs [$cm^{-2}\ s^{-1}$] produced in a cascade triggered by a 10^{20} eV proton as a function of the distance along the axes of the cascade (x).

be considered: (i) when an intergalactic magnetic field is present ($B_{IGM} \neq 0$); or (ii) when it is absent (e.g., inside the voids) ($B_{IGM} \simeq 0$). In the first case, synchrotron radiation will prevent the development of cascading due to the rapid draining of energy of the secondary electrons into low energy photons

$$E_\gamma \approx 2 \times 10^{11} \left(\frac{B_\perp}{10^{-9}G}\right)\left(\frac{E_e}{10^{20}eV}\right)^2 eV \qquad (1)$$

Moreover, for the $\approx 10^8$ photons produced, relativistic beaming and curvature of the electron's trajectory will reduce to only $\sim 2 \times 10^{-3}$ the number of photons that can be detected per event, per observer [5]. Thus, we must focus on the case for which $B_{IGM} \simeq 0$. In this case, the cascade is initiated by the interaction of an UHEP with a photon of the CMBR. Either a neutral or a charged pion may be produced and the initial energy is subsequently channeled to lower and lower energies through a $\gamma\gamma$-pair production, inverse Compton cycle [6], until no further γ-ray is produced when then the threshold for pair production due to interactions with the CMBR photons is reached. Thereafter, the interactions must involve higher background photons and the corresponding mean-free path, $\lambda_{\gamma\gamma}$, increases rapidly [see, e.g., 7-8].

As an example, Figs. 2-4 show the results for a typical cascade triggered by a 10^{20} eV UHEP while traversing a very low magnetic field ($B \lesssim 10^{-12}$ G) region of the IGM.

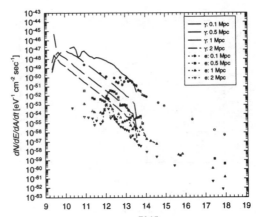

FIGURE 3. Spectra of γ-ray photons and electrons $[eV^{-1}\ s^{-1}\ cm^{-2}]$ produced in a cascade triggered by a 10^{20} eV-proton for different distances x.

CONCLUSIONS AND DISCUSSION

The previous results indicate that, as soon as the shower effectively develops (i.e., at $d \gtrsim 100$ kpc, from the π° decay), low energy electrons are produced. Consequently, an upper limit for the duration of the cascade at the detector can be calculated as the time delay between a 10^{12} eV electron and a γ-ray photon $[\Delta t \approx (1-\beta)d/c]$:

$$\Delta t \simeq 0.1 \left(\frac{d}{1Mpc}\right) s \qquad (2)$$

The results above also indicate that a cascade initiated by a single UHEP interaction with CMB photons should reduce to, at most, one photon of relatively low energy at the detector (see [5] for details). This renders the UHEP-CMB photon interaction practically unobservable.

The situation turns out to be different if one considers the background of γ-rays produced by the whole distribution of UHEP interacting with the CMBR

$$F_{obs}(r) = \frac{\pi \nu_{p,\gamma}}{4} \int_0^r \theta(r)^2 \Phi(r) \Delta t(r) r^2 dr \qquad (3)$$

Where $\nu_{p,\gamma}$ is the number of UHEP-γ_{CMB} interactions per unit time, per unit volume. Since the estimated flux of UHECR is $J(E > 10^{20})$ eV $\approx 3.3 \times 10^{-21}$

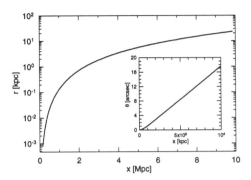

FIGURE 4. Radius (r) and aperture angle (θ) of the cascade in the laboratory reference frame as a function of the distance along the cascade axes (x).

$cm^{-2}\ sr^{-1}\ s^{-1}$ [9], $\nu_{p,\gamma} \approx 10^{-45} cm^{-3} s^{-1}$. Considering the volume of the local Universe within which the UHECR sources must be located ($d \lesssim 50$ Mpc), we derive a lower limit for the contribution of UHECR to the background diffuse γ-ray flux $F_{obs} \approx 10^{-10}\ cm^{-2}\ s^{-1}$. This value is about an order of magnitude smaller than the galactic diffuse background as estimated from our calculations and from the EGRET data [5]. Nonetheless, this contribution is comparable to the γ-ray diffuse background component due to blazers (see, e.g., Fegan, this conference [10]).

REFERENCES

1. Medina Tanco, G.A., Gouveia Dal Pino, E.M., and Horvath, J.E., *Astropart. Phys.* (in press) (1997).
2. Bloemen, H., *Ann. Rev. Astron. Astrophys.*, **27**, 469 (1989).
3. Sjostrand, T., *Computer Physics Commun.*, **82**, 74 (1994)
4. Shibata T. *ICRR-Report-343-95-9* (1995).
5. Medina Tanco, G.A., (in preparation) (1997).
6. Longair, M.S. *High Energy Astrophysics*, Camb. Univ. Press, 1991, vol. I., p.88.
7. Wowczyk et al., *J. Phys. A*, **5**, 1419 (1972).
8. Protheroe, R.J. and Johnson, P.A., *Astropart. Phys.*, **4**, 253 (1995).
9. Halzen, preprint: astro-ph/9704020.
10. Fegan, D.J., this conference (1997).

The Diffuse Galactic Continuum Observed with EGRET: Where's the Bump?

Jeff Skibo

E. O. Hulburt Center for Space Research, Code 7653, Naval Research Laboratory, Washington, DC 20375-5352

Abstract. It is shown that the spectrum of the diffuse Galactic emission observed with EGRET is consistent with being solely due to electron bremsstrahlung and displays no clear pion bump. However, the required electron spectrum is harder than that observed locally and it is necessary to confine the bulk of the electrons in dark molecular clouds to avoid excess inverse Compton emission. Futhermore, the resulting synchrotron radio spectrum is harder than that observed from the Galactic disk. Therefore, whereas the EGRET observations alone do not prove that nuclear cosmic rays pervade the Galaxy, it is difficult to reconcile all the evidence solely in terms of electrons.

INTRODUCTION

The Galactic plane is a prominent source of diffuse gamma ray emission at energies $\gtrsim 10$ MeV. There is general consensus that this emission is in most part produced by interactions of cosmic rays with ambient radiation and matter in the interstellar medium (ISM). If cosmic ray protons and electrons pervade the ISM with energy spectra and electron-to-proton ratio similar to that measured near earth, then the diffuse gamma ray emission from the Galactic disk would manifest a flattening or bump in the spectrum around 100 MeV. This feature results from the decay of neutral pions produced by interactions of the nuclear component of the cosmic rays with ambient interstellar gas and is expected to dominate over the electron bremsstrahlung continuum above 100 MeV [1]. Previous instruments operating at these energies, SAS-2: [2,3]; COS-B: [4–6], did not have sufficient sensitivity and energy resolution to resolve this feature.

Recent observations made with the EGRET instrument on the Compton Gamma Ray Observatory (CGRO) of the diffuse Galactic gamma ray emission reveal a spectrum which is inconsistent with the assumption that the cosmic ray spectra measured locally hold throughout the Galaxy [7,8]. The spectrum observed with EGRET below 1 GeV is in accord with, and supports, the assumption that the cosmic ray spectra and electron-to-proton ratio observed locally are uniform, however, the spectrum above 1 GeV, where the emission is supposedly dominated by π^0 decay, is harder than that derived from the local cosmic ray proton spectrum. After ruling out the possibilities that this discrepancy is due to instrumental effects, incorrect hadron physics, or point source contamination, Hunter et al. [8] conclude that the problem most likely lies with the cosmic ray proton spectrum and that a harder average proton spectrum would be in accord with the data. This conclusion has been corroborated in a recent reanalysis [9] of neutral pion production using modern event generation codes.

Historically, indirect evidence supporting the ubiquity of cosmic ray protons in the Galaxy was provided by elemental abundance measurements, which yield information on the average path length (grams cm^{-2}) traversed by nuclear cosmic rays. Assuming that the cosmic rays are initially of cosmic abundance and that most of the path length is traversed in the general ISM and not near the cosmic ray sources, then the mean spatial length traversed by a typical cosmic ray is on the order of a few thousand kpc. Clearly such particles sample a large volume of the Galaxy provided that their trajectories are not highly convoluted.

The only direct observational evidence that cosmic ray protons were not simply of local origin but existed elsewhere in the Galaxy was provided by the diffuse gamma ray emission observed from the Galactic disk. The emission above 1 GeV was thought to be largely of hadronic origin because the locally measured electron spectrum was too steep to produce the right gamma ray spectrum. The SAS-2 and COS-B data were adequately fit assuming that the cosmic ray spectra and proton-to-electron ratio measured locally held throughout the Galaxy. However, if the cosmic ray proton spectrum varies, as the EGRET observations imply, then it not unreasonable to suspect that the electron spectrum also varies. This warrants a reexamination of the question of whether the diffuse Galactic gamma ray emission can have a purely leptonic origin. Since the nonthermal radio synchrotron emission from the Galactic plane independently establishes the existence of cosmic ray electrons of the required energy, it seems natural to first exhaust this possibility before invoking the presence of a hadronic component. This approach has been taken before [10], where it was shown that the diffuse emission measured with COS-B at energies $\gtrsim$ 300 MeV, as well as the Galactic nonthermal radio emission are consistant with a purely leptonic origin. This paper expands upon this previous work using the new results obtained with EGRET.

RESULTS

In a simple thick target treatment the gamma ray production (photons per time per volume per energy) due to electron bremsstrahlung is given by

$$\frac{dN_\gamma}{dt\,dV\,d\epsilon}(\epsilon) = cn \int_\epsilon^\infty dE \, \frac{\frac{d\sigma}{d\epsilon}(\epsilon,E)}{\left|\frac{dE}{dt}(E)\right|} \int_E^\infty dE' \, Q(E') \qquad (1)$$

where E is the electron kinetic energy, ϵ is the photon energy, $\frac{d\sigma}{d\epsilon}(\epsilon,E)$ is the cross section differential in energy for bremsstrahlung production due to electron-ion [11], electron-atom [12] and electron-electron [13] interactions. The function $Q(E)$ is the electron injection rate per volume per kinetic energy assumed to be a power law, $Q(E) = Q_0 E^{-p}$, and $\frac{dE}{dt}(E)$ is the continuous energy loss rate for electrons due to ionization [14], Coulomb collisions, [15] bremsstrahlung [16], inverse Compton scattering and synchrotron emission [17] in a gas composed of H and He (10% by number) with ionization fraction

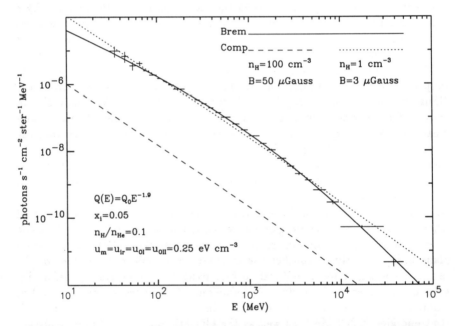

FIGURE 1. The average intensity of the diffuse Galactic gamma ray emission from the inner Galaxy, $-60° < \ell < 60°$, $-10° < b < 10°$ [8]. The curves are the bremsstrahlung and inverse Compton emission calculated for two sets of parameters which were chosen to yield identical bremsstrahlung spectra (solid curve). The inverse Compton spectra (dotted and dashed) are, however, different. The parameters in the lower left portion of the figure are common to both sets.

$x_i \equiv n_e/n_H$ and magnetic field B. The gamma ray emission produced by inverse Compton scattering of relativistic electrons with the microwave background, ambient infrared and optical photons is calculated according to the prescription given in reference [12].

The data in 1 represent the diffuse Galactic gamma ray emission observed with EGRET from the inner Galaxy averaged over $\pm 60°$ in Galactic longitude and $\pm 10°$ in latitude [8]. The curves are the bremsstrahlung and inverse Compton emission calculated for two sets of parameters. One set of parameters is representative of the general ISM, and the other of dark molecular clouds. The electron injection spectral indicies and normalizations were chosen so as to yield the same bremsstrahlung spectum (solid curve) in both cases. However, the inverse Compton spectra (dotted and dashed curves) are different for these cases. Whereas electron bremsstrahlung adequately describes the data in both cases, the inverse Compton emission in the general ISM case contributes at approximately the same level destroying the fit. It is difficult to reproduce the break in the gamma ray spectrum with the much more rigid Compton spectrum without invoking unnatural cutoffs in the electron spectrum. However, in the case were the gas to light ratio is significantly greater than that of the general ISM, the inverse Compton emission is suppressed. The presence of a pion bump would be most pronounced in the 30 - 500 MeV range. It is clear that gamma ray data alone do not require the addition of this component.

DISCUSSION

Although the diffuse Galactic gamma ray spectrum observed with EGRET does not in itself require the existence of cosmic ray protons and can be explained in terms of electrons alone, it is difficult to reconcile the requsite electron spectrum and its accompanying synchrotron radio spectrum with observations. The locally measured electron spectral index above 30 GeV, where the effects of solar modulation are negligible, is about 3.3 [18]. The electron spectrum in this analysis approaches the index 2.9 at these energies (radiative losses steepen the injection spectrum by one power). This alone should not be considered problematic because the EGRET data cannot be accommodated by the local cosmic ray spectra implying that the local cosmic ray spectra must be different from the Galactic average in any case [8,9].

However, it can be shown that the 10 MHz - 10 GHz synchrotron emission resulting from this electron spectrum is inconsistent with the radio observations. The spectrum of the nonthermal radio synchrotron emission is a complex issue, especially in the inner Galaxy where many thermal sources prevail. Recent surveys find radio spectral indicies ($dP/dV\,d\nu \propto \nu^{-\alpha}$) which vary over the sky in the range $0.3 \lesssim \alpha \lesssim 1$ and are generally steeper than 0.7 above a few hundred MHz. [19-22]. The radio spectral indicies calculated in this analysis are 0.25, 0.37, and 0.51 in the 10-100 MHz, 100-1000 MHz, and 1-10 GHz bands, respectively. This is generally flatter than the observed values. Furthermore, the need to suppress the inverse Compton emission by confining the bulk of the electrons to dark molecular clouds is inconsistent with the relatively large scale height of a few hundred pc for cosmic ray electrons inferred from radio maps [23]

I thank Dr. Martin Pohl for useful discussions.

REFERENCES

1. Stecker, F. W., *Cosmic Gamma Rays*, NASA SP 249 (1971)
2. Fichtel, C. E. et al., *ApJ* **198**, 163 (1975)
3. Hartman, R., Kniffen, D., Thompson, D., & Fichtel, C., *ApJ* **230**, 597 (1979)
4. Mayer-Hasselwander, H. A. et al., *A&A* **105**, 164 (1982)
5. Strong, A. W. et al., *A&A* **207**, 1 (1988)
6. Bloemen, J. B. G. M., *ARA&A* **27**, 469 (1989)
7. Bertsch, D. L. et al., *ApJ* **416**, 587 (1993)
8. Hunter, S. D. et al., *ApJ* **481**, 205 (1997)
9. Mori, M., *ApJ* **478**, 225 (1997)
10. Pohl, M. & Schlickeiser, R., *A&A* **252**, 565 (1991)
11. Koch, H. W. & Motz, J. W., *Rev. Mod. Phys.* **31**, 920 (1959)
12. Blumenthal, G. R. & Gould, R. J., *Rev. Mod. Phys.* **42**, 237 (1970)
13. Haug, E., *Zs. Naturforsch.* **30a**, 1099 (1975)
14. Longair, M. S., *High Energy Astrophysics: Volume 1*, Cambridge University press: Cambridge (1992)
15. Huba, J. D., *NRL Plasma Formulary*, Washington, DC: Naval Research Laboratory, NRL/PU/6790-94-265 (1994)
16. Ginzburg, V. L. & Syrovatskii, S. I., *The Origin of Cosmic Rays*, New York: McMillan (1964)
17. Rybicki, G. B. & Lightman, A. P., *Radiative Processes in Astrophysics*, New York: Wiley (1979)
18. Nishimura, J. et al., *21st Internat. Cosmic Ray Conf.* **3**, 213 (1991)
19. Webster, A. S., *MNRAS* **166**, 355 (1974)
20. Webber, W. R., Simpson, G. A. & Cane, H. V., *ApJ* **236**, 448 (1980)
21. Lawson, K. D., Mayer, C. J., Osborne, J. L., & Parkinson, M. L., *MNRAS* **225**, 307 (1987)
22. Reich, P. & Reich, W., *A&AS* **74**, 7 (1988)
23. Phillips, S. et al., *A&A* **103**, 405 (1981)

The pulsar contribution to the diffuse galactic γ-ray emission

M. Pohl[1,2], G. Kanbach[1], S.D. Hunter[3], B.B. Jones[4]

1 MPI für Extraterrestrische Physik, Postfach 1603, 85740 Garching, Germany;
2 Danish Space Research Institute, Juliane Maries Vej 30, 2100 København Ø, Denmark;
3 LHEA, GSFC, Code 662, NASA, Greenbelt, MD 20771;
4 W.W. Hansen Laboratory, Dept. of Physics, Stanford University, Stanford, CA 94305

Abstract. Here we investigate to what extent unresolved γ-ray pulsars contribute to the galactic diffuse emission, and further whether unresolved γ-ray pulsars can be made responsible for the excess of diffuse galactic emission above 1 GeV which has been observed by EGRET. Our analysis is based only on the properties of the six pulsars which have been identified in the EGRET data, and is independent of choice of a pulsar emission model.

We find that pulsars contribute very little to the diffuse emission at lower energies, whereas above 1 GeV they can account for 25% of the observed intensity in selected regions for a reasonable number of directly observable γ-ray pulsars (~12). While the excess above 1 GeV γ-ray energy is observed at least up to six or eight degrees off the plane, the pulsar contribution would be negligible there. Thus pulsars do significantly contribute to the diffuse galactic γ-ray emission above 1 GeV, but they can not be made responsible for all the discrepancy between observed intensity and model predictions in this energy range.

A recent analysis of the diffuse galactic γ-ray emission in the energy range of 30 MeV to 30 GeV [1] indicated that the spatial structure and total intensity of the emission observed by EGRET can be well understood as the result of interactions between cosmic rays and the interstellar medium in addition to a isotropic extragalactic background. However, at energies above 1 GeV the models predict only roughly 60% of the observed intensity.

The attempt of this paper is twofold. We want to provide constraints on the general contribution of the most likely input from discrete sources – pulsars – to the diffuse galactic γ-ray emission. We also want to find out whether or not unresolved pulsars can account for the observed excess at high γ-ray energies. Instead of using models as was done in previous studies [2–4] we will base our analysis solely on the properties of the six pulsars observed by EGRET. A detailed description of the method can be found elsewhere [5].

EGRET has up to now identified six pulsars by their light curves. The differential γ-ray photon spectra of these sources have been derived from pulsed analysis [6]. We will concentrate on nine energy bands spanning 50 MeV to 10 GeV, for which we have the observed photon flux $S_i(E_k)$ of pulsar i in energy band E_k. The basis of our modelling is the assumption that pulsars do not behave arbitrarily in their relation between the distance-normalized γ-ray intensity and the age, but that they follow a trend, a correlation between intensity and age. The best evidence comes from the data themselves: the χ^2 sums we obtain indicate that a correlation is an appropriate description of the data. The limited number of degrees of freedom forces us to use the most simple correlation model, a power-law:

$$S_m(t, E_k) = 10^{y_k} \left(\frac{t}{10^4 \text{ years}}\right)^{b_k} \text{ kpc}^2/\text{cm}^2/\text{sec}/\text{MeV} \qquad (1)$$

This relation can be fitted to the data on the basis of weighted least squares for all energy bins. The distance D_i and its uncertainty can be generally derived on the basis of the pulsar dispersion and rotation measure [7] and only for extremely nearby objects like Geminga by parallax measurements [8]. The age of pulsars can be estimated from the ratio of period and period derivative. The uncertainty of this measure of age is large, especially since pulsars often do not slow down solely by dipole radiation. We will use a factor of two for the age uncertainty δt_i, except for the Crab for which the true age is taken with a nominal uncertainty of 10%.

Three sources of uncertainty have to be considered in the fit: uncertainties in intensity, in distance, and in age. While the intensity error can be assumed to follow a Gaussian probability distribution, both the errors in age and distance enter with some power, so that their effective probability distributions are definitely not Gaussian. Furthermore, the way how age and distance estimates are derived lets us think that a Gaussian probability function is not a fair description of the actual error distribution of age and distance, respectively. We can account for these effects and still use the χ^2 method, if we assume that the error distribution for age and distance are Gaussians in the logarithm, i.e. the uncertainty is multiplicative rather than additive. Only in this case does any power of these parameters obey the same corresponding error distribution.

To account for the different uncertainty distributions in intensity on one side and age and distance on the other side, we have separated each argument χ_i of the χ^2-summation into three components $\chi_i^2 = \left(\chi_{i,1}^{-2} + \chi_{i,2}^{-2} + \chi_{i,3}^{-2}\right)^{-1}$, one for each source of uncertainty, where the components can be calculated in the system in which the uncertainty distribution may be taken as Gaussian. The best fit values and uncertainties of the parameters y_k and b_k of Eq.1 will serve as input for the calculation of the pulsar contribution to the diffuse galactic γ-ray emission.

For each source we can define a critical distance, up to which the source can be detected directly, and beyond which it would contribute to the diffuse

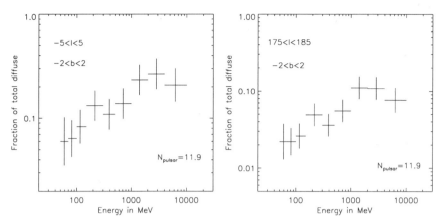

FIGURE 1. Here we show the fraction of the total diffuse intensity in direction of the Galactic Center (left) and galactic anticenter (right) which can be attributed to unresolved pulsars. The parameters to the model are given in the text. The error bars are derived by propagation of the parameter uncertainty in the pulsar model fit. The intensity due to pulsars scales almost linearly with the number of objects which are supposed to be seen as point sources, in this case 11.9 pulsars.

emission. We will assume that the Galaxy is a simple disk-like entity of radius r_h and half-thickness z_h, in which the line-of-sight through the disk determines the sensitivity threshold $F_c(l, b)$. The numbers r_h, z_h have been chosen such that on average the threshold corresponds to the flux of a 5σ source, respectively 4σ for $|b| > 10°$, in the summed data of Phases 1-4. The threshold in direction of the galactic poles would be $F_c(0, 90) = 7 \cdot 10^{-8}$ ph./cm²/sec, whereas in direction of the galactic center it would be $F_c(0, 0) = 4.8 \cdot 10^{-7}$ ph./cm²/sec. Then the critical distance can to be calculated as

$$X_{max}(l, b, t) = F_c^{-0.5}(l, b) \sqrt{\sum_{k=3}^{9} S_m(t, E_k) \delta E_k} \quad \text{kpc} \qquad (2)$$

where δE_k is the width of the energy bin E_k. The spatial distribution of pulsars is poorly determined. The distribution of radio pulsars indicates that there is a galactocentric gradient [7]. We may thus parametrize the normalized spatial distribution of pulsars in galactocentric cylinder coordinates (r,z) as

$$\rho(r, z) = 0.0435 \; \Theta(|z_c - z|) \; (z_c r_c^2)^{-1} \; \cosh^{-1}\left(\frac{r}{r_c}\right) \qquad (3)$$

where the cosh-term accounts for the galactocentric gradient and Θ is a step function. With τ_p^{-1} as the birth rate of γ-ray pulsars and ϵ as the fraction which radiates in our direction we can determine the number of directly detectable pulsars by integration over the line-of-sight, solid angle, and age

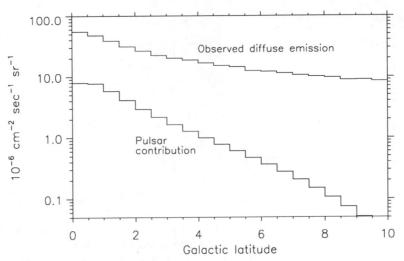

FIGURE 2. The latitude distribution of the diffuse emission above 1 GeV as observed by EGRET is much wider than that of the pulsar contribution, here shown as solid line for 12 directly observable pulsars. Both the data and the model have been averaged over longitude. Here we show the original observed intensity distribution compared to the psf-convolved model prediction. The statistical error of the observed emission is at the percent level and thus negligible.

$$N_{det} = \frac{\epsilon}{\tau_p} \int_0^{t_{max}} dt \oint d\Omega \int_0^{X_{max}} dx \, x^2 \, \rho(r,z) \tag{4}$$

as well as the diffuse emission of unresolved pulsars

$$I_{dif}(E_k) = \frac{\epsilon}{\tau_p \Omega} \int_0^{t_{max}} dt \int_\Omega d\Omega' \int_{X_{max}}^\infty dx \, \rho(r,z) \, S_m(t, E_k) \tag{5}$$

where x is the distance coordinate along the line-of-sight. Please note that any change in the total number of pulsars has similar impact on the number of directly observable pulsars as on the γ-ray intensity of unresolved objects. Thus the direct detections provide a strong constraint on the pulsar contribution to the diffuse galactic γ-ray emission and limit our choice of ϵ/τ_p. With Eq.5 we can now calculate the spectrum of unresolved pulsars and compare it to the spectrum of the total observed diffuse emission. This is shown for the Galactic Center direction and for the anticenter in Fig.1, where we plot the fraction of the total observed emission which is due to pulsars. Here we used for the spatial distribution of pulsars a radial scale length r_c=3.5 kpc and a half-thickness of z_c=0.15 kpc. The result changes little when we choose an age-dependent thickness which would account for older pulsars being more distant from the plane. The maximum age t_{max}, we have used $2 \cdot 10^6$ years, has only little influence on the result since it affects the number of directly observable objects N_{det} nearly as much as the predicted emission of unresolved pulsars I_{dif}.

The parameters used here ($\tau_p^{-1} = 0.01$ years^{-1} and $\epsilon = 0.15$) imply that around 12 pulsars should be detectable by EGRET. Since six are already identified this would mean that another six unidentified sources are actually pulsars. The small fraction of unresolved EGRET sources which shows pulsar-like or Geminga-like γ-ray spectra [9] argues strongly against the bulk of unidentified sources being pulsars. Also a substantial fraction of the unidentified EGRET sources in the galactic plane appears to be variable which makes the identification of *all* unidentified sources with pulsars even more unlikely [10]. In total, we think that around 12 directly observable pulsars, of which six are already identified, is a reasonable number, and that thus the calculated contribution of unresolved pulsars can be taken as serious estimate. Considering the integrated intensity above 100 MeV in this example pulsars would cause 13% of the observed intensity in the Galactic Center direction and about 5% in the anticenter direction. We have compared the latitude distribution of the observed diffuse emission above 1 GeV to what our model predicts for the pulsar contribution. The result is given in Fig.2 where we show the distribution of the diffuse intensity at energies above 1 GeV as averages over longitude. We find that pulsars contribute only very close to the plane where the line-of-sights are long, but the observed emission does fall off much less rapidly with latitude than the emission of unresolved pulsars. Also the observed spectral discrepancy above 1 GeV seems to extend to latitudes $|b| \geq 5°$. Thus there must be additional effects playing a rôle for the observed diffuse γ-ray emission above 1 GeV.

The EGRET Team gratefully acknowledges support from the following: Bundesministerium für Bildung, Wissenschaft, Forschung und Technologie, Grant 50 QV 9095 (MPE); and NASA Cooperative Agreement NCC 5-95 (SU).

REFERENCES

1. Hunter S.D. et al. 1997, ApJ, in press
2. Bailes M., Kniffen D.A. 1992, ApJ, 391, 659
3. Yadigaroglu I.-A., Romani R.W. 1995, ApJ, 449, 211
4. Sturner S.J., Dermer C.D. 1996, ApJ, 461, 872
5. Pohl M. et al. 1997, ApJ submitted
6. Fierro J.M. 1995, Ph.D. Thesis, Stanford University
7. Taylor J.H., Manchester R.N., Lyne A.G. 1993, ApJS, 88, 529
8. Caraveo P.A. et al. 1996, ApJ Letters, 461, L91
9. Merck M. et al. 1996, A&AS, 120, C465
10. McLaughlin M.A., et al. 1996, ApJ, 473, 763

The Total Cosmic Diffuse Gamma-Ray Spectrum from 9 to 30 MeV Measured with COMPTEL

S.C. Kappadath*, J. Ryan*, K. Bennett[†], H. Bloemen[§],
R. Diehl[||], W. Hermsen[§], M. McConnell*, V. Schönfelder[||],
M. Varendorff[||], G. Weidenspointner[||] and C. Winkler[†]

*Space Science Center, University of New Hampshire, Durham, NH 03824, U.S.A.
[†]Astrophysics Division, ESTEC, NL-2200 AG Noordwijk, The Netherlands
[§]SRON-Utrecht, 3584 CA Utrecht, The Netherlands
[||]Max-Planck-Institut für extraterrestrische Physik, 85740 Garching, Germany

Abstract. A preliminary COMPTEL Cosmic Diffuse Gamma-Ray (CDG) spectrum from 800 keV to 30 MeV was presented earlier at the 3rd Compton Symposium. The COMPTEL results represent the first significant detection of the CDG radiation in the 9 to 30 MeV range. Using high-latitude data from the first 5 years of the mission we have performed a new detailed measurement of the 9 to 30 MeV spectrum with finer energy binning. The new improved results are in good agreement with our previous estimates and are compatible with power-law extrapolations from higher energies. The measured 9–30 MeV spectra from the Virgo and South Galactic Pole observations are consistent with each other.

INTRODUCTION

One of the principle goals of the COMPTEL mission is the study of the Cosmic Diffuse Gamma (CDG) radiation. The COMPTEL instrument is ideally suited to measure the CDG radiation because of its large detection area, wide field-of-view and long exposure time. A *preliminary* COMPTEL CDG spectrum from 800 keV to 30 MeV has been presented earlier [1,2], see figure 1. The 2–10 MeV flux was lower than the pre-COMPTEL estimates [3–5] and showed no evidence for an MeV-bump, at least not at the levels claimed previously. The 2–30 MeV flux was compatible with power-law extrapolations from higher [6,7] and lower [8] energies. Only upper-limits were derived below ~2 MeV due to the presence of uneliminated background. A word of caution is appropriate due to the *preliminary* nature of the earlier COMPTEL results [1,2].

We have continued our investigations since the earlier results [1,2]. We have analyzed more data, discovered additional radioactive isotopes [9,10], made improvements in our response calculations and begun investigations of the systematic errors. In this paper we present a new detailed measurement of the 9 to 30 MeV CDG spectrum with finer energy binning. We also present a comparison between the CDG spectra derived from Virgo and South Galactic Pole (SGP) observations. In this work the CDG flux refers to the total γ-ray flux from high galactic latitudes. A detailed description of the COMPTEL instrument can be found in the calibration paper [11].

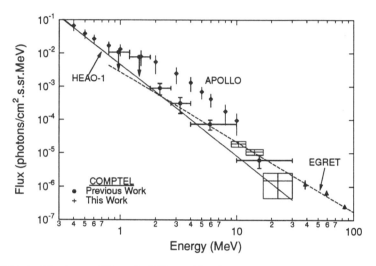

FIGURE 1. The latest (Virgo & SGP combined) 9–30 MeV CDG spectrum is shown together with the earlier COMPTEL results [1,2], the APOLLO measurements [5] and extrapolations from higher (EGRET) [6,7] and lower (HEAO-1) [8] energies.

OBSERVATIONS

In this work we have analyzed high-latitude observations from the first 5 years of the mission ($\sim$266 days of observations). Specifically, the data are all the Virgo observations ($b^{II} \sim 60°$) and most observations towards the SGP ($b^{II} < -50°$). The observed Virgo and SGP regions represent large areas towards the North and South Galactic Polar directions because of the wide COMPTEL field-of-view ($\sim$1.5 steradians) and the addition of data from adjacent pointings. The high-latitude observations minimize the γ ray contribution from the Galaxy. Also, the data are accumulated only during the periods when the Earth is outside the COMPTEL field-of-view. This eliminates any contamination from atmospheric γ rays.

DATA ANALYSIS

Since the CDG radiation is considered to be isotropic in space and constant in time, we expect no unique spatial or temporal signatures to distinguish the CDG radiation from the background radiation. The CDG measurement is made by first subtracting every instrumental background source and then attributing the residual flux to the CDG radiation.

In general the instrumental background can be decomposed into 'prompt' and 'long-lived' components. The prompt background is instantaneously produced by proton and neutron interactions in the spacecraft. Hence, it refers to the component that modulates with the instantaneous local cosmic-ray flux. The prompt background is assumed to vary linearly with the veto-scalar rates. The veto-scalar-rates are trigger-rates of the charge-particle shields (veto-domes) surrounding the main detectors and used in anti-coincidence [11]. Thermal neutron capture by the hydrogen in the upper detector produces a prompt 2.223 MeV photon ($\tau_{1/2}$ ~100 μs) which is well described by a linear function in veto-rate, hence we expect all prompt background to behave linearly with the measured veto-rate.

The long-lived background events are due to de-excitation photons from activated radioactive isotopes with long half-lives ($\tau_{1/2}$ >30 sec). Their decay rate is not directly related to the instantaneous cosmic-ray flux because of the long half-lives. The energy spectrum is used to determine the absolute contribution of each of the long-lived background isotopes [1].

In constructing the CDG spectrum, the first step is to determine the scattered γ-ray (forward-peak) count rates by explicitly fitting the Time-of-Flight (ToF) spectrum [1]. However, there are some additional background events in the forward-peak that originate in the upper part of the detector (both prompt and long-lived) that need to be accounted for. For each bin in total energy, the ToF-fitted count rates are ordered with the instantaneous veto-rates to construct veto-growth-curves (VGCs). In the absence of long-lived background (above ~4.2 MeV), the VGCs are fitted with a straight line to determine the count rate at zero veto-rate. Under the assumption that zero veto-rate corresponds to zero cosmic ray flux, the extrapolated flux at zero veto-rate is the desired CDG count rate. Below ~4.2 MeV, there are numerous long-lived activation lines whose behaviors are not accurately described by a linear VGC. Each of the radioactive isotopes must be accounted for individually. This includes identifying the isotopes, estimating their activities and subtracting their contributions to the VGC prior to correcting the prompt background. This is part of on-going work and results will be reported in future publications.

The CDG flux is then determined by deconvolving the resultant CDG count spectrum with the computed instrument response. The instrument response is determined from Monte Carlo simulations of a diffuse isotropic power-law source propagated through a detailed COMPTEL mass model.

RESULTS

Here we present the improved 9 to 30 MeV CDG spectrum. The 9-30 MeV range represents one of the optimum COMPTEL energy bands due to the high S/N ratio (>30%). The S/N decreases below 9 MeV, due to the prompt background dominating the 4-9 MeV band and due to the addition of the long-lived background component below ~4 MeV. Below 9 MeV, we presently feel the need for further investigations (especially of the systematic errors) before presenting the improved CDG spectrum.

The 9 to 30 MeV spectrum for the combined and separate Virgo and SGP observations are plotted in figures 1 and 2 respectively. The latest 9-30 MeV spectrum is consistent with our earlier results [1,2] and is compatible with power-law extrapolation from higher (EGRET) energies [6,7]. The COMPTEL results represent the first significant detection (7σ) of the 9-30 MeV CDG flux.

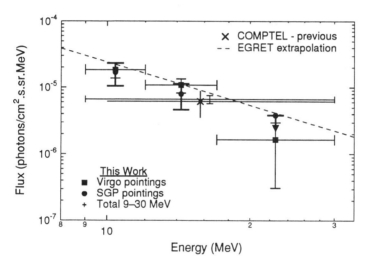

FIGURE 2. The CDG spectrum for the separate Virgo and SGP observations along with the total 9-30 MeV flux and the previous COMPTEL 10-30 MeV measurement [1].

The 9-30 MeV flux was computed for time- and pointing-separated subsets of the data (figure 3). They are consistent with a constant CDG flux (the null hypothesis of a constant CDG is rejected at the 68% level). Although the total 9-30 MeV flux from SGP is ~ 31% lower than the flux from Virgo pointings (figure 3), the null hypothesis of a uniform CDG is rejected only at the 71% level. The CDG spectra from the Virgo and the SGP observations (figure 2 and 3) are consistent with one another, even though the COMPTEL fluxes have not been corrected for the Galactic or point-source contribution.

Only *statistical* errors are plotted in the COMPTEL CDG measurements (figures 1, 2 and 3). Three major sources of *systematic* errors are: 1) the ToF-fit function, since the ToF-continuum [1] can be modeled with a quadratic or an exponential function, 2) the choice of veto-dome rates to generate the VGCs, since there are four independent veto-domes and 3) the COMPTEL effective-area calculations. We compute a total systematic error of ~15% for the 9–30 MeV flux result.

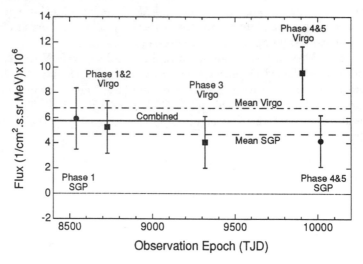

FIGURE 3. The time- and pointing-separated 9–30 MeV CDG flux together with the mean Virgo, mean SGP and combined 9–30 MeV flux.

Acknowledgements. This work is supported by NASA grant NAS5-26645, DARA grant 50 QV 90968 and the Netherlands Organization for Scientific Research (NWO).

REFERENCES

1. Kappadath, S.C. et al., 1996, A&AS, **120**, C619
2. Kappadath, S.C. et al., 1995, Proc. of 24th ICRC (Rome), **2**, 230
3. Schönfelder, V. et al., 1980, ApJ, **240**, 350
4. White, R.S. et al., 1977, ApJ, **218**, 920
5. Trombka, J.I. et al., 1977, ApJ, **212**, 925
6. Sreekumar, P. et al., 1997, ApJ, submitted
7. Kniffen, D.A. et al., 1996, A&AS, **120**, C615
8. Kinzer, R.L. et al., 1997, ApJ, **475**, 361
9. Oberlack, U. et al., 1997, these proceedings
10. Varendorff, M. et al., 1997, these proceedings
11. Schönfelder, V. et al., 1993, ApJS, **86**, 657

The Cosmic γ-Ray Background from Supernovae

K. Watanabe[†1], D.H. Hartmann[*], M.D. Leising[*], L.-S. The[*],
G.H. Share[⋆] and R.L. Kinzer[⋆]

[†]*NASA/GSFC code 660.1 Greenbelt, MD 20771*
[*]*Department of Physics and Astronomy
Clemson University Clemson, SC 29634-1911*
[⋆]*E.O. Hulburt Center for Space Research
Naval Research Laboratory, Washington D.C. 20375*

Abstract. The Cosmic γ-ray Background (CGB) near photon energies $E_\gamma \sim 1$ MeV may be partly, or even completely, due to γ-ray production in supernovae. In particular, γ-ray line emission from the decay chain $^{56}Ni \rightarrow ^{56}Co \rightarrow ^{56}Fe$ (847, 1238, 1770, 2599, 2030, 3250 keV) is the dominant source for the SN-induced CGB. Although iron synthesis occurs in all types of supernovae, SNIa contribute predominantly to the background mostly due to higher photon escape probabilities. Estimates of the global star formation history in the local universe yield a present-day rate density of $\simeq 3.7 \times 10^{-2} h^3 \; M_\odot \; Mpc^{-3} \; yr^{-1}$, but the star formation rate was about ten times higher at a redshift of order unity. The rate then decreases again, reaching the present-day value near redshifts $\simeq 5$. In addition to γ-rays from SNIa we also consider contributions to the CGB due to lines from SNII (such as ^{26}Al at 1.8 MeV, ^{44}Ti at 1.157 MeV, and ^{60}Co at 1.17 & 1.33 MeV). The γ-ray spectrum of Model W10HMM (Pinto & Woosley [8]) was time integrated to derive a template spectrum for Type II supernovae. For SNIa we also included the γ-ray continuum, using the W7 model (Nomoto *et al.* [5]), integrated over 600 days, as template. We discuss the various contributions of supernovae to the CGB, and emphasize the value of γ-ray observations in the MeV range as a potential tool for studies of cosmic chemical evolution.

INTRODUCTION

The CGB near photon energies $E_\gamma \sim 1 MeV$ may be partly, or even completely, due to γ-ray production in supernovae. In particular, γ-ray lines from the reaction chain $^{56}Ni \rightarrow ^{56}Co \rightarrow ^{56}Fe$ (847, 1238, 1770, 2599, 2030, 3250 keV) are the dominant sources of SN-induced CGB (Clayton & Silk 1969 [1], Clayton & Ward 1975 [2], The *et al.* [10]). Although iron synthesis occurs in all

---

[1]) also *USRA*

types of SN, SNIa contribute predominantly, because the product of SN rate and Fe yield, $(Rate \times yield)_{Ia} = R_{Ia}M_{Ia} = \dot{M}_{Ia} \gg \dot{M}_{II,Ib}$ ($R_{Ia} \leq 0.2R_{II,Ib}$ but $M_{Ni} \sim 0.5M_\odot$ in Ia and $\sim 0.075M_\odot$ in II). In addition, the massive envelopes in SNII do not allow many of the γ-rays to escape.

CGB Due to Lines from SNII Radioactivity

The ejection of some amount (M_{ej}) of radioactive material generates (assuming 1γ per decay) a total number of photons

$$N_\gamma = \frac{M_{ej}}{Am_u} = 1.2 \times 10^{53} A^{-1} M_{-4}, \tag{1}$$

where A is the mass number of the radioactive element, m_u is the atomic mass unit, and

$$M_{-4} = \frac{M_{ej}}{10^{-4} M_\odot}. \tag{2}$$

The supernovae rate is given by

$$R_{SNII} = \xi_{SNR} \dot{M}_\star \ yr^{-1}, \tag{3}$$

where $\dot{M}_\star$ is the star formation rate in $M_\odot yr^{-1}$, and ξ_{SNR} has units of $M_\odot^{-1}$. In our galaxy, $\dot{M}_\star = (1-10) M_\odot yr^{-1}$ and $R_{SNII} \sim 310^{-2} yr^{-1}$ (Timmes et al. [12]). Thus ξ_{SNR} is of order $10^{-2} M_\odot^{-1}$. Recent work on the global star formation history of the universe (Cole et al. [3]) suggsts a present-day star formation density

$$\dot{\rho}_\star(z=0) = \dot{\rho}_\star(0) \simeq 3.7 \times 10^{-2} h^3 \ M_\odot \ Mpc^{-3} \ yr^{-1}, \tag{4}$$

where h is related to the Hubble constant by $H_0 = 100 \ h \ kms^{-1} Mpc^{-1}$. The γ-ray production rate $\dot{n}_\gamma$ measured in photons $Mpc^{-3} \ yr^{-1}$ is thus given by

$$\dot{n}_\gamma = \dot{\rho}_\star(0) \xi_{SNR} N_\gamma. \tag{5}$$

The differential CGB flux can be expressed as

$$\frac{\partial F}{\partial \Omega}(\gamma cm^{-2} s^{-1} sr^{-1}) = F_0 C(z), \tag{6}$$

where the normalization factor is defined with Hubble length $L_H = c/H_0$ as

$$F_0 = (4\pi)^{-1} \dot{n}_\gamma L_H P_{esc} \simeq 3.53 \times 10^{-3} h^2 A^{-1} M_{-4} \xi_{SNR} P_{esc}, \tag{7}$$

where P_{esc} is the γ-ray escape probability from the supernova, and

$$C(z) = \int_0^\infty dz(1+z)^{-1} E(z)^{-1} \eta(z). \tag{8}$$

$E(z)$ is a function of z described in Peebles' 1993 [6] equation (13.3),

$$E(z) = \left[\Omega(1+z)^3 + \Omega_R(1+z)^2 + \Omega_\Lambda\right]^{1/2}. \tag{9}$$

For $\Omega_\Lambda = 0$, it simplifies to

$$E(z) = (1+z)(1+\Omega z)^{1/2}. \tag{10}$$

We use $\Omega = 1$ for the calculations presented here. The evolution of the star formation rate (SFR) with redshift based on cosmic chemical evolution sudies (Pei & Fall 1995 [7]; Lilly et al. [4]) is given as

$$\log(\eta(z)) = \begin{cases} 4\log(1+z) & \text{for } z \leq z_p \\ 4\log(1+z_p) - 0.25(z - z_p) & \text{for } z > z_p \end{cases}, \tag{11}$$

where $z_p \sim 1$. The global SFR was about 10 times higher at $z \sim 1$ than it is today.

CGB Due to γ-ray Continuum from SNII

Model W10HMM of Pinto & Woosley [8] [9] is integrated over time to calculate a template source function S(E) for the γ-ray continuum due to ^{56}Ni produced in SNII. The number of γ-rays per unit energy from SNII as a function of energy is given by

$$N_\gamma(E) = N_\gamma S(E) \quad (\gamma/keV). \tag{12}$$

The differential CGB flux is

$$\frac{\partial^2 F(E_\gamma)}{\partial E \partial \Omega}(\gamma cm^{-2} keV^{-1} s^{-1} sr^{-1}) = F_1 C_{con}(z), \tag{13}$$

where

$$F_1 = (4\pi)^{-1} \dot{n}_\gamma L_H \simeq 3.53 \times 10^{-5} h^2 A^{-1} M_{-4} \xi_0, \tag{14}$$

and

$$C_{con}(z) = \int_0^\infty dz E(z)^{-1} \eta(z) S(E_\gamma \times (1+z)). \tag{15}$$

ξ_0 is a scaling factor defined as $\xi_{SN} = \xi_0 10^{-2}$. Timmes et al. [12] investigated the star formation rate ($\dot{M}_\star^{MW}$) and the supernovae rate (R_{SNII}^{MW}) in the Milky Way utilizing the 1.8 MeV gamma-ray emission from radioactive ^{26}Al. From Eq.(3)

$$R_{SNII}^{MW} = \xi_0 10^{-2} \dot{M}_\star^{MW} yr^{-1} \tag{16}$$

we can thus calibrate the proportionality constant. Assuming Salpeter's IMF exponent, we see from Figure 1 of Timmes et al. [12] that $\xi_0 \sim 1$ for SNII and $\sim 1 \times 1/3$ for SNIa's (e.g. Tsujimoto et al. [13]). We take $M_{-4} = 7.5 \times 10^2$ for SNII (Pinto & Woosley [9]).

CGB Due to γ-ray Continuum & Positrons from SNIa

We calculate the contribution from SNIa which dominates in MeV region. The formalism of the calculation is the same as that of SNII. We neglect the time delay between SNIa and SNII for a given star formation history. We use the fully mixed version of W7 (Nomoto et al. [5]) of SNIa integrated in time over 600 days as the source function, $S(E)$, in Eq.(12). Note, The et al. [10] also use the same model. The mass in Eq.(14) is $M_{-4} = 5 \times 10^3$. The result of our calculation of the CGB contribution from SNIa agrees with the result of The et al. [10] multiplied by 2/3. This correction facor of 2/3 is the fraction of Fe from SNIa (Thielemann et al. [11]).

Many of the positrons in the Galaxy are probably produced by ^{56}Co decay in SNIa. Positron production in the decay of ^{56}Co occurs with a probability of $f_{\beta+} = 0.19$. One fourth of the positrons are in the ^1p_s state ($f_{^1p_s} = 1/4$, and annihilate with electrons giving 2 γ-rays of $E = m_e c^2 = 511$ keV. About $f_{esc} = 3\%$ of the γ-rays escape the SN environment. The remaining positrons are in the ^3p_s state resulting in the emission of three γ-rays, which produces a continuum in the 0 - 511 keV range. For $M_{56} \sim 0.5 M_\odot$ the number of positrons per SNIa is

$$N_{positron} = \frac{M_{56_{Co}}}{56 m_u} \ . \tag{17}$$

The number of 511 keV γs per SNIa is thus $N_{511} = N_{positron} \times f_{\beta+} \times f_{^1p_s} \times f_{esc} \times 2 \sim 3 \times 10^{52}$.

Summary

Our result agrees with previous work of The et al. [10]. Figure 1 shows that our estimated background flux is consistent with the recent CGB measurements with SMM (Watanabe et al.) [14]. This supports the notion that SNIa are the dominant source for the CGB in the MeV region. Accurate measurements of the CGB in this energy band provide a valuable constraint on theoretical models for γ-ray emission from supernovae and the cosmological history of the star formation rate.

REFERENCES

1. Clayton, D. D. and Silk, J., *ApJ*,**158**,L43(1969).
2. Clayton, D. D. and Ward, R. A., *ApJ*,**198**,241(1975).
3. Cole, S. et al. , *MNRAS*,**271**,781 (1994).
4. Lilly, S. J. et al. , *ApJ*, **460**,L1(1996).
5. Nomoto, K. et al. , *ApJ*, **286**, 644(1984).

6. Peebles, P.J.E., *Principles of Physical Cosmology*, NJ:Princeton University Press,1993.
7. Pei, Y. C. and Fall, S.M., *ApJ*, **454**, 69(1995).
8. Pinto, Philip A. and Woosley, S. E., *Nature*, **333**, 534(1988).
9. Pinto, Philip A. and Woosley, S. E., *ApJ*, **329**, 820(1988).
10. The, Lih-Sin *et al.* , *ApJ*, **403**, 32(1993).
11. Thielemann, F.-K. *et al.* , in *Supernovae* , ed. S Woosley, New York: Springer-Verlag,1991.
12. Timmes, F. X. *et al.* , *ApJ*, **479**, 760(1997).
13. Tsujimoto, T *et al.* , *MNRAS*, **277**, 945(1995).
14. Watanabe, K *et al.* , in Preparation.

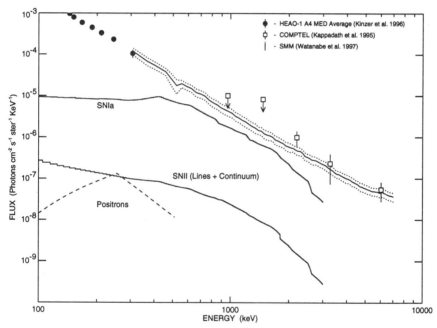

FIGURE 1. CGB from SNIa, SNII & positrons. We plot various observations of the CGB, including our latest analysis of the SMM data (Watanabe *et al.*) [14]. The dotted lines indicate the range of uncertainties in the best fit model for the SMM data (solid line). It is apparent that SNIa dominate the supernova-induced γ-ray background, and that the total emission from supernovae is consistent with the observed background. At energies below a few hundred keV contributions from Seyfert galaxies begin to dominate, and above 2-3 MeV the bulk of the CGB is probably due to blazar emission. Supernovae fill the gap between these classes of AGNs without the need for fine-tuning of any of the relevant parameters.

The γ-Ray and Neutrino Background and Cosmic Chemical Evolution

D. H. Hartmann*, K. Watanabe†, M. D. Leising*, L.-S. The*, and S. E. Woosley*

*Department of Physics and Astronomy
Clemson University Clemson, SC 29634-1911
†NASA/GSFC code 660.1 Greenbelt, MD 20771
*Board of Studies in Astronomy and Astrophysics
University of California, Santa Cruz, CA 96064

Abstract. The global star formation history of the universe and the associated supernova rate are now reasonably well known. Core collapse supernovae are copious sources of neutrinos, and produce an integrated neutrino flux at Earth of ~ 10 neutrinos/cm^2 s for each flavor. While terrestrial and extraterrestrial neutrino backgrounds exceed the supernova flux below ~ 15 MeV and above ~ 40 MeV, in the intermediate band the neutrino background from Type II/Ib supernovae may be detectable. Similarly, Type Ia supernovae produce the bulk of the extragalactic γ-ray background in the MeV range, while blazars dominate at higher energies and Seyfert galaxies at lower energies. The γ-ray and neutrino background probe cosmic chemical evolution.

I INTRODUCTION

The total solar neutrino flux at Earth is $\sim 10^{11}$ cm^{-2} s^{-1}, with a spectrum that truncates above ~ 15 MeV. At energies above ~ 40 MeV Cosmic Ray interactions in the atmosphere generate a dominating neutrino background. Between 20 and 30 MeV a low-background window exists (Koshiba 1992), through which one could discover the integrated neutrino background from stellar explosions (Guseinov 1966; Bisnovatyi-Kogan & Seidov 1984; Krauss, Glashow, & Schramm 1984; Dar 1985; Galeotti & Raiteri 1988). This background provides a trace of cosmic chemical evolution (Malaney & Chaboyer 1996; Malaney 1997), similar to the MeV γ-ray backgound, which is probably also due to cosmic supernova activity (Clayton & Silk 1969; Clayton & Ward 1975; The et al. 1993; Watanabe et al. 1997).

II THE NEUTRINO BACKGROUND

The total neutrino flux from past supernovae is estimated to be in the range 1–100 cm^{-2} s^{-1} (e.g., Woosley, Wilson, & Mayle 1986; Zhang et al. 1988; Totsuka 1992; Totani & Sato 1995; Totani, Sato, & Yoshii 1996(TSY)). New generation neutrino detectors, such as super-Kamiokande, may be able to detect this background (Totsuka 1992). The optimal energy range for detection (bounded by solar and atmospheric neutrinos) is $E_\nu = 20-30$ MeV. In this window the supernova neutrino flux is ~ 0.8 cm^{-2} s^{-1} (TSY); a challange to experimenters.

The contribution to the background is dominated by Type II/Ib supernovae, which radiate over 99% of the neutron star binding energy as thermal neutrinos (e.g., Woosley, Wilson, & Mayle 1986). Since these supernovae trace massive stars one can use the luminosity density in the blue band (or the H$_\alpha$ flux) as an indicator of the collapse rate. H$_\alpha$ observations, in conjunction with a Miller-Scalo Initial Mass Function (IMF), provide an estimate of the star formation rate in the present universe (Cole et al. 1994; Gallego et al. 1995)

$$\dot{\rho}_*(0) = (4 \pm 2)\ 10^{-2}\ h^3 \quad (M_\odot\ \text{Mpc}^{-3}\ \text{yr}^{-1})\ , \tag{1}$$

where h is the Hubble constant in units of 10^2 km s^{-1} Mpc^{-1}. Similar values follow from independent studies of galaxy luminosity functions from redshift surveys (e.g., Lilly al. 1996) and the Hubble Deep Field (Madau 1997; Connolly et al. 1997). The star-formation rate density can be converted to a supernova rate density by linear scaling. If the IMF is a universal function and if star formation in the Milky Way is representative of star formation in the universe, we scale a SFR of 3.7 M$_\odot$ yr^{-1} as $\zeta\ 10^{-2}$ SNe yr^{-1}. For SNII/Ib we use $\zeta = 2$ (Timmes, Diehl, & Hartmann 1997). The differential neutrino flux from cosmological core collapses is

$$\frac{\partial F_\nu}{\partial E_\nu} = 10^{-4}\ L_H\ \zeta\ h^3 \int_0^\infty dz\ E(z)^{-1} \eta(z)\ \Phi\left(E_\nu(1+z)\right)\ (\text{MeV Mpc}^2\ \text{yr})^{-1}, \tag{2}$$

where $L_H = c/H_0 = 3{,}000\ h^{-1}$ Mpc, $\eta(z) = \dot{\rho}_*(z)/\dot{\rho}_*(0)$ describes source evolution, Φ is the time integrated template spectrum, and

$$E(z) = (1+z)\ (1+\Omega z)^{1/2}\ . \tag{3}$$

The present-day star formation rate in the universe is significantly lower than it was during most of cosmic history (Pei & Fall 1995; Lanzetta, Wolfe, & Turnshek 1995; Timmes, Lauroesch, & Truran 1995; Malaney & Chaboyer 1996; Lu, Sargent, & Barlow 1996; Madau 1997; Connolly et al. 1997). From these studies we derive a schematic evolution function; $\text{Log}(\eta) = A\ \text{Log}(1+z)$ for redshifts less than $z_p \sim 1$, and $A \sim 4$. Beyond the peak, $\text{Log}(\eta)$ decreases linearly with redshift (slope B ~ -0.25 from Pei & Fall 1995).

To calculate the neutrino background we need an average integrated neutrino spectrum per event, Φ. The calculations of Woosley, Wilson, & Mayle (1986) suggest that the mean supernova spectrum resembles that of a 15 $M_\odot$ star (Totani & Sato 1995). We assume that the neutrino spectrum is thermal, and that the released binding energy is $E \sim 3 \ 10^{53} \ \beta$ ergs, split equally among six neutrino species. The electron neutrino temperatures obtained from theoretical models and observations of SN87A is $T_\nu \sim 4$ MeV. The integrated supernova background spectrum is

$$\frac{\partial F_\nu}{\partial E_\nu} \sim 5.6 \ h^2 \ \zeta \ \beta \ T_\nu^{-4} \ E_\nu^2 \ G(E_\nu) \quad (\nu \ \text{cm}^{-2} \ \text{s}^{-1} \ \text{MeV}^{-1}) \ , \qquad (4)$$

where

$$G = \int_0^\infty dz \ E(z)^{-1} \ (1+z)^2 \ \eta(z) \ [\exp(E_\nu(1+z)/T_0) + 1]^{-1} \ C_c \ . \qquad (5)$$

The function C_c takes into account that the neutrino temperature may decrease with time (see Hartmann & Woosley 1997 for details). Our "standard model" uses $h = 0.75$, $\Omega = 1$, $\Lambda = 0$, $\zeta = 2$, $A = 4$, $B = -1/4$, $z_p = 1$, $\beta = 1$, and $T_\nu(0) = 4$ MeV, and no cooling. This produces a total flux of 8 cm^{-2} s^{-1}, 5 times less that that of TSY. Most of the excess is due to their "burst" of star formation at early epochs. The flux in their 15–40 MeV band (0.8 cm^{-2} s^{-1}) is very similar to our value (see Hartmann & Woosley 1997 for details).

III THE GAMMA-RAY BACKGROUND

The cosmic gamma-ray background (CGB) near photon energies $E_\gamma \sim$ 1MeV may be partly, or completely, due to γ-rays from iron production in supernovae (Clayton & Silk 1969, Clayton & Ward 1975, The et al. 1993). SNIa dominate, because their product of rate and yield is larger, and also because massive envelopes in SNII inhibit γ-ray escape. The number of photons released from some amount (M_{ej}) of radioactive material is $N_\gamma = M_{ej}/Am_u = 1.2 \times 10^{53} A^{-1} M_{-4}$, where A is the mass number of the radioactive element, m_u is the atomic mass unit, and $M_{-4} = M_{ej}/10^{-4} M_\odot$. Given the cosmic star formation history, the γ-ray production rate is $\dot{n}_\gamma = \dot{\rho}_\star \ \xi_{SNR} N_\gamma$, and the differential background flux is

$$\frac{\partial F}{\partial \Omega}(\gamma \ \text{cm}^{-2} \ \text{s}^{-1} \ \text{sr}^{-1}) = F_0 \ C(z) \ , \qquad (6)$$

where

$$F_0 = (4\pi)^{-1} \dot{n}_\gamma L_H \simeq 3.53 \times 10^{-3} h^2 A^{-1} M_{-4} \ \xi_{SNR} \ , \qquad (7)$$

and

$$C(z) = \int_0^\infty dz (1+z)^{-1} E(z)^{-1} \eta(z). \qquad (8)$$

Estimating the emergent flux from shortlived radioactivity requires transport calculations, while longlived radioactivities (such as ^{44}Ti and ^{26}Al) produce photons that escape the ejecta without interactions (but contribute little to the flux because of their low abundances, Watanabe et al. 1997). ^{56}Co ($t_{1/2} = 77.12d$) decays to ^{56}Fe with line emission at 0.847 MeV (100%), 1.04 MeV (14%),1.24 MeV (68%), 1.77 MeV (16 %), 2.03 MeV (12 %),2.6 MeV (17%), and 3.24 MeV (12.5%). Compton scattering of these line photons generates a continuum spectrum below 3.24 MeV. For SNeII we use the radiation transport calculations for the W10HMM supernova model (Pinto & Woosley 1988) to determine the time integrated gamma-ray spectrum S(E). The number of γ-rays per unit energy is given by $N_\gamma(E) = N_\gamma S(E)(\gamma/keV)$. The differential CGB flux follows from

$$\frac{\partial^2 F(E_\gamma)}{\partial E \partial \Omega}(\gamma cm^{-2} keV^{-1} s^{-1} sr^{-1}) = F_1\, C_{con}(z)\,, \qquad (9)$$

where

$$F_1 = (4\pi)^{-1} \dot{n}_\gamma L_H \simeq 3.53 \times 10^{-5}\, h^2\, A^{-1}\, M_{-4}\, \xi_0\,, \qquad (10)$$

and

$$C_{con}(z) = \int_0^\infty dz E(z)^{-1} \eta(z)\, S(E_\gamma \times (1+z))\,. \qquad (11)$$

ξ_0 is a scaling factor defined as $\xi_{SN} = \xi_0 10^{-2}$. $\xi_0 \sim 1$ for SNII (Timmes et al. 1997) and $\sim 1/3$ for SNIa's (e.g. Tsujimoto et al. 1995). $M_{-4} = 7.5 \times 10^2$ for SNII. To determine the flux from SNIa's we use the fully mixed W7 model (Nomoto et al. 1984), integrated over 600 days. The average ejecta mass is assumed to be $M_{-4} = 5 \times 10^3$. We find a relative SNII contribution of $\sim$ 1%, confirming the results of The et al. (1993). The SNIa γ-ray background nearly matches the observed flux (see Watanabe et al. 1997), implying that supernovae are responsible for most of the CGB in the MeV regime.

IV DISCUSSION

The observed gamma-ray background can be explained as a superposition of three components. Seyfert galaxies dominate below a few hundred keV, blazar emission dominates above a few MeV, and the MeV range is due to photons from SNIa radioactivity ! In the case of neutrinos we focus on the 20–30 MeV window where solar or terrestrial backgrounds are low. While the total flux is ~ 10 cm^{-2} s^{-1}, the flux in the "window" is one order of magnitude lower. The current observational limit on the electron neutrino background is ~ 580 cm^{-2} s^{-1} in the 20–50 MeV band (Hirata 1991) and ~ 230 cm^{-2} s^{-1} in the 19–35 meV band (Zhang et al. 1988). Some enhancement of the detection probability is possible if $\nu_\mu \to \nu_e$ conversion occurs at a significant level (Totani & Sato 1996), because muon neutrinos have higher energies. If we can detect the supernova neutrino background and obtain high-quality measurements of the

MeV γ-ray background, a remarkable multi-particle tracer of cosmic chemical evolution would become available. The CGB already constrains the cosmic star formation history, and we anticipate the detection of the related neutrino flux by Super-Kamiokande, which can detect a flux of ~ 20 cm^{-2} s^{-1} during 5 years of operation (Nakamura et al. 1994). [This work was supported through NASA grants NAG 5-1578 and NAG 5-2843, and NSF grant AST 94-17161]

REFERENCES

1. Bisnovatyi-Kogan, S., & Seidov, Z. 1984, *Ann. N.Y. Acad. Sci.* 422, 319
2. Clayton, D. D., & Ward, R. A. 1975, ApJ 198, 241
3. Clayton, D. D., & Silk, J. 1969, ApJ 158, L43
4. Cole, S., et al. 1994, MNRAS 271, 781
5. Connolly, A. J., et al. 1997, astro-ph/9706255
6. Dar, A. 1985, Phys. Rev. Lett. 55, 1422
7. Gallego, J., et al. 1995, ApJ, 455, L1
8. Galeotti, P., & Raiteri, C. M. 1988, Astro. Lett. and Comm. 27, 49
9. Guseinov, O. K. 1966, AZh 43, 772
10. Hartmann, D. H., & Woosley, S. E. 1997, Astroparticle Physics, in press
11. Hirata, K. 1991, PhD thesis, U. of Tokyo, ICRR Report 239-91-8
12. Koshiba, M. 1992, Phys. Rep. 220, 229
13. Krauss, L. M., Glashow, S. L., & Schramm, D. N. 1984, Nature, 310, 191
14. Lanzetta, M. K., Wolfe, A. M., & Turnshek, D. A. 1995, ApJ 440, 435
15. Lu, L., Sargent, W. L. W., & Barlow, T. A. 1996, ApJ, in press
16. Lilly, S. J., Le Févre, O., Hammer, F., & Crampton, D. 1996, ApJ 460, L1
17. Madau, P 1997, astr-ph/9612157
18. Malaney, R. A., & Chaboyer, B. 1996, ApJ 462, 57
19. Malaney, R. A. 1997, astro-ph/9612012 preprint
20. Nomoto, K. et al., 1984, ApJ 286, 644
21. Nakamura, K., et al. 1994, in *Physics and Astrophysics of Neutrinos*, (Springer Verlag), eds. M. Fukugita and A. Suzuki, p. 249
22. Pei, Y. C., & Fall, S. M. 1996, ApJ 454, 69
23. Pinto, P. A. and Woosley, S. E. 1988, ApJ 329, 820
24. The, L.-S., Leising, M. D., & Clayton, D. D. 1993, ApJ 403, 32
25. Timmes, F. X., Lauroesch, J. T., & Truran, J. W. 1995, ApJ 451, 468
26. Timmes, F. X., Diehl, R., & Hartmann, D. H. 1997, ApJ, 479, 760
27. Totani, T., & Sato, K. 1995, Astroparticle Phys. 3, 367
28. Totani, T., & Sato, K. 1996, Int. J. Mod. Phys. D, 5, 519
29. Totani, T., Sato, K., & Yoshii, Y. 1996, ApJ 460, 303
30. Totsuka, Y. 1992, Rep. Prog. Phys. 55, 377
31. Tsujimoto, T et al. 1995, MNRAS 277, 945
32. Watanabe, K et al. 1997, these proceedings
33. Woosley, S. E., Wilson, J. R., & Mayle, R. 1986, ApJ, 302, 19
34. Zhang, W., et al. 1988, Phys. Rev. Lett. 61, 385

The contribution of blazars to the extragalactic diffuse γ-ray background

A. Mücke[1] & M. Pohl[2]

(1) MPI für extraterrestrische Physik, Postfach 1603, D-85748 Garching, Germany
(2) Danish Space Research Institute, Juliane Maries Vej 30, 2100 Copenhagen 0, Denmark

Abstract. We present results of a calculation of the blazar contribution to the extragalactic diffuse γ-ray background (EGRB) in the EGRET-energy range. Our model is based on the non-thermal emission processes known to be important in blazar jets, and on the unification scheme of radio-loud AGN. The background calculations place a lower limit of 5-16% AGN-contribution to the EGRB.

I INTRODUCTION

One of the most important results of the CGRO Energetic Gamma Ray Telescope (EGRET) in the field of extragalactic astronomy is the existence of an isotropic diffuse γ-ray background (EGRB) (e.g. Thompson et al. 1993).

With the recent identifications of over 60 blazars, the idea of a superposition of unresolved AGN as the origin of the EGRB becomes likely. The most common method to estimate the point source contribution from radio-loud blazars is to integrate the scaled radio luminosity function (e.g. Stecker et al. 1996). However, a simple direct luminosity correlation between radio and γ-rays is at least highly noisy if it exists at all (Mücke et al. 1997).

There exists strong evidence for relativistic beaming in EGRET-blazars (Dermer & Gehrels 1995) due to relativistic bulk motion of the emitting region. If this is the case, there must exist a population of sources with a jet pointing not towards us, which cannot be resolved and which contribute to the EGRB.

The purpose of this work is to calculate the integrated spectral contribution of unresolved flat-spectrum radio quasars (=FSRQs) and BL Lac objects to the EGRB in the EGRET energy range (∼30 MeV - 10 GeV) using a model which is based on the emission processes in blazars (Dermer & Schlickeiser 1993) and the unification scheme of radio-loud AGN (for a review see Urry & Padovani 1995). The most probable parent population of FSRQs are the Fanaroff-Riley II (=FR II) radio galaxies, while the parent population of BL

Lacs are the FR I radio galaxie. We use a Hubble constant of $H_0 = 75$ km s^{-1} Mpc^{-1} and $q_0 = 0.5$.

II THE MODEL

A Emission processes

The calculation of the source spectrum includes the inverse Compton process of soft photons from the accretion disk (EIC), the (homogeneous) synchrotron-self-Compton process (SSC) and the synchrotron process. We use the formalism of Dermer & Schlickeiser (1993) (DS) which implicitely assumes that energy losses due to EIC are greater than due to SSC or synchrotron emission. Furthermore, we assume that the initial e^--spectrum $N_e(\gamma)$ injected into the plasma stream follows a power law with $N_e \sim \gamma^{-2}$ between the limits $\gamma_1 < \gamma < \gamma_2$ for all blazars and evolves as the e^- lose their energy.

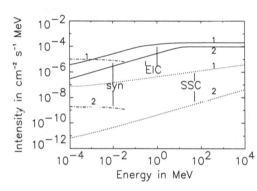

FIGURE 1. Time averaged EIC-, SSC- and synchrotron-spectra for instantaneous injection of e^- (1) and a continuous injection into a moving plasma blob (2). Source parameters are: bulk Lorentz factor $\Gamma = 10$, $\gamma_1 = 10^2$, $\gamma_2 = 10^6$, magnetic field B=2G, injection height $z_i = 300$ gravitational radii r_g, injected e^--energy $E_{inj} = 10^{45}$erg, viewing angle $\Theta = 5°$, radius of the outmoving blob $r_b = 30 r_g$, accretion disk luminosity $L_0 = 10^{44}$erg/s, source distance = 1.2 Gpc.

We consider two cases of electron injection into the jet: instantaneous injection at height z_i above the accretion disk and continuous injection into a moving plasma blob. The resulting e^--spectrum of the latter case is calculated by integrating the initial e^--spectrum over injection height z_i during the observation time Δt. For a low-energy cut-off $\gamma_1 > 1$ the equilibrium e^--spectrum follows a broken power law with $N_e \sim \gamma^{-2}$ for $\gamma < \gamma_1$ and $N_e \sim \gamma^{-3}$ for $\gamma > \gamma_1$. The resulting e^--spectrum for instantaneous e^--injection is discussed in DS. Fig.1 shows the time averaged EIC-, SSC- and synchrotron spectra of a typical EGRET-blazar for both cases. While the break in the EIC-spectrum of the single blob case is due to incomplete cooling (see DS), the spectral break for the case of a continuous flow is due to the break in the injection spectrum and might occur at MeV-energies. We also note that the SSC-spectra are harder than the EIC-spectra.

B   Modeling Log N-Log S

The measurement of the EGRB is based on the phase 1 to 3 summed skymaps (see Sreekumar et al., this proceedings). 70 detected point sources with a significance $\geq 4\sigma$ contribute to the source distribution Log N-Log S. Among them are 9 BL Lacs, 37 FSRQs and 24 so far unidentified sources. Though BL Lacs have on average lower flux values compared to FSRQs, their Log N-Log S-distributions do not differ significantly.

In order to estimate the AGN-contribution to the EGRB, we model the resolved Log N-Log S-function of each, FSRQs and BL Lacs, by Monte-Carlo simulations. The sources, each provided with two jets, are distributed homogeneously in an Euclidean universe with a redshift range of $z = 0.01 - 5$, and follow no density evolution. The viewing angles are distributed uniformly. A luminosity evolution $L \sim \exp(T(z)/\tau)$ (T(z)=look back time in units of Hubble time) and distribution $P(\Gamma) \sim \Gamma^{-\alpha}$ of bulk Lorentz factors Γ were used according to the unified picture ($\tau \simeq 0.3$ and $\alpha \simeq 4$ with $\Gamma = [5, 36]$ for BL Lacs/FR I (Urry et al. 1991, Stickel 1991); $\tau \simeq 0.2$ and $\alpha \simeq 2.3$ with $\Gamma = [5, 40]$ for FSRQs/FR II (Padovani & Urry 1992)). The injected e^--energy E_{inj} in BL Lacs relative to that in FSRQs is set to 0.2 in agreement with their mean relative luminosities. The parameters B=1G, $z_i = 100 r_g$, $r_b = 30 r_g$, $\Delta t = 10^6$ sec, luminosity of accretion disk photons $L_0 = 10^{44}$ erg/s are fixed for all models. We use the observed Log N-Log S-distributions of FSRQs and BL Lacs together with the relative numbers of resolved objects of each class ($19.6^{+6.5}_{-6.6}\%$ BL Lacs, $80.4^{+13.3}_{-13.2}\%$ FSRQs) to constrain our model and to adjust the free parameters E_{inj} and $\mathcal{N}_{BLLac}$ (=fraction of BL Lacs in the universe).

The calculations are carried out for models with an e^--spectrum between $\gamma_1 = 1$ and $\gamma_2 = 10^6$ (model 1), and $\gamma_1 = 100$ and $\gamma_2 = 10^4$ (model 2). Objects which are brighter than the sensitivity limit determine the resolved Log N-Log S while sources with flux values below that limit contribute to the EGRB. The unresolved blazar contribution to the EGRB then follows by summing up all unresolved AGN per solid angle of the whole survey.

We wish to point out that we have not taken into account any luminosity variations due to different amounts of injected e^--energies among FSRQs/BL Lacs. Furthermore, we have not considered any flux or spectral variability nor spectral differences between BL Lacs and FSRQs.

The observed log N-Log S-distributions of both FSRQs and BL Lacs were compared with the model distributions by a Kolmogorov-Smirnov-test (KS-test). We derive acceptable fits. The best-fit values of the free parameters can be found in Table 1. Though the best fit of the intrinsic number counts of BL Lacs $\mathcal{N}_{BLLac}$ lies between $20\% - 40\%$, $\geq 50\%$ BL Lacs can not be ruled out due to the uncertainty of the numbers of BL Lacs detected by EGRET so far.

We find that only 5-16% of the EGRB are due to unresolved AGN with roughly 30% unresolved BL Lacs. For a distribution of E_{inj} we expect that the unresolved blazar contribution to the EGRB increases. Therefore, the derived values should be considered as lower limits.

	Model 1		Model 2	
	single blob	cont.flow	single blob	cont.flow
$\mathcal{N}_{BLLacs}$	30%	30-40%	30%	30-40%
E_{inj}(erg)	$10^{43.6}$	$10^{43.7}$	$10^{43.2}$	$10^{43.4}$
$\frac{N_{res,BLLac}}{N_{res,tot}}$	19%	17-21%	16%	18-25%
P_{KS}	0.55	0.95-0.99	0.77	0.95
AGN contr.to EGRB	$\simeq 16\%$	$\simeq 6\text{-}7\%$	$\simeq 14\%$	$\simeq 5\text{-}7\%$
unres. BL Lacs	$\simeq 30\%$	$\simeq 28\text{-}41\%$	$\simeq 34\%$	$\simeq 34\%$

Table 1: The best-fit values $\mathcal{N}_{BLLacs}$ and E_{inj} with a goodness-of-fit probability P_{KS} tested with a KS-test between the observed and model Log N-Log S.

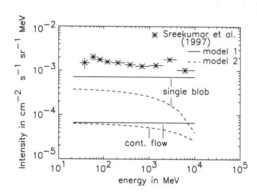

FIGURE 2. The EGRB-spectrum observed by EGRET (Sreekumar et al., this proceedings) together with the contribution of unresolved blazars estimated from model 1 ($\gamma_1 = 1$, $\gamma_2 = 10^6$) and model 2 ($\gamma_1 = 100$, $\gamma_2 = 10^4$) multiplicated with a factor of 7.

Fig.2 shows the background spectra of our best-fit models together with the observed EGRB-spectrum (Sreekumar et al., this volume). To avoid a cut-off at GeV-energies in the spectrum of unresolved blazars, the injection spectrum in the average source should provide electrons with Lorentz factors $\gamma > 10^4$.

III CONCLUSIONS AND DISCUSSIONS

We have modelled the source distribution of blazars to the EGRB without using any luminosity or flux relation between the γ-ray and any other spectral band. Instead, we base our model on physical parameters of blazars and the unification scheme of radio-loud AGN.

At least 5-16% of the EGRB can be explained by blazars which can not be resolved due to their misaligned jets and/or their large distances. Roughly 30% of those unresolved AGN are BL Lacs. With the stronger luminosity evolution of EGRET-AGN found by Chiang et al. (1995) the point-source contribution reaches about $\sim 21\%$. These values are lower limits since we have not considered any variations of luminosity due to different amounts of injected e^--energies. If we allow a power law distribution of injected energies by unresolved blazars.

The derived lower limits here can be compared with the results of a fluctuation analysis applied to the EGRET-EGRB (Willis 1996). It has been found that at least 5-14% of the EGRB are unresolved point-sources using the Chiang et al.- luminosity function together with the fluctuation constraints, and 4-15% of the EGRB has been predicted to be resolved in the decade of flux below the EGRET detection threshold on the basis of the fluctuation analysis alone. However, 100% source contribution to the EGRB can not be ruled out by this analysis. Therefore, our result here is still in agreement with the findings of a fluctuation analysis. Our model predicts a significant flattening of the Log N-Log S distribution in the next 2 decades below the EGRET threshold (see Fig.3) and the detection of roughly 100-300 more blazars with a telescope ~ 10 times more sensitive compared to EGRET.

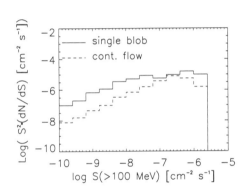

FIGURE 3. The predicted source distribution Log (S^2 (dN/dS))-Log S of EGRET-blazars for model 1: upper curve: case of single blob, lower curve: case of continuous flow. The main contribution to the EGRB comes from sources near the sensitivity limit.

REFERENCES

1. Chiang, J., Fichtel, C.E., von Montigny, C., Nolan, P.L., Petrosian, V. 1995, ApJ, 452, 156.
2. Dermer, C.D. & Gehrels, N. 1995, ApJ, 447, 103.
3. Dermer, C.D. & Schlickeiser, R. 1993, A&A, 416, 458.
4. Mücke, A., Pohl, M., Reich, P., et al. 1997, A&A, 320, 33.
5. Padovani, P. & Urry, M.C. 1992, ApJ, 387, 449.
6. Stecker, F.W. & Salamon, M.H. 1996, ApJ, 464, 600.
7. Stickel, M., Padovani, P., Urry, C.M. et al. 1991, ApJ, 374, 431.
8. Thompson, D.J., et al., 1993, ApJS, 86, 629.
9. Urry, C.M. & Padovani, P. 1991, ApJ, 371, 60.
10. Urry, C.M., Padovani, P. & Stickel, M. 1991, ApJ, 382, 501.
11. Urry, C.M. & Padovani, P., 1995, Pub.Astr.Soc.Pac., 107, 803.
12. Willis, T. 1996, PhD-thesis, University of Stanford, California.

Absorption of High Energy Gamma Rays by Interactions with Extragalactic Starlight Photons at High Redshifts

M.H. Salamon* and F.W. Stecker[†]

*Physics Department, University of Utah
Salt Lake City, UT 84112
[†]Laboratory for High Energy Astrophysics
NASA/Goddard Space Flight Center
Greenbelt, MD 02271

Abstract. We extend earlier calculations of the attenuation suffered by γ-rays during their propagation from extragalactic sources, obtaining new extinction curves for γ-rays down to 10 GeV in energy, from sources up to a redshift of $z = 3$.

The recognition that high energy γ-rays, propagating over cosmological distances, suffer electron-pair-producing interactions with photons from the extragalactic background radiation fields dates back to the 1960s [1–4]. The reaction $\gamma\gamma \to e^+e^-$ between a γ-ray of energy E and a background photon of energy ϵ can occur when the center-of-mass square energy s is above threshold, $s = 2E\epsilon(1 - \cos\theta) > 4m_e^2c^4$, where θ is the angle between the two photons' direction vectors. A γ-ray of energy E_{TeV} TeV therefore interacts only with background photons above a threshold energy $\epsilon_{\mathrm{thr}} \approx 0.3\mathrm{eV}/E_{\mathrm{TeV}}$. Since the number density of background photons decreases roughly as a power law in energy, most of the collisions occur near threshold. Thus, when estimating the extinction of 1 TeV gamma rays, it is the density of the infrared background which dominates; at 20 GeV, however, only UV photons near the Lyman limit can act as targets.

This mechanism was recently invoked [5] to explain why many EGRET blazars are not seen at $\sim$TeV energies by ground-based instruments such as Whipple, in spite of the fact that an extrapolation of the EGRET power-law spectra places them above the sensitivity limit of these ground-based detectors. The opacity τ seen by a γ-ray in its propagation from source to Earth is roughly $\tau \sim N\sigma_T d$, where N is the number of target soft photons

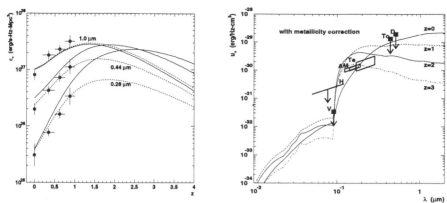

FIGURE 1. Left: The mean co-moving emissivity of stellar populations as a function of redshift for three different wavelengths, with (solid lines) and without (dashed lines) our metallicity correction added. The dashed lines are essentially a reproduction of the results of Ref. [11], and the observational data points are from the Canada-French redshift survey group [13]. Right: The computed co-moving background energy density as a function of wavelength for several redshifts. The data points shown are high galactic latitude detections or limits at redshift $z = 0$: A, Ref. [14]; D, Ref. [15]; H, Ref. [16]; M, Ref. [17]; Te, Ref. [18]; To, Ref. [19]; V, Ref. [20].

above threshold, σ_T is the Thompson cross section, and d is the distance to the source. For sources with redshift $z > 0.1$ (corresponding to most of the EGRET blazar sources), based on estimates of the diffuse IR background [6] the opacity is greater than unity for $\sim$ TeV γ-rays, making their ground-based detection unlikely.

With the advent of a new generation of ground-based instruments with anticipated γ-ray energy thresholds as low as 20 GeV, and with the likely future launch of GLAST [7] with sensitivity in the range $\sim$0.01 to 100 GeV, it is important to extend the opacity calculations down to the lowest relevant γ-ray energies. Although efforts along these lines have been made [8,9], very recent work on the evolution of star formation rates with redshift [10,11] justifies a new and more detailed calculation of γ-ray opacity.

The role played by the extragalactic starlight background (ESB) in the attenuation of γ-rays from extragalactic sources is defined in the exact expression for the γ-ray opacity τ,

$$\tau(E_0, z_e) = c \int_0^{z_e} dz \frac{dt}{dz} \int_0^2 dx \frac{x}{2} \int_0^\infty d\nu \, (1+z)^3 \frac{u_\nu(z)}{h\nu} \sigma_{\gamma\gamma}(s), \quad (1)$$

where E_0 is the *observed* γ-ray energy, z_e is the source redshift, $t(z)$ is the cosmic time, $x \equiv (1 - \cos\theta)$, θ is the angle between the photons' direction vectors, ν is the target photon frequency at redshift z, $u_\nu(z)$ is the photon energy density per unit frequency at redshift z, and $\sigma_{\gamma\gamma}$ is the Bethe-Heitler cross section for $\gamma\gamma \to e^+e^-$.

Apart from the uncertainty in cosmological parameters, the only unknown in the above equation is the ESB energy density $u_\nu(z)$. This can be determined if the mean emissivity per unit frequency, $\mathcal{E}_\nu(z)$, of starlight from galaxies is known:

$$u_\nu(z) = \int_z^{z_{max}} dz' \frac{dt}{dz} \mathcal{E}_{\nu'}(z') e^{-\tau_{cloud}(\nu,z,z')}. \quad (2)$$

Here z_{max} is the redshift for the turn-on of star formation, $\nu' = \nu(1+z')/(1+z)$, and the last factor accounts for the partial absorption of the starlight by intervening Lyα clouds during the ESB's propagation through intergalactic space [12].

The mean emissivity $\mathcal{E}_\nu(z)$ is the total stellar energy output per unit volume and frequency, averaged over all galaxies and proto-galaxies, at a given redshift. Consider a population of stars all born at the same instant, with an initial mass function (IMF) $\phi(M)dM \propto M^{-\alpha}dM$ ($\alpha = 2.35$ here). The total emission $S_\nu(T)$ from this population is the integral of the spectral energy output of each star (a function of its mass M, age T, and to a smaller extent its metallicity [21,22]) weighted by the IMF. As the age T of the population increases, $S_\nu(T)$ becomes redder, due to the shorter lifetimes of the bluer stars. The mean emissivity $\mathcal{E}_\nu(z)$ is then the convolution of $S_\nu(T)$ with the redshift-dependent stellar formation rate, $\dot{\rho}_s(z)$:

$$\mathcal{E}_\nu(z) = \mathcal{T}_{d,g}(\nu) \int_z^{z_{max}} dz' \frac{dt}{dz'} \dot{\rho}_s(z') S_\nu[T = t(z) - t(z')] \mathcal{L}(\nu, z'), \quad (3)$$

where $\mathcal{T}_{d,g}(\nu)$ is the probability that stellar photons of frequency ν will escape absorption by dust and gas in their parent galaxy, and $\mathcal{L}(\nu, z)$ is a frequency-dependent correction to S_ν which accounts for the increase in stellar metallicities with decrease in z.

Figure 1 shows our results for $\mathcal{E}_\nu(z)$ (Eq.3) and $u_\nu(z)$ (Eq.2). For $S_\nu(T)$ we have used the population synthesis models of Refs. [23,24,22]; for $\mathcal{L}(\nu, z)$ we have constructed an empirical correction function based on the work of Ref. [21]; the star formation rate $\dot{\rho}_s(z)$ comes from the beautiful analysis of Refs. [10,25]. (See Ref. [26] for more details.)

With Eq.1, the ESB opacity to γ-rays is calculated, and shown in Fig.2, both with and without the metallicity correction function $\mathcal{L}$ included. Given the uncertainties associated with $\mathcal{L}$ [26], the true opacities likely lie somewhere between the two sets of curves.

Figure 3 shows the effect of the ESB on γ-ray propagation from several blazars. Note that the spectral cutoffs occur at lower energies for blazars at higher redshifts, a distinctive signature which can discriminate this cutoff mechanism from intrinsic (intrasource) cutoff mechanisms. Also note that

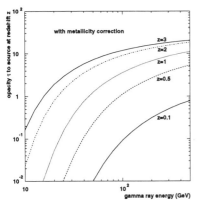
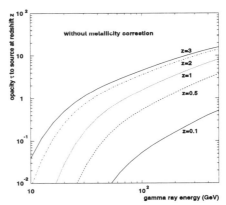

FIGURE 2. The opacity τ of the ESB to γ-rays as a function of γ-ray energy and source redshift z. Left: Curves calculated with the metallicity correction included. Right: Curves calculated without metallicity correction. The truth likely lies between these two sets of curves. We note that the opacities obtained here are independent of the value assumed for Hubble's constant (see Ref. [26] for details).

there is essentially no attenuation below 10 GeV, due to the sharp break in the energy density above the Lyman limit (Fig.1). Figure 3 also shows the beginning of the extinction of that component of the extragalactic γ-ray background above 20 GeV that is due to unresolved blazars [27]; this is compared with recent EGRET measurements of the extragalactic γ-ray background [28].

Acknowledgements: We thank M. Fall, M. Malkan, Y.C. Pei, P. Sreekumar, G. Worthey, and N. Wright helpful conversations and advice.

REFERENCES

1. Nishikov, A.I., *Sov. Phys., JETP* **14**, 393 (1962).
2. Gould, R.J. and Schreder, G.P., *Phys. Rev. Lett.* **16**, 252 (1966).
3. Stecker, F.W., *Ap.J.* **157**, 507 (1969).
4. Fazio, G.G. and Stecker, F.W., *Nature* **220**, 135 (1970).
5. Stecker, F.W., De Jager, O.C., and Salamon, M.H., *Ap.J.* **390**, L49 (1992).
6. Stecker, F.W. and De Jager, O.C., *Ap.J.* **476**, 712 (1997).
7. Bloom, E.D., *Sp. Sci. Rev.* **75**, 109 (1996).
8. Madau, P. and Phinney, E.S., *Ap.J.* **456**, 124 (1996).
9. Stecker, F.W. and De Jager, O.C., *Sp. Sci. Rev.* **75**, 413 (1996).
10. Pei, Y.C. and Fall, S.M., *Ap.J.* **454**, 69 (1995).
11. Fall, S.M., Charlot, S., and Pei, Y.C., *Ap.J.* **464**, L43 (1996).
12. Madau, P. and Shull, J.M., *Ap.J.* **457**, 551 (1996).
13. Lilly, S.J., et al., *Ap.J.* **460**, L1 (1996).

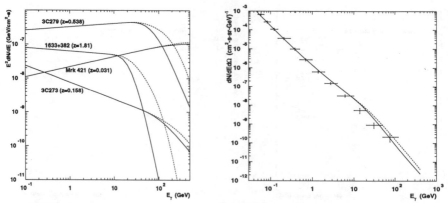

FIGURE 3. (Left: The attenuated power-law spectra of four prominent blazars. The solid (dashed) curves are calculated with (without) the metallicity correction factor. Right: Extragalactic γ-ray background spectrum from unresolved blazars, calculated for the EGRET point source sensitivity of 10^{-7} cm^{-2}s^{-1}; solid (dashed) line includes (does not include) metallicity correction, and data points are from EGRET [28].

14. Anderson, R.C., et al., *Ap.J.* **234**, 415 (1979).
15. Dube, R.R., Wickes, W.C., and Wilkinson, D.T., *Ap.J.* **232**, 333 (1979).
16. Holberg, J.B., *Ap.J.* **311**, 969 (1986).
17. Martin, C., Hurwitz, M., and Bowyer, S., *Ap.J.* **379**, 549 (1991).
18. Tennyson, P.D., et al., *Ap.J.* **330**, 435 (1988).
19. Toller, G.N., *Ap.J.* **266**, L79 (1983).
20. Vogel, S.N., et al., *Ap.J.* **441**, 162 (1995).
21. Worthey, G., *Ap.J.Suppl.* **95**, 107 (1994).
22. Charlot, S., Worthey, G., and Bressan, A., *Ap.J.* **457**, 625 (1996).
23. Bruzual, A.G. and Charlot, S., *Ap.J.* **405**, 538.
24. Charlot, S. and Bruzual, A.G., *Ap.J.* **367**, 126 (1991).
25. Fall, S.M. and Pei, Y.C., *Ap.J.* **402**, 479 (1993). (1993).
26. Salamon, M.H. and Stecker, F.W., submitted to *Ap.J.* (1997).
27. Stecker, F.W. and Salamon, M.H., *Ap.J.* **464**, 600 (1996).
28. Sreekumar, P., Stecker, F. W., and Kappadath, S. C., these proceedings.

Further COMPTEL observations of the region around GRO J1753+57: are there several MeV sources present?

O.R. Williams[4], K. Bennett[4], R. Much[4], V. Schönfelder[1],
W. Collmar[1], H. Bloemen[2], J.J. Blom[2], W. Hermsen[2], J. Ryan[3]

[1] *Max-Planck Institut für Extraterrestrische Physik, P.O. Box 1603, 85740 Garching, F.R.G.*
[2] *SRON-Utrecht, Sorbonnelaan 2, NL-3584 CA Utrecht, the Netherlands*
[3] *Space Science Center, Univ. of New Hampshire, Durham NH 03824, U.S.A.*
[4] *Astrophysics Division, ESTEC, P.O. Box 299, NL-2200 AG Noordwijk, the Netherlands*

Abstract.
GRO J1753+57 was first reported by COMPTEL as a very bright source in the range 1-3 MeV during an observation in November 1992. It was not detected in three other observations of the same region carried out in June 1991, December 1992, and March 1993 and it was not possible to identify a likely candidate at other wavebands which lay within the 3σ location contour.

Six further observations of the region have been made and after co-adding all the available data evidence has emerged for a more complex stucture than previously indicated. In the analysis presented here, we show that the observed signal can be successfully modelled by emission from a combination of two or more point sources which lie outside the original 3σ location contour of GRO J1753+57, at the positions of the blazar 4C 56.27, the EGRET quasar QSO 1739+522, and the unidentified EGRET source GRO J1837+59. All these sources can reasonably be expected to be γ-ray emitters.

However, uncertainties over the origin of the observed emission persist, in part due to the difficulty in unambiguously modelling COMPTEL data in the possible presence of several point sources. Moreover, it has previously been suggested that a diffuse galactic source related to high-velocity clouds may contribute to the observed emission in this region. This possibility would provide an alternative explanation for part of the observed signal and is not excluded by the analysis presented here, since extended emission is particularly difficult to image with COMPTEL.

INTRODUCTION

GRO J1753+57 was first reported by COMPTEL in the range 1-3 MeV during an observation in November 1992 [1]. The detection was consistent with a single, bright, variable point source. It was not detected in three other observations of the same region in June 1991, December 1992, and March 1993. Moreover, it was not possible to identify a convincing candidate at other wavebands within the 3σ location contour, despite a thorough search and follow up optical observations [2].

A total of 10 observations of the region have now been made. Co-adding all the available data and searching for a point source using a maximum likelihood-ratio method [4] results in the maximum likelihood-ratio map (MLM) shown in Fig. 1. The best fit position of GRO J1753+57 from the earlier analysis [1] is marked by a triangle. As has been previously shown [3] the feature visible in the 1-3 MeV range (which contains the greatest likelihoods) is not consistent with a single point source. This has significant consequences for the interpretation of this object since there are a number of interesting potential counterparts which lie inside the feature visible in Fig. 1, but which are outside the original 3σ location contour for a single point source. Namely:

- QSO 1739+522 an EGRET QSO at l,b=79.56,31.74
- GRO J1837+59 an unidentified Egret source at l,b=88.77,25.09
- 4C 56.27 a BL Lac object at l,b=85.7,26.1

MODELLING GRO J1753+57 USING SEVERAL POINT SOURCES

Modelling of COMPTEL data is complicated when several point sources are located within $sim 10°$. We have attempted to model the emission near GRO J1753+57 in terms of two sources, firstly by including sources at the locations of QSO 1739+522 and 4C 56.27 in the background model and secondly by including sources at the locations of QSO 1739+522 and GRO J1837+59 in the background model. The absence of significant residuals in either case show that either combination of sources can account for the observed signal i.e. we cannot distinguish between emission from 4C 56.27 and GRO J1837+59.

Fluxes from two nearby sources can be determined by simultaneously fitting two point spread functions (PSF) at the appropriate locations together with a background model. Fig. 2 shows the 1-3 MeV 'lightcurves' derived from simultaneous fits using PSFs at the locations of QSO 1739+522 and GRO J1837+59. The other standard COMPTEL energy bands (0.75-1.0 MeV, 3.0-10.0 MeV, 10-30 MeV) are omitted since they yield mainly upper limits. If sources at these locations give rise to the observed signal they seem to be variable. There is a tendancy that if one source has a high flux in a given observation, then the other source also has a high flux. This may indicate a systematic problem, or could merely be coincidental.

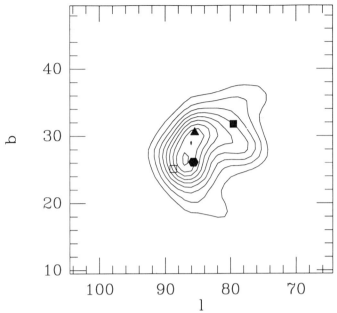

FIGURE 1. MLM of the 1-3 MeV range for all observations. Contours start at 10 with steps of 5. Locations of QSO 1739+522, GRO J1837+59, 4C 56.27 and the best fit position of GRO J1753+57 from the earlier analysis [1] are marked by a filled square, open square, hexagon, and triangle respectively.

Similar results have been obtained from simultaneous fits using PSFs at the locations of 4C 56.27 and QSO 1739+522. This is to be expected since we cannot distinguish between emission from 4C 56.27 and GRO J1837+59.

EGRET OBSERVATIONS OF GRO J1837+59

The EGRET observations of GRO J1837+59 reveal it as a credible candidate for some of the MeV emission from this region.

Fig. 2c shows the published EGRET data for GRO J1837+59 [5] to be compared with the COMPTEL 1-3 MeV light curve in Fig. 2a. During the observations in the period TJD 8943-8957, when COMPTEL sees a relatively high flux from this location, EGRET also observes a high flux. However, EGRET also observes a high flux in the period TJD 8406-8415, when COMPTEL sees no signal.

Fig. 3 compares the EGRET data from GRO J1837+59 in the period TJD 8943-8950 with the COMPTEL fluxes from the location of GRO J1837+59 in the period TJD 8943-8957. Within the statistical uncertainties, the COMPTEL fluxes are consistent with an extrapolation of the EGRET spectrum

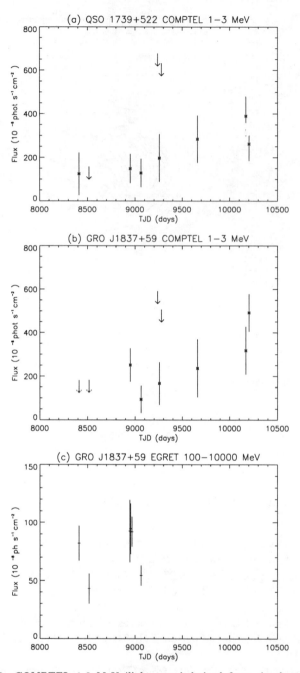

FIGURE 2. COMPTEL 1-3 MeV 'lightcurves' derived from simultaneous fits to (a) QSO 1739+522 and (b) GRO J1837+59. For comparison the EGRET lightcurve for GRO J1837+59 is shown in (c).

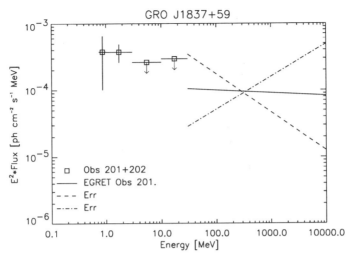

FIGURE 3. EGRET flux from GRO J1837+59 in the period TJD 8943-8950 [5] and the COMPTEL flux from GRO J1753+57 in the period TJD 8943-8957

CONCLUSIONS

In the analysis presented here we show that the observed signal can be successfully modelled by emission from a combination of two or more point sources which lie outside the original 3σ location contour of GRO J1753+57, at the positions of the blazar 4C 56.27, the EGRET quasar QSO 1739+522, and the unidentified EGRET source GRO J1837+59. All these sources can reasonably be expected to be MeV γ-ray emitters.

However, considerable uncertainties over the origin of the observed emission persist, in part due to the difficulty in unambiguously modelling COMPTEL data in the possible presence of several point sources. Moreover, it has previously been suggested that a diffuse galactic source related to high-velocity clouds may contribute to the observed emission in this region [3]. This possibility would provided an alternative explanation for at least part of the observed signal and is not excluded by the analysis presented here, since extended emission is particularly difficult to image with COMPTEL.

REFERENCES

1. Williams, O.R., et al., 1995 A&A 297, 21.
2. Carramiñana, A., et al., 1996 A&AS 120, 595.
3. Blom, J.J., et al. 1997, Proc. 2nd Integral Workshop, in press.
4. Bloemen, H., et al., 1994, A&AS 92, 419.
5. Thompson, D.J., et al., 1995, ApJS 101, 259.

Temporal and Spectral Studies of Unidentified EGRET High Latitude Sources

O. Reimer[1], D.L. Bertsch[2], B.L. Dingus[3], J.A. Esposito[2], R.C. Hartman[2], S.D. Hunter[2], B.B. Jones[4], G. Kanbach[1], D.A. Kniffen[5], Y.C. Lin[4], H.A. Mayer-Hasselwander[1], C. v.Montigny[7], R. Mukherjee[6], P.L. Nolan[4], P. Sreekumar[2], D.J. Thompson[2], W.F. Tompkins[4]

[1] *Max-Planck-Institut für Extraterrestrische Physik, 85740 Garching, Germany*
[2] *NASA/Goddard Space Flight Center, Code 661, Greenbelt, MD 20771, USA*
[3] *Physics Dept., Univerity of Utah, Salt Lake City, UT 84112, USA*
[4] *W.W.Hansen Exp. Research Lab, Stanford University, Stanford, CA 94305, USA*
[5] *Hampden-Sydney College, Hampden-Sydney, VA 23943, USA*
[6] *McGill University, Physics, Montreal, H3A 2T8, Canada*
[7] *Landessternwarte, 691117 Heidelberg, Germany*

Abstract.
Source distribution studies only provide meaningful information if the basic populations are carefully determined, especially if the statistics are limited. This necessitates restrictions from the approach of using the complete EGRET source catalog data. Considering various peculiarities of the catalogs a more uniform data set is defined for further source distribution studies. Here we focus on the temporal and spectral characteristic of the unidentified EGRET high latitude sources ($|b| > 10°$). Variability is addressed on the basis of flux histories with timescales of individual viewing periods, and spectra are determined for all sources based on the enhanced statistics from EGRET observations of Phase 1 to 4. The derived features are used to discuss whether the unidentified high latitude sources show similarities or differences from classes of already identified objects, or if no conclusive answer can be given.

THE EGRET POINT SOURCE CATALOGS

The most popular approach to population and correlation studies of EGRET point sources has been to use the EGRET point source catalogs. The study of the 2EG-catalog and its supplement (Thompson et al. [1], [2]) has become a common procedure for attempts to understand the nature of the unidentified EGRET sources by comparison with properties of known objects or object classes [3], [4], [5], [6], [7], [8]. The approach here is basically to take all catalog sources and divide them into subclasses by identifications, coordinates, physical properties etc. But various aspects from the compilation of the 2EG- and 2EGS-catalogs have to be taken into account in order to get an almost unbiased sample of sources as the fundamental starting point for source population studies. This becomes even more important and essential if corrections for the instrument response (exposure, detection sensitivity etc.) are introduced.

First, the 2EG- and 2EGS-catalogs consist of two different significance thresholds for detection, for $|b| < 10°$ of $\geq 5\sigma$ and for $|b| < 10°$ of $\geq 4\sigma$. If a separation study in galactic latitudes does not take this nonuniformity ("step") in the detectibility function into account, classes with different statistical significances and sensitivities for being a source will smear the characteristics of source distributions, e.g. Grenier 1997 [8]. It is difficult to balance this effect afterwards in order to conclude if source excesses are present or not.

Second, the criterion for including a source in the 2EG- and 2EGS-catalogs as detected is, if the significance criterion mentioned above is fulfilled in either a single viewing period, a combination of single viewing periods or the total superposition of all viewing periods. This means a source will be listed as detection if it fulfills the detectibility criterion in one single viewing period, even if it is well beneath that criterion in an analysis of the superposition of all viewing periods from phase 1+2 or phase 1+2+3. This becomes an essential point if one attempt to introduce an instrumental exposure correction in order to balance the uneven sky coverage of EGRET, e.g. Grenier 1997 [8]. Any use of the *total* exposure regarding sources in the catalogs below the sensitivity criterion for the appropriate *total* observation time is improper and will not correct for the nonuniformity.

HIGH LATITUDE SOURCE DISTRIBUTIONS

Noticing these peculiarities of the 2EG- and 2EGS-catalogs, a suitable subset of sources was extracted. It includes only sources above the $\geq 4\sigma$ significance criterion for the *total* superpositioned observations from phase 1 and 2. This is currently the only common available base for an appropriate exposure correction, because the 2EGS-catalog does not include the statistical significances of detections for the 2EG-sources for phase 1+2+3. The non-uniform sensitivity across $|b| = 10°$ for catalog sources is handled by setting the histogram bin size of the latitude distributions to $10°$ and restricting the interpretation of objects to either $|b| > 10°$ or $|b| < 10°$. A total of 66 sources will be used here as high latitude sources, 47 AGN, the LMC, and 18 unidentified sources. The 35 sources $|b| < 10°$ are only sketched for qualitative consideration.

The latitude distributions of different source classes meeting these selection criteria are shown in Fig. 1. In order to get a comparable visual impression the sources are always normalized to sources per steradian. No significant excess of unidentified sources at intermediate galactic latitudes $10° < |b| < 30°$ is indicated, although the enhancement in the Galactic plane is obvious.

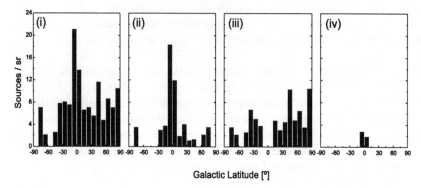

FIGURE 1. Latitude distributions of different source classes (i) all sources (ii) unidentified sources (iii) AGN (iv) PSR.

But this figure has to be completed by including the uneven instrument exposure and the structured galactic diffuse background. Mattox et al. [10] determined that the significance of detection s for an isolated EGRET point source could be expressed by $s \sim f\sqrt{e/bg}$, where f is the flux, e the exposure and bg the diffuse gamma ray background. Fig. 2 shows the latitude distribution of e, bg, and $\sqrt{e/bg}$, which is proportional to the detection significance for equally luminous sources .

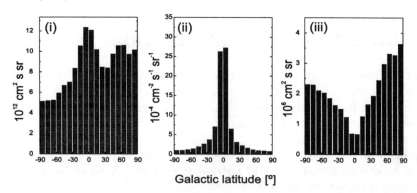

FIGURE 2. Latitude distributions of (i) exposure, (ii) diffuse background, (iii) $\sqrt{exposure/background}$, see text.

This picture suggests even more that one has to be careful in assessing source exesses in latitude profiles. The detection sensitivity is certainly lower in regions of enhanced diffuse emission, although longer sky coverage counteracts partly for the Galactic plane. For the high latitude sources rising gradients for the detectability are seen, with obviously better detection sensitivity at positive Galactic latitudes. This means an equally luminous point source would be more easily detected at high galactic latitudes, especially at *positive* high latitudes. Considering the slightly lower detection sensitivity at intermediate Galactic latitudes, limited evidence for an enhancement of unidentified sources is indicated.

VARIABILITY STUDIES

The Variability index V introduced by McLaughlin et al. [9] is used additionally for deciding if the unidentified high latitude sources show a specific feature with respect to other source classes. The unidentified high latitude sources tend to show a more non-variable behavior than AGNs, although a clear separation of source classes by only their variability seems impossible. Fig. 3 shows the Variability V for selected source classes.

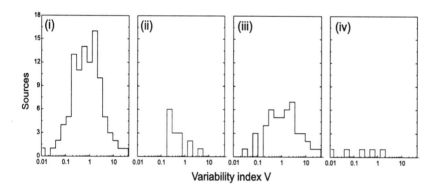

FIGURE 3. Variability index V for the following source classes (i) all sources (ii) all high latitude unidentified sources (iii) AGN (iv) PSR.

SPECTRAL STUDIES

A source spectrum is derived for all sources matching the selection criteria here based on the enhanced statistics of EGRET observations from phase 1 to 4. Flux values or upper limits for each of 10 energy intervals were determined.

These values were used to fit power law models to the energy spectra. The explicit spectra will be published in conjunction with the announced 3EG-catalog [11]; therefore here spectral indices were used only to investigate whether the unidentified high latitude sources show a characteristic feature with respect to other sources classes. In general, a range in the spectral indices as seen previously for high-energy γ-ray emitting objects is present. But in contrast to the distribution of the AGN, which show a nonuniform distribution centered near an index of -2, the distribution appears more equalized for the unidentified high latitude sources. The spectra of the few known γ-ray pulsars are harder compared to AGN. But as mentioned in Merck et al. for the Galactic sources [12], many of the unidentified high latitude sources have softer spectra than the known γ-ray pulsars. Fig. 4 shows the distribution of the spectral indices.

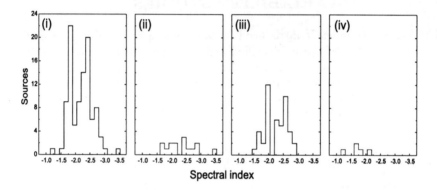

FIGURE 4. Spectral index α for the following source classes (i) all sources (ii) all high latitude unidentified sources (iii) AGN (iv) PSR.

REFERENCES

1. Thompson D.J., et al., 1995, ApJS, 101, 259
2. Thompson D.J., et al., 1996, ApJS, 107, 227
3. Kaaret P., Cottam J., 1996, ApJ. 462, L35
4. Kanbach G., et al., 1996, A&AS, 120, C461
5. Sturner S.J., et al. 1996, A&AS, 120, 445
6. Özel M.E., Thompson D.J., 1996, ApJ, 463, 105
7. Yadigaroglu I.-A., Romani R.W., 1997, ApJ, 476, 347
8. Grenier, I.A., Proceedings 2nd INTEGRAL Workshop, ESA SP-283 (1997)
9. McLaughlin M.A., et al., 1996, ApJ, 473, 763
10. Mattox J.R., et al., 1996, ApJ, 461, 396
11. Hartman, R.C., et al., in preparation
12. Merck M. et al., 1996, A&AS, 120, C465

Discovery of a non-blazar gamma-ray transient in the Galactic plane

M. Tavani[1], R. Mukherjee[2,3], J.R. Mattox[4], J. Halpern[1],
D.J. Thompson[5], G. Kanbach[6], W. Hermsen[7],
S.N. Zhang[8], R. S. Foster[9]

1. *Columbia Astrophysics Laboratory, Columbia University, New York, NY 10027.*
2. *USRA, NASA/GSFC, Code 610.3, Greenbelt, MD 20771.*
3. *Physics Dept., McGill University, 3600 Univ. St., Montreal, H3A 2T8, Canada.*
4. *Astronomy Dept., Boston University, 725 Commonwealth Avenue, Boston, MA 02215 .*
5. *Code 661, NASA Goddard Space Flight Center, Greenbelt, MD 20771.*
6. *MPE, Giessenbachstrasse, Garching bei Munchen, D-85748, Germany.*
7. *SRON-Utrecht, Sorbonnelaan 2, 3584 CA Utrecht, The Netherlands.*
8. *Universities Space Research Association, NASA/MSFC, Huntsville, AL, 35812.*
9. *Code 7210, Naval Research Laboratory, Washington, DC 20375.*

Abstract.
A new bright gamma-ray source was detected in June 1995 near the Galactic plane (GRO J1838-04). Subsequent EGRET observations did not detect significant flux from this source. The gamma-ray error box of GRO J1838–04 does not contain any spectrally-flat radio-loud source. GRO J1838–04 provides strong evidence for the existence of a *new* class of variable gamma-ray sources.

INTRODUCTION

Variable gamma-ray sources are particularly interesting among the ~40 unidentified gamma-ray sources concentrated near the Galactic plane [9,10]. The likelihood of detecting an active galactic nucleus (AGN) similar to those detected by EGRET at high latitudes is small in the Galactic plane. Furthermore, known gamma-ray pulsars show a constant gamma-ray flux (within uncertainties) and cannot explain the existence of variable sources. Detecting a variable gamma-ray source in the Galactic plane is therefore of great importance: it would imply the existence of a new class of gamma-ray emitters different from blazar-like AGNs and isolated pulsars. We studied the gamma-ray emission from the Galactic plane in search for transient sources between Galactic longitudes $l = 17°$ and $l = 32°$ during the CGRO Phase 4. This

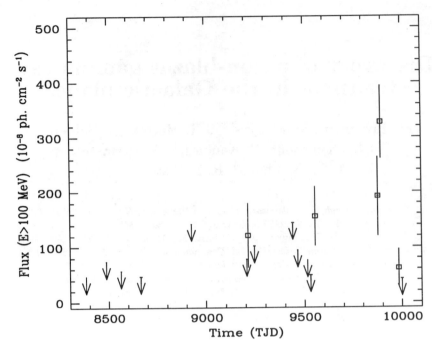

FIGURE 1. Time history of γ-ray flux detected by EGRET from GRO J1838–04. TJD is the truncated Julian date (JD), TJD=JD−2,440,000.0. 1σ flux errors and 2σ upper limits are reported, the upper limits being shown as downward arrows. The time interval is from April 1991 through early October 1995. EGRET data for VPs 421 and 422 have been combined. Figure adapted from ref. [7] with the inclusion of the last point obtained from public data of CGRO Cycle 5.

field contains interesting EGRET sources which are candidate for time variability, 2EG J1813−12 and 2EG J1828+01. During the viewing period (VP) 423 (June 20-30, 1995) a new gamma-ray transient was discovered. In the following, we briefly summarize the observations and discuss the implications of this discovery [7].

COMPTON GRO DETECTION OF GRO J1838–04

GRO J1838–04 is located near the Galactic plane in a field centered at Galactic coordinates $l = 27.31°$ and $b = +1.04°$ with an elongated error box (99% confidence) of major axis $\sim 1.4°$ and minor axis $\sim 0.8°$. The average γ-ray flux above 100 MeV for the whole VP 423 is $\Phi = (3.3 \pm 0.7) \cdot$

10^{-6} ph cm^{-2} s^{-1}. Fig. 1 shows the EGRET lightcurve of GRO J1838–04 since the beginning of the CGRO mission with the inclusion of a crucial Cycle 5 EGRET observation in October 1995 that failed to detect the source. The intensity level reached during the γ-ray flare had peak flux above 100 MeV of $(4.0 \pm 1.1) \cdot 10^{-6}$ ph cm^{-2} s^{-1} reached during the last 3.5-day interval of VP 423. The inferred γ-ray luminosity for isotropic emission corresponding to this peak flux is $L_\gamma \simeq 7.2 \times 10^{34} d_{kpc}^2$ erg s^{-1}, where d_{kpc} is the source distance in kpc. This is a very intense γ-ray transient with a γ-ray flux comparable with that of the Geminga pulsar [9] and of AGN flare peak intensities as for 3C 279 ($z = 0.538$) [1] and 0528+134 ($z = 2.07$) [4]. The EGRET spectrum above 30 MeV during the peak emission is consistent with a power-law of photon index 2.09 ± 0.18 [7]. Neither COMPTEL or BATSE detected significant high-energy emission during the gamma-ray flare or GRO J1838–04.

SEARCH FOR COUNTERPARTS

The Galactic plane near GRO J1838–04 was surveyed in the radio band at 20 cm by Helfand et al. [5]. No spectrally-flat bright radio source with blazar characteristics is within the 99% confidence error box of GRO J1838–04. The brightest radio source in the error box, 27.920+0.977 (source 'A' in Fig. 3 of ref. [7]) was determined to have a constant flux (~ 500 mJy at 2.2 GHz) and a steep radio spectral index $\alpha_r \simeq -1$ from data obtained at the Green Bank radio interferometer at 2.25 and 8.3 GHz during the period December 14 1995 - January 13 1996. This radio source certainly does not resemble the radio-loud blazars associated with gamma-ray sources as detected by EGRET at high Galactic latitudes [3,2]. Other radio sources in the GRO J1838–04 error box are weaker and unlikely to be blazar counterpart candidates.

Two radio pulsars are known in the error box, PSR B1831–04 and PSR B1834–04 with relatively small spindown luminosities ($10^{32.5}$ and $10^{33.1}$ erg s^{-1}, respectively [8]). The supernova remnant SNR27.8+0.6 is also in the error box [6]. No pulsar is associated with the remnant which appears to be of a center-filled plerionic type from its radio extended and core emission. Three X-ray sources in the ROSAT all-sky survey (RASS) database are within the 99% confidence error box of GRO J1838–04. None is associated with the radio sources of ref. [5]. More data are clearly necessary to obtain information on the GRO J1838–04 counterpart, and X-ray observations as well as optical and radio searches are under way.

DISCUSSION

The probability of finding a gamma-ray blazar as bright as GRO J1838–04 in the Galactic plane sky area within $\pm 2°$ of Galactic latitude was estimated to be $\sim 4 \times 0.4/4\pi \simeq 0.1$ [7]. At the γ-ray flux level near 10^{-6} ph cm^{-2} s^{-1}, there is no lack of sensitivity for EGRET observations near the Galactic plane. If radio-quiet AGNs also contributed to the gamma-ray emission, EGRET would have detected many more sources than observed. Therefore, GRO J1838–04 is not a bright γ-ray blazar of the kind usually detected by EGRET. If GRO J1838–04 is associated with an extragalactic source, it is highly improbable that a first detection of such a new γ-ray AGN would occur near the Galactic center.

A Galactic origin of GRO J1838–04 is not as yet supported by plausible counterparts. Several Galactic sources can be considered as plausible candidates for transient gamma-ray emission (see discussion of ref. [7]). However, there is no current evidence to support any specific Galactic model. More multiwavelength data are clearly needed to resolve this issue.

GRO J1838–04 may be the tip of the iceberg of a previously unrecognized class of transient gamma-ray sources, even though no other similar bright transient has been detected by EGRET near the Galactic plane (see the search in the anticenter region of ref. [11]). Its closeness to the Galactic plane helps discriminating against known types of γ-ray sources. Future high-sensitivity searches for gamma-ray transients supported by quick follow-up observations in the Galactic plane may discover more of these enigmatic sources and contribute to resolve the issue of their ultimate origin.

REFERENCES

1. Kniffen, D., et al., 1993, ApJ, 411, 133.
2. Mattox, J.R., et al., 1997, ApJ, 481, 95.
3. Montigny, C.V., et al., 1995, ApJ, 440, 525.
4. Mukherjee, R. et al., 1996, ApJ, 470, 831.
5. Helfand, D.J., et al., 1992, ApJS, 80, 211.
6. Reich, W., Furst, E. & Sofue, Y., 1984, A&A, 133, L4.
7. Tavani, M., et al., 1997, ApJ, 479, L109.
8. Taylor, J.H., Manchester, R.N. & Lyne, A., 1993, ApJS, 88, 529.
9. Thompson, D.J., et al., 1995, ApJS, 101, 259.
10. Thompson, D.J., et al., 1996, ApJS, 107, 227.
11. Thompson, D.J., et al., 1997, these Proceedings.

Searches for short-term variability of EGRET sources in the Galactic anticenter

D.J. Thompson[1], S.D. Bloom[2], J.A. Esposito[3], D.A. Kniffen[4], C. von Montigny[5]

[1] NASA/GSFC
[2] NASA/GSFC. NAS-NRC
[3] NASA/GSFC. USRA
[4] Hampden-Sydney
[5] Landessternwarte Heidelberg

Abstract. As part of a systematic search of the EGRET data base for short-term (less than one week) source variability, we have examined observations of the Galactic anticenter. The analysis reconfirms the short-term variability of quasar 0528+134 and the lack of significant variability of the Crab and Geminga pulsars. No other sources in the Galactic anticenter show convincing evidence of variability on 2-day time scales during CGRO Phases 1 and 2.

INTRODUCTION

In the energy range above 100 MeV, a number of blazars show short-term gamma-ray variability [1]. Gamma-ray pulsars by contrast show little variability [2]. At least one low-latitude, unidentified transient source without a flat-spectrum radio counterpart, GRO J1838-04 [3], varies significantly within a few days, suggesting a new class of gamma-ray sources. Finding short-term variability can therefore help determine the nature of unidentified EGRET sources, one of the key questions still unanswered in high-energy gamma-ray astrophysics:
- Variable + flat-spectrum radio source ⇒ possible blazar
- Variable, no radio counterpart ⇒ like GRO J1838-04
- Non-variable ⇒ possible pulsar.

METHOD

The analysis approach used is as follows:

1. Construct maps of EGRET data on 2-day intervals, using the energy range above 100 MeV.
2. Using the standard EGRET maximum likelihood program [4], calculate the flux or upper limit for known sources in the field of view. We use the second EGRET catalog source list [5].
3. After including all known sources, search the field of view for any new, possibly transient, sources.
4. Examine the data for evidence of variability.

In order to quantify the variability, we follow the approach of McLaughlin et al. [6]:

$$\chi^2 = \sum_{i=1}^{N_{vp}} (\frac{F(i) - \bar{F}}{\sigma(i)})^2, \qquad (1)$$

where $\bar{F}$ is the average flux in N_{vp} viewing intervals
and $\sigma(i)$ is the quadrature sum of the statistical and systematic uncertainties.

We choose the systematic uncertainty at 6%, following [6]. The systematic uncertainty may be larger [7].

The probability of finding a larger χ^2 than this by chance is

$$Q = 1 - P(\frac{N_{vp} - 1}{2}, \frac{\chi^2}{2}), \qquad (2)$$

where P is the incomplete gamma function.

The quantity $V = -\log Q$ describes the variability index:
$V = 1 \Rightarrow$ chance probability 10%
$V < 1 \Rightarrow$ no claim for variability
$V > 3 \Rightarrow$ strong claim for variability

RESULTS

We have examined the Galactic anticenter data from Phase 1 and Phase 2. Most known sources are not bright enough to be detected with significance greater than 3σ in a 2-day interval. Unless we find more than one 3σ detection, we can say nothing of consequence about variability. On 2-day time scales, only the Crab and Geminga pulsars are sufficiently bright that the systematic uncertainty affects the total uncertainty. For all other sources, the statistical uncertainty is completely dominant.

The figures show the results for: Crab/Geminga pulsars (Figure 1), quasar 0528+134 (Figure 2), and three unidentified EGRET sources (Figures 3, 4, and 5). Only 0528+134 provides convincing evidence for short-term variability. This result reconfirms the original analysis of Hunter et al. [8].

On average, each map yielded one excess with statistical significance of about 3σ in addition to the known sources. No excesses exceeding 4σ were found. The number of 3σ excesses is reasonably consistent with that expected from statistical fluctuations in the EGRET data [4]. As a result, we find no evidence for previously-unidentified transient sources in the data examined thus far.

CONCLUSION

Analysis of the early EGRET data from the Galactic anticenter finds no new evidence of variability on 2-day time scales. This study will continue.

REFERENCES

1. Bloom, S.D. et al., these proceedings (1997).
2. Fierro, J.M., PhD thesis, Stanford U. (1995).
3. Tavani, M. et al., *Ap.J.* **479**, L109 (1997).
4. Mattox, J.R. et al., *Ap.J.* **461**, 396 (1996).
5. Thompson, D.J. et al., *Ap.J.Supp.* **101**, 259 (1995).
6. McLaughlin, M.A. et al., *Ap.J.* **473**, 763 (1996).
7. Bertsch, D.L. et al., these proceedings (1997).
8. Hunter, S.D. et al., *Ap.J.* **409**, 134 (1993).

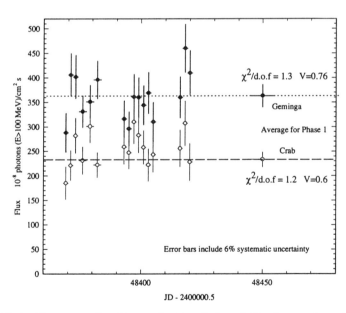

FIGURE 1. Short-term flux history for the Crab and Geminga pulsars during early EGRET observations.

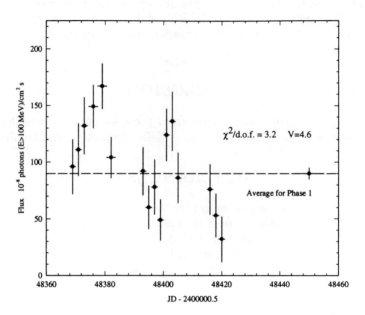

FIGURE 2. Short-term flux history for quasar 0528+134 during early EGRET observations.

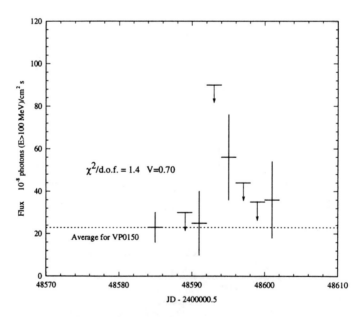

FIGURE 3. Short-term flux history for 2EGJ0323+5126

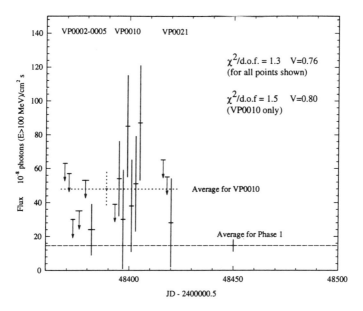

FIGURE 4. Short-term flux history for 2EGJ0506+3424

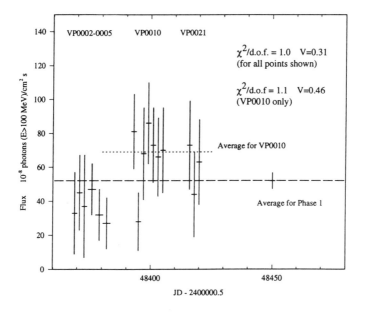

FIGURE 5. Short-term flux history for 2EGJ0618+2234

Short Time-scale Gamma-Ray Variability of Blazars and EGRET Unidentified Sources

S.D. Bloom[1], D.J. Thompson[2], R. C. Hartman[2],
C. von Montigny[3]

[1] NASA/GSFC. NAS/NRC Resident Research Associate
[2] NASA/GSFC.
[3] Landessternwarte Heidelberg

Abstract.
We have begun to examine the EGRET database for short term variations in the fluxes of blazars and the unidentified sources at high Galactic latitudes. We find that several AGN show previously unreported variability. Such variations are consistent with inverse Compton scattering processes in a shock propagating through a relativistic jet.

INTRODUCTION

Previous gamma-ray variability studies of EGRET detected blazars and unidentified sources have focused on long time-scale (weeks to years) variability [7], [6]. However, observations of bright flares, such as the outbursts of PKS 1622-297 [4] and 3C 279 ([2], [12]), clearly show variabilty on time-scales much less than a week. We therefore have set out to examine the EGRET data for variations on short time-scales (roughly, 2-14 days).

In general, observations of at least two weeks had been scheduled for most EGRET projects in order to detect weak sources. The standard counts maps were generated for the entire viewing period, and thus fluxes and spectral information determined from them are *averaged* properties over many days. Thus, any shorter time-scale variations would have been washed out. Since several blazars are known to vary dramatically on time-scales ~ 1 day-1 week at gamma-ray energies , we would expect that there are other blazars with similar, though probably less dramatic, variability properties. We have thus started an examination of EGRET data for such fluctuations. This process will also bring our attention to bright transients that only appear in part of one viewing period.

METHOD

In order to concentrate on blazars and potential blazars we have limited our study to viewing periods (VP's) centered on $b > |20°|$. A similar study on the Galactic anticenter region is reported on by Thompson et al [10]. Similar to Thompson et al [10], we have divided each viewing period into new two-day maps of > 100 MeV counts (accomplished for 35 VP's). A standard maximum likelihood analysis was performed on each map [3] to determine fluxes over two-day scales. An estimated systematic uncertainty of 6.5 % has been added to the statistical uncertainties [6]. We have then used a χ^2 analysis on fluxes and determined probability, P, that the data are consistent with a constant flux. The variability parameter of Mc Laughlin et al [6] is used to easily compare significance of variability among sources (and different VP's on the same source):

$$V \equiv |logP|$$

RESULTS

The blazar PKS 0208-512 (Figure 1) shows a factor of 2–3 increase over 3 weeks, followed by a rapid (2 day) factor of 3 decrease, and then another factor of 2–3 increase over the last week. The time history of the unidentified source 2EG J1835+5919 (Figure 2) is consistent with no variability over the 30 day period. However, inspection of Figure 2 reveals that there may have been a drop in flux over the last 10 days. The possible blazar 2EG J0220+4228 (0219+428) (Figure 3) shows a factor of 2–3 increase and decrease within one week. The blazar 0446+112 (Figure 4) undergoes a nearly factor of 4 decrease over one month, and nearly a factor of two within the last two weeks. No transients were found in any of the viewing periods studied.

DISCUSSION

The shortest time-scales of variability observed in this study are consistent with the model of Romanova and Lovelace [8]. They consider inverse Compton processes within a shock moving through a relativistic jet. A sudden acceleration of particles causes a brief synchrotron flare ($<< 1$ day) which will be accompanied by a simultaneous synchrotron self-Compton (SSC) flare over similar time-scales. However, since external inverse Compton models, such as the "mirror" model of Ghisellini & Madau [1], occur over larger spatial scales than the SSC models, the high energy flares will be delayed from the synchrotron flare and occur over several days. It is beyond the scope of this study to attempt any multiwavelength analysis and detailed theoretical modeling. In addition, we can not time-resolve the gamma-rays over < 1 day;

however, our future work will check for general consistency with these models. The relatively low amplitude and long duration of the flares in this study (as compared to, say, 1622-297 and 3C 279) favor external scattering models [8].

CONCLUSION

We have shown that three blazars undergo factors of 2–3 variability on time-scales $\sim$ 1 week. This variability is consistent with inverse Compton processes from a shock in a jet. The high amplitude variability seen within one week for 2EG J0220+4228 suggests that this source is much more likely to be related to the blazar 0219+428 than to the pulsar PSR J0218+4232 [11] at energies > 100 MeV. 2EG J1835+5919 shows only marginal short time-scale variability, and varies by less than a factor of two over longer time-scales [9]. There are no compelling radio counterparts to this source [5]. These results suggest that the source 2EG J1835+5919 is unlikely to be a blazar.

REFERENCES

1. Ghisselini, G. and Madau, P., *M.N.R.A.S.*, **280**, 67 (1996).
2. Hartman, R. C. *et al*, *Ap. J.* **461**, 698 (1996).
3. Mattox, J. R. *et al*, *Ap. J.* **461**, 39 (1996).
4. Mattox, J. R. *et al*, *Ap. J.* **476**, 692 (1997).
5. Mattox, J. R. *et al*, *Ap. J.* **481**, 95 (1997).
6. Mc Laughlin, M. A. *et al* , *Ap. J.* **473**, 763 (1996).
7. Mukherjee, R. *et al*, *Ap. J.*, in press (1997).
8. Romanova, M. M. and Lovelace, R. V. E. , *Ap. J.* **475**, 97 (1997).
9. Thompson, D. J. *et al*, *Ap. J. Suppl.* **101**, 259 (1995).
10. Thompson, D. J. *et al*, these proceedings (1997).
11. Verbunt, F. *et al* , *A.&A.*, **311**, 9 (1996).
12. Wehrle, A. E. *et al*, in preparation (1997).

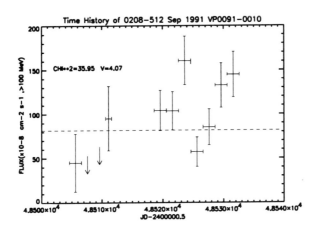

FIGURE 1. Short-term flux history for blazar PKS 0208-512

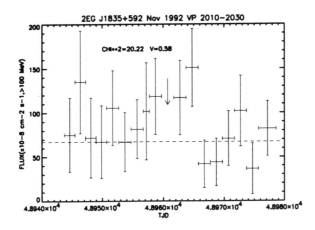

FIGURE 2. Short-term flux history for 2EG J1835+592

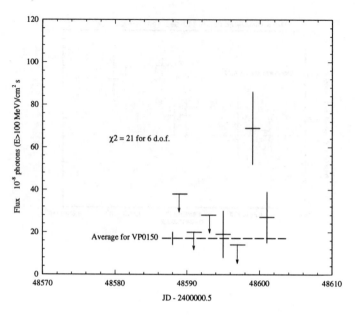

FIGURE 3. Short-term flux history for 2EG J0220+4228.

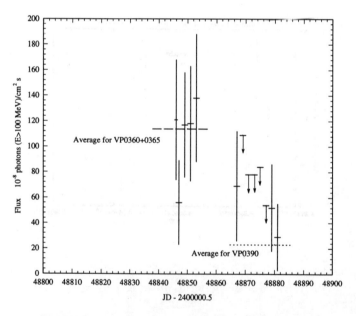

FIGURE 4. Short-term flux history for blazar 0446+112

Optical identification of EGRET source counterparts

A. Carramiñana[1], J. Guichard[1], K.T.S. Brazier[2], G. Kanbach[3], O. Reimer[3]

[1] *INAOE, Luis Enrique Erro 1, Tonantzintla 72840, MEXICO* [1]
[2] *Physics Department, Durham University, Durham 1DH 3LE, ENGLAND*
[3] *Max-Planck-Institut für Extraterrestrische Physik, Giessenbachstr D-85748, Garching, GERMANY*

Abstract. As part of a program for searching for Geminga-like candidates through the combination of γ-ray, X-ray and optical data, we have engaged ourselves in optical observations of selected ROSAT point sources. The strategy followed has been to survey the error boxes of the best ROSAT candidates in search of optical counterparts. If an optical source is present, as in the case of 2EG J2020+4026, an optical spectrum is acquired in order to ascribe the likelihood of the star been the X-ray and γ-ray source. More interestingly, in the case of the EGRET source 2EG J0008+73, inside the CTA 1 SNR, no optical counterpart was found and the limits imposed restrict the nature of the γ-ray source. In both cases, the most likely interpretation is that the γ-ray source is a Geminga-type object.

THE OBSERVATORIO ASTROFÍSICO GUILLERMO HARO

The Observatorio Astrofísico "Guillermo Haro" (OAGH) is located at 31° of Northern latitude, $\sim$13 kms North-West from the town of Cananea, in the state of Sonora. It is also at $\sim$ 85 kms SE of Mt Hopkins observatory, in Arizona. The 2.12 meter Ritchey-Chretien telescope is at 2480 metres above sea-level.

Two instruments have been devoted to this project:

1. the Landessternwarte Heidelberg Faint-Object Spectrograph Camera (LFOSC) which can perform imaging/photometry BVRI, and spectroscopy in two low resolution modes: (i) 4500Å to 9000Å coverage with

---

[1] through CONACYT grant 4142-E9404

~ 7.5 pixel^{-1} and (ii) 4000Å to 7000Å coverage with ~ 5.5 pixel^{-1}. The instrument is user friendly and switching between photometry and spectroscopy is straightforward. The field of view is 6×10 arcmin. It has capability for multiobject spectroscopy through slit-mask. The LFOSC instrument, whose performance is optimum in the R band, was designed and intensively used for identification and study of ROSAT sources.

2. the Boller & Chivens spectrograph has 10 dispersion modes, allowing a better spectral resolution than LFOSC. It is coupled to CCD Tektronix 1000 with good blue response

Before the end of 1997, a third mode of operation will be available, which will provide imaging/photometry in a smaller field of view than LFOSC, but with improved spatial resolution and response towards the blue and ultraviolet.

THE EGRET SOURCE 2EG J2020+4026

The γ-ray source 2EG J2020+4026 is inside the γ-Cygni SNR (G78.2+2.1). Using events with $E \geq 1$GeV photons the source location was pinned down to an 8 arcmin error box, containing only one ROSAT source, named RX J2020.2+4026. A Palomar Observatory Observatory plate of the location of the ROSAT source clearly shows a $m_v \sim 15$ star (Figure 1).

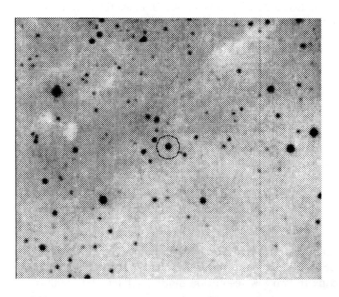

FIGURE 1. POSS image of the region around the ROSAT source RX J2020.2+4026. The $m_v \sim 15$ star is clearly seen inside the error box.

Optical observations of the RX J2020.2+4026 box were performed at the OAGH in October 3-5, 1995. Spectra were taken with the Boller & Chivens in the 3.2Å/pixel dispersion mode. The optical spectrum corresponds to a K0V star and shows no strong emission lines or signs of coronal activity, making its association with RX J2020.2+4026 uncertain.

The overall data is consistent with a Geminga type object. If the K0V star were RX J2020.2+4026, it is highly unlikely that it would correspond to 2EG J2020+4026. A more direct interpretation is that a Geminga-type neutron star is hidden in the ROSAT error box [2]. We note that this region is rather crowded, enhancing the possibility of source confusion.

OBSERVATIONS OF 2EG J0008+73

2EG J0008+73 is a bright EGRET source inside the SNR CTA1 (G119.5+10.2). Its location was pinned to a 15' error box (95% confidence) using $\geq$ 1 GeV γ-ray events. The point source RX J0007.0+7302 is inside EGRET error box, and independent X-ray studies indicate this is most probably a neutron star [5].

Optical observations of the RX J0007.0+7302 error box were made from the OAGH in Jan 1-5, 1997. Deep BVRI images were obtained with the LFOSC instrument. Total integration time over the three workable nights was $\gtrsim$ 2 hours per band.

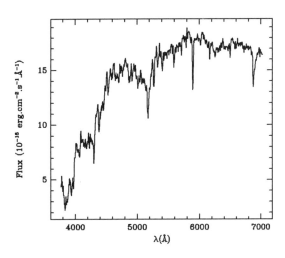

FIGURE 2. Optical spectrum of the star contained in the error box of RX J2020.2+4026.

The added images (Figure 3) show no point source brigther than $B \sim 23.1$, $V \sim 22.5$, $R \sim 22.6$, $I \sim 21.5$. A high coronal activity stellar counterpart for RX J0007.0+7302 is ruled out by these limits. A Bl Lac counterpart is unlikely but cannot be ruled out by these limits (requires down to $V \sim 24$). However, the lack of proven variability of the γ-ray source favours a galactic counterpart, and the overall data is consistent with a Geminga type object [3].

CONCLUSIONS

Optical studies of ROSAT sources associated with 2EG J2020+4026 and 2EG J0008+73 have identified these as good Geminga type candidates. Optical observations are a valuable link in multi-frequency studies aimed to identify Geminga-type objects.

REFERENCES

1. Bertsch D.L. et al 1992, Nature 357, 306
2. Brazier K.T.S. et al 1996, MNRAS 281, 1033
3. Brazier K.T.S. et al 1997, MNRAS submitted
4. Halpern & Holt 1992, Nature 357, 222
5. Slane P. et al 1997, ApJ in press

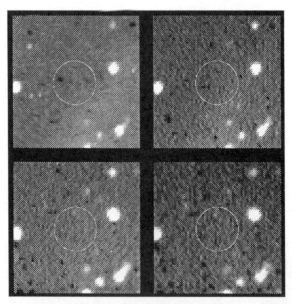

FIGURE 3. Deep BVRI images of the RX J0007.0+7302 error box. The 10" radius circle, centred on the ROSAT position, overestimates the actual error in location.

Possible Identification of Unidentified EGRET Sources with Wolf-Rayet Stars

R.K. Kaul and A.K. Mitra

Bhabha Atomic Research Centre, Nuclear Research Laboratory, Mumbai - 400 085, India

Abstract. We propose, on the basis of positional correlation, the possible association of 8 EGRET unidentified galactic plane sources with Wolf-Rayet stars having strong stellar winds. The probability of a chance positional association (within $\leq 1°$) of an EGRET sources with one of the 160 WR stars in the galaxy works out to be $\sim 10^{-4}$. We also discuss the likely origin of these gamma rays from progenitor particles accelerated by the strong Wolf-Rayet wind.

I INTRODUCTION

The second EGRET catalog (Thompson et al, 1995) contains 129 sources which were detected in the time-integrated EGRET exposures at energies $\geq$ 100 MeV. These include 6 pulsars, a solar flare, the LMC and 51 active galactic nuclei (AGN) as well as sources for which no definite identification with objects at other wavelengths is available yet. Of the 37 unidentified galactic plane sources, 10 are high confidence detections ($> 6\sigma$) while 27 are marginal detections ($< 6\sigma$). Mukherjee et al (1995) have reported that the unidentified galactic sources appear to have, on the average, greater luminosities than the pulsars detected by EGRET and conclude that pulsars can account for only a small fraction of the unidentified galactic sources, indicating that there is probably an additional galactic population of high energy gamma-ray sources whose nature is as yet unknown. Esposito et al (1996) have reported a statistically significant positional correlation between several unidentified EGRET sources and SNR's, while Kaaret and Cottam (1996) have claimed a statistically significant correlation between OB associations and unidentified EGRET sources in the galactic plane. Here we propose, again on the basis of positional association, that some of the unidentified EGRET sources at $|b| < 10°$ are associated with Wolf-Rayet (WR) stars with strong stellar winds which are capable of generating ≥ 100 MeV gamma-rays through non-thermal processes.

S.No.	EGRET Source	l	b	Associated WR star	l	b	Angular distance (degrees)
1.	2EGJ 1021-5835	284.45	-1.20	WR 19	283.89	-1.19	-0.56
				WR 20	284.51	-1.84	0.64
2.	2EG 1049-5847	287.63	0.40	WR 28	287.75	0.15	0.28
				WR 27	287.14	0.12	0.56
3.	2EGJ 1103-6106	290.30	-0.90	WR 38	290.57	-0.92	0.27
				WR 37	290.55	-1.05	0.29
				WR 39	290.63	-0.90	0.33
4.	2EGJ 1718-3310	353.31	2.48	WR 88	352.67	2.04	0.78
5.	2EGJ 1801-2312	6.73	-0.14	WR 105	6.52	-0.52	0.43
				WR 104	6.44	-0.49	0.45
6.	2EGJ 1825-1307	18.38	-0.43	WR 116	19.16	-0.32	0.79
				WR 114	17.54	-0.13	0.89
7.	2EGJ 2019+3719	75.46	0.60	WR 142	75.33	0.33	0.27
				WR 141	75.33	0.08	0.54
				WR 138	75.24	1.11	0.56
8.	2EGJ 2033+4112	80.19	0.66	WR 144	80.04	0.93	0.31
				WR 146	80.57	0.45	0.43
				WR 145	79.69	0.93	0.57

II EGRET-SOURCES AND WR STAR ASSOCIATIONS

We have compared the locations of the 37 unidentified EGRET sources with $|b| \leq 10°$, with the locations of the 160 WR stars listed in the catalogue of van der Hucht et al (1988). We find that 8 of these sources have one or more WR stars within 1° of the centroid of the EGRET error circle. More importantly, 4 of the gamma-ray sources have a WR star within ≤ 20 arcminute of the centroid of the EGRET error circle. The table above lists the EGRET sources with possible WR star counterparts within 1° of the center of the error circle, alongwith the galactic coordinates of the gamma-ray source and the associated WR star(s), the angular separation between the centroid of the EGRET error circle and the associated WR star and the distance of the WR star in kpc (wherever available).

In order to calculate the probability of a chance association between the EGRET-detected gamma-ray sources and the WR stars, we assume that the distribution of WR stars is random with respect to l and b. We divide the galactic plane bounded by $-10° \leq b \leq 10°$ into cells of $1° \times 1°$ size and calculate the probability of getting γ_1 objects out of a total of n_1 (say 1 out of 37 gamma-ray sources) and γ_2 objects out of a total of n_2 (say 1 out of the 160 WR stars) in one particular cell out of the N total cells. The probability may be written as

$$P_{\gamma_1,\gamma_2} = \left[C_{\gamma_1}^{n_1} \frac{(1-1/N)^{n_1-\gamma_1}}{N^{\gamma_1}} \right] \left[C_{\gamma_2}^{n_2} \frac{(1-1/N)^{n_2-\gamma_2}}{N^{\gamma_2}} \right] \quad (1)$$

This leads to $P_{(1,1)} \sim 10^{-4}$ and suggests an insignificant statistical correlation between WR stars and unidentified EGRET sources. Sturner and Dermer (1996) have also found a similar value ($\sim 4.5 \times 10^{-4}$) for the probability of chance alignment between 37 EGRET sources and 182 SNR's, assumed to be distributed randomly with respect to l and b in the galactic plane, from detailed computer simulations. Based on the above estimate of probability, one would expect only 1 WR star- gamma-ray source positional association by chance, as compared to 8 reported here. We, therefore, feel encouraged to claim that some of the associations reported here may have a physical basis, especially since ≥ 100 MeV gamma-ray generation in certain WR star systems has already been proposed on theoretical grounds.

III HIGH ENERGY GAMMA-RAY GENERATION IN WR STAR SYSTEMS

The Wolf-Rayet (WR) stars are very rare, hot, luminous objects whose spectra are dominated by broad (10^3 km s^{-1}) emission lines in the optical region. They are thought to be the remnants of initially massive stars that have lost their hydrogen-rich envelopes through mass transfer to a companion or by a strong stellar wind in the case of single stars.

Non-thermal radio emission from some WR stars, whose winds are typically ten times stronger than those of OB stars, has been reported through VLA observations (Becker and White, 1985). The possibility of the existence of relativistic electrons in the strong winds of WR stars, accelerated through the first-order Fermi acceleration by strong shocks has been proposed by White (1985) in order to explain the reported non-thermal radio emission. These relativistic electrons would also be expected to generate high energy gamma-rays through the synchrotron, bremsstrahlung or inverse-Compton interactions with the ambient magnetic and radiation fields (Pollock, 1987; Chen and White, 1994). The rate of cooling of the electrons due to inverse Compton process and synchrotron process is determined by the ratio of the photon and magnetic field densities :

$$\frac{U_{photon}}{U_{mag}} \simeq \frac{L_*}{4\pi r^2 (B^2/8\pi)c} \simeq 10^3 \; L_{39} \; R_{10}^{-2} \; B_*^{-2} \quad (2)$$

where L_* is the bolometric luminosity of the WR star and R_{10} is the radius of star in units of $10 R_\odot$. At least near the surface of the star, the value of this ratio can be $\sim 10^3$ for $L_{39} \equiv (L_*/10^{39} \; erg \; s^{-1}) = 1$ and $B_* \approx 1$ G. For such high values of ξ, the electron cooling will be totally dominated by IC process, and it can be shown that for typical shock acceleration rates, the maximum electron energy, as limited by the IC process, can easily be $E_{electron} \sim 10 \, GeV$.

The ratio of the integrated IC luminosity (i.e., $L_x + L_\gamma$) and the integrated synchrotron luminosity (L_{radio}) is also given by

$$\frac{L_x + L_\gamma}{L_{radio}} \sim \xi \sim 10^3 \qquad (3)$$

Since $L_{radio} \sim 10^{31} \, erg \, s^{-1}$, it is possible to have a value of $L_{x+\gamma} \sim$ few $10^{34} \, erg \, s^{-1}$, implying that a typical luminosity in the EGRET (≥ 100 MeV) band could be $L(> 100 \, MeV) \sim 10^{34} \, erg \, s^{-1}$. However, by noting the sensitivity of ξ on B_*, it is possible that such relatively high values of L_γ may be rather rare. On the other hand, for very favourable cases, it may be possible to have $L_\gamma(> 100 \, MeV) \geq 10^{35} \, erg \, s^{-1}$ too. Since, alongwith electron acceleration, the wind shock may also accelerate protons or ions (White, 1985), the bulk power of the wind for early type stars could lie between $\sim 10^{36} - 10^{37} \, erg \, s^{-1}$. The 0.4-4 keV X-ray luminosities seen from many hot stars ($10^{32} - 10^{33} \, erg \, s^{-1}$) indicate that the rate at which energy is deposited in wind shocks could be $\approx 10^{36} \, erg \, s^{-1}$ (Pollock, 1987). It is possible that a sizeable fraction of the energy dissipated by the shocks is used in accelerating ions by first order Fermi-acceleration near the shock front with $L_p \sim 10^{35} \, erg \, s^{-1}$. Although the WR winds are generally thin to γ-rays, they can offer an optical depth of few percent for the production of π^o via ion-wind collisions. Thus, under favourable cases, the ion generated γ-ray luminosity above 70 MeV's could be $\sim 10^{34} \, erg \, s^{-1}$ (Chen & White, 1991).

REFERENCES

1. Becker, R.H. and White, R.L., *ApJ*, **297**, 649 (1985)
2. Chen, W. and White, R.L., *ApJ Lett.*, **381**, L63 (1991)
3. Esposito, J. et al, *ApJ*, **461**, 820 (1991)
4. Fichtel, C.E. et al, *ApJS*, **94**, 551 (1994)
5. Kaaret, P. and Cottam, J., *ApJ Lett.*, **462**, L35 (1996)
6. Mukherjee, R. et al, 1995, *ApJ*, **441**, 61
7. Pollock, A.M.T., *A& A*, **171**, 135 (1987)
8. Sturner, S.J. and Dermer, C.D., *ApJ*, **461**, 872 (1996)
9. Thompson, D.J. et al, *ApJS*, **101**, 259 (1995)
10. van der Hucht, K.A. et al, *Sp. Sci. Rev.*, **28**, 227 (1988)
11. White, R.L., *ApJ*, **289**, 698 (1985)

Accreting Isolated Black Holes and the Unidentified EGRET Sources

Charles D. Dermer

*E. O. Hulburt Center for Space Research,
Code 7653, Naval Research Laboratory, Washington, DC 20375*

Abstract. Many of the unidentified >100 MeV EGRET γ-ray sources are conjectured to originate from a population of isolated $\approx$ 10-100 $M_\odot$ black holes accreting from the ISM. The low-latitude sources are mostly distant ($\sim$ several kpc) black holes accreting from high density molecular gas, whereas the medium latitude sources are black holes within a few hundred pc which accrete from the more tenuous local ISM. The γ-ray luminosity is assumed to be proportional to the Bondi-Hoyle accretion rate, and the nearby and distant sources have γ-ray powers of $\lesssim 10^{33}$ ergs s^{-1} and $\gtrsim 10^{35}$ ergs s^{-1}, respectively, in both cases far below the Eddington luminosity. The number of sources is compatible with estimates of black hole formation rates. In contrast to a pulsar origin for the unidentified sources, this model can account for γ-ray flux variability.

INTRODUCTION

One of the outstanding questions in gamma-ray astronomy is the nature of the >100 MeV sources which have not been shown to be members of established gamma-ray source populations such as rotation-powered pulsars or blazars. There are 96 unidentified EGRET sources in the Second EGRET Catalog [1] and Supplement [2], of which 40 are low-latitude ($|b| < 10°$) and 56 are medum/high-latitude ($|b| > 10°$) sources (see [3] for a review). The 100 MeV-5 GeV EGRET peak flux threshold ϕ_p for the identification of sources at medium and high latitudes is $\cong 10^{-7} \phi_{-7}$ ph cm^{-2} s^{-1} with $\phi_{-7} \approx 1$. At low latitudes, the peak flux threshold $\phi_{-7} \approx 4$ due to the increased background from the diffuse galactic γ-ray emissivity near the galactic plane.

A Gaussian fit to the latitude distributions of the low-latitude and medium/high-latitude unidentified sources yields standard deviations of $\sigma_b \cong$ 2° and $\cong 20°$, respectively (see, e.g., [4-7]). If the half-thickness or scale height of the underlying source population is $100 h_{100}$ pc, then simple arguments (e.g., [4]) indicate that the 100 MeV-5 GeV luminosities of the low and medium/high latitude unidentified source populations are

$$L_\gamma(\text{ergs s}^{-1}) \approx 7.5 \times 10^{31} \frac{\phi_{-7} h_{100}^2}{\sin^2 \theta} \cong \begin{cases} 2.5 \times 10^{35} (\phi_{-7}/4) h_{100}^2, & \text{if } |b| < 10° \\ 6 \times 10^{32} \phi_{-7} h_{100}^2, & \text{if } |b| > 10° \end{cases}$$
(1)

Here we argue that the spatial and luminosity distributions of the unidentified EGRET sources can be understood if high-energy gamma rays are produced by a population of isolated black holes which accrete from the ISM. The ~ 3 orders of magnitude difference between the luminosities of the two populations reflects the ratio of the gas densities of typical molecular clouds ($\gtrsim 10^2$ cm^{-3}) to that of the local ISM (≈ 0.1 cm^{-3}). Although some of the low-latitude unidentified EGRET sources are probably isolated neutron stars [7-9], supernova remnants [10,11], or blazars seen through the disk of the galaxy, this proposal for the remaining sources unifies the low and medium/high latitude unidentified sources into a single class of sources that accounts for the spatial distribution and can also account for time-variable sources.

ISOLATED BLACK HOLES IN THE GALAXY

A number of arguments supports the conjecture that the unidentified EGRET sources are black holes accreting from the ISM: (i) The distribution of the unidentified γ-ray sources is correlated with the ISM column density along the line of sight, with only a handful belonging to an isotropic source population [12]; (ii) variability of unidentified sources with a non-blazar origin implies a new class of high energy sources [13,14]; and (iii) the average gamma-ray spectral index α of the unidentified sources is $\gtrsim 2.0$ which, other than for the young 1000-year old Crab pulsar, significantly differs from the harder spectral indices of the known gamma-ray pulsars. This last point is illustrated in Fig. 1, where we compare the 100 MeV-2 GeV spectral indices of the six known EGRET gamma-ray pulsars [15,16] with the spectral indices in the EGRET energy range of 32 low-latitude unidentified sources [17].

Isolated black holes in the Galaxy are formed by the evolution of massive stars which explode as SN II. Using the initial mass function of Miller & Scalo [18] and a Galaxy age of 12×10^9 yr, we find that the total number of stars which have formed in the Milky Way with mass $M > 10\ M_\odot$ and $M > 25\ M_\odot$ are 2.0×10^8 and 2.3×10^7, respectively. Here we use a Galactic volume of 1.27×10^{11} pc^3, corresponding to a 15 kpc radius and a scale height of 90 pc. The total mass that has been processed through stars of mass $>10\ M_\odot$ and $>25\ M_\odot$ during the lifetime of the Milky Way is $3.3 \times 10^9\ M_\odot$ and $7.8 \times 10^8\ M_\odot$, respectively. Similar results are obtained using the Salpeter IMF.

In order to avoid chemical pollution in the galaxy, in particular, an overproduction of Oxygen, Twarog & Wheeler [19] argue that stars with masses $M \gtrsim 24$-$28\ M_\odot$ collapse directly into black holes with little expulsion of their

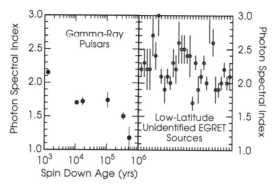

FIGURE 1. Comparison of photon spectral indices of the six EGRET-detected pulsars with the spectral indices of low-latitude unidentified sources [17].

envelope masses into the ISM. In view of the absence of a pulsar detection from SN 1987A, Brown & Bethe [20] argue that stars with masses between 10 and 25 $M_\odot$ produce 2 $M_\odot$ black holes, whereas stars which are initially more massive than 25 $M_\odot$ collapse into a black hole with nearly the same mass as the progenitor star. Detailed calculations by Timmes et al. [21] indicate that a 40 $M_\odot$ star will leave a remnant mass of $\cong$ 10 $M_\odot$, though fallback of the expelled debris could leave a more massive remnant. The optimistic estimates of Brown and Bethe indicate that the space density of > 25 $M_\odot$ black holes could be as large as 1 per 5500 pc^3, though the calculation in [21] suggests that this could be a substantial overestimate.

BONDI-HOYLE ACCRETION FROM THE ISM

It is not possible to fairly review the geography of the interstellar medium here, but it can be briefly summarized (see [22] and references therein). We are surrounded by "local fluff" within several pc of the Sun, with $n \sim 0.1$ cm^{-3} and $T \approx 8000$ K. Beyond is the "local bubble" which ranges from several pc to $\sim$ 100-200 pc. In the directions $120° \lesssim \ell \lesssim 200°$ and $270° \lesssim \ell \lesssim 15°$, the local bubble is composed of neutral 10^3-10^4 K hydrogen gas with a density $n_{\rm HI} \cong 0.07$ cm^{-3}. This gas occupies a 25% filling factor, and is embedded in a tenuous 10^6 K ionized gas of density $\cong 0.005$ cm^{-3} which produces the EUV/SXR background. The gas in the direction $200° \lesssim \ell \lesssim 270°$ to distances $\gtrsim 200$ pc is exceedingly dilute by comparison.

The locations and column densities of molecular clouds in the LISM have been mapped through CO observations and show that $\approx 4.0 \times 10^6$ $M_\odot$ of molecular gas are found within 1 kpc [23]. Molecular clouds have a fractal structure and a large range of densities, so it is an extreme oversimplification to assume that they have a uniform density. Nevertheless, we characterize the average density of molecular clouds by the value of $10^2 n_2$ cm^{-3}. This implies a volume-filling factor of molecular clouds in the Galaxy of $\approx 3 \times 10^{-3}/n_2$.

If stars with masses $\gtrsim 25 M_\odot$ collapse directly into black holes, then as many as 2×10^7 black holes with an average mass of $\approx 35 M_\odot$ could inhabit the Galaxy. The sound speed in $10^4 T_4$ K hydrogen gas is $c_s \simeq 1.2 \times 10^6 T_4^{1/2}$ cm s^{-1}, which is comparable to the transverse speeds of stars born with a 90 pc scale height. The Bondi-Hoyle mass accretion rate of $35 m_{35}$ Solar mass objects in the ISM is given by $\dot{M} = 4.6 \times 10^{14} \lambda m_{35}^2 n / v_6^3$ gm s^{-1}, where the eigenvalue $\lambda \cong 1$ and $v = 10^6 v_6$ cm s^{-1} is the quadratic sum of c_s and the speed of the object through the ISM gas. The emergent accretion luminosity

$$L(\text{ergs s}^{-1}) = \eta \dot{M} c^2 \cong 6 \times 10^{33} \; \lambda m_{35}^2 \eta_{-2} n[\text{cm}^{-3}] v_6^{-3} \tag{2}$$

(noting $\rho \approx 1.4 m_p n$), where $\eta = 0.01 \eta_{-2}$ represents the efficiency for converting the gravitational energy of the accreting matter into gamma radiation. Given the large uncertainties in η and the strong dependence of the luminosity on v, one sees that the implied luminosities from Bondi-Hoyle accretion in gases with densities of 0.1 cm^{-3} amd 10^2 cm^{-3} are in accord with the luminosities given in eq. (1).

The observed number of unidentified sources represents the product $N_{\text{bh}} \times \chi \times r$, where N_{bh} is the number of black holes in the Galaxy, χ is the volume-filling factor of gas, and r is the ratio of the volume of the Galaxy in which the sources are observed to the total Galactic volume. The 2° and 20° Gaussian standard deviations imply sampling distances of $\approx 3.0 h_{100}$ kpc and $\approx 0.3 h_{100}$ kpc for the low and medium/high-latitude sources, respectively. Using 15 kpc for the Galaxy's radius, our naive estimate for the population of $\gtrsim 25 M_\odot$ black holes therefore implies that there should be $\approx 2100 h_{100}^2$ and $2400 h_{100}^2$ low and medium/high latitude sources, respectively. If the number of massive black holes has been seriously overestimated by a factor of ≈ 100, as suggested in [21], then the observed number of unidentified EGRET sources is in rough agreement with the observations.

GAMMA RAYS FROM ISOLATED BLACK HOLES

The conjecture advanced in this paper unifies the low and medium/high-latitude sources, insofar as the low-latitude unidentified EGRET sources would naturally originate from the distant, high density molecular clouds that are found close to the Galaxy plane. In view of the large uncertainties in estimating the number of massive black holes in our Galaxy, the observed number of sources can be turned around to provide a constraint upon the birth rate of $\gtrsim 25 \; M_\odot$ black holes.

Accreting black hole sources are usually thought to be luminous X-ray sources. To the extent that bright X-ray sources are not coincident with the unidentified EGRET sources, this objection would argue against the hypothesis presented here. But our knowledge of black-hole accretion is generally obtained from luminous sources radiating at $\approx 0.01 - 1$ of the Eddington luminosity limit. The ratios of the γ-ray source luminosities to the Eddington luminosities for $\sim 35 M_\odot$ black holes are in the range $\sim 10^{-7}\text{-}10^{-4}$ where our understanding of accretion physics is based more upon theoretical studies than observations. This is the advection-dominated regime of accretion, and studies [24] of the Galactic Center black hole, which is also probably radiating at a low rate compared to its Eddington luminosity, indicate that the emergent luminosity becomes more strongly dominated by higher-energy radiation.

The apparently bimodal spatial distribution of the unidentified sources occurs because we can only detect nearby low luminosity sources and bright distant sources. The greater number of medium/high latitude unidentified sources in the ranges $120° \lesssim \ell \lesssim 200°$ and $270° \lesssim \ell \lesssim 30°$ is in qualitative agreement with the density distribution of the local ISM. Source variability is expected when black holes travel through regions of varying densities and accrete at variable rates.

REFERENCES

1. Thompson, D. J. et. al., *Ap. J. Supp.* **101**, 259 (1995).
2. Thompson, D. J. et. al., *Ap. J. Supp.* **107**, 227 (1996).
3. Mukherjee, R., Grenier, I. A., and Thompson, D. J., these proceedings.
4. Mukherjee, R. et. al., *Ap. J.* **441**, L61 (1995).
5. Kanbach, G. et. al., *Astron. Ap. Supp.* **120**, 461 (1996).
6. Özel, M.E., and Thompson, D. J., *Ap. J.* **463**, 105 (1996).
7. Kaaret, P., and Cottam, J., *Ap. J.* **462**, L35 (1996).
8. Yadigaroglu, I., and Romani, R. W., *Ap. J.* **467**, 347 (1997).
9. Sturner, S. J., and Dermer, C. D., *Ap. J.* **461**, 872 (1996)
10. Sturner, S. J., and Dermer, C. D., *Astron. Ap.* **281**, L17 (1995).
11. Esposito, J. A., et. al., *Ap. J.* **461**, 820 (1996).
12. Grenier, I. A., *Adv. Space Res.* **15**, 573 (1994).
13. Tavani, M. et. al., *Ap. J.* **479**, L109 (1997).
14. McLaughlin, M. A. et. al., *Ap. J.* **473**, 763 (1996).
15. Ramanamurthy, P. V. et. al., *Ap. J.* **447**, L109 (1995).
16. Thompson, D. J., et. al., *Ap. J.* **436**, 229 (1994).
17. Merck, M. et. al., *Astron. Ap. Supp.* **120**, 465 (1996).
18. Miller, G. E., and Scalo, J. M., *Ap. J. Supp.* **41**, 513 (1979)
19. Twarog, B. A., and Wheeler, J. C. *Ap. J.* **316**, 153 (1987).
20. Brown, G. E., and Bethe, H. A., 1994, *Ap. J.* **423**, 659 (1994).
21. Timmes, F. X., Woosley, S. A., and Weaver, T. A., *Ap. J.* **457**, 834 (1996).
22. Blaes, O., and Madau, P., *Ap. J.* **403**, 690 (1993).
23. Dame, T. M. et. al., *Ap. J.* **322**, 706 (1987)
24. Mahadevan, R., Narayan, R., and Krolik, J., *Ap. J.* **486**, in press (1997)

SEYFERT AND RADIO GALAXIES

Multi-Year BATSE Earth Occultation Monitoring of NGC4151

A. Parsons[*], N. Gehrels[*], W. Paciesas[†]

A. Harmon[‡], G. Fishman[‡], C. Wilson[‡], S. N. Zhang[°‡]

[*]*NASA/Goddard Space Flight Center*, [†]*University of Alabama at Huntsville*, [°]*USRA*,
[‡]*NASA/Marshall Space Flight Center*

Abstract. Because the Earth occultation technique allows BATSE to be used as an all-sky monitor, BATSE is uniquely able to provide continuous observations of NGC4151 over the entire lifetime of the Compton GRO mission. Continuous 2154 day light-curves for the Seyfert galaxy NGC4151 have been derived from Earth occultation analysis of CGRO BATSE data. Light-curves for the 20-70 keV and 70-200 keV bands are shown and found to have an average flux of 11 x 10^{-3} photons cm^{-2} s^{-1} and 2.3 x 10^{-3} photons cm^{-2} s^{-1} respectively. Evidence for intrinsic source variability is observed in both 20 - 70 keV and 70 - 200 keV bands with a fractional variation ratio measured at Fvar = 0.2.

INTRODUCTION

A continuous 2154 day (5.9 year) NGC4151 light-curve has been derived from Earth occultation measurements using the Burst And Transient Source Experiment (BATSE) on the Compton Gamma-Ray Observatory (CGRO). Here we describe the BATSE Earth occultation technique and present a preliminary light-curve for NGC4151. Because the Earth occultation technique allows BATSE to be used as an all-sky monitor, BATSE can provide continuous observations of NGC4151 over the entire lifetime of the Compton GRO mission. Broad band variability studies of emission from active galactic nuclei (AGN) can provide important clues to the source dynamics and geometry. Characteristic time scales for variability can be compared with those indicative of different physical processes such as the expected orbital, viscous, and light travel time scales. NGC4151, a nearby Seyfert 1.5 galaxy, has been extensively monitored in nearly all available energy bands. This AGN has been so well studied because it is one of the brightest extragalactic sources in 2-100 keV band and as such, is often a prime target for observation. Unfortunately, even though there is an extensive observational database for this source, many of its properties remain unclear. The spectral behavior of NGC4151 in the 2 - 30 keV band is characterized by a flux variation of a factor of 10 and a change in the spectral index α that is correlated with the flux. As reported in [10] the spectral index varies from $\alpha=1.3$ at lower x-ray fluxes to $\alpha=1.7$ at higher flux, with the time behavior of these variations consisting mainly of slow 1 day drifts with occasional larger amplitude flares. Like most of its fellow Seyfert galaxies, NGC4151's gamma-ray emission is typically observed to cut off at a few hundred keV. Hard x-ray measurements in the 2 - 200 keV range have failed to provide a consistent picture for the behavior of NGC4151. While some pre-CGRO balloon observations had detected flux that extended to MeV energies [7,8], the more recent SIGMA [3] and OSSE [4–6] observations display a spectral break below about 100 keV. Since 1991, OSSE has continued to observe this lower energy cutoff, and all OSSE NGC4151 observations to date indicate surprisingly little variation in either flux or spectral shape [4]. It is difficult to accommodate such little variability above 30 keV along with much greater variability at lower energies. Coincident 20 - 200 keV observations by BATSE are thus very important to bridge the gap in energy.

MSFC BATSE EARTH OCCULTATION TECHNIQUES

Using BATSE Earth occultation techniques, the flux of a source is determined from the difference in the BATSE LAD count rates as the source rises and sets behind the Earth's limb. With a full 4π steradian field of view, BATSE's eight LADs view all of the sky all of the tim and can continuously monitor the entire sky from 20 - 1900 keV. Although NGC4151 is the brightest Seyfert galaxy in the 20 - 200 keV energy range, it is still a relatively weak source for BATSE Earth occultation measurements. Data from many NGC4151 occultation steps must be averaged to achieve adequate signal-to-noise for a source flux measurement. The lower sensitivity of BATSE occultation measurements limits our minimum measured variation time scale to a few days.

Occultation steps are fit using the CONT data product which contains continuous raw count rates from each of the eight LADs. The 20 - 1900 keV CONT data is divided into 16 energy channels and has 2.048 s time resolution. The CONT data rates for each LAD are fit to source steps on top of an empirical quadratic background model. Model source spectra are convolved with the BATSE response matrices and fit to the observed occultation step rates to determine the source photon flux during a particular occultation step. To obtain an adequate signal-to-noise ratio for useful flux measurements, we must calculate a weighted average over many steps. As we increase the number of days included in our step averages, we necessarily reduce the error in the mean value at the expense of variability information on short time scales.

The base time unit for the BATSE NGC4151 light curves presented in this paper is 7 days We have chosen to fit the 20 - 200 keV data to an empirical power-law source spectrum with an exponential cutoff:

$$F(E) = F(E_0) \left(\frac{E}{E_0}\right)^{-1.6} \exp\left(\frac{E_0 - E}{E_{cutoff}}\right) \quad \frac{photons}{s \cdot cm^2 \cdot keV} \quad (1)$$

where $E_0, E_{cutoff} = 100$ keV.

Typically, more than one LAD will detect an occultation step; these multiple detector rates are fit simultaneously. The best fit model parameters are determined from a nonlinear least squares fit using the standard Levenberg-Marquardt algorithm [9]. A preliminary light curve is shown in Figure 1. Note that continuous flux histories are shown for both 20-70 keV and 70-200 keV energy ranges. These ranges were chosen to best match the BATSE channel energies. Note that the 70-200 keV fluxes have been scaled by a factor of 2. The errors shown in the data represent 1σ statistical errors in the fitted source flux.

The problem of contamination from bright sources that flare or are occulted at the same time as NGC4151 is illustrated in the raw step count rate history shown in Figure 2. Notice the outlying data point recorded by LAD #1 on TJD 8586.9. Since the fluxes are simultaneously fit to all illuminated detectors, contaminated step data in one of the detectors will skew the spectral fit and add large systematic errors to the affected flux measurements. Since NGC4151 is quite far off the galactic plane (75° galactic latitude), effects from interfering sources are fairly easily identified. A step is flagged and discarded if the relative count rates measured in the LADs are not compatible with the geometry of a single source in the direction of NGC4151. While some good occultation steps will certainly be discarded along with the contaminated steps, it is very unlikely that any contaminated data will be retained. As long as the rejection of good occultation steps is uncorrelated with the behavior of the source, the loss of good steps will result in a slight increase in statistical error but will have no systematic effect in our flux measurements.

VARIATION ANALYSIS

Is the measured flux distribution inconsistent with a constant flux? For both the 20-70 keV and 70-200 keV bands we calculate χ^2 for the flux measurements about a constant instrinsic source flux. The χ^2 about the weighted mean flux value of $(11\pm.13) \times 10^{-3}$ photons cm^{-2} s^{-1} for the 20-70 keV flux history is 632 for 307 degrees of freedom. The probability of measuring this χ^2 or larger by chance is less than 1×10^{-6}. Similarly, the χ^2 about the mean 70-200 keV flux $(2.3\pm.04) \times 10^{-3}$ photons cm^{-2} s^{-1} is 442 for 307 degrees of freedom. The probability of measuring this χ^2 or larger by chance is also less than 1×10^{-6} and thus we find that the 20-200 keV data is inconsistent with a constant source. Given that our data represent measurements from a varying source, we wish to compare the magnitude of the variations in the different energy ranges. We need to separate the variations from statistical measurement error from intrinsic source variations. To establish the intrinsic variability of the source we follow the lead of [1,2] where they remove the effect of measurement error by quadrature subtraction. Let σ_{var} represent the underlying unknown source variability, let σ_{data} represent the standard deviation of the measured fluxes and σ_{err} be the typical error in each flux datum. Since the measured error in the data is dominated by fluctuations in the BATSE instrument background, we may assume that the measurement errors in the BATSE fluxes are uncorrelated with the intrinsic NGC4151 flux. The total variation in the data may thus be written as $\sigma^2_{data} = \sigma^2_{var} + \sigma^2_{err}$. We may define the fractional variability F_{var} of the source as:

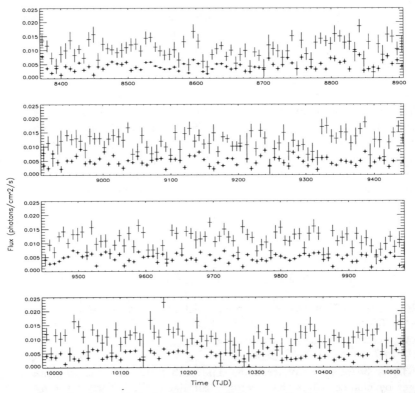

FIGURE 1. Continuous 2154 day (5.9 year) light curves for flux measurements in the 20-70 keV and 70-200 keV ranges. Note that the fluxes for the 70-200 keV band have been multiplied by a factor of 2 to better illustrate the variability.

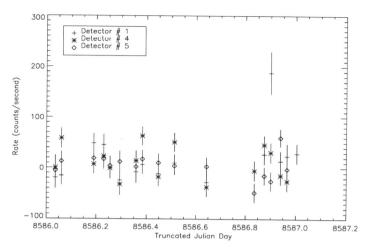

FIGURE 2. An example occultation step time series for LADs 1, 4 and 5 from TJD 8586 to 8587. Note the abrupt jump in the step count rate for LAD 1 at TJD 8586.9 which illustrates the effect of contamination from interfering sources.

$$F_{var} = \frac{\sigma_{var}}{\mu} = \frac{\sqrt{(\sigma_{data}^2 - \sigma_{err}^2)}}{\mu} \qquad (2)$$

For 20-70 keV and 70 - 200 keV energy bands, Fvar = 0.22 and 0.19 respectively. There does not appear to be much of a difference in the overall time behavior in the two bands. We are currently developing new statistical methods to better characterize the source variations on long time scales.

REFERENCES

1. Edelson, R.,*ApJ*, **401**, 516 (1992).
2. Edelson, R. A., et al.,*ApJ*, **470**, 364, (1996).
3. Finoguenov, A., et al., *Astronomy and Astrophysics*, **300**,101, (1995).
4. Johnson, et al., *MNRAS* (in preparation,) (1996).
5. Johnson, W. N., et al.,*Proceedings of the Second Compton Symposium*, ed. Fichtel, C. E., College Park, MD, 515,(1994).
6. Maisack, M., et al.,*ApJ. Let.*, **407**, L61, (1993).
7. Perotti, F., et al.,*Nature*, **282**,484,(1979).
8. Perotti, F., et al.,*ApJ*,**247**, L63, (1981).
9. Press, W. H., et al. *Numerical Receipes: The Art of Scientific Computing*, New York: Cambridge University Press,1989, ch 14, 702.
10. Yaqoob, T., et al.,*MNRAS*, **262**, 435, (1993).

Broad-band continuum and variability of NGC 5548

Paweł Magdziarz[1], Omer Blaes[2], Andrzej A. Zdziarski[3],
W. Neil Johnson[4] and David A. Smith[5]

[1] *Astronomical Observatory, Jagiellonian University, Orla 171, 30-244 Cracow, Poland*
[2] *Department of Physics, University of California, Santa Barbara, CA 93106*
[3] *N. Copernicus Astronomical Center, Bartycka 18, 00-716 Warsaw, Poland*
[4] *Naval Research Lab., Code 7651, 4555 Overlook Ave., SW, Washington, DC 20375-5352*
[5] *Department of Physics and Astronomy, University of Leicester, Leicester, LE1 7RH, UK*

Abstract.
We analyze a composite broad-band optical/UV/Xγ-ray spectrum of the Seyfert 1 galaxy NGC 5548. The spectrum consists of an average of simultaneous optical/*IUE*/*Ginga* observations accompanied by *ROSAT* and *GRO*/OSSE data from non-simultaneous observations. We show that the broad-band continuum is inconsistent with simple disk models extending to the soft X-rays. Instead, the soft-excess is well described by optically thick, low temperature, thermal Comptonization which may dominate the entire big blue bump. This might explain the observed tight UV/soft X-ray variability correlation and absence of a Lyman edge in this object. However, the plasma parameters inferred by the spectrum need stratification in optical depth and/or temperature to prevent physical inconsistency. The optical/UV/soft X-ray component contributes about half of the total source flux. The spectral variations of the soft-excess are consistent with that of the UV and argue that the components are closely related. The overall pattern of spectral variability suggests variations of the source geometry, and shows the optical/UV/soft X-ray component to be harder when brighter, while the hard X-ray component is softer when brighter.

INTRODUCTION

NGC 5548 is one of the brightest Seyfert 1 galaxies with optical/UV/Xγ continuum consisting of three characteristic components: the big blue bump, the soft-excess, and the hard X-ray thermal Comptonization continuum with the reflection hump. These components are known to be highly variable, but there have been no observing campaigns covering all of them simultaneously. The only simultaneous *IUE*/*Ginga* campaign (Clavel et al. 1992; Nandra et al. 1993) showed a tight correlation between the UV bump and hard X-ray component, thus supporting UV/X-ray reprocessing models. Recently, Marshall et al. (1997) have found a correlation between the UV and the soft X-ray component which suggests that the soft-excess also participates in the reprocessing. We analyze a composite broad-band optical/UV/Xγ spectrum in order to develop a continuum model and apply it to re-analyze the variability correlations in a simultaneous sub-set of available *IUE/Ginga* observations.

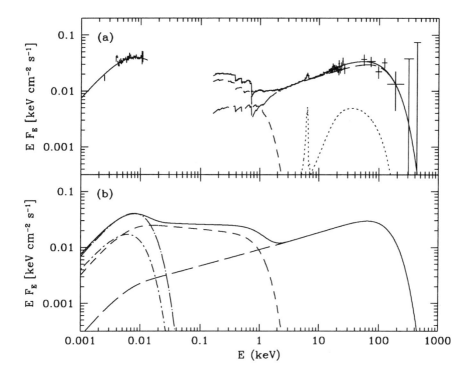

FIGURE 1. The intrinsic (i.e., after the effect of Galactic absorption at the soft excess and optical/UV reddening have been removed) broad-band composite optical/UV/Xγ spectrum of NGC 5548. The top panel shows the data and fitted model. The spectrum consists of an average over the simultaneous sub-set of optical/*IUE*/*Ginga* data supplemented with non-simultaneous average *ROSAT* and *GRO*/OSSE observations. The non simultaneous data were fitted with free normalizations and renormalized on the figure. The bottom panel shows a deconvolution of the spectrum into continuum components. The dot-short-dashed and the short-dashed curves show the disk and the soft-excess component respectively. The long-dash and the dotted (in top panel) curves show the thermal Comptonization coronal spectrum and the reflection component respectively. The solid curve gives the resulting model. The dot-long-dash curve shows the disk component fitted to the optical/*IUE* data alone. The energy scale is in the source frame.

BROAD-BAND SPECTRUM AND VARIABILITY

The optical/*IUE* and *Ginga* data come from the simultaneous sub set of the Jan. 1989–July 1990 campaign reported by Clavel et al. (1992) and Nandra et al. (1991) respectively. The *ROSAT* data is an average spectrum over a monitoring campaign of Dec. 1992–Jan. 1993 (Done et al. 1995). The *GRO*/OSSE data were compiled by McNaron-Brown et al. (1997, in preparation) from Phase 1 and 3 observations. We use the *Ginga* data from both the top-layer and mid-layer of the LAC detector. The mid-layer gives more effective area in the 10–20 keV range which is crucial for determining the Compton reflection component. That layer had previously been ignored due to problems with background subtraction.

Broad-band Continuum

We find the average spectra from non-simultaneous observations to be consistent within statistical discrepancies with that from the simultaneous campaign. However, the results have to be treated carefully since the averaging of the highly variable spectrum produces very wide confidence limits. The soft-excess component contributes significantly up to an energy $\sim$2 keV (Fig. 1a) which is marginally observed in the *Ginga* data. As was shown by Fiore, Matt, and Nicastro (1997), this component can not be simply explained by atomic processes. The *GRO*/OSSE spectra are consistent with a constant high energy cut-off at about 100 keV, and no variations of the spectral shape. This gives Comptonizing plasma parameters of kT_{HC} $\sim$50 keV and τ $\sim$2, consistent with those typical of Seyfert 1s (Zdziarski et al. 1997). Our re-analysis of the *ROSAT* campaign shows that the average data well constrain the soft-excess spectral index, $\Gamma=2.1^{+0.3}_{-0.2}$, but do not constrain the energy cut-off. This, however, is well established at $E_{SE}=0.6 \pm 0.1$ keV from *Ginga* observations. The average *IUE* data well constrain both the reddening of the spectrum E(B−V)=0.03 ± 0.01, and the maximum disk temperature kT_d=3.2 ± 0.2 eV. This argues against the high temperature disk (in agreement with Laor et al. 1997).

It is energetically possible for the entire blue bump (the dot-long-dash curve in Fig. 1b) to arise from reprocessing of the hard continuum assuming the X/γ source forms a patchy corona above the surface of the disk. However, the nature of the soft excess remains unclear in a such model and it is hard to explain the tight correlation in simultaneous *IUE–EUVE* observations (Marshall et al. 1997). On the other hand, extrapolation of the *ROSAT* soft X-ray power-law points exactly to the UV component, suggesting a Comptonization origin. Then the blue bump turns out to be dominated by Comptonization (the short-dash, and the dot-short-dash in Fig. 1b), and contains about half the total flux. However, Comptonizing the disk component by a hypothetical cold plasma requires an extreme optical depth ($\tau\sim30$), which needs to be explained by stratification of temperature and optical depth (Nandra et al. 1995).

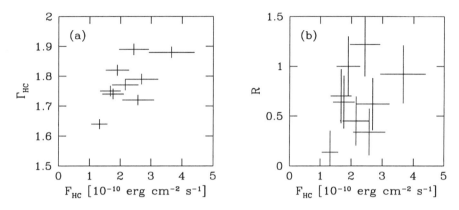

FIGURE 2. Correlation between the total flux emitted in the X/γ hard continuum with (a) photon spectral index, and (b) amount of reflection ($\Omega/2\pi$). When the source is brighter, it shows a softer spectral index and larger solid angle of cold matter intercepting the hard continuum.

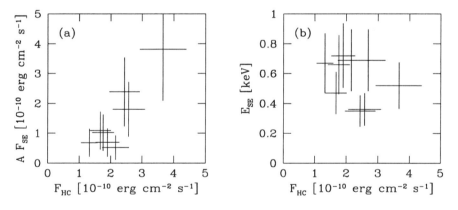

FIGURE 3. Correlation of (a) the total flux emitted in the soft-excess, and (b) the cut-off energy of the soft-excess component with the total flux emitted in the X/γ hard continuum. The total flux emitted in the soft excess is positively correlated with that from the hard continuum, while the cut-off energy of the soft excess remains constant. The soft excess is modeled by a cut-off power-law with the spectral index frozen at $\Gamma=2.2$ (cf. Walter & Fink 1993). The plotted F_{SE} flux is the soft excess flux multiplied by a model dependent constant, A.

Variability Correlations

The hard X-ray continuum shows significant correlation between the total flux changes and the spectral index and amount of reflection (Fig. 2a, b). Such a pattern of spectral variability suggests changes in the source geometry which produce more reflection when the source is brighter and softer. The total X/γ flux is correlated with the total optical/UV flux (cf. Nandra et al. 1993), but the energy balance is dependent on the spectral model assumed. The soft excess component required by the *Ginga* data seems to vary in a correlated fashion with the total flux emitted by the X/γ continuum (Fig. 3a). The cut-off energy of the soft-excess is consistent with being constant over the hard continuum changes (Fig. 3b). The above results suggest that the soft-excess component is related to reprocessing. Marshall et al.'s (1997) results suggest much higher variability amplitude in the EUV than in the UV. Hence one should expect variability in the soft-excess spectral index, which is consistent with the overall 'harder when brighter' variability pattern in the optical/UV, but opposite to the 'softer when brighter' pattern in the hard X/γ continuum.

CONCLUSIONS

We have shown that the soft-excess component in NGC 5548 may contribute to the energy reprocessing and its variability behavior is consistent with that of the big blue bump. The soft-excess is consistent with optically thick Comptonization, probably in a complex, stratified plasma. If the soft-excess is actually related to optically thick Comptonization, the entire big blue bump and the soft-excess may be dominated by the same component. This would explain both the tight *IUE–EUV* correlation (Marshall et al. 1997), and the absence of a Lyman edge in the UV spectrum (Kriss et al. 1997).

REFERENCES

1. Clavel, J., et al., *ApJ* **393**, 113 (1992).
2. Done, C., Pounds, K. A., Nandra, K., Fabian, A. C., *MNRAS* **275**, 113 (1995).
3. Fiore, F., et al., *MNRAS* **284**, 731 (1997).
4. Kriss, G., et al., *Emission Lines in Active Galaxies: New Methods and Techniques*, Peterson, B. M., et al., eds., IAU Coll. 159, in press
5. Laor, A., et al., *ApJ* **477**, 93 (1997).
6. Marshall, H. L., et al., *ApJ* **479**, 222 (1997).
7. Nandra, K., et al., *MNRAS* **248**, 760 (1991).
8. Nandra, K., et al., *MNRAS* **260**, 504 (1993).
9. Walter, R., and Fink, H. H., *AA* **274**, 105
10. Zdziarski, A. A., et al., *The Transparent Universe*, 2nd INTEGRAL Workshop, Winkler, Ch., et al., eds., ESA SP-382, p.373

Detection of a High Energy Break in the Seyfert Galaxy MCG+8-11-11

P. Grandi[1], F. Haardt[2],

G. Ghisellini[3], J. E. Grove[4], L. Maraschi[3], C. M. Urry[5]

[1] *Istituto di Astrofisica Spaziale, via E. Fermi 21, I-00044 Frascati (Roma)*, Italy
[2] *Dipartimento di Fisica, Università degli Studi di Milano, Milano, Italy*
[3] *Osservatorio Astronomico di Brera/Merate, Milano, Italy*
[4] *Naval Research Laboratory, Washington D.C., USA*
[5] *STScI, Baltimore, USA*

Abstract. We present the results from ASCA and OSSE simultaneous observations of the Seyfert 1.5 galaxy MCG+8-11-11 performed in August–September 1995. The ASCA data are well described by a hard power law ($\Gamma \sim 1.6 - 1.7$) absorbed by a column density slightly larger than the Galactic value. An iron line at 6.4 keV of EW$\sim$ 400 eV is also required by the data. The simultaneous OSSE data are characterized by a much softer power law with photon index $\Gamma = 3.0^{+0.9}_{-0.8}$, strongly suggesting the presence of a spectral break in the hard X/soft γ-ray band. In order to better investigate the complex hard X-ray spectrum of this source, a total spectrum has been produced by summing over all the observations of MCG+8-11-11 performed by OSSE so far. The joint ASCA+OSSE fit clearly showed a reflection component plus an exponential cutoff of the power law at $\sim$ 300 keV. Although the break could be due to downscattering of X-rays on cold e^+e^--pairs in a non-thermal plasma, the inferred low compactness of this source favours instead thermal or quasi-thermal Comptonization as the likely process of high energy radiation production.

INTRODUCTION

MCG+8-11-11 is a nearby (z=0.021) face-on spiral galaxy classified as a Seyfert galaxy of type 1.5. It is one of the brightest Seyfert galaxies in the X-rays showing flux variations by a factor 3–4 on time scale of years [1], and references therein). The 0.5-8 keV spectrum is described by a simple power law ($\Gamma = 1.6 - 1.9$) absorbed by Galactic Column density, N_H^{Gal}=2.03×10^{21} cm^{-2}, ([1]) It has never been observed by ROSAT and GINGA. It was detected in several balloon flights above 30 keV ([2], [3], [4], [5]) and is one of the four AGNs reported to show MeV emission ([3]). The GRO satellite confirmed

MCG+8-11-11 as an extremely intense AGN in hard X–rays/soft γ–rays, with data extending up to ∼ 200 keV. Notably COMPTEL and EGRET failed to detected the source at MeV/GeV energies ([6], [7]).

Here we present the ASCA and OSSE data that, for the first time, unambiguously reveal the presence of an exponential cutoff at ∼ 300 keV. A more extended discussion of the results can be found in [9].

LOW ENERGY SPECTRUM

The ASCA observations were splitted in two pointings of ∼ 10 ksec each on September 3 and 6 1995.

The X–ray flux of MCG+8-11-11 increased by ∼ 20% between the two ASCA observations. The count rates in the 0.6-9 keV range varied from 0.554±0.007 cts/s on September 3, to 0.669±0.008 cts/s on September 6 in SIS0. Both the observations are well described by an absorbed power law plus an iron line from cold material (see Table 1). In both epochs the value of N_H is slightly larger than the Galactic column density even if the systematic SIS overestimate of $N_H = 2 - 3 \times 10^{20}$ cm^{-2} is taken into account. The slightly different spectral slope between the two observations suggests a steepening of the spectrum with increasing luminosity (see Table 1), in agreement with the findings of [1].

Table 1: Low Energy Spectral Fits (0.6-9 keV)

Date	Γ	N_H 10^{21} cm^{-2}	E_{Fe} keV	σ_{Fe} keV	EW eV	χ^2 (d.o.f)
Sept. 3	1.57±0.04	$2.32^{+0.25}_{-0.23}$	$6.37^{+0.09}_{-0.06}$	$0.15^{+0.16}_{-0.07}$	392^{+129}_{-108}	526(508)
Sept. 6	$1.69^{+0.06}_{-0.04}$	$2.65^{+0.19}_{-0.12}$	$6.42^{+0.15}_{-0.12}$	$0.31^{+0.69}_{-0.13}$	401^{+171}_{-149}	594(558)

HIGH ENERGY SPECTRUM

MCG+8-11-11 was observed continuously by OSSE from 1995 August 22 to September 7.

The OSSE spectrum can be fitted with a simple power law in the 50-200 keV energy range ($\chi^2 = 4$ for 10 d.o.f). The photon index is extremely steep ($\Gamma = 3.0^{+0.9}_{-0.8}$) when compared to the ASCA slope, indicating the presence of a break in hard X-ray spectrum. The photon flux is $F_{50-150 keV} = 4.3 \times 10^{-4}$ photons cm^{-2} sec^{-1}.

In disagreement with the results from the HEAT balloon ([5]), no 112 keV line (interpreted as twice–scattered 511 keV annihilation line) was detected, although the OSSE and HEAT observations are very close in time (HEAT flew in July 1995), and similar continuum flux characterizes the two data sets. OSSE upper limit for the 112 keV line was obtained assuming a power law continuum and a gaussian line with a FWHM line width of 32 keV (as observed by [5]). The 95% confidence upper limit is $F_{112keV} = 9 \times 10^{-5}$ photons cm^{-2} s^{-1}, about a factor 10 below the HEAT value. As a further check, we also searched for a 170 keV (the once–scattered 511 keV line) feature in emission. As in the case of the lower energy line, only upper limits could be derived.

THE TOTAL SPECTRUM

MCG+8-11-11 was detected by OSSE in two other epochs in 1991 and 1993 ([8]). No significant luminosity variability appears between ours and previous observations. We therefore decided to combine all the available data to improve the S/N in the 50-200 keV energy range. The overall spectral shape of MCG+8-11-11 was then studied simultaneously fitting the "total" ASCA spectrum (from the two 1995 September observations) and the "total" OSSE spectrum (from all the three observations).

Three different models were tested: **1)** a power law model + an iron line, **2)** a power law model reflected by a cold disk plus an iron line, **3)** a power law with an exponential cut-off plus the reflection component plus an iron line.

All these models are reasonable fit to the data (see Table 2).

Table 2: Average Total Spectral Fits (0.6-200 kev)

Model[a]	Γ	N_H	E^b_{Fe}	σ^b_{Fe}	EW	R^c/E_{cut}	χ^2 (d.o.f)
		10^{21} cm^{-2}	keV	keV	eV	/keV	
(1)	$1.65^{+0.02}_{-0.03}$	$2.53^{+0.15}_{-0.13}$	6.43 ± 0.08	$0.28^{+0.10}_{-0.21}$	470^{+86}_{-97}		915 (871)
(2)	$1.71^{+0.03}_{-0.07}$	$2.73^{+0.18}_{-0.00}$	6.4(f)	0.3(f)	397^{+81}_{-73}	$0.50^{+0.35}_{-0.50}$	913(872)
(3)	1.73 ± 0.06	2.75 ± 0.22	6.4(f)	0.3(f)	230^{+222}_{-99}	$1.64^{+0.88}_{-0.78}$ 266^{+90}_{-68}	897(871)

[a]– (1) = Power Law + Line, (2) = Power Law + Line + Reflection, (3)=Cutoff Power Law + Line + Reflection)
[b]– (f)- the parameter is fixed to the indicated value
[c]– R is the ratio between the continuum and the reflection normalization

However the χ^2 significantly improves when a cut-off power law plus a reflection component is used to model the continuum (comparing model 2 and 3 $\Delta\chi^2 = 16$ is obtained). This is also evident in Figure 1, where the energy-density spectrum of the SIS0 and OSSE data is plotted.

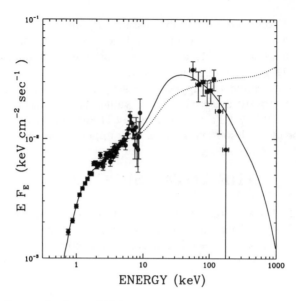

FIGURE 1. Energy density unfolded spectrum of the combined SIS0 and OSSE data. The solid line represents the power law plus reflection and gaussian line model. A simpler power law plus reflection and gaussian line model (dotted line) is also shown by comparison

DISCUSSION

The unambiguous detection of a break in the high energy spectrum of MCG+8-11-11 is suggestive of a thermal origin of the X-ray emission. The simplest thermal model assumes that the X-rays are produced via inverse Compton in a hot corona embedding a colder accretion disk. Soft thermal photons from the disk provide the main source of cooling for the hot electrons. The hard photons are an important source of heating for the disk which reprocesses them into soft photons ([10], [11]). In Comptonization models the X-ray spectral slope (Γ) is mainly determined by plasma temperature $\Theta = kT/mc^2$ and by the optical depth τ. Our measurements of the photon index and of the cutoff energy allow us to determine the physical parameters of the Comptonization region. From the E_{cut} value, a temperature $kT \simeq 130^{+50}_{-40}$ keV is inferred. Assuming plane parallel geometry, and a face-on line of sight, the derived temperature together with ASCA spectral index ($\Gamma = 1.6 - 1.7$ between 2-10 keV) leads to an estimate of the coronal optical depth of $\tau = 0.1 - 0.3$ (see Figures 1 and 2 in [12]), consistent with an extended corona picture (see Fig.2 in [12]). However, if the coronal electron temperature is in the lower part of the possible range of values (say $\lesssim 120$ keV), a photon starved source is probably required to obtain a spectrum as flat as observed. If this is the case, a structured corona ([13]; see also [14], and [15]), or a different geometry (e.g. [16]) must be invoked.

Our data seems also to indicate that pairs are not important in MCG+8-11-11. The observed 20% flux variation, if linearly extrapolated, leads to a doubling timescale $\Delta t_2 \simeq 15$ days, consistent with the earlier results [17], giving a compactness value of $\ell \sim 0.1$, where $\ell \equiv \frac{\sigma_T}{m_e c^3} L/c\Delta t_2$. On the other hand, if we translate the values of the coronal temperature into the maximum allowed compactness assuming a pair dominated plasma (see, e.g., Figure 3 of [18]), we find higher values, both in the case of an extended corona ($\ell \lesssim 600-3$) and of a hemispherical source ($\ell \lesssim 700-40$).

The spectral break could also be consistent with non thermal pair models, as discussed by [19]. But, in order to produce the observed break in the hard X–ray spectrum, non–thermal models require a large value of ℓ, not supported by variability results. We conclude that, at least in this weakly variable source, thermal models are preferred.

REFERENCES

1. Treves A., Bonelli G., Chiappetti L., Falomo R., Maraschi L., Tagliaferri G., Tanzi E.G., 1990, ApJ, 359, 98
2. Frontera F., Fuligni F., Morelli E., Ventura G., 1979, ApJ, 234, 477
3. Perotti F., Delle Ventura A., Villa G., Di Cocco G., Butler R.C., Carter J.N., Dean A.J., 1981, Nature, 292, 133
4. Perotti F., et al., 1990, A&A, 234, 106
5. Perotti F., Mattaini E., Quadrini E. , Santambrogio E., Bassani L., Stephen J.B., 1997, ApJ, 475, L89
6. Lin Y.C., et al., 1993, ApJ, 416, L53
7. Maisack M., et al., 1993, ApJ, 407, L61
8. Johnson W.N., et al., 1993, AAS, 97, 21
9. Grandi P. , Haardt F., Ghisellini G., Grove E., Maraschi L., Urry, C. M., 1997, ApJ Letter, submitted
10. Haardt F., Maraschi L., 1991, ApJ, 380, L51
11. Haardt F., Maraschi L., 1993, ApJ, 413, 507
12. Haardt F., Maraschi L., Ghisellini G., 1997, ApJ, 476, 620
13. Haardt F., Maraschi L., Ghisellini G., 1994, ApJ, 432, L95
14. Stern B., Poutanen J., Svensson R., Sikora M., Begelman M.C., 1995, ApJ, 449, L13
15. Poutanen J., Svensson R., 1996, ApJ, 470, 249
16. Dove J.B., Wilms J., Begelman M.C., 1997, ApJ, in press (astro-ph/9705108)
17. Done C., Fabian A.C., 1989, MNRAS, 240, 81
18. Svensson R., 1996, A &AS, 120, 475
19. Zdziarski A.A, Fabian A.C., Nandra K., Celotti A., Rees M.J., Done C., Coppi P.S., Madejski G.M., 1994, MNRAS, 269, L55

Compton Gamma-Ray Observatory Observations of the Nearest Active Galaxy Centaurus A

H. Steinle[1], K. Bennett[2], H. Bloemen[3], W. Collmar[1], R. Diehl[1], W. Hermsen[3], G.G. Lichti[1], D. Morris[4], V. Schönfelder[1], A.W. Strong[1], O.R. Williams[2]

[1] *Max-Planck-Institut für extraterrestrische Physik, Postfach 1603, 85740 Garching, Germany*
[2] *Astrophysics Division, ESTEC, 2200 AG Noordwijk, The Netherlands*
[3] *SRON-Utrecht, Sorbonnelaan 2, 3584 CA Utrecht, The Netherlands*
[4] *Space Science Center, University of New Hampshire, Durham, NH 03824, U.S.A.*
(CGRO-COMPTEL Collaboration)

Abstract. During the CGRO observations in the years 1991 to 1995, the region on the sky including the active radio galaxy Centaurus A, was in the wide field-of-view of COMPTEL and EGRET in 15 pointings of various durations. During some of these pointings, also OSSE made simultaneous observations of Cen A. BATSE, the fourth instrument on CGRO, continuously monitors this source. The analysis of the COMPTEL data obtained during these 15 pointings shows emission consistent with the position of Cen A, and spectra in two different emission states have been measured. The COMPTEL spectra in the energy range 0.75 - 30 MeV are consistent with the published Centaurus A spectra in the energy range 0.05 - 4 MeV from the OSSE instrument. For the higher emission state, a moderate steepening of the spectrum above 30 MeV is required to fit the EGRET data point. Thus, a spectrum of Cen A over 4 decades of energy is available for theoretical interpretation.

INTRODUCTION

The nearest active radio galaxy Centaurus A (Cen A, NGC 5128, PKS 1322-427) is at a distance of 3 - 4 Mpc [4,5]. Usually it is classified as a FR I type radio galaxy, as a Seyfert 2 object [2] and as a misdirected BL Lac type AGN [9]. Giant outer radio lobes extend to 10°, inner radio lobes co-align with the x-ray jet(s) as do optical filaments. From the complex optical appearance and subsequent simulations, Cen A is supposed to be the result of a recent merger between an elliptical and a spiral galaxy. The nucleus is hidden by a

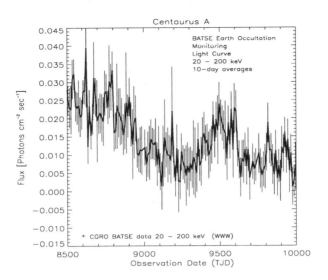

FIGURE 1. BATSE light-curve covering the time span from September 1991 to October 1995 in which all the observations used in this paper have been made. The photon flux in the energy band 20-200 keV averaged over 10 days is shown against the Truncated Julian Day number (TJD).

prominent dust layer. From investigations in the radio and optical range (jets, H II, stars), an inclination of the line-of-sight to the jet of 60° - 75° is derived [3,6]. Short term variability (days) is observed from radio- to γ-rays. Large variations in the hard X-ray/γ-ray luminosity have been observed in the past [1].

SPECTRA

Using the information provided by the BATSE instrument onboard CGRO, we defined the emission state of Cen A by comparing the measured hard X-ray flux (Figure 1) with the long term light curve of Cen A given in [1]. We found that we have observed Cen A in two different emission states: an intermediate state at the very first observation (VP 12; TJD 8546-8560) and in a low state in all following observations. These two states in hard X-rays are used here to also characterize the γ-ray state. We grouped the observations accordingly for the analysis. The hard X-ray flux varied by a factor 5 in the time interval covered by our observations, but the maximum flux was about a factor of 2 short of the historically observed highest flux [1].

By using the published OSSE spectra [7] and the EGRET data points (assuming the EGRET detection concerns Cen A [11,8]), we were able to derive spectra of both emission states covering the large γ-ray energy interval 50 keV to 1 GeV.

Intermediate Emission State Spectrum

During the first observation period (VP 12), the highest emission in hard X-rays of all CGRO observations was observed. However, compared to the historically observed emission states in hard X-rays, this is only an intermediate-high state [1].

We fitted the multi-instrument measurements between 50 keV and 1 GeV by a broken power-law model (with two breaks) of the form

$$I_1 \times (E/E_{b_1})^{-\alpha_1} \text{ for } E < E_{b_1},$$
$$I_1 \times (E/E_{b_1})^{-\alpha_2} \text{ for } E_{b_1} \leq E < E_{b_2},$$
and
$$I_2 \times (E/E_{b_2})^{-\alpha_3} \text{ for } E \geq E_{b_2}$$
(E_{b_1}, E_{b_2} are the energies in MeV at the breaks)

The resulting parameters are:

$$I_1 = (2.34 ^{+0.6}_{-0.6}) \times 10^{-2} \text{ } [Photons \text{ } cm^{-2} sec^{-1} MeV^{-1}]$$
$$I_2 = (5.4 ^{+24.8}_{-4.6}) \times 10^{-7} \text{ } [Photons \text{ } cm^{-2} sec^{-1} MeV^{-1}]$$
$$E_{b_1} = 0.15 ^{+0.03}_{-0.02} \text{ } [MeV]$$
$$E_{b_2} = 16.7 ^{+27.8}_{-16.3} \text{ } [MeV]$$
$$\alpha_1 = 1.74 ^{+0.05}_{-0.06}$$
$$\alpha_2 = 2.29 ^{+0.1}_{-0.1}$$
$$\alpha_3 = 3.29 ^{+0.7}_{-0.6}$$

Low Emission State Spectrum

For the low emission state, no fully simultaneous observations are available (from published data).

The OSSE observations are the sum of viewing periods 43, 215, 217, and 316. This data can be equally well fitted by a broken power-law model ($I_1 \times (E/E_b)^{-\alpha_1}$ for $0.05 E < E_b$, $I_1 \times (E/E_b)^{-\alpha_2}$ for $E_b \leq E < 1.0$; E in MeV, E_b = 0.14 MeV) and a power-law with exponential cut-off [7].

The COMPTEL observations are the sum of all 14 viewing periods (with Cen A in the field-of-view) of Phases I to IV/Cycle 4 except VP 12. A single power-law fits the data points very well: $I_c \times (E/E_{0_c})^{-\alpha_c}$ for $0.75 < E < 30$ (E in MeV, E_{0_c} = 1 MeV) with the parameters

$$I_c = (2.1 ^{+0.9}_{-0.9}) \times 10^{-4} \text{ } [Photons \text{ } cm^{-2} sec^{-1} MeV^{-1}]$$
$$\alpha_c = 2.6 ^{+0.8}_{-0.6}$$

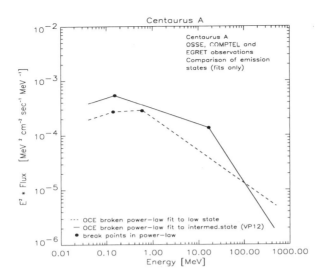

FIGURE 2. Comparison of the Centaurus A spectra measured in the two emission states. The break points in the power-law spectra are highlighted by the filled circles. The flux has been multiplied by E^2.

To connect to the OSSE fit, an additional (second) break at about 580 - 600 keV is necessary.

The extrapolated COMPTEL power-law fit exactly matches the EGRET data point. (The EGRET observations are the sum of viewing periods 207, 208, 215, and 217 (Phase II only); EGRET data point **not** used in fit).

The three resulting power-law indices for the whole energy range 50 keV to 1 GeV are:

$$\alpha_1 = 1.73 \quad \text{for } 0.05 < E < 0.14 \ (E \text{ in MeV})$$
$$\alpha_2 = 1.97 \quad \text{for } 0.14 < E < 0.59 \ (E \text{ in MeV})$$
$$\alpha_c = 2.6 \quad \text{for } 0.59 < E < 1000 \ (E \text{ in MeV})$$

γ-Ray Luminosity

The two measured spectra in the energy range 50 keV to 1 GeV can be integrated to derive the γ-ray luminosity in the two emission states. Assuming isotropical emission and a distance of 3.5 Mpc ($z = 0.0006$, $H_0 = 50$ km s^{-1} Mpc^{-1}, $q_0 = 0.5$), we obtain

$$L_\gamma^m = 5 \times 10^{42} \text{ erg s}^{-1}$$
$$L_\gamma^l = 3 \times 10^{42} \text{ erg s}^{-1}$$

for the intermediate and low emission state, respectively.

CONCLUSIONS

Centaurus A has been observed by all CGRO instruments several times during Phases I to IV/Cycle 4 in the years 1991 to 1995. Comparing the flux measured by BATSE with the long-term light-curve given in [1], two different emission states can be defined: an intermediate-high state in which the source was during the first part of Phase I and a low state, in which we measured Cen A during the remainder of the observations.

Spectra covering the energy range 50 keV to 1 GeV for both emission states have been measured. A hardening of the spectrum in the low emission state is possibly indicated. The derived isotropic luminosities of several 10^{42} erg s^{-1} for the covered energy range are only moderate when compared to other AGN, but one has to keep in mind, that Cen A is viewed from a very large angle with respect to the jet axis (60° - 75°) as deduced from radio and optical measurements [3,6].

Emission from Cen A up to high energies is detected: GeV - if the EGRET identification holds, but even in the COMPTEL data alone, Cen A is detected up to 30 MeV and the extrapolated power-law fits indicate emission at higher energies. This emission, together with a moderately steep spectral index in the high energy part of the power-law models, seem to rule out Compton-scatter models (as e.g. in [10]). The large angle between the jet and the line-of-sight would require that the measured high-energy photons are scattered by the same amount which is almost impossible in these models.

REFERENCES

1. Bond I.A., Ballet J., Denis M., et al., *Astron. & Astrophys.* **307**, 708-714 (1996).
2. Dermer Ch. D. and Gehrels N., *Ap.J.* **447**, 103-120 (1995).
3. Ebneter K. and Balick B., *P.A.S.P.* **95**, 675-690 (1983).
4. Harris G.L.H., Hesser J.E., Harris H.C., and Curry P.J., *Ap.J.* **287**, 175-184 (1984).
5. Hui X., Ford H.C., Ciardullo R., Jacoby G.H., *Ap.J.* **414**, 463-473 (1993).
6. Jones D.L., Tingay S.J., Murphy D.W., et al., *Ap.J.* **466**, L63-L65 (1996).
7. Kinzer R.L., Johnson W.N., Dermer C.D., et al., *Ap.J.* **449**, 105-118 (1995)
8. Nolan P.L., Bertsch D.L., Chiang J., et al., *Ap.J.* **459**, 100-109 (1996)
9. Morganti R., Fosbury R.A.E., Hook R.N., Robinson A., Tsvetanov Z., *M.N.R.A.S.* **256**, 1P-5P (1992)
10. Skibo J.G., Dermer C.D., and Kinzer R.L., *Ap.J.* **426**, L23-L26 (1994)
11. Thompson D.J., Bertsch D.L., Dingus B.L., et al., *Ap.J.S.* **101**, 259-286 (1995)

An Anisotropic Illumination Model of Seyfert I Galaxies

P.O. Petrucci*, G. Henri*, J. Malzac** and E. Jourdain**

*Laboratoire d'Astrophysique, Grenoble FRANCE
†Centre d'Etude Spatiale des Rayonnements, (CNRS/UPS), Toulouse FRANCE

Abstract. We develop a self-consistent model of Seyfert galaxies continuum emission. The high energy source is assumed to be an optically thin plasma of highly relativistic leptons ($e^+ - e^-$), at rest at a given height on the disk axis. Such a geometry is highly anisotropic, which has a strong influence on Compton process. Monte-Carlo simulations allow the superposition of a reflected component to the UV to X-ray spectrum obtained with our model, leading us to a first comparison with observations by fitting the high energy spectra of NGC4151 and IC4329a.

INTRODUCTION

We propose a new model, for Seyfert galaxies, involving a point source of relativistic leptons located above the disk (that could be physically realized by a strong shock terminating an aborted jet) emitting hard radiation by Inverse Compton (IC) process on soft photons produced by the accretion disk. The disk itself radiates only through the re-processing of the hard radiation impinging on it, i.e. we do not suppose any internal energy dissipation (cf Fig. 1). Such a geometry is highly anisotropic, which takes a real importance in the computation of IC process ([2], [3]). We treat both Newtonian and general relativistic cases [8], deriving a self-consistent solution in the Newtonian case. We have recently added Monte-Carlo calculations to take into account, in the high energy part of our synthetic spectra, the compton reflection component and the fluorescent iron line [9]. We present here the most important results supplied by the model.

ANGULAR DISTRIBUTION OF THE HOT SOURCE

It appears that the anisotropy of the soft photon field at the hot source level, leads to an **anisotropic Inverse Compton process**, with much more

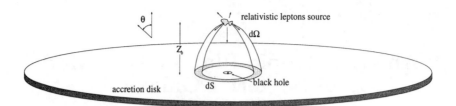

FIGURE 1. The general picture of the model. We have also drawn the trajectory of a beam of photons emitted by the hot source in a solid angle $d\Omega$ and absorbed by a surface ring dS on the disk.

radiation being scattered backward than forward. Such an anisotropic re-illumination could naturally explain the apparent X-ray luminosity, usually much lower than the optical-UV continuum emitted in the blue bump [10]. It can also explain the equivalent width observed for the iron line, which requires more impinging radiation than what is actually observed [7]. We plot in Figure 2 the angular distribution of the power emitted by the hot source in Newtonian metrics and for different values of the source height in Kerr metrics. It appears that the closer the source to the black hole is, the less anisotropic the high energy photon field is. This is principally due to the curvature of geodesics making the photons emitted near the black hole arrive at larger angle than in the Newtonian case.

DISK TEMPERATURE PROFILE

The radiative balance between the hot source and the disk allows to compute the temperature profile on the disk surface. It is, in fact, markedly different

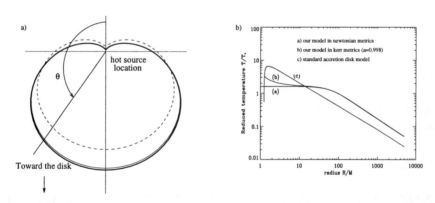

FIGURE 2. (a) Polar plots of $\dfrac{dP}{d\Omega}$ for $Z_0/M = 100$ (solid line) and $Z_0/M = 10$ (dashed line) in Kerr metrics with $a = 0.998$. The bold line corresponds to the Newtonian metrics. (b) Effective temperature versus r for $Z_0/M = 70$

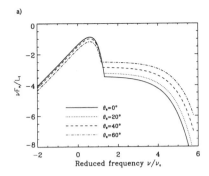
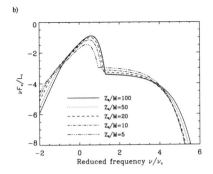

FIGURE 3. (a) Differential power spectrum for different values of the inclination angle in the Newtonian case. (b) Differential power spectrum for different values of Z_0 for the Kerr maximal case. We use reduced coordinates and logarithmic scales.

from "standard accretion disk model" as shown in Figure 2. Indeed, even if at large distances, all models give the same asymptotic behavior $T \propto R^{-3/4}$, in the inner part of the disk, it keeps increasing in "standard model" whereas, in our model, for $R \leq Z_0$, the **temperature saturates** around a characteristic value T_c. The differences between Newtonian and Kerr metrics come only from Gravitational and Doppler shifts, which are only appreciable for $R \leq 5M$.

INFLUENCE OF THE INCLINATION ANGLE

The overall UV to X-ray spectra can be deduced from this model. The bulk of the energy coming from the disk gives the well-known "blue-bump". On the other hand, in order to avoid run-away electrons, the relativistic particle distribution is supposed to be a **power law with an exponential cut-off** about 300 keV, giving the high energy part of the spectrum. One can see on Figures 3a Newtonian spectra for different inclination angles for $Z_0/M = 10$. For all inclination angles, the Kerr spectra are always weaker in UV and brighter in X-ray than the Newtonian ones. These effects are much less pronounced for high Z_0/M values because the emission area is much larger, and thus is less affected by relativistic corrections.

INFLUENCE OF THE HOT SOURCE HEIGHT

Figure 3b shows the overall spectrum, for different values of Z_0/M in Kerr metrics for $\theta = 0°$. The relativistic effects become important for values of Z_0/M smaller than about 50. They produce a variation of intensity lowering the blue-bump and increasing the hard X-ray emission. The change in the UV range is due to the transverse Doppler effect between the rotating disk and the observer, producing a net red-shift. In the X-ray range, the variation is due to the high energy dependence on Z_0/M (cf Figure 2). The observed **X/UV ratio** can then be strongly altered by these effects. Quantitatively, the luminosity ratio between the maximum of the blue-bump and the X-ray plateau goes from $\simeq 300$ in the Newtonian case, to $\simeq 10$ for $Z_0/M = 3$.

SCALING LAWS

With some further assumptions, the model predicts scaling laws quite different from the standard accretion models. If one assumes a constant high energy cut-off (possibly fixed by the pair production threshold) and a constant solid angle subtended by the hot source, then the following mass scaling laws apply:

$$T_c = constant$$
$$L_c \propto M^2$$

The first equality could explain the weak variations of the blue bump component, even though the luminosity L ranges over 6 orders of magnitude from

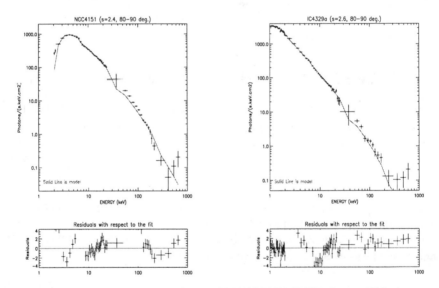

FIGURE 4. The data are deconvolued ROSAT/GINGA/OSSE data published respectively in [4] and [11]. Note that in both spectra we added a fictive data point around 40 kev with large error bars to join GINGA and OSSE data. The parameters are the electron index s, the high energy cut-off ν_0, the viewing angle θ and the absorber column density N_h. In these fits the only free parameter is N_h, while the others are kept at a convenient value. For both fits $\nu_0 = 100 kev$, θ=80-90 deg. The best fit value for N_h is 1.210^{23} for NGC4151 and 4.610^{21} for IC4329a. The reduced χ^2 are respectively 12.0 and 5.92

source to source [10]. The second one is in agreement with the results of [1] where they find in a sample of Seyfert I and quasars a correlation between the mass and the luminosity under the form $L \propto M^\beta$ with $\beta = 1.8 \pm 0.6$.

THE REFLECTION COMPONENT

We considered an infinite disc of neutral matter with solar abondances (opacities given by [6]), and hydrogen column density of $10^{25} cm^{-2}$. We found that about 10 % of the total energy emitted toward the disc is reflected and does not contribute to the disc heating. However, the reflection component has important effects on the shape of the observed spectrum. Due to the strong anisotropy of the primary source, the ratio of the reflected to the primary component can be very high. The spectrum strongly depends on the viewing angle θ with respect to the disc normal. For nearly edge on inclinations, the reflection contribution is weak. As the inclination decrease the reflection component increase [5]. Conversely, the primary spectrum decreases with decreasing θ. So that, for a face on inclination, reflection dominates the hard X-ray spectrum. A rough comparison to IC4329a and NGC 4151 spectra shows that our model is reasonably close to the data if the orientation is nearly edge-on (cf. fig 4). We can note that the anisotropic illumination model predicts a weak UV component for high inclination angles which is in qualitative agreement with the observed UV to soft X-ray luminosity ratio in these two objects.

Acknowledgment : We are very grateful to A. Zdziarski for providing us the NGC4151 and IC4329a data.

REFERENCES

1. Collin-Souffrin S., Joly M., *Disks and Broad line regions*, In: Duschl W.J., Wagner S.J. (eds) Physics of AGN. Springer-Verlag, 1991, p. 195
2. Ghisellini G., George I.M., Fabian A.C., Done C., *MNRAS* **248**, 14, (1991).
3. Henri, G., Petrucci, P.O., *A&A* in press
4. Madejski G.M. et al, *ApJ* **438**, 672 (1995)
5. Magdziarz P. , Zdziarski A.A., *MNRAS* **273**, 837 (1995)
6. Morrisson R., McCammon D.,*ApJ* **270**, 119 (1983)
7. Nandra K., GeorgeI.M., Mushotzky R.F., Turner T.J., Yaqoob T., *ApJ* **477**, 602 (1997)
8. Petrucci, P.O., Henri, H., *A&A* in press
9. Pounds, K. A., Nandra, K., Stewart, G. C., George, I. M., Fabian, *Nature* **344**, 132 (1990)
10. Walter R., Fink, H.H., *A&A* **274**, 105 (1993)
11. Zdziarski A.A., Johnson W.N., Magdziarz P., *MNRAS* **283**, 193 (1996)

Scattered Emission and the X-γ Spectra of Seyfert Galaxies

J. Chiang, C. D. Dermer and J. G. Skibo

E. O. Hulburt Center for Space Research, Code 7653, Naval Research Laboratory, Washington DC 20375-5352

Abstract. We have developed a Monte Carlo code in order to compute the emergent X-ray and γ-ray spectra for different geometries and orientations of Seyfert galaxies. Our calculation includes reflection off a cold, optically thick accretion disk, obscuration and reflection by a neutral torus, and reprocessing of the radiation by an ionized cone of material located within the opening angle of the torus. We assume a standard spectral form for the central X-γ continuum emission, and we attempt to fit the broad band spectra of Seyfert galaxies NGC 4945, NGC 4151 and IC 4329A using data obtained by the *Ginga*, *ROSAT*, *ASCA* and OSSE instruments. These three objects span the range of Seyfert classifications, and by fitting their spectra we infer the torus geometries surrounding their central X-γ sources. In particular, we find that the strength of the reflection hump can be a useful diagnostic for constraining the covering factor of the disk-torus system as seen by the central continuum source.

INTRODUCTION

The standard picture of Seyfert unification consists of a central massive ($M_{BH} \gtrsim 10^7 M_\odot$) black hole, an accretion disk, a broad emission line region, an obscuring torus surrounding these regions, and a warm ionized medium lying above the disk-torus-BLR system perhaps within the observer's line-of-sight. This scenario was motivated by the discovery of broad emission lines in the polarized flux of the Seyfert 2 galaxy NGC 1068 [1]. In this object, it is believed that an optically thick torus of material blocks our direct line-of-sight to the broad line region and central continuum source, while high altitude ionized material above the torus but within its opening angle reflects the broad line emission into our line-of-sight so that it can be detected in polarized light. This observation led to the unified picture in which the disparity in the spectral characteristics between Seyfert 1 and Seyfert 2 galaxies can be attributed to orientation effects. In subsequent observations, Miller & Goodrich [10] have seen broad lines in the polarized light of several other Seyfert 2s, confirming this general picture. In addition, observations of cones of [OIII]λ5007 emission in several Seyfert objects, including NGC 1068 [13], suggest collimation which may be due to a torus geometry. However, this inference is confounded by the fact that NGC 4151 also possesses such a cone and yet, from optical/UV observations of its broad line region, it does not appear to have an obscuring torus [4].

If Seyfert 1 and Seyfert 2 objects possess similar central sources of continuum radiation, then the differences in their apparent spectra should tell us something about the nature of the obscuring material surrounding their inner regions. Conversely, reasonable physical assumptions about this surrounding environment should allow us to infer something about the intrinsic spectra of these objects. In the X-γ energy range ($\sim$ 1 keV–1 MeV), it is believed that thermal Comptonization processes produce the intrinsic central radiation. The spectra from these models are typically parameterized by an exponentially cut-off power-law with photon spectral indices $\sim$ 1.9 and e-folding energies in the range 300–500 keV [14]. For Seyfert 2 objects, photoelectric absorption by the obscuring material at energies $\lesssim$ 10 keV substantially reduces the soft X-ray flux relative to that found in Seyfert 1s. In both types of objects, neutral material in the accretion disk and torus reprocesses some of the thermal Comptonization emission and produces strong iron Kα emission at 6.4 keV and a "reflection" continuum hump in the range 10–100 keV.

Analyses of X-γ data obtained by the *Ginga*, *ROSAT*, *ASCA* and *OSSE* instruments have fit the observed spectra from specific objects by parameterizing the various components of absorbing and scattering material in terms of absorbing columns, covering factors, and reflection components. Aside from considering inclination angles in the reflection component fits and "complex" absorbers, these analyses largely disregard the specific geometry of the material which is modifying the intrinsic source of radiation (cf. [12]). In contrast, several authors have performed detailed Monte Carlo calculations of the expected emission from a Seyfert nucleus enshrouded by a geometrically and optically thick torus and have examined the contribution of the torus to the reflection component and the iron Kα emission as a function of inclination angle and torus optical depth [7,5,9]. Using simpler geometries consisting of slabs and spheres, Jourdain & Roques [6] have attempted to fit the hard X-ray spectra of NGC 4151, NGC 4507 and the upper limits of NGC 1068. However, to our knowledge, there have not been any Monte Carlo calculations which use the full torus geometry to *fit* the X-γ spectra for any specific object.

MONTE CARLO CALCULATIONS

We have developed a three-dimensional Monte Carlo code in order to calculate the effect of the geometry described above on the nuclear X-γ emission. We have attempted to fit the spectra of several Seyfert galaxies in an effort to bridge the gap between the Monte Carlo calculations and their application to individual objects. This enables us to make more specific statements about the feasible geometries of the torus environment for the various objects we consider.

The geometry consists of a torus with circular cross-section, an optically thick, geometrically thin accretion disk, and twin cones of fully ionized material above and below disk (Figure 1). Somewhat above and below the disk we place sources of Xγ radiation intended to mimic the spectra from a thermal Comptonizing plasma. Presently, the spectrum of these sources is taken to be an exponentially cut-off power-law of the form $E^{-\gamma}\exp(-E/E_c)$ where the energy spectral index is in the range $\gamma \simeq 0.8$–1.0 and the e-folding energy is $E_c \gtrsim 100$ keV. These values are consistent with intrinsic spectra inferred for Seyfert 1s (see e.g., [12,15]). We use the full Klein-Nishina cross-section to calculate the Compton scattering and the cross-sections of Bałucinska-Church & McCammon [2] and abundances of Morrison & McCammon [11] to model the photoelectric absorption.

We have created XSPEC table files of our Monte Carlo results for various ranges of parameters, e.g., spectral index, high altitude cone scattering depth, and inclination. Using these tables we have fit the X-γ spectra of IC 4329A and NGC 4151 using the data from [8] and [15], respectively. The results of our fits are shown in Figure 2. Both of these fits require that the opening angle of the torus be $\theta_T \gtrsim 75°$ in contrast to a geometrically thick torus with $\theta_T \sim 45°$ expected in the standard model. The reason for this large opening angle is due primarily to the weakness of the reflection component in the hard X-rays in the 8–30 keV range. With a geometrically thick torus, the covering factor, $R \equiv \Omega/2\pi$, by cold material of the central continuum will be substantially greater than unity.

We have also fit by-eye the published unfolded X-γ spectrum of NGC 4945 [3]. The results are shown in Figure 3. Although a full spectral fitting procedure will eventually be necessary, from our crude fits we find evidence for a much weaker reflection component in this object as well, in contrast to that expected for a geometrically thick torus. In the 2–20 keV *Ginga* band, the shape of the absorbed continuum is not well modeled by our calculation, particularly above 10 keV where the model spectrum is systematically and significantly lower than the observed spectrum despite the very large uncertainties in the data.[1] If we translate the spectrum upwards to match the data in this band, we will introduce a discrepancy in the 7–10 keV *Ginga* band where the model will then over-predict the observed flux. This over-prediction can be traced again to the presence of a strong reflection hump due to the large disk-torus covering factor. The contribution of the reflection hump can be clearly

---

[1] We note that the three observations comprising these data were taken at three different times (OSSE: 1994 November; *ASCA*: 1993 August; *Ginga*: 1990 July), and variability may account for some of the discrepancy in the 10–20 keV *Ginga* band.

seen in the spectrum scattered into our line-of-sight by the high altitude cone (dashed histogram in Figure 3).

This work was performed while J.C. held a NRC-NRL Research Associateship. C.D. and J.S. were supported by the Office of Naval Research.

REFERENCES

1. Antonucci, R. R. J, & Miller, J. S. 1985, ApJ, 297, 621
2. Bałucinska-Church, M., & McCammon, D. 1992, ApJ, 400, 699
3. Done, C., Madejski, G. M., & Smith, D. A. 1996, ApJ, 463, L63
4. Evans, I. N., et al. 1993, ApJ, 417, 82
5. Ghisellini, G., Haardt, F., & Matt, G. 1994, MNRAS, 267, 743
6. Jourdain, E., & Roques, J. P. 1995, ApJ, 440, 128
7. Krolik, J. H., Madau, P., & Życki, P. T. 1994, ApJ, 420, L57
8. Madejski, G. M., et al. 1995, ApJ, 438, 672
9. Matt, G., Brandt, W. N., & Fabian, A. C. 1996, MNRAS, 280, 823
10. Miller, J. S., & Goodrich, R. W. 1990, ApJ, 355, 456
11. Morrison, R. & McCammon, D. 1983, ApJ, 270, 119
12. Nandra, K. & Pounds, K. A. 1994, MNRAS, 268, 405
13. Pogge, R. 1988, ApJ, 328, 519
14. Zdziarski, A. A., Johnson, W. N., Done, C., Smith, D., & McNaron-Brown, K. 1995, ApJ, 438, L63
15. Zdziarski, A. A., Johnson, W. N., & Magdziarz, P. 1996, MNRAS, 283, 193

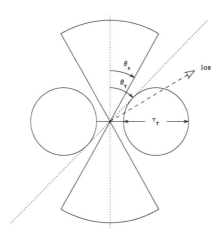

FIGURE 1. Schematic of the geometry used in our Monte Carlo calculation. The sources of intrinsic nuclear radiation (not shown) are located along the symmetry axis just above and below the accretion disk.

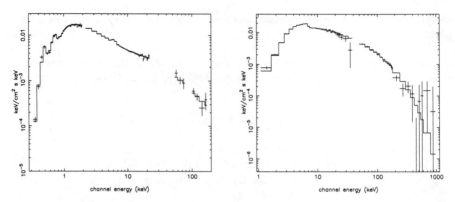

FIGURE 2. *Left*: Model calculations for IC 4329A. The absorption below ~ 1 keV is likely due to a photoionized absorber along our line-of-sight producing edge features in OVII and/or OVIII. The data are *ROSAT*, *Ginga*, and OSSE observations described in [8]. *Right*: Model calculations for NGC 4151. The data are *Ginga* and OSSE observations from 1991 June described in [15].

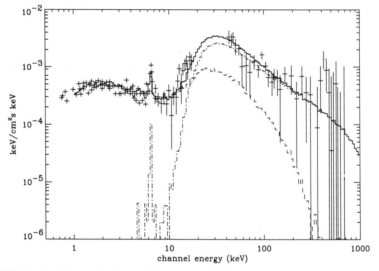

FIGURE 3. Model calculation of the X-γ spectrum of NGC 4945. A geometrically ($\theta_T = 45°$) and optically thick ($N_{H,T} = 10^{25}\,\mathrm{cm}^{-2}$) torus is assumed. The data are the *ASCA/Ginga*/OSSE observations shown in Figure 1 of [3]. The solid histogram is the emergent spectra; the dashed histogram is the part scattered into the line-of-sight by the high altitude ionized cone; and the dot-dashed histogram is the part emerging from the torus.

Pair models revivified for high energy emission of AGNs

Gilles Henri* and Pierre-Olivier Petrucci*

*Laboratoire d'Astrophysique de l'Observatoire de Grenoble
BP 53
F 38 041 Grenoble Cedex 9 France

Abstract. We argue that although the high energy emission of radio-loud and radio-weak AGNs are markedly different, they all exhibit remarkable features (either spectral break or upper cut-off) around or just below MeV energy. The critical energy is correlated with the expected Doppler factor determined by radio characteristics of these objects. This supports the idea that high energy emission is closely linked to pair production, with absorption or annihilation features. In the case of radio-loud blazars, gamma-gamma absorption explains remarkably well the position and the magnitude of the spectral break, with very few ad-hoc assumption on the particle distribution. In the case of Seyfert galaxies, a static non thermal source, associating pair production and reacceleration, can explain the main features of high energy spectra. A self-consistent model associating anisotropic Inverse Compton process and disk reillumination predicts unambiguously the X/UV ratio, which is highly orientation dependant. We discuss some implications of this model for the understanding of present observations.

INTRODUCTION

The detection by CGRO of more than 60 AGNs above 100 keV has revealed that the various classes, determined by their radio properties, are also markedly different in the gamma-ray range [1]. Whereas a subset of the blazar class can be extremely luminous above 1 MeV, with a spectrum extending at least up to 10 TeV for a few BL Lacs [2], all the radio-quiet Seyfert galaxies show a cut-off around 100 keV. The radio galaxy Cen A has also a spectral break in this range, but recent COMPTEL observations show that its spectrum extends above 1 MeV, making it a intermediate case between radio-quiet and radio-loud objects.
Although the very different shapes of the SED indicate some fundamental difference in the emission mechanism, one can note a remarkable feature common to all spectra: the spectral index is always abruptly changing at some critical energy whose value is ranging from about 100 keV to 10 Mev, that

is close to MeV within a factor 10. All Seyfert spectra are exponentially cut-off around 100 keV (the precise value depends on the importance of the reflected component). The blazars are characterized by a spectral break at a few MeV, some of them, the so-called MeV blazars ([3]), being occasionnaly exceptionnally bright around this energy. In the case of the radio-galaxy Cen A, a spectral break has been detected around 100 keV, but higher energy emission has been also detected by COMPTEL. One can thus find a correlation between the break energy and the Doppler factor $\mathcal{D}$ associated with the relativistic motion. The blazars, with a relativistic jet pointing towards the observer, have probably a Doppler factor of the order of 10. Cen A jet, on the other side, lies most probably near the plane of the sky. The expected transverse Lorentz factor is thus smaller than 1. In both cases, the observed break is compatible with $\mathcal{D} \times \nabla \prime\prime$ keV, the expected value for Doppler-shifted pair absorption. Radio-quiet objects, where no relativistic motion has been detected so far, have probably Doppler factor around 1, but the physics may be intrinsically different. Nevertheless, the observed break is also not too far from the canonical value of 511 keV. In the following sections, we present more detailed argument showing how the high energy emission could arise from a compact, pair-dominated, non-thermal source in *any* AGN class.

PAIR PRODUCTION IN A RELATIVISTIC JET

General model

The case of pair production in a relativistic jet has been studied in detail in Marcowith et al. [4], in the frame of the two-flow model [5]. In this model, a powerful magnetized $p^+ - e^-$ jet is assumed to be emitted from an accretion disk at mildly (non necessarily highly) relativistic velocity, $v \sim 0.5c$. It generates strong MHD turbulence, capable of accelerating some non thermal electrons at high random Lorentz factor, $\bar{\gamma} \sim 10^5$. By Inverse Compton (IC) process on the UV photons from the accretion disk, as is usually considered, these electrons can produce gamma-rays. If the source is compact enough, i.e. sufficiently dense and close to the black hole, these gamma-rays can produce at turn $e^+ - e^-$ pairs, which can be reaccelerated in-situ to generate a pair cascade. The best site for the pair plasma formation is the inner core of the jet, which should be void of protons due to the inner minimal radius of stability of an accretion disk orbiting around a black hole. Thus, in this region, there is no threshold for electron acceleration due to absorption of Alfvèn waves by ambient protons, and freshly created pairs can be continuously reaccelerated as soon as they are formed by 2nd order Fermi process. Also the bulk Lorentz factor is *not* a free parameter in this model: it is fixed by the radiation pressure of the accretion disk, to the equilibrium value for which light aberration makes the net flux vanish in the plasma frame. This value is approximately

$$\gamma_b \simeq (z/r_g)^{1/4}.$$

Stationary solutions

With these processes taken into account, it is possible to solve the continuity equations for particles and photons along the jet, taking as free parameters: the jet radius $R(z)$, the disk luminosity L_{UV}, the inclination angle i, and the spectral index of the particle distribution, that depends on the ill-known acceleration mechanism. To start the integration of differential continuity equations, one needs also a supplementary boundary condition, that can be taken as the value of $\tau_{\gamma\gamma}(1MeV)$ at the point where $\tau_T = 1$. However, the final results depend only weekly on this parameter. The main results are the following :
- it is possible to find stationnary solutions corresponding to the formation of a optically thick plasma, verifying simultaneously at some distance z_0

$$\tau_T = \sigma_T n_e R \simeq 1 \qquad (1)$$

and

$$\tau_{\gamma\gamma}(1MeV) \simeq 1. \qquad (2)$$

The distance z_O depends mainly on disk luminosity, that controls the soft photon density, and on the jet radius. It is of the order of $10^2 r_g$ for a near Eddington accreting object with $R = 10 r_g$
- the behaviour of this solutions is the following: before z_0, one has a low density of pairs and a high soft photon density. The plasma is optically thin to pair creation at low energy, but the photon of highest energy are absorbed to create new pairs. The pair density grows gradually along the jet; the pair creation opacity grows accordingly, and more and more photons are absorbed. Around z_0, gamma absorption and pair creation are maximum, and the Thomson optical depth becomes of the order of unity . Then most photons are IC scattered: the soft photon density drops exponentially. Gamma-ray luminosity, pair creation and $\tau_{\gamma\gamma}$ are all decreasing. The pair density reaches a plateau, until annihilation makes it decrease again on a longer scale. For each photon energy $\epsilon m_e c^2$, there is a gamma photosphere where $\tau_{\gamma\gamma}(\epsilon) = 1$. after which a photon can freely escape. The gamma-ray source is thus concentrated around z_0, which can explain the fast variability observed in many objects.
- due to differential absorption of gamma-rays, there is a spectral break around $0.511 \times \mathcal{D}$ MeV, where $\mathcal{D}$ is the Doppler factor. This spectral break has nothing to do with the cooling of particles. It may be larger than 0.5, as observed in many blazars. In fact, the model predicts an approximate simple relationship between the X-ray and gamma-ray energy spectral indexes : $\alpha_\gamma \approx 2\alpha_x$

- for all solutions, the MeV luminosity in the jet frame is approximately constant, depending only on the jet radius through the condition (2) (note however that the apparent luminosity depends strongly on the viewing angle through the Doppler factor). The spectra can vary by a change in the spectral index s. In this case, one should observe a variation at high energy, in the EGRET range, but high luminosities would be correlated with low spectral index (flatter spectra) such that the extrapolations to the MeV range should approximately intersect. Such a behaviour has indeed been observed in some objects as 3C273 [6].
- as there is a copious pair production, a broad annihilation feature is expected, leading to at leats some MeV excess or bump. As the Doppler boosting factor is not the same for annihilation and for IC process, the ratio between annihilation component and IC process depends on the inclination angle. A strong MeV bump can be observed in some cases, explaining the so-called "MeV-blazars" [3].
- there are also optically thin solution where the condition (2) is never satisfied. They correspond to a lower pair density and thus a lower luminosity. The spectral break occurs at higher energy. These states could describe the quiescent states of blazars, whereas the flaring states are best described by optically thick solutions.

HIGH ENERGY EMISSION OF SEYFERTS

Let us now turn to the case of radio-quiet Seyfert galaxies. As combined results of OSSE, COMPTEL and EGRET have revealed it, high energy spectra of radio-quiet objects are very different from those of blazars, with a typical spectrum cut-off below a few 10^2 keV. However, once again, this cut-off is close to the pair production threshold: it is thus conceivable that it is related to pair creation process. Furthermore, the relationship between radio-quiet and radio-louds objects is still unexplained. We propose a model in which the high energy source of Seyferts is realized by a strong shock terminating an aborted MHD jet very close to the black hole. These shocks accelerate relativistic particles that produce also gamma-rays, but contrarily to blazars, pairs cannot escape relativistically because they are not channelled by the ambient MHD structure. Although pair models proposed some years ago by various authors don't seem to fit the observational data, mainly because they predict too strong an annihilation line, it may be that some variants could actually fit them. Pair cration models with reacceleration as proposed in [7] predict a natural high energy cut-off to avoid pair run-away. As pair production is limited by this cut-off, no strong annihilation line is emitted.

Non thermal models have also very interesting consequences on the angular distribution of the emitted radiation. Inverse Compton process in the strongly anisotropic photon field of the accretion disk will produce mainly backscattered radiation. This is could well explain the strong reillumination observed in many Seyferts by the Compton reflection component and the fluorescent Fe Kα line. It is in fact possible to construct a self-consistent model [8] where the disk emission is entirely due to reprocessing of high energy radiation, produced itself entirely by Inverse Compton scattering by a relativistic pair source located at some height above the disk. Angular distribution of the X-ray radiation and temperature profiles of the disk are then univoquely determined. In this case, the apparent X/UV ratio is a function of inclination angle. In general, for small inclination angles (disks seen almost face-on), this ratio is smaller than 1, due to the strong anisotropy of IC process.It can be as low as 0.012 for $i = 0$. This can explain the "X-ray" deficit plaguing the illumination model. The Seyfert galaxies with relatively high X/UV ratio, on the other side, would be seen at comparatively high inclination angle. A strong reflection component is also predicted [9].

This model predicts also different scaling laws that the standard accretion disk models. If all distances scale like the mass of the central objects, one can show that the following scaling laws apply for the central temperature and total luminosity [8]:

$$T_c \propto M^0 \approx constant \qquad (3)$$
$$L_t \propto M^2. \qquad (4)$$

These relationships seem closer to what is observed in statistical samples than the predictions of standard accretion theories [10], [11].

REFERENCES

1. Dermer C.D., and Gehrels N.,*ApJ*, **447**, 103 (1995).
2. Punch M., et al., *Nat*, **358**, 477 (1992).
3. Blom, J. J., Bennett, K., Bloemen, H., et al., *A&A* **298**, L33 (1995).
4. Marcowith, A., Henri, G., and Pelletier, G., *MNRAS* **277**, 681 (1995).
5. Sol, H., Pelletier, G., and Asséo, E., *MNRAS* **237**, 411 (1989).
6. von Montigny, C., Aller, H.,Aller, M., et al., *ApJ* **483**, 161 (1997).
7. Done, C., Ghisellini, G., and Fabian, A. C., *MNRAS* **245**, 1 (1990).
8. Henri, G. and Petrucci, P.O. ; Petrucci, P.O., and Henri, G.,*A&A* in press (1997).
9. Petrucci, P.-O., et al., *these proceedings*, poster 14-8 (1997).
10. Walter, R., Fink, H.H., *A&A*, **274**, 105 (1993).
11. Collin-Souffrin S., and Joly M., in *Duschl W.J., Wagner S.J. (eds) Physics of Active Galactic Nuclei*. Springer-Verlag, 1991, p. 195

Big Blue Bump And Transient Active Regions in Seyfert Galaxies

Sergei Nayakshin[†] and Fulvio Melia[*] [1]

[†] *Physics Dept., The University of Arizona, Tucson AZ 85721*
[*] *Physics Dept. & Steward Observatory, The University of Arizona, Tucson AZ 85721*

Abstract. An important feature of the EUV spectrum (known as the Big Blue Bump, hereafter BBB) in Seyfert Galaxies is the narrow range in its cutoff energy E_c from source to source, even though the luminosity changes by 4 orders of magnitude. Here we show that if the BBB is due to accretion disk emission, then in order to account for this "universality" in the value of E_c, the emission mechanism is probably optically thin bremsstrahlung. In addition, we demonstrate that the two-phase model with active regions localized on the surface of the cold disk is consistent with this constraint if the active regions are very compact and are highly *transient*, i.e., they evolve faster than one dynamical time scale.

INTRODUCTION

The UV to soft X-ray spectrum of many Active Galactic Nuclei (AGNs) may be decomposed into a non-thermal power-law component and the so-called Big Blue Bump (BBB), which cuts off below about 0.6 keV (e.g., [17]). The observed spectral shape of the bump component in Seyfert 1's hardly varies, even though the luminosity L ranges over 6 orders of magnitude from source to source. [18] concluded that the cutoff energy E_c of the BBB is very similar in different sources whose luminosities vary by a factor of 10^4. [18] pointed out that if the variations in the ratio of the soft X-ray excess to UV flux from one object to another are interpreted as a change in the temperature of the BBB, then this change is smaller than a factor of 2.

Early theoretical work on the BBB spectrum focused on the role of optically thick emission from the hypothesized accretion disk surrounding the central engine (e.g., [15,11,4,10,12]). An alternative model, in which the BBB is interpreted as thermal, optically thin free-free radiation, has been proposed by

---

[1]) Presidential Young Investigator.

[1,3,6,2]. In these studies, the incident X-ray intensity is always assumed to be stationary in time. However, the most recent work on the physics of the high-energy sources suggests that a likely origin for the illuminating X-rays are magnetic flares above the surface of the cold accretion disk (e.g., [8,13]). In this paper, we attempt to identify which of the various models for the BBB emission can account for the observed near-independence of E_c on the AGN luminosity. Our main goal is to determine if a viable mechanism can arise as a result of X-ray illumination of the disk by a *transient magnetic flare*.

THE RADIATION FLUX AND EMISSION MECHANISMS

At least initially, the accretion rate is $\dot{M} \sim M^2$, where M is the black hole mass, but this constitutes a runaway process in the sense that $L/L_{Edd} \propto t$, where t is the time, and L_{Edd} is the Eddington luminosity. When $L \to L_{Edd}$, the outward radiation pressure presumably suppresses the inflow, with the effect that L saturates at the value $\sim L_{Edd} \propto M$. As a statistical average, we thus expect that $L \propto M$.

In view of this, let us next examine how the various different emission mechanisms fare in their prediction of the BBB cutoff energy $E_c(L)$. For any radiation process, the flux F scales as L over the emitting area, which itself scales as M^2. Thus, in general we expect that $F \sim L^{-1}$. The blackbody flux is $F_{bb} = \sigma T^4$, where T is the effective temperature, and so $T \sim L^{-1/4}$. Thus, when L varies by 4 orders of magnitude, it is expected that T ought to itself vary by a factor of 10. This is not consistent with the observations discussed above.

The disk also may radiate as a 'modified blackbody', for which the flux is then given by $F_{mb} \sim 2.3 \times 10^7 T^{9/4} \rho_d^{1/2}$ erg cm^{-2} s^{-1}, where ρ_d (in g cm^{-3}) is the disk mass density and T is in Kelvins. For the likely situation of a radiation-dominated disk, $\rho_d \sim L^{-1}$, and so $T \sim L^{-2/9}$, which again is not consistent with the data.

Optically thin bremsstrahlung, on the other hand, produces a flux

$$F_{ff} = \varepsilon_{ff}\, d = 6.1 \times 10^{20} T^{1/2} \rho^2 d \text{ erg cm}^{-2}\text{ s}^{-1}, \qquad (1)$$

where ε_{ff} (erg cm^{-3} s^{-1}) is the free-free emissivity, and d is the geometrical thickness of the emitting region. Thus, since d presumably scales as $R_g \equiv 2GM/c^2 \propto L$, T is independent of L because $F \sim L^{-1}$.

TIME-INDEPENDENT X-RAY ILLUMINATION

It is well known that an active region (AR) radiating X-rays above the cold disk will produce a reflected component and a reprocessed UV spectrum due to the absorbed X-ray flux [7,9]. The characteristic Thomson optical depth τ_T at which the incident X-rays are absorbed or scattered to lower energies is of order a few. The problem has also been addressed by, e.g., [14,22,5,16]. As far as the UV portion of the spectrum is concerned, the photoionized reflection models produce a temperature that is either well below [16] the observed value ~ 60 eV, or one that is strongly dependent on M (for example, T changes by about a factor of 2 for a change in M by a factor of 10 in Figure 2 of [14]). Physically, this is explained by the fact that in the stationary case the thermal equilibrium in the whole disk below the X-ray emitting region is established, and thus the characteristic temperature of the UV emission is representative of the disk itself rather than the reflection process. The arguments given in §2 then show that the undesirable correlation between T and L ensues. We also note that calculations of the static X-ray reflection/reprocessing cannot be simply extended to the reflection of X-rays from such dynamic processes as magnetic flares, whose presence seem to be necessary in order to explain the hard X-ray spectrum of Seyferts (e.g., [13]).

TIME-DEPENDENT X-RAY ILLUMINATION OF THE DISK

In the following, the principal distinction between this and the time-independent case is the type of equilibrium established during a flare. The X-ray skin, being a very small fraction of the total thickness of the disk will adjust very quickly to a quasi-equilibrium with the incident X-radiation. The rest of the material below the skin will be out of equilibrium due to the fact that the time scale required to establish such a state is far longer than the flare lifetime.

The compactness l of the AR (here assumed to be where the magnetic flare occurs) is defined according to $l \equiv F_x \sigma_T \Delta R_a / m_e c^3$, where ΔR_a is the typical AR size. The value of l is expected to be rather high (~ 100, [20]). The incident X-ray flux F_x can be deduced from l and ΔR_a:

$$F_x \equiv \frac{l m_e c^3}{\Delta R_a \sigma_T} = 3.6 \times 10^{17} \text{erg cm}^{-2} \text{sec}^{-1} \frac{l_2}{\Delta R_{13}}, \qquad (2)$$

where σ_T is the Thomson cross section, $l_2 \equiv l/100$ and $\Delta R_{13} \equiv \Delta R_a / 10^{13}$ cm. The scaling of ΔR_a is based on the expectation that it should be of order the accretion disk scale height (H_d).

When this flux turns on, the radiation ram pressure on the surface of the disk greatly exceeds the equilibrium thermal pressure from within. The X-ray skin therefore gains momentum and an inward plow phase is initiated that very quickly slows down the density wave. The density itself is expected to increase in the X-ray skin until the reprocessed UV free-free emissivity balances the incident X-ray flux, at which point,

$$\rho = 2.8 \times 10^{-7} \, \text{g cm}^{-3} \frac{l_2}{\Delta R_{13}} T_5^{-1/2} (\tau_x/3)^{-1}, \tag{3}$$

where $T_5 \equiv T/10^5$ K. If the X-ray skin were to contract further, the UV emissivity would exceed the incident radiation flux which clearly violates energy conservation. Comparing this with the gas density ρ_d in a cold accretion disk, we see that the latter is smaller than ρ by 2-3 orders of magnitude.

Also, a quasi-equilibrium between internal pressure $2(\rho/m_p)kT$ of the compressed gas and the incident X-ray flux will be established. The radiation energy flux in the Eddington approximation is $F_e = -c/3 \, (du_{\text{rad}}/d\tau)$, where u_{rad} is the radiation energy density and τ is the total optical depth. To find the total compressional force F_{rad} acting on the absorbing/reflecting layer, we integrate over τ:

$$F_{\text{rad}} = 1/c \int_0^{\tau_x} d\tau \, F_e = (1/3) \left[u_{\text{rad}}(0) - u_{\text{rad}}(\tau_x) \right]. \tag{4}$$

In a steady state situation, we expect $F_{\text{rad}} = 0$. At the other extreme, when all the incident X-ray flux is re-radiated back to the corona, $u_{\text{rad}}(\tau_x) = 0$ and $u_{\text{rad}}(0) = 2\sqrt{3} F_x/c$ (the skin is then just a mirror reflecting the incident momentum flux). The pressure equilibrium condition is therefore $2(\rho/m_p)k_b T = 2/3^{1/2} A [F_x/c]$, where k_b is Boltzmann's constant, and the unknown parameter $0 < A < 1$ reflects our ignorance of the specific details in this layer. The X-ray flux drops out of the equation when we include the energy balance condition $F_x = \varepsilon_{\text{ff}} d = F_{\text{UV}}$, in which F_{UV} is given by Equation (1). It is this cancellation of the exact value of F_x that leads to the mass-invariance of the temperature. Assuming we know the exact value of A (our guess is that it is probably between 1/2 and 1), we therefore infer a unique value for T:

$$T = 3.8 \times 10^5 \, \text{K} \left(A\tau_x/\sqrt{3} \right)^2. \tag{5}$$

The validity of this treatment rests on the assumption that the intrinsic disk flux is negligible compared to the local X-ray flux from the active region, for otherwise the compressional effects will not work to produce the required UV flux and BBB temperature. This condition is certainly met when $l \gg 1$. [22] have shown that the quality of the X-ray data are high enough to distinguish between reflection from weakly ionized/neutral gas and ionized gas if the latter has ionization parameter ξ larger than about 200. Since the data are adequately fitted with a neutral reflector, we then require that the ionization parameter of the gas should not exceed 200. Since the gas density in the X-ray skin is higher than in the static models, our estimated value for ξ in the X-ray skin is always $\lesssim$ few tens, and thus is consistent with observations. On the other hand, static reflection is characterized by a gas density smaller by typically 2-3 orders of magnitude, and therefore the ionization parameter is too high for this reflection.

DISCUSSION AND CONCLUSIONS

We have produced a simple explanation for the relatively universal value of the BBB temperature in the face of large variations in luminosity. In effect, one could define a so-called "bremsstrahlung temperature" T_{brems} as the temperature at which the radiation pressure of the bremsstrahlung photons in an optically thin plasma is equal to the gas pressure. Equation (5) gives a value of T that is close to T_{brems}. Our derived temperature compares favorably with the cutoff energy range ($\sim$ 36-80 eV) observed in the sample of Zhou et al.(1997).

This work was partially supported by NASA grant NAG 5-3075.

REFERENCES

1. Antonucci, R., & Barvainis, R. 1988, *ApJ*, **332**, L13.
2. Barvainis, R. 1993, *ApJ*, **412**, 513.
3. Barvainis, R., & Antonucci, R. 1990, BAAS, **22**, 745
4. Czerny, B., & Elvis, M. 1987, *ApJ*, **321**, 305.
5. Czerny, B., & Zycki, P. T. 1994, *ApJ*, **431**, L5.
6. Ferland, G. J., et al. 1990, *ApJ*, **363**, L21.
7. Guilbert, P., & Rees, M.J. 1988, *MNRAS*, **233**, 475.
8. Haardt F., et al. 1994, *ApJ*, **432**, L95.
9. Lightman, A.P., & White, T.R. 1988, *ApJ*, **335**, 57.
10. Laor, A., & Netzer, H. 1989, *MNRAS*, **238**, 897.
11. Malkan, M., & Sargent, W.L.W. 1982, *ApJ*, **254**, 22.
12. Mushotzky, R.F., et al. 1993, *Annu. Rev. Astron. Astrophys.*, **31**, 717
13. Nayakshin, S., & Melia, F. 1997, these proceedings
14. Ross, R.R., & Fabian, A.C. 1993, *MNRAS*, **261**, 74.
15. Shields, G.A. 1978, Nature, 272, 706
16. Sincell, M.W., & Krolik, J.H. 1997, *ApJ*, **476**, 605S.
17. Walter, R., & Fink, H.H. 1993, A&A, **274**, 105
18. Walter, R., et al. 1994, A&A, **285**, 119
19. White, T.R., Lightman, A.P., and Zdziarski, A.A. 1988, *ApJ*, **331**, 939.
20. Zdziarski, A.A. et al. 1996, A&AS, 120, 553
21. Zhou et al. 1997, *ApJ (Letters)*, **475**, L9.
22. Zycki et al. 1994, *ApJ*, **437**, 597.

Magnetic Flares and the Observed $\tau_T \sim 1$ in Seyfert Galaxies

Sergei Nayakshin[†] and Fulvio Melia[*] [1]

[†]*Physics Dept., The University of Arizona, Tucson AZ 85721*
[*]*Physics Dept. & Steward Observatory, The University of Arizona, Tucson AZ 85721*

Abstract. We here consider the pressure equilibrium during an intense magnetic flare above the surface of a cold accretion disk. Under the assumption that the heating source for the plasma trapped within the flaring region is an influx of energy transported inwards with a group velocity close to c, e.g., by magnetohydrodynamic waves, this pressure equilibrium can constrain the Thomson optical depth τ_T to be of order unity. We suggest that this may be the reason why $\tau_T \sim 1$ in Seyfert Galaxies. We also consider whether current data can distinguish between the spectrum produced by a single X-ray emitting region with $\tau_T \sim 1$ and that formed by many different flares spanning a range of τ_T. We find that the current observations do not yet have the required energy resolution to permit such a differentiation. Thus, it is possible that the entire X-ray/γ-ray spectrum of Seyfert Galaxies is produced by many independent magnetic flares with an optical depth $0.5 < \tau_T < 2$.

INTRODUCTION

Many of the observational characteristics of X-ray/UV spectra of Seyfert galaxies fit within the currently popular two-phase patchy accretion disk-corona model [2–4,8,9], in which the X-ray emitting region is located within 'active regions' (AR) above the disk. Recent calculations by [9] show that the Thomson optical depth τ_T in Seyfert galaxies must be close to unity. [4] suggested that the ARs may be magnetic flares occurring above the accretion disk's atmosphere and showed that their compactness l may be quite high (~ 30), so that pairs can be created. The explanation for the observed value of $\tau_T \sim 1$ based on the pair equilibrium condition relies on the assumption that the particles are confined to a rigid box, so that no pressure constraints

---

[1]) Presidential Young Investigator.

need to be imposed. This is unphysical for a magnetic flare where the particles are free to move along the magnetic field lines.

The 'universal' X-ray spectral index of Seyfert Galaxies suggests that the emission mechanism is thermal Comptonization with a y-parameter close to one [2,3]. Observations of Seyfert Galaxies point to a compactness parameter $\sim 1 - 100$ [8]. One can show that the radiation pressure dominates over the gas pressure in Seyferts if the plasma has single temperature and $l \gg 1$. There must be an agent that energizes the particles to enable them to radiate at this high rate. We foresee two possibilities for the nature of this 'agent': (i) the gravitational field, and (ii) an external flow of energy into the system.

Insofar as the first possibility is concerned, there does not appear to be a scale that sets τ_T to have a value of 1. For example, in standard accretion disk theory, the inner radiation pressure-dominated regime has an optical depth that depends on several parameters, such as the accretion rate and the α-parameter. It is even less obvious why τ_T should be ~ 1 in the gas pressure dominated regimes since there the pressure has no reference to τ_T at all.

In the case of magnetic field, the dynamic portion of the field supplies a "ram" pressure that is related in a known way to its energy density. If the magnetic energy flux into the X-ray emitting region is known, this also constrains the inwardly directed momentum flux (the compressional force) into the system. Thus, the compressional force exerted on the active region by the magnetic field is expected to correlate with the source luminosity. What makes this useful in terms of setting the optical depth of the system is that a similar correlation exists between the luminosity and the outwardly directed radiation pressure in the emitting region.

PRESSURE EQUILIBRIUM

Let us first suppose that the X-ray source is a sphere with Thomson optical depth τ_T, and that the energy is supplied radially by magnetohydrodynamic waves. The waves carry an energy density ε and propagate with velocity v_a. For definitiveness, we assume that these are Alfvén waves, in which case the momentum flux that enters the X-ray source is $(1/2)\varepsilon$. The magnetic energy of the Alfvén waves is in equipartition with the oscillating part of the particle energy density, and so we can estimate the gas pressure as being of the same order as the ram pressure of the oscillating part of the magnetic field, i.e. $(1/2)\varepsilon$. Finally, we assume that all of the wave energy and momentum are absorbed by the source.

The energy equilibrium for the AR is then given by $F_r = \varepsilon v_a$, whereas in pressure equilibrium $P_r \simeq \tau_T F_r/c \simeq \varepsilon$. Dividing the latter equation by the former, one obtains for the equilibrium Thomson optical depth:

$$\tau_T \simeq \frac{c}{v_a} \qquad (1)$$

This value does not depend on luminosity. [5] show that the Alfvén velocity is close to c for typical conditions in an accretion disk.

Suppose now that the geometry is not perfectly spherically symmetric, and that instead the Alfvén waves can enter the X-ray source through an area A_a, but the radiation leaves through an area $A_r \gtrsim A_a$, which is plausibly just the total area of the AR. In this case, since the energy balance is now $F_r A_r = \varepsilon v_a A_a$, the equilibrium τ_T is changed to

$$\tau_T \simeq \left(\frac{c}{v_a}\right)\left(\frac{A_r}{A_a}\right). \qquad (2)$$

The magnetic field energy density of the flare is likely to be a fraction of the underlying disk energy density and the typical size ΔR_a of the flare is expected to be of the order of the disk scale height [1,4,5]. Now, the confinement of the plasma inside the flare, and the observed condition $l \gg 1$, require that $B^2/8\pi \gg P_r \gg P_g$. The tube is thick (meaning that its cross sectional radius is of the order of its length), since the pressure in the disk's atmosphere is insufficient to balance the tube's magnetic field pressure. Also, we may assume that sideways expansion of the flux tube is stopped due to the magnetic field tension. We then need to consider the pressure equilibrium along the magnetic field lines only. The magnetic waves propagate upwards along the magnetic flux tube and heat the particles. These in turn produce X-radiation by up-scattering the UV radiation from the disk. The radiation pressure from the AR is pushing the gas along the magnetic lines, i.e., downwards to the disk. This downward direction of the radiation pressure arises naturally in a two-phase model (unlike the situation within the accretion disk) since here most of the energy is released above the disk's atmosphere [5]. The balance of radiation pressure with the magnetic ram pressure then sustains the AR optical depth as discussed above. Since A_r/A_a is probably of order $\sim$ one to a few, and with $v_a \sim c$, we therefore expect

$$\tau_T \sim 1 - 2. \qquad (3)$$

The lowest values of the equilibrium τ_T can be reached because A_r in this equation is not necessarily the total area of the source – some of the X-ray flux can be reflected by the underlying disk and re-enter the AR. Some of this re-entering flux can be parallel to the incoming Alfvén energy flux, and thus the effective A_r is smaller than the full geometrical area of the source. Furthermore, we have assumed a one-temperature gas, and have neglected the gas pressure in our calculation. In addition, protons may account for a sizable fraction of the total pressure in the AR, which then leads to a reduction in τ_T.

RANGE IN τ_T PERMITTED BY OBSERVATIONS

[9] produced a fit of the average $Ginga$/OSSE spectrum of Seyfert 1 galaxies. They found that the radial optical depth of the ARs is $\tau_T \sim 1$. Here, we

will examine whether the Seyfert spectrum can be due to a combination of spectral components from flares with different τ_T, but the same y-parameter (set arbitrarily at 1.3). A constant y-parameter is a natural consequence of the fixed geometry of the flare [2].

We first compute the spectrum from flares with a range of Thomson optical depths assuming that they all have the same luminosity. We then convolve these spectra with a Gaussian probability distribution that a flare occurs with τ_T. The composite spectrum $F(E)$ (in energy/sec/keV) is

$$F(E) = \int_0^\infty d\tau_T \ \exp\left[-\frac{(\tau_T - \tau_0)^2}{\tau_\sigma^2}\right] F(E, \tau_T), \qquad (4)$$

where $F(E, \tau_T)$ is the spectrum from a single flare with τ_T. We take $\tau_0 = 1.14$ and adopt several values of τ_σ = const to represent the possible spread in τ_T between different flares. The individual spectra are computed assuming a slab geometry using an Eddington frequency-dependent approximation for the radiative transfer, using both the isotropic and first moments of the exact Klein-Nishina cross section [7]. Although this geometry is clearly different from that of a realistic flare, our point here is to test the possibility of co-adding spectra with different τ_T.

Figure 1 shows the results of our calculation for $\tau_0 = 1.14$ and two values of τ_σ: 0.7 and 1.5 (short-dashed and long-dashed curves correspondingly). The spectrum from a single flare (solid curve) with $\tau_T = 1.14$ is also shown for comparison. It can be seen that the plot for $\tau_\sigma = 0.7$ is hardly distinguishable from that for $\tau_T = 1.14$. Moreover, these curves differ the most above 100 keV, where the OSSE data typically have error bars larger than this difference (see, e.g., Fig. 1 in [9]). On the basis of this simple test, we would expect that Seyfert spectra may be comprised of contributions from many ARs encompassing a range $(0.5 - 2)$ of τ_T. This conclusion is very important for the magnetic flare model, since it is otherwise difficult to see how different flares could produce exactly the same τ_T.

CONCLUSIONS

We have considered the consequences of imposing a pressure equilibrium on the active regions of Seyfert Galaxies under the assumption that the emission arises within energetic magnetic flares above the surface of a cold disk. We showed that if the energy is supplied to the X-radiating plasma by the influx of some energy source with a group velocity $\sim c$, then τ_T probably falls within the range $1-2$. The current X-ray/γ-ray observations are consistent with this range of Thomson optical depths. We conclude that magnetic flares remain a viable explanation for the spectra observed in Seyfert Galaxies. Alternative explanations, based on a gravitational confinement of the ARs, appear to be incapable of accounting for the observed universality in the value of τ_T.

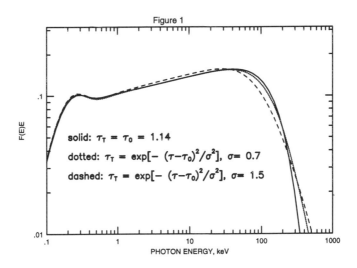

Figure 1

ACKNOWLEDGMENTS

This work was partially supported by NASA grant NAG 5-3075.

REFERENCES

1. Galeev, A. A., Rosner, R., & Vaiana, G. S., 1979, *ApJ*, **229**, 318.
2. Haardt F. & Maraschi L., 1991, *ApJ*, **380**, L51.
3. Haardt F. & Maraschi L., 1993, *ApJ*, **413**, 507.
4. Haardt F., et al., 1994, *ApJ*, **432**, L95.
5. Nayakshin & Melia 1997a, submitted to ApJ, also astro-ph/9705254
6. Nayakshin & Melia 1997b, ApJL, in press
7. Nagirner, D. J., & Poutanen, J, 1994, Astrophys. Space Phys., **9**, 1
8. Svensson, R. 1996, A&AS, 120, 475
9. Zdziarski, A.A., Gierlinski, M., Gondek, D., & Magdziarz, P. 1996, A&AS, 120, 553

Physical Constraints for The Active Regions in Seyfert Galaxies

Sergei Nayakshin[†] and Fulvio Melia[*] [1]

[†]*Physics Dept., The University of Arizona, Tucson AZ 85721*
[*]*Physics Dept. & Steward Observatory, The University of Arizona, Tucson AZ 85721*

Abstract. We discuss various physical constraints on the Active Regions (AR) in Seyfert 1 Galaxies. We show that a viable model that can account for these constraints is one in which the ARs are magnetically confined and 'fed'. In addition, the large compactness parameter ($l \gg 1$) required to explain the "unique" X-ray index of these sources, in conjunction with the physical conditions needed to make the optical depth close to unity, suggests that the magnetic energy density in the AR should be comparable to the equipartition value in the accretion disk, and that it should be released in a flare-like event. We argue that the constraints can be satisfied if the high-energy emission occurs within magnetic flares above the surface of the disk.

INTRODUCTION

Haardt & Maraschi (1991, 1993) showed that if most of the energy is dissipated in a hot corona overlying a cold accretion disk, then the resulting spectra naturally explain many of the observed features in Seyfert Galaxies. In particular, the hardening of the spectra above about 10 keV [10] and a broad hump at ~ 50 keV (e.g., [21]) are explained by reflection of the hard X-rays from the cold disk. The direct component of the spectrum, a power-law with spectral index $\alpha_{2-18} = 1.95 \pm 0.15$ [10], is produced via inverse Compton upscattering of the soft photons from the cold disk.

However, observationally, the hard X-ray luminosity, L_h, can be a few times smaller than the luminosity, L_s, in the soft UV-component. This is inconsistent with the uniform two-phase disk corona model, because the latter predicts about the same luminosities in both X-rays and UV (due to the fact that all UV radiation arises as a consequence of reprocessing of the hard X-ray flux, which is about equal in the upward and downward directions). To overcome

---

[1]) Presidential Young Investigator.

this apparent difficulty, Haardt, Maraschi & Ghisellini (1994) [HMG94] introduced a patchy disk-corona model. The model assumes that the X-ray emitting region consists of separate 'active regions'(AR), independent of each other. In this case a part of the reprocessed as well as intrinsic radiation from the cold disk escapes to the observer directly, rather than entering the ARs, thus allowing for a greater ratio of L_s/L_h. Recently, Stern et al. (1995) and Poutanen & Svensson (1996) carried out state of the art calculations of the radiative transport of the anisotropic polarized radiation and showed that the model is very robust in its predictions.

Galeev, Rosner and Vaiana (1979) showed that the physical conditions in the accretion disks around black holes are such that the magnetic field is likely to grow to its equipartition value. This magnetic field is then transported to the surface of the disk by buoyancy where its energy is released in a flare-like event. This approach is, in a sense, superior to the traditional spectral modelling approach, in that it attempts to include all the relevant physics self-consistently (e.g., [20,13,19]).

It is imperative that these two classes of models find a common ground of self-consistency. In particular, in the spectral modelling, the mechanism by which the gravitational energy of the cold disk is transported to the optically thin corona is not specified. It is *assumed* that some process can provide the needed electron heating, and often a reference is made to magnetic fields. Moreover, no hydrostatic equilibrium is imposed, and instead particles are artificially kept in a box. HMG94 showed that the compactness can be high enough during the active phase if one assumes that the entire magnetic field energy is transferred to the particles during a few light-crossing time scales. Recent findings by Nayakshin & Melia (1997; see the two accompanying papers in these proceedings) support the magnetic flare model and motivate us here to attempt to collect the various pieces of the puzzle.

PHYSICAL CONSTRAINTS ON THE ACTIVE REGIONS

The compactness parameter l is defined according to $l \equiv F_\gamma \sigma_T \Delta R_a / m_e c^3$, where F_γ is the radiation energy flux at the top of the AR and ΔR_a is the typical size of the AR. Notice that this defines the *local* compactness, which characterizes the local properties of the plasma, unlike the global compactness $l_g \equiv L\sigma_T/R'm_e c^3$ (where L is the total luminosity of the source and R' is the typical size of the region that emits this luminosity). the latter should be compared to the observed compactness rather than the former. Depending on the ratio of the total active area ΔS_a covered by the ARs to R'^2, the local compactness can be either larger or smaller than the global (i.e., observed) one.

A large local compactness is strongly preferred in current pair-dominated two-phase models (e.g., [18,22]). To give the right spectra, τ_T should be relatively large (~ 1). In the context of the pair-dominated two-phase model, the only mechanism for fixing the optical depth is by pair equilibrium, and thus one needs $l \gg 1$, when numerous pairs are created. In addition, radiation mechanisms put their own limitations on the local compactness. Fabian (1994) shows that in order for the Compton emissivity to dominate over the bremsstrahlung one, the compactness of the plasma should be larger than $l \sim 0.04 \Theta^{-3/2}$, where Θ is the electron temperature in units of $m_e c^2/k_B$. For the typical $\Theta \sim 0.2$, this requires that $l \gtrsim 0.5$.

Geometry, Confinement and LifeTime

The geometry of localized X-ray sources above the disk (e.g., [7,18,22]) requires the active regions to be transient on a disk thermal time scale $t_{\rm th} \sim \alpha^{-1} t_{\rm h}$, where $t_{\rm h} = H_d/c_s$. An integral assumption of the two-phase model is that the internal disk emission is negligible compared with the X-ray flux of the AR, at least during the active phase [14]. Assuming that a fraction ~ 1 of the total energy content of the surface area of the disk immediately below the AR is transferred into the AR, the time scale for the release of this energy should then be much shorter than $t_{\rm th}$, during which the internal disk energy is released as radiation. Otherwise the localized ARs actually produce a *steeper* spectrum than a full corona.

The plasma in the ARs should be confined during the active phase. Gravitational confinement does not work here due to several reasons. First, the *locally* limited Eddington compactness l is at most $\sim 50/(1+2z)$ for ARs in a roughly semi-spherical shape, where z is the positron number density n_+ over that of protons n_p, while the relatively large Thomson optical depth $\tau_T \equiv \sigma_T n_p (1+2z) \sim 1$ obtained by Zdziarski et al. (1996) requires a compactness of a few hundred (if no magnetic field is involved and particles are in a rigid box). Second, there is no mechanism for counter balancing side-ways expansion of the plasma. Therefore, since there seem to be no other reasonable possibility for the AR plasma confinement, we conclude that it is the magnetic field that provides the confining pressure. Any mechanism will fail to confine the plasma during a time larger than about one dynamical time scale for the disk, since adjacent points with slightly different radii are torn apart on this time scale due to the disk differential rotation. Thus, confinement of the plasma requires the lifetime to be shorter than one hydrostatic time scale.

To be consistent with the two-phase accretion disk-corona model (and observations) and physics, one needs very short lived phenomena to occur above the disk atmosphere. In fact, the whole evolution of the AR should happen faster than the disk hydrostatic time scale. To confine the plasma with high compactness parameter $l \gg 1$, one needs mechanisms other than gravitational confinement. This points to a magnetic flare as the most likely candidate.

MAGNETIC FLARES AND ACCRETION DISKS

Galeev, Rosner & Vaiana (1979) showed that magnetic flares on the surface of accretion disks are more than likely to be created, since internal dissipative processes are ineffective in limiting the growth of magnetic field fluctuations. As a consequence of buoyancy, magnetic flux should be expelled from the disk to a corona, consisting of many magnetic loops, where the energy is stored. It has also been speculated that, just as in the Solar case, the magnetically confined, loop-like structures (which we shall collectively call magnetic flares) carry the bulk of the X-ray luminosity. The X-rays are assumed to be created by upscattering of the intrinsic disk emission. A number of Solar magnetic flare workers have elaborated on this subject since (e.g., [9,1,2,20,19].

We will now assume that by some process the magnetic field energy is transferred to the particles. The magnetic field is limited by its equipartition value in the midplane of the disk. The size of the AR, ΔR_a, is of the order of the disk scale height H_d. Let us assume that the the energy transfer to the particles occurs on the time scale t_l equal to the light crossing time H_d/c times some number $b \gtrsim$ a few. We will also assume that the flare occurs at 6 gravitational radii, where most of the bolometric luminosity is produced. Using the results of [17], we obtain:

$$l \lesssim 400 \frac{\mathcal{L}}{\alpha b} \frac{\varepsilon_m \Delta R_a}{\varepsilon_d H_d} \qquad (1)$$

where ε_d is the midplane energy density, ε_m is the magnetic energy density $\lesssim \varepsilon_d$. HMG94 suggested that plausible values for b and α are 10 and 0.1, respectively. We can also assume that $\varepsilon_m \sim 0.1\varepsilon_d$.

[11] have shown that if the compactness parameter $l \gg 1$ during energetic magnetic flares, it seems likely that the optical depth of the plasma trapped inside the magnetic flux tube is in the range $0.5 - 2$. They also showed that it is possible that the X-ray spectra from Seyfert Galaxies consist of many independent spectra from sources that have a Thomson optical depth in the range $0.5 - 2$. As pointed out by [7], the spectra from magnetic flares will be similar to the time-independent spectra from static regions with the same compactness and optical depth. Thus, calculations of emergent spectra from static Active Regions, preferably in the shape of hemispheres, apply for magnetic flare spectra as long as their lifetime is $\gtrsim$ a few life-crossing times of the source.

Recent work by [22] suggests that the optical depth of roughly unity is explained by a high compactness of the emitting region ($\sim$ several hundred). However, this observations can also be explained by invoking pressure balance (see the accompanying paper in these proceedings). In that event, the plasma consists of the protons to a large extent. We think that pairs are only marginally important for the flares, since their optical depth seems to be relatively large (~ 1.2), which requires a very large ($\gtrsim 300$) compactness. If protons are present, then a much smaller compactness will do, as long as it is $\gg 1$. Finally, notice that pair (see [16]) runaway is not possible for magnetic flares, since the plasma can expand along magnetic field lines.

CONCLUSIONS

We have shown that the physical constraints on the localized active regions in the context of the two-phase corona-accretion disk model unambiguously require that the active regions be magnetically 'fed' and confined. In view of the large compactness required to explain the observed optical depth ~ 1, this further implies that the magnetic energy should be released in a flare-like event with a duration $\sim$ few light-crossing times for the AR. It also implies that the magnetic field energy density in the flare region should be a fair fraction of the equipartition value in the *midplane* of the accretion disk. Provided these two conditions are met, it seems very plausible that magnetic flares can explain the current observations of Seyfert Galaxies quite well, with the overall spectrum being a composite of contributions from many different flares that produce similar spectra as long as $l \gg 1$.

REFERENCES

1. Burm, H. 1986, A&A, **165**, 120
2. Burm, H. & Kuperus, M. 1988, A&A, **192**,165
3. Fabian, A. C. 1994, ApJS, 92, 555
4. Galeev, A. A., Rosner, R., & Vaiana, G. S., 1979, *Ap. J.*, **229**, 318.
5. Haardt F. & Maraschi L., 1991, *Ap. J.*, **380**, L51.
6. Haardt F. & Maraschi L., 1993, *Ap. J.*, **413**, 507.
7. Haardt F., et al., 1994, *Ap. J.*, **432**, L95.
8. Iwasawa K., et al., 1996, *M.N.R.A.S.*, **282**, 1038.
9. Kuperus, M., & Ionson, J. 1985, A&A, 148, 309
10. Nandra, K., & Pounds, K. A., 1994, *M.N.R.A.S.*, **268**, 405.
11. Nayakshin & Melia 1997a, see this volume (" Phys. Constr.")
12. Nayakshin & Melia 1997b, see this volume (" Magnetic Flares")
13. van Oss, R.F., van den Oord, G.H.J., & Kuperus, M. 1993, A&A, **270**, 275
14. Poutanen, J, & Svensson, R. 1996, *Ap. J.*, **470**, 249.
15. Stern et al. 1995, *Ap. J.*, **449**, L13.
16. Svensson, R. 1982, *Ap. J.*, **258**, 335.
17. Svensson, R., & Zdziarski, A. 1994, *Ap. J.*, **436**, 599.
18. Svensson, R. 1996, A&AS, 120, 475
19. Volwerk, M., van Oss, R.F., & Kuijpers, J., 1993, A&A, **270**, 265
20. de Vries, M., & Kuijpers, J., 1992, A&A, **266**, 77
21. Zdziarski, A. A. et al. 1995, *Ap. J.*, **438**, L63.
22. Zdziarski, A.A. et a. 1996, A&AS, 120, 553

Are Gamma-ray Bursts related to Active Galactic Nuclei?

Javier Gorosabel and Alberto J. Castro-Tirado

LAEFF-INTA, P.O. Box 50727, 28080 Madrid, Spain.

Abstract. We study the possible correlation of a selected sample of 340 gamma-ray bursts that can be localized to ≤ 7 deg^2 with different subdivisions of the Véron & Véron-Cetty compilation of quasars-AGNs. The intention is to confirm whether the bursts are related to radio-quiet quasars as it was claimed by Schartel et al., who found a 96.4% confidence level correlation between radio-quiet quasars and a subsample of 134 events from the BATSE 3B catalogue. We find a 90% correlation between polygonal GRBs (IPN GRBs) and radio-quiet quasars, however the correlation vanishes for circular GRBs (BeppoSAX and WATCH GRBs). On the other hand, a 97.6% correlation is found between BL lac objects and circular GRBs. We discuss the possible connection between the results of this study and those obtained by Schartel et al. 1996.

1 INTRODUCTION

Gamma–ray bursts (GRB) remain one of the most elusive mysteries in high energy astrophysics due to the lack of knowledge of the distance scale. However, the recent optical spectrum of GRB970508 may finally settle the issue of the distance scale in favour of the cosmological models, al least for some of the bursts [6]. This view is strengthened by the announced correlation between BATSE bursts and radio-quiet quasars [7], (hereafter SC96). SC96 reported that GRBs seem to be related to radio-quiet quasars on the basis of a 96.4% confidence level correlation between a subsample of 134 GRBs from the BATSE 3B catalogue and a sample of 7146 radio-quiet quasars. Our study aims at verifying the results obtained by SC96 using an improved sample of GRBs.

2 METHOD

With the intention of comparing the angular characteristics of a sample of 340 well-localized bursts and different subsamples of quasars-AGNs (§2.1), we have performed simulations of random sets of GRBs in order to obtain the distributions for the positional coincidences (§2.2) and the correlation estimate (§2.3).

2.1 The γ - ray bursts and quasars-AGNs samples

Our selected sample of 340 GRBs can be divided into two groups of 311 polygonal and 29 circular error boxes each. The circular error boxes are based on the positions provided by *BeppoSAX* and by the WATCH instruments on the *Granat* and *Eureca* satellites [1]. A considerable fraction of the polygonal error boxes have been obtained from different compilations [2,3,5,4]. The mean area of the 340 error boxes in our sample is 1.02 deg^2 in contrast to 9.0 deg^2 for the sample used by SC96. On the other hand we used the latest (7th) version of the "Catalogue of quasars and Active Nuclei" [8]. This catalogue is divided into three sections -quasars, BL Lac objects and AGNs. In addition to this division, we have considered certain subsamples depending on their spectral characteristics. Hereafter, the quasars-AGNs subsamples will be represented by the label $l = \{1, 2, .., 18\}$. The 18 samples are listed in Table 1.

2.2 Simulations of GRB sets and positional coincidences

One of the most important points is the estimation of the background level of positional coincidences, i. e., the expected value for chance coincidences in order to obtain a comparison with the real coincidences. For this purpose, 1500 sets of 340 GRBs were created (hereafter j=$\{1,2,..,1500\}$). Each simulated set was obtained as a random rotation in the galactic plane of the original GRB set. By definition, a positional coincidence occurs when there is a GRB error box containing at least one source. This definition is based on the fact that any observed burst can be associated to a single counterpart. So, the number of coincidences of the real GRBs with the 18 subsamples can be calculated, and labelled as C_0^l and P_0^l, $l = \{1, 2, .., 18\}$, respectively. Similarly, C_j^l and P_j^l, $l = \{1, 2, .., 18\}, j = \{1, 2, .., 1500\}$, were calculated. They represent the number of coincidences between the 1500 simulated GRBs sets and the 18 quasars-AGNs subsamples. The mean value defines a background level that can be compared to the real coincidences C_0^l and P_0^l, $l = \{1, 2, .., 18\}$, respectively. Let us define $PC^l(\%)$ and $PP^l(\%)$ as the fractional number of simulated GRBs sets for which $C_j^l < C_0^l$ and $P_j^l < P_0^l$.

2.3 The correlation estimate

We remark that the probability of finding the GRB–related object, becomes smaller as one moves away from the center of the GRB error box. In order to consider this effect, a normalized probability distribution was fitted for each error box. Thus, for the j-th simulated GRB set, $j=\{1,2,..,1500\}$, and for the l-th quasar-AGN subsample, $l=\{1,2,..,18\}$, two quantities (hereafter quoted as correlations) T_j^l and M_j^l were introduced. T_j^l and M_j^l give the correlation of the polygonal error boxes and the circular error boxes belonging to the j-th GRB catalogue with the l-th quasar-AGN subsample. T_j^l can be expressed by the following expression:

$$T_j^l = \sum_{k=1}^{311}\sum_{i=1}^{N^l} t_{ijk}^l \text{ being } t_{ijk}^l = \begin{cases} \frac{-\ln(1-s)}{a_{jk}} \exp(((\frac{x_{ijk}^l}{X_{jk}^l})^2 + (\frac{y_{ijk}^l}{Y_{jk}^l})^2)\ln(1-s)) & \text{if the quasar-AGN is in the GRB error box.} \\ 0 & \text{if the quasar-AGN is outside the GRB error box.} \end{cases}$$

with $s=0.9973$, X_{jk}^l and Y_{jk}^l the 3σ dimensions of the k-th polygon belonging to the l-th simulated catalogue along the principal axes, x_{ijk}^l and y_{ijk}^l the distances from the i-th quasar-AGN of the l-th subsample to the principal axes of the k-th polygon belonging of the j-th set, and a_{jk} the area of the k-th polygon of the j-th set. M_j^l was obtained substituting t_{ijk}^l by a simple Gausssian-like function $m_{ijk}^l = \frac{-\ln(1-s)}{\pi r_{jk}^2} \exp(\left(\frac{d_{ijk}^l}{r_{jk}}\right)^2 \ln(1-s))$. On the other hand, the same method was applied to the true GRB set, which provided 18 values of the correlation T_0^l and M_0^l, $l=\{1,2,..,18\}$. The distributions for T_j^l and M_j^l, $l=\{1,2,..,18\}$, $j=\{1,2,..,1500\}$, were calculated in order to detect an excess in the value of T_0^l and M_0^l.

3 RESULTS AND DISCUSSION

As it is displayed in Table 1, for BL Lac (all) objects 94.7% of the simulations reveal less coincidences than the real circular GRBs. This excess is not noticeable in polygonal boxes which show a value of 25.8%. The excess is also detected by the correlation function (Table 2, 97.6% excess). These results do not agree with those obtained by SC96, who found only a 13.5% correlation using a 134 BATSE GRB sample. This disagreement can be easily explained taking into account that BATSE did not detect any of the two WATCH GRBs containing BL Lac objects.

On the other hand we detect an excess of radio-quiet quasars in polygonal error boxes (89.7% in the coincidences and 90.0% in the correlation function), although this possible connection is not detected in circular error boxes. Considering that SC96 used a BATSE GRB sample and the fact that a great number of polygonal error boxes used in this study are based on positions provided by BATSE-IPN, such results could agree with those obtained by SC96. On the other hand, only three of the 29 circular GRBs were observed by BATSE. This fact could explain the lack of radio-quiet quasars in circular GRBs. A possible reason could be due to the different sensitivity of the

Main Class	Subsample	N^l	C_0^l	$<C_j^l>$	$PC^l(\%)$	P_0^l	$<P_j^l>$	$PP^l(\%)$
QUASAR	all	8609	3	3.93	19.5	29	24.17	82.4
QUASAR	radio-quiet	7146	2	2.44	29.2	23	17.57	89.7
QUASAR	radio-loud	1377	2	1.66	48.8	7	8.24	25.2
QUASAR	highly polarized	72	0	0.12	0.0	0	0.43	0.0
BL Lac	all	220	2	0.39	94.7	1	1.41	25.8
BL Lac	confirmed	93	1	0.15	86.0	0	0.53	0.0
BL Lac	highly polarized	76	1	0.14	86.1	0	0.52	0.0
BL Lac	radio selected	119	1	0.20	81.4	0	0.83	0.0
BL Lac	X-ray selected	82	0	0.12	0.0	1	0.47	62.3
AGN	AGN	1553	2	1.32	46.5	9	9.39	39.3
AGN	Seyfert 1	888	1	1.06	32.9	3	5.65	8.8
AGN	Seyfert 2	496	0	0.72	0.0	4	3.13	61.4
AGN	Prob. Seyfert	97	2	0.11	99.2	1	0.63	52.1
AGN	LINER	71	0	0.09	0.0	1	0.49	55.6
AGN	radio-quiet	1346	2	1.52	58.5	8	8.12	42.8
AGN	radio-loud	166	0	0.28	0.0	0	1.21	0.0
	AG	1165	1	0.57	54.3	10	4.46	99.1
	Nucl. H II	116	0	0.12	0.0	0	0.53	0.0

TABLE 1. positional coincidences. N^l represents the number of sources contained in the l-th subsample. C_0^l and P_0^l are the number of coincidences of the circular and polygonal error boxes belonging to the real GRB set with the l-th subsample of quasars-AGNs. $<C_j^l>$ and $<P_j^l>$ represent the mean value of the coincidences of the circular and polygonal error boxes belonging to the simulated GRB set with the l-th subsample of quasars-AGNs. $PC^l(\%)$ and $PP^l(\%)$ represent the fractional number of simulated GRBs sets for which $C_j^l < C_0^l$ and $P_j^l < P_0^l$.

experiments, as WATCH is sampling the strongest bursts and BATSE is also detecting a fainter population. It must be stated that the sample of GRBs and the number of coincidences are too small to suggest that there are two populations. This point can be better addressed when new and smaller error boxes will become available in the near future. There are two subsamples that also shows an important correlation; the subsample called AG and the probable Seyfert galaxies. The first one include all the AGN objects without any morphological type assigned whereas the second one, comprise AGNs that probably are Seyfert galaxies (but are not confirmed yet). Therefore, both subsamples include an unclear family of objects that can not be classified. Thus, any conclusion about this excess seems unreliable. For the other classes of AGNs and quasars no excess coincidences above random expectation were found.

REFERENCES

1. Castro-Tirado, A., Brandt, S., Lund, N., Lapshov, I., Terekhov, O. & Sunyaev,

R. 1994, AIP Conf. Proc. 307, p. 17.
2. Golenetskii, S. V., Guryan, Yu. A., Dumov, G. B., Dyatchkov, A. V., Panov, V. N., Khavenson, N. G. & Sheshin, L. O. 1986, Ioffe Technical Institute, St, Petersbourg, preprint 1026
3. Hurley, K. 1994, AIP Conf. Proc. 307, p. 29.
4. Hurley, K. 1997, private communication.
5. Lund, N. 1995, APSS, 231, 217
6. Metzger et al. 1997, Nature 387, 878.
7. Schartel, N., Andernach, H., Greiner, J. 1996, A&A accepted. Astro-ph 9612150.
8. Véron-Cetty M.P., Véron P., 1996, A catalogue of quasars and active galactic nuclei (7th edition), ESO Scientific Report, (in press).

Main Class	Subsample	N^l	M_0^l	$\langle M_j^l \rangle$	$PM^l(\%)$	T_0^l	$\langle T_j^l \rangle$	$PT^l(\%)$
QUASAR	all	8609	1.49	4.73	35.4	79.31	52.29	86.7
QUASAR	radio-quiet	7146	0.89	3.49	39.9	75.03	41.88	90.0
QUASAR	radio-loud	1377	0.60	0.88	64.3	3.63	7.89	42.9
QUASAR	highly polarized	72	0.00	0.05	0.0	0.0	0.28	0.0
BL Lac	all	220	1.36	0.14	97.6	0.05	1.71	48.5
BL Lac	confirmed	93	1.26	0.07	98.2	0.0	0.31	0.0
BL Lac	highly polarized	76	0.09	0.06	93.1	0.0	0.28	0.0
BL Lac	radio selected	119	1.26	0.06	98.6	0.0	0.65	0.0
BL Lac	X-ray selected	82	0.00	0.04	0.0	0.05	1.08	57.7
AGN	AGN	1553	0.46	0.99	54.5	4.57	8.98	43.6
AGN	Seyfert 1	888	0.10	0.56	53.9	3.81	5.07	62.1
AGN	Seyfert 2	496	0.00	0.36	0.0	0.74	2.50	51.5
AGN	Prob. Seyfert	97	0.35	0.09	96.1	0.01	0.81	75.0
AGN	LINER	71	0.00	0.06	0.0	0.00	0.43	66.3
AGN	radio-quiet	1346	0.45	1.83	62.7	4.48	8.58	49.9
AGN	radio-loud	166	0.00	0.13	0.0	0.00	2.73	0.0
	AG	1165	0.10	0.52	65.5	22.45	6.98	95.8
	Nucl. H II	116	0.00	0.03	0.0	0.0	0.87	0.0

TABLE 2. correlations. N^l represents the number of sources contained in the l-th subsample. M_0^l and T_0^l are the correlation of the circular and polygonal error boxes belonging to the real GRB set with the l-th quasars-AGNs subsample. $<M_j^l>$ and $<T_j^l>$ represent the mean value of the correlation of the circular and polygonal error boxes belonging to the simulated GRB set with the l-th subsample of quasars-AGNs. $PM^l(\%)$ and $PT^l(\%)$ represent the fractional number of simulated GRBs sets for which $M_j^l < M_0^l$ and $T_j^l < M_0^l$.

BLAZARS

Evidence for γ-Ray Flares in 3C 279 and PKS 1622-297 at $\sim$10 MeV

W. Collmar[1], V. Schönfelder[1], H. Bloemen[2], J.J. Blom[2], W. Hermsen[2], M. McConnell[3], J.G. Stacy[3], K. Bennett[4], O.R. Williams[4]

[1] *Max-Planck-Institut für Extraterrestrische Physik, 85740 Garching, Germany*
[2] *SRON-Utrecht, Sorbonnelaan 2, 3584 CA Utrecht, The Netherlands*
[3] *University of New Hampshire, IEOS, Durham NH 03824, USA*
[4] *Astrophysics Division, SSD/ESA, NL-2200 AG Noordwijk, The Netherlands*

Abstract. The EGRET experiment aboard the Compton Gamma-Ray Observatory (CGRO) has observed at energies above 100 MeV strong gamma-ray flares with short-term time variability from the gamma-ray blazars 3C 279 [1] and PKS 1622-297 [2]. During these flaring periods both blazars have been detected by the COMPTEL experiment aboard CGRO at photon energies of $\sim$10 MeV, revealing simultaneous γ-ray activity down to these energies. For both cases the derived fluxes exceed those measured in previous observations, and 3C 279 shows an indication for time variability within the observational period. Both sources show evidence for 'hard' MeV spectra. In general the behaviour of both sources at γ-ray energies is found to be quite similar supporting the conclusion that the underlying physical mechanism for both γ-ray flares might be the same.

INTRODUCTION

The EGRET experiment aboard CGRO has detected more than 60 blazar-type AGN [3] thereby greatly widening the field of extragalactic γ-ray astronomy. Most of them are observed to be time variable and several sources showed remarkable flares. During the last two years the two most intense flares along the whole EGRET mission have been observed from the sources 3C 279 [1] and PKS 1622-297 [2], which occured on top of an already high γ-ray flux level.

The COMPTEL experiment [4], measuring 0.75-30 MeV γ-rays, has detected 8 of these $\sim$60 EGRET blazars [5]. Among them are 3C 279 and PKS 1622-297. In this paper we report first results on these sources for the time periods for which these flares have been observed by EGRET at energies above 100 MeV.

OBSERVATIONS AND DATA ANALYSIS

The γ-ray flare events have been observed during a three week observation of the Virgo sky region from January 16 to February 6, 1996 for 3C 279 in CGRO Cycle 5 during the viewing periods (VPs) 511.0 and 511.5, and for PKS 1622-297 during a four week observation towards the Galactic Center region from June 6 to July 10, 1995 in CGRO Cycle 4 covering the VPs 421-423.5.

We have applied the standard COMPTEL maximum-likelihood analysis method (e.g. [6]) to derive detection significances, fluxes, and flux errors of γ-ray sources in the four standard COMPTEL energy bands (0.75-1 MeV, 1-3 MeV, 3-10 MeV, 10-30 MeV), and a background modelling technique which eliminates in a first approximation source signatures but preserves the general background structure [7]. For the 10-30 MeV range the improved COMPTEL data cuts [8], increasing its' sensitivity, have been applied. To derive source fluxes, several sources located in the surrounding sky region (e.g. 3C 279 and 3C 273 in Virgo) have been simultaneously fitted in an iterative procedure, leading to a simultaneous determination of the fluxes of several potential sources and a background model which takes into account the presence of sources. For the analysis of PKS 1622-297 a diffuse emission model has been included in the fitting procedure as well.

RESULTS

3C 279

The blazar 3C 279 is detected by COMPTEL with a significance of $\sim 4\sigma$ during this observational period of three weeks on Virgo in CGRO Cycle 5 (Fig. 1). The observed flux level in the 10-30 MeV band is the highest ever detected. This is the first redetection of 3C 279 by COMPTEL for energies above (>3 MeV) since 1991, when the blazar showed another γ-ray flare observed simultaneously by EGRET and COMPTEL ([9,10]). Subdividing the three week period into the individual VPs 511.0 (two weeks) and 511.5 (one week) reveals evidence for a flux jump by roughly a factor of 4 (Fig. 2) within 10 days. The flux value of 3C 279 measured in VP 511.5 is the largest ever observed by COMPTEL. The significance that the two fluxes are different is 2.6σ, and represents the shortest time varibility yet observed by COMPTEL from any blazar. This rise in flux is consistent with the EGRET observations at energies above 100 MeV. During VP 511.0 3C 279 was redetected by EGRET at a high flux level and rose up to the largest value ever in VP 511.5 [1].

The spectral analysis shows positive evidence for the source only at energies above 3 MeV. Together with the upper limits derived at the lower energy bands, this indicates a 'hard' (photon spectral index $\alpha < 2$) energy spectrum at MeV energies (Fig. 3).

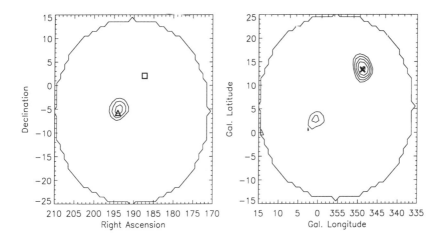

FIGURE 1. COMPTEL 10-30 MeV skymaps (detection significances) of the Virgo region (left) and the galactic center (right) for the relevant time periods. The contour lines start at a detection significance of 3.0σ with a step of 0.5σ assuming χ_1^2-statistics for known sources. The locations of 3C 279 (Δ), 3C 273 ($\square$), and PKS 1622-297 (**X**) are indicated.

PKS 1622-297

The blazar PKS 1622-297 is detected with a significance of $\sim 5\sigma$ during the four week pointing towards the Galactic Center in CGRO Cycle 4 (Fig. 1). This is the first detection of this blazar by COMPTEL as was the case for EGRET [2]. The COMPTEL light curve, even though the flux variations are not statistically significant, follows the general trend reported by EGRET at energies above 100 MeV. There is evidence for the source during all four individual VPs showing MeV flaring of PKS 1622-297 for at least one month (Fig. 2). The largest flux value is observed during VP 423.0 consistent with the time period of the major flare observed by EGRET. A flux drop by a factor of ~ 2.5 between the two last VPs, again consistent in trend with EGRET, is a hint for MeV variability as well.

The spectral analysis shows that the source is mainly detected in the highest COMPTEL energy band (10-30 MeV), which, together with the upper limits derived at lower energies, indicates a 'hard' (photon spectral index $\alpha < 2$) spectrum on average (Fig. 3). However, we like to point out that the source is located just above the plane near the galactic center region, which is a difficult region for quantitative analysis, especially for energies below 3 MeV, due to the diffuse MeV emission of the Galaxy. Although a diffuse emission model was included in the analysis procedure, the presented spectral results should be considered as preliminary.

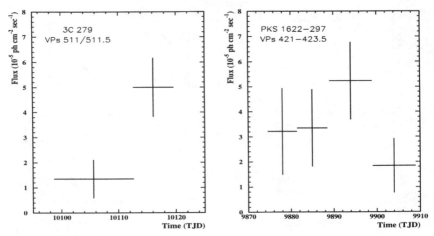

FIGURE 2. Light curves in the 10-30 MeV range of 3C 279 (left) and PKS 1622-297 (right). The fluxes are given for the individual CGRO VPs. The error bars are 1σ. The flux change of 3C 279 by a factor of 3.7 within 10 days is the shortest time variability yet observed from any blazar by COMPTEL. The fluxes measured during these flaring intervals are the largest ever observed from any blazar by COMPTEL in this energy band.

SUMMARY

We have reported first results of COMPTEL observations of two blazars, 3C 279 and PKS 1622-297, for time periods for which EGRET (>100 MeV) observed the two strongest γ-ray flares ever occuring on top of an already high flux level. Although there are differences in detail, the general MeV behaviour of both sources resembles each other surprisingly accurately. Both sources are detected during these periods of high γ-ray activity at high γ-ray energies, which by itself demonstrates simultaneous MeV-flaring activity. Note, that PKS 1622-297 is detected for the first time by COMPTEL. Both sources are only detected at the highest COMPTEL energies. This, together with the upper limits derived at the lower energies, leads to evidence for hard (photon spectral index $\alpha < 2$) MeV spectra. The 10-30 MeV flux follows for both sources the general flux trend as seen by EGRET with evidence for time variability in the case of 3C 279 and a hint in the case of PKS 1622-297. These similarities support the conclusion that the underlying physical mechanism for γ-ray activity is the same for both sources.

Detailed COMPTEL analyses, concentrating on subsets of these observations are in progress to derive informations on possible time-shifts between the high-energy EGRET and the low-energy COMPTEL γ-ray emission. Especially PKS 1622-297 is a promising candidate for such investigations because

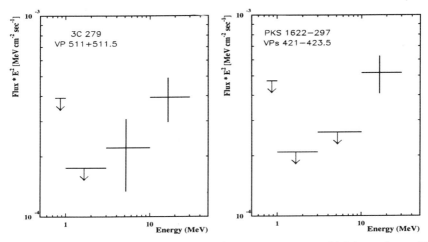

FIGURE 3. Energy spectra of 3C 279 (left) and PKS 1622-297 (right) are shown. The spectra, averaged over the whole 3- and 4 week observations of both blazars, are presented in a differential flux × E^2 representation. The error bars are 1σ and the upper limits are 2σ. Both sources are mainly detected at the upper COMPTEL energy bands, which, together with the upper limits at lower energies, indicates a 'hard' MeV spectrum for both cases.

the observation covers a time period of high γ-ray activity in which a flux spike is observed. For 3C 279 the observations are only available at the rising part of the flare. It remains unclear whether the top of the MeV emission is covered by COMPTEL.

REFERENCES

1. Wehrle A., Pian E., Maraschi L., et al., 1997, *ApJ* submitted
2. Mattox J., Wagner S.J., Malkan M., et al., 1997, *ApJ* **476**, 692
3. Kanbach G., 1996, Proc. of the 'Workshop on Gamma-Ray Emitting AGN'; eds. J.G. Kirk et al., Heidelberg, Germany, p. 1
4. Schönfelder V., Aarts H., Bennett K., et al., 1993, *ApJS* **86**, 657
5. Collmar W., 1996, Proc. of the 'Workshop on Gamma-Ray Emitting AGN'; eds. J.G. Kirk et al., Heidelberg, Germany, p. 9
6. de Boer H., Bennett K., Bloemen H., et al., 1992, In: *Data Analysis in Astronomy IV*, eds. V. Di Gesu et al., New York, USA, p. 241
7. Bloemen H., Hermsen W., Swanenburg B.N., et al., 1994, *ApJS* **92**, 419
8. Collmar W., Wessolowski U., Schönfelder V., et al., 1997, these Proceedings
9. Kniffen D.A., Bertsch D.L., Fichtel C.E., et al., 1993, *ApJ* **411**, 133
10. Williams O.R., Bennett K., Bloemen H., et al., 1995, *A&A* **298**, 33

EGRET Observations of PKS 0528+134 from 1991 to 1997

R. Mukherjee,[a] D. L. Bertsch,[b] S. D. Bloom,[b] B. L. Dingus,[c] J.
A. Esposito,[b] R. C. Hartman,[b] S. D. Hunter,[b] G. Kanbach,[d] D.
A. Kniffen,[e] A. Kraus,[f] T. P. Krichbaum,[f] Y. C. Lin,[g] W. A.
Mahoney,[h] A. P. Marscher,[i] H. A. Mayer-Hasselwander,[d] P. F.
Michelson,[g] C. von Montigny,[j] A. Mücke,[d] P. L. Nolan,[g] M.
Pohl,[k] O. Reimer,[d] E. Schneid,[l] P. Sreekumar,[b] H. Teräsranta,[m]
D. J. Thompson,[b] M. Tornikoski,[m] E. Valtaoja,[m] S. Wagner,[j] A.
Witzel[f]

a McGill University, Physics, Montreal, H3A 2T8, Canada[1]
b NASA/GSFC, Code 661, Greenbelt, MD 20771
c Dept. of Physics, University of Utah, Salt Lake City, UT 84112
d Max-Planck-Institute für Extraterrestrische Physik, D-85748 Garching, Germany
e Hampden-Sydney College, P. O. Box 862, Hampden-Sydney VA 23943
f Max-Planck-Institute für Radioastronomie, D-53121 Bonn, Germany
g W. W. Hansen Expt. Physics Lab., Stanford Univ., Stanford, CA 94305-4085
h JPL, California Institute of Technology, Pasadena, CA 91109
i Dept. of Astronomy, Boston University, Boston, MA 02215
j Landessternwarte-Koenigstuhl 69117 Heidelberg, Germany
k Danish Space Research Institute, 2100 Copenhagen O, Denmark
l Northrop Grumman Corporation, Mail Stop A01-26, Bethpage, NY 11714
m Metsahovi Radio Research Station, SF-02540 Kylmala, Finland

Abstract.
The compact, radio-loud quasar PKS 0528+134 has been one of the brightest active galactic nuclei (AGN) detected by EGRET. The flux history of the source including data up to March 1997, as well as the spectra measured during the flaring and non-flaring states is presented. Multiwavelength observations of PKS 0528+134 are also presented. A detailed relativistic SSC jet model agrees well with the multiwavelength spectrum; the data, however are insufficient to discriminate between this and other emission models.

---

[1] USRA Research Scientist, NASA/GSFC, Code 610.3

INTRODUCTION

EGRET (Energetic Gamma Ray Experiment Telescope) detected high energy γ-rays from PKS 0528+134 for the first time during the early pointings of CGRO from April to June, 1991 [1]. Since then PKS 0528+134 has been observed several times and is one of the brightest active galactic nuclei (AGN) detected by EGRET. It is located near the Galactic anticenter at $l = 191.°37$, $b = -11.°01$ ($\alpha = 5^h28^m6^s.8$, $\delta = +13°29'42''$ J2000), has a redshift of $z = 2.07$ and a mean optical brightness of $m_v = 20$. In this article we report on the Phase 1 to Cycle 6 observations of PKS 0528+134 made by EGRET.

FLUX HISTORY

Table 1 lists the dates of observation and viewing periods during which PKS 0528+134 was within 30° of the EGRET instrument axis. The integrated flux above 100 MeV, as well as the significance of detection of PKS 0528+134 in each VP is listed in Table 1. The following consecutive viewing periods with similar pointing directions were added together in order to improve the significance of detection: 0.2, 0.3, 0.4, and 0.5; 36.0 and 36.5; 321.1 and 321.5.

Fig. 1 shows the history of the γ-ray flux of PKS 0528+134 above 100 MeV for all the EGRET observations to date. The highest flux from the source was detected in 1993, March 23 – 29, when the γ-ray flux was comparable to that of the Crab and nearly a factor of 3 higher than the quiescent flux. A χ^2 test of the data indicates a probability $P < 10^{-15}$ that the flux variations are consistent with a constant flux. Defining the variability index of a blazar as $V = -\log P$ [2], we get a value of $V > 15$ for PKS 0528+134. Comparing

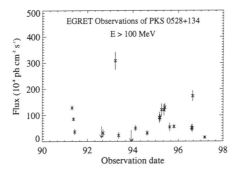

FIGURE 1. Flux of γ-rays above 100 MeV from PKS 0528+134 over the period 1991 April to 1997 March (Phases 1 through Cycle 6). 2σ upper limits are shown as downward arrows.

TABLE 1. EGRET observations of PKS 0528+134 from 1991 to 1997

Viewing Period	Observation Dates	Flux $\times 10^{-7}$ ph cm^{-2} s^{-1}	Significance σ	Inclination Angle of Source
0.2-0.5	1991 Apr 22 – May 07	12.9 ± 0.9	20.3	8.°0
1.0	1991 May 16 – 30	8.5 ± 0.8	13.5	6.3
2.1	1991 Jun 08 – 15	3.6 ± 1.1	3.8	5.1
36.0+36.5	1992 Aug 11 – 20	< 5.5^a	-	22.0
39.0	1992 Sep 01 – 17	3.2 ± 1.4	2.6	23.9
213.0	1993 Mar 23 – 29	30.8 ± 3.5	13.6	9.1
221.0	1993 May 13 – 24	2.3 ± 1.2	2.2	6.4
310.0	1993 Dec 01 – 13	< 4.0^a	-	15.7
321.1+321.5	1994 Feb 08 – 17	4.9 ± 1.2	5.0	12.9
337.0	1994 Aug 09 – 29	3.2 ± 1.0	3.6	13.5
412.0	1995 Feb 28 – Mar 07	9.1 ± 2.1	5.8	13.1
413.0	1995 Mar 07 – 21	9.0 ± 1.3	9.3	7.7
419.1	1995 Apr 04 – 11	12.1 ± 2.4	6.6	17.4
419.5	1995 May 09 – 23	12.0 ± 2.2	7.5	20.9
420.0	1995 May 23 – Jun 06	13.0 ± 1.6	11.2	9.8
426.0	1995 Aug 08 – 22	5.5 ± 1.6	4.0	8.5
502.0	1995 Oct 17 – 31	5.7 ± 0.8	8.5	0.9
526.0	1996 Jul 30 – Aug 13	5.2 ± 1.2	5.4	8.3
527.0	1996 Aug 13 – 20	5.1 ± 1.6	4.3	9.3
528.0	1996 Aug 20 – 27	17.3 ± 2.1	11.7	12.1
616.1	1997 Feb 18 – Mar 18	1.1 ± 0.5	2.7	0.0

a: 2σ upper limit

this value of V to that of the other blazars detected by EGRET [3], we note that PKS 0528+134 is one of the four most variable blazars with one of the highest weighted average fluxes, ever to be detected by EGRET. In fact PKS 0528+134 is one of the few blazars seen by EGRET that has flared regularly. Strong variations in flux were observed in 1991, 1993, 1995, and 1996. During the recent Cycle 6 observations (Feb-Mar 1997), the source was in one of its lowest states.

SPECTRUM

The background-subtracted γ-ray spectrum of PKS 0528+134 was determined by dividing the energy range 30 MeV - 10 GeV into 10 intervals, and estimating the number of source photons in each of the bins. The data were fit to a single power-law model using $F(E) = k(E/E_0)^{-\alpha}$ ph cm^{-2} s^{-1} MeV^{-1}, where where $F(E)$ is the flux, E is the energy, α is the photon spectral index, k is the coefficient, and E_0 is the energy normalization factor. The spectrum of the source was determined for those viewing periods where the source was detected at $> 6\sigma$ in the > 100 MeV analysis. Table 2 shows the results of the

TABLE 2. EGRET spectral analysis for PKS 0528+134

Viewing Period	Spectral Index (α)	$k \times 10^{-9}$ ph cm^{-2} s^{-1} MeV^{-1}	E_0 MeV	χ^2/n_f
0.2-0.5	2.27 ± 0.07	3.26 ± 0.19	199	1.53
1.0	2.44 ± 0.14	2.29 ± 0.20	171	0.83
213.0	2.21 ± 0.11	8.24 ± 0.76	200	0.78
413.0	2.21 ± 0.16	0.92 ± 0.15	255	1.07
419.5	2.43 ± 0.21	2.87 ± 0.51	189	0.84
420.0	2.37 ± 0.13	2.48 ± 0.29	206	1.19
502.0	2.32 ± 0.16	1.16 ± 0.17	235	0.55
528.0	2.44 ± 0.37	2.73 ± 0.76	159	0.19
Sum	2.56 ± 0.17	1.87 ± 0.20	150	0.56

spectral analysis for PKS 0528+134 in the different viewing periods. "Sum" corresponds to the sum of viewing periods 2.1, 36.0, 36.5, 39.0, 310.0, 321.1, 321.5, and 337.0, when the source was weak.

The power per log frequency interval emitted by PKS 0528+134 is shown in Fig. 3 for three different epochs. Included in the figure are observations made during and around the flare in 1993 March, in 1992 September (when PKS 0528+134 was in a low γ-ray state), as well as some preliminary data obtained during the recent campaign in 1997 February. There was no evidence for intra-day variability in the optical/IR observations during the 1997 February campaign.

DISCUSSION AND CONCLUSIONS

EGRET results [6,3] have shown that, in the radio to γ-ray multiwavelength spectra of blazars, the power in the γ-ray range equals or exceeds the power in the infrared-optical band. The high γ-ray luminosity of the blazars suggests that the emission is likely to be beamed and, therefore, Doppler boosted into the line of sight. The models that are used to explain the emission mechanism in blazars are the leptonic jet models: (a) SSC or Synchrotron self-Compton model [7], (b) ERC or external radiation Compton model [8,9] and the hadronic jet model: (c) PIC or proton-initiated-cascade model [10]. The 1993 March data for PKS 0528+134 were fit to the relativistic jet SSC code of Marscher and Travis [5], which is shown as a solid curve in Fig. 3 [4]. While the fit demonstrates that a jet emitting SSC radiation is a viable model for explaining the general characteristics of the multiwaveband emission from PKS 0528+134, it is not possible to discriminate between this and other emission models based on the spectrum alone. For instance, the low-state data of PKS 0528+134 (1994 August) was fit quite well with the ERC model [11]. Similarly, the SSC, ERC, and PIC models have all been shown to fit the

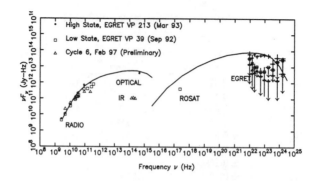

FIGURE 2. Multiwavelength spectrum of PKS 0528+134 during three different epochs. The solid line is a model fit synchrotron self-Compton spectrum from a relativistic jet to the 1993 data. 2σ upper limits are shown as downward arrows.

multiwavelength spectrum of 3C 273 rather well [12], and both the SSC and ERC models have yielded reasonable fits to the spectrum of 3C 279 [13]. The 1997 February data are preliminary, and analysis is currently under way.

In conclusion, the limited data that we have on PKS 0528+134 prevent us from being able to distinguish between the different theoretical models on the basis of the spectra alone. A future high-energy γ-ray mission should play a key role in resolving the physics of these powerful objects.

R. Mukherjee acknowledges support from NASA Grant NAG5-3696 and would also like to thank Prof. D. Hanna and the High Energy Physics group at McGill University for their hospitality.

REFERENCES

1. Hunter, S. D., et al., *ApJ* **409**, 134 (1993).
2. McLaughlin, M. A., et al., *ApJ* **473**, 763 (1996).
3. Mukherjee, R., et al., *ApJ* accepted **490** (1997).
4. Mukherjee, R., et al., *ApJ* **470**, 831 (1996).
5. Marscher, A. P. & Travis, J. P., *A&AS* **120**, 537 (1996).
6. von Montigny, C., et al., *ApJ* **440**, 525 (1995).
7. Bloom, S. D., & Marscher, A. P., *ApJ* **461**, 657 (1996).
8. Dermer, C. D., & Schlickeiser, R., *ApJS* **90**, 945 (1994).
9. Ghisellini, G., & Madau, P., *MNRAS* **280**, 67 (1996).
10. Mannheim, K., & Biermann, P. L., *A&A* **53**, L21 (1992).
11. Sambruna, R. M., et al., *ApJ* **474**, 639 (1997).
12. von Montigny, C., et al., *ApJ* **483**, 161 (1997).
13. Hartman, R. C., et al., *ApJ* **461**, 698 (1996).

Imaging Analysis of PKS0528+134 During Its Flare with A Direct Demodulation Technique

S.Zhang, T.P.Li, M.Wu, W.Yu

*High Energy Astrophysics Lab.,Institute of High Energy Physics,
P.O.BOX 918-3,100039,Beijing,China*

Abstract. The quasar PKS0528+134 was observed to flare by EGRET/CGRO during the viewing period 213 (VP213). COMPTEL data from VP213 have been reanalyzed by using the direct demodulation method. Images of the source are obtained at energies above 3 MeV and the time variability in the energy range 10-30 MeV is consistent with the result of EGRET. There is no spectral break at energies above 3 MeV during the flare.

INTRODUCTION

The quasar PKS0528+134 was first detected at γ-ray energies by EGRET as one of the brightest active galactic nuclei (AGN)[1]. Along with the sources Crab and Geminga, it located near Galactic anticenter at l=191.37°, b=-11.01°. The redshift of this quasar is about 2.07. Following the detection of PKS0528+134 by EGRET, Collmar[2,3] analyzed the COMPTEL data and discovered the source at low energy γ-ray. They obtained some preliminary results from data of the viewing periods 0 and 1 (VP0 and VP1): The source was detected mainly at energies above 10 MeV; The detections and non-detections during different viewing periods follow the trend seen by EGRET; A spectral break exists in the upper part of the COMPTEL energy range at energies between 10 and 30 MeV (for VP0 and VP1). PKS0528+134 was seen to flare during viewing period 213 (VP213) by EGRET and γ-rays ($\geq$ 30 MeV) were detected at a level approximately three times greater than the observed intensity in earlier observations[4]. However, comparison of the EGRET and COMPTEL results from the data of VP213 has not been reported because of the poor statistics at energies below 30 MeV. Considering the direct demodulation method[5-9] owns the ability of reconstructing objects from incomplete and noisy data, we have reanalyzed the COMPTEL data from VP213 using this method. We report in this paper the main results of PKS0528+134 observed by COMPTEL during VP213.

OBSERVATION AND ANALYSIS METHOD

From the beginning of the all-sky survey of COMPTEL PKS0528+134 has been in the field of view several times. During the observation of March 23 – 29,1993 (VP213) the source was located about 9.0° to the COMPTEL pointing direction. We used a direct demodulation method to process this observation data.

The direct demodulation method has already been successfully applied to scan observation data of slat collimator telescopes, e.g. the scan imaging for Cyg X-1 by the balloon-borne hard X-ray collimated telescope HEAP-4[10] and the reanalyzing result for EXOSAT-ME Galactic plane survey[11]. It can also be applied to analyze observation data from rotating modulation telescope[12], coded-mask aperture telescopes[8] and other types (e.g. COMPTEL) of telescope. The process of direct demodulation contains two steps: First, the continual intensity distribution of background are produced by iteratively solving the modulation equation under continual constraints. Second, the intensity of object sky can be obtained by solving the modulation equation again under constraints of the produced continual background. After the subtraction of the continual background, the real intensity distribution of the sources in the object sky and the uncontinuous portion of the observed background are left for the final reporting. Gauss-Seidel iteration method is adopted in this paper and the error estimations are derived by the bootstrap technique.

RESULTS

In the four energy ranges of COMPTEL VP213, images of PKS0528+134 are obtained only at the energies above 3 MeV (Figure 1(a) and 1(b)). The nebula and pulsar Crab and the quasar PKS0528+134 are resolved exactly in these two skymaps. The flux of Crab is $(3.57\pm0.68)\times10^{-4}$ ph cm^{-2} s^{-1} and $(9.57\pm2.57)\times10^{-5}$ ph cm^{-2} s^{-1} in energy ranges 3-10 and 10-30 MeV. Flux of the PKS0528+134, with significances of about 2.79σ and 2.54σ in energy ranges 3-10 and 10-30 MeV respectively, is weaker than that of the Crab (see Table 1). The excesses of the structure in the skymaps may reflect the uncontinuous portion of the background in the observed data.

In Table 1, results of VP0 and VP1 for COMPTEL are from Collmar[3]. Results of VP0,VP1 and VP213 for EGRET are from Mukherjee[13]. Results with symbol(*) are obtained with the direct demodulation method. Table 2 shows the time variability of PKS0528+134 observed by COMPTEL[2,3] and by EGRET[13] covering three viewing periods. The intensity ratios of low

TABLE 1. Flux of PKS0528+134

VP	EGRET $\geq$ 30 MeV Flux (10^{-7} ph cm^{-2} s^{-1})	COMPTEL 10-30 MeV Flux (10^{-5} ph cm^{-2} s^{-1})	COMPTEL 3-10 MeV Flux (10^{-5} ph cm^{-2} s^{-1})
0	12.9±0.9	3.2±1.2	4.0±2.8
1	8.5±0.8	2±1.2 (1.95±1.56*)	≤ 4.3
213	30.8±3.5	7.71±3.03*	16.6±5.95*

TABLE 2. Intensity ratios of PKS0528+134

Intensity ratio	EGRET $\geq$ 30 MeV	COMPTEL 10-30 MeV	COMPTEL 3-10 MeV
VP1/VP0	0.66±0.08	0.63±0.44	≤ 1.08
VP213/VP0	2.39±0.32	2.41±1.31	4.15±3.26
VP213/VP1	3.62±0.54	3.86±2.76	≥ 3.86

state to low state (VP1/VP0) and flare state to low states (VP213/VP0 and VP213/VP1) observed by COMPTEL in energy range 10-30 MeV are consistent with that observed by EGRET at energies above 30 MeV. However, this consistence is hard to be seen in the lower energy range 3-10 MeV.

The COMPTEL spectral results together with the simultaneously measured EGRET spectrum of PKS0528+134[13] during VP1 and VP213 are shown in Figure 2(a) and Figure 2(b). The COMPTEL spectrum of the source during VP1 is from Collmar[3]. The spectral break in energy range 10-30 MeV observed during VP1 of the low state (see Figure 2(a)) does not appear in the observed spectrum during VP213 of the flare state (see Figure 2(b)) even down to the energy of 3 MeV.

CONCLUSION

The quasar PKS0528+134 is weaker than the Crab at energies below 30 MeV. The time variability observed by COMPTEL in energy range 10-30 MeV is similar to the results of EGRET. The common feature of the spectral break for the blazar-type quasar detected by COMPTEL[2,3] does not appear for PKS0528+134 at least at energies above 3 MeV during its flare.

ACKNOWLEDGEMENTS

The authors are grateful to Dr. W.Collmar and Dr. Y.C.Lin for helpful discussions and data assistance.

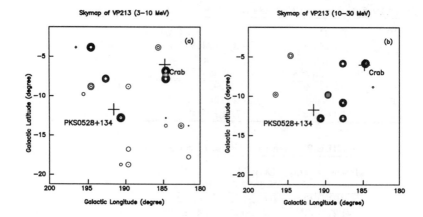

FIGURE 1. The direct demodulation maps of COMPTEL data from VP213. Contours start from f_1 with steps increment factor k from the first step Δc and the contour f_n is taken as $((1-k^{n-1})/(1-k))\Delta c + f_1$, where k,$\Delta c$,$f_1$ are 1.2, 9×10^{-7} ph cm^{-2} s^{-1}, 1.1×10^{-6} ph cm^{-2} s^{-1} in Figure 1(a) and 1.2, 5×10^{-7} ph cm^{-2} s^{-1}, 7×10^{-7} ph cm^{-2} s^{-1} in Figure 1(b). Contours number n is 6 in both Figures.

REFERENCES

1. Hunter S.D., et al.,*ApJ* **409**, 134 (1993)
2. Collmar W., et al.,*Proc. AIP Conference* **304**,659 (1994)
3. Collmar W., et al., *Proc. of XXXII ICRC* **1**, 168 (1993)
4. Sreekamar P., et al.,*IAU Circ.* No.5753 (1993)
5. Li T.P. and Wu M., A direct method for spectral and image restoration, in: Worrall D.M., Biemesderter C. & Barnes J., eds., *Astronomical Data Analysis Software and System I.*, A.S.P. Conf. Ser. **25**, 229(1992)
6. Li T.P. and Wu M., *ApSS* **206** 91 (1993)
7. Li T.P. and Wu M., *ApSS* **215**, 213 (1994)
8. Li T.P., *Exper. Astron.* **6**, 63 (1995)
9. Li T.P., in: Hu W.R., eds., *Space Science in China* 25 (1996)
10. Lu Z.G., et al., *Nucl. Instr. and Meth. in Phys. Res. Sec. A* **362**, 551 (1995)
11. Lu F.J, Li T.P., Sun X.J., Wu M. and Page C.G., *A&AS* **115**, 395(1996)
12. Chen Y., Li T.P. and Wu M.,Direct Demodulation technique for Rotating Modulation Collimator Imaging, *Exper. Astron.*, (1996)(submitted)
13. Mukherjee R., et al., *ApJ* **470**, 831 (1996)

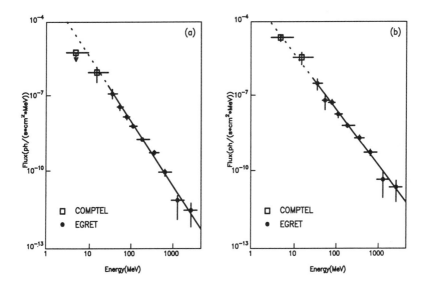

FIGURE 2. The spectra of PKS0528+134 observed by COMPTEL and EGRET during VP1 (Figure 2(a)) and VP213 (Figure 2(b)).

First Results of an All-sky Search for MeV-emission from Active Galaxies with COMPTEL

J. G. Stacy[*], J. M. Ryan[*], W. Collmar[†], V. Schönfelder[†], H. Steinle[†], A. W. Strong[†], H. Bloemen[‡], J. J. Blom[‡], W. Hermsen[‡], O. R. Williams[§], and M. Maisack[¶]

[*]*Space Science Center, University of New Hampshire, USA*
[†]*Max Planck Institute for Extraterrestrial Physics, Garching, Germany*
[‡]*SRON-Utrecht, Utrecht, The Netherlands*
[§]*Astrophysics Division, ESA/ESTEC, Noordwijk, The Netherlands*
[¶]*University of Tübingen, Germany*

Abstract.

We present first results of a systematic search through COMPTEL all-sky maps for evidence of MeV gamma-ray emission from active galaxies. All-sky maximum-likelihood and flux maps have been produced from publicly-available COMPTEL datasets for individual CGRO viewing periods for the 4.5-year period covering Phases 1 to 4 of the CGRO mission (1991-1995), in four standard energy bins spanning the sensitive range of COMPTEL (0.8-30 MeV). From these all-sky maps we have extracted statistical likelihoods of source emission and corresponding fluxes or upper limits for ~150 candidate objects (primarily active galaxies) lying at high Galactic latitudes $|b| > 10°$. Included in these results is a compilation of COMPTEL fluxes and/or limits for the unidentified high-latitude sources of the the Second EGRET Catalog. Our scientific objective is to derive the global properties of high-energy AGN in the medium-energy gamma-ray regime, and to provide cumulative fluxes or limits in the COMPTEL energy range for all high-energy AGN of interest. We describe our mapping and flux-extraction procedure and present a first sample of the results obtained in the form of representative figures. These cumulative results will ultimately be compared with studies of AGN for individual viewing periods, and used in statistical investigations of the various sub-classes of active galaxies.

SCIENTIFIC MOTIVATION

The Imaging Compton Telescope (COMPTEL) aboard the *Compton Gamma Ray Observatory* (CGRO) is sensitive to medium-energy gamma radiation from 0.8 to 30 MeV. Measurements with COMPTEL effectively bridge the gap between the hard x-ray and the high-energy gamma-ray regimes. As a wide-field, imaging instrument COMPTEL has carried out the first comprehensive survey of the sky at MeV-energies [1,2]. The medium-energy gamma-ray portion of the spectrum is known to be of prime importance in the study of the broadband properties of a number of classes of astrophysical sources. For active galactic nuclei (AGN), in particular, the power per natural logarithmic frequency interval (νF_ν) is known to peak in the MeV region of the spectrum. Spectral breaks are typically required to join observations spanning several decades in energy around the MeV gamma-ray band [3]. Clearly, medium-energy observations carried out with COMPTEL will play a significant role in constraining the many theoretical models proposed to explain high-energy emission from a variety of astrophysical sources [2].

In the present study we have carried out a systematic search through COMPTEL composite all-sky maps covering the first four phases of the CGRO mission (1991-1995) for evidence of MeV gamma-ray emission from known or suspected gamma-ray sources, particularly those detected in neighboring energy bands by the CGRO/EGRET and OSSE instruments. All-sky maximum-likelihood and flux maps have been produced from standard-processing COMPTEL datasets for individual CGRO viewing periods, in four standard energy bins spanning the sensitive range of COMPTEL (0.8-30 MeV). From these all-sky maps we have extracted statistical likelihoods of source emission and corresponding COMPTEL fluxes or upper limits for an extensive list of target objects, including the unidentified sources of the most recent EGRET catalog. Our scientific objective is to derive the global properties of gamma-ray sources in the medium-energy gamma-ray regime, and to provide cumulative fluxes or limits in the COMPTEL energy range for all high-latitude sources of general interest.

COMPTEL COMPOSITE ALL-SKY MAPS

We have developed IDL-based software routines to produce composite COMPTEL all-sky maps, using as input standard-processing COMPTEL maximum-likelihood skymaps (MLMs) by CGRO viewing period. A composite COMPTEL all-sky map is constructed in the following manner: After user specification of an output coordinate system (either Galactic, or right ascension and declination (B1950 or J2000)) and projection option ("flat" rectangular, Aitoff, or north or south polar), a reference list of individual COMPTEL MLMs by energy band and viewing period is read, and individual

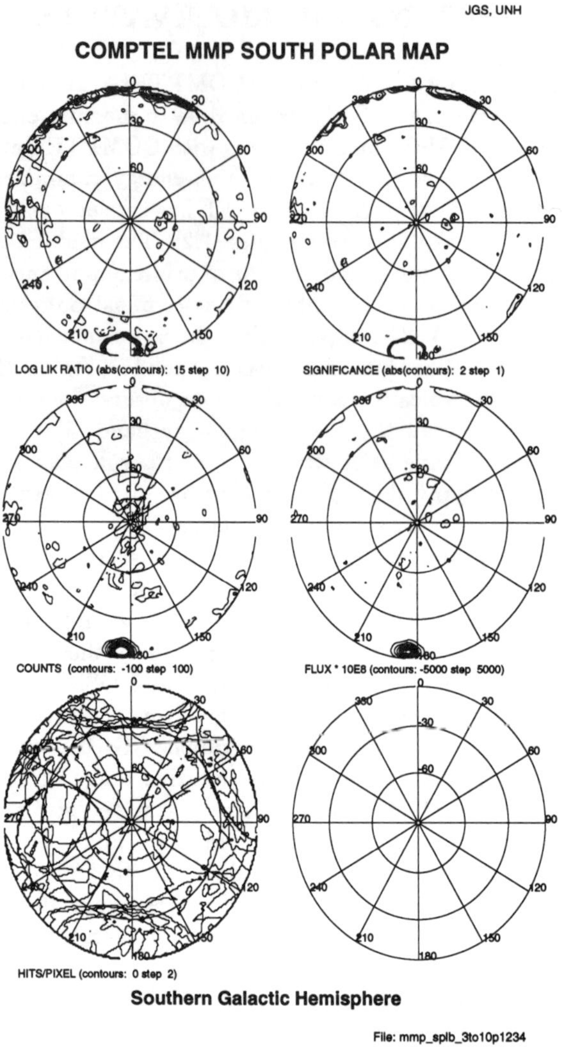

FIGURE 1. Sample layers of a composite COMPTEL "all-sky" map (in this case a polar projection covering the southern Galactic hemisphere) combining data for CGRO Phases 1 through 4 over the energy range 3 to 10 MeV. Top left: sum of the log-likelihood ratios of the 109 input maps; top right: significance (in σ) of the log-likelihood values, assuming χ^2_{N+2} statistics, where N equals the number of input maps contributing to a given pixel of the composite map; middle left: counts; middle right: flux ($\times 10^8$ ph cm^{-2} s^{-1}); bottom left: "hits/pixel" (i.e., number of input maps contributing to a given pixel); bottom right: reference grid in Galactic coordinates. The prominent source at the bottom of individual maps is the Crab pulsar.

MLM files are retrieved in succession from archive. For each MLM dataset the coordinate system of the input map is determined, and its region of overlap calculated with respect to the output all-sky map. Each input MLM map is then processed, pixel by pixel, with weighted values accumulated and assigned to the appropriate pixel in the output map, adopting a bicubic spline interpolation. Once the last input MLM map is processed, a final pass is made through the array containing the output composite map, and final weighted-mean counts, fluxes, and errors are computed.

Output composite COMPTEL skymaps (termed "MMP" datasets) are written as multilayer FITS files. The layers of the output skymap file correspond essentially to those of the input MLM files (i.e., log-likelihood ratio, counts, counts error, flux, flux error), with an additional layer corresponding to a measure of the significance of the summed log-likelihood ratio, based on the number of input maps contributing to a given output pixel, and with the number of "hits per pixel" also stored as a separate layer of the output map (see Figures 1 and 2).

A separate suite of IDL software tools has also been developed to extract statistical likelihood, source significance, and flux values from the COMPTEL composite all-sky maps described above, for user-specified input source lists. In this process, a user simply provides an input list of target objects by position. The flux-extraction routine then reads the composite all-sky maps of interest and computes average output values within one-pixel radius of the specified source positions. Recommended COMPTEL team-standard corrections (time-of-flight, livetime, etc. [4]) are applied to obtain final fluxes and significances of detection.

COMPTEL AGN SURVEY

Using the composite COMPTEL all-sky maps we have carried out a systematic search for sources of MeV gamma-ray emission at high Galactic latitudes. All-sky maps were produced in north and south Galactic polar projection for CGRO Phases 1 through 4, in the four standard energy ranges of COMPTEL (0.75–1, 1–3, 3–10, and 10–30 MeV). Statistical likelihoods, significances of potential source detections, and associated fluxes, errors, or upper limits, have been derived for an extensive target list of ∼150 objects at $|b| > 10°$, consisting of AGN and other known but unidentified high-latitude gamma-ray sources [5].

At present, we continue to assess the nature of a number of marginally significant statistical excesses, both point-like and extended in appearance. A full compilation of these results for all of our candidate search objects will soon appear in the COMPTEL source catalog [6]. Ultimately, these cumulative results will be compared with source studies for individual CGRO viewing periods, and used in statistical investigations of source properties by class.

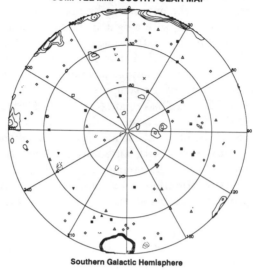

FIGURE 2. Enlarged view of the "significance" (or "confidence of detection," in σ) layer of the composite COMPTEL map covering the southern Galactic hemisphere combining data for CGRO Phases 1 through 4 over the energy range 3 to 10 MeV (see Figure 1). Symbols indicate the positions of known or candidate gamma-ray sources (at $|b| > 10°$).

The COMPTEL project is supported by the German government through DARA grant 50 QV 90968, by NASA under contract NAS 5-26646, and by the Netherlands Organization for Scientific Research (NWO).

REFERENCES

1. Schönfelder, V., et al., *ApJS.* **86**, 657 (1993).
2. Schönfelder, V., et al., *A&A Suppl. Ser.* **120**, C13 (1996).
3. Collmar, W., et al., *A&A.* in press (1997).
4. Diehl, R., *"COMPTEL Data Analysis Standards,"* COMPTEL Internal Report COM-MO-DRG-MGM-231.9 (3 March 1996).
5. Thompson, D. J., et al., *ApJS.* **101**, 259 (1995).
6. Schönfelder, V., et al., in preparation (1997).

Variability time scales in the gamma-ray blazars using structure function analysis

Giridhar Nandikotkur*, P. Sreekumar[†], D. A. Carter-Lewis*

Department of Physics and Astronomy
Iowa State University Ames, IA 50011
[†] *USRA,Goddard Space Flight Center, NASA Greenbelt, MD 20771*

Abstract. About 50 active galactic nuclei (AGN) have been detected by the Energetic Gamma Ray Experiment Telescope (EGRET) on board the Compton Gamma Ray Observatory since its launch in 1991. A majority of them are flat spectrum quasars and a few of them are BL Lacertae (BL Lac) objects. Some of the current models predict different time scales for variability in various regions of the electro-magnetic spectrum. Using structure function analysis, we are examining the variability time-scales in a few of the sources in the gamma-ray region. The existence of a maximum time-scale of correlation is being investigated in each source. The possible existence of multiple time-scales in single sources is also being explored. We present the preliminary results of an analysis of data on 3C279 in the 100 MeV-10000 MeV region.

INTRODUCTION

The EGRET instrument aboard the Compton Gamma Ray Observatory has reported the detection of more than 50 blazars from the cumulative observations since the launch in 1991 (von Montigny et.al. [1]; Mukherjee et.al. [2]). In general, the gamma-ray luminosity of these sources dominates that at other wavelengths (see [1]). The high-energy emission is often explained as arising from inverse Compton scattering of relativistic electrons either from the synchrotron photons (SSC models)(Maraschi, Ghisellini, & Celloti [3]; Bloom & Marscher [6]) or external soft photons from either the accretion disk or broad-line region (ERC models) (Dermer & Schlickeiser [4]; Sikora, Begelman & Rees [9]; Blandford & Levinson [5]). A majority of the gamma-ray blazars show evidence for time-variability, however, a select few have been observed numerous times and have clearly exhibited high and low states of emission. Assuming that the bulk of the observed variability is intrinsic to the source, it is generally believed that shocks propagating in these jets give rise to much of the observed intensity variations. Accurate estimates of the time-scales for these variations is crucial in distinguishing between these models.

The goal of this investigation is to examine using structure function analysis, the archival EGRET data on a small set of strong gamma-ray blazars, and to identify the time-scales for variability that are present in these sources.

STRUCTURE FUNCTION ANALYSIS

The first order SF is defined as (Simonetti, Cordes, & Heeschen [8]): $D_f^1(\tau) = <[f(t+\tau) - f(t)]^2>$ and is estimated using the expression

$$D_f^1(\tau) \equiv \frac{1}{N(\tau)} \sum_{t=1}^{N-\tau} (f(t+\tau) - f(t))^2 \qquad (1)$$

the average being taken over all measurements separated by lag τ and $N(\tau)$ being the number of such pairs. Here f(t) and f(t+τ) represent the flux at time t, and (t+τ) respectively.

For a stationary process, the structure function has a simple dependence on the more-widely-used autocorrelation function ($\rho(\tau)$) given by the expression: $D_f^1(\tau) = 2\sigma^2(1 - \rho(\tau))$. A plot of the structure function ($D_f^1(\tau)$) vs. the lag (τ) for an "ideal" process consists of a plateau at small lags, and a monotonic rise to an asymptotic value at large lags (see fig 1 in Hughes, Aller & Aller [7].) The lower plateau at shorter lags gives a measure of the fluctuations in the measurement noise ($2\sigma_{noise}^2$), while the plateau at large lags is equal to $2\sigma_{signal}^2$. The nature of the rise between the two plateaus yields information about the process that gives rise to the observed variations. The structure function at large lags is not a good representation of the sample because of the fewer number of pairs that contribute towards its computation. We use the following empirical rule is used while calculating the structure function: $N(\tau) \geq 30$ and $\tau \leq T/2$ where T is the duration of observation.

For the simplest system involving a single process (without periodicity), the value of the lag where the second plateau begins, gives the maximum time scale of correlation (T$_{max}$). A periodicity or pseudo-periodicity in the flux will show up as an oscillation in the structure function whereas a significant *global* rising trend in the dataset shows up as a sharp rise in the structure function at larger lags.

One of the principal motivations for choosing this method was its suitability to a typical EGRET data set which consists of a limited number of observations from the source and each observation providing limiting counting statistics. However, with a careful choice of the lag and the bin-size parameters, for a well-observed and bright EGRET source (e.g., 3C279), one can obtain more than 30 pairs for lags that are equal to half of the time period of observation. The method is not extremely sensitive to gaps in the data set which allows some flexibility in combining viewing periods that are very close to each other. Absolute knowledge of the diffuse background is not necessary because of the differencing property of the method, if we assume that the EGRET background is fairly constant over time (a reasonable assumption since the bulk of the diffuse emission comes from the cosmic-ray interactions in the inter-stellar medium.) Traditional time series methods can introduce a substantial bias in the analysis and have other problems of windowing, aliasing etc., and sometimes, one runs into the problems of infinite autocovariances. However, the structure function can still exist in such cases. The SF method is however, sensitive to extreme values of the data points (this is especially the case when the source is flaring). But robust methods of estimation can handle such situations where one assigns smaller weights to the extreme points.

The structure function of a large number of processes obeys a power law at small lags whose dependence can be expressed as $D_f^1(\tau) \simeq \tau^{2H}$. The parameter H is an indication of the nature of memory associated with the processes and the distribution of power across the frequencies. If $0 < H < \frac{1}{2}$, then the process has a short memory and if $\frac{1}{2} < H < 1$, then the process has a long memory. $H = \frac{1}{2}$ implies that the observations are uncorrelated. The slope of $\text{Log}(D_f^1(\tau))$-$\text{Log}(\tau)$ plot gives the value of 2H.

Thus the parameters of interest during the structure function analysis would be, the maximum time-scale of correlation and the slope in the $\log(D_f^1(\tau))$-$\log(\tau)$ space. These quantities can be compared from source to source or from one class of sources to the other.

RESULTS & CONCLUSION

We have performed the structure function analysis of the archival data for 3C279 which is one of the brightest blazars that has been detected by EGRET. The preliminary results for viewing period 3.0 (June 1991) are being presented. The plot of structure function ($D_f^1(\tau)$) vs. the lag (τ) has been examined for the existence of any structure different from that produced by a steady source. The structure function could indicate spurious features that arise from variations not associated with the source, such as those due to a non-uniform exposure. In an attempt to filter out such features, the structure function was evaluated at two background locations as well. Only features in the SF that are not reproduced in the SFs from background points were subjected to further analysis.

Figure 1 shows the variation of flux with time for viewing period 3.0 which was a 13 day observation period. The flux has been calculated at intervals of 0.2 days. The structure function plot is shown in figure 2. Two distinct time scales are seen. The first one is at two days and there is a second time scale at~ 4.5-5 days. Since it is a 13-day observation period, both the time scales are less than half of the time period of observation and are thus statistically

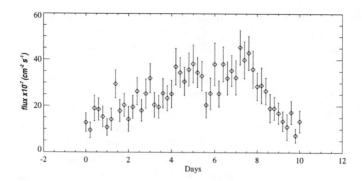

FIGURE 1. Flux Plot for the 3C279 for Viewing period 3.0

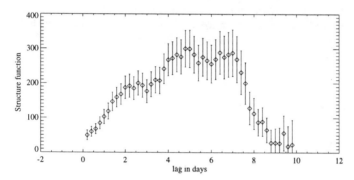

FIGURE 2. Structure Function Plot for the 3C279 for VP 3.0

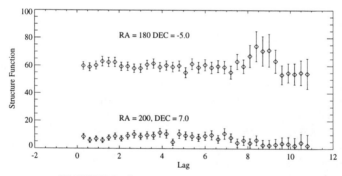

FIGURE 3. Structure Function Plot for two background points

significant. The existence of a third plateau might indicate the interplay of two different processes that give rise to the observed variability. The log-log plot of the structure function revealed a slope of 0.7 (H = 0.35) which indicates the presence of short-range effects. The structure functions of the background points used to study the effects not associated with the source, are shown in figure 3. The structure function values for the point RA=180, DEC=-5.0, have been arbitrarily shifted by 50 units for clarity. There is no structure in the background as indicated by the absolutely flat structure function.

Exhaustive analysis of the public archival data is being carried out to investigate the presence of structure in other viewing periods of observation. It might be early to conclude that the time scale of 4.5-5 days is the maximum time-scale of correlation of the source but it is certainly the case for viewing period 3.0.

REFERENCES

1. von Montigny et.al., *ApJ* **440**, 525 (1995).
2. Mukherjee et. el. , *ApJ* , In press (1997).
3. Maraschi, Ghisellini, & Celloti, *ApJ* **397**, L5 (1992).
4. Dermer & Schlickeiser , *ApJS* **90**, 945 (1994).
5. Blandford & Levinson, *ApJ* **441**, 79 (1995).
6. Bloom & Marscher, *ApJ* **461**, 657 (1996).
7. Hughes, Aller & Aller , *ApJ* **396**, 469 (1992).
8. Simonetti J. H., Cordes Jim, & Heeschen , *ApJ* **296**, 46 (1985).
9. Sikora, Begelman & Rees , *ApJ* **421**, (1994).

A spectral study of gamma-ray emitting AGN

M. Pohl[1,2], R.C. Hartman[3], P. Sreekumar[3], B.B. Jones[4]

1 MPI für Extraterrestrische Physik, Postfach 1603, 85740 Garching, Germany;
2 Danish Space Research Institute, Juliane Maries Vej 30, 2100 København Ø, Denmark;
3 LHEA, GSFC, Code 662, NASA, Greenbelt, MD 20771;
4 W.W. Hansen Laboratory, Dept. of Physics, Stanford University, Stanford, CA 94305

Abstract. In this paper we present a statistical analysis of the γ-ray spectra of flat-spectrum radio quasars (FSRQ) compared to those of BL Lacs. The average spectra and possible systematic deviations from power-law behaviour are investigated by summing up the intensity and the power-law fit statistic for both classes of objects. We also compare the time-averaged spectrum to that at the time of γ-ray outbursts.

The spectrum of the average AGN is softer than that of the extragalactic γ-ray background. It may be that BL Lacs, which on average have a harder spectrum than FSRQs, make up the bulk of the extragalactic background.

We also find apparent cut-offs at both low and high energies in the spectra of FSRQs at the time of γ-ray outbursts. While the cut-off at high energies may have something to do with opacity, the cut-off at low energies may be taken as indication that the γ-ray emission of FSRQs is not a one component spectrum.

I THE AVERAGE SOURCE

We have first analysed the summed EGRET data for Phases 1 to 3 (May 1991 to October 1994). At energies above 100 MeV 148 point sources with likelihood test statistic TS $= 2(\ln \lambda_0 - \ln \lambda)$ of at least 9, of which 44 are coincident with AGN known to emit γ-rays in the EGRET range. These 44 AGN divide into 33 FSRQ and 11 BL Lacs [1].

By adding the observed spectra of the AGN we obtain the intensity one would see as diffuse background if the sources were unresolved. This can be compared to the spectrum of the diffuse extragalactic γ-ray background [2,3]. It is surprising to see that the average γ-ray spectrum of all AGN is softer than that of the extragalactic γ-ray background by $\delta s = 0.17 \pm 0.045$. In Fig.1 we plot the ratio of the average AGN spectrum to that of the extragalactic background, which should be a constant if both were compatible. A Fischer-Snedecor test indicates that this ratio is better described by a linear relation

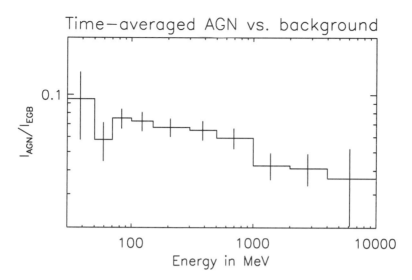

FIGURE 1. The ratio of the average intensity of all observed AGN to that of the extragalactic diffuse γ-ray background. This ratio is not compatible with a constant and thus the observed AGN would, if they were unresolved, give a softer diffuse emission than we observe in the background. This implies that the background can not be the superposition of unresolved AGN with the same characteristics as the observed objects.

than by a constant with 3.5 σ significance. If unresolved AGN are responsible for the extragalactic background, they can thus not have the same spectral characteristics as the resolved ones. We can also compare the average FSRQ spectrum to that of BL Lacs (see Fig.2). The average γ-ray spectrum of FSRQs is softer than that of BL Lacs by $\delta s = 0.12 \pm 0.08$. The difference in spectral index is not very significant which is mainly due to the large uncertainties in the average BL Lac spectrum. This does not imply that in single viewing periods BL Lacs have always harder spectra than FSRQ. In fact we see a remarkable spread of spectral indices for both classes of objects when individual viewing periods are considered [4]. But the spectrum of the average in both object classes appears to be different, and therefore they would contribute with different spectral characteristic to the diffuse γ-ray background. It may be that the extragalactic background is predominantly produced by BL Lacs with a harder spectrum than that of the average resolved AGN. We have also searched for systematic deviations from power-law behaviour. For each AGN we fit a power-law spectrum to the data. We can sum the fit statistic in the ten energy bands

$$\chi_{tot} = \frac{1}{\sqrt{N}} \sum_{i=1}^{N} \chi_i \quad , \quad \chi = \frac{I_{mod} - I}{\delta I} \qquad (1)$$

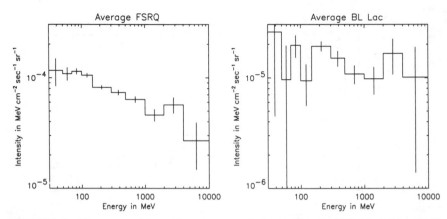

FIGURE 2. The summed intensity spectrum of the 33 identified FSRQ (left) and 11 identified BL Lacs (right) in the total EGRET allsky data above 100 MeV. The error bars are derived by Gaussian error propagation of the individual uncertainty measures. The spectrum of FSRQ is significantly softer than that of the diffuse background, while that of BL Lacs is consistent with it.

There are no significant deviations from power-law behaviour in the average spectra of FSRQ and BL Lacs.

II THE SPECTRA AT PEAK

Here we consider only the variable sources when they were at flare state. We are left with a total of 28 FSRQ and only 6 BL Lacs [1]. When repeating the same analysis as in the preceding chapter, we get no change in the results for BL Lacs. This is to a large extent due to the limited statistic in the remaining sample.

For FSRQ on the other hand we find that the spectrum at peak is significantly harder than that of the time-averaged emission (see Fig.3). The change in slope is $\delta s = 0.18 \pm 0.06$, and a Fischer-Snedecor test indicates a significance of $4\,\sigma$. We also obtain an indication for intensity deficits both at low energies and at high energies. This is also seen in the summed power-law fit statistic. We have thus repeated the power-law fits under the constraint that the fit is based on the energy band between 70 MeV and 4 GeV and then extended to calculate the true deviations in the outer energy bands. The result is shown in Fig.4. At energies below 70 MeV there is a deficiency of intensity compared to power-law behaviour with total statistical significance of $3.0\,\sigma$ while at high energies above 4 GeV we observe an intensity deficit with $2.6\,\sigma$ significance. Please note that here we have searched for deviations from power-law behaviour in individual source spectra. The significance of our finding does not imply similar significance for structure in the average spectrum of sources

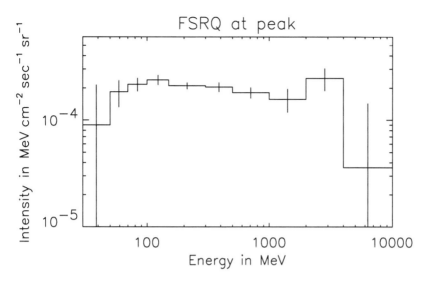

FIGURE 3. The summed intensity spectrum of the 28 variable FSRQ at the time of peak flux. The error bars are derived by Gaussian error propagation of the individual uncertainty measures. The peak spectrum of FSRQ is significantly harder than that of the time-averaged emission of FSRQ as displayed in Fig.2.

as shown in Fig.3. This result is stable with respect to the choice of sources. We have omitted the sources which have less than 6σ significance at the time of flare and the outcome remains unchanged. We have also included secondary flares and again the result is unchanged. We have further tested the reliability of our method by Monte-Carlo simulations and we have found no significant systematics in the distribution of the variable χ. Even accounting for calibration uncertainties at low γ-ray energies the statistical uncertainties are much larger than the systematical uncertainties.

Most FSRQ show a spectral break at MeV energies. It is however questionable whether such an extended spectral turnover is sufficient to account for the observed deficit below 70 MeV, which is a factor 10 higher in energy than the typical break energy. We prefer to interpret the result in the sense that the γ-ray spectrum of FSRQs is not a one component spectrum, but rather the superposition of different emission processes. Detailed simulations show that a low energy cut-off in the injection spectrum of radiating electrons can account for the observed behaviour [5].

The fact that we see the high-energy cut-off only at flare states, when photon densities are high, points at opacity effects as cause. In case of backscattered accretion disk photons the opacity will sharply increase at a few GeV. However, correlations between optical depth and the flux at a few 100 MeV will occur only when the γ-ray outburst is caused by an increased flux of target photons.

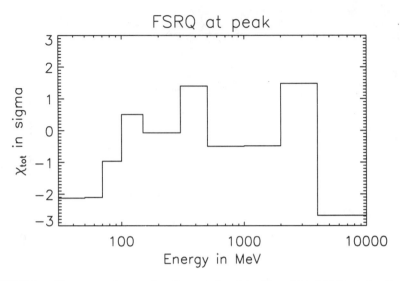

FIGURE 4. The summed power-law fit statistic of the 28 variable FSRQ at the time of their peak flux level. Here the fit is based only on the data between 70 MeV and 4 GeV, i.e. omitting the outer energy bands for which a deviation was suspected. The total statistical significance of the spectral breaks is 3.0 σ at low energies and 2.6 σ at high energies.

One may also think of photon-photon pair production on the high energy end of the self produced synchrotron spectrum. Here a correlation with the flux level can be naturally explained. Simulations show that at least in simple geometries the synchrotron-self-Compton component tends to swamp the high energy end of the synchrotron spectrum (Böttcher, Pohl and Schlickeiser, in prep.), so that there is no natural reason to let this effect become important at a few GeV γ-ray energy. Though opacity effects seem to be involved it is not yet clear where the target photons are supposed to come from. Further simulations may help to understand the cut-offs better.

The EGRET Team gratefully acknowledges support from the following: Bundesministerium für Bildung, Wissenschaft, Forschung und Technologie, Grant 50 QV 9095 (MPE); and NASA Cooperative Agreement NCC 5-95 (SU).

REFERENCES

1. Pohl M. et al. 1997, A&A in press
2. Kniffen D.A et al. 1996, A&AS 120, C615
3. Sreekumar P. et al. 1997, ApJ submitted
4. Mukherjee R. et al. 1997, ApJ submitted
5. Böttcher M., Schlickeiser R.: 1996, A&A, 306, 86

EGRET Observations of PKS 2005−489

Y.C. Lin[1], D.L. Bertsch[2], S.D. Bloom[3], B.L. Dingus[4],
J.A. Esposito[3], S.D. Hunter[2], B.B. Jones[1], G. Kanbach[5],
D.A. Kniffen[6], H.A. Mayer-Hasselwander[5], P.F. Michelson[1],
C. von Montigny[7], R. Mukherjee[8], A. Mücke[5], P.L. Nolan[1],
M.K. Pohl[9], O.L. Reimer[5], E.J. Schneid[10], P. Sreekumar[3],
D.J. Thompson[2], W.F. Tompkins[1]

[1] *W. W. Hansen Exp. Phys. Lab., Stanford Univ., Stanford, CA 94305*
[2] *NASA/Goddard Space Flight Center, Code 661, Greenbelt, MD 20771*
[3] *Univ. Space Res. Assoc., NASA/GSFC, Code 610.3, Greenbelt, MD 20771*
[4] *Phys. Dept., Univ. of Utah, Salt Lake City, UT 84112*
[5] *Max-Plack-Institut für extraterrestrische Physik, Postfach 1603, 85740 Garching, Germany*
[6] *Dept. of Phys. and Astr., Hampden-Sydney College, Hampden-Sydney, VA 23943*
[7] *Landessternwarte Heidelberg-Königstuhl, 69117 Heidelberg, Germany*
[8] *Dept. of Phys., McGill Univ., Montreal, Que, H3A 2T8, Canada*
[9] *Danish Space Research Institute, Juliane Maries Vej 30, 2100 Copenhagen O, Denmark*
[10] *Northrop Grumman Corporation, Mail Stop A01-26, Bethpage, NY 11714*

Abstract. PKS 2005−489, at z = 0.071, is one of the closest BL Lacertae objects, bright in radio to X-ray wavebands. It has been in EGRET's field of view a total of six times within an aspect angle of 30° from Phase 1 to Phase 3 (1991 May to 1994 October), none afterwards. But the total EGRET exposure to this region of the sky was still rather low. PKS 2005-489 was listed in the First EGRET Catalog at the significance $\sqrt{TS}$ = 4.3, in the context of the maximum likelihood analysis adopted by the EGRET team as part of the standard EGRET software package. Later when the EGRET data were reanalyzed with revised software and improved diffuse model to generate the Second EGRET Catalog, the detection significance of PKS 2005−489 was found to be at $\sqrt{TS}$ = 3.8, slightly below the $\sqrt{TS}$ = 4.0 cut line to qualify for an entry in the Second EGRET Catalog. The status of EGRET results on PKS 2005−489 at this time is presented in this paper.

INTRODUCTION

The bright radio source PKS 2005−489 was identified as a BL Lacertae object by Wall et al. (1986) [1], who also measured the redshift at 0.071. It has since been known to be a strong source in all wavebands from radio to X-ray frequencies. Thus PKS 2005−489 ranks as one of the closest and brightest BL Lac objects in the whole sky. Recently Sambruna, Urry, Ghisellini, & Maraschi (1995) [2] gave a summary of the observational results of PKS 2005−489 from radio frequencies to high-energy gamma rays together with discussions of the relevance of these results to the beaming jet models of blazars. The broadband spectrum of PKS 2005−489 in Sambruna et al. (1995) [2] is reproduced here as Figure 1.

EGRET OBSERVATIONS

EGRET has observed PKS 2005−489 a total of six times within an aspect angle of 30° from Phase 1 to Phase 3, but none afterwards. In the first EGRET

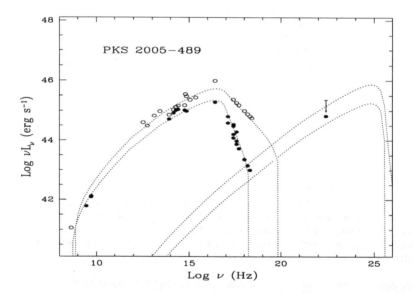

FIGURE 1. Spectral Energy Distribution (SED) of PKS 2005-489 in Sambruna, Urry, Ghisellini, & Maraschi (1995). Data are taken from the literature, but are not simultaneous. Dotted curves are theoretical fits in the paper.

Catalog (Fichtel et al. 1994) [3], PKS 2005−489 is listed as one of the EGRET-detected sources with detection significance of $\sqrt{TS} = 4.3$, in the context of the maximum likelihood analysis adopted by the EGRET team as part of the standard EGRET software package (Mattox et al. 1996) [4]. Later when the EGRET data were reanalyzed with revised software and improved diffuse model to generate the Second EGRET Catalog (Thompson et al. 1995) [5], the detection significance of PKS 2005−489 was found to be at $\sqrt{TS} = 3.8$, slightly below the $\sqrt{TS} = 4.0$ cut line to qualify for an entry in the Second EGRET Catalog. PKS 2005−489 falls in one of the several regions in the sky where the total EGRET exposure is considerably lower than other sky areas. This could be the reason why this source has not been detected by EGRET at higher significance levels. But PKS 2005−489 remains a viable EGRET source candidate, as can be seen in the following.

In Table 1, we list the six viewing periods and the measured photon fluxes or the 2-σ flux upper limits, as well as the average photon fluxes calculated for the summed data of Phase 1 (1991 May to 1992 November) and for Phase 1 plus Phase 2 (1991 May to 1993 September), all for energy E > 100 MeV. Furthermore, we tabulate in Table 2 the corresponding photon fluxes of Phase 1 and of Phase 1 plus Phase 2 for E > 1 GeV. The photon fluxes and upper limits in Table 1 are further plotted in Figure 2.

DISCUSSIONS

Several salient features stand out from the results in Tables 1 and 2.

TABLE 1. EGRET Photon Fluxes of PKS 2005−489 for E > 100 MeV.

Viewing period	Viewing angle deg	Date	Photon flux 10^{-7} cm^{-2} s^{-1}	$\sqrt{TS}$
VP0050	29.99	1991/07/12 − 26	< 2.8[a]	
VP0350	15.07	1992/08/06 − 11	2.66±1.66	2.0
VP0380	15.07	1992/08/27 − 09/01	2.18±1.15	2.4
VP0420	14.13	1992/10/15 − 29	0.61±0.58	1.2
VP2090	8.37	1993/02/09 − 22	< 1.2[a]	
VP3230	22.12	1994/03/22 − 04/05	< 1.5[a]	
Phase 1		1991/05/16 − 1992/11/17	1.31±0.46	3.4
Phases 1 & 2		1991/05/16 − 1993/09/07	0.82±0.31	3.0

[a] 2-σ upper limit

TABLE 2. EGRET Photon Fluxes of PKS 2005−489 for E > 1 GeV.

Viewing period	Viewing angle deg	Date	Photon flux 10^{-7} cm^{-2} s^{-1}	$\sqrt{TS}$
Phase 1		1991/05/16 − 1992/11/17	0.32±0.15	3.9
Phases 1 & 2		1991/05/16 − 1993/09/07	0.25±0.10	4.4

(1) The flux excess at the position of PKS 2005−489 was measurable in three of the six viewing periods in Table 1. The other three observations were either too brief or too far from the instrument axis, resulting in poor exposures and large flux upper limits. VP2090 might be an exception to this situation, but the flux upper limit in VP2090 is still consistent with other flux values.

(2) The summed data produced more significant flux excesses than individual viewing periods. This is an indication that flux excesses also exist in other viewing periods with viewing angles larger than 30°.

(3) What is most important is the fact that the fluxes for E > 1 GeV are more significant than the corresponding E > 100 MeV fluxes. This is an indication that this source is brighter at higher energies relative to the background level. Comparing Tables 1 and 2, we see that the fluxes for E > 1 GeV are larger than one tenth of the corresponding E > 100 MeV flux by more than 2σ error margin. The photon spectrum should thus be flatter than E^{-2}. A direct determination of the photon spectrum is not possible due to the small amount of collected source photons.

Thus it appears that, even though PKS 2005−489 failed to make to the Second EGRET Catalog (Thompson et al. 1995) [5], it remains a serious EGRET source candidate and may very well be readily detectable at TeV energies by the facilities in the southern hemisphere.

After the discovery of gamma radiation of blazars, by CGRO and by ground-based TeV facilities, the theoretical studies of the emission mechanisms of

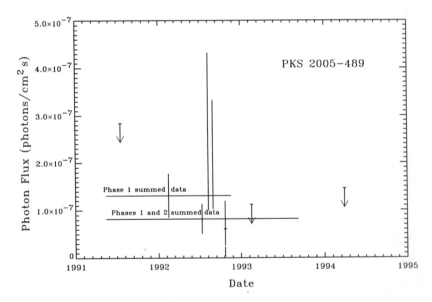

FIGURE 2. Light curve of PKS 2005−489 for E > 100 MeV in EGRET observations.

blazars seem to have converged to the general picture (See, e.g. Urry 1996; Ghisellini & Maraschi 1996) [6] [7] that the broadband spectrum of a blazar consists of two general components: (a) a lower-energy component which is believed to be the result of synchrotron radiation of a beam of relativistic particles, most likely electrons although theories with proton beams can also be made to explain the observed data, and which peaks, in the νF_ν plot, in the optical to EUV region; (b) a higher-energy component which is likely the result of inverse Compton scattering of the same beam of relativistic particles on some ambient field of soft photons and which peaks in the GeV-TeV region. Furthermore, if the peak of the lower-energy component is in the optical region, the higher-energy component will peak at the GeV region. Strong EGRET flux can be expected but TeV photons may not be produced significantly. On the other hand, if the peak of the lower-energy component is in the EUV or low-energy X-ray region, then the higher-energy component will peak in the TeV region. This may produce a signidicant TeV source, but EGRET flux will be small or invisible. Many of the bright high-latitude EGRET sources tend to support one half of this picture. What we need next is more examples in the TeV region. It is by now a very well-known fact that Mrk 421 is a relatively dim EGRET source but a very prominent TeV source. The only other TeV blazar Mrk 501 has barely produced some tangible flux excess in the EGRET data (Hartman 1997) [8]. To make meaningful constraints on theoretical models, we should have positive EGRET fluxes, not just an upper limit. At the same time, the EGRET flux should not be too strong as dictated by the theoretical idea that MeV/GeV fluxes of TeV sources are likely to be on the weak side. Thus PKS 2005−489 forms a very rare find in this regard: peak of the lower-energy component at EUV and bright in this region, moderate but positive EGRET flux excess with photon spectrum flatter than E^{-2}.

REFERENCES

1. Wall, J. V., et al. 1986, MNRAS, 219, 23.
2. Sambruna, R. M., Urry, C. M., Ghisellini, G., & Maraschi, L. 1995, ApJ, 449, 567.
3. Fichtel, C. E., et al. 1994, ApJS, 94, 551.
4. Mattox, J. R., et al. 1996, ApJ, 461, 396.
5. Thompson, D. J., et al. 1995, ApJS, 101, 259.
6. Urry, C. M. 1996, Proc. Blazar Continuum Variability, A.S.P. Conf. Ser. Vol. 110, p.391
7. Ghisellini, G., & Maraschi, L. 1996, Proc. Blazar Continuum Variability, A.S.P. Conf. Ser. Vol. 110, p. 436
8. Hartman, R. C. 1997, private communication

Whipple Observations of BL Lac Objects at E > 300 GeV

M. Catanese*, P.J. Boyle†, J.H. Buckley§, A.M. Burdett‡, J. Bussóns Gordo†, D.A. Carter-Lewis*, M.F. Cawley†, D.J. Fegan†, J.P. Finley‖, J.A. Gaidos‖, A.M. Hillas‡, F. Krennrich*, R.C. Lamb**, R.W. Lessard‖, C. Masterson†, J.E. McEnery†, G. Mohanty*, J. Quinn†,§, A.J. Rodgers‡, H.J. Rose‡, F.W. Samuelson*, G.H. Sembroski‖, R. Srinivasan‖, T.C. Weekes§, J. Zweerink* [1]

*Dept. of Physics and Astronomy, Iowa State University, Ames, IA 50011
†National University of Ireland
§F.L. Whipple Observatory, Harvard-Smithsonian CfA, P.O. Box 97, Amado, AZ 85645
‡Dept. of Physics, Univ. of Leeds, Leeds, LS2 9JT, Yorkshire, England, UK
‖Dept. of Physics, Purdue Univ., West Lafayette, IN 47907
**Space Radiation Laboratory, California Institute of Technology, Pasadena, CA 91125

Abstract. Two BL Lacertae objects have been detected at TeV energies using the atmospheric Čerenkov imaging technique. The Whipple Observatory γ-ray telescope has been used to observe all the BL Lacertae objects in the northern hemisphere out to a redshift of ∼0.1, of which two (BL Lacertae and W Comae) are EGRET sources. We report preliminary results of the observations including the tentative detection of very high energy emission from a third BL Lac object, 1ES 2344+514.

INTRODUCTION

With the detection of very high energy (VHE, $E > 300$ GeV) emission from the two BL Lacertae objects (BL Lacs), Markarian 421 (Mrk 421) [1] and Mrk 501 [2], the Whipple collaboration began a survey of nearby BL Lacs to search for VHE emission. A collection of such sources could lead to constraints on γ-ray emission models and also an estimate of the density of extragalactic background IR light through its attenuation of the VHE γ-ray spectra [3,4].

BL Lacs are members of the blazar class of active galactic nucleus (AGN), the only type of AGN detected above 100 MeV. BL Lacs are particularly

---

[1] This research is supported by grants from the U.S. Department of Energy and NASA, by PPARC in the UK and by Forbairt in Ireland.

promising candidates for VHE emission because of two features of their emission spectra. First, BL Lacs have weak or no emission lines in their optical spectra which may indicate that they have less γ-ray absorbing material near the source [5]. Second, in models where the high energy emission is produced by inverse Compton (IC) scattering of low energy photons by the same electrons which produce the synchrotron emission [6–9], the extension of the synchrotron emission of X-ray selected BL Lacs (e.g., Mrk 421 and Mrk 501) into the X-ray waveband implies a higher maximum γ-ray energy than for radio-loud BL Lacs (e.g., W Comae) and flat spectrum radio quasars (e.g., 3C 279) where the synchrotron emission ends in the optical to UV range.

Table 1 lists the objects observed in our BL Lac survey. We have limited our search to $z \lesssim 0.1$ to reduce the loss of γ-ray flux by absorption on background IR light. We applied a two-part approach to the survey. First, we selected a few promising candidates for long exposures to search for low-level VHE γ-ray emission, such as was observed with Mrk 501 in its initial detection [2]. Second, the remaining objects were observed for less total time, but the observations were spread out over a long time period in order to maximize the chance that we might see an episode of high emission, such as has been seen in Mrk 421 [10] and Mrk 501 [11]. This latter approach led to the likely detection of a third VHE-emitting BL Lac, 1ES 2344+514.

OBSERVATIONS AND ANALYSIS

The VHE observations reported in this paper were made with the atmospheric Čerenkov imaging technique [12] using the 10m optical reflector located at the Whipple Observatory on Mt. Hopkins in Arizona (elevation 2.3 km). A camera consisting of 109 photomultiplier tubes (3°.0 field of view) is mounted in the focal plane of the reflector and records images of atmospheric Čerenkov radiation from air showers produced by γ-rays and cosmic rays [13]. The energy threshold of the observations reported here is 350 GeV.

Čerenkov light images from γ-rays are typically smaller and more elliptical than background hadronic images and they are preferentially oriented toward the source location. The basic γ-ray selection was based on the Supercuts criteria [14], but some modifications were implemented to account for recent changes to the telescope (see [15] for details).

The results reported in this paper use a *Tracking* analysis (cf., [2]) wherein events from on-source pointings whose orientations do not point toward the object's direction are used to determine the background level. A large collection of off-source observations and non-detected objects other than BL Lacs were combined to estimate the factor which converts the off-source events to a background estimate. If no significant emission is seen from a candidate source, a 99.9% confidence upper limit on the rate is calculated using the method of Helene [16]. The count rates or upper limits are converted to integral fluxes

TABLE 1. BL Lac Observations

Object	z	Type[a]	Observ. Epoch	Exp. (hrs)	Excess	Max. Daily Exc.	Flux (Crab)[b]	Flux[c]
1ES 2344+514	0.044	X	1995/96	20.5	5.3σ	6.0σ	0.16±0.03	1.7±0.3
			20-12-95	1.8	6.0σ		0.63±0.11	6.6±1.1
			1996	32.1	1.6σ	2.1σ	<0.081	<0.85
Markarian 180	0.046	X	1996	20.9	-0.4σ	0.6σ	<0.105	<1.10
1ES 1959+650	0.048	X	1996	3.7	0.3σ	1.2σ	<0.128	<1.34
3C 371	0.051	R	1995	12.7	0.6σ		<0.179	<1.88
I Zw 187	0.055	X	1996	2.3	0.6σ	1.2σ	<0.150	<1.58
1ES 2321+419	0.059?	X?	1996	6.4	-1.3σ	0.3σ	<0.106	<1.11
BL Lacertae[d]	0.069	R	1995	39.1	-1.0σ	0.5σ	<0.06	<0.63
1ES 1741+196	0.083	X	1996	8.8	-1.7σ	0.6σ	<0.046	<0.48
W Comae	0.102	R	1996	16.6	-0.4σ	1.4σ	<0.056	<0.59

[a] Indicates whether the object is radio selected (R) or X-ray selected (X).
[b] Flux, or upper limit, is expressed in units of the Crab Nebula flux.
[c] Integral fluxes or upper limits are quoted above 350 GeV in units of $10^{-11} \text{cm}^{-2}\text{s}^{-1}$. Flux upper limits are at the 99.9% confidence level.
[d] Results from [19].

by expressing them as a multiple of the Crab Nebula count rate and then multiplying that fraction by the Crab Nebula flux, $I(> 350 \text{ GeV}) = 1.05 \times 10^{-10}$ photons cm^{-2} s^{-1} [17]. This procedure assumes that the Crab Nebula VHE γ-ray flux is constant, as 7 years of Whipple Observatory data indicate [17], and that the object's photon spectrum is identical to that of the Crab Nebula between 0.3 and 10 TeV, $dN/dE \propto E^{-2.5}$ [17], which may not be the case.

RESULTS

In Table 1 we present the results of the BL Lac observations. With the exception of 1ES 2344+514, no object shows evidence of VHE emission. In particular, the EGRET sources W Comae [18] and BL Lacertae [19] are not detected despite long exposures. Also, only 1ES 2344+514 shows evidence of short term activity.

Most of the excess from 1ES 2344+514 during 1995 comes from an apparent flare on 1995 December 20 (see Figure 1). Without the flare, the excess for the 1995/96 season is 3.7σ which, if due to γ-rays, corresponds to a flux of 0.11±0.03 times the Crab Nebula flux. We find a non-statistically significant excess from this object in 1996 perhaps indicating that the average flux level dropped below the telescope sensitivity limit, as occasionally happens with Mrk 421 [20]. We consider the detection tentative because we see no conclusive evidence for a consistent signal nor is there independent confirmation of a high state for this object (e.g., from X-ray observations) during the flare.

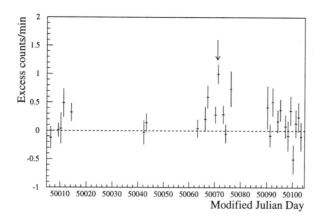

FIGURE 1. The light curve for γ-ray observations of 1ES 2344+514 between Oct. 1995 and Jan. 1996. The flare on 20-12-1995 is indicated by the arrow.

CONCLUSIONS

Both Mrk 421 and Mrk 501 have energy spectra, expressed as νF_ν, with peaks of comparable amplitude at X-ray and γ-ray energies [20,11]. If the X-ray selected BL Lacs in this sample are similar, an upper limit below the X-ray power output would indicate some reduction of the VHE flux at the source or en-route to Earth. Our current limits are not below the lowest measured X-ray fluxes for these objects. Further observations will be needed to clearly establish whether these objects emit VHE γ-rays at the expected levels.

For W Comae and BL Lacertae, the Whipple upper limits are well below the power output at EGRET energies (see Fig. 2 and [19]). This is consistent with expectations from IC emission models. However, the VHE observations are not contemporaneous, so it may also be that the γ-ray emission was in a low state when these objects were observed at the Whipple Observatory.

Finally, the detection of 1ES 2344+514, if confirmed, would be consistent with the other detected VHE blazar sources: it is a nearby, X-ray selected BL Lac (the third closest known) whose γ-ray emission is variable.

REFERENCES

1. Punch, M., et al., *Nature* **358**, 477 (1992).
2. Quinn, J. et al., *ApJ* **456**, L83 (1996).
3. Gould, J. R., & Schréder, G. P., *Phys. Rev.* **155**, 1408 (1967).
4. Stecker, F. W., de Jager, O. C., and Salamon, M., *ApJ* **415**, L71 (1993).
5. Dermer, C. D., & Schlickeiser, R., *ApJS* **90**, 945 (1994).
6. Königl, A., *ApJ* **243**, 700 (1981).

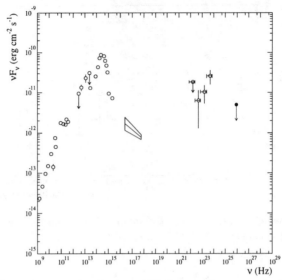

FIGURE 2. Spectral energy distribution of W Comae. Shown are the VHE upper limit (filled circle) and archival data (open circles and box) [18,21–26].

7. Bloom, S. D., & Marscher, A. P., *ApJ* **461**, 657 (1996).
8. Dermer, C. D., Schlickeiser, R., & Mastichiadis, A., *A&A* **256**, L27 (1992).
9. Sikora, M., Begelman, M. C., and Rees, M. J., *ApJ* **421**, 153 (1994).
10. Gaidos, J. A., et al., *Nature* **383**, 319 (1996).
11. Catanese, M., et al., *ApJ Letters*, submitted (1997).
12. Cawley, M. F., & Weekes, T. C., *Exp. Astron.* **6**, 7 (1995).
13. Cawley, M. F., et al., *Exp. Astron.* **1**, 173 (1990).
14. Reynolds, P. T., et al., *ApJ* **404**, 206 (1993).
15. Catanese, M., et al., in *Towards a Major Atmospheric Detector-IV*, ed. M. Cresti, 335 (1996).
16. Helene, O., *Nucl. Instr. Meth.* **212**, 319 (1983).
17. Hillas, A. M., et al., in preparation (1997).
18. Sreekumar, P., et al., *ApJ* **464**, 628 (1996).
19. Catanese, M., et al., *ApJ* **480**, 562 (1997).
20. Buckley, J. H., et al., *ApJ* **472**, L9 (1996).
21. Lamer, G., Brunner, H., and Staubert, R., *A&A* **311**, 384 (1996).
22. Edelson, R., et al., *ApJS* **83**, 1 (1992).
23. Cruz-Gonzalez, I., and Huchra, J. P., *AJ* **89**, 441 (1984).
24. Impey, C. D., and Neugebauer, G., *AJ* **95**, 307 (1988).
25. Gear, W. K., et al., *MNRAS* **267**, 167 (1994).
26. Owen, F. N., Spangler, S. R., and Cotton, W. D., *AJ* **84**, 351 (1980).

Multiwavelength Observations of Markarian 421

J.H. Buckley*, P. Boyle[†], A. Burdett[‡], J. Bussóns Gordo[†], D.A. Carter-Lewis[§], M. Catanese[§], M.F. Cawley[†], D.J. Fegan[†], J.P. Finley[‖], J.A. Gaidos[‖], A.M. Hillas[‡], F. Krennrich[§], R.C. Lamb[**], R.W. Lessard[‖], C. Masterson[†], J. McEnery[†], G. Mohanty[§], J. Quinn[†,*], A. Rodgers[‡], H.J. Rose[‡], F. Samuelson[§], G.H. Sembroski[‖], R. Srinivasan[‖], T.C. Weekes*, J. Zweerink[§] [1]

F.L. Whipple Observatory, Harvard-Smithsonian CfA, P.O. Box 97, Amado, AZ 85645
[†]*National University of Ireland*
[‡]*Dept. of Physics, Univ. of Leeds, Leeds, LS2 9JT, Yorkshire, England, UK*
[§]*Dept. of Physics and Astronomy, Iowa State University, Ames, IA 50011*
[‖]*Dept. of Physics, Purdue Univ., West Lafayette, IN 47907*
[**]*Space Radiation Laboratory, California Institute of Technology, Pasadena, CA 91125*

Abstract. We report on TeV γ-ray observations of Mrk 421 made with the Whipple Observatory's 10 m telescope over the period 1994 to 1996 and summarize the results of contemporaneous multiwavelength observations. These observations indicate short variability timescales (<1 hr) and correlations in the TeV γ-ray, X-ray, extreme UV, and optical flux. The broadband spectrum for Mrk 421 determined from data taken in three epochs is presented and implications of these data on the physical parameters in the jet are discussed.

INTRODUCTION

The Whipple 10 m telescope has detected >300 GeV γ-rays from two BL Lac objects: Mrk 421 and Mrk 501. The core-dominated radio emission and highly variable broad-band (radio to X-ray) continuum emission of these objects is thought to result from beamed synchrotron radiation from relativistic jets viewed at a small angle to the jet axes [1]. A second distinct spectral component is observed at energies above the synchrotron cutoff. This high energy emission may result from synchrotron self-Compton (SSC) emission [2], Compton up-scattering of external accretion disc photons (EC models) [3,4] or cascades produced by very high energy protons (PIC models) [5]. Here we describe how correlated multiwavelength observations constrain these models.

---

[1] This research is supported by grants from the U.S. Department of Energy and NASA, by PPARC in the UK and by Forbairt in Ireland.

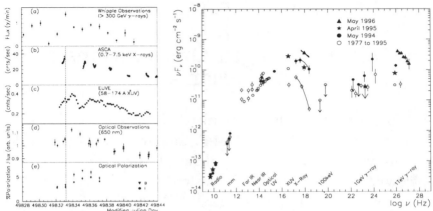

FIGURE 1. Left; γ-ray, X-ray, extreme UV and Optical light-curves for Mrk 421 during the period 1995 April 20 to May 5 [7]. Right; Spectral energy distribution of Mrk 421.

OBSERVATIONS AND DISCUSSION

The Whipple γ-ray telescope employs a 10 m optical reflector to image Čerenkov light from air showers. By making use of the distinctive differences in the shower images, a γ-ray signal can be extracted from the large background of hadronic showers. The large effective area ($\approx 5 \times 10^8$ cm^2) enables sensitive measurements of the $E > 300$ GeV (VHE) emission for hard-spectrum sources.

First evidence for variability in the TeV emission from Mrk 421 and for a correlation with the X-ray emission was obtained during a TeV flare in 1994 May [6]. Longterm monitoring of Mrk 421 subsequently revealed that the γ-ray emission is best characterized by a succession of rapid flares with a baseline level below the Whipple sensitivity limit [7].

Multiwavelength observations taken in 1995 (Fig. 1a, [7]) indicate correlations of the γ-ray and X-ray data and a correlation with the longer wavelength data when a shift of ~1 day is introduced relative to the γ-ray emission. Polarization measurements suggest variations in the synchrotron component of the optical emission. The variability amplitude is comparable for the X-ray and TeV emission, but larger in the X-ray (~400%) than the XUV (~200%) or the optical band (~20%).

In its most dramatic flare on 1996 May 7, Mrk 421 reached a flux ten times that of the Crab Nebula, with a doubling time of ≈1hr and a subsequent decay by a factor of ~30 in <24hr. A second flare observed eight days later reached about half this level, with doubling and decay timescales of ≈15 min [8] (Fig. 2a). One day prior to the TeV flare, the RXTE 2-10 keV X-ray flux [9] was found to be nearly twice that previously recorded during intense flares [10]. Optical data obtained during this period do not show a simple correlation with the TeV emission (Fig. 2b [11]), but are consistent with a delayed flare with a relative amplitude similar to the 1995 April optical flare. The data also show a 0.5 magnitude increase of the overall optical level perhaps due to the

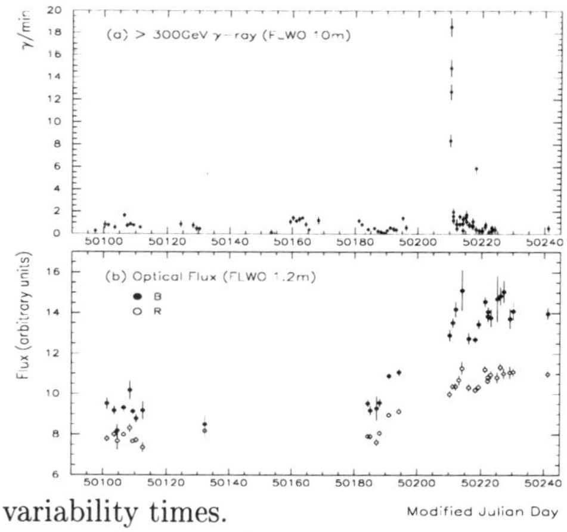

FIGURE 2. Whipple >300 GeV γ-ray and optical flux (arbitrary units) of Mrk 421 including the flare events of 1996 May 7 and 1996 May 15. The optical data was obtained by relative CCD photometry with the Whipple Observatory 1.2m telescope using standard R and B filters [11].

superposition of flares with long variability times.

Fig. 1b shows the spectral energy distribution (SED) derived from these data. The peak flux and spectrum of the 2-10 keV emission [9] together with the Whipple VHE spectrum derived from data taken on 1996 May 7 [12] are indicated by filled triangles. The stars show data taken during the flare in 1995 April. The filled circles are from measurements taken in the period around 1994 May 15 [6]. For reference the open circles show archival data from 1977-1994 including the OSSE upper limits [13] and EGRET data [6,7].

The correlated variability of the TeV and X-ray emission and apparent spectral roll-over in these bands are naturally explained if the same population of electrons produces the X-ray emission by synchrotron radiation and the TeV emission by inverse-Compton (IC) scattering. The observation of spectral hardening in the X-ray emission during flares [10,9] and increasing amplitude of variability between the UV and X-ray wavebands [7] indicates that variations in the synchrotron spectrum are due to an increase in the maximum electron energy [6] or the Doppler factor. A lack of significant spectral variability in the TeV range [12] could result from the termination of the VHE spectrum by intrinsic or external absorption by γ-γ pair production [12].

We assume that the emission is beamed with a Doppler factor $\delta = \Gamma^{-1}(1-\beta\cos\theta)^{-1}$ where Γ and β are the bulk Lorentz factor and velocity of the jet, and θ is the angle between the line of sight and the jet axis [1]. In the following discussion, the particle Lorentz factor γ and the magnetic field B are given in the comoving frame with B in units of Gauss. The observed synchrotron energy E_{syn} and IC energy E_C are in keV and TeV respectively.

For $\delta > 1$, the measured luminosity L_{syn} is boosted by $\sim \delta^3$ compared with the intrinsic luminosity, and the limit on the size of the emission region is relaxed $R < ct_{\text{var}}\delta$. This results in a reduction of the opacity for pair production $\tau_{\gamma\gamma} \propto \sigma_T L_{\text{syn}}(E_s)/R$ (where $E_s = 2(m_e c^2)^2/E_\gamma$ is the threshold energy). As δ is increased, the waveband corresponding to E_s moves from the optical up through the UV. The observed correlations of the TeV with the optical–UV emission and short variability timescales ($t_{\text{var}} \approx 1$ day) in the data of 1995

April imply that the low energy photons and the VHE emission are produced in the same compact region. From the condition $\tau_{\gamma\gamma} < 1$, one derives a lower limit on the Doppler factor of this region $\delta \gtrsim 5(t_{\text{var}}/10^5 \text{ sec})^{-1/5.4}(F_U/10 \text{ mJy})^{1/5.4}$ where F_U is the optical U-band flux [14]. During the 1996 May 7 flare, evidence for an optical/γ-ray correlation is less compelling. Using $F_U = 15.8$mJy (from data taken on May 7 after a galaxy light subtraction [11]) and assuming that $\sim 20\%$ of this optical flux is produced in the same region as the TeV flux, a lower limit of $\delta \gtrsim 8$ is derived.

The maximum synchrotron energy $E_{\text{syn,max}} = 1.2 \times 10^{-11} \delta \gamma_{e,\text{max}}^2 B$ determined from the X-ray data can be used to derive the magnetic field in the emission region. If the VHE emission arises from IC scattering, then regardless of the nature of the seed photons, the kinematic limit $\delta \gamma_{e,\text{max}} > E_{C,\text{max}}/m_e c^2$ applies, where $E_{C,\text{max}}$ is the maximum observed γ-ray energy. Thus one obtains the limit $B \lesssim 2.2 \times 10^{-2} E_{\text{syn,max}} \delta E_{C,\text{max}}^{-2}$ [14]. For the 1996 May 7 flare $E_{C,\text{max}} > 5$ TeV [12] and $E_{\text{syn,max}} \approx 5$ keV [9] giving $B \lesssim 0.005 \delta$ G.

B is also bounded from below since the cooling time for electrons producing the X-ray variations $t_{e,\text{cool}} \approx 8 \times 10^8 \delta^{-1} \gamma_e^{-1} [(1+\eta)B]^{-2}$s [2] must be less than the variability timescale t_{var}. Substituting $\gamma_e = 3 \times 10^5 (E_{\text{sync}}/\delta B)^{1/2}$ and assuming equal contributions of synchrotron and IC losses (i.e., $\eta \approx 1$) $t_{e,\text{cool}} = 1.3 \times 10^3 B^{-3/2} \delta^{-1/2} E_{\text{syn}}^{-1/2}$. From the X-ray data [9] we estimate a decay time of $t_{\text{var}} \approx 8$ hr at $E_{\text{syn}} \approx 3$ keV. The condition $t_{\text{var}} > t_{e,\text{cool}}$ then implies $B > 0.09 \delta^{-1/3}$. In order for the upper and lower limits to be satisfied, one obtains an independent limit on the Doppler factor [15] of $\delta > 9$. For $\delta \approx 10$ one obtains $B \approx 0.04$ G. However if the Compton-drag limit ($\delta \lesssim 10$ [16]) does not apply for a BL Lac (with a faint accretion disk) then $\delta \gg 10$ is possible.

For EC models, the Comptonized photons must have energies < 0.1 eV for scattering to occur in the Thomson regime and to avoid significant attenuation by pair production [4]. Such low-energy external photons could come from thermal IR produced by dust [4] or from the emission peaked in the far-IR predicted for the ion-supported torus of a low accretion rate AGN [17]. The shortest variability timescales for Mrk 421 imply an emission region with a dimension of $\approx 40 R_S$ [14] (where R_S is the gravitational radius of a $10^8 M_\odot$ black hole) within the $\sim 50 R_S$ extent of the hot torus [17] and compatible with the geometry for the EC mechanism [3].

It has been suggested that the acceleration of electrons to energies $\gtrsim 5$ TeV is problematic [18,19], but these calculations make tacit assumptions about the form of the power spectrum of magnetic field fluctuations and about the shock velocity. For supernova remnants, there is direct observational evidence for electron shock acceleration to ~ 100 TeV in an ambient field of $B \sim 10^{-5}$ G [20]. A maximum energy $E_{e,\text{max}} \lesssim 60 B^{-1/2} (\beta_S/0.1)^2$ TeV is determined by equating the loss time $t_{e,\text{cool}}$ and the acceleration time $\tau_{\text{acc}} \gtrsim 10(\beta_s/0.1)^{-2} E_{\text{TeV}} B^{-1}$ sec (in the Bohm limit) [20]. It is unclear how large the shock velocity β_S is compared with typical values of $\beta_S \approx 3 \times 10^{-3}$ for SNRs, or how the acceleration process is altered for an oblique relativistic shock, however values of $E_{e,\text{max}} \gtrsim$

50 TeV are not ruled out.

In PIC models [5] electrons and protons accelerated in the same shocks produce the synchrotron and high energy emission respectively. Interactions of extremely high energy ($> 10^{16}$ eV) protons with low energy synchrotron photons results in the production of pions; γ-ray emission is produced in the subsequent electromagnetic cascades. Variability timescales of the γ-ray emission are determined either by the proton or the low energy electron synchrotron (UV and longer wavelength) cooling times. The cooling time for protons is much longer than that of electrons $t_{p,cool}/t_{e,cool} \approx 3 \times 10^7 \gamma_e/\gamma_p$ since $\sigma_{p\gamma} \approx 10^{-3}\sigma_T$ and since synchrotron cooling is inefficient for protons [19]. Thus large magnetic fields, high Doppler factors and high proton Lorentz factors are required to reproduce the observations for Mrk 421, and the best fit to the SED gives $B=40$ G, $\gamma_p = 6 \times 10^7$, and $\delta = 31$ [21]. However, for data taken in May 1994, a magnetic field of $B \approx 0.2$ G is derived from the ASCA X-ray data (with no input from the VHE data) by comparing the relative time lags between low and high energy X-rays to differences in their cooling times $t_{e,cool} \propto B^{-3/2}\delta^{-1/2}E_{syn}^{-1/2}$ [10]. The large disparity with the PIC value for B implies that the observed synchrotron flares do not come from the same environment as the cascade emission, making it difficult for the present PIC models to account for observed optical-X-ray/TeV correlations.

REFERENCES

1. Blandford, R. D., & Königl, A., *ApJ* **232**, 34 (1979).
2. Königl, A., *ApJ* **243**, 700 (1981).
3. Dermer, C. D., Schlickeiser, R., & Mastichiadis, A., *A&A* **256**, L27 (1992).
4. Sikora, M., Begelman, M. C., and Rees, M. J., *ApJ* **421**, 153 (1994).
5. Mannheim, K., *A&A* **269**, 67 (1993).
6. Macomb, D. J., et al., *ApJ* **449**, L99 (1995).
7. Buckley, J. H., et al., *ApJ* **472**, L9 (1996).
8. Gaidos, J. A., et al., *Nature* **383**, 26 (1996).
9. Schubnell, M.S., et al., these proceedings, (1997).
10. Takahashi, T., et al., *ApJ* **470**, L89 (1996).
11. McEnery, J.E., Buckley, J.H., in *Proc. of the Conference: Blazars, Black Holes and Jets*, Girona, in press (1997).
12. Zweerink, J., et al., *ApJ*, submitted (1997).
13. McNaron-Brown, K., et al., *ApJ* **451**, 575 (1995)
14. Buckley, J. H., et al., *Advances in Space Research*, in press (1997).
15. Dermer, C.D., private communication (1997).
16. Melia, F., & Königl, A., *ApJ* **340**, 162 (1989).
17. Rees, M. J., et al., *Nature* **295**, 17 (1982).
18. Biermann, P.L. & Strittmatter, P.A., *ApJ* **322**, 643 (1987).
19. Mannheim, K., in *Proc. of the 32nd Moriond Meeting*, Les Arcs, in press (1997).
20. Reynolds, S.P., *ApJ* **459**, L13 (1996).
21. Mannheim, K., *Sp.Sc.Rev.* **75**, 331 (1996).

Observation of Strong Variability in the X-Ray Emission from Markarian 421 Correlated with the May 1996 TeV Flare

Michael Schubnell[1]

Randall Laboratory of Physics
University of Michigan, Ann Arbor, Michigan 48109

Abstract. We observed the BL Lac object Markarian 421 with the X-ray satellite RXTE and the Whipple Air Cerenkov Telescope during a two week correlated X-ray/gamma-ray campaign in May 1996. Two dramatic outbursts with extremely rapid and strong flux variations were observed at TeV energies during this period. The X-ray emission in the 2-10 keV band was highly variable and reached a peak flux of 5.6×10^{-10} erg cm^{-2} s^{-1}, a historic high. Similar behavior was observed for the TeV emission. In contrast to earlier near-simultaneous X-ray/gamma-ray observations of Mrk 421, the variability amplitude is much larger at TeV than at X-ray energies. This behavior is expected in Synchrotron Self-Compton models.

INTRODUCTION

Present blazar research focuses on the question of how relativistic jets are formed and how particles are accelerated to energies beyond a TeV. To find an answer to this question we are trying to understand the very basic properties of the jet, i.e. particle population, velocities, total energy, and magnetic field strength. The non-thermal blazar emission from radio to UV is generally thought to be synchrotron radiation beamed from a relativistic jet viewed at small angles. In the case of Markarian 421, the synchrotron emission dominates up to X-ray energies. A second component, observed above the synchrotron break, is typically relatively flat and extends to X-ray and sometimes to gamma-ray energies, as in the case for Markarian 421.

In the context of current theoretical models invoked to explain the broadband energy spectrum of Markarian 421 and other similar X-ray selected BL Lac objects, correlated observations can be used to understand the physics of

---

[1] Support for this work was provided in part by NASA contract NAG 5-3264

the electron population believed to be the progenitor of the high energy emission. The energy spectrum near the spectral break around a few keV may give a handle on the maximum electron energy in the system. Mufson et al. (1990) report a two-component X-ray spectrum with a steepening tail above 4.5 keV. This suggests the contribution of two different components to the X-ray emission: synchrotron radiation from the highest energy electrons and inverse-Compton scattered photons from the lowest energy electrons. In such a scenario, the X-ray band plays a crucial role in the study of the broadband emission [12].

Markarian 421 has been monitored previously and has shown rapid variability on short time scales. Detailed studies based on EXOSAT observations showed significant variation in the 0.5-10 keV band on time scales from several hours to several days [5]. This strongly suggests that typical X-ray high states are relatively short lived. Near coincident flaring in the X-ray and TeV gamma-ray emission was observed in 1994 and in 1995 [1,6] which prompted us to propose extensive observations with the newly launched X-ray satellite, XTE, and the Whipple telescope.

OBSERVATIONS

The X-ray observations discussed here were carried out with the PCA (Proportional Counter Array) onboard the Rossi X-Ray Timing Explorer [2]. Between 1996 May 4 and May 21, the PCA was pointed on average 4 times every day at Markarian 421. A typical observation resulted in a net exposure of 600 seconds.

The individual X-ray spectra were fitted with a single power-law model with energy spectral index α, $F(E) = F_0 E^{-\alpha}$, with the absorbing column N_H fixed at the Galactic value ($1.5 \times 10^{20} cm^{-2}$) [3]. We found that the simple power law function describes the data well in the energy range between 2 and 10 keV if we allow for a 2% systematic error for energies between 3.5 and 6 keV. This additional systematic error accounts for small uncertainties in the channel-to-energy conversion which was not finalized at the time of this analysis. Alternately, we also have applied a single power-law model and allowed for a varying column density. While this resulted in slightly improved χ^2 values for fits between 2 keV and 10 keV, it also resulted in N_H values ≈ 20 times larger than the Galactic value. A broken power-law model, successfully applied to ASCA data [10], did not fit any of the spectra well.

RESULTS

During the observation period two very remarkable outbursts were observed at TeV energies (May 7 and May 15). The May 7 flare is the most intense flare recorded at TeV energies [4]. During the course of 2 hours of observing, the gamma-ray emission increased steadily from 4 times to about 30 times the average flux. The second TeV flare (May 15) was less pronounced but shows an even shorter doubling time of ≈ 20 minutes.

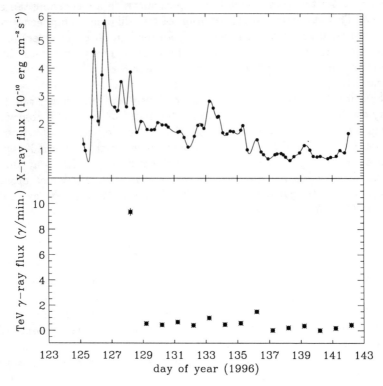

FIGURE 1. The X-ray (2-10 keV) and TeV (> 300 GeV) emission from Markarian 421 during May 1996. We included a fitted cubic spline function as a guidance for the eye.

The time history of the May 1996 X-ray and TeV observations is shown in figure 1. The X-ray light curve during the high state between day 125 and day 129 shows very rapid quasi-periodic variations with similar rise and fall times of the order of several hours. This behavior is an indication that geometrical effects may need to be considered in order to explain the observed variability. The 2-10 keV flux was high throughout the observing period compared to previous measurements (see [9] for a compelation of the long term variability of Markarian 421). On May 5 (day 126) the 2-10 keV flux increased to 5.6×10^{-10} $erg\ cm^{-2}s^{-1}$, brighter than all previous observations.

The TeV high state on May 7 (day 128) clearly occurred during strong activity in the X-ray band. Restricted by the bright moon, TeV observations could not be obtained prior to day 128. This leaves some ambiguity in determining the exact time delay between TeV and X-ray high state. However, the close coincidence of the TeV and X-ray flare suggests that both arise from the same electron population in the jet.

An interesting conclusion can be drawn from comparing the relative amplitudes of the flare states in both bands. While previous flare observations [1,6] claim a comparable amplitude in the variability, during this observation, the flux at TeV energies rose a factor of ≈ 30 above the quiescent level, compared to a factor of ≈ 5 for the keV flux. This is expected in Synchrotron Self-Compton (SSC) models where the X-rays are the seed photons that are Comptonized to produce the TeV flux. Simultaneous optical data taken with the 48 inch optical telescope on Mt. Hopkins did not show an increase in the flux during this period [7]. This also confirms previous observations where the flux at longer wavelengths remained relatively constant during strong X-ray/TeV flares.

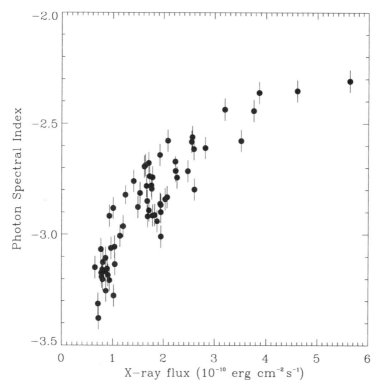

FIGURE 2. Correlation between the observed 2-10 keV flux and the photon spectral index α (A single power-law function with index α, such that the flux $\Phi(E) = \Phi_0 E^{-\alpha}$, describes the spectra well).

The variation in the observed flux shows a strong correlation with the photon spectral index (figure 2). A spectral hardening is observed during a phase of increased emission. This can be explained by the injection of an electron component more rapid than typical synchrotron cooling time scales [11].

SUMMARY

We detected strong variability in the 2-10 keV emission from Markarian 421 during observations carried out in May 1996, parallel to the TeV observations by the Whipple Collaboration. The correlated variability of X-ray and TeV emission strongly supports models which involve synchrotron radiation and the Inverse Compton (IC) process to describe the spectral energy distributions of blazars. In the case where the seed photons for the IC process are the synchrotron photons (SSC models), the IC flare amplitude is expected to be proportional to the square of the synchrotron flare amplitude. The presented data suggest this scenario, in which, at least in the case of Markarian 421, the variability in the TeV luminosity (IC) is caused by variability of the X-ray photons (synchrotron). The quasi-periodicity in the X-ray emission, with similar time scales in the rise and fall times of the individual flares indicate that geometric effects may have to be considered to understand the perturbations which lead to the observed variability.

ACKNOWLEDGMENTS

The author thanks the Whipple Collaboration for cooperation concerning the TeV data, and J. Lochner, G. Rohrbach, and A. Rots of the RXTE Guest Observer Facility for providing excellent technical support. I also thank C. Akerlof, R. Sambruna, and M. Urry for helpful discussions.

REFERENCES

1. Buckley, J. H., et al., *ApJ* **472**, L9 (1996).
2. Bradt, H. V., Rothschild, R. E., & Swank, J. H., *A&A* **97**, 355 (1993).
3. Elvis, M., Lockman, F. J., & Wilkes, B. J., *AJ* **97**, 777 (1989).
4. Gaidos, J. A., et al., *Nature* **383**, 319 (1996).
5. George, U. M., Warwick, R. S., & Bromage, G. E. *MNRAS* **232**, 793 (1988).
6. Macomb, D. J., et al., *ApJ* **449**, L99 (1995).
7. McEnery, J., et al., to appear in proc. 25th ICRC (1997).
8. Mufson, S. L., et al., *ApJ* **354**, 116 (1990).
9. Takahashi, T., et al., Mem. Soc. Astron. Ital., 67, 533 (1996).
10. Takahashi, T., et al., *ApJ* **470**, L89 (1996).
11. Tashiro, M., et al., *PASJ* **47**, 131 (1995).
12. Urry, C. M., et al., *STSI preprint* 1017 (1996).

The Energy Spectrum of Mrk 421

Frank Krennrich

*Physics & Astronomy Department, Iowa State University, Ames, IA, 50011
for the Whipple Collaboration[1]*

Abstract.
Gamma-ray emission from Mrk 421 between 300 GeV and 5 TeV is reported here with emphasis on the energy distribution. The energy spectrum of Mrk 421 is presented here for data taken during a single night of observation on May 7 1996 when the gamma-ray rate reached the highest level ever recorded by a ground-based γ-ray telescope. We also present data from large zenith angle observations to search for high energy gamma-rays in the 5 TeV range taken during 3 nights of observation in June 1995. Both data sets show no evidence for a cutoff in the spectrum up to energies of 5 TeV.

INTRODUCTION

The detection of Mrk 421 at $E \geq 500$ GeV [1] has caused interest for two reasons: the emission process at the source can be severely constrained and photons can be used to probe the interaction with the infrared background radiation. AGN models that involve the production of gamma-rays via inverse Compton scattering of low energy "seed" photons from high energy electrons in the inner region of a relativistic jet have difficulties explaining multi-TeV emission. In the case of Mrk 421 a cutoff at the source is expected in the TeV region [2], but no terminal energy value for the γ-ray emission has been given. Other models such as the proton-blazar [3] model do predict γ-ray emission well beyond the 10 TeV-regime. It is therefore important to extend

---

[1] D.A. Carter-Lewis, M.A. Catanese, G. Mohanty, F. W. Samuelson, J. Zweerink, *Iowa State University*; J.H. Buckley, T.C. Weekes, *Fred Lawrence Whipple Observatory, Harvard-Smithsonian CfA, P.O. Box 97, Amado, AZ 85645-0097*; M.F. Cawley, *Physics Department, St.Patrick's College, Maynooth, County Kildare, Ireland*; P.J. Boyle, J. Bussóns-Gordo, D.J. Fegan, C. Masterson, J.E. McEnery, J. Quinn, *Physics Department, University College, Dublin 4, Ireland*; J.P. Finley, J.A. Gaidos, R.W. Lessard, G.H. Sembroski, R. Srinivasan, *Department of Physics, Purdue University, West Lafayette, IN 47907*; A.C. Burdett, S.M. Bradbury, A.M. Hillas, A.J. Rodgers, H.J. Rose, *Department of Physics, University of Leeds, Leeds, LS2 9JT, Yorkshire, England, UK*; R.C. Lamb *Space Radiation Laboratory, California Institute of Technology*

the γ-ray observations to higher energies to distinguish between models in which electrons or protons are the dominant particle species. In addition TeV γ-ray observations from Mrk 421 can be used to derive constraints on the density of the extragalactic starlight. The physical process is: $\gamma\gamma \to e^+e^-$, the absorption of the high energy radiation through pair production off optical or near infrared photons [4]. The detection of a cutoff in the energy spectra of AGN could indicate either the detection of the IR background or a turnover intrinsic to the source. The non-detection of a cutoff improves upper limits on the IR density significantly and constrains AGN models severely. Here we present observations of the BL Lacertae object Mrk 421 using two different techniques to search for γ-ray emission in the multi-TeV energy range.

First we show the energy spectrum applying two standard methods ([7]; [8]) of deriving energy spectra to data taken during the 'big flare' on May 7 1996. We also present observations using the so-called large zenith angle technique to search for the highest energies in the spectrum of Mrk 421 ([10]).

THE ENERGY SPECTRUM

The Whipple collaboration has developed two methods to derive energy spectra from TeV γ-ray observations. Method 1 uses a selection of γ-ray images after modifying "supercuts" to apply to a wide range of energies. The energy calibration of the instrument relies on laboratory optical measurements. The energy is estimated for individual events which are then binned ([7]) to form a spectrum. Method 2 using a spherical selection window for γ-ray images employs a different energy calibration method by using the images from cosmic-ray induced air showers. The energy is not individually estimated for each event and simulations are used to match the observed size spectrum (size = number of photoelectrons contained in an image) by varying the index of a simulated γ-ray spectrum ([8]) until it fits the observed spectrum.

Those two methods have been applied to data from observations of the Crab Nebula showing a consistent result on the energy spectral index and flux of the Crab Nebula between 400 GeV and 5 TeV ([8]). Both methods have been applied to the observations of Mrk 421 during the big flare.

The data from the 'big flare' on May 7 1996 is suitable for deriving an energy spectrum because it provides rich statistics and a high signal to noise ratio. The γ-ray flux of Mrk 421 reached up to 15 γ/minute on May 7 ([6]) which is about 10 times the steady flux of the Crab Nebula. The big flare observations on May 7 1996 were carried out to search for sub-hour scale time variability and exclusively ON-source runs were recorded. For the spectral analysis the background estimate is derived from 5 OFF-source runs carried out during different nights of the 1995/96 observation season. The variance of the 5 different OFF-source runs was used to estimate the systematic error of the background measurement. The total live time of the ON-source observations is

108 minutes. Figure 1 showing the differential flux as a function of the primary energy, shows no evidence for a cutoff. A preliminary spectrum derived from the data ([11]) can be expressed as follows:

$$F(E) = (2.24 \pm 0.12 \pm 0.7) \times 10^{-6} \left(\frac{E}{TeV}\right)^{-2.56\pm 0.07 \pm 0.1} \left(\frac{photons}{s\, m^2\, TeV}\right) \quad (1)$$

LARGE ZENITH ANGLE OBSERVATIONS

The second technique presented here using data from June 1995 utilizes the large zenith angle atmospheric Čerenkov technique ([9]; [12]). The data reported here are taken at zenith angles between 45°-60°. Air showers developing at large zenith angles have their shower maximum further away from the detector and therefore, the Čerenkov light is spread out over a larger area at ground level. Consequently the detector collection area increases and, as a result, large-zenith angle observations have increased statistics at higher energies. Since the Čerenkov photon density at observation level decreases with increasing zenith angle, the energy threshold increases. This causes the energy region covered by the observations to shift to significantly higher energies relative to standard small-zenith angle observations. The energy threshold is defined here so that more than 90% of the events which pass a corresponding size cut have energies above the stated energy threshold. This definition actually quotes a threshold 30% lower than the standard definition which corresponds to the energy where the detection rate has a maximum([13]; [7]) and is therefore more conservative in searching for the highest energies. The increase in collection area and energy threshold has been determined by Monte Carlo simulations and has been tested for consistency with observations of the Crab Nebula ([10]). The analysis to suppress the hadronic background efficiently has been developed using simulation results and shown to be more sensitive (by a factor of 1.66) ([10]) than using standard 'Supercuts' ([5]) which are optimized for small zenith angle observations.

HIGH ENERGY EMISSION OF MRK 421

Observations of Mrk 421 at large zenith angles were carried out during June 1995. For the analysis of the multi-TeV emission we have chosen 4 runs when the γ-ray rate was above 1.5 γ/minute. This data collected in 112 minutes of observation at 45° - 50° and 55° - 60° zenith angle has been analyzed at different energy thresholds 2 TeV, 4 TeV, 5 TeV and 8 TeV (see Table 1).

Figure 2 showing the α-distribution (orientation angle of image relative to the γ-ray source) for events with energies E$\geq$ 5 TeV exhibits an excess of 25 events with a significance of 5σ. The numbers at different energy thresholds

TABLE 1. Number of excess events from June 20, 21, 28 1995.

Energy Threshold (TeV)	Excess Events	Significance (σ)
2.0±0.6	109	9.3
4.0±1.2	41	6.0
5.0±1.5	25	5.0
8.0±2.4	10	3.1

are consistent with a power-law spectrum and no evidence for a cutoff is present in the data.

CONCLUSIONS

We have derived γ-ray spectra from Mrk 421 in the energy range 400 GeV - 8 TeV. Two different data sets using two different techniques have been used. The standard spectrum analysis using data from the big flare (May 7 1996) shows a single power-law distribution with no evidence for a cutoff up to at least 5 TeV. Large zenith angle observations show a similar extension of the high energy emission with a clear detection of γ-rays up to $E \geq 5$ TeV (5.0 σ) and evidence for $E \geq 8$ TeV (3.1 σ) photons.

REFERENCES

1. Punch M., et al., *Nature* **358**, 477 (1992).
2. Sikora M., Begelman M.C., Rees M.J., *ApJ* **421**, 153 (1994).
3. Mannhein K., *A & A* **269**, 67 (1993).
4. Gould R.J., & Schrèder G., *Phys. Rev.* **155**, 1408 (1967).
5. Reynolds P.T., et al., *ApJ* **404**, 206 (1993).
6. Gaidos J.A., et al., *Nature* **383**, 319 (1996).
7. Mohanty G., et al., *Astroparticle Physics* (submitted).
8. Hillas A.M., et al., *ApJ* (in preparation).
9. Krennrich F., et al., *Towards a Major Atmospheric Čerenkov Detector-IV*, ed. M. Cresti 161 (1995).
10. Krennrich F., et al., *ApJ* **481**, 758 (1997).
11. Zweerink J., PhD Thesis, Iowa State University (1997).
12. Tanimori T., et al., *ApJ* **429**, L61 (1994).
13. Weekes T.C., *Nuovo Cimento B* **35**, 95 (1976).
14. Rodgers A.J., private communication (1997).

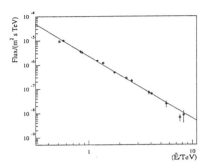

FIGURE 1. The differential flux as a function of the primary energy for the 'big flare' observation on May 7 1996 derived with method 1 (filled circles; Zweerink (1997)) and method 2 (empty circles; Rodgers (1997)).

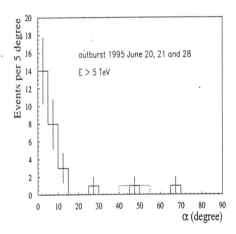

FIGURE 2. The α distribution for the data of June 20, 21 and 28 of a high state of emission at an energy threshold of 5 ± 1.5 TeV. The energy threshold has been defined to ensure that more than 90% of those events are above 5 ± 1.5 TeV (Krennrich et al. 1997).

Study of the Temporal and Spectral Characteristics of TeV Gamma Radiation from Mkn 501 During a State of High Activity by the HEGRA IACT Array

F.Aharonian[1], A.Akhperjanian[2], J.Barrio[3,4], K.Bernlöhr[1], J.Beteta[4],
S.Bradbury[3], J.Contreras[4], J.Cortina[4], A.Daum[1], T.Deckers[5], E.Feigl[3],
J.Fernandez[3,4], V.Fonseca[4], A.Fraß[1], B.Funk[6], J.Gonzalez[4], V.Haustein[7],
G.Heinzelmann[7], M.Hemberger[1], G.Hermann[1], M.Heß[1], A.Heusler[1],
W.Hofmann[1], I.Holl[3], D.Horns[7], R.Kankanian[1,2], O.Kirstein[5], C.Köhler[1],
A.Konopelko[1], H.Kornmayer[3], D.Kranich[3], H.Krawczynski[1,7], H.Lampeitl[1],
A.Lindner[7], E.Lorenz[3], N.Magnussen[6], H.Meyer[6], R.Mirzoyan[3,4,2], H.Möller[6],
A.Moralejo[4], L.Padilla[4], M.Panter[1], D.Petry[3], R.Plaga[3], J.Prahl[7], C.Prosch[3],
G.Pühlhofer[1], G.Rauterberg[5], W.Rhode[6], R.Rivero[6], A.Röhring[7],
V.Sahakian[2], M.Samorski[5], J.Sanchez[4], D.Schmele[7], T.Schmidt[6], W.Stamm[5],
M.Ulrich[1], H.Völk[1], S.Westerhoff[6], B.Wiebel-Sooth[6], C.A.Wiedner[1],
M.Willmer[5], H.Wirth[1] (HEGRA collaboration)

[1] *Max-Planck-Institut für Kernphysik, D-69029 Heidelberg, Germany*
[2] *Yerevan Physics Institute, 375036 Yerevan, Armenia*
[3] *Max-Planck-Institut für Physik, Föhringer Ring 6, D-80805 Munich, Germany*
[4] *Facultad de Ciencias Fisicas, Universidad Complutense, E-28040 Madrid, Spain*
[5] *Universität Kiel, Inst. für Kernphysik, D-42118 Kiel, Germany*
[6] *BUGH Wuppertal, Fachbereich Physik, D-42119 Wuppertal, Germany*
[7] *Universität Hamburg, II. Inst. für Experimentalphysik, D-22761 Hamburg, Germany*

Abstract. During the period March-April 1997, a high flux level of TeV γ-rays was observed from Mkn 501, using the HEGRA stereoscopic system of four imaging atmospheric Cherenkov telescopes (IACTs). The almost background-free detection of γ-rays with a rate exceeding in average $100\,\mathrm{h}^{-1}$, coupled with good energy resolution of the instrument $\leq 25\%$ allowed a study of the flux variation on time scales between 5 min and days, and measurements of the differential energy spectrum of the source for selected periods. Here we briefly discuss the results of observations with emphasis on the measurements of temporal and spectral characteristics of the source at 3 subsequent periods in April when the average flux of the source showed a remarkable increase of the flux from about one to eight times the flux observed from the Crab Nebula.

INTRODUCTION

Among the few TeV cosmic γ-ray sources are two BL Lac objects, Mkn 421 and Mkn 501 [1] – highly variable AGNs without strong emission lines, but showing a strong nonthermal (synchrotron) component of radiation from radio to hard X-ray wavelengths [2]. Interestingly, Mkn 501 has not yet been detected at MeV/GeV energies, and Mkn 421 is one of the weakest γ-ray emitting AGNs detected by the EGRET [3]. On the other hand, TeV γ-rays have not been detected from much more powerful radio loud AGNs and quasars which are strong GeV γ-ray emitters. Generally, this is in agreement with theoretical expectations that nearby BL Lac objects rather than quasars are most likely to be detected in TeV γ-rays (see e.g. [4,5]). The ASCA and Whipple discovery of correlated flares of Mkn 421 in keV and TeV energy bands supports the models which assume that both components originate in relativistic jets due to synchrotron and inverse Compton radiation of a single population of relativistic electrons [6]. In practice, this implies that the TeV radiation, combined with X-ray observations is likely to be a crucial channel for informing us about nonthermal processes in BL Lacs. TeV observations of distant extragalactic objects can address another important issue – to set meaningful limits on the diffuse extragalactic background at infrared and optical wavelengths (see [1] and references therein).

The first measurements of TeV radiation from Mkn 501 in 1995/1996 had shown a flux level significantly below the Crab flux [1]. However, in early March 1997, the Whipple group communicated the observation of strong TeV γ-ray emission from Mkn 501 which has increased by an order of magnitude, compared with the period of 1995/1996.

OBSERVATIONS

The HEGRA collaboration is continuously monitoring Mkn 501 with six imaging telescopes located on Canary Island La Palma since March'97. The results obtained by the $5\,\mathrm{m}^2$ telescope CT1, which operates in the 'stand alone' mode, will be published elsewhere [7].

The observations of Mkn 501 by the partially completed stereoscopic system of HEGRA IACTs consisting of 4 telescopes, CT3–CT6 [8], are described in [9]. The telescopes have mirrors with $8.5\,\mathrm{m}^2$ area and 5 m focal length and are equipped with 271-pixel photomultiplier cameras with $0.25°$ pixel size and $\approx 4.3°$ effective field of view. The recent observations of TeV γ-rays from the Crab Nebula by the same CT3/4/5/6 telescope system [8], combined with Monte Carlo calculations, provide good understanding of the performance of the instrument in its current status, namely: energy threshold $\approx 500\,\mathrm{GeV}$, angular resolution as good as $0.1°$, energy resolution $\leq 25\%$, minimum detectable flux of 0.3 Crab for 1 h observation of a point source.

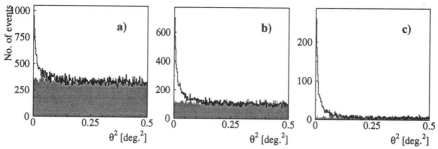

FIGURE 1. Distribution $dN/d\theta^2$ of events in the square of the angle θ relative to the direction to the source. The shaded histogram shows the background. (a) before shape cuts, (b) after loose shape cuts, and (c) after tight shape cuts.

The presented Mkn 501 data sample comprises observations from March 15/16 to April 13/14 with a total exposure time of 26.7 hours. The image analysis, reconstruction of the shower axis, and determination of the shower energy are described elsewhere [8,9]. Fig. 1 shows the $dN/d\theta^2$ distributions of the reconstructed shower directions for all events which triggered at least two telescopes. Remarkably, due to the good angular resolution as well as the strong suppression of hadronic showers at the trigger level, a clear γ-ray signal is visible already in the raw data (Fig.1a). Application of the image shape cuts (γ-rays generate narrower and more regular images compared with the images of hadronic showers) provides further suppression of the background caused by cosmic rays. To maintain the high efficiency of acceptance for γ-ray events, below we use very loose cuts (Fig.1b) which reduce the background by a factor of 3, while the number of events in the peak remains almost unchanged. Finally, at the $\approx 50\%$ expense of signal statistics, the background can be almost eliminated by applying tight shape cuts [8] (Fig.1c). The reconstructed position of the source is consistent with position of Mkn 501 within 0.01°.

TIME VARIABILITY AND ENERGY SPECTRA

To investigate time variability, data are grouped in different time bins. Fig.2a shows the detection rate on a night by night basis for the whole data set. While the rate decreases by $\approx 60\%$ during the first 9 nights, it increases by a factor of 3.5 during April 9/10 and again by a factor 4 during April 12/13. Fig.2b gives a closer view on the period from April 12 to April 14 in 5 min intervals. While one could interpret a sub-hour structure into these data, they are not inconsistent with a constant flux within each of these 3 nights.

The stereoscopic HEGRA IACT system with its energy resolution of about 25% enables detailed study of the differential energy spectrum of γ-rays, if the photon statistics exceeds several hundreds. This allows us to measure not

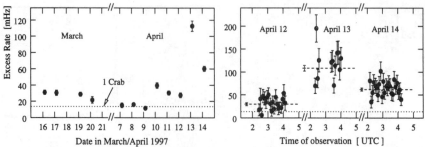

FIGURE 2. Detection rate of Mkn 501 on a night by night basis at zenith angles $\leq 30°$ (a) for the whole data set and in 5 min. intervals (b) for the last 3 nights. The dashed lines indicate the average per night, the dotted line shows the Crab detection rate.

only the spectrum on average for the whole March-April period (approximately 2500 γ-ray events detected at zenith angles up to 45°), but also for shorter periods, from 1 to 3 nights of observations depending on the flux level. Fig.3 shows the energy spectra of Mkn 501 measured during 3 periods of source activity in April : (i) Apr 7-10, (ii) Apr 11-12; (iii) Apr 13. The combination of ≥ 1 subsequent (similar) days in one period provides reasonable statistics for an evaluation of the differential spectra of the source at 3 different average flux levels (see Fig.2). Remarkably, in the case of a strong flare of the source on Apr 13 the spectrum can be derived even in a few hours of continuous observations. In Fig.3 we show the measured fluxes between 1 and 10 TeV. In fact, the statistics of γ-rays detected at both lower and higher energies formally is sufficient for flux estimates also below 1 TeV and above 10 TeV. However, due to a strong energy dependence of γ-ray detection efficiencies at lower energies and possible saturation effects at very high energies, the quantitative evaluation of the spectra in these energy bands is still in progress. In Fig.3 only statistical errors are shown. The systematic errors on the flux caused by different reasons, like the accuracy of calculations of the collection area and determination of the energy of γ-rays, as well as an uncertainty in the absolute energy calibration, currently are estimated as large as 45%. Within such uncertainties, the fluxes shown in Fig.3 can be fitted by a 'conventional' power-law spectrum, although we may notice a tendency of a gradual steepening of the observed spectra, especially for the statistically rich ones corresponding to the whole March-April dataset and to the flare in Apr 13.

The observations of Mkn 501 by the HEGRA IACT system allow a study of the time variation of the energy spectrum of the source based on the temporal behavior of the hardness ratios in counting rates, which are rather insensitive to the uncertainties in the collection areas. The hardness ratios $r_1 = R(\leq 1\,\text{TeV})/R(1 - 2.5\,\text{TeV}) = 0.310;\ 0.266;\ 0.274$, $r_2 = R(1 - 2.5\,\text{TeV})/R(2.5 - 5\,\text{TeV}) = 1.70;\ 1.71;\ 2.21$, and $r_3 = R(2.5 - 5\,\text{TeV})/R(5 - 10\,\text{TeV}) = 2.26;\ 2.80;\ 2.32$ respectively for the Apr10-12, Apr 13, and to-

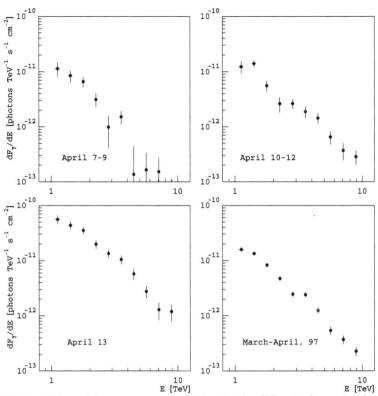

FIGURE 3. Differential energy spectra of Mkn 501 for different observational periods measured by the HEGRA IACT system. Only statistical errors are shown.

tal March-April datasets do not show, within the statistical uncertainties of about ±20%, significant variation of the spectral shape of Mkn 501 during the whole period of observations, whereas the amplitude of the absolute flux does vary during these periods by almost 1 order of magnitude.

REFERENCES

1. Weekes, T.C. et al., Review Paper; these Proceedings (1997).
2. Urry, C.M., Padovani, P., *PASP* **107**, 803 (1995).
3. von Montigny C. et al., *Ap.J* **440**, 525 (1995).
4. Dermer, C., Schlickeiser, R., *Ap.J. Suppl.* **90**, 945 (1994).
5. Ghisellini, G, Maraschi, L., *ASP Conf. Ser.*, in press (1997).
6. Takahashi, T. et al., *Ap.J* **470**, L89 (1996).
7. Lorenz, E. et al., these Proceedings (1997).
8. Daum, A. et al., *Astroparticle Physics*, submitted (1997).
9. Aharonian, F.A. et al., *A & A Letters*, submitted (1997).

Multiwavelength Observations of a Flare from Markarian 501

M. Catanese*, S.M. Bradbury‡, A.C. Breslin‖, J.H. Buckley§,
D.A. Carter-Lewis*, M.F. Cawley‖, C.D. Dermer†, D.J. Fegan‖,
J.P. Finley**, J.A. Gaidos**, A.M. Hillas‡, W.N. Johnson†, F.
Krennrich*, R.C. Lamb‡‡, R.W. Lessard**, D.J. Macomb††, J.E.
McEnery‖, P. Moriarty‖‖‖‖, J. Quinn‖,§, A.J. Rodgers‡, H.J.
Rose‡, F.W. Samuelson*, G.H. Sembroski**, R. Srinivasan**,
T.C. Weekes§, J. Zweerink* [1]

*Dept. of Physics and Astronomy, Iowa State University, Ames, IA 50011
‡Dept. of Physics, Univ. of Leeds, Leeds, LS2 9JT, Yorkshire, England, UK
‖National University of Ireland
§F.L. Whipple Observatory, Harvard-Smithsonian CfA, P.O. Box 97, Amado, AZ 85645
†E.O. Hurlburt Ctr for Space Res., Code 7650, NRL, Washington, D.C. 20375
**Dept. of Physics, Purdue Univ., West Lafayette, IN 47907
‡‡Space Radiation Laboratory, California Institute of Technology, Pasadena, CA 91125
††NASA/Goddard Space Flight Center, Code 662, Greenbelt, MD 20771
‖‖‖‖Dept. of Physical Sciences, Regional Technical College, Galway, Ireland

Abstract. We present multiwavelength observations of Markarian 501 (Mrk 501) during a flare in 1997 April. Evidence of correlated variability is seen in observations above 350 GeV taken with the Whipple Observatory very high energy (VHE) γ-ray telescope, 50 keV - 10 MeV data from OSSE, and quicklook results from the All-Sky Monitor (ASM) of *RXTE*. EGRET did not detect Mrk 501. Short term optical correlations are not conclusive but the U-band flux observed with the 1.2m telescope of the Whipple Observatory was 10% higher than in 1997 March. The average energy output of Mrk 501 appears to peak in the 2 keV to 100 keV range. The VHE γ-ray energy output is somewhat lower but more variable. These results imply an extension of the synchrotron emission in Mrk 501 to at least 100 keV, the highest observed in a blazar.

INTRODUCTION

Mrk 501 was discovered as a γ-ray source at $E > 300$ GeV by the Whipple collaboration [1] and recently confirmed by the HEGRA collaboration [2]. Like the other TeV-emitting AGN, Mrk 421, Mrk 501 exhibits day-scale variability in its VHE γ-ray emission [1], but prior to 1997, the γ-ray emission from Mrk 421 was generally observed to have a higher mean flux, be more

---

[1] This research is supported by grants from the U.S. Department of Energy and NASA, by PPARC in the UK and by Forbairt in Ireland.

frequently variable, and exhibit higher amplitude and shorter time-scale flares [3,4]. EGRET has never detected Mrk 501 [5].

Here, we present Whipple Observatory VHE γ-ray and optical, OSSE, and EGRET observations of Mrk 501 during a flare in 1997 April. We also include quicklook X-ray results from the ASM. These observations show evidence of a correlation among the wavebands covered by the above experiments, except EGRET which detected no emission from Mrk 501. More details of these results are described elsewhere [6].

OBSERVATIONS AND ANALYSIS

The VHE observations were made with the atmospheric Čerenkov imaging technique [7] using the 10 m optical reflector of the Whipple Observatory [8]. Observations of Mrk 501 in 1997 between April 7 and April 19 were taken nightly for a total exposure of 19.9 hours. The mean flux above 350 GeV during this period was $(16.4\pm3.9)\times10^{-11}$ photons cm^{-2} s^{-1}, which is ~1.6 times the flux of the Crab Nebula. The flux ranged from a low of $(4.9\pm1.8)\times10^{-11}$ photons cm^{-2} s^{-1} on April 19 to a high of $(40.5\pm9.6)\times10^{-11}$ photons cm^{-2} s^{-1} on April 16. The latter is the highest flux recorded from Mrk 501 by the Whipple Observatory. No evidence of hour-scale variability was found within this data set.

Optical observations were taken April 7 – 15 with the 1.2m telescope of the Whipple Observatory using standard UBVRI filters. The 12' full field CCD frames were analyzed using relative photometry with a 6" aperture applied to Mrk 501 and several comparison stars in the field of view. The U-band fluxes (see Fig. 1d) are expressed in arbitrary units with no galaxy light subtraction. These results are described in detail elsewhere [9].

CGRO observed Mrk 501 April 9 – 15 as the result of a Target of Opportunity initiated in response to the reported high level of VHE γ-ray activity [10]. The OSSE [11] data taken during this 6 day period yielded a very strong detection of Mrk 501 (see Table 1). The daily flux observed by OSSE varied by over a factor of 2, peaking on April 13 (see Fig. 1). Mrk 501 has not previously been observed by OSSE, so no comparison to previous flux levels can be made. The average 50 – 150 keV flux detected from Mrk 501 is higher than the emission seen by OSSE from any other blazar [12] except for the highest recorded state of 3C 273 [13]. The average of the OSSE data for this 6 day observation is well fit by a single power law in the energy range 0.05 – 2.0 MeV of the form:

$$N(E) = (1.8 \pm 0.1) \times 10^{-2} \left(\frac{E}{0.1 \text{MeV}}\right)^{-2.08\pm0.15} \text{ photons cm}^{-2} \text{ s}^{-1} \text{ MeV}^{-1}. \quad (1)$$

There was no statistically significant variation in the power law index with time for the 6 daily average OSSE measurements.

TABLE 1. OSSE flux measurements for Mrk 501

	0.05-0.15 MeV	0.15 MeV - 0.50 MeV	0.5 - 3.5 MeV
Flux[a]	24.2±1.3	2.0±0.3	<0.2

[a] Flux units are 10^{-3} photons cm^{-2} s^{-1} MeV^{-1}. The upper limit is 2σ.

The EGRET [14,15] observational data were analyzed with maximum likelihood techniques [16], taking into account the contributions of isotropic and Galactic diffuse emission [17] and other bright sources near the position of Mrk 501. The observations taken between April 9 and April 15 indicate an excess of 1.5σ, which corresponds to a 2σ upper limit of $I(>100\text{ MeV}) < 3.6 \times 10^{-7}$ photons cm^{-2} s^{-1}.

RESULTS

Figure 1 shows the light curves, plotted as daily flux levels, for the contemporaneous observations of Mrk 501. The average flux level in the U-band in 1997 March is also included in the figure to indicate the significant ($\gtrsim$10%) increase in flux between March and April. An 11 day rise and fall in flux is evident in the VHE and ASM wavebands. The rise of this variation is present in the OSSE data and may also be present in the optical data, although the variation is small (at most 6%) due in part to the galaxy light contribution. The VHE, OSSE and ASM data all show evidence of a peak in the emission on April 13. The ratio of the fluxes between April 13 and April 9 are 4.2, 2.6, 1.7, and 1.01 for the VHE, OSSE, ASM, and U-band emission respectively.

In Figure 1 we show the spectral energy distribution, plotted as the power per logarithmic bandwidth, νF_ν, versus frequency, ν. The average flux for the period from April 9 to April 15, the duration of the *CGRO* ToO, is indicated by the filled circles. The OSSE data for this plot have been divided into 8 energy bands between 50 and 470 keV. No excess emission is seen above 470 keV so those data are not plotted. A peak in the power output occurs within the 2-100 keV range and the VHE γ-ray power output is much higher than at EGRET energies but somewhat below that in the 2-100 keV range.

DISCUSSION

For Mrk 501, like Mrk 421 [3,25], the VHE γ-rays and the soft X-rays vary together, the energy budget at X-rays and γ-rays is comparable, there is evidence of correlated optical variability, and the cutoff in the synchrotron emission occurs at much higher energies than in radio-selected BL Lac objects or flat-spectrum radio quasars. However, Mrk 421 has a large deficit in energy output between 1 keV and 50 keV [12,3] while for Mrk 501 there is instead a

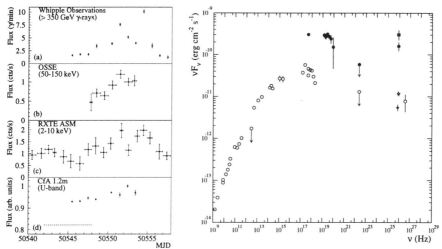

FIGURE 1. *Left:* (a) VHE γ-ray, (b) 50 keV - 150 keV, (c) 2-10 keV, and (d) U-band light-curves of Mrk 501 for the period April 2 (MJD 50540) to April 20. The dashed line in (d) indicates the average U-band flux in March. *Right:* The spectral energy distribution of Mrk 501. Contemporaneous data taken April 9 – 15 (filled circles), archival data (open circles) [18–24,2], the mean VHE γ-ray flux in 1995 (filled star) [1] and in 1996 (open star), and the maximum VHE γ-ray flux, detected on April 16, (filled square) are shown.

peak in the power output at ∼100 keV, and the deficit has shifted to between 1 MeV and 350 GeV. Also, the variability amplitude for the VHE γ-rays is larger than the X-ray and OSSE variations, whereas Mrk 421 shows comparable amplitude variations between VHE γ-rays and X-rays [3,25].

A likely explanation of the OSSE detection is that the synchrotron emission in Mrk 501 extends to 100 keV. The correlated variability between the hard X-rays and the VHE γ-rays implies that the same nonthermal electrons responsible for the two synchrotron emission produce VHE γ-rays by Compton scattering of low energy photons [26–29]. With these assumptions and following the procedure outlined by Catanese et al. [6], we obtain an expression for the allowed magnetic field strength of the emission region:

$$[E_{\rm obs}({\rm eV})\delta \Delta t_{\rm obs}^2({\rm days})]^{-1/3} \lesssim B \lesssim 2E_{\rm syn}(100~{\rm keV})\delta/[E_{\rm C}({\rm TeV})]^2 \quad (2)$$

where $\Delta t_{\rm obs}$ is the decay timescale of the emission observed at energy $E_{\rm obs}$, δ is the Doppler factor of the jet, B is the magnetic field in Gauss, $E_{\rm syn}$ is the synchrotron cutoff energy, and $E_{\rm C}$ is the maximum energy of the inverse Compton photons. If the lower limit in equation 2 is found to exceed the upper limit when δ is set equal to 1, then beaming is implied. The *RXTE* ($E_{\rm obs}$ = 2000 eV) flux varies on time-scales at least as short as one day, implying that $B \gtrsim 0.08\delta^{-1/3}$ G. If the high-energy spectrum of Mrk 501 continues without break to $\gtrsim 5$ TeV, then beaming is implied.

$\gamma\gamma$ transparency arguments also provide lower limits to δ. The > 350 GeV γ-rays interact with the X-ray photons in the extreme relativistic regime of pair production (i.e., $E_{\rm VHE}E_{\rm X} \gg m_e^2 c^4$), which has been treated by Dermer & Gehrels [31]. One finds that $\tau_{\gamma\gamma} \lesssim 0.1$ for this interaction, so no beaming is required. Improved limits to the Doppler factor for either of the above tests will be obtained if short time scale correlations are detected between VHE γ-rays and EUVE or optical photons. Also, if the spectrum observed with OSSE extends beyond 0.511 MeV, $\gamma\gamma$ transparency arguments can be applied to this γ-ray emission directly [12].

REFERENCES

1. Quinn, J. et al., *ApJ* **456**, L83 (1996).
2. Bradbury, S. M., et al., *A&A* **320**, L5 (1997).
3. Buckley, J. H., et al., *ApJ* **472**, L9 (1996).
4. Gaidos, J. A., et al., *Nature* **383**, 319 (1996).
5. Thompson, D. J., et al., *ApJS* **101**, 259 (1995).
6. Catanese, M., et al., *ApJ Letters*, submitted (1997).
7. Cawley, M. F., & Weekes, T. C., *Exp. Astron.* **6**, 7 (1995).
8. Cawley, M. F., et al., *Exp. Astron.* **1**, 173 (1990).
9. Buckley, J. H., & McEnery, J. E., in preparation (1997).
10. Breslin, A. C., et al., *IAUC*, 6596 (1997).
11. Johnson, W. N., et al., *ApJS* **86**, 693 (1993).
12. McNaron-Brown, K., et al., *ApJ* **451**, 575 (1995).
13. McNaron-Brown, K., et al., *ApJ* **474**, L85 (1996).
14. Kanbach, G., et al., *Space Sci. Rev.* **49**, 69 (1988).
15. Thompson, D. J., et al., *ApJS* **86**, 629 (1993).
16. Mattox, J. R., et al., *ApJ* **461**, 396 (1996).
17. Hunter, S. D., et al., *ApJ* **481**, 205 (1997).
18. Gear, W. K., et al., *MNRAS* **267**, 167 (1994).
19. Owen, F. N., Spangler, S. R., & Cotton, W. D., *AJ* **84**, 351 (1980).
20. Impey, C. D., & Neugebauer, G., *AJ* **95**, 307 (1988).
21. Kotilainen, J. K., et al., *MNRAS* **256**, 149 (1992).
22. Edelson, R., et al., *ApJS* **83**, 1 (1992).
23. Singh, K. P., & Garmire, G. P., *ApJ* **297**, 199 (1985).
24. Staubert, R., et al. *A&A* **162**, 16 (1986).
25. Macomb, D. J., et al., *ApJ* **449**, L99 (1995).
26. Maraschi, L., Ghisellini, G., & Celotti, A., *ApJ* **397**, L5 (1992).
27. Bloom, S. D., & Marscher, A. P., *ApJ* **461**, 657 (1996).
28. Dermer, C. D., Schlickeiser, R., & Mastichiadis, A., *A&A* **256**, L27 (1992).
29. Sikora, M., Begelman, M. C., & Rees, M. J., *ApJ* **421**, 153 (1994).
30. Marscher, A. P., & Gear, W. K., *ApJ* **298**, 114 (1985).
31. Dermer, C. D., & Gehrels, N., *ApJ* **447**, 103 (1995).

Recent Observations of γ-rays above 1.5 TeV from Mkn 501 with the HEGRA 5 m² Air Čerenkov Telescope

D. Kranich, E. Lorenz[1], D. Petry for the HEGRA Collaboration

Max-Planck-Institut für Physik, Föhringer Ring 6, 80805 München, Germany

Abstract. Since February 1997 the BL Lac object Mkn 501 is in a "high state" of γ-ray emission. The HEGRA collaboration has studied Mkn 501 with their air Čerenkov telescopes on La Palma. Here we report on observations with the 5 m² telescope (threshold ≈ 1.5 TeV) operated in a stand alone mode. We observed a rapidly varying flux between 0.5 to 6 times of that from the Crab Nebula. On average a Mkn 501 flux of $(2 + 1.3 - 0.5) \times 10^{-11} \text{cm}^{-2}\text{s}^{-1}$ has been determined. The spectrum extends at least up to 10 TeV with an integral power law coefficient of 1.8 ± 0.2 and seems to be steeper than in 1996.

INTRODUCTION

The AGN Mkn 501 is known to be a VHE γ-ray emitter [1,2] while it has not been observed by EGRET in the HE domain above 100 MeV. It is a variable source with a significant increase in intensity since its discovery by [1] in 1995. Since about February 1997 the source is in a 'high state' of emission showing strong variability [3]. The HEGRA collaboration has studied Mkn 501 since early March with their air Čerenkov telescopes (ACT). The ACTs are part of a detector complex for the study of cosmic rays over an energy range between ≈ 500 GeV and 10^{17} eV. The installation is located on the Roque de los Muchachos, La Palma, Islas Canarias (28.8° N, 17.8° W, 2200 m a.s.l.). In total, 6 ACTs (5 of 8.5 m² and one with 5 m² mirror area) are in operation. Four of the 8.5 m² ACTs were operated in the so-called stereo-mode while the 5 m² ACT (hereafter called CT1) was operated in a stand-alone mode. Here we report on observations with CT1 while the results from the stereo system are reported elsewhere at this symposium and first data are already submitted for publication [4]. Compared to the system, CT1 has a higher threshold (1.5

---
[1] corresponding author

vs. ≈ 0.5 TeV) and only a 3.25° camera of 127 pixels, but was available for longer observation time starting at larger zenith angles. Also, the DAQ has a larger dynamic range for pulse height recording and observations can be carried out under the presence of moon light [5]. Technical details of CT1 are given in [6].

DATA SAMPLE AND ANALYSIS

Since the 9th of March (MJD 50517) the source Mkn 501 has been observed continuously whenever weather conditions permitted it (≈ 30 % of time lost). During periods of moonlight, observations were carried out with reduced photomultiplier (PM) HV resulting in a 30-60 % higher threshold. Only during the days of full moon and the source being closer than 30° to the moon the measurements had to be stopped because PM anode currents exceeded safe limits. Observation times were typically 2-5 hours per night with raw-data trigger rates of 0.5 - 0.9 Hz (zenith angle dependent). The trigger condition was a coincidence of 2 pixels (of 127) exceeding each 17 photoelectrons (PE). After MJD 50568 the threshold was lowered to 13 PE by increasing the HV by 6 % resulting in a raw- data trigger rate of up to 1.8 Hz.

The data were recorded entirely in the so-called tracking mode. The filtering and processing of the data and background determination followed the prescription given in [2]. After applying some filter cuts to remove accidental

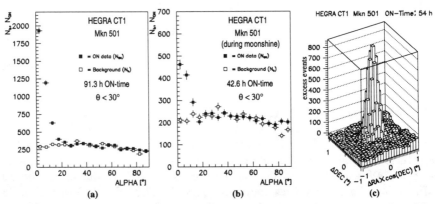

FIGURE 1. (a) The ALPHA distribution for on-source data taken during moonless night and background after all cuts except that on ALPHA. The angle ALPHA describes the orientation of the shower image and is expected to be small for γ-showers from the direction of the assumed source-position. At ALPHA < 10° there are 2554 excess events (40.4 σ). (b) like (a) but for the data taken at times when the moon was above the horizon. At ALPHA < 10° there are 455 excess events (12.4 σ). (c) Excess event distribution obtained by varying the assumed source position by up to ±1° along the RA and DEC axes (see text).

triggers etc., standard Hillas image parameters were calculated. For the selection of γ-ray candidates modified Supercuts, so-called dynamical supercuts [7] were applied. These cuts vary with energy (equivalent parameter: SIZE), shower impact parameter (equivalent parameter: DIST) and the zenith angle θ. The cuts were at first optimised with Monte Carlo (MC) data and then fine-tuned on previously recorded data from the Crab Nebula.

Fig. 1a and b show the ALPHA distributions for the total data with $\theta < 30°$ which was used to produce the light curve shown in figure 3 together with the background determined from previously recorded off-source data. Fig. 1c shows a "lego plot" of excess events versus assumed source position. The excess is centered within 0.02° (= tracking error of CT1) of the nominal Mkn 501 position and has an angular spread of $\sigma = (0.10 \pm 0.02)°$. For a short period, CT1 was pointed 0.3° away from the source and the data showed the excess with the same shift and identical spread. The significance of the daily signal varied between 1 and 15 σ depending on the flux and threshold while the excess rate varied between 2 and 60 h^{-1}. The background, however, remained stable (see fig. 2).

RESULTS

The rates were converted into a flux taking into account the collection area (determined by Monte Carlo simulations), change of threshold due to θ, increase of threshold due to the different HV settings and the dynamical supercuts (loss of γ-ray events of $\approx$ 40 %). Details will be given in a forthcoming paper. In order to assess the variability of the source, we present here the lightcurve for the data at $\theta < 30°$ with an average threshold around 1.7 TeV (Fig. 2). The threshold rises with zenith angle and with reduced HV therefore the flux data from the concerned observations were extrapolated to 1.5 TeV by using a power law coefficient of -1.8 (see below). The complete analysis of the data up to $\theta = 60°$ will be presented elsewhere. The daily flux showed large variations sometimes exceeding the CRAB flux by up to a factor 6. We determine an average flux (MJD 50517 to MJD 50608) of

$$F_{\text{Mkn501,March-June1997}}(E > 1.5\text{TeV}) = (2 + 1.3 - 0.5) \times 10^{-11} \text{cm}^{-2}\text{s}^{-1}$$

where the errors given are mainly systematic. The comparable flux in 1996 was [2] $F_{\text{Mkn501,March-August1996}}(E > 1.5\text{TeV}) = (2.3(\pm 0.4)_{\text{stat}}(+1.5 - 0.6)_{\text{syst}}) \times 10^{-12} \text{cm}^{-2}\text{s}^{-1}$ and the Crab Nebula flux [2] $F_{\text{Crab,1996}}(E > 1.5\text{TeV}) = (7.7(\pm 1.0)_{\text{stat}}(+4.6 - 1.9)_{\text{syst}}) \times 10^{-12} \text{cm}^{-2}\text{s}^{-1}$.

From MJD 50548 to 50554 (April 9 - 15, 1997) EGRET was pointed to the direction of Mkn 501 but no evidence for γ-ray emission above 100 MeV has been found from the preliminary analysis [8] with an upper limit in the range of the previous one of $1.7 \times 10^{-7} \text{cm}^{-2}\text{s}^{-1}$ while we measured during the same time a mean flux of $(3 + 1.9 - 0.8) \times 10^{-11} \text{cm}^{-2}\text{s}^{-1}$. This indicates that

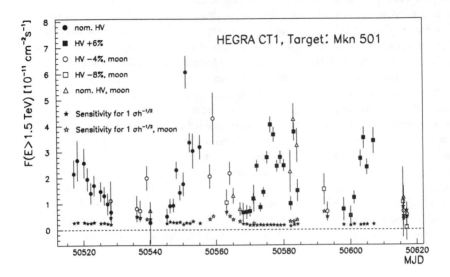

FIGURE 2. The preliminary daily flux measurements from March to June 1997 from the data at zenith angle $\theta < 30°$. The different symbols indicate the different data taking conditions. Also indicated is the sensitivity, i.e. the flux corresponding to a γ-rate which is equal to the squareroot of the background rate in h^{-1}; this gives a measure of the stability of the general observing conditions.

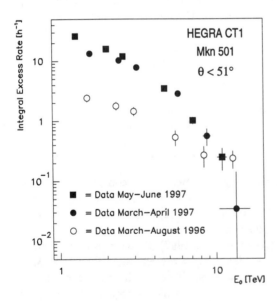

FIGURE 3. Integral γ-candidate excess rates vs. energy threshold E_0 for all data taken in 1996 (220 h on-time) and 1997 (54 h on-time in the March-April dataset and 80 h in the May-June dataset) at zenith angles $\theta < 51°$. The bars are the statistical errors. The threshold was raised by applying a zenith angle dependent cut on SIZE. The zenith angle distributions of the three datasets up to 51° are similar.

the spectrum must be very hard with an integral power law index between 100 MeV and 1.5 TeV < 1. In Fig. 3 we show the average integral γ-excess rates as a function of energy thresholds E_0 for the first and the second half of the current observation period and for the data from 1996 (the 1996 data published in [2] were reanalysed using the dynamical supercuts which resulted in a 6.8 sigma signal with 494 excess events). These rates can be turned into fluxes by dividing by the appropriate expected rate which has to be determined from Monte Carlo studies. Preliminary investigations show that up to about 10 TeV the 1997 spectrum can be described by a power law with an integral coefficient of 1.8 ± 0.2 which is within the errors compatible with the slope of 1.58 ± 0.51 of the 1996 data but the convergence of the rates at larger energies seems to indicate that the spectrum in 1997 is steeper than it was in 1996 and that the flux at energies above 10 TeV is essentially unchanged. This needs further investigations. Both the 1996 and 1997 spectra are incompatible with a large cosmological IR background calculated by [9] (and references therein) although a single source observation is not sufficient to rule out all possible scenarios.

As Mkn 501 was also observed by the Whipple and CAT groups (see their contributions to this symposium) and both locations are at different longitude it is expected that soon nearly continuous data will be available covering a long period. The ongoing observations of Mkn 501 at energies from radio waves to ≈ 10 TeV will provide a unique opportunity to study the multi-wavelength behaviour of BL Lac objects.

Acknowledgements

We thank T. C. Weekes for the early information on the increase in activity of Mkn 501. The support of the German BMBF and Spanish CICYT is gratefully acknowledged. Also we thank the Instituto de Astrofisica de Canarias for the use of the site and the provision of excellent working conditions.

REFERENCES

1. Quinn J., et al., *ApJ* **456**, L83 (1996).
2. Bradbury S.M., Deckers T., Petry D., et al., *A&A* **320**, L5 (1997).
3. Breslin, A.C., et al., IAU Circular No. 6592 (1997).
4. Aharonian, F., et al., preprint astro-ph/9706019
5. Raubenheimer C., et al., *Observation of the Crab Nebula in the presence of moonlight*, contributed paper, 25th ICRC, Durban, (1997).
6. Mirzoyan R., et al., *NIM A* **351**, 513 (1994).
7. Kranich, D., and Petry, D., contributed paper, Workshop "Towards a major atmospheric Cherenkov Detector V", South Africa, August 1997.
8. Kanbach, G., (EGRET) private communication
9. Stecker, F.W., and de Jager, O.C., preprint astro-ph/9608072

BeppoSAX Monitoring of the BL Lac Mkn 501

E. Pian[1,2], G. Vacanti[3], G. Tagliaferri[4], G. Ghisellini[4],

L. Maraschi[5], A. Treves[6], C. M. Urry[1], F. Fiore[7], P. Giommi[7],

E. Palazzi[1], L. Chiappetti[8], R. M. Sambruna[9]

[1] *STScI, 3700 San Martin Drive, Baltimore MD 21218*
[2] *ITESRE-CNR, via P. Gobetti 101, I-40129 Bologna, Italy*
[3] *ESA ESTEC, Space Science Department, Astrophysics Division, Postbus 299, NL-2200 AG Noordwijk, The Netherlands*
[4] *Osservatorio Astronomico di Brera, Via Bianchi 46, I-22055 Merate (Lecco), Italy*
[5] *Osservatorio Astronomico di Brera, Via Brera 28, I-20121 Milan, Italy*
[6] *Department of Physics, University of Milan at Como, Via Lucini 3, I-22100 Como, Italy*
[7] *SAX SDC, Via Corcolle 19, I-00131, Rome, Italy*
[8] *IFCTR-CNR, Via Bassini 15, I-20133 Milan, Italy*
[9] *NASA/Goddard Space Flight Center, Greenbelt, MD 20771*

Abstract.

The BL Lac object Mkn 501 was observed with the BeppoSAX satellite on 7, 11, and 16 April 1997 during a phase of high activity at TeV energies, as monitored with the Whipple, HEGRA and CAT Cherenkov telescopes. Over the whole 0.1-200 keV range the spectrum was exceptionally hard ($\alpha \leq 1$, with $F_\nu \propto \nu^{-\alpha}$) indicating that the X-ray power output peaked at (or above) ~ 100 keV. This represents a shift of at least two orders of magnitude with respect to previous observations of Mkn 501, a behavior never seen before in this or any other blazar. The overall X-ray spectrum hardens with increasing intensity and, at each epoch, it is softer at larger energies. The correlated variability from soft X-rays to the TeV band points to models in which the same population of relativistic electrons produces the X-ray continuum via synchrotron radiation and the TeV emission by inverse Compton scattering of the synchrotron photons or other seed photons.

INTRODUCTION

Mkn 501, one of the closest (z=0.034) BL Lacertae objects, and one of the brightest at all wavelengths, was the second source, after Mkn 421, to be detected at TeV energies by the Whipple and HEGRA observatories [1,2]. Historically its spectral energy distribution (νF_ν) resembles that of BL Lac objects selected at X-ray energies (or HBL [3]), having a peak in the EUV-soft X-ray energy band. Accordingly, the 2–10 keV spectra observed so far were relatively steep, with energy spectral indices α larger than unity ($F_\nu \propto \nu^{-\alpha}$). From the EXOSAT data base the hardest X-ray spectrum of Mkn 501, observed in one of its brightest states, had a spectral index of 1.2 ± 0.1 [4]. *Einstein* measured in two occasions spectral indices consistent with values smaller than 1 within the large errors [5].

Mkn 501 was observed with BeppoSAX [6] over a period of $\sim$10 hours on each 7, 11, and 16 April 1997, during a multiwavelength campaign involving ground-based TeV Cherenkov telescopes (Whipple, HEGRA and CAT), plus other satellites, CGRO (EGRET and OSSE), RXTE, ISO and optical telescopes. in this paper we briefly report the BeppoSAX results and their implications for blazar models, deferring to a separate paper [7] for an exahustive presentation.

OBSERVATIONS, ANALYSIS, AND RESULTS

Event files for the LECS (0.1–10 keV), and the three MECS (1.5–11 keV) experiments were linearized and cleaned with SAXDAS; light curves and spectra were accumulated for each pointing with the SAXSELECT tool, using 8.5 and 4 arcmin extraction radii for the LECS and the MECS, respectively, that provide more than 90% of the fluxes. For the background subtraction we used the files accumulated from blank fields available from the BeppoSAX Science Data Center (SDC) public ftp site. Source+background and background spectra were accumulated for each of the four PDS units (15-300 keV), using the XAS software package. The source was significantly detected up to the highest energy channels. Each net spectrum was binned in energy intervals to reach a signal-to-noise ratio larger than 20, up to 150 keV. The grouped spectra from the four units were then coadded for each pointing.

Spectral analysis was performed with the XSPEC 9.01 package, using the response matrices released by the SDC. For each observation, the LECS and MECS spectra have been jointly fit with both a simple and a broken power-law model. The latter yields a better representation of the data, and a fitted N_H closer to the Galactic value [8]. We then fixed N_H at the Galactic value, and determined under this assumption the best-fit parameters for the broken power-law model. The spectral indices below and above the break energy of $\sim$2 keV, respectively, are 0.63 ± 0.04, 0.91 ± 0.02 (7 Apr), 0.64 ± 0.04, 0.80 ± 0.02 (11 Apr), 0.40 ± 0.04, 0.59 ± 0.02 (16 Apr). Single power-law fits to the PDS data in the 20–200 keV range give a good representation of the spectra. The spectral indices are 0.98 ± 0.13 (7 Apr), 0.79 ± 0.09 (11 Apr), 0.84 ± 0.04 (16 Apr). At each epoch the overall X-ray spectrum is softer at higher energies and in each band hardens with increasing intensity (see Figure 1).

DISCUSSION

In Figure 1 the unfolded and unabsorbed X-ray spectra from the first and last BeppoSAX observations are compared with previously observed X-ray spectra [4] and with data in the radio [9], mm [10–13], far-infrared [14], optical [15,16], UV [17]; TeV-rays [1,2,18] (squares). Nearly simultaneous Whipple TeV data [19] are indicated as filled circles, while the open circle (13 April) and the TeV fitting curve along with its 1-σ confidence range (15-20 March) are from the HEGRA experiment [20].

The present X-ray observations imply a dramatic hardening of the spectral energy distribution in the medium X-ray band and an increase of the (apparent) bolometric luminosity of a factor ≥ 20 with respect to previous epochs. The really striking feature is that the peak of the power output is found to *shift* in energy by a factor ≥ 100. Moreover for the first time in any object the peak is observed to *occur in the hard X-ray range*, definitely above $\sim 50 - 100$ keV. Since in the optical the source was nearly normal [21] the change of the spectral energy distribution seems to be confined to energies greater than ~ 0.1 keV, as also indicated by the apparent pivot of the three BeppoSAX spectra.

The overall continuity of the X-ray spectra presently observed with previous UV and soft X-ray measurements suggests that the X-ray emission constitutes the high energy end of the synchrotron component and thus that its peak frequency (i.e., where $\alpha = 1$) increased by more than two orders of magnitude with respect to previous observations of Mkn 501.

The TeV emission also brightened by more than a factor of 5 in the first two weeks of April, with the most intense TeV flare peaking on 16 April [19], the date of the last BeppoSAX observation. However, unlike the X-ray spectrum, the TeV spectrum was rather steep ($\alpha \simeq 1.5$) and did not show noticeable temporal variation [20,22].

The synchrotron self-Compton [23–25] model explains naturally the correlated flaring at X-ray and TeV energies if the energy distribution of the emitting electrons, $N(\gamma)$, changes at the highest energies. In fact the highest energy electrons produce X-rays via synchrotron and the TeV radiation via inverse Compton scattering. Due to the very high electron energies involved, the scattering cross section for energetic photons is reduced by the Klein-Nishina effect and only photons below the Klein-Nishina threshold ($h\nu \leq mc^2/\gamma$) are effectively upscattered. Therefore a) the inverse Compton flux does not vary more than the synchrotron flux, as it would normally be observed; b) the peak of the inverse Compton power shifts in frequency less than the synchrotron peak when the critical electron energy changes.

The shift of $\sim$2–3 orders of magnitude in the frequency ν_S of the synchrotron peak cannot be ascribed to either a variation of Doppler factor δ or mag-

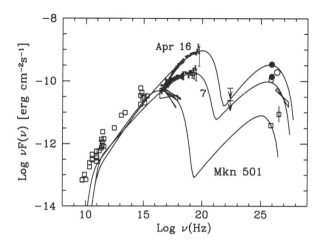

FIGURE 1. Spectral energy distribution of Mkn 501. See text for the data points. The solid lines indicate fits with a one-zone, homogeneous synchrotron self-Compton model: in the "quiescent state" the electrons are continuously injected with a power-law distribution ($\propto \gamma^{-2}$) between $\gamma_{min} = 2 \times 10^4$ and $\gamma_{max} = 4 \times 10^5$. For the fit to the 7 April spectrum the injected electron distribution ($\propto \gamma^{-1.4}$) extends from $\gamma_{min} = 1$ and $\gamma_{max} = 4 \times 10^6$. For the fit to the 16 April spectrum, the injected electron distribution ($\propto \gamma^{-1.1}$) extends from $\gamma_{min} = 1$ and $\gamma_{max} = 8 \times 10^6$. For all models, the beaming factor is $\delta = 10$ and the magnetic field $B \sim 0.5$ Gauss.

netic field B alone (enormous variations would be demanded, since $\nu_S \propto B\delta$). Therefore a real change in power is implied. Assuming that the variation is only due to a change in the electron energy, this must have increased by roughly a factor of $\sim$10–30. The corresponding shift of the inverse Compton peak is expected to be a factor of $\sim$10–30, in the linear Klein-Nishina regime. Since the cooling time of these very high energy electrons is rather short and the synchrotron peak did not move back to the quiescent position during at least 10 days, a mechanism of continuous particle injection is required.

Three simple, one-zone synchrotron self-Compton models along these lines are shown in Figure 1 for the quiescent state and the 7 and 16 April states of Mkn 501. The $N(\gamma)$ distribution has been found self-consistently solving the continuity equation including continuous injection of relativistic particles, radiative losses, electron-positron pair production and taking into account the Klein-Nishina cross section [26]. The parameters of the fits are given in the figure caption. A variation of the maximum energy of the emitting electrons, together with an increased luminosity and a flattening of the injected particle distribution can describe the observed spectra quite well. We have also assumed that the region where the variable and dominant X-ray and TeV flux is produced is more compact than the region responsible for the "quiescent" spectrum, but characterized by the same beaming factor and magnetic field.

REFERENCES

1. Quinn, J., et al. 1996, ApJ, 456, L83
2. Bradbury, S. M., et al. 1997, A&A, 320, L5
3. Padovani, P., & Giommi, P. 1995, ApJ, 444, 567
4. Sambruna, R. M., et al. 1994, ApJS, 95, 371
5. Urry, C. M., Mushotzky, R. F., & Holt, S. S. 1986, ApJ, 305, 369
6. Boella, G., et al. 1997, A&AS, 122, 299
7. Pian, E., et al. 1997, ApJL, submitted
8. Elvis, M., Lockman, F. J., & Wilkes, B. J. 1989, AJ, 97, 777
9. Gear, W. K., et al. 1994, MNRAS, 267, 167
10. Steppe, H., et al. 1988, A&AS, 75, 317
11. Wiren, S., Valtaoja, E., Teräsranta, H., & Kotilainen, J. 1992, AJ, 104, 1009
12. Lawrence, A., et al. 1991, MNRAS, 248, 91
13. Bloom, S. D., & Marscher, A. P. 1991, ApJ, 366, 16
14. Impey, C. D., & Neugebauer, G. 1988, AJ, 95, 307
15. Véron-Cetty, M.-P., & Véron, P. 1993, ESO Scientific Report No. 13, 1
16. Burbidge, G., & Hewitt, A. 1987, AJ, 93, 1
17. Pian, E., & Treves, A. 1993, ApJ, 416, 130
18. Weekes, T. C., et al. 1996, A&AS, 120, 603
19. Catanese, M., et al. 1997, ApJ, submitted
20. Aharonian, F., et al. 1997, A&A, submitted
21. Buckley, J., & McEnery, J. 1997, in preparation
22. Aharonian, F., et al. 1997, in Proc. of the 4th Compton Symposium, Williamsburg, 27-30 April 1997, in press
23. Jones, T. W., O'Dell, S. L., & Stein, W. A. 1974, ApJ, 188, 353
24. Ghisellini, G., Maraschi, L., & Dondi, L. 1996, A&AS, 120, 503
25. Mastichiadis, A., & Kirk, J. G. 1997, A&A, 320, 19
26. Ghisellini, G. 1989, MNRAS, 236, 341

Multiwavelength Observations of the February 1996 High-Energy Flare in the Blazar 3C 279

A. E. Wehrle[1], E. Pian[2], C. M. Urry[2], L. Maraschi[3], G. Ghisellini[4], R. C. Hartman[5], G. M. Madejski[5], F. Makino[6], A. P. Marscher[7], I. M. McHardy[8], J. R. Webb[9], G. S. Aldering[10], M. F. Aller[11], H. D. Aller[11], D. E. Backman[12], T. J. Balonek[13], P. Boltwood[14], J. Bonnell[5], J. Caplinger[15], A. Celotti[16], W. Collmar[17], J. Dalton[12], A. Drucker[12], R. Falomo[18], C. E. Fichtel[5], W. Freudling[17], W. K. Gear[19], N. Gonzalez-Perez[20], P. Hall[21], H. Inoue[6], W. N. Johnson[22], M. R. Kidger[20], R. I. Kollgaard[23], Y. Kondo[5], J. Kurfess[22], A. J. Lawson[8], B. McCollum[5], K. McNaron-Brown[22], D. Nair[9], S. Penton[24], J. E. Pesce[2], M. Pohl[17], C. M. Raiteri[25], M. Renda[12], E. I. Robson[19,26], R. M. Sambruna[5], A. F. Schirmer[13], C. Shrader[5], M. Sikora[35], A. Sillanpää[27], P. S. Smith[28], J. A. Stevens[12], J. Stocke[24], L. O. Takalo[27], H. Teräsranta[29], D. J. Thompson[5], R. Thompson[15], M. Tornikoski[29], G. Tosti[30], P. Turcotte[12], A. Treves[31], S. C. Unwin[32], E. Valtaoja[29], M. Villata[25], S. J. Wagner[33], W. Xu[1], A. C. Zook[34]

[1] *Infrared Processing Analysis Center, MC 100-22, California Institute of Technology, Pasadena, CA 91125*
[2] *Space Telescope Science Institute, 3700 San Martin Drive, Baltimore, MD 21218*
[3] *Osservatorio Astronomico di Brera, Via Brera 28, I-20121 Milan, Italy*
[4] *Osservatorio Astronomico di Brera, Via Bianchi 46, I-22055 Merate (Lecco), Italy*
[5] *NASA/Goddard Space Flight Center, Greenbelt, MD 20771*
[6] *ISAS, 3-1-1, Yoshinodai, Sagamihara, Kanagawa 229, Japan*
[7] *Department of Astronomy, Boston University, 725 Commonwealth Avenue, Boston, MA 02215*
[8] *Department of Physics, University of Southampton, Southampton SO9 5NH, United Kingdom*
[9] *Department of Physics, Florida International University, University Park, Miami, FL*

[10] *University of Minnesota, Department of Astronomy, 116 Church St., SE Minneapolis, MN 55455*
[11] *University of Michigan, Physics and Astronomy, 817 Dennison Building, Ann Arbor, MI 48109*
[12] *Franklin & Marshall College, Physics & Astronomy Department, P.O. Box 3003, Lancaster, PA 17604-3003*
[13] *Colgate University, Department of Physics & Astronomy, 13 Oak Dr., Hamilton, NY 13346-1398*
[14] *1655 Main St., Stittsville, Ontario, Canada K2S 1N6*
[15] *IUE Data Analysis Center, NASA/GSFC, Greenbelt, MD 20771*
[16] *ESTEC, Astroph. Div. Space Sci. Dept., Postbus 299, 2200 Noordwijk, The Netherlands*
[17] *Max Planck-Institut für Extraterrestrische Physik, Giessenbachstrasse, D-85740 Garching bei München, Germany*
[18] *Osservatorio Astronomico di Padova, Via Osservatorio 5, I-35122 Padova, Italy*
[19] *Centre for Astrophysics, University of Central Lancashire, Preston, PR1 2HE, United Kingdom*
[20] *Instituto de Astrofísica de Canarias, E-38200, La Laguna, Tenerife, Spain*
[21] *Steward Observatory, University of Arizona, Tucson AZ 85721*
[22] *Naval Research Lab., 4555 Overlook Av., SW, Washington, DC 20375-5352*
[23] *Department of Astronomy and Astrophysics, The Pennsylvania State University, University Park, PA 16802*
[24] *University of Colorado, JILA, Campus Box 440, Boulder, CO 80309-0440*
[25] *Osservatorio Astronomico di Torino, Strada Osservatorio 20, I-10025 Pino Torinese, Italy*
[26] *Joint Astronomy Center, 660 N. Aohoku Place, University Park, Hilo, HI 96720*
[27] *Tuorla Observatory, Tuorla 21500 Piikkiö, Finland*
[28] *NOAO/KPNO, N. Cherry Avenue, P.O. Box 26732, Tucson, AZ 85926*
[29] *Metsähovi Radio Research Station, 02540 Kylmala, Finland*
[30] *Osservatorio Astronomico, Università di Perugia, I-06100 Perugia, Italy*
[31] *SISSA/ISAS, Via Beirut 2-4, I-34014 Miramare-Grignano (Trieste), Italy*
[32] *MS 306-388, Jet Propulsion Laboratory, California Institute of Technology, 4800 Oak Grove Drive, Pasadena, CA 91109*
[33] *Landessternwarte, Heidelberg-Königsstuhl, D-69117 Heidelberg, Germany*
[34] *Pomona College, Department of Physics & Astronomy, 610 College Ave., Claremont, CA 91711-6359*
[35] *Copernicus Astronomical Center, Polish Academy of Science, Warsaw, Poland*

Abstract.

We report CGRO, RXTE, ASCA, ROSAT, IUE, HST and ground-based observations of a large flare in 3C 279 in February 1996. X-rays and $\gamma-$ rays peaked simultaneously (within one day). We show simultaneous spectral energy distributions prior to and near the flare peak. The $\gamma-$ ray flare was the brightest ever observed in this source.

In February 1996, the blazar 3C 279 flared dramatically in X- and γ-rays towards the end of a three-week pointing by CGRO. The intensity of the flare was a factor of $\sim$3 and $\sim$10 in X- and γ- rays, respectively. It is the strongest γ-ray flare ever observed for this object. During one short period of 8 hours, the γ-ray flux increased by a factor of 2.6.

Our collaboration carried out observations with CGRO, RXTE, ASCA, ROSAT, IUE, HST, ISO and many ground-based observatories. The most striking result is that the RXTE light curve showed an outburst which was simultaneous with the CGRO-EGRET flare within the temporal resolution of $\sim$1 day. The optical-UV light curves, which were not as well-sampled during the high energy flare, exhibited smaller variations (factor of $\sim$2) and less obvious correlation. The flux at millimetric wavelengths was quite high, near an historical maximum. ISO observations were obtained in non-standard modes with the photometer and camera; they will be presented elsewhere (Barr et al, in preparation). Lightcurves are shown in Figure 1. A full description is presented in Wehrle et al. [1].

We show (Figure 2) simultaneous spectral energy distributions of 3C 279 prior to and near to the flare peak. In the simplest class of synchrotron self-Compton model, the peak of the gamma ray energy distribution should vary as the square of the height of the optical-IR peak. Thus, in a log-log plot of the broad-band energy distribution (Fig. 2), the displacement of the γ-ray peak would be twice that of the broad optical-IR peak. Figure 2 shows that the γ-ray variation is in fact larger than this, which poses problems for the simplest emission models.

A Note on Previous and Current Campaigns:

Hartman et al. [2] describe the assembly of nearly contemporaneous multiwavelength data, centered on CGRO and GINGA observations, during the June 1991 γ-ray flare [3]. Our first campaign, in which the CGRO, ROSAT, IUE and ground-based observations were coordinated in advance, found the source in a low state during Dec. 1992-Jan. 1993 [4]. The data reported here, using scheduled observations with CGRO, ROSAT, and IUE, independent observations made with HST, performance verification phase observations by ISO, Target Of Opportunity observations with ASCA and RXTE plus many ground-based observatories, are from the second of four campaigns. Our third (coordinated) and fourth (ad hoc) campaigns took place in Dec. 1996-Jan. 1997 and June 1997, involving CGRO, RXTE, ISO and several ground-based observatories.

We are grateful to the satellite and telescope schedulers for making the campaigns possible. A. Wehrle, S. Unwin, E. Pian, and W. Xu acknowledge support from the NASA Long Term Space Astrophysics Program. M. Urry, E. Pian, and J. Pesce acknowledge support from NASA grants NAG8-1037,

NAG5-2538, and NAG5-3138. M.F. Aller and H.D. Aller acknowledge support from NSF grant AST-9421979.

REFERENCES

1. Wehrle, A. E. et al. *(submitted to Ap.J.)* (1997).
2. Hartman, R.C., Collmar, W., von Montigny, C., and Dermer, C.D. 1997 (these proceedings)
3. Hartman, R.C. et al. *Ap.J.* **461**, 698 (1996).
4. Maraschi, L. et al. *Ap.J.* **435**, 91 (1994).

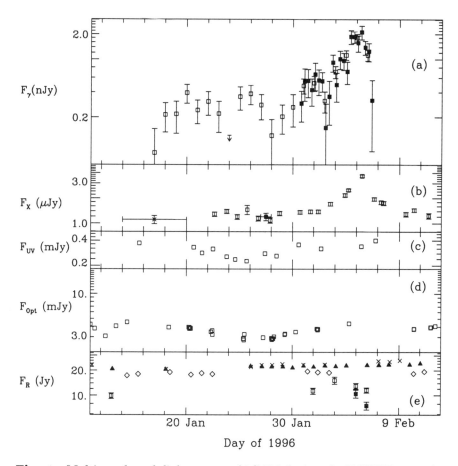

Fig. 1 – Multiwavelength light curves of 3C 279 during the EGRET campaign (1996 January 16 - February 6): *(a)* EGRET fluxes at >100 MeV binned within 1 day (open squares) and 8 hours (filled squares); *(b)* X-ray fluxes at 2 keV: besides the XTE data (open squares), the isolated ASCA (filled square) and ROSAT-HRI (cross) points are reported with horizontal bars indicating the total duration of the observation; *(c)* IUE-LWP fluxes at 2600 Å, reduced by a factor of 0.44; *(d)* optical data from various ground-based telescopes in the R (open squares). *(e)* JCMT photometry at 0.45 mm (filled squares) and 0.8 mm (open squares), radio data from Metsähovi at 37 GHz (crosses) and 22 GHz (filled triangles), and from UMRAO at 14.5 GHz (open diamonds). Errors, representing 1-σ uncertainties, have been reported only when they are bigger than the symbol size.

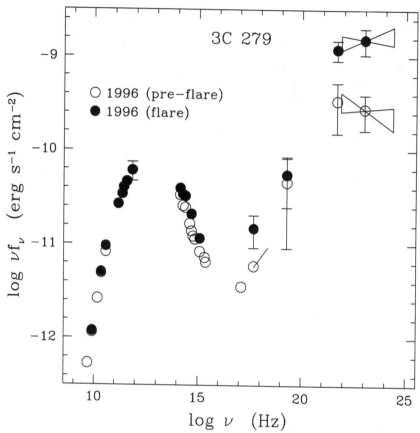

Fig. 2 – Radio-to-γ-ray energy distribution of 3C 279 in pre-flare (open dots) and flaring state (filled dots) in January-February 1996. The UV, optical and near-IR data have been corrected for Galactic extinction. The slope of the ASCA spectrum ($\alpha_\nu = 0.7$) has been reported normalized at the RXTE point closest in time. The EGRET power-law spectra referring to the 16-30 January and 4-6 February periods are shown, normalized at 0.4 GeV. Errors have been reported only when they are bigger than the symbol size.

Radio to γ-Ray Observations of 3C 454.3: 1993-1995

M. F. Aller[1], A. P. Marscher[2], R. C. Hartman[3], H. D. Aller[1], M. C. Aller[1], T. J. Balonek[4], M. C. Begelman[5], M. Chiaberge[6], S. D. Clements[7], W. Collmar[8], G. De Francesco[6], W. K. Gear[9], M. Georganopoulos[2], G. Ghisellini[6,10], I. S. Glass[11], J. N. González-Pérez[12], P. Heinämäki[13], M. Herter[14], E. J. Hooper[15], P. A. Hughes[1], W. N. Johnson[16], S. Katajainen[13], M. R. Kidger[12], A. Kraus[17], L. Lanteri[6], G. F. Lawrence[18], G. G. Lichti[8], Y. C. Lin[19], G. M. Madejski[20,21], K. McNaron-Brown[22], E. M. Moore[2], R. Mukherjee[3,23], A. D. Nair[7], K. Nilsson[13], A. Peila[6], D. B. Pierkowski[4], M. Pohl[8], T. Pursimo[13], C. M. Raiteri[6], W. Reich[17], E. I. Robson[24], A. Sillanpää[13], M. Sikora[25], A. G. Smith[7], H. Steppe[26], J. Stevens[24], L. O. Takalo[13], H. Teräsranta[27], M. Tornikoski[27], E. Valtaoja[27], C. von Montigny[3,14], M. Villata[6], S. Wagner[14], R. Wichmann[14], and A. Witzel[17]

[1] *Dept. of Astronomy, Dennison Bldg., U. Michigan, Ann Arbor, MI 48109*
[2] *Dept. of Astronomy, Boston University, 725 Commonwealth Ave., Boston, MA 02215*
[3] *NASA/Goddard Space Flight Center, Code 661, Greenbelt, MD 20771*
[4] *Dept. of Physics & Astronomy, Colgate U., Hamilton, NY 13346-1398*
[5] *U. Colorado, JILA, Box 440, Boulder, CO 80309-0440*
[6] *Osservatorio Astronomico di Torino, I-10025 Pino Torinese, Italy*
[7] *Astronomy Dept., U. Florida, S. S. R. B., Gainesville, FL 32611*
[8] *Max-Plank-Institut für extraterrestrische Physik, D-85740, Garching, Germany*
[9] *Royal Observatory, Blackford Hill, Edinburgh, EH9 3HJ, Scotland*
[10] *Osservatorio di Brera-Merate, V. Bianchi, 46, Merate, Italy*
[11] *SAAO, PO Box 9, Observatory 7935, South Africa*
[12] *Instituto de Astrofísica de Canarias, E-38200 La Laguna, Spain*
[13] *Tuorla Observatory, University of Turku, SF-21500, Piikkiö, Finland*
[14] *Landessternwarte Heidelberg-Königstuhl, D-69117 Heidelberg, Germany*
[15] *U. Arizona, Steward Observatory, Tucson, AZ 85721-0001*
[16] *NRL, Code 7651, 4555 Overlook, SW, Washington, DC 20375-5352*
[17] *Max-Planck-Institut für Radioastronomie, D-53121, Bonn, Germany*
[18] *Dept. of Physics, U. Minnesota, Minneapolis MN 55455*

[19] HEPL, Stanford Univ., Stanford, CA 94305
[20] NASA/Goddard Space Flight Center, Code 666, Greenbelt MD 20771
[21] Dept, of Astronomy, U. Maryland, College Park, MD 20742
[22] Dept. of Physics & Astronomy, George Mason U., Fairfax, VA 22030
[23] Universities Space Research Association
[24] Joint Astronomy Centre, 660 N. Aohoku Place, Hilo HI 96720
[25] N. Copernicus Ast. Center, PL-00-716, Warszawa, Poland
[26] Instituto de Radioastronomia Milimetrica, E-18012, Granada, Spain
[27] Metsähovi Radio Research Station, FIN-02540 Kylmälä, Finland

Abstract. Results from a multiwaveband study of the γ-ray-bright compact radio source 3C 454.3 are presented.

GOALS

As part of a program to study the emission from γ-ray bright AGNs, we present broadband observations of the QSO 3C 454.3 obtained in a worldwide coordinated effort during CGRO cycles 3-5. The purpose of the project was to search for correlated, possibly time-delayed, activity in the radio through X-ray regions using long-term data with sufficient sampling to define the variability, and to obtain simultaneous data across the electromagnetic spectrum at bright phase, in order to test and constrain proposed emission mechanisms. 3C 454.3 was chosen for study because of its extreme γ-ray brightness (photon flux $> 10^{-6}/\text{cm}^{-2}\text{s}^{-1}$ for E$>$ 100MeV) in August 1992, making it a likely candidate for future bright-phase detections. The period discussed here includes 3 OSSE detections, and 3 bright and 2 faint ($<2\sigma$) EGRET detections.

RESULTS

3C 454.3 has been monitored in the optical and cm-wavelength bands for several decades; these data show continuous activity with typical timescales of the order of months in the optical and of the order of years at cm-wavelengths, with peak-to-trough flux variations by a factor of ≤ 2.5; a discrete correlation function analysis of radio-optical activity through 1992 found past events to be uncorrelated (Clements et al. 1995). The present study includes groundbased and satellite observations during 1993-1995, and a 16-day period of intensive monitoring during November-December 1995 chosen to coincide with CGRO VP 507. Table 1 lists the observatories which participated in the program. Table 2 summarizes all EGRET detections and includes both photon flux, F, and flux density, S, at 10^{23} Hz, from modeling using the tabulated γ.

A goal of the program was to identify time-correlated broadband activity during EGRET detection periods. Figure 1 shows the data for selected spec-

TABLE 1. Contributing Observatories and Spectral Coverage

Observatory	Instrument	Region	Band
CGRO	EGRET	γ-ray	0.03 - 10 GeV
	COMPTEL	γ-ray	0.75 - 30 MeV
	OSSE	hard X-ray	50 - 150 keV
ROSAT	PSPC or HRI	soft X-ray	0.5 - 2 keV
Foggy Bottom Obs.	16 in	optical	VRI
Heidelberg	70 cm	optical	BR
Rosemary Hill Obs.	0.76 m	optical	BVRI
Steward Obs.	61 in	optical	BVRI
Torino	1.05 m	optical	BVR
Tuorla Obs.	1.03 m	optical	V
Teide Obs.	82 cm	optical	BVRI
Teide Obs.	1.5 m	IR	JHK
SAAO Sutherland	1.9 m	IR	JHK
Mt. Lemmon	1.5 m	IR	JHK
UKIRT	3.8 m	IR	JHK
JCMT	15 m	mm/sub-mm	150, 231, 273, 375 GHz
SEST	15 m	mm	90, 230 GHz
OVRO	mm array	mm	90-115 GHz
Effelsberg	100 m	mm/cm	1.7, 2.7, 4.8, 5.0
			8.4, 10.6, 32 GHz
Metsähovi	13.7 m	mm/cm	22.2, 36.8 GHz
UMRAO	26 m	cm	4.8, 8.0, 14.5 GHz

TABLE 2. EGRET Data for 3C 454.3

VP	Start Date	End Date	F	ΔF	S pJy	ΔS	σ	γ	$\Delta\gamma$
19.0	92/01/23	92/02/06	82	9	139	15	13.5	2.24	0.11
26.0	92/04/23	92/04/28	93	24	157	41	5.4		
28.0	92/05/07	92/05/14	26	14	43	24	2.1		
37.0	92/08/20	92/08/27	124	19	210	32	10.1	2.13	0.17
320.0	94/03/08	94/03/15	29	10	49	17	3.4		
327.0	94/05/17	94/05/24	35	11	60	19	3.8	2.01	0.26
336.0	94/08/01	94/08/04	< 60		< 102		1.7		
410.0	95/01/24	95/02/14	49	7	83	12	10.2	2.28	0.13
507.0+	95/11/28	95/12/14	< 23		< 39		1.9		

1) VP 507.0+ denotes the result for VPs 507.0+507.5. 2) F is the flux ($E > 100$ MeV) in 10^{-8} photons cm^{-2}s^{-1}; when $\sigma < 2.0$, 95% ULs are given for F and S. 3) γ is the photon number spectral index F(E)$\propto$E$^{-\gamma}$.

tral regions. 3C454.3 was active in all bands during this time period, but no clear pattern of correlated time-delays from band-to-band could be identified. The global behavior in the millimeter-through-centimeter wavelength range does show time-delays characteristic of a diverging, partially-opaque flow, but there is no clear correlation between the radio and the optical variability. We find no 1:1 correlation between EGRET and OSSE detection periods and rises or declines in either the radio or the optical light curves.

During the campaign period (28 November 1995 - 14 December 1995) the EGRET-detected γ-ray flux was $<1/5$ of its historical maximum, and the flux was relatively low in other bands. Over this period the flux in the centimeter and soft X-ray bands decreased by $\sim 10\%$. Short-term peak-to-peak fluctuations of $\sim 30\%$ dominated the optical flux behavior. Although the

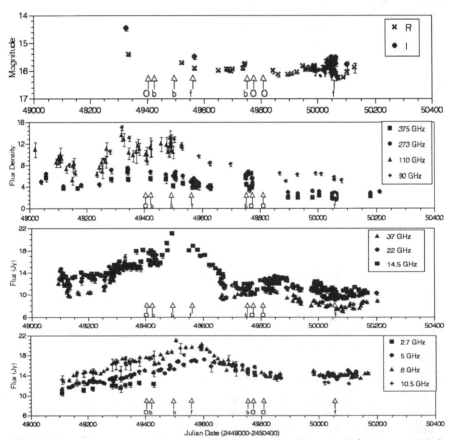

FIGURE 1. From top to bottom: Light curves in the optical (red magnitudes), submm/mm, mm/cm, and cm spectral bands. EGRET detections are marked by b (bright) and faint (f) along the abscissa. OSSE detections are marked by O.

source was too weak at MeV energies to search for variability correlated with lower frequency activity during the campaign, the simultaneous broadband data obtained during this period allowed us to construct a spectrum for this faint detection epoch which is shown in Figure 2; for comparison we show results from bright phase VP 410 which are based on the sparse simultaneous observations.

Our data show that the source was active across the spectrum during CGRO detections, but there is no well-defined correlation between these detection times and the phases seen in the radio-through-optical activity. The observed range of variability in the radio-to-soft X-ray bands is much smaller than in the γ-ray band, and this must be addressed by viable emission models. X-ray data with sufficient sampling to define the variability and simultaneous observations across the electromagnetic spectrum at key time periods are clearly needed for properly understanding the emission mechanism in this and other γ-ray-bright AGNs.

REFERENCES

1. Clements, S. D, Smith, A. G., Aller, H. D., & Aller, M. F., *AJ* **110**, 529 (1995).

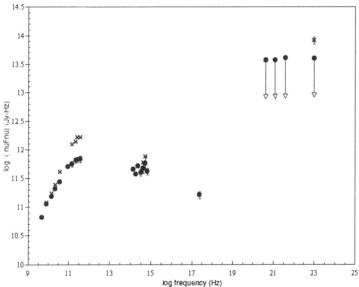

FIGURE 2. Spectral energy distributions based on simultaneous data during the campaign period VP 507 (circles) and during bright phase VP 410 (crosses). For VP 507 the error bars indicate the range of the variability when known; error estimates for individual observations are smaller than the size of the symbol.

Multi-wavelength Radio Monitoring of EGRET Sources and Candidates

P.G. Edwards, J.E.J. Lovell

Institute of Space and Astronautical Science, Japan

R.C. Hartman

NASA Goddard Space Flight Center, USA

M. Tornikoski

Metsähovi Radio Research Station, Finland

M. Lainela

Turku University Observatory, Finland

P.M. McCulloch

University of Tasmania, Australia

B.M. Gaensler, R.W. Hunstead

University of Sydney, Australia

Abstract. From several southern hemisphere radio-wavelength monitoring programs, we have compiled multi-wavelength observations of a number of both EGRET-detected sources and sources with similar radio-properties that have not been detected by EGRET. The radio observations were made at 0.8, 2, 5, 8, 90 and 230 GHz, though not all sources were monitored at all wavelengths. The data are compared with EGRET data to the end of Cycle 4 (October 1995) in an effort to confirm previously reported correlations between activity at radio wavelengths and EGRET detections, and to investigate whether the non-EGRET-detected sources were in radio- and gamma-ray-quiet phases during this period.

INTRODUCTION

EGRET-detected AGN have been identified at high statistical confidence with strong, compact, flat-spectrum radio sources [1]. However, not all radio sources sharing these properties have been detected above 100 MeV. An understanding of why this is the case will be of great value in determining the means by which these gamma rays, which dominate the energy budget of the EGRET AGN, are produced.

Reich et al. have examined the radio state of AGN detected by EGRET [2]. With the exception of the strongest radio sources (e.g. 3C273 and 3C279) Reich et al. found enhanced radio emission between 2.7 and 230 GHz with a delay of up to several months relative to the EGRET detection.

Valtaoja and Teräsranta refined this analysis in a comparison of Metsähovi 22 and 37 GHz radio flux density monitoring and EGRET phase 1 observations [3]. They concluded that highly polarized quasars (HPQ) are much more likely to be detected by EGRET than other types of AGN, and that gamma-rays are detected from these objects when they are in the rising or peak stages of a millimetre wavelength flare.

Mattox et al. note that EGRET is detecting $\sim$10% of blazars, but that a 'duty cycle' of $\sim$30% is observed for strong EGRET sources [1]. This duty cycle is too large to explain the fact that only one-tenth of the blazars are detected, unless there is strong variability on timescales longer than $\sim$1 year.

Given the suggested correlations between source activity at radio and gamma-ray wavelengths, a close study of the long-term variability of EGRET source and candidates at radio wavelengths is timely. Studies to date have largely been limited to the northern hemisphere. We have collected data from several southern hemisphere monitoring programs to examine the variability of a selection of southern EGRET sources and a group with similar radio characteristics that EGRET has not, to date, detected.

The specific questions that we seek to answer are whether EGRET detects AGN during periods of increased activity at radio wavelengths, whether the sources EGRET has not detected show different behaviour, and whether there any correlations between gamma ray flux and contemporaneous radio flux density at any wavelength.

The seventeen southern hemisphere sources included in this study are listed in Table 1 with classifications from a preliminary literaure search. Due to space restrictions, in this paper we present the results for one source in the sample, though discuss the preliminary conclusions of this study more generally.

OBSERVATIONS

The 843 MHz data are part of an on-going program being conducted with the Molonglo Observatory Synthesis Telescope [4]. The 2.3 and 8.4 GHz ob-

TABLE 1. The full sample of sources in this program.

Source name	EGRET source	Redshift	Classification
PKS 0208−512	E	1.003	QSO HPQ
PKS 0438−436		2.852	QSO HPQ
PKS 0454−234		1.009	BLLac HPQ
PKS 0454−463		0.858	QSO LPQ
PKS 0521−365	E	0.055	Ngal BLLac HPQ
PKS 0537−441	E	0.894	BLLac OpVar
PKS 0637−752		0.654	QSO
PKS 1127−145	E	1.187	QSO
PKS 1424−418		1.522	QSO HPQ
PKS 1510−089	E	0.360	QSO HPQ OpVar Sy1
PKS 1514−241		0.042	Ngal BLLac
PKS 1622−253	E	0.786	Gal LPQ
PKS 1730−130	E	0.902	QSO OpVar LPQ
PKS 1741−038		1.054	QSO HPQ
PKS 1921−293		0.352	QSO OpVar HPQ
PKS 2005−489		0.071	QSO OpVar HPQ
PKS 2052−474	E	1.489	BLLac LPQ

servations were made as part of a series of observations with the University of Tasmania's 26 m telescope [5]. The 90 and 230 GHz observations were made with the SEST telescope [6]. The data are compared with EGRET data to the end of Cycle 4 (October 1995).

VLBI observations of many of the sources in this sample, which are of great importance in understanding the nature of these objects on the parsec-scale, have been presented elsewhere (e.g. [7–10] and references therein).

PKS 2005−489

PKS 2005−489 is a highly polarized quasar. Although not a robust EGRET detection, it has been detected at marginal significance on several occasions [11]. The radio monitoring of this source, shown in Figure 1, is the least of all the sources in this survey.

The radio flux densities are plotted with the same linear scale to facilitate comparison between frequencies. No statistically compelling detection of the source has been made by EGRET and so open circles are plotted corresponding to the periods when the source was within 30° of the instrument axis.

The 90 GHz flux density has increased by a factor of 2 over 5.5 years. The 230 GHz data, while clearly undersampled, indicates that shorter timescale variability may also be a feature of this source. The error bars are determined from the rms of the observations, with an additional uncertainty due to pointing and calibration uncertainties added in quadrature [6]. Interestingly, the rapid increase at 230 GHz seen in 1992 coincides with the three EGRET pointings to this area of the sky which resulted in the sources inclusion as a marginal detection in the first EGRET catalog [12]. A subsequent reanalysis, and phase two observations, decreased the likelihood that the source was emitting gamma-rays at these energies (see [11] for more details).

CONCLUSIONS

The results for other sources in this study will be presented in full elsewhere. The radio monitoring data we have are relatively sparse – particularly so for PKS 2005-489.

In general, the EGRET sources in this sample show significant activity at 90 and 230 GHz. Although a suggestion of weak gamma ray emission from PKS 2005-489 coincides with a period of rising millimetre flux density, in general EGRET detections do not appear to be confined only to periods when the 90 GHz flux density is increasing.

The sources not detected by EGRET are similarly variable at millimetre wavelengths, and a number of the sources have been within the EGRET field of view during such radio flares but not detected at gamma ray energies. Particularly notable is the repeated non-detection of PKS 1921-293 during an extended period when it was very bright at radio wavelengths.

The lack of discernible difference in the total flux density and its variability between the two, admittedly small, groups, in this sample indicates either that radio flux density levels are not reliable indicators of gamma ray activity, or that, for at least some sources, other factors are more dominant in determining whether the source is detectable by EGRET.

REFERENCES

1. Mattox, J.R. et al., *ApJ* **481**, 95 (1997).
2. Reich, W. et al., *A&A* **273**, 65 (1993).
3. Valtaoja, E. and Teräsranta, H., *A&A* **297**, L13 (1995).
4. Hunstead, R.W. and Gaensler, B.G., *Extragalactic Radio Sources*, eds R. Ekers et al., Dordrecht: Kluwer, p. 103 (1996).
5. King, E.A. et al., in preparation (1997).
6. Tornikoski, M. et al., *A&ASupp* **116**, 157 (1996).
7. Tingay, S.J. et al., *ApJ* **464**, 170 (1996).
8. Tingay, S.J. et al., these proceedings (1997).
9. Shen, Z.-Q. et al., *AJ*, in press (1997).
10. Shen, Z.-Q. et al., in preparation (1997).
11. Lin, Y.C. et al., these proceedings (1997).
12. Fichtel, C.E. et al., *ApJSupp* **94**, 551 (1994).

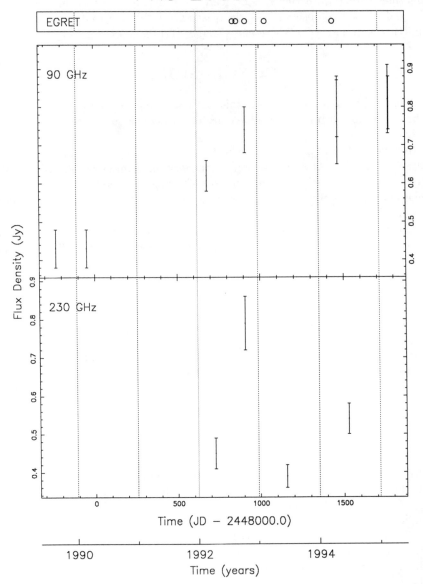

FIGURE 1. Flux density variation of PKS 2005–489 at 90 and 230 GHz. Open circles denote EGRET pointings which included the source within the field of view.

VLBI Observations of Southern Hemisphere Gamma-Ray Loud and Quiet AGN

Tingay, S.J. and Murphy, D.W.

California Institute of Technology, Jet Propulsion Laboratory (MS238-332)
4800 Oak Grove Drive, Pasadena, CA 91109
email: tingay@hyaa.jpl.nasa.gov; dwm@casa.jpl.nasa.gov

Edwards, P.G.

Institute of Space and Astronautical Science
3-1-1, Yoshinodai, Sagamihara-shi, Kanagawa 229, Japan
email pge@orihime.isaslan1.isas.ac.jp

Costa, M.E., McCulloch, P.M. and Lovell, J.E.J.

University of Tasmania
GPO Box 252C, Hobart, Tasmania 7001, Australia

Jauncey, D.L., Reynolds, J.E., and Tzioumis, A.K.

CSIRO, Australia Telescope National Facility
PO Box 76, Epping, NSW 2121, Australia

Preston, R.A., Meier, D.L., and Jones, D.L.

California Institute of Technology, Jet Propulsion Laboratory (MS238-332)
4800 Oak Grove Drive, Pasadena, CA 91109

Nicolson, G.D.

Hartebeesthoek Radio Astronomy Observatory
PO Box 443, Krugersdorp 1740, Transvaal, South Africa

CP410, *Proceedings of the Fourth Compton Symposium*
edited by C. D. Dermer, M. S. Strickman, and J. D. Kurfess
© 1997 The American Institute of Physics 1-56396-659-X/97/$10.00

Abstract. We have observed, using very long baseline interferometry (VLBI), some EGRET-identified AGN in the Southern Hemisphere: PKS 0208−512; PKS 0521−365; PKS 0537−441; PKS 1127−145; PKS 1622−253; PKS 1622−297; PKS 1730−130; and PKS 1908−201 as well as some prominent, compact southern radio sources which are not gamma-ray loud: PKS 0438−438; PKS 0637−752; and PKS 1921−293. We are searching for radio wavelength properties on parsec-scales which can distinguish between gamma-ray loud and gamma-ray quiet AGN. Such properties might include relativistic beaming indicators such as parsec-scale jet proper motions and radio core brightness temperatures, as well as parsec-scale to kiloparsec-scale jet bend angles.

INTRODUCTION

The EGRET instrument, aboard the *Compton Gamma-Ray Observatory*, has identified, with high confidence, over 40 compact extragalactic radio sources as sources of > 100 MeV gamma-rays (Mattox et al. [3]). The gamma-ray luminosities of some of these sources are exceedingly high, up to 10^{48-49} ergs s^{-1}, equivalent to the Eddington limit for a $10^{10} M_\odot$ black hole, if isotropic emission is assumed.

The radio sources identified by EGRET are generally found to have superluminal jets on the milliarcsecond scale and high radio core brightness temperatures, in excess of the nominal inverse Compton limit. In light of their gamma-ray and radio properties the most commonly favored paradigm for the EGRET-identified radio sources is as follows: the gamma-ray emission region lies deep within the base of a relativistic jet which is directed close to our line of sight. The beamed radio emission from the jet produces high brightness temperature radio cores and superluminal jet motions, and the gamma-ray emission is beamed to give high apparent luminosities.

It would seem worthwhile, therefore, to search for differences between the EGRET-identified radio sources and gamma-ray quiet radio sources, based upon relativistic beaming models, by probing their jets with the highest resolution possible, using VLBI.

RELATIVISTIC BEAMING INDICATORS

pc-scale jet speeds

We have compared the apparent pc-scale jet speeds of the EGRET-identified and gamma-ray quiet radio sources in the compilation of Vermeulen & Cohen [9] and have found no significant evidence of any difference [6].

Radio core brightness temperatures

We have derived radio core brightness temperatures, and thus estimated lower limits for the jet Doppler factors, for the sources we have observed thus far at frequencies between 1.6 and 8.4 GHz (those mentioned in the abstract) (Tingay et al. [7]). We have also compared data from the literature for EGRET-identified and gamma-ray quiet radio sources, most notably from the space VLBI results of Linfield et al. [1,2] and the ground based 22 GHz VLBI observations of Moellenbrock et al. [4].

Even though Moellenbrock et al. find EGRET-identified radio sources to have (on-average) higher brightness temperatures than gamma-ray quiet radio sources, our results and the results of Linfield et al. show a wide range in brightness temperature which is apparently independent of whether or not the source has been identified in gamma-rays. This indicates that there is no one-to-one correspondence between beamed radio emission and gamma-ray emission, at least among the most prominent sources.

Jet bending

Finally, we have made a study of jet bending in the 42 EGRET-identified radio sources listed by Mattox et al. and a sample of 26 gamma-ray quiet, flat spectrum radio sources derived from the complete sample of Pearson & Readhead [5], following the suggestion of von Montigny et al. [10] that strong flat-spectrum radio sources which are gamma-ray quiet may preferentially have bent or misaligned jets, causing the gamma-ray emission to appear Doppler dimmed. We have investigated the frequency and degree of jet bending on pc and kpc scales, as well as pc-scale to kpc-scale misalignments. Statistical tests show no evidence that the jet bending characteristics of radio sources can be associated with the presence or absence of gamma-ray emission. Full results of this investigation will be presented elsewhere (Tingay et al. [8]).

CONCLUSIONS

Even though relativistic beaming is likely to play an important role for the appearance of gamma-ray sources, it does not - to first order, and for the quantity and quality of data available thus far - seem to explain why some flat-spectrum radio sources have strong gamma-ray emission and many others do not. At the least it seems that there is no clear one-to-one relationship between gamma-ray emission and beamed radio emission.

REFERENCES

1. Linfield, R.P. et al. 1989, ApJ, 336, 1105

2. Linfield, R.P. et al. 1990, ApJ, 358, 350
3. Mattox, J.R. et al. 1997, ApJ, 481, 95
4. Moellenbrock, G.A. et al. 1996, AJ 111, 2174
5. Pearson, T.J. & Readhead, A.C.S. 1988, ApJ 328, 114
6. Tingay, S.J. et al. 1996, ApJ 464, 170
7. Tingay, S.J. et al. 1997a, in preparation
8. Tingay, S.J. et al. 1997b, in preparation
9. Vermeulen, R.C. & Cohen, M.H. 1994, ApJ 430, 467
10. von Montigny, C. et al. 1995 A&A, 299, 680

VLBA Monitoring of Three Gamma-Ray Bright Blazars: AO 0235+164, 1633+382 (4C 38.41) & 2230+114 (CTA 102)

W. Xu & A. E. Wehrle

Infrared Processing and Analysis Center, MC 100-22, Jet Propulsion Laboratory, California Institute of Technology, Pasadena, California, 91125, USA

A. P. Marscher

Astronomy Dept., Boston University, Boston, Massachusetts, 02215, USA

Abstract. We present VLBA observations of three gamma-ray bright blazars: 0235+164, 1633+382 (4C38.41) and 2230+114 (CTA102). 0235+164 was unresolved at 22 and 43 GHz during the period between Nov. 1993 and Aug. 1996. The brightness of the core varied significantly. 1633+382 (4C38.41) has a sharp leading edge in its jet. The jet edge moved approximately at a constant velocity of $(9.9 \pm 0.7)\ h^{-1}c$ between March 1994 and May 1995. The velocity is much faster than that of $(6.1 \pm 1.1)\ h^{-1}c$ measured by Barthel et al. (1995). This may indicate that a component is moving at a faster speed. 2230+114 (CTA 102) produced several new components between Feb. 1995 and Aug. 1996. These components moved at apparent velocities of $(12 - 21)\ h^{-1}c$. The components at ~ 1.7 mas and ~ 8 mas seem to be stationary.

I INTRODUCTION

We have been monitoring a representative sample of gamma-ray bright blazars with VLBA at 22 and 43 GHz since 1993, as part of multiwaveband campaigns. The ultimate goals of these campaigns are to determine the origin(s) of the gamma-ray emission and the structure of the innermost region of blazars. The VLBA monitoring can determine whether the gamma-ray flares are associated with production of superluminal knots. Measurement of the proper motion of the knots will provide an estimate of the bulk Lorentz factor. These information is critical for discriminating between theoretical models.

We present VLBA observations of three gamma-ray bright blazars: 0235+164, 1633+382 (4C38.41) and 2230+114 (CTA102) in this paper.

II   0235+164

0235+164 is a BL Lac object ($z = 0.94$) seen through foreground galaxies. It is one of the most violently variable objects. It has shown optical variability with an amplitude of five magnitudes (e.g. Stickel et al. 1993). Its flux density at high radio frequencies (15, 22 and 37 GHz) decayed nearly 95% from the peak in late 1992 within four years (Aller, private communication; Terasranta, private communication). The rapid variability of 0235+164 may be due to microlensing events produced by the foreground galaxies (Stickel et al. 1993).

0235+164 shows a dominant core and some faint extended structure on kpc scales (Murphy et al. 1993), but is unresolved at pc scales (Wehrle et al. 1992). We observed 0235+164 at three epochs at 22 GHz, and at five epochs at 43 GHz between Nov. 1993 and Aug. 1996. It was unresolved during our monitoring program, but its brightness varied significantly.

III   1633+382 (4C38.41)

1633+382 is among the first group of AGNs detected by EGRET. It is identified with a quasar at redshift $z = 1.814$. Strong variations at radio and optical wavebands have been detected (e.g. Barthel et al. 1995). 1633+382 has a core-dominated triple morphology on kpc scales at radio frequency (Murphy et al. 1993). It has a core-jet structure on pc scales. The jet bends through ~90 degrees in a region between ~10 mas and ~30 mas from the core (Pearson & Readhead, 1988; Polatidis et al. 1995). Barthel et al. (1995), based on three observations at 5 GHz over a period of 7 years (from 1979.25 to 1986.89), measured a proper motion of (0.16 ± 0.03) mas/yr, which corresponds to an apparent velocity of (6.1 ± 1.1) $h^{-1}c$ ($H_0 = 100$ h $km/s/Mpc$, $q_0 = 0.5$)

We made seven epochs of observations at 22 GHz with VLBA from March 1994 to May 1996. The maps (Fig. 1) show that the jet has a sharp leading edge at 22 GHz and that during our observation, the jet "edge" moved at a constant speed while maintaining roughly the same brightness. The jet "edge" moved (0.56 ± 0.04) mas over 2.17 years, yielding an apparent velocity of (9.9 ± 0.7) $h^{-1}c$. The core varied dramatically during our monitoring. It changed from ~1 Jy in early 1994 to ~2 Jy in early 1995, then back to ~1.3 Jy in 1996. The change of the flux density in the core accounted for most of the change in the total flux density.

The apparent velocity which we measured is much faster than that reported by Barthel et al. (1995). This may indicate that 1633+382 produced a faster component over the last decade.

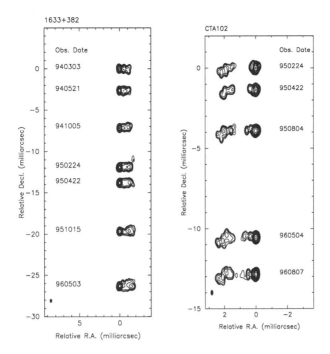

Figure 1. (left) VLBA monitoring of 1633+382 at 22 GHz. Restoring beam is FWHM 0.40 x 0.25 mas, p.a. 0°. Contours are in factors of 2. The lowest contour is 4.7 mJy/beam.

Figure 2. (right) VLBA monitoring of 2230+114 at 43 GHz. Maps are rotated by 48° clockwise. Restoring beam is FWHM 0.30 x 0.15 mas, p.a. 0°. Contours are in factors of 2. The lowest contour is 3.3 mJy/beam.

IV   2230+114 (CTA 102)

2230+114 is identified with a quasar at redshift $z = 1.037$. It is a peculiar object. Bååth (1987) reported that the components detected at 932 GHz separated at an apparent velocity of $(18 \pm 4)\ h^{-1}c$. Wehrle & Cohen (1989) found no detectable motion at 5 GHz, with an upper limit 0.5 mas/yr. Recently, Rantakyrö et al. (1996) combined VLBI images at different frequencies and concluded that the apparent motion appears to increase with distance from the core, from 0 c close to the core to $(15 \pm 6)\ c$ at $\sim$10 mas from the core.

We observed 2230+114 at 43 GHz at 5 epochs between Feb. 1995 and Aug. 1996 (Fig. 2), and at 22 GHz at 4 epochs between May 1994 and April 1995 (Figs. 3 and 4). The two outer components, at $\sim$8 mas and $\sim$1.7 mas southeast to the core respectively, have complex structure and seem to be stationary compared to previous observations (Rantakyrö et al. 1996; Wehrle & Cohen 1989). However, the structure within $\sim$1 mas of the core changed

TABLE 1. Apparent Motion of Components in 2230+114

Components	Timespan (years)	Angular Distance Traveled (mas)	Proper Motion (mas/yr)	Apparent Velocity (h^{-1} c)
C1	0.36	0.28	0.77	~21
C2	0.81	0.34	0.42	~12
C3	0.20	0.12	0.60	~17

dramatically. It is clear that several new components were produced between Feb. 1995 and Aug. 1996.

Our result indicates that components near the core move at high velocities (12 - 21 c) (Table 1), with no evidence of motion in the region between ~1.5 mas and ~10 mas from the core. It contradicts with the suggestion by Rantakyrö et al. (1996) that components accelerate with distance. We suggest that the jet in 2230+114 moves along a helical path, and that the two stationary components are where the jet curves towards the line of sight. The jet bending at ~10 mas from the core (Wehrle & Cohen 1989) is consistent with this scenario.

ACKNOWLEDGMENTS

This work was supported partly by the NASA Long Term Space Astrophysics Program and and the NASA Gamma Ray Observatory Guest Investigator Program. WX is supported by the National Research Council Associateship Programs. The National Radio Astronomy Observatory is a facility of the National Science Foundation, operated under a cooperative agreement by Associated Universities, Inc.

REFERENCES

1. Bååth, L. B., *Superluminal Radio Sources*, eds. Zensus & Pearson, p206 (1987).
2. Barthel et al., *ApJ*, **444**, L21 (1995).
3. Murphy, D. W., Browne, I. W. A., & Perley, R. A. *MNRAS*, **264**, 298 (1993).
4. Pearson, T. J., & Readhead, A. C. S., *ApJ*, **328**, 114 (1988).
5. Polatidis, A. G., Wilkinson, P. N., Xu, W., Readhead, A. C. S., & Pearson, T. J. *ApJS*, **98** , 1 (1995).
6. Rantakyrö, F. T., Bååth, L. B., Dalacassa, D., Jones, D. L., & Wehrle, A. E. *A&A*, **310**, 66 (1996).
7. Stickel, M., Fried, J. W., & Kuhr, H, *A&AS*, **98**, 393 (1993).
8. Wehrle, A. E., & Cohen, M., *ApJ*, **346**, L69 (1989).
9. Wehrle et al. *ApJ*, **391**, 589 (1992).

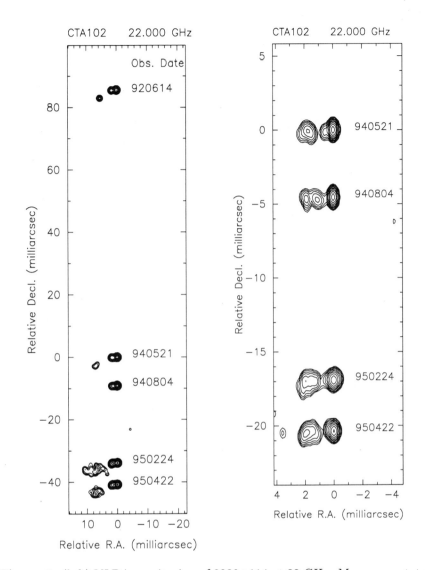

Figure 3. (left) VLBA monitoring of 2230+114 at 22 GHz. Maps are rotated by 48° clockwise. Restoring beam is FWHM 1.0 x 1.0 mas, p.a. 0°. Contours are in factors of 2. The lowest contour is 4.0 mJy/beam. The image of 920614 is reconstructed from the Gaussian model published by Rantakyrö et al. (1996).

Figure 4. (right) VLBA monitoring of 2230+114 at 22 GHz. Maps are rotated by 48° clockwise. Restoring beam is FWHM 0.6 x 0.3 mas, p.a. 0°. Contours are in factors of 2. The lowest contour is 3.7 mJy/beam.

Coordinated Millimeter-wave Observations of Bright, Variable Gamma-ray Blazars with the Haystack Radio Telescope

J. Gregory Stacy*, W. Thomas Vestrand*, and
Robert B. Phillips†

*Space Science Center, University of New Hampshire, USA
†MIT Haystack Observatory, Westford, Massachusetts, USA

Abstract.
We present results of an ongoing program to monitor at millimeter wavelengths a select sample of bright, variable gamma-ray blazars with the 37-m Haystack radio telescope [1]. Our primary objective during the 1996–1997 observing season was to follow the blazars 3C 279 and PKS 0528+134, each the subject of intense multiwavelength observing campaigns in parallel with observations carried out with the *Compton* Gamma Ray Observatory (CGRO). Our secondary objective was to monitor those active galactic nuclei (AGN) identified as particularly strong candidates for gamma-ray flares based on their CGRO detection history. Quasi-weekly monitoring sessions were carried at frequencies of 43 and 86 GHz with the Haystack telescope using new beam-switching instrumentation. In particular, we have applied a promising new technique (termed "multiple drift scanning") that has been demonstrated to reliably and repeatedly measure continuum source fluxes to levels of a few hundred milliJanskys at 43 GHz, and to ~1-1.3 Jy at 86 GHz. We describe our observing program and summarize the results obtained to date. In brief, the majority of sources monitored during the 1996-1997 observing season, though bright, have remained relatively steady in their millimeter emission.

SCIENTIFIC MOTIVATION

The detection of highly-variable gamma-ray emission from active galactic nuclei (AGN), primarily blazars, with the EGRET experiment aboard the *Compton* Gamma Ray Observatory (CGRO) [1-3] has opened a new era of

---

[1] Radio astronomy at the Haystack Observatory of the Northeast Radio Observatory Corporation (NEROC) is supported by the National Science Foundation.

study of these objects. The rapid variability observed in gamma-rays (∼days, [1,2]) indicates that the high-energy emission arises from regions very near the central engine of the galaxy, where jet formation and the acceleration of relativistic plasma occurs. The gamma-rays constitute a new direct probe of the "inner-jet" region, heretofore unobservable, even by VLBI techniques, in quasars and related objects.

As emphasized by numerous investigators [4,5], coordinated multiwavelength observations of flares from blazars, when combined with the high-energy data, offer the prospect of determining the physical structure and properties of the inner-jet region, and will provide the critical diagnostic measurements necessary for discriminating between the various theoretical models (primarily Compton-scattering models) put forward to explain the broadband emission from blazars. A key ingredient of such multiwavelength campaigns is the ability to measure the time-delays, between wavebands, of the brightness variations occurring during a flare. The relative order and delay of these frequency-dependent variations differ according to the predictions of the various models.

We have been participating in a large, international effort to follow the brightest and most variable of the gamma-ray AGN detected by the CGRO/EGRET instrument, in hopes of observing some in active or flaring states. Recent studies suggest a specific connection between the gamma-ray and mm-wave emission in blazars [4,6,7]. High-energy flaring activity in the gamma-ray blazar 3C 279 in January 1996, for example, has been reported with correlated emission in lower-frequency bands [8]. These observations demonstrate the feasibility of measuring the frequency-dependent time delays of flare emission from blazars, and the particular value of such measurements in the millimeter (and shorter) wavebands, where opacity effects are minimized. In fact, elevated mm-wave fluxes are now used as a major criterion for declaring a CGRO target-of-opportunity observation, if seen to exceed specified threshholds.

OBSERVING PROGRAM

Quasi-weekly observations, weather permitting, were carried out with the Haystack radio telescope from December 1996 through April 1997. Primary monitoring observations were conducted at a frequency of 43 GHz (7 mm), with retuning to 86 GHz (3 mm) under exceptional conditions. Our allocated LST range covered the period from $\sim 6^h - 15^h$, with occasional extensions to higher and lower times. Sources are observed when possible both rising and setting at an elevation corresponding to the optimum gain of the Haystack telescope (at ∼40 degrees elevation). Planets and bright sources (e.g., 3C 273) are used for pointing checks. Planets (primarily Mars) are used as flux calibrators, with observations of the near-steady quasar 3C 286 [9] as an independent

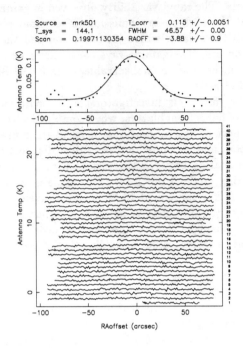

FIGURE 1. Sample multiple drift scans of MRK 501 taken at 43.12 GHz (7 mm) on 23 April 1997 (TJD 10561) during the CGRO target-of-opportunity observations of VP 617.8. The measured source antenna temperature of 0.115 K corresponds to a gain-corrected intensity of ∼0.9 Jy.

check on instrument stability and performance.

To minimize the effects of the Haystack radome, beam-switched continuum measurements are carried out with a somewhat "retro" observing technique, termed Multiple Drift Scanning (or MDS, [10]). In the MDS technique, the telescope is positioned ahead of the source of interest, collecting data as the target drifts through the antenna beam. The antenna is then "leap-frogged" ahead of the moving target, and the data-collection sequence repeated as necessary to achieve the desired signal-to-noise (see Figure 1). The accumulated drift-scan data are added together and fit by a gaussian profile corresponding to the beam-size of the telescope. The resulting source antenna temperature is converted to a source flux after calibration. The MDS technique has been demonstrated to reliably and repeatedly measure continuum source fluxes to levels of a few hundred milliJanskys at 43 GHz, and to ∼1–1.3 Jy at 86 GHz. Further, the MDS method is estimated to be ∼2.5 times more sensitive per unit time compared to the more traditional ON-OFF method often used in determining continuum source strength [10]. Further enhancements to the Haystack continuum setup this past observing season included: 1) a higher beam-switching rate (10 Hz vs 5 Hz), and 2) a larger effective bandwidth (~240 MHz) for data collection.

PRELIMINARY SOURCE FLUXES

Preliminary source fluxes measured at 43 GHz for a selection of targets observable from Haystack are shown in Figure 2. In brief, the majority of sources, though bright, have remained relatively steady or have been declining in their millimeter emission over the course of the 1996-1997 observing season. A fuller, more complete report of these observations is in preparation.

In parallel with these CGRO-related efforts, extragalactic sources are also being monitored regularly at 86 GHz with the Haystack telescope as part of a separate program to identify bright sources for potential VLBI measurement at 3-mm wavelength with the Coordinated Millimeter Very-long-baseline Array (CMVA). These monitoring results at 86 GHz are publicly accessible and posted at the CMVA site on the World Wide Web, at http://dopey.haystack.edu/cmva/CMVA.html .

The radio measurements described here are intended to complement, and will ultimately be combined with, the results of a number of related observing programs being carried out by us and numerous other investigators in different wavebands covering the entire spectrum.

We gratefully acknowledge the assistance of the scientific and technical staff of the Haystack Observatory, making efficient observations possible. This work was supported in part by NASA grant NAG 5-3671 to the University of New Hampshire.

REFERENCES

1. Mukherjee, R., et al., *ApJ*, submitted (1997).
2. von Montigny, C., et al., *ApJ.* **440**, 525 (1995).
3. Fichtel, C., et al., *ApJS.* **94**, 551 (1994).
4. Marscher, A. P., and Bloom, S. D., *The Second Compton Symposium (AIP Conf. Proc. 304)*, eds. C. E. Fichtel, N. Gehrels, and J. P. Norris, New York: AIP, 1994, p. 572.
5. Dermer, C. D., and Schlickeiser, R., *Science.* **257**, 1642 (1992).
6. Valtaoja, E., and Teräsranta, H. *A&A.* **297**, L13 (1995).
7. Reich, W., et al., *A&A.*, **273**, 65 (1993).
8. Wehrle, A. E., et al., AAS/HEAD meeting, San Diego, CA (1996).
9. Ott, M., et al., *A&A.* **284**, 331 (1994).
10. Barvainis, R. E., in preparation (1997).

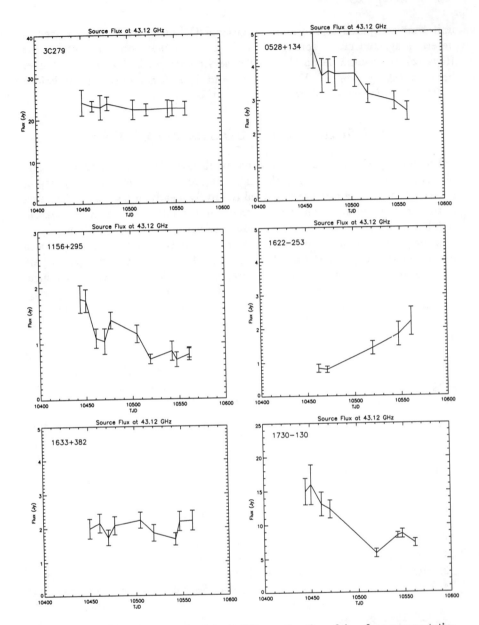

FIGURE 2. Preliminary source fluxes at 43 GHz as a function of time for a representative sample of the gamma-ray AGN observed with the Haystack telescope. *Note the varying vertical flux scale.* With the possible exception of 1622-253, all sources showed relatively steady or declining emission levels during the 1996-1997 observing season.

The Burst Activity of Millimeter Wavelengths Compared to Gamma-Activity of AGN

Harri Teräsranta

Metsähovi Radio Research Station
02540 Kylmälä
Finland

Abstract. From our more than 15 years long monitoring of AGN at 22 and 37 GHz we have selected 50 best sampled sources. The flux on of the source has been devided to two parts, stationary component, which corresponds to the minimum flux observed and the outburst co mponent. We have calculated the energy-ratios for the burst and stationary component. This ratio is highest for the sources classified as HPQ, BL Lac and EGRET detected sources. The ratio for LPQ's is significantly lower. The highest burst-ratio sources s how also super-compact structure in VLBI maps.

INTRODUCTION

The long term continuum monitoring of AGN with the Metsähovi radio telescope has during the 17 years of operation formed an unique database of nearly 35000 observations at 22 and 37 GHz. We have selected 50 best sampled sources from our monitoring sample of flat spectrum northern AGN [1]. Sources were classified according to the optical properties in HPQ (highly polarized quasars) and LPQ (low polarized quasars), as well as to BL Lac type objects. From our continuum monitoring at 22 and 37 GHz we have estimated the minimum flux densities to all of those 50 sources and devided the energy of the source emitted at these frequencies to two parts, the quiet level (at minimum flux levels) and the outburst energy. The ratio of these energies we call the burst ratio. From these 50 sources selected to this study, 18 has been detected with EGRET according to the second EGRET catalog [2]. This study is to investigate whether the burst ratio at radio wavelengths and the gamma radiation in AGN are somehow connected.

THE BURST-RATIO

We have set the lowest value from the flux record as the quiet level. To smoothen the daily variations a weekly mean value was used for the analysis. From the flux(time) plots the energy is proportional to flux * dt, so the burst ratio is the ratio between the areas of burst emission and stationary emission:

$$BR = \frac{E_b}{E_q} \quad (1)$$

In Figure 1 is demonstrated the deviation to burst and quiet energies for 3C 273 at 22 GHz.

The areas were integrated numerically from the flux tables. In Table 1. are presented the burst ratios at 22 and 37 GHz as well as the source classification. A bigger burst ratio would mean a higher portion of the energy coming from the outburst components and could also be a measure of the source compactness as a large quiet level might partly come from far outside the core (diffuse radiation).

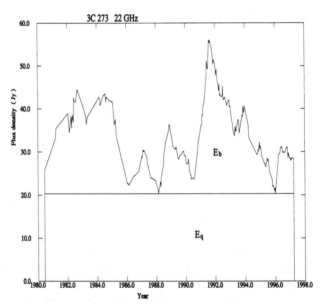

FIGURE 1. The division for burst and quiet energies.

CLASSIFICATION OF THE SOURCES

The HPQ/LPQ classification is set according to the optical polarization level, a source with a polarization rate larger or equal to 3.0% is named a

TABLE 1. Burst ratios for the monitored sources at 22 and 37 GHz and the source classification.

Source	BR22	BR37	BR22/BR37	type
OC 012	1.785	2.318	0.770	BLO/HPQ
0109+224	1.529	1.604	0.953	BLO/HPQ
DA 55	4.202	3.528	1.191	QSO/HPQ
0149+218	0.560	1.273	0.440	QSO
0202+149	0.797	0.673	1.184	QSO/HPQ/EGR
0234+285	0.925	0.715	1.294	QSO/HPQ/EGR
AO 0235+164	6.973	4.565	1.527	BLO/HPQ/EGR
3C 84	0.989	2.161	0.458	GAL
NRAO 140	1.147	1.118	1.026	QSO/LPQ
NRAO 150	0.946	1.376	0.688	QSO
OA 129	1.311	1.643	0.798	QSO/HPQ/EGR
3C 120	0.652	0.569	1.146	GAL
0458-020	1.331	1.005	1.324	QSO/HPQ/EGR
0528+134	2.012	1.106	1.819	QSO/LPQ/EGR
DA 193	0.535	0.437	1.224	QSO/LPQ
OH 471	0.509	0.688	0.740	QSO/LPQ
PKS 0735+17	1.208	1.367	0.884	BLO/HPQ/EGR
0736+017	1.703	1.892	0.900	QSO/HPQ
OI 090.4	1.122	2.481	0.452	BLO/HPQ
0804+499	2.825	2.397	1.179	QSO/HPQ/EGR
OJ 287	2.915	3.081	0.946	BLO/HPQ
4C 39.25	1.971	1.813	1.087	QSO/LPQ
0953+254	0.326	0.282	1.156	QSO/LPQ
OL 093	0.903	1.106	0.816	QSO/HPQ
4C 29.45	1.517	1.192	1.273	QSO/HPQ/EGR
ON 231	3.976	4.868	0.817	BLO/HPQ/EGR
3C 273	0.724	0.937	0.773	QSO/LPQ/EGR
3C 279	0.683	1.004	0.680	QSO/HPQ/EGR
1308+326	4.443	4.564	0.973	BLO/HPQ
1413+135	4.860	11.873	0.409	BLO/HPQ
OQ 530	2.461	1.635	1.505	BLO/HPQ
OR 103	1.381	2.391	0.578	QSO/HPQ
1510-089	1.686	2.558	0.659	QSO/HPQ/EGR
4C 14.60	0.818	2.391	0.342	BLO/HPQ
1611+343	0.622	1.124	0.553	QSO/LPQ/EGR
4C 38.41	1.042	1.005	1.037	QSO/LPQ/EGR
OS 562	1.911	1.343	1.423	QSO/LPQ
3C 345	1.517	1.658	0.915	QSO/HPQ
1652+398	0.666	0.850	0.784	BLO/HPQ
1739+522	2.074	3.902	0.532	QSO/HPQ/EGR
1741-03	1.173	1.177	0.997	QSO/HPQ
1749+096	3.647	4.032	0.905	BLO/HPQ
2005+40	0.769	0.768	1.001	QSO
OX 057	0.532	0.409	1.301	QSO/LPQ
2145+067	0.382	0.477	0.801	QSO/LPQ
BL Lac	1.770	1.984	0.892	BLO/HPQ

TABLE 1. (cont.) Burst ratios for the monitored sources at 22 and 37 GHz and the source classification.

Source	BR22	BR37	BR22/BR37	type
2201+315	0.832	1.158	0.718	QSO/LPQ
3C 446	1.231	1.342	0.917	BLO/HPQ
CTA 102	0.310	0.570	0.544	QSO/HPQ/EGR
3C 454.3	0.941	0.958	0.982	QSO/HPQ/EGR

HPQ, all other LPQ. From two sources, 0149+218 and 2005+40 were found no observation for the polarization, those sources were put to the LPQ group. There is not yet a known optical counterpart for NRAO 150 and it was thus left out from the classification based on the polarization properties. Also the Galaxies 3C 84 and 3C 120 were omitted from the analysis. As the polarization rate is variable, it is likely that some of the sources currently classified as LPQ will end to the HPQ group in the future when more polarization observations are available. The sources classified as BL Lac's were according to [3].

DIFFERENCES BETWEEN GROUPS

From the sources in the study, 32 were HPQ, 15 were LPQ, 15 were BL Lac objects and 18 were detected with EGRET during the phases 1-3. Some of the groups overlap, as only HPQ and LPQ are exclusive to each other. Most of the EGRET sources are HPQ. A clear difference can be seen at least between the HPQ and LPQ groups, the former have a generally larger burst ratio. The largest burst ratios are still found in the BL Lac group. The EGRET sources are mostly HPQ and thus their distribution is similar. The small number of sources in the groups does not give a sufficient statistics to make difference between groups, however, a simple T-test was performed and the propabilities of the groups coming from the same parent population are given in Table 2. When comparing the burst ratios at 22 and 37 GHz, the differences between groups were small.

Many sources with high burst ratio are found to be ultracompact, according to VLBI observations, like OJ 287, ON 231 and 1413+135 [4]. The extended radio luminosity was below detection limits in 1413+135 and 1749+096 [5].

DISCUSSION

Most of the EGRET detected AGN seem to have a high burst-ratio and many of those are also classified as HPQ. Still some of the sources active at gamma-rays show relatively no variations at radio and mm- wavelengths during the gamma outburst. A properly sampled monitoring effort at a higher

TABLE 2. Probabilities that the source groups are from the same population.

groups	frequency	P
BLO-EGR	22GHz	0.135
BLO-EGR	37GHz	0.047
LPQ-BLO	22GHz	0.0018
LPQ-BLO	37GHz	0.0027
HPQ-BLO	22GHz	0.234
HPQ-BLO	37GHz	0.249
LPQ-EGR	22GHz	0.075
LPQ-EGR	37GHz	0.031
HPQ-LPQ	22GHz	0.0092
HPQ-LPQ	37GHz	0.010
HPQ-EGR	22GHz	0.507
HPQ-EGR	37GHz	0.239

frequency (90 GHz) might show whether these gamma-outbursts are related to mm-outbursts, but just fail to reach the lower frequencies, like 22 and 37 GHz. The high burst-ratios could be an indication of a high compactness in the source, as well as a higher probability to detect gamma-radiation from the source. More optical polarization observations would be needed for a few sources like 0528+134, to see whether they still can be classified as LPQ.

REFERENCES

1. Teräsranta H. et al., A&AS 94, 121, 1992.
2. Thompson D. J. et al., ApJS 101, 259, 1995.
3. Burbidge G. and Hewitt A.: "BL Lac objects and rapidly variable QSOs – an overview", in Variability of Blazars, eds. E. Valtaoja and M. Valtonen (Cambridge, Cambridge University Press), p. 4-38, 1992.
4. Perlman E. and Stocke J., AJ 108, 56, 1994.
5. Punsly B., ApJ 473, 152, 1996.

Relationships between radio and gamma-ray properties in active galactic nuclei

Anne Lähteenmäki*, Harri Teräsranta*, Kaj Wiik*
and
Esko Valtaoja[†]

*Metsähovi Radio Research Station, Helsinki University of Technology, Finland
[†] Tuorla Observatory, University of Turku, Finland

Abstract. We compare the Metsähovi 22 and 37 GHz total flux density variations of active galactic nuclei with EGRET Phase 1+2+3 gamma-ray data. There is a connection between the type of the source, the phase of the radio flare, the associated radio variability brightness temperature and the gamma-ray emission. It seems that the gamma emission originates within the same shocks which produce the synchrotron flares in radio —induced by the synchrotron self-Compton mechanism.

INTRODUCTION

The close connection between radio and gamma-ray emission in EGRET-detected sources indicates that the production of gamma-rays is somehow tied to the existence of a synchrotron-emitting radio jet. No radio-quiet AGN have been detected with EGRET and the majority of the detected sources can be classified as compact, flat-spectrum, 'blazar'-type radio sources. Most sources are detected only intermittently, and a number of sources with apparently similar radio properties have never been seen with EGRET. Time variability must therefore be a major ingredient in models for gamma-ray production in AGN. Based on the analysis of EGRET Phase 1 data and Metsähovi radio monitoring data we have earlier proposed that detectable amounts of gamma-rays are seen mainly in quasars with high optical polarization (a subset of 'blazar'-type quasars), and only when high radio frequency flux is rising. The most simple explanation is that gamma-rays are produced by the SSC mechanism in the same shocks which produce the radio outbursts [3,4]. We have extended our analysis to include later EGRET and Metsähovi 22 and 37 GHz monitoring data. We have looked at the average properties of each source, comparing the average Phase 1+2+3 gamma-ray flux (or, in the case of non-detected Metsähovi radio sources, Phase 1 upper limits) with a number of average radio properties. We have also looked at the radio state of each source at the time of each EGRET pointing towards it.

The almost twenty years of continuum monitoring of active galaxies in Metsähovi has resulted in an extensive database at 22 and 37 GHz. By using this high frequency total flux density variation data we have calculated the intrinsic variability brightness temperature $T_{b,var}$ and the Doppler boosting factor $D = (T_{b,var}/T_{lim})^{1/3}$ for each source (almost 80 sources out of the total of $\sim$ 130 sources). The phase ϕ of the radio flare was also determined. For statistical analysis we used two methods: Spearman correlation to find out whether the gamma-ray flux of the detected sources depends on any of the source properties and the Kruskal-Wallis test to find out significant differences between detected and nondetected sources.

AVERAGE PROPERTIES

We looked at the following average properties of each source:

1. The average Phase 1+2+3 EGRET gamma-ray flux [2], and in the case of nondetections, the Phase 1 upper limits [1].

2. The average radio properties: the redshift, the superluminal expansion speed, the observed 22 and 37 GHz fluxes S_{ave} and variability timescales, the associated ever observed highest brightness temperature, the Doppler boosting factor, the intrinsic Lorentz factor Γ and the viewing angle θ (calculated from D and Γ).

Figure 1 shows the EGRET detection versus classification. Two features are apparent: High polarization quasars (HPQs) are much more likely to be detected than any other AGN and the high detection rate of 'blazars' is entirely due to HPQs (BL Lacs have a low detection rate compared even to 'non-blazar' quasars (low polarization quasars, LPQs). The detection probability depends only on $T_{b,var}$ (or D) (2, $P_{Kruskal-Wallis} = 0.0285$). Thus sources with largest amounts of Doppler boosting are strong gamma-ray emitters. The average gamma-ray flux for the detected sources depends most strongly on $T_{b,var}$ ($P_{Spearman} = 0.0010$) and it also possibly depends on the average radio flux $S(ave)$ ($P_{Spearman} = 0.032$). Since $D = D(\Gamma, \theta)$, the average gamma-ray flux for the detected sources also depends on Γ and θ ($P_{Spearman} = 0.086$ for Γ and 0.023 for θ). A small viewing angle may be more decisive for gamma emission than a large Lorentz factor (i.e. large flow speed).

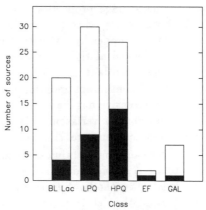

FIGURE 1. The fraction of gamma-ray detected sources for different classes of objects.

POINTINGS

We also looked at the radio state of each source at the time of each EGRET pointing (189 cases all together) towards it. We have studied each detection separately and in the case of nondetections, we have looked at the radio state during the Phase 1 pointing giving the lowest upper limit.

Detection probability during a single EGRET pointing depends most clearly on $T_{b,var}$ (or D) and the phase of the associated radio flare ($P_{Kruskal-Wallis} =$ 0.0016 for $T_{b,var}$ and 0.0031 for ϕ,3). For the rising part of the flare (phase $0 \longrightarrow 1$) we found 27 detections and 49 nondetections and for the falling part

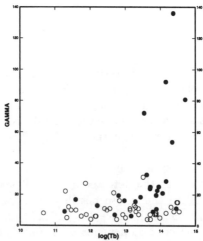

FIGURE 2. Average gamma-ray flux vs. brightness temperature. Filled, detections; open, upper limits.

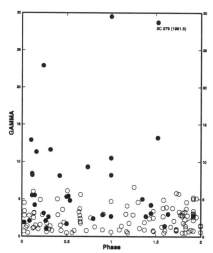

FIGURE 3. Observed gamma-ray flux vs. radio flare phase

(phase 1 ⟶ 2) only 9 detections and 61 nondetections. Detection probability also seems to depend on the strength of the radio flare, S_{max} ($P_{Kruskal-Wallis}$ = 0.0290, 4). Also the strength of the detected gamma-ray emission depends on $T_{b,var}$ and S_{max}. It does not correlate with phi ($P_{Spearman}$ = 0.023 for $T_{b,var}$, 0.008 for S_{max} and 0.281 for phi). Radio flares are often overlapping and without further information it is not possible to determine to which flare the gamma-ray emission is related to. Thus many of the detections during the falling phase may be related to another flare rising at the same time (c.f. 3C 279, 3). In our analysis we have consistently used values for the strongest flare at the time of each pointing.

SUMMARY

The most probable gamma-ray AGN to be detected is a highly polarized quasar with an ongoing and rising high frequency radio flare and a large associated variability brightness temperature (or Doppler factor). The time averaged gamma-ray flux increases with the estimated amount of Doppler boosting in the source. The gamma-ray flux during an individual EGRET pointing also increases with $T_{b,var}$ but also with the maximum amplitude of the outburst, S_{max}. Therefore the gamma-ray properties of AGN are closely related both to the average radio properties as well as to the radio properties of individual outbursts (shocks in the jet) growing at the time of the EGRET pointing. Such correlations cannot be explained unless the gamma emission originates within the same shocks which produce the synchrotron flares at radio, IR and higher frequencies. The found statistical connections support the

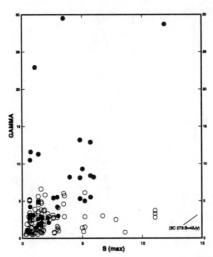

FIGURE 4. Observed gamma-ray flux vs. radio flare peak flux

SSC framework, and in our opinion are sufficient to eliminate most —if not all— of the proposed alternative models for gamma-ray production.

REFERENCES

1. Fichtel, C.E. et al. *ApJS* **94**, 551 (1994).
2. Thompson, D.J. et al. *ApJS* **107**, 227 (1996).
3. Valtaoja, E. and Teräsranta. H., *A & A* **297**, L13 (1995).
4. Valtaoja, E. and Teräsranta. H., *A & AS* **120**, 491 (1996).

Fast Variations of Gamma-Ray Emission in Blazars

Stefan J. Wagner*, Corinna von Montigny, and Martin Herter

swagner @ lsw.uni-heidelberg.de
Landessternwarte, Königstuhl, 69117 Heidelberg, Germany

Abstract. The largest group of sources identified by EGRET are Blazars. This sub-class of AGN is well known to vary in flux in all energy bands on time-scales ranging from a few minutes (in the optical and X-ray bands) up to decades (radio and optical regimes). In addition to variations of the gamma-ray flux between different viewing periods, the brightest of these sources showed a few remarkable gamma-ray flares on time-scales of about one day, confirming the extension of the "Intraday-Variability (IDV)" phenomenon into the GeV range. We present first results of a systematic approach to study fast variability with EGRET data. This statistical approach confirms the existence of IDV even during epochs when no strong flares are detected. This provides additional constraints on the site of the gamma-ray emission and allows cross-correlation analyses with light curves obtained at other frequencies even during periods of low flux.
We also find that some stronger sources have fluxes systematically above threshold even during quiescent states. Despite the low count rates this allows explicit comparisons of flare amplitudes with other energy bands.

Introduction

Most of the identified sources in the energy band studied with EGRET are Blazars. One of the defining criteria of this class is pronounced variability in the optical and radio bands on time-scales of weeks to months [13]. Later it was realized that these sources often show strong and variable X-ray emission as well. During the last decade it became clear that variations can be traced down to much shorter time-scales in all energy bands. Rapid changes "IDV" [16] can now be traced down to the shortest time-scales which can be probed in all of the different wavelength regimes. In the radio domain variations have been found to occur within less than two hours [11], [7], [5], [15], in the optical and near-IR regimes variations on time-scales of hours [18], [10], [14] down to a few minutes [15] have been reported. In the X-ray regime variations have been reported to be even faster [1], [12], [18].

Blazars also turned out to be the dominant class of sources in the GeV regime. It was soon realized that they vary from one EGRET viewing period to another (von Montigny, et al., this meeting). If observed during a particularly bright state, individual objects were shown to vary on shorter time-scales as well (e.g. [6]). With typical recorded flux-densities of about a photon per hour it is impossible to probe the regime of very fast variations (< hours) which can be tackled in other energy bands. Statistical approaches enable investigations of the fastest variation that *can* be studied. They also allow comparisons throughout the entire range of time-scales with those in other regimes of the electromagnetic spectrum.

Rapid Gamma-ray Flares

Already one of the first EGRET pointings towards a Blazar found the source 3C 279 in a state of enhanced activity. Detailed analysis revealed variations on time-scales of a few days with high significance [6], [3]. Comparable variations were found in other sources, including, e.g., PKS 0528+134 [4], PKS 1406-076 [17], PKS 1633+382 [8], and PKS 2251+158 [2]). Hunter et al. [4] illustrated that the variability of PKS 0528+134 was persistent over at least eight weeks with several maxima rather than a singular, well defined peak. This is comparable to the variations of the synchrotron emission of Blazars in general, as seen e.g. in the optical regime [16]. In an attempt to constrain the shortest time-scales, Mattox et al. [9] analysed the brightest flare of any gamma-emitting Blazar, PKS 1633-29, by binning the light curve with a variable width in time such that the individual bins had comparable errors. This revealed time-scales as short as about 6 hours in the observers frame.

In order to study the statistical properties of fast variations we analysed the EGRET observations of a larger set of Blazars on time-scales of about one

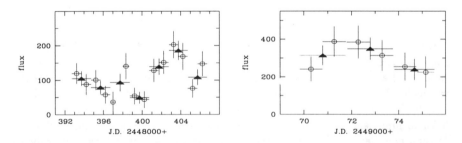

FIGURE 1. Light curves of PKS 0528+134 with a binning of one (open circles) and two days (full triangles) during two different viewing periods (flux is given in $10^{-8}\ \gamma\ cm^{-2}\ s^{-1}$). While χ^2 tests indicate significant variations in the left panel (not in the right one), a two day binning is too poor to reveal the shortest time-scales.

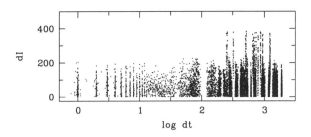

FIGURE 2. Differences in intensity of PKS 0528+134 between any two epochs as a function of the difference in time log(dt/1 day) between the measurements.

day. All observations of a selected source available from the EGRET archive were taken and binned into one-day maps. This binning was chosen to remain sensitive to the fastest changes seen in EGRET data so far. Rebinning to longer time-scales could be performed without any significant degradation in accuracy. Examples of light curves binned into intervals of one and two days, respectively, are shown in Figure 1.

The entire data set of every well-observed source can be studied in over 100 one-day maps. This is a sufficiently large data base to carry out statistical studies, comparable to the structure function analysis used in temporal studies at other wave-bands. A similar method has been used by [18] to study IDV in 0716+714 in the radio and optical bands.

Statistical Analyses

All light curves of a source, binned in one-day intervals (as shown in Figure 1), are combined and used to create structure functions. In order to determine characteristic temporal changes in flux, every combination of two measurements $I(t_i)$, $I(t_j)$ is used to derive one measurement $dI = I(t_i)-I(t_j)$ at $dt = (t_i-t_j)$. The entire data set of PKS 0528+134 is given in Figure 2. The structure function as defined by [18] was derived from the set of dI(dt) by averaging a) equal bins in dt, b) equal bins in log(dt), and c) bins of equal numbers of measurement points dI(dt). With the small number of measurements, the uncertainty introduced by choosing a particular binning and phase may be significant. Figure 3 illustrates the resulting structure functions, derived by using the three different ways of binning. Each panel gives four curves, representing shifts of the phase by 25 % of the width of the bins.

Since the exposure of different one-day maps changed dramatically, the errors on the dI(dt) varied significantly. This is taken into account in the computation of the structure functions by individual weights but is not illustrated with error bars in Figure 2 for clarity. Instead we present another version of Figure 2 where the dI(dt) are given in terms of their significance dS(dt) =

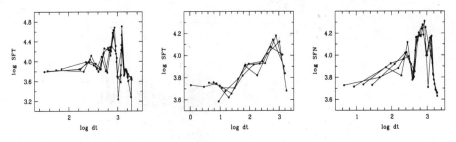

FIGURE 3. Structure functions of PKS 0528+134 derived from bins of dt = 100 days (left panel, phase offsets of 25 days), $\log(dt/1\ \text{day}) = 0.4$ (central panel, phase offsets of $\log(dt/1\ \text{day}) = 0.1$), and n = 800 (right panel, phase offset of dn = 200).

$dI(dt)/\sigma_{dI(dt)}$ (Figure 4, left panel). The right panel compares the histogram of dS(dt) for dt ~ 1 day with the normal distribution expected for a steady source with random errors, which clearly illustrates the excess at large dS.

Interpreting the results shown in Figure 4 directly as evidence for the statistical occurrence of variability relies on an accurate determination and propagation of errors. We consistently chose conservative errors.

We tested different approaches by either fixing the positions of the sources or leaving them as free parameters and by fixing the background or leaving it as free parameters. Despite subtle differences, we arrived at consistent results.

As an alternative approach we treated other sources within the same pointing directions. If errors are not estimated correctly, one may still regard the least variable source within any field of view as constant and derive the actual spread in errors from the dispersion of those sources with least variability. This will allow the identification of variable sources even if there are unknown contributions to the total errors as long as constant sources exist. Although gamma-bright pulsars may be considered as being constant on time-scales much longer than their pulsation periods and much shorter than cooling time-

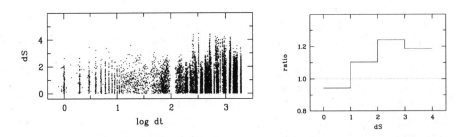

FIGURE 4. *Left*: dS(dt) distribution - differences in intensity between any two points, normalised to their significance. *Right*: Ratio of actual dS(dt) (for dt ~ 1 day) for PKS 0528+134 to those of the normal distribution for a constant source. There is a clear excess of differences at the 2-3 Sigma level, indicating variability on a statistical level.

scales, we found significant differences between the structure functions of the Crab Pulsar and Geminga. These results will be presented separately.

Results and Discussion

We computed structure functions of several gamma-bright Blazars using light curves obtained from maps with a temporal binning of one day. IDV is detected in the EGRET data on a significant level even during those observations when the sources are not exceptionally bright. Gamma-ray IDV has a high duty cycle and is not confined to individual flares. This is similar to the variability seen at lower photon energies, and permits cross-correlation studies even if IDV remains at low amplitudes. Comparing the fluxes of Blazars during their low states with those derived at fixed but "empty" positions in the same one-day maps, we find a statistical indication for PKS 0528+134 and 0716+714 to have a steady flux about 1 sigma above the background. This is consistent with previous upper limits but confirms that EGRET data are sensitive enough to reach even the lower envelope of the range in flux densities.

REFERENCES

1. Feigelson, E.D., Bradt, H., McClintock, J., et al. 1986. *Ap.J.*, 302, 337–351.
2. Hartman, R.C., Bertsch, D.L., Dingus, B.L., et al. 1993. *Ap.J.Lett*, 407, L41.
3. Hartman, R.C., Webb, J.R., Marscher, A.P., et al. 1996. *Ap.J.*, 461, 698.
4. Hunter, S.D., Bertsch, D.L., Dingus, B.L., et al. 1993. *Ap.J.*, 409, 134–138.
5. Kendziora-Chudcer, L., Jauncey, D.L., Wieringa, M., Walker, M.A., Nicolson, G.D., et al., in *IAU Coll. 164: Radio Emission from Galactic and Extragalactic Compact Sources*, G. Taylor, J. Wrobel, & A. Zensus (Eds.), 1997, in press.
6. Kniffen, D.A., Bertsch, D.L., Fichtel, C.E., et al. 1993. *Ap.J.*, 411, 133–136.
7. Kraus, A., Ph.D. thesis, University of Bonn, 1997.
8. Mattox, J.R., Bertsch, D.L., Chiang, J., et al. 1993. *Ap.J.*, 410, 609–614.
9. Mattox, J.R., Wagner, S.J., Malkan, M., et al. 1997. *Ap.J.*, 476, 692.
10. Miller, H.R. & Noble, J.C. 1996. in *Blazar Continuum Variability*, H.R. Miller, J.R. Webb, & J.C. Noble (Eds.), ASP. Vol. 110, pp. 17–30.
11. Quirrenbach, A., Witzel, A., Qian, S.-J., et al. 1989. *A.& A.*, 226, L1–L4
12. Remillard, R.A., Grossan, B., Bradt, H.V., et al. 1991. *Nature*, 350, 589–592.
13. Stein, W.A., in *Pittsburgh Conference on BL Lac Objects*, A.M. Wolfe (Ed.), Pittsburgh University, 1978, pp. 1–21.
14. Takalo, L.O., Sillanpää, A., Pursimo, T., et al. 1996, *A.& A. Suppl.*, 120, 313.
15. Wagner, S.J. 1997. in *IAU Coll. 164: Radio Emission from Galactic and Extragalactic Compact Sources*, J. Wrobel, & A. Zensus (Eds.), in press.
16. Wagner, S.J. & Witzel, A. 1995. *Ann. Rev. Astron. Astrophys.*, 33, 163–197.
17. Wagner, S.J., Mattox, J.R., Hopp, U., et al. 1995. *Ap.J. Lett.*, 454, L97–L100.
18. Wagner, S.J., Witzel, A., Heidt, J., et al. 1996, *A.J.*, 111, 2187–2211.

A $z = 2.1$ quasar as the optical counterpart of the MeV source GRO J1753+57

A. Carramiñana[1], V. Chavushyan[2,3], J. Guichard[1]

[1] INAOE, Luis Enrique Erro 1, Tonantzintla 72840, MEXICO
[2] Special Astrophysical Observatory, Zelenchuk, RUSSIA
[3] Byurakan Astrophysical Observatory, Byurakan, ARMENIA

Abstract. The COMPTEL source GRO J1753+57 is probably the most extreme case of a MeV peaked spectrum. Because of its spectrum and variability, this source, still undetected by EGRET, was initially suspected to be a MeV blazar. However, no Bl Lac or quasar inside the 2σ COMPTEL error is known. A first search for an optical counterpart provided unlikely candidates.

During this first study, an optical counterpart was found for the flat-spectrum radio source 1755+578, but no spectroscopic identification could be made. In more recent campaigns, we have obtained optical spectra for this object, one with the 6-m telescope at Zelenchuk, which indicate that this radio source is a quasar with a redshift of 2.1. Given its radio and optical characteristics, the quasar 1755+578 is the most likely counterpart of GRO J1753+57, suggesting this is indeed a very bright MeV blazar.

THE COMPTEL SOURCE GRO J1753+57

GRO J1753+57 was discovered serendipitously by [12] in COMPTEL data recorded in November 92 in a flaring state. Its flux decayed afterwards by a factor of ~ 2 in a week and finally went below detection threshold. Its was marginally detected ~ 100 days later at a flux level $\lesssim 5$ the flare maximum.

The γ-ray spectrum of this souce is also particular, although not unique. It was clearly seen between 1 MeV and 8 MeV, but remained undetected at higher energies. The ($E^2 * dN/dE$) distribution peaked in the (1.75-3.0) MeV window defined by [12]. This spectrum is similar of that observed in two blazars, PKS0208-512 and GRO J0506-612, where the MeV emission surpasses GeV emission by more than one order of magnitude. These objects have been named "MeV blazars". The non detection by EGRET of GRO J1753+57 indicates that its MeV spectrum dominates the overall γ-ray emission as much as in the established "MeV blazars".

Because of its variability and spectral properties, GRO J1753+57 was initially suspected to be another MeV blazar. However, no obvious candidate was identified in the COMPTEL error box by [12], leading to a search for an optical counterpart [5].

THE SEARCH FOR AN OPTICAL COUNTERPART OF GRO J1753+57

Following the direction set by [12], we selected radio loud objects, X-ray sources and NGC objects inside a $2° \simeq 2\sigma$ radius around the COMPTEL location given by [12]. No QSO/AGN from the Veron & Veron catalogue [11] is located in this region, the completness limit been at $V \sim 17$. Figure 1 shows 10 NGC objects and 9 radio bright sources around the MeV source. Our search provided three candidates [5]:

1. NGC 6474: an active galaxy and X-ray source, whose optical spectrum makes it an interesting object by itself. It is most probably a Seyfert, which would make the physical association unlikely.

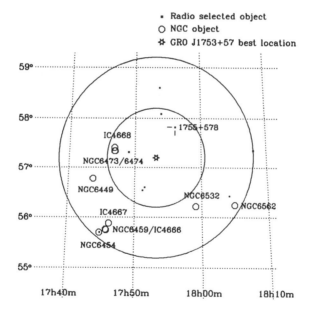

FIGURE 1. Diagram of the region around GRO J1753+57. The circles are centred on the best COMPTEL location for the object and have 1° and 2° radii respectively, corresponding to 1σ and 2σ boxes respectively.

2. NGC 6454: elliptical galaxy and also radio loud flat-spectrum object. Detected in X-rays by the Einstein. However its stellar component dominates the optical emission and has only mild nuclear activity.

3. 1755+578, the best radio candidate: loud and flat spectrum. An optical object was identified with the 2.1m telescope of INAOE's "Guillermo Haro" observatory at Cananea, but its optical spectra had too low signal-to-noise to provide spectral identification. We measured (8/3/96) were: $B = 19.22\pm0.16$, $V = 18.55\pm0.18$, $R = 18.09\pm0.20$ and $I = 17.54\pm0.18$. Variability in time scales of ~ 1 day was searched for, with null results. The study was inconclusive.

Given the difficulty of assigning the identification of GRO J1753+57 to NGC6474 or NGC6454, further studies of GRO J1753+57 were mandatory.

1755+578 AS GRO J1753+57

A first long integration spectrum was taken in June 1995 with the LSW LFOSC spectrograph at the Cananea 2.1m observaroty. A single emission line was detected at $\lambda \lesssim 5000$Å but no identification was possible. More recently, (feb 97) a better spectrum was obtained with the long-slit spectrograph of the 6m telescope at SAO under mediocre weather conditions. This spectrum, show in Figure 2, is dominated by the strong Lyα emission at the edge of the spectral range, SiIV/OIV] (1400Å), CIV (1549Å) and CIII] (1909Å) in emission redshited by $z = 2.101$.

The identification of 1755+578 as a quasar makes it the prime candidate to be the optical counterpart of GRO J1753+57. It is a broad line QSO, with radio loud, flat spectrum emission. VLA maps have provided evidence for jets [9]. We note that no public X-ray data exist for this object. The non-detection by EGRET translates in a 95% confidence upper limit $\sim 8 \times 10^{-8}\,\text{cm}^2\,\text{s}^{-1}$ for $E_\gamma > 100\,\text{MeV}$ (see Fig. 4 of [14]).

Assuming isotropy, and using a broken power-law fit to the data given by [12], one obtains the following luminosities for GRO J1753+57:

- $L(0.75\text{MeV} - 30.0\text{MeV}) \simeq 2.69 \times 10^{49}\,\text{erg s}^{-1}$ during outburst;

- $L(E_\gamma > 100\text{MeV}) < 3.5 \times 10^{47}\,\text{erg s}^{-1}$, averaged;

- $L_V \simeq 1.3 \times 10^{46}\,\text{erg s}^{-1}$, most probably during quiescence.

It is almost certain that the MeV emission is highly beamed. If one assumes $L_\gamma \approx L_V$, the beaming factor is $\sim 10^{-3}$. Beaming is also required for transparency arguments. If we assume the optical emission to be isotropic and ~ 0.1 Eddington limit, as the luminosity in the Lyα line is certainly much larger, the central mass is $M_* \approx 1.0 \times 10^9\,M_\odot$, with a Schwarszchild radius of about 10^4 light-seconds. The γ-ray density near the object is then $\sim 2 \times 10^{14}\,\text{cm}^{-3}$, assuming $E_\gamma \sim 2\,\text{MeV}$, and taking $\sigma_{\gamma\gamma} \sim (\pi/2)r_e^2$, one obtains a photon mfp for pair production of just 14 light-seconds. Beaming is required to avoid this contradiction. Note that because of its high redshift, a more detailed analysis has to be made in the rest frame of the source. Here we referred to photons *after* they have been emitted.

GRO J1753+57 AND THE MEV BLAZARS

Although this identification would make GRO J1753+57 the third "MeV blazar" so far, there are now several extragalactic sources whose observed γ-ray flux is dominated by the MeV component to different extends (Table 1). All are high polarization quasars, except for PKS0528+134 and 3C273, which are low polarization QSO. We included the unidentified COMPTEL source GRO J1040+48, for which a high velocity cloud interpretation also exist [4].

Ratio of fluxes for the MeV bright extragalactic sources have been calculated using EGRET photon fluxes and spectral indexes given in [14] and fitting broken power-laws to the COMPTEL data. Special remarks above some of these objects:

- GeV faint, PKS 0506-612 appears in the first EGRET catalogue but not in the second one. It is the preferred counterpart of the EGRET, altough another possibility is PKS 0522-611.

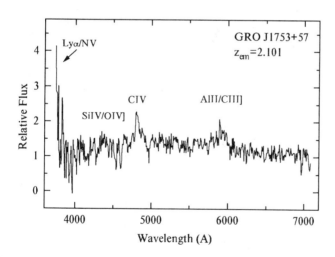

FIGURE 2. Spectrum of 1755+578 obtained with the long-slit spectrograph of the SAO 6m telescope.

TABLE 1. The MeV bright blazars

Source	z	Observed $F(0.75\text{-}30\text{MeV})/F(E\geq 100\text{ MeV})$	Ref.
GRO J1040+48?	?	≥ 120	[15]
PKS 0506-612	1.093	$\gtrsim 80$	[1]
GRO J1753+57	2.101	≥ 80	[12]
3C273	0.158	~ 8	[13]
CTA 102	1.037	~ 4	[2]
PKS 0208-512	1.003	~ 3	[3]
3C454.3	0.859	~ 1.5	[2]
PKS 0528+134	2.06	~ 1	[6]

- PKS0208-512 is bright for both COMPTEL and EGRET, its spectrum extending to several GeV.
- CTA102 and 3C454.3 are sometimes classified as Bl Lac objects.
- the COMPTEL flux for PKS0528+134 has been taken as $\sim 15\%$ Crab in the 3-10MeV and 10-30MeV bands, according to [6].

One of the models proposed for MeV blazars [8] produces a blue shifted annihilation line. Given that the broken power-law spectral fit of GRO J1753+57 peaks at about 2.75 MeV, and considering its redshift, the annihilation $e^{\pm}$ would need to be redshifted by $\Gamma \approx 11$ to fit the observations.

REFERENCES

1. Bloemen et al 1995, A&A 293, L1
2. Blom et al 1995, A&A 295, 330
3. Blom et al 1995, A&A 298, L33
4. Blom J.J. et al. 1997, A&A 321, 288
5. Carramiñana et al 1996, A&A Sup Ser 120, 595
6. Collmar et al 1993, Proc 1st Compton Symp, AIP Conf 280, 483
7. Fichtel C.E. et al. 1994, ApJSS 94, 551
8. Henri G., Pelletier G. & Roland J. 1993, ApJ 404, L41
9. Taylor G.B. et al. 1994, ApJSS 95, 345
10. Taylor G.B. et al. 1996, ApJSS 107, 37
11. Véron-Cetty M.P. & Véron P. 1991, *A catalogue of quasars and active nuclei*, ESO
12. Williams et al 1995, A&A 297, L21
13. Williams et al 1995, A&A 298, 33
14. Thompson D.J. et al. 1995, ApJSS101, 259
15. Iyudin et al 1996, A&A 311, L211

ASCA Observations of Blazars and Multiband Analysis

Tadayuki Takahashi,[1] Hidetoshi Kubo,[1] Greg Madejski,[2] Makoto Tashiro[3] and Fumiyoshi Makino[1]

[1] *Institute of Space and Astronautical Science, Sagamihara, Kanagawa, Japan*
[2] *Laboratory for High Energy Astrophysics, Code 662, NASA GSFC, MD 20771, U.S.A*
[3] *Department of Physics, University of Tokyo, Bunkyo-ku, Tokyo, Japan*

Abstract. We present results of multiband analysis for 18 blazars observed with the X-ray satellite *ASCA*, half of which were also observed contemporaneously with *EGRET* instrument onboard Compton Gamma-ray Observatory (CGRO). Our multi-band analysis suggests that the magnetic field is $0.1 \sim 1$ gauss. With these values of B, we estimate γ_b is $10^3 \sim 10^4$ for quasar-hosted blazars (QHBs), and 10^5 for X-ray selected BL Lac objects (HBLs).

INTRODUCTION

Multi-frequency studies are very important tools in the identification of major emission processes in blazars. Recent studies show that the overall spectra have at least two pronounced components in $\nu F(\nu)$: the low energy peak (LE) and the high energy peak (HE) (see, e.g., [1]). The peak of the LE component is distributed over 4 decades of frequency: $10^{13} - 10^{17}$ Hz. The HE component, on the other hand, peaks in the γ-ray band, in the MeV - to - GeV range, but in the case of BL Lacs, it sometimes extends to the TeV range. The local power-law shape and the smooth connection of the entire radio - to - UV/soft X-ray spectrum, as well as the relatively high level of polarization observed from radio to the UV implies that the emission from the LE component is most likely produced via the synchrotron process of relativistic particles radiating in magnetic field. The HE component is generally thought to involve Comptonization of lower energy photons by the same particles that radiate the LE component. In general, in quasar-hosted blazars (QHBs), the LE component peaks at lower peak frequencies than it does in the X-ray selected BL Lac objects (HBLs) [2].

It is very intriguing to study inter-relations between processes that are responsible for emission in both components. Since the launch of *ASCA*, we have carried out extensive multiband campaigns of blazars [3–6]. In this paper, we study the relation between the LE component and the HE component, in terms of the peak frequencies (ν_{LE} and ν_{HE}) and the ratio of the luminosity of these components (L_{HE}/L_{LE}).

ASCA OBSERVATIONS

With satellite for X-ray astronomy, *ASCA*, we observed 18 blazars, of which 10 were also observed contemporaneously with *EGRET* as parts of multiband campaigns [7]; such simultaneous observing campaigns are essential for these highly variable objects. The *ASCA* data clearly show that the X-ray spectra of HBLs – these with high $\nu_{LE} \sim 10^{16} - 10^{17}$ Hz – are the softest, with the power law energy index $\alpha \sim 1 - 1.5$, and they form the highest observable energy tail of the LE peak. QHBs – which have lower $\nu_{LE} \sim 10^{13} - 10^{14}$ Hz – have the hardest X-ray spectra ($\alpha \sim 0.7$); those are *not* on the extrapolation of the LE peak, but are consistent with the lowest observable energy end ("onset") of the HE peak. For radio selected BL Lac objects (RBLs), the X-ray spectra are intermediate, and in a few cases are concave.

MULTIBAND ANALYSIS

Although it is almost certain that the process responsible for the HE component is the Comptonization of lower energy photons, the origin of these "seed photons" is still unresolved. The leading models propose either the synchrotron photons produced internally in the jet (the Synchrotron-Self Compton, or the SSC model; see, e.g., [8]) or photons external to the jet (the External Radiation Compton or the ERC model; see, e.g., [9,10])

In the following analysis, we assume that *both* SSC and ERC processes may take place in blazars, especially for QHBs in which the γ–ray flux strongly dominates the radiative output. In order to estimate the luminosity due to SSC emission, we assumed a simple, homogeneous model, in which photons are produced in a sphere of radius R with a constant magnetic field B. We considered the radiation by a single population of relativistic electrons, with a broken power-law in energy distribution, and a break point at γ_b, where γ is the Lorentz factor of electrons (similar to e.g. [11]). We also assume that the radiation spectrum of the LE component peaks at a frequency corresponding to that radiated by the electrons with γ_b. The peak frequency of the synchrotron component (ν_{sync}) is then given by:

$$\nu_{sync} = 1.2 \times 10^6 \gamma_b^2 B \delta \text{Hz} \quad (1)$$

Therefore the wide distribution of ν_{LE} observed in blazars could be attributed to the distribution of either γ_b or B or δ. If the electron energy is still in the

Thomson regime, the expected peak of the SSC component is $\nu_{SSC} = \frac{4}{3}\gamma_b^2 \nu_{sync}$
The ratio of the luminosity of the SSC component L_{SSC} to the synchrotron luminosity L_{sync} is $\frac{L_{SSC}}{L_{sync}} = \frac{u_{sync}}{u_B}$, where $u_{sync} = \frac{L_{sync}}{4\pi R^2 c \delta^4}$ is the energy density of the synchrotron photons, and $u_B = B^2/8\pi$ is the energy density of the magnetic field.

To check the validity of the assumption that the observed HE component is solely due to SSC emission, we calculated the beaming factor (δ) which is obtained from the equation, using L_{sync}, ν_{sync}, L_{SSC}, and ν_{SSC}:

$$L_{ssc} = 1.6 \times 10^{12} \left(\frac{L_{synch}^2}{cR^2 \nu_{synch}^4 \delta^2} \right) \nu_{ssc}^2 \text{erg/s} \qquad (2)$$

The peak frequency and the luminosity of both the LE and HE components are calculated from a third-order polynomial fit to the spectra in $\nu F \nu$. Since we only have upper limits for γ-ray emission for 6 out of the 18 blazars, we use 12 blazars only for the multiband analysis. The upper limit of the size R can be estimated from the observed time variability (Δt) given by $R < c\Delta t \delta/(1+z)$. We also calculate values with a fixed value of $R = 0.01$ pc.

With the assumption that the SSC process dominates the HE peak – or ν_{HE} and L_{HE} are equal to ν_{ssc} and L_{ssc} – the values of δ calculated for 7 blazars are in the range of $\sim 5 - 10$. These values are consistent with the superluminal expansion results [12], and the limits obtained from arguments involving the γ-ray opacity [13]. However the analysis for 5 QHBs and RBLs shows the values of δ to be much too large, sometimes exceeding $\delta \sim 100$. This suggests that an additional emission mechanism – such as the ERC process – contributes significantly in the γ-ray regime.

Since the synchrotron radiation is observed from blazars, the SSC component should exist at some level. The fact that QHBs have a hard X-ray spectra, with $\alpha \sim 0.7$ – which are *not* on the extrapolation of the synchrotron optical/UV continuum – suggests that the main X-ray emission of the QHBs is due to a separate emission process. Also, the spectral indicies for some QHBs in the γ-ray band are clearly harder than those obtained in the X-ray band [7]. We thus adopt the SSC process as the dominant mechanism of the production of X-rays. We then estimate L_{ssc} in the $\log(\nu) - \log(\nu F \nu)$ space by the following method. We assume that it is equal to or lower than the extrapolation of the *ASCA* spectrum, but higher than the highest *ASCA* value in the energy band of *ASCA*. As shown in Fig. 1, we further constrain L_{ssc} by assuming $R = c\Delta t\delta/(1+z)$ and $5 < \delta < 20$, using the formula in Eq. 2. The value of L_{ssc} in the center of this region is shown for the 5 blazars with "excessive" values of δ in Fig. 2(a). The values for the other 7 blazars are calculated by assuming $L_{SSC} = L_{HE}$.

Once we obtain L_{SSC}, we can calculate the strength of the magnetic field B as follows:

$$B = 0.27 \left(\frac{R}{10^{-2}\text{pc}}\right)^{-1} \left(\frac{\delta}{10}\right)^{-2} \sqrt{\left(\frac{L_{sync}}{10^{46}\text{erg/s}}\right)\left(\frac{L_{sync}}{L_{ssc}}\right)} \quad \text{gauss} \quad (3)$$

γ_b is then calculated using Eq.1,

From our analysis, B is inferred to be $0.1 \sim 1$ gauss (see Fig. 2(b)) for $\delta=10$, an intermediate value. With these values of B, we estimate γ_b is $10^3 \sim 10^4$ for QHBs, and 10^5 for XBLs (Fig. 2(c); $\delta=10$). This difference of γ_b between the two sub-classes of blazars suggests that the relativistic electrons are accelerated to higher energies in blazars with higher LE peak frequencies (e.g. HBLs) than in those with lower LE peak frequencies (e.g. QHBs). Alternatively, since B and γ_{el} depend on the value of δ, the results might be affected by changing δ. But the effect of varying δ is at most a factor of 4 in B and a factor of 1.4 in γ_b for $\delta \sim 5-20$, which is small compared with the difference of γ_b. Although the larger value of δ would be a way to obtain a larger γ_b, the larger δ also leads to ae higher ν_{sync} (Eq. 1), which is not consistent with the lower ν_{sync} for the higher γ_b objects in our results.

For QHBs, where a dense external radiation fields exist, the ERC emission most likely contributes as much or even more significantly than the SSC emission in γ-ray band (see, e.g., [14]). Therefore, our analysis indicates that the difference of γ_{el} is related to the large photon density in QHBs as compared with that of XBLs; Compton upscattering of these photons may well provide the observed GeV flux. It should be noted that the TeV photons have been observed only from XBLs which turned out to have higher γ_{el}.

SUMMARY

Our results show that the non-thermal emission from blazars, observed from radio to GeV/TeV γ-rays, is radiation of very energetic particles via both synchrotron and Compton processes. While the overall spectra of all blazars over this wide range of energies appear similar from one object to another, the X-ray regime is an important band, where the the emission due to both processes overlaps: it is either the high energy end of the synchrotron emission or the lowest observable energy end of the Comptonized spectrum. The *ASCA* data allow investigation of the details of the emission and estimation of the fundamental physical parameters, such as B and γ_{el}.

REFERENCES

1. von Montigny, C., et al., *ApJ*, **440**, 525 (1995).
2. Sambruna, R. M., Maraschi, L., & Urry, C. M. , *ApJ*, **463**, 444 (1996).
3. Madejski, G.,et al., *ApJ*, **459**, 156 (1996).
4. Takahashi, T., et al., *ApJ*, **470**, L89 (1996).
5. Takahashi, T., et al., in *All-sky X-ray Observations in the Next Decade* (1997).

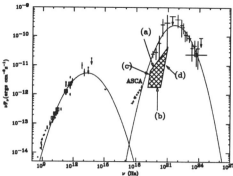

FIGURE 1. Multiband spectrum of CTA102. Hatched region corresponds to the estimated range of ν_{ssc} and L_{ssc} (a) a linear extrapolation of the ASCA spectrum, (b) the highest ASCA value, (c) (ν_{ssc}, L_{ssc}) for $\delta=5$, and (d) (ν_{ssc}, L_{ssc}) for $\delta=20$.

6. Kubo, H. et al., preprint (1997).
7. Kubo, H., PhD thesis, Univ. of Tokyo (1997).
8. Konigl, A. 1981, *ApJ*, **243**, 700
9. Dermer, C. D., Schlickeiser, R., & Mastichiadis, A. 1992, *A&A*, **256**, L27
10. Sikora, M., Begelman, M., & Rees, M., *ApJ*, **421**, 153 (1994).
11. Inoue, S. and Takahara, F., *ApJ*, **463**, 555 (1996).
12. Vermeulen, R. C., & Cohen, M. H., *ApJ*, **430**, 467 (1994).
13. Dondi, L., & Ghisellini, G., *MNRAS*, **273**, 583 (1996).
14. Sikora, M., Madejski, G., Moderski, R., & Poutanen, J., *ApJ*, 484 108 (1997).

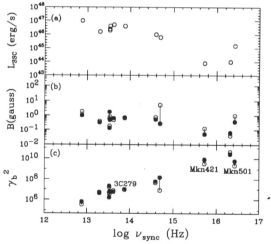

FIGURE 2. The distribution of parameters. (a) γ-ray luminosity (L_{SSC}), (b) magnetic field (B), and (c) the electron Lorentz factor γ_e as a function of the peak frequency of the LE component. The size R is estimated from the observed time variability (open symbol). We also plot values calculated with $R = 0.01$ pc (filled symbol).

Spectral modelling of gamma-ray blazars

M. Böttcher, H. Mause and R. Schlickeiser

Max-Planck-Institut für Radioastronomie, Postfach 20 24, D - 53 010 Bonn, Germany

Abstract. We present model calculations reproducing broadband spectra of γ-ray blazars by a relativistic leptonic jet, combining the EIC and the SSC model. To this end, the evolution of the particle distribution functions inside a relativistic pair jet and of the resulting photon spectra is investigated. Inverse-Compton scattering of both external (EIC) as well as synchrotron photons (SSC) is treated using the full Klein-Nishina cross-section and the full angle-dependence of the external photon source. We present model fits to the broadband spectra of Mrk 421 and 3C279 and the X-ray and γ-ray spectrum of PKS 1622-297. We find that the most plausible way to explain both the quiescent and the flaring states of these objects consists of a model where EIC and SSC dominate the observed spectrum in different frequency bands. For both Mrk 421 and 3C279 the flaring states can be reproduced by a harder spectrum of the injected pairs.

INTRODUCTION

The discovery of many blazar-type AGNs (e. g., Hartman et al. 1992) as sources of high-energy γ-ray radiation, dominating the apparent luminosity, has revealed that the formation of relativistic jets and the acceleration of energetic charged particles, that generate nonthermal radiation, are key processes to understand the energy conversion process. Emission from relativistically moving sources is required to overcome γ-ray transparency problems implied by the measured large luminosities and short time variabilities (for review see Dermer & Gehrels 1995).

Dermer, Miller & Li (1996) have recently inspected the acceleration of energetic electrons and protons in the central AGN plasma by comparing the time scales for stochastic acceleration with the relevant energy loss time scales. Their results demonstrate that with reasonable central AGN plasma parameter values low-frequency turbulence can energize protons to TeV and PeV energies where photo-pair and photo-pion production are effective in halting the acceleration (Sikora et al. 1987, Mannheim & Biermann 1992). Once the accelerated protons reach the thresholds for the latter processes they will generate plenty of secondary electrons and positrons of ultrahigh energy which are now injected at high energies into the acceleration scheme. The further fate of the secondary particles depends strongly on whether they find themselves in a compact environment set up by the external accretion disk, or not. Secondary particles within the photosphere will initiate a rapid electromagnetic cascade which has been studied by e.g. Mastichiadis & Kirk (1995), which might even lead to runaway pair production and associated strong X-ray flares (Kirk & Mastichiadis 1992), and/or due to the violent effect of a pair catastrophy (Henri, Pelletier & Roland 1993) ultimately lead to an explosive event and the emergence of a relativistically moving component filled with energetic electron-positron pairs.

It is the purpose of the present investigation to follow the time evolution of the relativistic electrons and positrons as the emerging relativistic blob moves out. We generalize the approach used in earlier work (Dermer & Schlickeiser 1993, Dermer, Sturner & Schlickeiser 1996) by accounting for Klein-Nishina effects and including external inverse-Compton scattering as well as synchrotron self-Compton scattering self-consistently.

SIMULATIONS OF THE TEMPORAL EVOLUTION OF PARTICLE AND PHOTON SPECTRA

We assume the relativistic pairs to be isotropically distributed in a spherical volume of radius R_B, located at height z_i above the accretion disk plane. The minimum Lorentz factor of the pairs at the time of injection is expected to be in the range of the threshold value of the primary protons' Lorentz factor for photo-pair production. We find this threshold at $\gamma_{1\pm} \sim 10^3$. The pair distribution above this cutoff basically reflects the acceleration spectrum of the protons which can extend up to $\gamma_{2\pm} \sim 10^6$. The lack of a significant cutoff at high photon energies due to γ-γ pair production in the EGRET spectra of many γ-ray blazars and in particular the detection of TeV γ-rays from Mrk 421 suggests that a new jet component must be produced/accelerated outisde the γ-ray photosphere for TeV photons. We find this photosphere due to the interaction of γ-rays with accretion disk radiation to be located around $z \approx 5 \cdot 10^{-3} M_8^{-1}$ pc where M_8 is the mass of the central black hole in units of $10^8\, M_\odot$.

The blob is assumed to move outward perpendicularly to the accretion disk plane with velocity $c\beta_\Gamma = c\sqrt{1 - 1/\Gamma^2}$. The cooling mechanisms we take into account are inverse-Compton scattering of accretion disk photons (EIC), synchrotron and synchrotron-self-Compton (SSC) losses to arbitrarily high order. For comparison to observations, we calculate the time-averaged emission, since the integration times of present-day γ-ray observing instruments are much longer than the cooling time-scales resulting from our simulations.

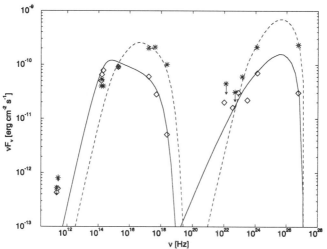

FIGURE 1. Fit to the broadband spectrum of Mrk 421 in its quiescent state (squares; solid line) and during the May 1994 flare (stars; dashed line)

MODEL CALCULATIONS

We used our code to fit the spectra of three γ-ray blazars, namely the flat-spectrum radio quasar 3C279, the BL-Lac object Mrk 421 and the quasar PKS 1622-297. Throughout our calculations, we assumed $H_0 = 75$ km s^{-1} Mpc^{-1}, $q_0 = 0.5$, and $\Lambda = 0$. The fits to Mrk 421, as shown in Fig. 1, are discussed in detail in Böttcher et al. (1997).

The quasar 3C279

The quasar 3C279 was observed in a broad-band campaign by Hartman et al. (1996) during an outburst in 1991 June and in a more quiescent state in 1991 September – October. For the latter phase, unfortunately, no simultaneous observations in the infrared – optical – UV band were available. The typical flare timescale of this object is several days, which restricts the size of the emitting region to $R_B \lesssim D \cdot 10^{17}$ cm where $D = (\Gamma [1 - \beta_\Gamma \cos\theta_{obs}])^{-1}$ is the Doppler factor. Our fits to the quiescent and the flaring phase of 3C279 are shown in Fig. 2. The parameters for the quiescent state were

$$\gamma_{1\pm} = 200 \qquad \gamma_{2\pm} = 7 \cdot 10^4$$
$$s = 2.5 \qquad n = 3\,\text{cm}^{-3}$$
$$R_B = 3 \cdot 10^{17}\,\text{cm} \qquad B = 0.1\,\text{G}$$
$$L = 10^{46}\,\text{erg s}^{-1} \qquad M = 5 \cdot 10^8\,M_\odot$$
$$z_i = 6 \cdot 10^{-2}\,\text{pc} \qquad \Gamma = 15$$
$$\theta_{obs} = 2^0$$

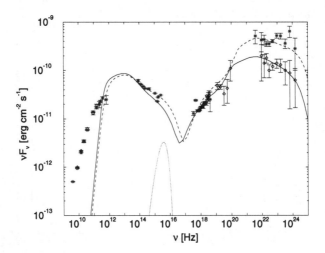

FIGURE 2. Fit to the broadband spectrum of 3C279 in its quiescent state (squares; solid line) during 1991 September – October, and during the flare of 1991 June (stars; dashed line). Dotted curve: accretion disk spectrum

where $\gamma_{1,2\pm}$ are the cutoffs of the initial pair distributions, s is the initial spectral index of the particle spectra $(n(\gamma) \propto \gamma^{-s})$, z_i is the height of the injection site above the accretion disk plane, L is the total luminosity of the accretion disk and M is the masss of the central black hole.

The flaring state could be reproduced by a harder injection spectrum with $s = 2.4$, $\gamma_{1\pm} = 300$ and $\gamma_{2\pm} = 8 \cdot 10^4$.

The quasar PKS 1622-297

Recently, Mattox et al. (1997) have reported an intense γ-ray flare of the quasar PKS 1622-297 during which the X-ray and the γ-ray spectrum can be represented by two different power-laws. Our model calculation, which is illustrated in Fig. 3, demonstrates that the combined SSC/EIC model is well suited to reproduce such a two-component spectrum.

The parameters chosen for the fit shown in Fig. 3 are:

$$\begin{aligned}
\gamma_{1\pm} &= 400 & \gamma_{2\pm} &= 7 \cdot 10^4 \\
s &= 2.7 & n &= 420\,\text{cm}^{-3} \\
R_B &= 10^{17}\,\text{cm} & B &= 0.08\,\text{G} \\
L &= 10^{46}\,\text{erg s}^{-1} & M &= 10^8\,M_\odot \\
z_i &= 2.5 \cdot 10^{-2}\,\text{pc} & \Gamma &= 15 \\
\theta_{obs} &= 4^0
\end{aligned}$$

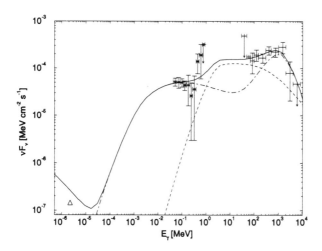

FIGURE 3. Fit to the X-ray and γ-ray spectrum of PKS 1622-297 during the flare reported by Mattox et al. (1997); the optical data point is from non-simultaneous observations and does not correspond to the flare state of PKS 1622-297. Dashed: EIC, dot-dashed: SSC

REFERENCES

1. Böttcher, M., Mause, H., & Schlickeiser, R., *A&A*, in press (1997)
2. Dermer, C. D., and Gehrels, N., *ApJ* **447**, 103 (1995)
3. Dermer, C. D., Miller, J. A., and Li, H., *ApJ* **456**, 106 (1996)
4. Dermer, C. D., & Schlickeiser, R., *ApJ* **416**, 458 (1993)
5. Dermer, C. D., Sturner, S. J., & Schlickeiser, R., *ApJ*, **109**, 103 (1997)
6. Hartman, D., et al., *ApJ* **385**, L1 (1992)
7. Hartman, R. C., Webb, J. R., Marscher, A. P., et al., *ApJ* **461**, 698 (1996)
8. Henri, G., Pelletier, G. & Roland, J., *ApJ* **404**, L41 (1993)
9. Kirk, J. G., & Mastichiadis, A., *Nature* **360**, 135 (1992)
10. Mannheim, K., & Biermann, P. L., *A&A* **253**, L21 (1992)
11. Mastichiadis, A., & Kirk, J. G., *A&A* **295**, 613 (1995)
12. Mattox, J. R., Wagner, S. J., Malkan, M., et al., *ApJ* **476**, in press (1997)
13. Sikora, M., Kirk, J., T., Begelman, M. C., et al., *ApJ* **320**, L81 (1987)

Modelling the rapid variability of blazar emission

J. G. Kirk and A. Mastichiadis

Max-Planck-Institut für Kernphysik, Postfach 10 39 80, D-69029 Heidelberg

Abstract. A homogeneous, synchro-self-Compton model (SSC) is used to model the spectra of gamma-ray emitting blazars. The constraints imposed by the rapid variability of the X-rays on a time scale of 1000 s can be met, but require a very large bulk Lorentz factors. These are not consistent with the bulk Lorentz factors implied by observations of apparent super-luminal motion in such sources. A model is also presented in which the target photons for the inverse Compton scattering stem from an external black body source. During a flare, this model predicts a lower relative amplitude of variation in the TeV regime as compared with the X-ray regime. It also gives rise to a steep spectrum in the TeV range. Finally, by varying the electron injection stochastically, the typical patterns of time variation of emission at different wavelengths are presented.

SYNCHROTRON-SELF-COMPTON MODELS

In previous work [1] we have shown that a homogeneous region containing magnetic field and relativistic electrons can reproduce the observed spectrum of blazars over a large wavelength range, whilst at the same time allowing time variations on the scale of roughly 1 day. A fit to observations of Mkn 421 is shown in Fig. 1. The radiation processes which dominate in our source model are synchrotron emission (in the radio to X-ray range) and inverse Compton scattering of the synchrotron photons (in the gamma-ray range). The source is assumed to be pervaded by a magnetic field of uniform strength, and is moving at a small angle θ to the line of sight with a bulk Lorentz factor of Γ. The computations involve the numerical solution of the time-dependent, spatially averaged kinetic equations for electrons and photons. The flaring behaviour in Fig. 1 is modelled as a sudden increase of the maximum Lorentz factor to which the injected electron population extends.

Seven independent parameters are needed to determine a stationary spectrum in this model. They are the Doppler-boosting factor $\delta = [\Gamma(1-\beta\cos\theta)]^{-1}$ (where β is the bulk velocity of the source), the size of the source R, the magnetic field B, the mean time during which particles are confined in the source

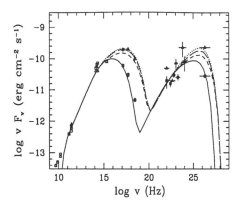

FIGURE 1. Model of the flare of Mkn 421 observed in 1994, [2,3]. The light crossing time of the source allows variations on a timescale of $R/(c\delta) \approx 1$ day. The curves show the spectrum 1, 2 and 3 crossing times after a sudden increase of γ_{max}.

TABLE 1. Parameters of the model fits to Mkn 421

	δ	R (cm)	B (mG)	ct_{esc}/R	ℓ_e	s	γ_{max} (in flare)
Fig. 1	15	4.7×10^{16}	70	3.3	1.9×10^{-5}	1.67	2×10^5 (1×10^6)
Fig. 2	60	1.9×10^{15}	130	13.2	1.6×10^{-6}	1.67	7×10^4 (3.6×10^5)
Fig. 3	22.5	6.8×10^{16}	100	1.7	2.6×10^{-6}	1.67	1.3×10^5 (6.5×10^5)

t_{esc}, and three parameters determining the injected relativistic electron distribution: its luminosity, or compactness ℓ_e, the spectral index s and the maximum Lorentz factor γ_{max}. Note that it is necessary to include a finite escape time from the radiating region (which can also be thought of as an adiabatic loss time) in order to account for the rather flat spectra seen in the radio to infra-red region – these photons must be radiated by particles which have not been allowed to cool significantly. The quiescent state and flare shown in Fig. 1 have parameters given in Table 1, where all quantities are given as measured in the rest-frame of the source.

Recently, very rapid variations in the TeV flux of this object have been reported [4]. Shorter timescales can be achieved within the model, principally by changing the Doppler boosting factor δ. In Fig. 2 a fit is shown in which the light-crossing time of the source has been reduced to 1000δ s. However, the escape time t_{esc} needed is significantly longer, and the magnetic field larger. This means that the variability timescale of a flare is dominated by the time taken to cool or escape from the source, rather than by the light-crossing time. The parameters of the model are given in Table 1. Another problem with this and similar fits is that the implied values of the bulk Lorentz factor lie substantially above those indicated by observations of apparent superluminal

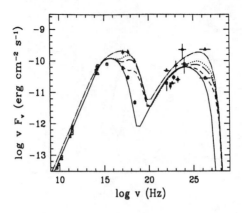

FIGURE 2. The best fit for the flare of 1994, with a variability timescale of $R/(c\delta) \approx 1000\,\text{s}$. The curves show the spectrum 1, 3, and 10 crossing times after a sudden increase of $\gamma_{\max}$, as well as the final steady state.

motion [5]. A possible solution lies in allowing for a non-spherical emitting region e.g., a laminar region behind a shock front [6,7].

EXTERNAL PHOTONS

As an alternative to the synchrotron self-Compton models, we have considered the effect of introducing an external source of soft, black-body photons. These dominate over the synchrotron photons as targets for inverse Compton cooling when their energy density exceeds that of the magnetic field. However, their effect is diminished for very high energy electrons, because of the fall-off of the Klein-Nishina cross-section with increasing energy. In a flare, the synchrotron self-Compton model predicts an increase in the gamma-ray flux by a factor which is proportional to the square of the relative increase in the synchrotron flux. If external photons dominate as targets, the increases should be linearly related at low energies, and the high energy gamma-rays should react even less sensitively in the Klein-Nishina regime. To illustrate this effect, we have chosen, in the model displayed in Fig. 3, external photons of temperature $0.5\,\text{eV}$ – the lowest consistent estimate for photons which probably originate in broad line clouds. The resulting fit to the quiescent state of Mkn 421 is reasonable. The flare, on the other hand, fails to reach the required amplitude in the GeV–TeV range because of the Klein-Nishina fall-off. A further problem is that the predicted spectrum in the TeV range is very steep. It is difficult to account for the observation of photons of energy greater than a few TeV [8–10] in this model.

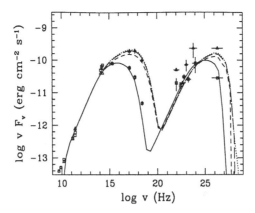

FIGURE 3. The predicted spectrum of Mkn 421, assuming energy losses are dominated by inverse Compton scattering on photons which originate outside of the source and have a black-body temperature of 0.5 eV. During a flare, the IC component increases in direct proportion to the increase in X-ray luminosity, provided the scattering is purely Thomson. In our case, however, there is a contribution from the Klein-Nishina regime, which reduces the IC response. The spectrum cuts off steeply above roughly 1 TeV

VARIABILITY

To emphasise the features of multi-wavelength variability in our model we plot in Fig. 4 the results of a stochastic variation of the injection rate of relativistic electrons. Three timescales are important in this respect: the light-crossing time of the source, the synchrotron cooling time and the escape time from the source. The light curves at lower frequency are clearly smoother than those at high frequencies, which is simply a result of the increased cooling time. However, there is also a pronounced memory effect – the IR flux rises through almost the whole time period plotted, whereas the X-ray flux, which closely tracks the injected electrons undergoes a sequence of three almost equally intense flares with very little build up of the underlying emission. Rapid, small amplitude changes in the UV and IR fluxes are seen with timescale much shorter than the appropriate cooling time. These are caused by the sudden switch-on and off of the electron injection, which is assumed to extend over a crossing time, and the slightly longer timescale on which electrons emitting at these frequencies leak out of the source. Despite this, there is not always a simple one-to-one correlation between an X-ray flare and its IR counterpart. Clearly visible in Fig. 4 is the 'soft-lag' characteristic of synchrotron cooling. This phenomenon is predicted to occur when the electron acceleration rate is much faster than the cooling rate [11]– as assumed in the calculations presented here – and has been observed within the X-ray band in

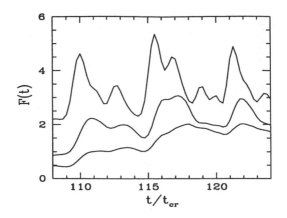

FIGURE 4. The X-ray, UV and IR fluxes predicted from the homogeneous model assuming stochastic fluctuations in the injection rate of relativistic electrons. In this episode of high level X-ray emission, injection produces an almost immediate response at all frequencies. However, at UV and IR wavelengths, there is a longer term increase in the level of emission, as those electrons injected at higher energy slowly cool down and start to emit more strongly in these bands. The highest energy electrons have a cooling time roughly equal to the crossing time, so that in the UV, the flux responds on timescales of up to about 30 crossing times.

Mkn 421 [12] as well as other sources [13,14].

REFERENCES

1. Mastichaidis A., Kirk J.G. 1997a A&A 320, 19
2. Macomb D.J., et al. 1995 ApJ 449, L99
3. Macomb D.J., et al. 1996 ApJ 459, L111 (Erratum)
4. Gaidos J.A. et al. 1996 Nature 383, 319
5. Vermeulen R.C., Cohen M.H. 1994 ApJ 430, 467
6. Kirk J.G. 1997 in Proceedings of the Cracow Workshop on 'Relativistic Jets in AGN'.
7. Mastichaidis A., Kirk J.G. 1997b, in preparation
8. Krennrich F. et al. 1997 ApJ 481, 758
9. Bradbury S.M. et al. 1997 A&A 320, L5
10. Aharonian F. et al. 1997 submitted to A&A
11. Kohmura Y. et al. 1994 PASJ 46, 131
12. Takahashi T. et al. 1996 ApJ 470, L89
13. Sembay S. et al. 1993 ApJ 404, 112
14. Tashiro M. et al. 1995 PASJ 47, 131

OVERVIEWS, SURVEYS, AND MISCELLANEOUS

BeppoSAX Overview

Luigi Piro*
on behalf of the BeppoSAX team [1]

*Istituto Astrofisica Spaziale, C.N.R. Via E. Fermi 21 00044 Frascati, Italy

Abstract. The combination of the broad band narrow field instruments, the wide field cameras and gamma-ray burst monitor along with the operational capabilities of the whole BeppoSAX mission provide an unprecedented opportunity to study the X-ray variable sky. This presentation will focus on two main topics: the broad band spectra and variability of AGN and the capabilities of the mission to observe transient phenomena, presenting some key results obtained on gamma-ray bursts, in particular the discovery of X-ray afterglows. Further results on galactic sources are illustrated by other contributions by team members.

INTRODUCTION

The X-ray satellite SAX, named BeppoSAX after launch in honour of Giuseppe (Beppo) Occhialini, is characterized by an unprecedented combination of broad band coverage (from 0.1 to 300 keV), relatively large area and low background, good energy resolution, and imaging capabilities (resolution of about 1 arcmin) in the range of 0.1-10 keV: the sensitivity of the scientific payload and the balanced performances of the low energy and high energy instruments allows the exploitation of the full band of BeppoSAX for sources as weak as 1 mCrab. This capability, in conjunction with the presence of wide field instruments primarily aimed at discovering transient phenomena, which could then be observed with the broad band instruments, provides an unprecedented opportunity to study the broad band behaviour of several classes

---

[1] The BeppoSAX team is composed by scientists from: Istituto Astrofisica Spaziale (IAS), C.N.R., Frascati and Unita' GIFCO Roma; Istituto di Fisica Cosmica ed Applicazioni Informatica (IFCAI) and Unita' GIFCO, Palermo; Istituto Fisica Cosmica e Tecnologie Relative (IFCTR), CNR and Unita' GIFCO, Milano; Istituto per le Tecnologie e Studio Radiazioni Extraterrestri (ITeSRE), C.N.R., Bologna and Universita' di Ferrara; Space Research Organization of the Netherlands (SRON), The Netherlands; Space Science Department (SSD), ESA, Noordwijk, The Netherlands; BeppoSAX Science Data Center, ASI, Rome; BeppoSAX Science Operation Center, N.Telespazio, Rome

of X-ray sources. The time taken from a Target of Opportunity alert to the acquisition of the target with the broad band instruments can be as short as few hours.

The broad band capability is provided by a set of instruments co-aligned with the Z axis of the satellite, Narrow Field Instruments (hereafter NFI) and composed by:

- MECS (Medium Energy Concentrator Spectrometers): a medium energy (1.3-10 keV) set of three identical grazing incidence telescopes with double cone geometry [1,2] with position sensitive gas scintillation proportional counters in their focal planes [3].

- LECS (Low Energy Concentrator Spectrometer): a low energy (0.1-10 keV) telescope, identical to the other three, but with a thin window position sensitive gas scintillation proportional counter in its focal plane [4].

- HPGSPC, a collimated High Pressure Gas Scintillation Proportional Counter (4-120 keV, [5]).

- PDS, a collimated Phoswich Detector System (15-300 keV, [6])

Access to large regions of the sky ($\sim$ 3200 degree^2) with an angular resolution of 5'(FWHM) in the range 2-30 keV is provided by:

- two coded mask proportional counters (Wide Field Cameras, WFC, [7]), perpendicular to the axis of the NFI and pointed in opposite directions. The field of view is 20 deg square (FWHM), although sources can be detected with less sensitivity in a field of 40 deg $\times$ 40 deg in total. Source location accuracy should be of $\sim$ 2 arcmin, depending on the source statistics.

Finally, the anticoincidence scintillator shields of the PDS (GRBM [8]) will be used as a gamma-ray burst monitor in the range 60-600 keV with a fluence greater than about 10^{-6} erg cm^{-2} and with a temporal resolution of about 1 ms.

More details on the mission and its instruments can be found in [9,10], in the special session devoted to BeppoSAX of the SPIE Vol. 2517 and on line at: *http://www.sdc.asi.it*.

BROAD BAND SPECTROSCOPY OF AGN WITH THE NFI

In the Science Verification Phase (SVP) carried out from July to September 1996 more than 30 targets were observed; Some of the results obtained on galactic sources are presented in this conference [11–14,22]. Four bright AGN were included in the SVP: the Seyfert 1 galaxies MCG-60-30-15, NGC 4151, the quasar 3C273 and the blazar PKS2155-304. Several other AGN are being observed in the AO phase. Hereafter a brief review of the first results.

Warm absorbers

ASCA results have demonstrated that absorption features by highly ionized matter are common in Seyfert spectra (e.g. [15]). The MCG-6-30-15 BeppoSAX low energy spectrum [16] is consistent with a warm absorber scenario where highly ionized species of Oxygen and Neon are responsible for the most prominent features below 1 keV. However a comparison of light curves below and above 1 keV displays a pattern of variability that cannot be easily explained by simple photoionization models, indicating a non-equilibrium condition. Similar absorption features are starting to be detected in the spectrum of other kind of AGN, including objects where beaming should play an important role in the X-ray emission. Previous observations of PKS2155-304 [20] showed an absorption structure consistent with resonant Ly-α absorption from OVIII. This feature is distinctly detected in the BeppoSAX observation [21]. Furthermore, the long BeppoSAX observation of 3C273 has shown, for the first time, a narrow absorption feature around 0.6 keV [17], which is consistent with the same emission process, provided it takes place in a matter with a bulk velocity of $\sim 60000 km\ s^{-1}$ towards the observer. This can be associated with the jet of 3C273.

Hard components

Along with the measurements of the reflection component in bright Seyfert galaxies (MCG-6-30-15: [27], NGC 4151: see next section), the very low and stable background of the PDS [22] has allowed the detection of hard tails in other AGN. In the case of NGC 1068 [23] it has been the first detection reported so far, and has allowed to confirm the *Compton thick dominated* scenario of a sub class of seyfert 2 galaxies, where the X-ray emission above a few keV is dominated by reflection from the inner region of a torus. A similar evidence is possibly present in NGC 7674 [24]. Another new result is the detection of a hard tail in PKS2155-304 [22,21], which is in this case attributed to the inverse Compton component expected in the SSC scenario.

Weak AGN

Several AO programs concentrating on weak (i.e. $<\ mCrab$) AGN are being carried out. The very low and stable background afforded by the MECS allows a good determination of the two-component spectrum of Narrow Line

Seyfert 1 galaxies [25] and a 100% detection rate of optically selected Seyfert 2 galaxies [26].

NGC 4151: a case study of the spectral complexity of AGN

One of the observational challenge in the study of emission-line AGN is to disentangle the different contribution of the several features present in the spectrum. A significant example demonstrated by BeppoSAX is the case of MCG-6-30-15, where the red tail of the iron line, attributed to the relativistic motion of an accretion disk around a MBH, may be in part produced by the extension to lower energies of a strong reflection component [27].

The X-ray spectrum of NGC 4151 is the most complex observed so far in AGN, being characterized by narrow and broad spectral features from soft to hard X-rays [28–30]. It is then the best candidate to verify the unique capability of BeppoSAX to disentangle spectral features over the 0.1-200 keV energy range.

In fig.1 we show the spectrum of NGC 4151 as observed by BeppoSAX in the fist part of the observation of July 96, when the source was in a low state. Exposure times were $\sim$ 7 ksec for the LECS and $\sim$ 25 ksec for the MECS, HPGSPC and PDS. In order to underline the several broad and narrow features overimposed on the continuum, we have normalized the spectrum to that of 3C273, which has a similar intrinsic power law slope and is almost featureless (but for the absorption edge and a soft excess [17]).

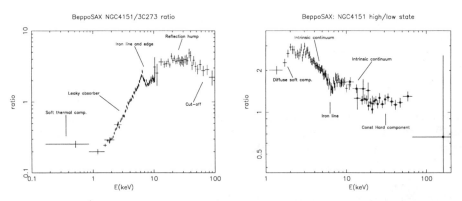

FIGURE 1. BeppoSAX observation of NGC 4151. In the left panel the ratio of the spectrum of this object to that of 3C273 underline the presence of several broad and narrow features that appear as deviations from a constant line. The right panel is the ratio of the high to low states observed in July 96 few days apart and shows the complex spectral variability present in this object

A key information on the origin of these components is furtherly provided by their (different) time evolution and their response to the variation of the intrinsic continuum. The spectral variability study furthermore allows to discriminate between alternative modellizations that provide equally good description of the spectrum.

From this point of view NGC 4151 provides a instructive example. During the July observation the source underwent an increase of the 2-10 kev flux by a factor of 2 in about 2 days. In fig.1 (right panel) we show the ratio of the spectra taken in the high state (when the LECS was not operating) to the low state (the one utilized in the left panel). The presence of *complex* spectral variability is evident. There is a general trend of the spectrum to get steeper when the source is brighter. This is likely due to a change of the slope of the intrinsic continuum. However there are at least three remarkable features that deviate from this trend (that, in the log - log scale of the fig. would be represented by a straight line): a) a constant component, influencing the spectrum up to about 3 keV, produces the bent observed around that energy. This is likely produced by the scattering of the nuclear continuum on an electron cloud, in the same scenario suggested for Seyfert 2 galaxies. b) The iron line does not follow the variation of the ionizing continuum, indicating that the material is at distances greater than about 2 light days. This evidence indicates that the bulk of the line is not produced in an AD. c) Above about 10 keV the ratio flattens, indicating the presence of a hard less-variable component, most likely reflection from a thick medium.

FROM THE γ-RAY BURST TO ITS AFTERGLOW

For more than twenty years after their discovery the origin of γ-ray bursts has remained one of the great mysteries in Astrophysics. This has been largely due to the difficulty in identifying a counterpart. Accurate position have only been provided in the past for a few dozens GRB's [18,19], several days after the event.

The BeppoSAX mission represents a big step forward, in that it combines instrumental and operational capabilities that allow the detection (by the GRBM) and positioning of GRB's to within a few arcmin (by the WFC, see fig. 2) in less than 2-4 hours. This in turn allows to carry out fast (within 6-8 hours after the event) follow up observations with the NFI (fig. 3), that provide sub arcmin positioning. This procedure has led to the discovery of the X-ray afterglow in GB970228 [31], followed by a similar detection in GB970402 [32] and GB970508 [33].

Quick dissemination of the coordinates by the BeppoSAX team has allowed fast observations by ground based facilities in other wavelenghts, with the discovery of optical transients associated to GB970228 [34] and GB970508 [35]. In the latter case the first spectroscopic observation of such an event [36]

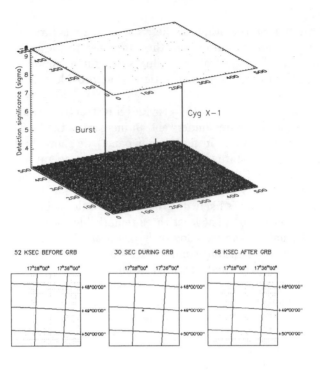

FIGURE 2. WFC images of GB960720: the upper panel shows the full field of view of the WFC (40x40 deg) during the burst, with Cyg X-1 close to the edge. The lower panel is the sequence of images centered on the GRB: 50 ksec before, 30 s during the burst, 50 ksec after the burst

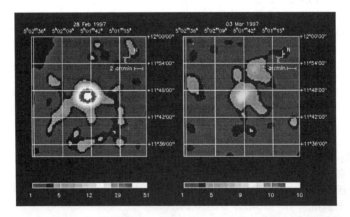

FIGURE 3. GB970228: MECS images of the afterglow observed on Feb.28, 8 hours after the burst (left image). On Mar.3 (right image) the source had decreased by a factor of 20

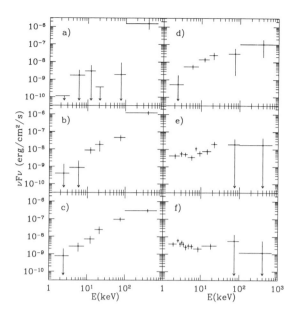

FIGURE 4. GB960720: time evolution of the spectrum in νF_ν: a) to d): 1 s step; e): from 4 to 8 s; f): from 8 to 17 s

strongly argues for an extragalactic origin of GRB.

Not less important is the capability of BeppoSAX to perform X to gamma-ray spectral measurements of GRB's, combining the GRBM and WFC spectra. This provides strong constraints to radiative mechanisms of GRB's.

In fig.4 we show the very fast time evolution of the spectrum of GB960720 [37]: in a few seconds the energy at which most of the luminosity is emitted shifts from the MeV region to the keV region. Furthermore, the spectrum in the initial phase of the event betrays the presence of an optically thick source rapidly evolving in a thin configuration. No other class of sources in the universe shows such a fast and extreme evolution. More theoretical work is needed to account for these results.

REFERENCES

1. Citterio O. et al. 1985, SPIE Proc. 597, 102
2. Conti G. et al. 1994, SPIE Proc. 2279, 101
3. Boella, G., Chiappetti, L., et al. 1997b, A&AS, 122, 327
4. Parmar, A.N., et al. 1997, A&AS, 122, 309
5. Manzo, G., et al. 1997, A&AS, 122, 341
6. Frontera, et al. 1997, A&AS, 122, 357
7. Jager, R., et al. 1997, A&AS, in press

8. Costa E. et al., 1997, Adv. Space Res., submitted
9. Piro L., Scarsi L. & Butler R.C., 1995, SPIE Proc. 2517, 169
10. Boella, G., Butler, R.C., et al. 1997a, A&AS, 122, 299
11. Dal Fiume D. et al. 1997, this conference
12. Kuulkers E. et al. 1997, this conference
13. Orlandini, M. et al. 1997, this conference
14. Cusumano G. et al. 1997, this conference
15. Reynolds C. S. 1997, MNRAS, in press
16. Orr, A. et al. 1997, A&A, in press
17. Grandi P., et al. 1997, A&A, in press
18. Atteia, J.-L. et al. 1987, ApJS, 64, 305.
19. Boer, M. et al. 1994, in Gamma Ray Bursts, eds. G.J. Fishman - J.J. Brainerd - K. Hurley, (New York, AIP Conf. Proceedings), 307, 458
20. Madejsky G.M. et al., 1991, ApJ, 370, 198
21. Giommi, P., et al. 1997, A&A, in press
22. Frontera, et al. 1997, this conference
23. Matt G., et al. 1997, A&A, submitted
24. Malaguti G. et al., 1997, this conference
25. Comastri A., et al. 1997, A&A, submitted
26. Salvati M., et al., 1997, A&A, in press
27. Molendi S., et al. 1997, A&A, submitted
28. Perola G.C. et al. 1986, ApJ 306, 508
29. Warwick R.S., Done C. & Smith D.A. 1995, MNRAS 275, 100
30. Zdziarski, A. A., Johnson, W. N. & Magdziarz P. 1996, MNRAS 283, 193
31. Costa E., et al. 1997, Nature, 387, 783
32. Piro L. et al. 1997, IAU circ. 6617
33. Piro L. et al. 1997, IAU circ. 6656
34. Van Paradijis J., et al 1997, Nature, 386, 686
35. Djorgovski S.G. et al 1997, Nature, 387, 876
36. Metzger M.R. et al 1997, Nature, 387, 879
37. Piro L. et al. 1997, A&A, in press

Initial Results from the High Energy Experiment *PDS* aboard *BeppoSAX*

F. Frontera[1,2], D. Dal Fiume[1], E. Costa[3], M. Feroci[3],
M. Orlandini[1], L. Nicastro[1], E. Palazzi[1],
G. Zavattini[2] and P. Giommi[4]

[1] *Istituto Tecnologie e Studio Radiazioni Extraterrestri, CNR, 40126 Bologna, Italy*
[2] *Dipartimento di Fisica, Università di Ferrara, 44100 Ferrara, Italy*
[3] *Istituto Astrofisica Spaziale, CNR, 00144 Frascati*
[4] *BeppoSAX Science data Center, Roma, Italy*

Abstract. The high energy experiment *PDS* is one of the Narrow Field Instruments aboard the X-ray astronomy satellite *BeppoSAX*. It covers the energy band from 15 to 300 keV. Here we report results on its in-flight performance and observations of galactic and extragalactic X-ray sources obtained during the Science Verification Phase of the satellite: in particular Crab, Cen X-3, 4U1626-67 and PKS2155-305.

INTRODUCTION

The Phoswich Detection System (*PDS*) is one of the Narrow Field Instrument aboard the X-ray astronomy satellite *BeppoSAX*, a program of the Italian Space Agency (ASI) with Dutch participation [1]. A detailed description of the instrument can be found in [2].

The *PDS* was designed to operate in the hard X-ray range from 15 to 300 keV and to perform high sensitivity spectroscopic and temporal studies of celestial X-ray sources, in particular their continuum emission. Classes of sources accessible to *PDS* include High Mass X-ray Binaries (HMXRB), Low Mass X-ray Binaries (LMXRB), Am Her-type sources, supernova remnants (in particular Crab-like sources) and Active Galactic Nuclei (AGN).

In this paper we report on functional performance of *PDS* and on some scientific results obtained from the observation of different classes of X-ray sources performed during the Science Verification Phase (SVP) of the *BeppoSAX* satellite.

IN-FLIGHT FUNCTIONAL PERFORMANCE

BeppoSAX was launched from Cape Canaveral with an Atlas-Centaur rocket on April 30, 1996. Its orbit is almost equatorial (3.9°) at an altitude of 600 Km. During the *BeppoSAX* Commissioning Phase (1 May–30 June 1996), the *PDS* was switched on and tested. Since then all subsystems continue to properly operate with very good performance.

The anticoincidence (AC) shields provide a reduction of the background level by about a factor two with respect to the level obtained with the phoswich technique alone. In addition, the AC system strongly decreases the background modulation along the *BeppoSAX* orbit.

The background level B of the *PDS* is the lowest obtained thus far with high energy instruments at satellite orbits, specially at energies beyond 100 keV. In the 15–300 keV band, B is about 2.0×10^{-4} Cts cm^{-2} sec^{-1} keV^{-1}: this is $\sim 70\%$ of that published in the *BeppoSAX* handbook [3]. In 100–300 keV B is a factor 3 lower than that expected. The background modulation along a single *BeppoSAX* orbit and on one day time scale is about 20%, while longer term variations (*e.g.*, build-up effects) are negligible.

We evaluated the systematic error in the background subtraction introduced by the rocking collimator technique, by computing the background level variation between the ON- and OFF-source positions in 32 ksec observing time. We obtained 0.17 ± 0.09 mCrab in the 15–300 keV energy band, corresponding to a 5σ instrument sensitivity of 0.9 mCrab. This has to be compared to the value of 0.5 mCrab, if only Poisson statistics is taken into account.

SCIENTIFIC RESULTS

We will discuss some scientific results obtained with *PDS* during the *BeppoSAX* SVP, that show the actual spectral and timing capability of the instrument in flight.

Crab Nebula

The source has been observed for calibration purposes two times, in September 1996 (obs.I) and in April 1997 (obs.II). The spectral deconvolution makes use of a response function derived from a Monte Carlo code that describes the interaction of a photon beam with the *PDS* instrument, complemented by the ground calibrations [4].

By fitting the total spectrum of Crab with a single power law model we obtained a photon index α consistent for both observations: $\alpha = 2.119 \pm 0.002$ for obs.I and $\alpha = 2.112 \pm 0.003$ for obs.II. The normalization parameters at 1 keV are 8.60 ± 0.06 and 8.57 ± 0.04 for obs.I and obs.II, respectively. The fitting was not so good: the reduced χ^2 was 2.89 (77 dof) for obs.I and 3.09

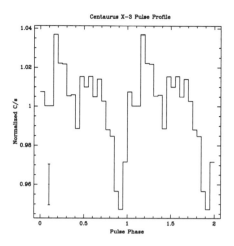

FIGURE 1. *BeppoSAX/PDS* pulse profile of the 4.84 sec X-ray binary pulsar Cen X-3.

(67 dof) for obs.II. These high χ^2 are partly due to a break in the Crab high energy spectrum also observed by other groups [5], and partly to wiggles in the count rate spectrum between 30 and 60 keV, whose instrumental origin is under investigation. These systematic effects are estimated to be less than 5% in the residuals and less than 1% in the spectral index reconstruction.

We can conclude, from these results, that the photon index is consistent with previous results on the source spectrum [5], while the normalization parameter is about 10% lower than the extrapolation of the power law spectrum measured with the MECS instrument [6] aboard *BeppoSAX*. It is however consistent with published results on Crab [5] within their uncertainties.

Centaurus X-3

This classical X-ray pulsar, belonging to the class of HMXRBs, was observed for ~ 6000 sec during a low intensity level (30 mCrab) at high energies. The source was clearly detected up to about 50 keV. The hard X-ray spectrum of the source is consistent with an optically thin thermal bremsstrahlung with $kT = 8.2 \pm 0.2$ keV and normalization parameter 0.66 ± 0.03. The plasma temperature and flux level are consistent with those measured by HEAO-1/A4 [7]. The *PDS* pulse profile of the pulsar ($P_P \sim 4.84$ sec) is characterized by a dip (see Fig. 1). It appears different from the average pulse profiles previously reported [7]. An investigation on the origin of this different behavior is under way.

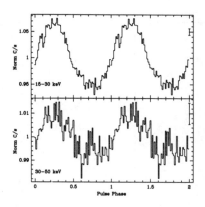
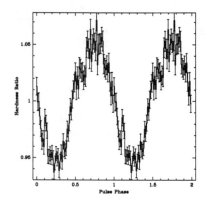

FIGURE 2. a): 4U1626–67 pulse profiles in two energy bands as observed by *PDS*. b): Hardness ratio between the 30–50 and 15–30 keV pulse profiles.

4U1626–67

This X-ray pulsar ($P_P \sim 7.7$ sec) is one of the few LMXRBs that show pulsed emission. It was detected with *PDS* up to 50 keV in ~ 70 ksec exposure time. The spectral analysis is in progress, while the pulse profiles of the pulsar in two hard X-ray ranges are shown in Fig. 2a. As can be seen, 4U1626–67 exhibits a sinusoidal shape and a peculiar feature: the hardness ratio between 30–50 keV and 10–30 keV pulse profiles turns out to be anti-correlated with the intensity profile (Fig. 2b).

PKS2155–304

PKS2155–304 is one of the strongest BL Lac objects in the 2–10 keV energy band. It is the first time the source has been simultaneously observed in a broad-energy band (0.1–300 keV), crucial for studying the relationship among different emission components. The source was clearly detected by *PDS* up to 100 keV (exposure time ~ 100 ksec). The ratio between the source spectrum and the Crab spectrum is shown in Fig. 3. It is apparent that the source spectrum is softer than Crab below 10–20 keV, while it is harder or similar to the Crab spectral slope above 10–20 keV. An extended paper on the *BeppoSAX* results on this source can be found elsewhere [8].

The most probable interpretation of the hard X-ray component is Inverse Compton radiation associated to a Self-Synchro Compton (SSC) process, while the lower energy component is synchrotron radiation.

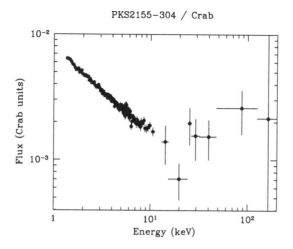

FIGURE 3. Observed ratio between PKS2155-305 and Crab spectra. Data from both MECS and *PDS* were used.

CONCLUSION

The SAX/PDS experiment is performing as designed and shows a flux sensitivity that is in agreement or better than that given in the *BeppoSAX* handbook [3].

REFERENCES

1. Boella, G. et al., A&AS, **122** 299 (1997)
2. Frontera, F. et al., A&AS, **122**, 357 (1997)
3. Piro, L. et al., *BeppoSAX* Handbook issue 1.0 (1995)
4. Zavattini, G. et al., to be presented at SPIE symposium (1997)
5. Bartlett, L.M., PhD Thesis, The University of Maryland (1994)
6. Boella, G. et al., A&AS, **122**, 327 (1997)
7. Howe, S.K. et al., ApJ, **272**, 678 (1983)
8. Giommi, P. et al., A&A, submitted (1997)

The CFA Batse Image Search (CBIS) as used for a Galactic plane survey

D. Barret[1], J.E. Grindlay[2], P.F Bloser[2], G.P. Monnelly[2]
B.A. Harmon[3], C.R. Robinson[4] & S.N. Zhang[4]

[1] *Centre d'Etude Spatiale des Rayonnements, CNRS-UPS, Toulouse, FRANCE*
[2] *Harvard Smithsonian Center for Astrophysics, Cambridge, USA*
[3] *Marshall Space Flight Center, Huntsville, USA*
[4] *Universities Space Research Association, Huntsville, USA*

Abstract. The CFA BATSE IMAGE SEARCH (CBIS) system allows automated analysis of large sets of BATSE images. CBIS is being used to conduct a deep survey of the Galactic plane to search for faint transients ($\sim$ 100 mCrab) in the archival data [2]. Furthermore, together with the earth occultation technique [3], CBIS is now applied to a real time imaging survey of the plane to discover new transients and monitor known sources. The CBIS processing is made of two sequential steps: First, the images are scanned to search for excesses. Second, the positions of all these excesses are cross-correlated to produce a list of candidate sources. In this paper, we will detail the algorithms of the scanner and the cross-correlator.

INTRODUCTION

As part of a collaboration between the BATSE team at Marshall, Harvard University and now CESR, we have initiated a survey of the Galactic plane to search for faint transients in the Galaxy. The scientific motivation for the survey and its initial results are described in [1], [2]. Basically the overall goal of this extended survey is to constrain the total number of black hole systems in the Galaxy, and constrain the number of black hole versus neutron star systems. This is required to constrain theories of massive star evolution and and formation of binaries. In this paper, we shall focus only on the technique; namely the CFA Batse Imaging Search (CBIS) system that we are currently using. First, we will discuss in some extents the BATSE imaging technique when used for a survey, and then describe each step of the CBIS processing.

THE BATSE IMAGING TECHNIQUE

The BATSE imaging technique developed by [4], [5], [6] allows to map the entire sky with an angular resolution of $\sim 1°$, a location accuracy of $\sim 0.1-0.5°$ (depending on the source brightness and location), and a sensitivity of ~ 50 mCrab for a 15 day integration time. Let us first summarize the basic features of the technique, and discuss some issues related to a survey.

The Earth Occultation Transform Imaging

When the limb of the earth sweeps through a given region of the sky (say the field of view of the image, FOV), the detectors record changes in the counting rate. The portion of the limb in the FOV is approximated as a straight line at a given projection angle. The location and orientation of each line is known to better than $0.1°$; several occultations will produce as many lines parallel to each other. The counting rate between adjacent lines when differentiated becomes indicative of the intensity of sources integrated along the line. This signal is used as input to the Radon transform which inverts the data to produce an image. As the projection angles are not completely sampled, a Maximum Entropy Method (MEM) is used to reconstruct the final image contours (see [5] for more details).

Some key issues to conduct a survey using the BATSE imaging technique

- **Removal of bright sources:** As we are interested by searching weak sources, bright sources must be removed first from the images. What we mean by bright sources are the persistent and/or highly variable bright hard X-ray sources such as the Crab, Cyg X-1 and 4U1700-37, plus the transient sources during outburst (e.g. GROJ0422+32). The occultation flux is computed simultaneously for all the sources to be removed. For a given source, if the significance of the fitted flux, averaged over one day in one detector, is larger than 3σ, then the step is removed from that detector. This increases the sensitivity provided that only (and all) bright sources are properly removed. However, the occultation step is fitted with a generic step profile, such that residuals will show up in the images due to the mismatch between the generic and actual step profiles. These residuals will appear at a position slightly offset from the bright source position (less than $1°$). In addition, residuals will appear if the source is viewed by several detectors, and the significance of the fitted flux (which is less than 3σ in a single detector) becomes significant when all the detectors are combined together.

- **Missing Images:** Given the CGRO orbit parameters, sources with declination angles between -40° and 40° will always be imaged. Sources near the poles have about 30% less exposure. Images could also be missing because of data gaps which result from incomplete earth occultation coverage of the entire FOV of the image. For the Galactic plane survey in its present form [2], we have found that about 10% of the images are missing. This means that as we are checking candidates for time consecutive detections, CBIS has to keep track of which and when images are missing. This complicates significantly the code for selecting the final candidates (see below).

- **Field of View (FOV):** As the size of the field of view increases, potential distortions of the images appear due to approximation of the Earth's limb projected on the sky as a straight line [4]. These distortions cause the reconstructed positions to be offset from actual source locations. For crowded regions, such as most of the Galactic plane, it is therefore suggested to use small FOV (typically $\sim 10°$). Over an image, only the inner part is reliable for studies, as most of the excesses located in the outer parts are produced by interfering sources nearby. These excesses will disappear by moving the center of the FOV. This means that in a survey, any given location should be imaged within the center part of at least one image, and there should be good overlap between adjacent images. In practice we have found that only excesses with offset angles from the center of the FOV less than 0.8 times the radius of the image should be considered.

A survey implies an automated analysis system

For the survey as currently implemented [1] [2], the galactic plane from l=0 to 360 (b= $\pm 12°$) is covered by $90 \times 12° \times 12°$ images (Fig. 1). This ensures a relatively good overlap between images. As for the survey we wish the technique to be sensitive to 100 mCrab transient events with durations of a few weeks, the integration time of the images was choosen to be 15 days. A new set of 90 images is made every 5 days. Assuming that we are searching in only one energy band (i.e 20-50 keV), to process 6 years of data, we thus have an initial set of ~ 40000 images to search. Given this number, it is obvious that an automated system is needed to analyze the entire image set. We have thus decided to develop a scanner which looks through all the images and returns the positions of each excess found. *This is the BATSE Image Automated Scanner (BIAS).*

BATSE IMAGE AUTOMATED SCANNER (BIAS)

A peak in the image has more or less an elongated gaussian shape. Its shape depends mainly on the difference between the projection angle of the

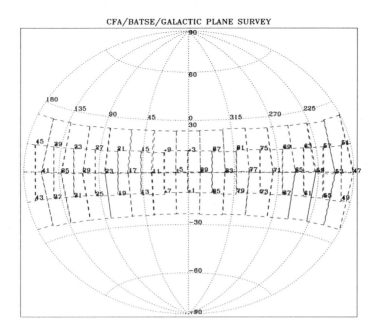

FIGURE 1. Positions of the fields used for the current BATSE Galactic Plane survey. For clarity in the figure, only the fields with an odd number have been represented.

rising and setting limbs. It thus varies with time. For example, if the rising and setting limbs are roughly parallel, then the excess will be elongated in the direction of the limb. As the shape of an excess varies with time, we decided to search for excesses without making any assumptions about the point spread function of the technique.

The Maximum Entropy Method (MEM) used in the reconstruction smoothes the large scale structures of the images, and tends to concentrate the flux towards the brightest spots of the images. This means that on average, a very large number of pixels have very low values and can be excluded from the analysis. It is worth noting that as a consequence of the MEM used, a peak in the image contours is not related to the true source intensity but instead indicative of the relative intensity of sources within the field. This means that we had to chose arbitrarily a low flux threshold (i_{thresh}; typically 0.1 in pixel flux units) below which the pixels in the images are not considered by the scanner.

A feature in the image consists of a group of pixels above i_{thresh}. The smoothed shape of the excess implies that all these pixels are connected (i.e. they all have one neighbor in the group). However, at the tail of the wings of the excess (i.e close to i_{thresh}), fluctuations will cause pixels to be disconnected from the group. For that reason, the BIAS requests for a pixel to belong to a

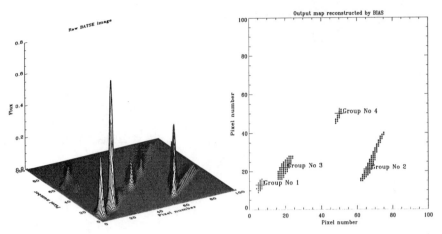

FIGURE 2. Raw BATSE image (**left**) and the "cleaned" map as reconstructed by the BIAS (**right**). Four groups of pixels are found. The cross indicates the position of the maximum of each group. Group no 2 is in fact associated with the X-ray burster 4U1608-52.

group that it has a neighbor in the group at a distance less than r_{crit}. With this assumption, *the scanner simply determines the number of distinct groups in the image (see Fig. 2)*.

Within a group, as the pixel with the largest flux is the most likely to be associated with the true source position; the scanner returns the position (α, δ) of the peak. The flux associated with the excess is not the peak flux but the integrated flux of all the pixels in a group above i_{thresh}.

BATSE IMAGE SOURCE SEARCHER (BISS)

When the scanner has run on a large set of images, the output is a large set of positions (see Fig. 3). For the assumed parameters (i_{thresh}=0.1, r_{crit}=3) the averaged number of excesses per image is ~ 1. Suppose now that an unknown transient source underwent a 20 day outburst (say the flux from the source remained above the sensitivity threshold for 20 days). The location of this source is such that it is imaged with good sensitivity in two fields. What will happen the way the survey is conducted? The BIAS will pick up 8 excesses in 4 time consecutive images in 2 fields. These 8 excesses should be associated with consistent positions (α, δ) (i.e. with averaged distance to the mean position less than the angular resolution of the technique).

This example demonstrates the need for an algorithm which is able to cross-correlate the list of BIAS positions to search for groups of positions which might be associated with real sources. *This is the BATSE Image Source Searcher (BISS)*. The philosophy behind the BISS is essentially the same as

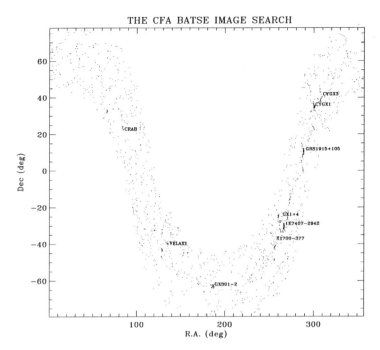

FIGURE 3. The dots indicate the raw positions found by the BIAS after scanning of 850 days of Galactic plane survey images. Residuals from bright sources like Cyg X-1 are clearly visible. These positions are input to the BISS which will locate all the clusters of positions (like the one around GX301-2).

for the BIAS, except that the BIAS groups pixels in images whereas the BISS groups positions in sky coordinates (α, δ). The way BISS works can be described as follows:

For each BIAS position, BISS computes the number of positions within a given search radius (r_{search}, typically 1-2°). If the total number of positions in the search radius circle is larger than 2, then a mean position can be derived for the group. Starting from this new position, BISS recomputes the number of surrounding positions until no additional BIAS positions are found within r_{search} of the current mean. It is worth pointing out that this method maximizes the number of positions around a mean starting from one initial BIAS position. This means that the algorithm has to run on all the BIAS positions as it is sensitive to where it starts (in other words BIAS positions found around a mean cannot be discarded for the subsequent processing). Now starting from this mean position, it is possible to make a second order optimization by moving the current mean along a discrete grid around the starting mean with a grid step small compared to the angular resolution of the technique. This makes the search more sensitive, as in the first method the iterations are based on the actual BIAS positions. However the grid search is much more time consuming that in practice it cannot be used for large datasets.

At the end of the first iteration step, BISS returns a list of mean positions, each being associated with a list of raw BIAS positions. Then, the list of mean positions is reordered in decreasing number of associated BIAS positions. The next step consists of merging groups which have BIAS positions in common. At the end of this step, each BIAS position previously involved in at least one group, now belongs to one group and one only. The merged groups are now defined by a new mean position and a new list of BIAS positions.

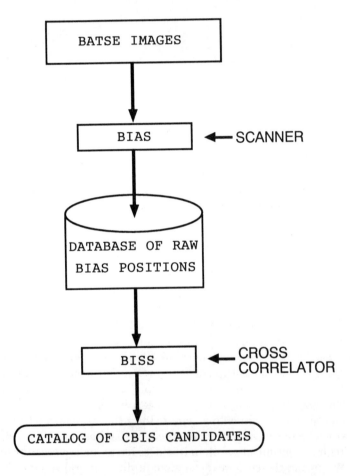

FIGURE 4. The CBIS architecture. The BIAS and BISS are coded in IDL. The database is also an IDL database.

WHAT IS A CBIS CANDIDATE ?

From the BISS output, a catalog of candidate sources can be computed. For each BISS candidate, we have a list of all the BIAS positions involved, and for each of these, the information identifying the image from which it was derived, namely, the start and stop time, the field number, the energy band.

The next step consists of removing as many unreal candidates as possible. As said above, a transient event in the data will appear as time consecutive detections in one (or more) fields and multiple detections in contiguous fields. To select the final list of candidates, we proceed in two steps: First, the candidate source must be associated with a minimum number of BIAS positions which depends on the integration time of the images, the time overlap between two consecutive images, and space overlap between fields. Second, given the exposure time and the location of the candidate with respects to the field coverage, we apply stronger criteria such as a minimum number of time consecutive detections, simultaneous detections in all continguous fields in which the candidate source is imaged with good sensitivity (its offset from the center of the field has to less than 0.8 times the radius of the image).

Some additionnal tests can be performed as checking whether the candidate is close to a bright source, or whether it remains when BISS is run with different r_{search}. Once the candidate source has survived all these tests, then we proceed with the standard analysis (generation of light curve, spectrum, checks on the limb geometry, etc...). This analysis gives the last word on the reality of a source. Figure 4 summarizes the whole CBIS architecture.

CONCLUSIONS

The CBIS system is now operational. All the hard X-ray sources that have been discovered or detected with the routine occultation analysis [3] have also been seen by CBIS [2]. This gives us strong confidence that the CBIS algorithms are working properly. There are several ways the whole system can be improved to conduct a survey. First to remove the systematics, the field coverage needs to be optimized to ensure a good and uniform coverage of the whole region to be surveyed. It has become clear that the coverage used for the current Galactic plane survey ($90 \times 20° \times 20°$ fields) is not optimum, and we are expecting to move to $420 \times 10° \times 10°$ fields. Also, we have to improve the way bright sources are removed before the images are made. This affects the sensitivity of the search in wide regions around bright sources. Finally one could also improve the way the CBIS candidates are selected but this requires optimization of the field coverage first. With these improvements, undoubtedly the main scientific goal of the survey which is the search for faint transients in the Galaxy, should be reachable [2]. The application of CBIS for real time data will also help the BATSE team to discover new sources that would have been missed otherwise. The efforts that have been put in the developement of CBIS should be soon rewarded.

REFERENCES

1. Grindlay, J.E et al.,Proc. of the 3rd CGRO Symposium, **120-4**, 145 (1996).
2. Grindlay, J.E et al., The INTEGRAL Meeting, ESA-SP **382**, 551 (1997).
3. Harmon, B. A, et al., *The Compton Obs. Sci. Workshop*, NASA, **CP3137**, 69 (1992)
4. Zhang, S. N. et al., *Nature*, **336**, 245 (1993).
5. Zhang, S. N. et al., *IEEE*, **41(4)**, 1313 (1994).
6. Zhang, S. N. et al., *Experimental Astronomy*, **Vol6, No 4**, 57 (1995).

TeV Gamma Ray Emission from Southern Sky Objects and CANGAROO Project

T.Kifune[1], S.A.Dazeley[2], P.G.Edwards[3], T.Hara[4],
Y.Hayami[5], S.Kamei[5], R.Kita[6], T.Konishi[7], A.Masaike[8],
Y.Matsubara[9], Y.Matsuoka[9], Y.Mizumoto[10], M.Mori[11],
H.Muraishi[6], Y.Muraki[9], T.Naito[12], K.Nishijima[13],
S.Ogio[5], J.R.Patterson[2], M.D.Roberts[1], G.P.Rowell[2],
T.Sako[9], K.Sakurazawa[5], R.Susukita[14], A.Suzuki[7],
R.Suzuki[5], T.Tamura[15], T.Tanimori[5], G.J.Thornton[2],
S.Yanagita[6], T.Yoshida[6], and T.Yoshikoshi[1]

[1] *Institute for Cosmic Ray Research, University of Tokyo, Tanashi, Tokyo 188, Japan*
[2] *Department of Physics and Mathematical Physics, University of Adelaide, South Australia 5005, Australia,* [3] *Institute of Space and Astrophysical Science, Sagamihara 229, Japan,* [4] *Faculty of Management Information, Yamanashi Gakuin University, Kofu 400,* [5] *Department of Physics, Tokyo Institute of Technology, Tokyo 152,* [6] *Department of Physics, Ibaraki University, Mito 310,* [7] *Department of Physics, Kobe University, Kobe 637,* [8] *Department of Physics, Kyoto University, Kyoto 606,* [9] *Solar-Terrestrial Environmental Laboratory, Nagoya University, Nagoya 464-01,* [10] *National Astronomical Observatory of Japan, Tokyo 181,* [11] *Department of Physics, Miyagi University of Education, Sendai 980,* [12] *Department of Earth and Planetary Physics, University of Tokyo, Tokyo 113,* [13] *Department of Physics, Tokai University, Hiratsuka 259,* [14] *Institute of Physical and Chemical Research, Wako 351-01* [15] *Department of Engineering, Kanagawa University, Yokohama 221, Japan*

Abstract. We report recent results of the CANGAROO Collaboration on very high energy gamma ray emission from pulsars, their nebulae, SNR and AGN in the southern sky. Observations are made in South Australia using the imaging technique of detecting atmospheric Cerenkov light from gamma rays higher than about 1 TeV. The detected gamma rays are most likely produced by the inverse Compton process by electrons which also radiate synchrotron X-rays. Together with information from longer wavelengths, our results can be used to infer the strength of magnetic field in the emission region of gamma rays as well as the energy of the progenitor electrons. A description of the CANGAROO project is also given, as well as details of the new telescope of 7 m diameter which is scheduled to be in operation within two years.

INTRODUCTION

A firm foundation for VHE (Very High Energy) gamma ray astronomy [1] was laid with the use of the imaging Čerenkov technique to detect a signal from the Crab by the WHIPPLE group [2]. Detection of VHE emission from PSR B1706−44 soon followed by the 3.8 m imaging Čerenkov telescope of the CANGAROO (Collaboration of Australia and Nippon(Japan) for a GAmma Ray Observatory in the Outback) Project, which commenced operation in 1992. Gamma rays from this pulsar had been discovered by CGRO EGRET soon after its launch. The 3.8 m telescope, located in the southern hemisphere, has the advantage of being able to study many Galactic objects near the Galactic center, while a number of Čerenkov imaging telescopes that have commenced operation in the northern hemisphere concentrate mainly on observing the Crab nebula and nearby AGNs.

THE CANGAROO PROJECT

The 3.8 m diameter telescope [3] is located near Woomera, South Australia (at 136°E and 31°S, 160 m above sea level). The camera consists of 256 photomultiplier tubes with a total field of view of about 3° diameter, enabling a fine angular resolution to construct images of Čerenkov photons. The threshold energy of detectable gamma rays from the zenith was estimated to be 1∼2 TeV before November 1996. The recoating of the mirror of the telescope with aluminium has since reduced the threshold energy by a factor of 2.

Another telescope of 10 m diameter is scheduled to commence operation in 1998, to exploit the region of ∼ 100 GeV energies. A radio antenna design is used for the construction of the main body of the telescope, which has an alt-azimuth mount. The light collecting mirrors attached to the parabolic antenna are spherical with 80 cm diameter and made of carbon reinforced plastic material which has the merit of having less weight than glass. We will start with a total area equivalent to a 7 m diameter mirror. The focal length is 8 m, and a camera consists of 512 photomultiplier tubes of Hamamatsu R4124 of 1 ns rise time. The diameter of the tube is 13 mm with the use of light guide to collect photons across 0.12° per tube.

RESULTS OF CANGAROO OBSERVATION

Galactic Objects

The galactic objects CANGAROO has observed are listed in Table 1. In addition to the Crab nebula [4], positive evidence has been obtained on PSR B1706-44 [5] and the Vela pulsar nebula [6].

TABLE 1. List of Galactic Objects of CANGAROO Observation

objects	integral flux (10^{-12} cm-2 s^{-1})	threshold (TeV)	observation time (hrs)	references
Crab	0.8	7	61	[4], [7]
PSR B1706−44	8	∼1	84	[5]
Vela	2.9 (nebula)	2.5	119 + (28)$^{(a)}$	[6]
	<1.4 (pulsar position)	2.5	119 + (28)$^{(a)}$	[6]
PSR B1055−52	<0.95	2	69	[8]
PSR B1758−23	< 2	∼1	48	[9]
PSR B1259−63	∼ 4 ?	∼3	38 + (>51)$^{(a)}$	[10]
PSR B1509−58	< 2	2	50 + (>31)$^{(a)}$	[10]
SN 1006	a hint of a signal	∼2	35 + (>30)$^{(a)}$	

(a) data taken in 1997 with analysis yet to be done

Crab: The Crab is seen at zenith angles of 53° − 60° near its culmination from the observation site in Australia. The low elevation causes an increase of threshold energy but with an increase of detection area [4], enabling us to enjoy excellent sensitivity at ∼ 10 TeV energies. The energy spectrum observed extends to as high as ∼ 50 TeV without an apparent change of slope from the power law spectrum in 300 GeV to ∼ 1 TeV region [7].

EGRET pulsars; PSR B1706−44 and Vela pulsar:
These two EGRET pulsars are found to be VHE gamma ray emitters, but there is no evidence of VHE emission from PSR B1055−52 [8]. The signals from PSR B1706−44 [5] and the Vela [6] show no modulation with the pulsar spin period, which suggests that pulsar nebula is, like the Crab, responsible for the VHE emission. The ratio of the observed VHE to X-ray luminosity of the nebulae are larger than in the case of the Crab. The regime of synchrotron and inverse Compton processes of relativistic electrons to radiate X and gamma rays then indicates that the magnetic field in the nebulae is about two orders of magnitude weaker than the Crab nebula. VHE gamma rays are spatially peaked as offset to the south-east direction from the Vela pulsar by about 0.13°, and the spatial size of emission appears broader than the point spread function. The contribution from the pulsar position is given as the upper limit from the pulsar position in Table 1. We expect that our 1997 data with a reduced gamma ray threshold energy will be useful to infer the VHE spectrum. More knowledge with better accuracy, particularly on the energy spectrum both in X-ray and VHE bands, are required to restrict parameters of emission models and to infer the structure of the nebulae.

Other sources: Observations have not been made yet for objects such as young, short period pulsars or X-ray binaries, which are not detected by EGRET so far but on which extensive VHE efforts have been made [12].

Among the binary pulsars, PSR B1259−63 is a peculiar object that has a highly eccentric orbital motion around a giant companion Be star. A VHE signal of $\sim 4\sigma$ was detected near the time of the previous periastron in 1994, and confirmation will be sought in the data from around the next periastron in May 1997.

Supernova remnants are another target of prime importance for VHE gamma ray study. One of the unidentified EGRET sources is coincident with the SNR W28 which is possibly associated with PSR B1757−24. Evidence for point source emission was not detected for either W28 or PSR B1758−23 [9], though we can not exclude VHE emission, as appears in the earlier data [11], from the vicinity of the center of EGRET error circle which is apparently shifted from W28 towards the giant molecular cloud M20. Searches remain necessary for the unknown position of VHE emission possibly extended in the complex system of SNR, molecular cloud and pulsar. PSR B1509−58 is also embedded in a complex structure of plerion activity, and a considerable amount of data has been accumulated. An opportunity for VHE gamma rays to show direct evidence of shock acceleration of relativistic particles at the supernova shell is provided by shell type supernovae, such as SN1006 from which the ASCA satellite detected non-thermal X-rays to suggest progenitor electrons up to ~ 100 TeV [14].

Active Galactic Nuclei

No TeV signal has been detected from nearby southern AGNs; Cen A; PKS 0521−36 (z=0.055); PKS 2316−42 (z=0.055); PKS 2005−49 (z=0.071); EXO 0423−08 (z=0.039) etc. [13]. The upper limits are around $\sim 1 \times 10^{-12}$ cm^{-2} s^{-1} at 2 TeV, which lies near the quiescent Whipple flux levels of the northern BL Lac objects which are VHE sources, *i.e.*, Mrk 421 and 501.

DISCUSSIONS AND SUMMARY

The CANGAROO detections are summarized in Fig. 1. The vertical axis is the product F·E (in the unit of erg cm^{-2} s^{-1}) of integral photon flux F and threshold energy E, and the horizontal axis the observation time used for analysis. The detection sensitivity is near 10^{-12} erg cm^{-2} s^{-1} which corresponds to a luminosity of 10^{32} erg s^{-1} at a distance of 1 kpc. The significance of detection limited by statistical fluctuations is shown by the dotted line (Fig. 1), which is normalized to the detection of 100 gamma rays at 1 TeV over 20 hrs. The preliminary result on PSR B1259−63 is marked by "B". "S" indicates the flux due to the inverse Compton process expected from SN1006 when the magnetic field is as weak as the interstellar value. Loose upper limits on PSR B1509−58 and on the Vela pulsar position are given due to a limited sensitivity of the current technique of imaging Čerenkov photons to spatially extended emission.

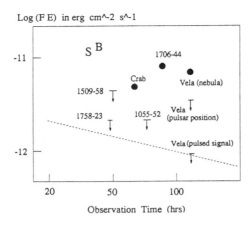

FIGURE 1. Energy flux observed by CANGAROO.

VHE gamma rays have a unique role to provide information about the high energy nature of these objects. There are an increasing number of interesting Galactic objects; unidentified EGRET sources, pulsar nebulae and SNRs of non-thermal X-rays, objects with relativistic jets etc., and CANGAROO will continue to enjoy a 'hard time' in choosing the targets for observation.

The project is supported by a Grant-in-Aid for Scientific Research of the Japan Ministry of Education, Science, Sports and Culture, and also by the Australian Research Council.

REFERENCES

1. See review by Weekes, T.C. et al., in this *Proc. 4th CGRO Symp.* (1997).
2. Weekes, T.C. et al., *Ap. J.* **342**, 379 (1989).
3. Hara, T. et al., *Nucl. Inst. Meth. Phys. Res.* **332**, 300 (1993).
4. Tanimori, T. et al., *Ap. J. Lett* **429**, L61 (1994).
5. Kifune, T. et al., *Ap. J. Lett* **438**, L91 (1995).
6. Yoshikoshi, T. et al., submitted to *Ap. J. Lett.* (1997).
7. Sakurazawa, K. et al., to appear in *Proc. 25th Int. Cosmic Ray Conf. (Durban)*, (1997).
8. Susukita, R., *Ph D Thesis, Kyoto University* (1997).
9. Mori, M. et al., *Proc. 24th Int. Cosmic Ray Conf. (Rome)*, **2**, 491 (1995).
10. Sako, T. et al., to appear in *Proc. 25th Int. Cosmic Ray Conf. (Durban)*, (1997).
11. Kifune, T. et al., *Proc. 23rd Int. Cosmic Ray Conf. (Calgary)*, **1**, 444 (1993).
12. Kifune, T., *Space Science Review* **75**, 31 (1996).
13. Roberts, M.D. et al., to appear in *Proc. 25th Int. Cosmic Ray Conf. (Durban)*, (1997).
14. Koyama, K. et al., *Nature* **378**, 255 (1995).

Saturated Compton Scattering Models for the Soft Gamma-Ray Repeater Bursts

I. A. Smith, E. P. Liang, A. Crider, D. Lin

Department of Space Physics and Astronomy, Rice University, MS-108, 6100 South Main, Houston, TX 77005-1892

M. Kusunose

Kwansei Gakuin University, Uegahara Ichiban-cho, Nishinomiya 662, Japan

Abstract. The Soft Gamma-Ray Repeaters (SGR) are sources of brief intense outbursts of low energy gamma rays. Most likely they are a new manifestation of neutron stars. In this paper, we explore Compton scattering models for the bursts. We use a BATSE SGR 1900+14 burst to show that Comptonization by a thermal plasma alone is unable to explain the shape of the burst spectrum, but that adding a small but significant fraction of non-thermal plasma produces an acceptable fit.

INTRODUCTION

The Soft Gamma-Ray Repeaters (SGR) are a small but highly unusual class of sources of brief intense outbursts of low energy gamma rays. There are three known SGR. Two of them (SGR 1900+14 and SGR 1806-20) are in our Galaxy, while the third (SGR 0525-66) is generally believed to be in the LMC. It appears that they are new manifestations of neutron stars. However, accounting for the wide variety of multiwavelength observations as well as the bursts themselves has produced no clear picture for these sources. See Smith (1997) for the most recent review.

The SGR burst spectra are one of the features that distinguish them as a separate class of objects: they are harder than in X-ray bursters, but softer than in gamma-ray bursters. The burst spectra are usually similar from burst to burst. The apparent exception was the well known 1979 March 5 event from SGR 0525–66, whose initial spectrum may be consistent with those of gamma-ray bursters [2].

Many possible physical mechanisms have been directly fitted to the burst data to determine whether they are allowed. An optically thin thermal bremsstrahlung can usually fit the spectra above ~ 20 keV, e.g. Figure 2 of Kouveliotou et al. (1993). However, there is a definite turn-over below this in the SGR 1806–20 bursts seen by ICE [1].

In this paper, we explore Compton scattering models for the bursts. Fenimore et al. (1994) showed that Comptonization by a thermal plasma of a black body of injected soft photons gave acceptable though not spectacular fits to the ICE bursts of SGR 1806–20. A problem with this model is that the spectrum falls rapidly beyond the peak. This could not be carefully tested with the ICE data, that had limited statistics above 100 keV. Here we use a BATSE SGR 1900+14 burst [3] to show that a thermal plasma alone is unable to explain the shape of the spectrum, but that adding a small but significant fraction of non-thermal plasma produces an acceptable fit.

COMPTON SCATTERING MODEL

We used a Monte Carlo code similar to the one used in Liang et al. (1997) to simulate gamma-ray burst spectra. However, while the old code only included non-thermal electrons, the new one includes both non-thermal and thermal electrons. (We now use this hybrid code for all of our GRB simulations too.)

For all the runs presented here, the geometry is a uniform sphere of particles with a uniform injection of soft thermal photons of temperature 5.11×10^{-4} keV. (The results are insensitive to the temperature, provided it is small). A large optical depth $\tau = 50$ is chosen to produce the roll-over, assuming this is a common feature of SGR spectra. Larger values of τ produce steeper slopes below the peak. The non-thermal electrons have a power law distribution with power law index $p = 7$ up to a cut-off Lorentz factor $\gamma_{\max} = 10$. (The results are insensitive to these choices, provided p is large). The intersection of this power law with the thermal electron distribution determines the lower Lorentz factor $\gamma_{\min}$ of the non-thermal electrons.

Figure 1 compares some representative runs with a BATSE SGR 1900+14 burst. The temperature of the thermal electrons, and the fraction of particles that are thermal versus non-thermal are the only parameters that change in the runs presented. It can be seen that the best fit is found when a $\sim$ 10% component of power-law electrons is added to the thermal plasma. The spectrum for the 99% thermal run drops too rapidly above the peak.

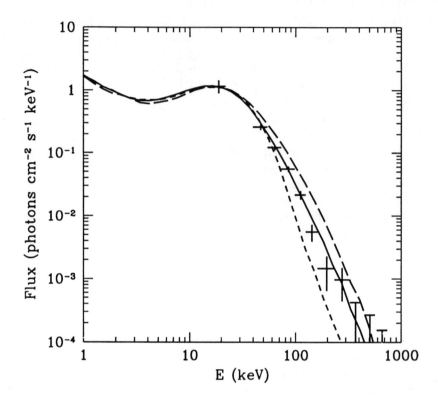

FIGURE 1. Saturated Compton scattering models compared to a BATSE SGR 1900+14 burst. Short dashed curve: 99% thermal electrons, $kT_e = 9$ keV, 1% non-thermal. Solid curve: 90% thermal electrons, $kT_e = 4$ keV, 10% non-thermal. Long dashed curve: 80% thermal electrons, $kT_e = 1$ keV, 20% non-thermal. The data points are for the only BATSE SGR 1900+14 burst that had sufficient data to derive a spectrum (Figure 2 of Kouveliotou et al. 1993). This illustrates that to explain the spectrum of this burst using a Compton scattering model, a small but significant fraction of non-thermal plasma is required.

CONCLUSIONS

To explain the spectrum of the BATSE SGR 1900+14 burst using a Compton scattering model, a small but significant fraction of non-thermal plasma is required. This shows that this SGR 1900+14 burst agrees with the result that was indicated by the ICE bursts from SGR 1806–20.

In the near future, we will be able to thoroughly test the Compton scattering model because SGR 1806-20 has been particularly active recently with many bursts seen by BATSE and RXTE [4,5]. Of the ~ 30 BATSE triggers, several were visible above 100 keV, making them very valuable for our spectral fitting. In addition to the large bursts seen by BATSE, numerous small bursts were seen by RXTE to come in groups with burst rates as high as one per minute, and with peak fluxes about 2 orders of magnitude fainter [5]. These will provide important information about the spectra at lower energies, and the spectra of fainter bursts. From these bursts we will be able to extract the relevant parameters for each burst (or group of bursts), e.g. τ and T_e, and determine the changes from burst to burst. This will provide essential information for any physical model of the sources.

ACKNOWLEDGEMENTS

This work was supported by NASA grants NAG 5-1547, NAG 5-2772, and NAG 5-3824 at Rice University.

REFERENCES

1. Fenimore, E. E., Laros, J. G., & Ulmer, A. 1994, ApJ, 432, 742
2. Fenimore, E. E., Klebesadel, R. W., & Laros, J. G. 1996, ApJ, 460, 964
3. Kouveliotou, C., et al. 1993, Nature, 362, 728
4. Kouveliotou, C., et al. 1996, IAUC, 6501
5. Kouveliotou, C., et al. 1996, IAUC, 6503
6. Liang, E., Kusunose, M., Smith, I. A., & Crider, A. 1997, ApJ, 479, L35
7. Smith, I. A. 1997, in The Gamma-Ray Universe Revealed by CGRO, to appear

The GRB 970111 error box 19-hours after the high energy event

Alberto J. Castro-Tirado[1], Javier Gorosabel[1], Nicola Masetti[2], Corrado Bartolini[2], Adriano Guarnieri[2], Adalberto Piccioni[2], Jochen Heidt[3], Tomas Seitz[3], Edouard Thommes[4], Christian Wolf[4], Enrico Costa[5], Marco Feroci[5], Filippo Frontera[6,7], Danielle Dal Fiume[7], Luciano Nicastro[7], Eliana Palazzi[7] and Niels Lund[8]

[1] *LAEFF-INTA, P.O. Box 50727, 28080 Madrid, Spain.*
[2] *Dipartimento di Astronomia, Università di Bologna, Italy.*
[3] *Landessternwarte Heidelberg, Heidelberg, Germany.*
[4] *Max Plank Institute fuer Astronomie, Heidelberg, Germany.*
[5] *Istituto di Astrofisica Spaziale, CNR, Frascati, Italy.*
[6] *Dipartimento di Fisica, Università di Ferrara, Italy.*
[7] *Istituto Tecnologie e Studio Radiazioni Extraterrestri, CNR, Bologna, Italy.*
[8] *Danish Space Research Institute, Copenhagen, Denmark.*

Abstract. We present the results of the follow-up observations performed 19 hours after the Gamma–Ray Burst GRB 970111, detected by BeppoSAX, BATSE/CGRO and Ulysses. The gamma-ray fluence of this burst is seven times larger than that of GRB 970228, the first burst for which an optical counterpart was found. However, no counterpart was detected for GRB 970111 to a limiting magnitude of B $\sim$ 23 and R $\sim$ 22.6.

1 INTRODUCTION

The main problem for the resolution of the γ–ray burst (GRB) mystery has been the lack of knowledge of the distance scale [1]. Today it is widely accepted that the finding of counterparts provides some clues about their distances and perhaps, their nature. With the advent of the Italian-Dutch BeppoSAX satellite, it has been possible for the first time, to perform multiwavelenth deep searches few hours after the event, in order to detect delayed transient emission that could be asociated with the gamma-ray event.

Here we present the results of the follow-up observations performed 19 hours after the Gamma–Ray Burst GRB 970111. The gamma-ray fluence of this burst is seven times larger than that of GRB 970228, the only burst for which an optical counterpart has been presumably found so far[2,3].

GRB 970111 was detected as a 50-s three-peaked Gamma-Ray Burst on January 11, 1997 by the Wide Field Camera of the X-ray satellite BeppoSAX [4]. The burst was located at $\alpha(2000.0) = 15^h28^m24^s$, $\delta(2000.0) = 19°40'.0$ (error box radius = 10'). Soon after the event (17.8 h), SAX was reoriented towards the burst location, and two faint X-ray sources were reported within the GRB error box [5], hereafter labelled as 'a' and 'b'. Source 'a' was resolved by ROSAT into two objects [6]. 67-h after the event, Guarnieri et al. [7], found that no object in the X-ray error boxes or surroundings showed remarkable optical variations. In the meantime, and taking into account the detection of the burst by both the GRB detector on Ulysses and BATSE on CGRO, the location was constrained to the intersection of an annulus and the SAX error box [8], and it was found that only BeppoSAX source 'a' lied within it. This object was detected also as a non-variable radio source [9], coincident with one of the three sources detected by Rosat on 16 January [10]. However, due to a misalignment noticed for the BeppoSAX Wide Field Cameras [11], the reported GRB error box had to be shifted by 4'.2 from its original position; and a new region seven times smaller than the former was derived when combining the SAX and IPN data [12].

OBSERVATIONS

Images covering a large field of view (16' diameter) were obtained on 12 January, starting 19 hours after the gamma-ray event, at the 2.2-m telescope at the German-Spanish Calar Alto Observatory (equipped with CAFOS). Fortunately enough, the refined position was just within our images. We also used the 1.5-meter telescope (equipped with BFOSC) at the Bologna Astronomical Observatory, on January 14-15, 17 and 31, on February 14, 17-18 and on March 5 and 13. B and R filters were employed in all images. Table 1 reports the complete list of observations.

Images of the PG 1047+003 sequence [13] and NGC 4147 were obtained, in order to calibrate the field of GRB 970111. After the standard cleaning procedure for bias and flat field, the frames were processed with the DAOPHOT II package [14] and the *ALLSTAR* procedure inside MIDAS or with simple aperture photometry if the objects were too faint to be detected with DAOPHOT II. Then, the fields were calibrated with the standard Landolt stars quoted above. A comparison among frames acquired on the same night in different bands and performed with a simple FORTRAN code allowed us to determine the $B - R$ colors of more than 300 objects within the SAX error box for GRB 970111, of which about 100 lie within the IPN/SAX error box.

The same procedure was applied to frames collected on different nights but in the same filter to search for the variable stars of the field.

Date of 1997	Filter	Frames	Exp. times (min)	Telescope
Jan. 12	B,R	1,1	25;10	Calar Alto 2.2m
Jan. 14	B,R	1,1	20	Loiano 1.5m
Jan. 17	B,R	1,2	45;15	Loiano 1.5m
Feb. 10	B,R	1,1	25;10	Calar Alto 2.2m
Feb. 11	R	1	10	Calar Alto 2.2m
Mar. 5	R	2	30	Loiano 1.5m

TABLE 1. Journal of the GRB 970111 observations.

3 RESULTS AND DISCUSSION

No objects within the entire IPN/BeppoSAX error box showed any remarkable blue excess (we found that no $B - R$ was lower than 0.7). Being the field at fairly high galactic latitude ($b^{II} = 53°$) the interstellar absorption should not be high, so, the values of the color indices quoted above should not be very different from the unabsorbed ones. Moreover, no fading object was detected, between the first set of images and the later data. Any fading was ≤ 0.2 mag for objects with B $\leq$ 21, R $\leq$ 20.8 and ≤ 0.5 mag for those down to B = 23, R = 22.6 [15]. Further observational details will be published elsewhere [16].

Several models of GRBs predict that the flux density F_ν is proportional to $\nu^{1/3}$ below a critical synchroton frequency ν_m [17-19]. Taking into account the γ–ray fluence at maximum, which was few times 10^{-5} erg cm^{-2} [20], the simple extrapolation of the spectrum leads to a flux of ~ 0.2 mJy in the optical band (~ 18 mag), which was certainly not observed (the upper limit is 23 mag or ~ 0.002 mJy). This could indicate some kind of absorption by intervening matter close to the source.

REFERENCES

1. Fishman, G. J. & Meegan C.A., 1995, Ann. Rev. Astron. Astrophys., 33, 415.
2. Groot, P. J. et al. 1997, IAU Circ. 6584.
3. Guarnieri, A. et al. 1997a, Astron. and Astrophys. (in press).
4. Costa E., Feroci M., Piro L. et al., 1997, IAU Circ. 6533.
5. Butler R.C., Piro L., Costa E. et al., 1997, IAU Circ. 6539.
6. Voges W., Boller T., Greiner J., 1997, IAU Circ. 6539.
7. Guarnieri A., Bartolini C., Piccioni A. et al., 1997, IAU Circ. 6544.
8. Hurley K., Kouvelioutou C., Fishman G., Meegan C., 1997a, IAU Circ. 6545.
9. Frail D.A., Kulkarni S.R., Nicastro L. et al., 1997, IAU Circ. 6545.
10. Frontera F., Costa E., Piro L. et al., 1997, IAU Circ. 6567.
11. in't Zand J., Heise J., Hoyng P. et al., 1997, IAU Circ. 6569.
12. Hurley K., Kouvelioutou C., Fishman G., Meegan C., van Paradijs J., 1997b, IAU Circ. 6571.
13. Landolt A.U., 1992, AJ, 104, 340.
14. Stetson P.B., 1987, PASP, 99, 191.

15. Castro–Tirado, A, J., Gorosabel, J., Heidt, J. et al. IAU Circ. 6598.
16. Gorosabel, J. et al. 1997, Astron. and Astrophys. (in preparation).
17. Katz, J. I. 1994, Astrophys. J. 432, L107.
18. Mészáros, P. M. and Rees, M. J. 1993a, Astrophys. J. 405, 278.
19. Mészáros, P. M. and Rees, M. J. 1993b, Astrophys. J. 418, L59.
20. Fishman, G. E. 1997, talk presented at the Elba Workshop on Latest results on GRBs, 26-27 May, Elba (Italy).

The Duration - Photon Energy Relation In Gamma-Ray Bursts and Its Interpretations

Demosthenes Kazanas, Lev G. Titarchuk & Xin-Min Hua

Code 661, Goddard Space Flight Center, Greenbelt, MD 20771

Abstract. Fenimore et al. (1995) have recently presented a very tight correlation between the spectral and the temporal structure in Gamma Ray Bursts (hereafter GRBs). In this note we present two simple models which can account in a straightforward fashion for the observed correlation. These models involve: (a) The impulsive injection of a population of relativistic electrons and their subsequent cooling by synchrotron radiation. (b) The impulsive injection of monoenergetic high energy photons in a medium of Thompson depth $\tau_T \sim 5$ and their subsequent downgrading in energy due to electron scattering. We present arguments for distinguishing between these two models from the existing data.

INTRODUCTION

Fenimore et al. (1995) have presented a very well defined correlation between the average duration of the subpulses usually present in the time profile of a given burst, or the duration of the entire burst in the absence of subpulses, and the energy channel of observation. They have found that the average pulse width, as determined either by the autocorrelation function of the entire GRB or by specific fits of the time profiles of individual subpulses in an given profile, is well fitted by a power law in the energy E at which the observation is made; more specifically they have found that $\Delta \tau \simeq AE^{-0.45 \pm 0.05}$, where A is a proportionality constant. The above reference thus made quantitative an effect that had been noticed earlier by Fishman et al. (1992) and Link, Epstein & Priedhorsky (1993) who indicated that, in general, the various peaks in the GRB time profiles are shorter and better defined at higher energies. This unique piece of information may, therefore, serve as a stepping stone in probing the physical processes associated with the radiation emission in GRBs and maybe the entire GRB phenomenon.

I THE MODELS

It is worth noting that such a relation precludes from the outset the possibility the observed spectra are the result of up–Comptonization of soft photons by a thermal, hot electron population; in this case the the higher energy photons are those which would have spent longer time within the hot, scattering medium, leading to a correlation between the burst duration $\Delta \tau$ and the photon energy E of opposite sense to that observed.

Consider now the impulsive injection of relativistic particles at a given volume and assume the emission to be optically thin. If synchrotron or inverse Compton losses dominate the evolution of the electron distribution, then the loss rate per electron is $\dot\gamma \propto B^2\gamma^2$, where γ is the Lorentz factor of a given electron and B the magnetic field (if the losses are dominated by inverse Compton, then B^2 should be replaced by the ambient photon energy density). Therefore, the characteristic electron life time is $\Delta \tau \sim \gamma/\dot\gamma \propto \gamma^{-1}$. Since the characteristic energy of synchrotron (or inverse Compton) emission is $E \propto \gamma^2 \epsilon_0$ (ϵ_0 is a characteristic energy), expressed in terms of E rather than γ, the time duration is $\Delta \tau \propto E^{-1/2}$.

Consider, alternatively, the impulsive injection of monoenergetic high energy photons of energy E_0 in a medium of Thomson depth $\tau_T \gg 1$ and of electron temperature $kT \ll E_0$. Photons of energy E suffer fractional energy loss of order $\Delta E/E = (1-\mu)E/m_e c^2$ per collision, where μ is the cosine of the photon scattering angle, or in terms of the photon wavelength λ, $\Delta\lambda = (1-\mu)$. Averaged over solid angle, $\langle \Delta\lambda \rangle = 1$. So after n scatterings, $\langle \lambda_n \rangle = n + \lambda_0$ where λ_0 is the original wavelength (omitted for $n > 1$). In the diffusion regime, the dispersion of photon wavelengths is also propotional to the number of scatterings. Hence the following relation between the number of scatterings or time (in units of scattering time): $\Delta \tau \propto \lambda^2 \propto E^{-2}$. For a finite medium, this diffusion process lasts for a number of scatterings $n \propto \tau_T^2$,

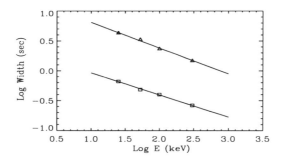

FIGURE 1. The burst duration – photon energy correlation as given in Fenimore et al. (1995)

thus the maximum width (in wavelegth) of the escaping photons would be $\Delta\lambda \propto \tau_T$. In this case, however, the precise law between the pulse width and the energy depends on the photon source distribution within the scattering medium, as it will be exhibited in more detail in the next section. Nonetheless, the qualitative correlation between the energy and duration of pulses will always be as that outlined above.

Relativistic Electron Cooling

Consider the simplest case of time-dependent synchrotron radiation by a population of non-thermal electrons: Relativistic electrons of Lorentz factor γ, with a power law distribution of index Γ i.e. $Q_e(\gamma) = q\gamma^{-\Gamma}$, are injected in a volume of tangled magnetic field B at an instant $t = t_0$ and are left to cool by emission of synchrotron (on inverse Compton radiation). The electron distribution function $N(\gamma, t)$ in time is given by

$$N(\gamma, t) = \begin{cases} q\gamma^{-\Gamma}(1 - \beta\gamma t)^{\Gamma-2} & \text{for } \gamma < 1/\beta t \\ 0 & \text{for } \gamma > 1/\beta t \end{cases} \quad (1)$$

Given the electron distribution function $N(\gamma, t)$, the emitted spectrum can be calculated by convolving $N(\gamma, t)$ with the synchrotron emissivity $\epsilon(\gamma, \nu)$. However, the form of the emerging spectrum can be easily calculated using the δ-function approximation to the single electron emissivity, i.e. $\epsilon(\gamma, \nu) \propto \delta(\nu - \gamma^2 \nu_c)$, where $\nu_c \sim 4 \times 10^6 B(G)$ Hz is the cyclotron frequency.

Figure 2 shows the synchrotron emission as a function of time at a set of frequencies, spaced logarithmically, with each curve corresponding to a frequency $\times 10$ smaller than the previous one. The values of the parameters used in this

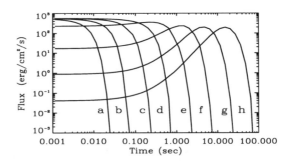

FIGURE 2. The synchrotron emission at individual frequencies as a function of time, for the relativistic cooling electron model. The frequencies corresponding to each curve are: (a) 10^{24} Hz, (b) 10^{23} Hz, (c) 10^{22} Hz, (d) 10^{21} Hz, (e) 10^{20} Hz, (f) 10^{19} Hz, (g) 10^{18} Hz, (h) 10^{17} Hz.

calculation are $\Gamma = 2$, the minimum and maximum electron Lorentz factors $\gamma_m = 10^6$ and $\gamma_M = 10^9$ and $B \sim 30$ Gauss. It is apparent in this figure that the burst duration is inversely proportional to the square root of the radiation frequency, in accordance with the relation reported by Fenimore et al.(1995). It is also apparent in this figure that, for the particular form of electron injection considered in this example (i.e. impulsive injection), there is no time lag in peak emission between frequencies in the range $\nu > \gamma_m(t_0)^2 \nu_c$, i.e. for frequencies higher than the synchrotron frequency of the lowest energy electrons at the time of injection. Time lags in peak emission develop only for frequencies smaller than that above, resulting from the "cooling" of the lowest energy electrons to energies $\gamma_m(t) < \gamma_m(t_0)$.

Energy Downgrading by Electron Scattering

Consider alternatively the injection of photons of energy $E \sim m_e c^2$ near the center of a cold ($T_e \ll E$), spherical (for simplicity) electron cloud of Thomson depth $\tau_T \sim 5$. As the photons random walk out of the cloud, they also lose energy in collisions with the ambient electrons. A monoenergetic photon pulse at the center of the cloud spreads, therefore, both in energy and in time. The duration of the pulse for photons of a given energy is given by the convolution

$$J_u(\lambda) = \sum_{n=0}^{\infty} \Psi_n(\lambda) P_u(n) \qquad (2)$$

where $\Psi_n(\lambda)$ is the photon distribution after n scatterings. Therefore the photon spectrum after time u will have the form

$$J_u(\lambda) = \frac{1}{\sqrt{2\pi\sigma_\lambda^2(u)}} exp\left[-\frac{(\lambda - m_\lambda^u)^2}{2\sigma_\lambda^2(u)}\right], \qquad (3)$$

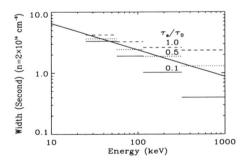

FIGURE 3. The The pulse width - energy relation for the downscattering model for different values of the τ_s/τ_0 ratio.

which represents a gaussian of width u about $\lambda - \lambda_0$. This is simply given by the width of the gaussian in equation (6), which considering the definitions of $\sigma_\lambda^2(u)$ and m_λ^u, yields $\lambda - \lambda_0 \simeq 2\sqrt{2u}$, i.e. that $\Delta\tau \propto E^{-2}$, as suggested by our heuristic arguments.

To remedy the situation we have produced a more complex, realistic model. This model consists of two concentric spheres; it is assumed that cold electrons are uniformly distributed over the entire volume of the larger sphere of radius R_0. The smaller sphere, of radius R_s, represents the volume within which the high energy photon sources are located and the photon injection takes place. The distribution of photon sources is assumed to be uniform within the volume of radius R_s, thus introducing an additional parameter in the problem namely the ratio of the radii of these two spheres $R_s/R_0 = \tau_s/\tau_0$. It is also assumed that the photon injection takes place impulsively at $t = 0$, at energies ~ 1 MeV. For the treatment of this, more detailed model we have used a Monte–Carlo code developed by one of us (XMH; see Hua & Titarchuk 1995).

In Figure 3 we present the widths in time associated with photons escaping frorm the scattering cloud at several energy bands, as obtained by our Monte-Carlo calculation; the energy bands were chosen in the same fashion as in the analysis of Fenimore et al. (1995). The widths of the pulses as a function of the energy are given for three values of the ratio τ_s/τ_0, namely 1.0, 0.5 and 0.1. The photon injection was assumed to be monoenergetic at $E_0 = 1$ MeV and the total Thomson depth of the source was taken to be $\tau_0 = 10$. The value $\tau_s/\tau_0 = 0.5$ provides good fit to the relation found by Fenimore et al.(1995), indicated by the straight solid line.

II CONCLUSIONS

We have presented above two models which can account for the recently discovered correlation of the GRB duration (or the duration of subpulses within a given burst) as a function of the photon energy. Both models appear to be able to reproduce the observed power law dependence of $\Delta\tau$ on E of this relation, as well as the general form of the observed GRB spectra.

REFERENCES

1. Fenimore, E. E., et al. 1995, ApJ, 448, L101
2. Hua, X. M. & Titarchuk, L. G., 1995, ApJ,449, 188
3. Link, B., Epstein, R.I., & Priedhorsky, W.C. 1993, ApJ, 408, L81
4. Schaeffer, B. E. et al. 1994, ApJS, 92, 285

FUTURE MISSIONS AND INSTRUMENTATION

IBIS: The Imaging Gamma-Ray Telescope on Board INTEGRAL

Pietro Ubertini
on behalf of the IBIS Consortium

Istituto di Astrofisica Spaziale CNR - Via E. Fermi 23, 00044 - Frascati, Italy

Abstract. INTEGRAL (International Gamma Ray Astrophysics Laboratory) will be the follow up on the succesful high energy missions CGRO and GRANAT. The Scientific goal of INTEGRAL is to address the fine spectroscopy with imaging and accurate positioning of celestial X-ray emission as well as large scale diffuse emission studies. These achievement will be possible by means of a high spectral resolution Spectrometer (SPI) and a high angular resolution Imager (IBIS) supplemented by an X-ray monitor and an Optical Monitoring Camera. INTEGRAL, with 2 years nominal lifetime possibly extended to 5 years, will be an observatory-class mission. The observing time will be divided into a General Programme, open to the scientific community with scientific targets selected by a peer review committee, and a Core Programme which is guaranteed time for the INTEGRAL Science Working Team (ISWT).

INTRODUCTION

IBIS, (Imager on-Board INTEGRAL Satellite), is an essential element of the mission with its powerful diagnostic capabilities of fine imaging source identification and spectral sensitivity to both continuum and broadened lines. It will observe, simultaneously with the other instruments on board INTEGRAL, celestial objects of all classes ranging from the most compact galactic systems to extragalactic objects. IBIS features a coded aperture mask and a novel large area multilayer detector which utilizes both Cadmium Telluride and Cesium Iodide elements to achieve the fine angular resolution <12 arcmin, over a wide energy range (15 keV to 10 MeV) and good resolution spectroscopy (6%-7% 0.1-1 MeV) required to satisfy the mission's imaging objectives. A large international consortium, composed by 14 Institutes belonging to 9 countries is committed to develop, build and fly this instrument.

SCIENTIFIC RATIONALS

The results obtained in the last half decade in the gamma ray range have provided a new picture of the high energy sky. Of particular relevance are the data collected with both CGRO and GRANAT. The first one provids a large collecting area from keV to GeV but moderate (degree level) imaging capaility; conversely, GRANAT featurs a lower sensitivity with very good imaging capability up to a few MeV. The complementarity of these two missions have clearly shown the need to combine in a single instrument both sensitivity and imaging in order to perform a forward step in the understanding of the high energy sky. IBIS has been designed considering these requirements as essential. More recently, the must for high angular capability to disentangle the spectra of neighbouring sources in crowded regions such as the Galactic Center has been re-enforced by the unique results of the Wide Field Cameras on board the BeppoSAX (Bazzano el al., 1997a, 1997b, in 't Zand et al., 1997a), capable to study and monitor at the same time as much as 32 sources in the Galactic Central radian. Such an achievement is due to the capability of the position sensitive X-ray detector operated with a coded mask, providing high localisation accuracy over a wide field of view (Jager et al., 1997). IBIS will permit to extend imaging to the gamma ray domain assuring good spectral resolution over a wide energy range. This will overcome source confusion limits allowing to measure detailed spectra from different types of systems, as just demonstrated by the WFCs at energies up to ∼30 keV (Bazzano et al, 1997c, Heise, 1997, in 't Zand et al., 1997b, Ubertini et al., 1997a) and at lighter energy by SIGMA (Goldwurm et al., 1994, Vargas et al. 1997).

FIGURE 1. IBIS detector/hopper view.

INSTRUMENT DESCRIPTION

The basic concept of the IBIS gamma ray detector is to have two pixellised arrays: a matrix of 128x128 CdTe elements as a front layer coupled with a matrix of 64x64 CsI bars. This set up provides a wide spectral range and high sensitivity spectroscopic capability. The detector aperture is restricted by a thin tungsten passive shield, covering the distance between the mask and the Position Sensitive Detector (PSD), and is shielded in all other directions by an active Bismuth Germanate (BGO), 20 mm thick, Veto system, as shown in Figure 1. The imaging is obtained projecting on the PSD a shadowgram provided by the use of a 15 mm thick tungsen based a coded mask positioned 3100 mm above the top detector plane. Sky images, spectrally resolved, will be obtained with standard deconvolution processes (Caroli et al., 1987). The two detector layers allow the paths of photons to be tracked in three dimensions as they scatter and interact with more than one element/plane. In this way the detector is operated as a photoelectric detector at lower energies and as a Compton detector/telescope at higher energy thus improving the signal-to-noise ratio and in turn the high energy sensitivity.

The CdTe Array. The Cadmium Telluride (CdTe) is used in order to achieve a good spatial and spectral resolution in the lower end of the detector energy range. In fact the possibility to have detectors thickness of a few mm optimises their use in the low energy domain. The IBIS CdTe, with 2 mm thick chrystals, will provide a 50% efficiency at 150 keV and will assure a good detection sensitivity down in the X-Ray domain at energy lower than 15 keV. The CdTe spectroscopic capability and variation with energy is relatively well represented with the relationship: $\Delta E/E = 2E^{-0.7}E$ in keV (Lebrun et al. 1995). This result is an upper limit to the spectral resolution. In fact, the optimum operative temperature for these detectors is around 0 °C (both for the FWHM and the peak to valley ratio), while the data in the figure has been collected at room temperature ($\sim 20°$).

The CsI Layer. The CsI array is a matrix composed by 64x64 elements, 30 mm thick, that is placed 100 mm underneed the CdTe layer in order to achieve a good stopping power up to several MeV, till maintaining a high spatial resolution and good spectral capability. The large sensitive area, exceeding 3000cm^2, is designed for optimal performance at 511 keV. The CsI layer is divided, as the CdTe one, in eight rectangular modules of 512 detector elements each. This design provides a high degree of modularity having the CsI modules the same cross-sectional shape as those of the CdTe ones. The design and construction of the detector elements have been studied in order to optimize the low energy threshold, resolution, and signal-to-noise ratio (Labanti et al., 1996, Di Cocco et al., 1996).

The spectroscopic capabilities of the CsI array at different energies are summarised as follows: $\Delta E/E$ (FWHM) %: 9.5 at 511 keV, 7.7 at 662 keV and 5.2 at 1275 keV.

The Veto Shield. The sides and rear of the detector planes are surrounded by an active Bismuth Germanate (BGO) veto shield consisting in a 5 side BGO array made out of 16 independent BGO slabs read-out by 32 photomultipliers (PMT). The high density and mean Z of BGO ensures that a thickness of 20 mm is very efficient to reduce the detector background due to leakage through the shielding of Cosmic Diffuse Background and gamma-rays produced in the spacecraft. The 16 BGO crystals are viewed by two PMT which provide veto signals to the detector electronics. Analog signals from the Veto and Calibration System are collected, after preamplification, in Veto Electronics Boxes and used by the detector layer electronics for event rejection or tagging. Particular care was devoted to obtain a detection lower thresholds for the BGO array down to 50 keV, in order to optimise the rejection of BGD events. In Figure 2 are shown preliminary results obtained from a Development Model of a BGO unit. As can be seen the lower threshold is close to the required 50 keV.

The Passive Collimating System and the Shielding Tube. The current collimating system of IBIS is made of two subsystems mechanically independent based on tungsten as passive material. The first one is an hopper structure on the top of the ISGRI layer that is a truncated pyramid with an height of 550 mm and a profile that ideally joins the detector perimeter to the active mask sides. The second one is a structure made of four walls (two vertical and two inclined) closing the aperture down to the hopper level. This set up was chosen in order to reduce as much as possible the Cosmic X-Ray Background flux up to 150-200 keV within the limited mass constraint (Natalucci and Caroli, 1996).

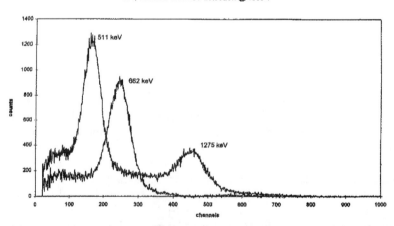

FIGURE 2. Preliminary spectral results of a slab of BGO 340x75x20 mm (DM) seen by a single 3 inches photomiltiplier. The line are 511, 662 and 1275 keV.

The Coded Mask Assembly. The Coded Mask is an array of squared tungsten elements (densimet D18), 15 mm thick, providing 80% transparency at 20 keV and 70% opacity (minimum) at 3 MeV. The basic pattern is a 53x53 Modified URA (MURA). The Mask assembly is a square graphite honeycomb sandwich with tungsten elements embedded inside to obtain the required alignment (better than a fraction of mm) and optical characteristics. The mask will be aligned on ground by using laser/optical prism assemblies. In flight the alignment will be guaranteed by the structure of the PLM, which will provide high rigidity and low thermal expansion. A proper Mechanical Ground Support Equipment (MGSE) and jigs will be manufactured to keep the mask in the correct position during ground calibration planned at instrument level. The transparency of the IBIS mask support structure for on-axis and off-axis sources (19 degrees) is respectively: 68% and 52% at 15 keV, 80% and 70% at 20 keV and 88% and 84% at 50 keV.

The on Board Calibration System. On-board calibrations are necessary in order to monitor and control the instrument in flight performances. The response function of the instrument will be refined through comparison of predicted performances obtained by modelling and ground calibration with in-orbit measurements. The calibration system will use a ^{22}Na radioactive source that produces, for each single disintegration, an opposed pair of 511 keV photons (100%), and one 1.275 MeV photon (96.6%). This set up will permit the measurement and check of the quantum efficiency as a function of energy and time, the monitoring of the gain changes and energy resolution as a function of energy and time.

Field of View and Point Source Location Accuracy (PSLA). The imaging capability over a wide field of view is a key parameter for IBIS and for the INTEGRAL Mission as a whole. In order to ensure a full compatibility of the IBIS field of view with the Spectrometer one allowing for joint observations of any region of the sky, we have choosen a FOV of 9x9 degree Fully Coded, 19x19 degree FWHM and 29x29 degree Zero Response. Such an extended FOV will also enable a better sky coverage increasing the number of detectable sources, and allow a monitoring as frequent as possible of sources, which is mandatory due to the high variability of the X-gamma ray sky. The angular resolution of the telescope is 12 arcmin. Simulations have shown that assuming no error in pointing axis reconstruction and/or other systematic effects, IBIS will locate in the sky a strong pointlike source (30 sigma) with an accuracy of better than 30 arcsec. This value is substantiated by the actual results obtained with the SIGMA telescope on the Galactic Centre (Goldwurm, 1995).

THE OBSERVING PROGRAMME

INTEGRAL is an observatory-class mission lead by the European Space Agency with the payload complement instrumentation provided by four European cosortia, including USA contributions. The observatory has a two years nominal life time, to be eventually extended to five years (Winkler, 1997). The observing time will comprise a General Programme and a Core Programme (Gehrels et al., 1997). The General Programme will constitute 65% of the time of the first year and will increase to 75% in later years. It will be open to the scientific community, with observations selected from peer-reviewed proposals. The Core Programme will constitute the rest of the time and will be dedicated to key investigations devoted to regular surveys of the Galactic Plane and deep sky fields. The Core Programme is planned and carried out by the INTEGRAL Science Working Team, chaired by the INTEGRAL Project Scientist and will be dedicated to three key investigations: Deep observation of the Central Radian of the Galaxy, the Galactic Plane Survey and Transient monitoring, by means of weeky scans, and pointings at selected sources and fields.

The Central Radian of the Galaxy. One of the main scientific objectives of the deep survey of the central radian of the Galaxy is to map and resolve at arcminute level the continuum emission of the galactic ridge and, at the same time, map line emission from nucleosynthetic radioiosotopes. It will also be possible to perform deep imaging and spectroscopic studies of the Galactic Centre region resolving isolated gamma-ray point source with an accuracy of < one arcminute and provide spectral information with outstanding resolution of ~2 keV. At least 90 sources known as X and gamma ray emitters are contained in this region at low energy (<10 keV). Spectra are available for about 40 objects associated with X-ray pulsars, black hole candidates, X-ray sources in Globular clusters showing burster behaviour and transient. In Figure 3 are shown the persistent and transient sources in the GC region amenable to study within a single IBIS pointing. In the Figure 4 is shown a reconstructed image of a simulated observation of the Galactic Centre, in the

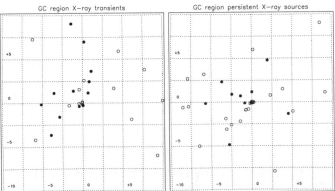

FIGURE 3. Galactic Centre Region Transients and Persistents Sources Distribution. Dots indicate sources emmitting at energy > 30 keV (From Bazzano et al., 1997d).

low energy band (40-80 keV). The image shows the 30% sensitivity FOV of the telescope, with an exposure time of about 2 days. The sources intensity were simulated using the results obtained with the SIGMA telescope (Goldwurm et al., 1997). The inner rectangular box represents the SIGMA telescope FOV, with similar sensitivity of ~30%.

Frequent Galactic Plane Scans. One of the expected outcome of the INTEGRAL Mission will be the regular survey of the Galactic Plane. The two goals of these regular scans are the production of a time and spatially resolved maps of the Galaxy over a wide energy band (15 keV up to 5 MeV), even if with a modest exposure, and the search for sources in outburst in order to trigger target of opportunity observations if warranted. As part of this study a number of Galactic Transients will be detected and possibly deep follow-on observations provided by means of pre-planned target of opportunity (see Ubertini et al., 1997b, for more details).

Pointings at Selected Sources. In order to complement the Galactic Central radian deep exposure and the Galactic Plane scans a few dedicated observations will be performed on gamma-ray sources of particular astrophysical interest to solve key open issues. Because of the limited amount of time available for these investigations in the Core Programme few of them will have a long exposure, typically 10^6s while few of them could be shorter snap-shots of about 10^5s on particularly bright sources amenable for studies at lower energies (~100-500 keV).

An example. IBIS will be particularly suited for the study and definition of the high energy spectral shape of AGNs; we expect that over 200 Seyfert type objects will be visible by IBIS above the 5 sigma level up to 100 keV in a typical 1 day exposure; 20 objects can be detected up to 500 keV thus allowing spectral studies to be performed on quite large sample of sources

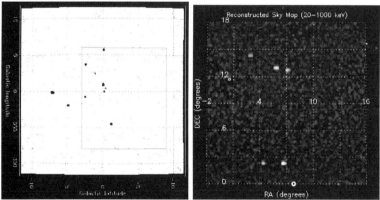

FIGURE 4. a, simulated observation of the Galactic Centre; b, simulated observation of the region containing 3C273 (50-150 keV band).

(Di Cocco et al., 1997). At higher energies the main investigations will be focussed on Blazars whose spectra extends to MeV energies, with the main aim to determin the spectral characteristics. In order to achieve the required sensitivity typically 10^6 sec observations are necessary and we expect to detect a number of serendipitous sources from each exposure. In the Figure 4b is shown a simulated IBIS observation of the extragalactic sky region (12 x 12 degree) containing the high energy source 3C273 in the 50-150 keV band: a number of sources of various classes (Seyfert 1, 2, QSOs) are visible above 10σ (Malizia et al., 1997).

Acknowledgements

The author acknowledge the IBIS Consortium, the System Team and the INTEGRAL ISWT. A particular thanks to Mr. S. Di Cosimo for the careful editing of the manuscript.

REFERENCES

1. Bazzano, A., et al, *Proocedings of the 4th CGRO Symposium* 1997a
2. Bazzano, A., et al, , *Proocedings of the 4th CGRO Symposium* 1997b
3. Bazzano, A., et al., IAU6597, 1997c
4. Bazzano, A., et al., ESA SP-382, 261, 1997d
5. Caroli, E., et al., *Spa. Sci. Rev.* Vol. 45, 3-4, p. 349, 1987
6. Di Cocco, G., et al., SPIE, Vol. 2806, p. 280, 1996
7. Di Cocco, G., et al., *Proceedings of the XXV ICRC Symposium* Durban, 1997
8. Gehrels, N., et al., ESA SP-382, 587, 1997
9. Goldwurm, A., et al., *Nature* V.371, p.589, 1994
10. Goldwurm, A., *Experimental Astronomy* Vol. 6, p. 9, 1995
11. Goldwurm, A., et al., *Proocedings of the 4th CGRO Symposium* 1997
12. Jager, R., et al., *A&A* 1997, in Press
13. Heise, J., IAU 6606, 1997
14. in 't Zand, J.J.M., et al., *Proocedings of the Potsdam Symposium* June 18-20, 1997a
15. in 't Zand, J.J.M., et al., IAU 6618, 1997b,
16. Labanti C., et al., SPIE, Vol. 2806, p. 269, 1996
17. Lebrun, F., et al., SPIE, Vol. 2586, p. 258, 1996
18. Malizia, A., et al., ESA SP-382, 439, 1997
19. Natalucci L., and Caroli, E., SPIE, Vol. 2806, p. 289, 1996
20. Ubertini, P., et al., IAU 6668, 1997a
21. Ubertini et al., *Workshop for ASM and GRB mission in X-ray band* Riken, 1997b in press
22. Vargas, M., et al., ESA SP-382, 129, 1997
23. Winkler, C., ESA SP-383, 573, 1997.

SPI: A High Resolution Imaging Spectrometer for INTEGRAL

B. J. Teegarden[1], J. Naya[2], H. Seifert[2], S. Sturner[2], G. Vedrenne[3], P. Mandrou[3], P. von Ballmoos[3], J. P. Roques[3], P. Jean[3], F. Albernhe[3], V. Borrel[3], V. Schonfelder[4], G. G. Lichti[4], R. Diehl[4], R. Georgii[4], P. Durouchoux[5], B. Cordier[5], N. Diallo[5], J. Matteson[6], R. Lin[7], F. Sanchez[8], P. Caraveo[9], P. Leleux[10], G. K. Skinner[11], P. Connell[11]

[1] *NASA, Goddard Space Flight Center, Greenbelt, MD 20771, USA*
[2] *NASA/USRA, Goddard Space Flight Center, Greenbelt, MD 20771, USA*
[3] *CESR, F-31029 Toulouse, France*
[4] *MPE, D-85740 Garching, Germany*
[5] *CEA/Saclay, SAP, F-91191 Gif-sur-Yvette, France*
[6] *Univ. of California, San Diego, CA 92093, USA*
[7] *Univ. of California, Berkeley, CA 94720, USA*
[8] *I.F.I.C., Univ. of Valencia, 46100 Burjassot, Spain*
[9] *IFC del CNR, 20133 Milano, Italy*
[10] *Univ. of Louvain, B-1348 Louvain-la-Neuve, Belgium*
[11] *Univ. of Birmingham, Birmingham B15 2TT, UK*

Abstract. SPI (Spectrometer for INTEGRAL) is a high spectral resolution gamma-ray telescope using cooled germanium detectors that will be flown on board the INTEGRAL mission in 2001. It consists of an array of 19 closely-packed germanium detectors surrounded by an active bismuth germanate (BGO) anti-coincidence shield. The instrument operates over the energy range 20 keV to 8 MeV with an energy resolution of 1-5 keV. A tungsten coded-aperture mask located 1.7 m from the detector array provides imaging over a 15° fully-coded field-of-view with an angular resolution of $\sim 3°$. The point source narrow-line sensitivity is estimated to be $3\text{-}7 \times 10^{-6}$ ph cm^{-2} s^{-1} over most of the range of the instrument (E > 200 keV) for a 10^6 s observation. With its combination of high sensitivity, high spectral resolution and imaging, SPI will improve significantly over the performance of previous instruments such as HEAO-3, OSSE, and Comptel. It can be expected to take a major step forward in experimental studies in nuclear astrophysics. The SPI instrument is being developed under the auspices of the European Space Agency by a large international team of scientists and engineers in both Europe and the United States.

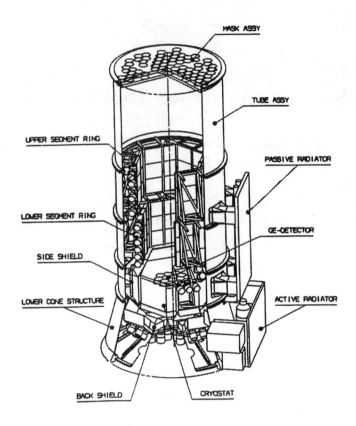

FIGURE 1. Spectrometer for INTEGRAL (SPI).

GENERAL DESCRIPTION

The Spectrometer for INTEGRAL (SPI) is a new telescope that incorporates the best current technology for fine spectroscopy in the gamma-ray region (for more detail see [1,2]). It is one of the two major instruments on the ESA INTEGRAL mission. INTEGRAL is the second of the ESA Medium-Class Missions (M2) and is scheduled for launch in the first half of 2001. It will be the successor to the NASA Compton Gamma-Ray Observatory (CGRO) and will make a significant step forward in gamma-ray astronomy particularly in the areas of fine spectroscopy and fine imaging. SPI employs an array of 19 high-resolution cooled germanium detectors in a close-packed hexagonal array (see Fig. 1). The energy range is 20 - 8000 keV, and over this range the energy resolution varies from $\sim$ 1 to 5 keV FWHM. The detectors are cryogenically cooled by two dual Stirling cycle refrigerators to a nominal operating temperature of 85K. In addition to providing a precise determination of the photon energy the Ge detector array also acts as an imaging device. The positional information comes from the knowledge of which detector was impacted, so the camera has only 19 pixels. Nevertheless, it can provide valuable information on the spatial distribution of the gamma-ray emission. A tungsten coded-aperature mask is located 1.7 m from the detector plane. Images are formed by deconvolving the shadow pattern cast by this mask on the detector plane. The fully-coded field-of-view of the telescope is 15° and the angular resolution is ~3° FWHM. Extensive simulations of the imaging performance have been carried out at the U. of Birmingham.

The Ge detector array is surrounded by an active anti-coincidence shield made of bismuth germanate (BGO) scintillator (see Fig. 1). This hi-Z material is ideally suited to suppressing the strong flux of background radiation generated in both the SPI instrument and in the spacecraft. For long observations SPI is background-limited, and it is critical to maintain the overall instrument background at the lowest possible level. Further suppression of the internal Ge detector background is obtained through the use of on-board pulse shape discrimination (PSD).

Extensive calculations and modelling of the SPI background have been performed [3–5]. Based on these simulations, the predicted narrow-line sensitivity for SPI is $\sim 2 \times 10^{-5}$ ph cm^{-2} s^{-1} at 511 keV (note that this value is high due to the present of an instrumental background line at 511 keV) and $\sim 5 \times 10^{-6}$ ph cm^{-2} s^{-1} at 1 MeV for a 10^6 sec observation. These sensitivities are typically a factor of 10 - 50 times better than the narrow-line sensitivities of prior instruments.

SPI is co-aligned on the INTEGRAL bus (a derivative of the XXM bus) with the other principal telescope, IBIS [6], and the two monitor instruments, JEM-X [7] and OMC [8]. More detail on these instruments and the overall mission can be obtained at http://astro.estec.esa.nl/SA-general/Projects/Integral/integral.html. The baseline mission scenario for INTEGRAL is to be launched on a Proton vehicle into a 48-hr orbit with an apogee of 75000 km and a perigee of 46000 km. In this orbit there will be very little loss of data due to passage through the radiation belts or earth occulation. In contrast the typical CGRO observational duty cycle is 35%. The spacecraft is constrained to point within 90°± 40° of the sun at the beginning of life and 90°± 30° at the end of life. Nominal mission lifetime is 2 years with a likely extension to 5 years. Final assembly, integration and test of the SPI experiment will take place at CNES, Toulouse, France. The following sections discuss various SPI subsystems in more detail.

DETECTOR/CRYOSTAT SUBSYSTEM

Each Ge detector is encapsulated in its own hermetic enclosure to protect it from contamination and to ease integration and replacement, if necessary. The lightweight capsules are constructed from aluminum and are sealed by electron beam welding after installation of the Ge detector. The 19 detector capsules are contained within a cryostat which serves to thermally isolate them from the ambient environment so that they can be cooled to the required 85 K operating temperature. The detectors are mounted to a cold-plate that is connected via a cold-finger to the Stirling cycle coolers that are located outside the BGO shield. The cryostat itself is thermally isolated from the surrounding shield and is operated at a temperature of 210 K to decrease the thermal load on the cold plate and Ge detectors. An on-board annealing capability (up to 100 C) is included to remove radiation damage effects if necessary. The cold plate, cold-finger, and most of the cryostat are all constructed from beryllium. This is because beryllium produces much less background (particularly at 511 keV) than all other materials suitable for cryostat construction. Also it introduces only a small amount of absorption at the 20 keV instrument threshold. A preamp for each Ge detector is mounted to the cryostat immediately behind the detector array cold plate. The detector/cryostat assembly will be fabricated and integrated at CESR and CNES, Toulouse, France.

BGO ANTICOINCIDENCE SHIELD

The detector/cryostat array is surrounded by an active BGO anticoincidence shield. The shield serves both to suppress instrument background (due to both internal and external sources) and to limit the SPI field-of-view. The mean pathlength in BGO traversed by a photon that impacts the detector array is ~5 cm. It was found through extensive modelling [4,5] that a minimum in the overall 511 keV background was obtained for a shield thickness of ~ 5 cm. This is as a result of the competition between externally-produced background which goes inversely with shield thickness and background due to neutron production in the shield itself which is roughly proportional to shield thickness. The shield is constructed of 90 different BGO modules. Each module is viewed by two redundant PMTs. Shield dead-time is a prime consideration. Extensive simulations and laboratory measurements have been made and the design of the electronics has been optimized to minimize this dead time. The internal housing of the shield is constructed from a low-Z light-weight carbon composite to minimize the amount of background producing inert material inside the shield. The shield subsystem will be fabricated, integrated and tested at MPE/Garching in collaboration with DASA.

CODED APERTURE MASK

The coded aperture mask is constructed from 3-cm thick hexagonal tungsten blocks mounted on a graphite-epoxy honeycomb plate that has a high transparency to gamma-rays. The tungsten blocks are >90% opaque over the entire SPI energy range. The mask pattern has been optimized for the SPI geometry and has 120° rotational symmetry. The mask elements are the same size as the detector modules (6 cm). With a mask/detector separation of 1.7 m the angular resolution of SPI will be $\sim$ 3° FWHM. The mask subsystem is the responsibility of the University of Valencia, Spain.

ELECTRONICS

Each detector has its own associated electronics that includes preamp, shaping amp, discriminators and ADC. The detector signals are digitized to 15 bit accuracy and time-tagged to an accuracy of $\sim$ 100 μsec. In the normal mode the pulse height, time tag and PSD information for every accepted event (i.e. those the do not trigger the anticoincidence shield) are transmitted to the ground. Multiple-detector events are also recorded and transmitted. These constitute as much as 30% of the total event rate at higher energies. In addition the bad events (those that trigger the anticoincidence shield) are accumulated into 19 spectra (1 for each detector). These spectra are periodically ($\sim$30 min) transmitted to the ground and will provide useful information to better define the in-flight calibration. In addition to the pulse-height and time tag information, various housekeeping parameters (count rates, temperatures, voltages, mode flags) are included in the telemetry stream. The analog front-end electronics is being developed at CESR/Toulouse, the digital front-end electronics at CEN/Saclay and the digital processing electronics at CNES/Toulouse.

Additional electronics (PSD) is included to aid in suppresssion of internally-produced detector background. In the energy range 0.2-1.5 MeV the dominant component of the background results from β-decays in the Ge detectors induced primarily by secondary neutrons. These β-decays go primarily into the ground state (i. e. no accompanying photon). Since the range of the β in Ge is short (<few mm) the energy deposition of these events is localized to a small region. This is in contrast to the primary photons that constitute the signal. Most of these photons undergo multiple compton scattering before either loosing all of their energy or departing the detector. The current pulse generated by an interaction in the Ge detector depends on the location and multiplicity of the interactions in that detector. By analyzing this pulse (with the PSD) one can distinguish between single- and multiple-site events. It is estimated that the background in the range 0.2-1.5 MeV can be reduced by as much as a factor of 10. The PSD electronics is being developed jointly by the Univ. of Calif., San Diego, the Univ. of Calif., Berkeley, and the Lawrence

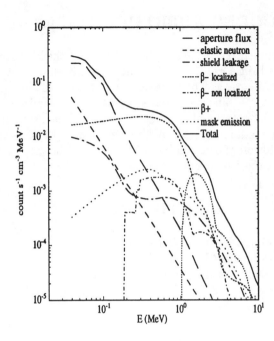

FIGURE 2. SPI instrumental background components.

Berkeley Labs.

BACKGROUND AND SENSITIVITY

Extensive simulations of the SPI background have been performed using primarily the GEANT code developed for analyzing particle interactions in high-energy physics experiments. The results of these simulations have been used in many important instrument design trade studies (e. g. shield thickness, active vs. passive mask, cryostat materials, PSD performance). Every attempt has been made to optimize the SPI design to obtain the lowest possible background and highest possible sensitivity. The various components of the SPI background are shown in Fig. 2 [4,5]. Below 100 keV the aperture flux from the diffuse cosmic x-ray background dominates. From $\sim$ 200 keV to $\sim$ 1.5 MeV the dominant component is β^--decay. It is this component that will be suppressed by the PSD. Above 1.5 MeV the strongest contribution is β^+-decay, but with significant contributions also from the tungsten mask and from shield leakage. In Fig. 3a [4,5] the SPI efficiency vs. energy is shown with and without event selection. The most important event selection is that due to PSD which causes a loss of efficiency due to its requirement

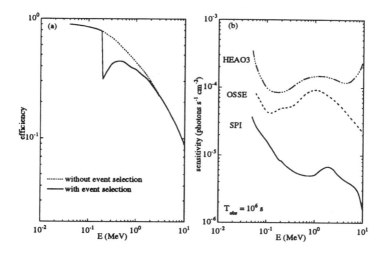

FIGURE 3. a) Efficiency vs. energy with and without event selection. b)SPI narrow-line sensitivity compared with other instruments.

for single-site interactions. In Fig. 3b [4,5], the SPI narrow-line sensitivity is shown as derived from the background given in Fig. 2 and the efficiency given in Fig. 3a. SPI is also compared with OSSE and HEAO-3 (the only previous comparable high resolution spectrometer). Typically SPI will be 20-30 times more sensitive to narrow lines than HEAO-3, and it will open an entirely new region of observational phase space in gamma-ray astronomy.

IMAGING

In the energy range of SPI the signal-to-background ratio will in general be small (typically a few to 10%). Knowledge of the background is critical to producing high quality gamma-ray images and spectra. This knowledge will be obtained primarily by using a dithering technique where the spacecraft pointing direction will varied in a pre-determined fashion. The typical pattern will be a 5 x 5 grid of pointings about the target direction each separated by 2°. Extensive imaging simulations have been performed for SPI. Most recently these simulations have used the GSFC-developed MGEANT code (http://lheawww.gsfc.nasa.gov/docs/gamcosray/legr/integral) to calculate individual photon interactions in the instrument and to produce event lists containing the energy losses in each of the SPI detectors. This has been done for a variety of different types of astrophysical gamma-ray sources (e.g. point sources of various spectra and positions, diffuse 511 keV, diffuse 1809 keV). These event lists were then fed into

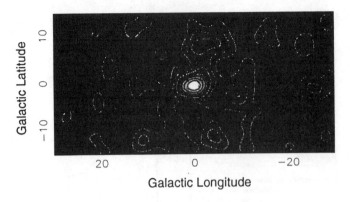

FIGURE 4. Simulated SPI map of OSSE 511 keV measured emission.

the Univ. of Birmingham-developed CAPTIF image processing package (http://www.sr.bham.ac.uk/~phc/CAPTIF) and simulated sky maps were generated. An example of such a map of the diffuse 511 keV emission is given in Fig. 4. The input data used was the measured OSSE 511 keV distribution [9,10]. The simulation reproduces quite well the essential features of the OSSE map. For more details on these simulations and other results see Connell et al [11].

REFERENCES

1. Lichti, G.G. et al., *SPIE Conference Proceedings*, Vol. 2806, ed. B. D. Ramsey & T. A. Parnell (1996).
2. Mandrou P. et al., *Proceedings of 2nd INTEGRAL Workshop*, ESA Publication Division, SP-382, St. Malo, p. 591 (1996).
3. Naya, J. et al., *Gamma-ray and Cosmic-Ray Detectors, Techniques and Missions*, ed. B.D. Ramsey & T. A. Parnell, Proc. SPIE, vol 2806, p. 457, (1996).
4. Jean, P. et al., in *Gamma-ray and Cosmic-Ray Detectors, Techniques and Missions*, ed. B.D. Ramsey & T. A. Parnell, Proc. SPIE, vol 2806, p. 457, (1996).
5. Jean, P. et al., *Proceedings of 2nd INTEGRAL Workshop*, ESA Publication Division, SP-382, St. Malo, p. 635 (1996).
6. Ubertini P. et al., *Proceedings of 2nd INTEGRAL Workshop*, ESA Publication Division, SP-382, St. Malo, p. 599 (1996).
7. Westergaard, N. J. et al., *Proceedings of 2nd INTEGRAL Workshop*, ESA Publication Division, SP-382, St. Malo, p. 605 (1996).
8. Gimenez A. et al., *Proceedings of 2nd INTEGRAL Workshop*, ESA Publication Division, SP-382, St. Malo, p. 613 (1996).
9. Purcell, W. R. et al., *Proceedings of 2nd INTEGRAL Workshop*, St. Malo, p. 67 (1996).
10. Cheng, L.-X. et al., *ApJ* **481**, L43-47 (1997).

11. Connell, P. H. et al., these proceedings (1996).

The spectral line imaging capabilities of the SPI germanium spectrometer on INTEGRAL

G. K. Skinner*, P. H. Connell*, J. Naya[†], H. Seifert[†],
S. Sturner[†], B. J. Teegarden[†] and A. W. Strong[‡]

*School of Physics and Astronomy, University of Birmingham,
Edgbaston, Birmingham, B15 2TT,
[†]NASA Goddard Spaceflight Center, Greenbelt, MD 20771 USA
[‡]Max-Planck Institut für extraterrestrische Physik, Garching, Germany

Abstract.
The SPI spectrometer on Integral will offer the high spectral resolution possible with germanium detectors, together with the capability of imaging both point and extended sources. To assess the performance of the instrument both a GEANT Monte Carlo model and ray-tracing techniques are being used to produce simulated data streams. These are being analysed with independent software systems representative of those which may be used with flight data. In this way the advantages and feasibility of various approaches to the reconstruction and deconvolution problem can be investigated. Example results obtained using a number of techniques, including linear and maximum entropy methods as well as some newly developed techniques, are presented.

Introduction

The SPI germanium imaging spectrometer, to fly on the ESA INTEGRAL mission in 2001, is described in more detail elsewhere [4,6,11]. Briefly it consists of 19 Germanium detectors in a heavy (576 kg) BGO active shield with a coded mask having 63 opaque elements made of 3 cm thick Tungsten which offer $\sim 3°$ resolution over a fully coded field of view of 15° diameter (40° to zero response).

The instrument is particularly designed for studies of gamma-ray line emission. Most known sources of such emission are diffuse, which necessitates a high sensitivity to extended emission, a requirement which dictates the wide field of view and leads to the choice of the angular resolution. The ambiguities inevitable in attempting to image ~200 resolvable pixels within the field of view with only 19 detector elements are resolved by using multiple pointings. The analysis of the data in these circumstances requires new techniques which are being developed through simulations and trials as illustrated here.

Simulations

In order to reveal any effects of simplifying assumptions made in the data analysis it is important to include in the data simulation stage as many as possible of the effects that occur in a real instrument. This is done by using a detailed model of the instrument in a GEANT Monte Carlo simulation to generate a simulated data stream.

The results obtained by running the analysis procedures using as input the simulated datastreams produced by Monte Carlo runs can be compared with the assumed distribution of emission (both spatial and energy) which went into the simulations. They also be compared with results obtained with much simpler 'ray-tracing' simulations.

The intention in this work has been to keep the simulation and reconstruction stages as independent as possible in order to provide the maximum level of consistency checking. For example event data streams simulated by Monte Carlo runs at GSFC are transferred to the Birmingham group in a format similar to that in which flight data will eventually be telemetered and are analysed there by independent image reconstruction software.

The examples illustrated here use both simulation by the ray tracing approach (*e.g.* Figure 2) and by full Monte Carlo simulations (*e.g.* Figure 3).

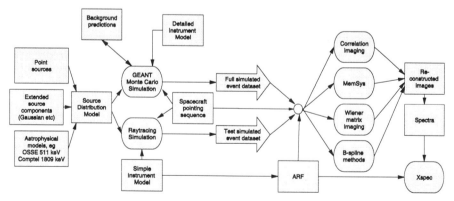

FIGURE 1. The system for simulation and for image reconstruction from simulated data

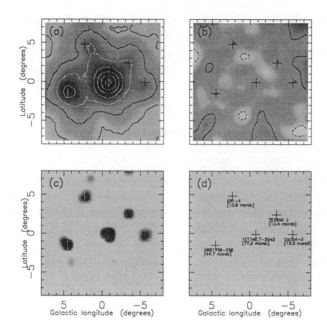

FIGURE 2. Reconstructions from simulated data of a field of point sources. Source field: Galactic centre sources seen by Sigma in the energy range 70–150 keV, as in (d), simulated by ray tracing. Spacecraft pointing sequence: 169 pointings of 20 minutes on a 2° pitch grid $=2 \times 10^5 s$. Reconstruction: (a) correlation mapping (b) correlation mapping with IROS (iterative removal of sources). 5 sources have been fitted and removed from the positions indicated by dots. Residual noise is < 1 millicrab rms (c) B-spline method. Note how noise manifests itself as occasional peaks towards the edge and the corners where the exposure is lowest.

Image reconstruction methods

All of the methods being investigated for reconstructing images require a knowledge of the response of the instrument to radiation of all possible energies from different directions. This is more than simply a description of the shadowing by the mask. For example, the shield has significant transparency in the upper part of the energy range and there are off-axis effects, shadowing by components outside the instrument and energy dependencies to be taken into account. This information is stored in a common format - the ARF (Figure 1). The ARF is large and complex to compute so it is important to evaluate the effects of making simplifying assumptions.

The image reconstruction techniques investigated include:

Correlation methods lead to poor resolution and significant sidelobes when used directly. An example is in Figure 2(a). For point sources they can be combined with iterative removal of sources [3], *e.g.* Figure 2(b).

Matrix inversion is subject to the well known instabilities and noise amplification but can be stabilised by Wiener-like matrix techniques similar to that described by Rideout et al [8] (Figure 3(a)).

Maximum Entropy techniques are being explored using the Memsys 5 package [5] (not illustrated here).

Spline methods using B-splines are proving very successful. B-splines [9] are basis functions which have the properties that the value, gradient and curvature are all zero outside a well defined region. By solving for the amplitude of each of a set of B-spline on a regular grid which gives the best χ^2 fit to the observed data, an image is obtained which has all of the smooth well-defined properties of a splined surface. Examples are shown in Figures 2(c) and 3(b).

Methods can be combined, for example in principle, as we will show elsewhere [1], the B-spline method can be combined with the Weiner matrix approach [8] while achieving a considerable improvement in computational efficiency. In Maximum Entropy, positivity is implicit. With others methods it can be imposed as a constraint. This makes the response non-linear, but reduces the number of parameters used to describe the source field (nodes where the intensity would be negative are effectively removed). We note that the method can be extended by also allowing a higher density of nodes in regions where there is strong highly structured emission (including point sources). In this form it can become a variant of the pixon method [2] in which a an entropy based on the likelihood and the number of parameters is optimised.

REFERENCES

1. Connell, P.H., In preparation (1997).
2. Dixon, D.D., et al., *Ap. J. Suppl. Ser.* **120** 683 (1996).
3. Hammersley, A.P., Ponman, T.J., and Skinner, G.K., *Nucl. Intrum. Meth.* **A311**, 585 (1992).
4. Lichti, G.G., et al, in *Proceedings of the SPIE conference, Denver,1996* SPIE 217 (1996).
5. Strong, A.W. *Experimental Astronomy* **6/4**, 97 (1995).
6. Mandrou, P., in *Proceedings of the 2nd INTEGRAL Workshop, St Malo, 1996* ESA **SP-382,** 591 (1997).
7. Purcell, W. R. This volume.
8. Rideout, R. and Skinner, G.K., *Astron. Astrophys. Suppl. Ser.* **120**, 589 (1996).
9. Schoenberg, I.J., and Whitney, A. *Trans. Amer. Math. Soc.* **74**, 246 (1953).
10. Skinner, G.K., et al, in *Proceedings of the 2nd INTEGRAL Workshop, St Malo, 1996* ESA **SP-382**, 591, (1997).
11. Teegarden, B.J., et al, This volume.

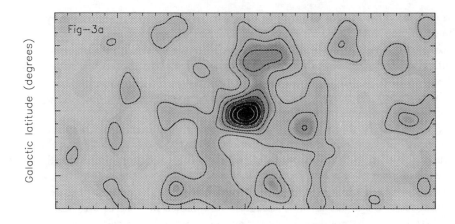
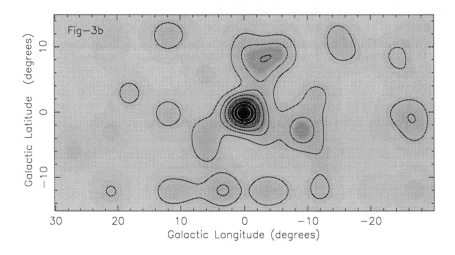

FIGURE 3. Simulations of response to 511 keV emission from the galactic centre, using data from a GEANT simulation assuming the distribution found by Purcell using OSSE data and with image reconstruction by (a) Modified Wiener and (b) B-splines

The IBIS view of the galactic centre: INTEGRAL's imager observations simulations

P. Goldoni*, A. Goldwurm*, P. Laurent*, F. Lebrun*

CEA/DSM/DAPNIA, SAp CEA-Saclay F-91191 Gif-sur-Yvette FRANCE

Abstract. The Imager on Board Integral Satellite (IBIS) is the imaging instrument of the INTEGRAL satellite, the hard-X/soft-gamma ray ESA mission to be launched in 2001. It provides diagnostic capabilities of fine imaging (12' FWHM), source identification and spectral sensitivity to both continuum and broad lines over a broad (15 keV–10 MeV) energy range. It has a continuum sensitivity of $2\ 10^{-7}$ ph cm^{-2} s^{-1} at 1 MeV for a 10^6 seconds observation and a spectral resolution better than 7 % at 100 keV and of 6 % at 1 MeV. The imaging capabilities of the IBIS are characterized by the coupling of the above quoted source discrimination capability with a very wide field of view (FOV), namely 9° × 9° fully coded, 29° × 29° partially coded FOV.
We present simulations of IBIS observations of the Galactic Center based on the results of the SIGMA Galactic Center survey. They show the capabilities of this instrument in discriminating between different sources while at the same time monitoring a huge FOV. It will be possible to simultaneously take spectra of all of these sources over the FOV even if the sensitivity decreases out of the fully coded area. It is envisaged that a proper exploitation of both the FOV dimension and the source localization capability of the IBIS will be a key factor in maximizing its scientific output.

THE IBIS TELESCOPE

The IBIS telescope [10] is a coded mask imaging system based on a 1.5 cm thick tungsten mask of 95 × 95 elements of 1.1 × 1.1 cm^2, designed from a replication of a 53 × 53 MURA basic pattern [3] which gives it a high angular resolution ($\approx$ 12') over a wide field of view (FOV) (9° × 9° fully coded, 29° × 29° partially coded at zero response). The IBIS detection system is composed of two planes, an upper layer made of 16384 squared CdTe pixels (ISGRI) and a lower layer made of 4096 CsI scintillation bars (PiCsIT). This system enables high sensitivity continuum spectroscopy (E/ΔE>10) and a wide spectral range (15 keV – 10 MeV).

The simulation we performed are for the moment limited to the ISGRI upper layer [5]. The ISGRI pixels are 4× 4 mm^2, 2 mm thick crystals of Cadmium Telluride, a semiconductor operating at ambient temperature, providing a spectral resolution of about 8% at room temperature. The 128×128 = 16384 pixels are arranged in 8 modules separated by dead zones 2 pixel wide needed by the mechanical structures which sustain the detector plane. The sensitivity loss caused by dead zones is not large, however the absence of sensitive elements in the detector plane must be properly taken into account during deconvolution procedures.

CODED MASK IMAGING

In the X/gamma-ray domain, source localization can be achieved through the use of coded mask imaging systems [1]. These systems represent the most performant way to achieve source localization in this energy domain. In recent years this was demonstrated by the fast localization of X-ray Novae by the SIGMA coded mask telescope [9]and by the first localization of the X counterpart of a gamma ray burst thanks to the Wide Field Cameras of the BeppoSAX satellite [2] .
The incoming radiation is modulated by a mask with opaque and transparent elements and then recorded by a position sensitive detector. In this way it

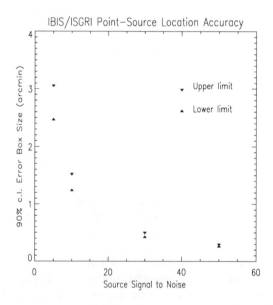

FIGURE 1. *IBIS/ISGRI Point Source Location Accuracy vs. signal to noise. Values are comprised between triangles*

is possible to simultaneously measure source and background thus avoiding the on/off technique usually employed in gamma-ray astronomy which is very sensitive to background variability and observing conditions. The angular resolution of such system is defined by the angle subtended by one hole at the detector which is 12' for IBIS. However point source location accuracy will also depend on the ratio R between mask element size and detector pixel size and will be proportional to the source signal to noise (S/N) ratio. The positional error can be easily computed for the case of integer ratios R as a function of S/N [6]. ISGRI/IBIS pixels subtend an angle of $\sim$ 4.6' and therefore R is equal to 2.4, we thus expect to reach a position accuracy which is between the values obtained for R=2 and for R=3. These values for a number of S/Ns are reported in Fig. 1. The instrument's extended field of view is divided in two parts, the fully coded field of view (FCFOV), where each source will project a complete (shifted) basic pattern onto the detector plane, and the partially coded FOV (PCFOV) where source flux will be only partially modulated by the mask. Telescope's sensitivity will depend on the amount of modulation and therefore it will be constant in the FCFOV and decreasing with angle from telescope axis in the PCFOV. This is shown in Fig. 2.

Sky images are produced through an algorithm which is basically a balanced correlation between detector image and mask pattern (see [6]). Reconstructed sky images however are affected by large amount of coding noise in form of

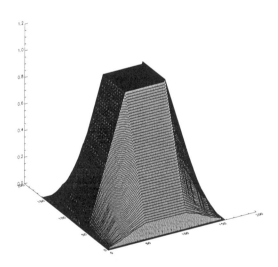

FIGURE 2. *3D image of the IBIS/ISGRI normalized sensitivity over the whole field of view.*

ghost peaks and extended modulation due to the sources in the FC or PC FOVs. An iterative source cleaning procedure must therefore been applied in the data analysis to correctly deconvolve sky images. Such procedures have been developed by our group and succesfully used for the data analysis of the SIGMA coded mask telescope images [6] and now modified and adapted to simulated IBIS images. Indeed the simulations will allow us to test and improve the data analysis techniques.

IMAGING SIMULATIONS OF THE GALACTIC BULGE

The Galactic Bulge is the zone of the sky with the highest source density in the X/gamma-ray domain. A typical INTEGRAL Galactic Bulge observation will last about 10^5 s, i.e. a day. We simulated such an observation in the energy range 40-80 keV assuming sources are at their average brightness level as measured with the SIGMA/GRANAT telescope [8] during its 7 year long galactic center survey ([7,9]) and adding a contribution also from 4U 1700–37, a hard X-ray high mass binary, at a level about 1/2 lower than the peak flux value detected by SIGMA [4]. We then applied sky image reconstruction and cleaning algorithm as described in the previous section.
Results of the simulation (Fig. 3) show the expected imaging performances of the IBIS telescope. All sources including the faint Terzan 1 ($\approx$ 8 mCrabs) are clearly detected at more then 5 sigma level with the brighter (the "microquasar" 1E 1740.7-2942) at more than 40 sigma. The sources are well separated even in the very center of the galaxy where source confusion problems may arise. It can be seen that 4U 1700–37 is also easily detected by the IBIS instrument, which proves the possibility of taking a high significance spectrum even if the source is at more than 11° from pointing axis. The possibility to monitor such huge field gives to IBIS a very high potential for search and high precision ($< 1'$) localization of high-energy brigth transient sources like X-ray novae and gamma-ray bursts. Moreover thanks to the wide field and the galatic plane survey program, it will be possible to detect serendipitous, faint sources like Ti^{44} lines from hidden supernovae in the galactic plane.

REFERENCES

1. Caroli E. et al., *Space Sci. Rev..* **45**, 349 (1987)
2. Costa E. et al., *Nat.* **387**, 743 (1997)
3. Gottesman S. R. and Fenimore E., *Applied Optics.* **28**, 20 (1989).
4. Laurent, P. et al., *A.&A.* **260**, 237 (1992)
5. Lebrun F. et al., Proc. of "Imaging in High Energy Astronomy", Eds. L. Bassani and G. DiCocco, Kluwer Acad., The Netherlands (1995)
6. Goldwurm A., *Exper. Astron.* **6** 9 (1995)

7. Goldwurm A. et al., *Nature*, **371** 589 (1994)
8. Paul J. et al., *Adv. Space Res.*, **11**(8), 289 (1991)
9. Vargas M. et al., Proc. of 2nd INTEGRAL Workshop, *ESA SP*-382, 129 (1997)
10. Ubertini P. et al., *SPIE*, **2806**, 24 (1996)

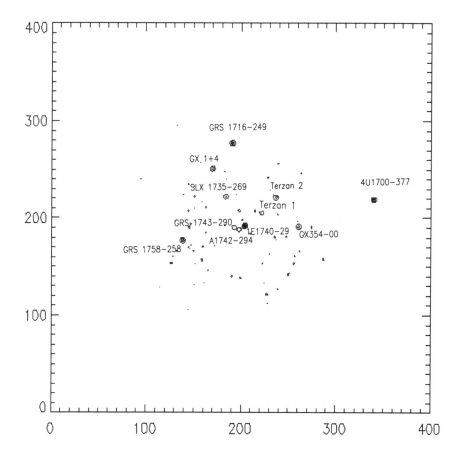

FIGURE 3. *5-level contour image of the complete ISGRI FOV of the Galactic Center in the 40-80 keV band. Units on axis are sky pixels (about 4.6 arcminutes see text). Note the clear detection of the 4U1700-377 flare at more than 11° from the pointing direction*

Can the INTEGRAL-Spectrometer SPI detect γ-ray lines from local galaxies?

R. Georgii, R. Diehl, G.G. Lichti and V. Schönfelder

Max-Planck Institut für extraterrestrische Physik
Giessenbachstr., D-85740 Garching, Germany

Abstract. The spectrometer SPI on board of INTEGRAL, the future γ-ray mission of ESA, is designed to measure γ-ray lines with a high energy resolution and a high sensitivity. It will detect γ-ray lines of long-living isotopes such as ^{26}Al from extragalatic γ-ray sources. In the case of a positive detection important constraints could be set on the yields of those isotopes in supernovae and on supernova rates in the observed galaxies. An estimation of the expected observation times for the 1809 keV γ-ray line of ^{26}Al for different galaxies, mainly galaxies of the local group, is presented. The most promising candidate for such a detection of the 1809 keV γ-ray line is the Large Magellanic Cloud (LMC).

INTRODUCTION

γ-ray line spectroscopy is a powerful tool for tracing the nucleosynthesis processes in our galaxy. Especially the detection of ^{26}Al in the interstellar medium (ISM) [1] enabled us to set important constraints on theoretical models describing the nucleosynthesis in stars and supernovae (see [2] for a recent review). The detection of the 1809 keV line from other galaxies would narrow those limits even further and would support our understanding of star formation and evolution. Especially the limits on the yield of ^{26}Al sources and the supernova rate of galaxies could be constrained by such a detection. The aim of this work is to estimate the observation time for SPI for ^{26}Al from different galaxies in dependence on uncertainties in astrophysical and instrumental parameters.

ESTIMATION OF THE ^{26}Al FLUX FROM GALAXIES

Assuming all the ^{26}Al is being produced in supernovae of type Ib or II, i.e. typically an aluminum mass of $y_{al} = 10^{-4} M_\odot$ per supernova is being ejected in the ISM, assuming a constant supernova rate f of a galaxy during the half-life time τ_{al} of ^{26}Al (about 1 Myr), the number N_{26} of ^{26}Al nuclei in the ISM can be found to be

$$N_{26} = \frac{M_\odot \cdot y_{al} \cdot b_r}{m_{al}} \cdot \frac{\tau_{al}}{\tau_{mean}} . \quad (1)$$

Here $M_\odot$ denotes the solar mass, b_r the branching ratio for the 1809 keV decay (about 0.97), m_{al} the atomic mass of ^{26}Al and $\tau_{mean} = 1/f$ the mean time between the occurrence of supernovae in a galaxy. Together with the distance D the flux Φ_{al} in the 1809 keV line can be estimated to be

$$\Phi_{al} = \frac{N_{26}/\tau_{al}}{4\pi \cdot D^2} . \quad (2)$$

In Table 1 estimated fluxes using (2) are given for several galaxies, assuming the production sites to be point like structures compared to the spatial resolution of SPI (2.5°) [3]. Table 1 shows that with the estimated narrow line sensitivity for SPI, $S_{line} = 5 \cdot 10^{-6}$ s^{-1}cm^{-2} at 1 MeV for 10^6 sec [4], we can expect to observe the 1809 keV line in members of the local group only, especially the Large Magellanic Cloud (LMC) and maybe the Small Magellanic Cloud (SMC).

TABLE 1. The estimated fluxes for the 1809 keV line for different galaxies. The values for the supernova rates f are taken from [3] and [5]. The estimated observation time for a 2σ detection with SPI is given.

Object	Type	Distance D [kpc]	SN rate f [yr^{-1}]	^{26}Al flux Φ_{al} [s^{-1}cm^{-2}]	Estimated time [yr]
M33	Sc	840.0	0.0062	$1.04 \cdot 10^{-8}$	$2.89 \cdot 10^3$
M31	Sb	770.0	0.0083	$1.7 \cdot 10^{-8}$	$1.14 \cdot 10^3$
SMC	Ir	58.8	0.0011	$3.7 \cdot 10^{-7}$	2.2
LMC	Ir	50.1	0.0045	$2.1 \cdot 10^{-6}$	0.07[a]
NGC 6822	Ir	520.0	0.0004	$1.7 \cdot 10^{-8}$	10.2
NGC 253	Starburst	3100.0	0.1	$1.2 \cdot 10^{-8}$	$2.0 \cdot 10^3$
NGC 4946	Sc	3100.0	0.01	$1.2 \cdot 10^{-9}$	$2.0 \cdot 10^3$

[a] (≈ 26 days)

ESTIMATION OF THE OBSERVATION TIME WITH SPI

The INTErnational Gamma-RAy Laboratory (INTEGRAL) is a medium-size mission (M2) of ESA with a scheduled launch in April 2001. One of the two main instruments is the SPectrometer on Integral (SPI), which is mainly designed for high-resolution spectroscopy of gamma-ray lines. It consists of a cooled Ge-detector array with a typical energy resolution of 2 keV at 1.33 MeV and will measure in an energy range of 20 keV to 8 MeV. This instrument is an excellent candidate for the detection of ^{26}Al and all the following calculations will be performed with its instrumental parameters.

The sensitivity S of a detector, defined as the limiting flux of an observation is given by the formula

$$S = \frac{n \cdot \sqrt{N+B}}{X} \quad , \qquad (3)$$

where n defines the significance of the observation, N the number of source counts, B the number of background counts and X the exposure, i.e. the product of the effective area A_{eff} of the detector times the effective observation time t_{obs}.

B is the product of b, the background in the detector per keV, second and volume, A the area, h the height and ΔE the energy resolution of the detector times 1.5 as a heuristic factor accounting for the correct inclusion of the background [6]. N is the product of the exposure X and the source flux Φ_{source}. In the background-dominated case, i.e. $B >> N$, we can substitute in equation (3) the sensitivity with the source flux, thus leading to the following formula for the observation time

$$t_{obs} = n^2 \cdot \frac{(\Phi_{source} \cdot A_{eff} + b \cdot A \cdot a \cdot h \cdot \Delta E)}{(A_{eff} \cdot \Phi_{source})^2} \qquad (4)$$

Using the values $A = 500$ cm^2, $A_{eff} = 92$ cm^2, $h = 7$cm, $\Delta E = 2.1$ keV [7], $a = 21/19$ being a correction for a coded mask telescope and $b = 1.7 \cdot 10^{-3}$ MeV^{-1} cm^{-2} sec^{-1} [6] for SPI, the observation times listed in the last column of Table 1 can be expected for a 2 σ observation of the 1809 keV line. Due to uncertainties in the astrophysical parameters as the ^{26}Al yield and the supernova rate of a galaxy these numbers can change significantly. In Figure 1 a contour plot shows the variation of the observation time in days for LMC in dependence of those two parameters. There is also an uncertainty in the determination of instrumental parameters as the energy resolution of the Ge-detector array and the background event rate. Especially the latter one can be wrong up to a factor of two, thus giving considerable changes in the observation time. This is due to the fact, that the background rate is determined via Monte-Carlo simulations using only a very preliminary mass model for SPI.

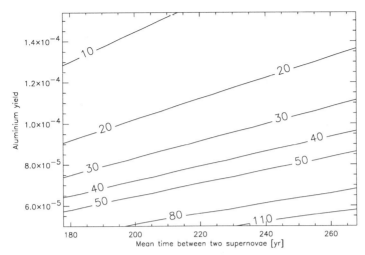

FIGURE 1. The observation time in days for LMC in dependence of the two astrophysical parameters, the aluminum yield y_{al} and the mean time between two supernovae τ.

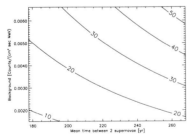

FIGURE 2. The observation time in days for a 2σ detection in LMC in dependence of the background event rate b and the mean time between two supernovae τ_{mean}. An aluminum yield of $10^{-4} M_\odot$ was assumed.

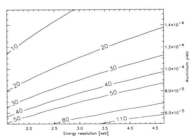

FIGURE 3. The observation time in days a 2σ detection in LMC in dependence of the aluminum yield y_{al} and the energy resolution ΔE. A supernova rate of 0.45 per century was assumed.

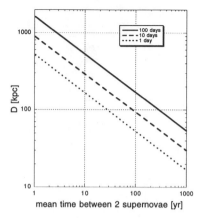

FIGURE 4. The distance for a 2σ detection of the 1809 keV line versus the mean time between two supernovae τ_{mean} for 3 different observation times (1 day, 10 days, 100 days).

In Figure 2 and Figure 3 contour plots of the observation time in days for LMC are shown for different combinations of astrophysical and instrumental parameters. In Figure 4 the distance for a 2σ detection of the 1809 keV line is plotted versus the supernova rate assuming a standard aluminum yield of $10^{-4} M_\odot$ and the instrumental parameters for SPI.

CONCLUSIONS

The 1809 keV line measurement of ^{26}Al from the LMC appears a realistic prospect of the INTEGRAL mission with the SPI instrument. This holds even taking into account the uncertainties in the supernova rate and the aluminum yield of the candidate sources and even further accounting for instrumental uncertainties from the final design of SPI. There might also be a chance to detect ^{26}Al from the SMC. The chances to observe the 1809 keV line from M31 or M33 or even from galaxies outside of the local group are very low.

REFERENCES

1. Mahoney,W.A., Ling, J.C., Wheaton, Wm.A., Jacobson, A.S., *APJ* **286**, 578, (1984).
2. Prantzos, N., Diehl, R., *Phys. Report* **267**, 1, (1996).
3. Timmes, F.X. and Woosley, S.E., *APJ* **481**, L81, (1997.)
4. Winkler, C., *Exp. Astron.*, **6**, 85, (1995).
5. Tammann,G., *Supernovae*, Les Houches, 1990, p. 27.
6. Pierre, J., PhD thesis, Toulouse, 1996, p. 53.
7. Lichti, G.G., et al., Proceedings of SPIEs anual meeting, Denver, 1996, p. 222.

Contribution of passive materials to the background lines of the spectrometer of *INTEGRAL* (SPI)

N. Diallo[1], B. Cordier[1], M. Collin[2], F. Albernhe[3]

[1] *CEA/DSM/DAPNIA/SAP, CEA Saclay, 91191 Gif sur Yvette, France.*
[2] *CNRS/Centre d'Etude Spatiale et des Rayonnements, Toulouse, France.*
[3] *CEA Bruyères-le-Châtel, Service de Physique Nucléaire, BP 12, 91180 Bruyères-le-Châtel, France.*

Abstract. The passive materials inside the SPI BGO shield represent a non negligible fraction of the SPI detection unit total mass, containing many iron-rich materials, if we except the berrylium. The bombardment of the passive materials by the cosmic rays and by their secondaries in the BGO shield produce radioactive unstable nuclei which in turn decay and possibly emit a γ-ray. In this paper we will compute the contribution of various passive materials inside the shield to the SPI background lines. This study will able us then to determinate for each line of astrophysical interest which passive materials should be avoided while conceiving a γ-ray spectrometer, as a potential emitter of this line. It has also enabled us to deduce the less noisy material to be choosen when technological trade off occur.

INTRODUCTION

INTEGRAL is an ESA's high-energy astrophysics mission to be launched into a high perigee orbit early in the next decade. One of the two mission's main telescopes is the gamma-ray spectrometer SPI. The spectrometer SPI is dedicated to high-resolution γ-ray line astrophysics in the energy range 20 keV to 8 MeV, with an energy resolution in the few parts-per-thousand range and with a sensivity of $(2\text{-}5) \times 10^{-6} \gamma/\text{s/cm}^2$. The only detectors which can achieve the required energy resolution under flight conditions are cooled Ge detectors. Therefore the central element of SPI are 19 large-volume Ge crystals which are arranged in an hexagonal geometry. These 19 Ge crystals are housed in a cryostat which cools them down to $\sim 85K$. The Ge-detectors array is surrounded (with the exception of the field of view) by a massive coded-aperture mask mounted 170 cm above the detector plane. The main

TABLE 1. Main scientific targets of the *INTEGRAL*/SPI mission.

Nuclide	Astrophysical site	Energy (keV)
^{7}Be	Galactic novae	478
^{16}O*	Orion complex	6129
^{26}Al	Galactic novae, supernovae, Wolf Rayet	1809
^{22}Na	Galactic novae	1274
^{44}Ti	CAS A, SN 1987A, galactic supernovae	68 78 1157
^{56}Co	Galactic & extra-galactic supernovae	847 1238
^{60}Fe	Galactic supernovae	1173 1332

scientific objectives are summarised in Table 1. Fore more details on the SPI telescope see Lichti at al. (1996).

In order to achieve the sensivity requirements, $(2\text{-}5) \times 10^{-6} \gamma/\text{s}/\text{cm}^2$ around 1 MeV, it is important to decrease the background. However, in space, the telescope will be continually irradiated by the CR protons. These protons passing through the heavy BGO anti-coincidence shield will generate a large population of secondary neutrons. Finally, a population of mixed protons and neutrons will irradiate continuously the detectors and the the passive materials inside the anti-coincidence shield. This irradiation produces unstable nuclides which will decay emitting β particles ans γ-ray photons with half lives ranging from less than one second to a few months and years. In the spectrometer activation we can distinguish two cases: the detectors activation and the passive materials activation. The unstable nuclides created inside the Ge detectors will contribute to the background via their β emission. The continuum spectra produced by these physical processes has been calculated by Jean et al. (1997) and Diallo et al. (1997). In the order hand, the unstable nuclides created in the pasive materials inside the anti-coincidence shield will contribute to the background via their γ-ray emisssion. Unfortunately in this last case, the created nuclides are able to produce background lines at the same energy than the scientific targets summarised in Table 1. In this study we propose then to use an hadronic Monte Carlo code in order to estimate the effects of the passive materials on the sensivity. We will compute the background contribution of each material at the specific energies presented in Table 1.

COMPONENTS OF THE CALCULATION

The cosmic ray spectrum

The orbits proposed for the *INTEGRAL* mission are both highly eccentric. In the case of the Proton launcher, the satellite will not cross the active fraction of radiation belts, and in absence of solar flare accelerated particles, the background photons will be essentially induced by interactions of CR (mainly protons in the 500 Mev to few GeV range) with the spectrometer materials. The incident CR proton spectrum used in our simulations is given by Meyer (1969) and the calculations are performed for an isotropic flux at solar maximum (2 protons/s/cm^2).

The TIERCE code

In general, the hadronic interactions triggered by energetic CR in the detector are very difficult to simulate with good accuracy due to the complexity of the vast number of nuclear processes involved combined with the incompleteness of the tabuled nuclear data. The TIERCE code developed at CEA Bruyères-le-Châtel enables us to make these computations with a good accuracy. TIERCE is a Monte-Carlo code which transport hadrons, electrons and γ. It is one of the few available to simulate correctly the spallation reactions. TIERCE includes the High Energy Transport Code (HETC) for energetic hadrons and the Monte-Carlo N-particle MCNP code for the low energy neutrons (< 20 MeV). HETC is a nucleon/meson transport code which uses an intranuclear cascade and evaporation model. MCNP uses multigroup scattering and absorption cross-section data (Bersillon et al. 1996). TIERCE includes also a geometrical definition module and a decay module link to a decay library. In our application it allows us to define a SPI mass model, to bombard this model with high energy protons, to compute the secondary particles spectra in all the volumes, to generate unstable nuclides, to follow the decays and to compute the emitting β and γ spectra. This code permits to compute the decay of a volume at a given time for a continuous excitation source: one can then estimate the spectrometer radio-activity evolution versus the running time in orbit.

The SPI hadronic numerical model

The detector array of the hadronic numerical model is composed of 19 hexagonal (6cm corner to corner × 7cm height) Ge volumes. Each detector is encapsulated in its Aluminium container. The thickness of each capsule is 2mm around the Ge detector. The 19 detector modules are enclosed in a Be cryostat supported by a Be cold plate. The camera is surrounded by a 5.5 cm thickness hexagonal anti-coincidence shield, made of BGO. In this model we do not simulate the upper anti-coincidence rings and the mask. The numerical model is immersed in an isotropic cosmic proton flux, and we compute the resulting proton and neutron spectra present inside the BGO shield. The passage through the BGO blocks induces a decrease of the high energy protons and a strong increase of the low energy protons. The maximum of the proton spectrum is shifted to lower energy (150 MeV). This phenomenon is due to the ionisation energy losses and the creation of secondary protons by spallation reactions. In addition to the protons, the passage through the shield creates a large amount of neutrons. Above 1 MeV, the neutron spectrum is distributed like a power law with an index of 1.75. In the thermal neutron range the shape is simular to a plateau. The proton and neutron spectra present inside the BGO shield are represented in Figure 1 and Figure 2.

The passive materials

Table 2 summarises the nature and the mass of passive materials foreseen inside the anti-coincidence shield. All these passive materials constitute the different camera elements like cryostat, capsules, plates, screws, electronic pre-amplifiers... Of course, this matter is spatially distributed inside the BGO shield. In order to compute the radioactivity yields of these different materials, we propose to irradiate them with the inner cryostat particle population computed by our hadronic model. The radioactivity yields are calculated for a continuous CR bombardment of six months.

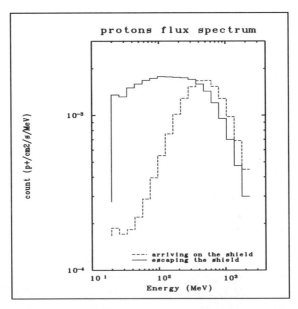

FIGURE 1. Proton flux spectrum irradiating the BGO and Proton flux spectrum inside the BGO shield irradiating the camera.

RESULTS

In Table 3, the γ-ray yield of the different materials is given at the astrophysical target energy. The yields are expressed in γ production per second and per gram of material multiplied by a factor of 10^{-5}. We see immediately that the use of steel pollutes the 846 keV (^{56}Co) and 1157 keV (^{44}Ti) lines. In the same way, the use of Titanium (Ta6V) pollutes the 1157 keV (^{44}Ti) line and the use of PTFE the 6129 keV (^{16}O*) line. Concerning Aluminium and steel, the effects of the different blendings seem to be marginal. In Table 4, tacking into account the total mass of each material, we represent for each scientific target the total amount of γ-ray lines emitted inside the BGO shield. Of course this approach is only qualitative while we do not simulate the γ-ray transport. We simply multiplied the yield value by the mass value. Evidently the calculation of the detected γ-ray lines is more complex, because the materials are spatially distributed. This computation implies to take into account the different solid angle versus the detectors and and the auto-absorption of each material. We attend to do this work in a next future by using the CERN GEANT Monte Carlo software. However, with this qualitative approach we note that the ^{16}O*, ^{44}Ti and ^{56}Co lines are the more polluted astrophysical targets. On the contrary the ^{26}Al line seems to be little disturbed by the passive materials activation.

In Table 5, we represent the positron production in the different passive mate-

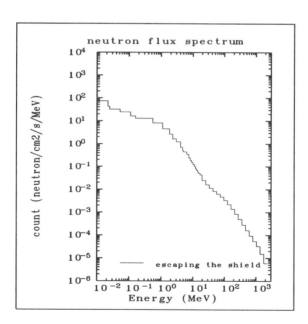

FIGURE 2. Neutron flux spectrum inside the BGO shield irradiating the camera.

TABLE 2. Passive materials present inside the BGO anti-coincidence shield.

Material	Composition in % of mass	density (g/cm^3)	Mass (g)
Al5083	Al 92 Mg 4.5 Mn 0.7 Si 0.4 Fe 0.4 Zn 0.2 Ti 0.1	2.66	2033
Au4G	Al 93 Cu 4.2 Mg 0.8 Si 0.8 Mn 0.7 Fe 0.7 Zn 0.2	2.76	994
BeO	Be 99 O 1 Fe 0.1	1.84	15359
Brass	Cu 63 Zn 37	8.45	284
Ceramic	Ti 60 O 40	4.05	782
CuBe 2	Cu 98 Be 2	8.25	1454
Constantan	Cu 55 Ni 45	8.90	128
Steel 1810	Fe 72 Cr 18 Ni 10	7.90	2506
Steel C18.8	Fe 74 Cr 18 Ni 8	7.90	1676
Steel Z6CNU 17.04	Fe 78 Cr 17 Ni 4	7.95	1976
PTFE	F 76 C 24	2.20	1132
Ta 6V	Ti 90 Al 6 V 4 O 0.02	4.43	1451
TE 630	Si 38 O 30 C 16 Ca 5.6 B 1.3 H 1.5	1.85	1746
Vespel	C 69 O 21 N 7 H 3	1.43	104

TABLE 3. γ-ray lines yields at specific energies of the passive materials inside the SPI BGO shield after 6 months on orbit (units are in $10^{-5}\gamma/s/g$).

Energy (keV)	477	847	1157	1173	1238	1274	1332	1809	
Al5083	0.8	0.1	2.2	0.0	0.0	21.5	0.0	2.4	2.3
Au4G	8.6	3.3	3.3	0.1	0.2	17.5	2.1	2.2	1.4
BeO	6.7	0.3	0.1	0.0	0.0	0.0	0.0	0.0	0.8
Ceramic	4.7	0.0	166.9	0.0	0.0	1.2	0.0	0.0	35.2
CuBe 2	0.0	16.1	30.4	4.2	9.0	0.6	25.0	0.0	0.1
Constantan	0.0	34.0	48.0	4.3	22.4	0.6	27.6	0.0	0.0
S 1810	0.1	164.0	68.6	0.5	5.3	1.2	0.9	0.0	0.4
S C18.8	0.0	169.7	64.5	0.4	4.9	1.0	1.0	0.0	0.4
S Z6 1704	0.0	146.5	61.0	0.2	3.0	1.1	0.5	0.0	0.2
PTFE	11.5	0.0	0.0	0.0	0.0	0.0	0.0	0.0	704.5
Ta 6V	0.5	0.0	296.6	0.0	0.0	4.4	0.0	0.3	1.0
Vespel	20.6	0.0	0.0	0.0	0.0	0.0	0.0	0.0	12.7

rials, these values can give us an idea of the 511 kev line production. Evidently, as the others γ-ray lines, we must take into account the mass distribution of the passive materials inside the cryostat to determinate with good accuracy the 511 keV line detected in the Ge.

REFERENCES

Bersillon, O. et al., in 2nd Conference on Accelerator-Driven Transmutation technologies and Applications, Kalmar (1996).

Diallo, N., Collin, M., Bersillon, O. et al., in ESA SP-382, The Transparent Universe, ESA Publications Division, 631 (1997).

TABLE 4. Total amount of γ-ray lines emitted at specific energies by the passive materials inside SPI BGO shield after 6 months on orbit (units are in $10^{-2}\gamma/s$).

Energy (keV)	477	847	1157	1173	1238	1274	1332	1809	
Al5083	1.7	0.2	4.5	0.0	0.0	43.8	0.0	5.0	5.0
Au4G	8.5	3.2	3.3	0.1	0.3	17.5	2.0	2.2	1.4
BeO	103.5	5.9	2.4	0.0	0.0	0.2	0.0	0.0	13.0
Ceramic	3.7	0.0	130.0	0.0	0.0	0.9	0.0	0.0	27.5
Cu Be	0.2	23.4	44.2	6.2	13.01	0.9	36.2	0.0	0.2
Constantan	0.0	4.3	6.1	0.5	2.9	0.0	3.5	0.0	0.0
S 1810	0.3	410.7	171.9	1.1	13.4	3.1	2.3	0.0	1.2
S C18.8	0.1	284.4	108.2	0.7	8.2	1.7	1.7	0.0	0.7
S Z6 1704	0.1	289.6	120.4	0.4	5.9	2.1	1.0	0.0	0.5
PTFE	13.0	0.0	0.0	0.0	0.0	0.0	0.0	0.0	797.5
Ta 6V	0.7	0.0	430.4	0.0	0.0	6.3	0.0	0.4	1.4
Vespel	2.1	0.0	0.0	0.0	0.0	0.0	0.0	0.0	1.3

TABLE 5. Positron production in the passive materials.

Material	Mass (g)	$\beta^+/s/g\ 10^{-3}$	β^+/s
Al5083	2033	5.04	10.25
Au4G	994	5.42	5.39
BeO	15359	0.13	1.99
Ceramic	782	7.58	6.24
CuBe 2	1454	14.26	20.74
Constantan	128	11.89	1.52
Steel 1810	2506	4.37	10.95
Steel C18.8	1676	4.10	6.88
Steel Z6CNU 17.04	1976	3.46	6.85
PTFE	1132	8.37	9.48
Ta 6V	1451	6.40	9.29
Vespel	104	8.12	0.84

Jean, P., in *Etude des performances et modélisation de spectromètre gamma pour l'astrphysique nucléaire*, Thèse de l'Université Paul Sabatier Toulouse (1997).

Lichti, G.G., Sch:onfelder, V., Diehl, R. et al., SPIE 2806,217 (1996).
Meyer, J.-P., Ann. Rev. Astron. Astrophys., 7, 1 (1969).

MGEANT—A Generic Multi-Purpose Monte-Carlo Simulation Package for Gamma-Ray Experiments

Helmut Seifert[*,†], Juan E. Naya[*,†], Steven J. Sturner[*,†], and Bonnard J. Teegarden[*]

[*]*NASA/Goddard Space Flight Center, Greenbelt Maryland 20771*
[†]*Universities Space Research Association*

Abstract. We present a generic multi-purpose Monte-Carlo simulation package, based on GEANT and the CERN Program Library, which is appropriate for gamma-ray astronomy, and allows the rapid prototyping of a wide variety of detector systems. Instrument specific geometry and materials data can simply be supplied via input files, and are independent of the main code. The user can select from a standard set of event generators and beam options (implemented as "plug-ins") which would be needed in an astrophysical or instrument calibration context. The philosophy of this approach is to facilitate the implementation of new software add-ons and changes in the instrumental setup. This is especially useful for projects which involve a large group of developers and collaborators. MGEANT has been successfully used to perform background calculations and to generate the instrument response for the *WIND*/TGRS instrument, and is currently being used to study the background, sensitivity, response, and imaging capabilities of the *INTEGRAL*/SPI.

INTRODUCTION

MGEANT, developed by at the Low Energy Gamma Ray (LEGR) Group at NASA/GSFC, is based on the GEANT program which is supported by the Application Software Group, Computing and Networks Division, CERN Geneva, Switzerland. GEANT is an extensive system of general purpose, detector description, and simulation tools which are used in the design, optimization, and testing of complex high energy physics experiments, the development of associated analysis programs, and the interpretation of experimental data. The program is well suited for this purpose since it is able to simulate the transport of elementary particles through an experimental setup, and to represent the setup and the particle trajectories graphically. GEANT has been used extensively at NASA/GSFC, and in particular has been used by the LEGR Group to model the Transient Gamma Ray Spectrometer (TGRS) on the *WIND* spacecraft [1], the Gamma-Ray Imaging Spectrometer (GRIS) [2], as well as new prototype detectors. Currently, GEANT is used to study the *INTEGRAL*/SPI [3], and to support the development of data analysis software for this instrument.

While GEANT offers the framework and the raw tools for experimental setup, event simulation, and visualization, it is the responsibility of the user to code all the relevant subroutines describing the experimental environment, to assemble the appropriate program segments and utilities into an executable program, and to compose the appropriate data cards which control the execution of the program. Although seemingly straightforward and simple, using GEANT correctly and efficiently generally involves a long and steep learning curve. Also, the process of setting up a simulation of a particular instrument and/or experiment requires sometimes difficult and/or nonintuitive programming steps. Often, even simple modifications to an existing setup require changes in the program code itself, which may be difficult to perform by a user who is not the author of the program. To alleviate these problems, the idea was to develop from GEANT a tool which, on one hand, would be specific to gamma ray astronomy and instrument development, but would also be generic and flexible enough to allow for a rapid and simple implementation of a wide variety of experiment configurations. For the program itself, a modular, "object oriented" approach was chosen. Rather than hardcoding specific instrument geometries and materials, these data are supplied via input files. Finally, event generators and beam options are implemented as "plug-ins" which, for efficiency, are included during the program compile and link phase. The available set of plug-ins is constantly being expanded and, it is the hope, eventually will also incorporate contributions from the user community outside NASA/GSFC.

OVERVIEW OF MGEANT

MGEANT is being developed using the same tools which are used for the releases of the CERN Program Library, i.e. CVS, imake, makedepend, gmake, and the C Preprocessor. The package can be downloaded from the LEGR WWW site http://lheawww.gsfc.nasa.gov/docs/gamcosray/legr/mgeant as an Imake project with source code (C and FORTRAN) and header files, and an Imakefile and associated configuration files for installation. Certain configurations of MGEANT also require the CFITSIO package, which is available from the HEASARC at http://heasarc.gsfc.nasa.gov/docs/software/fitsio.

From the distribution, MGEANT simulators with many different properties can be built. For example, an interactive version may be useful for visualizing both the experimental setup and the particle transport, while a non-interactive version may be more efficient for simulations in production mode. The user may choose among several output formats and, depending on the application, may select from different hadronic interaction interfaces. Also, a variety of beam generators (plane wave/astrophysical point source; calibration source/point source at a finite distance; isotropic beam; diffuse emission map) and spectrum models (monoenergetic/multi-line spectrum; exponential; power-law, broken power-law, or power-law with exponential cut-off; gamma-ray burst model; cosmic ray spectrum for solar minimum or maximum; user spectrum) are available. Finally, apart from simulations where the relative instrument orientation and source location are fixed, an astrophysical observing program with different instrument pointings (e.g. a scan of the Galactic plane) can be simulated as well.

SIMULATIONS WITH MGEANT

TGRS (launched on the *WIND* spacecraft in Nov. 1994) is designed to perform spectroscopy of bright gamma-ray bursts and solar flares in the $\sim$20-8000 keV energy range, with a resolution which is 5-30 times better than that of earlier-generation detectors. TGRS consists of a single, large, high-purity Ge crystal, which is radiatively cooled to $\sim$85 K to minimize the effects of radiation damage. Using an on-board occulter, TGRS is also ideally suited for continuous long-term monitoring of sources close to the ecliptic, such as the Crab and the Galactic Center. MGEANT was used on *WIND*/TGRS for modelling the instrument background (Figure 1), and for the generation of the detector response matrices for spectral deconvolution (Figure 2).

The spectrometer SPI, one of the four instruments on INTEGRAL (International Gamma-Ray Astrophysics Laboratory), is to perform high resolution spectroscopy ($E/\Delta E = 500$) of celestial gamma-ray sources in the energy range 15-10000 keV. Integral will be launched in 2001, with a projected lifetime of 2-5 years. In the case of SPI, MGEANT was used to study the continuum background due to induced prompt reactions (γ diffuse cosmic; cosmic ray p and e^- interactions, including secondaries). Work in progress is the study of delayed reactions (β decays, e^- capture). Studies of background lines concentrate on the lines from decays of isotopes in the passive material. Although the line emissivity for a selected passive material currently must be supplied by the user, in the future these emissivities will be automatically calculated within MGEANT. First results of studies of the SPI imaging capabilities are presented elsewhere in these Proceedings [4]. For deconvolving the spectrum of point sources currently several methods are investigated. In the most favourable scheme, the Imaging Procedure succeeds in separating the spatial from the spectral response, making generation of a response matrix for spectroscopy very simple. MGEANT is used for these studies both to generate the simulated

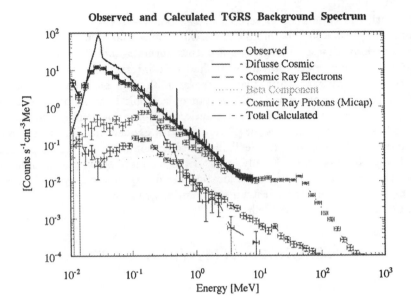

FIGURE 1. MGEANT calculations of the TGRS instrument background.

data sets, and to generate the responses (Figure 3).

REFERENCES

1. Owens, A., et al., *Space Sc. Rev.* **71**, 273 (1995)
2. Tueller, J., et al., *Nuclear Spectroscopy of Astrophysical Sources*, New York: AIP Press, 1988, pp 439–443
3. Mandrou, P., et al., *The Transparent Universe*, Proceedings 2nd INTEGRAL Workshop, 1996, pp 591–598
4. Skinner, G., et al., *These Proceedings*

FIGURE 2. The response matrix shown above was generated with MGEANT for GRB 950822, which had an angle of incidence of ~46° with respect to the detector normal. This matrix is covers ~20–15000 keV, with a binning of ~1 keV below ~477 keV, and ~10% above ~477 keV.

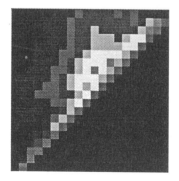

FIGURE 3. The first SPI spectral response matrix for an on-axis point source. The energy range 20–8192 keV is divided into 21 logarithmic bins.

A Small Scan Angle-Dependent Background Systematic in Non-Standard OSSE Observations

J. D. Kurfess[1], K. McNaron-Brown[2], W.R. Purcell[3],
R. L. Kinzer[1], & W. N. Johnson[1]

[1] *E. O. Hulburt Center for Space Research*
Naval Research Laboratory, Washington DC 20375
[2] *George Mason University, Fairfax, VA 22030*
[3] *Northwestern University, Evanston, IL 60208*

Abstract. We have used the large set of OSSE data acquired during AGN observations and high galactic latitude mapping observations to study scan-angle dependent background. A small scan-angle dependent background is observed in continuum bands and is adequately characterized by a sinusoidal function of scan angle. The effect is not significant for observations which use the standard OSSE background offsets or for gamma ray lines like the 511 keV positron annihilation line. It can be significant for non-standard observations of sources near the OSSE sensitivity limit which use large or asymmetrical background offsets or one-sided offsets.

INTRODUCTION

The standard OSSE observational technique uses offset background pointings acquired at ±4.5° from the source position. This technique implicitly assumes that the background is either independent of or varies linearly as a function of scan angle. Any non-linear scan angle dependence coupled with the use of large or asymmetrical background offsets or one-sided background offsets may produce a scan-angle background systematic.

OSSE mapping and AGN observations covering a wide range of scan angles have been used to study potential scan angle-dependent background effects. It has been found to be an important effect in the analysis of mapping observations. It can also be important for other observations where large or asymmetrical background offsets are used and when the flux is near the OSSE limiting sensitivity. This applies to many galactic plane observations and to mapping low-energy continuum emission in the halo around the galactic center region. It also applies to several other observations where non-standard pointing strategies had to be applied, including observations of Cassiopeia A (The et al. 1996) and Orion (Murphy et al. 1996). It will be important to assess this scan-angle dependent background following the recent reboost of GRO to higher altitudes (and a higher background environment).

OSSE OBSERVATIONS AND DATA ANALYSIS

The OSSE instrument consists of four nearly identical detector systems (Johnson et al. 1993). Each detector has a passive tungsten collimator which defines a 3.8° × 11.4° field-of-view (FOV). Each detector also has an independent single-axis drive system which can orient the detector through a total angular range of 192°. The detectors are located in pairs on the spacecraft such that detectors 1 and 2 have clear FOVs from −51° to 90° from the S/C Z-axis and detectors 3 and 4 have clear FOV from −11° to 141° from the S/C Z-axis. This is shown in Figure 1.

The standard analysis of OSSE data has been described by Purcell et al. (1992) and Johnson et al. (1993). For each 2-minute source observation, an estimated background is determined for each energy channel by fitting a quadratic function (in time) to the 3 or 4 available background observations which were acquired within 6 minutes before or after the source observation. These background observations are typically acquired alternately at scan angles of +4.5° and −4.5° relative to the source location to reduce any potential scan angle dependence of the local background environment. The quadratic

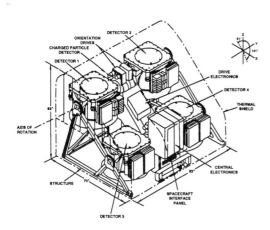

Figure 1. OSSE general configuration. The detectors can be rotated though an angular range of 192° as shown. The exterior thermal shield and support structure have been removed in this view and are indicated in outline.

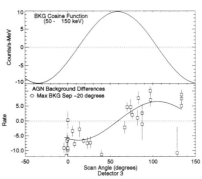

Figure 2. (top) Scan angle background (50–150 keV) determined from OSSE observations for Detector 3. (bottom) AGN "Background Differences" vs. scan angle for large background offsets. The solid line is the background residual expected from the model shown in top panel.

fits to the background observations provide an adequate correction for temporal variation in the background rates. However, the spatial dependence relative to the S/C is assumed to be absent (or linear with symmetric offsets) in standard analyses. The total spectrum from a source observation over a typical viewing period of 14 days is obtained by summing the individual 2-minute background-subtracted spectra, with appropriate normalization for livetime and angular response.

OSSE mapping observations differ significantly from standard observations. Rather than having a single, well-defined source position as in standard observations, mapping observations typically consist of sequences of eight unique pointing positions for each detector and cover a scan angle range of $\sim 15°$ − $25°$. There are no unique "source" and "background" positions (see Dixon et al. 1997). Instead, each pointing position is treated sequentially as a "source" position, with others as "background" positions. Background estimation is performed as described above for standard OSSE observations. Specifically, the backgrounds used for each "source" position are taken to be the $\pm$ 1 and $\pm$ 3 2-minute intervals relative to the source observation. The same process is performed for each of the pointing positions in the observation sequence, and the individual two-minute difference spectra are summed by detector and by pointing position. Each spectrum represents the observed spectrum at one scan angle measured relative to several other scan angles. In the absence of sources in the fields, and any systematic effects, the difference spectra for each scan angle should be consistent with zero. Any scan angle dependent background effect would result in non-zero residuals in the differences for one or more of the summed difference spectra. The individual difference spectra are statistically correlated because spectra were used as both "source" and "background" estimates. These correlations are calculated and included in propagating the uncertainties when summing the difference spectra.

High galactic latitude mapping observations (Virgo, south galactic pole, north ecliptic pole) have provided evidence for an approximately sinusoidal scan angle-dependent background throughout most of the 0.05 − 10 MeV energy range. The magnitude of this component over the full scan angle range is typically about 1% of the overall background. Figure 2(a) shows the sinusoidal background component term determined from analysis of the mapping data in the 50 − 150 keV energy range for OSSE detector 3. This sinusoidal component is significant at the 15 − 20σ level. At energies between 150 keV and 1 MeV, the significance of the effect is much smaller.

Given this background component, one can anticipate that the OSSE offset pointing to determine backgrounds will produce a systematic effect which depends on the scan angle and the size of the background offsets. For example, for the background component shown in Figure 2(a), source observations taken at a scan angle of 60-70 degrees (the peak of the sinusoidal background component for detectors 3 and 4) will have background offsets which produce slightly lower count rates. This results in a systematically positive source flux in the absence of a real source. As the magnitude of the background offsets are increased, the systematic error also increases in a non-linear fashion.

The validity of this effect has been tested by using OSSE AGN observations. By comparing the spectra of the two background positions (normally on opposite sides of the source location) the AGN background data confirm the scan angle-dependent background effect. By treating each background position as a "source" and using the AGN position as a reference, the relative count rates in the two background positions can be determined. The difference between these two background spectra, referred to as "background differences", results in a spectrum which has essentially removed the original AGN source component. Figure 2(b) shows that a small scan angle-dependent residual in these background differences is observed, and is in good agreement with the differences that are predicted from the mapping observations. Note that Figure 2(b) represents the differential of the scan angle dependent background component shown in Figure 2(a).

ANALYSIS OF MAPPING AND AGN DATA

The combination of the AGN "background difference" and mapping processing provided a set of background subtracted spectra, each representing the spectrum measured at one scan angle relative to one or more other scan angle positions. The data also cover a wide range of scan angles, enabling a sensitive study of the shape and intensity of any scan angle-dependent background effect. Scan-angle dependence for specific lines is, in general, small so we focus on the scan-angle dependence for continuum observations in this paper. The data from each continuum band were fitted separately with an analytic model for the shape of the scan angle-dependent background as a function of OSSE scan angle. The analytic model was used to determine the expected residual background for each of the difference spectra based on the specific scan angles used as "source" and "background". Various analytic models were investigated, and adequate fits were obtained using a simple cosine model in which the amplitude, phase and period were free parameters in the fit. Until the cause of this instrumental systematic is better understood, this cosine model is assumed to be an approximate description for the scan angle-dependent background. The model fits were performed using linear χ^2-minimization methods and included the covariance in the data.

Some of the mapping data also included significant responses to one or more gamma ray sources in the mapped region (e.g., 3C 273, and M87/NGC4388 in the Virgo mapping data; see Dixon et al. 1997). In these cases, the responses for the relevant sources were determined and the intensities of the sources included as extra free parameters in the model fitting process. Figure 3 shows examples of the systematic effect for several observational configurations and compares the magnitude of the effect with the OSSE sensitivity (2σ sensitivity for a 5×10^5 second observation) shown by the bold solid lines. The systematic effect is shown as a function of energy for observations taken near the Z-axis. The figure shows the absolute magnitude of the systematic at the Z-axis for symmetrical background offsets of $\pm 4.5°$ and $\pm 10.5°$, as well as for a one-sided background observation with offsets $-4.5°$ and $-7.0°$.

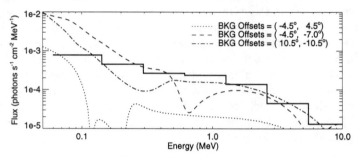

Figure 3. Comparison of the expected scan angle background systematic and the OSSE 2σ sensitivity (histogram) for Z-axis observations with several different background strategies.

CONCLUSIONS

The nominal OSSE background offsets produce a systematic that is small compared to the OSSE sensitivity at all energies. This is valid for standard observations ($\pm 4.5°$ backgrounds) at all scan angles. Therefore, the scan angle-dependent background is not a significant concern for standard OSSE observations. However, larger or asymmetric background offsets can produce systematic effects which are not negligible compared to the OSSE sensitivities. Single-sided backgrounds (where two background positions are used, but both are on the same side of the source) can generate a systematic effect at a level of 3σ or higher. A procedure, $SCAN\_CORRECTION$, will be available in the OSSE IGORE data analysis system to calculate a corrected spectrum for any OSSE observation. A more detailed description of the scan-angle dependent background will be available soon as part of the IGORE documentation, and will be the topic of a more detailed publication.

REFERENCES

1. Dixon, D., et al. 1997 *Astrophys. J.* **484** 891.
2. Johnson, W.N., et al., 1993 *Astrophys. J. Supp.* **86** 693.
3. Murphy, R. J., et al., 1996 *Astrophys. J.* **473** 990.
4. Purcell, W.R., et al., 1992 in *"The Compton Observatory Science Workshop,"* NASA Conference Publication 3137, p. 15.
5. The, L.-S., et al. 1996 *Astron. and Astrophys. Supp.* **120**, 357.

A time dependent Model for the activation of COMPTEL

Martin Varendorff*, Uwe Oberlack*, Georg Weidenspointner*,
Roland Diehl*, Rob van Dijk[†], Mark McConnell[††] and James Ryan[††]

*Max-Planck-Institut für extraterrestrische Physik, D-85740 Garching, Germany
[†]ESTEC, Space Science Department, Keplerlaan 1, NL-2201 AZ Noordwijk, Netherlands
[††]University of New Hampshire, Morse Hall, Durham, N. H., 03824, USA

Abstract. The structure of the CGRO satellite is irradiated by cosmic rays and trapped particles from radiation belts. These incident particles produce radioactive nuclei in nuclear reactions with the satellite structure. Most of the radiation dose can be attributed to the passages through the South Atlantic Anomaly. The incident particle flux on the COMPTEL instrument is estimated from the event rate of a plastic scintillation detector. This event rate is modeled with a Neural Network simulation. The increase of the event rate during SAA passages is taken as a measure for the amount of induced radioactivity. A Neural Network Model is used to derive the buildup of radioactive nuclei in the instrument over the first five years of the mission. Measurements of the internal ^{22}Na- and ^{24}Na-activity are used to estimate the proton flux in the SAA. The result is consistent with earlier measurements and models.

INTRODUCTION

The signal to background ratio is small for Compton telescopes in astrophysical experiments. Therefore a good understanding of the background is essential. The goal of this study is to provide a good measure for the activation of the COMPTEL instrument, which is helpful for the construction of background models.

Instrument:
A detailed description of COMPTEL onboard CGRO can be found in Schönfelder et al. (1993). COMPTEL consists of two sets of detector modules. An incoming gamma-ray photon is Compton scattered in the upper D1 detector (liquid scintillator) and assumed to be totally absorbed in the lower D2 detector (NaI). Two calibration units ("Cal-units") are located between the two detection layers at the side of the instrument opposite each other (Snelling

et al., 1986). The Cal-unit is comprised of a ^{60}Co doped plastic scintillation wafer, viewed by two 1/2 inch PMTs. The primary use of the Cal-unit is to tag the gamma-rays from the decay of ^{60}Co.

Input data:
The count rate of the plastic scintillator inside Cal-unit B varies considerably with respect to several parameters. Figure 1 shows the variation of the count rate of Cal-unit B with time. The effect of the satellite reboost around TJD 9300 from about 330 km above earth to about 440 km is clearly visible. The dependance on solar cycle is shown in Figure 2. The black dots represent measurements at the beginning of the mission around solar maximum; the gray dots, near solar minimum in the same altitude range.

Neural Network Simulation:
A Neural Network Simulation was chosen to model the cal-unit rates as a function of the orbital parameters, because most of the functional dependencies are unknown and because the parameter space is unevenly filled. Measurement noise and data gaps prevent use of the raw data itself as a model for the activation of the instrument.

A Neural Network Simulation Model with 7 input neurons, and 2 hidden layers à 10 neurons and one output neuron was built. The activation $o_{(j,n)}$ of the neuron j in layer n equals the sigmoidal function F of the weighted sum of the

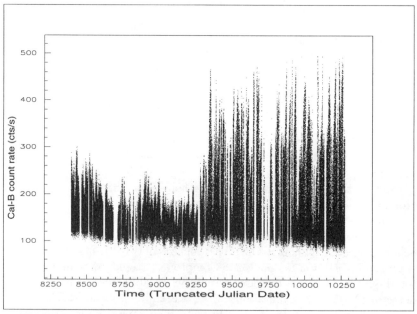

FIGURE 1. Measured count rates of Cal-unit B as a function of time.

activation from all N neurons in layer n-1 plus a bias b_j:

$$o_{(j,n)} = F(\sum_{i=1}^{N(n-1)} w_{ij} \cdot o_{(i,n-1)} + b_j), \quad \text{with} \quad F(x) = \frac{1}{1+e^{-x}}$$

Each input neuron presents one of the seven input parameters as its activation to the network. The input values are propagated through the network using the formulas above. The output of the last neuron represents the model value corresponding to the input values. The following model parameters were chosen: height above earth, geographic longitude and latitude, time since launch (to include solar cycle variations), and the orientation of the satellite relative to the propagation direction (azimuth, declination), to account for asymmetries in the incoming particle flux. To take care of the discontinuity of the azimuth angle at 360°/0°, the sine and the cosine of this angle are used as input parameters. During the training of the network using the backpropagation algorithm, the weights w_{ij} and the biases b_j are optimized, minimizing the deviation between model and data. A similar approach was made for the Neural Network simulation of the background count rate of COMPTEL (Varendorff et al., 1996).

Measurement of the cascade lines:
At the decay of ^{22}Na a 1.275 MeV photon and β^+-particle are emitted simultaneously. The annihilation of the positron (after its deceleration) produces

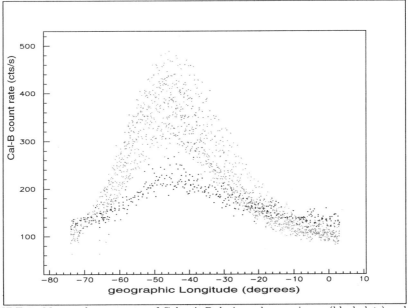

FIGURE 2. Measured count rates of Cal-unit B during solar maximum (black dots) and solar minimum (gray dots).

two 511 keV photons moving in opposite directions. One of the 511 keV photons may hit the upper detector (D1), while the 1.275 MeV photon (from the same original decay) hits the lower detector (D2). The complementary process (1.275 MeV photon in D1, 511 keV photon in D2) is not detected due to the threshold of the D2 detector (650 keV). The measured apparent time of flight of such an event is only slightly smaller than the time of flight of a photon which is Compton scattered in D1 and absorbed in D2. After the decay of ^{24}Na a photon at 1.369 MeV and a photon at 2.754 MeV are emitted simultaneously. A detailed description of the fitting of the cascade lines (^{22}Na and ^{24}Na) using spectra of the D2 detector can be found in Oberlack et al., 1997.

RESULTS

Model data comparison:

The number of nuclei generated as a function of time is then assumed to be proportional to the value produced by the activation model. This is then folded together with the corresponding half life times of the isotopes. A comparison of the measurement of the activation of the COMPTEL instrument and the model is shown in Figure 3 (each dot represents one observation period). The scaling factor from the measured Cal-unit count rates to the number of ^{22}Na- and ^{24}Na-nuclei as well as the initial number of those nuclei are taken from a fit of the model to the measured data. The scatter in ^{22}Na is caused by the statistical and systematic error of the fit to the data. The activition in ^{24}Na with a half life time of only 15 hours shows much more scatter than the activation in ^{22}Na. This is caused by the different orientations of the satellite relative to the anisotropic incident particle flux. The strong increase of the cascade lines after reboost around TJD 9300 is clearly visible in both plots.

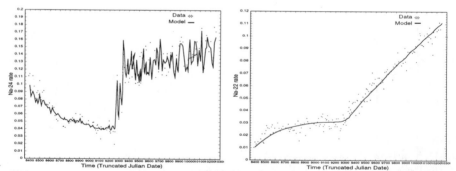

FIGURE 3. Comparison of the measured ^{24}Na- (left) and ^{22}Na-activity (right) with the predicted values from the Neural Network Model. The difference between the two plots is determined by the drastic difference in isotope lifetimes (15 h for ^{24}Na, 2.6 years for ^{22}Na).

The recent reboost to a hieght of about 517 km will increase the combined background from ^{22}Na and ^{24}Na by a factor of 2 to 3 in the next 1 to 2 years.

Estimation of the average incident proton flux:

The radioactive isotope ^{22}Na is produced by the interaction of energetic protons with Al-atoms in the structure of the satellite. The measured time of flight distribution of the events from the ^{22}Na decay shows, that mainly events from the structure near the upper detector (D1) contribute to the signal. The Al-mass around D1 is approximately 125 kg with a thickness in the range of 1 cm to 2 cm. The average daily proton fluence is then estimated by comparing the daily increase in the number of proton produced nuclei with an estimate of the efficiency of the production of these nuclei by protons. To simplify the calculation, all the Al mass was put in a plate of 2 cm thickness and irradiated by protons following the spectral shape of a model calculation of the orbital proton spectrum (Stassinopoulos, 1981). At the beginning of the mission during solar maximum at an orbital altitude of 440 km a daily production of $1.7 \cdot 10^7$ nuclei of ^{22}Na was derived from measurements of the ^{22}Na count rate. A daily proton fluence ($E > 100$ MeV) of $2.3 \cdot 10^5$ protons/cm^2 can be inferred from those numbers. For an altitude of 462 km a daily fluence of $5 \cdot 10^5$ protons/cm^2 was predicted by Dyer et al. (1994).

Acknowledgements

We would like to thank the MITRE Corporation for providing the Aspirin/MIGRAINES software package. The COMPTEL project is supported by the German government through DARA grant 50 QV 90968, by NASA under contract NASS-26645 and by the Netherlands Organisation for Scientific Research (NWO).

REFERENCES

1. Schönfelder, V., Aarts, H., Bennett, K., et al., 1993, Astrophysical Journal Suppl. **86**, 657.
2. Snelling, M., et al., 1986, Nuclear Instruments & Methods, **A248**, 545-549
3. Oberlack, U., et al., "A New Background Model for COMPTEL 1.8 MeV Data", these Proceedings, Poster 20-5.
4. Varendorff, M., Forrest, D., McConnell, M., and Ryan, J., 1996, A&A Suppl., **120**, 699-702.
5. Stassinopoulos, E. G., "GRO charged particle radiation study", NASA Report X-601-81-38, Table 3.
6. Dyer, C. S., Truscott, P. R., Evans, H. E., Hammond, N., Comber, C., Battersby, S., 1994, IEEE Transactions on Nuclear Science, Vol. **3**, 438.

Statistical analysis of COMPTEL maximum likelihood-ratio distributions: evidence for a signal from previously undetected AGN

O.R. Williams[4], K. Bennett[4], R. Much[4], V. Schönfelder[1],
J.J. Blom[2], J. Ryan[3]

[1] *Max-Planck Institut für Extraterrestrische Physik, P.O. Box 1603, 85740 Garching, F.R.G.*
[2] *SRON-Utrecht, Sorbonnelaan 2, NL-3584 CA Utrecht, the Netherlands*
[3] *Space Science Center, Univ. of New Hampshire, Durham NH 03824, U.S.A.*
[4] *Astrophysics Division, ESTEC, P.O. Box 299, NL-2200 AG Noordwijk, the Netherlands*

Abstract.
The maximum likelihood-ratio method is frequently used in COMPTEL analysis to determine the significance of a point source at a given location. In this paper we do not consider whether the likelihood-ratio at a particular location indicates a detection, but rather whether distributions of likelihood-ratios derived from many locations depart from that expected for source free data. We have constructed distributions of likelihood-ratios by reading values from standard COMPTEL maximum-likelihood ratio maps at positions corresponding to the locations of different catagories of AGN. Distributions derived from the locations of Seyfert galaxies are indistinguishable, according to a Kolmogorov-Smirnov test, from those obtained from "random" locations, but differ slightly from those obtained from the locations of flat spectrum radio loud quasars, OVVs, and BL Lac objects. This difference is not due to known COMPTEL sources, since regions near these sources are excluded from the analysis. We suggest that it might arise from a number of sources with fluxes below the COMPTEL detection threshold.

INTRODUCTION

The maximum likelihood-ratio method can be applied to COMPTEL data to determine the significance of a point source at a given location [1]. In this paper we do not consider whether the likelihood-ratio at a particular location indicates a detection, but rather whether distributions of likelihood-ratios derived from many locations depart from that expected for source free data.

Maximum likelihood-ratio distributions were generated by taking positions either from the EGRET AGN search list or from the EGRET source catalogue [2] and reading values at those positions from the standard COMPTEL maximum-likelihood ratio maps. We have divided the EGRET AGN search list into the various sub-catagories shown in Table 1. Data from observation periods up to, and including 523, have been used.

To ensure our distributions are not dominated by known objects, locations within 15° of the Crab and of the galactic plane have been discarded, together with locations within 6° of other known MeV sources.

A Kolmogorov-Smirnov (KS) test can be used to compare two measured distributions. It returns the probability that the two distributions are drawn from the same underlying population i.e. a low probability indicates different distributions. In the absence of any sources, the distributions of likelihoods from any class of objects should be compatible so we can use the KS test to search for evidence of a signal in any distribution.

In order to use the KS test to search for distributions containing signal a source-free distribution is required. Such a distribution has been created by taking our total search list of AGN, changing the sign of the galactic latitude and looking up the likelihood at that "mirrored" point.

Generating a source-free distribution by randomly choosing locations on the sky was *not* successful: its use indicated a "signal" from all the catagories of AGN, including e.g. Seyfert galaxies which have not been detected with COMPTEL. This problem may arise as a consequence of the non-isotropic nature of the AGNs in the search list.

Some of the regions studied have (unphysical) "negative" fluxes. The likelihood values from such regions are indicated by multiplying the true value by -1. These "negative" likelihoods are then treated in the same way as positive ones because only the measured distribution is important and not whether an individual measurement represents an upper limit or a detection.

PRELIMINARY RESULTS

Table 1 shows the probabilities obtained by comparing the 3845 flux measurements from the "mirror" distribution with distributions drawn from various classes of sources. The number of measurements in each distribution is shown in square brackets. The first two columns gives results for the entire AGN search-list (i.e. for all catagories of AGN) and the EGRET source catalogue distributions. The strongest evidence (0.004) for a signal is in the 10.0-30.0 MeV band AGN search list distribution - although 8 trials have been performed (which decreases the significance to 0.05).

The AGN search list has been sub-divided into four different catagories of sources. Columns 4-7 in Table 1 show the probabilities obtained using distributions derived from these different AGN classes. We note that the possible signal in the 10-30 MeV band, seen in the undivided AGN distribution, is largely contributed by the radio-loud quasars.

TABLE 1. KS tests of likelihoods: Probability that the "mirror" distribution and distributions from different source catagories arise from the same underlying population.

Energy (MeV)	All AGN classes [4647][1]	EGRET sources [695]	SyG[2] [941]	RLQ[3] [3007]	XBL[7] [452]	RG[8] [222]	FSRQ[4]+ XBL+RBL[5] [3062]	SyG+RG +SSRQ[6] [1424]
0.75-1	0.679	0.668	0.089	0.729	0.744	0.265	0.394	0.308
1-3	0.491	0.541	0.443	0.708	0.072	0.066	0.861	0.351
3-10	0.657	0.038	0.638	0.328	0.480	0.175	0.018	0.543
10-30	0.004	0.297	0.188	0.005	0.121	0.665	0.003	0.142

[1] [#]: number in the sample
[2] SyG: Radio quiet quasars and Seyfert galaxies
[3] RLQ: Radio loud quasars and BL Lacs (i.e. FSRQ, RBL, and SSRQ)
[4] FSRQ: Flat spectrum radio loud quasars
[5] RBL: Radio selected BL Lacs
[6] SSRQ: Steep or intermediate spectrum radio loud quasars
[7] XBL: X-ray selected BL Lacs
[8] RG: Radio galaxies

Finally, column 8-9 shows the AGN search list divided into two catagories: the first consisting of objects from classes which have shown emission in the γ-ray region (flat spectrum radio quasars and x-ray or radio selected BL Lac objects) and the other of objects from classes which have not shown emission in the γ-ray region (Seyfert galaxies, steep spectrum radio quasars and radio galaxies). We see not only evidence for a signal in the 10.0-30.0 MeV band for the known γ-ray emitting classes, but also marginal evidence in the 3-10 MeV band. Care must be taken when interpreting these probabilities since we are performing extra trials on the data, however we do have an *a priori* knowledge which leads us to divide the sources into these catagories.

LIKELIHOOD DISTRIBUTIONS

The results section shows a possible "signal" in the 3-10 MeV and 10-30 MeV from classes of AGN known to emit γ-rays. Fig. 1 shows the observed and predicted likelihood distributions in the 10-30 MeV band for the radio loud quasar and "mirror" distributions. The observed values are in reasonable agreement with the predicted values, but there is no obvious sign of any signal, i.e. the observed values for the radio loud quasars do not exceed those for the "mirror" distributions.

Fig. 1 shows that the observed "signal" does not arise from a few sources with a relatively high likelihood, since such sources would produce a visible excess. It must therefore arise from a larger number of sources with low likelihoods. Fig. 2 shows the *difference* in all four energy bands between the integrated likelihood distributions of the combined blazar plus flat spectrum

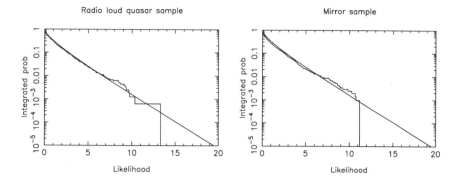

FIGURE 1. Observed and predicted distributions in the 10.0-30.0 MeV band for the radio loud quasar and "mirror" samples.

radio quasar sample and the "mirror" sample, after scaling to correct for the different number of observations. The numerical difference is shown in the plots on the left and the significance of that difference is shown in the plots on right. There is an excess of ∼100 events at likelihoods between 1 and 5 in the 10-30 MeV band and a smaller effect in the 3-10 MeV bands.

CONCLUSIONS

The analysis presented here demonstrates the potential of this method for studying likelihood distributions from classes of AGN with COMPTEL rather than clearly detected sources. The "mirror" method of generating a "source-free" distribution provides consistent results but is limited by statistics, however generating a "source-free" background by selecting random locations on the sky gives unsatisfactory results.

Preliminary results from the KS analysis shows some evidence for a "signal" in the 10-30 MeV band, and perhaps also the 3-10 MeV band, related to radio loud AGN. The "signal" does not arise from known COMPTEL sources. The data is consistent with ∼100 source observations having measured likelihoods of between 1 and 6, although the uncertainties are large. The number of individual sources implied by these results could be factor of 10 or more less than this, since all locations on the sky have been repeatedly viewed by COMPTEL.

REFERENCES

1. Thompson, D.J., et al., 1995, ApJS 101, 259.
2. de Boer, H., et al., 1992 in: *Data Analysis in Astronomy IV*, ed. M.A. Albrecht & D. Egret, Kluwer Academic Publishers, p.79.

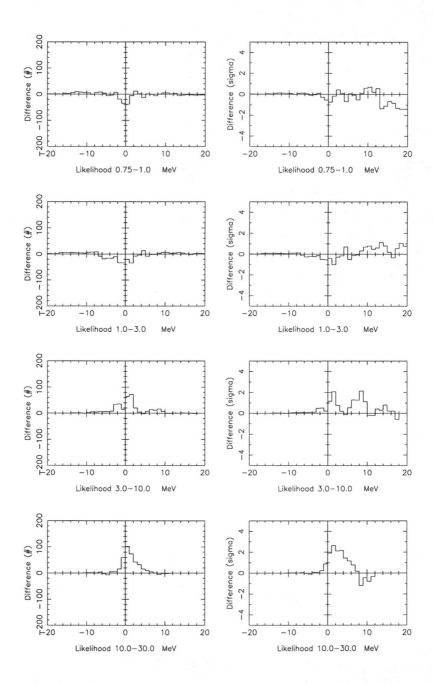

FIGURE 2. Differences between the integrated likelihood distributions of the FSRQ+XBL+RBL and "mirror" samples. The plots on the left show the numerical difference and those on the right show the difference in σ

Improved COMPTEL 10-30 MeV Event Selections for Point Sources from Inflight Data

W. Collmar[1], U. Wessolowski[1], V. Schönfelder[1], G. Weidenspointner[1], C. Kappadath[2], M. McConnell[2], K. Bennett[3]

[1] *MPI für Extraterrestrische Physik, Germany*
[2] *University of New Hampshire, USA*
[3] *Astrophysics Division, ESA/ESTEC, The Netherlands*

Abstract. After several years in orbit the COMPTEL experiment aboard the COMPTON Gamma-Ray Observatory has collected a substantial amount of data from the MeV sky. We have used the inflight event data collected from the Crab, which is the brightest point source at MeV energies, to optimize our event selections for point sources. For the COMPTEL 10-30 MeV range we have derived a set of improved parameter selections, which leads to a reduction of background events and - at the same time - increases the number of source events, resulting in an obvious improvement in the signal-to-background ratio for point sources. Due to a revised cut on the PSD parameter a background reduction of ~23% occurs. A narrowing and shifting of the TOF window results in a further slight (~2%) reduction of background, however, also in a slight increase (~5%) of source events. The revised event cuts improve significances of 'real' point sources in imaging analysis typically by ~1σ.

MOTIVATION

The COMPTEL experiment [1] aboard the Compton Gamma-Ray Observatory (CGRO) measures 0.75-30 MeV γ-rays. COMPTEL is the first Compton telescope exploring this difficult energy range from a satellite platform. The sensitivity of COMPTEL - like any other observing instrument - depends on the rate of background events, which is strongly influenced by event selections, i.e. cuts on several recorded event parameters like e.g. the energies on the upper and lower COMPTEL detectors, the time-of-flight (TOF) between the upper and lower detectors, and the pulse-shape discrimination (PSD) to discriminate between γ-rays and neutrons.

During the first year of the CGRO mission the *standard* COMPTEL event selections had been derived - neglecting any energy dependence - by using some early inflight data from the Crab, which is the brightest point source at MeV energies. These selections have been modified only slightly over the years. Now, after several years in orbit, much more inflight data on the Crab are available, providing better statistics to investigate the observed COMPTEL event distributions more deeply. This is especially important for the energy bands (e.g. above 10 MeV), for which the event rate is small and therefore only limited event statistics was available several years ago.

METHOD

The basic idea of our analysis is to use the Crab source signature in COMPTEL event space as signature for celestial (non-background) events and optimize the parameter cuts on this signature. For the presented analysis we have overlayed 42 days of Crab observations. To derive quantitative results we fitted a simple model to the so-called Angular Resolution Measure (ARM) spectra, which display the angular offset between an event circle and the Crab location (Fig. 1a). The model consists of a first gaussian representing the source (Crab) contribution, and a second gaussian on top of a constant to describe the background contribution. The ARM spectra are generated in a looping procedure for all kinds of combinations of event parameter cuts and are fitted with the model. For each fit we calculated the source and background events as well as the signal-to-background ratio and the source significance for several windows around the best-fit source position. To cross-check our results we applied two supplementory analysis methods: 1) background subtraction in event parameters, and 2) visualization of the selected event distributions. Consistency of the different methods was required for final results.

Finally, the modified event selections are checked by applying a complete three-dimensional maximum-likelihood and maximum-entropy data analysis, including standard background generation and the full COMPTEL response.

RESULTS

The analysis has started for all four standard COMPTEL energy ranges (0.75-1 MeV, 1-3 MeV, 3-10 MeV, 10-30 MeV), and the work is still ongoing. Positive results, however, have been derived so far only for the 10-30 MeV range. The event parameters of interest investigated are PSD, TOF, the earth horizon angle ζ, the Compton scatter angle $\overline{\varphi}$, and the so-called cutoff rigidity for cosmic rays. The rigidity dependence on source significances had not been investigated previously, and for PSD only a general restriction was applied to cut out neutron events. Table 1 gives the standard and optimized cuts on these event parameters for the 10-30 MeV range.

The two parameter sets of Table 1 differ mainly in two respects: 1) the narrowing of the PSD window results in a decrease of ~23% of background

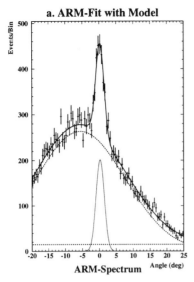
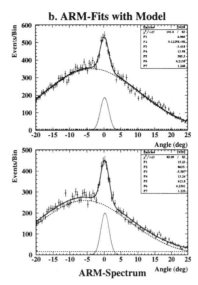

FIGURE 1. (a) Fit of an 10-30 MeV ARM spectrum with our model. The solid line shows the best fit of our model to the data. The dotted line represents the Crab (source) contribution, and the two dashed lines symbolize the individual (constant, gaussian) background components. The total background is the sum of the two dashed lines.
(b) Two selected ARM fits on the same data are shown. The ARM-spectra have been generated by applying the standard (upper panel) and revised (lower panel) selections. The number of events ('entries') decreases significantly while, in contrast, the amplitude for the Crab events (fit parameter 'P5') increases slightly.

events by cutting off the tail of the neutron event distribution but keeping the source events, and 2) the shifting and narrowing of the TOF window results in a further slight decrease ($\sim 2\%$) of background events, however, at the same time, in an increase of source events by $\sim 5\%$ (Fig. 1b). The change of ζ from $5°$ to $0°$ is of minor importance, and for the selections on the other two event parameters no improvement could be found. According to our ARM fits the new selections improve the signal-to-background ratio from 38% to 50%, and the Crab significance from 15.4σ to 18.3σ for a $\pm 2\sigma$-window around the best-fit position.

The optimized event cuts yield also improved results in COMPTEL imaging analysis, because the 3-dimensional dataspaces are filled with less background and more source events. Figure 2, as an example, shows two maximum-likelihood images of a three week observation of the Virgo region generated by applying the standard and the revised event cuts. The detection significance of the γ-ray blazar 3C 279, which showed a strong γ-ray flare ([2,3]) during this time period, increases from $\sim 3\sigma$ to $\sim 4\sigma$, and the new map appears to be

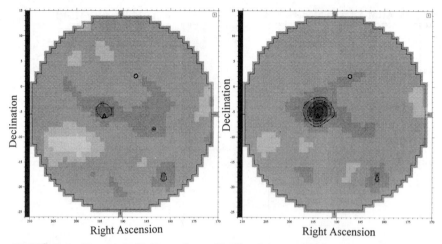

FIGURE 2. Two 10-30 MeV maximum-likelihood images of the Virgo region with an observation time of three weeks (CGRO VPs 511+511.5). The left one was generated with standard selections while the right one was generated with the optimized selections. Both images are displayed on the same greyscale code and contour lining. The contour lines start at a detection significance of 2.5σ with a step of 0.5σ assuming χ_1^2-statistics for known sources. The increase in significance ($\sim 3\sigma$ to $\sim 4\sigma$) of the quasar 3C 279 ($\triangle$) is evident.

less noisy. To check whether an improvement for all-sky maps is possible as well, we generated an all-sky maximum-entropy map for the sum of all data up to October '96 (Phase 1 to Cycle 5) using the revised event selections (Fig. 3). This map is considered to be an improvement compared to the standard maps (e.g. [4]), because it is generally 'cleaner' and point sources are visible more clearly, like the blazar PKS 1622-297 (l,b: $349°,13°$) for example.

SUMMARY

We have started to revisit the COMPTEL event selection criteria by taking advantage of the increased amount of inflight data. The improved event

TABLE 1. COMPTEL standard and optimized event selections (10-30 MeV) for the different parameters of interest

Selections	PSD channel #	TOF channel #	ζ [°]	$\overline{\varphi}$ [°]	Rigidity
standard	0 - 110	115 - 130	> 5	> 4	all
optimized	65 - 85	113 - 126	> 0	> 4	all

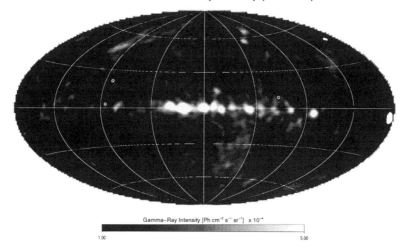

FIGURE 3. COMPTEL 10-30 MeV all-sky maximum-entropy map generated with the optimized event selections and centered at the Galactic Center. Diffuse emission from the galactic plane is obvious. In addition, there are several point sources clearly visible, like e.g. the Crab (l,b: $185°,-6°$) and Vela (l,b: $264°,-3°$) pulsars, and the blazars 3C 279 (l,b: $304°,57°$), 3C 273 (l,b: $290°,64°$), and PKS 1622-297 (l,b: $349°,13°$).

statistics offers the opportunity for a detailed study of the event parameter distributions as function of energy, and therefore for improved data selections. First positive results have been achieved for the highest COMPTEL energy range (10-30 MeV), although the optimal selections might not yet be reached. A reduction in background events of $\sim 25\%$ and - at the same time - an increase of source events by $\sim 5\%$ has been derived which 'converts' to an improvement of $\sim 1\sigma$ for 'real' point sources in standard maximum-likelihood imaging analysis. In addition, the maps generally look 'cleaner' because the amplitude of possible artifacts decreases. Investigations of other energy ranges yielded no significant improvements so far, however, work is still in progress. The overall goal is to increase the scientific return from COMPTEL by sensitivity improvements due to optimal event selections.

REFERENCES

1. Schönfelder V., Aarts H., Bennett K., et al., 1993 *ApJS* **86**, 657
2. Wehrle A., Pian E., Maraschi L., et al., 1997, *ApJ* submitted
3. Collmar W., Schönfelder V., Bloemen H., et al., 1997, these Proceedings
4. Strong A., Diehl R., Oberlack U., et al., 1997, Proc. 2nd Integral Workshop, St. Malo, 1996, ESA SP-382, 533

Earth Occultation Technique with EGRET Calorimeter Data above 1 MeV

Brenda L. Dingus[1], D. L. Bertsch[2], and E. J. Schneid[3]

[1] *University of Utah, Salt Lake City, UT 84112, U.S.A.*
[2] *NASA/Goddard Space Flight Center, Greenbelt, MD 20771, U.S.A.*
[3] *Northrop-Grumman, Bethpage, NY 11714, U.S.A.*

Abstract. The technique of earth occultation has produced many exciting results from the BATSE data. We examine the possibility of using this technique on the Total Absorption Shower Calorimeter (TASC) of EGRET. The TASC has an effective area of a few 1000 cm^2 and is 8 radiation lengths deep. Spectra from 1-200 MeV are collected every 33 sec and the rate at 4 energies is monitored every 2 sec. The detector is unshielded and uncollimated so the background is large. The statistical error on the background measurements require several days of exposure to detect the Crab at the lowest energies. Longer exposures would be needed due to systematic errors in determining the background. However, the wide field of view (the effective area is nearly 1000 cm^2 even through the back of the spacecraft) could be used to monitor variability and confirm fluxes of sources such as the black hole candidates, Cyg X-1 and GRO J0422+33.

EGRET'S TOTAL ABSORPTION SHOWER CALORIMETER (TASC)

The TASC is a 76 cm x 76 cm x 8 radiation lengths NaI(Tl) detector (Thompson, et al. 1992). The effective area for gamma-ray detection is reduced by the spacecraft material that surrounds it, so an EGS Monte Carlo is used with the spacecraft mass model to calculate the effective area vs energy and for different incidence directions. This effective area is over 1000 cm^2 for a large solid angle which even includes through the back of the spacecraft. For comparison, the effective area at 1 MeV of OSSE is 300 cm^2 and of Comptel is 40 cm^2.

However, the TASC is unshielded and uncollimated, so the background is very large. The background rates for different energy ranges are plotted in Figure 1 for a typical time interval of a few hours early in the mission.

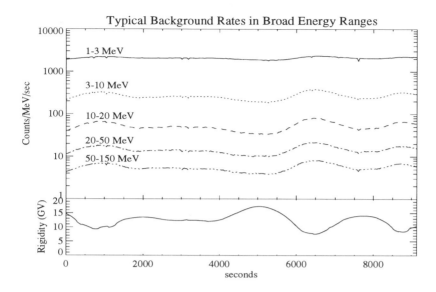

FIGURE 1. Background rates and rigidity vs. time for a typical, several hour interval early in the mission.

The rates vary with solar activity and with rigidity, especially at the highest energies. At lower energies the rates depend also on the time since the SAA due to activation. Spectra from 1-200 MeV are collected every 32.768 sec, and the rate above the 4 energies of approximately 1, 2.5, 7, and 20 MeV is monitored every 2.048 sec.

PREVIOUS TASC OBSERVATIONS

The TASC has proven very useful in measuring the spectrum of transient events. The simplest analysis is to measure the spectrum of gamma-ray bursts. The background is obtained from a linear fit to a few minutes before and after the burst. One of the weaker gamma-ray bursts detected by EGRET, 940301, is observed up to nearly 10 MeV in a single 32.768 sec interval, and the spectrum is measured to be $0.26 \pm 0.02 (E/MeV)^{-2.48\pm0.10}$ photons/(cm^2 sec MeV) (Schneid et al. 1995).

The TASC has also been used to search for delayed emission from gamma-ray bursts. The burst of 940217 had > 30 MeV emission lasting for nearly 2 hours as detected by the spark chambers. The TASC flux was measured during the last 34 minutes of the 2 hour interval as $\sim 10^{-3}$ photons/(cm^2 sec MeV) (Hurley et al. 1994). This analysis required using ± 15 and ± 30 orbits from the delayed time interval to calculate the background. These orbits correspond to almost the same orbital position and, therefore, similar rigidity. However, using background from such a long time from the observation does add errors due to the variability of the charged particle background. This is a very weak detection, but the flux measured is comparable to the flux from the Crab. However, a Comptel observation of this interval yielded upper limits which are near the values we measured.

The spectra and time evolution of MeV lines from solar flares have also been studied with the TASC. Again, the analysis required using ± 15 and ± 30 orbits from the flare to obtain the background due to the long duration of the solar flare. Two flares were detected in the TASC at 70 degrees off axis (Dingus et al. 1995) and the 4 June 1991 flare was detected through the back of the spacecraft (Schneid et al. 1996).

EARTH OCCULTATION WITH THE TASC

The sensitivity of the TASC will depend on the ability to subtract the background. We are looking into the two methods used on BATSE data. One method described by Harmon et al. (1993) at MSFC does a quadratic fit to the background using a fixed time interval before and after the occultation step. The duration of time interval used in the TASC analysis will have to be determined. We may be able to use the 4 energy levels of housekeeping rates to better estimate the variability. However, these rates are compressed for telemetry and at the highest rates(i.e. lowest energies) significant information is lost. Skelton et al. (1993) at JPL has another technique which models the background as a function of rigidity and other parameters.

A rough indication of the sensitivity can be calculated from the background rates and effective areas. Let S be the number of source gamma-rays detected and B be the number of background particles and photons. Then

$$\frac{S}{\sqrt{B}} = \frac{F \times A \times t}{\sqrt{N \times t}}$$

where F is the integral flux from the source over a specified energy range, A is the effective area, t is the time, and N is the number of background counts (see Figure 1). The time can be expressed as $t = 32.768$ sec $\times$ (0.3 $\times$ live time) $\times N_{int} \times (30 \times N_{days})$ since the TASC collects spectra for 32.768 sec with a typical live fraction of 30%. N_{int} is the number of 32.768 time intervals over which the background is estimated, and $30 \times N_{days}$ is the number of occultation steps. Using the Crab flux of $3 \times 10^{-3}(E/MeV)^{-2.3}$ photons/(cm^2 sec MeV) and $N_{int} = 5$ (similar to that used by BATSE), N_{days} can be calculated for a 3 sigma detection (i.e. $S/\sqrt{B} = 3$) as shown in Table 1. The energy ranges and Nint will be optimized by studying the data. Also, we may choose to exclude occultations with exceptionally high background, such as times near the South Atlantic Anomaly. The TASC analysis will be simplified by the considerably fewer sources above 1 MeV. It is not unusual for BATSE to have another source occultation occuring in the time interval when the background is being estimated. However, the TASC analysis will undoubtably be more difficult due to the large and variable background. The previous calculation does not consider any of these systematics.

WORK IN PROGRESS

We are first studying and optimizing the technique on the Crab. The Crab has been viewed for many weeks at a wide variety of angles off axis. Once the technique has been proven to work on the Crab, we will look at the black hole candidates detected by Comptel – Cyg X-1 (McConnell et al 1994) and GRO J0422+32 (van Dijk et al 1995). These two sources are chosen because they have fluxes comparable to the Crab at 1 MeV. Comparisons will be made between the flux measured by Comptel and of the earth occultation technique using EGRET. We can look at these sources and other sources at times when they are not within the field of view of Comptel. For example, Cyg X-1 has been directly opposite the field of view several times and the TASC has considerable effective area through the back of the spacecraft.

REFERENCES

Dingus, B.L. et al. 1994 AIP Conf. Proc. 294, 177
Harmon, B.A. et al. 1993 AIP Conf. Proc. 304, 314
Hurley, K. et al. 1994 Nature, 372, 652
Li, H., Kusunose, M., Liang, E.P., 1996 ApJ, 460, L29
McConnell, M. et al. 1994 ApJ, 424, 933
Schneid, E.J. et al. 1995 ApJ, 453, 95
Schneid, E.J.et al. 1996 A&A Supp., 120, 299
Skelton, R.T. et al. 1993 AIP Conf. Proc. 304, 1189
Thompson, D.J. et al. 1993, ApJ S, 86, 629
van Dijk, R. et al. 1995, A&A, 296, L33

TABLE 1. Fewest number of days to observe a 3 sigma detection of the Crab. Actual time for detection will be larger because only statistical errors in the background are used.

Energy Range	# of Days
1-3 MeV	1.4
3- 10 MeV	6.1
10-20 MeV	64
20-50 MeV	270
50-150 MeV	3840

Maximum-Entropy analysis of EGRET data

M. Pohl[1,2] for the EGRET collaboration, and A.W. Strong[1]

1 MPI für Extraterrestrische Physik, Postfach 1603, 85740 Garching, Germany;
2 Danish Space Research Institute, Juliane Maries Vej 30, 2100 København Ø, Denmark

Abstract. EGRET data are usually analysed on the basis of the Maximum-Likelihood method [1] in a search for point sources in excess to a model for the background radiation (e.g. [2]). This method depends strongly on the quality of the background model, and thus may have high systematic uncertainties in region of strong and uncertain background like the Galactic Center region.

Here we show images of such regions obtained by the quantified Maximum-Entropy method. We also discuss a possible further use of MEM in the analysis of problematic regions of the sky.

I   THE METHOD

Scientific data analysis means using our data to make inferences about various hypotheses. However, what the data give us is the likelihood of getting our specific data set as a function of the hypothesis. To reverse the conditioning Bayes' theorem can be used which invokes a prior probability distribution for the hypotheses as additional input. The prior probability distribution can be fairly simple, for example a scaling of a known behaviour in a system to its subsystem.

The consequence of this is that the "best" hypothesis can be derived by maximising the entropy (for an extensive introduction see [3])

$$S(h) = \sum_{i=1}^{L}(h_i - m_i - h_i \log(\frac{h_i}{m_i})) \qquad (1)$$

under the constraint of a likelihood argument. Here h_i is the hypothesis in pixel i and m_i is our prior expectation.

An implementation of this method is given by the MEMSYS5-package, which further allows us to derive a posterior probability bubble and thus the uncertainty of the "best" hypothesis.

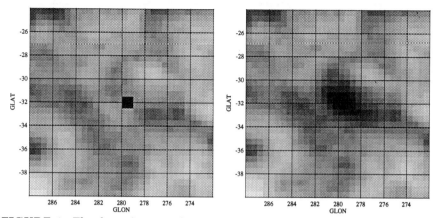

FIGURE 1. The sky region around LMC as seen by EGRET. The grey scale is linear in integrated intensity above 100 MeV. The image on the right side is the result of a Maximum-Entropy deconvolution with a model of the galactic diffuse emission as prior expectation, whereas on the left side we see the image resulting when a point source at the position of LMC is added to the prior expectation. Though the right image seems to indicate a substantial spatial extent of the LMC emission, the data seem to be compatible with LMC being 1° in size or less.

II THE INFLUENCE OF THE PRIOR EXPECTATION

From the definition it is clear that the prior expectation of the analyst has a strong influence on the result. This implies also that without information on what kind of prior was used the results of a MEM run have to be taken with care.

It is often argued that MEM is advantageous for the analysis of diffuse emission or extended structure. In Fig.1 we show the region around the Large Magellanic Cloud (LMC) twice. One image is derived without having LMC included into the prior expectation, and the second is the result of adding a point source to the prior expectation based the flux and best position of LMC [4]. Though the prior distribution is different only in an area of 1 square degree, the final image is influenced over roughly 20 square degrees. This implies that a) the appearance of LMC depends strongly on the prior expectation, and b) any source of unknown extent, which is not accounted for in the prior distribution, may result in a fuzz of a few degrees extent. So with MEM we do find sources, but it is hard to distinguish point sources from structure like that in the sky distribution of galactic diffuse emission.

A similar effect is seen in crowded regions like the Galactic Center, as shown in Fig.2. Here we have restricted ourselves to γ-ray energies above 300 MeV so that the smaller extent of the point-spread function can be used to advantage.

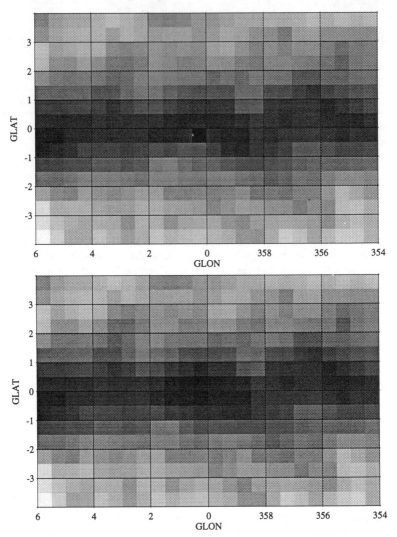

FIGURE 2. The galactic center region at energies above 300 MeV displayed in a linear grey scale. The upper image is derived with a model of the galactic diffuse emission and a galactic center point source in the prior distribution, whereas for the lower image only the model of diffuse emission has been used to construct the prior distribution. As in Fig.1 the appearance of the Galactic Center source depends strongly on our expectation.

The second EGRET catalogue [4] lists a highly significant source right at the position of the Galactic Center, but notes potential problems of the likelihood analysis in this region. In the MEM analysis we again see a strong influence of the prior model on the resulting image, however on smaller scale. This is due both to our restriction to higher energies and to the better statistic than in case of LMC.

When we now artificially increase the flux of the galactic center source in the data, but still use the catalogue flux in our prior distribution, then the additional intensity will be spatially distributed like in the case of no point source in the prior, i.e. the lower panel of Fig.2.

III DISCUSSION

We have shown that on small scales the MEM images tend to be influenced strongly by the prior expectation. Thus point sources can not be easily distinguished from structure in the diffuse emission. Also, in single viewing periods the prior distribution is likely to be dominant of the likelihood statistic and thus the final image. Thus MEM can not be used in a simple way to deduce the light curve of point sources in regions of high confusion level. However, what we can do is to use MEM to improve the background model for the likelihood analysis of known point sources in crowded regions of the γ-ray sky.

Despite our concerns about the limited reliability of individual structure in the MEM output, it is still true that MEM produces images which are highly reliable on larger scales. As an example we show an allsky image in Fig.3. Please note that the grey scale here is logarithmic and spans more than two orders of magnitude in integrated intensity above 100 MeV. The point sources stick out very clearly. None of the point sources is in the prior distribution, so that all of them have an extent of a few degrees in the image.

The EGRET Team gratefully acknowledges support from the following: Bundesministerium für Bildung, Wissenschaft, Forschung und Technologie (BMBF), Grant 50 QV 9095 (MPE); NASA Cooperative Agreement NCC 5-93 (HSC); NASA Cooperative Agreement NCC 5-95 (SU); and NASA Contract NAS5-96051 (NGC).

REFERENCES

1. Mattox J.R. et al. 1996, ApJ, 461, 396
2. Hunter S.D. et al. 1997, ApJ, 481, 205
3. Gull S.F., Skilling J.: 1984, IEEE Proceedings 131(F), 646
4. Thompson D.J. et al. 1995, ApJS, 101, 259

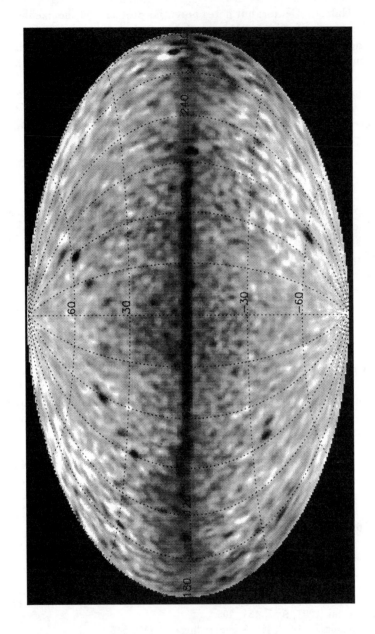

FIGURE 3. Maximum-Entropy image of the whole sky in Aitoff projection. The grey scale is logarithmic in integrated intensity above 100 MeV.

Non-parametric estimates of high energy gamma-ray source distributions

D. D. Dixon*, E. D. Kolaczyk**, J. Samimi[†], and M. A. Saunders[††]

*Institute of Geophysics and Planetary Physics
University of California
Riverside, CA 92521
**Department of Statistics
University of Chicago
Chicago, IL
[†]Sharif University
Tehran, Iran
[††]Systems Optimization Laboratory
Stanford University
Stanford, CA

Abstract. We discuss the application of non-parametric techniques to EGRET data, to estimate source intensity distributions. The methods used are model independent, or they use the model as the null hypothesis in a non-parametric testing framework. Two particular techniques are applied, both employing wavelets. TIPSH denoising operates directly on the data, recovering the statistically significant residual based on a null hypothesis. Basis Pursuit Inversion also accounts for the instrumental response, and estimates the "unblurred" source distribution. In addition, the Basis Pursuit Inversion framework allows us to estimate point and diffuse contributions separately.

MOTIVATION

The analysis of EGRET data is complicated both by the "blurring" due to the point spread function (PSF), and by the photon limited nature of the data, leading to Poisson fluctuations. To date, estimates of point and diffuse source distributions have been carried out largely via parametric Maximum Likelihood (ML) fitting [1], using model components for the Galactic and extra-Galactic diffuse γ-ray emission, as well as one or more point sources. Source detection is carried out by translating the point source component over the field-of-view (FOV) and searching for a position which yields a sufficiently large likelihood ratio.

The other side of the statistical estimation coin is *non-parametric*, where we have no particular model, but rather try to estimate directly the functional form of the underlying flux distribution. Nominally, this implies the solution of an intregral equation and thus is independent of any particular model. Practical applications generally involve some discretization, imposing a "model" of sorts, albeit a very large one. Nearly always, practical applications also require some sort of *regularization* to deal with the effects of noise. As such, there is always some tradeoff between the statistical bias and variance of the solution, for any particular choice of regularization; one of the main programs in the development of these techniques is to reduce the bias at a given variance level, but one can never eliminate it. Non-parametric methods should therefore be viewed not as alternatives to parametric methods, but rather as complementary.

For example, the parametric ML technique used for point source detection in EGRET data suffers some difficulties. An obvious problem occurs if the model is inaccurate, e.g., the diffuse model contains errors, or if there are multiple "unknown" sources in the FOV. Another is the *multiple tests problem* [2]. If the FOV does contain multiple sources, the detection and subtraction of the first (and presumably strongest) source alters the statistical distribution, since this estimate is not independent of the dataset. As a result, the detection of subsequent sources will in reality not be done at the stated levels of statistical significance [3]. A suitable non-parametric approach avoids these difficulties by estimating the locations and strengths of all the sources in the FOV. This result can then be used to build a model for parametric ML, which can generate error boxes on positions, flux estimates, significances, etc.

TIPSH DENOISING

"Denoising" refers to the process of estimating the noise-free data without removing instrumental effects. The TIPSH algorithm [5,6] accomplishes this using the Haar wavelet transform in a translationally invariant framework [4]. With the Haar wavelet, which is essentially a step function, we may calculate exact threshholds for Poisson distributed data. The translationally invariant framework then allows us to average out artifacts due to the non-smooth wavelet morphology.

The particular variant of TIPSH that we have applied to EGRET data is a recent extension of that described in [5,6]. This extension allows us to postulate some count distribution as a null hypothesis and recover the "significant" residuals. "Significance" here is controlled by the selection of a p-value cutoff. The p-value is the sum of the probability mass function above some cutoff implied by the Poisson distribution under the null hypothesis. Specification of the p-value then allows us to calculate the threshholds for the wavelet coeffcients. Below this threshhold, the coefficient is considered consistent with the

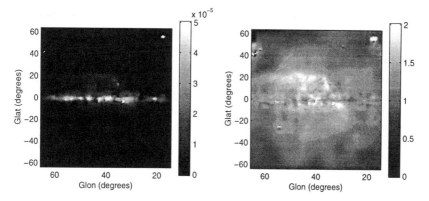

FIGURE 1. (a) Denoised residual, from EGRET data, $E > 1$ GeV, through VP 429.0 (units in photons cm^{-2} s^{-1} MeV^{-1} sr^{-1}). Though it is difficult to see in grayscale, there is a low-level extended halo apparent around the Galaxy. (b) Ratio of denoised data to null hypothesis (diffuse model). Note that the ratio is *larger* at higher latitudes, tending to be closer to 1 along the Galactic plane. Values have been truncated in both figures, to reduce the dynamic range and make relevant features clearer.

null-hypothesis, and discarded. Above the threshhold, the coefficent is deemed "significant" and kept in the final result. We chose a p-value threshhold (formally called the significance level) of 10^{-6} for this analysis, corresponding to a 2% false detection rate.

For the analysis presented here, we have used a null hypothesis consisting of the expected diffuse emission. The two components of this model are the Galactic diffuse model distributed with the LIKE software package (obtainable from the Compton Observatory Science Support Center), and the scaled exposure factor to represent the expected photon distribution from an isotropic component. The scaling factor was obtained by fitting the exposure factor to the data in otherwise empty portions of the sky. The data used are coadded EGRET observations through VP 429.0, for $E > 1$ GeV.

Results are shown in Figure 1. The denoised residual (data minus null hypothesis) is shown in Figure 1a. The maximum flux has been truncated, to reduce the dynamic range and aid in seeing features. Several point sources are clearly visible, as well as other excesses along the Galactic plane. Also apparent is a larger scale "halo". Comparison with simulated data generated from the null hypothesis indicates that much of the structure in Figure 1 is "significant", though with such non-parametric techniques, it is difficult to assess this quantitatively. The largest errors in the denoised simulations were 2×10^{-5}, and largely confined to the smallest scale wavelets. Figure 1b shows the ratio between the denoised data and the hypothesis, again truncated to reduce the dynamic range. It is interesting to note that apart from some known point sources [7] and localized deviations (possible point source candidates), the ratio along the plane is in the neighborhood of 1–1.5, while it is considerably larger at higher latitudes. In [8] it is noted that the Galactic diffuse model underpredicts the total Galactic flux. Using TIPSH, we can now estimate *where* the underprediction occurs, and find that it is not so much along the plane, but rather at higher latitudes (though of course, we here only refer to the *ratio*, and not the total flux differences on and off the plane). Some of these features are discussed elsewhere in this volume [9].

BASIS PURSUIT INVERSION

Here, we attack the more difficult problem of inverting the instrument response. Basis Pursuit Inversion (BPI) is motivated by the technique of Basis Pursuit [10], where one attempts to find a compact representation of a signal from an overcomplete dictionary of basis "atoms", using the vector 1-norm of the solution as the discriminant. The nominal Basis Pursuit formulation is easily extended to the inverse problem, where our dictionary now consists of basis atoms for the image convolved with the instrument response. Mathematically, we write the problem as

$$\min_a L(a, d) + c^T a, a \geq 0, Fa \geq 0, \qquad (1)$$

where a is a vector of coefficients representing the image, d is the data, $L(a, d)$ is some measure of likelihood or goodness-of-fit, and c is a constant vector. The last constraint enforces non-negativity on the image fluxes, with F transforming from the problem basis to the pixel basis. We have replaced the vector 1-norm of Basis Pursuit with the term $c^T a$, allowing greater flexibility and the potential for including prior information.

In analyzing EGRET data, we have used the knowledge that the source distribution consists of two components: point-like and diffuse. Since point-like sources are highly localized, our basis dictionary includes δ-functions (i.e., pixels) as an image basis. The diffuse component is expected to show some degree of spatial correlation, so we have also included wavelets (which compactly encode spatially correlated signals). We thus expect point source emission to be represented by δ-functions, and diffuse emission by wavelets, allowing the two components to be separated. A preliminary result is shown in Figure 2, where we used $L(a, d) = ||RFa - d||^2$, with R being the EGRET response matrix.

The current performance of BPI for EGRET data, while impressive, is still not entirely satisfactory. In our previous attempts, the basic formulation was to use some quadratic function (e.g., mean squared error or χ^2) for $L(a, d)$, and then modify c to attempt to include prior information, account for Poisson statistics, etc. Though this allowed us to stay in the computationally simpler linear regime, and treat the problem as a perturbed linear program [10], we

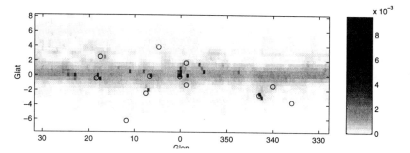

FIGURE 2. BPI reconstructed EGRET data from Phases 1–3, $E > 100$ MeV (units in photons cm^{-2} s^{-1} MeV^{-1} sr^{-1}). Reconstructed point sources appear as single bright pixels. 2nd Catalog sources with significance $> 5\sigma$ are shown. Note that there are some differences; in source confused regions, such as near the Galactic center, we might expect this, due to inaccuracies in the model used for EGRET source analysis. The absence of reconstructed sources off the Galactic plane is an artifact of the Poisson nature of the data.

found it difficult to achieve satisfactory suppression of artifacts and accurate source detection over the *entire* image. We now believe this difficulty arises from the Poisson nature of the data, and the inability of a quadratic function to capture the distributional characteristics at low count levels. We are therefore implementing a scheme based on the Poisson log-likelihood for $L(a,d)$, which we believe will significantly improve the source detection accuracy.

ACKNOWLEDGEMENTS

This work was supported by NASA Grant NAG5-3666. DDD thanks Dieter Hartmann for suggesting the possibility of a high energy Galactic halo.

REFERENCES

1. Mattox, J. R., *et al.*, ApJ, **461**, 396 (1996).
2. Worsley, K. J., Adv. in Appl. Probab., **27**, 943 (1995).
3. Mattox, J. R., AIP Conf. Proc. ed. James Matthews, **220** (1991).
4. Donoho, D. L. and Coifman, R. R., in *Wavelets and Statistics*, Antoniadis, A. and Oppenheim, G. (eds.), Springer-Verlag (1995).
5. Kolaczyk, E. D., Technical Report, Department of Statistics, The University of Chicago (1996).
6. Kolaczyk, E. D., ApJ, **483**, to appear (1997).
7. Thompson, D. J. *et al.*, ApJ Suppl., **101**, 259 (1995).
8. Hunter, S. D. *et al.*, ApJ, **481**, 205 (1997).
9. Hartmann, D. H. *et al.*, this proceedings (1997).
10. Chen, S. S., Donoho, D. L., and Saunders, M. A., SISC, to appear (1997).

Development of Gas Micro-Structure Detectors for Gamma-Ray Astronomy

S.D. Hunter[*], S.V. Belolipetskiy[†], D.L. Bertsch[*], J.R. Catelli[*],
H. Crawford[§], W.M. Daniels[*], P. Deines-Jones[*], J.A. Esposito[*],
H. Fenker[‡], B. Gossan[§], R.C. Hartman[*], J.B. Hutchins[†], J.F.
Krizmanic[*], V. Lindenstruth[§], M.D. Martin[†], J.W. Mitchell[*],
W.K. Pitts[†], J.H. Simrall[†], P. Sreekumar[*], R.E. Streitmatter[*],
D.J. Thompson[*], G. Visser[§], & K.M. Walsh[†]

[*] *NASA/Goddard Space Flight Center, Greenbelt, MD 20771*
[†] *University of Louisville, Louisville, KY 40292*
[‡] *Thomas Jefferson National Accelerator Facility, Newport News, VA 23606*
[§] *University of California-Berkeley, Berkeley, CA 94720*

Abstract. Large area gas micro-structure detectors are being developed for the next generation high-energy gamma-ray telescope as part of NASA's SR&T program to support new technologies. These low-cost detectors are produced by laser micromachining of metalized polyimide films layered on carbon fiber composite substrates. This integrated detector and support design reduces the detector complexity and associated assembly costs.

Accomplishments to date include testing of a 32 channel ASIC for the frontend electronics and integration of functional hardware into prototype detectors for tests of the FPGA readout system and event display software.

I PROJECT DESCRIPTION

Development of several new technologies for the next generation, high-energy gamma-ray telescope are being funded as part of the NASA SR&T program. Our collaboration was formed to develop large area, gas micro-structure detectors (GMSDs), which are proportional chambers produced using modern microfabrication techniques. The goal of this three year program is to fabricate a prototype telescope with an active area of approximately 50 cm×50 cm. Advances in laser micromachining make it possible to fabricate very large area detectors (up to 1 m×1 m), at reasonable cost, for this next generation telescope. These large area, low cost detectors enable the design of a gamma-ray track imager with many layers of very thin (≤0.01 radiation length) converter foils. Such an imager would consequently have very high angular resolution. For example, a track imager with 50 tracking layers, position resolution of 200 μm, and an active area similar to the GLAST baseline instrument would have at least twice the sensitivity of the baseline instrument at $E_\gamma \geq 300$ MeV due to a reduction in the width of the point-spread-function [1]. Below 300 MeV, the improvement in sensitivity could be even greater.

Large area GMSD detectors offer the additional advantages of reducing the mechanical complexity and the number of readout channels per layer, which translates into higher reliability and lower cost. The largest practical size for a GMSD tracking detector is determined by the gas gain and the capacitance per sense element. The calculated capacitance is ~ 40 pF/m, and we have achieved a gas gain of 6000. This confirms that a detector with 1 m sense elements will be sensitive to single primary ionization electrons when read out with our low-power, low-noise, front-end ASIC electronics. A 2 m $\times$ 2 m telescope could be constructed as a 2 $\times$ 2 array of 1 m $\times$ 1 m detectors, with all the electronics and interconnects on the periphery of the detector array. In addition, a 1 m $\times$ 1 m detector has slightly more active area (because the detector is one piece and has no dead area or support walls) than sixteen 25 cm $\times$ 25 cm detectors yet requires only 1/4 as many readout channels per layer. For a fixed instrument power budget, 4 times as many tracking layers can, therefor, be incorporated into the tracker design allowing thinner convertor foils and hence better angular resolution or alternatively, larger detector area with comparable resolution.

Laser micromachining will be used to fabricate the GMSD anode and cathodes onto separate layers of metalized polyimide film, which have been bonded to a multi-layer carbon-fiber composite sheet, to provide mechanical support and alignment. The lead foil convertor material is sandwiched into the carbon fiber lay-up, providing an integrated assembly of detector and convertor. The GMSD electric field configuration, Figure 1, is similar to the high-gain CAT, GEM and MICROMEGAS structures described in the recent literature [3–5]. The active volume of the GMSD is determined by a third electrode placed a few millimeters above the cathode plane. This "drift" electrode generates an electrostatic field in which the ionization electrons drift to the anode. This design is expected to have much higher gain and better stability than the more familiar microstrip gas chamber (MSGC) designs [6].

II CURRENT STATUS

We have constructed functional 10 cm $\times$ 10 cm MSGCs with 270 anodes on a 400 μm pitch. The detectors are operated with the cathodes at negative high voltage and the anodes at ground. The ASIC amplifier and readout electronics are mounted to polyimide circuit boards, and connected to the detector with

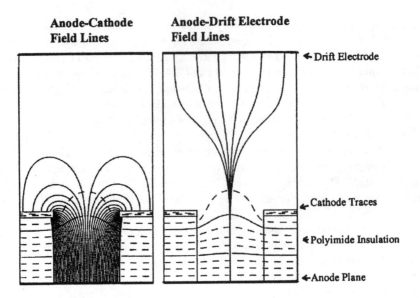

FIGURE 1. Schematic cross-sectional view of a GMSD detector showing the electric field lines from the cathode traces (left pannel) and from the drift electrode (right pannel) to the anode plane [2].

FIGURE 2. Assembly of 10 cm×10 cm MSGC detector planes in a stack.

wire bonds.

We have developed a low-power, 32 channel ASIC combining front-end charge amplification, shaping amp, discriminator, and sparse read-out for these detectors. Functional prototypes, with a power consumption of less than 200 μW per channel, have been produced through the MOSIS process. Each hit channel produces a logic high signal. These are OR'ed together inside each ASIC, producing a signal that indicates at least one channel in this ASIC was hit. The OR output from 8 ASICs is again OR'ed into what we call a major OR, indicating that at least one channel in the group of 8 ASICs (256 channels in total) was hit. Coincidences of the major OR signals from different planes are used to form a trigger for the instrument. When a trigger is issued the signals from each channel's discriminator are latched into a register in each ASIC. When a read signal is issued, this register is shifted to a second (readout) register and the ASIC is reset to accept another event. If there is another event while the first event is being read out, the signals are latched into a second register and the system goes dead until the first event read, whereupon the second register is shifted to the readout register and the system is ready for another event. Readout is accomplished in less than 50 μs. In the event of a series of hits that do not meet all the criteria for a triggered event, the ASICs are sent a clear signal to reset all discriminators. The clear process takes less than 1 μs to complete. In this manner the detector is able to self trigger at rates up to 20 kHz.

The readout itself is accomplished by cycling through the ASICs. The first ASIC with hit channels is read out, one channel at a time, until it lowers its hit flag. Readout skips to the next ASIC with hit channels and continues until all hit channels have been read. For the prototype instrument, the data stream consists of a series of 16 bit words that signify the address of each hit channel (14 bits or 16k channels max) and 2 bits indicating cluster size (1 to 4 channels per word). The readout and data formatting are controlled by an FPGA for each set of 8 ASICs. Data is simply pushed onto a FIFO from which it is taken by the DAQ CPU until the FIFO is emptied.

III FUTURE GOALS

Now that functional prototype components have been demonstrated, the next steps are to produce large area GMSDs for a prototype of a space flight instrument and to further test and refine the basic building blocks of the tracking system. We are now concentrating upon the development of the composite materials, laser micromachining techniques, and assembly steps required to produce a large integrated detector and support structure.

Like the EGRET spark chambers, GMSDs are gas detectors and therefore gas purity and instrument lifetime have been raised as concerns for the use of GMSDs on a future gamma-ray mission. However, unlike the EGRET spark chambers, GMSDs operate without producing sparks, and the associated generation of a large number of free radicals, in the gas. To eliminate gas contamination, we are fabricating the GMSDs from materials (e.g. polyimide) which are known to be clean, i.e. have low outgassing. Further, the GMSD gas mixture contains only noble gases, e.g. Ar and Xe, the purity of which can be maintained by continuous exposure to a passive getter. In contrast the EGRET gas contains ethane (C_2H_4). Thus, gas cleanliness and instrument lifetime is considered to be much less of an issue for the GMSD design than for EGRET. Long-duration operation will be verified in a test chamber, which is nearing completion.

Production of micro-machined GMSD's is scheduled to begin at the end of 1997. In keeping with the modular design of the instrument, the ASIC and FPGA chips for the 50 cm × 50 cm detectors will be bump bonded to circuit boards, which will then be mounted to traces on the detectors. This process, which offers significant cost savings compared to wire bonding, is being developed in collaboration with an industry partner. The ASIC design for these prototype detectors is nearly complete. Pending final testing, the large number required for the prototype tracker will be purchased. Detailed Monte Carlo simulations are in progress to optimize the detector pitch, converter thickness, and number of layers, within the total power and mass constraints of a space flight instrument, to provide the maximum angular resolution, sensitivity, and scientific return.

REFERENCES

1. Hunter, S.D., Hartman, R.C., Streitmatter, R.E. 1996, in Exploring the Astrophysics of Extremes: Workshop on the Next Generation High-Energy Gamma-Ray Telescope, Goddard Space Flight Center
2. Keith Solberg, Indiana University, Cyclotron Facility
3. Giomatris, Y., et al. 1996, Nucl. Instrum. Methods, A376, 29
4. Bouclier, R., et al. 1997, IEEE Nucl. Sci. Symposium, Anaheim, CA (also CERN-PPE/96-177 and CERN-PPE/97-32)
5. Bartol, F., et al. 1996, J. Phys. III (France), 6, 337
6. Peskov, V., Ramsey, B., Kolodziejczak, J., & Fonte, P. 1997, Nucl. Inst. Methods Phys. Res., in press

The Design of a 17 m ⊘ Air Cerenkov Telescope for VHE Gamma Ray Astronomy above 20 GeV

E. Lorenz for the MAGIC Telescope Design Group

Max-Planck-Institut für Physik
Föhringer Ring 6
D-80805 München, Germany

Abstract. The 20-300 GeV energy range is up to now inaccessible to gamma ray astronomy. Here we report on a design of a 17 m ⊘ air Cerenkov telescope, dubbed MAGIC telescope, which will have a threshold of 20 GeV (trigger threshold < 10 GeV), a high collection area of > 10^5 m² (100 GeV) and a high gamma/hadron separation power. The hardware investments are estimated to be around 3.5 M$ while it would take 2.5-3.5 years for the construction.

INTRODUCTION

Gamma ray (shortcut γ) astronomy is at present the only well established method to explore the relativistic universe. Due to technical limitations the energy range between 20 GeV, the upper limit of contemporary γ satellites, and 300 GeV, the lower limit of large air Cerenkov telescopes (ACT), is up to now inaccessible to research. Satellites are flux limited due to their small collection area of typically 0.1 m². On the other hand 20 GeV electromagnetic air showers contain still enough high momentum electrons above the Cerenkov threshold such that ACTs should be able to detect them provided the collected signal is large enough and can be discriminated against the night sky background light (NSB). Here we will report on a novel ACT design with the necessary increase in sensitivity.

THE PHYSICS GOALS

The main research targets of the new ACT will be:

a) the study of active galactic nuclei (AGN) up to $z \approx 2.8$. Measurements will also allow one to set stringent constraints on the existence and size of the (up to now unquantified) infrared (IR) background.

b) the systematic study of possible galactic γ emitters such as supernova remnants (SNR), plerions, X-ray binaries, unidentified EGRET sources etc. which will hopefully lead to the identification of the main sources of cosmic radiation up to about 10^{15} eV in our galaxy.

c) testing of gamma ray bursts (GRB) in the new energy window. EGRET recently reported a GRB associated with γs up to $\approx$ 20 GeV [1]. A telescope like MAGIC with a $> 10^5$ times larger collection area should be able to provide an answer to the 'galactic halo' vs. cosmological distance scenario.

d) search for exotics such as for the lightest SUSY particle etc.

e) overlap between satellite and ACT observations.

STEPS TO LOWER THE THRESHOLD

Using the presently best telescope, the Whipple installation [2], as a yardstick the following changes are considered:

a) Increase of the mirror area from 74 to 234 m^2 (= collector of $\approx$ 17 m $\oslash$).

b) Use of photosensors with a GaAsP photocathode with about 45 % quantum efficiency (QE) extended to 650 nm [3]. Low energy showers or showers at large zenith angles are stopped high up in the atmosphere and a sizeable fraction of the UV- and blue Cerenkov light is lost due to Rayleigh scattering. A gain of 3 (8-10 at large zenith angles) should be possible over conventional bialkali photocathodes.

c) Photosensors of $\approx$ 100% photoelectron collection and an excess noise factor $\approx$ 1.

d) Use of optimised light catchers, Winston cones, in front of the PMs for $\approx$ 100% light collection and shielding against off-angle background light.

e) Use of an isochronous (parabolic shaped) mirror in order to minimize the gating time for pulse height recording and 0.1° pixels. We expect 1.5-3 photoelectrons/pixel from the NSB for $\tau \approx$ 8 nsec.

f) Two additional goals are to operate the telescope in the presence of moon light and to have basically a full sky coverage while conserving a low threshold. In both cases a red extended sensor is of importance to conserve a low threshold.

THE TELESCOPE

The telescope is modelled after a 17 m solar collector [4]. The main mirror support dish consists of a three layer space frame made from carbon fiber-epoxy tubes for low weight and stiffness. A fundamental requirement is that the inertia of the telescope must be low in order to reposition it for GRB

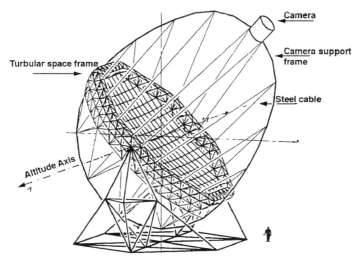

FIGURE 1. A model for the 17 m ⌀ MAGIC telescope

searches within 30-60 sec to any position on the sky. A finite element analysis of the frame shows that deformations can be held below 3.5 mm against the nominal curvature at any position for a combined frame- and mirror weight of less than 9 tons. Fig. 1 shows a computer generated image of the telescope. The telescope has a tessellated mirror with a basic element size of 50 × 50 cm. The elements are lightweight sandwich aluminium panels with internal heating for prevention of dew and ice deposits. A high quality reflecting surface with a surface roughness of < 10 nm is achieved by diamond turning. A preproduction series showed high optical quality with a focal spot size of typically 6 mm diameter. We plan to use a novel scheme for mirror adjustments and small corrections during telescope turning in order to counteract small residual deformations of the 17 m frame. Always four mirror elements will be preadjusted on a lightweight panel together with a switchable laser pointer. The panel can be tilted by two stepping motors under control of a videocamera that compares on demand the actual laser spot position on a screen with the nominal one. A prototype works successfully. The telescope will have a 3.6° ⌀ camera with a pixel size of 0.1° in the central region of 2° ⌀ and a coarser one of 0.2° in the outer part. As photon detector we intend to use novel hybrid PM from INTEVAC with a high QE (45%) red extended GaAsP photocathode combined with an avalanche diode as secondary amplification element [3]. At present the photosensor is considered as the critical path and it might be necessary to use a conventional PM camera from one of the HEGRA telescopes as a start-up replacement (threshold $\approx$ 35 GeV). The camera will be connected to the electronics ground station by 100 m optical fibers working in the analog mode for the transfer of the fast PM signals [5]. Signals will be

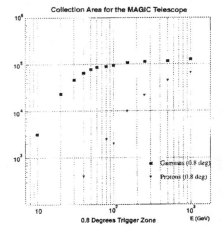 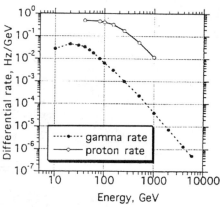

FIGURE 2. Collection area as a function of E

FIGURE 3. Differential counting rates as function of E

digitised by ≥ 250 MHz, 8-bit F-ADCs. This allows for noise minimisation, good timing measurements, buffering for the multilevel trigger system and eventually to combine more telescopes for quasi-stereo observations. All the novel components are either already tested or in use in other research fields. We estimate a hardware price of about 3.5 M$ and a construction time of 2.5-3.5 years.

SOME PERFORMANCE DATA

MC simulations show that the telescope has a trigger threshold (= maximum differential counting rate) of ≈ 10 GeV, i.e., a threshold for high quality data around 20 GeV and a rather large collection area in zenith position plateauing to $\approx 10^5$ m^2 (at ≈ 100 GeV) when using a trigger area of $1.6°$ ⌀ in the camera. Opening the trigger area to the full camera diameter increases the collection area to $> 3 \times 10^5$ m^2 for TeV signals. Fig. 2 shows the collection area as a function of energy E while fig. 3 shows the differential rate for a hypothetical gamma source with an integral flux of 10^{-11}/cm^2 sec at 1 TeV and a slope -1.7 together with the charged cosmic background. For large zenith angle observations the collection area will increase considerably but at the expense of a higher threshold. The quality factor using only image shape parameters rises from about 3 in the sub-100 GeV region to at least 8 above 1 TeV. Simulations using new γ/h separation algorithms are ongoing. In table 1 a coarse performance comparison with other projects is shown. Compared to satellite-borne γ detectors the collection area of MAGIC is at least a factor 10^5 times larger while the investment would be only in the range of 1-5%. Also the telescope could be built in much shorter time.

TABLE 1. Light flux sensitivity, thresholds and mirror area of MAGIC and some contemporary telescopes

Telescope	minimal photon flux [photons/m^2]	threshold	collector [m^2]	comments
AIROBICC [6]	4000	12 TeV	50 × 0.125	wide angle detector
HEGRA CT1 [7]	220	1.5 TeV	5	≥ 100 photoel./image
HEGRA CT3-6	150	700 GeV	8.4	≥ 100 photoel./image
CAT [8]	35 (?)	300 GeV (?)	18	≥ 30 photoel./image
WHIPPLE	35	300 GeV	74	≥ 300 photoel./image
WHIPPLE 98	16	100 GeV	74	≥ 100 photoel./image
VERITAS [9]	16/tel. ?	60 GeV (?)	9 × 74	
MAGIC	1.1	20 GeV	234	≥ 80 photoel./image

CONCLUSIONS

In summary it seems to be possible to construct a telescope with a low threshold of 20 GeV, high γ/h separation power, a large collection area, nearly full sky coverage and potential to measure in the presence of moon light. The telescope has the potential to discover and study a few hundred new γ sources. The telescope should be able to overlap with satellite observations and bridge the gap between the keV - 20 GeV energy range covered by satellites and the range of contemporary ACTs above 300 GeV.

ACKNOWLEDGEMENTS

Herewith I would like to thank my colleagues from the MAGIC design group for provision of the information and many fruitful discussions. Part of the development work for component studies has been supported by the German BMBF and the Spanish CICYT.

REFERENCES

1. K. Hurley et al.; Nature **372** (1994) 652
2. M.F. Cowley and T.C. Weekes; Exp. Astron. **6** (1996) 7
3. S. Bradbury et al.; Nuc. Inst. Meth. **A387** (1997) 45
4. Schlaich, Bergermann und Partner: Solar Power Plant with a Membrane Concave Mirror. Company report. Hohenzollernstr. 1; D-70178 Stuttgart.
5. A. Karle et al.; Nuc. Inst. Meth. **A387** (1997) 274
6. A. Karle et al.; Astopart. Phys. **3** (1995) 321
7. R. Mirzoyan et al.; Nuc. Inst. Meth. **A351** (1994) 513
8. B. Degrange et al.; Proc. Int. Workshop: Towards a Major Atmospheric Cerenkov Detector II (1993) 235 and 305
9. The VERITAS Proposal; T. Weekes et al. (1996)

Monte Carlo Simulations of the Timing Structure of Cherenkov Wavefronts of Sub-100 GeV Gamma Ray Air Showers

D.R. Peaper*, C.L. Gottbrath*, M.P. Kertzman* and G.H. Sembroski[†]

*DePauw University, Greencastle, Indiana 46135
[†]Purdue University, W.Lafayette, Indiana 47907

Abstract. We present a Monte-Carlo study of the phenomenology of Cherenkov light wavefronts from low energy gamma ray induced air showers. Experimentally the measurements of the spatially distributed arrival times of the wavefronts of the Cherenkov light of gamma ray air showers have been used to extract the directions of the showers. This has mainly been done for > 500 GeV showers using a conical fit to the timing structure of the wavefront. This directionality is then used to contribute to the rejection of background showers (mainly hadron induced showers) which arrive isotropically. Investigation of the arrival times of simulated Cherenkov photons from gamma ray induced air showers of energies 100 GeV and below reveals that there is greater variation in the morphology of the wavefronts than at the higher energies and that the fitting of simple conical functions to determine arrival directions may no longer be appropriate. We demonstrate that the detailed structure of the wavefront of these low energy gamma ray showers is primarily determined by the height distribution of the emitting cascade particles. Preliminary work suggests a correlation between the shape of the wavefront and the height of shower-maximum.

I INTRODUCTION

We first present an analytical investigation of the expected structure of the wavefront timing. This is followed by a study of the morphology of Monte-Carlo generated showers (A description of the Monte-Carlo air shower simulation methods used are given in [1]) which are then compared with the analytical results.

II ANALYTICAL ANALYSIS

Zenith Emission Tracks The amount of time it takes a visible photon moving in a direction that is at an angle Θ to the vertical, traveling from an emission altitude Z to the observatory altitude Z_{obs} is

$$T(Z, Z_{obs}) = \frac{1}{c \cdot cos(\Theta)} \left\{ (Z - Z_{obs}) + \frac{\eta_0}{\rho_0} \cdot \int_{Z_{obs}}^{Z} \rho(z) dz \right\} \quad (1)$$

where $\eta(z)$ is related to the atmospheric index of refraction $n(z) = (1 + \eta(z))$ and is related to the atmospheric density $\rho(z)$ by $\eta(z) = \eta_0 \cdot \left(\frac{\rho(z)}{\rho_0}\right)$ where η_0 and ρ_0 are the values at sea level. We are interested in the timing as a function of the distance x of the impact from the origin which is defined as the point where the original emitting particle will hit. We define $d \equiv (Z - Z_{obs})$ as the vertical distance the photon travels. We then have $cos(\Theta) = \frac{d}{\sqrt{d^2 + x^2}}$ and since are interested in the timing of the wavefront relative to the time of impact of the v=c emitting particle we subtract the time the emitting particle will take to reach altitude Z_{obs}. The last integral in Equation 1 is the amount of matter (gm/cm^2) a photon traverses going vertically from altitude Z to altitude Z_{obs} and we can define g(Z) to be the depth in gm/cm^2 at altitude Z. T becomes:

$$T(Z_{obs}, d) = \sqrt{d^2 + x^2} \cdot \left[\frac{1}{c} + \frac{\eta_0}{\rho_0} \cdot \left(\frac{g(Z_{obs}) - g(Z_{obs} + d)}{c \cdot d} \right) \right] - \frac{d}{c \cdot cos(\Theta_i)} \quad (2)$$

where we have made allowance for the possibility of the emitting particle moving in a non-vertical direction with a direction Θ_i. If we define

$$\alpha = \left[\frac{1}{c} + \frac{\eta_0}{\rho_0} \cdot \left(\frac{g(Z_{obs}) - g(Z_{obs} + d)}{c \cdot d} \right) \right] \text{ and } \beta = \frac{d}{c \cdot cos(\Theta_i)} \quad (3)$$

then Equation 2 can be rearranged as:

$$\frac{(T + \beta)^2}{d^2 \alpha^2} - \frac{x^2}{d^2} = 1 \quad (4)$$

This is the equation of a hyperbola. We show in Figure 1 the timing structure for various emission altitudes for a v=c vertical ($\Theta_i = 0$) emitting particle.

Inclined Emission Tracks Consider a v=c particle inclined from zenith by angle $\Theta_i \neq 0$. The timing structure is made up of point sources along this emission track which again give hyperbolas as specified in Equation 4. Now however the horizontal locations of the emission points are displaced by $x_{step} = d \cdot \sqrt{1 - cos(\Theta_i)^2}$ as the vertical altitude d of the emission points changes. In Equation 4 the quantity x is replaced by $x - x_{step}$. Further, we can define a timing correction, $T_{plane} = \frac{x}{c} \cdot \sqrt{1 - cos(\Theta_i)^2}$ which makes a plane

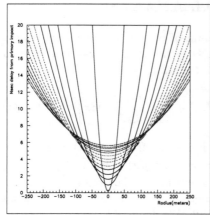
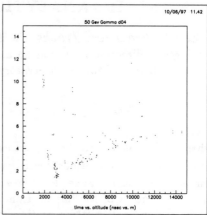

FIGURE 1. Theoretical arrival times of photons at an altitude of 1300m from point sources at various altitudes starting at 1500m and stepping by 1000m.

FIGURE 2. Corelation between the arrival time and altitude of emission of the Cherenkov photons from a 50 GeV Gamma Ray shower impacting in an area 15m x 13m at the core position.

wave moving at speed v=c down the emission track appear to 'simultaneously' impact the ground at all points x. Equation 4 becomes:

$$\frac{\left(T + \beta - \frac{x}{c} \cdot \sqrt{1 - cos(\Theta_i)^2}\right)^2}{d^2 \alpha^2} - \frac{x - x_{step}^2}{d^2} = 1 \tag{5}$$

The cross term $T \cdot x$ can be eliminated by a rotation of axis in the x-t plane. This results again in an equation for a hyperbola but with the symmetry axis slightly rotated in the x-t plane.

We now look at the shape of the wavefront at points transverse to the direction of inclination, ie. along the y axis. At a particular altitude d the emission point is offset by x_{step}. The timing curve is a hyperbola which is symmetric in rotation about its z axis. The transverse y axis going through the origin slices up through this offset form. The y axis timing is equivalent to the timing at points $y = \sqrt{x^2 - x_{step}^2}$. Substituting this y for x in Equation 4 gives:

$$\frac{(T + \beta)^2}{d^2 \alpha^2 \left(1 - \frac{x_{step}^2}{d^2}\right)} - \frac{x^2}{d^2 \left(1 - \frac{x_{step}^2}{d^2}\right)} = 1 \tag{6}$$

which is again a hyperbola.

Discussion For inclined emission tracks the timing structure is independent of inclination. All inclinations show evidence of timing focusing. For zenith emission tracks this focusing occurs in an annulus which has a inner radius of 80m and a width of about 80m. The timing width of the light pulse at the center of the annulus (radius=120 m) is less then .5 ns. Plots of Equations 5 and 6 for an inclined emission track indicate that the annulus now has the shape an ellipse. The long axis of the ellipse has an inner radius that goes like $80m/cos(\Theta_i)$. The transverse axis has grown to 100m. Across the annulus we can also see that the envelope of the timing has a conical behavior spatially.

A second observation is that the timing structure at the center of a shower is a good measurement of the altitude of emission distribution. From Equation 2 we can see that at x,y=0 ie at $\Theta = 0$, the timing structure for $\Theta_i = 0$, is:

$$T_{origin}(Z_{obs}, d) = \frac{\eta_0}{\rho_0} \cdot \frac{(g(Z_{obs}) - g(Z_{obs} + d))}{c} \qquad (7)$$

Thus Figure 1 and Equation 7 indicate that at the core of a shower the shape of the timing pulse reflects the structure of the altitude of emission distribution of the shower.

III SHOWER MORPHOLOGY

The timing structure of the Cherenkov light produced by real gamma ray generated air showers will differ from that of a single track of isotropically emitting point sources. In an atmospheric cascade, the number of Cherenkov emitting particles varies with altitude. The shower particles have a distribution of directions due to various processes including multiple scattering, bending in the Earth's magnetic field, and any change in direction due to the process which created the particle. Further, the emission along the particle track is directed according to the Cherenkov emission angle, and is definitely not isotropic. Finally, due to the statistical nature of the shower development the exact distribution of particles varies considerably, and the shower to shower fluctuations will be quite large. Even with these factors contributing to the complexity of real air showers, the timing and shape of the wavefront are well described by our simple model.

Equation 7 shows that the timing structure of the shower core reflects that distribution of the altitudes of emission. Figure 2 shows a scatter plot of arrival times for photons arriving within a 15m x 13m area centered at the shower core vs. the altitude of emission for those same photons. This figure incdicates that the timing distribution is correlated to the altitude of emission. Note that this is true only near the shower core. Due to the focusing effect mentioned earlier, we would not expect the envelope of the timing structure to be much affected at the timing annulus radius.

IV SUMMARY, CONCLUSIONS AND RECOMMENDATIONS

Pulse Shape at Center Measures Altitude of Emission Distribution The distribution of arrival times of the Cherenkov light at the center of a shower is highly correlated with the altitude of emission distribution of light in a shower. This distribution is also expected to be highly correlated with the particle density with altitude distribution in the shower.

Timing Focusing Occurs Independently of Shower-maximum. The concentration of the arrival time of light in a zenith shower occurs in an annulus about the shower center. This focusing creates a wavefront that is to first order conical in shape and is independent of both shower energy and the altitude of shower-maximum.

Inclined Showers Have Timing Properties Simular to Zenith Showers The Cherenkov light from inclined showers displays the same characteristics seen in zenith showers. The center pulse shape is a good measure of the altitude of emission distribution. Focusing of the timing of the light occurs in an annulus about the center of the shower. For inclined showers this annulus becomes elliptical in shape with its long axis being in the vertical plane of the inclination.

Measurement of the Edge of the Light Pulse at the Annulus A 0.5 ns or better arrival time measurement is needed for those detectors measuring the annulus of a shower. This gives the needed accuracy to fit a cone to the annulus to get a good estimate of the shower direction and the location of the shower center.

Measurement of the Timing Pulse Shape Pulse shape measurement at the center of a shower with at least 1 ns resolution and 10 ns duration is recommended. This allows for a good estimation in the altitude of shower-maximum for a particular shower which is critical for accurate shower energy determination. Monte-Carlo results indicate that a 1 ns resolution should be adequate to determine the altitude of shower-maximum to within $\pm$ 1 Km.

Fitting Strategy for Determination of the Shower Direction We suggest that for the best possible determination of the direction of a shower that the procedure of a fitting a cone to the timing information be only done to those detectors in the annulus. Including timing information from the center of the shower will only reduce the accuracy of the fit due to the dependency of the timing at the center of a shower to the distribution of emission altitudes.

REFERENCES

1. Kertzman, M.P. and Sembroski, G.H., *Nucl. Inst. Meth.* **A**, 343, (1994),pp. 629-643.

The University of Durham Mark 6 VHE Gamma Ray Telescope

P. M. Chadwick, M. R. Dickinson, N. A. Dipper, J. Holder,
T. R. Kendall, T. J. L. McComb, K. J. Orford, S. M. Rayner,
I. D. Roberts, S. E. Shaw and K. E. Turver

Department of Physics, University of Durham, South Road, Durham DH1 3LE, UK

Abstract. The operation of the University of Durham Mark 6 atmospheric Čerenkov telescope is discussed. The telescope has been used to detect gamma rays at energies ≥ 150 GeV and to achieve good discrimination between gamma ray and hadron initiated showers, using both conventional imaging and novel fluctuation measures. The telescope was commissioned in 1995 and a description of its operation is presented. Verification of the performance during observations of PSR B1706-44 is described.

INTRODUCTION

The Mark 6 telescope is located at Narrabri, N.S.W., Australia. It consists of three flux collectors, each of area 42 m², focal length 7.0 m and focal ratio $f/1.0$ placed on a single alt-azimuth mount. The left and right flux collectors each have a detector consisting of 19 close-packed hexagonal PMTs at their focus, the pixel size being 0.5°. The detector at the focus of the central mirror is an imaging camera consisting of 91 1" circular PMTs arranged in a hexagonal close-packed array and surrounded by 18 2" circular PMTs. The pixel size of the 1" PMTs is 0.25°. The telescope uses a novel 3-fold spatial plus 4-fold temporal coincidence system and is triggered by requiring a coincidence between a corresponding pair of PMTs in the left and right detectors viewing the same area of sky plus any adjacent 2 of the 7 PMTs in the camera also viewing that region of sky, all in a 10 ns interval. This unique triggering system allows reliable detection of gamma rays at low energies, and provides an additional γ-ray discriminant to the conventional imaging parameters.

The design and construction of the telescope has been described [1–3]. It has been in routine operation for more than 2 years; this paper describes aspects of the performance of the telescope.

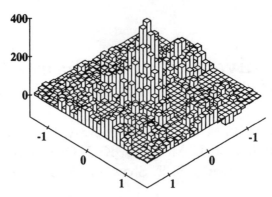

FIGURE 1. *The pointing accuracy of the Mark 6 telescope demonstrated with data taken during observations of PSR B1706-44. The vertical axis shows the number of excess events, and the horizontal axes show the offset from the position of PSR B1706-44 in degrees.*

OPERATION AND PERFORMANCE

The attitude of the telescope is measured in two ways. Shaft encoder positions are recorded for each event to 14-bit accuracy. This gives a measurement of the telescope's pointing to a relative accuracy of ± 0.022°. In addition, a coaxial optical CCD camera is mounted on the telescope. The output of this CCD camera is continuously monitored by microcomputer, which measures the position and brightness of a guide star within its 2° × 2° field. This information is integrated into the data stream on an event by event basis; guide stars of magnitude $m_v \leq 6$ can be employed, with absolute position sensing better than ± 0.03°.

Gamma ray candidate events from PSR B1706-44 were selected on the basis of shape alone. These events were then subjected to an analysis in which the source was been assumed to be at a matrix of positions in the field of view of the telescope's camera [4]. For each false source position, the values of the pointing angle *ALPHA* for the events have been calculated and the selection criterion *ALPHA* ≤ 25° applied. We show in Figure 1a plot of the significance of the source detection as a function of assumed source position. The relative accuracy of the Mark 6 telescope pointing is found to be ± 0.02°.

The Telescope Trigger

The Mark 6 telescope uses a unique 3-fold spatial plus 4-fold temporal coincidence system, which is an improvement over that previously reported [1]. An event trigger is initiated whenever a signal is detected from a pair of PMTs at similar positions in the left/right detectors. The central flux collector, viewed by the 'camera', completes the event trigger when any 2 of the group of 7 camera PMTs which cover the same area of sky as the left/right PMTs which have responded also produces a signal, provided that these 2 camera PMTs are adjacent.

The event trigger is achieved using a hardware first-level trigger with a decision time of $\sim$ 10 ns. The most sensitive element of the system at present is the central flux collector/camera combination; work on increasing the sensitivity of the left/right detectors continues. This system allows reliable, but low efficiency, detection of light ($\sim$ 40 photons m^{-2}) corresponding to gamma rays with energy as low as 150 GeV.

Stability of Performance

The Mark 6 telescope is equipped with a comprehensive telescope performance monitoring system. The PMTs' anode currents and noise rates are recorded throughout observations, their relative gains are measured using the laser calibration system and the digitizer pedestals are measured using 'null' events. In addition, environmental conditions such as wind speed, atmospheric pressure, air temperature are monitored.

PMT performance is affected by temperature, and to ensure stability throughout observations all PMT detector packages are equipped with servo-controlled heating systems. During a typical 4 hour observation, the digitizer pedestals and relative gains of the PMTs are measured to vary by < 3%.

The radiator in a ground based gamma ray detector is the atmosphere; few attempts have been made to measure the clarity of the radiating atmosphere, other than to carefully monitor the count rate of the telescope. Ideally, a measure of atmospheric clarity independent of the telescope is required. In addition to the brightness of the guide star (see above), we measure the far infra-red (FIR) emission (8 − 14 μm) for a 2° × 2° region of the sky covering the field of view of the gamma ray telescope. The presence of cloud or particulate matter (typical of an inversion layer) in the radiating region leads to an increase in the measured radiation temperature. Figure 2 shows strong the correlation between 8−14 μm radiation temperature and telescope count rate. The advantages of this system are its immediate response to changes in atmospheric conditions and its independence. Small changes in the atmospheric conditions have been detected using FIR for which no change in telescope performance was noted.

Energy Calibration

The estimation of the energy threshold for the detection of gamma rays by an atmospheric Čerenkov telescope is a difficult task. The following is a

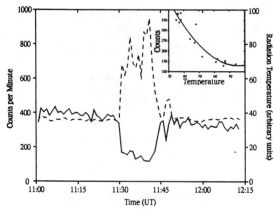

FIGURE 2. *FIR radiation temperature (dotted line) and count rate (solid line) during a night's observation with the Mark 6 telescope. The inset shows the measured relationship between radiation temperature and count rate*

preliminary approach to the problem of estimating the energy threshold of the Mark 6 telescope; a more rigorous investigation is in progress.

Simulations of the photon yield from 40000 cosmic rays were generated from a circular field of view 2° in radius and out to a maximum impact parameter of 250 m have been completed. The simulations were performed for a telescope inclined at 20° to the zenith and presented to a model of the response of the telescope. This telescope model was optimised to account for the observed cosmic ray background and was then used with simulations of gamma ray air showers to give the expected threshold energy. 50000 gamma ray showers were generated with energies ranging from 100 to 10^5 GeV for a source with a power law spectrum of index -2.4 (the measured value of the differential spectral index of the Crab nebula in the VHE range [5]).

The effective collecting area for an atmospheric Čerenkov telescope depends on the gamma ray energy, the Čerenkov light pool size and the triggering probability. Figure 3a shows the effective area for the Mark 6 telescope as a function of energy for gamma rays from a point source. The threshold energy of an atmospheric Čerenkov telescope may be defined as the energy at which the differential gamma ray flux is a maximum. This is illustrated in Figure 3b, which predicts a threshold for gamma rays of ~ 300 GeV. The systematic error in the determination of the energy threshold is conservatively estimated as $\pm 50\%$. It should also be noted that the values derived here are the threshold energy and mean effective area assuming 100% gamma ray retention during event selection.

The performance of the Mark 6 telescope has been verified using data from an observation of PSR B1706-44 [4]. These have shown that, for events corre-

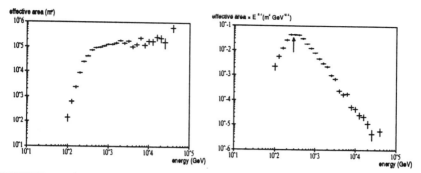

FIGURE 3. a). The effective collecting area for gamma ray detection calculated for the Mark 6 telescope. b). The simulated differential detection rate for gamma rays as a function of energy threshold for the Mark 6 telescope.

sponding to γ-ray energy ≥ 300 GeV, conventional imaging techniques enhance the γ-ray signal and provide a 7σ detection from ~ 10 hours of data, after rejection of $> 99\%$ of background events.

CONCLUSIONS AND FUTURE DEVELOPMENTS

The University of Durham Mark 6 telescope was deployed on the Narrabri site in July 1994 and has collected data since March 1995. In its present configuration, the telescope has a gamma ray imaging energy threshold of ~ 300 GeV, and is capable of detecting gamma rays of energy 150 GeV. Further modest reductions in the energy threshold will be possible.

ACKNOWLEDGEMENTS

This work was funded by the U.K. Particle Physics and Astronomy Research Council. We are grateful to the University of Sydney for the lease of the site at Narrabri. The assistance of the staff of the Physics Department, University of Durham, in the construction of the telescope is gratefully acknowledged.

REFERENCES

1. Chadwick, P. M. et al., *Proc. 24th Int. Cosmic Ray Conf.* **3** 416 (1995).
2. Chadwick, P. M. et al., *Towards a Major Atmospheric Cerenkov Detector IV*, ed. M Cresti, 301 (1995).
3. Chadwick, P. M. et al., *Astron. Astrophys. Suppl. Ser.* **120** 657 (1996).
4. Chadwick, P. M. et al., this conference.
5. Vacanti, G. et al., *Astrophys. J.* **377** 467 (1991).

SOLAR TOWER ATMOSPHERIC CHERENKOV EFFECT EXPERIMENT (STACEE) FOR GROUND BASED GAMMA RAY ASTRONOMY

D. Bhattacharya[1], M.C. Chantell[2], P. Coppi[3], C.E. Covault[2],
M. Dragovan[2], D.T. Gregorich[4], D.S. Hanna[5], R. Mukherjee[5],
R.A. Ong[2], S. Oser[2], K. Ragan[5], O.T. Tümer[1], D.A. Williams[6]

[1] *Institute of Geophysics and Planetary Physics, University of California, Riverside, CA 92521*
[2] *Enrico Fermi Institute, University of Chicago, Chicago, IL 60637, USA*
[3] *Department of Astronomy, Yale University, New Haven, CT 06520, USA*
[4] *Department of Physics and Astronomy, California State University, LA, Los Angeles, CA 90032, USA*
[5] *Department of Physics, McGill University, Montreal, Quebec H3A 2T8, Canada*
[6] *Santa Cruz Institute for Particle Physics, University of California, Santa Cruz, CA 95064, USA*

Abstract. The STACEE experiment is being developed to study very high energy astrophysical gamma rays between 50 and 500 GeV. During the last few years this previously unexplored region has received much attention due to the detection of sources up to about 10 GeV by the EGRET instrument on board the CGRO. However, the paucity of detected sources at ~ 1 TeV indicates that fundamental processes working within these sources and/or in the intergalactic space are responsible for the cutoff in the photon spectra of the EGRET sources. The cutoff or the spectral change of these sources can be observed with ground-based Cherenkov detectors with a very low threshold. The use of large arrays of mirrors at solar power facilities is a promising way of lowering the threshold. Using this concept a series of tests were conducted at the National Solar Thermal Test Facility (NSTTF) at Sandia National Laboratories (Albuquerque, NM) with a full size prototype of the STACEE telescope system. The tests show that STACEE will be capable of meaningful exploration of the gamma-ray sky between 50 and 500 GeV with good sensitivity.

Introduction

The cosmic sources detected by the EGRET instrument above 30 MeV [1] can be broadly categorized under AGNs, galactic pulsars and unidentified sources. The spectra of many of these sources extend up to 10 GeV–above which the effective area of the instrument becomes too small to achieve sensitive flux detections. The bolometric power of the EGRET AGNs is often dominated by gamma-ray luminosity; a few of them have shown variability on the timescale of a day during their flaring state [2]. Most of the blazar gamma-ray models involve some kind of beaming, with one class of models considering accelerated electrons producing gamma-ray photons through the inverse Compton process and the other class considering protons generating gamma-rays through nuclei interaction and photomeson production. However, many of the EGRET AGNs remain undetected in the TeV region, even though their extrapolated fluxes are well above the sensitivities of imaging Cherenkov telescopes. The Whipple Observatory has obtained only upper limits for 3C 279, one of the brightest EGRET AGNs (see Figure 1). However, another AGN, Mrk 421, is detected at very high energy and also at EGRET's energy range. This is despite the fact that the flux of Mrk 421 measured by EGRET is much smaller than that from 3C 279. The measurement of an intrinsic break of the spectrum in the unexplored region could constrain theoretical models of gamma-ray generation within AGNs.

It is possible that because photons from 3C 279 (redshift = 0.54) have to travel a much greater distance than Mrk 421 (redshift = 0.031) they are strongly attenuated by pair production against the intergalactic infrared background. The intergalactic infrared radiation field is sensitive to rates of early star formation because much of the energy in starbursts escapes in the infrared [3]. Hence, measuring the spectra from 10 to 500 GeV for several AGN at different redshifts has the potential not only to determine whether infrared absorption is taking place, but also to quantify the infrared background and constrain cosmological models.

In the case of the Crab pulsar, STACEE can achieve two objectives: it can verify whether a hardening of the unpulsed spectrum occurs above a few GeV as the Compton component becomes more important than synchrotron radiation [4]. The second objective involves the detection of the spectral turnover of the pulsed emission; pulsed emission has not been detected at TeV energies. In Figure 1, we show the projected sensitivity of STACEE. The Crab, a strong and steady source, should produce an 8σ excess in one hour. Extrapolations for the EGRET spectra for ten AGNs indicate that they should be detected ($> 5\sigma$) by STACEE in 50 hours or less of observation. Similarly, STACEE can expect to see nine of the EGRET unidentified sources and two supernova remnants seen by EGRET.

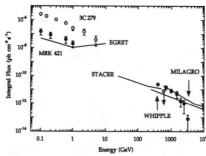

FIGURE 1. *Projected sensitivity of STACEE and other contemporary experiments. The STACEE and Whipple curves correspond to the 5σ sensitivity for 50 hrs of observations, the typical time devoted to a single source in one observing year. Milagro curve is for 5σ sensitivity obtained within a year. The spectra of 3C 279 and Mrk 421 as obtained by EGRET and Whipple are also shown.*

STACEE Concept and Prototype Performance

The achievable energy threshold of an atmospheric Cherenkov telescope, E_{th}, is proportional to the night sky photon background Φ_{bkg}, field-of-view of the telescope Ω, trigger gate width τ, and inversely proportional to the mirror collection area A and efficiency ϵ for converting a photon striking a heliostat into a photoelectron in its PMT [5]: $E_{th} \propto \sqrt{\frac{\Phi_{bkg}\Omega\tau}{\epsilon A}}$. Using a large number of solar heliostat mirrors a big detection area can be achieved to obtain low energy thresholds [6]. Cherenkov light from the showers in the atmosphere is reflected by the heliostats to a secondary mirror on the central receiving tower. The secondary focuses light onto a photomultiplier camera, with light from each heliostat directed to a separate photomultiplier tube. To verify the feasibility of the concept we have built and tested a fully functional STACEE prototype. The prototype included a secondary telescope, PMT camera, and data acquisition electronics for 8 channels (and subsequently for 16 channels). Tests were conducted at the NSTTF site at Sandia during August, 1996, October 1996, May 1997, and June 1997. For the 1997 tests sixteen heliostats were used. Initial development of the project and much of the test results are described in [7], [8], [9] and [10]. This paper gives results from only 1996 tests.
Site suitability and prototype description : For the August and October 1996 tests, we used eight heliostats in coincidence. Using UBV photometry on several stars, optical transmission of the atmosphere as a function of the zenith angle was measured [8]. The measurements agree with an ideal model atmosphere to within 3%. The Cherenkov signal must be detected against a photon background comprising air glow, starlight and scattered light from artificial sources. The background photon flux is calculated by measuring the currents in the PMTs. The current from each PMT due to background is

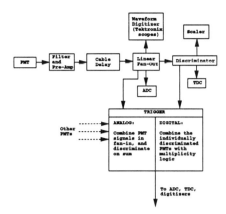

FIGURE 2. *Schematic of the electronics setup used for the prototye tests (one channel of eight).*

given by $I = \Phi_{bkg} e G \Omega \epsilon A$, where I is the PMT current in Amperes, e is the charge of the electron in Coulombs and G is the PMT gain. We measured the ambient flux to be $4.3 \pm 0.9 \times 10^{12}$ ph m^2 s^{-1} sr^{-1}, about twice the flux at a typical dark mountain site. This background level is adequate for astronomical observations. The heliostat pointing stability, focusing properties and reflectivity are also measured and found to be suitable for the experiment [8].

Figure 2 shows a schematic for the data acquisition system. We measured the trigger rate as a function of discriminator threshold. A digital trigger condition of any four out of eight PMTs was used. Figure 3 shows the digitized pulse height spectrum for 35 minutes of coincidence data for cosmic ray air showers arriving from the zenith. A fit to the falling edge of the spectrum yields a power law index of -2.7 $\pm$ 0.1 (statistical error only), which agrees well with the known differential index of the cosmic ray spectrum in this energy range. This demonstrates that we are in fact triggering on Cherenkov light from cosmic ray showers. We have also tested the efficiency of the system for different canting angles of the heliostats with respect to the assumed location of the shower maximum of air showers in the atmosphere.

Energy Threshold : We observed cosmic ray air shower events arriving from the zenith at a rate of 5.1 Hz using the digital trigger configuration of the STACEE prototype. The background accidental rate was 0.2 Hz. The observed rate equals the product of the cosmic ray shower flux above the threshold, effective collection area and acceptance solid angle of the instrument. The effective area and FOV were calculated from Monte Carlo simulations, and the energy threshold was estimated from the cosmic ray trigger rate.

An alternative method was also employed to obtain the threshold. The measured photoelectron yields were compared to the expected photon densities on the ground as a function of primary energy. From Figure 3 we define a minimum pulse height needed to trigger a PMT. This can be related to an

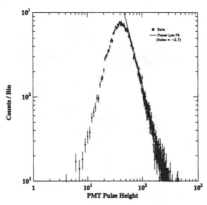

FIGURE 3. *Pulse height spectrum obtained from observations of air showers at the zenith.*

equivalent number of photoelectrons and the corresponding number of photons striking the heliostat using the average PMT quantum efficiency and the collection efficiency of the optics. Dividing by the projected area of a heliostat we find the photon density on the ground necessary to trigger a channel. From Monte Carlo calculations, we can estimate the lowest energy γ ray that will produce a high enough photon density to trigger the experiment. Both techniques yield the same result to within statistical uncertainties (for a detailed description of this calculation see [8]). The energy threshold for cosmic ray air showers is approximately 300 GeV. This yields an effective threshold for the gamma-ray showers of around 75 GeV.

These tests have demonstrated that the STACEE concept is suitable for gamma-ray astronomy. We have already performed tests with 16 heliostats in coincidence; future work will employ up to forty-eight heliostats and several secondary mirrors. We also hope to undertake simultaneous observations in conjunction with space-based detectors for high energy gamma rays and ground-based telescopes operating in the TeV region to provide contiguous spectral coverage of cosmic sources.

REFERENCES

1. Thompson, D. J. et al.: 1995, *ApJ*, **101**, 259.
2. Hartman, R. C. et al.: 1997, *AIP Proc.*, this issue.
3. Stecker, F. W. & de Jager: 1997, O. C., *ApJ*, **476**, 712.
4. de Jager O. C.& Harding, A.: 1991, *Proc. of 22nd ICRC, Dublin*, **1**, 572.
5. Weekes, T. C.: 1988, *Phys. Reports*, **160**, 1 (1988).
6. Tümer, O. T. et al.: 1990, *Nucl. Phys. B (Proc. Suppl.)*, **14A**, 351.
7. Ong, R. A. et al.: 1996, *Astroparticle Physics*, **5**, 353.
8. Chantell, M. C. et al.: 1997, *submitted to Nucl. Inst. Meth.*, astro-ph/9704037.
9. Williams, D. A. et al.: 1997, *ICRC Proc.*, OG 10.3.10.
10. Covault, C. E. et al., 1997: *ICRC Proc.*, OG 10.3.34.

On the Potential of the HEGRA IACT Array

Felix A. Aharonian* (HEGRA collaboration)

*Max-Planck-Institut für Kernphysik, D-69029 Heidelberg, Germany

Abstract. The status of the HEGRA array of five imaging atmospheric Cherenkov telescopes is briefly discussed. Our current understanding of the telescopes based on Monte Carlo studies and recent stereoscopic observations of TeV γ-rays from the Crab Nebula, allows the following conclusions on the performance of the instrument: effective energy threshold $E_{\text{eff}} \simeq 500\,\text{GeV}$; angular resolution $\delta\theta \simeq 0.1°$; energy resolution $\Delta E/E \leq 25\%$; minimum detectable flux for 100 h observations of point sources $J_{\text{min}} \simeq 25\,\text{mCrab}$; minimum detectable energy fluence of transient $\leq 1\,\text{h}$ flares of TeV γ-rays $S_{\text{min}} \simeq 5 \cdot 10^{-8}\,\text{erg/cm}^2$.

STATUS

The HEGRA collaboration is close to complete an array of five imaging atmospheric Cherenkov telescopes (IACTs) located on the Roque de los Muchachos, Canary Island La Palma (28.8° N, 17.9°). The excellent optical conditions and good infrastructure provided by the Instituto de Astrophysica de Canarias make the area a prime site for ground-based γ-ray astronomy.

The telescope array, primarily designed for stereoscopic observations of γ-radiation at energies above several hundred GeV [1], is formed by five identical IACTs - one at the center, and four others at the corners of a 100 m by 100 m square area. The multi-mirror reflector of each telescope has an area $8.5\,\text{m}^2$. Thirty 60 cm diameter front aluminized and quartz coated spherical mirrors with focal length of about 5 m are independently installed on an almost spherical frame of an alt-azimuth mount. The point spread function of the reflector is better than 10 arcminutes. Each telescope is equipped with a 271-channel camera [2] with a pixel size of about 0.25° which results in the telescope's effective FoV$\approx 4.3°$. The digitization of the PMT pulses is performed with a 120 MHz FADC system. The system trigger demands *at least two* fired pixels in each of *at least two* telescopes.

A partially completed system 'CT3/4/5/6' consisting of four telescopes, is taking data in the stereoscopic mode since 1996 [3]. The last nominal telescope of the system will be available after replacement of the 61-channel camera of the second (prototype) HEGRA telescope CT2 by a standard 271-pixel high resolution camera. We plan to run, after the test and calibration work, the whole telescope array as a *single detector* at the end of 1997.

The basic concept of the HEGRA IACT array is the *stereoscopic approach* based on simultaneous detection of air showers by ≥ 2 telescopes, which allows unambiguous and precise reconstruction of the shower parameters on an *event-by-event* basis, superior rejection of hadronic showers, and effective suppression of the background light of different origins (night sky background, local muons, etc.). The recent observations of the Crab Nebula [3] and Mkn 501 [4] by the 'CT3/4/5/6' system strongly support the theoretical expectations concerning the features of the stereoscopic imaging technique [5] and confirm the predictions for the performance of the HEGRA IACT array [1]. These data demonstrate that (1) the accuracy of reconstruction of the direction of *individual* γ-ray primaries can be as good as $0.1°$, (2) the suppression of hadronic showers at the hardware and software levels can essentially exceeds factor of 100, and (3) the energy threshold of the telescopes operating in coincidence mode can be reduced by $\approx 50\%$. In particular the confinement of the γ-ray signal in the angular region $\sim 0.1°$, in addition to the significant suppression of hadronic showers at the *trigger* level, provides '4σ-per-1 h' signal from the Crab, already before the application of 'shape' cuts to the shower images. The 'shape' cuts allow further, by a factor of 100, suppression of the CR background which results in only ≈ 1 (!) background event/h in the 'signal domain', while the number of γ-ray events exceeds 20 events per 1 h [3].

PERFORMANCE

The *energy domain* of the instrument, for the given spectrum of γ-rays, is characterized by the detection rate of primaries in the energy interval $\Delta E \approx E$, i.e. by the function $E \cdot dR/dE$, where the differential detection rate dR/dE is determined by the shower collection area and the spectrum of primary particles. In Fig.1 we present detection rates for 4 different spectra of primary γ-rays from point sources, as well as for isotropic fluxes of CR protons and electrons. The maximum of the photon statistics is achieved at energies between 500 GeV (for steep spectra) and 1 TeV (for flat spectra). For power-law spectra from a point source, with a flux 10^{-11} ph/cm^{2}s above 1 TeV, the total detection rates vary between $70\,h^{-1}$ ($\alpha = 2$) to $150\,h^{-1}$ ($\alpha = 3$). In the case of the Crab Nebula we expect, after application of 'shape' cuts to the shower images at different viewing angles, and parametrized as functions of the image amplitude and distance to the shower core [6], about $50\,h^{-1}$ γ-rays on top of the CR contamination with a rate of only few events per 1 hour.

Since these fluxes are detected at background-free conditions, the HEGRA IACT array with it's energy resolution $\leq 25\%$ allows, for an observational

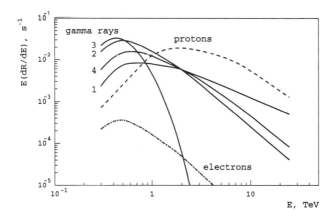

FIGURE 1. Detection rates of γ-rays from point sources. For comparison the rates for the CR protons and electrons are also shown. The rates correspond to showers which trigger the 5-IACT system operating in the '2-telescope coincidence' mode, and are contained in the 'source domain' $\theta \leq 0.15°$. Curves 1 and 2 correspond to power-law spectra of photons $dJ/dE \propto E^{-\alpha}$ with $\alpha = 2$ and 3 respectively; the curve 3 corresponds to power-law spectrum with exponential cutoff: $dJ/dE \propto E^{-2} \exp(-E/250\,\text{GeV})$; curve 4 corresponds to the inverse Compton spectrum of the Crab. The γ-ray spectra are normalized to an integral flux above 1 TeV: 10^{-11} ph/cm²s for curves 1,2,4, and 10^{-12} ph/cm²s for curve 3.

time ~ 100 h, good spectroscopic measurements of strong TeV sources in a broad energy region from 0.3 TeV to ≥ 30 TeV. Furthermore, the excellent angular resolution ≤ 0.1° enables a study of the spatial distribution of γ-ray production regions in such strong sources on *arcminute* scales. Therefore, *the detailed study of angular and spectral characteristics of established TeV sources is considered as one of the important objectives of the HEGRA IACT array.*

The *second* feature of the HEGRA IACT array is its high sensitivity for search for new TeV sources with fluxes down to the level of ≈25 mCrab. This program is planned to be realized in 2 general modes:

Survey Mode

The superior background rejection by the HEGRA telescopes allows a search for TeV γ-rays from point-like sources with fluxes ≥ 0.1 Crab at almost background-free conditions. This allows drastic reduction of the observation time allocated to individual targets, and opens a unique possibility for effective *sky surveys* with an observatio time ≈ 1 h per 2° × 2° region at a sensitivity which for $t_{\text{obs}} \leq$ few hours is determined essentially by the photon statistics ('at least 10 detected γ-rays'):

$$J_{\min} \simeq 0.2 \, (t_{\text{obs}}/1\,\text{h})^{-1} \, \text{Crab}.$$

In particular, this implies that all *persistent* Galactic TeV emitters with apparent luminosity $L_\gamma (\geq 0.5\,\text{TeV}) \simeq 1.5 \cdot 10^{33} (d/1\,\text{kpc})^2$ erg/s would be revealed by a survey of the Galactic Disk, an essential part of which ($10° \leq l \leq 250°$) is effectively observable from La Palma at zenith angles $\leq 50°$.

It is important to note that the individual telescopes of the HEGRA system, operating in the 'stand alone' (mono) mode, have energy threshold of about 750 GeV and sensitivity which allows detection of a signal from the Crab with a statistical significance $\simeq 2.5\sigma\,(t/1\,\text{h})^{1/2}$. Thus the '$5\sigma$' detection of faint sources at the flux level of 0.2 Crab by a single telescopes of the system requires exposure time of about 100 h, while only 1 h observation time would be sufficient for detection of such fluxes by the telescope system which operates in the 'stereo mode'. This clearly demonstrates the principally new level of the imaging technique which can be gained by implementation of the stereoscopic approach.

The HEGRA IACT array is an effective instrument in the search for possible *transient* or *burst-like* sources. In particular, the minimum detectable fluence $S_\gamma \simeq 5 \cdot 10^{-8}$ erg/cm^2 during the burst time ≤ 1 h would allow detection of possible VHE counterparts of GRBs, if only a small part of the total energy of the burst were released in TeV photons. Moreover, the HEGRA IACT array can explore new types of highly sporadic nonthermal phenomena from different classes of astrophysical objects which may appear well below the sensitivities of satellite-borne detectors at MeV and GeV energies.

Target Mode

The minimum detectable fluxes of faint VHE sources at the level of ≤ 0.1 Crab are determined essentially by the signal-to-noise ratio $S/\sqrt{(B)}$. In particular, the detection of a γ-ray signal at the 5σ level

$$J_{\min} \simeq 0.05\,(t_{\text{obs}}/25\,\text{h})^{-1/2}\,\text{Crab}.$$

Note that in the *target* mode the accumulation of 25 h requires typically 1 or 2 weeks of continuous observations during moonless and cloudless nights. Since in some cases the observation time of a target of special interest could be increased up to 100 h, the minimum detectable flux may reach a level of 25 mCrab which corresponds to the *energy flux* above 500 GeV approximately $(1.5 - 2) \cdot 10^{-12}$ erg/cm^{2}s. This implies, in particular, that all blazars emitting γ-rays at energies $E' \geq 50/\delta_{10}$ GeV (in the frame of the jet; $\delta_{10} = \delta_j/10$ is the jet's Doppler factor), should be detected by the HEGRA IACT array if the intrinsic γ-ray luminosity of the source exceeds $L'_\gamma \simeq 3 \cdot 10^{39}(z/0.1)^2$ erg/s ($H_0 = 75$ km/s Mpc).

Although from the point of view of the nonthermal energy budget the search for γ-rays from distant blazars with redshifts $z \geq 0.1$ could be easily justified, the effective energy threshold of the HEGRA around 500 GeV cannot guarantee success in the case of such distant sources due to possible absorption of TeV γ-rays in the intergalactic photon fields. Nevertheless it should be emphasized that the HEGRA IACT array can provide reasonable sensitivities also below 500 GeV. Indeed, conservative estimates show that long-term ($t_{obs} \sim 100\,h$) observations should enable detection of a flux around 300 GeV as low as $f_E \simeq 10^{-11}\,erg/cm^2 s$, which extends the extragalctic distance scales available for the HEGRA up to 1 Gpc.

Extended Sources

The HEGRA IACT array with its superior γ/hadron separation efficiency, in combination with its high γ-ray detection rates, allows an effective study of moderately extended sources with angular radius $\phi \simeq 1°$. The minimum detectable flux, integrated over the angular size of the source,

$$J_{min} \simeq 0.15\,(t_{obs}/100\,h)^{-1/2}\,Crab,$$

is sufficient to probe the energy budget of SNRs in accelerated cosmic ray protons and nuclei down to $10^{49}\,erg$, as well as to study the π^0-decay and inverse Compton components of the diffuse radiation of the Galactic Disk at the flux level of $10^{-9}\,cm^{-2}s^{-1}ster^{-1}$.

SUMMARY

The HEGRA telescope array perhaps can be considered as the first (scaled-down) model of the next generation of very large IACT arrays. Therefore it is difficult to overestimate the importance of the recent observations of the Crab Nebula by the HEGRA telescopes, which convincingly confirm the Monte Carlo predictions concerning the potential of the stereoscopic approach, and give an optimistic view for the forthcoming '100 GeV' threshold telescope arrays like HESS and VERITAS. On the other hand, even with its relatively high energy threshold, the HEGRA IACT array already demonstrates very good performance. This assures that this instrument will serve as one of the most powerful tools of TeV γ-ray astronomy during the next several years.

REFERENCES

1. Aharonian, F. in Proc. Towards a Major Atmospheric Cherenkov Detector-II, Calgary (ed. R.C.Lamb), 81 (1993)
2. Hermann, G., in Proc. Towards a Major Atmospheric Cherenkov Detector-IV, Padova (ed. M.Cresti), 396 (1995).
3. Daum, A. et al., *Astroparticle Physics*, submitted (1997).
4. Aharonian, F. et al., *A & A Letters*, submitted (1997).
5. Aharonian, F. et al., *Astroparticle Physics* 6, 343 (1997).
6. Konopelko, A., in Proc. Towards a Major Atmospheric Cherenkov Detector-IV, Padova (ed. M.Cresti), 373 (1995).

A Site for Čerenkov Astronomy in the White Mountains of California

John R. Mattox & Steven P. Ahlen

Boston University, Boston, MA 02215

Abstract. We have identified a site at an elevation of 3,900 meters on the Barcroft Plateau in the White Mountains of California which would be a good site for Čerenkov γ-ray astronomy. The atmospheric conditions are conducive. There is ample infrastructure. And the site is large enough to accommodate an array of telescopes which at this altitude is expected to provide impressive performance.

INTRODUCTION

The performance of an array of imaging Čerenkov γ-ray telescopes (Ahlen and Marin 1996) can be significantly enhanced by locating it at a higher site. The energy threshold of a Čerenkov telescope is inversely proportional to the intensity of the Čerenkov light reaching the detector for a specific event. Therefore, the energy threshold is proportional to D^2, where D is the distance from the detector to the region where the Čerenkov light is produced. The elevation of maximum Čerenkov light production is energy dependent and varies between showers, but for a 100 GeV γ-ray, it averages about 11 km. Therefore, if a Čerenkov observatory were built at 3900 m elevation, the energy threshold could thus be expected to be 54% of the energy threshold of the same facility at 1300 m, the elevation of the Whipple Base Camp where the VERITAS array is being constructed. This would not cause the source count rate at zenith to change substantially (assuming a power law spectrum with a photon index of −2 for the source) because the effective area of the detector is also proportional to D^2. Because low-energy cosmic-ray protons produce relatively little atmospheric Čerenkov light, a reduced energy threshold is very desirable.

Another ≈10—20% gain in energy threshold at 3900 m elevation can be expected due to the decreased attenuation of Čerenkov light because of the decreased atmospheric pressure and reduced concentrations of aerosols in the air. There is no corresponding decrease in effective area. Furthermore, sources can be observed over a larger fraction of the sky. Observation can be made from 3900 m at zenith angles as large as $\approx 40°$ before the distance to the maximum Čerenkov light production exceeds that of a source observed from 1300 m at a zenith angle of $\approx 0°$.

Thus, we suggest that the benefits for Čerenkov γ-ray astronomy of a high altitude site are compelling — well worth the inconvenience of working at this altitude. For this site, the logistics appear to be manageable. Human performance is somewhat impaired at 3900 m, and some people are quite uncomfortable. However, most people can readily adapt to living and working at this altitude (Cudabeck 1984).

THE WHITE MOUNTAINS

Fritz Zwicky brought the White Mountains to the attention of the Office of Naval Research (Zwicky 1950). Zwicky wrote, "As far as a site for observation of celestial objects and of the optical phenomena in the upper atmosphere is concerned, I consider White Mountain Peak the best on the North American continent."

White Mountain Peak is located 32 km northeast of the town of Bishop, CA (see Figure 1). The elevation is 4,342 m. In 1948, the Navy established a research facility at Crooked Creek, 16 km south of White Mountain Peak (see Figure 1) at an elevation of 3,094 m. This facility was used initially to test the high-altitude operation of infrared sensors. In 1950, the University of California was contracted by the Navy to operate the White Mountain facility. In 1951, the Barcroft Facility was established between Crooked Creek and White Mountain Peak at an elevation of 3,801 m; and in 1955 a research hut was erected on the summit. We have investigated the feasibility of establishing a Čerenkov gamma-ray observatory on a plateau 0.8 km north of the Barcroft Facility at an elevation of 3,886 m, see Figure 1. This plateau is 1 km northeast of Mt. Barcroft. We refer to it henceforth as the Barcroft Plateau.

For a natural mountainous region, the Barcroft Plateau is very level. A central 200 m by 250 m rectangular region is level to within several meters. And, it drops off by less than 13 meters at the perimeter of a 300 m by 450 m rectangular region. Before making detailed plans, it would be appropriate to conduct a high resolution GPS survey of the Barcroft Plateau (using differential GPS). This data could be used with a GIS program to choose mirror locations.

In 1978, the Navy transferred all White Mountain facilities to the University of California which continues to operate then as the White Mountain Research Station (WMRS). See http://orpheus.ucsd.edu/wmrs/. In addition to the three high-altitude sites described above, WMRS now includes a facility 6 km east of Bishop, at an elevation of 1,234 m.

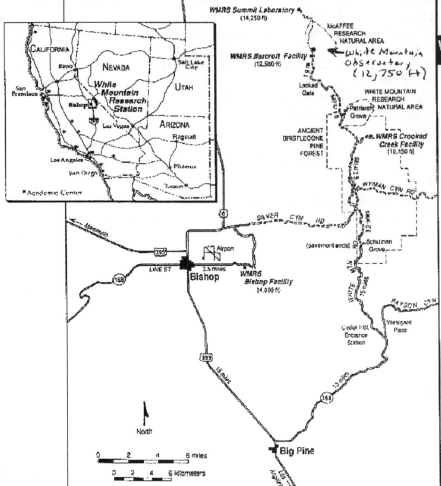

Figure 4. Location of the White Mountain Research Station Facilities

FIGURE 1. A map of the White Mountain region reproduced from the WMRS 1995 annual report. The proposed site for an atmospheric Čerenkov γ-ray observatory is drawn in.

THE POTENTIAL OF THE WHITE MOUNTAINS FOR OBSERVATION OF OPTICAL PHENOMENA

In the 1970's, the White Mountain Range was very seriously considered by UC Berkeley as the site for the 10 meter telescope which was eventually called the Keck telescope. The White Mountain Range was expected to be as good as Mauna Kea in Hawaii for astronomy, but Mauna Kea was chosen because of considerations of ease of winter access.

The White Mountain Ridge runs north-south. The western edge of the ridge falls off abruptly into the Owens valley. The Sierra Nevada Range also runs north-south on the opposite side of the Owens valley. Thus, prevalent western winds travel for 45 to 75 km without obstruction before reaching the White Mountain Ridge. It is expected that air turbulence due to passage over the Sierra Nevada Range will be largely dissipated before reaching the White Mountains. Therefore, good seeing is expected. Indeed, measurements made by UC Berkeley astronomers in the 1970's in collaboration with M. Walker at the summit of White Mountain Peak found a seeing disk smaller than $1''$ for at least part of 55% of the 64 nights for which observations were made. Unfortunately, the method did not discern the actual size of the seeing disk when it was smaller than $1''$— the diffraction limit of the 6" telescope they used. The average seeing disk at White Mountain Peak was smaller than $1''$ on 36% of nights. This compares very favorably with Kitt Peak for which Walker measured an average seeing disk smaller than $1''$ on only 15% of 253 nights. Of course, seeing is not a consideration for a Čerenkov gamma-ray facility.

The prevalent western winds also imply that the White Mountain Range is in the precipitation shadow of the Sierra Nevada Range. Indeed, the crest of the White Mountains receive 1/3 of the winter snow fall of the crest of the Sierra Nevada. We were able to reach the WMRS Barcroft Facility by snowmobile on 1/9/97 - right after a 100 year storm brought severe flooding to much of California to the north and west. We expect that winter operation of our proposed facility is quite practical (although we grant that it will be more difficult than Mauna Kea).

The precipitation shadow of the Sierra Nevada Range also causes the air at White Mountain to be very dry. The UC Berkeley group reported that sunrise measurements of precipitable water vapor pressure on White Mountain Peak yielded a median value of 1.30 mm in the summer, and 0.72 mm during the rest of the year.

The sky above the Barcroft Plateau is very dark. It is surrounded by a great expanse of lightly developed, and undeveloped land. The Sierra Mountains and the Owens valley to the east are sparsely populated because there is very little land which can be developed. The mountains extending to the north and south for 100 km in each direction are completely undeveloped. And, the desert regions of Nevada to the east are very lightly developed. The nearest population concentration is the city of Bishop, 30 km to the south-west. The population is only 4000, and it is completely shielded by Mt. Barcroft. The brightest source of artificial light visible from the Barcroft Plateau is the city of Las Vegas, 300 km to the south-east. It contributes substantially to the sky brightness near the horizon in that direction, but does not affect the sky brightness away from the horizon. We made a sky brightness measurement on the evening of 1/6/1997 at Schulman Grove, 20 km south of the Barcroft Plateau (see Figure 1). Our result for a zenith pointing was not inconsistent with the natural sky brightness of 22 magnitudes per square arc second.

The White Mountains offers the prospect of operating an observatory without substantial inefficiency due to cloud cover. The WMRS Barcroft Facility was manned year round from the early 1950's through the early 1970's and weather observations were systematically made. Summer weather in the White Mountains is substantially better than winter weather because high pressure air in the eastern pacific impedes the eastward flow of Pacific air. The best weather is in the fall. The 20 year average for the summer months (June, July, August, and September) was a clear sky or only scattered clouds (which will allow for Čerenkov observation) 88% of the time, and a clear sky, 58% of the time. For the other months, a clear sky or scattered clouds was observed 70% of the time, and a clear sky, 36% of the time. Thus, we can expect to loose 30% of potential observing time to cloud cover in non-summer months, and only 12% during summer months.

Substantial winds occur in the White Mountains. Therefore careful consideration must be given to wind during the design of mirrors to be placed there. From a study of extensive wind data (see http://bu-ast.bu.edu/~mattox/wmo.html for details) we expect that mirrors designed to operate in winds of velocity up to 44 knots could be used at least 90% of the time in the winter, and 99% of the time in the summer.

Also, we expect that a mirror which can withstand a wind speed of 150 knots in a stowed position without damage will suffice. A definitive upper limit on the maximum wind-speed which has occured during the last 20 years on the Barcroft Plateau can be obtained through a careful analysis of the 18' Ash dome which has stood there since 1978 without wind damage. We plan to install a high quality anemometer on the Barcroft Plateau before the winter of 97/98 and begin to acquire extensive wind-speed data. This will allow us to refine our expectations.

The White Mountains are visited by severe lightning storms during the summer months. The storms usually occur in the afternoon, and skys clear by early evening, so observations will usually not be impaired. Several lightning rods could be erected near a new facility to prevent direct lightning strikes to equipment. Fiber-optic signal cables between the mirror/detector complexes and control room could further mitigate possible damage.

Infrastructure

The only necessary component of infrastructure which is currently missing on the Barcroft Plateau is a high band-width Internet connection, and that could easily be accomplished with a microwave link (see http://bu-ast.bu.edu/~mattox/wmo.html). The value of WMRS facilities, equipment and furnishings is conservatively estimated to be in excess of $15 M. One can drive a standard rental car to the Barcroft Plateau during the summer and fall. WMRS has built an underground powerline from Crooked Creek to the Barcroft station. Up to 150 kva is available for our project at the standard power utility price. Also, the WMRS offers well equipped workshops and skilled artisans who can work on new projects.

The Crooked Creek and Barcroft Facilities both provide comfortable living conditions — hot meals and dormitory-style sleeping accommodations. We anticipate that observers who are still accommodating to the Barcroft altitude will drive 16 km down the mountain to sleep at the Crooked Creek Facility 700 meters lower than Barcroft.

The Crooked Creek and Barcroft Facilities are currently closed during the winter months because there is insufficient potential research activity to justify the cost of winter operations. However, both are equipped for winter operations, and we expect that the addition of Čerenkov γ-ray astronomy to other potential winter activities at Barcroft would enable winter operations to commence. The WMRS has the necessary equipment to keep the road to Crooked Creek plowed through the winter. They also have a snowcat which could be used to transport personnel and equipment from Crooked Creek to Barcroft during the winter months.

We are in contact with local pilots who could provide year-round air transport for personnel between the Barcroft airport and Bishop with light airplanes. The trip down would take $\approx$ 8 minutes, and the trip up, $\approx$ 20 minutes. During the winter, the planes can be equipped with ski-wheels to use a packed snow landing strip at Barcroft. Ground transport would always be available, and would be used instead of aircraft during inclement weather inorder to maintain a large safety margin in the use of aircraft.

REFERENCES

1. Ahlen, S.P., and Marin, A., 1996, SPIE Proceedings 2806, p. 64.
2. Cudabeck, D.D., 1984, PASP, 96, 463
3. Zwicky, F., 1950, Final Report, Office of Naval Research Contract Nonr-24421, Jan. 1950, Data on the Suitability of the White Mountains of California as an Observatory Site.

Simulation of HEAO 3 Background

B. L. Graham*, B. F. Phlips[†],
R. A. Kroeger[‡], and J. D. Kurfess[‡]

*George Mason University, Fairfax, VA
[†] USRA, Washington, DC
[‡] Naval Research Laboratory, Washington, DC

Abstract. A Monte Carlo technique for modeling background in space-based gamma-ray telescopes has been developed. The major background components included in this modeling technique are the diffuse cosmic gamma-ray flux, the Earth's atmospheric flux, and decay of nuclei produced by spallation of cosmic rays, trapped protons and their secondaries, the decay of nuclei produced by neutron capture, and the de-excitation of excited states produced by inelastic scattering of neutrons. The method for calculating the nuclear activation and decay component of the background combines the low Earth orbit proton and neutron spectra, the spallation cross sections from Alice91 [2], nuclear decay data from the National Nuclear Data Center's (NNDC) Evaluated Nuclear Structure Data File (ENSDF) database [3], and three-dimensional gamma-ray and beta transport with Electron Gamma-ray Shower version 4 (EGS4) [4] using MORSE combinatorial geometry. This Monte Carlo code handles the following decay types: electron capture, β^-, β^+, meta-stable isotope and short lived intermediate states, and isotopes that have branchings to both β^- and β^+. Actual background from the HEAO 3 space instrument are used to validate the code.

INTRODUCTION

Calculations of the radioactive background in γ-ray detectors has been a challenging problem since before the first balloon flight of a γ-ray detector [1]. Reliable calculation of the background rate is essential for future missions because the sensitivity of a γ-ray telescope is background limited. A Monte Carlo simulation code has been developed that models the background rate due to the decay of spallation products in space based γ-ray telescopes. The measured background for the High Resolution Gamma Ray Spectrometer aboard HEAO 3 was used to validate this code. This model includes the geometry and composition of the instrument, the incident proton and neutron fluxes on the components of the detectors, the cross-sections for spallation, inelastic scattering and capture, the branching ratios in the decay scheme for the unstable spallation products, the half-lives of the unstable spallation products, and the Monte Carlo transport code that propagates γ-rays and βs. The Monte Carlo code does not propagate α-particles, protons, neutrons or model α-decays.

MODEL

A mass model, which describes the geometry and composition, was made for the components of HEAO 3. HEAO 3 consisted of four Ge co-axial detectors in a CsI anti-coincidence shield. Each of the Ge detectors was 5.4 cm in diameter and 4.5 cm thick. The detectors were held in a thin aluminum cup. These were all contained in an aluminum cryostat. There was an aluminum cold plate between the detectors and a silver cold finger. There also was a stainless steel baseplate (nominally composed of 10% Nickel, 20% Chromium and 70% Iron). The cryostat was surrounded on all sides by a 6.6 cm thick CsI shield, with holes in the upper lid of the shield which collimated the field of view to $\sim 30°$ [5]. The decays of isotopes produced by the neutron and proton activation of Ge, Al, CsI, Steel and Ag were modeled. Components of the spacecraft outside the shield were not included in the mass model for this work.

The activation of the detectors and surrounding materials was due to a flux mostly composed of secondary neutrons and protons produced by cosmic rays and trapped protons incident on the spacecraft. In this work, the neutron and proton flux used was calculated for an orbit of 500 km at an inclination of 44° [5]. The proton and neutron flux used was scaled according to the scaling method used by Gehrels [6] which was based on data from the balloon instrument, Low Energy Gamma-ray Spectrometer (LEGS) [7].

SIMULATION

The cross-sections for spallation from neutrons and protons incident on the detectors, shields and support structures were taken from Alice91 [2]. These cross-sections and the neutron and proton flux were evaluated over the energy range of 1 to 250 MeV, thus primary cosmic-ray protons ($>$ few GeV) are not included in this model. The reported errors on the Alice91 spallation cross-sections was less than 30%. The NNDC's Evaluated Nuclear Data File (ENDF) database contains measured cross-sections for neutron capture and inelastic scattering of neutrons to excited states. The model used the γ-ray spectrum from the Earth's atmosphere as measured by the Solar Maximum Mission (SMM) [8]. This spectrum was folded through both the open collimator response and the response to shield leakage. A ^{40}K calibration source was modeled inside the cryostat and was fitted to the measured data only at the 1.46 MeV line. The cosmic diffuse γ-ray background was from the HEAO 1 A4

[9]. This flux is also folded through the open collimator response and the response to shield leakage. The ENSDF database was used for β branchings and half-lives for β-unstable isotopes, and for energy levels and gamma-ray branching ratios for decay products, as well as the half-lives of intermediate states of decay products ($> 1\mu s$). Also included in NNDC databases were the x-ray fluorescence yields and energies of the characteristic x-ray(s) for each isotope [3].

EGS4 is a Monte Carlo transport code that was used for β and γ transport [4]. Included in EGS4 are routines for the following photon interactions: photo-electric absorption; Compton scattering and pair production, as well as routines for the scattering of electrons; ionization energy losses; Moller, Mott and Bhabha scattering. The combinatorial geometry package MORSE-CG was used, allowing for full three-dimensional propagation of βs and γs.

A routine was designed and implemented for initializing EGS4 to model a nuclear decay, one decay at a time. The βs and γ-rays produced for a given isotope in a single decay follow a single series of branchings in a cascade to the ground state of the decay product for each decay modeled via Monte Carlo. This cascade is produced starting from the β-unstable spallation product isotope and ends at the ground state of the decay product. At the β-unstable spallation product a β branch or Electron Capture (EC) energy or Internal Transition (IT) is chosen from the branchings available (or appropriate) for that spallation product isotope via Monte Carlo selection from the branching ratios. If a β branch was selected then the endpoint energy for that branch was used to determine the energy of the β from the appropriate β-spectrum [11,12]. Then the β endpoint energy, or the EC energy, was subtracted from the ΔQ value for the decay of this β unstable spallation product to determine the energy level of the decay product for the start of the γ-ray cascade. At each energy level of the decay product a γ-ray was randomly selected from the possible γ-rays for that energy level according to their branching ratios. The next energy level was determined by subtracting the energy of the γ-ray selected from the value of the previous energy level until the ground state of the decay product is reached.

For each β-unstable isotope that was a spallation product, a number of decays were modeled. For every decay the energy deposited by each interaction in the detectors or the shield was summed for each detector and the shield. The energy resolution was applied to the summed energy for each active element. Then a histogram was made of all decays of a spallation product where the energy deposited in the detector was above the detector threshold, and the energy deposited in the shield was below the shield's threshold. These histograms were summed together with appropriate normalization for the decay rate of each isotope and summed with the cosmic diffuse γ-ray flux and the Earth's atmospheric γ-ray flux, and elastic neutron flux and was compared to the measured background rate.

The production rate for each β-unstable isotope j is

$$R_{prod_j} = \sum_i \int F_n(E)\sigma_{n_{j,i}}(E)a_i\rho_i\frac{N_o}{A_i}VdE + \sum_i \int F_p(E)\sigma_{p_{j,i}}(E)a_i\rho_i\frac{N_o}{A_i}VdE \tag{1}$$

where i is a stable isotope in volume V, a_i is its fractional abundance, ρ_i the density, N_o is Avogadro's number, A_i is the atomic weight, $\sigma_{n_{j,i}}$ and $\sigma_{p_{j,i}}$ are the cross-sections for neutron and proton spallation to β-unstable isotope j from isotope i, and F_p and F_n is the cosmic-ray proton and neutron flux [7]. The fractional abundance a_i was calculated for the isotopic fraction for each element in each material.

A daily South Atlantic Anomaly (SAA) exposure of 220 minutes per day was used. An assumption of 10 SAA passages per day was used resulting in an average of 22.0 minutes for each SAA passage. The detectors on HEAO 3 were turned on starting from 600 s after an SAA passage until the beginning of the next SAA passage.

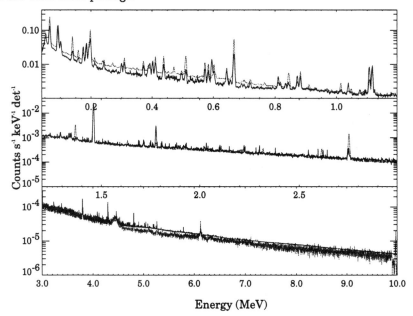

FIGURE 1. The sum of all components included in this model (gray curve) versus the measured background rate from HEAO 3 (black curve)

Figure 1 shows the background spectrum modeled for HEAO 3 summed over all components and compared to the measured data taken from the first 50 days of operation [5,10]. Included in the model were all unstable isotopes with production >0.01% of the most abundantly produced from Ge (56 isotopes), all unstable isotopes with production >1.0% of the most abundantly produced from Al (11 isotopes), steel (33 isotopes), CsI (36 isotopes), and Ag (19 isotopes).

CONCLUSIONS

The Monte Carlo model for HEAO 3 captures most of the major line features and shape of the continuum background. Agreement is good throughout the background spectrum suggesting that the essential physics is being modeled correctly. In the continuum, the data and model agree to within 45% of each other. The difference can be partially attributed to the errors of ~30% in the spallation cross-sections in Alice91 and to the uncertainties in the proton and neutron flux. There are 133 lines observed in the data from HEAO 3. Forty-seven of these lines are not observed in the model, of which 15 are from Bi, U, Pb, Tl and Th contaminates (not modeled) and 7 lines in the HEAO 3 data that were not identified [10]. There are no lines in the model that are not also found in the data within the statistics of comparision. Eighty-six lines are observed in the model that agree with lines in the data. Most of the lines not apparent in the model are weak lines, thus may be absent due to limited statistics.

The HEAO 3 simulation has been used to validate the Monte Carlo Model. With a few minor exceptions that are still being investigated, the model performs well. The future plans for this model is to apply this model to more complicated instruments such as INTEGRAL and a proposed high resolution Compton telescope.

This work was supported by the Office of Naval Research, ONR, work request # N0001497WX40001, task area # LR0340643.

REFERENCES

1. L. E. Peterson *J. Geophys. Res.* **68**, 979 (1963).
2. Blann and Vonach, *Phys. Rev* **28**, 1475 (1983).
3. National Nuclear Data Center (NNDC), Brookhaven National Laboratory, Brookhaven, NY.
4. W.R. Nelson et.al. **SLAC-265** (Stanford University), (1985).
5. W. Mahoney et.al. *NIM* **179**, 363 (1980).
6. N. Gehrels, *NIM A* **313**, 513 (1992).
7. N. Gehrels, *NIM A* **239**, 324 (1985).
8. J. Letaw, et.al. *J. Geophys. Res.* **94**, 1221 (1989).
9. L. E. Peterson private communication.
10. W. Wheaton et.al. *High Energy Radiation Background in Space*, American Institute of Physics, 304 (1989).
11. K. Seigbahn *Alpha, Beta and Gamma-Ray Spectroscopy*, 1337-1339 (1965).
12. Evans *Atomic Physics*, McGraw-Hill, 548-552 (1955).

Activation of Gamma Detectors by 1.2 GeV Protons

J. L. Ferrero[1], C. Roldán[1,2], I. Arocas[2], R. Blázquez[2], B. Cordier[3], J. P. Leray[3], F. Albernhe[4], V. Borrel[4]

[1] Instituto de Ciencia de Materiales. [2] Departamento de Física Aplicada, Universitat de València, Av. Doctor Moliner 50, 46100 Burjassot, Valencia, Spain.
[3] CEA, DAPNIA Sap CE-Saclay. Orme des Merisiers Bat 709. 91191 Gif-sur-Yvette, France.
[4] Centre d'Etude Spatial des Rayonnements, 9 Avenue Col. Roche. BP 4346, Toulouse, France.

Abstract. Radioactivity induced by cosmic rays in the elements of space gamma-ray detectors was simulated under irradiation with 1.2 GeV protons. Radioactivity decay data from one minute to one month after irradiation were analyzed for Cd, Te, CsI, and BGO targets. In this paper we present the identification of unstable γ emitters isotopes activated in the samples, the temporal evolution of the 511 keV γ line and the calculation of some isotopes half life and yields of production through the evolution in time of their γ lines.

INTRODUCTION

Previous works [1,2] have demonstrated that induced radioactivity from energetic protons present in cosmic radiation is a major source of background in gamma ray telescopes. A consequence of the bombarding of protons is the production of unstable isotopes through spallation mechanisms and subsequent γ, β⁻ and β⁺ emissions. This contribution to the background is perhaps the most difficult to remove because of the delayed gamma emission.

In order to measure the induced radioactivity and identify the unstable isotopes generated in elements entering the gamma detectors we have simulated the cosmic protons with protons beams delivered by an accelerator and irradiated several samples of materials constituting gamma detectors.

We present the results of an experiment designed to identify the short and long-lived spallation products induced in Cd, Te and CsI samples by protons of 1.2 GeV. Temporal evolution of the 511 keV γ line for Cd, Te, CsI and BGO samples are shown. Finally, determination of some isotopes half-life and production yields are calculated.

EXPERIMENTAL METHOD

The experiment was performed at the Saturne Laboratory at Saclay, France. Small samples of Cd, Te, CsI and BGO were irradiated by a 1.2 GeV beam delivered by a synchrotron with an average flux of $\approx 10^9$ p· s^{-1}· cm^{-2}. The Te sample was prepared putting powder of tellurium inside a methacrylate cube of 1 cm side. The characteristics of the irradiated samples are shown in Table 1.

The total accumulated doses of protons are shown in Table 1 and they have been calculated by measuring the intensity of the 1.275 MeV γ-line from ^{22}Na having a branching ratio of 99.95% and a half life of 2.602 years. This isotope was produced in aluminum foils of 1x1x0.5 cm^3 by the ^{27}Al(p, 3p3n)^{22}Na reaction. These foils backing the samples and both have been irradiated at the same time [1,3].

The induced radioactivity in the samples was measured in situ 1 hour and 2 hours after irradiation at Saclay, and 1 week and 1 month after irradiation in the Environmental Radioactivity Laboratory at the University of Valencia, Spain. First and second hour measurements were carried out using a 40% HPGe and a 60% HPGe detectors, respectively. Longtime measurements, 1 week and 1 month after irradiation, were carried out using a 40% HPGe detector in a low background environment using a 5cm thick Pb shield. The time lapsed between the end of the irradiation and the beginning of the measurement and the measurement time intervals are shown in Table 2.

TABLE 1. Characteristics of irradiated samples.

Sample	Volume (cm^3)	Mass (g)	Dose (x 10^{10} protons)
CsI	1 x 1 x 1	9.51	5.5 ± 1.5
Te	1 x 1 x 1	2.99	0.91 ± 0.04
BGO	π·(0.75)^2·0.1	1.35	33 ± 4
Cd	1 x 1 x 0.4	3.52	2.4 ± 0.6

TABLE 2. Measurements time intervals. **A**: Time interval from the stop of the irradiation to the start of the measurements. Acquisition time intervals after irradiation during **B**: first hour, **C**: second hour, **D**: one week, **E**: one month.(m: minute, s: second, h: hour, d: day)

SAMPLE	B	C*	D*	E*	F*
Cd	1m 31s	1m(2), 2m(2),4m(2),8m(1),12m(1),18 m(2)	1h(1)	1d(1)	3d(1)
Te	1m 27s	2m(5),10m(2),30m(1)	1h(1)	1d(1)	3d(1)
CsI	1m 30s	1m(2), 2m(2),4m(2),8m(1),12m(1),18 m(2)	1h(1)	1d(1)	3d(1)
BGO	8m 13s	1m(2), 2m(2),4m(2),8m(1),12m(1),18 m(2)	1h(1)	1d(1)	3d(1)

* Figures between ( ) refer to number of spectra. Spectra were acquired sequentially.

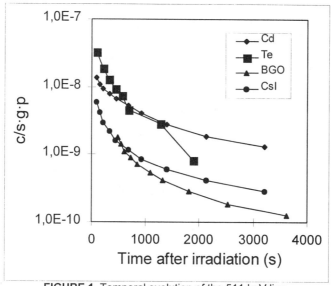

FIGURE 1. Temporal evolution of the 511 keV line.

RESULTS AND DISCUSSIONS

Annihilation Peak Evolution

Most of the unstable isotopes created in the samples are β^+ emitters. The resulting positrons are annihilated inside the detector or the surrounding material and 2 γ of 511 keV are emitted. Since β^+ emitters are in general short-lived the temporal evolution of the 511 keV γ line will provide us a good information about the behaviour of short-lived radioisotopes produced in the samples.

Evolution of the 511 keV line for Cd, Te, CsI and BGO samples is shown in Figure 1. In the x-axis it appears the time after irradiation in seconds, and the y-axis represents counts per second obtained from the peaks in the spectra, normalized to gramme of sample and dose of protons.

Tellurium data have been corrected by a factor coming from the activation of the methacrylate ($C_5H_8O_2$) surrounding the sample. Proton irradiation on ^{12}C produces ^{11}C (β^+ emitter, half-life=20.4 m) by the $^{12}C(p,pn)^{11}C$ reaction (σ=28.1 mb)[5]. This means that there will be an additional activity which must be subtracted from the peak areas in the tellurium spectra. Because of the half-life of ^{11}C this effect it will be only significant in the first hour after irradiation measurements.

Figure 1 give us information about the contribution to the gamma background in the space detectors. We can conclude that CsI and BGO would be a better choice for constituting space gamma detectors because of the low counting rates observed in both samples. On the other hand, Cd and Te show a higher activity of β^+ emitters, being more important the tellurium contribution to the background, in spite of the data correction.

Isotope Identification

The isotope identification was performed on the gamma spectra obtained for the samples. The criteria used for isotope identification were: (i) for a given isotope, all its important gamma rays should be observed and identified; (ii) the measured gamma-ray line intensities should correspond to the expected branching ratio; (iii) if an isotope is seen in different measurements, its observed intensities should follow the expected half-life.

Most of the peaks on the spectra were consistent with a specific radioisotope produced by proton irradiation or with natural radioactivity (which is observed in the background spectrum recorded by the Ge detector). Only a few peaks of low statistics were not identified in some of the spectra. Because of the large number of isotopes associated with each line, unambiguous identification was possible by observing other emission lines from the isotope in accordance with the expected branching ratio and half-life.

A total of 101 radionuclides from Cd, 110 from Te, and 175 from CsI have been identified. In some cases the metastable and the ground states have been observed. The atomic numbers of the identified isotopes ranging between Z=31 and Z=51 for the Cd sample, between Z=36 and Z=53 for the Te sample and between Z=31 and Z=56 for the CsI sample. The half-life of the identified isotopes ranging from 34 seconds to 120 days for the Cd sample, from 36 seconds to 115 days for the Te sample and from 23 seconds to 154 days for the CsI samples. Most of these isotopes decay via β^+ and ε (electron capture) and a minor number of isotopes decay via IT (internal transition) and β^-. The complete information about the identified isotopes is available from the first author of this paper.

Calculations of Some Isotopes Half-Life. Production Yields.

Once we were sure that an isotope was present in the irradiated sample we could study the evolution in time, through some spectra, of one of its gamma lines to calculate its half-life.

TABLE 3. Half-life and production yield of the identified isotopes ^{88}Nb, ^{114}Sb and ^{118}In obtained from experimental data.(m: minute, g: gramme, p: proton)

Isotope	Spectral line (keV)	Sample	Half-life (Theory)[4]	Half-life (Exp)	Production Yield (10^{-5} g^{-1} p^{-1})
^{88}Nb	1057.00	Cd	14.3 / 7.8* m	11.0 ± 0.3 m	See text
^{88}Nb	1082.60	Cd	14.3 / 7.8* m	11.1 ± 0.3 m	See text
^{114}Sb	1299.89	Te	3.49 m	3.7 ± 0.2 m	2.07 ± 0.14
^{114}Sb	887.62	Te	3.49 m	3.7 ± 0.2 m	2.8 ± 0.3
^{114}Sb	887.62	CsI	3.49 m	2.84 ± 0.06 m	1.8 ± 0.5
^{118}In	1050.75	Te	4.4 m	4.1 ± 0.2 m	1.75 ± 0.13

*metastable.

The slope of the linear fitting of the logarithm of the peak area (in counts per second) of each line in the spectra versus the time after irradiation (in seconds) will provide us a value of the corresponding isotope half life. If we extrapolate the decay line to t=0 seconds, we will obtain the counts per second of the isotope just after the irradiation. Correcting this value with the branching ratio of the γ-line and with the detector efficiency at the suitable energy, we will obtain the activity of the isotope. We obtain the production yield (number of isotopes produced by proton irradiation) by dividing this activity by its disintegration constant (this is possible the identified isotopes are not part of a decay chain).

Half-life and isotope production yield of three selected isotopes ^{88}Nb, ^{114}Sb and ^{118}In, that are not part of a decay chain, are calculated as an example and are indicated in Table 3. In this table, we show the calculation of the half-life for the isotope ^{88}Nb through two of its lines when the ground and excited states are both present. For this reason we have not calculated the production yield. The value of the half-life obtained is between the metastable and ground state half-life. For the isotope ^{114}Sb we calculate the decay of two different lines of the same isotope from the Te sample, obviously we have obtained the same half life in both cases (3.7±0.2 m), and this result is coherent with its tabulated half-life value of 3.49 m [4]. However, the half-life obtained for ^{114}Sb 2.84±0.06 m, in the CsI sample, differs from the previous ones. In the three cases the production yield is similar, this means that the production of ^{114}Sb is approximately the same in Te than in CsI. In general, the calculated half-life differ from tabulated data in less than 7%; however for ^{114}Sb from CsI sample 18% of discrepancy was observed.

CONCLUSIONS

Samples of Cd, Te, CsI and BGO was irradiated with energetic protons of 1.2 GeV in order to determine the nature of the induced radioactivity. For the Ce, Te and CsI samples, proton-induced radionuclides with a half-life greater than 30 seconds have been identified. Temporal evolution of the 511 keV gamma line has been determined for all samples and the results shown that CsI and BGO samples presents counting rates one order of magnitude lower than Cd and Te samples. The half-life and production yields of some isotopes have been calculated from the temporal evolution of their gamma lines presents in the collected spectra. Results shown a good concordance with tabulated data.

ACKNOWLEDGEMENTS

J.L.F., C.R., I. A. and R.B. acknowledge the DGICYT (Project n° PB95-1082A) for financial support.

REFERENCES

[1] J. Ferrrero et al., *Astron. Astrophys. Suppl. Ser.* **120**, 669-671 (1996).
[2] J. A. Ruiz et al., *Astrophys. J. Suppl.* **S 92(2)**, 683 (1994).
[3] R. Bodemann et al., *Nucl. Instr. Meth. Phys. Res.* **B42**, 76 (1989).
[4] E. Browne and R. B. Firestone, *Table of Radioactive Isotopes*, New York: John Wiley & Sons, 1986.
[5] J. Tobailem, C.H. de Lassus St. Genies, 1975 and 1981, *additif No2* a la CEA-N-1466 (1) (1975), CEA-N-1466 (5) (1981).

PARTICIPANT LIST
THE FOURTH COMPTON SYMPOSIUM
WILLIAMSBURG, VIRGINIA APRIL 27-30, 1997

Aharonian, Felix
MPI f. Kernphysik
aharon@fel.mpi-hd.mpg.de

Aller, Hugh
University of Michigan
haller@umich.edu

Aller, Margo
University of Michigan
margo@astro.lsa.umich.edu

Arzoumanian, Zaven
Cornell University
arzouman@spacenet.tn.cornell.edu

Bandyopadhyay, Reba
Oxford University
rmb@astro.ox.ac.uk

Baring, Matthew
NASA/GSFC
baring@lheavx.gsfc.nasa.gov

Barnes, Sandy
USRA/NASA/GSFC
barnes@grossc.gsfc.nasa.gov

Barret, Didier
CESR, Toulouse
barret@cesr.cnes.fr

Barrett, Paul
USRA/NASA/GSFC
barrett@compass.gsfc.nasa.gov

Barthelmy, Scott
USRA/NASA/GSFC
scott@lheamail.gsfc.nasa.gov

Bazzano, Angela
IAS/CNR
angela@saturn.ies.fro.cnr.it

Bennett, Kevin
ESTEC/ESA
kbennett@astro.estec.esa.nl

Bertsch, David
NASA/GSFC
dlb@mozart.gsfc.nasa.gov

Bhattacharya, Dipen
Univ. of California, Riverside
dipen@tigre@ucr.edu

Bignami, Giovanni
ASI/Roma
bignami@asizou.asi.it

Blazquez, Raquel
Valencia University

Bloemen, Hans
SRON-Utrecht
H.Bloemen@sron.run.nl

Bloom, Steve
NASA/GSFC
bloom@egret.gsfc.nasa.gov

Boettcher, Markus
MPIfR, Bonn
mboett@mipfr-bonn.mpg.de

Bower, Geoffrey
University of California, Berkeley
gbower@astro.berkeley.edu

Brazier, Karen
University of Durham
karen.brazier@dur.ac.uk

Bridgman, William, T.
USRA/NASA/GSFC
bridgman@grossc.gsfc.nasa.gov

Briggs, Michael
University of Alabama, Huntsville
briggs@gibson.msfg.nasa.gov

Buchholz, James
California Baptist College
jbuchholz@earthlink.net

Bunner, Alan
NASA Headquarters
alan.bunner@hq.nasa.gov

Caravro, Patrizia
IFC/Milano
pat@irctr.mi.cnr.it

Carraminana, Alberto
INAOE, Mexico
alberto@inaoep.mx

Castro-Tirado, Alberto
LAEFF-INTA
ajct@laeff.esa.es

1653

Catanese, Michael
Iowa State University
catanese@egret.sao.arizona.ed

Chakrabarty, Deepto
MIT
deepto@space.mit.edu

Chaty, Sylvain
CEA Saclay
chaty@discovery.saclay.cea.fr

Chen, Andrew
Columbia University
awc@astro.columbia.edu

Chen, Wan
NASA/GSFC
chen@rosserv.gsfc.nasa.gov

Chen, Xingming
Lick Observatory
chen@ucolick.org

Cheng, LingXiang
University of Maryland
lxcheng@astro.umd.edu

Chernenko, Anton
Space Research Institute
anton@cgrsmx.ili.rssi.ru

Chernyakova, Maria
Astro Space Center LPI
masha@sigma.iki.rssi.ru

Cheung, Cynthia
NASA/GSFC
cynthia.cheung@gsfc.nasa.gov

Chiang, James
Naval Research Laboratory
chiang@asse.nrl.navy.mil

Clayton, Donald
Clemson University
clayton@gamma.phys.clemson.edu

Cline, Thomas
NASA/GSFC
cline@lheavx.gsfc.nasa.gov

Collmar, Werner
Max-Planck Inst.
wec@mpe-garching.mpg.de

Connell, Paul Henry
Birmingham University
phc@stor.sr.bham.ac.uk

Coppi, Paolo
Yale University
coppi@blazar.astro.yale.edu

Corbel, Stephane
CEA Saclay
corbel@discovery.saclay.cea.fr

Cordes, James
NAIC/Cornell University
cordes@spacenet.tn.cornell.edu

Crider, Anthony
Rice University
acrider@spacsun.rice.edu

Cui, Wei
MIT
cui@space.mit.edu

Cusumano, Giancarlo
IFCAI CNR
cusumano@ifcai.pa.cnr.it

Dal Fiume, Daniele
TESRE/CNR Bologna
daniele@tesre.bo.cnr.it

Daugherty, Joseph
UNC Asheville
daughtery@cs.unca.edu

Davis, Stanley
NASA/GSFC
davis@grossc.gsfc.nasa.gov

Deal-Giblin, Kim
University of Alabama, Huntsville
deal@gibson.msfc.nasa.gov

Del Sordo, Stefano
IFCAI-CNR
delsordo@ifcai.pa.cnr.it

Dermer, Charles
Naval Research Laboratory
dermer@osse.nrl.navy.mil

Diallo, Nene
CEA Saclay
diallo@integral.saclay.cea.fr

Diehl, Roland
Max-Planck Inst.
rod@mpe-garching.mpg.de

Dieters, Stefan
UAH/MSFC

Digel, Seth
NASA/GSFC
digel@gsfc.nasa.gov

Dingus, Brenda
University of Utah
dingus@mail.physics.utah.edu

Dixon, David
Univ. of California, Riverside
dixon@agouti.ucr.edu

Dogiel, Vladimir
P.N. Lebedev Physical Inst.

Dorman, Lev
IZMIRAN, TECHNION, UNAM

Doron, Ted
Chemical Abstracts Service
tdoron@cas.org

Dubath, Pierre
ISDC
pierre.dubath@obs.unige.ch

Durouchoux, Philippe
Saclav-Service & Astrophysique
durouchouse@sapyzg.saclay.cea.fr

Edwards, Philip
ISAS
pge@vsop.isas.ac.jp

Eikenberry, Stephen
Harvard-Smithsonian
seikenberry@cfa.harvard.edu

Esposito, Joseph
USRA/NASA/GSFC
jae@egret.gsfc.nasa.gov

Fargion, Daniele
University of Italy
Fargion@Roma1.infn.it

Fegan, David
Univ. College, Dublin
djfegan@ferdia.ucd.ie

Fender, Robert
University of Sussex
rpf@star.maps.susx.ac.uk

Feroci, Marco
IAS-CNR
feroci@saturn.ias.fra.cnr.it

Finger, Mark
USRA/NASA/MSFC
finger@batse.msfc.nasa.gov

Finley, John
Purdue University
finley@purdd.physics.purdue.edu

Fishman, Gerald
NASA/MSFC
fishman@ssl.msfc.nasa.gov

Focke, Warren
NASA/GSFC
Warren.B.Focke.1@gsfc.nasa.gov

Ford, Eric
Columbia University
eric@astro.columbia.edu

Frail, Dale
NRAO
dfrail@nrao.edu

Gail, Bill
Ball Aerospace Corp.
bgail@ball.com

Galvan, Edward
New Mexico State University
bmcnamar@nmsu.edu

Gehrels, Neil
NASA/GSFC
gehrels@gsfc.nasa.gov

Georgii, Robert
MPI Extra Terrestrische Physik
rog@mpe-garching.mpg.de

Gierlinski, Marek
Jagiellonian University
gier@camk.edu.pl

Godwin, Linda
NASA/JSC

Goldoni, Paolo
SAP/CEA-Saddy
paolo@discovery.saddy.ced.fr

Gorosabel, Javier
LAEFT-INTA
jgu@laexff.esa.es

Gottbrath, Chris
DePauw University

Gouveia Dal Pino, Elisabete
University of Sao Paulo
dalpino@jet.iagusp.usp.br

Graham, Bradley
Naval Research Laboratory
graham@osse.nrl.navy.mil

Grandi, Paola
IAS/CNR
grandi@alphasax2.ias.fra.cnr.it

Greiner, Jochen
AIP Potsdam
jgreiner@aip.de

Grenier, Isabelle
Centre d'etudes de Saclay
isabelle.grenier@cea.fr

Gros, Maurice
CEA Saclay
maurice.gros@saclay.cea.fr

Grove, J. Eric
Naval Research Lab
grove@osse.nrl.navy.mil

Gruber, Duane
Univ. of California, San Diego
dgruber@ucsd.edu

Gursky, Herbert
Naval Research Laboratory
gursky@ssd0.nrl.navy.mil

Gwinn, Carl
Univ. of California, Santa Barbara
cgwinn@physics.ucsb.edu

Hallum, Jeremy
Boston University
jhallum@bu-ast.bu.edu

Harding, Alice
NASA/GSFC
harding@twinkie.gsfc.nasa.gov

Harris, Michael J.
USRA/NRL
harris@osse.nrl.navy.mil

Harrison, Thomas
New Mexico State University
tharriso@nmsu.edu

Hartman, Bob
NASA/GSFC
rch@egret.gsfc.nasa.gov

Hartmann, Dieter
Clemson University
hartmann@grb.phys.clemson.edu

Heindl, William
Univ. of California, San Deigo
wheindl@mamacass.ucsd.edu

Helfer, Larry
University of Rochester
pany@tsepiot.pas.rochester.edu

Henri, Gilles
Observatoire de Grenoble
henri@obs.ujf-grenoble.fr

Hermsen, Willem
SRON Utrecht
W.Hermsen@sron.ruu.nl

Hernanz, Margarita
IEEC/CSIC
hernanz@ieec.fcr.es

Howard, William
USRA
whoward@usra.edu

Hua, Xin-Min
NASA/GSFC
hua@rosserv.gsfc.nasa.gov

Hui, Li
Los Alamos National Lab
hli@lanl.gov

Hunter, Stanley
NASA/GSFC
sdh@egret.gsfc.nasa.gov

Hurley, Kevin
Univ. of California, Berkeley
kherley@sunspot.ssl.berkeley.edu

Jaffe, Tess
HSTX/GSFC
jaffe@rosserv.gsfc.nasa.gov

Jauncey, David
Australia Telescope
djauncey@atnf.csiro.au

Jean, Pierre
CESR, Toulouse
jean@cesr.cnes.fr

Johnson, Neil
Naval Research Laboratory
johnson@osse.nrl.navy.mil

Jones, Brian
Stanford University
bbjones@egret7.stanford.edu

Jung, Gregory
USRA/NRL
jung@osse.nrl.navy.mil

Kaaret, Philip
Columbia University
kaaret@astro.columbia.edu

Kaluzienski, Lou
NASA Headquarters

Kanbach, Gottfried
Max-Planck Inst.
gok@mpe-garching.mpg.de

Kappadath, S.
University of New Hampshire
s.cheenu.kappadath@unh.edu

Kaspi, Victoria
MIT
vicky@space.mit.edu

Kawai, Nobuyuki
RIKEN Institute
nkawai@postman.riken.go.jp

Kazanas, Demosthenes
NASA/GSFC
kazanas@lheavx.gsfc.nasa.gov

Kertzman, Mary
DePauw University

Kifune, Tadashi
University of Tokyo
tkifune@icrr.u-toyko.ac.jp

King, Edward
Australia Telescope
eking@atnfcsiro.au

Kinzer, Robert
Naval Research Laboratory
kinzer@osse.nrl.navy.mil

Kirk, John
Max Planck Institut Kernphysik
kirk@boris.mpi-hd.de

Kotani, Taro
RIKEN Institute
kotani@riken.go.jp

Kouveliotou, Chryssa
USRA/NASA/MSFC
kouveliotou@batse.msfc.nasa.gov

Kozlovsky, Benzion
NASA/GSFC
bzk@pair.gsfc.nasa.gov

Krennrich, Frank
Iowa State University
frank@egret.sao.arizona.edu

Kretschmar, Peter
Integral Science Data Centre
peter.kretschmar@obs.unige.ch

Krimm, Hans
Hampden-Sydney College
hansk@pulsar.hsc.edu

Kroeger, Richard
Naval Research Laboratory
kroeger@osse.nrl.navy.mil

Kuiper, Lucien
SRON Utrecht
L.Kuiper@sron.ruu.nl

Kurczynski, Peter
University of Maryland
kraken@rosserv.gsfc.nasa.gov

Kurfess, James
Naval Research Laboratory
kurfess@osse.nrl.navy.mil

Kuulkers, Erik
University of Oxford
erik@astro.ox.ac.uk

Lahteenmaki, Anne
Metsahovi Radio Res. Station
alien@kurp.fi

Lamb, Richard
Caltech
lamb@srl.caltech.edu

Langston, Glen
NRAO
glangston@nrao.edu

Leeber, Dawn
New Mexico State University
dleeber@nmsu.edu

Leising, Mark
Clemson University
leising@nova.phys.clemson.edu

Lessard, Rodney
Purdue University
lessard@purdd.physics.purdue.edu

Leventhal, Marvin
University of Maryland
ml@astro.umd.edu

Lewin, Walter
MIT
lewin@space.mit.edu

Liang, Edison
Rice University
liang@spacsun.rice.edu

Lichti, Giselher
MPI Extra Terrestrische Physik
grl@mpe-garching.mpg.de

Lin, Ying-chi
Stanford University
lin@egret0.stanford.edu

Ling, James
Jet Propulsion Laboratory
jling@jplsp.jpl.nasa.gov

Lorenz, Eckart
MPI-Physics
ecl@hegrat.mppmu.mpg.de

Macomb, David
USRA/NASA/GSFC
macomb@cosmic.gsfc.nasa.gov

Madejski, Greg
NASA/GSFC

Magdziarz, Pawel
Jagiellonian University
pavel@camk.edu.pl

Mahoney, William
Jet Propulsion Laboratory
wam@heag4.jpl.nasa.gov

Majmudar, Deepa
Columbia University
dpm@astro.columbia.edu

Malzac, Julien
CESR (CNRS/UPS)
malzac@cest.cnes.fr

Mandrou, Pierre
CESR-Toulouse
mandrou@cesr.cnes.fr

Martin, Michael
University of Louisville
mmarting@zeno.physics.louisville.edu

Massaro, Enrico
Instituto Astronomico - Roma
massaro@astrm2rm.astro.it]

Matteson, Jim
Univ. of California, San Diego
jmatteson@ucsd.edu

Mattox, John
Boston University
mattox@bu.edu

Matz, Steven
Northwestern University
s-matz@nwu.edu

Mayer-Hasselwander, Hans
Max-Planck Inst.
hrm@mpe-garching.mpg.de

McBreen, Brian
University College Dublin
bmcbreen@ollamh.ucd.ie

McCollough, Michael
USRA/NASA/MSFC
mccollough@bowie.msfc.nasa.gov

McConnell, Mark
University of New Hampshire
mark.mcconnell@unh.edu

McLaughlin, Maura
Cornell University
mclaughl@spacenet.tn.cornell.edu

McNamara, Bernie
New Mexico State University
bmcnamar@nmsu.edu

Medina-Tanco, Gustavo
Royal Greenwich Observatory
gmt@ast.cam.ac.uk

Melia, Fulvio
University of Arizona
melia@physics.arizona.edu

Messina, Daniel
NRL/SFA, Inc.
messina@osse.nrl.navy.mil

Milne, Peter
Clemson University
pmilne@astro.phys.clemson.edu

Mirabel, Felix
CEA Saclay
mirabel@discovery.saclay.cea.fr

Mitra, Abhas
Nuclear Research Laboratory
nrl@magnum.barc.ernet.in

Miyaji, Shigeki
Chiba University
miyaji@c.chiba-u.ac.jp

Mori, Masaki
Miyagi University of Education
m-mori3@ipc.miyakyo-u.ac.jp

Morris, Daniel
University of New Hampshire
dmorris@comptel.sr.unh.edu

Moscoso, Michael
University of Texas
mdm@astro.as.utexas.edu

Moskalenko, Igor
MPE Garching
imos@mpe-garching.mpg.de

Mukherjee, Reshmi
USRA/McGill Univ.
muk@hep.physics.mcgill.ca

Muslimov, Alex
NASA/GSFC
muslimov@lheal.gsfc.nasa.gov

Nagel, Steve
NASA/JSC

Nandikotkur, Girikhar
Iowa State University
giridhar@iastate.edu

Narayan, Ramesh
Harvard-SAO
rnarayan@cfa.harvard.edu

Naya, Juan
USRA/NASA/GSFC
naya@fgrs2.gsfc.nasa.gov

Nayakshin, Sergei
University of Arizona
serg@physics.arizona.edu

Norris, Jay
NASA/GSFC
norris@grossc.gsfc.nasa.gov

Oberlack, Uwe
Max-Planck Inst.
ugo@mpe-garching.mpg.de

Obrebski, Tina
Naval Research Laboratory
tina@osse.nrl.navy.mil

Olive, Jean-Francois
CESR, Toulouse
olive@sigma-o.cesr.cnes.fr

Orlandini, Mauro
TESRE/CNR Bologna
orlandini@tesre.bo.cnr.it

Paciesas, William
UAH/MSFC
william.paciesas@msfc.nasa.gov

Palmer, David
USRA/NASA/GSFC
palmer@lheamail.gsfc.nasa.gov

Palumbo, Giorgio
Bologna Uniiversity
gqcpalumbo@astbo3.bo.astro.it

Parizot, Etienne
CEA-Seclay
parizot@cea.fr

Parsons, Ann
NASA/GSFC
parsons@lheamail.gsfc.nasa.gov

Peaper, David
DePauw University
dpeaper@depauw.edu

Pentecost, Elizabeth
USRA
lpenteco@usra.edu

Petrucci, Pierre-Olivier
Observatoire de Grenoble
petrucci@obs.ujf-grenoble

Phlips, Bernard
USRA/NRL
phlips@osse.nrl.navy.mil

Pian, Elena
STSci
pian@stsci.edu

Piro, Luigi
IAS/CNR
piro@alpha.ias.fra.cnr.it

Pitts, Karl
University of Louisville
kpitts@zeno.physics.louisville.edu

Pohl, Martin
Max-Planck Inst.
mkp@mpe-garching.mpg.de

Poutanen, Juri
Uppsala Astron. Observ.
juri@astro.uu.se

Prince, Thomas
Caltech
prince@srl.caltech.edu

Purcell, William
Northwestern University
w-purcell@nwu.edu

Ramaty, Reuven
NASA/GSFC
ramaty@pair.gsfc.nasa.gov

Ray, Alak
NASA/GSFC
ark@olegacy.gsfc.nasa.gov

Reimer, Olaf
Max-Planck Inst.
olr@mpe-garching.mpg.de

Rephaeli, Yoel
Stanford University
yoelr@memsch.stanford.edu

Reynolds, John
ATNF/CSIRO
jreynold@atnf.csiro.au

Robinson, Craig
USRA/NASA/MSFC
robinson@ssl.msfc.nasa.gov

Rothschild, Richard
Univ. of California, San Diego
rothschild@ucsd.edu

Rubin, Bradley
RIKEN Institute
rubin@crab.riken.go.jp

Ryan, James
University of New Hampshire
jryan@unh.edu

Ryde, Felix
Stockholm Observatory
felix@astro.su.se

Said, Slassi-Sennou
Univ. of California, Berkeley
slassi@ssi.berkeley.edu

Saito, Yoshitaka
RIKEN Institute
saito@crab.riken.go.jp

Salamon, Michael
University of Utah
salamon@cosmic.physics.utah.edu

Santangelo, Andrea
IFCAI-CNR
andrea@ifcai.pa.cnr.it

Schlickeiser, Reinhard
MPI Radioastronomie
rschlickeiser@rmpifr.bonn.mpg.de

Schonfelder, Volker
Max-Planck Inst.
vos@mpe-garching.mpg.de

Schubnell, Michael
University of Michigan
schubnel@umich.edu

Scott, Matthew
USRA/NASA/MSFC
scott@gibson.msfc.nasa.gov

Seifert, Helmut
USRA/NASA/GSFC
seifert@lheamail.gsfc.nasa.gov

Sembroski, Glenn
Purdue University
sembroski@purdd.physics.purdue.edu

Shapiro, Maurice
University of Maryland

Share, Gerald
Naval Research Laboratory
share@osse.nrl.navy.mil

Shrader, Chris
USRA/NASA/GSFC
shrader@grossc.gsfc.nasa.gov

Sikora, Marek
N. Copernicus Astron. Center
sikora@camk.edu.pl

Silva, Allison
New Mexico State University
bmcnamar@nmsu.edu

Skibo, Jeff
Naval Research Laboratory
skibo@osse.nrl.navy.mil

Skinner, Gerry
University of Birmingham
gks@star.sr.bham.ac.uk

Smith, David
Univ. of California, Berkeley
dsmith@ssl.berkeley.edu

Smith, Ian
Rice University
ian@spacsun.rice.edu

Sreekumar, P.
USRA/NASA/GSFC
sreekumar@gsfc.nasa.gov

Srinivasan, Radhika
Purdue University
srinivasan@purdd.physics.purdue.edu

Stacy, Greg
University of New Hampshire
greg.stacy@unh.edu

Starrfield, S.
Arizona State University
sumner.starrfield@asu.edu

Staubert, Ruediger
University of Tuebingen
staubert@astro.uni-tuebingen.de

Stecker, Floyd
NASA/GSFC
stecker@lheavx.gsfc.nasa.gov

Steinle, Helmut
MPI f. Extrat. Physik
hcs@mpe-garching.mpg.de

Stollberg, Mark
UAH/MSFC
stollberg@gibson.msfc.nasa.gov

Strickman, Mark
Naval Research Laboratory
strickman@osse.nrl.navy.mil

Strohmayer, Tod
USRA/NASA/GSFC
stroh@pcasrv1.gsfc.nasa.gov

Strong, Andrew
Max-Planck Inst.
aws@mpe-garching.mpg.de

Sturner, Steven
USRA/NASA/GSFC
sturner@tgrosf.gsfc.nasa.gov

Subramanian, Prasad
George Mason University
psubrama@gmu.edu

Swank, Jean
NASA/GSFC
swank@pcasun1.gsfc.nasa.gov

Takahashi, Tadayuki
ISAS
takahasi@astro.isas.ac.jp

Tatischeff, Vincent
NASA/GSFC
tatische@lhea1.gsfc.nasa.gov

Tavani, Mario
Columbia University
tavani@astro.columbia.edu

Teegarden, Bonnard
NASA/GSFC
bonnard@lheamail.gsfc.nasa.gov

Terasranta, Harri
Metsahovi Radio Res. Station
hte@alpha.hut.fi

The, Lih-Sin
Clemson University
lishin@astro.phys.clemson.edu

Thompson, Dave
NASA/GSFC
djt@egret.gsfc.nasa.gov

Timmes, Frank
Univ. of California, SC
fxt@burn.uchicago.edu

Tingay, Steven
Jet Propulsion Lab
tingay@hyaa.jpl.nasa.gov

Titarchuk, Lev
NASA/GSFC
titarchuk@lheavx.gsfc.nasa.gov

Tompkins, Bill
Stanford University
billt@leland.stanford.edu

Tueller, Jack
NASA/GSFC
tueller@gsfc.nasa.gov

Turver, Keith
Durham University
k.e.turver@durham.ac.uk

Tzioumis, Tasso
ATNF, CSIRO
atzioumi@atnf.csiro.au

Ubertini, Pietro
IAS-CNR
uberotini@alpha1.ias.fra.cnr.it

Ulmer, Melville
Northwestern University
m-ulmer@nwu.edu

Urry, C. Megan
STSci
cmu@stsci.edu

Valtaoja, Esko
Turku University
esko.valaoja.utu.fi

van der Hooft, Frank
University of Amsterdam
vdhooft@astro.uva.nl

van der Meulen, Roel
SRON Utrecht
vdmeulen@sron.ruu.nl

van Dyk, Rob
ESTEC/ESA
rvdijk@estec.esa.ni

van Paradijs, Jan
Univ. of Alabama, Huntsville
jvp@astro.uva.nl

Varendorff, Martin
Max-Planck Inst.
mgv@mpe-garching.mpg.de

Vedrenne, Gilbus
CESR, Toulouse

Vestrand, Tom
University of New Hampshire
tom.vestrand@unh.edu

Vilhu, Osmi
University of Helsinki
osmi.vilhu@helsinki.fi

von Ballmoos, Peter
CESR, Toulouse
pvb@cesr.cnes.fr

von Montigny, Corinna
LSW Heidelberg
cvmontig@lsw.uni-heidelberg.de

Wagner, R.
Ohio State University
rmw@lowell.edu

Wagner, Stefan
LSW Heidelberg
swagner@lsw.uni-heidelberg.de

Wang, John
University of Maryland
jcwang@astro.umd.edu

Watanabe, Ken
USRA/NASA/GFSC
watanabe@grossc.gsfc.nasa.gov

Weekes, Trevor
Whipple Observatory
tweekes@cfa.harvard.edu

White, Stephen
University of Maryland
white@astro.umd.edu

Wietfeldt, Fred
NIST
few@rrdstrad.nist.gov

Wiik, Kaj
Metsahovi Radio Res. Station
kah.wiik@hut.fi

Williams, Grant
Clemson University
ggwilli@hubcap.clemson.edu

Williams, Owen
ESTEC/ESA
owilliam@estec.esa.nl

Wilms, Jorn
IAA Tubingen
wilms@astro.uni.turbingen.de

Wilson, Robert
NASA/MSFC
wilson@gibson.msfc.nasa.gov

Winkler, Christoph
ESA/ESTEC
cwinkler@estsa2.estec.esa.nl

Woods, Peter
Univ. of Alabama, Huntsville
peter.woods@msfc.nasa.gov

Xu, Wenge
Caltech
wx@ipac.caltech.edu

Yu, Wenfei
Inst. of High Energy Physics
wenfei@gibson.msfc.nasa.gov

Yusef Zadeh, Farmad
Northwestern University
zadeh@nwu.edu

Zdziarski, Andrzej
N. Copernicus Astron. Center
aaz@camk.edu.pl

Zhang, S. Nan
USRA/NASA/MSFC
zhang@ssl.msfc.nasa.gov

Zychi, Piotr
University of Durham
piotr.zycki@durham.ac.uk

Author Index

A

Aharonian, F., 361, 1397
Aharonian (HEGRA collaboration), F. A., 1631
Ahlen, S. P., 1636
Akhperjanian, A., 1397
Akyüz, A., 1137
Albernhe, F., 1535, 1559, 1647
Aldering, G. S., 1417
Allen, G. E., 1089
Aller, H. D., 1417, 1423
Aller, M. C., 1423
Aller, M. F., 1417, 1423
Applegate, J. H., 1167
Arnett, D., 1119
Arocas, I., 1647

B

Backman, D. E., 1417
Bailes, M., 583, 602
Balonek, T. J., 1417, 1423
Bandyopadhyay, R., 892
Baring, M. G., 171, 638, 1157
Barret, D., 75, 697, 724, 734, 952, 1498
Barrio, J., 1397
Bartolini, G., 1516
Bazán, G., 1119
Bazzano, A., 719, 729
Begelman, M. C., 1423
Belolipetskiy, S. V., 1606
Bennett, K., 537, 542, 583, 588, 829, 967, 1084, 1099, 1198, 1218, 1243, 1298, 1341, 1582, 1587
Bernlöhr, K., 1397
Bertsch, D. L., 509, 1248, 1346, 1371, 1592, 1606
Beteta, J., 1397
Bhattacharya, D., 1137, 1147, 1626
Bignami, G. F., 387, 573
Biller, S., 558
Blaes, O., 1288
Blanco, P. R., 897, 1089, 1147
Blázquez, R., 1647

Bloemen, H., 249, 537, 829, 967, 1074, 1084, 1099, 1109, 1114, 1218, 1243, 1298, 1341, 1356
Blom, J. J., 1243, 1341, 1356, 1582
Bloom, S. D., 1257, 1262, 1346, 1371
Bloser, P. F., 697, 724, 734, 952, 1498
Boirin, L., 724
Boltwood, P., 1417
Bonnell, J., 1417
Borrel, V., 957, 1535, 1647
Böttcher, M., 1473
Bouchet, L., 957
Boyle, P. J., 558, 592, 1142, 1376, 1381
Bradbury, S., 1397
Bradbury, S. M., 1142, 1402
Brazier, K. T. S., 588, 597, 1267
Breslin, A. C., 1402
Bridgman, W. T., 977
Briggs, M. S., 687
Brinkmann, W., 922
Buccheri, R., 542
Buckley, J. H., 558, 592, 1142, 1376, 1381, 1402
Buckley, T., 665, 675
Burdett, A. C., 1142
Burdett, A. M., 558, 592, 1376, 1381
Burger, M., 878
Bussóns Gordo, J., 558, 1142, 1376, 1381
Bykov, A. M., 249, 1074

C

Campbell-Wilson, D., 937
Caplinger, J., 1417
Caraveo, P. A., 387, 568, 573, 1535
Carramiñana, A., 583, 588, 597, 1267, 1462
Carter-Lewis, D. A., 558, 592, 1142, 1361, 1376, 1381, 1402
Case, G., 1137
Cassé, M., 1059
Castro-Tirado, A. J., 1333, 1516
Catanese, M. A., 558, 592, 1142, 1376, 1381, 1402
Catelli, J. R., 1606

Cawley, M. F., 558, 592, 1142, 1376, 1381, 1402
Celotti, A., 1417
Chadwick, P. M., 612, 1621
Chakrabarty, D., 773
Chantell, M. C., 1626
Charles, P. A., 892
Chaty, S., 917
Chavushyan, V., 1462
Chen, X., 995
Cheng, L. X., 902, 1012
Chernenko, A., 653
Chernyakova, M. A., 822
Chiaberge, M., 1423
Chiang, J., 1308
Chiappetti, L., 1412
Churazov, E., 957
Clements, S. D., 1423
Cline, T. L., 1007
Cocchi, M., 719, 729
Collin, M., 1559
Collmar, W., 307, 829, 863, 1243, 1298, 1341, 1356, 1417, 1423, 1587
Colombo, E., 592
Connell, P., 1535
Connell, P. H., 1544
Connors, A., 407, 542, 583
Contreras, J., 1397
Coppi, P., 1626
Corbel, S., 912, 932, 937
Cordes, J. M., 633
Cordier, B., 1535, 1559, 1647
Cortina, J., 1397
Costa, M. E., 1433, 1493, 1516
Covault, C. E., 1626
Crary, D. J., 803, 947
Crawford, H., 1606
Crider, A., 868, 1512
Cui, W., 813, 839, 854
Cusumano, G., 553, 758, 793

D

Dal Fiume, D., 553, 758, 793, 1493, 1516
Dalton, J., 1417
Dame, T. M., 1188
D'Amico, N., 602
Daniels, W. M., 1606

Daum, A., 1397
Dazeley, S. A., 1507
Deal, K., 687
Deckers, T., 1397
De Francesco, G., 1423
de Gouveia Dal Pino, E. M., 1203
Deines-Jones, P., 1606
de Jager, O. C., 171
Del Sordo, S., 758, 793
Dermer, C. D., 271, 307, 977, 1044, 1152, 1275, 1308, 1402
Diallo, N., 1535, 1559
Dickinson, M. R., 612, 1621
Diehl, R., 218, 542, 967, 1074, 1084, 1099, 1104, 1109, 1114, 1198, 1218, 1298, 1535, 1554, 1577
Dieters, S., 617
Digel, S. W., 1188
Dingus, B. L., 407, 1248, 1346, 1371, 1592
Dipper, N. A., 612, 1621
Dixon, D. D., 1039, 1137, 1601
Dobrinskaya, J., 932
Dogiel, V. A., 1069
Done, C., 962
Dorman, L. I., 1172, 1178, 1183
Dotani, T., 844, 922
Dove, J., 849
Dragovan, M., 1626
Drucker, A., 1417
Durouchoux, Ph., 887, 912, 932, 937, 1535
Dyachkov, A., 957

E

Ebisawa, K., 623, 844, 922
Edwards, P. G., 1428, 1433, 1507
Eikenberry, S. S., 547
Ellison, D. C., 1157
Esin, A., 887
Esposito, J. A., 1248, 1257, 1346, 1371, 1606

F

Falomo, R., 1417
Fazio, G. G., 547

Fegan, D. J., 361, 558, 592, 1142, 1376, 1381, 1402
Feigl, E., 1397
Fender, R. P., 798, 813, 932, 937
Fenker, H., 1606
Fernandez, J., 1397
Feroci, M., 758, 1493, 1516
Ferrero, J. L., 1647
Fichtel, C. E., 436, 1417
Filippenko, A. V., 932
Finger, M. H., 57, 617, 739, 748, 773, 778, 803, 947
Finley, J. P., 558, 592, 1142, 1376, 1381, 1402
Fiore, F., 1412
Fishman, G. J., 509, 687, 927, 947, 952, 1012, 1283
Fletcher, S., 1099
Focke, W., 854
Fonseca, V., 1397
Ford, E. C., 697, 703, 734
Foster, R. S., 813, 922, 1253
Fraß, A., 1397
Freudling, W., 1417
Freyberg, M. J., 1069
Frontera, F., 758, 1493, 1516
Funk, B., 1397

G

Gaensler, B. M., 1428
Gaidos, J. A., 558, 592, 1142, 1376, 1381, 1402
Galvan, E., 665, 675
Gautier, T. N., 912
Gear, W. K., 1417, 1423
Geckeler, R. D., 753
Gehrels, N., 3, 524, 902, 1007, 1012, 1283
Georganopoulos, M., 1423
Georgii, R., 1109, 1535, 1554
Gerard, E., 892
Ghigo, F. D., 813
Ghisellini, G., 1293, 1412, 1417, 1423
Giarrusso, S., 553, 793
Gierliński, M., 844
Gilfanov, M., 887, 957
Giommi, P., 1412, 1493
Glass, I. S., 1423

Goldoni, P., 957, 1549
Goldwurm, A., 957, 1549
Gómez-Gomar, J., 1125
Gonzalez, J., 1397
González-Pérez, J. N., 1417, 1423
Gordo, J., 592
Gorosabel, J., 1333, 1516
Gossan, B., 1606
Gottbrath, C. L., 1616
Gotthelf, E., 1027
Graham, B. L., 1642
Grandi, P., 1293
Gregorich, D. T., 1626
Greiner, J., 907
Grenier, I. A., 394, 1157, 1188
Grindlay, J. E., 122, 697, 724, 734, 952, 1498
Grove, J. E., 122, 788, 1079, 1293
Gruber, D. E., 744, 753, 788, 897, 1089
Guainazzi, M., 793
Guarnieri, A., 1516
Guichard, J., 1267, 1462

H

Haardt, F., 1293
Hall, P., 1417
Halpern, J. P., 568, 1253
Hanna, D. S., 1626
Hara, T., 1507
Harding, A. K., 39, 607, 638, 648
Harmon, A., 878, 922, 1283
Harmon, B. A., 122, 141, 687, 697, 713, 734, 773, 813, 834, 839, 873, 927, 947, 952, 1498
Harris, M. J., 418, 1007, 1079, 1094
Harrison, T. E., 665, 670, 675, 679, 942
Hartman, R. C., 307, 1248, 1262, 1346, 1366, 1417, 1423, 1428, 1606
Hartmann, D. H., 1039, 1147, 1223, 1228
Haustein, V., 1397
Hayami, Y., 1507
Heidt, J., 1516
Heikkila, C. W., 665
Heinämäki, P., 1423
Heindl, W. A., 744, 788, 854, 897, 1089
Heinzelmann, G., 1397
Heise, J., 719, 729

Hemberger, M., 1397
Henri, G., 1303, 1313
Hermann, G., 1397
Hermsen, W., 39, 537, 542, 583, 588, 829, 967, 1074, 1084, 1099, 1109, 1114, 1218, 1243, 1253, 1298, 1341, 1356
Hernanz, M., 1125
Herter, M., 1423, 1457
Heß, M., 1397
Heusler, A., 1397
Higdon, J. C., 912, 1089
Hillas, A. M., 558, 592, 1142, 1376, 1381, 1402
Hjellming, R. M., 141, 813
Hofmann, W., 1397
Holder, J., 612, 1621
Holl, I., 1397
Hooper, E. J., 1423
Horns, D., 1397
Hua, X.-M., 122, 858, 982, 987, 1520
Hughes, P. A., 1423
Hunstead, R. W., 1428
Hunter, S. D., 192, 1188, 1213, 1248, 1346, 1371, 1606
Hurley, K. H., 407, 1007
Hutchins, J. B., 1606

I

Illarionov, A. F., 822
Inoue, H., 922, 1417
in't Zand, J., 719, 729
Isern, J., 1125
Iyudin, A., 1109

J

Jahoda, K., 692, 844, 1089
Jauncey, D. L., 1433
Jean, P., 1535
Johnson, W. N., 283, 844, 1079, 1288, 1402, 1417, 1423, 1572
Jones, B. B., 783, 1213, 1248, 1366, 1371
Jones, D. L., 1433
José, J., 1125

Jourdain, E., 881, 957, 1303
Jung, G. V., 708, 1079

K

Kaaret, P., 697, 703, 734
Kamae, T., 628
Kamei, S., 1507
Kanbach, G., 597, 1213, 1248, 1253, 1267, 1346, 1371
Kankanian, R., 1397
Kappadath, S. C., 344, 1218, 1587
Kaspi, V. M., 583, 602
Katajainen, S., 1423
Kaul, R. K., 1271
Kawai, N., 628, 922
Kazanas, D., 122, 982, 987, 1520
Kendall, T. R., 612, 1621
Kendziorra, E., 788
Kertzman, M. P., 1616
Khavenson, N., 957
Kidger, M. R., 1417, 1423
Kifune, T., 361, 1507
Kinzer, R. L., 192, 1079, 1193, 1223, 1572
Kirk, J. G., 1478
Kirstein, O., 1397
Kita, R., 1507
Kniffen, D. A., 524, 1248, 1257, 1346, 1371
Knödlseder, J., 1104, 1109, 1114
Köhler, C., 1397
Kolaczyk, E. D., 1039, 1601
Kollgaard, R. I., 1417
Kommers, J., 687
Kondo, Y., 1417
König, M., 763
Konishi, T., 1507
Konopelko, A., 1397
Kornmayer, H., 1397
Kotani, T., 922
Kouveliotou, C., 96, 687, 947
Kozlovsky, B., 1049
Kranich, D., 1397, 1407
Kraus, A., 1346, 1423
Krawczynski, H., 1397
Krennrich, F., 558, 592, 1142, 1376, 1381, 1402
Krennrich, for the Whipple Collaboration, F., 563, 1391

Kretschmar, P., 788
Kreykenbohm, I., 788
Krichbaum, T. P., 1346
Krizmanic, J. F., 1606
Kroeger, R. A., 141, 1642
Krolik, J. H., 972
Kubo, H., 1467
Kuiper, L., 537, 542, 583, 588, 829
Kuleshova, N., 957
Kunz, M., 744
Kurfess, J. D., 509, 708, 1079, 1147, 1193, 1417, 1572, 1642
Kusunose, M., 868, 1512
Kuulkers, E., 683

Lindenstruth, V., 1606
Lindner, A., 1397
Ling, J. C., 858
Lingenfelter, R. E., 1089
Litterio, M., 643
Lorenz, E., 1397, 1407
Lorenz, E. for the MAGIC Telescope Design Group, 1611
Lovell, J. E. J., 1428, 1433
Lu, F. J., 578
Lund, N., 1516
Luo, C., 878, 992
Lyne, A., 588

L

Lähteenmäki, A., 1452
Lainela, M., 1428
Lamb, R. C., 558, 592, 1142, 1376, 1381, 1402
Lammers, U., 683
Lampeitl, H., 1397
Lanteri, L., 1423
Lattanzi, M. G., 573
Laurent, P., 957, 1549
Lawrence, G. F., 1423
Lawson, A. J., 1417
Leahy, D. A., 748
Lebrun, F., 1549
Leeber, D. M., 942
Leising, M. D., 163, 1017, 1022, 1104, 1147, 1223, 1228
Leleux, P., 1535
Leonard, D. C., 932
Leray, J. P., 1647
Lessard, R. W., 558, 592, 1142, 1376, 1381, 1402
Leventhal, M., 208, 902, 1012
Lewin, W. H. G., 687, 947
Li, T. P., 578, 1351
Liang, E. P., 461, 868, 873, 878, 932, 992, 1512
Lichti, G. G., 542, 1084, 1109, 1298, 1423, 1535, 1554
Lin, D., 868, 1512
Lin, R., 1535
Lin, Y. C., 283, 783, 1248, 1346, 1371, 1423

M

Macomb, D. J., 1402
Madejski, G. M., 283, 1417, 1423, 1467
Magdziarz, P., 1288
Magnussen, N., 1397
Mahoney, W. A., 912, 1346
Maisack, M., 788, 1356
Majmudar, D., 1167
Makarov, V. V., 573
Makino, F., 1417, 1467
Malzac, J., 1303
Manchester, R. N., 583, 602
Mandrou, P., 1535
Mandzhavidze, N., 1054
Maraschi, L., 1293, 1412, 1417
Marscher, A. P., 1346, 1417, 1423, 1437
Martin, M. D., 1606
Martini, P., 892
Masaike, A., 1507
Masetti, N., 1516
Mason, P. A., 665, 670, 675, 679
Massaro, E., 553, 643
Massone, G., 573
Masterson, C., 558, 592, 1142, 1376, 1381
Mastichiadis, A., 1478
Matsubara, Y., 1507
Matsuoka, M., 922
Matsuoka, Y., 1507
Matteson, J., 1535
Mattox, J. R., 568, 1152, 1253, 1636
Matz, S. M., 407, 808
Mause, H., 1473

Mayer-Hasselwander, H. A., 1248, 1346, 1371
McCollough, M. L., 813, 834
McCollum, B., 1417
McComb, T. J. L., 612, 1621
McConnell, M. L., 122, 542, 829, 967, 1084, 1099, 1198, 1218, 1341, 1577, 1587
McCulloch, P. M., 1428, 1433
McEnery, J. E., 558, 592, 1142, 1376, 1381, 1402
McHardy, I. M., 1417
McLaughlin, M. A., 633
McNamara, B. J., 665, 670, 675, 679, 942
McNaron-Brown, K., 1079, 1417, 1423, 1572
Medina Tanco, G. A., 1203
Meegan, C., 407
Meier, D. L., 1433
Melia, F., 1000, 1318, 1323, 1328
Meyer, H., 1397
Michelson, P. F., 783, 1346, 1371
Mignani, R., 573
Milne, P. A., 1017, 1022
Mineo, T., 553
Mioduszewski, A. J., 813
Mirabel, I. F., 141, 892, 902, 917
Mirzoyan, R., 1397
Misra, R., 1000
Mitchell, J. W., 1606
Mitra, A. K., 818, 1271
Mizumoto, Y., 1507
Mohanty, G., 558, 592, 1142, 1376, 1381
Möller, H., 1397
Monnelly, G. P., 1498
Moore, E. M., 1423
Moralejo, A., 1397
Morfill, G. E., 1069
Morgan, E. H., 897, 907
Mori, M., 623, 768, 1507
Moriarty, P., 592, 1402
Morris, D. J., 1084, 1109, 1114, 1298
Moskalenko, I. V., 863, 881, 1162
Moss, M., 932
Much, R. P., 537, 542, 829, 1243, 1582
Mücke, A., 1233, 1346, 1371
Mukherjee, R., 394, 1253, 1346, 1371, 1423, 1626
Muller, J. M., 719, 729

Muraishi, H., 1507
Muraki, Y., 1507
Murphy, D. W., 1433
Murphy, R. J., 17, 1079
Muslimov, A., 648

N

Nagase, F., 922
Nair, A. D., 1423
Nair, D., 1417
Naito, T., 1507
Nandikotkur, G., 1361
Narayan, R., 461, 887
Natalucci, L., 719, 729
Navarro, J., 602
Naya, J., 1535, 1544
Naya, J. E., 1567
Nayakshin, S., 1318, 1323, 1328
Nevalainen, J., 887
Nicastro, L., 553, 793, 1493, 1516
Nicolson, G. D., 1433
Nilsson, K., 1423
Nishijima, K., 1507
Nolan, P. L., 783, 1248, 1346, 1371
Nomoto, K., 1089
Nowak, M., 849

O

Oberlack, U., 1104, 1109, 1114, 1577
Ogio, S., 1507
Olive, J.-F., 724
Ong, R. A., 1626
Oosterbroek, T., 683, 758
Orford, K. J., 612, 1621
Orlandini, M., 758, 793, 1493
Osborne, J. L., 612
Oser, S., 1626
Owens, A., 683

P

Paciesas, W. S., 283, 697, 713, 734, 803, 813, 834, 839, 927, 952, 1283
Padilla, L., 1397
Palazzi, E., 758, 1412, 1493, 1516

Palmer, D., 1007
Panter, M., 1397
Parizot, E. M. G., 1059, 1064
Parmar, A. N., 553, 683, 758
Parsons, A., 1283
Patterson, J. R., 1507
Paul, J., 1059
Peaper, D. R., 1616
Peila, A., 1423
Pelling, M. R., 897, 1089
Penton, S., 1417
Perryman, M. A. C., 573
Pesce, J. E., 1417
Petrucci, P. O., 1303, 1313
Petry, D., 1397
Petry, D. for the HEGRA Collaboration, 1407
Phillips, R. B., 1442
Phlips, B. F., 854, 1642
Pian, E., 1412, 1417
Piccioni, A., 1516
Pierkowski, D. B., 1423
Piraino, S., 793
Piro, L., 793
Piro, L. on behalf of the BeppoSAX team, 1485
Pitts, W. K., 1606
Plaga, R., 1397
Pohl, M. K., 449, 1034, 1213, 1233, 1346, 1366, 1371, 1417, 1423
Pohl, M. for the EGRET collaboration, 1596
Pooley, G. G., 813
Poutanen, J., 887, 972
Prahl, J., 1397
Preston, R. A., 1433
Prince, T. A., 57
Prosch, C., 1397
Pühlhofer, G., 1397
Purcell, W. R., 208, 708, 902, 1027, 1079, 1193, 1572
Pursimo, T., 1423

Q

Quinn, J., 558, 592, 1142, 1376, 1381, 1402

R

Ragan, K., 1626
Raiteri, C. M., 1417, 1423
Ramaty, R., 449, 1007, 1049, 1054
Rauterberg, G., 1397
Ray, A., 607
Rayner, S. M., 612, 1621
Reich, W., 1423
Reimer, O. L., 597, 1248, 1267, 1346, 1371
Remillard, R. A., 907
Renda, M., 1417
Rephaeli, Y., 271
Revnivtsev, M., 957
Reynolds, J. E., 1433
Reynolds, S. J., 1157
Rhode, W., 1397
Rivero, R., 1397
Roberts, I. D., 612, 1621
Roberts, M. D., 1507
Roberts, M. S. E., 783
Robinson, C. R., 713, 734, 813, 834, 922, 927, 1498
Robson, E. I., 1417, 1423
Rodgers, A., 1381
Rodgers, A. J., 558, 592, 1142, 1376, 1402
Rodriguez, L. F., 141, 902
Röhring, A., 1397
Roldán, C., 1647
Roques, J.-P., 957, 1535
Rose, H. J., 558, 592, 1142, 1376, 1381, 1402
Rothschild, R. E., 744, 753, 788, 897, 1089
Rowell, G. P., 1507
Rubin, B. C., 713, 778, 927, 947
Rupen, M. P., 141, 813
Ryan, J. M., 17, 537, 542, 829, 967, 1084, 1099, 1109, 1114, 1198, 1218, 1243, 1356, 1577, 1582
Ryde, F., 972

S

Sahakian, V., 1397
Saito, Y., 628
Sako, T., 1507

Sakurazawa, K., 1507
Salamon, M. H., 1238
Sambruna, R. M., 1412, 1417
Samimi, J., 1039, 1601
Samorski, M., 1397
Samuelson, F. W., 558, 592, 1142, 1376, 1381, 1402
Sanchez, F., 1535
Sanchez, J., 1397
Sandhu, J. S., 602
Santangelo, A., 758
Saunders, M. A., 1601
Schandl, S., 763
Schirmer, A. F., 1417
Schlickeiser, R., 449, 1473
Schmele, D., 1397
Schmidt, T., 1397
Schneid, E. J., 1346, 1371, 1592
Schönfelder, V., 509, 542, 583, 588, 829, 863, 967, 1069, 1074, 1084, 1099, 1109, 1114, 1198, 1218, 1243, 1298, 1341, 1356, 1535, 1554, 1582, 1587
Schubnell, M., 1386
Scott, D. M., 617, 739, 744, 748, 778, 803
Segreto, A., 553, 758, 793
Seifert, H., 1007, 1535, 1544, 1567
Seitz, T., 1516
Sembroski, G. H., 558, 592, 1142, 1376, 1381, 1402, 1616
Shahbaz, T., 892
Share, G. H., 17, 1079, 1223
Shaw, S. E., 612, 1621
Sheth, S., 878
Shibata, S., 628
Shrader, C. R., 3, 328, 892, 1417
Sikora, M., 494, 1417, 1423
Sillanpää, A., 1417, 1423
Silva, A., 665
Simrall, J. H., 1606
Skibo, J. G., 449, 977, 1044, 1152, 1208, 1308
Skinner, G. K., 1535, 1544
Smale, A., 724
Smith, A. G., 1423
Smith, D. A., 962, 1288
Smith, D. M., 208, 902, 1012
Smith, I. A., 110, 868, 932, 1512
Smith, M. J. S., 719, 729
Smith, P. S., 1417

Song, L. M., 578
Sood, R. K., 932, 937
Sparks, W. M., 1130
Spencer, R. E., 937
Sreekumar, P., 436, 768, 1248, 1346, 1361, 1366, 1371, 1606
Srinivasan, R., 558, 592, 1142, 1376, 1381, 1402
Stacy, J. G., 1341, 1356, 1442
Stamm, W., 1397
Stappers, B. W., 602
Stark, M. J., 692
Starrfield, S., 1130
Staubert, R., 744, 753, 763, 788
Stecker, F. W., 344, 1238
Steinle, H., 283, 542, 829, 1298, 1356
Stelzer, B., 753
Steppe, H., 1423
Stevens, J. A., 1417, 1423
Stocke, J., 1417
Stollberg, M. T., 803
Streitmatter, R. E., 1606
Strickman, M., 607
Strohmayer, T. E., 692, 773
Strong, A. W., 192, 537, 542, 829, 1074, 1099, 1109, 1114, 1162, 1198, 1298, 1356, 1544, 1596
Sturner, S. J., 1152, 1535, 1544, 1567
Sun, X. J., 578
Sunyaev, R., 957
Susukita, R., 1507
Suzuki, A., 1507
Suzuki, R., 1507
Swank, J. H., 692, 724, 854, 897, 995, 1089

T

Taam, R. E., 995
Tagliaferri, G., 1412
Takahashi, T., 1467
Takalo, L. O., 1417, 1423
Takeshima, T., 922
Tamura, T., 1507
Tanaka, Y., 922
Tanimori, T., 1507
Tashiro, M., 1467
Tatischeff, V., 1049, 1054
Tavani, M., 75, 697, 734, 922, 1253

Teegarden, B. J., 1007, 1535, 1544, 1567
Templeton, M., 665, 670, 675
Teräsranta, H., 1346, 1417, 1423, 1447, 1452
Thaddeus, P., 1188
The, L.-S., 1022, 1147, 1223, 1228
Thommes, E., 1516
Thompson, D. J., 39, 394, 1248, 1253, 1257, 1262, 1346, 1371, 1417, 1606
Thompson, R., 1417
Thornton, G. J., 1507
Timmes, F. X., 218
Tingay, S. J., 1433
Titarchuk, L. G., 477, 982, 987, 1520
Tompkins, W. F., 783, 1248, 1371
Tornikoski, M., 1346, 1417, 1423, 1428
Tosti, G., 1417
Treves, A., 1412, 1417
Truran, J. W., 1130
Tserenin, N., 957
Tueller, J., 902, 1012
Tümer, O. T., 1626
Turcotte, P., 1417
Turver, K. E., 612, 1621
Tzioumis, A. K., 937, 1433

U

Ubertini, P. on behalf of the IBIS Consor 1527
Ubertini, P., 719, 729
Ueda, Y., 922
Ulmer, M. P., 39, 633
Ulrich, M., 1397
Unwin, S. C., 1417
Urry, C. M., 1293, 1412, 1417

V

v.Montigny, C., 1248
Vacanti, G., 1412
Valtaoja, E., 1346, 1417, 1423, 1452
van der Hooft, F., 947
van der Klis, M., 947
van der Meulen, R. D., 537, 542, 1074
van Dijk, R., 829, 967, 1099, 1577
van Paradijs, J., 96, 617, 687, 947
Varendorff, M., 542, 1218, 1577

Vargas, M., 887, 957
Vaughan, B. A., 849
Vedrenne, G., 1535
Vestrand, W. T., 768, 1442
Vilhu, O., 887
Villata, M., 1417, 1423
Visser, G., 1606
Völk, H., 1397
von Ballmoos, P., 1109, 1114, 1535
von Montigny, C., 307, 1257, 1262, 1346, 1371, 1423, 1457

W

Wagner, R. M., 892
Wagner, S. J., 1346, 1417, 1423, 1457
Wallyn, P., 912
Walsh, K. M., 1606
Waltman, E. B., 813
Wang, J. C. L., 658
Watanabe, K., 1223, 1228
Webb, J. R., 1417
Weekes, T. C., 361, 558, 592, 1142, 1376, 1381, 1402
Wehrle, A. E., 328, 1417, 1437
Weidenspointner, G., 1218, 1577, 1587
Wessolowski, U., 1109, 1587
West, M., 558
Westerhoff, S., 1397
Wheaton, Wm. A., 858
White, N. E., 922
Wichmann, R., 1423
Wiebel-Sooth, B., 1397
Wiedner, C. A., 1397
Wiescher, M. C., 1130
Wiik, K., 1452
Williams, D. A., 1626
Williams, O. R., 1243, 1298, 1341, 1356, 1582
Willmer, M., 1397
Wilms, J., 753, 788, 849
Wilson, C. A., 617, 687, 773, 834, 1283
Wilson, R. B., 739, 748, 773, 778, 803
Winkler, C., 542, 1109, 1114, 1218
Wirth (HEGRA collaboration), H., 1397

Witzel, A., 1346, 1423
Wolf, C., 1516
Woods, P., 687
Woosley, S. E., 1089, 1228
Wu, M., 578, 1351

X

Xu, W., 1417, 1437

Y

Yamaoka, K., 922
Yanagita, S., 1507
Yoshida, T., 1507
Yoshikoshi, T., 1507

Yu, W., 578, 734, 1351
Yusef-Zadeh, F., 1027

Z

Zavattini, G., 1493
Zdziarski, A. A., 283, 844, 1288
Zhang, S., 578, 1351
Zhang, S. N., 141, 697, 713, 734, 813, 834, 839, 873, 878, 922, 927, 952, 1253, 1283, 1498
Zook, A. C., 1417
Zweerink, A., 558
Zweerink, J., 592, 1142, 1376, 1381, 1402
Zych, A., 1137
Życki, P. T., 962